MANTLE CONVECTION

THE FLUID MECHANICS OF ASTROPHYSICS AND GEOPHYSICS

A series edited by PAUL H. ROBERTS
University of California, Los Angeles, USA

Volume 1 SOLAR FLARE MAGNETOHYDRODYNAMICS
Edited by E. R. Priest

Volume 2 STELLAR AND PLANETARY MAGNETISM
Edited by A. M. Soward

Volume 3 MAGNETIC FIELDS IN ASTROPHYSICS
Ya. B. Zeldovich, A. A. Ruzmaikin, D. D. Sokoloff

Volume 4 MANTLE CONVECTION: Plate Tectonics and Global Dynamics
Edited by W. R. Peltier

Volume 5 DIFFERENTIAL ROTATION AND STELLAR CONVECTION: Sun and Solar-type Stars
G. Rüdiger

MANTLE CONVECTION
Plate Tectonics and Global Dynamics

Edited by
W. R. PELTIER
University of Toronto
Toronto, Ontario, Canada

GORDON AND BREACH SCIENCE PUBLISHERS
New York London Paris Montreux Tokyo Melbourne

Gordon and Breach Science Publishers

Post Office Box 786
Cooper Station
New York, New York 10276
United States of America

Post Office Box 197
London WC2E 9PX
England

58, rue Lhomond
75005 Paris
France

3-14-9, Okubo
Shinjuku-ku, Tokyo
Japan

Private Bag 8
Camberwell, Victoria 3124
Australia

Library of Congress Cataloging in Publication Data

Mantle convection.

(Fluid mechanics of astrophysics and geophysics ; v. 4)
Bibliography: p.
Includes index.
1. Earth—Mantle. I. Peltier, W. R. II. Series.
QE509.M2618 1989 551.1′16 88-34720
ISBN 0-677-22120-7

Contents

Introduction to the Series

Anyone who surveys the shelves of a science library will soon convince himself that fluid mechanics has taken more than its share of the rapid growth of scientific publication following World War II. And within fluid mechanics, publications of astrophysical and geophysical interest have enjoyed increasing prominence. There are a number of obvious reasons for this, ranging from the academic to the practical. Prominent is the long-term aim of understanding and partially controlling the Earth's atmopshere, with implications that range from the environmental one of minimizing the spread of harmful industrial wastes to the humanitarian one of maximizing agricultural yields. No less pressing in these days of developing ocean resources is the need to understand ocean movements, how they can be harnessed to quench man's thirst for energy, how they affect oceanic life and the harvest man takes from the sea. Seismology and vulcanology are of scarcely less practical significance to many populous areas, and they owe their existence to crustal stresses set up by the motion of an underlying mantle. Beneath the mantle, which is often modelled by a highly viscous fluid, lies the Earth's core in whose internal flows lies the cause of the Earth's magnetism.

Like geomagnetism, most topics in astrophysical fluid mechanics are hard to model in the laboratory. Some are principally of academic interest. Exceptions are the interaction of the solar wind with the Earth's external magnetic field and magnetic storms and their ionospheric consequences on radio reception. The number of topics in fluid mechanics of academic interest in astrophysics is legion, and includes the study of stellar winds in general, the motion of gas clouds and H II regions, the pulsation of stars, their rotation, their magnetism, their formation by gravitational collapse, the propagation of density waves in galaxies, the maintenance of galactic magnetic fields, pulsar magnetospheres, supernova explosions, the motion of planetary atmospheres, and so forth. Some of these subjects have an old and respected history, but many arose only in the recent past. Sometimes the impetus for the interest in a topic has come from discoveries made during space exploration or from orbiting observatories but often it can be understood as part of a natural progression; for example, as the classic problem of the structure and evolution of a static spherically symmetric star becomes increasingly well solved, what more natural than to examine the 'exceptional' departures: the pulsation of a star, its distortion by rotation, tides, or magnetism?

A thread common to many of these topics is the fluid mechanics of rotating stratified (and perhaps electrically conducting) fluids. Not surprisingly, concepts useful in large-scale atmospheric dynamics and oceanography find parallels in solar, planetary and cosmical physics. A scientist in one of these disciplines may well find himself reading, comprehending, and perhaps being helped by a paper in another. It was with this idea in mind that the publishers encouraged me in 1977 to broaden the scope of the journal *Geophysical Fluid Dynamics* to include papers of astrophysical interest. The response has been so favorable that they urged me to start a series of books to complement it.

This is the fourth volume in that series on *The Fluid Mechanics of Astrophysics and Geophysics*, and follows books on *Solar Flare Magnetohydrodynamics*, *Stellar and Planetary Magnestism*, and *Magnetic Fields in Astrophysics*. The format of the third book was the traditional one, in which authors give their own personal account of an entire field. The first two volumes were collections of reviews and articles that the editors welded into unified state-of-the-art accounts by adding supplementary material of their own. This is also the format of the present book.

The past twenty years have witnessed a remarkable revolution in man's understanding of the surface on which he lives, the crust, and the mantle underlying it. Not only has the fluid mechanist met exciting new challenges elucidating unfamiliar dynamical behaviors, but the geophysicist has developed new insights into processes at work in the earth as a whole: the motions of crustal plates, their relation to mantle convection, the "D"-layer, the topography of the core-mantle boundary, the thermal and dynamical interaction of core and mantle, and so forth. That the material was available for an exciting book could hardly be doubted, but was the time ripe for a survey? Was there not a real danger that a book would be seriously dated during the few months necessary for its production? I was overjoyed when Professor Peltier opined that the subject of mantle convection had reached a plateau, which could be meaningfully surveyed. I am grateful to him and his team of experts for spending the time and effort necessary to bring this book to fruition; I am sure they would have preferred to have devoted themselves to advancing their subjects by research! But students in the area will undoubtedly value their scientific altruism highly. The book will, I am certain, be judged a worthy successor to the three previous volumes of the series.

P. H. Roberts

CHAPTER 1

Mantle Convection and Plate Tectonics: The Emergence of Paradigm in Global Geodynamics

W.R. Peltier†

Department of Physics, University of Toronto
Toronto, Ontario, Canada, M5S 1A7

1. Introduction

The revolution in the Earth sciences which began in earnest during the decade of the 1960s was based upon the increasingly widespread acceptance of the idea of continental drift—that is the notion that continental masses have suffered large horizontal displacements with respect to one another throughout geological time. This revolution occurred in large part because of the explosion of knowledge which was then taking place in our understanding of the ocean floor and on the basis of which a new version of the drift hypothesis, which came to be called seafloor spreading, was rapidly accepted (Hess, 1962; Dietz, 1961). This idea, and later refinements and elaborations of it, now constitutes the guiding principle in terms of which most research on the large scale dynamical state of the planet is organized.

At the heart of this paradigm is the understanding that the Earth's mantle (roughly its outer 50 percent by radius) must behave as a viscous fluid when it is subjected to a shear stress of sufficient temporal duration. To the extent that this rheological ansatz is correct, it follows that if the mantle is subjected

†John Simon Guggenheim Foundation Fellow.

to a radial temperature gradient that is sufficiently in excess of adiabatic then it will become convectively unstable and be sustained in its motion by the continuous conversion of gravitational potential energy to kinetic energy that is effected by the hydrodynamic buoyancy force. The observed spreading of the seafloor away from mid-oceanic ridges is understood to constitute a surface manifestation of such deep-seated mantle convection as is the horizontal convergence of material which is occurring in deep oceanic trenches. This and other aspects of the pattern of surface motions associated with convection has been described within the framework of the set of ideas that has come to be called "plate tectonics" (e.g., Morgan, 1968; McKenzie and Parker, 1967). In this picture the surface of the Earth is seen as being divided into a relatively small number of quasi-rigid patches called "plates" which interact with each other only at their boundaries and in so doing conspire to produce all of the dramatic events which are responsible for continuously modifying the planetary surface (e.g., earthquake activity, volcanic activity, mountain building, etc.). Continents are often believed to comprise large scale chemical heterogeneities within the plates which are simply carried passively by them over the Earth's surface. Although the practitioners of plate tectonics do concern themselves with the issue of the origin of the forces which "drive" the plates (e.g., Forsyth and Uyeda, 1975), their literature has tended to address this issue by attempting to discover motive power in forces labelled "ridge-push" and "trench-pull" that are derivative of the plates themselves rather than arising, as is the case in a fluid whose motion is sustained by thermal convection, in the buoyancy forces focussed in the hot upwelling and cold downwelling plumes in which lateral heterogeneity of density is concentrated.

Our goal in this book is to describe the way in which various avenues of investigation have begun to produce a reasonably coherent modern theory of the thermal convection process which does in fact explain not only the surface motions evident in plate tectonics but also the structure of the planet's gravitational field and many of its other large scale characteristics. Although this dynamical theory remains incomplete in many crucial respects, the observational database which it is being developed to explain has very recently been enormously expanded by the application of various techniques of seismic tomography to "image" the lateral heterogeneity of density throughout the volume of the mantle. These direct, though somewhat "blurred," images of the convective circulation promise to revolutionize our understanding of mantle dynamics in as radical a fashion as the seafloor observations of the 1960s revolutionized our view of surface kinematics. The time, therefore, is propitious, we believe, to compile in one place the main experimental and theoretical results upon which our present understanding of the dynamical state of the Earth's mantle is based. However, because considerable controversy remains on many of the most fundamental issues, we have elected to attempt to produce this compilation in the form of a series of essays (which comprise the chapters of this book) by a group of scientists who have each been especially active in this area of geophysical research, and who have been selected to represent most of the major subdisciplines whose unique contributions will be required to produce a truly satisfactory understanding of the mantle convection process. In the following subsection of this introductory chapter I have attempted to provide a brief summary of the most important observational results upon which our present understanding is based. A brief précis of this book which attempts both to consolidate and to extend this understanding, will be found in Section 3.

2. Evidence for a Mantle General Circulation

Although even the earliest geographers were impressed with the complimentary nature of the main features of the coastlines of Africa and South America, it remained for Bullard *et al.* (1965) to describe the goodness of fit statistically and this was done subsequent to the advance of Hess' hypothesis of seafloor spreading. Figure 1.1, from Dietz and Holden (1970), illustrates the original super-continent of Pangea which is assumed to have existed prior to its breakup, subsequent to which the individual continental fragments were dispersed to their present locations. That the present arrangement of continental blocks may indeed have been effected by the breakup of Pangea was first established using paleomagnetic methods by S.K. Runcorn (1956) and E. Irving (1956). Reconstruction of the paleo pole-paths for Europe and for North America showed that they were coincident until the Triassic period (about 200 million years ago), after which they diverged, an event interpreted as coinciding with continental breakup.

2.1 Paleomagnetic Evidence

Paleomagnetic methods have in fact played a dominant role in establishing the validity of the spreading hypothesis. In 1963, Laurence Morely [unpublished, but see his historical account of the attempt at publication (1987)] and Vine and Mathews (1963), in the course of studies of the pattern of magnetic anomalies over an ocean ridge, independently realized that the patterns observed could be understood in terms of Hess' theory of seafloor

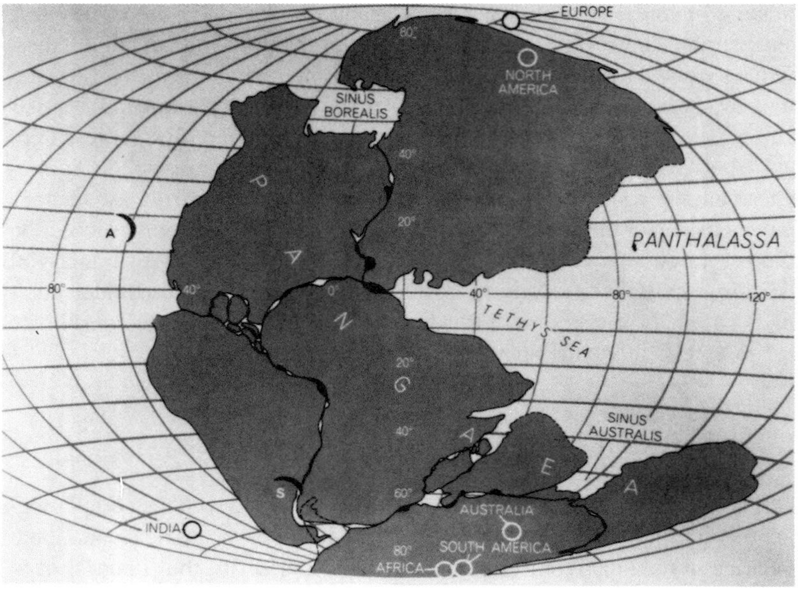

Figure 1.1. Reconstruction of the supercontinent Pangea as it may have looked 200 million years ago. The relative positions of the continents are based upon numerical best fits (from Dietz and Holden, 1970).

spreading. An example of the characteristic magnetic anomaly patterns which obtain over oceanic ridges is shown on Figure 1.2 which is from the classic paper of Hertzler *et al.* (1966) who were then working at the Lamont-Doherty Geological Observatory of Columbia University, whose scientists played such an important role in this crucial era of seafloor exploration. Morely and Vine and Mathews proposed that such characteristic and beautifully symmetric patterns were produced when hot material, rising from the mantle beneath the ridge crest, was cooled below the Curie point, acquiring in the process an induced magnetization in the local Earth's magnetic field. Since the Earth's predominantly dipolar magnetic field undergoes essentially random reversals of polarity through geological time, the material reaching the surface is "normally" or "reversely" magnetized depending upon the polarity of the field at the time it is cooled below the Curie point. As this material spreads in either direction away form the ridge axis the symmetric pattern of magnetic anomalies is generated. The analogy has often been

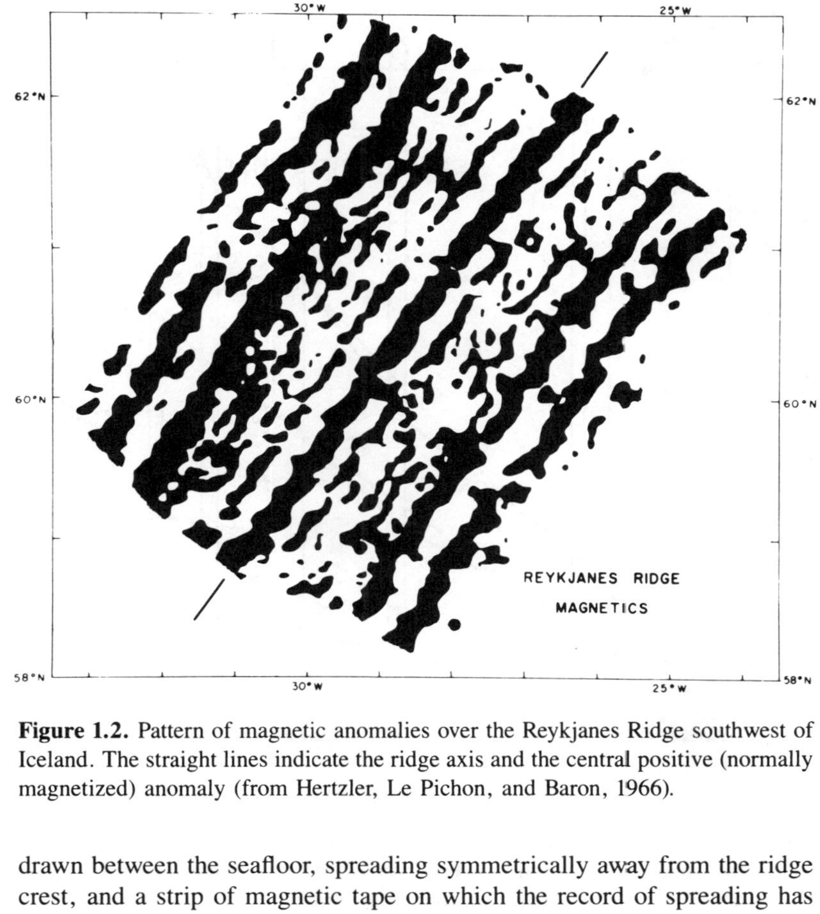

Figure 1.2. Pattern of magnetic anomalies over the Reykjanes Ridge southwest of Iceland. The straight lines indicate the ridge axis and the central positive (normally magnetized) anomaly (from Hertzler, Le Pichon, and Baron, 1966).

drawn between the seafloor, spreading symmetrically away from the ridge crest, and a strip of magnetic tape on which the record of spreading has conveniently been recorded for our subsequent reading.

Since the age of each of the alternate reversed and normal magnetic anomalies which symetrically parallel the ridge axis may be determined rather accurately by appeal to the timescale of geomagnetic reversals established on land, it is possible to deduce the rate of spreading of the seafloor which is characteristic of each segment of the globe encircling pattern of ridges. The establishment of the geomagnetic reversal timescale through analyses of the land based record from volcanically active regions by Cox *et al.* (e.g., 1967) in fact constituted a fundamental cornerstone of the logical ediface that grew into the modern "theory" of plate tectonics. Figure 1.3, from the paper by Vine (1966), illustrates examples of spreading rates

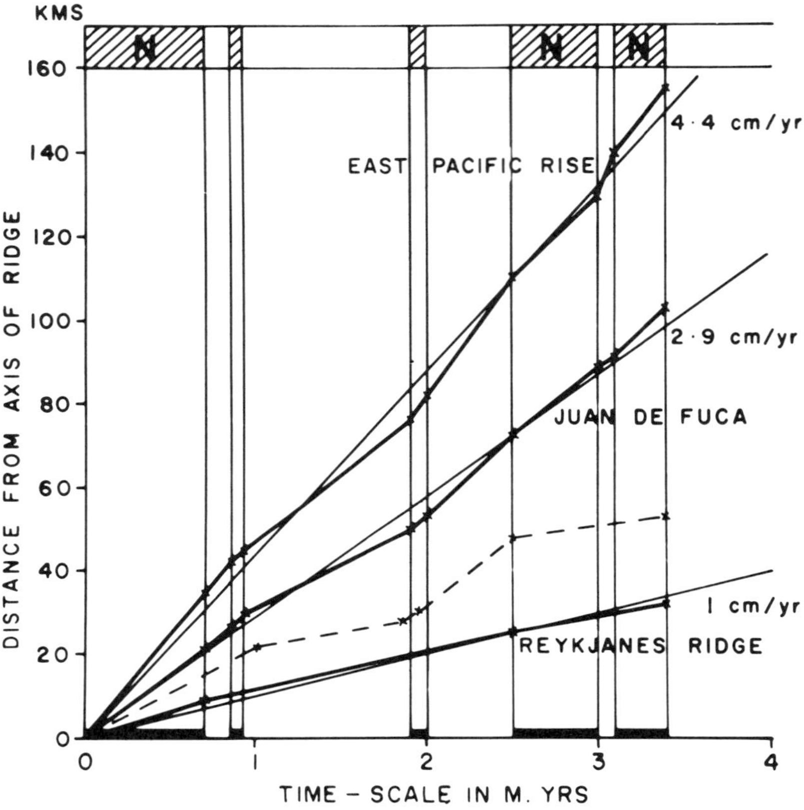

Figure 1.3. Inferred normal and reversed boundaries in the oceanic crust plotted against the time scale of geomagnetic reversals for various spreading ridges. The slopes are the local spreading rates (from Vine, 1966).

inferred in this manner for three different mid-oceanic ridge segments which range from 1 cm/yr for the Reykjanes Ridge segment of Figure 1.2 to 4.4 cm/yr for the East Pacific Rise. The spreading rate for the Juan de Fuca Ridge of the west coast of Canada is an intermediate one of 2.9 cm/yr.

2.2 Seismic Evidence

Just as the methods of paleomagnetism were crucial in establishing that hot mantle-derived material is continuously rising under and spreading horizontally away from the axes of mid-oceanic ridges, so the methods of seismology proved crucial in establishing the existence of the required complimentary regions in which cold material was elsewhere sinking back into the interior in the fashion characteristic of a thermally forced circulation. Prior to the advance of the seafloor spreading hypothesis, Benioff (1954) had in fact suggested that the previously identified deep ocean trenches were regions where old-cold ocean-floor was being down-thrust into the mantle along what are now called Benioff zones which are defined by the internal locus of deep seismicity. Figure 1.4 shows Benioff's original data for the Kurile-Kamchatka region illustrating the geometry of the process that is now referred to as the "subduction" of oceanic lithosphere. In fact it remained until the occurrence of the Alaska earthquake of 1964 before truly convincing evidence of the validity of Benioff's suggestion was collected. This evidence was presented in important papers by Plafker (1965) and by Stauder and Bollinger (1966). The analysis of the earthquake source mechanism as a function of depth along Benioff zones using first motion data (e.g., Isacks and Molnar, 1971) has played an important role in the development of mantle convection models. The cessation of seismicity below a depth of 650–670 km, where a discontinuity of the seismic parameters occurs, and the compressive nature of the deep focal mechanisms, have been interpreted as implying that the cold oceanic slab does not penetrate below this depth. As will be discussed in detail in Chapter 3 of this book, however, this idea has not stood the test of time, thus undermining what has been one of the main lines of observational evidence that has been invoked to support the notion that the convective circulation of the upper mantle does not penetrate the lower.

On the basis of global distributions of seismicity such as discussed by Isaacs *et al.* (1968), it was recognized that the system of ridges and trenches, and their important interconnections via the transform faults discovered by J. Tuzo Wilson (1965), was continuous over the Earth's surface. Sinuous features comprising these three types were seen to divide the surface of the Earth into a relatively small number of "plates," the interiors of which were relatively devoid of seismicity. One example of a fairly recently compiled map of global seismicity is shown in Figure 1.5. Figure 1.6 shows a corresponding map of plate boundaries on which is also superimposed symbols denoting the various geological processes which are characteristic of each boundary segment. The idea of "plate tectonics" is basically to describe these geological processes, which include mountain building, volcanism, and seismicity as consequences of the interactions among this relatively small number (12) of plates. It is supposed in the context of this kinematic

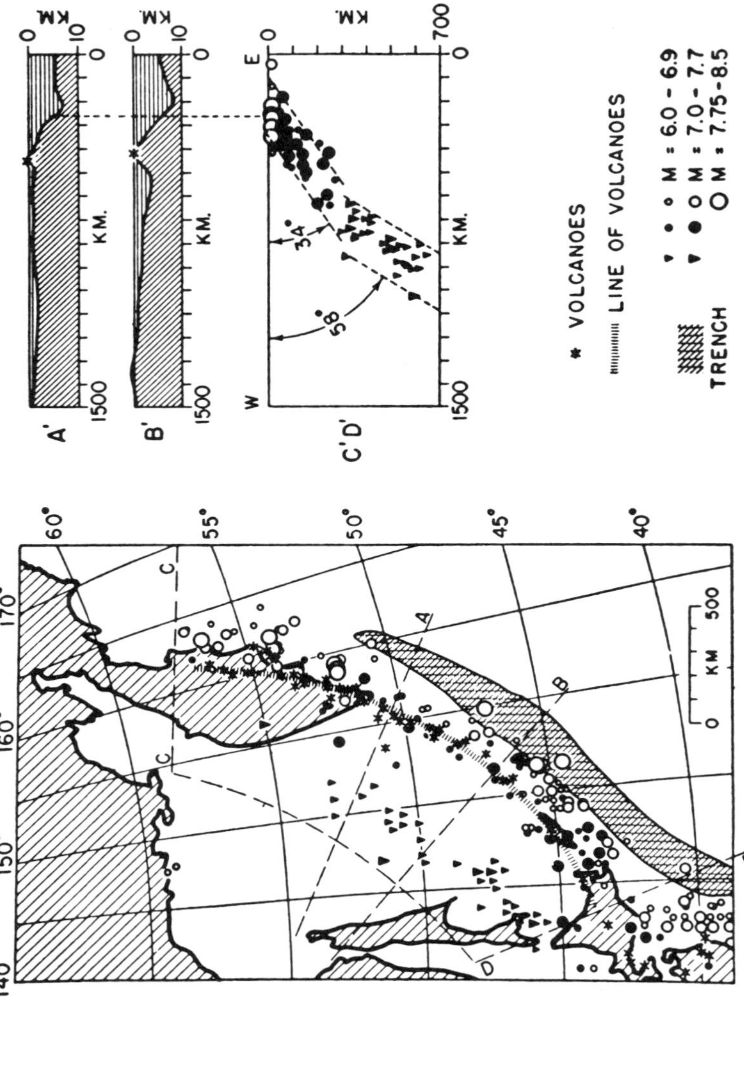

Figure 1.4. Location map and composite depth cross section of Kurile-Kamchatka earthquake sequences (from Benioff, 1954).

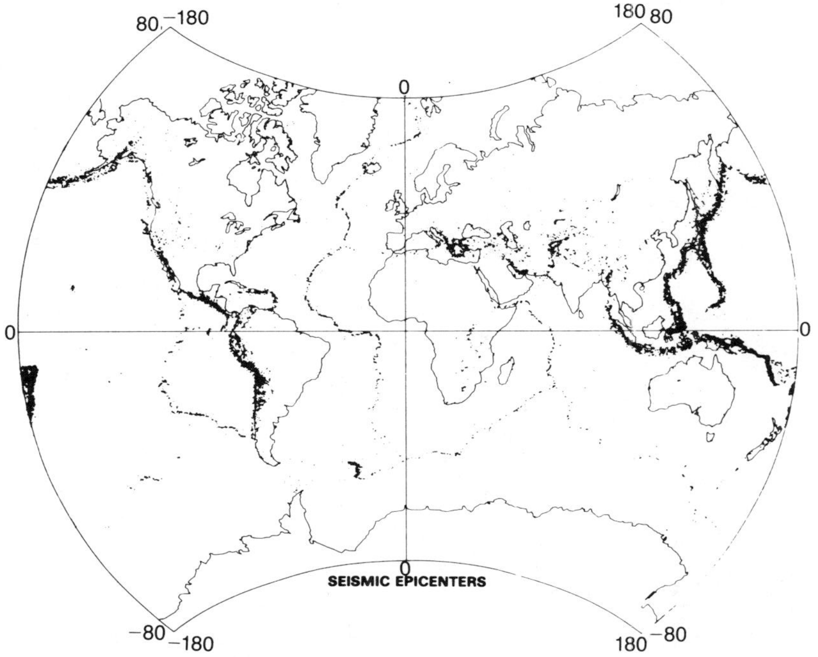

Figure 1.5. Global map of seismic epicenters for all earthquakes in the magnitude range 4.5–5.5 with depths 0–700 km. Note the correlation between seismicity and plate boundaries. (Based upon NOAA data tape, Van der Grinten projection, Goddard Space Flight Center, 1978).

theory that all interactions between the plates occur along the boundaries which separate them, which each have one of the three characteristic forms: spreading ridges, deep ocean trenches (subduction zones), and transform faults. Within the context of this "theory" neither the existence of the plates nor the reason for their relative motion is explained. Although the approximately 1:1 relation between convection cells and surface plates envisioned by Hess would appear to provide a natural explanation for both the existence of the plates and their relative motion, this is not a view that is universally accepted.

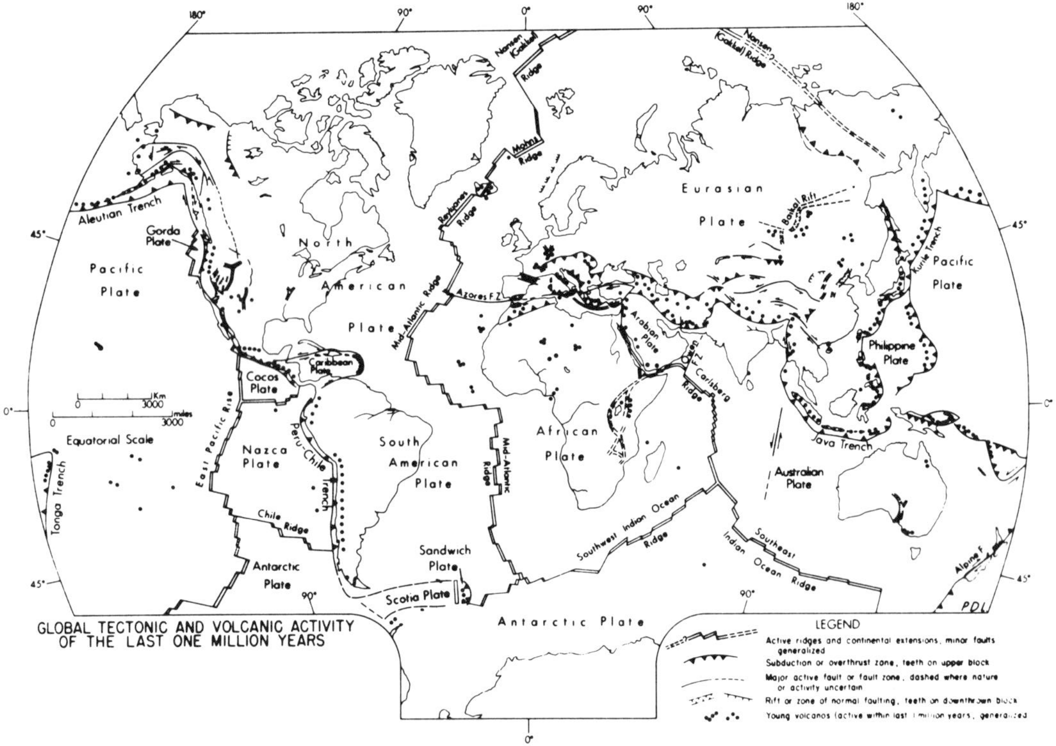

Figure 1.6. Global map of plate boundaries and of tectonic and volcanic activity. (Goddard Space Flight Center, 1978, based upon The Physical World,® 1975 by the National Geographic Society, Van der Grinten projection, P.D. Lowman Jr.).

2.3 Thermal and Gravitational Evidence

Besides this information from paleomagnetism and seismology, which has been crucial to the development of the ideas of plate tectonics, there are other geophysical data which are proving equally useful in constraining the new generation of mantle convection models which are now being developed to provide a dynamical explanation of these kinematic observations. One of the most interesting of these datasets consists of the observed variations of heat flow and oceanic bathymetry as a function of ocean floor age away from mid-oceanic ridges. Examples of such data from Parsons and Sclater (1977) are shown in Figures 1.7 and 1.8 for bathymetry and heat flow respectively. When plotted as a function of the square root of the age of the lithosphere, both data fall essentially on straight lines although there is a systematic departure form this characteristic behavior, particularly for bathymetry, for ages in excess of about 70 million years. McKenzie and Sclater (1968) showed that the characteristic $t^{1/2}$ variation was expected on the basis of a simple model consisting of a plate (the lithosphere), of constant thickness, spreading at a uniform rate and subject to constant temperature at its end (the ridge crest). In their analysis no justification was provided for the assumption of constant lithospheric thickness and no explicit hypothesis was made regarding the reason for the spreading. Such partial analyses are in fact characteristic of much of the earliest work that sought to address the origin of the surface motions characteristic of plate tectonics. A standard assumption in this work has been that mid-ocean ridges are passive in the sense that they occur simply where the plates decide to fracture under the influence of a large scale stress field originating "elsewhere." As the plates are pulled apart by the action of this stress field, hot material rises to fill the void. The alternative assumption, and one which is born out by the results from seismic tomographic imaging of the Earth's upper mantle (see Chapters 9 and 10) is that ridge crests are the locations at which impact occurs between hot upwelling thermal plumes and the Earth's surface. In this alternative view the oceanic ridges are clearly active features which define the surface traces of the locations of hot upwelling, just as the oceanic trenches define the complementary regions of surface downwelling. That this alternative view is fully compatible with bathymetry and heat flow data was first established by Jarvis and Peltier (1982) who employed a simple model of high Rayleigh number convection in a box to show that the observed square root of age variation was entirely explicable if the ridges were associated with the surface stagnation points of rising thermal plumes. Their analysis, which is reviewed in detail in Chapter 7 of this book, also demonstrated that if the oceanic lithosphere were assumed to constitute the cold thermal boundary layer of the large scale

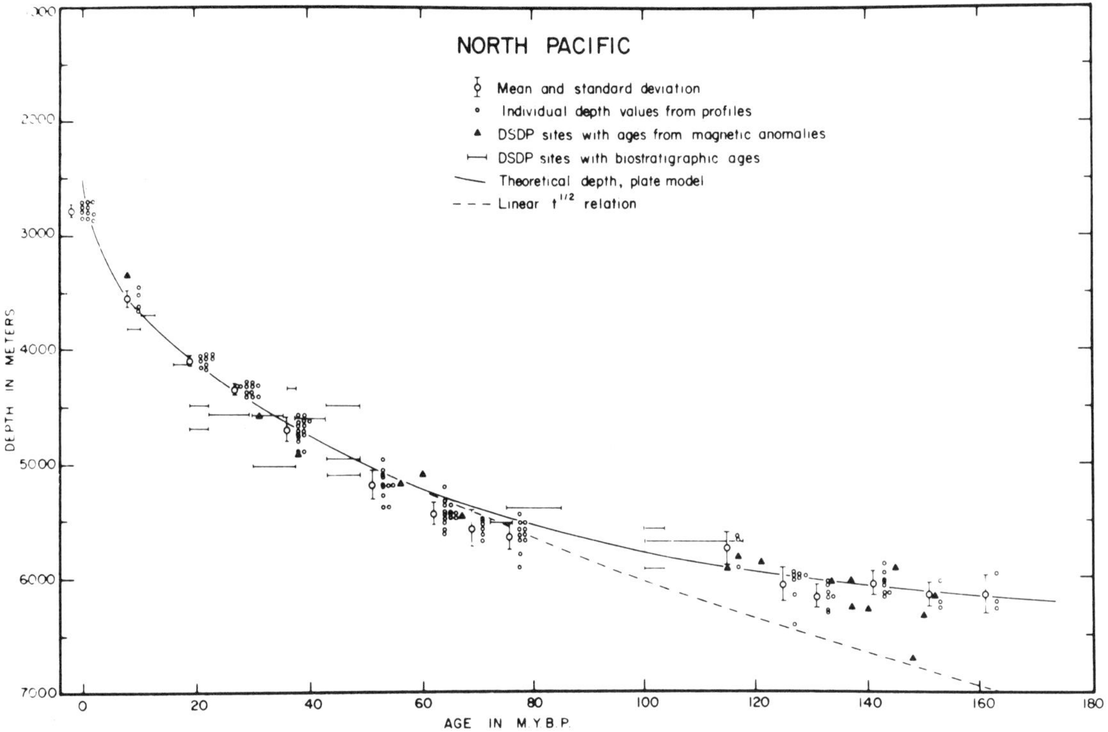

Figure 1.7. Plot of individual depth measurements used in calculating mean depths and depths at DSDP sites for the North Pacific (from Parsons and Sclater, 1977): I: mean and standard deviation; ○ individual depth values from profiles; Δ DSDP sites with ages from magnetic anomalies; ⊢—⊣: DSDP sites with biostratigraphic ages; ———: theoretical depth, plate model; – – –: linear $t^{1/2}$ relation.

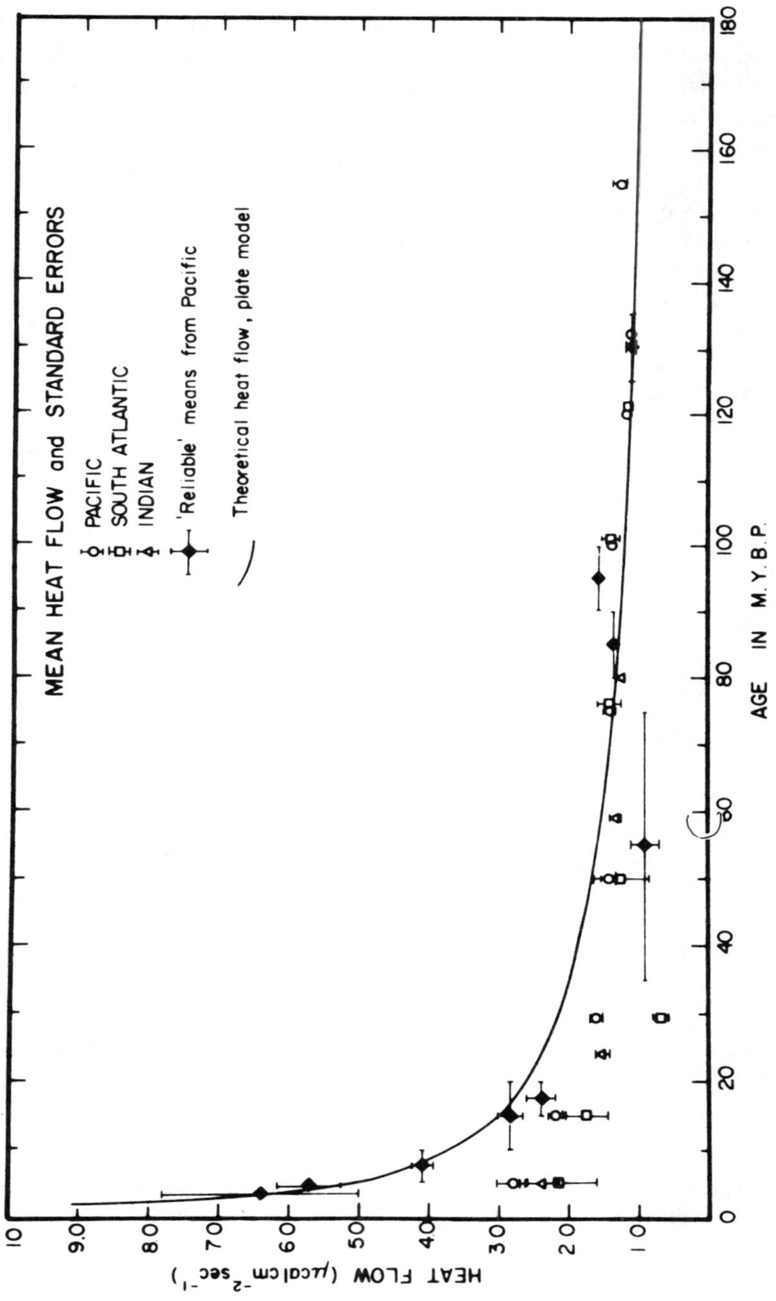

Figure 1.8. Heat flow as a function of age of the ocean floor indicating decay as $t^{1/2}$ (from Parsons and Sclater 1977): ○, Pacific; □, South Atlantic; Δ, Indian; ◆: "reliable" means from Pacific; ——: theoretical heat flow, plate model.

convective circulation in the mantle, then the observed variations of heat flow and bathymetry could be employed to constrain the extent to which the surface heat flow could be due to internal radioactive heating of the mantle as opposed to heat loss from the core.

In the vicinity of individual oceanic trenches the surface gravitational acceleration is observed to exhibit variations which are as characteristic of these regions of cold downwelling as heat flow data are of regions of hot upwelling. Perhaps the most striking feature in this field is observed in the form of an anomaly in vertical acceleration along a profile normal to the axis of the trench. Figure 1.9 illustrates such a profile across the Tonga Trench which shows data from Talwani *et al.* (1961) upon which are superimposed the results of model calculations due to Griggs (1972). The modelling demonstrates that the characteristic form of the observed anomaly cannot be fitted by topography alone, but is easily explained by the existence beneath the trench of a slab of cold high density material dipping at an angle equal to

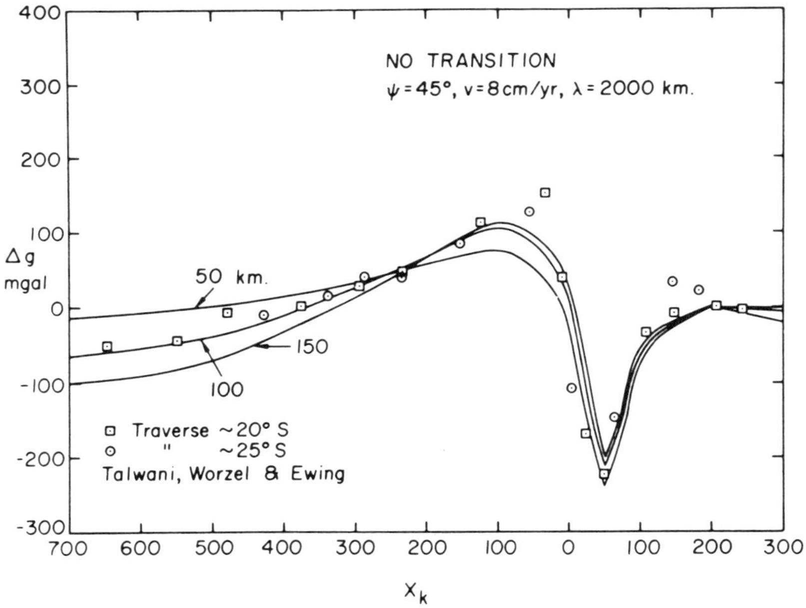

Figure 1.9. Gravity anomaly in milligals as a function of distance from the trench axis. A model with a dipping cold slab under the trench fits the data (from Griggs, 1972); no transition, $\lambda = 45°$, $v = 8$ cm/y, $\lambda = 2000$ km, □ traverse 20°S, ○ transverse 25°S.

that of the local seismic Benioff zone. Models such as these were introduced originally by McKenzie (1969) and Minear and Toksöz (1970). The form of the gravity anomaly over the Tonga trench is similar to those found in other locations. In contrast to their success in reconciling heat flow and bathymetry data from the vicinity of ridge crests, simple convection models have yet to adequately explain the gravity anomalies in the vicinity of trenches. Some discussion of this issue will be found in Chapters 9 and 10 of this book.

Even on a global scale the structures of the surface heat flow and gravitational fields have many features which show them to be strongly linked to surface plate kinematics and thus to the mantle convection process. Figure 1.10 illustrates the first of these two respective fields in terms of the GEM 10B model of geoid heights consisting of spherical harmonic constituents up to degree and order 32 of Lerch *et al.* (1979). The field shown has the dynamically inferred flattening of Nakiboglu (1982) removed. Similarly Figure 1.10b illustrates the global model of surface heat flow of Chapman and Pollack (1980) which consists of spherical harmonic constituents up to degree and order 18. The main features observable in the large scale gravitational field are the pronounced geoid highs associated with the subduction zones, whereas the main features revealed by the heat flow map are the extraordinary high heat flows associated with the two fastest spreading ridges on the Earth's surface, namely the East Pacific Rise off the west coast of South America, and the Indian Ridge which cuts through the center of the Indian Ocean and south of Australia. These global fields are clearly central to the mantle convection process and will have to be explained in detail by any successful model.

3. Conclusions and Discussion of the Outline of This Book

The brief review above should serve to adequately motivate the separate and more detailed contributions which form the individual chapters to follow. The reader interested in a more complete account of the geophysical analyses which were crucial to establishing the plate tectonics paradigm will find the compilation of original papers by Cox (1973) very useful. The focus of the ensuing chapters is upon the problem of constructing the dynamical theory which must eventually replace the descriptive kinematic theory which is embodied in the ideas of plate tectonics. All that we have learned in the process of constructing this kinematic theory is most succinctly presented in Figure 1.11 on which is shown the global pattern of surface motions accord-

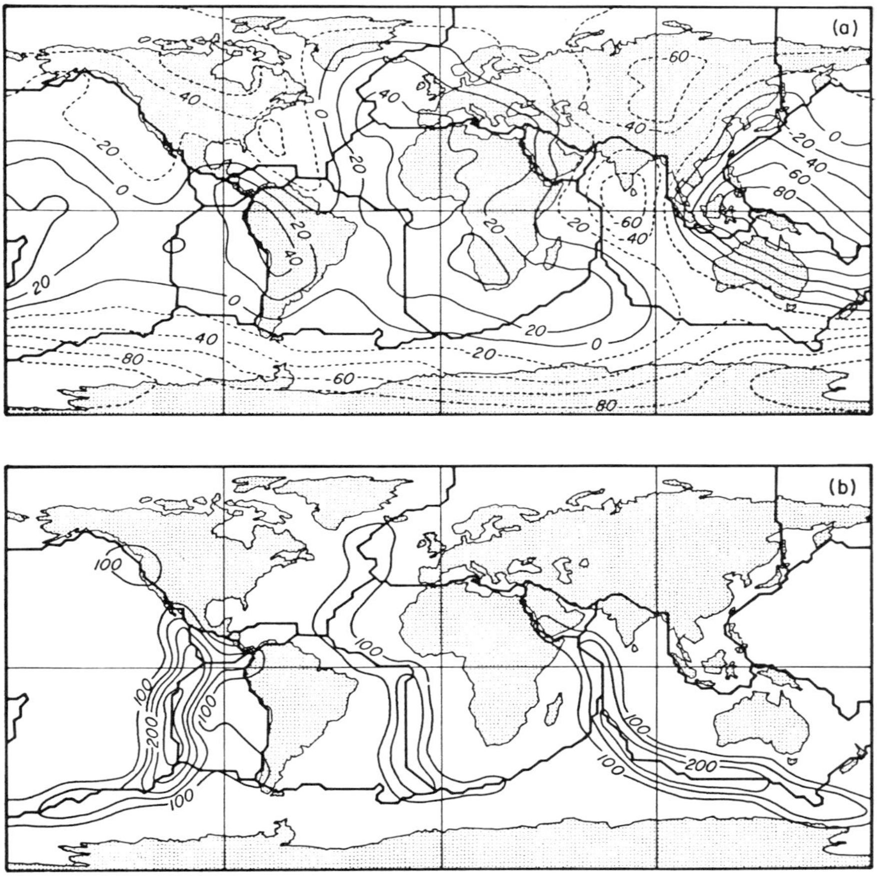

Figure 1.10. a, Geoid height anomaly in meters from the GEM 10B model of Lerch *et al.* (1979) corrected by removal of the dynamically inferred flattening of Nakiboglu (1980). Solid contours are positive anomalies, dashed contours are negative anomalies. Contour interval is 20 m. b, Surface heat flow map of Pollack and Chapman (1981). Contours in mill-watts per square meter with contour interval of 50 mWm^{-2}.

ing to the paleomagnetically and seismically constrained reconstruction of Minster and Jordan (1978). This absolute motion model has been constrained by the assumption that surface "hot spots" like Hawaii define an absolute frame of reference. What we have yet to learn is precisely how this pattern of surface motions is created, through what source of energy it is maintained

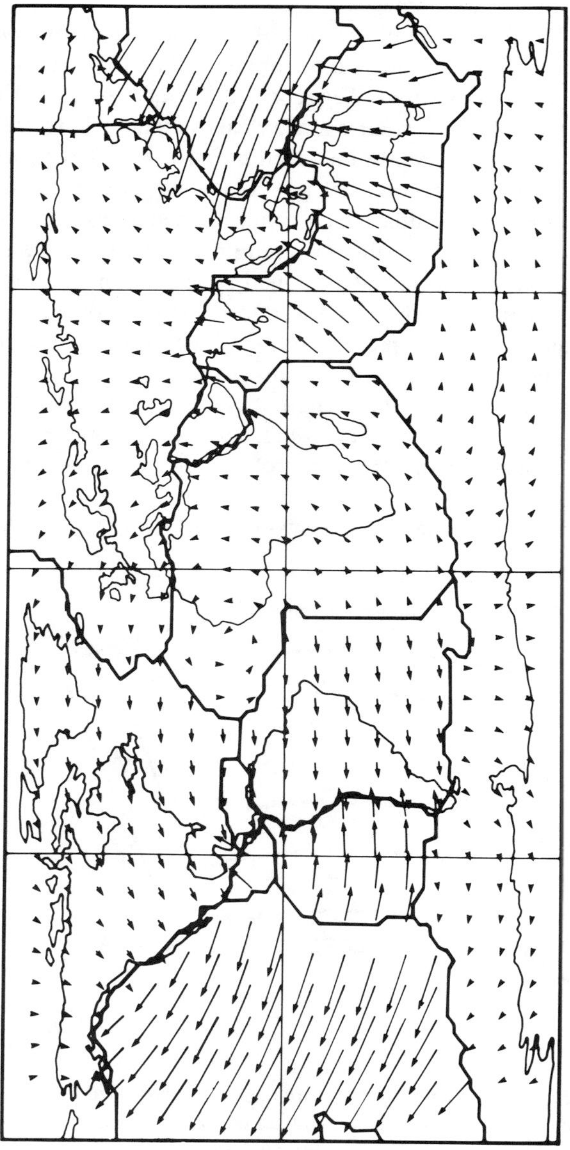

Figure 1.11. Surface plate velocities from the reconstruction of Minster and Jordan (1978) relative to the "hot spot" frame. Plate boundaries are shown as heavy solid lines.

against frictional dissipation, and how the surface motions are coupled to motions in the planets' deep interior.

In the view of Hess (1962) the connection between the surface motions and those in the Earth's deep interior could be simply understood in terms of the convection hypothesis of Alfred Weggener (1929) and Aurther Holmes (1931). Hess view of this connection is made clear in Figure 1.12 (from Hess, 1962) in which he has apparently imagined an intimate relation between ridges and hot upwellings and trenches and cold downwellings and that the circulation in the Earth's mantle which connects them consists of simple convection cells, each of which fills the entire region between the Earth's surface and the core-mantle boundary. One of the main lines of enquiry which links the chapters of this book together is the issue of the extent to which this simple view may provide an adequate reconciliation of the previously described observations. For many historical reasons, which are discussed at length hereafter, this simple view has, until recently, been considered untenable. That it is a view which deserves to be reconsidered is quite strongly suggested by the many lines of evidence which are discussed in what follows.

The book comprises ten main chapters which are themselves divided into

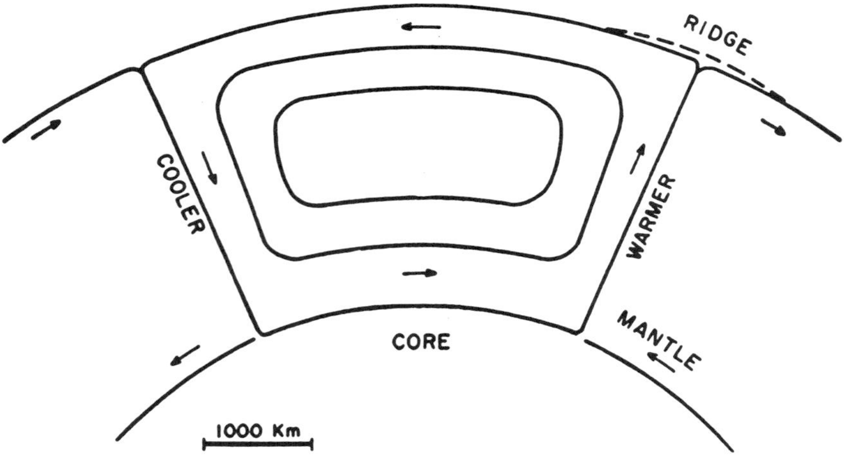

Figure 1.12. Sketch of the mantle general circulation as envisioned by Hess (1962). Convection currents fill the entire mantle and are of the "whole mantle" style with cold downwellings and hot upwellings extending from the core-mantle boundary to the Earth's surface.

two main sections. In the first of these sections, following an introduction to the basic concepts of thermal convection in Chapter 2, is contained analyses of the physical state of the Earth's mantle from four different perspectives. Chapter 3 is devoted to a review of recent results from seismology which is focussed squarely upon the issue of what can be said about the plausibility of "whole mantle convection" (the simple view of Hess) on the basis of seismic travel time data from earthquakes located in the vicinity of downgoing slabs. These data appear to require that the anomalously cold material of the slabs penetrates to a depth well below the 670 km discontinuity and thus support the whole mantle convection scenario. In Chapter 4 the physical and chemical states of the mantle are discussed from the point of view of the high pressure physics and chemistry of the dominant mineral phases of which the Earth's mantle is constituted. This analysis demonstrates that, although the 420 km and 670 km seismic discontinuities which bracket the mantle transition zone may be simply explicable in terms of equilibrium phase transformations, some difficulty appears to remain in explaining why they are as sharp as seismology requires them to be, a result which clearly leaves open the issue as to whether some chemical contribution to the discontinuities may be required. Of course, if such chemical contrast were required, then Hess' simple view of the circulation would be incorrect as convection which was of the whole mantle style would so mix the system that no such chemically distinct horizon(s) could be maintained. Chapter 5 is devoted to a detailed reappraisal of the extent to which geochemical data may be brought to bear upon the issues of mantle mixing and crustal growth, particularly using the Rb-Sr and Nd-Sm isotopic systems. An important conclusion which appears to follow from this review is that the early interpretations of these isotopic data, as requiring the existance of chemically distinct mantle reservoirs with the volumes of depleted and pristine material very nearly equal to the volumes of the upper and lower mantle, are rather more ambiguous than the initial analyses suggested, the result being that the data do not in fact provide strong (if any) support for a "layered mantle" hypothesis (the antithesis of the Hess view). The final chapter in Section 1, Chapter 6, considers the mantle from the point of view of its rheology through examination and interpretation of the various signatures of the process of glacial isostatic adjustment. These data strongly suggest that the Earth's mantle is very nearly isoviscous and therefore most probably in a state of vigorous whole mantle convection of the sort envisioned by Hess.

The second section of this book comprises Chapters 7 through 11 and consists of material explicitly dealing with hydrodynamic modeling of the mantle convection process. Beginning with the simplest possible isochemical and isoviscous models of mantle convection, Chapter 7 provides a very detailed discussion of the extent to which specific geophysical observations may be reconciled in this simplest possible dynamical scenario. In Chapter 8 considerable physical complexity is added to the two dimensional plane layer geometry of the models of Chapter 7, physical complexity in the form of phase transitions, chemical discontinuities, strong pressure and temperature variations of viscosity, and non-Newtonian behavior. These analyses all have a strong bearing upon the whole mantle convection–layered mantle convection debate. Chapter 9 provides a first glimpse of the way in which the new results from global seismic tomographic imaging are beginning to revolutionize work on models of mantle convection. In this chapter a detailed discussion is provided of the construction of one such "tomograph" and the depth dependent lateral heterogeneity of mantle density which can be inferred from it is employed in the context of a simple, yet three dimensional and spherical, mantle convection model to analyze the extent to which such models are able to reconcile global geoid height anomalies such as those illustrated in Figure 1.10a. Chapter 10 describes further work based upon similar tomographic models which addresses the issue of whether the surface velocity field itself (shown on Figure 1.11) may also be explicable in terms of similarly first order convection models. The result of this analysis will be construed as the first successful direct test of the validity of the convection hypothesis itself since we shall show that the new seismic data enable a reasonable first order prediction of the poloidal component of the surface velocities. Also contained in this chapter is an initial discussion of results from a new spherical axi-symmetric model of high Rayleigh number mantle convection which reveals some novel features of the circulation which are due specifically to its spherical geometry. The final chapter in the book, Chapter 11, addresses the issue of the impact which mantle convection must have had upon the thermal evolution of the Earth (and other terrestrial planets) and the crucial role which it must thereby play in the energetics of the process whereby the Earth's magnetic field is generated through magnetohydrodynamic processes in the core.

4. References

Benioff, H., 1954, Orogenesis and deep crustal structure: Additional evidence from seismology. *Bull. Geol. Soc. Am.*, **65**, 385–400.

Bullard, E.C., J.E. Everett, and A.G. Smith, 1965, The fit of the continents around the Atlantic. *Philos. Trans R. Soc. London, Ser. A.* **258**, 41–51.

Chapman, D.S., and H.N. Pollack, 1980, Global heat flow: Degree 18 spherical harmonic representation. *EOS Trans. AGU*, **61**, 683.
Cox, A., R.R. Doell, and G.B. Dalrymple, 1963, Geomagnetic polarity epochs and Pleistocene geochronometry. *Nature*, **198**, 1049–1051.
Cox, A., 1969, Geomagnetic reversals. *Science*, **163**, 237–245.
Cox, A., 1973, *Plate Tectonics and Geomagnetic Reversals*. San Francisco: W.H. Freeman and Company.
Dietz, R. S., 1961, Continent and ocean basin evolution by spreading of the sea floor. *Nature*, **190**, 854–857.
Dietz, R.S., and J.C. Holden, 1970, Reconstruction of Pangea: Breakup and dispersion of continents, Permian to present. *J. Geophys. Res.*, **75**, 4939–4955.
Forsyth, D.W., and S. Uyeda, 1975, On the relative importance of the driving forces of plate motion. *Geophys. J. R. Astron. Soc.*, **43**, 163–200.
Griggs, D.T., 1972, The sinking lithosphere and the focal mechanisms of deep earthquakes. In Robertson, E.C. (ed.), *The Nature of the Solid Earth*. New York: McGraw-Hill, pp. 361–384.
Hess, H.H., 1962, History of ocean basins. In A.E.J. Engel, H.L. James, and B.F. Leonard (eds.), *Petrologic Studies: A Volume in Honour of A.F. Buddington*. New York: Geol. Soc. Amer., pp. 559–620.
Holmes, A., 1931, Radioactivity and Earth movements. *Trans. Geol. Soc. Glasgow,* **18** (Part 3), 559–606.
Irving, E., 1956, Paleomagnetic and paleoclimatological aspects of polar wandering. *Pure Appl. Geophys.,* **33,** 23–41.
Isacks, B., and P. Molnar, 1971, Distribution of stresses in the descending lithosphere from a global survey of focal-mechanism solutions of mantle earthquakes. *Rev. Geophys. Space Phys.,* **9**, 103–174.
Isacks, B., J. Oliver, and L. R. Sykes, 1968, Seismology and the new global tectonics. *J. Geophys. Res.,* **73**, 5855–5899.
Jarvis, G.T., and W.R. Peltier, 1982, Mantle convection as a boundary layer phenomenon. *Geophys. J. R. Astron. Soc.,* **68**, 389–427.
Lerch, F.J., S.M. Klosko, R.E. Laubscher, and C.A. Wagner, 1979, Gravity model improvement using GEOS 3 (GEM 9 and 10). *J. Geophys. Res.,* **84**, 3897–3916.
McKenzie, D.P., 1969, The relation between fault plane solutions for earthquakes and the directions of the principle stresses. *Seismol. Soc. Amer. Bull.,* **59**, 591–601.
McKenzie, D.P., and W.J. Morgan, 1969, Evolution of triple junctions. *Nature,* **224**, 125–133.
McKenzie, D.P., and R.L. Parker, 1967, The North Pacific: An example of tectonics on a sphere. *Nature,* **216**, 1276–1280.
McKenzie, D.P., and J.G. Sclater, 1968, Heat flow inside the island arcs of the northwestern Pacific. *J. Geophys. Res.,* **73**, 3173–3179.
Minear, J.W., and M.N. Toksöz, 1970, Thermal regime of a downgoing slab and new global tectonics. *J. Geophys. Res.,* **75**, 1397–1419.
Minster, J.B., and T.H. Jordan, 1978, Present day plate motions. *J. Geophys. Res.,* **83**, 5331–5354.

Morely, L.W., 1986, Early work leading to the explanation of the banded geomagnetic imprinting of the ocean floor. *EOS Transactions AGU,* **36**, 665–666.

Morgan, W.J., 1968, Rises, trenches, great faults and crustal blocks. *J. Geophys. Res.,* **73**, 1959–1982.

Nakiboglu, S.M., 1982, Hydrostatic theory of the Earth and its mechanical implications. *Phys. Earth Planet Int.,* **28**, 302–311.

Parsons, B., and J.G. Sclater, 1977, An analysis of the variation of ocean floor bathymetry and heat flow with age. *J. Geophys. Res.,* **82**, 803–827.

Plafker, G. 1965, Tectonic deformation associated with the 1964 Alaskan earthquake. *Science,* **148**, 1675–1687.

Runcorn, S.K., 1956, Paleomagnetic comparisons between Europe and North America. *Proc. Geol. Soc. Can.,* **8**, 77–85.

Stauder, W., and G.A. Bollinger, 1965, The S-wave project for focal mechanism studies—earthquakes of 1963. Air Force Off. Sci. Res., Grant AF-AFOSR 62-458, Rep.

Talwani, M., 1961, A review of marine geophysics. *Mar. Geol.,* **2**, 29–80.

Vine, F.J., 1966, Spreading of the ocean floor: New evidence. *Science,* **154**, 1405–1415.

Vine, F.J., and D.H. Mathews, 1963, Magnetic anomalies over oceanic ridges. *Nature,* **199**, 947–949.

Wegener, A., 1929, *Die Enstehung der Kontinente und Ozeane*. Braunschweig: Vieweg. 4th Ed. Transl., 1929, as *The Origin of Continents and Oceans*. London: Methuen. Reprinted, 1966, by Dover.

Wilson, J.T., 1965, A new class of faults and their bearing on continental drift. *Nature,* **207**, 343–347.

CHAPTER 2

Fundamentals of Thermal Convection

F.H. Busse

Institute of Geophysics and Planetary Physics
University of California at Los Angeles, Los Angeles, California

1. Introduction

By thermal convection we are referring to the motions induced by thermal buoyancy in a layer of fluid heated from below or heated by internal heat sources. Thermal convection is a ubiquitous phenomenon in nature which occurs on a large variety of scales from centimeter sized convection cells in a puddle of water cooled by evaporation to the giant cells of the sun extending over several 10^5 km. One of the most fascinating convection processes occurs in the solid parts of some planets including the Earth. Although solid material does not ordinarily move like a fluid over large scales in space and in time, fluid like motions do indeed occur in the Earth's mantle. The by now well documented phenomenon of sea floor spreading and continental drift is the evidence for such motions at the Earth's surface.

Apart from its geophysical and astrophysical applications, convection in a fluid layer heated from below has received much attention in recent years because it represents the simplest form of hydrodynamic instability. As long as the horizontal extent of the layer is sufficiently large compared to its vertical dimension the physical conditions are homogeneous and isotropic with respect to the horizontal directions. Because of this property and because of the simple source of energy, thermal convection has become a favored subject for the study of the nonlinear evolution of instabilities and the transition to turbulent fluid flow. The relative simplicity of buoyancy induced motion seems to be reflected even in the turbulent state. The examples of mesoscale convection in the Earth's atmosphere or of solar granulation and supergranulation exhibit an amount of regularity and structure which is surprising in view of the huge Reynolds and Rayleigh numbers associated with those phenomena. Convection is thus a prime example for the coherent structures in turbulent fluids which have been attracting much attention in recent years.

The outline of a theory of convection given in this chapter does not include a discussion of the fully developed state of turbulent convection. Because of the high Prandtl number and the relatively modest Rayleigh numbers characterizing convection in the Earth's mantle, mixing length theory and other semi-empirical descriptions of turbulent convection do not seem to be particularly useful in the context of mantle convection. The outline will instead emphasize results which can be derived rigorously from basic equations and which can be compared quantitatively with results of laboratory experiments. Experience has shown that these results can often be extrapolated to geophysical and astrophysical applications even though empirical eddy diffusivities must sometimes be used in place of the molecular diffusivities relevant on the laboratory scale.

Before entering the mathematical formulation of problems of convection, it is useful to consider the energy balance which characterizes the instability of a static layer heated from below. An upper bound on the potential energy which can be released per unit time by a fluid parcel moving with the vertical velocity V is given by $\rho_0 \beta \Delta T g V$ where g is the acceleration of gravity, ΔT is the temperature difference across the layer and where a linear dependence of the density ρ on the temperature T has been assumed

$$\rho = \rho_0(1 - \beta(T - T_0)). \tag{1.1}$$

The parcel does not retain its original density, however, as it moves up or down since thermal conduction opposes the distortion of the isotherms caused by fluid motion. The characteristic velocity of isotherms returning to their equilibrium position in the absence of fluid motion is given by κ/d where κ is the thermometric conductivity and d is a typical length scale which we shall identify with the layer thickness. Restricting the attention to small velocities V—nonlinear effects complicate the argument for larger values of V—we conclude that only the fraction Vd/κ of the density difference across the layer enters a realistic expression for the potential energy release per unit time,

$$L_P = C_1 \rho_0 \beta \Delta T g V^2 d/\kappa. \tag{1.2}$$

C_1 is a numerical constant which will depend on the details of the motion and the boundary conditions. Convection can occur if L_P exceeds the energy L_D dissipated per unit time by viscous friction. Using $\rho_0 \nu V/d^2$ as an estimate for the friction-term in the Navier-Stokes equations we find

$$L_D = C_2 \rho_0 \nu V^2/d^2, \tag{1.3}$$

where again details of the problem enter the numerical factor C_2. The condition $L_P \geq L_D$ for instability can be written in the form

$$\mathrm{R} \equiv \frac{\beta \Delta T g d^3}{\kappa \nu} \geq \frac{C_2}{C_1} \equiv \mathrm{R}_c, \tag{1.4}$$

where the non-dimensional expression on the left hand side of the inequality is called the Rayleigh number honoring Lord Rayleigh (1916) who first derived a mathematical theory of convection in a layer heated from below. The lowest possible ratio C_2/C_1 for given physical conditions of the layer is called the critical value of the Rayleigh number, R_c, or critical Rayleigh number. The reader unfamiliar with fluid mechanical terminology should note the inconsistent usage in that non-dimensional parameters as well as their critical values are called "numbers."

The Rayleigh number is the fundamental parameter in the theory of convection. In experiments on the laboratory scale only one other non-dimensional parameter is needed to characterize the effect of fluid properties on convective motion. This parameter is the ratio ν/κ and is called the Prandtl number. Other parameters are important in the description of geophysical and astrophysical convection processes and will be mentioned in connection with the basic equations.

2. Basic Equations

The equations describing motions in a chemically homogeneous fluid find their most general form in the principles of conservation of mass, momentum and energy,

$$\partial_t \rho + \partial_j \rho u_j = 0, \tag{2.1}$$

$$\partial_t \rho u_i + \partial_j \rho u_j u_i = \rho g_i + \partial_j \mathsf{S}_{ij}, \tag{2.2}$$

$$\partial_t \rho E + \partial_j \rho u_j E = \mathsf{S}_{ij} \partial_j u_i + \rho Q - \partial_j h_j, \tag{2.3}$$

where u_i is the velocity vector, S_{ij} is the stress tensor, E is the internal energy, h_i is the heat flux vector and Q is the rate of internal heating per unit volume. The body force per unit mass g_i will be identified with gravity in the following. The symbol ∂_i denotes the differentiation with respect to the coordinate x_i of a Cartesian system of coordinates. Repeated indices in any term imply the summation over those indices. ∂_t denotes the differentiation with respect to time. The stress tensor can be separated into its isotropic and deviatoric parts

$$\mathsf{S}_{ij} = -p\delta_{ij} + \tau_{ij}. \tag{2.4}$$

For a Newtonian fluid the latter part is given by

$$\tau_{ij} = \mu(\partial_i u_j + \partial_j u_i - \tfrac{1}{3}\delta_{ij}\partial_k u_k), \tag{2.5}$$

while the pressure p in general differs from the thermodynamic equilibrium pressure p_e according to the relationship

$$p = p_e - \tilde{\mu}\partial_k u_k. \tag{2.6}$$

μ and $\tilde{\mu}$ are called the dynamic and bulk viscosity, respectively. Since convection motions exhibit little divergence, $\partial_k u_k \approx 0$, the difference $p - p_e$ is usually neglected. Using relationship (2.6) we write the energy equation (2.3) in the form

$$(\partial_t + u_j\partial_j)E + p_e(\partial_t + u_j\partial_j)\rho^{-1} = C_p(\partial_t + u_j\partial_j)T - \beta T \rho^{-1}(\partial_t + u_j\partial_j)P_e$$
$$= \frac{1}{\rho}\partial_j \lambda \partial_j T + Q + \Phi + \frac{\tilde{\mu}}{\rho}(\partial_k u_k)^2, \tag{2.7}$$

where the left hand side describes the change of the entropy per unit mass times the temperature T as experienced by a moving fluid parcel. We have assumed that the heat flux is proportional to the negative temperature gradient with the conductivity λ as the constant of proportionality. This approximation is certainly valid for the molecular transport of heat and also applies for the radiative heat flux inside the earth. The dissipation function Φ is defined by

$$\Phi = \tau_{ij}\partial_j u_i/\rho, \tag{2.8}$$

and is positive in general. C_p is the specific heat at constant pressure and β is the thermal expansivity. The energy equation (2.7) remains valid for non-Newtonian fluids if the term $\tilde{\mu}\partial_k u_k$ is replaced by a suitable expression for $p - p_e$. Although non-Newtonian rheology is undoubtedly relevant to flow in the Earth's mantle, we shall restrict the attention to the special Newtonian case on which almost all work on mantle convection has been based.

The equations discussed so far must be supplemented by an equation of state

$$\rho = \rho(p_e, T), \tag{2.9}$$

and by expressions for the material properties such as μ and λ in order to provide a complete set of equations. These complete equations are not very useful, however, because they describe many other phenomena besides convective motions. Solutions will be dominated by rapid changes in time owing to compressional waves rather than by the slow changes in time expected of convection motions. To rectify this situation the anelastic approximation has been introduced (Ogura and Phillips, 1962; Gough, 1969) in which the density ρ is replaced by the density $\bar{\rho}$ of some static reference state except in the gravity term of equation (2.2). The anelastic approximation requires that the convection velocity is small compared to the velocity of sound, a condition which of course is extremely well satisfied in the Earth's mantle. Equations based on the anelastic approximation have been applied to the problem of mantle convection by Jarvis and McKenzie (1980). But since the equations are rather complicated they will not be reproduced here.

In contrast to convection zones in stars which typically extend over many scale heights, the thickness of the Earth's mantle is certainly much less than the density scale height, i.e., the distance over which an isentropic density distribution varies by a factor e. It thus seems likely that the variation of density with depth has only a minor influence on convection in the earth mantle and that the Boussinesq approximation provides a reasonable basis for theoretical investigations. The Boussinesq approximation follows from two assumptions. First, a constant density is assumed except in the gravity term of the equation of motion and, secondly, the pressure dependence of the density in this term is neglected in comparison with the temperature dependence. This latter assumption can be expressed mathematically if the equation of state (2.9) is written in the form

$$\rho = \rho(T_s, p_s)[1 - \beta(T - T_s) + K_T^{-1}(p - p_s)], \tag{2.10}$$

where T_s, p_s refer to an isentropic hydrostatic reference state and K_T denotes the isothermal modulus of bulk compressibility. Since typical values $T - T_s$ are given by the amount ΔT_{pot} by which the temperature difference across the convection layer exceeds the isentropic difference and since a typical value $p - p_s$ is given by $\rho g \beta \Delta T_{\text{pot}} d$, the condition of negligible dependence of the density on $p - p_s$ can be written in the form

$$\frac{\rho g d}{K_T} = \frac{g d \beta}{C_v \Gamma} << 1, \tag{2.11}$$

where Γ is the Grüneisen parameter defined by $\Gamma = \beta K_T/\rho C_v$. Because Γ is of the order unity, the limit (2.11) expresses the property that the energy gained by exchanging two fluid parcels at the top and bottom of the convection layer is small compared to the energy gained by cooling the lower parcel by the amount ΔT_{pot}. In the case of the Earth's mantle the left hand side in relationship (2.11) can be estimated to be about 0.5 for whole mantle convection which indicates that effects of compressibility influence, but do not dominate the dynamics of convection. Condition (2.11) also insures that the change of density throughout the layer owing to pressure variations is small. The condition of a nearly constant density is thus satisifed if the condition

$$\beta \Delta T << 1 \tag{2.12}$$

is assumed.

In its narrow definition the Boussinesq approximation requires that all material properties other than the density are also constant. But this latter assumption is quite independent from the basic assumptions mentioned above and problems such as the influence of a strong temperature dependence of the viscosity can be treated well within the context of the wider Boussinesq approximation in the sense of conditions (2.11) and (2.12). It is only for reasons of mathematical simplicity that the narrow definition of a Boussinesq fluid will be adopted in the following unless indicated otherwise.

Before continuing the discussion of the general equations it is of interest to consider the special steady solution of (2.2), (2.7) describing the static state,

$$\nabla p = \rho \mathbf{g} \equiv -\boldsymbol{\lambda}\rho_0[1 - \beta(T - T_0)]\, g_0 f_1, \tag{2.13a}$$

$$\kappa\nabla^2 T = -Q \equiv -Q_0 f_2. \tag{2.13b}$$

The unit vector $\boldsymbol{\lambda}$ opposite to the direction of gravity has been introduced and g_0 and Q_0 are constants. The existence of a static solution $\mathbf{u} \equiv 0$ requires that the functions f_1 and f_2 depend only on the coordinate $\boldsymbol{\lambda}\cdot\mathbf{x}$ in the direction of the unit vector $\boldsymbol{\lambda}$. Small deviations from this condition introduce slow circulations which eventually can be taken into account as perturbations. But in the following the existence of a steady static solution will be assumed.

Using d, d^2/κ and $\nu\kappa/\beta g_0 d^3$ as scales for length, time and temperature, respectively, we introduce as new dimensionless variables the deviation Θ of the temperature and the deviation ΠP of the pressure from their respective static distributions. The Boussinesq approximation of the basic equations (2.1), (2.2), (2.7) can then be written in the dimensionless form

$$\nabla\cdot\mathbf{v} = 0, \tag{2.14a}$$

$$\nabla^2\mathbf{v} + \boldsymbol{\lambda}\Theta f_1 - \nabla\Pi = \mathrm{P}^{-1}(\mathbf{v}\cdot\nabla\mathbf{v} + \partial_t\mathbf{v}), \tag{2.14b}$$

$$\nabla^2\Theta + \boldsymbol{\lambda}\cdot\mathbf{v}\,(\mathrm{R}f_3 + \mathrm{R}_Q f_4) = \mathbf{v}\cdot\nabla\Theta + \partial_t\Theta, \tag{2.14c}$$

where $\mathbf{v}$ is the dimensionless velocity vector. The functions f_3 and f_4 depend only on the coordinate $\boldsymbol{\lambda}\cdot\mathbf{x}$ and are determined by the static solution of the problem. In addition to the heat source distribution they reflect in general the geometrical configuration of the layer. f_1 may not only describe variations of gravitational force but also variations of the thermal expansivity in which case the parameter β describes its mean value. In view of the applications of the theory, plane layers and spherical shells are the primary examples for which the functions f_n will be specified below.

The dimensionless parameters are the Rayleigh number R, the heat source Rayleigh number R_Q and the Prandtl number P which are defined by

$$\mathrm{R} \equiv \beta(\Delta T - \Delta T_a)g_0 d^3/\nu\kappa, \quad \mathrm{R}_Q \equiv \beta Q_0 g_0 d^5/\nu\kappa\lambda, \quad \mathrm{P} = \nu/\kappa. \tag{2.15}$$

The isentropic temperature difference ΔT_a across the layer is given by

$$\Delta T_a = -\int (\beta T/C_p)\mathbf{g}\cdot d\mathbf{x},$$

where the integral is extended between top and bottom of the layer, in other words $\Delta T - \Delta T_a = \Delta T_{\text{pot}}$. Although we have been referring implicitely to a layer in which the temperature gradient of the static state typically exceeds in magnitude the isentropic value, this need not to be the case throughout the layer. Problems of convection layers with substantial stably stratified portions are of importance in astrophysical and geophysical applications and are included in the above formulation. But the Rayleigh number may be negative in that case and another definition may be more appropriate. The dissipation function Φ does not appear in equation (2.14c) because of assumption (2.11). Since the release of potential energy and its conversion into viscous dissipation is small compared to the heat transported by convection, Φ is negligible in the Boussinesq approximation. The dimensionless parameter of relationship (2.11) is therefore also sometimes called the dissipation number. For a discussion of the effects of finite Φ in problems of mantle convection see Hewitt *et al.* (1975).

The equations (2.14) must be supplemented by boundary conditions for the velocity vector $\mathbf{v}$ and the temperature field Θ. The no-slip condition at a rigid boundary and the stress-free condition at a free surface are commonly used. Assuming the origin of the position vector on the lower boundary, these conditions are given by

$$\mathbf{v} = 0 \qquad \text{at } \boldsymbol{\lambda}\cdot\mathbf{x} = 0, 1, \tag{2.16a}$$

for rigid boundaries, and by

$$\mathbf{v}\cdot\boldsymbol{\lambda} = 0, \quad \boldsymbol{\lambda} \times (\boldsymbol{\lambda}\cdot\nabla)\mathbf{v} + \boldsymbol{\lambda}\cdot(\boldsymbol{\lambda} \times \nabla)\mathbf{v} = 0 \quad \text{at } \boldsymbol{\lambda}\cdot\mathbf{x} = 0, 1, \tag{2.16b}$$

for stress-free boundaries. A combination of the conditions (2.16) is also possible, of course. The continuity of the temperature and the heat flux requires

$$\Theta = \Theta_e, \qquad \boldsymbol{\lambda}\cdot\nabla\Theta = \gamma_e\boldsymbol{\lambda}\cdot\nabla\Theta_e \qquad \text{at } \boldsymbol{\lambda}\cdot\mathbf{x} = 0, 1, \tag{2.17}$$

where Θ_e is the change in the exterior temperature distribution introduced by convection and γ_e is the ratio between exterior and interior heat conductivity. Often the limit $\gamma_e \rightarrow \infty$ is assumed in which case condition (2.17) reduces to

$$\Theta = 0 \qquad \text{at } \boldsymbol{\lambda}\cdot\mathbf{x} = 0, 1. \tag{2.18}$$

In discussing the boundary conditions we have assumed that the distortion by the convection flow of the bounding surfaces is negligible. This problem is examined in section 3.4 where it will be shown that neglecting the distortion is justified in the limit (2.11).

A simplification of the analysis of equations (2.14) can often be obtained by using the representation

$$\mathbf{v} = \nabla \times (\nabla \times \boldsymbol{\lambda}\Phi) + \nabla \times \boldsymbol{\lambda}\psi, \tag{2.19}$$

for the velocity field $\mathbf{v}$. That (2.19) is a general representation for arbitrary solenoidal vector fields $\mathbf{v}$ has been shown in the spherical case, $\boldsymbol{\lambda} = \mathbf{x}/|\mathbf{x}|$, by Backus (1958) and in the planar case of a constant vector $\boldsymbol{\lambda}$ by Warner (1972).

In the following the spherical and the planar case of equations (2.14) will be discussed in special cases.

3. Linear Theory of Convection

3.1. Convection in Planar Layer

In a planar layer it is appropriate to introduce a Cartesian system of coordinates (x, y, z) with the z-coordinate in the direction of $\boldsymbol{\lambda}$. Gravity and the temperature gradient of the static state are assumed to be constant throughout the layer, $f_1 = f_3 = 1$, and internal heat sources will be neglected at this point, $\mathrm{R}_Q = 0$. In that case the following linear equations for steady convection flow of infinitesimal amplitude are obtained from equations (2.3),

$$\nabla \cdot \mathbf{v} = 0, \tag{3.1a}$$

$$\nabla^2 \mathbf{v} + \boldsymbol{\lambda}\Theta - \nabla\Pi = 0, \tag{3.1b}$$

$$\nabla^2\Theta + \boldsymbol{\lambda}\cdot\mathbf{v}\mathrm{R} = 0. \tag{3.1c}$$

After taking the Laplacian of the z-component of the $(\text{curl})^2$ of equation (3.1b) we can eliminate the temperature Θ and obtain a single equation for the vertical velocity component v_z,

$$(\nabla^6 - R\Delta_2)v_z = 0, \tag{3.2}$$

where Δ_2 is the horizontal Laplacian, $\Delta_2 \equiv \partial^2/\partial x^2 + \partial^2/\partial y^2$. Because of the homogeneity of the problem with respect to the x, y-coordinates, equation (3.2) admits separable solutions,

$$v_z = w(x,y)\, f(z), \tag{3.3}$$

with

$$\Delta_2\, w(x,y) + \alpha^2\, w(x,y) = 0. \tag{3.4}$$

For stress-free boundaries of nearly infinite conductivity the simple conditions

$$v_z = \frac{\partial^2}{\partial z^2} v_z = \frac{\partial^4}{\partial z^4} v_z = 0 \quad \text{at } z = 0,\ 1 \tag{3.5}$$

are obtained which permit a solution $f(z)$ in terms of trigonometric functions,

$$f_n(z) = \sin n\pi z, \qquad n = 1, 2, \ldots \tag{3.6a}$$

The corresponding value of the Rayleigh number is given by

$$R_n = ((n\pi)^2 + \alpha^2)^3/\alpha^2. \tag{3.6b}$$

Only solutions associated with the smallest value of the Rayleigh number are of physical interest. The absolute minimum of expression (3.6b) is reached for $n = 1$, and is called the critical value of the Rayleigh number,

$$R_c = 27\pi^4/4 \text{ corresponding to } \alpha_c = \pi\sqrt{2}. \tag{3.6c}$$

In the case of rigid boundaries an analytic expression for $f(z)$ can also be obtained (Pellew and Southwell, 1940); but because of its more complicated form numerical solutions are usually preferred (Reid and Harris, 1958). The boundary conditions for v_z are given by

$$v_z = \frac{\partial}{\partial z} v_z = \left(\frac{\partial^2}{\partial z^2} - \alpha^2\right)^2 v_z = 0 \qquad \text{at } z = 0, 1, \tag{3.7}$$

and critical value of the Rayleigh number is considerably higher than in the stress-free case,

$$R_c = 1707.76 \text{ corresponding to } \alpha_c = 3.117. \tag{3.8}$$

The critical wavenumber α_c indicates that the horizontal extent of the circulating motion is comparable with the depth of the layer. In the stress-free case with less restricting boundaries the horizontal extent is somewhat larger.

Figure 2.1 shows the dependence on α of the Rayleigh number for the lowest mode for both cases of boundary conditions. The minimum at a finite value α_c reflects the property that narrow convection cells corresponding to high values of α are ineffective because of the strong heat exchange between up- and downgoing fluid. For very large scale motions the vertical component becomes small in comparison to horizontal component and the release of potential energy is less efficient. This latter effect can be compensated by a decrease of thermal dissipation in the case of thermally insulating boundaries. In the limit of $\gamma_e \to 0$ the last condition in the boundary conditions (3.5) and (3.7) must be replaced by $\partial\left(\frac{\partial^2}{\partial z^2} - \alpha^2\right)^2 v_z/\partial z = 0$ with the result that the minimizing value α_c tends to zero with γ_e for both rigid and stress-free boundary conditions. This limit is of interest because it permits a solution of equations (3.1) in terms of polynomials for the z-dependence (Jakeman, 1968)

$$f(z) = z(z - 1)(z^2 - z - 1), \quad R_c = 120 \quad \text{for stress-free boundaries} \tag{3.9a}$$

$$f(z) = z^2(z - 1)^2 \quad , R_c = 720 \quad \text{for rigid boundaries} \tag{3.9b}$$

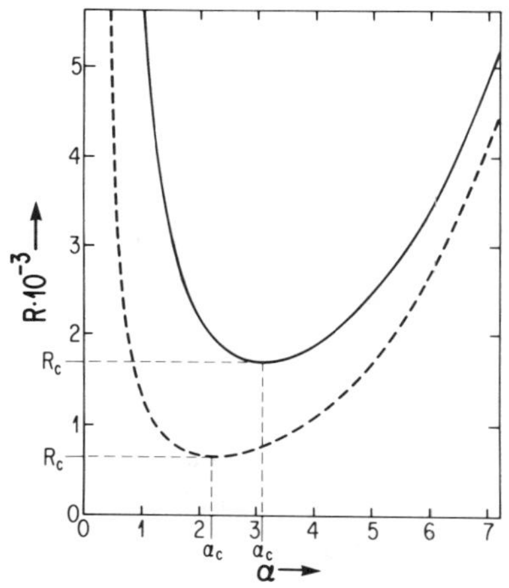

Figure 2.1. The Rayleigh number R for onset of convection as a function of the wavenumber α in the case of rigid (solid) and stress-free boundaries (dashed).

For arbitrary positive values γ_e the critical values of R and α increase monotonically between the cases (3.9) for $\gamma_e = 0$ and the cases (3.6c) and (3.8) in limit $\gamma_e = \infty$ (Sparrow *et al.*, 1964).

To get a feeling for the conditions required for the onset of convection in a laboratory apparatus it is useful to inspect Table 2.1 where the depth is indicated which is required for different fluids to achieve the critical Rayleigh number (3.8) with a temperature difference of 1°C.

3.2. Energy Stability Analysis

The assumption that the onset of convection can be described by equations (3.1) can be justified rigorously by considering the energy integrals of equations (2.14). By multiplying equation (2.14b) by $Rc^2\mathbf{v}$ and equation

Table 2.1. Properties of fluids in convection experiments at 20°C.

	Mercury	Air	Water	Glycerin
Prandtl number P	0.027	0.66	6.94	11,460
Depth d [cm] needed for R_c = 1708 for $T_2 - T_1$ = 1°C	0.773	2.54	0.497	3.47

(2.14c) by Θ, adding the result and averaging it over the planar or spherical fluid layer we obtain the following balance

$$\frac{1}{2}\frac{d}{dt}[\mathrm{RP}^{-1}c^2\langle \mathbf{v}\cdot\mathbf{v}\rangle + \langle\Theta\Theta\rangle]$$

$$= -\mathrm{R}c^2\langle|\nabla\times\mathbf{v}|^2\rangle - \langle|\nabla\Theta|^2\rangle + \mathrm{R}c\langle\Theta\boldsymbol{\lambda}\cdot\mathbf{v}\, f_5\rangle, \quad (3.10)$$

where the angular brackets indicate the average and the definition

$$f_5 = (c^2 f_1 + f_3 + \mathrm{R}_Q f_4/\mathrm{R})/c \quad (3.11)$$

has been introduced. The positive constant c is arbitrary at this point and will be determined later. We want to find the lowest value R_E of R for which the balance (3.10) can be satisfied with vanishing left hand side. By admitting all fields $\mathbf{v}$ and Θ that satisfy equation (2.14a) and the boundary conditions, but not necessarily the equations (2.14b) and (2.14c), we insure that R_E provides a lower bound for all Rayleigh numbers R for which non-decaying convection of arbitrary amplitude is possible. After introducing new variables according to the definitions

$$\mathsf{S}^2 = \mathrm{R}; \qquad \hat{\mathbf{v}} = c\mathsf{S}\mathbf{v}, \quad (3.12)$$

the problem can be reduced to the determination of the minimum S_E of the variational functional

$$\mathsf{S}(\hat{\mathbf{v}},\Theta) \equiv \{\langle|\nabla\Theta|^2\rangle + \langle|\nabla\times\hat{\mathbf{v}}|^2\rangle + \tau\nabla\cdot\hat{\mathbf{v}}\}\langle\boldsymbol{\lambda}\cdot\hat{\mathbf{v}}\Theta f_5\rangle^{-1}; \quad (3.13)$$

where τ has been introduced as a Lagrange multiplier in order to accommodate the constraint (2.14a). The Euler-Lagrange equations corresponding to an extremal value of the functional (3.13) are given by

$$\nabla\cdot\hat{\mathbf{v}} = 0, \quad (3.14a)$$

$$\nabla^2\hat{\mathbf{v}} + \tfrac{1}{2}\nabla\tau + \mathsf{S}\boldsymbol{\lambda}\Theta\,\tfrac{1}{2}f_5 = 0, \quad (3.14b)$$

$$\nabla^2\Theta + \mathsf{S}\boldsymbol{\lambda}\cdot\hat{\mathbf{v}}\,\tfrac{1}{2}f_5 = 0. \quad (3.14c)$$

In the special case when the ratio

$$(f_3 + \mathrm{R}_Q f_4/\mathrm{R})/f_1 \quad (3.15)$$

is equal to a positive constant it is possible to show that equations (3.14) are identical to equations (2.14) with vanishing right hand side. After setting c^2 equal to (3.15) and reversing the transformation (3.12) this identity becomes evident. Since the critical Rayleigh number R_c for steady infinitesimal convection is thus equal to R_E there cannot exist subcritical finite amplitude convection in this case (Sorokin, 1953). Another consequence is that onset of convection in the form of oscillatory motion cannot occur at a Rayleigh number less than R_c. This latter property is sometimes called the principle of exchange of stabilities.

When the ratio (3.15) is not equal to a constant the lowest eigenvalue S_E of equations (3.14) provides a lower bound for the Rayleigh number at which non-decaying convection can exist. By maximizing S_E as a function of c the optimal bound based on the balance (3.10) can be found. For applications of this procedure in the case of convection in spherical shells see Joseph and Carmi (1966).

3.3. Numerical Methods

Only in exceptional cases it is possible to obtain explicit analytical expressions for the solutions of the linear equations obtained from the general equations (2.14) by neglecting the right hand side. Two methods, the Galerkin method and the "shooting" method, are commonly employed to obtain numerical approximations to the exact solutions.

In the Galerkin method the dependent variables are expanded in terms of a complete system of orthogonal functions which satisfy the appropriate boundary conditions. In the case of rigid boundaries with infinite conductivity, for example, Φ, Θ are assumed in the form

$$\Phi = w(x,y)\,\Sigma a_n \phi_n(z) \tag{3.16a}$$

$$\Theta = w(x,y)\,\Sigma b_n \sin n\pi z, \tag{3.16b}$$

where the functions $\phi_n(z)$ are the solutions of the boundary value problem

$$\frac{\partial^4}{\partial z^4}\phi_n(z) = \lambda_n^4 \phi_n(z) \text{ with } \phi_n = \frac{\partial}{\partial z}\phi_n = 0 \text{ at } z = 0,\ 1,$$

and are given by

$$\phi_n(z) = \begin{cases} \dfrac{\sinh \lambda_n z}{\sinh \frac{1}{2}\lambda_n} - \dfrac{\sin \lambda_n z}{\sin \frac{1}{2}\lambda_n} & \text{for even } n \\[2ex] \dfrac{\cosh \lambda_n z}{\cosh \frac{1}{2}\lambda_n} - \dfrac{\cos \lambda_n z}{\cos \frac{1}{2}\lambda_n} & \text{for odd } n \end{cases} \tag{3.17a}$$

where the values λ_n are determined as solutions of the transcendental equations

$$\begin{aligned} \coth \tfrac{1}{2}\lambda_n - \cot \tfrac{1}{2}{}_n &= 0 \qquad \text{for even } n \\ \tanh \tfrac{1}{2}\lambda_n + \tan \tfrac{1}{2}\lambda_n &= 0 \qquad \text{for odd } n \end{aligned} \tag{3.17b}$$

For numerical values of α_n see an appendix in Chandrasekhar's (1961) book. After multiplying equation (3.1b) by $\sin m\pi z$ and equation (3.1c) by $\phi_m(z)$ and integrating them over the interval $0 \leqslant z \leqslant 1$ a system of linear equations for the unknown coefficients a_n, b_n is obtained. The condition that the determinant of the coefficient matrix vanishes determines the possible values of R for a given value of α. Curves such as that shown in Figure 2.1 are obtained by computing the lowest value of R as a function of α.

For the actual computation the number of equations must be truncated, of course. The solution for R converges well for low values of the truncation parameter N which denotes the number of coefficients a_n or b_n taken into account. But the Galerkin method is not an efficient way to obtain results of high accuracy. On the other hand it can be readily generalized for the solution of the nonlinear equations as will be discussed in Section 5 of this chapter. The shooting method is highly efficient way to obtain accurate solutions; but it is restricted to systems of linear equations, since it is based on the superposition principle.

The shooting method overcomes the problem of unknown boundary conditions at the starting point of the forward integration, $z = 0$, by generating as many independent solutions as there are unknown boundary conditions. If there are, for example, $2N$ linear first order differential equations

$$y_n' = f(y_1, \ldots, y_{2n}, z) \qquad \text{for } n = 1, \ldots, 2N, \tag{3.18}$$

but only N boundary conditions

$$\sum_{i=1}^{2N} \gamma_{ni} y_i = 0 \text{ at } z = 0, \qquad n = 1, \ldots, N, \tag{3.19}$$

then N independent solutions $y_n^{(m)}$ can be obtained by requiring, in addition to (3.19), that the boundary conditions

$$\sum_{i=1}^{2N} \beta_{ni} y_i^{(m)} = \delta_{nm} \text{ at } z = 0, \qquad n = 1, \ldots, N \tag{3.20}$$

be satisfied. The coefficients β_{ni} are arbitrary except for the property that the determinant of the combined $2N \times 2N$ matrix $(\gamma_{ni}, \beta_{ni})$ must not vanish. After integrating equations (3.18) for $y_n^{(m)}$ by the Runge-Kutta method, for instance, the general solution can be written in the form

$$Y_n = \sum_{m=1}^{N} c_m y_n^{(m)} \qquad n = 1, \ldots, 2N,$$

and the coefficients c_m can be chosen to satisfy the remaining N boundary conditions at $z = 1$,

$$\sum_{i=1}^{2N} \eta_{ni} y_i = 0 \text{ at } z = 1, \qquad n = 1, \ldots, N \tag{3.21}$$

These boundary conditions yield N homogeneous equations for the unknown c_m which are solvable only if the determinant of the coefficient matrix vanishes,

$$\det |A_{nm}| = 0 \text{ with } A_{nm} \equiv \sum_{i=1}^{2N} \eta_{ni} y_i^{(m)} \tag{3.22}$$

Since the coefficients A_{nm} are functions of R for a given value of α, so is the determinant. All possible values of R can thus be obtained as zeros of the determinant using a Newton-Raphson iteration method. Since computations based on the method outlined here can be performed easily by small computers, there is no need to present here the variety of solutions describing the onset of convection for different boundary conditions.

3.4. Dynamic Deflections of the Boundary

In formulating the boundary conditions (3.5) it has been assumed that the position of the boundary is not significantly affected by the convection motion. In this section we examine this assumption. Starting with the solution (3.3), (3.6a) for stress-free boundaries we investigate the dimensionless deflection ζ of the surface as a perturbation. If ζ does indeed turn out to be a small quantity in comparison with other convection effects its neglection will be justified.

The basic condition at the free surface is that normal as well as tangential components of the stress tensor vanish. The condition for the tangential component is already taken into account in (3.5); the condition for the normal component of the dimensionless stress tensor S_{ij} can be written in the form

$$0 = n_i n_j \mathsf{S}_{ij} \approx \mathsf{S}_{zz} = g(z-1)d^3/\kappa^2 - \Pi \mathrm{P} + 2\mathrm{P}\partial_z v_z \quad \text{at } z = 1 + \zeta, \tag{3.23}$$

where the difference between the normal vectors $\mathbf{n}$ and $\boldsymbol{\lambda}$ can be neglected because of the smallness of the deflection. The quantities Π and $\partial_z v_z$ can be expressed in terms of $\partial_z \Theta$ after the divergence of equation (3.1b) has been taken and equation (3.1c) has been used,

$$\zeta = -\beta \Delta T [2(\pi^2 + \alpha^2) + \mathrm{R}/(\pi^2 + \alpha^2)]\, \mathrm{R}^{-2} \partial_z \Theta|_{z=1}. \tag{3.24}$$

We have made use of the definition (1.4) for R and the expressions (3.6b) in the case $n = 1$. It is obvious from relationship (3.24) that ζ is indeed small and that the effects of deflection are negligible within the Boussinesq approximation. If, however, the convection layer is overlain by another fluid layer of only slightly lower density ρ_1 then ζ would be increased by a factor $\rho_0(\rho_0 - \rho_1)^{-1}$ in which case it may have to be taken into account in the formulation for the boundary condition (3.5).

The result (3.24) can be used to make another point relevant to the problem of mantle convection. Since Θ vanishes at the boundary, the deflection ζ is positive over regions of upwelling and negative over regions of downwelling. The corresponding excess or defect of mass is opposite in sign to the change in the density of the fluid layer. Using the solution

$$\Theta = w(x,y)\ \Theta_0 \sin \pi z \tag{3.25}$$

of equations (3.1) the two contributions, δm_1 and δm_2, to the horizontal variation in mass of the fluid layer can be evaluated,

$$\delta m_1 = \rho_0 \zeta d = \rho_0 \beta \Delta T \pi \Theta_0 w(x,y) \mathrm{R}^{-1}(\pi^2 + 3\alpha^2)/(\pi^2 + \alpha^2)^2 \tag{3.26a}$$

$$\delta m_2 = -\rho_0 \beta \Delta \mathrm{T} \mathrm{R}^{-1} \int_0^1 \Theta dz = -\beta \Delta T \rho_0 \Theta_0 2 w(x,y)/\pi \mathrm{R} \tag{3.26b}$$

The contribution δm_2 from the interior has the opposite sign of the contribution of δm_1 from the upper boundary and exceeds it in absolute value. But when the deflection of the lower boundary is taken into account which is equal to δm_1 in the simple model considered here, then the various contributions nearly cancel, at least in the long wavelength limit $\alpha \to 0$. The nearly compensating influences of boundary deflection and lateral density variations are the reason that gravity anomalies can not readily be correlated with convection flow in the Earth's mantle.

3.5. Onset of Convection in Spherical Shells

Long before the concept of continental drift and plate tectonics became generally accepted, the idea that large scale thermally driven flows exist in the Earth's mantle had been entertained. Chandrasekhar (1962) mentions this idea in his book as the main motivation for the mathematical analysis of convection in spherical shells. The relevant chapter in that book can still serve as a comprehensive review of the linear theory. In this section only a brief outline of the results of linear theory is given in order to provide the foundation for the discussion of the nonlinear problem in Section 4.3.

The basic equations (2.14) have been written in such a way that they describe both convection in plane layers and in spherical shells. Usually the following three gravity distributions are assumed:

I. Gravity distribution of a uniform sphere $\qquad f_1 = r(1 - \eta)$

II. Constant gravity $\qquad f_1 = 1$

III. Gravity with negligible contribution from fluid shell $\qquad f_1 = r^{-2}(1 - \eta)^{-2} \qquad (3.27)$

The functions f_1 of the distance r from the center are normalized such that g_0 is defined as the gravity at the surface of the sphere; η is the ratio of inner to outer radius. The case II is not really consistent with the assumptions of a Boussinesq fluid. But since the gravity in Earth's mantle is approximately constant and since it can be closely modelled by an appropriate choice of the densities of the core and the fluid shell, case II is a popular choice.

The general solution for pure thermal conduction with constant thermal conductivity determines the functions f_3 and f_4,

$$f_3 = \eta(1 - \eta)^{-2} r^{-2},$$

$$f_4 = -\eta(\eta + 1)/[6r^2(1 - \eta)^3] + r/3. \qquad (3.28)$$

In addition to the cases of vanishing heat sources in the fluid and vanishing ΔT,

$$[A]\text{: } \mathrm{R}_Q = 0; \qquad [B]\text{: } \mathrm{R} = 0, \qquad (3.29a)$$

the case of a temperature gradient which is linear in r is of special interest,

$$[C]\text{: } \mathrm{R}_Q = \mathrm{R}\, 6(1 - \eta)/(1 + \eta), \qquad (3.29b)$$

since it permits the extension to the case of a uniformly heated fluid sphere with $\eta = 0$.

After introducing the representation (2.19) for the velocity field and taking the radial component of the curl curl of equation (2.14b) we obtain as linear equations for steady convection

$$\nabla^4 L_2 \Phi/r - L_2 \Theta f_1/r = 0, \qquad (3.30a)$$

$$\nabla^2 \Theta + (\mathrm{R} f_3 + \mathrm{R}_Q f_4)\, L_2 \Phi/r^2 = 0, \qquad (3.30b)$$

where L_2 is the negative Laplacian on the unit sphere, i.e., in spherical coordinates (r, θ, ϕ)

$$L_2 = -(\sin\theta)^{-1} \frac{\partial}{\partial\theta}\left(\sin\theta \frac{\partial}{\partial\theta}\right) - (\sin\theta)^{-2} \frac{\partial^2}{\partial\phi^2}. \qquad (3.31)$$

Equations (3.30) permit separable solutions

$$\Phi = w(\theta,\phi)\, h(r), \qquad \theta = w(\theta,\phi)\, g(r), \qquad (3.32)$$

where w satisfies the equation

$$L_2 w = l(l + 1)w \qquad \text{with } l = 1, 2, \ldots \tag{3.33}$$

In general, solutions for $h(r)$ and $g(r)$ can be obtained only numerically. At the boundaries stress-free conditions

$$f = f'' = g = 0 \qquad \text{at } r = \eta(1 - \eta)^{-1} \text{ and } r = (1 - \eta)^{-1}, \tag{3.34}$$

or no slip conditions

$$f = f' = g = 0 \qquad \text{at } r = \eta(1 - \eta)^{-1} \text{ and } r = (1 - \eta)^{-1}, \tag{3.35}$$

or a combination of both are used. In Table 2.2 selected results for various combinations of gravity distributions (3.27) and static temperature distributions are displayed in dependence on the radius ratio η. It is obvious from the data that the critical Rayleigh number approaches the values (3.6c) and (3.8) of the plane layer case as η tends to unity. The critical values for an inner stress-free boundary and outer rigid boundary becomes the same as in the reversed case in the limit $\eta \rightarrow 1$; the corresponding value for a plane layer is 1100.65 according to Reid and Harris (1958). The minimizing value of l follows approximately the relationships

$$l \approx (1 + \eta)(1 - \eta)^{-1}\, 3.117/2 \text{ for rigid boundaries} \tag{3.36a}$$

$$l \approx (1 + \eta)(1 - \eta)^{-1}\, \pi/\sqrt{8} \text{ for free boundaries} \tag{3.36b}$$

which can be obtained by simple extrapolation from the plane layer limit. The correspondence between the data of the spherical case and the corresponding plane layer limit could be made even closer if an average value of the gravity instead of the surface value g_0 would be used in the definition of the Rayleigh number. But even without this adjustment the close correspondence between the onset of convection in spherical shells and plane layers is

Table 2.2. Critical Rayleigh numbers as a function of the radius ratio η

$\eta =$	0.1	0.2	0.3	0.4	0.5	0.6	0.8
IC ff		1-1281	2-1110	2-988	3-902	4-834	10-735
IC rr	1-3812	2-3138	2-2867	3-2537	4-2337	6-2173	13-1914
IC rf	1-1509	2-1527	2-1375	3-1348	4-1305	5-1243	12-1175
IC fr	1-3471	2-2883	2-2347	3-2048	4-1807	5-1591	12-1311
IA ff	1-3248	1-1907	2-1451	2-1133	3-979	4-872	
IIIC ff	1-179	1-229	2-283	2-317	3-367	4-419	
IIC ff					3-679		

Compiled from Chandrasekhar (1962), Geiger and Busse (1981), and Zebib *et al.* (1983).
The integer preceding the Rayleigh number and separated from it by a hyphen gives the value of l for which the Rayleigh number reaches a minimum. The symbols in the left columns indicate the combination of conditions (3.27) and (3.29); rf indicates a lower rigid and an upper stress-free boundary.

clearly evident. In finishing the discussion of Table 2.2 we note that cases IA and IIIC give rise to the same equation after Θ or Φ have been eliminated from the system (3.30) except that the Rayleigh number in the case IIIC is smaller by the factor $(1 + \eta)\eta/2$ than in the case IA.

Among the different combinations of gravity and temperature distributions there are two distinguished ones which satisfy the condition that expression (3.15) is independent of r. Because of this property the cases IC and IIIA lead to self-adjoint systems of equations (3.30). Only for these cases does the principle exchange of stabilities apply according to the proof of Section 3.2. But there is no indication that onset of convection in the form of oscillatory motion occurs in any of the other possible cases. On the other hand, subcritical finite amplitude onset of convection must be expected in general for those formulations of equations (3.30) which are not self-adjoint as will be discussed in detail in Section 4.3.

4. Weakly Nonlinear Theory of Convection

4.1. Convection Patterns

Until now we have bypassed the problem of the horizontal pattern of convection that is realized when a layer heated from below becomes unstable. Both the infinitely extended plane layer and the spherical shell are characterized by the isotropy and homogeneity with respect to the horizontal dimensions, and the eigenvalue R of the linear problem is thus degenerate. Because of the finite size of a spherical shell the degeneracy of spherical convection is finite, while the degeneracy becomes infinite in the ideal limit of an infinite plane layer. It is characteristic of nonlinear problems that the degeneracy of solutions of the corresponding linearized problems is in part removed by the effects of finite amplitude. But in general some degeneracy still persists and a stability analysis is required to isolate one or more solutions that are physically realizable. These two aspects of the problem of preferred convection patterns will now be discussed in the case of a plane layer.

The general solution of equation (3.4) can be written in the form

$$w(x,y) = \sum_{m=-N}^{N} c_m \exp\{i\mathbf{k}_m \cdot \mathbf{x}\}, \tag{4.1}$$

where the horizontal vectors $\mathbf{k}_m$ are arbitrary except for the property

$$|\mathbf{k}_m| = \alpha, \qquad \mathbf{k}_m = -\mathbf{k}_{-m}.$$

To insure that (4.1) is a real expression the condition $c_{-m} = c_m^+$ must be imposed where the superscript + indicates the complex conjugate. It is convenient to use the normalization condition

$$\sum_{m=1}^{N} |c_m|^2 = 1/2. \tag{4.2}$$

The summation limit N may tend to infinity in which case expression (4.1) represents a general almost periodic function which satisfies equation (3.4).

The weakly nonlinear limit of equations (2.14) can be investigated in terms of an expansion in powers of the amplitude ε of convection. Using the representation (2.19) of the velocity field we write

$$\begin{aligned} \Phi &= \varepsilon\Phi_1 + \varepsilon^2\Phi_2 + \varepsilon^3\Phi_3 + \ldots \\ \Theta &= \varepsilon\Theta_1 + \varepsilon^2\Theta_2 + \varepsilon^3\Theta_3 + \ldots \\ \mathrm{R} &= \mathrm{R}_0 + \varepsilon\mathrm{R}_1 + \varepsilon^2\mathrm{R}_2 + \ldots . \end{aligned} \tag{4.3}$$

There is not need to consider the function ψ since it does not enter the problem until the order ε^3 (Schlüter *et al.*, 1965). In the limit of infinite Prandtl number, which will be assumed in the following, the vertical component of vorticity

$$\boldsymbol{\lambda}\cdot\nabla \times \mathbf{v} = -\Delta_2\psi$$

vanishes exactly as does the radial component in the analogous case for spherical convection. Assuming $f_1 = f_3 = 1$, $\mathrm{R}_Q = 0$ as in the case of equations (3.1) we obtain the nonlinear equations for steady convection flow

$$\nabla^4\Delta_2\Phi - \Delta_2\Theta = 0, \tag{4.4a}$$

$$\nabla^2\Theta - \mathrm{R}\Delta_2\,\Phi = \nabla\partial_z\Phi\cdot\nabla\Theta - \nabla^2\Phi\partial_z\Theta. \tag{4.4b}$$

The solution in the order ε is given for stress-free boundary conditions (3.5) by expressions (3.3) and (3.6) for $n = 1$,

$$\begin{aligned} \Phi_1 &= \alpha^{-2}\sin\pi z \sum_{m=-N}^{N} c_m \exp\{i\mathbf{k}_m\cdot\mathbf{s}\}, \\ \Theta_1 &= (\pi^2+\alpha^2)^2\Phi_1, \\ \mathrm{R}_0 &= (\pi^2+\alpha^2)^3/\alpha^2, \end{aligned} \tag{4.5}$$

Although $\alpha = \alpha_c \equiv \pi/\sqrt{2}$ will usually be assumed there is no need to specify α at this point. In the order ε^2 the equations

$$\nabla^4\Delta_2\Phi_2 - \Delta_2\Theta_2 = 0, \tag{4.6a}$$

$$\nabla^2\Theta_2 - R_0\Delta_2\Phi_2 = \nabla\partial_z\Phi_1\cdot\nabla\Theta_1 - \nabla^2\Phi_1\partial_z\Theta_1 + R_1\Delta_2\Phi_1, \quad (4.6b)$$

must be solved. This inhomogeneous linear system of equations is solvable if and only if the inhomogeneous part is orthogonal to all solutions of the adjoint homogeneous problem. The linear system (3.1) is self-adjoint for the usual kinds of boundary conditions. But if the expression (3.15) is not constant the property of self-adjointness is lost and the adjoint homogeneous problem must be solved in general before the solvability condition can be evaluated. In the present case all terms on the right hand side of (4.6b) except for the last one are antisymmetric with respect to $z = ½$ and thus orthogonal to the solutions of the linear problem. The solvability condition is therefore satisfied with $R_1 = 0$ and the solution Θ_2, Φ_2 can be obtained in the form

$$\Phi_2 = -\sin 2\pi z \sum_{n,m=-M}^{N} c_n c_m H(\mathbf{k}_n\cdot\mathbf{k}_m) \exp\{i(\mathbf{k}_n + \mathbf{k}_m)\cdot\mathbf{x}\}, \quad (4.7a)$$

$$\theta_2 = -\sin 2\pi z \sum_{n,m=-M}^{N} c_n c_m (4\pi^2 + 2\alpha^2 + 2\mathbf{k}_n\cdot\mathbf{k}_m)^2 H(\mathbf{k}_n\cdot\mathbf{k}_m) \exp\{i(\mathbf{k}_n + \mathbf{k}_m)\cdot\mathbf{x}\}. \quad (4.7b)$$

with

$$H(\mathbf{k}_n\cdot\mathbf{k}_m) \equiv \frac{(\alpha^2 - \mathbf{k}_n\cdot\mathbf{k}_m)(\pi^2 + \alpha^2)^2\alpha^{-4}\pi/2}{(4\pi^2 + 2\alpha^2 + 2\mathbf{k}_n\cdot\mathbf{k}_m)^3 - 2R_0(\alpha^2 + \mathbf{k}_n\cdot\mathbf{k}_m)}$$

The equations in the order ε^3

$$\nabla^4\Delta_2\Phi_3 + \Delta_2\Theta_3 = 0, \quad (4.8a)$$

$$\nabla_2\Theta_3 - R_0\Delta_2\Phi_3 = \nabla\partial_z\Phi_2\cdot\nabla\Theta_1 + \nabla\partial_z\Phi_1\cdot\nabla\Theta_2 - \nabla^2\Phi_2\partial_z\Theta_1 - \nabla^2\Phi_1\partial_z\Theta_2 + R_2\Delta_2\Phi_1 \quad (4.8b)$$

do not have to be solved explicitly since the solvability conditions yield the constraints for the coefficients c_n and the expression for R_2, which are the main goals of the weakly nonlinear analysis. By multiplying (4.8a) by $2R_0 \sin\pi z \exp\{-i\mathbf{k}_p\cdot\mathbf{x}\}$ and (4.8b) by $2(\pi^2 + \alpha^2)^2 \sin\pi z \exp\{-i\mathbf{k}_p\cdot\mathbf{x}\}$, adding the equations and averaging them over the fluid layer we obtain the following relationship

$$0 = -R_2 c_p(\pi^2 + \alpha^2)^2\alpha^{-2} + \sum_{m,n,l} (4\pi^2 + 2\alpha^2 + 2\mathbf{k}_n\cdot\mathbf{k}_m)^2 H(\mathbf{k}_n\cdot\mathbf{k}_m)(\mathbf{k}_l\cdot(\mathbf{k}_n + \mathbf{k}_m) + 2\alpha^2)$$
$$(\pi^2 + \alpha^2)^2\pi(4\alpha^2)^{-1}\delta(\mathbf{k}_l + \mathbf{k}_n + \mathbf{k}_m - \mathbf{k}_p)c_m c_n c_l \quad \text{for } -N \leq p \leq N. \quad (4.9)$$

For reasons of symmetry equations (4.9), (4.2) are solved by

$$|c_1|^2 = \ldots = |c_N|^2 = (2N)^{-1}, \tag{4.10}$$

whenever the $\mathbf{k}_n$ vectors are distributed in a regular fashion, i.e., the angle between any two adjacent vectors is the same. Of particular interest are the solutions (4.10) corresponding to $N = 1, 2, 3$. In the cases $N = 1$ and $N = 2$, the undetermined phases of the complex coefficients c_m reflect the freedom of translations of any solution in the horizontal plane. Only a single physically distinct solution is given by (4.10) in these cases. For $N = 1$ it is called the roll solution, while the solution for $N = 2$ is known as square pattern solution as shown in Figure 2.2. In the case $N = 3$ a one-dimensional manifold of distinct solutions is described by (4.10) since the number of undetermined phases exceeds the number of degrees of freedom for translations in the horizontal plane. Of physical importance are the two hexagon solutions

$$l\text{-hexagons corresponding to } c_1c_2c_3 = (2N)^{-3/2}, \tag{4.11a}$$

$$g\text{-hexagons corresponding to } c_1c_2c_3 = -(2N)^{-3/2}, \tag{4.11b}$$

but the possibility that other solutions, such as the regular triangle solution with $c_1c_2c_3 = i(2N)^{-3/2}$ (Gorkov, 1957) may be important in some yet undiscovered physical context cannot be excluded. A detailed mathematical discussion of patterns within a hexagonal lattice has been given recently by Buzano and Golubitsky (1983) and by Golubitsky *et al.* (1984).

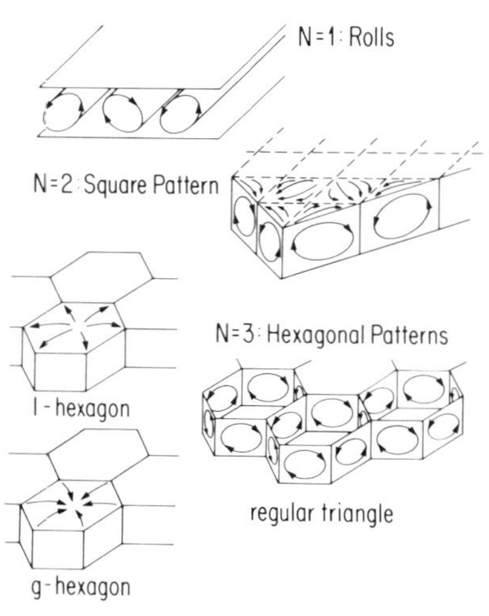

Figure 2.2. Patterns of steady convection.

A physical property of special interest is the enhancement of heat transport across the fluid layer due to convection. When the temperature difference ΔT between the boundaries is fixed, the integration of the horizontally averaged equation (2.14c) yields

$$\frac{\partial}{\partial z}\overline{\Theta} = \overline{v_x\Theta} - \langle v_z\Theta\rangle, \tag{4.12}$$

where the bar indicates the horizontal average and the angular brackets indicate the average over the entire fluid layer. Equation (4.12) remains valid even in the case of time-dependent convection, as long as the averaged quantities remain constant in time. Since v_z vanishes at $z = 0,1$ the heat flux is carried by conduction only at the boundaries. In the case $f_1 = f_3 = 1$, $R_Q = 0$ considered in this section, the Nusselt number, which is defined as the ratio between the heat transport in the presence of convection and the heat transport without convection, can be written in the form

$$\mathrm{Nu} \equiv (\mathrm{R} + \langle v_z\Theta\rangle)/\mathrm{R}. \tag{4.13}$$

The weakly nonlinear analysis described above permits us to evaluate the quantity

$$\Lambda \equiv d\langle v_z\Theta\rangle/d\mathrm{R}|_{\mathrm{R}=\mathrm{R}_0} = \langle -\Delta_2\Phi_1\Theta_1\rangle/\mathrm{R}_2. \tag{4.14}$$

Λ describes the factor by which the slope of the heat flux as a function of the Rayleigh number changes after the onset of convection. For the three simple solutions discussed above we find from equations (4.9)

$$N = 1: \quad \Lambda_{\text{rolls}} = 2 \tag{4.15a}$$

$$N = 2: \quad \Lambda_{\text{squares}} = 2[1 + 8\chi_0^2(32\chi_0^3 - 1)^{-1}(1 + \alpha^2/\pi^2)^{-1}]^{-1}, \tag{4.15b}$$

$$N = 3: \Lambda_{\text{hexagons}} = 2[1 + 24\chi_1^2(64\chi_1^3 - 1)^{-1}(1 + \alpha^2/\pi^2)^{-1} + 8\chi_2^2(64\chi_2^3 - 3)^{-1}(1 + \alpha^2/\pi^2)^{-1}/3]^{-1}, \tag{4.15c}$$

where the definitions

$$\chi_0 \equiv (\pi^2 + \alpha^2/2)(\pi^2 + \alpha^2)^{-1}, \qquad \chi_1 \equiv (\pi^2 + \alpha^2/4)(\pi^2 + \alpha^2)^{-1},$$
$$\chi_2 \equiv (\pi^2 + 3\alpha^2/4)(\pi^2 + \alpha^2)^{-1}$$

have been used. Expressions (4.15) have been derived by Malkus and Veronis (1959) who together with Gorkov (1957) initiated the study of nonlinear convection. It is obvious from expressions (4.15) that rolls have a higher heat transport than squares and that the heat transport of the latter exceeds that of hexagons. This relationship continues to hold in the case of finite Prandtl number and in the case of rigid boundaries. For details see the papers of Malkus and Veronis (1959), Schlüter *et al.* (1965) and Busse (1967a).

Since there are an infinite number of possible steady solutions available according to relationship (4.10) which describes only a fraction of the full manifold of solutions, a stability analysis must be added to distinguish one or more physically realizable forms of convection. The fact that convection rolls are most effective in transporting heat suggest that they also excel in the property of stability. It can indeed be demonstrated by an analysis similar to that given above that they represent the only stable solution among all steady solutions (Schlüter *et al.*, 1965). This confirms, in this particular case, the Malkus hypothesis of the preference of the solution of maximum transport (Malkus, 1954b). In general, however, this hypothesis is not valid. Many cases are known where the convection solution of maximum heat transport is unstable. At higher Rayleigh numbers, for example, the heat transport of the square pattern solution exceeds that of the roll solution in the case of rigid boundaries, but the squares remain unstable (Frick *et al.*, 1983).

The results discussed so far are representative for equations which are symmetric about the median plane of the layer, such as equations (4.4). Even when stress free boundaries are assumed on one side and no slip conditions on the other side, rolls continue to be the preferred solution (Busse, 1967a). The thermal boundary conditions, however, may affect the preferred pattern of convection. In the limit $\gamma_e \approx 0$ the square pattern solution is preferred (Busse and Riahi, 1980; Proctor, 1981) and the changeover from rolls to squares appears to occur in the neighborhood of $\gamma_e \approx 1$, depending on various secondary parameters (Jenkins and Proctor, 1984; Riahi, 1984).

4.2. Asymmetry Effects

The fact that R_1 vanishes in the analysis of the preceding section could have been predicted on the basis of the coincidence of linear stability and energy stability result as demonstrated in section 1.32. A non-vanishing R_1 implies the existence of subcritical finite amplitude convection since the direction of motion can always be chosen such that $\varepsilon R_1 < 0$. This possibility is excluded as long as the linear equations can be derived from the energy functional (3.13). In this section an example of the more general class of problems will be considered, which are either not self-adjoint in the linear limit or exhibit nonlinear terms with a symmetry different from that of the advection terms on the right hand side of equations (2.14).

When the fluid layer has a uniform distribution of heat sources the function f_4 is a linear function of z. In this section we assume

$$f_4 = z - \tfrac{1}{2}. \tag{4.16}$$

In the limit R = 0 the choice (4.16) for f_4 corresponds to a parabolic temperature profile in the static state with equal temperatures on both boundaries. In the following the opposite limit

$$\delta \equiv \mathrm{R}_Q << 1$$

will be assumed and the solution will be represented in terms of a double power series,

$$\Phi = \sum_{\nu=0,\mu=1}^{\infty} \varepsilon^{\mu}\delta^{\nu}\Phi_{\mu}^{(\nu)};$$

$$\theta = \sum_{\nu=0,\mu=1} \varepsilon^{\mu}\delta^{\nu}\theta_{\mu}^{(\nu)}; \qquad \mathrm{R} = \sum_{\nu=0,\mu=0} \varepsilon^{\mu}\delta^{\nu}\mathrm{R}_{\mu}^{(\nu)}. \tag{4.17}$$

The basic equations to be solved differ from the system (4.4) only by the addition of the term $f_4\delta\Delta_2\Phi$ on the right hand side of the heat equation. Starting with solution (4.5) we obtain in the order $\varepsilon^1\delta^1$ the system of equations

$$\nabla^4\Delta_2\Phi_1^{(1)} - \Delta_2\Theta_1^{(1)} = 0, \tag{4.18a}$$

$$\Delta^2\Theta_1^{(1)} - \mathrm{R}_0^{(0)}\,\Delta_2\Phi_1^{(1)} = f_4\Delta_2\,\Phi_1^{(0)} + \mathrm{R}_0^{(1)}\Delta_2\Phi_1^{(0)}. \tag{4.18b}$$

Because f_4 is antisymmetric with respect to the plane $z = \tfrac{1}{2}$, the solvability is trivially satisfied with $\mathrm{R}_0^{(1)} = 0$ and the solution can be written in the form

$$\Phi_1^{(1)} = h(z)\Phi_1^{(0)}; \qquad \Theta_1^{(1)} = \nabla^4\Phi_1^{(1)} \tag{4.19}$$

where $h(z)$ is a function which is antisymmetric in $z - \tfrac{1}{2}$. In the order $\varepsilon^1\delta^2$ the solvability condition yields a non-trivial result. A negative value $\mathrm{R}_0^{(2)}$ is obtained (Krishnamurti, 1968a) which indicates a decrease of the critical value of the Rayleigh number for onset of convection.

A more interesting result is obtained in the order $\varepsilon^2\delta^1$ of the problem,

$$\nabla^4\Delta_2\Phi_2^{(1)} - \Delta_2\Theta_2^{(1)} = 0 \tag{4.20a}$$

$$\nabla^2\Theta_2^{(1)} - \mathrm{R}_0^{(0)}\Delta_2\Phi_2^{(1)} =$$

$$f_4\Delta_2\Phi_2^{(0)} + \nabla\frac{\partial}{\partial z}\Phi_1^{(1)}\cdot\nabla\Theta_1^{(0)} + \nabla\frac{\partial}{\partial z}\Phi_1^{(0)}\cdot\nabla\Theta_1^{(1)} -$$

$$\nabla^2\Phi_1^{(1)}\frac{\partial}{\partial z}\Theta_1^{(0)} - \nabla^2\Phi_1^{(0)}\frac{\partial}{\partial z}\Theta_1^{(1)} + \mathrm{R}_1^{(1)}\Delta_2\phi_1^{(0)}. \tag{4.20b}$$

The solvability conditions obtained in the same manner as in the case of relationships (4.9) yield

$$R_1^{(1)} c_p = B \sum_{n,m} \delta(\mathbf{k}_n + \mathbf{k}_m + \mathbf{k}_p) c_n c_m \text{ for all } p, \tag{4.21}$$

where B is a positive number.

These equations are satisfied with $R_1^{(1)} = 0$ for any distribution of $\mathbf{k}_n$-vectors which does not include two vectors which enclose an angle of 60° between them. If the latter possibility occurs all vectors $\mathbf{k}_n$ must enclose the angle of 60° with two other vectors. The simplest solution of this kind is the hexagon solution. The two solutions (4.11) satisfy conditions (4.21) with

$$\sqrt{3/2}\, R_1^{(1)} = \begin{cases} B & \text{for } l\text{-hexagons,} \\ -B & \text{for } g\text{-hexagons.} \end{cases} \tag{4.22}$$

Other solutions in the form (4.10) with $N = 3$ do not satisfy conditions (4.21) since $c_1 c_2 c_3$ is not real for those solutions.

The asymmetry effect represented by the term of the order δ not only affects the relationship between Rayleigh number and amplitude ε of convection, but also influences the stability of competing convection patterns. In Figure 2.3 the parabolic curves based on approximate formula

$$R - R_c \approx \varepsilon\delta R_1^{(1)} + \varepsilon^2 R_2^{(0)} \tag{4.23}$$

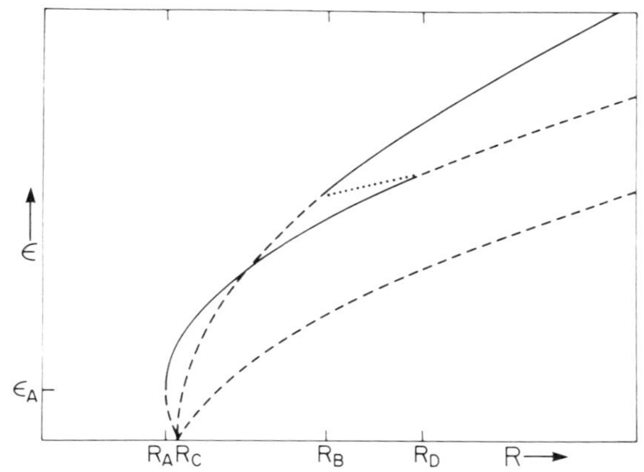

Figure 2.3. The dependence of the amplitude ε of convection on the Rayleigh number R for rolls, l- and g-hexagons. The respective solutions are unstable along the broken parts of the curves.

are shown for the three solutions of interest with the dashed parts indicating the regimes where the respective solutions are unstable. For details of the stability analysis see Busse (1967a). The case of internal heating, $\delta > 0$, has been assumed; for internal cooling, $\delta < 0$, the roles of l-hexagon and g-hexagon solutions would be reversed. There are a number of features displayed in Figure 2.3 which are worth emphasizing.

1. Except in the exactly symmetric case a hexagonal pattern of convection is preferred at the onset of convection. Experiments such as those of Silveston (1958) show the transition to roll convection as the Rayleigh number is increased.
2. There exists a range of Rayleigh numbers where two solutions are stable at the same time. This property gives rise to an hysteresis effect in that the hexagonal convection cells are not replaced by rolls until the increasing Rayleigh number exceeds R_D while the rolls persist for $R > R_B$ as the Rayleigh number is decreased. A similar hysteresis loop involving the hexagonal solutions and the state of no motion exists for $R_A \leq R \leq R_C$.
3. For small amplitudes of convection, $\varepsilon \leq \varepsilon_A \equiv |R_1^{(1)}|/R_2^{(0)}$, there exists no stable form of steady convection. This property is of little consequence as long as the Rayleigh number is the externally prescribed parameter. But in geophysical as well as in astrophysical applications of convection theory, the heat flux rather than the temperature difference must be regarded as given parameter. In that case the interesting prospect of time dependent convection arises. As shown by Busse (1967b), convection in the form of nonlinear relaxation oscillations may occur.

The properties discussed above are summarized in Figure 2.4 which shows the stability of convection patterns as a function of the asymmetry parameter $B\delta$. Because of the general nature of the perturbation theory result, it is only through the quantity B that any asymmetric property enters the theory of steady convection as well as the stability theory. Different kinds of asymmetries in the convection layer as for example those introduced by the temperature dependence of the material properties give rise to the same result. The temperature dependence of viscosity is usually the most important cause of asymmetry in material properties. Its importance for the preference of hexagonal convection cells was recognized already by Graham (1933). Tippelskirch (1956) did some imaginative experiments on convection in liquid sulphur to demonstrate the reversal of motion in the hexagonal cells as the temperature dependence of viscosity changes sign. This concept was supported theoretically by the analyses of Palm (1960) and Segel and Stuart (1962). A systematic stability analysis was given by Busse (1962, 1967a)

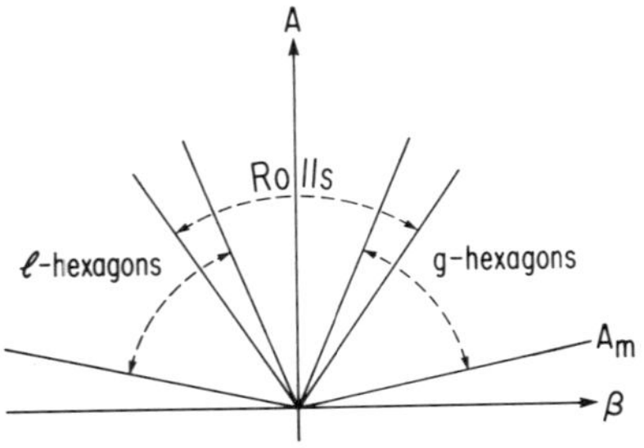

Figure 2.4. Stability range of convection in the form of rolls and hexagons as a function of the amplitude $A = |\varepsilon|$ and the asymmetry parameter $\beta = \delta B$.

who also included the temperature dependence of other material properties. For recent quantitative experimental investigation see Ahlers (1980) and Walden and Ahlers (1981).

The case of an internally heated layer used as an example in this section is most easily realized in the laboratory by a constant rate of change of the mean temperature of the convection layer. Krishnamurti (1968 a,b) reports both theoretical and experimental results for this problem. In contrast to the work by others (see Section 5.7) Krishnamurti finds l-hexagons in a layer with decreasing mean temperature which corresponds to the internally heated case. This discrepancy was first pointed out by Roberts (1967) and may be traced to an incorrect sign in expression (14) of Krishnamurti (1968a). There still remains, however, the question whether the sign of motion in the pattern observed by Krishnamurti (1968b) could have been misinterpreted.

4.3. Patterns of Convection in Spherical Shells

Equations (3.33) and (3.4) have the common property that they admit a large manifold of potential convection patterns. But the degeneracy of the eigen solutions of the linear problem (3.30)

$$\Phi = f(r)w_l(\theta,\phi) \equiv f(r) \sum_{m=-l}^{l} c_m Y^m(\theta,\phi) \tag{4.24}$$

is finite because of the finite domain of the spherical shell in contrast to the infinite domain of the plane layer. In the representation (4.24) spherical harmonics Y_l^m are assumed in the form

$$Y_l^m = \hat{P}_l^m(\cos\theta)\exp\{im\phi\},$$

where the functions $\hat{P}_l^m$ differ from the commonly used associated Legendre polynomials by a normalization factor,

$$\hat{P}_l^m(z) \equiv [(2l + 1)(l - m)!/(l + m)!]^{1/2} P_l^m(z). \tag{4.25}$$

The factor has been chosen to insure that the average of $[\hat{P}_l^m]^2$ over the sphere equals unity.

Because coordinates have been introduced in the representation (4.24) the analysis of different patterns of convection is more difficult than in the planar case. Solutions which can be transformed into each other by a rotation on the sphere cannot readily be distinguished from solutions which describe physically distinct convection flows. We shall separate these two sources of the multiplicity of solutions by referring to orientational degeneracy and to pattern degeneracy. This point can be elucidated by considering the case $l = 1$ of solution (4.24). Since they are two degrees of freedom of rotation on the sphere, expression (4.24) describes only a single physically distinct convection flow. Assuming the normalization condition

$$\sum_{m=-l}^{l} |c_m|^2 = 1, \tag{4.26}$$

we find that the freedom in the choice of the coefficients c_n is equivalent to the freedom of translations on the sphere. In the case $l \geq 2$, however, there are $2l$ arbitrary coefficients c_m only two of which reflect the orientational degeneracy while the remainder corresponds to the pattern degeneracy of the problem.

The problem of possible convection pattern is much richer in the spherical case since every degree l of the spherical harmonics must be considered separately. Only for low integers l have solutions been investigated in the literature and questions about the symmetry of the bifurcating steady solutions pose some interesting mathematical problems (see, for example, the papers by Chossat, 1979; Sattinger, 1980; Golubitsky and Schaeffer, 1982; and Ihrig and Golubitsky, 1984). But there are a few results of a more general nature which we shall explore by introducing the expansion (4.3). Starting with the general expression (4.24) for Φ_1 we consider the solvability condition for the equations of the order ε^2. We note that the solvability conditions (4.21) in the planar case can be interpreted as the Euler–Lagrange equations for the extremum value $L \equiv R_1^{(1)} B^{-1}$ of the functional

$$\mathcal{L}(w) \equiv \langle www \rangle / \langle ww \rangle^{3/2}, \tag{4.27}$$

where w is a general function of the form (4.1). Similarly we find for the spherical case that R_1 is proportional to the extremum value $L^{(l)}$ of the functional

$$\mathcal{L}^{(l)}(w_l) \equiv \langle w_l w_l w_l \rangle / \langle w_l w_l \rangle^{3/2}, \tag{4.28}$$

where the angular brackets indicate the average over the spherical convection layer. The formulation (4.28) is quite general since it can be shown (Busse, 1975) that the θ,ϕ-derivatives in the nonlinear terms on the right hand side of equations (2.14) can be eliminated by partial integrations. We also note that in contrast to the case discussed in Section 4.1 self-adjointness of the linear operator is not assumed. Because of the spherical symmetry of the problem the solutions of the linear problem adjoint to system (3.30) can be written in the same form (3.32) except that the r-dependence will differ in general.

From the general expression (4.28) it can be concluded immediately that R_1 must vanish for odd l. Because of the property of spherical harmonics

$$Y_l^m\,(\pi-\theta,\ \phi+\pi) = (-1)^l\ Y_l^m(\theta,\phi), \tag{4.29}$$

the integral over $(w_l)^3$ always vanishes for odd l. But for even l there exists for any solution w_l corresponding to a local extremum of (4.28) another solution, $-w_l$, with the same value of R_1, but with opposite sign. It thus transpires that convection solutions of the form (4.24) with odd l resemble the roll and square pattern solutions of the plane layer case. When l is even a close similarity with the two hexagon solutions (4.11) can be seen. These similarities allows us to draw the following general conclusions which we shall state without entering into a detailed mathematical discussion.

1. Whenever the minimizing value R_c obtained from the linear problem (3.30) corresponds to an odd l and differs little from the value R_0 for $l+1$ or $l-1$, then the contribution εR_1 in the expansion (4.3) may lead to the subcritical finite amplitude onset of a convection pattern with the even degree $l+1$ or $l-1$. In other words, convection patterns which are symmetric in their radial motion on opposite sides on the sphere have a slight edge over patterns for which a reversal of motions leads to the same pattern, but translated on the sphere according to relationship (4.29) for odd l.
2. Hysteresis effects at the onset of convection and the associated possibility of time dependent convection in the case of fixed heat flux can be expected for convection solutions with even l.

Only when an energy balance of the form (3.10) can be derived from the basic equations and when the ratio (3.15) is constant, can $R_1 = 0$ be concluded for even l. In that case as in the case of odd l the solvability conditions of order ε^3 of the problem must be considered. Fortunately the resulting equations for the coefficients c_n can still be discussed without reference to the radial dependence on the problem. But the nonlinear relationships between the coefficients c_n become sufficiently complicated that only the cases $l = 2, 3$ have been treated in detail (Busse and Riahi, 1983). Here we confine ourselves to stating the results which so far have been obtained:

$l = 1$: a single solution exists; there is no pattern degeneracy.
$l = 2$: two solutions exist corresponding to the axisymmetric pattern.
$l = 3$: three solutions exist; the solution with tetrahedronal symmetry is preferred in general. The axisymmetric solution is always unstable.
$l = 4$: four solutions exist corresponding to the axisymmetric pattern and the pattern of octahedronal symmetry. The latter type of solutions exhibit a large value of $|R_1|$ and one of them is thus preferred.
$l = 6$: eight solutions exist corresponding to four different patterns. The two solutions of dodecahedronal symmetry exhibit the largest value of $|R_1|$ and one of them is thus preferred.

The preferred pattern of convection in spherical shells for $\lambda = 2, 3, 4, 6$ are displayed in Figure 2.5. It is interesting to note that axisymmetric solutions are not preferred for $l > 2$, even though they exist for arbitrary l as solutions of the basic equations. This property may be due to the latitudinal inhomogeneity of convection cells which characterizes all axisymmetric solutions for $l \geq 3$. On the other hand, those solutions reflecting the symmetries of the Platonic bodies are characterized by equal cells of motion which closely resemble hexagonal cells in the plane and are therefore preferred at least for small amplitudes of convection. It should be noted that the cube has the same symmetry as the octahedron and that the symmetries of the isocahedron and the dodacahedron are equivalent as is evident from the patterns shown in Figure 2.5.

The high degree of symmetry of the convection patterns displayed in Figure 2.5 suggests that the corresponding motions will remain stable when the Rayleigh number increases much beyond the critical value. But there is also the possibility that hybrid patterns could become important which can only be described by a combination of spherical harmonics of odd and even degree l. The "tennis ball" pattern, for example, can be described as a sum of spherical harmonics of degrees 1 and 2. A three cell pattern which basically arises from a combination of spherical harmonics of degrees 2 and 3 has been found in numerical simulations of nonlinear convection in spherical shells by Baumgardner (1983). Even axisymmetric convection patterns with more than two cells can become stable at higher Rayleigh numbers (Zebib *et al.* 1983) where the pattern is no longer determined by a single harmonic. It must be expected that there coexist even more stable steady solutions in the case of spherical shells than in the plane layer case for

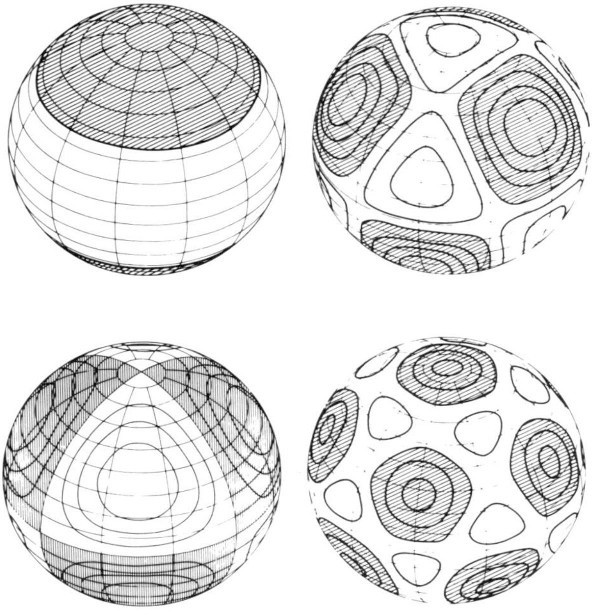

Figure 2.5. Preferred patterns of convection in spherical shells for $l = 2, 3, 4$, and 6. The motion is either rising or descending in the shaded areas.

which the coexistence of rolls and hexagons is shown in Figures 2.3 and 2.4. The possibilities of multiple steady states are only beginning to be explored and interesting results can be expected from future research in this area.

5. Strongly Nonlinear Convection at Infinite Prandtl Number

5.1. Numerical Solutions

The fact that convection in the form of two-dimensional convection rolls is stable for Rayleigh numbers close to the critical value for the most commonly assumed boundary conditions has been a boon for the numerical analysis of convection. Two-dimensional numerical integration schemes could be applied with the expectation that physically relevant information will be obtained. In fact, two-dimensional convection has become the favored testing ground for newly developed schemes for the numerical integration of boundary value problems. Only recently have numerical solutions for three-dimensional problems of convection become feasible. In this section we shall focus on some older research in the two-dimensional problem that is particularly relevant to high Prandtl number fluids.

We consider the nonlinear equations (4.4) as in the discussion of the weakly nonlinear problem. The limit of infinite Prandtl number P simplifies the problem considerably in that the linear equation of motion can be solved exactly for a given temperature distribution. If, for example, the Fourier expansion

$$\Theta = \sum_{\lambda,\nu} b_{\lambda\nu} \exp\{i\lambda\alpha x\} \sin \nu\pi z \tag{5.1}$$

is assumed, the solution of equation (4.4a) can be written in the form

$$\Phi = \sum_{\lambda,\nu} b_{\lambda\nu} \exp\{\mathrm{i}\lambda\alpha x\}\; v_\nu(\lambda\alpha,z), \tag{5.2}$$

where

$$v_\nu(\lambda\alpha,z) \equiv [(\nu\pi)^2 + (\lambda\alpha)^2]^{-2} \sin \nu\pi z \tag{5.3a}$$

in the case of stress-free boundary conditions (3.5), while

$$v_\nu(\lambda\alpha,z) \equiv [(\nu\pi)^2 + (\lambda\alpha)^2]^{-2} \sin \nu\pi z + \frac{\nu\pi}{[(\lambda\alpha)^2 + (\nu\pi)^2]^2}$$

$$\frac{\tfrac{1}{2}\sinh \lambda\frac{\alpha}{2} \cosh \lambda\alpha\zeta - \zeta\cosh \lambda\frac{\alpha}{2} \sinh \lambda\alpha\zeta}{-\sinh \lambda\frac{\alpha}{2} \cosh \lambda\frac{\alpha}{2} - \lambda\frac{\alpha}{2}} \quad \text{for odd } \nu,$$

$$\frac{\tfrac{1}{2}\cosh \lambda\frac{\alpha}{2} \sinh \lambda\alpha\zeta - \zeta\sinh \lambda\frac{\alpha}{2} \cosh \lambda\alpha\zeta}{\sinh \lambda\frac{\alpha}{2} \cosh \lambda\frac{\alpha}{2} - \lambda\frac{\alpha}{2}} \quad \text{for even } \nu, \tag{5.3b}$$

with $\zeta \equiv z - \tfrac{1}{2}$ is obtained in the case of rigid boundary conditions (3.7). After multiplying equation (4.4b) by $\exp\{-i\varkappa\alpha x\} \sin \mu\pi z$ and averaging the result over the fluid layer we obtain a nonlinear system of algebraic equations for the unknowns $b_{\lambda\nu}$ which can be solved by a Newton-Raphson iteration method. Because of the symmetry of the problem several simplifications are possible. Any periodic solution of equations (4.4) can be shifted such that Θ becomes a symmetric function in x with real coefficients $b_{\lambda\nu}$. In addition it is found that the equations (4.4) admit a subclass of solutions for which the $b_{\lambda\nu}$ vanish whenever $\lambda + \nu$ is odd. Since this class includes the solutions obtained in the weakly nonlinear limit of the problem, the attention is usually restricted to solutions with even $\lambda + \nu$. This symmetry can also be expressed in spatial coordinates,

$$\Theta\,(x,z - \tfrac{1}{2}) = -\Theta(\pi\alpha^{-1} - x, \tfrac{1}{2} - z). \tag{5.4}$$

There are other two-dimensional periodic forms of convection, for example those corresponding to two convection rolls on top of each other. But these solutions have found little attention since there is no reason to expect that they are physically realizable.

The summation indices in expressions (5.1) and (5.2) run through the intervals $-\infty < \lambda < \infty$, $0 < \nu < \infty$. The system of algebraic equations for the unknowns $b_{\lambda\nu}$ must thus be truncated before a numerical solution can be obtained. A convenient and economic prescription in terms of computer time is given by

$$b_{\lambda\nu} \equiv 0 \text{ for } \lambda + \nu > N_T. \tag{5.5}$$

If N_T is chosen sufficiently high the physically relevant properties of the solution are expected to become essentially independent of N_T. In the solutions of equations (4.4) for stress-free boundaries (Veronis, 1966; Straus, 1972) and for rigid boundaries (Busse, 1967c) it was found that reasonable convergence is achieved when the convective heat transport changes by less than about 1 percent when N_T is replaced by $N_T - 2$. Results shown in the above mentioned papers indicated that $N_T = 10$ yields reliable results up to R $= 2 \cdot 10^4$ according to this criterion.

One of the goals of numerical computations is the determination of the dependence of the Nusselt number on the Rayleigh number. But this task is complicated by the property that the convective heat transport varies with the wavenumber α. The wavenumber maximizing the heat transport increases with R whereas a decrease is observed in experiments. Thus the power laws which are used to fit the numerical data at high Rayleigh numbers are usually given for a fixed wavenumber α. For stressfree conditions the power law

$$\text{Nu} = 1.96(\text{R}/\text{R}_c)^{1/3} \tag{5.6}$$

appears to fit the data for $\alpha = \alpha_c = \pi/\sqrt{2}$. The expression (5.6) has been given by Moore and Weiss (1973) who did their computations for the Prandtl number of water, P = 6.8. But the numerical data of Straus (1972) for infinite Prandtl number and the data of Veronis (1966) for P = 6.8 and P = 100 agree reasonably well with the power law (5.6). For rigid boundaries the computations of Busse (1967c) and more recent ones done in connection with work of Frick *et al.* (1983) can be represented by

$$Nu = 1.77(R/R_c)^{0.225}. \tag{5.7}$$

Since the accuracy of the numerical results decreases as the Rayleigh number increases to a few times 10^5, the exponent in formula (5.7) is not in serious disagreement with the exponent ⅕ deduced from boundary layer theory (see Section 5.4). From a physical point of view, formulas (5.6), (5.7) are of doubtful value because they are based on the questionable assumption that convection rolls retain the critical value of their wavenumber. Moreover, the computations of Moore and Weiss (1973) indicate that a transition to oscillatory convection has occurred by the time the Rayleigh number has reached 100 times its critical value. In the case of rigid boundaries a transition to three-dimensional convection occurs at about 13 times the critical value as will be discussed in the next section. Such an instability also occurs in the case of one free and one rigid boundary for which Kvernvold (1979) found the fit

$$Nu = 1.6(R/R_c)^{0.3} \tag{5.8}$$

from his computational results which extend up to Rayleigh numbers of the order $50 \cdot R_c$.

One of the most interesting results of the numerical computations at high Rayleigh number is the development of thermal boundary layers and of the plume like structure of the temperature field in the up- and down-moving flow. This is already evident at fairly moderate Rayleigh numbers as shown in Figure 2.6. At higher Rayleigh numbers the interior of each convection roll becomes even more isothermal while boundary layer and plume structures become thinner. The mushroom like spreading of the hot and cold plumes as they meet the opposite boundary is responsible for the property that the gradient of the horizontally averaged temperature actually reverses in the interior of the layer. This reversal appears to be most pronounced for high Prandtl numbers and for low values of α, where more space is available for the spreading plume. A typical profile of the mean temperature is shown in Figure 2.7. A typical picture of the streamlines of convection rolls is shown in Figure 2.8a. The buoyancy of the hot (cold) plume draws the streamlines into the lower left (upper right) corner and creates a maximum of the vorticity at the rigid boundary as shown in Figure 5.8b. This property contrasts with the opposite distortion caused by inertial effects at Prandtl numbers of the order unity or lower in which case the vorticity maximum is shifted towards the lower right (upper left) corner (Clever and Busse, 1981).

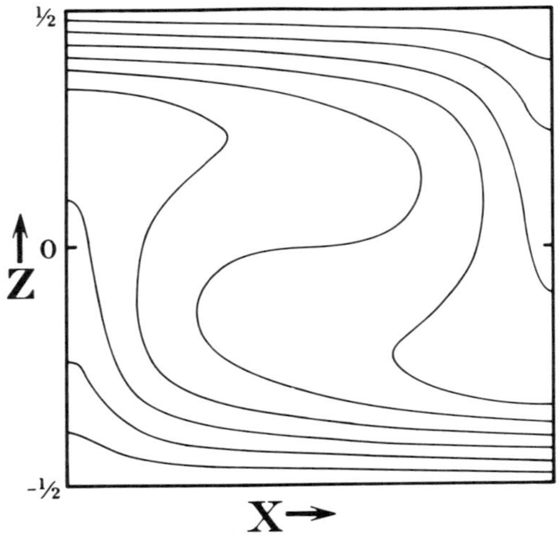

Figure 2.6. The isotherms of a convection roll with $\alpha = 3.117$ for $R = 2 \cdot 10^4$, $P = \infty$ in the presence of rigid upper and lower boundaries.

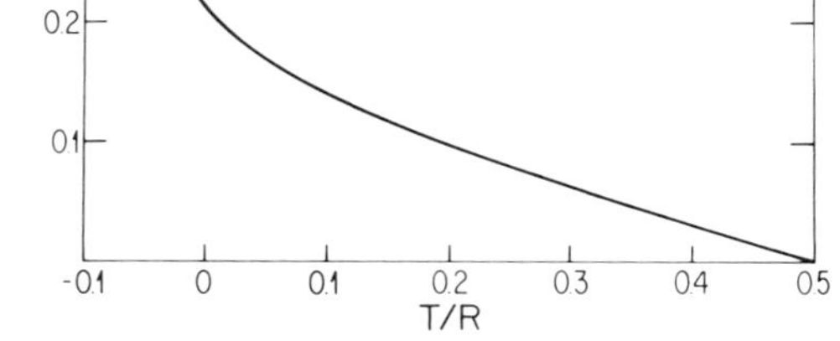

Figure 2.7. The variation of the mean temperature for $\alpha = 3.117$, $R = 2.2 \cdot 10^4$, $P = \infty$. Only the lower half of the profile is shown which is antisymmetric with respect to the median plane of the layer.

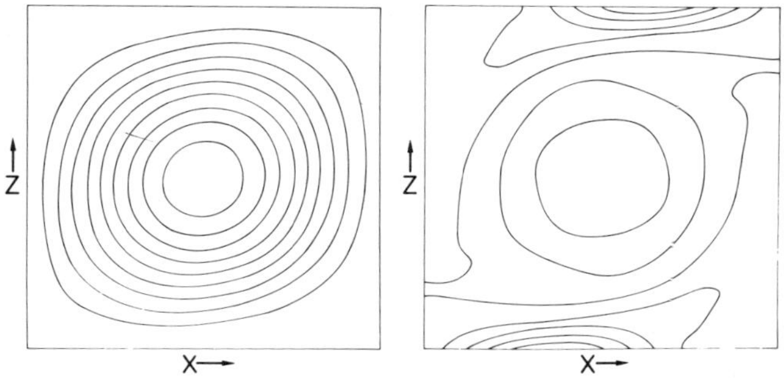

Figure 2.8. (a), The streamlines of a convection will be in the presence of rigid boundaries for $\alpha = 3.117$, $R = 2 \cdot 10^4$, $P = \infty$. (b), Lines of constant vorticity for the same parameter values as in (a).

The relatively thick thermal boundary layer near the rigid boundary suggests that it may become unstable owing to the same mechanism that leads to the onset of convection in a static fluid layer. A rough analysis supports this expectation. Using the dimensionless thickness δ of the thermal boundary layer as indicated in Figure 2.9, the dimensionless heat transport H can be written in the form

$$H \approx R/2\delta, \tag{5.9}$$

since the temperature gradient alone must carry the heat flux near the boundary. The criterion for the instability of the thermal boundary layer

$$R\delta^3/2 > R_c \tag{5.10}$$

can be expressed in terms of the heat transport H according to relationship (5.9),

$$H < R(R/2R_c)^{1/3}/2 \text{ for instability.} \tag{5.11}$$

Instability of the thermal boundary layers must thus be expected whenever the Nusselt number increases with a power of less than one-third of the Rayleigh number. Convection with rigid boundaries should be unstable according to criterion (5.11), but convection with stress-free boundaries should remain stable with respect to this mechanism. The detailed stability analysis presented in Section 5.2 confirms these expectations. The simple arguments presented above do not take into account the stabilizing effect of the flow of the basic convection roll. But since the instability can occur in the

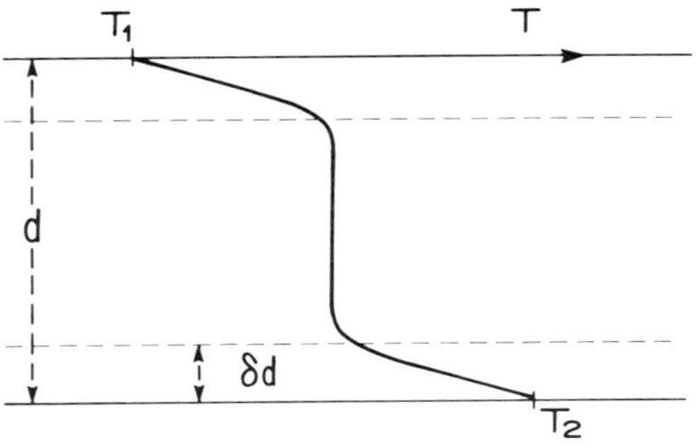

Figure 2.9. Sketch of the profile of the horizontally averaged temperature in a convection layer. δ is in dimensionless thickness of the thermal boundary layers.

form of rolls at right angles to the given rolls, the stabilizing effect is relatively small. It is well known that the onset of longitudinal rolls is not affected by the presence of a plane parallel shear (Deardorff, 1965).

The numerical studies of two-dimensional convection rolls that have been made for various physical conditions are too numerous to be listed here. A particularly extensive study is that of McKenzie *et al.* (1974). Some others will be mentioned in later sections of this chapter. A numerical analysis of nonlinear axisymmetric convection in spherical shells based on an extension of the method described in this section has been performed by Zebib *et al.* (1980, 1983) and Schubert and Zebib (1980).

5.2. Instabilities of Convection Rolls

In view of the large manifold of steady solutions that exist in nonlinear problems such as that of convection in a fluid layer heated from below, it is important to test the stability of numerically obtained solutions. The computational effort and expense that is required to test the stability with respect to three-dimensional disturbances of infinitesimal amplitude is actually not significantly different from the computational cost of generating the steady two-dimensional solution. The stability analysis also provides valuable information about the three-dimensional or time-dependent convection solution that replaces the two-dimensional solution in the case of instability. In the following these features will be exhibited in the discussion of a few examples.

The equations for infinitesimal disturbances $\tilde{\Phi}$, $\tilde{\Theta}$ superimposed onto the steady solutions Φ, Θ of equations (4.4) are given by

$$\nabla^4\Delta_2\tilde{\Phi} - \Delta_2\tilde{\Theta} = 0 \tag{5.12a}$$

$$\nabla^2\tilde{\Theta} - R\Delta_2\tilde{\Phi} = \nabla\partial_z\tilde{\Phi}\cdot\nabla\Theta + \nabla\partial_z\Phi\cdot\nabla\tilde{\Theta} - \nabla^2\tilde{\Phi}\partial_z\Theta - \nabla^2\Phi\partial_z\tilde{\Theta} + \sigma\tilde{\Theta}, \tag{5.12b}$$

where a time dependence of the form $\exp\{\sigma t\}$ has been assumed for the disturbances. Equations (5.12) together with boundary conditions of the form (3.5) or (3.7) constitute a linear boundary value problem with σ as eigenvalue. The task is to find the eigenvalue σ with maximum real part. When the latter is positive the steady solution is unstable; otherwise it is stable with respect to infinitesimal disturbances. Because of its linearity, equation (5.12a) can be solved exactly. Using the general representation

$$\tilde{\Theta} = \{\sum_{\lambda,\nu} \tilde{b}_{\lambda\nu} \exp\{i\lambda\alpha x\} \sin \nu\pi z\} \exp\{idx + iby + \sigma t\}. \tag{5.13a}$$

We find

$$\tilde{\Phi} = \{\sum_{\lambda,\nu} b_{\lambda\nu} \exp\{i\lambda\alpha x\} v_\nu([(\nu\alpha + d)^2 + b^2]^{1/2}, z)\} \exp\{idx + iby + \sigma t\}. \tag{5.13b}$$

The ansatz (5.13a) is general with respect to the y- and t dependence, because equations (5.12) do not depend explicitly on y or t. With respect to the x-dependence the generality of expression (5.13a) is guaranteed by the Floquet theory of differential equations with periodic coefficients. After multiplying equation (5.12b) by $\sin \mu\pi z \exp\{-i[(\kappa\alpha + d)x + by] - \sigma t\}$ and averaging it over the fluid layer we obtain a linear homogeneous system of equations for the unknowns $\tilde{b}_{\lambda\nu}$. The system is solvable if and only if the determinant of the coefficient matrix vanishes. The eigenvalues σ are determined as the zeros of the determinant for given values of R, α, b and d. The sign of the maximum of the real part of σ as a function of b and d decides the question of stability as has already been mentioned above.

The boundary of the region of stable convection rolls in the R, α-plane obtained by this method is shown in Figure 2.10 for the case of rigid boundaries. For low Rayleigh number the stability of rolls is limited towards low values of α by the zig-zag instability and towards high α by the cross roll instability. The main purpose of these instabilities is to change the wavenumber of the rolls to a more advantageous value near α_c. The zig-zag instability accomplishes this most elegantly by a wavy deformation of the roll pattern. Since the rolls are lengthened by the process, their effective wavenumber increases as shown in Figure 2.11a. The zig-zagging pattern of rolls developing from this instability has given it its name. Experimental observations of the zig-zag instability are shown in the paper of Busse and Whitehead (1971).

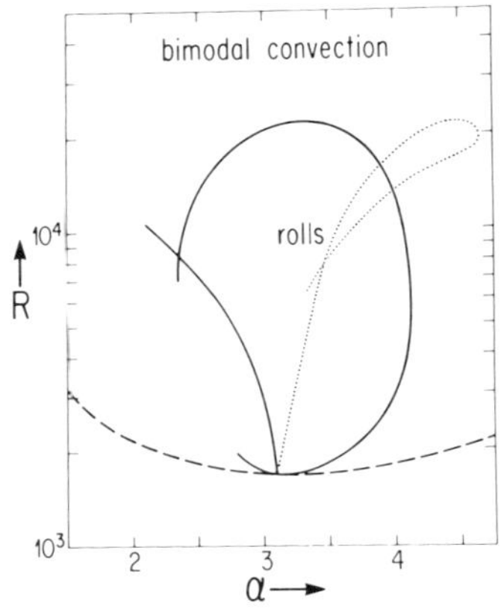

Figure 2.10. Region of stable convection rolls as a function of the Rayleigh number R and the wavenumber for rigid boundaries (after Busse, 1967c). The dashed line indicates the Rayleigh number for onset of convection and the dotted line describes the wavenumber b of crossroll disturbances growing near the stability boundary. The stability boundary bounding the region of stable rolls towards small values of α for low R is caused by the zig-zag instability.

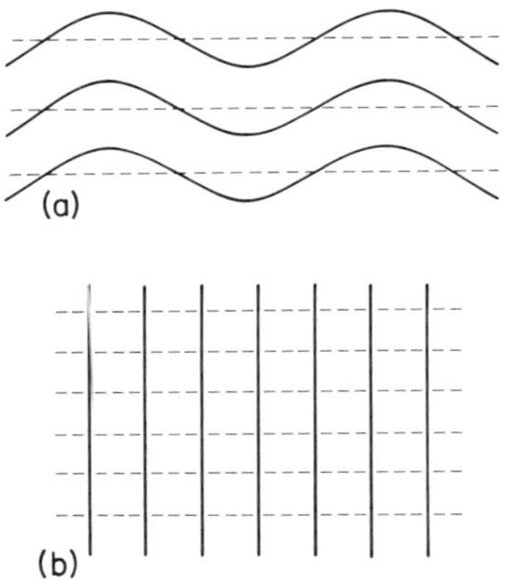

Figure 2.11. (a), Zig-zag and (b), cross-roll instabilities. Dashed lines indicate the original roll pattern.

A mechanism as simple as that of the zig-zag instability does not exist at high wavenumbers α. The region of stable rolls is thus more extensive for $\alpha > \alpha_c$ than for $\alpha < \alpha_c$ at low Rayleigh numbers. The cross-roll instability generates new rolls with a wavenumber close to α_c in the direction perpendicular to the given rolls as shown in Figure 2.11b. The perpendicular direction is preferred because it minimizes the interaction with the given rolls which have a stabilizing influence. The distortion of the profile of the mean temperature owing to the convective heat transport prevents the cross-roll instability from growing in the stable region. But for wavenumbers α sufficiently different from α_c the distortion is weakened and cross-roll disturbances can grow. The symmetry with respect to α_c is quite evident in Figure 2.10.

For Rayleigh numbers above 10^4 the character of the cross-roll instability changes. The wavenumber b of the strongest growing cross-roll disturbances increases and when the Rayleigh number exceeds a value of the order $1.5 \cdot 10^4$ the wavenumber b intersects the stability boundary as seen in Figure 2.10. This implies that the cross-roll disturbances can not develop into a pattern of stable two-dimensional rolls. Instead a three-dimensional form of convection is generated known as bimodal convection. As indicated schematically in Figure 2.12, this form can be described as a row of short wavelength rolls superimposed at a right angle onto the basic roll pattern. The short wavelength and the slightly stronger motion of the secondary rolls near the boundaries indicate that the thermal gradients at the boundaries provide the energy for the secondary motion, just as has been suggested by the heuristic discussion of the criterion (5.11) for instability.

Bimodal convection is a steady form of convection and remains stable for a considerable range of Rayleigh numbers if the Prandtl number is sufficiently high. Numerical solutions (Frick *et al.*, 1983) have shown that the heat transport is increased significantly by the onset of the secondary rolls in agreement with the experimental measurements which go back to Malkus (1954a). Visual observations have been made only more recently after the instability theory became available. They are in good agreement with the theoretical predictions (Krishnamurti, 1970a; Busse and Whitehead, 1971).

As expected, no transition to bimodal convection is found in the case of

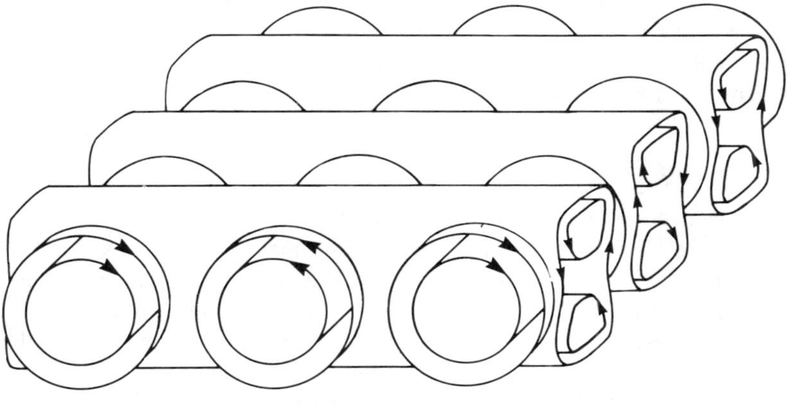

Figure 2.12. Schematic sketch of bimodal convection.

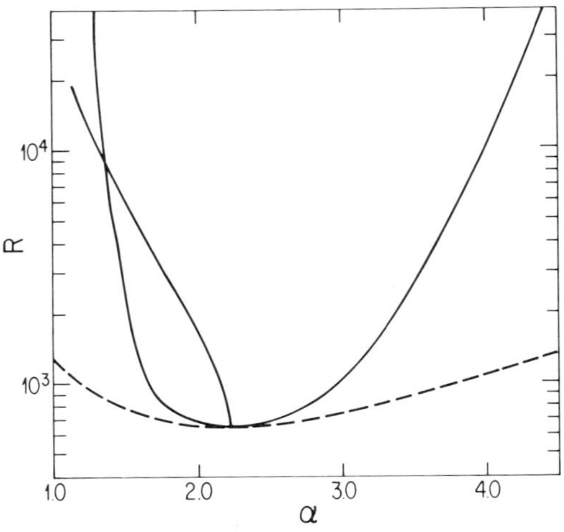

Figure 2.13. Region of stable convection rolls as a function of the Rayleigh number R and the wavenumber α for stress-free boundaries (after Straus, 1972; and Bolton and Busse, 1984).

stress-free boundaries (Straus, 1972). As shown in Figure 2.13, the cross-roll instability continues to bound the regime of stable rolls toward small as well as large values of α, but it does not close the region of stable rolls towards high Rayleigh numbers. The situation is somewhat complicated owing to the fact that at high but finite Prandtl numbers, rolls are unstable with respect to the skewed varicose instability (Busse and Bolton, 1984; Bolton and Busse, 1985). But since the growthrate of this instability vanishes in the limit of infinite Prandtl number it becomes unimportant in that limit and has not been taken into consideration in Figure 2.13.

5.3. Mean-Field Theories

The desire to obtain solutions for much higher Rayleigh numbers than is possible by direct numerical integration of the nonlinear equations (2.14) has motivated researchers to look for simplifying approximations which allow the retention of the major part of the nonlinearity while simplifying the mathematical analysis significantly. Malkus (1954b) introduced the idea that the fluctuating components of the nonlinear interaction terms on the right hand side of (2.14) can be neglected in first approximation, such that the only nonlinear effect acting on a convection mode is the change of the mean temperature profile owing to the convective heat transport. In this simplest form the mean-field approximation is also known as the quasi-linear approximation. Using the standard assumption $f_1 = f_3 = 1$, $R_Q = 0$ for a plane convection layer the quasi-linear approximation can be obtained from equation (4.4) by replacing Θ on the right hand side by $\overline{\Theta}$ which is given by (4.12),

$$\nabla^4\Delta_2\phi - \Delta_2 \Theta = 0, \tag{5.14a}$$

$$\nabla^2\Theta - R\Delta_2\Phi = \Delta_2\Phi(\overline{\Delta_2\Phi\Theta} - \langle\overline{\Delta_2\phi\Theta}\rangle). \tag{5.14b}$$

Note that the assumption of infinite Prandtl number is not necessary in deriving these equations. Since there is no mean velocity field, equations (5.14) remain unchanged in the case of finite Prandtl number. But because the neglection of fluctuating terms is a less severe assumption in the limit of infinite Prandtl number, it can be expected that the solutions of (5.14) resemble most closely the solutions for large Prandtl number convection. A Prandtl number dependence can eventually be included, when a hexagonal interaction of the type considered in Section 4.2 is taken into account in addition to the interaction with the mean field (Roberts, 1966; Gough *et al.*, 1975).

Equations (5.14) admit simple solutions of the form (3.3), (3.4) for Φ and Θ. Herring (1963, 1964) obtained those solutions numerically for stress-free and rigid boundary conditions. Asymptotic expressions for free boundaries have been derived by Howard (1965) and for rigid boundaries by Roberts (1966) and Stewartson (1966). The free boundary result, Nu $=2.83\ (R/R_c)^{1/3}$, compares well with the numerical result (5.6) and the result (5.26a) of the boundary layer analysis to be discussed below. For rigid boundaries the result Nu $\propto R^{1/5}$ obtained for a fixed wavenumber α also exhibits the same order of magnitude as the numerical result (5.7) and the expression (5.26b) of the boundary layer theory. But when the wavenumber α_m maximizing the heat transport is used, the relationship Nu $\propto R^{3/10}$ is obtained. This latter relationship may be more typical for experimental measurements of turbulent convection at high Rayleigh number. The high value of the wavenumber α_m, however, which increases proportionally to $R^{1/4}$, does not have any correspondence in the measured spectrum of convection. Moreover, in his later paper Herring (1964) found that solutions characterized by a single value of α are unstable at high Rayleigh number and that solutions with two different values of α seemed to be preferred. We shall return to this point below.

While the mean field approximation leads to mathematically convenient equations, it is difficult to eliminate the arbitrariness inherent in this approach. These shortcomings have led Howard (1963) to a different approach in which the fields Φ, Θ are determined which maximize the heat transport. Instead of obeying the basic equations such as equations (4.4) the extremalizing fields, Φ, Θ must satisfy only the Euler-Lagrange equations of a variational problem and the conditions at the rigid boundaries. The Euler-Lagrange equations share with equations (5.14) the mean-field character, i.e., they are quasi-linear. Howard (1963) obtained extremalizing solutions with a single wavenumber α; extremalizing solutions with an arbitrary number of different α, so-called mutli-α-solutions, were obtained by Busse (1969). The asymptotic upper bound derived from these latter solutions is independent of the Prandtl number and yields

$$Nu \leq 0.031 \cdot R^{1/2}. \tag{5.15}$$

Chan (1971) restricted the attention to the limit of infinite Prandtl number and took equation (5.12a) into account as an additional constraint. He thus was able to arrive at the asymptotic upper bound

$$Nu \leq 0.152 R^{1/3} \tag{5.16}$$

for the heat transport at infinite Prandtl number. He also showed that the mean-field equations (5.14) are closely related to the Euler-Lagrange equations of the variational problem and that the multi-α-solutions of equations (5.14) give rise to a maximum heat transport obeying the same power law as the upper bound (5.16) but with a smaller numerical coefficient.

While the upper bounds (5.15) and (5.16) may be much too high to provide realistic estimates for the heat transport by convection at asymptotically high Rayleigh numbers, the corresponding extremalizing solutions exhibit a number of physically interesting features. In fact, the transition from a single-α-solution to a double-α-solution occurs in a similar way as the transition to bimodal convection discussed in the preceding section. Other properties such as the dependence of the fluctuating temperature on the distance from the boundary and the profile of the mean temperature throughout the convection layer also show many realistic features. By adding more constraints derived from the basic equations it should be possible to obtain even more realistic extremalizing solutions. But the analytic and mathematically rather attractive multi-α-solutions probably must be replaced by numerically generated solutions as the variational problems become more complex.

5.4. Boundary Layer Theory

The boundary layer analysis that has been helpful in solving mean-field equation of the form (5.14) can also be applied directly to solve the basic equations (2.14). Not all problems of convection are accessible to boundary layer analysis; but the standard equations (4.4) give rise to two-dimensional solutions that exhibit a boundary layer structure for asymptotically high Rayleigh numbers as indicated schematically in Figure 2.14. Rising hot and descending cold plumes that are beginning to assume a boundary layer character can be seen in Figure 2.6 at moderate Rayleigh numbers. The much stronger horizontal thermal boundary layers have already been discussed in connection with Figure 2.8.

For the boundary layer analysis it is convenient to recast the basic equations (2.14) into the form

$$\nabla^2 \mathbf{v} - \boldsymbol{\lambda}\,\hat{\Theta} - \nabla\Pi = 0, \tag{5.17a}$$

$$\nabla^2 \hat{\Theta} - \mathbf{v}\cdot\nabla\,\hat{\Theta} = 0, \tag{5.17b}$$

where

$$\hat{\Theta} \equiv \Theta - \mathrm{R}z \tag{5.17c}$$

is the deviation of the temperature from the average value of the layer. As before we have assumed $f_1 = f_3 = 1$, $\mathrm{R}_Q = 0$, $\mathrm{P} = \infty$ and have neglected the time derivative since we are interested in steady solutions. Because the analysis will be restricted to two-dimensional solutions, a stream function ψ can be used to represent the velocity field,

$$\mathbf{v} = \nabla \times \mathbf{j}\psi, \tag{5.18}$$

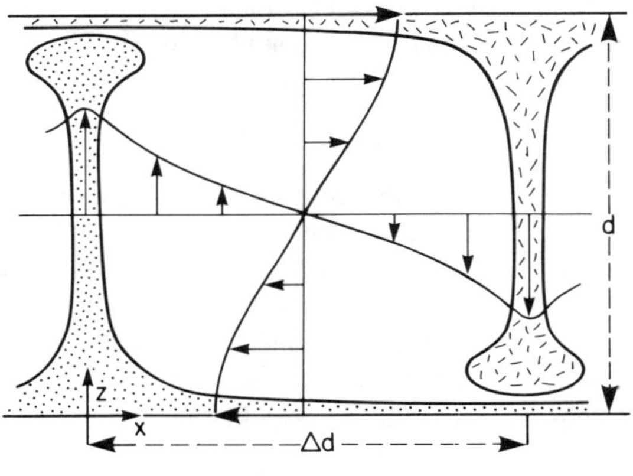

Figure 2.14. Sketch of thermal plumes and thermal boundary layers assumed in the boundary layer theory of convection rolls.

where **j** is the unit vector in the y-direction. The boundary layer analysis is based on an isothermal interior, $\hat{\Theta} \approx 0$, where the stream function must satisfy the equation

$$\nabla^4 \hat{\psi} = 0, \tag{5.19}$$

which follows from the y-component of the curl of the equation of motion (5.14a). The top and bottom thermal boundary layers make a negligible contribution to the curl of the buoyancy force, and the boundary conditions which must be applied to the solution of equation (5.19) are thus the same as before,

$$\hat{\psi} = \partial^2_{zz} \hat{\psi} \qquad \text{at } z = 0,1 \tag{5.20a}$$

for stress-free boundaries and

$$\hat{\psi} = \partial_z \hat{\psi} \qquad \text{at } z = 0,1 \tag{5.20b}$$

for rigid boundaries. The rising hot plume and the descending cold plume which are the distance Δ apart in dimensionless units provide the buoyancy to drive the convection flow. Since the pressure can not vary much across the thin plumes the buoyancy is balanced solely by the viscous stress as indicated in Figure 2.14. By integrating equation (5.17a) across the plume of width $2\delta_p$ at $x = 0$ we obtain

$$-\partial_x v_z|_{x=\delta_p} = \frac{1}{2} \int_{-\delta_p}^{\delta_p} \tilde{\Theta}\, dx = \frac{1}{2} [v_z|_{x=0}]^{-1} \mathrm{Nu}\, R\Delta. \tag{5.21}$$

The last equality follows from the property that each plume transports across the layer the heat flux collected from the boundary area of width Δ. Relationship (5.21) gives rise to the nonlinear boundary condition for $\hat{\psi}$,

$$\partial^2_{xx} \hat{\psi} \cdot \partial_x \hat{\psi} = \mp \tfrac{1}{2} \mathrm{Nu}\, R\Delta \qquad \text{at } x = (\tfrac{1}{2} \mp \tfrac{1}{2})\Delta, \tag{5.22}$$

which implies singularities at the corners of rectangular region. These corners must be analyzed carefully in the detailed theory (Roberts, 1979) but they do not have to concern us here. A second relationship between the magnitude of the vertical velocity and the heat transport NuR is obtained from equation (5.17b) in the thermal boundary layers above the hot plume and below the cold plume. By neglecting x-derivatives in comparison with z-derivatives in those boundary layers we obtain

$$\partial^2_{zz} \tilde{\Theta} = v_z \partial_z \tilde{\Theta}. \tag{5.23}$$

Using the Taylor expansion

$$v_z \approx (z - 1)\partial_z v_z|_{z=1} \qquad \text{for } 1 - z \approx 0(\delta) \qquad (5.24a)$$

suggested by condition (5.20a) in the case of stress-free boundaries and the expansion

$$v_z = (z - 1)^2 \partial_{zz}^2 v_z|_{z=1} \qquad \text{for } 1 - z \approx 0(\delta) \qquad (5.24b)$$

in the case of rigid boundaries we conclude from equation (5.23) that the thickness δ of the horizontal thermal boundary layers is given by

$$\delta^{-2} \approx \partial_z v_z|_{z=1} \qquad \text{for stress-free boundaries,} \qquad (5.25a)$$

and by

$$\delta^{-3} \approx \partial_{zz}^2 v_z|_{z=1} \qquad \text{for rigid boundaries.} \qquad (5.25b)$$

Relationships analogous to (5.25), for the corner at $x = 0$, $z = 1$, hold, of course, for the corner at $x = \Delta$, $z = 0$. Since $\delta^{-1} \approx \mathrm{Nu}/2$ according to (5.9) and since all the derivatives of v_z appearing in above formulas are of the same order of magnitude as v_z itself, relationships (5.21) and (5.25) yield

$$\mathrm{Nu} \approx \mathrm{R}^{1/3} \qquad \text{for stress-free boundaries,} \qquad (5.26a)$$

and

$$\mathrm{Nu} \approx \mathrm{R}^{1/5} \qquad \text{for rigid boundaries.} \qquad (5.26b)$$

The detailed analysis that is required for the determination of the dependence of the Nusselt number on the aspect ratio Δ of the convection roll is too complicated to be presented here. The problem of matching horizontal and vertical boundary layers in the corner regions is still not completely solved in spite of the considerable theoretical work that has been devoted to this subject. Robinson (1965) and Turcotte and Oxburgh (1967) derived boundary layer theories valid for $\mathrm{P} = \infty$ in the case of stress-free boundaries. These theories have been improved by G.O. Roberts (1979) who also derived the correct asymptotic law for rigid boundaries. A detailed analysis of the problem with an application to mantle convection has been given by Olson and Corcos (1980). While the general magnitude of the heat flux agrees well with that obtained from high Rayleigh number computations, the dependence on the aspect ratio Δ is much stronger in the analytical expressions of the boundary layer theory. A recent analysis by Zimmermann (1983) suggests that neglected higher order terms in the corners may be responsible for this discrepancy. Other discrepancies between numerical results and asymptotic theory have been pointed out in the recent comparison study undertaken by Jarvis and Peltier (1982).

5.5. Experimental Observations

Experiments on high Rayleigh number convection in a high Prandtl number fluid face the difficulty that the Prandtl number increases with the viscosity while the thermal diffusivity of liquids is typically of the order 10^{-7} to 10^{-6} m^2/sec or above this range in the case of metals. In order to reach high Rayleigh numbers, the dimensions of the laboratory convection layer must thus be increased. Layers of 100 cm in the horizontal extent and with a thickness of the order of 10 cm are typically used (Whitehead and Parsons, 1978). The conditions for the high Prandtl number regime of convection are not quite as stringent, however, as they sometimes have been stated. Moore and Weiss (1973) find on the basis of their numerical computations that the high Prandtl number regime is achieved for

$$P \gtrsim (R/R_c)^{2/3} \tag{5.27a}$$

in the case of stress-free boundaries. This criterion can be interpreted as the condition that the Reynolds number of convection is less than of order unity. Using relationships (5.6) and (5.25a) for a Reynolds number defined by $P\partial_z v_z|_{z=1}$, we recover criterion (5.27a). Using the analogous procedure in the case of rigid boundaries we find as an approximate definition of the high Prandtl number regime

$$P \gtrsim (R/R_c)^{3/5}. \tag{5.27b}$$

There are a number of experimental investigations well within this range with the experiment of Whitehead and Parsons achieving the highest Rayleigh numbers. Best understood is the range of relatively moderate Rayleigh numbers below the transition to bimodal convection. Modern techniques of velocity and temperature measurements have allowed a detailed comparison between observations and theoretical predictions. The work of Dubois and Bergé (1978) and Dubois *et al.* (1978) finds as good an agreement between measurements and the data of the small amplitude perturbation theory and the theory of Busse (1967c) as can be expected. A similarly detailed comparison between experimental observations and theoretical computations for convection in a box has been done by Bühler *et al.* (1979) and Oertel (1982). Because of the small aspect ratio of the box, the influence of the side walls had to be taken into account in the latter study, which required three-dimensional computations.

Bimodal convection, which replaces two-dimensional convection above R = 22,600, has been less well studied. The additional wavenumber in third dimension enlarges the parameter space and the three-dimensional nature of the convection together with the well developed thermal boundary layers provides a considerable challenge for the numerical analyst. The computations of Frick *et al.* (1983) are based on an extension of the Galerkin method described in Section 5.1 and describe bimodal convection over a fairly typical range of the parameter space. (Note that the two surfaces shown in Figure 3 of that paper are identical and should not have been drawn as intersecting surfaces.) The stability analysis of bimodal convection included in the work of Frick *et al.* shows reasonable agreement with the experimental observations of Whitehead and Chan (1976). In this latter study the two basic wavenumbers, α_1 and α_2, of bimodal convection were varied in a systematic way with the method of controlled initial conditions and the stability of the steady convection flow was tested. For certain combinations of the wavenumbers, as for example $\alpha_1 = 4.3$ and $\alpha_2 = 5.2$, bimodal convection appears to be a stable steady solution according to Whitehead and Parsons (1978) up to Rayleigh numbers of the order of $8 \cdot 10^5$, the highest reached in their experiment.

Whitehead and Parsons report on a different form of steady convection, called square pattern convection, which could easily be realized with controlled initial conditions at high Rayleigh numbers, but which was unstable at Rayleigh numbers below 10^5. This form of convection is not just a form of bimodal convection with equal wavenumbers, $\alpha_1 = \alpha_2$, which is indicated by the property that the typical wavenumber is much higher than that of the basic rolls in bimodal convection. The distinction between square pattern convection and bimodal convection is clearly evident from the theoretical analysis of Frick *et al.* (1983) which shows that square pattern convection transports even more heat than bimodal convection. In spite of this property square pattern convection was shown to be unstable in the range up to Rayleigh numbers of the order $7 \cdot 10^4$ for which reliable results could be obtained. Evidently square cells finally become competitive for $R \gtrsim 10^5$.

The typical flow at Rayleigh numbers of the order of 10^5 and above is neither bimodal convection nor square pattern convection. Instead spoke pattern convection is usually realized in experiments started with random initial conditions. At Rayleigh numbers of the order 10^5 it is difficult to separate spoke pattern convection from a randomly distorted pattern of bimodal cells. But as shown in the photographs of Whitehead and Parsons, the spoke centers become more pronounced once R exceeds $1.5 \cdot 10^5$. Spoke pattern convection is more clearly distinguished from other forms of convection in lower Prandtl number fluids outside the regime (5.27b) where its wavelength is much larger than in the high Prandtl number regime. The characteristic feature that motions crossing the convection layer from one thermal boundary layer to the other occur mainly in the spoke centers appears to be common to both the low and the high Prandtl number versions of spoke pattern convection.

The measurements of the Nusselt number in high Prandtl number fluids exhibit a power law relationship which exceeds the asymptotic dependence (5.26b) for two-dimensional convection by a considerable margin. Rossby (1969) represents his measurements for silicone oil with P = 200 in the form

$$\mathrm{Nu} = 0.18\mathrm{R}^{0.281}, \tag{5.28a}$$

or equivalently

$$\mathrm{Nu} = 1.46(\mathrm{R}/\mathrm{R}_c)^{0.281}. \tag{5.28b}$$

The relationship corresponds well to typical values for bimodal convection according to the computations of Frick *et al.* (1983). There is thus no evidence that spoke pattern convection increases significantly the efficiency of the convective heat transport beyond that of bimodal convection. Rossby's data do not go much beyond the Nusselt number 10 and more measurements at high Rayleigh numbers may be needed to establish an accurate relationship of the form (5.28). Eventually compositional instead of thermal convection may fulfill this goal. Convection in an isothermal layer with salty water above fresh water (see, for example, Green and Foster, 1975) is characterized by a Prandtl number of the order 500 and permits the realization of very high Rayleigh number. Another possibility realized in an electrochemical redox cell has recently been proposed by Ward and Leblanc (1984).

5.6. Transition to Time Dependent Convection

One of the most intriguing problems of thermal convection is the onset of time dependence in high Prandtl number fluids heated from below. According to the experiments of Whitehead and Parsons it appears to be possible that stable steady solutions exist in the limit $R \to \infty$ as long as the inequality (5.27b) is satisfied. Since these special solutions are usually not realized when experiments are started from random initial conditions, there is no contradiction implied if all experiments of the latter kind show the onset of time dependence well within the regime (5.27b). The onset of time dependence occurs typically over a broad range of Rayleigh numbers. First an oscillatory motion is noticeable at a spot where the convection pattern is particularly inhomogeneous. More spots exhibiting oscillations are observed as the Rayleigh number increases. But only at much higher Rayleigh number does every convection cell participate in some kind of oscillatory motion. The Rayleigh number at which an oscillating spot can first be noticed is hard to determine, but it appears to increase with Prandtl number (Busse and Whitehead, 1974). This property is controversial since Krishnamurti (1970b, 1973) published data indicating a Prandtl number independent onset of oscillations at $R \approx 6 \cdot 10^4$ for $P \gtrsim 50$. Willis and Deardorff (1970) did not observe oscillations below $R \approx 10^5$ for $P = 57$ and Whitehead and Parsons (1978) did observe only a single isolated oscillating spot in this large aspect ratio layer with $P = 8600$ when they increased the Rayleigh number to $1.5 \cdot 10^5$. Clearly further research is needed to understand the onset and nature of oscillations in high Prandtl number fluid.

Since oscillations are observed well within the regime (5.27b) they cannot be caused by the momentum advection terms which are responsible for the onset of time dependence in convecting fluids of low and moderate Prandtl number (Busse, 1972; Clever and Busse, 1974). A definitive theory of oscillations in high Prandtl number convection is still missing. But a mechanism proposed by Howard (1966) seems to capture some of the basic properties of the observed oscillatory convection.

Howard envisions a periodic breakup of the thermal boundary layer. The breakup phase corresponds to a rapidly growing instability as a result of which a parcel of fluid or "thermal" emerges from the boundary layer. It is followed by a recovery phase dominated by conduction in which the thermal boundary grows until it again becomes unstable. Using definition (5.17c) the latter phase can be described by the solution

$$\hat{\Theta} = \frac{1}{2} R \frac{2}{\sqrt{\pi}} \int_{z/2\sqrt{t}}^{\infty} e^{-\zeta^2} \, d\zeta \tag{5.29}$$

of the conduction equation

$$\frac{\partial^2}{\partial x^2} \hat{\Theta} = \frac{\partial}{\partial t} \hat{\Theta} \tag{5.30}$$

near the lower boundary. At the time t^* the thickness

$$\delta^* = \sqrt{\pi t^*} = (\hat{\Theta}^{-1} \partial \Theta / \partial z|_{z=0})^{-1} \tag{5.31}$$

of the thermal boundary layer has increased sufficiently that the criterion for stability

$$R\delta^{*3} < R_\delta \approx 10^3 \tag{5.32}$$

is violated and a rapid eruption of a blob of hot fluid from the thermal boundary layer ensues. The criterion (5.32) must not be interpreted as a condition for static stability. Instead it describes the condition under which the growth of any instability is not rapid enough to compete effectively with the growth of the thermal boundary layer itself. Once (5.32) is violated, a superexponential growth occurs, as can be seen in the numerical computations by Foster (1965). Howard's assumption that the breakup phase is short compared to the recovery phase thus seems to be well justified for fluids with $P \gtrsim 1$. According to relationships (5.31), (5.32) the period of the process is given by

$$t^* \approx \frac{1}{\pi}(R/R_\delta)^{-2/3}, \tag{5.33}$$

which agrees quite well with the observations of Rossby (1966), Krishnamurti (1970b) and Busse and Whitehead (1974). There is a different interpretation possible, however, since the period of a fluid parcel carried around in a convection roll decreases with $R^{-2/3}$ in the case of stress-free boundaries or $R^{-3/5}$ in the case of rigid boundaries according to relationships (5.21), (5.26). In fact, Krishnamurti's observations clearly indicate the circulation of hot or cold blobs of fluid in a convection roll as the cause of oscillations. The two interpretations do not necessarily contradict each other, of course, since the circulating thermal blob may well serve as the trigger of the process envisioned by Howard.

The instability of the thermal boundary layer has been used before in Sections 5.1 and 5.2 to explain the onset of bimodal convection and the question of the relationship with the present discussion arises. The basic mechanism is indeed the same. But in the case of only slightly curved convection rolls a transition to a nearly steady form of convection is possible, while in less well ordered patterns a steady secondary flow apparently can not be achieved. This property may explain why oscillations in high Prandtl number fluids appear first in the most strongly non-uniform part of a convection pattern and why oscillatory convection is associated with the relatively inhomogeneous spoke pattern.

5.7 Convection in an Internally Heated Layer

Internal heating by radioactive elements in the earth's mantle is a major contributor to the buoyancy force driving mantle convection. For this reason convection in an internally heated layer has received much attention among theoretical and experimental geophysical fluid dynamicists. At low amplitudes of convection the problem does not differ much from other cases of convection with asymmetric properties. As has been discussed in Section 4.2 g-hexagons must be expected at the onset of convection and hysteresis effects associated with subcritical finite amplitude convection may be noticed in experiments. But the perturbation theory of Section 4.2 provides little guidance when the asymmetry is large or when the Rayleigh number exceeds many times its critical value.

A simple consideration shows that a boundary layer structure such as that discussed in Section 5.4 cannot be expected for internally heated convection. The convection cell of Figure 2.14 is characterized by the property that only the outermost part participates in the heat transport and in the dynamics. The fluid in the interior of the convection roll does not experience temperature variations and plays only a passive role in the dynamics of convection. These statements remain basically valid even when the simple solution of Figure 2.14 is replaced by a three-dimensional time-dependent structure such as spoke pattern convection. In the internally heated layer, however, each parcel of fluid must reach the cooled boundary in order to accomplish an effective heat transport, and a passive role of some parts of the fluid layer is not possible. Typically only plumes of cool fluid form and the circulation is closed by a broad upwelling towards the cold upper boundary. The two-dimensional numerical computations of Peckover and Hutchinson (1974) and of McKenzie *et al.* (1974) show this effect clearly at high Rayleigh numbers; but it is also evident in the three-dimensional computations carried out by Thirlby (1970) at more moderate Rayleigh numbers. An experimental demonstration of the asymmetry between up- and downflow is shown in the paper of De La Cruz-Reyna (1976).

Among the many possibilities of boundary conditions for internally heated convection, two cases have been studied in considerable detail, both experimentally and theoretically. Case A is characterized by an insulating bottom boundary, while in case B both boundaries are kept at the same constant temperature,

$$\text{case A:}\quad \mathrm{R} = \mathrm{R}_Q/2, \quad \partial_z\Theta = 0 \text{ at } z = 0, \quad \Theta = 0 \text{ at } z = 1,$$

$$\text{case B:}\quad \mathrm{R} = 0,\ \Theta = 0 \text{ at } z = 0,\ 1.$$

In the theoretical analysis of equations (2.14) $f_1 = f_3 = 1, f_4 = z - ½$ are set and no-slip boundaries are usually assumed in both cases corresponding to rigid plates in the experimental realizations. Tritton and Zarraga (1967) initiated the experimental study of convection in an internally heated layer by using an electrolytically conducting fluid through which a horizontal current is forced. This experiment stimulated the theoretical analysis of case A by Roberts (1967) which is based on a mean-field approximation but which includes the horizontal interaction of the form (4.21), i.e., it is capable of distinguishing between rolls and hexagonal forms of convection. The main discrepancy between theoretical calculations and experimental observations occured in the wavenumber of convection cells. Tritton and Zarraga found a strong decrease of the latter while the theory predicted a slight increase. Schwidersky and Schwab (1971) later showed that the temperature dependence of the electrical conductivity interacts quite strongly with the dynamics of the convection flow. The nonuniformity of the electric current leads to a widening of the convection cells and gives rise to convection rolls aligned with the direction of the current at high Rayleigh numbers.

The mean-field approximation used by Roberts (1967) is not suitable for the analysis of the relative stability of rolls and hexagons. Later computations without this approximation by Tveitereid and Palm (1976) showed good agreement with Nusselt numbers calculated by Roberts, but indicated large differences in the stability properties. According to Tveitereid and Palm g-hexagons are stable from the neighborhood of the critical value of the Rayleigh number up to 15 times that value which is as high as the computations have been carried out. This finding agrees with the more recent experimental observations by De La Cruz-Reyna (1976) and Carrigan (1982, 1985). In order to avoid the difficulties caused by a horizontal current, these experimenters generated a nearly uniform internal heat source distribution by a vertical current. The horizontal electrodes cause some inconvenience in visualizing the convection pattern; but this inconvenience appears to be unavoidable in achieving high Rayleigh number convection driven by nearly uniform heating. An interesting observation made by Carrigan (1985) is that the direction of flow in polygonal cells changes at Rayleigh numbers of about 40 times the critical value. The sheet-like streamers of cold fluid seen by Carrigan are also observed in high Rayleigh number experiments with a water layer cooled from above (Spangenberg and Rowland, 1961; Katsaros, 1978) which basically resemble convection in an internally heated layer. This form of convection may well be a modification of spoke pattern convection in which the sheet-like spokes dominate over the spoke collecting centers.

Patterns of convection and onset of time dependence in internally heated layers are sensitive to the boundary conditions. Convection observed in case B differs significantly from case A which has been discussed so far. Kulacki and Goldstein (1972) failed to see any steady convection in their experimental realization of case B, but their interferometric measurements did not permit a good visualization of the pattern of convection. The theoretical analysis of case B by Tveitereid (1978) indicates that stable *g*-hexagons are possible up to 3.6 times the critical value of the Rayleigh number. Above that value no stable steady convection solution appears to exist. The main difference between cases A and B is the strongly stably stratified region near the bottom boundary in the latter case. This region supports internal waves and may be responsible for the onset of time dependence at relatively low Rayleigh numbers.

In view of the variety of phenomena that occur in internally heated convection layers a more complete analysis of equations (2.14) is desirable. An initial effort in this direction has been made by Clever (1979) who solved equations (2.14) in the case of rigid conducting boundaries for two-dimensional rolls and analyzed their stability as a function of the parameters R and R_Q. Not surprisingly rolls are found to be stable only within the region $R_Q < 8R$, $R \lesssim 2.5 \cdot 10^4$ of the parameter space.

6. Special Effects

6.1. Convection with Strongly Temperature Dependent Viscosity

The strong variation of the effective viscosity in the earth's mantle is probably the single most important cause of differences between mantle convection and convection in ordinary laboratory fluids. It is not surprising then that convection in a fluid layer with strongly temperature dependent viscosity has received much attention among theoretical and experimental geophysical fluid dynamicists. A number of numerical computations of two-dimensional convection with various assumptions about the temperature or depth dependence of the viscosity have been done (Foster, 1969; Torrance and Turcotte, 1971; Houston and DeBremaecker, 1975; DeBremaecker, 1977; Kopitzke, 1979; Schmeling and Jacoby, 1981). They have usually been motivated by particular aspects of the phenomenon of mantle convection; some of these studies will be discussed in later chapters of this book. In this section the attention will be restricted to more general properties of the problem.

The influence of small variations of the viscosity with temperature has already been discussed in Section 4.2. While the perturbation theory yields explicit analytical expressions for the effect of the temperature dependence of various material properties, it is clearly inadequate if the variation of a property throughout the convection layer exceeds the order of magnitude of its mean value. The general effect of larger variations appears to be that the range of stability of hexagonal convection is extended to higher Rayleigh numbers than predicted by the perturbation theory while the range of subcritical finite amplitude convection turns out to be much smaller. This effect is generally seen in experiments (Hoard *et al.*, 1970) and also applies, by the way, in the case of internal heating where the perturbation theory can be compared with the three-dimensional computations (Tveitereid and Palm, 1976). Qualitatively new effects appear only when the maximum ratio r of viscosities in the layer reaches a value of the order four or more as will be discussed below.

The onset of convection in the form of infinitesimal disturbances can be determined quite easily even for large viscosity variations. A variety of different dependences of the viscosity on temperature have been discussed by Stengel *et al.* (1982). It is common usage to express all variations relative to the viscosity value corresponding to the temperature given by the arithmetic mean of the boundary temperatures of the layer. With this definition the critical value of Rayleigh number varies relatively little over a large range of r. In the realistic case of an exponential dependence on temperature the critical Rayleigh number first increases then decreases with increasing r. In the case of a linear dependence on temperature the critical Rayleigh number decreases monotonically (Busse and Frick, 1985) with growing r. This ultimate decrease of the critical Rayleigh number reflects the onset of convection in the low viscosity boundary layer near the hot surface.

At finite amplitudes the confinement of the convection to a sublayer near the hot boundary becomes even more pronounced. The sublayer is nearly isothermal because the flux of heat across it is carried by convection. The remainder of the fluid layer is nearly motionless and a steep temperature gradient is required to conduct the heat flux across it. Laboratory measurements by Richter *et al.* (1983) in a layer with values of r up to 10^5 show this phenomenon quite clearly. In view of this dramatic change from the constant viscosity case it is surprising that global properties such as the Nusselt number change relatively little. It was first noticed by Booker (1976) in his experimental study that the heat flux is about the same as in a constant viscosity layer with the same Rayleigh number if the definition introduced above is used. The agreement becomes even better if the Rayleigh number is measured in units of its critical value as determined by linear theory (Booker and Stengel, 1978). Thus the Nusselt number law in form (5.28b) applies to convection with viscosity variations within the accuracy of measurements. This result has been confirmed and extended up to $r = 10^5$ by Richter *et al.* (1983). It should be remembered that all available fluids with large viscosity variations exhibit a nearly exponential dependence of the viscosity on temperature. Extensions of the empirical relationship (5.28b) to fluids with different viscosity-temperature dependence may not be possible. Boundary layer analyses for the case of an exponential temperature dependence of viscosity in extension of the simple model discussed in Section 5.4 have been presented by Morris and Canright (1984) and by Fowler (1985). Their findings are in general agreement with the observations of Richter *et al.* (1983).

Another area in which variations of viscosity corresponding to values for r of the order 10 appear to have little effect is the stability of convection rolls. Richter (1978) did experiments with r up to 20 and found that both rolls and hexagonal convection cells could be realized over a large range of Rayleigh numbers depending on initial conditions. This result agrees qualitatively with the overlapping stability regions of rolls and hexagons shown in Figure 2.3. Using controlled initial conditions Richter extended the experiments of Busse and Whitehead (1971) on the stability of rolls to the case of strongly varying viscosity and found relatively small changes. The main deviation from the theoretical results of Figure 2.10 is that the transition to bimodal convection occurs at a lower Rayleigh number. This observation is readily explained by the low viscosity in the lower thermal boundary layer which is the main source of buoyancy for the secondary rolls.

At viscosity ratios r of the order 10 the new planform of square convection cells was discovered by Oliver and Booker (1983). Square cell convection can be found in the parts of the parameter region where both rolls and hexagons are stable as shown in Figure 2.15 taken from the recent extensive experimental study of White (1985). For values of r larger than 30 squares replace rolls entirely and they appear to become the dominant mode of convection for larger values of r. But it is not clear whether there exists a stability boundary with increasing r where hexagonal convection cells are unstable with respect to square cell disturbances.

The predominance of square cells at high values of r cannot be explained on the basis of the small amplitude theory of Sections 3.1. The square pattern convection shown in Figure 2.2 exhibits symmetry in up- and down-motions and does not resemble the square cells observed by Oliver and Booker (1983) and White (1988). The latter cells show rising motion in the interior and more narrow regions of descending motion on all sides of the square. The relationship between the two types of square cells can be understood when the square pattern of Figure 2.16a and its pendant generated by higher harmonies are superimposed. In constant viscosity fluids the pattern of Figure 2.16b also exists but it possesses different vertical symmetry. Only with strong asymmetries present in the fluid layer do the vertical dependence of the basic convection pattern Figure 2.16a and that of its higher harmonic overlap significantly and give rise to the highly asymmetric square cells seen in the experiments.

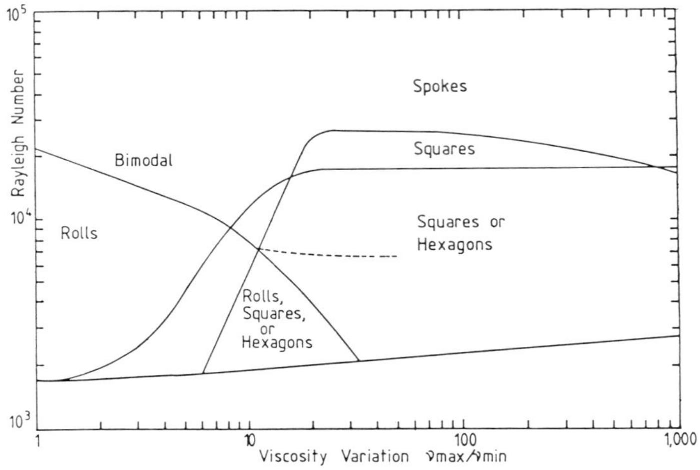

Figure 2.15. Observations of convection patterns as a function of the Rayleigh number and the ratio r of the viscosity at the upper and lower boundaries (from White, 1982). Lyle's Golden Syrup was used as the convection fluid.

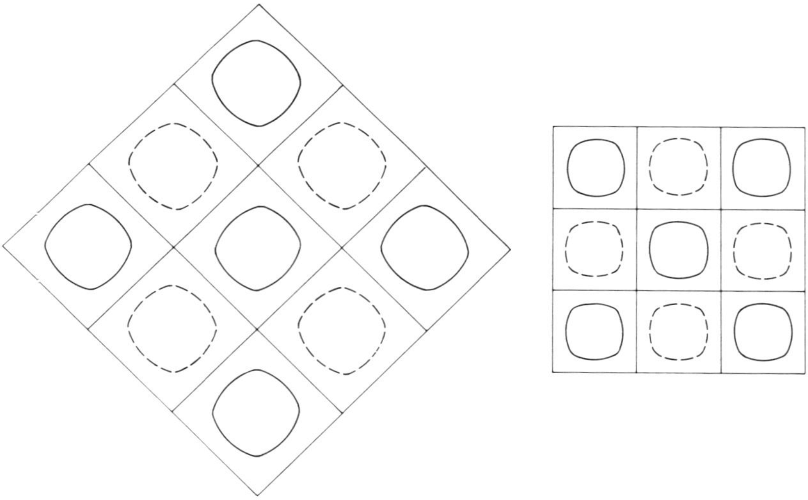

Figure 2.16. Superposition of two square patterns symmetric in up-(solid line) and down-moving (dashed line) convection flow yields a pattern of strongly rising motion surrounded by a square boundary of descending flow. The left pattern corresponds to the basic **k**-vectors, $\mathbf{k}_1$ and $\mathbf{k}_2$, the right pattern is generated by $\mathbf{k}_1 + \mathbf{k}_2$ and $\mathbf{k}_1 - \mathbf{k}_2$.

An analysis of finite amplitude square cell convection has been done recently by Busse and Frick (1985). For mathematical convenience a linear dependence of the viscosity on temperature has been assumed in that work which prevents a quantitative comparison with experimental observations. But qualitatively the agreement is good. There is a region in the R-r-space where both rolls and squares are stable and the squares exhibit the kind of asymmetry discussed above. Although square cells do not appear to be much more efficient than rolls in transporting heat, they are well adjusted to the asymmetric properties of the convection layer caused by strong viscosity variations. In this respect squares even exceed hexagons because the ratio of boundary area to volume of the cell is lower for the latter kind of pattern.

When the investigations of convection in the presence of strong viscosity variations were started it was hoped that the results might show increased width of the convection cells which could be used to explain the relatively large wavelength of mantle convection. Theoretical computations (Davies, 1977) and especially the experimental observations mentioned above show the opposite effect. Because the effective depth of convection is decreased, its wavelength also decreases. The stagnant high viscosity upper part of the convection layer, of course, does not simulate very well the elastic and rheological properties of the lithosphere. Additional assumptions must be introduced to permit plate like motions. We refer to Jacoby and Schmeling (1982) for a discussion of this problem. Another aspect that must be kept in mind is the question as to what extent temperature dependent viscosity simulates the much more complicated rheological laws that undoubtedly apply to mantle material. Investigations in this direction (Parmentier *et al.*, 1976; Parmentier and Morgan III, 1982; Christensen, 1983) have been confined to the two-dimensional problem, but they show surprisingly small differences in comparison with the corresponding case of Newtonian viscosity. The experimental study by Liang and Acrivos (1970) on convection in non-Newtonian fluids has come to the same conclusion; only the Nusselt number showed a slight increase over the corresponding Newtonian case.

6.2. Convection in a Fluid Layer with Phase Changes

It has long been realized that phase changes of mantle material are likely to have a strong influence on the dynamics of mantle flow. In this section we do not intend to enter the discussion of realistic models for phase changes occurring in the mantle. Instead we shall consider the additional terms that have to be taken into account in the basic equations and their main effects on the dynamics of convection.

We restrict the attention to univariant phase transition and assume that the two phases do not differ in their material properties except that the phase of higher density ρ_2 exceeds that density ρ_1 of the other phase by a small amount $\Delta\rho$. The Clausius-Clapeyron relationship yields

$$(dT/dp)_c = T\Delta\rho/q\rho_1\rho_2, \tag{6.1}$$

where q is the latent heat released in the transition from the lighter to the denser phase and where the subscript c indicates that the derivative must be taken along the line of coexistence of the two phases in the p-T-space. This line can be described by

$$f(p,T) = 0. \tag{6.2}$$

We shall use the convention that f is a decreasing function of the pressure for a given temperature. According to Le Chatellier's principle this implies that $f < 0$ for the denser phase.

The effect of the phase transition can be incorporated into the energy equation (2.7) by replacing the constant specific heat C_p and the coefficient of expansion β by

$$c_p \equiv C_p + q\delta(f)\partial f/\partial T, \tag{6.3a}$$

$$-(\partial\rho/\partial T)_p/\rho \equiv \beta + \rho_0^{-1}\Delta\rho\delta(f)\ \partial f/\partial T, \tag{6.3b}$$

respectively, where ρ_0 is the mean of ρ_1 and ρ_2 and δ denotes the Dirac δ-function. In the linearized equation of state (2.10) the phase transition introduces an additional term proportional to $\Delta\rho$,

$$\rho = \rho(T_s, p_s)[1 - \beta(T - T_s) + K_T^{-1}(p - p_s)] - \Delta\rho\delta(f)\left[\left(\frac{\partial f}{\partial T_s}\right)(T - T_s) + \left(\frac{\partial f}{\partial p_s}\right)(p - p_s)\right]. \tag{6.4}$$

In the limit (2.11) of the Boussinesq approximation

$$\left|\frac{p - p_s}{p_s}\right| << \left|\frac{T - T_s}{T_s}\right|$$

holds and the terms proportional to $p - p_s$ in expression (6.4) can be neglected. Even with this simplification the fully nonlinear problem of convection through a phase transition is a complex one and most studies therefore have been restricted to small disturbances of the static state.

The equations (3.1) for steady infinitesimal convection in the case $f_1 = f_3 = 1$, $R_Q = 0$ assume the following form if the two phases are separated by the surface $z = z_0$ in the static state,

$$\nabla \cdot \mathbf{v} = 0, \tag{6.5a}$$

$$\nabla^2 \mathbf{v} + \boldsymbol{\lambda}\Theta(1 + \delta(z - z_0)\mathrm{P}_q) - \nabla\Pi = 0, \tag{6.5b}$$

$$\nabla^2\Theta + \boldsymbol{\lambda}\cdot\mathbf{v}(\mathrm{R} + \mathrm{R}_\alpha s_{12}\delta(z - z_0)) = 0. \tag{6.5c}$$

The three dimensionless parameters are defined by

$$\mathrm{P}_q = \frac{q\Delta\rho}{|q|\pi_0\beta\Delta T}\left|1 - \left(\frac{dT}{dp}\right)_c\left(\frac{dT_s}{dp_s}\right)^{-1}\right|^{-1}, \tag{6.6a}$$

$$\mathrm{R}_\alpha = \frac{|q|g\beta d^3}{C_p\nu\kappa} \tag{6.6b}$$

$$s_{12} = \left(\frac{dT_s}{dp_s} - \left(\frac{dT}{dp}\right)_c\right)\left|\frac{dT_s}{dp_s} - \left(\frac{dT}{dp}\right)_c\right|^{-1}. \tag{6.6c}$$

In the derivation of equations (6.5) the relationships

$$\left(\frac{dT}{dp}\right)_c = \frac{-\partial f/\partial p}{\partial f/\partial T} \tag{6.7}$$

$$\delta(f) = \delta\left[\left.\frac{\partial f}{\partial z}\right|_{z_0}(z - z_0)\right] = \delta(z - z_0)\left|\frac{\partial f}{\partial z}\right|_{z_0}^{-1} \tag{6.8}$$

have been used. Equations (6.5) differ from those derived by Busse and Schubert (1971) only in that negative values of q are admitted and a different scale for the temperature is used. It is assumed that the adiabatic temperature difference ΔT_a used in the definition (2.15) of R does not include the effect of the phase transition.

As long as q is positive, i.e., for exothermic phase transitions, the onset of convection is similar to that in a single phase fluid. A negative value of s_{12} has the same effect as a subadiabatic temperature distribution. When s_{12} becomes positive, i.e., when the denser phase overlies the lighter phase, an unstable configuration arises. In both cases the distortion of the interface caused by latent heat effects is in the direction of the flow through the interface as can be seen from the expression

$$\zeta = \Theta(1 - (dT/dp)_c/(dT_s/dp_s))^{-1}\nu\kappa/g\beta\Delta T d^3 \tag{6.9}$$

for the dimensionless distortion ζ. The situation is quite different for endothermic phase transitions with $q < 0$ where only the case $s_{12} > 0$ must be considered if the attention is restricted to static temperature fields which increase towards higher pressure. The dynamic effect of the boundary distortion is now opposite to that of thermal expansion and instability can occur in the case R < 0 even though the light phase is on top of the dense phase.

Table 2.3 summarizes the effects discussed above. Note that the sign of ζ/v_z determines whether the interface distortion is in the direction of flow through the interface or opposite to it. The results agree with those displayed in Figures 1 and 2 of Schubert *et al.* (1975) in the cases of a dense phase overlying a lighter phase, $q \cdot s_{12} < 0$. Both, the exothermic and the endorthermic version, are likely to have applications in the Earth's mantle. The olivine-spinel transition is known to be exothermic, the spinel-post spinel transition is likely to be endothermic. Quantitative estimates of dynamic effects of these phase transitions have been made by Schubert and Turcotte (1971), Peltier (1973), Schubert *et al.* (1975) and others. Nonlinear effects have been considered by Turcotte and Schubert (1971), Richter (1973) and Olson and Yuen (1982). Another property that must be taken into account in realistic models of mantle convection is the bivariant or multi-variant nature of phase transitions in the mantle (see, for example, Schubert *et al.*, 1975). We refer to later chapters in this book for a more detailed discussion of aspects of this problem.

The case of the denser phase overlying the lighter phase is not likely to occur in the Earth's mantle because of its inherent instability. It was pointed out by Busse and Schubert (1971), however, that dissipative effects can in principle stabilize a dense phase on top of a lighter phase. This system thus is qualitatively different from the system of a dense fluid overlying an immiscible lighter fluid which is always unstable owing to the Rayleigh-Taylor instability. A geophysical application of the case $q \cdot s_{12} > 0$ has been discussed by Schubert and Straus (1980) in considering the occurrence of water above steam in geothermal areas. Because of an incorrect assumption about

Table 2.3. Effects of small values of $|q|$ and $|P_q|$ on the stability of the static state with $dT_s/dp_s > 0$.

		Exothermic case		Endothermic case
		$q > 0,\ s_{12} < 0$	$q > 0,\ s_{12} > 0$	$q < 0,\ s_{12} > 0$
Effect of latent heat		Stabilizing	Destabilizing	Destabilizing
Effect of boundary distortion	$R < 0$	Stabilizing	Stabilizing	Destabilizing
	$R > 0$	Destabilizing	Destabilizing	Stabilizing
Effect of latent heat on boundary distortion		$\zeta/v_z > 0$	$\zeta/v_z > 0$	$\zeta/v_z > 0$
Effect of temperature advection on boundary distortion		$R\zeta/v_z < 0$	$R\zeta/v_z > 0$	$R\zeta/v_z > 0$

the distortion of the phase boundary, the schematic diagrams in their paper disagree with the results of Table 2.3.

6.3. Multi-Component Convection

The evidence from isotopic data for several reservoirs of mantle material (DePaolo, 1981) has stimulated the interest in problems of convection in a fluid with several chemically distinct components. For the simplest case of two immiscible fluids of equal properties but slightly different mean density the solution (3.6) can be used in the case $n = 2$ to describe the onset of convection. In addition to equal properties the condition that the thickness $d/2$ of the lighter fluid is equal to that of the heavier fluid underneath must be satisfied. This idealized case is far too special to serve as a basis for a comprehensive discussion of double layer convection. In general a parameter space of high dimensions must be considered since the ratios of about all material properties for the two fluids enter the problem as additional dimensionless parameters. Some cases in which simplifying assumptions can be made have been discussed by Busse (1981). Among the new features of double layer convection we mention the inducement of circulation by viscous coupling in a stable layer by the onset of convection in the neighboring layer. When the viscosity ratio is sufficiently different from unity, thermal coupling is also possible with a reversed sense of the induced circulation. For a discussion of experimental observations of two layer convections we refer to Nguyen and Kulacki (1980). The problem has been considered under more special assumptions in connection with the problem of mantle convection by several authors; see, for example, Richter and Johnson (1974), Olson (1981), Richter and McKenzie (1981) and Spohn and Schubert (1982).

An interesting phenomenon can occur when two immiscible fluids in a layer heated from below have about the same mean density, but differ in their coefficients of expansion such that the range of density of fluid B throughout the layer includes the range of density of fluid A. Besides convection in two or more layers, a mode of convection appears to be realizable in which fluid B envelops parcels of fluid A. Figure 2.17 shows two possible modes of convection in the special case when the mean densities of the two fluids are identical. The configuration of the upper sketch corresponds to a minimum of the potential energy of the static state. The lower drawing depicts the mode of maximum heat transport for a given Rayleigh number when viscosity and heat conductivity of the two fluids are comparable. Laboratory studies of this form of convection have apparently not yet been done, because it is difficult to avoid the complicating influence of surface tension.

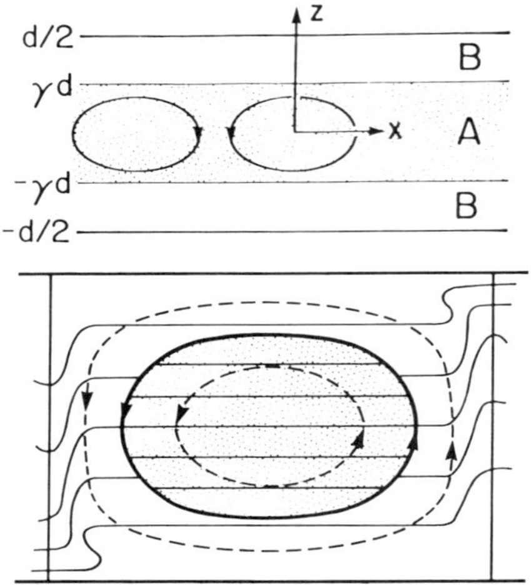

Figure 2.17. Configurations of two-dimensional convection in a layer with two fluids, A and B, with the same average density. Fluid A (stippled area) has a lower coefficient of thermal expansion (after Busse, 1983).

6.4. Buoyancy Driven Motions in a Stably Stratified Layer

Convection induced by an unstable density stratification is not the only possible cause of motions in planetary mantles. Another cause of flow is the "thermal blanket"—or "crustal thickness"—instability which occurs when a fluid layer of high viscosity μ^* and low heat conductivity λ^* overlies a layer of lower viscosity μ but with higher heat conductivity λ. There exists a static solution of the problem in which a uniform heat flux traverses both layers from below. But since it is assumed that the temperature gradient is subadiabatic in both layers, the static solution corresponds to a stable density stratification.

The instability occurs in the form of growing oscillations, i.e., as overstability, in the following way. An initial thickening of the upper layer impedes the heat flux from below and causes a rise of the temperature in the lower layer. The buoyancy force generates an ascending motion whose viscous stresses act on the upper layer and cause it to thin. On the other hand a thickening of the upper layer occurs over descending motions. When the time lag with which the temperature excess under the thinning upper layer reverses corresponds to about a half period of the oscillation, a mutual reinforcement between motion and temperature perturbation becomes possible, and the amplitude of the oscillation can grow exponentially as long as nonlinear effects are negligible. The criterion for instability can be written in the form

$$\left(\frac{\lambda}{\lambda^*} - 1\right)\left(\frac{\omega}{2\kappa}\right)^{1/2} d - 3\left(\frac{\kappa}{2\omega}\right)^{12} D - (\rho - \rho^*)gd/4\omega\mu^* > 0, \qquad (6.10a)$$

where the frequency ω of the oscillation must satisfy the relationship

$$\omega^2 \approx H\left(1 - \frac{\lambda^*}{\lambda}\right)\rho^*\kappa\beta g/4\mu^*\lambda^*. \qquad (6.10b)$$

It has been assumed that the thickness d of the upper layer is small in comparison with the thickness D of the lower layer and that the distortion of the interface is large in comparison with the distortion of the upper boundary of the upper layer. The other symbols have their usual meaning: κ is the thermal diffusivity of the lower layer, ρ is its density and β its thermal expansivity; H is the heat flux density in the upward direction. For details we refer to the discussion of this instability by Busse (1978). It should be noted that in the expression for C on p. 8 of that paper $D\lambda^2$ should be replaced by $D^{-3}\lambda^2$. It is shown in the same paper that internal heat sources in the upper layer have the same effect as a lower conductivity. Plots of the solution in the phase of maximum motion and in the phase of maximum distortion of the upper layer are shown in Figure 2.18. The solution derived by Busse (1978) actually corresponds to the limit when the temperature gradient equals the

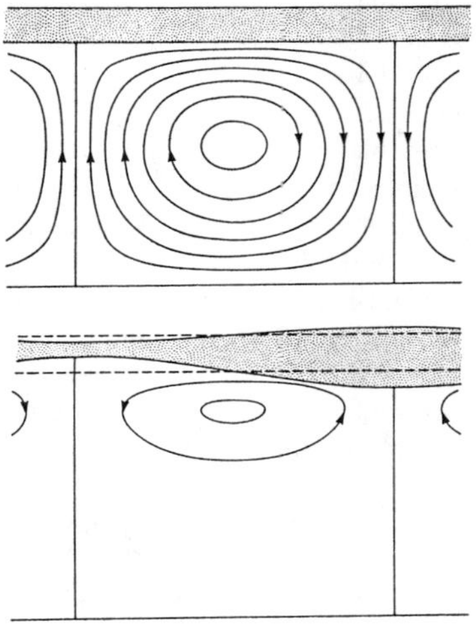

Figure 2.18. Streamlines of the "thermal blanket" instability in the form of a standing wave. The upper picture shows the phase when the distortion of the upper high viscosity layer (stippled) vanishes, the lower picture shows the phase of maximum distortion (after Busse, 1978).

adiabatic gradient in the lower layer; this assumption, which is also used for expressions (6.10), is not essential for the analysis.

The onset of the thermal blanket instability as overstability is of potential importance for mantle flow because of the time dependence it introduces. It can serve as a simple model for a periodic breakup and convergence of continental lithosphere in the form of supercontinents. Detailed models which combine the thermal blanket mechanism with buoyancy driven convection are likely to yield some interesting modes of nonlinear time dependent motion.

Note added in proof: This chapter was written at the end of 1983. Only minor updates have been done since that time.

7. References

Thermal Convection Bibliography

Ahlers, G., Effect of departures from the Oberbeck-Boussinesq approximation on the heat transport of horizontal convecting fluid layers. *J. Fluid* **98**, 137–148, (1980).

Backus, G.E., A class of self sustaining dissipative spherical dynamos. *Ann. Phys.* **4**, 372–447 (1958).

Baumgardner, J.R., A three-dimensional finite element model for mantle convection, Ph.D. Thesis, University of California at Los Angeles (1983).

Bolton, E.W. and F.H. Busse, Stability of convection rolls in a layer with stress-free boundaries. *J. Fluid Mech.* **146**, 115–125 (1985).

Booker, J.R., Thermal convection with strongly temperature-dependent viscosity. *J. Fluid Mech.* **76**, 741–754 (1976).

Booker, J.R., and K.L. Stengel, Further thoughts on convective heat transport in a variable-viscosity fluid. *J. Fluid Mech.* **86**, 289–291 (1978).

Bühler, K., K.R. Kirchartz, and H. Oertel, Jr., Steady convection in a horizontal fluid layer. *Acta Mechanica* **31**, 155–171 (1979).

Busse, F.H., Das Stabilitätsverhalten der Zellularkonvection bei endlicher Amplitude, Dissertation, University of Munich, 1962. English translation by S.H. Davis, Rand Rep. LT-66-19, Rand Corp., Santa Monica, California.

Busse, F.H., The stability of finite amplitude cellular convection and its relation to an extremum principle. *J. Fluid Mech.* **30**, 625–649 (1967a).

Busse, F.H., Non-stationary finite amplitude convection. *J. Fluid Mech.* **28**, 223–239 (1967b).

Busse, F.H., On the stability of two-dimensional convection in a layer heated from below. *J. Math. Phys.* **46**, 140–150 (1967c).

Busse, F.H., On Howard's upper bound for heat transport by turbulent convection. *J. Fluid Mech.* **37**, 457–477 (1969).

Busse, F.H., The oscillatory instability of convection rolls in a low Prandtl number fluid. *J. Fluid Mech.* **52**, 97–112 (1972).

Busse, F.H., Patterns of convection in spherical shells, J. Fluid Mech. **72**, 67–85 (1975).

Busse, F.H., A model of time-periodic mantle flow. *Geophys. J. Roy. Astr. Soc.* **52**, 1–12 (1978).

Busse, F.H., On the aspect ratio of two-layer mantle convection. *Phys. Earth Plan. Int.* **24**, 320–324 (1981).

Busse, F.H., Multiple solutions for convection in a two component fluid. *Geophys. Res. Letters* **9**, 519–522 (1982).

Busse, F.H. and E.W. Bolton, Instabilities of convection rolls with stress-free boundaries near threshold. *J. Fluid Mech.* **146**, 115–125 (1984).

Busse, F.H. and H. Frick, Square pattern convection in fluids with strongly temperature dependent viscosity. *J. Fluid Mech.* **150**, 451–465 (1985).

Busse, F.H. and N. Riahi, Nonlinear convection in a layer with nearly insulating boundaries. *J. Fluid Mech.* **96**, 243–256 (1980).

Busse, F.H. and N. Riahi, Patterns of convection in spherical shells II. *J. Fluid Mech.* **123**, 283–302 (1982).

Busse, F.H. and G. Schubert, Convection in a fluid with two phases. *J. Fluid Mech.* **46**, 801–912 (1971).

Busse, F.H. and J.A. Whitehead, Instabilities of convection rolls in a high Prandtl number fluid. *J. Fluid Mech.* **47**, 305–320 (1971).

Busse, F.H. and J.A. Whitehead, Oscillatory and collective instabilities in large Prandtl number convection. *J. Fluid Mech.* **66**, 67–79 (1974).

Buzano, E. and M. Golubitsky, Bifurcation on the hexagonal lattice and the planar Bénard problem. *Trans. Roy. Soc. (London)* **A308**, 617–677 (1983).

Carrigan, C.R., Multiple-scale convection in the Earth's mantle: A three-dimensional study. *Science* **215**, 965–967 (1982).

Carrigan, C.R., Convection in an internally heated, high Prandtl number fluid: A laboratory study. *Geophys. Astrophys. Fluid Dyn.* **32**, 1–21 (1985).

Chan, S.-K., Infinite Prandtl number turbulent convection. *Stud. Appl. Math.* **50**, 13–49 (1971).

Chandrasekhar, S., *Hydrodynamic and Hydromagnetic Stability*. Oxford: Clarendon Press (1961).

Chossat, P., Bifurcation and stability of convective flows in a rotating or not rotating spherical shell. *SIAM J. Appl. Math.* **37**, 624–647 (1979).

Christensen, U., Convection in a variable viscosity fluid: Newtonian versus power-law rheology. *Earth Planet Sci. Lett.* **64**, 153–162 (1983).

Clever, R.M., Heat transfer and stability properties of convection rolls in an internally heated fluid layer. *J. Appl. Math. and Physics (ZAMP)* **28**, 585–597 (1977).

Clever, R.M. and F.H. Busse, Transition to time-dependent convection. *J. Fluid Mech.* **65**, 625–645 (1974).

Clever, R.M. and F.H. Busse, Low Prandtl number convection in a layer heated from below. *J. Fluid Mech.* **102**, 61–74 (1981).

Davies, G.F., Whole-mantle convection and plate tectonics. Geophys. J. Roy. Astr. Soc. **49**, 459–486 (1977).

Deardorff, J.W., Gravitational instability between horizontal plates with shear. *Phys. Fluids* **8**, 1027–1030 (1965).

DeBremaecker, J.-C., Convection in the earth's mantle. *Tectonophysics* **41**, 195–208 (1977).

De La Cruz-Reyna, S., The thermal boundary layer and seismic focal mechanisms in mantle convection. *Tectonophysics* **35**, 149–160 (1976).

DePaolo, D.J., Nd isotopic studies: Some new perspectives on Earth structure and evolution. *EOS, Trans. Amer. Geophys. Union* **62**, 137–140 (1981).

Dubois, M. and P. Bergé, Experimental study of the velocity field in Rayleigh-Bénard convection. *J. Fluid Mech.* **85**, 641–653 (1978).

Dubois, M., C. Normand, and P. Bergé, Wavenumber dependence of velocity field amplitude in convection rolls—theory and experiments. *Int. J. Heat Mass Transfer* **21**, 999–1002 (1978).

Foster, T.D., Stability of a homogeneous fluid cooled uniformily from above. *Phys. Fluids* **8**, 1249–1257 (1965).

Foster, T.D., Convection in a variable viscosity fluid heated from within. *J. Geophys. Res.* **74**, 685–693 (1969).

Fowler, A.C., Fast thermoviscous convection. *Stud. Appl. Math.* **72**, 189–219 (1985).

Frick, H., F.H. Busse, and R.M. Clever, Steady three-dimensional convection at high Prandtl number. *J. Fluid Mech.* **127**, 141–153 (1983).

Golubitsky, M. and D. Schaeffer, Bifuraction with 0(3) symmetry including applications to the Bénard problem. *Comm. Pure Appl. Math.* **35**, 81–111 (1982).

Golubitsky, M., J.W. Swift, and E. Knobloch, Symmetries and pattern selection in Rayleigh-Benard convection. *Physica* **10D**, 246–276 (1984).

Gorkov, L.P., Stationary convection in a plane liquid layer near the critical heat transfer point. Sov. Phys. J.E.T.P. **6**, 311–315 (1957).

Gough, D.O., The anelastic approximation for thermal convection. *J. Atmos. Sci.* **26**, 448–456 (1969).

Gough, D.D., E.A. Spiegel, and J. Toomre, Modal equations for cellular convection. *J. Fluid Mech.* **68**, 695–719 (1975).

Graham, A., Shear patterns in an unstable layer of air. *Phil. Trans. Roy. Soc.* **A232**, 285–96 (1933).

Green, L.L. and T.D. Foster, Secondary convection in a Hele Shaw cell. *J. Fluid Mech.* **71**, 675–687 (1975).

Herring, J.R., Investigation of problems in thermal convection. *J. Atmos. Sci* **20**, 325–338 (1963).

Herring, J.R., Investigation of problems in thermal convection: Rigid boundaries. *J. Atmos. Sci.* **21**, 277–290 (1964).

Hewitt, J.M., D.P. McKenzie, and N.O. Weiss, Dissipative heating in convective flows. *J. Fluid Mech.* **68**, 721–738 (1975).

Hoard, C.Q., C.R. Robertson, and A. Acrivos, Experiments on the cellular structure in Bénard convection. *Int. J. Heat Mass Transfer* **13**, 849–855 (1970).

Houston, M.H., and J.-C. DeBremaecker, Numerical models of convection in the upper mantle. *J. Geophys. Res. 80*, 742–751 (1975).

Howard, L.N., Heat transport by turbulent convection. *J. Fluid Mech.* **17**, 405–432 (1963).

Howard, L.N., Boundary layer treatment of the mean field equations. *Woods Hole Oceanographic Inst. GFD Notes, no. 65–51,* **I**, 124–126 (1965).

Howard, L.N., Convection at high Rayleigh number. In H. Görtler (ed.), *Proc. 11th Internat. Congress of Appl. Mech.*, Munich (1964), Berlin: Springer, pp. 1109–1115 (1966).

Ihrig, E. and M. Golubitsky, Pattern selection with 0(3) symmetry. *Physica* **13D**, 1–13 (1984).

Jacoby, W.R., and H. Schmeling, On the effects of the lithosphere on mantle convection and evolution. *Phys. Earth Planet. Inst.* **29**, 305–319 (1982).

Jakeman, E., Convective instability in fluids of high thermal diffusivity. *Phys. Fluids* **11**, 10–14 (1968).

Jarvis, G.T. and D.P. McKenzie, Convection in a compressible fluid with infinite Prandtl number. *J. Fluid Mech.* **96**, 515–583 (1980).

Jarvis, G.T. and W.R. Peltier, Mantle convection as a boundary layer phenomenon. *Geophys. J. Roy. Astr. Soc.* **68**, 386–424 (1982).

Jenkins, D.R. and M.R.E. Proctor, The transition from roll to square cell solutions in Rayleigh-Bénard convection. *J. Fluid Mech.* **139**, 461–471 (1984).

Joseph, D.D. and S. Carmi, Subcritical convection instability, Part 2—Spherical shells. *J. Fluid Mech.* **26**, 769–777 (1966).

Katsaros, N. B., Turbulent free convection in fresh and salt water: Some characteristics revealed by visualization. *J. Phys. Oceanogr.* **8**, 613–626 (1978).

Kopitzke, U., Finite element convection models: comparison of shallow and deep mantle convection, and temperature in the mantle. *J. Geophys.* **46**, 97–121 (1979).

Krishnamurti, R., Finite amplitude convection with changing mean temperature, Part 1. Theory. *J. Fluid Mech.* **33**, 445–455 (1968a).

Krishnamurti, R., Finite amplitude convection with changing mean temperature, Part 2. An experimental test of the theory. *J. Fluid Mech.* **33**, 457–463 (1968b).

Krishnamurti, R., On the transition to turbulent convection, Part 1. The transition from two to three-dimensional flow. *J. Fluid Mech.* **42**, 295–307 (1970a).

Krishnamurti, R., On the transition to turbulent convection, Part 2. The transition to time-dependent flow. *J. Fluid Mech.* **42**, 309–320 (1970b).

Krishnamurti, R., Some further studies on the transition to turbulent convection. *J. Fluid Mech.* **60**, 285–303 (1973).

Kulacki, F.A. and A.A. Emara, Steady and transient thermal convection in a fluid layer with uniform volumetric energy sources. *J. Fluid Mech.* **83**, 375–395 (1977).

Kulacki, F.A. and R. J. Goldstein, Thermal convection in a horizontal fluid layer with uniform energy sources. *J. Fluid Mech.* **55**, 271–287 (1972).

Kvernvold, O., Rayleigh-Bénard convection with one free and one rigid boundary. *Geophys. Astrophys. Fluid Dyn.* **12**, 273–294 (1979).

Liang, J.F. and A. Acrivos, Experiments on buoyancy driven convection in non-Newtonian fluid. *Rheol. Acta* **9**, 447–455 (1970).

Malkus, W.V.R., Discrete transitions in turbulent convection. *Proc. Roy. Soc. (London)* **A225**, 185–195 (1954a).

Malkus, W.V.R., The heat transport and spectrum of thermal turbulence. *Proc. Roy. Soc. (London)* **A225**, 196–212 (1954b).

Malkus, V.W.R. and G. Veronis, Finite amplitude convection. *J. Fluid Mech.* **4**, 225–260 (1958).

McKenzie, D.P., J.M. Roberts, and N.O. Weiss, Convection in the Earth's mantle: Towards a numerical simulation. *J. Fluid Mech.* **62**, 456–538 (1974).

Moore, D.R., and N.O. Weiss, Two-dimensional Rayleigh-Bénard convection. *J. Fluid Mech.* **58**, 289–312 (1973).

Morris, S. and D. Canright, A boundary-layer analysis of Bénard convection with strongly temperature-dependent viscosity. *Phys. Earth Planet. Int.* **36**, 355–373 (1984).

Nguyen, A.-T. and F.A. Kulacki, Convection in multi-layer systems. In P.A. Davis and S.K. Runcorn (eds.), *Mechanisms of Continental Drift and Plate Tectonics*. New York: Academic Press. pp 259–266 (1980).

Oertel, H., Thermal instabilities. In J. Zierep and H. Oertel, Jr. (eds.), *Convective Transport and Instability Phenomena*. Karlsruhe: G. Braun, pp. 3–24 (1982).

Ogura, Y. and N.A. Phillips, Scale analysis of deep and shallow convection in the atmosphere. *J. Atmos. Sci.* **19**, 173–179 (1962).

Oliver, D.S. and J.R. Booker, Planform of convection with strongly temperature-dependent viscosity. *Geophys. Astrophys. Fluid Dyn.* **27**, 73–85 (1983).

Olson, P., Mantle convection with spherical effects. *J. Geophys. Res.* **86**, 4881–4890 (1981).

Olson, P. and G.M. Corcos, A boundary layer model of mantle convection with surface plates. *Geophys. J. Roy. Astr. Soc.* **62**, 192–220 (1980).

Olson, P., and D.A. Yuen, Thermochemical plumes and mantle phase transitions. *J. Geophys. Res.* **87**, 3993–4002 (1982).

Palm, E., On the tendency towards hexagonal cells in steady convection. *J. Fluid Mech.* **8**, 183–192 (1960).

Parmentier, E.M. and J. Morgan III, Thermal convection in non-Newtonian fluids: Volumetric heating and boundary layer scaling. *J. Geophys. Res.* **87**, 7757–7762 (1982).

Parmentier, E.M., D.L. Turcotte, and K.E. Torrance, Studies of finite amplitude non-Newtonian thermal convection with application to convection in the earth's mantle. *J. Geophys. Res.* **81**, 1839–1846 (1976).

Peckover, R.S. and I.H. Hutchinson, Convective rolls driven by internal heat sources. *Phys. Fluids* **17**, 1369–1371 (1974).

Pellew, A. and R.V. Southwell, On maintained convective motion in a fluid heated from below. *Proc. Roy. Soc. (London)* **A126**, 312–343 (1940).

Peltier, W.R., Penetrative convection in the planetary mantle. *Geophys. Fluid Dyn.* **5**, 47–88 (1973).

Proctor, M.R.E., Planform selection by finite-amplitude thermal convection between poorly conducting slabs. *J. Fluid Mech.* **113**, 469–485 (1981).

Rayleigh, Lord, On convection currents in a horizontal layer of fluid, when the higher temperature is on the under side. *Phil. Mag.* **32**, 529–546 (1916).

Reid, W.H. and D.L. Harris, Some further results on the Bénard problem. *Phys. Fluids* **1**, 102–110 (1958).

Richter, F.M., Convection and the large-scale circulation of the mantle. *J. Geophys. Res.* **78**, 8735–45 (1973).

Richter, F.M., Experiments on the stability of convection rolls in fluids whose viscosity depends on temperature. *J. Fluid Mech.* **89**, 553–560 (1978).

Richter, F.M. and G.E. Johnson, Stability of a chemically layered mantle. *J. Geophys. Res.* **79**, 1635–1639 (1974).

Richter, F.M. and D.P. McKenzie, On some consequences and possible causes of layered mantle convection. *J. Geophys. Res.* **86**, 6133–6142 (1981).

Richter, F.M., H.-C. Nataf, and S.F. Daly, Heat transfer and horizontally averaged temperature of convection with large viscosity variations. *J. Fluid Mech.* **129**, 173–192 (1983).

Roberts, G.O., Fast viscous Bénard convection. *Geophys. Astrophys. Fluid Dyn.* **12**, 235–272 (1979).

Roberts, P.H., Convection in horizontal layers with internal heat generation theory. *J. Fluid Mech.* **30**, 33–49 (1967).

Roberts, P.H., On non-linear Bénard convection. In R. Donnelly, R. Hermann, and I. Prigogine (eds.), *Non-Equilibrium Thermodynamics, Variational Techniques, and Stability*. Chicago: University of Chicago Press, pp. 125–162 (1966).

Robinson, J.L., A boundary layer convection cell, Woods Hole Oceanographic Institution Report No. 65—51, 92–112 (1965).

Rossby, H.T., An experimental study of Bénard convection with or without rotation. Ph.D. Thesis, Massachusetts Institute of Technology (1966).

Rossby, H.T., A study of Bénard convection with and without rotation. *J. Fluid Mech.* **36**, 309–335 (1969).

Sattinger, D.H., Selection mechanisms in symmetry breaking phenomena. In R. Rautmann (ed.), *Approximation Methods for Navier-Stokes Equations*. pp. 494–504, *Lecture Notes in Mathematics*, vol. 771 (1980).

Schlüter, A., D. Lortz, and F. Busse, On the stability of steady finite amplitude convection. *J. Fluid Mech.* **23**, 129–144 (1965).

Schubert, G. and J.M. Straus, Gravitational stability of water over steam in vapor-dominated geothermal systems. *J. Geophys. Res.* **85**, 6505–6512 (1980).

Schubert, G. and D.L. Turcotte, Phase changes and mantle convection. *J. Geophys. Res.* **76**, 1424–1432 (1971).

Schubert, G., D.A. Yuen, and D.L. Turcotte, Role of phase transitions in a dynamic mantle. *Geophys. J. Roy. Astr. Soc.* **42**, 705–735 (1975)

Schubert, G. and A. Zebib, Thermal convection in an internally heated infinite Prandtl number fluid in a spherical shell. *Geophys. Astrophys. Fluid Dyn.* **15**, 65–90 (1980).

Schmeling, H. and W.R. Jacoby, On modelling the lithosphere in mantle convection. *J. Geophys.* **50** 89–100 (1981).
Schwiderski, E.W. and H.J.A. Schwab, Convection experiments with electrolytically heated fluid layers. *J. Fluid Mech.* **48**, 703–719 (1971).
Segel, L.A. and J.T. Stuart, On the question of the preferred mode in cellular thermal convection. *J. Fluid Mech.* **13** 289–306 (1962).
Silveston, P. L., Wärmedurchgang in waagerechton Flüssigkeitsschichten. *Forsch. Ing. Wes.* **24**, 29–32, 59–69 (1958).
Sorokin, V.S., Variational method in the theory of convection (in Russian). *Prikl. Mat. Mekh.* **17**, 39–48 (1953).
Spangenberg, W.W. and W.R. Rowland, Convection circulation in water induced by evaporative cooling. *Phys. Fluids* **4**, 783–800 (1961).
Sparrow, E.M., R.J. Goldstein, and V.H. Johnson, Thermal instability in a horizontal fluid layer: Effect of boundary conditions and non-linear temperature profile. *J. Fluid Mech.* **18**, 513–528 (1964).
Spohn, T. and G. Schubert, Modes of mantle convection and the removal of heat from the earth's interior. *J. Geophys. Res.* **87**, 4682–4696 (1982).
Stengel, K.C., D.S. Oliver, and J.R. Booker, Onset of convection in a variable-viscosity fluid. *J. Fluid Mech.* **120**, 411–431 (1982).
Straus, J.M., Finite amplitude doubly diffussive convection. *J. Fluid Mech.* **56**, 353–374 (1972).
Stewartson, K., Asymptotic theory in limit $R \to \infty$. In R. Donnelly, R. Hermann, and I. Prigogine (eds.), *Non-Equilibrium Thermodynamics, Variational Techniques, and Stability.* Chicago: University of Chicago Press, pp. 158–162 (1966).
Thirlby, R., Convection in an internally heated layer. *J. Fluid Mech.* **44**, 673–693 (1970).
Tippelskirch, H., Über Konvektionszellen insbesondere in flüssigem Schwefel. *Beitr. Phys. Atmos.* **29**, 37–54 (1956).
Torrance, K.E. and D.L. Turcotte, Thermal convection with large viscosity variations. *J. Fluid Mech.* **47**, 113–125 (1971).
Tritton, D.J., and M.N. Zarraga, Convection in horizontal layers with internal heat generation, Experiments. *J. Fluid Mech.* **30**, 21–31 (1967).
Turcotte, D.L. and E.R. Oxburgh, Finite amplitude convection cells and continental drift. *J. Fluid Mech.* **28**, 29–42 (1967).
Turcotte, D.L. and G. Schubert, Structure of the olivine-spinel phase boundary in the descending lithosphere. *J. Geophys. Res.* **76**, 7980–7987 (1971) (see also correction in vol. **77**, 2146).
Tveitereid, M., Thermal convection in a horizontal fluid layer with internal heat sources. *Int. J. Heat Mass Transfer* **21**, 335–339 (1978).
Tveitereid, M. and E. Palm, Convection due to internal heat sources. *J. Fluid Mech.* **76**, 481–400 (1976).
Veronis, G., Large-amplitude Bénard convection. *J. Fluid Mech.* **26**, 49 (1966).
Walden, R.V. and G. Ahlers, Non-Boussinesq and penetrative convection in a cylindrical cell. *J. Fluid Mech.* **109**, 89–114 (1981).

Ward, III, W.J. and O.H. LeBlanc, Jr., Rayleigh-Benard convection in an electrochemical redox cell. *Science* **225**, 1471–1473 (1984).

Warner, W.H., Poloidal and toroidal potentials for solenoidal fields. *J. Appl. Math. Phys. (ZAMP)* **23**, 221 (1972).

Willis, G.E. and J.W. Deardorff, The oscillatory motions of Rayleigh convection. *J. Fluid Mech.* **44**, 661–672 (1970).

White, D.B., Experiments with convection in a variable viscosity fluid. Ph.D. Thesis, Cambridge University (1982).

White, D.B., The planforms and onset of convection with a temperature-dependent viscosity. *J. Fluid Mech.* **191**, 247–286 (1988).

Whitehead, J.A. and G.L. Chan, Stability of Rayleigh-Bénard convection rolls and bimodal flow at moderate Prandtl number. *Dyn. Atmosph. Oceans* **1**, 33–49 (1976).

Whitehead, J.A. and B. Parsons, Observations of convection at Rayleigh numbers up to 760,000 in a fluid with large Prandtl number. *Geophys. Astrophys. Fluid Dyn.* **9**, 201–217 (1978).

Zebib, A., G. Schubert, J.L. Dein, and R.C. Paliwal, Character and stability of axisymmetric thermal convection. *Geophys. Astrophys. Fluid Dyn.* **23**, 1–42 (1983).

Zebib, A., G. Schubert, and J. Straus, Infinite Prandtl number thermal convection in a spherical shell. *J. Fluid Mech.* **97**, 257–277 (1980).

Zimmermann, G., Boundary layer solution for the Rayleigh-Bénard convection problem, Report. Max-Planck-Institut für Strömungsforschung (1983).

CHAPTER 3

Seismic Imaging of Boundary Layers and Deep Mantle Convection

Thomas H. Jordan,[†] Arthur L. Lerner-Lam,[‡] and Kenneth C. Creager[§]

[†]Department of Earth, Atmospheric and Planetary Sciences, Massachusetts Institute of Technology, Cambridge, Massachusetts 02139

[‡]Lamont-Doherty Geological Observatory, Palisades, New York 10964

[§]Geophysics Program, University of Washington, Seattle, Washington 98195

1. Introduction

Although much has been recently learned about the surface manifestations of convection in the Earth's mantle and few geophysicists would dispute its existence, there has not yet developed a consensus on the answers to some of the most fundamental questions regarding the nature of this convection. A vigorous debate is currently focused on the vertical extent of the convecting layer comprising the lithospheric plates. According to one school, this layer is limited from below by a chemical discontinuity at a depth of approximately 700 km; according to another, it extends over most if not the entire depth of the mantle. Geophysical thinking has been divided along these lines for the last dozen or so years, and it is now widely appreciated that further progress on mantle dynamics will be inhibited until one or the other of these models (or both) can be convincingly excluded.

It is perhaps less appreciated that the question of stratified versus whole-mantle convection was raised nearly two decades before the advent of plate tectonics. In September, 1950, a UNESCO-sponsored colloquium was convened in Hershey, Pennsylvania, to consider "Plastic Flow and Deformation Within the Earth" (Gutenberg *et al.*, 1951). The participants included Vening Meinesz, Griggs, and Hess, among others. They were by-and-large faithful to the mobilistic theories which sought to explain surface tectonics in terms of convective processes in the deep interior. Their faith was shaken by Francis Birch, who presented a paper concerning the inhomogeneity of the mid-mantle transition zone. He concluded that, "the layer between 200 and 800 km may be interpreted as a region in which the transition to the high-pressure form takes place, probably with a change in chemical composition as well . . . To assume unimpeded circulation in this region is to ignore its definitely transitional nature." In the vigorous discussion following his report, J.H.F. Umbgrove summarized the quandry very succinctly: "If chemical change is involved, we have to give up the hypothesis of convection currents passing through this zone. If it is only a change of phase, the problem remains whether Birch's results flatly oppose the hypothesis of convection currents over the entire 2900-km sector of the mantle, or whether there is a way out of the apparent difficulty. *A positive answer to this question seems crucial with regard to any further discussion of convection currents*" (emphasis added). It was in this context that the Dutch geophysicist P.P. Bijlaard, then at Cornell, submitted the diagram reproduced in Figure 1.1 and commented, "It might be possible that, as a consequence of differences in temperature, convection currents (I) between depths of 900 and 2900 km would occur, which might induce secondary currents (II) in the higher

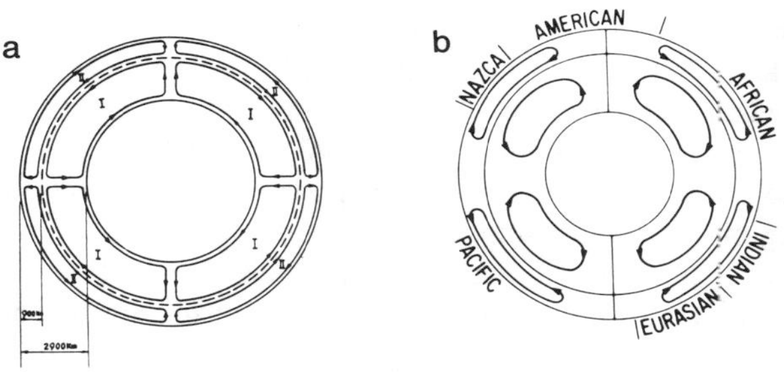

Figure 1.1. Two schematic models of stratified mantle convection. Model (a) was proposed by P.P. Bijlaard three and a half decades ago (Gutenberg *et al.*, 1951) and (b) is from a recent paper by Busse (1983).

regions of the mantle." Figure 1.1 juxtaposes Bijlaard's scheme with a similar diagram of stratified, quadrupolar convection extracted from a recent paper by Busse (1983). The comparison provides one measure of how far our thinking on this subject has evolved in the ensuing three-and-a-half decades.

Convection currents of the sort illustrated in Figure 1.1 involve lateral temperature contrasts of several hundred degrees over scale lengths of thousands of kilometers, and these contrasts will induce heterogeneity in the speeds of seismic waves. At conditions appropriate to the deep mantle, the logarithmic variation of seismic velocities with temperature ranges from about $-5 \times 10^{-5}\,°C^{-1}$ for compressional waves to $-1 \times 10^{-4}\,°C^{-1}$ for shear waves, so that the magnitude of the velocity heterogeneity will be several percent, regions of higher temperature having lower wave speeds. If the large-scale flow within the mantle were as well organized as hypothesized in Figure 1.1, a strong quadrupolar signature of this convective pattern should be present in the aspherical variations of wave speeds, especially in the transition zone comprising the boundary layers separating the two systems, and a correlative signature should also exist in the quadrupolar components of the geopotential field. In fact, Busse's sketch in Figure 1.1 was motivated by the recent discoveries by Dziewonski *et al.* (1977) and Masters *et al.* (1982) of a quadrupolar components of mantle heterogeneity with just these properties.

The results of Dziewonski *et al.* (1977) and Masters *et al.* (1982) do not, unfortunately, discriminate satisfactorily between models of shallow and deep circulation, nor do the considerable refinements in the mapping and interpretation of large-scale aspherical structure made in the subsequent studies of Woodhouse and Dziewonski (1984), Nataf *et al.* (1984, 1986), Dziewonski (1984), and Hager *et al.* (1985). A particular problem is a "resolution gap" in the ability of present methods to image the mid-mantle transition zone, which has thus far prevented the detailed global mapping of lateral heterogeneity in this critical region. Nevertheless, these studies mark the coming-of-age of three-dimensional structural seismology, which holds the promise of providing a much refined picture of mantle convection.

This chapter purports to review some of the contributions seismology has made to the long-standing problems of mantle dynamics. The Earth is a complex body with elastic and anelastic properties that are anisotropic and heterogeneous on many scales. Extracting useful information about this complexity from inaccurate and often woefully insufficient data sets derived by eyeballing or numerically processing a bunch of wiggly lines is a technically difficult endeavor—some would say an art form—and its practitioners have developed a healthy skepticism regarding any physical interpretations extrapolated from the results. As in any actively moving field, conflicting interpretations by reputable groups of investigators abound, bewildering as much as informing the non-specialists. On the other hand, an historical perspective suggests that, once the smoke has cleared, many conclusions which now appear controversial will be seen as established or refuted by papers published years ago. Recognition of these facts breeds a certain timidity among reviewers, especially those who wish to affect a disinterested stance. Aware of these difficulties, we have abandoned any attempt to write an objective, comprehensive and balanced account of the subject, and have chosen to concentrate on a few topics where our opinions are well informed (or at least well formed) by direct exposure to seismological data.

We begin our review with a discussion of plate structure in Section 2, placing particular emphasis on the boundary layer configurations of the continents. The existence of deep-seated "roots" beneath the continental cratons, only recently a very controversial subject, has now been confirmed by a variety of seismic methods. We infer that the average thickness of the plates in these regions is substantially greater than allowed by the simple thermal boundary models of classical plate tectonics. These cratonic structures appear to be dynamically controlled by a chemical boundary layer in the uppermost mantle with large lateral variations in thickness. Section 3 addresses the fate of subducted oceanic lithosphere, around which many dynamical issues revolve. We summarize the information on deep-focus seismic zones, concentrating on recent high-resolution images of near-source structure obtained by residual-sphere analysis. These images indicate the penetration of descending slabs into the lower mantle beneath a number of western Pacific subduction zones and, therefore, the existence of significant mass flux across the 650-km discontinuity. On the basis of this evidence, we conclude that the hypothesis of rigorous stratification of convection into an upper-mantle system and a lower-mantle system of differing compositions is not consistent with the available seismological data. The data on the large-scale structure of the mantle beneath the plates is reviewed in Section 4. We discuss the global models of aspherical heterogeneity obtained by the techniques of body-wave and surface-wave tomography and some of the questions they pose for dynamical models. In particular, we present the evidence for low-wavenumber, large-amplitude heterogeneity in the vicinity of the core-mantle boundary derived from core-penetrating *PKP* waves and argue that this heterogeneity is most plausibly explained by the existence of chemical, as well as thermal, boundary layers at this interface.

2. Structure of the Tectonic Plates

In the simplest model of plate tectonics, the plates are themselves the upper thermal boundary layer of the mantle convection system (Turcotte and Oxburgh, 1967), deriving their rigidity and ability to transmit deviatoric stress from the cold, strong mechanical boundary layer forming their outer parts (Elsasser, 1969). The thermal boundary layer theory of plate structure has enjoyed great success in explaining the large-scale physiography and evolution of the major ocean basins, as well as many of their smaller scale features. Attempts have been made to integrate the continents into this thermal boundary layer model, but these have been less successful.

Continents are the most obvious manifestation of lateral heterogeneity in the Earth and the most dynamically important component of a surficial chemical boundary layer (Figure 2.1). Whereas the thin basaltic crust of the oceans is continually being recycled into the mantle via the subduction process, the thicker, silica-rich crust of the continents remains buoyantly trapped at the surface. The disposition of continental crust thus exerts considerable control on plate kinematics (Isacks *et al.*, 1968; McKenzie, 1969). Moreover, crustal thickness is a significant parameter in determining the overall rheology of the lithosphere and consequently the strain field within the plates caused by the forces exerted at plate boundaries (Isacks *et al.*, 1968; McKenzie, 1969; Vink *et al.*, 1984). It has been conjectured that these recognized crustal properties are sufficient to explain the various peculiarities of continental tectonics visible at present or evident in the geological record, including the diffuse nature of continental seismicity, the

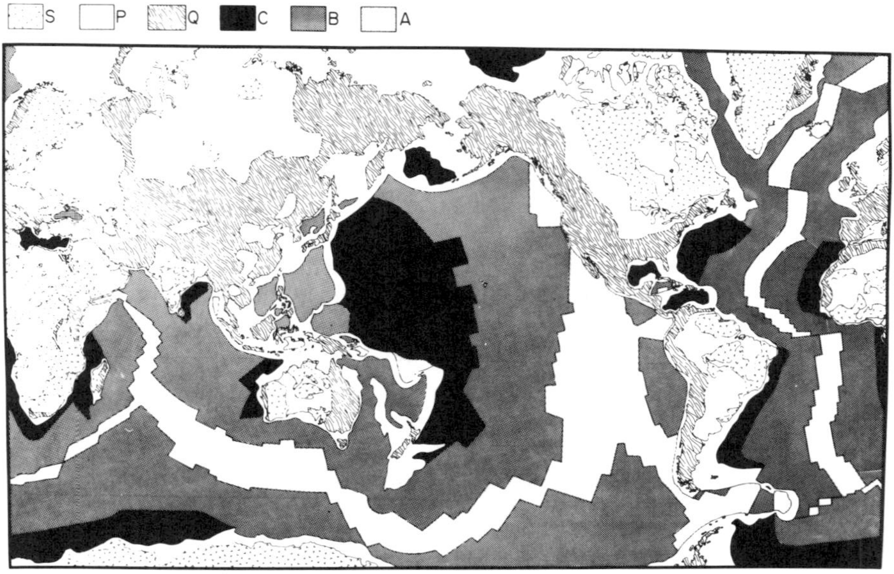

Figure 2.1. Mercator projection of the globe between 75°N and 70°S showing Jordan's (1981a) tectonic regionalization GTR1. Oceanic crust is partitioned into three age regions: *A*, young ocean (<25 My); *B*, intermediate-age ocean (25–100 My); and *C*, old ocean (>100 My). Subaerial continental crust is divided into three regions based on generalized tectonic behavior during the Phanerozoic: *S*, Precambrian shields and platforms; *P*, Phanerozoic platforms; and *Q*, Phanerozoic orogenic zones and magmatic belts. White areas are regions of submerged continental or transitional crust, including continental margins, island arcs, and oceanic plateaus adjoining continental crust.

complexity of continental plate boundaries and continent-continent interactions, the tendency for continental masses to collect into "supercontinents," the age distribution of continental rocks, and so forth. According to the conventional theory of plate tectonics, the thermal, mechanical and chemical properties of the lithospheric plates below a depth of 100 km or so are largely independent of crustal type and can be couched in a unified model of boundary layer evolution without endowing special properties to the subcontinental mantle or special dynamics to continental drift (Crough and Thompson, 1976; Kono and Amano, 1978; Sclater *et al.*, 1980, 1981). The conventional model of plate structure is illustrated in Figure 2.2.

It has become clear from seismological studies, however, that the conti-

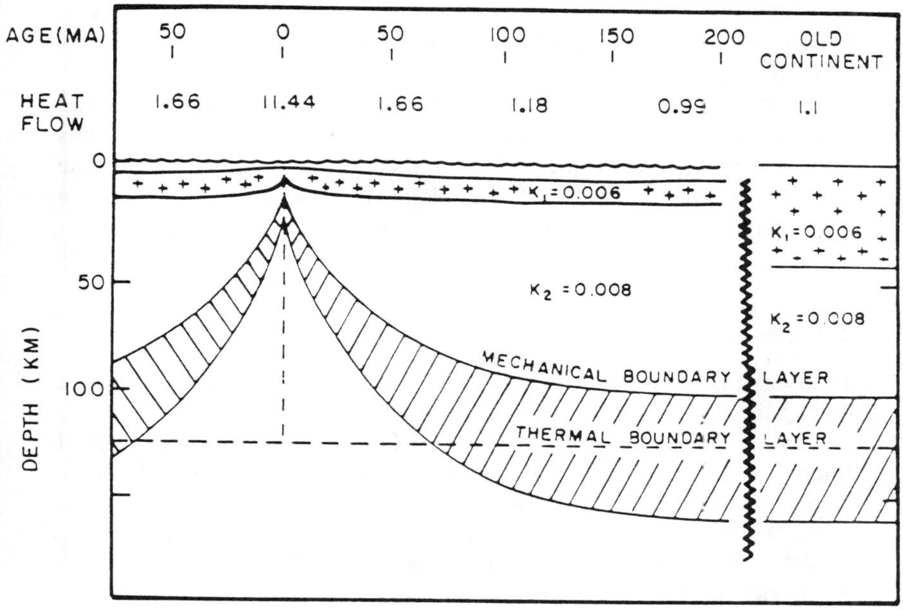

Figure 2.2. The unified thermal boundary-layer model of plate structure proposed by Sclater *et al.* (1981, Figure 13). In this model, old oceans and old continents have similar sub-crustal thermal structures and a characteristic tectospheric thickness of about 125 km. Heat flow is in units of 1 μcal cm^{-2}s^{-1} (42 mW/m^2).

nental cratons are underlain by an extensive layer of anomalous mantle material, geographically variable and up to several hundred kilometers thick which translates with the continents during the course of plate motions. Since it was first proposed ten years ago (Jordan, 1975a), the hypothesis that the plates in old continental regions are, on average, much thicker than even the oldest oceanic plate has been controversial. In a review of seismological and other evidence, Anderson (1979) dismissed the thick-plate hypothesis, stating, "Continental roots extend no deeper than about 150–200 km . . . Oceanic and continental geotherms converge above about 200 km and become less steep than the melting gradient at greater depth." Sclater *et al.* (1980, 1981) presented a comprehensive analysis of plate thermal structure in both continental and oceanic regions, and concluded that "the range in possible geotherms at depths below 100 to 150 km under continents and oceans overlaps and that the thermal structure beneath an old stable continent is indistinguishable from that beneath an old ocean were it at equilibrium. Oceans and continents are part of the same thermal system" (Sclater *et al.*, 1981, p. 11535).

These assessments notwithstanding, simple boundary-layer cooling models such as the one illustrated in Figure 2.2 are no longer tenable. In this section we review the evidence for thick continental plates, emphasizing some of the more controversial seismological issues but also summarizing other geophysical, geochemical and geological data on deep continental structure.

2.1. General Considerations

In discussing plate structure, we shall find it convenient to discriminate among the three types of boundary layers:

Thermal boundary layer (TBL). The TBL is defined by the thermal thickness L_T of the plate, equal to the depth where the geotherm reaches, or comes close to reaching, the mantle adiabat. Heat transfer within the TBL is dominated by conduction, and the thermal gradients are super-adiabatic.

Mechanical boundary layer (MBL). The MBL is defined by the effective mechanical thickness L_M of the TBL for long-term loads; that is, the depth to an isotherm T_M such that the deviatoric stresses at a given (small) strain rate are less than some specified (small) value for $T > T_M$ (McNutt, 1984). The relationship between L_M and L_T has been investigated by numerical calculations employing realistic upper-mantle rheologies (Bodine *et al.*, 1981), and the flexure data from both oceans and continents have been shown to be consistent with $T_M \approx 550$–600°C (McNutt, 1984; Kusznir and Karner, 1985).

Chemical boundary layer (CBL). In a layered, convecting body with a chemical composition and evolutionary history as complicated as the Earth's, it can be expected that the mechanisms of mass transport will, over time, build up compositionally distinct boundary layers of intermediate density at the major chemical transitions. At the Earth's surface, the most obvious manifestation of a CBL is the continental and oceanic crust, but this CBL also involves the uppermost mantle. The primary process for the major-element modification of the upper mantle is the extraction of basaltic melt, and the residuum left behind by this process has a lower density than the parental rock (O'Hara, 1975; Boyd and McCallister, 1976). Compositional variations in upper-mantle density are most easily characterized in terms of a normative density $\hat{\rho}$, defined to be density of a mineral assemblage equilibrated at mantle conditions (e.g., within the garnet stability field) but referenced to standard pressure and temperature (Jordan, 1979a). The surficial CBL is defined by the thickness L_C of the crust and any subjacent mantle whose normative density has been significantly reduced by the extraction of a basaltic melt.

Because conduction will dominate over advection only when the mass motions are very small, the TBL must remain effectively coherent during plate motions. From a kinematical point of view, the TBL defines the plate, or lithosphere in the conventional terminology of plate tectonics. Traditionally, however, "lithosphere" denotes the strong outer shell of the Earth capable of withstanding large deviatoric stresses over geologically long periods of time (Barrell, 1914; Daly, 1940). To maintain the distinction between these two types of boundary layers—one kinematical, one dynamical—Jordan (1975a) proposed to adapt Elsasser's (1969) term "tectosphere" to designate the TBL, reserving lithosphere for the MBL. This usage will be followed here.

2.2. Oceanic Tectosphere as a Simple Thermal Boundary Layer

In the oceans, a CBL comprising a basaltic crust and a subjacent, chemically complementary layer of dunitic and harzbugitic residua and cumulates is emplaced at the spreading centers (Ringwood, 1969; Kay *et al.*, 1970). The crust has an average thickness of about 6–8 km and the residual layer about 25–30 km, implying $L_C \approx 35$ km. This ophiolitic sequence translates away from the ridge crests with little modification, whereas a conductively cooled TBL and an associated MBL thicken with age. At small ages, the temperature profile of an oceanic plate decays according to a cooling half-space model; in particular, the TBL thickness increases as the square-root of age:

$$L_T = a_T t^{1/2} \tag{2.1}$$

If we define L_T as the depth where the temperature comes within, say, 3 percent of the mantle adiabat (e.g., Parsons and McKenzie, 1978), then $a_T \approx$ 15 km My$^{-1/2}$. The TBL thickness at 70 My is thus about 125 km. The cooling half-space model predicts that the water depth (corrected for variations in crustal loading) should increase as the square-root of time, which it is observed to do in regions of post-Mesozoic age (Davis and Lister, 1974; Trehu, 1975). At ages greater than 70 My or so, however, the water depth increases less rapidly than predicted, levelling towards an asymptotic value about 3500 m below the rise crests (Sclater *et al.*, 1975; Parsons and Sclater, 1977). A number of explanations have been put forward to explain this asymptotic behavior, including radioactivity in the upper mantle (Forsyth, 1977), shear-strain heating (Schubert *et al.*, 1976), phase changes (Wood and Yuen, 1984), and advection of heat from depth by large-scale circulation (Jarvis and Peltier, 1980), small-scale instabilities at the base of the plate (Richter and Parsons, 1975; Parsons and McKenzie, 1978) or localized upwellings beneath hotspots (Heestand and Crough, 1981). With the exception of phase changes, all of these mechanisms involve retarding the rate of TBL growth by upward heat flux, so that L_T attains an asymptotic value of the order of one hundred kilometers. Such an interpretation implies that the thermal evolution of the oceanic tectosphere is better described by a cooling plate model of the sort suggested by Langseth *et al.* (1966) and developed by McKenzie (1967) and Sclater and Francheteau (1970) than by a cooling half-space model. From the bathymetry data, Parsons and Sclater (1977) estimate the asymptotic thickness of the plate to be 125 ± 10 km.

Geoid anomalies provide an even more sensitive discriminant between the cooling half-space and cooling plate models than does bathymetry. Half-space cooling implies a geoid anomaly that decays linearly with age at a constant rate of −0.15 m/My (Haxby and Turcotte, 1978), whereas the plate model of Parsons and Sclater (1977) predicts an asymptoic flattening of the anomaly beginning at about 40 My (Parsons and Richter, 1980; Sandwell and Schubert, 1980). Because subtectospheric heterogeneities dominate the geoid at long wavelengths, this flattening is difficult to observe directly (Sandwell and Schubert, 1980), but it does express itself in the variation of short-wavelength anomalies across fracture zones (Detrick, 1981). The fracture-zone anomalies observed on the GEOS 3 and SEASAT altimeters are consistent with cooling plate models having asymptotic values of L_T ranging from 70–125 km (Detrick, 1981; Sandwell and Schubert, 1982; Cazenave *et al.*, 1983; Cazenave, 1984; Watts *et al.*, 1985). The Parsons-Sclater value lies at the upper end of this range.

The observations of flexural rigidity provide direct estimates of the effective elastic thickness of the oceanic lithosphere, which in turn can be related to L_M through an assumed rheology (Beaumont, 1979; Bodine *et al.*, 1981; McNutt, 1984). Bodine *et al.* (1981) have demonstrated that the relaxation for typical loads is complete within 1–10 My, and L_M thus depends primarily on L_T at the time of loading. Therefore, at small ages we expect

$$L_M = a_M t^{1/2} \qquad (2.2)$$

where a_M is a constant. Plate flexures observed on the outer rise of trenches and those measured from young volcanic loads show a systematic increase in the effective elastic thickness with the square-root of age, although there is much scatter (Watts *et al.*, 1980; McNutt and Menard, 1982). The outer-rise data are more suitable for estimating the constant in equation 2.2, because the loading data from mid-plate volcanoes appear to be perturbed by the vulcanism (McNutt, 1984). Correcting for plate-curvature effects, McNutt (1984) obtains a_M = 3.8–4.2 km $My^{-1/2}$, corresponding to the 550–600°C isotherms of the cooling half-space model. The flexture data in the older oceanic regions are consistent with the cooling plate model having an asymptotic value of L_M equal to approximately 40 km (McNutt, 1984). Hence, the thickness of the MBL in the ocean basins is only about one-third the TBL thickness.

The plate structure of a thermally mature ocean basin is schematically represented in Figure 2.3. The parameters of the oceanic adiabat are T_0 = 1300°C and $(\partial T/\partial P)_S$ = 0.5°C/km. The value of the average potential (zero-pressure) temperature T_0 is particularly important, since it sets the temperature scale for convection in the upper mantle. Early estimates of T_0 varied widely, ranging from a low of 550°C (McKenzie, 1967) to nearly 1500°C (Sclater and Francheteau, 1970). Jordan (1975a) considered various types of evidence, including empirical geothermometry on mantle minerals and models of basaltic differentiation, to arrive at the estimate used here. Solomon (1976) criticized this choice, but it has been supported by subsequent studies (Forsyth, 1977; Parsons and Sclater, 1977). The TBL thickness in Figure 2.3 is 150 km, somewhat larger than Parsons and Sclater's (1977) estimate and the asymptotic values derived from the fracture-zone offsets observed on altimeter records, though not significantly so. An MBL thickness of 60 km is obtained from the depth to the 600°C isotherm. A CBL with L_C = 50 km is defined by the deviation of the normative density profile below a mantle reference assemblage, here taken to be Ringwood's (1966) pyrolite ($\hat{\rho}$ = 3.40 Mg/m^3). Iron and alumina depletions of the mantle part of the CBL are assumed to reduce the normative density by 0.06 Mg/m^3, equivalent to the removal of about 20 mol % of olivine basalt from pyrolite (Jordan, 1979a).

2.3. Seismological Constraints on the Structure of the Oceanic Tectosphere

Seismic refraction experiments utilizing chemical explosions have elucidated the structure of the crust and uppermost mantle beneath the ocean basins. The models derived from such experiments are generally consistent with ophiolite model of the oceanic CBL (Christensen and Salisbury, 1975; Salisbury and Christensen, 1978; Spudich and Orcutt, 1980). In particular,

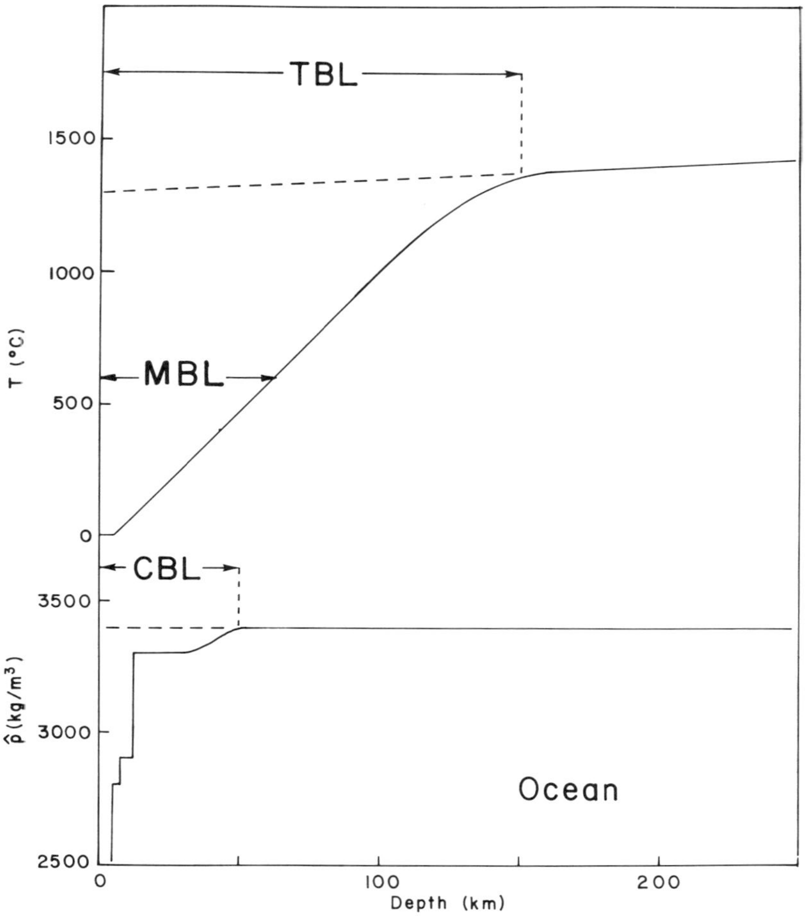

Figure 2.3. Tectospheric structure of a thermally mature ocean basin, showing the depth extent of its thermal, mechanical and chemical boundary layers.

the compressional velocities of the uppermost mantle show a strong (~7 percent) azimuthal variation, the highest velocities being parallel to the spreading direction at the time of crustal formation (Hess, 1964; Raitt *et al.*, 1969; Morris *et al.*, 1969; Keen and Barrett, 1971; Shor *et al.*, 1973; Bibee and Shor, 1976; Shearer and Orcutt, 1985). This azimuthal anisotropy has been observed in ophiolites and is attributable to the tectonic alignment of olivine and pyroxene crystals during the emplacement of the residual harzburgitic layer at spreading centers (Christensen and Salisbury, 1979; Christensen and Lundquist, 1982; Nicolas and Violette, 1982).

For shallow seismic sources in the oceans, the first-arriving compressional wave out to distances on the order of 2000 km is the P_n phase, a complex, high-frequency wavetrain propagating in the uppermost mantle (Sutton and Walker, 1972; Walker, 1977; Sereno and Orcutt, 1985). Measurements of P_n apparent velocities in the Pacific Basin east of Japan using ocean-bottom seismometers show significant azimuthal anisotropy out to distances as great as 1000 km (Shimamura and Asada, 1983). The fast axis of the anisotropy is perpendicular to the local orientation of the magnetic anomalies, not parallel to the present plate motion, indicating that it was frozen in at an earlier stage of plate evolution and is thus tectospheric. Because the velocity gradient is small in the uppermost mantle, the depth sampled at this distance is not well constrained, but the tectospheric anisotropy appears to extend below the CBL, perhaps to depths as great as 70–100 km (Shimamura, 1984).

It is difficult to study the structure of the oceanic upper mantle below the TBL by conventional marine refraction techniques, because the P_n phase obscures the later arrivals from deeper horizons. To obtain constraints on thermal evolution of the oceanic tectosphere, seismologists have resorted to the use of dispersive surface waves, of both the Rayleigh (*P-SV*) and Love (*SH*) type. The former are polarized in the plane of propagation, and the latter normal to it. At a given frequency, the speeds of these waves are observed to increase systematically with the average crustal age along the propagation path, and the variations of this increase with frequency can be used to construct age-dependent models of the oceanic upper mantle. Unfortunately, surface-wave studies are hampered by the limited distribution of seismic sources and receivers in the interiors of the oceanic plates, which restricts the power of the data to resolve lateral variations in structure, and by the limited range of frequencies over which the dispersion can be measured along pure oceanic paths (0.005–0.1 Hz), which restricts the power of the data to resolve radial variations in structure. Moreover, the whole business is complicated by the existence of upper-mantle anisotropy.

Seismological models of the oceanic upper mantle are usually parameterized by three layers: a high-velocity "lid," a low-velocity "channel," and a "sub-channel" (e.g., Kanamori and Press, 1970; Leeds *et al.*, 1974). The lid is sometimes referred to as the "seismological lithosphere," although its relationship to either the MBL or the TBL depends on the details of a rheological model (Anderson and Minster, 1980). In particular, the lid-channel boundary, which appears to be quite sharp, may be controlled by the intersection of the geotherm with the mantle solidus (Anderson and Sammis, 1970; Forsyth, 1977). We denote the depth to this boundary by L_S.

Nearly all of the detailed studies of the age-dependent properties of the oceanic upper mantle have employed surface-wave paths confined to the Pacific Basin. Leeds *et al.* (1974) and Leeds (1975) obtained isotropic models of the Pacific from the age variation of Rayleigh-wave phase velocities; in their models, L_S increased from 12 km at 5 My to 116 km at 150 My. Motivated by the cooling half-space model of the oceanic TBL, Yoshii (1975) adopted the functional form

$$L_S = a_S t^{1/2} \tag{2.3}$$

and showed that Rayleigh-wave group velocities are consistent with $a_S = 7.5$ km My$^{-1/2}$. Forsyth (1975) measured the dispersion of both Love and Rayleigh waves propagating over paths in the eastern Pacific to constrain the anisotropy, as well as the age dependence, of the Pacific and Nazca plates. He found that the azimuthal variation of the surface-wave velocities is small, but that polarization anisotropy (i.e., anisotropy with a vertical axis of symmetry; transverse isotropy) is required to satisfy both data types. In his models, L_S increases rapidly to about 60 km at 10 My and thereafter changes slowly, attaining about 70 km at 50 My. Schlue and Knopoff (1977, 1978) also investigated anisotropy. They were able to satisfy a large set of Love and Rayleigh data with models in which the only anisotropy is a transversely isotropic low-velocity channel and the only lateral variation an increase in L_S with age. Their models approximate equation (2.3) with $a_S = 10$ km My$^{-1/2}$. They identified L_S as the top of a region of partial melting and associated the polarization anisotropy with the distribution of melt in cracks preferentially oriented by a shear-flow gradient in the channel.

Yu and Mitchell (1979) and Mitchell and Yu (1980), on the other hand, concluded that the differences in Love and Rayleigh dispersion are best modelled by an anisotropic lid overlying an isotropic low-velocity channel. They obtained $L_S = 110$ km for $t > 100$ My. Their procedures corrected the surface-wave data for frequency-dependent anelastic dispersion (Liu *et al.*, 1976), which had been ignored in the previous studies.

Isotropic velocity models are specified by two elastic parameters. The analysis of polarization anisotropy employed by Forsyth (1975), Schlue and Knopoff (1977), and Mitchell and Yu (1980) allowed for differences in the propagation speeds of horizontally and vertically polarized shear waves, v_{SH} and v_{SV}, but assumed isotropic compressional velocities, thus approximating a transversely isotropic material with three elastic parameters (Schlue and Knopoff, 1978). In general, however, a transversely isotropic material has five elastic parameters: the two shear-wave velocities; two compressional-wave velocities, v_{PH} and v_{PV}; and a fifth parameter η, specifying the wave speeds at angles intermediate to the vertical and horizontal (Anderson, 1966). Anderson and Regan (1983) and Regan and Anderson (1984) have argued that three-parameter models of oceanic anisotropy are inadequate, and they have obtained by trial-and-error structures satisfying the surface-wave data using the full five-parameter formulation. Their models have a high-velocity, nearly isotropic lid that thickens with age and an anisotropic low-velocity channel with age-dependent velocities. In these models, L_S reaches a maximum thickness of only 50 km, about half that obtained in previous work and comparable to L_M values obtained from flexural studies. Anderson and Regan (1983) argue that the thicknesses of both the lid and the MBL may be controlled by effects other than temperature, such as mineralogy, crystal orientation or partial melting.

Figure 2.4a shows the values of L_S for three surface-wave models—the isotropic structure of Leeds *et al.* (1974) and Leeds (1975), the pseudo-anisotropic structure of Schlue and Knopoff (1977), and the fully parameterized transversely isotropic structure of Anderson and Regan (1983)—and it compares them with the MBL and TBL thicknesses from the Parsons-Sclater plate model. The seismological estimates vary substantially, but all three fall between L_M and L_T for $t < 70$ My. The corresponding isotherms calculated from the cooling half-space model range from about 650°C to 1100°C. The higher value is consistent with the proposition that L_S is defined by the intersection of the oceanic geotherm with the mantle solidus (Forsyth, 1977), whereas the lower value is not.

The discrepancies among the velocity structures illustrate the nonuniqueness of seismological models derived from real data sets. Anderson and Regan (1983) argue that their models are superior because they adopt a more general parameterization. It appears, however, that the values of L_S given by the Anderson-Regan model are a factor of two lower primarily because the shear velocities in their low-velocity channel are as much as 0.3 km/s higher than the other two models. Anderson and Regan (1983) and Regan and Anderson (1984) did not attempt to demonstrate that models with greater values of L_S (and correspondingly lower channel velocities) cannot satisfy the same data sets. Indeed, given the increased flexibility of their five-parameter formulation, the availability of such models is quite plausible. Until the problems of nonuniqueness receive more attention, any conclusions on the physical interpretation of L_S cannot be considered definitive.

Although the ability of the surface-wave data to constrain localized properties such as L_S is limited, they do fix the age dependence of the travel times

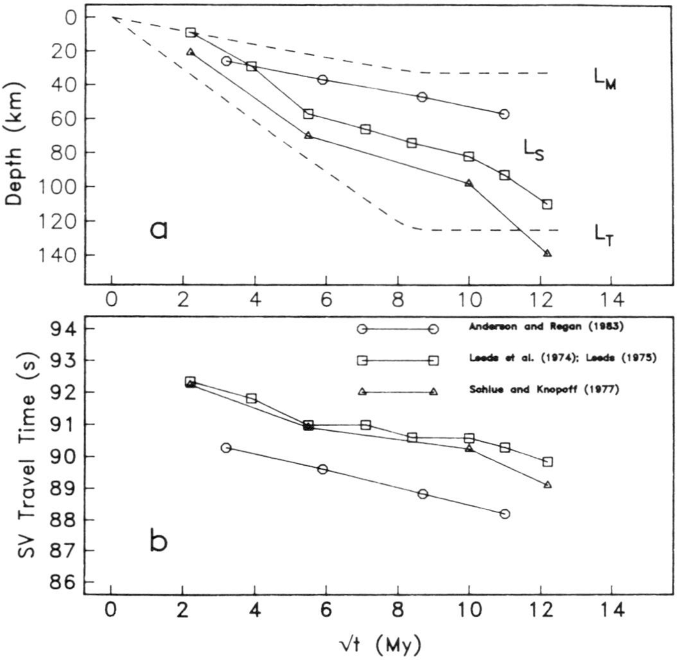

Figure 2.4. (a), Estimates of seismic lid thickness, L_S, versus the square root of crustal age from three surface-wave models: the isotropic structures of Leeds *et al.* (1974) and Leeds (1975) (squares), the pseudo-anisotropic structure of Schlue and Knopoff (1977) (triangles), and the fully-parameterized transversely isotropic structure of Anderson and Regan (1983) (circles). MBL and TBL thicknesses are from the Parsons and Sclater (1977) plate model. (b), Vertical one-way *SV* travel times through the upper 400 km of the three surface-wave models, showing square-root-of-age behavior.

of shear waves propagating vertically through the upper few hundred kilometers of the oceanic upper mantle. The travel time between depths of z_0 and z_1 is

$$\tau_S(z_0,z_1) = \int_{z_0}^{z_1} v_{SV}{}^{-1}\, dz \tag{2.4}$$

Figure 2.4b plots the vertical travel times through the upper 400 km derived from the three surface-wave models of Figure 2.4a. This functional varies linearly with the square-root of age for all three models over the entire range of the observations (5–150 My). The travel times from the Anderson-Regan structure differ from the other two by a constant offset of about −1.4 s, due primarily to differences in the shear velocities below 220 km, where the fundamental-mode data have poor resolution. However, the changes in travel time with age are nearly identical in all three models, having a slope of 0.27 s $My^{-1/2}$. This slope agrees with direct observations of vertical shear-wave travel times derived from ScS_2-ScS differential travel times (Sipkin and Jordan, 1976) (see Figure 2.5) and S-wave source anomalies (Duschenes and Solomon, 1977).

The fact that the vertical shear-wave travel times continue to decay as the square-root of age beyond 100 My has significant implications for the thermal and convective structure of the oceanic upper mantle. As discussed in the previous section, there is considerable non-seismological evidence that oceanic TBL evolves according to the cooling plate model with $L_T \leq 125$ km in old ocean basins. The values of L_S in Figure 2.4a conform more with the cooling half-space model, but given their non-uniqueness, it appears that structures consistent with both the surface-wave data and the cooling plate hypothesis are feasible. The vertical travel times in Figure 2.4b do indicate, however, that the average temperature of the oceanic upper mantle in the upper few hundred kilometers does not become asymptotically constant, but continues to decrease as the square-root of time. That this should be so is suggested by the appreciable flow of heat out of the mantle (~50 mW/m^3) in even the oldest ocean basins (Hager and O'Connell, 1981). If small-scale convection in the upper mantle beneath old ocean basins is the mechanism which regulates the asymptotic value of plate thickness, as hypothesized by Richter and Parsons (1975) and Parsons and McKenzie (1978), then this observation implies that there is a secular cooling of the sub-tectospheric mantle at ages greater than the onset of TBL instability (70 My in the Parsons-McKenzie model). This secular cooling will, in turn, contribute to driving a large-scale flow, which we believe must involve the lower mantle (see Section 3).

2.4. The Deep Structure of Continents

Like the oceans, the continents show a systematic decrease in surface heat flow and elevation and an increase in TBL thickness with age, provided that this age is defined not as the time since the original formation of the crust but as the interval since the last major thermal or orogenic event (Polyak and Smirnov, 1968; Sclater and Francheteau, 1970; Sclater *et al.*, 1981). The thermal reactivation and subsequent conductive relaxation of a continental TBL have been used to explain a wide variety of epeirogenic phenomena,

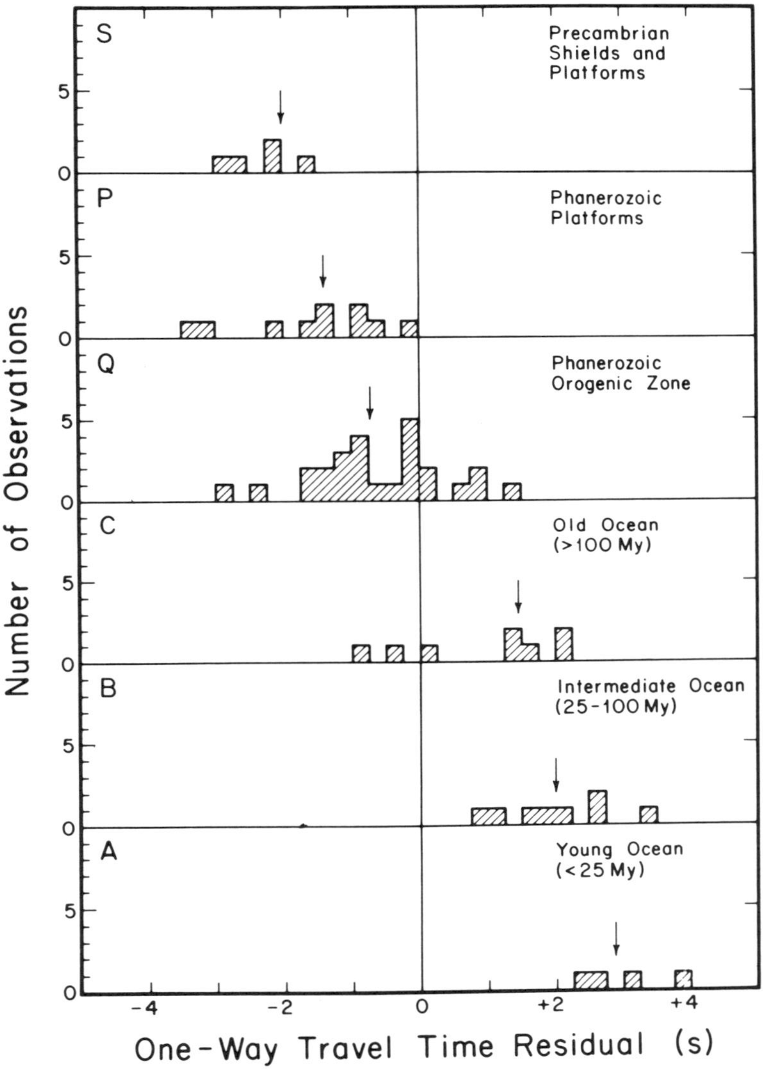

Figure 2.5. Histograms of Sipkin and Jordan's (1976) ScS_2-ScS differential travel-time residuals with surface-reflection points classified according to the tectonic regionalization of Figure 2.1. Residuals are relative to Jeffreys-Bullen tables, have been corrected for source depths, ellipticity, station elevations, and elevations of surface reflection points, and have been normalized to one-way times. Arrows indicate medians for each tectonic region.

including continental margin subsidence (Sleep, 1971; Steckler and Watts, 1978), the formation of continental basins (Sleep, 1971; Sleep and Snell, 1976; Haxby *et al.*, 1976; McKenzie, 1978; Sclater and Christie, 1980), and plateau uplift (Crough, 1979). Within the late Proterozoic and Phanerozoic orogenic belts peripheral to the cratons, where most of the large vertical motions have occurred, the parameters of continental TBL growth are compatible with those observed for the oceans; in particular, the asymptotic thickness of the epicontinental TBL is similar to the oceanic value (Sleep, 1971; Steckler and Watts, 1978; Sclater and Christie, 1980).

These facts have motivated Sclater *et al.* (1980, 1981) to extend to the continents the model for TBL evolution advanced by Parsons and McKenzie (1978) for the ocean basins (Figure 2.2). Continents are viewed as superficial crustal edifices that modify the strength of the MBL and, owing to their buoyancy, play an important role in plate tectonics, but they do not correlate with profound differences in tectospheric structure. The TBL configuration of old continents is postulated to be similar to that of old ocean basins: following rejuvenation, both oceanic and continental TBLs grow to an asymptotic thickness of about 125 km, which is regulated in steady-state by small-scale instabilities at the base of the MBL. The ancient cratons are those portions of the continental masses that have escaped any major thermal reactivation and collisional orogenesis since the Precambrian. This model implies that the temperature and composition, and thus the elastic properties, of the TBL underlying old continents should be, on the average, the same as those of an old ocean basin. Significant local variations may occur due to the thermal fluctuations associated with small-scale convection, but these should not be a strong function of crustal type.

The hypothesis that old continents and old oceans have similar TBL structures can be tested by seismological experiments. Seismologists have known for some time that the average shear velocity in the upper mantle beneath the continental cratons is considerably greater than the averages for oceanic and orogenic provinces (Brune and Dorman, 1963; MacDonald, 1963; Toksöz and Anderson, 1966; Knopoff, 1972), but the depth extent of these variations has been uncertain. Nearly all of the early work on continent-ocean heterogeneity was based on measurements of the phase and group velocities of fundamental-mode surface waves. In the frequency band typically observed over relatively pure oceanic and continental paths (0.005–0.10 Hz), the fundamental-mode data alone have poor resolving power below about 200 km, and they are particularly insensitive to structural variations in the 200–400 km depth range. Dziewonski (1971), for example, was able to satisfy a large, global set of fundamental-mode Rayleigh-wave data with models where the only significant differences between the continental and oceanic structures were confined above 200 km.

Okal (1977) re-examined the implications of the surface-wave data using an augmented data set and concluded that, "Rayleigh-wave phase velocities for continental areas (both shield and mountainous regions) fall within the range of oceanic models, and the difference between average oceanic and continental velocities is on the same order of magnitude as the variation within the oceanic plates due to age. This is incompatible with Jordan's models of strong, deep lateral heterogeneities between oceans and continents." Okal did not attempt a resolving power analysis; he simply compared his data with the dispersion calculated from published models (in fact, the dispersion curves he attributed to Jordan's (1975b) models are incorrect). A more precise statement is that the fundamental-mode Rayleigh-wave phase velocities *cannot resolve* continent-ocean differences much below about 200 km.

An additional constraint on continent-ocean heterogeneity was introduced by Sipkin and Jordan (1975, 1976), who estimated the lateral variations in the shear velocities integrated over the entire upper mantle from the travel times of ScS_n phases. (ScS_n reflects n times from the core-mantle interface and $n-1$ times from the free surface.) They showed that the differences observed in ScS_n travel times were much larger than predicted by Dziewonski's (1971) regionalized models of continent-ocean heterogeneity. They argued that, since the surface-wave data fix the average velocities in the upper 200 km or so, there must exist substantial lateral variations below this depth that correlate with crustal type. Jordan (1975a) applied Parker's (1972) quadratic programming technique to obtained a lower bound of 310 km on the depth extent of continent–ocean heterogeneity from two numbers, Sipkin and Jordan's (1975) average ScS time difference and Kanamori's (1970) difference in the average phase velocities of 190-s Love waves. Figure 2.5 gives histograms of Sipkin and Jordan's (1976) ScS_2-ScS residuals for each of the six crustal types in the tectonic regionalization of Figure 2.1.

In a parallel study of ScS_2-ScS differential travel times, Okal and Anderson (1975) argued that old continents and old oceans have similar residuals and concluded that the differences between them could be confined above 200 km. The discrepancy between the Sipkin-Jordan and Okal-Anderson results was used by Sclater *et al.* (1980) to argue against the thick-plate model. However, Sipkin and Jordan (1980) obtained a much larger set of ScS_m-ScS_n differential travel times from the newly installed digital networks. They found a one-way time difference averaging $+2.6 \pm 0.6$ s between western Pacific and continental paths, consistent with the median difference of +3.0 s between old oceans (Region C) and stable continents (Regions S and P) obtained from Figure 2.5.

The strong correlation between average upper-mantle shear velocities and crustal type is critical to the interpretation that most of the lateral variation observed in the ScS_n times is caused by tectospheric, not sub-tectospheric, structure (Jordan, 1975a). Figure 2.6 illustrates this correlation for ScS_2-ScS differential travel times from paths crossing northern Eurasia and the Arctic Basin. These differential times show a spectacular monotonic variation of more than 8 s in two-way travel time (4 s in one-way time) across the Phanerozoic fold belts and continental margin separating the high-velocity Siberian shield from the low-velocity Arctic Basin. This observation supports the hypothesis that the transition from the thick tectosphere of the cratons to the thin tectosphere of the oceans typically occurs in the Phanerozoic mobile belts surrounding the cratons.

Additional evidence for the correlation of vertical travel times with surface geology is provided by S-wave station anomalies. Station anomalies are derived by averaging the residuals observed at a particular receiver over many seismic sources to reduce the perturbations from heterogeneities far from the station (Doyle and Hales, 1967). Wickens and Buchbinder (1980) have complied a set of 42 station anomalies for North America, and their contoured anomaly map is shown in Figure 2.7. The smallest S-wave vertrical travel times are found for stations on the Canadian Shield and the largest in the Basin and Range Province of the western United States, the total variation being in excess of 5 s. This observation is an indication that the Basin and Range is underlain by a high-temperature, oceanic-type mantle structure (Doyle and Hales, 1967; Knopoff, 1972; Grand and Helmberger, 1984a), as required by heat-flow studies (Lachenbruch and Sass, 1980).

The strongest gradients in the S-wave station anomalies coincide with the Phanerozoic fold belts that bound the North American craton, consistent with the tectonic relationships observed in the ScS_2-ScS differential travel times of Figure 2.5. In general, Precambrian shields and platforms (Region S of Figure 2.1) are faster than the stable continental areas covered by platform sediments during the Phanerozoic (Region P), which in turn are faster than the Phanerozoic magmatic belts and orogenic zones (Region Q). On the average, all continental crustal types are underlain by mantle faster than even the oldest oceanic areas (Region C), although there is considerable overlap between the individual observations for Regions Q and C (Figure 2.5). These travel-time differences cannot be explained by, say, differences in the distribution of upwelling and downwelling convective plumes between continents and oceans; the correlation with long-term (~600 My) tectonic behavior of

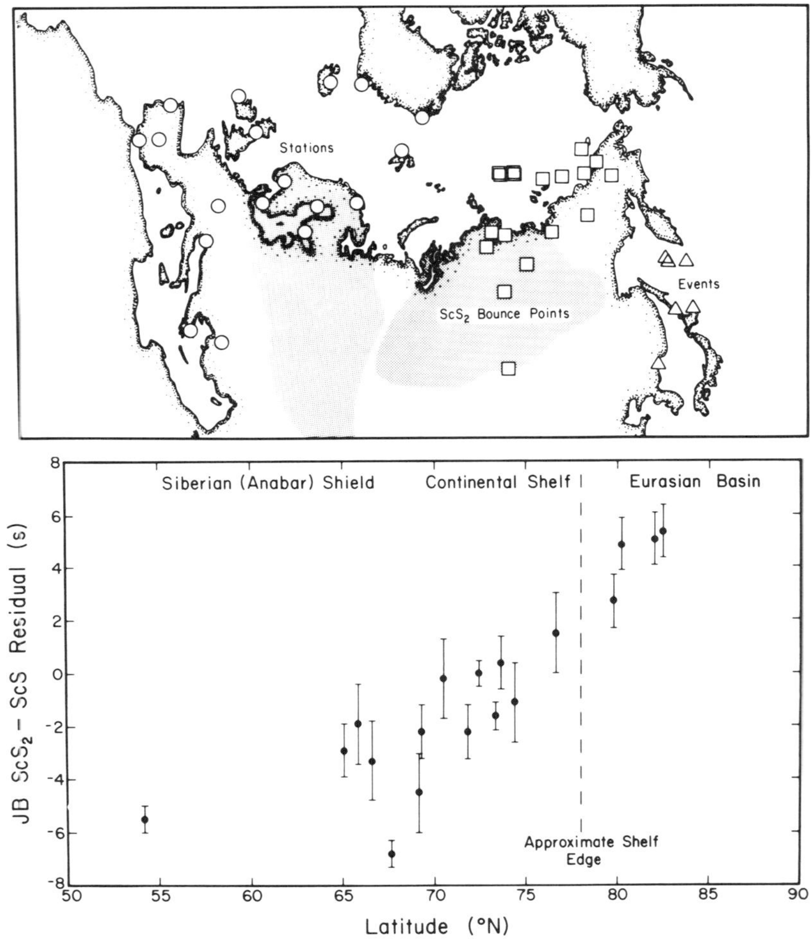

Figure 2.6. Transition from the high-velocity Siberian shield to the low-velocity Arctic ocean basin observed in the ScS_2-ScS differential travel times of Sipkin and Jordan (1976). Top panel: Map of northern Eurasia showing the distribution of sources (triangles), receivers (circles), and ScS_2 surface reflection points (squares). Stippling identifies stable continent (regions S and P of Figure 2.1). Projection is azimuth equidistant about a point in the Sea of Okhotsk. Bottom panel: ScS_2-ScS residuals plotted against latitude of the surface reflection points. Residuals are two-way times with respect to the Jeffreys-Bullen tables, corrected for elevations of surface-reflection points. The coherent increase of the residuals with latitude is indicative of tectospheric thinning from shield to ocean basin across a quiescent margin of the sort illustrated in Figure 2.10.

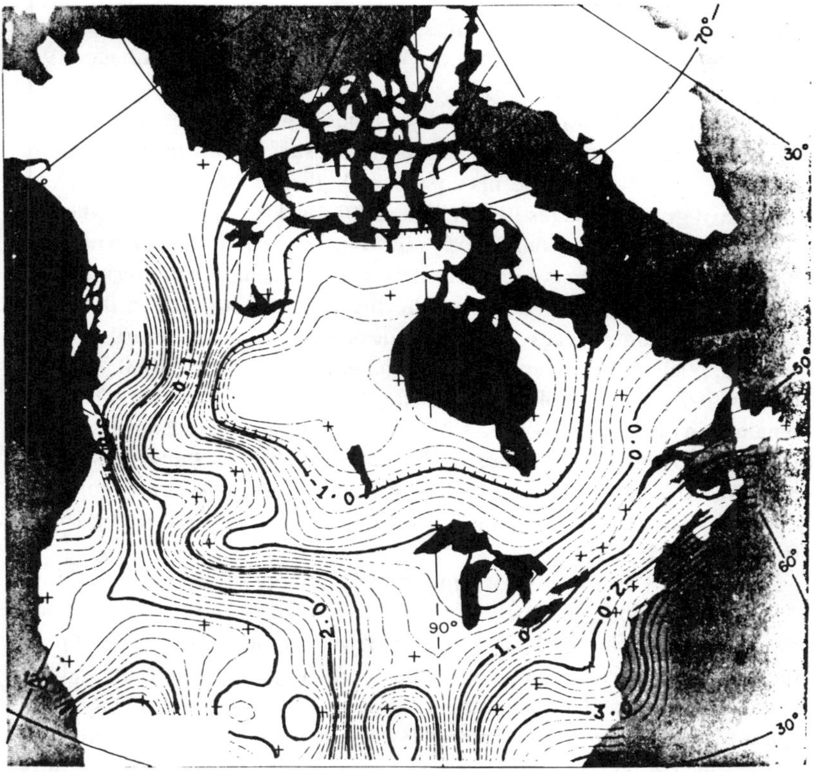

Figure 2.7. *S*-wave station anomalies for North America from Wickens and Buchbinder (1980). Light contours are spaced at 0.2-s intervals in one-way vertical travel time; heavy contours are at 1-s intervals. Crosses indicate locations of stations used to produce the map. The total variation in station anomaly is in excess of 5 s, and the pattern of variation is very coherent with the tectonic regionalization of Figure 2.1. The coherent increase of the residuals away from the Canadian shield across the quiescent North Atlantic continental margin is indicative of tectospheric thinning of the sort illustrated in Figure 2.10.

the crust is just too good, and the correlation with other indicators of the mantle flow regime (e.g., the geoid) is too poor (Jordan, 1975a). Instead, they require profound variations in tectospheric structure. As repeatedly emphasized by Jordan (1975a, 1978, 1979b, 1981b) and Sipkin and Jordan (1976, 1980), the vertical travel-time data in combination with fundamental-mode surface-wave data are sufficient to reject unified TBL models of the sort proposed by Crough and Thompson (1976), Kono and Amano (1978) and Sclater *et al.* (1980, 1981).

This inference is fully supported by a series of new high-resolution studies of lateral heterogeneity using higher-mode surface waves and multiply reflected body waves. The dispersion of higher-mode Rayleigh waves is capable of giving good resolution in the 200–400 km depth range, where the fundamental-mode data only poorly constrain the structure, but the measurement of this dispersion is complicated by the interference of the various mode groups. Frequency-wavenumber filtering methods have been developed which can separate successfully the mode branches out to the fourth overtone when applied to seismometer arrays having apertures of 1000 km or so (Nolet, 1975, 1977; Nolet and Panza, 1976; Cara, 1978). Data has been obtained in the band 0.01–0.05 Hz by these techniques over Eurasian (Nolet, 1975, 1977) and Pacific (Cara, 1979) paths. Inversions show substantially greater (>0.2 km/s) shear velocities beneath Eurasia in the depth range 100–200 km; differences as large as 0.1 km/s persist to depths exceeding 250 km (Cara *et al.*, 1980).

The interval on the seismogram corresponding to the group arrival times of the higher modes lies between the direct *S* phase and the fundamental-mode Rayleigh wave. Advances in numerical techniques now make it possible to obtain relatively complete synthetic seismograms for this interval from spherically symmetric structures. To take advantage of this capability, Lerner-Lam and Jordan (1983) have developed an inversion procedure which derives the path-averaged properties of the crust and upper mantle by fitting the observed time-domain waveforms with synthetic waveforms. Because the complex interference patterns among the higher-mode wave groups are modelled directly without attempting to separate the spectral properties of individual higher modes, the method is less susceptible to bias than mode-separation techniques, and good resolution of path-averaged structures can be obtained to depths of 500 km or so with smaller data sets (Gee *et al.*, 1985). Lerner-Lam and Jordan (1983) applied their procedure to the study of the Eurasian continent and Pacific Ocean, using source and receiver distributions similar to those employed by Cara (1979) and Cara *et al.* (1980). They obtained continental and oceanic models with shear-velocity differences to depths on the order of 400 km, consistent with the previous studies.

The waveform-inversion technique of Lerner-Lam and Jordan (1983) has been recently used to study the structural differences between northern Eurasia and the western Pacific (Lerner-Lam and Jordan, 1987). Isotropic shear-velocity models for these two regions are compared in Figure 2.8a.

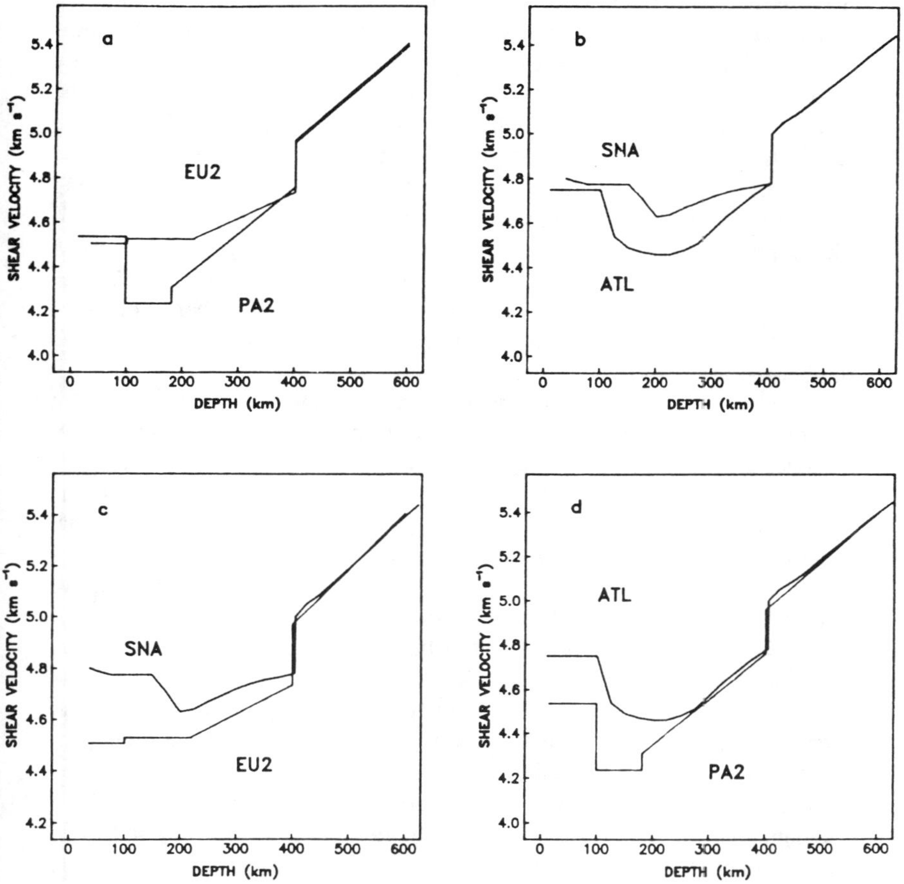

Figure 2.8. (a), Isotropic *SV*-velocity models for northern Eurasian (EU2) and the western Pacific (PA2) from the higher-mode waveform inversions of Lerner-Lam and Jordan (1987). (b), Isotropic *SH*-velocity models for the Canadian Shield (SNA) and the western Atlantic (ATL) from the multiple-*S* body-wave study of Grand and Helmberger (1984a, 1984b). (c), Comparison of EU2 and SNA. (d), Comparison of PA2 and ATL.

EU2 is a structure derived from Kuril-Japan earthquakes recorded by a vertical-component receiver array in western Europe. Although the paths sample a variety of structures in northern Eurasia, from the tectonically active region east of the Verkhoyansk Suture Zone to the Baltic Shield, the waves propagate mostly over stable continent (Regions S and P of Figure 2.1). PA2 is obtained from Kuril-Japan sources recorded by stations in the southwestern Pacific, with paths almost entirely over oceanic crust having ages greater than 100 My (Region C). PA2 features a 100-km thick lid, a well-developed low-velocity channel, and a relatively steep gradient in the *S* velocities between the channel and a discontinuity at 400 km depth. EU2 has no channel and small gradients down to 400 km. EU2 has higher *S* velocities than PA2 above 370 km. Both provide excellent fits to the fundamental and higher-mode waveforms.

Of course, the existence of models with differences to some maximum depth z^* does not imply that such differences must exist. The models are not unique, and all we can really assert is that extending structural differences down to the level z^* is consistent with the data; that is, we cannot reject the null hypothesis that $z_{\max} \leq z^*$, where $z_{\max}$ is an upper bound on the actual depth of continent-ocean variations. However, if no models with differences confined shallower than z^* can be found that fit the data adequately, then we have grounds for rejecting the null hypothesis in favor of the alternative $z_{\max} > z^*$. Lerner-Lam and Jordan (1987) have applied an inversion technique they call "squeezing" to demonstrate that the data used to derive EU2 and PA2 are inconsistent with $z_m < 220$ km. The higher-mode data alone require deep structural differences between old continents and old oceans independent of any inferences based on vertical travel times.

The waveforms on the seismogram between the direct *S* phase and the fundamental-mode surface waves can be described in terms of the interference of higher modes or, equivalently, as the superposition of various body waves, of which the multiply reflected *S* waves are the most prominent. The modal decomposition used by Lerner-Lam and Jordan (1983, 1987) is more convenient for the *P-SV* system, whereas the body-wave decomposition is more convenient for *SH*. Grand and Helmberger (1984a, 1984b, 1985) have applied a body-wave analysis to the study of *SH*-polarized *S, SS* and *SSS* waves in the distance range 10° to 80° and have obtained models satisfying the travel times and waveforms. Figure 2.8b compares their *SH*-velocity structures for stable North America (SNA) and the western Atlantic Ocean (ATL). The models are identical below the 400-km discontinuity, but SNA is faster than ATL above this depth. Grand and Helmberger inverted their data by trial-and-error and have not yet attempted any quantitative resolving power analysis, but it is clear that their data have a resolution of structural details below 200 km that is at least comparable, if not superior, to the *P-SV* data used by Lerner-Lam and Jordan (1983, 1987) and Gee *et al.* (1985).

Therefore, both the *P-SV* and *SH* waveform-modelling techniques demonstrate that the mantle beneath stable continents has higher shear-wave velocities than beneath the old oceans to at least 200–250 km in depth and perhaps to 400 km, consistent with the inferences from vertical travel-time data of Sipkin and Jordan (1975, 1976, 1980). We note, however, that significant discrepancies remain among the three data sets. Figure 2.8c compares SNA with EU2, and Figure 2.8d compares ATL with PA2. Although all four models are essentially identical below the 400-km discontinuity, the *SH*-polarized data give substantially higher shear velocities in the uppermost mantle. The differences between ATL and PA2 exist primarily in the upper 200 km, but SNA is faster than EU2 all the way to 400 km. The SNA shear velocities decrease with depth between 150 and 200 km, forming a low-velocity channel, whereas EU2 has no such channel. This particular difference may be due to a lack of resolving power in one or both data sets, but the large (0.16 km/s) average velocity difference between SNA and EU2 above 400 km is not.

The differences may reflect true regional variations, at least in part, but we are skeptical that an explanation based entirely on path differences can account for the discrepancies. Grand and Helmberger (1985) have modelled the transition from the active foldbelts of central Asia to the Russian Platform using their multiple-*S* technique, and, averaged over the uppermost mantle, the *lowest* velocities they find are still higher than EU2, even though the paths used in deriving EU2 traverse primarily the high-velocity shields and platforms of northern Eurasia. In fact, the vertical *S*-wave travel time through EU2 is actually greater than through ATL, in conflict with the ScS_n and station-anomaly observations. We suspect that the discrepancies evident in Figures 2.8c and 2.8d are diagnostic of polarization anisotropy. The sign of the velocity differences is consistent with other observations of polarization anisotropy in the upper mantle (e.g., Dziewonski and Anderson, 1981; Anderson and Regan, 1983), and similar model discrepancies have been noted by Cara *et al.* (1980) and Leveque and Cara (1983) using the dispersion of fundamental and higher-mode Rayleigh and Love waves measured by spatial filtering. A definitive statement must await the treatment of both *P-SV* and *SH* data sets from common paths by the same inversion technique.

Discrepancies also exist between the vertical *S*-wave travel times calculated from the models and the ScS_n observations. Despite the large continent-ocean differences in both sets of models, the vertical time for PA2 (suitably corrected for water depth) is only 1.5 s greater than EU2, and that for ATL is only 1.8 s greater than SNA. In comparison, the observed ScS_2-ScS difference between old oceans (Region C) and Precambrian shields and platforms (Region S) exceeds +3 s (Figure 2.5). The models in Figure 2.8 are isotropic structures, and their vertical travel times may be biased by the existence of polarization anisotropy. It is unlikely, however, that they overestimate the average structural contrasts between the continental cratons and old ocean basins, or the depth to which these contrasts persist.

2.5. The Importance of the Continental Chemical Boundary Layer

The seismological data reviewed in Section 2.4 imply that the stable parts of the continents are underlain by a TBL whose average thickness exceeds 220 km and most likely reaches 400 km, at least locally beneath the old cratons. These data clearly exclude the thin-TBL model of Figure 2.2. The most straightforward alternative is some variant of the cooling half-space model; that is, a thick TBL is grown by the long-term conductive decay of the continental thermal profile, unlimited by small-scale convective instability below the MBL (Chapman and Pollack, 1977). A cooling half-space model, like the cooling plate model, ascribes only a superficial character to the CBL, although it can play an important kinematical role. Because the continental CBL is buoyant, the thermal ages of the continents are typically much greater than the oceans, thus accounting for the correlation between tectospheric thickness and crustal type. The cratons are simply those areas that have been tectonically stabilized by the growth of a very thick TBL and have remained undisrupted by large-scale convective activity and collisional orogenesis.

Indeed, a continued increase in L_T according to equation 2.1 for 700 My is sufficient to grow a TBL about 400 km in thickness, which would satisfy the seismological data. At least three major objections can be raised against the cooling half-space model, however:

1. The local Rayleigh number of the sub-MBL portion of such a thick TBL is large ($>10^4$), indicating that it would be convectively unstable (Jordan, 1975a; Parsons and McKenzie, 1978).
2. To maintain isostatic balance, the growth of a thick TBL would require subsidence of the continents, just as it does in the oceans. The cratons should thus be covered with several tens of kilometers of sediments, which is not observed (Jordan, 1978; Oxburgh and Parmentier, 1978; Poupinet and de Voogd, 1981).

3. The geoid anomalies for thick TBLs with near-surface compensation are substantial (Haxby and Turcotte, 1978; Parsons and Richter, 1980), so that the cratons should be evident in the geoid, which they are not (Jordan, 1975a).

It is clear from seismology that there are substantial lateral variations in the thickness of the continental TBL, but it is equally clear from arguments (1)–(3) that the compensation of these variations does not occur in the crust, as required by simple cooling models.

To circumvent these difficulties, a more complex model of the tectosphere has been developed which assigns a major role to an augmented CBL beneath the continental cratons (Jordan, 1975a, 1978, 1981b). Its basic premise is that the mass excesses associated with cold, thick, cratonic TBLs are locally compensated by variations in the composition of the upper mantle. That is, below the MBL, the normative density of the continental tectosphere is less than the normative density of the oceanic upper mantle at the same depth. The continental TBL is thus stabilized against convective disruption by a thick CBL. This scenario can be inferred directly from the geophysical observations, independent of any petrological model (Jordan, 1975a).

The existence of a thick CBL beneath the cratons is required by the petrological data as well. The mechanism for chemical compensation of the continental TBL is basalt depletion. Mantle mineralogy in the depth range 70–400 km approximates that of a garnet lherzolite (Ringwood, 1975, Yoder, 1976), and the subtraction of basalt from a garnet lherzolite leaves a residue less dense than the parental material (O'Hara, 1975; Boyd and McCallister, 1976; Green and Liebermann, 1976). Peridotite nodules from kimberlite pipes in cratonic areas are observed to have major-element compositions depleted in Al_2O_3, FeO and CaO relative to compositions derived for the oceanic upper mantle (Ringwood, 1966; Nixon *et al.*, 1973). The normative density differences calculated from these compositions are sufficient to account for the thermal differences inferred by geophysical methods (Jordan, 1978, 1979a). For example, the average continental garnet lherzolite obtained from kimberlite xenoliths by Jordan (1978) has a normative density 1.3 percent less than Ringwood's (1966) pyrolite, which is sufficient to offset a continent-ocean temperature difference of several hundred degrees. These xenoliths appear to have equilibrated at depths of 150–200 km and temperatures of 1000–1200°C (Boyd, 1973; Finnerty and Boyd, 1984). In fact, oceanic temperatures at ths depth are inferred to be about 300°C higher (Figure 2.3). Other quantitative tests of the basalt depletion hypothesis have been made by Jordan (1978, 1979a) using the xenolith data.

The temperature and normative density profiles of a hypothetical continental craton are sketched in Figure 2.9 assuming $L_T = 225$ km, $L_C = 200$ km and $L_M = 90$ km. These values are consistent with the major-element data from kimberlite xenoliths (Jordan, 1979a; Boyd *et al.*, 1985), as well as a number of geophysical observations, including heat flow and heat production data (Sclater *et al.*, 1980; Nicolayson *et al.*, 1981) and flexural studies (Kusznir and Karner, 1985).

We note, however, that values of L_T and L_C adopted for this diagram probably underestimate those typical for the old, stable cratons, where the seismological data almost surely require $L_T > 300$ km. The temperature profile in Figure 2.9 was computed assuming the heat production of the sub-crustal tectosphere is negligible, an assumption often made in heat-flow analyses (e.g., Sclater *et al.*, 1980) but one which is likely to be incorrect. Jordan (1981b) has pointed out that a thicker TBL is required if the heat production of the sub-crustal CBL is relatively high, as indicated by the evidence for metosomatic enrichment of the CBL in large-ion lithophile (LIL) elements (see below). A TBL thicker than 225 km can also be inferred directly by comparing subcontinental temperatures derived by empirical geothermometry on mantle xenoliths with oceanic temperatures extrapolated along the mantle adiabat (Jordan, 1975a, 1978).

A continental tectosphere approaching 400 km in thickness is illustrated in Figure 2.10, which shows the transition from a cratonic to an oceanic upper-mantle structure across a typical quiescent continental margin. In this hypothetical model, the largest gradients in L_T occur beneath the Phanerozoic platforms and fold belts peripheral to the cratons. Transitions of this type are found between the Canadian Shield and the Atlantic Ocean at the latitude of Boston and along the northern margin of Eurasia, as is evident from the vertical travel-time data of Figures 2.6 and 2.7.

2.6. Implications for Mantle Evolution and Dynamics

The hypothesis that the continental cratons are underlain by a thick TBL stabilized by thick, basalt-depleted CBL has a number of far-reaching (and thus testable) implications and ramifications regarding mantle evolution and dynamics. We include here a brief catalog of some of the more obvious consequences of the model, most of which have been discussed previously by Jordan (1978, 1979b, 1981b):

Advective thickening as a mechanism for tectospheric growth. If the CBL controls the thickness of the continental tectosphere, then the TBL cannot grow by simple conductive cooling. In fact, unless there is a mechanism which either steadily depletes the continental upper mantle in its basaltic

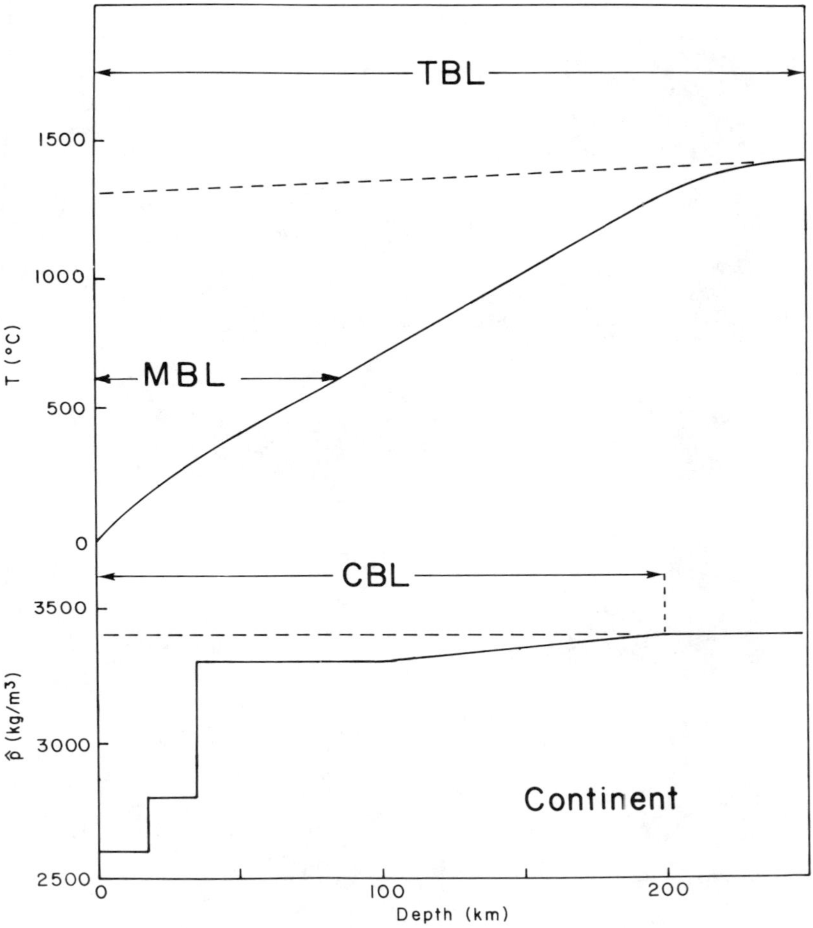

Figure 2.9. Tectospheric structure of an old continental craton. As discussed in the text, the thickness of the TBL is a lower bound to average values inferred for most cratons.

constituents or adds depleted material to the base of the tectosphere, the continental tectosphere beneath the stable cratons cannot grow progressively with time at all. Impressive outpourings of continental flood basalts do occasionally occur, but the volume of the mantle depleted by these episodes is actually quite small (on the order of 10^6 km^3 in a major episode such as the

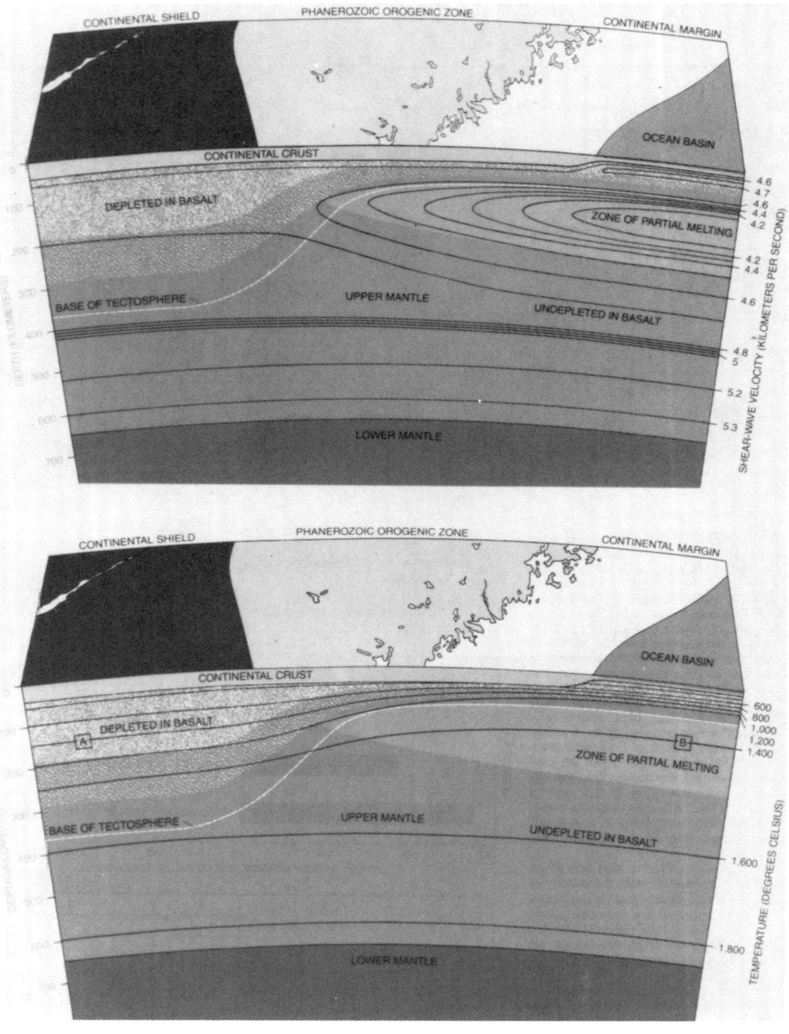

Figure 2.10. Hypothetical model of the transition in upper-mantle structure from a stable continental craton to an old ocean basin across a quiescent margin, from Jordan (1979c). The top panel shows shear velocities contoured in 0.2-km/s increments. The bottom panel shows temperature contoured in 200°C increments.

Miocene flooding of the Columbia River Basin in the northwestern U.S.). In comparison, the volume of depleted mantle beneath Regions S and P of Figure 2.1 is on the order of 10^{10} km^3, assuming an average CBL thickness of 200 km. In round numbers, the amount of basalt missing from the subcontinental upper mantle is about twice the present-day volume of the oceanic crust and a little less than half the total volume of the continental crust, orders of magnitude greater than the integrated volume of basalt erupted in continental regions since their stabilization in late-Archean, early-Proterozoic time.

Oxburgh and Parmentier (1978) have suggested that the CBL beneath the cratons could increase progressively by the addition of depleted mantle rising diapirically from the oceanic CBL. Seismic imaging of descending slabs in the northwestern Pacific by residual-sphere techniques (Section 3.4) shows no evidence of slab disruption by such a mechanism, although it is certainly possible that a significant amount of basalt-depleted material from the oceanic CBL has been transferred to the continents over the course of geological history. We are skeptical, however, that progressive underplating of the cratons coupled with thermal decay of the TBL is a viable model for tectospheric growth, because it is likely that these processes would operate with significantly different spatial and temporal scales and, hence, the balance between chemistry and temperature would be difficult to maintain.

The mechanism we prefer is advective thickening of the continental tectosphere during major episodes of orogenic compression, such as those which accompany the formation of supercontinents. The advective-thickening hypothesis is the foundation of the evolutionary model proposed by Jordan (1978, 1981b). Vulcanism along the active continental margins (and to a much lesser extent in intracontinental rifts) generates buoyant, refractory mantle resistant to convective recycling. Augmenting this material is any depleted peridotite transferred from the oceanic CBL by lateral accretion across subduction zones or by diapirism from descending slabs. Dispersed regions of buoyant mantle and their superjacent (but not necessarily chemically complementary) crustal columns are amalgamated during the course of plate motions and eventually swept up and accreted to the major continental masses. Major compressive events at convergent plate boundaries, particularly the violent episodes of continent-continent collision, further consolidate and thicken both the CBL and the TBL by downward advection (Figure 2.11). Competing against these constructive processes are the destructive, double-diffusive instabilities associated with tectospheric heating. These instabilities can lead to thinning of the CBL by continental extension, occasionally initiating Atlantic-type drift. Ultimately, however, perhaps

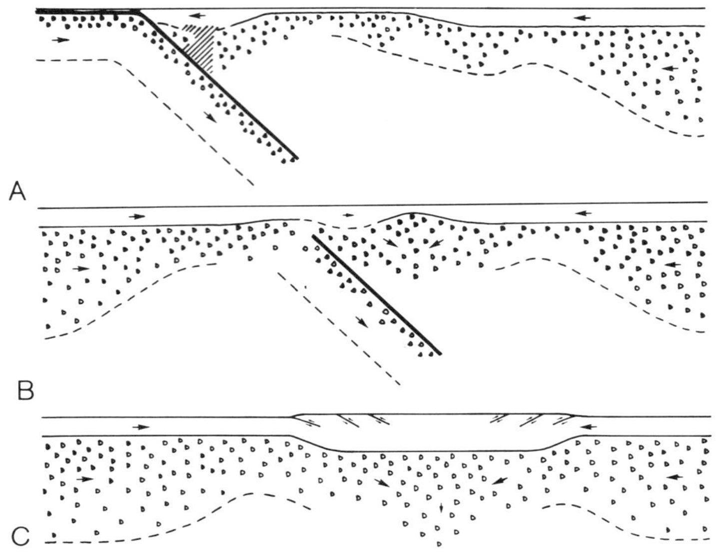

Figure 2.11. Schematic illustration of the formation of thick, cool, basalt-depleted continental tectosphere by advective thickening. The thin, inhomogeneous CBL accreted along an active continental margin (a) is consolidated and thickened by continent-continent collision (b) and subsequent compressive orogenesis (c). Dotted pattern is mantle of low normative density depleted in Fe and Al relative to Mg; cross-hatching shows region of melting beneath an active island arc, and arrows indicate relative motions.

only after repeated orogenic cycles, a thick, cohesive and relatively stable tectosphere—a continental craton—is formed.

Age of the continental tectosphere. This hypothetical scenario for tectospheric development of the continents conforms to actualistic principles and illustrates that a thickened CBL is a plausible consequence of the Wilson cycle. However, the model requires that the primary episodes of basalt extraction precede cratonization. Hence, the CBL structure of an ancient craton as regards major elements (though not necessarily minor ones) must date from its stabilization. Jordan (1981b) emphasized that this is a major, testable prediction of the tectospheric model. He pointed out that the data from South Africa are consistent with this prediction. In particular, he noted Kramers (1979) work on sulphide inclusions in diamonds, which indicated the inclusions were closed to *Pb* in excess of 2000 My ago.

In a recent, more definitive study of southern African diamond inclusions, Richardson *et al.* (1984) have obtained *Sm-Nd* and *Rb-Sr* model ages of 3200–3300 My for sub-calcic garnets encapsulated by diamonds. Peridotitic inclusions in diamonds (Gurney *et al.*, 1979) show residual major-element compositions characteristic of the entire kimberlite nodule suite (Mathais *et al.*, 1970), which provides strong evidence that the CBL extends to depths exceeding 150–200 km beneath the Kaapvaal craton. Although Richardson *et al.* (1984) do not comment on the tectospheric model, their age data are consistent with diamond formation during an episode of advective thickening of the CBL prior to the stabilization of Kaapvaal craton in the late Archean (Hunter, 1974).

If a thick, basalt-depleted CBL formed at this time, then a thick TBL must have also, since the low-density CBL would not otherwise be stable. This consequence of the tectospheric model has been confirmed by the application of geothermometry to monomineralic silicate inclusions in South African diamonds, which show equilibration temperatures of 900–1200°C (Boyd *et al.*, 1985). These temperature are similar to the range observed in the granular garnet lherzolite xenoliths, but the former were frozen in at the time of diamond formation (presumably of Archean age), whereas the latter were equilibrated in the Cretaceous. These data are consistent with a variety of other evidence (see Richter (1985) for a summary) that the geotherm beneath the cratons was not much different than it is today. Estimates of subcontinental temperatures in the Archean calculated from convection models invoking simple TBL cooling (oceanic-type tectosphere) invariably give much higher temperatures and imply significant melting, inconsistent with these observations (Davies, 1979; Richter, 1985).

The continental CBL as a geochemical reservoir. The cold, refractory CBL of the continental tectosphere constitutes a particularly stable reservoir for large-ion lithophile (LIL) chemical species. Peridotite nodules from kimberlite pipes with residual major-element compositions have generally had their minor-element compositions significantly altered by subsequent metasomatic events (Harte *et al.*, 1975; Menzies and Rama Murthy, 1980), providing direct evidence that the CBL has been impregnated by volatiles and highly differentiated liquids. The ubiquitous and diachronous emplacement of kimberlites in cratonic regions indicates that some of this flux of LIL-rich material may originate in the deep mantle and progressively modify the CBL after its advective thickening, but the isotopic compositions of diamond inclusions suggests that significant LIL enrichment occurred beneath South Africa in the interval between basalt depletion and diamond crystallization (Richardson *et al.*, 1985). In the context of the evolutionary model discussed above, the most likely site for the metasomatic alteration is the mantle wedge above the descending slab, the metasomatic agent being the fluids released from the slab itself. This leads to a scenario where subduction drives the initial differentiation of the CBL as well as its subsequent LIL enrichment prior to advective thickening; continued LIL enrichment may then proceed, but presumably at a much lower rate.

Most geochemical mass-balance calculations (e.g., Jacobsen and Wasserberg, 1979; O'Nions *et al.*, 1979) have ignored the sub-crustal CBL of the continents as a geochemical reservoir for LIL constituents. If it is as important as the considerations given here and by Jordan (1978, 1981b) suggest, then any conclusions based on these mass-balance calculations must be viewed with some skepticism, particularly those regarding the volume of the mantle which has been differentiated to produce the continents.

Implications for mantle heat flow. Metasomatism of the CBL introduces heat-producing elements to the continental tectosphere that can contribute significantly to the surface heat flux. Because of the alteration by the host kimberlite, it is difficult to infer precisely the heat production directly from the xenolith data, but Jordan (1981b) estimated that the total contribution of the sub-crustal CBL could be as high as 0.15 mW/m^2, about half of the heat flux at the base of the crust. The inference that the heat production of the CBL is high can be made independently by requiring that the conductive geotherm of the continental TBL satisfy the low (1000–1200°C) temperatures in the 150–200 km depth range obtained by empirical geothermometry (Jordan, 1975a). This conclusion is inconsistent with the thermal models derived by Sclater *et al.* (1980, 1981), who assumed that the heat production below the upper crust is negligible. (They were able to obtain overlapping bounds between oceanic and continental geotherms only because they made this assumption; if significant heat production is included below the upper crust, much thicker TBLs are required for old continents (Vitorello and Pollack, 1980).)

If the heat production of the continental CBL is as high as these arguments would suggest, then the heat flux at the base of the TBL beneath cratons is estimated to be on the order of 15 mW/m^3, less than half the oceanic value. The difference in sub-tectospheric continental and oceanic heat flow must be maintained by a large-scale component of mantle circulation, presumably one involving the participation of the lower mantle.

Tectospheric structure and cratonic stability. It is clear from the seismological data presented in Section 2.4 that there are large variations in the thicknesses of the continental TBL. According to the tectospheric model, these should be correlated with variations in CBL thickness. Probably the best single measure of these variations is τ_S, the verical travel time of shear waves through the upper mantle. The data on τ_S given in Figures 2.5–2.7 clearly indicate that the tectosphere is thickest under those parts of the continents that have remained tectonically stable during the Phanerozoic (Regions S and P of Figure 2.1) and thinnest under the Phanerozoic mobil belts and orogenic zones. This correlation suggests strongly that cratonic stability is regulated by variations in the thickness of the continental tectosphere: repeatly active mobile belts and orogenic zones mark regions where a thick CBL never stabilized, whereas the Archean cratonic nuclei are those areas where a deep CBL was stabilized at an early date. This would explain why the present-day distinction between cratons and mobile belts has, in many cases, persisted since mid-Proterozoic time (Watson, 1976) and why continents tend to break apart where they have broken before. Moreover, it suggests that the fundamental transition from Katarchean (> 3000 My B.P.) permobility to Proterozoic linear (geosynclinal and mobil belt) tectonics (Windley, 1977) marks the time when the area occupied by a stable, advectively thickened CBL became significant.

3. Fate of the Subducted Slab

Aside from the plate motions themselves, the most visible manifestation of convection in the Earth's mantle is the subduction of oceanic tectosphere beneath island arcs and active continental margins. Plate tectonics postulates that the surface area of the Earth remains constant in time, so that the total area created by sea-floor spreading must equal the area destroyed by subduction. The total area flux calculated from present-day plate motions amounts to 2.9 km^2/yr (Minster and Jordan, 1980). Since the typical thickness of a mature plate is on the order of 100 km or so (Parsons and Sclater, 1977), the subduction process involves about 300 cubic kilometers of oceanic lithosphere every year, which is sufficient to cycle the entire mantle through the plate system once every 3000 My. Nearly all of the plate consumption is concentrated along the oceanic trenches ringing the Pacific Basin (Figure 3.1). Not surprisingly, it is the seismological study of these subduction zones which tells us most about the depth extent of plate-tectonic convection.

3.1. Earthquakes in Subduction Zones

Subduction zones are marked by the depth extent of their seismic activity, which includes most of the world's earthquakes. Seismologists traditionally

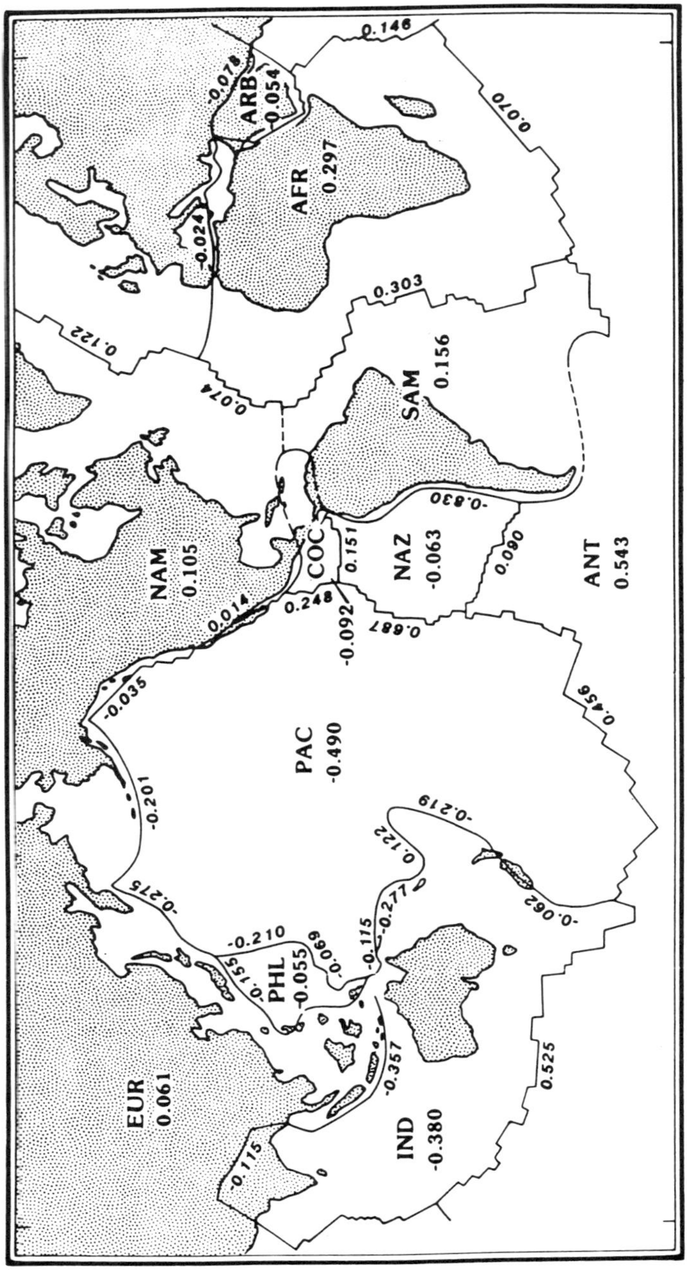

Figure 3.1. Global map showing rates of plate creation (positive values) and destruction (negative values) in units of km^2/yr for individual plates and sections of plate boundaries, from Garfunkel *et al.* (1986, Figure 3).

classify an earthquake's hypocental depth h as shallow-focus ($h \leq 70$ km), intermediate-focus ($70 < h \leqslant 300$ km), or deep-focus ($h > 300$ km) (Gutenberg and Richter, 1954). Although the historical record for some subduction zones shows only shallow seismicity (e.g., the Cascades and southern Chile), most are characterized by earthquakes with foci considerably deeper than 70 km. The first evidence for seismicity below the lithospheric shell came from observations of anomalously early core-refracted (*PKP*) arrivals at distant seismographic stations (Turner, 1922) and anomalously large differences between the travel times of the direct compressional (*P*) and shear (*S*) phases at nearby stations (Wadati, 1928). The global distribution of 141 deep earthquakes was mapped by Turner (1930), revealing nearly all of the intermediate-focus and deep-focus seismic zones known today. As travel-time tables were extended to account for focal depth (Wadati and Masuda, 1933), and depth-sensitive phases, such as the near-focus surface reflections *pP* and *sP*, were recognized and incorporated into earthquake-location procedures, hypocenter determinations became sufficiently precise to resolve the seismicity into planar features dipping at steep, but variable, angles to depths as great as 700 km (Wadati, 1935; Gutenberg and Richter, 1938).

There was, of course, much speculation on the physical mechanism responsible for these deep seismic zones. It was recognized that earthquakes of intermediate and deep focus generally have characteristics similar to shallow-focus earthquakes and could be explained by slip on faults (Leith and Sharpe, 1936), and there was an awareness that the driving forces responsible for this faulting might be those associated with convection currents and continental drift (Wadati, 1935; Leith and Sharpe, 1936; Holmes, 1965; Vening Meinesz, 1954). Benioff (1949, 1954) and Honda and Masatsuka (1952) concluded that deep seismic zones mark huge fault planes along which oceanic upper mantle is being thrust under the island arcs and continental orogenic zones, but they did not appeal to convection as a mechanism.

It was not until three decades after they were first delineated that the Wadati-Benioff zones were explained as the loci of slabs of oceanic lithosphere subducting into the mantle (Oliver and Isacks, 1967; Isacks *et al.*, 1968). Most of the seismic energy released in subduction zones is by shallow-focus, reverse-faulting earthquakes on the interface between the converging plates, and the slip vectors of these events are consistent with the relative plate motions (McKenzie and Parker, 1967; Isacks *et al.*, 1968). However, detailed focal-mechanism studies demonstrated that the seismicity below about 70 km is not caused by slip on an interface between two major blocks, as postulated by Benioff and Honda, but represents fracturing within the cold, brittle core of the descending slab (Isacks *et al.*, 1968; Isacks and Molnar, 1969). From the point of view of mantle convection, deep seismicity thus marks the locus of the detached thermal boundary layer as it recycles into the mantle (Turcotte and Oxburgh, 1968). The recognition of this fact was one of the final steps in the formulation of the plate-tectonic theory.

Since the establishment of modern global seismic networks in about 1960, it has been possible to locate routinely earthquakes having magnitudes larger than about five with a precision usually less than 20 km. These networks, in combination with local and regional stations, have been used to delineate the seismicity of descending slabs and to determine the focal mechanisms of many of the larger earthquakes in nearly all of the world's subduction zones. Among the numerous constraints on the subduction process established by these studies, two have tended to dominate discussions regarding the fate of the lithospheric slab: (1) seismicity abruptly terminates at depths of 650–700 km, and (2) deep slabs are deforming in response to compression oriented down the dip of the seismic zone.

World seismicity as a function of depth is plotted in Figure 3.2, taken from a recent compilation by Vassiliou *et al.* (1984). The number of earthquakes above a given magnitude level decays exponentially to about 300 km, where there is a seismicity minimum, followed by a resurgence in activity from 500 to 700 km (Leith and Sharpe, 1936; Sykes, 1966; Isacks *et al.*, 1968). No earthquake has been reliably located below about 700 km, and the statistical uncertainties of the hypocenters are consistent with a sharp cutoff a few tens of kilometers less than this depth (Stark and Frohlich, 1985).

Deep-focus seismic zones invariably involve old, cold oceanic lithosphere being subducted at relatively fast rates, whereas zones lacking deep-focus earthquakes are characterized by young, relatively hot lithosphere and/or lower rates. This suggests that the maximum depth of earthquakes corresponds to a critical temperature above which slab material is too hot and weak to support the stresses needed for brittle failure (Isacks *et al.*, 1968; McKenzie, 1969; Griggs, 1972). For a particular seismic zone, this depth then defines an "assimilation time" given by the ratio of the length of the zone to its subduction rate, which should correlate with the plate's "cooling time," i.e., its crustal age prior to subduction. Such a correlation is in fact observed, the former being about one-tenth the latter (Molnar *et al.*, 1979; Wortel, 1982; Shiono and Sugi, 1985).

The concept of a critical temperature does not provide a satisfactory explanation for the sharpness of the deep seismicity cutoff, however. The Mariana and Japan seismic zones, for example, both have deep-focus events

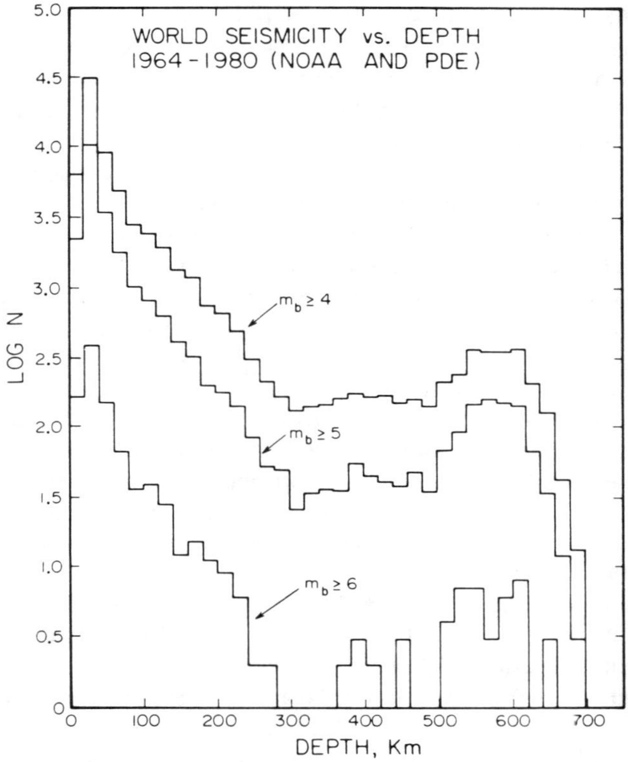

Figure 3.2. Total number of earthquakes recorded worldwide for the period 1964–80 plotted in 20-km depth intervals, from Vassiliou *et al.* (1984, Figure 1). The three curves give cumulative values about the indicated body-wave magnitude.

and involve old slabs, but the ratios of their assimilation to cooling times differ by nearly a factor of two. Moreover, it offers no explanation for the seismicity minimum at 300–500 km and the resurgence of earthquake activity at greater depths, which is a feature of individual subduction zones as well as the worldwide average (Leith and Sharpe, 1936; Sykes, 1966; Richter, 1979; Vassiliou *et al.*, 1984).

The global seismicity cutoff nearly coincides with a rapid increase in seismic wave velocities at an average depth of 650–670 km (King and Calcagnile, 1976; Given and Helmberger, 1980; Burdick, 1981; Walck,

1984a, 1984b), and the two are thought to be causally related (e.g., Richter, 1979; Stark and Frohlich, 1985). This 650-km (or 670-km) discontinuity is the deepest of two such high-gradient features in the mid-mantle transition zone (Niazi and Anderson, 1966; Johnson, 1967), both identified with phase changes in major silicate minerals, of which olivine is thought to be the most abundant (Ringwood, 1975; Weidner, 1985). At approximately 650 km, olivine is expected to transform from its γ phase to a solid solution of magnesiowüstite and perovskite, involving the displacement of silicon cations from four-fold to six-fold coordination (Liu, 1976b; Bell *et al.*, 1979; Ito and Yamada, 1982; Jeanloz, this volume). A phase transition in the pyroxene-garnet component of the mantle to a perovskite structure occurs near the same depth and also involves a large ($\sim$10 percent) increase in density (Liu, 1976a; Yagi *et al.*, 1979; Ito and Yamada, 1982).

The 650-km discontinuity has become the conventional demarcation between the upper and lower mantle. The basic question posed by its association with the seismicity cutoff is whether or not it presents a barrier to slab penetration. It could act as a barrier if it were a transition to a compositionally denser lower mantle, rather than (or in addition to) a simple phase change (Richter and Johnson, 1974; Liu, 1979; Jeanloz and Thompson, 1983), or if the phase transition at 650 km had a large, negative Clapeyron slope (Schubert *et al.*, 1975; Christensen and Yuen, 1984) or marked a jump to a higher viscosity (McKenzie and Weiss, 1975). Such a scenario is consistent with the model of stratified mantle convection preferred by many workers (e.g., McKenzie *et al.*, 1974; DePaolo and Wasserburg, 1976; Jeanloz and Richter, 1979; O'Nions *et al.*, 1979; Jacobsen and Wasserberg, 1979; Richter and McKenzie, 1981; Anderson, 1981; Lees *et al.*, 1983; McKenzie, 1983; Giardini and Woodhouse, 1984).

The hypothesis that deep slabs encounter resistance to penetration below about 650 km is indicated both by the higher levels of energy release at the base of deep seismic zones and by the mechanisms of faulting observed during this energy release. In the far field at low frequencies, source mechanisms of deep earthquakes are characterized by a second-order moment tensor (Gilbert, 1971; Randall, 1971) whose principal axes are thought to be controlled by the deviatoric stress field associated with the rupture (Isacks and Molnar, 1969; Knopoff and Randall, 1970). Although other mechanisms, including isotropic compression due to phase changes, have been postulated in the source processes of deep-focus events (Benioff, 1963; Knopoff and Randall, 1970; Gilbert and Dziewonski, 1975), most appear to conform to the simplest model of faulting, a dislocation described by a double couple (Honda and Masatsuka, 1952; Honda, 1962), or for some of the larger deep-focus multiple events, a series of such dislocations distributed in space and time (Sasatani, 1980).

In an important pair of papers, Isacks and Molnar (1969, 1971) obtained first-motion, double-couple solutions for a number of intermediate-focus and deep-focus earthquakes in subduction zones and noted several systematic features in the orientations of their principal axes:

1. For a particular continuous piece of slab, either the compressional axes or tensional axes tend to cluster in the down-dip direction.
2. Within the intermediate-focus zone, slabs are in a state of
   a. down-dip tension if there are no deep-focus earthquakes,
   b. down-dip tension if there is an aseismic gap between the intermediate- and deep-focus zones, and
   c. down-dip compression if the seismicity is continuous into the deep-focus zone.
3. Within the deep-focus zone, descending slabs are in a state of down-dip compression.

Generalization (1) motivates the hypothesis that the earthquakes are located in the cold thermal core of the slab and controlled by the dynamics of its flow into the mantle. Generalization (2a) is consistent with the idea that shallow slabs sink under their own weight into a relatively low-viscosity asthenosphere, whereas (2c) and (3) suggest that slab penetration is resisted by a higher viscosity mesosphere at the base of the seismic zone. (2b) then implies that an aseismic gap is indicative of a gap in the slab itself, or at least in its ability to transmit down-dip deviatoric stress.

Vassiliou (1984) has studied the statistics of a large number of focal mechanisms of intermediate- and deep-focus events and confirms generalizations (1) and (3), but shows that, although most intermediate-focus earthquakes in the highly active Tonga seismic zone are in a state of down-dip compression, intermediate-focus events in all other slabs are more consistent with down-dip extension, regardless of the depth to the seismicity cutoff or the existence of an aseismic gap. Therefore, conclusion (2c) does not generally apply.

Isacks and Molnar's original interpretation in terms of an Earth mechanically layered into a lithosphere, asthenosphere and mesosphere (Daly, 1940) is now outmoded, since the arguments for a high-viscosity lower mantle (e.g., McKenzie, 1966) or at least an inpenetrable boundary layer immediately below the 650-km discontinuity (McKenzie and Weiss, 1975) appear to be incorrect (Cathles, 1975; Peltier, 1976, 1980; Sammis *et al.*, 1977). Nevertheless, their observations continue to have a major impact on the dynamical modeling of slab penetration. In discussing the generalizations given above, for example, McKenzie and Richter (1976, pp. 73–74) conclude, "The simplest explanation for this behavior is that the slab meets great resistance to its motion at 600 kilometers and is unable to penetrate below 700 kilometers . . . the convective circulation of which the plates are a part is confined to a region 700 km deep."

That deep-focus zones are in down-dip compression has been confirmed by many subsequent studies (e.g., Richter, 1979), including recent moment-tensor inversions using extensive digital data bases (Giardini, 1984; Vassiliou, 1984). The idea that penetration into the lower mantle is resisted at 650 km has also been corroborated by detailed investigations of the Tonga deep-focus zone, the world's most seismically active. The slab in this region appears to be highly contorted (Sykes, 1966), and it is cut by planes of deformation delineated by the distribution of precisely located hypocenters, as well as the focal mechanisms of specific earthquakes (Billington and Isacks, 1975; Billington, 1980; Giardini and Woodhouse, 1984, 1986). Hypocenters cluster along one or both of the nodal planes suggesting that, like shallow-focus earthquakes, the intermediate- and deep-focus events represent slip on a plane. Giardini and Woodhouse (1984) conclude that, "Both the moment tensor solutions and the seismicity patterns indicate thickening and bottoming out of the descending material, and suggest that subduction experiences great resistance in penetrating the 670-km discontinuity . . . We observe horizontal displacements of the seismically active portion of the slab, and infer that material moves laterally rather than vertically at the base of the deep seismic zone. This favours the view that subduction does not continue into the lower mantle, although we clearly cannot infer the fate of subducted material after it ceases to be seismically active."

3.2. Thermal Structure of Subduction Zones

A key piece of evidence for the continuity of the subducted slab was the observation by Oliver and Isacks (1967) that high-frequency compressional and shear waves propagating up the Wadati-Benioff zone in the Tonga-Kermadec island arc do so without much energy loss, whereas those propagating in the mantle wedge above the zone are severely attenuated. This effect can be quantified in terms of the seismic quality factor Q, which is proportional to the ratio of the potential energy stored in a wave cycle to that irreversibly lost by internal friction. They obtained path-averaged Q values for the slab that were several times those for the wedge, a fact explained by the relatively cold temperature of the subducted lithosphere. They also demonstrated that the travel times along ray paths through the seismic zone are significantly less than those through the ambient mantle, again as expected for a cold slab. Similar observations had been published for the Japan seismic zone by Japanese seismologists using their own local networks (see review by Utsu, 1971). These and many other studies of upward propagating waves (Mitronovas and Isacks, 1971; Suyehiro and Sacks, 1979; Hubert and Frohlich, 1981) and downward propagating waves (Davies and McKenzie, 1969; Jacob, 1972; Sleep, 1973) have established that lithospheric slabs above deep-focus zones have seismic velocities averaging 5–10 percent higher than the surrounding mantle. This velocity contrast implies a temperature differential on the order of 1000°K.

The temperature structure of the oceanic lithosphere as a function of its age is reasonably well described by simple thermal-boundary-layer models (Parsons and Sclater, 1977). Since the age of the plate can be determined directly from its magnetic signature seaward of the trench, the temperatures in the subducting plate are known near the surface. By combining this initial condition with the subduction velocity and the geometry of the seismic zone, one can calculate the thermal history of the descending slab, subject to certain assumptions regarding thermodynamic parameters and heat production of the slab and the kinematics of flow in the vicinity of the slab (Hasebe *et al.*, 1970; Minear and Toksöz, 1970; Andrews and Sleep, 1974). Simple plate models indicate that large temperature contrasts between the slab and ambient mantle should exist at depths of 700 km (McKenzie, 1969; Griggs, 1972); lithosphere which has cooled down for 100 My cannot thermally reequilibrate with the hot mantle in the 10 My or so it takes to reach the base of the seismic zone. This conclusion is unmodified by adding the effects of radioactive heating (Toksöz *et al.*, 1973) and phase changes (Schubert *et al.*, 1975). The importance of shear-strain heating has been more controversial. By assuming strain heating to be very large, Minear and Toksöz (1970) and Toksöz *et al.* (1971) were able to construct slab models that thermally equilibrated at depths on the order of 700 km. However, the high levels of strain heating used in their calculations do not appear to be physically reasonable, either on the basis of rheology (Turcotte and Schubert, 1973; Andrews and Sleep, 1974; Bird, 1978) or energy balance (Griggs, 1972; Creager, 1984).

Thermal models of descending slabs have proven to be successful in predicting the variation of seismic wave velocities between the slab interior and the ambient mantle. For this purpose, the relationship between lateral velocity anomaly and lateral temperature anomaly is usually linearized and some value of the effective thermal coefficient of velocity, $\partial v_p/\partial T$, is specified. By calibrating the velocity structure of the slab using travel times from the nuclear explosion LONGSHOT, Sleep (1973) obtained an empirical value for this coefficient of −0.9 m/s/°K. A 1000°K thermal anomaly at a depth of a few hundred kilometers thus yields a velocity contrast of about 10 percent, which is roughly consistent with the observations. (As discussed below, more recent estimates suggest Sleep's value is about 50 percent too high.) Three-dimensional wave-speed models constructed in this manner have been used to study the perturbing effect of slab structures on *P*-wave travel times (Toksöz *et al.*, 1971; Sleep, 1973; Creager and Jordan, 1984, 1986a), *P*-wave amplitudes (Sleep, 1973), fault-plane solutions from *P*-wave first motions (Engdahl *et al.*, 1977), and earthquake locations (Engdahl *et al.*, 1977; Fujita *et al.*, 1981). The velocity gradients in these models are sufficiently high that the ray-bending effects are large (Toksöz *et al.*, 1971) and an accurate computation of travel times requires the use of three-dimensional ray tracing algorithms, such as those discussed by Julian and Gubbins (1977).

3.3. Additional Arguments Against Aseismic Extensions of Slabs

If the slab penetrates the mantle as a simple detached thermal boundary layer, unimpeded and without disruption, then the thermal-history calculations show that the large thermal contrasts of old lithosphere subducted at typical plate speeds (50–100 km/My) will persist to great depths. These thermal anomalies should be marked by zones of high seismic velocities along aseismic extensions of the slabs below the deepest earthquakes.

To image a velocity anomaly beneath a seismic source requires the use of teleseismic body phases, downward propagating waves which travel to great distances from the epicenter. A basic diagram employed for this purpose is the *residual sphere*, defined by Davies and McKenzie (1969) to be the graph of travel-time residuals on the unit focal sphere: the difference between the observed travel time and the time computed from a standard spherically symmetric Earth model is plotted at the take-off angle and azimuth of the ray path as it leaves the source. These authors made *P*-wave residual-sphere diagrams for shallow-focus events in subduction zones to display the teleseismic signature of high-velocity lithospheric slabs. Mitronovas and Isacks (1971) looked for a systematic variation in teleseismic *P*-wave residuals as a function of azimuth for deep-focus Tonga events; they found no evidence for an aseismic extension of the slab but suggested that the best way to answer this question would be to analyze the residual spheres of deep-focus earthquakes. A number of subsequent searches for evidence of deep penetration using residual-sphere techniques were negative, as indicated by the following quotations:

"For deep-focus earthquakes, events 8–12 (in Tonga), no clear pattern of travel-time anomalies is evident. The residuals for events 8 and 10 suggest that the seismic zone has a slightly higher velocity than the surrounding mantle, but until more data are available, this conclusion must remain tentative." (Toksöz *et al.*, 1971, p. 1131)

"We applied the method of Davies and McKenzie (1969) to the Tonga region with a more extensive data set than that used by Toksöz *et al.* (1971), but, like them, we were unable to find convincing evidence for anomalies associated with the slab below 300 km. The errors in reported travel-time data plus incomplete coverage of the focal sphere combine to make the method of Davies and McKenzie too crude a technique for the definition of slab anomalies as a function of depth in most regions." (Sondergeld *et al.*, 1977, p. 537)

"Plots of travel-time residuals on the focal sphere (of earthquakes in nearly every subduction zone) failed to reveal characteristically large negative residuals which are taken as evidence of source anomalies for shallow earthquakes at island arcs (see, e.g., Toksöz *et al.*, 1971). Thus, source anomalies, if present at all in deep-focus travel-time data, can safely be ignored." (Sengupta and Julian, 1978, p. 192)

Not only did the application of the Davies-McKenzie technique to large sets of teleseismic P-wave travel times by several groups of investigators fail to reveal aseismic extensions of slabs, it did not confirm the large temperature contrasts in deep-focus zones predicted by the thermal history calculations.

These negative results might be explained by either of two effects. First, the thermal equilibration time of the slab, and thus the velocity contrast of deep slab material, could be significantly overestimated by simple subduction models. One mechanism for decreasing the equilibration time is chemically induced instability in the upper mantle (Figure 3.3a). There are significant density variations within the slab associated with the compositional stratification of the oceanic lithosphere: a harzburgitic residuum of basaltic differentiation presumably occupies a layer ~30 km thick subjacent to the oceanic crust, which will have a normative density 1–2 percent less than undepleted mantle (Green and Liebermann, 1976; Oxburgh and Parmentier, 1977; Jordan, 1979a). Oxburgh and Parmentier (1977) have discussed the possibility that this low-density layer might detach itself from the slab and rise diapirically through the upper mantle. If so, the cold thermal core of the slab would be advectively disrupted.

A second possibility is that the velocity contrast of deep slab material is as large as the thermal calculations predict, but that the lateral velocity gradi-

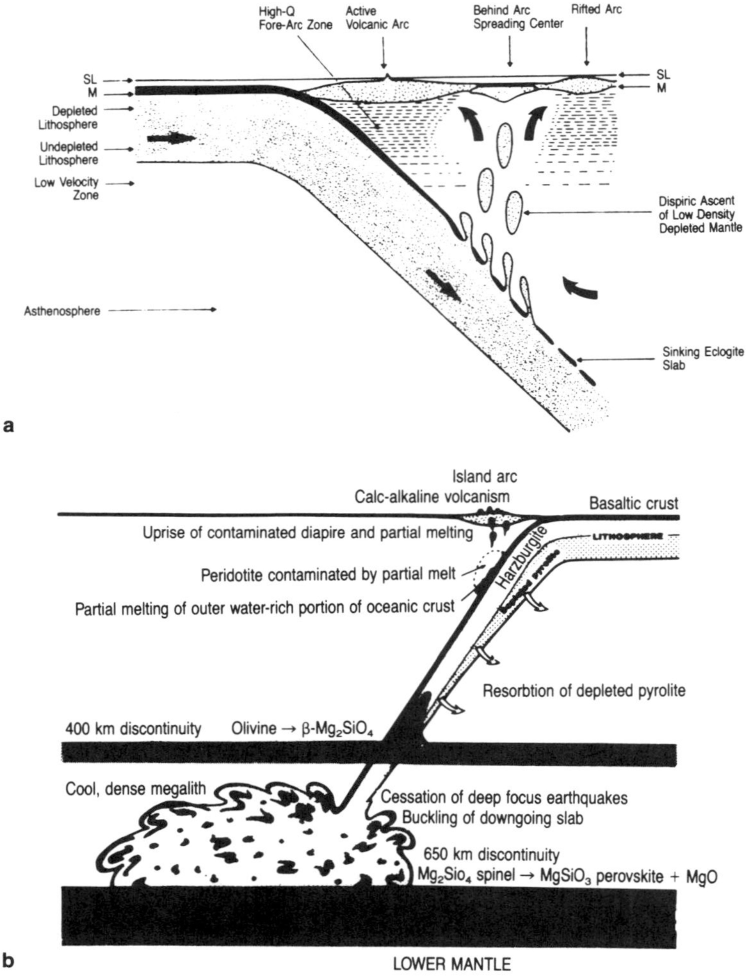

Figure 3.3. Cartoons depicting three possible fates of the subducted slab. Panel (a) shows how the descending slab might be disrupted by convective instability owing to its compositional stratification (from Oxburgh and Parmentier, 1977, Figure 4). Panel (b) depicts Ringwood's (1982, Figure 9) "megalith" model, where the subducted lithosphere buckles to form a melange at the 650 km seismic discontinuity. Panel (c) is a schematic illustration of isotherms for subduction flow in a compositionally stratified mantle (from Anderson, 1981); convection below 650 km is thermally coupled to convection above this discontinuity, so that descending plumes in the lower mantle are aligned beneath descending slabs in the upper mantle.

c 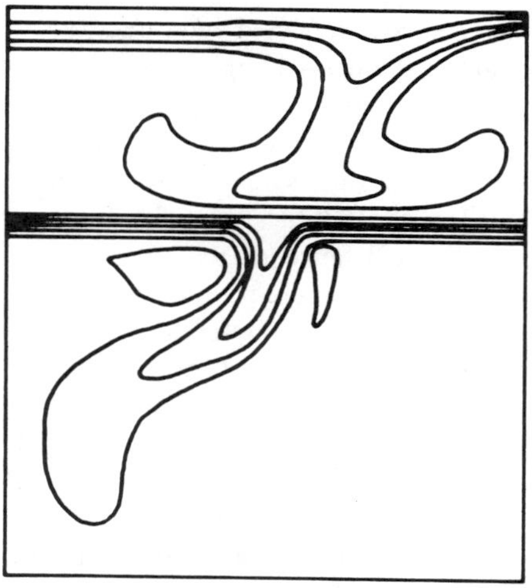

Figure 3.3. *Continued*

ents are reduced by the horizontal flow of slab material away from the seismically active regions. One scenario for the fate of the slab consistent with this possibility is Ringwood's (1982) "megalith" hypothesis, illustrated in Figure 3.3b, where the buckling of former oceanic crust and the harzburgite component of the subducted oceanic lithosphere forms a large melange at the 650-km discontinuity. The seismic velocity gradients in the deep-focus zones would also be decreased, of course, if the slab were simply entrained in an upper-mantle convection system bounded by a phase change, a viscosity increase, and/or a compositional discontinuity at 650 km (Figure 3.3c) As discussed previously, this hypothesis is consistent with the deformation of the slab inferred from the focal mechanisms of most deep-focus earthquakes (Isacks and Molnar, 1971) and, in Tonga at least, the distribution of seismicity (Giardini and Woodhouse, 1984).

3.4. Evidence for Aseismic Extensions

It now appears that near-source anomalies associated with deep slab structure do indeed exist, but the residual-sphere methodology employed in the studies quoted above was inadequate to resolve them. Evidence for deep-seated, high-velocity anomalies in the vicinity of subduction zones began to accumulate about ten years ago. Jordan and Lynn (1974) used the arrival-time differences between core-reflected and direct seismic phases—*PcP-P* and *ScS-S* differential travel times—to map a localized, high-velocity anomaly in the lower mantle beneath the Caribbean. They associated this feature with a convective downwelling marked by the Middle America Trench. On the basis of these results, as well as some observations by Weichert (1972) and Engdahl (1975), Jordan (1975b) argued for deep-mantle circulation of subducted material. The existence of the Caribbean anomaly has been corroborated by Lay (1983) and more recently in a tomographic analysis of multiple-*S* times by Grand (1987). Fitch (1975, 1977) applied a multiple-event technique to the analysis of travel times from earthquakes beneath Peru, Tonga and the New Hebrides, and showed that the *in situ* velocity contrasts at the levels of the deepest events are not much diminished from the 5–10 percent range inferred for shallower depths, thereby confirming the predictions of the thermal-history calculations. Frohlich and Barazangi (1980) examined the travel times of core-reflected phases from deep-focus Tonga events and concluded they were consistent with high-velocity anomalies beneath the sources.

The first definitive study directly imaging slab penetration into the lower mantle was Jordan's (1977) residual-sphere analysis of the 29 Jan 1971 deep-focus earthquake beneath the Sea of Okhotsk (h = 540 km). The clean waveforms from this event and the good station distribution afforded by its location in the Kuril-Kamchatka seismic zone yielded the spectacular signature of a high-velocity, slab-like feature extending to a depth of about 1000 km with a strike and dip consistent with the geometry of the seismic zone (Figure 3.4, bottom panel). Several aspects of Jordan's analysis differed from previous residual-sphere studies. Rather than using *P*-wave travel times from standard bulletins, he hand-picked the travel times of *S* and *ScS* phases directly from the seismograms. Because these shear waves propagate slower than compressional waves, their travel-time perturbations owing to mantle heterogeneities are a factor of 3–4 larger, and the signal-to-noise ratio (SNR) is correspondingly higher. The amplitude of the slab anomaly shown in the lower diagrams of Figure 3.4 has a peak-to-peak amplitude of about eight seconds, much larger than any timing uncertainties. The global network of seismic stations gave an azimuthal distribution superior to that available for the more active deep-focus zone in Tonga. By incorporating the travel times of the core-reflected *ScS* phase, he was able to improve the residual-sphere coverage at small take-off-angles, allowing excellent sampling of the quadrapolar anomaly pattern diagnostic of a steeply dipping, aseismic slab. To increase the SNR of this anomaly pattern, he performed two operations on the raw residuals not made in previous residual-sphere studies: the travel

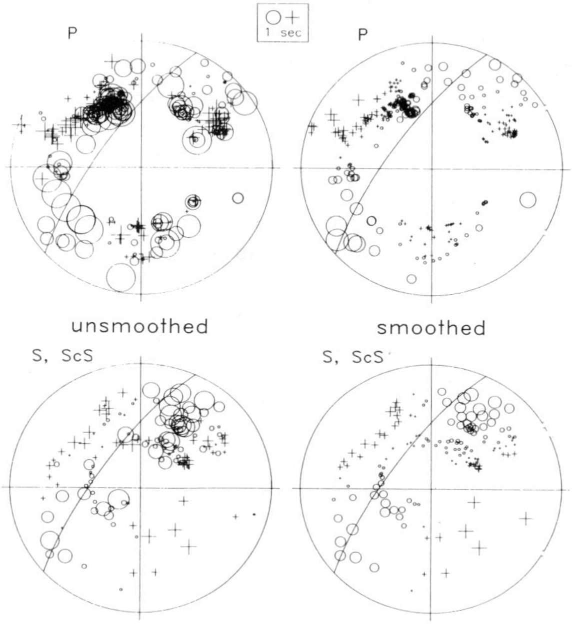

Figure 3.4. Observed residual spheres for *P*-times (top row), and *S* and *ScS* times (bottom row) of the 29 Jan. 1971 deep-focus earthquake beneath the Sea of Okhotsk. The former were derived from the ISC data base by Creager and Jordan (1984), whereas the latter were hand-picked by Jordan (1977). Residual spheres are equal-area, lower-hemisphere projections, circumscribed by the locus of rays with take-off angles of 60° from vertical. Circles represent negative residuals, crosses are positive; symbol size is proportional to residual magnitude. Symbol scale shown in box is 1 second for *P* times and 3.5 seconds for *S* times. Lines represent the plane tangent to the slab model shown in Figure 3.7 (cross-section AA′). In the left column, the times have been corrected for ellipticity and station corrections, and *P* residuals have been computed with respect to the relocated hypocenters. In the right column, these corrected residuals have been smoothed using the stochastic low-pass filter of Creager and Jordan (1986). The smoothed residual spheres are dominated by negative residuals corresponding to rays propagating in the plane of the slab, indicative of slab penetration into the lower mantle.

times were first corrected for known large-scale heterogeneities, in particular the upper-mantle variations beneath the recording stations described by "station anomalies." (A station anomaly is the mean travel-time residual observed for many events at a given station relative to a network average.) The residuals were then smoothed by a filter specifically formulated to enhance low-wavenumber signals discretely sampled on the unit focal sphere, thereby averaging over random reading errors and higher wavenumber inhomogeneities far from the source region.

For the high-quality *S*-wave data obtained by Jordan (1977), smoothing was hardly necessary, since the slab signature completely dominates the unfiltered residuals (Figure 3.4, bottom left). Unfortunately, reading such travel times from analog seismograms is tedious, and his work was limited to a single event. *P*-wave travel times, on the other hand, are readily available for all teleseismically observed earthquakes, since they are compiled by agencies such as the International Seismological Centre (ISC) to locate hypocenters. Timing and reading errors lower the SNR of these data significantly, and the combination of these errors with the poor coverage obtainable in many subduction zones such as Tonga proved to be a major source of frustration in early attempts to identify deep slabs by residual-sphere methods, as is clear from the comments quoted above.

However, in a follow-up study of the 29 Jan 1971 Okhotsk event and other deep-focus earthquakes in the Kuril-Kamchatka seismic zone, Means and Jordan (1980) were able to derive *P*-wave residual spheres from ISC data which had clear signatures of a deep slab (Figure 3.4, top left). Encouraged by these results, we have developed more sophisticated algorithms for processing ISC *P* times to obtain residual-sphere images of near-source heterogeneity (Creager and Jordan, 1984, 1986a). The application of these new techniques to deep-focus earthquakes in the Kuril-Kamchatka, Japan, Izu-Bonin and Mariana seismic zones has revealed aseismic extensions of subducted slabs along the entire northwest margin of the Pacific plate.

Travel-time residuals with respect to a spherically symmetric Earth model are produced not only by near-source structure, but by heterogeneities integrated along the entire ray path, as well as by random observational errors and mislocation of the hypocenter. We can take advantage of recent work on aspherical Earth structure to correct for the inhomogeneities away from the source that may corrupt the slab's residual-sphere signature. Our procedures adjust the travel times for hydrostatic ellipticity, near-receiver structure as expressed in station anomalies (Dziewonski and Anderson, 1983), and large-scale lower-mantle heterogeneity derived from global inversions of *P* times (Dziewonski, 1984). Of course, smaller scale heterogeneities not associated with the slab and not predicted by current aspherical Earth models do indeed exist, but their spectrum on the residual sphere is shifted to higher wavenumbers relative to near-source heterogeneities of equal spatial scale. To damp out these high-wavenumber components, we follow Jordan's (1977) approach of low-pass filtering the residual sphere. As illustrated in Figure 3.4 (top), this operation considerably enhances the SNR of near-source structure relative to unmodeled inhomogeneities and random observational errors.

Travel-time residuals are computed with respect to some assumed location of the source, usually one calculated using a radial velocity model and thus biased in some unknown way by the aspherical structure of the Earth. Our algorithm adjusts the hypocenter and origin time of each earthquake to fit the corrected residuals in a least-squares sense, which is equivalent to orthogonalizing the residual-sphere anomaly pattern with respect to its degree-zero and degree-one spherical harmonics (Davies and McKenzie, 1969; Creager and Jordan, 1984). This procedure, like any projection, introduces additional bias but has the advantage that the net mislocation effect can then be precisely specified by a known linear operator.

Because of the mislocation effects, as well as the considerable distortion of the residual sphere by ray-path bending in the vicinity of the focus (Toksöz *et al.*, 1971), the interpretation of the residual-sphere data is greatly facilitated by the comparison of the observed patterns with theoretical graphs generated from three-dimensional models of near-source heterogeneity. We obtain the latter from simple thermal models of the descending slab. Using the techniques developed by Toksöz *et al.* (1973) and Sleep (1973), the two-dimensional temperature field of the slab is computed assuming an initial, pre-subduction thermal profile and a kinematically prescribed velocity field, which includes entrained mantle flow in the vicinity of the slab and induced corner flow beneath the back-arc basin. A three-dimensional velocity model is constructed by assuming the lateral velocity contrasts are proportional to the thermal contrasts (Sleep, 1973) and rotating the resulting structure about an arc pole (Engdahl *et al.*, 1977). Theoretical residual spheres are computed by tracing rays through the three-dimensional velocity model for the same station set sampled by the actual data. We can then account for the mislocation bias by projecting the theoretical residual spheres with respect to the location parameters using the identical operator applied to the data. The bias caused by smoothing the observed residuals is taken into account by applying the same low-pass filter to the theoretical residuals.

Examples of the results obtained by Creager and Jordan (1986a) for a representative set of intermediate-focus and deep-focus events in the north

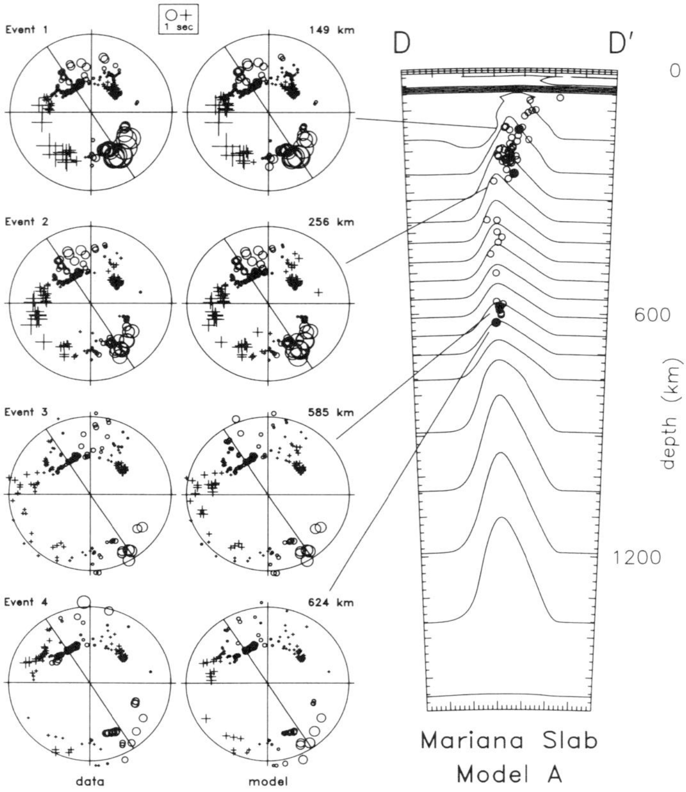

Figure 3.5. Observed (left) and theoretical (center) residual spheres for *P* waves of four intermediate- and deep-focus earthquakes in the Mariana slab (from Creager and Jordan, 1986). Slab Model A, used to compute theoretical diagrams, is shown in cross-section *DD′* (right); no vertical exaggeration. Plotting conventions are identical to Figure 3.4. Lines through NW and SE quadrants represent planes tangent to the slab model, which has a vertical dip below 200 km and a local strike of 147°. *P*-wave velocities on model cross-section are contoured at 0.25-km/s intervals; ambient values are radial model of Herrin *et al.* (1968). Velocity perturbations were calculated from a thermal model characterized by slab penetration to 1350 km, assuming $\partial v_p/\partial T = -0.5$ m/s/°K. Circles are projections of well-recorded ISC hypocenters within 150 km of section *DD′* in Figure 3.7 about an arc pole at 20°N, 90°E. Theoretical residual spheres were computed by shooting rays through the 3-D slab structure obtained by rotating the 2-D model about this arc pole. The residual-sphere diagrams for both the data and model have been orthogonalized with respect to location parameters and smoothed according to the procedures described in the text.

west Pacific are presented in Figures 3.5 and 3.6. In every case, the data are well fit by simple models of the descending slab having very few free parameters. The strike and dip of the slab above the seismicity cutoff are generally constrained to within 5°–10° by the distribution of well-located hypocenters (some of which are projected onto the model cross-sections in Figure 3.7), and these are consistent with the slab geometry inferred directly from the residual-sphere anomalies of intermediate-focus events.

With the exception of Japan, the strikes and dips of the aseismic extensions obtained from the residual-sphere patterns of deep-focus events are in similar agreement with those extrapolated from the deep-focus seismicity. The simplest geometry is in the Marianas, where the seismically active part of the slab below 200 km and its aseismic extension below 650 km are nearly vertical, yielding very consistent quadrapolar anomaly patterns (Figure 3.5). The amplitude of the residual-sphere anomaly decays by about a factor of two between the shallowest and deepest events shown in Figure 3.5, but this decay is due almost entirely to the increase in ambient seismic velocities through the mantle transition zone, and not to any substantial decay of the velocity contrasts—and thus thermal contrasts—within the slab (Creager and Jordan, 1986a). (From 200 to 700 km, the compressional-velocity contrast between the cold thermal core of the slab and the ambient mantle is reduced in the model from 0.49 km/s to 0.43 km/s, only 15 percent. By Fermat's Principle, the travel-time perturbations are approximately proportional to this contrast divided by the squared inverse of the ambient velocity. The latter varies by a factor of two over this depth interval.)

In Kuril-Kamchatka, the geometry is complicated by considerable variations in the dip of the seismic zone, both along strike and with depth. The residual-sphere anomaly patterns from the intermediate-focus earthquakes are very different from the deep-focus ones, implying a slab dip of about 20° less (Jordan, 1977; Creager and Jordan, 1984). This change of dip or "kink" occurs rather abruptly at about 500 km along the entire length of the slab, and in the northern part of the zone, it is evident in both the seismicity of the region and the orientation of the compressive axes of focal mechanisms (Veith, 1974; Strelitz, 1975; Sasatani, 1976).

The residual-sphere data for Japan require a similar change in dip at about the same depth, varying from about 30° or less in the seismic zone shallower than 500 km to greater than 50° below the seismicity cutoff near 600 km (Creager and Jordan, 1986a). Unlike in Kuril-Kamchatka, the cutoff is too shallow to permit adequate resolution by the seismicity alone, and there is no evidence for it at all in the focal mechanisms, which have compressive axes dipping parallel to the shallower seismicity (Yoshii, 1983; Giardini, 1984).

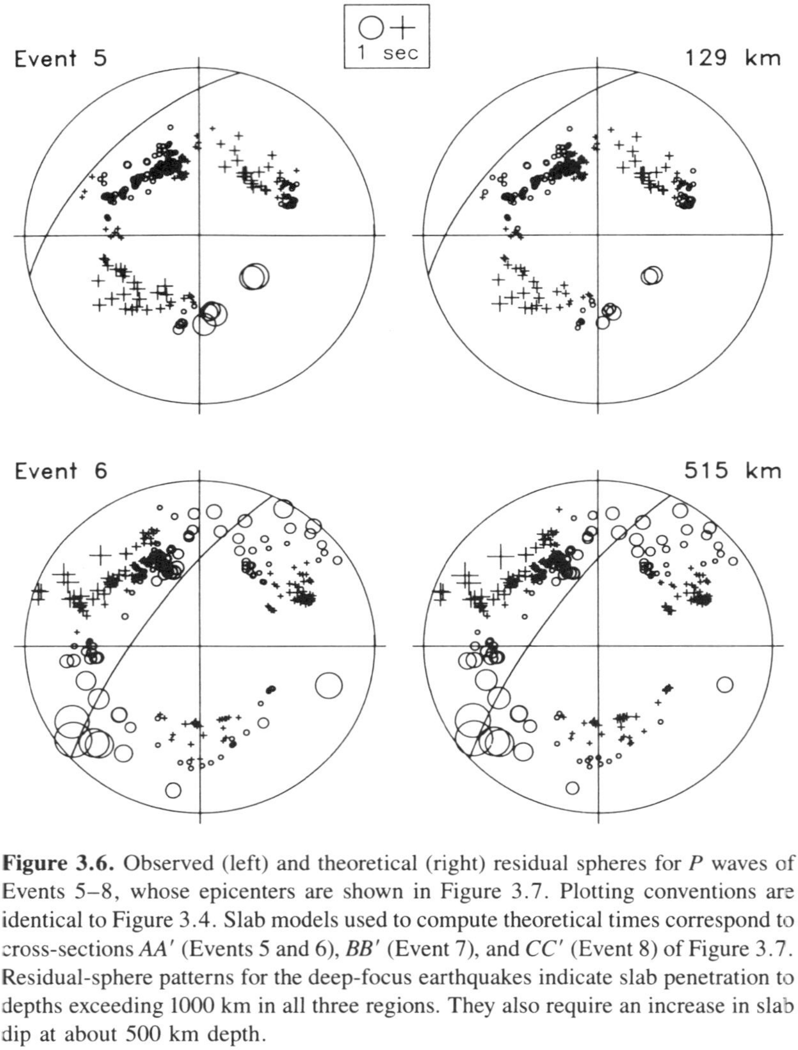

Figure 3.6. Observed (left) and theoretical (right) residual spheres for *P* waves of Events 5–8, whose epicenters are shown in Figure 3.7. Plotting conventions are identical to Figure 3.4. Slab models used to compute theoretical times correspond to cross-sections *AA′* (Events 5 and 6), *BB′* (Event 7), and *CC′* (Event 8) of Figure 3.7. Residual-sphere patterns for the deep-focus earthquakes indicate slab penetration to depths exceeding 1000 km in all three regions. They also require an increase in slab dip at about 500 km depth.

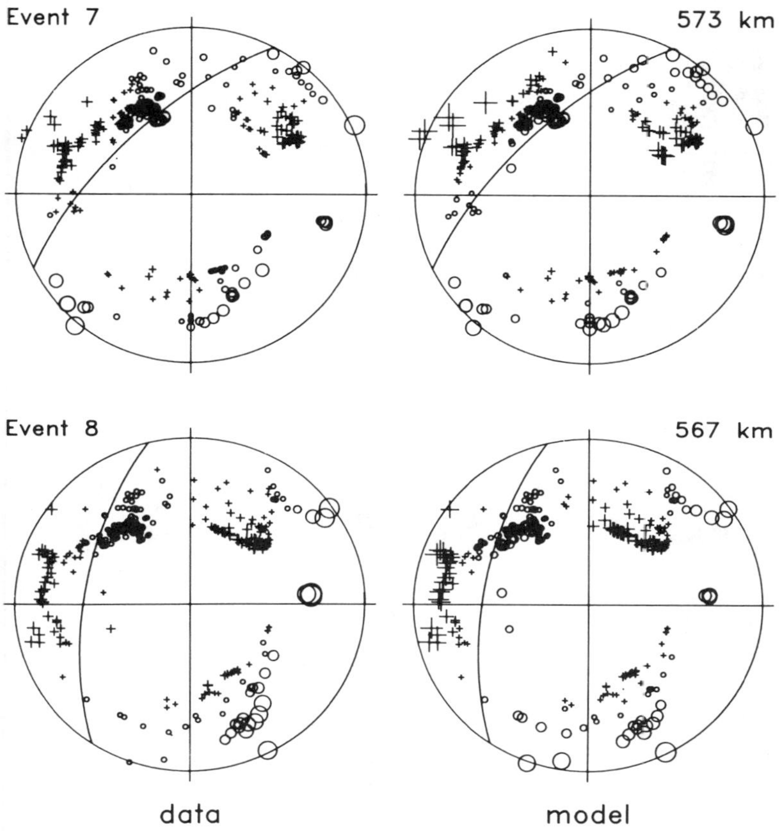

Figure 3.6. *Continued*

Although the dynamical significance of these kinks is not yet understood, it appears that the flow of material in the northwest Pacific subduction zones tends to steepen with depth, becoming nearly vertical at the base of the upper mantle.

The velocity anomaly is scaled to the thermal anomaly by an effective thermal coefficient of velocity, $\partial v_p/\partial T$. We have calibrated this parameter using the anomaly amplitudes observed for the intermediate-focus events, which are not sensitive to the slab extent below the seismic zone (Creager and Jordan, 1986a). The calibration procedure yields $-0.4 \geq \partial v_p/\partial T \geq -0.6$ m/s/°K for the models shown in Figure 3.7. In these models, the effects of

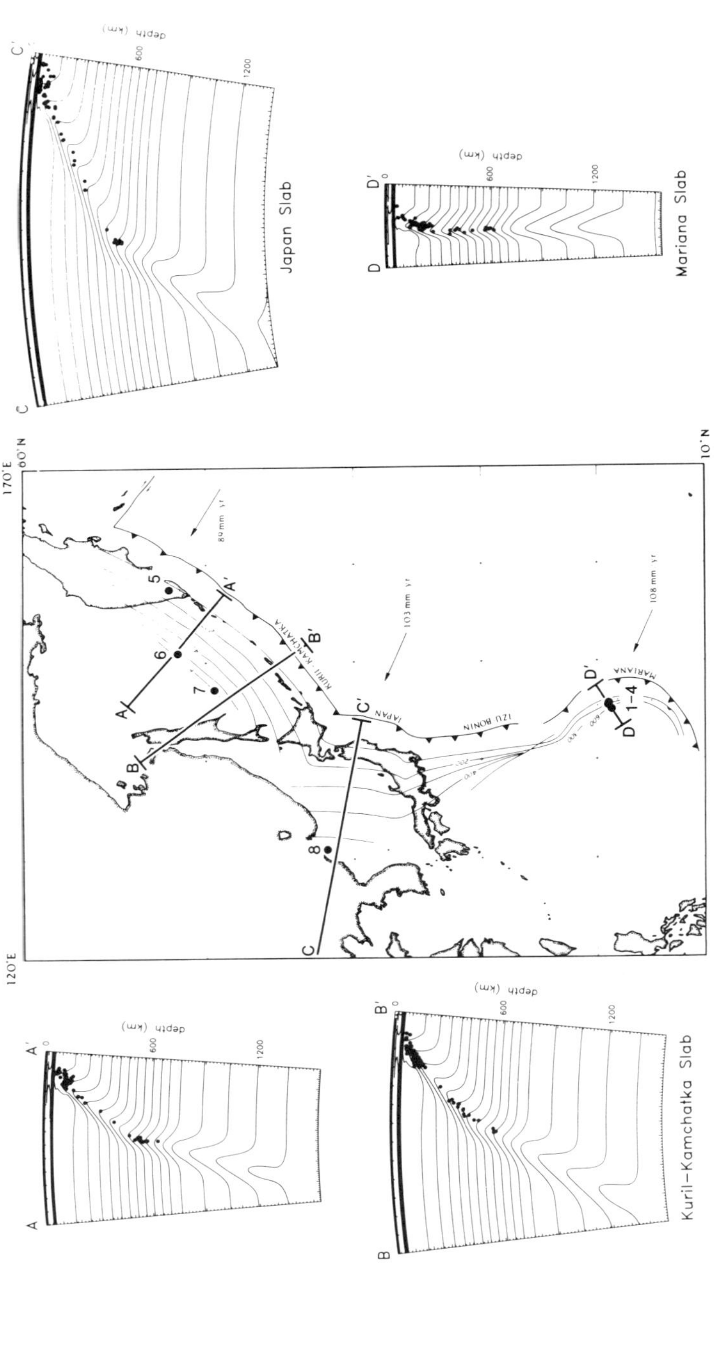
Japan Slab
Mariana Slab
Kuril–Kamchatka Slab
depth (km)
0
600
1200
A
A'
B
B'
C
C'
D
D'
170°E
60°N
120°E
10°N
KURIL–KAMCHATKA
JAPAN
IZU BONIN
MARIANA
103 mm yr
108 mm yr
1–4
5
6
7
8

phases changes on the velocity structure of the slab are not included; incorporating them reduces $\partial v_p/\partial T$ by 10–20 percent, yielding an average value of about -0.4 m/s/°K (Creager and Jordan, 1986a). Though somewhat lower than the estimate of -0.9 m/s/°K obtained by Sleep (1973) for the Aleutian slab (which would be reduced if this slab extends beyond the seismicity cutoff, as indicated by Spencer and Engdahl (1983)), our empirical thermal coefficients are consistent with recent laboratory measurements of $\partial v_p/\partial T$ on mantle minerals at high temperatures (Sumino *et al.*, 1977, 1983; Sumino, 1979; Suzuki *et al.*, 1983), as well as the value inferred from seismic and geoid modeling of large-scale lower-mantle structure (Hager *et al.*, 1985). This consistency lends credibility to the modeling assumptions. We note that it implies that $\partial v_p/\partial T$ does not vary substantially between the upper and lower mantle.

With the geometry and velocity contrast of the slab specified by the seismicity and by the travel times from intermediate-focus earthquakes, the extent of the slab is constrained by the times from the deepest events. All of the slab structures displayed in Figure 3.7 are characterized by penetration depths of about 1400 km—twice as deep as any earthquake—and from the good fits these models provide to the data in Figures 3.5 and 3.6 (as well as a number of other deep-focus events not shown here), it can be concluded that penetration depths of this order are certainly consistent with the *P*-wave travel times. To explore more rigorously the constraints the data place on the extent of the slab, the dependence of the residuals on slab length must be considered in detail. If the extent below the hypocenter is small, the amplitude of the residual-sphere anomaly is approximately proportional to this length. If it is large, however, the ray paths will turn out of the slab before they reach its terminus, and arbitrarily increasing the length of the slab will not affect the residual pattern. This phenomenon is referred to as "saturation" of the residual sphere, and the minimum level at which saturation is achieved, as measured by variance reduction, is termed the "saturation depth" (Creager and Jordan, 1986a). Obviously, the saturation depth is a *lower bound* to the actual penetration depth of the slab.

Every deep-focus earthquake we have examined in our work thus far appears to be saturated with respect to slab length (Creager and Jordan, 1986a). For the Marianas and Izu-Bonin, the estimate of the saturation depth is about 1000 km. The saturation depth increases to 1200 km or so for Kuril-Kamchatka, where deeper sources are available and the sharp velocity gradients associated with the upper slab interface are better sampled by the global network. Japan yields an intermediate value of about 1100 km. In the context of our modeling assumptions, slab penetration to at least these depths is needed to obtain fits as good as those shown in Figures 3.6 and 3.7; slab penetration to arbitrarily greater depths is consistent with the data, although not required.

The lower bound of 1000–1200 km on the penetration depths of northwest Pacific slabs inferred from the residual-sphere data appears to be robust with respect to plausible, unmodeled complications in the near-source veloicty structure. A number of these, including the effects of anisotropy, phase changes and variations in slab width, have been considered in detail by Creager and Jordan (1986a). The models shown in Figure 3.7 do not include the localized velocity anomalies associated with the major phase transitions near 400 and 650 km depth, for example, but incorporating them explicitly into the velocity structure only increases the lower bounds on penetration depths. Furthermore, because of ray-bending effects, the anomaly patterns are surprisingly insensitive to slab width, and the residuals for the deep-focus events cannot be explained by shorter but wider slab structures (see Fischer *et al.*, 1988).

3.5. Slab Penetration into the Lower Mantle

The penetration of simple, detached thermal boundary layers to depths greater than 1000 km can explain the near-source travel-time anomalies from deep-focus earthquakes in the northwest Pacific, but given the substantial implications this hypothesis has for models of mantle convection, it is important to inquire whether alternate interpretations are possible. Specifically, are our results compatible with the stratification of the Earth's solid shell into an upper-mantle and a lower-mantle convection system having little or no mass flux across the 650-km discontinuity?

The spatial relationships between upwelling and downwelling plumes in a two-layer convecting system will involve both the temperature contrasts and stresses across the interface separating the layers, as well as its dynamic topography. In situations where the stress coupling dominates the interaction, downwelling plumes in the upper layer will be associated with divergent flows on both sides of the boundary and, hence, upwelling plumes in the lower (McKenzie and Richter, 1981). This is the relationship usually envisaged in cartoons of two-layer mantle convection; e.g., Figure 1.1. To the extent that we can expect upwellings in the lower mantle to have lower than average seismic velocities, the convective patterns implied by this relationship are clearly inconsistent with the residual-sphere data, which show high seismic velocities directly beneath the seismic zones. If, however, the flow in the vicinity of the interface is dominated by the temperature contrasts, say because the two systems are mechanically decoupled by a low-viscosity boundary layer, then regions along the interface cooled by downwelling within the upper layer might be the sites of downwelling in the lower layer as well. Models with these features have been the subject of a recent numerical study by Daly and Parsons (1984), who find that alignment can be accomplished in two dimensions, providing the viscosity across the boundary increases by a factor of 10–1000. Slab-like features beneath the deep-focus zones imaged by our techniques may not be themselves lithosphere penetrating from above, but represent downwellings in a separate, lower-mantle convection system induced by the existence of cold slabs in the upper mantle.

A full test of this alternative will no doubt require three-dimensional calculations of stratified convection in a mantle with a realistic temperature structure and rheology, a formidable task. We can argue, however, that the rigorous constraints placed on such a model by the residual-sphere data tend to decrease its plausibility (Creager and Jordan, 1984, 1986a):

1. The P-velocity contrasts observed in the lower-mantle extensions of slabs have the same magnitude as their upper-mantle contrasts, and $\partial v_p/\partial T$ is observed to be a weak function of depth; the temperature contrasts in the lower mantle must therefore be nearly the same as in the upper mantle—i.e., on the order of 900°K. The thermal boundary layer at the top of the lower mantle in a two layer convecting system is expected to have a contrast of only about 300°K (Jeanloz and Richter, 1979), which is insufficient to explain the high-velocity zones beneath the deep-focus earthquakes.
2. The seismic data establish severe constraints on the geometry of the lower-mantle flow. In particular, the slab-like features below 650 km lie along the direct extrapolations of the upper-mantle slabs, matching both their strike and dip. Across the 650-km boundary, the strikes cannot differ by more than about 10°, and the features cannot be laterally offset normal to their strike by as much as, say, 100 km without destroying the fits to the residual-sphere data shown in Figures 3.5 and 3.6. Moreover, their along-strike extents appear to be very similar. For example, the residual-sphere patterns for deep events in the northern parts of the Japan and Kuril-Kamchatka seismic zones show asymmetries well modeled by the along-strike terminations predicted by the subduction geometry. Therefore, the lower-mantle flow pattern must match a non-steady upper-mantle flow pattern very precisely in both geographical coordinates over a region with a highly contorted slab geometry (where strikes vary by 70°).
3. Even supposing that a tightly coupled system with sufficient temperature contrasts could be constructed, the thermal gradients in the vicinity of the

650-km interface would be distorted by the horizontal flow of material both above and below it. A substantial decrease in the horizontal temperature gradient normal to the slab strike might be expected throughout a region whose vertical extent equals the combined thicknesses of the thermal boundary layers; say, 100 km or more. The travel-time data from deep-focus events are very sensitive to the lateral gradients in this depth interval, and they show no evidence of this distortion, nor does the seismicity. Indeed, the upper surface of what appears to be a continuous slab is well defined by the sharp transition from positive to negative residuals observed for many of the deep earthquakes (Figure 3.6).

The other seismological evidence pertaining to the subduction zones of the northwest Pacific is consistent with deep penetration. As elsewhere, the focal mechanisms of deep-focus earthquakes in this region imply down-dip compression of the slab (Katsumata and Sykes, 1969; Isacks and Molnar, 1971; Stauder and Maulchin, 1976; Yoshii, 1983), indicative of some resistance to penetration. The kinematical implications of this dynamical statement are not straightforward, however. A number of authors have shown that a relatively small increase in the viscosity at 650 km, an order of magnitude or so, is sufficient to engender down-dip compression in the deep-focus zone (O'Connell, 1977; Richter, 1979; Vassiliou *et al.*, 1984). On the basis of their extensive calculations, Vassiliou *et al.* (1984) conclude that "whether the 670 km barrier is a chemical discontinuity or a viscosity jump is not resolvable by the seismicity data alone. The stress patterns—both orientation and magnitude—are explainable, to first order, with either a chemical discontinuity or a viscosity contrast."

Down-dip compression of the slab in the 300–700 km depth range may also be caused by the forces associated with the elevation or depression of upper-mantle phase boundaries within the slab. A phase transition with a Clapeyron slope γ will be displaced from its unperturbed radius by a distance $\delta r = \gamma\, \delta T/(dP/dR)$, where δT is the temperature anomaly of the cold slab and dP/dr is the ambient pressure gradient; the displacement is positive (up) for an exothermic transition (positive γ), and negative (down) for an endothermic transition (negative γ). The Clapeyron slopes for the transformations associated with the two major seismic discontinuities, determined by laboratory experiments at the appropriate temperatures and pressures, are $\gamma_{400} =$ 3.5 MPa/°K (Suito, 1977) and $\gamma_{650} = -2$ MPa/°K (Ito and Yamada, 1982). The value of γ_{650} is the more uncertain, but Ito and Yamada's recent experimental work confirms earlier estimates (Ahrens and Syono, 1967; Bassett and Ming, 1972) and theoretical arguments (Navrotsky, 1980) that the transition is endothermic. The corresponding deflections of the phase boundaries are +0.10 km (up) and −0.05 km (down) per degree of thermal anomaly, respectively. From the thermal models of the northwest Pacific slabs derived to match the residual-sphere data, we infer that the regions of anomalous phase should extend on the order of 100 km above the 400-km discontinuity and 50 km below the 650-km discontinuity, if the transitions occur at their equilibrium positions. These perturbed phase boundaries very nicely bracket the region of anomalously intense deep-focus seismicity observed over the interval 300–700 km (Figure 3.2), and both will act to compress the slab between these depths.

The potential influence of phase transitions on slab stresses in the mid-mantle transition zone was recognized by Isacks and Molnar (1971) in their original discussion of focal-mechanism systematics, and calculations of the forces involved have been made by Turcotte and Schubert (1971) and Schubert *et al.* (1975). The large uncertainty in the nature and Clapeyron slope of the 650-km transition (Jackson *et al.*, 1974; Anderson, 1976; Jeanloz and Thompson, 1983) and the possibility that the 400-km transition may be displaced downward from its equilibrium position in the slab by kinetic effects (Sung and Burns, 1976a, 1976b; Sung, 1979; Poirier, 1981) prevent any definitive assessment of this mechanism, but the rather large negative value of γ_{650} implied by Ito and Yamada's (1982) experiments makes more plausible the hypothesis that the forces related to the phase transitions control both the seismicity levels and stress orientations of deep-focus earthquakes, at least in the relatively simple deep seismic zones of the northwest Pacific.

The fate of the Tonga slab is more controversial. It is the most seismically active slab, accounting for half of the deep-focus earthquakes reported in the ISC catalogue, and is highly contorted relative to the seismic zones of the Northwest Pacific. Giardini and Woodhouse (1984) interpreted the seismicity and focal mechanisms as evidence against penetration. When examined in detail, however, these data imply the existence of shear on nearly horizontal planes, with material deep in the slab moving southward relative to shallower material. As pointed out by Giardini and Woodhouse (1986), Tonga-Kermadec is the only major subduction zone where both the subducting and overriding plates have a significant (~50 km/My) along-strike component of velocity in the hotspot reference frame (Minster and Jordan, 1980, Figure 5). That is, the slab beneath Tonga is apparently being dragged northward through the mantle, so that its deformation may be controlled, at least in part, by this along-strike motion.

S-wave travel-time observations of Frohlich and Barazangi (1980) and our own analysis of *P*-wave residual spheres (Fischer *et al.*, 1986) are consistent with an aseismic, high-velocity zone beneath the deep-focus Tonga earthquakes. The poor distribution of seismic stations relative to these events limits the resolution of slab geometry, but Fischer *et al.* (1986) have shown that the observed ISC *P* times are well modeled by thermal slab models consistent with the seismicity. Using the thermal coefficient of velocity obtained for northwest Pacific slabs, they obtain saturation depths on the order of 1000 km, which they interpret as a lower bound on the depth of slab penetration beneath Tonga.

There are other regions where the distribution of seismicity suggests the horizontal motion of slabs near 650 km. The seismic zone has a sigmoidal curvature in the northern part of the Izu-Bonin Arc and beneath the Banda Sea (Isacks and Molnar, 1971; Cardwell and Isacks, 1978). The most extreme case is the deep-focus zone beneath the Fiji Plateau, which when projected normal to the New Hebrides Arc appears as a long, flat slab resting on the 650-km discontinuity (Isacks and Molnar, 1971; Pascal *et al.*, 1973). This seismicity appears to comprise two separate episodes of subduction, however, the southernmost events being associated with the current subduction in the New Hebrides Arc and the northern set with the now defunct Vitiaz Trench (Chase, 1971; Pascal *et al.*, 1978). When the northern set is projected on a plane normal to the Vitiaz Trench, the argument for horizontal motion of the slab at 650 km becomes less clear. In general, regions of upward curvature to the deep seismic zone occur where there are significant along-strike changes in the slab geometry; e.g., in the transition from shallow, westward-directed subduction beneath Japan to steep, southwestward-directed subduction beneath Izu-Bonin, or in the tight curvature of the Banda Sea. Given these complications and the resistance to subduction through 650 km implied by the endothermic nature of the phase transitions at this level, the association of these features with a general inability of slabs to penetrate through 650 km seems tenuous.

Indeed, the overall geometry of subduction obtained from vertical cross-sections through most deep seismic zones would appear to imply that the flow of mantle material is directed downward at 650 km. This impression has been quantified in a kinematical study of large-scale mantle flow by Hager and O'Connell (1978) and Hager *et al.* (1983). They find that the dips of the seismic zones are much better matched by the streamlines associated with models of whole-mantle convection than with models where the circulation is restricted above 650 km. Hager and O'Connell's inferences have been rejected because they incorrectly equate the instantaneous particle trajectories given by streamlines with the instantaneous locus of the slab given by seismicity (Garfunkel *et al.*, 1986), but their qualitative conclusions are in agreement with the seismic evidence.

3.6. Implications for Mantle Convection and the Nature of the 650-km Discontinuity

Lithospheric slab penetration into the lower mantle to depths exceeding 1000 km provides a simple and complete explanation of the near-source travel-time anomalies observed for deep-focus earthquakes in the northwest Pacific and Tonga, and it is consistent with the constraints on the subduction process in most other deep-focus seismic zones. The excellent fits achieved by uncomplicated thermal-boundary layer models to the complicated residual-sphere patterns shown in Figures 3.5 and 3.6 cannot be ignored in discussions of mantle dynamics. In our opinion, the seismological evidence is sufficiently strong that models of mantle convection that do not conform with this behavior should be rejected, or at least severely downweighted in any hierarchy of working hypotheses.

Rigorous stratification of the mantle into a two-layer, convecting system with a boundary at 650 km is one such model. Our results imply the flux of mantle material across this boundary must be large. If we consider only the Pacific subduction zones from the Marianas to Tonga and adopt a nominal plate thickness of 125 km (Parsons and Sclater, 1977), the slab flux alone, exclusive of any entrained flow, amounts to 100 km^3/yr (cf. Figure 3.1). If we extend this calculation to include all oceanic subduction zones and therefore the entire plate flux, we obtain approximately 300 km^3/yr. At this rate, the volume of the upper mantle is thrust into the lower mantle about once every billion years. Entrainment could increase the flux severalfold. Exchange rates of this magnitude cannot be easily reconciled with models which isolate the entire lower mantle as a primitive geochemical reservoir (e.g., O'Nions *et al.*, 1979; Jacobsen and Wasserberg, 1979).

Observations of high-frequency, high-amplitude $P'P'$ phases reflected from the bottom side of the 650-km discontinuity have been used as evidence that the velocity jump at this depth is caused by an abrupt change in major-element chemistry, rather than a change of phase (Lees *et al.*, 1983). As discussed by Creager and Jordan (1984, 1986a), the data on the aseismic extensions of slabs do not favor this hypothesis. The P-velocity increase at 650 km is approximately 4 percent (Given and Helmberger, 1980; Burdick, 1981; Walck, 1984b), which is about equal to the difference between the average velocity in the interior of the slab and the ambient upper mantle due to the thermal contrasts. If the entire increase at 650 km were caused by a compositional change, then the velocity contrast between the slab interior and the ambient lower mantle would be reduced by 4 percent, assuming the slab has an upper-mantle composition. On the other hand, the residual-sphere data do not allow the slab velocity contrast in the lower mantle to be significantly less than in the upper mantle, so we would then be forced to require that $\partial v_p/\partial T$ increases by a factor of two across the 650-km transition. As discussed above, both laboratory data (Sumino *et al.*, 1977, 1983; Sumino, 1979; Bonczar and Graham, 1982; Suzuki *et al.*, 1983) and direct comparisons of our results with observations of large-scale lower-mantle heterogeneities (Hager *et al.*, 1985) argue against an increase of this magnitude.

We should emphasize that these arguments restrict only the chemical change manifested by an increase in seismic velocities at 650 km. Compositional changes which, say, increase density but decrease compressional velocity (e.g., an increase in iron content (Bonczar and Graham, 1982)) are not limited by our data, insofar as they are consistent with deep slab penetration. Christensen and Yuen (1984) have investigated a range of time-dependent convection models and conclude that a compositional density jump of a few percent would not impede penetration through the 650-km discontinuity, nor would an endothermic phase change with a Clapeyron slope on the order of that derived by Ito and Yamada (1982) (-2 MPa/°K). These hypothetical properties are not excluded by slab penetration, but they could have significant dynamical effects, especially on convection at higher wavenumbers. One can envisage, for example, a mantle where the anomalous properties of this transition impede the average vertical flux across 650 km relative to other levels, although not the penetration of old lithosphere. Such a system might have a structure very different from simple "whole-mantle" convection.

Our current methods are unable to resolve aseismic extensions of slabs much below about 1000 km, which is only one-third the depth of the mantle. Slab penetration to greater depths is consistent with our data, though speculative, so we cannot exclude the existence of stratified convection with boundary layers at this depth or below. Given the slab geometry shown in Figure 3.7, however, the plausibility of convective stratification in the upper half of the mantle must be questioned. Northwest Pacific slabs all have aseismic extensions whose dip angles are greater than or equal to the dips of the seismic zones, so that their net curvature is concave downward. The change in dip with depth is particularly noteworthy in the cases of Kuril-Kamchatka and Japan, which are characterized by an abrupt change from shallow to steep dips at about 500 km. The dynamical significance of these transition-zone kinks is not understood, but their kinematical implications are important. Garfunkel (1975) and Garfunkel *et al.* (1986) have argued that slab migration is generally retrograde (i.e., opposite to the subduction direction) at rates of 10–25 km/Ma and, therefore, that migrating slabs do not separate mantle convection cells, but instead are entrained in the circulations beneath the overriding plates. If true, then the motions of slab particles near the base of the convecting layer should be concave upward. The fact that slabs tend to dip more steeply below the seismic zone suggests that the convecting layer is thicker than 1000 km.

4. Structure Beneath the Plates

A three-dimensional map of seismic wave speeds in the mantle will expose the structure of convection beneath the plates, just as a shadow-graph reveals the structure of convection in a laboratory tank. Unlike the clear view obtained in laboratory situations, however, the seismological pictures of mantle convection are strongly distorted by the heterogeneous structure of the plates themselves, and the resolution is limited by the irregular distributions of sources and receivers. The interpretation is further complicated by the need to separate the effects of the temperature variations from compositional heterogeneity and mineralogical anisotropy. Nevertheless, the seismic imaging of interior structure should be able to map a variety of convective features, including the upwelling plumes needed to balance downward plate flux, small-scale convection in the upper mantle beneath the plates, thermal and chemical boundary layers at the core-mantle interface, topography dynamically imposed on the mantle phase transitions, any internal boundary layers separating chemically stratified regions of the mantle, and, of course, the downwellings driven by descending slabs of oceanic lithosphere.

As reviewed in Section 3, the delineation of features in the last category has relied upon the constraints on slab geometry provided by deep seismicity and the understanding of slab structure derived from thermal-boundary-layer theory. Models of mantle convection are not sufficiently advanced to predict the geographic distribution, depth extent, or even the characteristic horizontal dimensions of the other features, so that seismologists must start from scratch, detecting the structures and resolving them from noise, as well as determining their average properties. Any feasible presentation of a structural model must involve a limited number of parameters, and both discrete-block grids and orthogonal-function expansions have been employed. In the latter, a perturbation to some spherically symmetric reference Earth model $\mathbf{m}_0(r)$ is represented as a sum over fully normalized surface spherical harmonics:

$$\mathbf{m}(r,\theta,\phi) = \mathbf{m}_0(r) + \sum_{s=0}^{\infty} \sum_{t=-s}^{+s} \delta\mathbf{m}_s^t(r) Y_s^t(\theta,\phi) \tag{4.1}$$

As written here in terms of the complex surface harmonics Y_s^t (Edmonds, 1960), the coefficients $\delta\mathbf{m}_s^t$ are complex-valued n-tuples of functions satisfying the symmetry conditions $\delta\mathbf{m}_s^{-t} = (-1)^t \delta\mathbf{m}_s^{t*}$, so that there are $2s + 1$ independent, real-valued n-tuples per degree. Horizontal resolution is limited by truncating the expansion at some maximum wavenumber s_{max}. Vertical resolution is limited either by truncating an expansion in vertical wavenumber or discretizing the radial coordinate.

A more fundamental choice involves which functions to include in $\delta\mathbf{m}$. For an isotropic, elastic perturbation, n equals 3, corresponding to variations in the P velocity, S velocity, and density: $\delta\mathbf{m} = (\delta v_p, \delta v_s, \delta\rho)$. However, velocity anisotropy is potentially important, especially near the surface, and additional functions must be introduced to account for anisotropic heterogeneities (e.g., Crampin *et al.*, 1984; Tanimoto and Anderson, 1985). In the simple case of hexagonal symmetry, for example, five more functions are needed (three elastic parameters and two orientation parameters). Presently available data sets are inadequate to resolve the details of such complex models, so that most workers take the perturbations to be isotropic, assuming anisotropic effects can be averaged out. If this assumption is false, the resulting models can be biased.

Ideally, all of the information on the seismogram should be used to construct three-dimensional Earth models. Although significant progress has been made on the problem of inverting the entire seismogram for Earth structure (Dziewonski and Steim, 1982; Lerner-Lam and Jordan, 1983; Woodhouse and Dziewonski, 1984), this is not possible. Simply put, much of the information on the seismogram—an example would be the amplitudes of high-frequency body waves—is too sensitive to the small-scale details of source and Earth structure not included in feasible model parameterizations. For this reason, most studies have relied on the phase information derived from coherent wave groups. The basic data functionals containing this information are the travel times of body waves, the phase velocities of surface waves, and the apparent center frequencies of free-oscillation multiplets. These data functionals satisfy variational principles (those of Fermat and Rayleigh), which allow their dependence on model perturbations to be linearized and, just as important, show how they average higher wavenumber heterogeneity. In this section we will be concerned with the constraints placed by these data on low-wavenumber aspherical variations, and our discussion will be focused on global models, rather than regional studies.

4.1. Upper Mantle

Long-period seismograms from large, shallow-focus earthquakes are dominated by the successive passages of fundamental-mode surface waves propagating along the great circles connecting the sources and receivers. Each group can be identified by its wave type (R for PSV-polarized Rayleigh waves; G for SH-polarized Love waves) and orbit number n. Groups traveling the minor arc between source and receiver are designated by odd orbit numbers ($n = 1, 3, 5, \ldots$), whereas those traveling the major arc have even orbit numbers ($n = 2, 4, 6, \ldots$). By observing the phase differences between two groups of the same type and orbital direction on the same seismogram (e.g., R_1 and R_3, or G_2 and G_6), one can obtain great-circle averages of the local phase velocities that are insensitive to the source properties and instrument response (Satô, 1958). Although they can be very precisely measured down to frequencies of 5 mHz or so, these averages constrain only the coefficients in equation (4.1) with s even, corresponding to an "even-parity" perturbation that is symmetric under reflection through the center of the Earth (Backus, 1964). To first order, the odd-parity heterogeneity (s odd) does not affect the phase difference between successive orbits.

Measurements of great-circle phase velocities were used to construct the first global models of upper-mantle heterogeneity (Toksöz and Anderson, 1966; Kanamori, 1970; Dziewonski, 1971; Okal, 1977). The Earth's surface was divided into a number of tectonic regions within which the structure was presumed to be laterally homogeneous. Dispersion curves for the individual tectonic regions were isolated by a regression of the phase-velocity data against the fraction of the path length traversing each structural type, and these curves were inverted for regional velocity profiles. Toksöz and Anderson's (1966) original study used a tectonic regionalization comprising only three crustal types (stable continents, ocean basins, and orogenic zones), whereas Okal's (1977) analysis incorporated seven (shields, mountain belts, trenches and marginal basins, and four oceanic regions based on crustal age). The six-bin regionalization GTR1 (Jordan, 1981a) is shown in Figure 2.1.

Adopting an *a priori* regionalization eliminates the even-parity ambiguity intrinsic to the great-circle data and reduces the unknowns to a manageable number, but it also precludes the detection and analysis of heterogeneity not correlated with surface tectonics. The importance of sub-tectospheric variations on the dispersion of long-period waves was first quantified in a free-oscillation study by Masters *et al.* (1982). At very low frequencies (less than, say, 5 mHz), the seismogram is most conveniently represented as a sum over discrete normal modes, which for a spherically symmetric Earth are collapsed into degenerate multiplets. The effect of aspherical heterogeneity is to split these multiplets (Woodhouse and Dahlen, 1978). Because the multiplet spectra are broadened by attenuation, the individual spectral lines are difficult to resolve, and the primary effect of asphericity is to shift the apparent center frequency of a multiplet away from its spherically symmetric value by an amount that is linearly related to the heterogeneity (Jordan, 1978b). This frequency shift can be accurately measured (Silver and Jordan, 1981) and, at high enough wavenumber, can be expressed as an integral of the local eigenfrequency around the great-circle connecting the source and receiver (Jordan, 1978b). Hence, the apparent center frequencies of fundamental-mode standing waves provide the same kind of information as the great-circle dispersion of fundamental-mode traveling waves; in particular, they are sensitive to only even-parity heterogeneity.

Using the newly installed IDA network of digital seismometers, Masters *et al.* (1982) made center-frequency measurements on nearly 4,000 spheroidal-mode multiplets, sampling 557 paths in the frequency range 1.4–5.2 mHz. Automated processing of digital records permitted a dramatic increase in the quantity and quality of the data (consider that Okal's (1977) study of just a few years earlier was based on only 29 paths). The free-oscillation data compiled by Masters *et al.* (1982) indicated a strong quadrupolar ($s = 2$) component of mantle heterogeneity. The heterogeneity correlated positively with the geoid but only weakly with Jordan's (1981b) GTR1 regionalization (Figure 2.1); the degree-two model gave a total variance reduction of 61 percent, compared with only 24 percent obtained from the regionalization. Based on the scaling of the frequency shifts with wavenumber, Masters *et al.* (1982) argued that this heterogeneity is predominantly sub-tectospheric, and their preferred model localized it in the mantle transition zone between about 400 and 650 km depth. They speculated that it may be related to a large-scale component of mantle convection.

The large amount of digital seismic data which became available in the first part of the present decade also motivated systematic studies of fundamental-mode dispersion using traveling-wave techniques. Nakanishi and Anderson (1983) obtained great-circle phase velocities in the 3–10-mHz band for 198 Love-wave paths and 253 Rayleigh-wave paths. They confirmed the existence of the $s = 2$ heterogeneity discovered by Masters *et al.* (1982) but, following Kawakatsu (1983) and Souriau and Souriau (1983), interpreted this feature as the degree-two component of trench-related heterogeneity, possibly concentrated at shallow (<200 km) depths.

If the location, mechanism and time function of an earthquake are known, phase velocities can be measured over the major and minor arcs connecting the source to the receiver, as well as over full great-circle paths. Nakanishi and Anderson (1984) used this single-station method to recover information on both the even-parity and odd-parity components of upper-mantle structure. They constructed maps of phase-velocity variations for Love and Rayleigh waves out to degree and order six from an analysis of about 200 paths. The Nakanishi-Anderson fundamental-mode data set has been inverted for anisotropic, laterally heterogeneous models of the upper mantle by Nataf *et al.* (1984, 1986). Tanimoto and Anderson (1985) analyzed a somewhat larger data set for lateral heterogeneity and azimuthal anisotropy.

The most complete analysis of upper-mantle aspherical structure published thus far is the study by Woodhouse and Dziewonski (1984; see also Dziewonski and Woodhouse, 1987). Employing a waveform-inversion technique based on high-frequency asymptotic approximations, they inverted some 2000 seismograms over 870 paths for an isotropic aspherical model of shear-velocity perturbations complete to degree and order eight ($s_{max} = 8$). Radial structure was represented by expanding each harmonic coefficient $\delta\mathbf{m}_s^t(r)$ as a cubic polynomial in r, so that their parameterization comprised a total of 243 scalar quantities. They presented two models: M84A, for which the data were uncorrected for lateral variations in crustal structure, and M84C, for which the data were corrected for a two-region (continent-ocean) crust. Although the seismograms used by Woodhouse and Dziewonski (1984) are dominated by the large-amplitude groups of the fundamental Love and Rayleigh waves, their waveform inversion scheme does attempt to fit the body-wave and higher-mode signals to frequencies as high as 22 mHz, and the resolving power of their data should be significantly better than data sets comprising fundamental modes alone.

Figure 4.1 shows the M84C structure averaged over the upper 400 km of the mantle, obtained by integrating the S-wave travel-time perturbations along vertical ray paths from 40 km to 400 km. As noted by Woodhouse and Dziewonski (1984), the structure in this depth interval is dominated by the high velocities associated with continental cratons and the low velocities corresponding to actively spreading mid-ocean ridges; i.e., tectospheric heterogeneity of the sort discussed in Section 2. In an attempt to separate tectospheric from sub-tectospheric heterogeneity, we have projected the perturbation in Figure 4.1 onto the GTR1 regionalization. The six geographical functions of GTR1 were filtered to degree eight prior to performing the projection (Lerner-Lam and Jordan, 1987). The resulting map (Figure 4.2) shows the part of the M84C travel-time perturbation that is correlated with large-scale surface tectonics and therefore plausibly attributable to tectospheric structure. The average one-way S-time anomalies obtained by this procedure are given for the individual regions in Table 3.1. The S anomalies in the table are generally consistent with recent studies of regional upper-

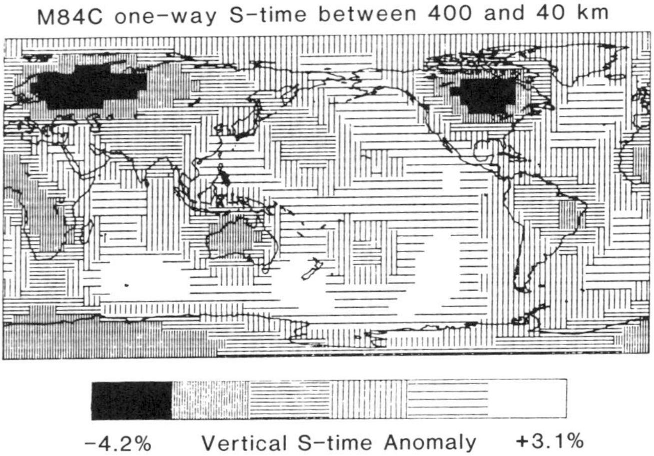

Figure 4.1. Degree-eight spherical structure averaged over upper 400 km of the mantle, obtained by integrating the S-wave travel-time perturbations from Woodhouse and Dziewonski's (1984) M84C model along vertical ray paths from 40 km to 400 km. Darker areas are faster regions.

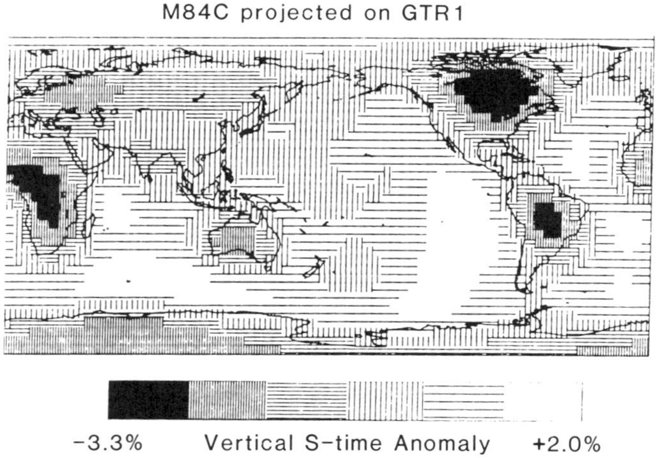

Figure 4.2. Projection of the M84C travel-time anomaly map onto the tectonic regionalization GTR1 shown in Figure 2.1. Spherical harmonic expansions of the six geographical functions of GTR1 were truncated at degree eight prior to performing the projection. Darker areas are faster regions.

Table 3.1.

Region	Description	Areal fraction	*S* anomaly (s)
A	Young oceans (<25 My)	.13	1.4
B	Intermediate-age oceans (25–100 My)	.35	0.4
C	Old oceans (>100 My)	.13	0.0
Q	Phanerozoic orogenic zones	.22	−0.2
P	Phanerozoic platforms	.10	−1.3
S	Precambrian shields and platforms	.07	−2.7

mantle structure derived by high-resolution waveform modeling. For example, the vertical *S*-time difference in the depth interval 40–400 km calculated from Grand and Helmberger's (1984a, 1984b) models ATL (western Atlantic) and SNA (stable North America) is +2.3 s, which compares with the C − (P + S)/2 difference of +2.0 s computed for M84C. The latter value is essentially identical to the difference obtained from Lerner-Lam and Jordan's (1987) models PA2 (western Pacific) and EU2 (northern Eurasia), discussed in Section 2.4.

Figure 4.3 displays the difference between Figures 4.1 and 4.2. Removing the components of M84C correlated with the GTR1 regionalization reduces

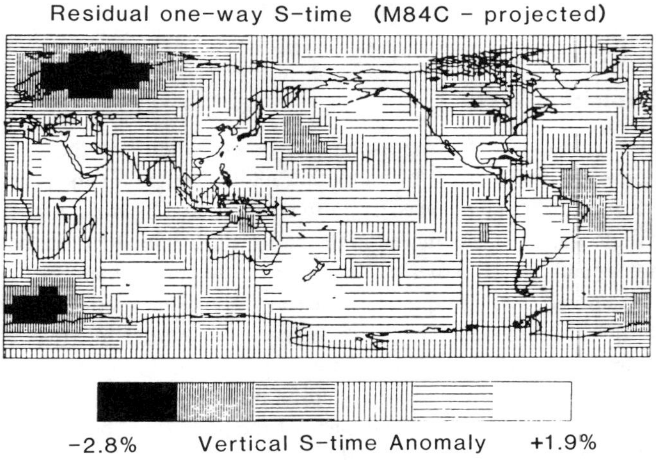

Figure 4.3. Degree-eight pattern obtained by subtracting Figure 4.2 from Figure 4.1. Darker areas are faster regions.

the amplitude of the pattern by about a factor of two and shifts it to higher wavenumber. To the extent that Figure 4.2 adequately represents the thermal boundary layer structure of the plates, this subtraction should produce an estimate of sub-tectospheric heterogeneity. However, the ability of the Woodhouse-Dziewonski data set to resolve low-amplitude, short-wavelength features has not been demonstrated. At high wavenumbers, the approximations used in their waveform-inversion procedure are less valid (Jordan, 1978b; Woodhouse and Girnius, 1982). Moreover, since M84C was parameterized by a spherical harmonic expansion truncated at $s_{max} = 8$ without tapering, some of the energy in the high-wavenumber oscillations may result from sidelobe ringing. Given these caveats and the uncertainties associated with stripping off plate structure, any interpretation of Figure 4.3 in terms of convective processes is risky. For example, a dominant feature on the residual map is the fairly large negative travel-time anomaly centered on the Ural Mountains and Barents Sea. Much of this area is classified in GTR1 as an orogenic zone (Region Q) and therefore is assigned the fairly low velocity derived by averaging it with other tectonically active areas, whereas it may actually have a plate structure more characteristic of continental shields. Some of the positive features, such as the Red Sea anomaly discussed by Nakanishi and Anderson (1983) and Woodhouse and Dziewonski (1984), may be manifestations of convective upwellings, but these features are rather small, localized anomalies. We conclude that at the present resolution of global seismic studies, lateral heterogeneity in the upper 400 km of the mantle is dominated by tectospheric structure.

4.2. Transition Zone

The same is not true for the aspherical heterogeneity of the mid-mantle transition zone. In Figure 4.4 we plot the M84C vertical S-time perturbation for the 400–650-km layer. The structure is dominated by two antipodal high-velocity anomalies, one centered in the westernmost Pacific near the equator and one beneath South America. The degree-two component of this structure is compared with the quadrupolar transition-zone model of Masters *et al.* (1982) in Figure 4.5. The Y_2^0 coefficient is smaller in M84C, but the models are otherwise very similar in both pattern and amplitude. Since the Woodhouse-Dziewonski inversion method provides good constraints on the odd-parity as well as even-parity parts of the aspherical heterogeneity, we conclude that both of the equatorial high-velocity regions are genuine features. In particular, it appears that the South American anomaly can not be explained as simply a "far-side" artifact of shallow trench-related hetero-

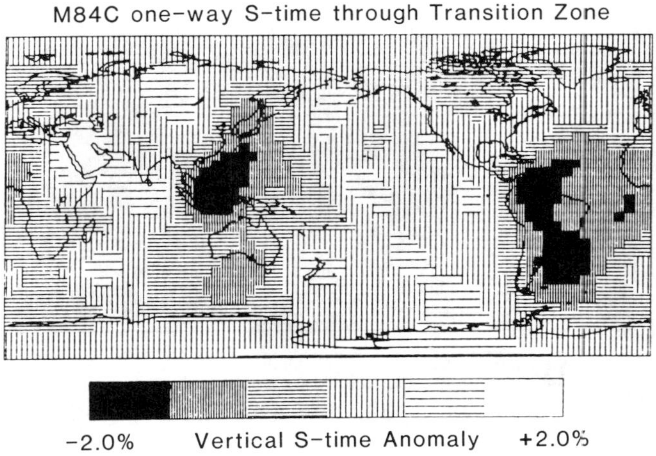

Figure 4.4. Degree-eight spherical structure averaged over the transition zone, obtained by integrating the *S*-wave travel-time perturbations from Woodhouse and Dziewonski's (1984) M84C model along vertical ray paths from 400 km to 650 km. Darker areas are faster regions.

geneity localized in the western Pacific, as advocated by Kawakatsu (1983), Souriau and Souriau (1983) and Nakanishi and Anderson (1983).

The high transition-zone velocities are most likely related to the downward flux of cold material in subduction zones. The western Pacific anomaly shown in Figure 4.5 encompasses trenches which account for about half of the global plate flux back into the mantle (Figure 3.1), whereas the source of cold material for the South American feature is presumably the slab subducted in the Peru-Chile trench. This hypothesis is supported by the fact that the transition-zone anomalies correlate positively with the degree-two geoid (Masters *et al.*, 1982). Moreover, Forte and Peltier (1987) have calculated the surface divergence of the tangential velocity field described by Minster and Jordan's (1978) AM1-2 plate-motion model, and they find that the quadrupolar component of this field is very similar to the Masters *et al.* (1982) anomaly pattern (Figure 4.6). At this wavenumber, the zones of convergence correspond to the high-velocity, high-density regions and the zones of divergence to low-velocity, low-density regions.

The dominance of the second-degree harmonic in the low-wavenumber spectrum of transition-zone heterogeneity implies some sort of large-scale organization of the flow through this region. Busse's (1983) proposal for

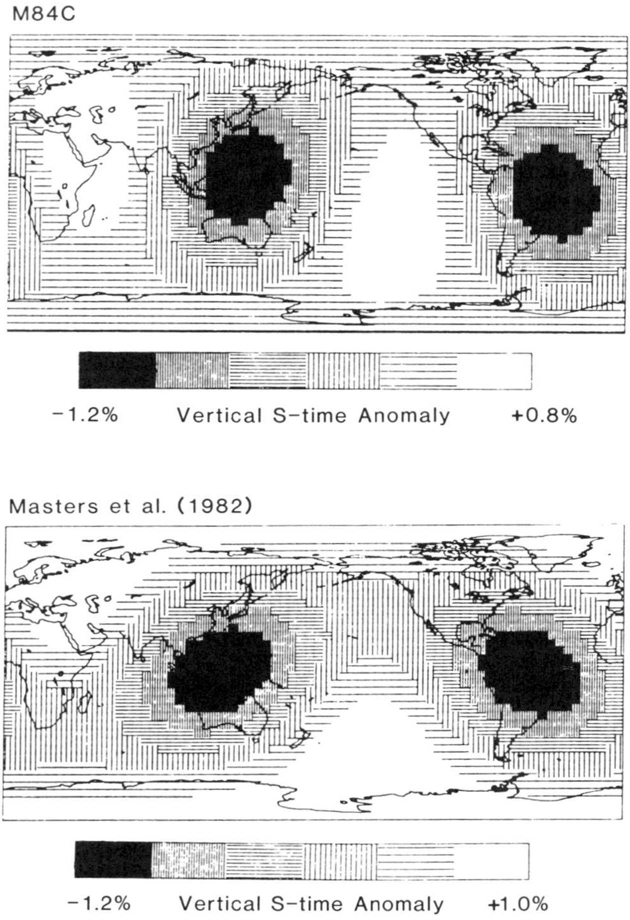

Figure 4.5. Comparison of degree-two transition-zone structure derived from Woodhouse and Dziewonski's (1984) M84C model (upper panel) with the transition-zone model of Masters *et al.* (1982) (lower panel). *S*-wave travel-time perturbations computed by integrating models along vertical ray paths from 400 km to 650 km. Darker areas are faster regions.

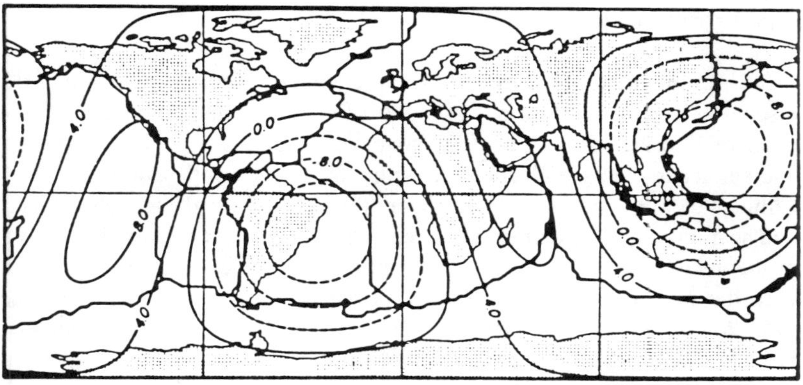

Figure 4.6. Degree-two component of surface divergence computed by Forte and Peltier (1987) from Minster and Jordan's (1978) plate-motion model. The close correspondence of convergence (dashed contours) with the high-velocity regions of Figure 4.5 suggests that the transition-zone anomalies seen in the seismic data are dominated by cold downwellings in the large-scale flow.

opposing quadrupolar convection patterns in a chemically stratified mantle, illustrated in Figure 1.1b, is one possible scheme. The most obvious way to discriminate this kind of model from those involving deeper circulation of upper-mantle material is to examine the relationships among structures on both sides of the 650-km discontinuity. For subduction zones with deep-focus seismicity, this can be done locally using the residual-sphere techniques described in Section 3. The results provide convincing evidence that the old, cold slabs continue across the discontinuity.

The nature of the flow through the transition zone on a larger, more global scale remains unknown, however, and additional information regarding seismic heterogeneity in the vicinity of 650 km is badly needed. Unfortunately, mapping the variations of seismic wave speeds at this depth is difficult. In the frequency range of typical surface-wave experiments (>5 mHz), the ability of the data to resolve structure below about 400 km is poor. Fundamental-mode free-oscillation data of the sort used by Masters *et al.* (1982) are affected by inhomogeneities throughout the mantle, but they provide only averages over broad depth ranges. The detailed mapping of the interface between the upper and lower mantle will require extensive sets of body-wave and higher-mode data. Progress in this direction has been recently reported by Grand (1987) and Dziewonski and Woodhouse (1987).

4.3. Lower Mantle

The lower mantle of the Earth, Bullen's region D, extends from the seismic discontinuity at a depth of about 650 km to the core-mantle boundary at a depth of 2886 km. There is some evidence for small discontinuities or at least anomalous gradients at several radii within the lower mantle (Johnson, 1969; Chinnery, 1969), especially in its upper 400 km (Muirhead and Hales, 1980) and near its base (Lay and Helmberger, 1983a, 1983b). But throughout most of the lower mantle, the radial variation of seismic velocities is consistent with the near-adiabatic compression of a homogeneous material (Birch, 1952; Jordan and Anderson, 1974). It has long been appreciated by seismologists that, at low wavenumbers, the RMS amplitude of lateral heterogeneity in the central part of the lower mantle is much smaller than in the upper mantle. This is not surprising from the point of view of mantle convection. To the extent that the seismological structure is largely determined by thermobaric variations owing to high Rayleigh number convection, the heterogeneity in the mid-mantle is expected to be dominated by downwelling plumes from the transition zone and upwelling plumes from the core-mantle boundary, giving rise to a relatively "white" horizontal wavenumber spectrum. In contrast, the boundary layers should have "red" spectra, with power concentrated at the low wavenumbers corresponding to the characteristic horizontal scales of the boundary layer flow (Jarvis and Peltier, 1986).

The most obvious manifestations of downwellings are the slab anomalies discussed in Section 3, and there is now good evidence that these account for much of the aspherical structure in the depth range 700–1700 km. The first detailed study of a velocity anomaly in the lower mantle related to convective downwelling was by Jordan and Lynn (1974), who compiled *PcP-P* and *ScS-S* differential travel times at North American stations for two deep-focus earthquakes in South America. By using differential times they were able to reduce the effects of upper-mantle heterogeneity and isolate the signature of a high-velocity region at depths of 800–1400 km beneath the Caribbean, which Jordan and Lynn (1974) and Jordan (1975b) associated with subduction at the Middle American trench. The anomaly has a horizontal (east-west) scale of 500–1000 km and an estimated velocity contrast of 1 percent or so, implying a temperature contrast on the order of 200°C.

Additional evidence for the Caribbean anomaly and other localized features in the lower mantle was obtained in a more comprehensive travel-time study by Lay (1983). More recently, Grand (1987) inverted an extensive set of travel times from multiply reflected *S*-waves to image of the mantle beneath North and Central America. In his maps, the Caribbean anomaly extends downward to at least 1700 km and northward beneath the United States, in a position roughly coincident with the locus of subduction during the early opening of the Atlantic. The analysis of regional travel times by Lay (1983) and global inversions of body-wave data by Dziewonski and Woodhouse (1987) indicate that this high-velocity feature continues around the periphery of the Pacific into Eurasia. Although these studies have not yet attained sufficient horizontal resolution to delineate the precise relationship of the high-velocity anomaly to subducted slabs, they do corroborate the evidence for deep slab flux presented in Section 3.

Jordan and Lynn (1974) found that the ratio of S-time perturbations to P-time perturbations for the Caribbean anomaly is about 3.5:1. The same ratio was obtained for lower-mantle slab anomalies in the northwest Pacific by Creager and Jordan (1984), and a similar value is required to satisfy the P and S data pertaining to larger scale variations in the lower mantle (Dziewonski and Woodhouse, 1987; Anderson, 1987). This ratio is close to that computed from S and P station anomalies, which sample upper-mantle inhomogeneities (Doyle and Hales, 1967; Hales and Herrin, 1972), but it is larger than the one derived from laboratory measurements of temperature derivatives. If the lower-mantle velocity perturbations are caused solely by temperature variations, then the travel-time data imply

$$(\partial \ln v_s/\partial T)/(\partial \ln v_p/\partial T) \approx 2 \tag{4.2}$$

The ratio calculated for candidate mantle mineralogies from the laboratory data of Anderson *et al.* (1968) is about 40 percent less. Hales and Doyle (1967) and Hales and Herrin (1972) suggested that, in the case of the station anomalies, the discrepancy between the seismic and laboratory estimates of $(\partial \ln v_s/\partial \ln v_p)$ could be indicative of partial melting in the upper mantle. Such a mechanism is much less plausible (though not completely excluded) for lower-mantle heterogeneity. Anderson (1987) points out that, whatever the mechanism, (4.2) indicates that anharmonic effects are more important in shear than in compression.

Although S-waves have been useful in delineating local and regional anomalies breath subduction zones, most of what is known about the aspherical structure of the lower mantle on a global scale comes from the analysis of P-wave travel times. In a pioneering study, Julian and Sengupta (1973) demonstrated that the scatter in P times from deep-focus earthquakes increases considerably beyond about 80°, implying that the amplitude of the lateral variations grows as the core-mantle boundary is approached. This hypothesis was confirmed by Sengupta and Toksöz (1976), who inverted 1490 P times and 314 S times for a lower-mantle model parameterized by velocity perturbations in 1400 blocks (individual block dimensions were 10° × 10° × 500-km). The sparse distribution of sources and receivers and the relatively small number of times severely limited the resolving power of Sengupta's data set, however.

Dziewonski *et al.* (1977) improved the sampling of lower-mantle strucutre considerably by using data from bulletins of the International Seismological Centre (ISC). The ISC compiles *P* times from a global network of several thousand stations for routine hypocenter determination, and over 1000 of these stations have reported enough teleseismic data to allow the estimation of station anomalies (Poupinet, 1979; Dziewonski and Anderson, 1983). Dziewonski *et al.* (1977) inverted nearly 700,000 *P* times for a model of the lower mantle comprising 120 blocks, each having approximate dimensions of 36° × 60° × 750-km. They did not relocate events or correct for station anomalies. The analysis of the ISC data was refined by Dziewonski (1984) using a time-term algorithm that iteratively perturbed lower-mantle structure, event locations, and station anomalies. He represented lower-mantle velocity perturbations to angular degree 6 and radial degree 4 using the same orthogonal-function expansion employed in the upper-mantle study of Woodhouse and Dziewonski (1984).

The compressional-velocity models of Dziewonski *et al.* (1977) and Dziewonski (1984) show a significant correlation with the geoid at angular degrees 2 and 3. Hager *et al.* (1985) have investigated this correlation by computing the dynamic response of a viscously stratified mantle (Richards and Hager, 1984). Their analysis assumes the density perturbations driving the flow are proportional to the seismically observed velocity perturbations, described by the lower-mantle models of Dziewonski (1984) and Clayton and Comer (1983). (Clayton and Comer's unpublished model is a discrete-block structure derived by a tomographic inversion of ISC *P* times.) Hager *et al.* (1985) correct the observed geoid for slab-related density anomalies not resolved by the global seismic models (Hager, 1984). They obtain a good match between the theoretical and observed geoids for mantle models with no chemical stratification, but they infer a viscosity increase by a factor of 10–30 at a depth of about 650 km, consistent with Hager's (1984) conclusion that an increase of 30–100 is needed to match slab-related geoid anomalies. Their calculations account for the dynamically supported topography at the free surface and the core-mantle boundary, with their preferred degree-3 model predicting total topographic amplitudes of about 1.4 km and 3 km, respectively.

The dynamic-response calculations published by Hager *et al.* (1985) do not account for the effects of upper-mantle and transition-zone structure on the slab-corrected geoid, nor do they allow for lateral variations in viscosity, which may be necessary to establish that average viscosity increases substantially across the mid-mantle transition zone. Hager's hypothesis should be testable by seismic methods, however, since it has considerable implications for the kinematics of mantle flow through this region. For example, the distortion of descending slabs owing to any viscosity increase must be small enough to satisfy the constraints imposed by the residual-sphere data discussed in Section 3. Preliminary calculations by Fischer *et al.* (1988) indicate that the compressive thickening of northwest Pacific slabs normal to their strike cannot be much more than a factor of three or so during their transit through the depth interval 500–800 km.

On a more global scale, the resolution gap between the upper-mantle and lower-mantle modeling techniques is again a problem. In Dziewonski's (1984) model L02.56, the power in all harmonics is highest at 650-km discontinuity and decreases rapidly with depth, reaching a minimum near 2000 km before rising towards the core-mantle boundary. The increase at the base of the lower mantle appears to be real, but as Dziewonski (1984) notes, the peak at the top of the lower mantle may be due to the inadequacy of the source and station corrections to account for the structure above 650 km. Problems with these corrections may explain why the seismic models of Dziewonski (1984) and Clayton and Comer (1983) do not agree very well with each other or correlate well with the geoid above $s = 3$. At higher wavenumbers, the dynamic response comparisons will require seismic structures that are continuous through the transition zone. Even at the lowest wavenumbers, such self-consistent structures may be necessary to properly interpret the geoid constraints.

4.4. Core-Mantle Boundary

At the base of the mantle, just above the core-mantle boundary (CMB), lies Bullen's Region D″, where the vertical gradients in the seismic velocities are small and the horizontal gradients are large relative to the mid-mantle. Early workers invoked compositional gradients—e.g., an increase in iron content with depth—to account for the anomalous seismic velocity profile in D″ (Bullen, 1950), but modern explanations of these gradients have tended to focus on the role of a thermal boundary layer (TBL) at the base of the mantle. A well-developed TBL is expected if there is significant heat flow across the CMB (Jones, 1977; Jeanloz and Richter, 1979; Elsasser *et al.*, 1979; Doornbos, 1983). Because of the destabilizing effects of a strongly temperature-dependent viscosity, flow at the base of the mantle is expected to be time-dependent, with hot, light material accumulating in pools and venting upward in thermal jets (Yuen and Peltier, 1980; Stacey and Loper, 1983; Loper and Stacey, 1983). The strong lateral temperature gradients associated with this flow should therefore give rise to seismically detectable aspherical structure.

In this scenario, the magnitude of the seismic heterogeneity in D″ is governed by the average temperature drop ΔT_m across a boundary layer whose maximum thickness we take to be h_m. The total variation in vertical P-wave travel time is then

$$|\delta\tau_m| \leq 2v_p^{-2}(\partial v_p/\partial T)h_m\Delta T \tag{4.3}$$

A simple parabolic boundary layer analysis can be used to scale this quantity with the heat flux from the core, F, and material properties in D″ (Jeanloz and Richter, 1979):

$$h_m\Delta T_m \propto (\nu F/g\alpha\rho C_p)^{1/2} \tag{4.4}$$

Estimates based on lower-mantle kinematic viscosities of $\nu = 10^{18}$ m^2/s are $h_m \sim 100$ km and $\Delta T_m \sim 500°$ (Jeanloz and Richter, 1979; Elsasser *et al.*, 1979). Adopting a temperature coefficient of $\partial v_p/\partial T = -0.5$ m/s/°K, inferred for the lower mantle from the slab-related anomalies and geoid-correlated velocity perturbations discussed in Section 3, we obtain $|\delta\tau_m| \leq 0.3$ s.

Aspherical heterogeneity does indeed increase near the base of the mantle (Julian and Sengupta, 1973), and the magnitude of the increase obtained in recent lower-mantle P-wave models is commensurate with the predictions of the thermal boundary layer analysis. In the lower 200 km of Dziewonski's (1984) L02.56 structure, for example, the peak-to-peak variation in vertical P time is about 0.2 s, an order of magnitude higher than within a comparable layer a thousand kilometers above the CMB. However, evidence is accumulating which indicates that the structure of the CMB may be more complicated than either the P-wave or simple TBL models suggest. The amplitude decay of short-period P-waves into the core shadow (Ruff and Helmberger, 1982) and the waveforms of *SH* waves at shorter distances (Lay and Helmberger, 1983a,b) imply that the radial velocity gradients within D″ are large and variable, perhaps indicative of compositional layering. Moreover, strong aspherical effects have been observed in many types of seismic waves which interact with the CMB, including P_{diff}, *PnKP, SnKS, PcP* and *ScS* phases (see reviews by Jordan (1979b), Doornbos (1983), and Lay (1987)), and it is not clear that thermally induced velocity differences are sufficient to explain the large amplitude and broad wavenumber spectrum of the observed variations.

As in the crust and uppermost mantle, the lateral structure at or near the CMB appears to span the spectrum from microscale (<30 km) to large scale (>3000 km) heterogeneity (Jordan, 1979b). Since the density jump across the CMB exceeds that at the free surface (4.4 Mg/m^3 vs. 3.3 Mg/m^3), it is plausible that convection in the mantle and core has deposited chemically differentiated material at the CMB to produce compositionally distinct layers of intermediate density, similar in some respects to the surficial chemical boundary layer (CBL) discussed in Section 2. To explain the seismic data, it has been postulated that interactions with the convective flows above and below the CMB cause variations in the thickness of the CBL, forming continent-like rafts of material with anomalous elastic properties (Jordan, 1979b; Ruff and Anderson, 1980).

This hypothesis has recently received support from a tomographic study of CMB structure using the travel times of core-penetrating *PKP* waves (Creager and Jordan, 1986b). The existence of large-scale aspherical structure somewhere below the lower mantle is required by the anomalous splitting of the low-degree free oscillations sensitive to core structure (Masters and Gilbert, 1981; Ritzwoller *et al.*, 1986; Giardini *et al.*, 1987) and anomalous travel times of PKP_{DF} (Poupinet *et al.*, 1983; Morelli *et al.*, 1986; Creager and Jordan, 1986b), but the form and distribution of this structure is as yet uncertain. The models proposed by the various groups have very different distributions of heterogeneity in the inner core, outer core, and lower mantle. Although aspherical heterogeneity of the outer core can account for core-mode splitting (Ritzwoller *et al.*, 1986), the low viscosity of core fluid makes inhomogeneity of the required magnitude extremely unlikely (Stevenson, 1987). Based on extensive observational work and modeling studies, seismologists at Harvard University have shown that much of the core-mode splitting and PKP_{DF} travel-time dispersion can be explained by an inner core that is both aspherical and anisotropic (Woodhouse *et al.*, 1986; Morelli *et al.*, 1986). However, lateral heterogeneity of large amplitude and horizontal scale is also observed in the travel times of PKP_{AB} (Creager and Jordan, 1986b) and PKP_{BC} (Morelli and Dziewonski, 1987), two branches which do not interact with the inner core.

Preliminary maps of CMB structure have been produced by Morelli and Dziewonski (1987) using PKP_{BC} and PcP travel times and by Creager and Jordan (1986b) using PKP_{AB} and PKP_{DF}. The data sets are disjunct, the modeling procedures are very different, and there are discrepancies between the models at even the lowest wavenumbers. Nevertheless, the amplitudes of the vertical P-wave travel-time variations through the CMB are about the same; both yield a peak-to-peak value on the order of one second. Creager and Jordan's vertical resolution is very limited, although they argue that their data are not well described by a model with aspherical heterogeneity confined to D″, indicating that either CMB topography and/or a core-side boundary layer are involved. On the other hand, because they employ the mantle-side reflection *PcP*, Morelli and Dziewonski are able to distinguish CMB topography from boundary-layer structure. They find that, when plotted on the CMB, the *PcP* residuals are anticorrelated with those of PKP_{BC}, which is diagnostic of a structure dominated by the topography. The topography needed to explain the Morelli–Dziewonski data set is on the order of 12 km in total amplitude. If the Creager-Jordan travel-time map is interpreted in terms of CMB topography alone, an amplitude of nearly 16 km is required.

It appears unlikely that topography of this magnitude and scale can be dynamically supported by mantle convection. Hager *et al.* (1985) have estimated the dynamic topography of the CMB from lower-mantle seismic models and the geoid, and their preferred model, which is characterized by whole-mantle convection with a ten-fold increase in viscosity at the 650-km discontinuity, has a peak-to-peak amplitude of only 3 km, a factor of four or five too small to account for the travel times. Moreover, their predicted topography does not show a very good correlation with either the Morelli-Dziewonski or Creager-Jordan maps. Hence, even if the magnitude of the dynamically supported topography could be increased over Hager *et al.*'s estimate by increasing the viscosity gradient in the mantle, satisfying the seismic data would remain difficult.

These considerations motivate the alternate hypothesis that CMB heterogeneity is associated with one or more CBLs trapped by the density contrast at the CMB. The primary advantage of the CBL hypothesis is that it provides a mechanism to induce large velocity variations which are not necessarily correlated positively with density variations. Although Birch's law (Birch, 1961) enforces such a correlation among silicates of roughly equal atomic weight, the admixture of iron, for example, increases density but *decreases* seismic velocities. It is plausible, therefore, to imagine localized areas of the CMB occupied by material of distinct chemistry, temperature and seismic velocities which has very little density constrast with surrounding areas. Such a scenario avoids the need for unrealistic dynamical configurations to support the aspherical heterogeneity.

Given the preliminary nature of the seismic models and the differences in the CMB structures they imply, it is premature to enter into too detailed speculation regarding the properties and behavior of CMB boundary layers.

Although the basic argument that the large magnitude and rich spectrum of CMB heterogeneity requires some sort of chemical mechanism appears to be sound, we do not know to what extent the material of anomalous composition is distributed above the CMB or below it. Because of the mantle's high viscosity, Region D″ is the preferred site for large-scale chemical inhomogeneity, but the possibility of lateral variations within a thin core-side boundary layer cannot be excluded, especially if the CMB has significant topography. We also have very few constraints on the composition and formation history of the CMB layering. We can envisage at least four mechanisms: (1) the CMB could have inherited a distinct chemistry from the early differentiation of the Earth; (2) iron-rich material could be deposited at the base of the mantle by plate-tectonic convection; (3) an alloy-rich slag could accumulate at the top of the core from solidification of the inner core; and, (4) a layer of intermediate density could form by chemical reaction between the mantle and core. Anderson (1975) and Ruff and Anderson (1980) have discussed possibility (1) in terms of a refractory (Ca, Al)-rich assemblage resulting from inhomogeneous accretion of the Earth. Knittle and Jeanloz (1986) discuss experimental evidence for possibility (4).

5. Conclusions

Boundary layers at the surface and core-mantle interface appear to dominate the large-scale, three-dimensional variation of seismic wave speeds in the Earth's interior. In both cases, the seismological data imply a magnitude of heterogeneity difficult to satisfy by simple thermal boundary layer models and appear to require the existence of substantial chemical boundary layers. The mantle part of the surficial CBL resides as thick ($>$200 km) accumulations beneath the stable continental shields and platforms. It has been depleted in iron and aluminum relative to magnesium by the extraction of basaltic components, and the low normative density resulting from this depletion locally compensates the low temperatures of the cratonic tectosphere, stabilizing it against convective disruption. The most plausible model for the formation of thick TBLs beneath the cratons is not conductive decay, but advective thickening by continent-continent collision. The low heat flow into the base of the continental tectosphere inferred from thermal modeling and the tendency of continents to be gregarious indicates that the configuration of the continental CBL is intimately coupled to the large-scale convective flow driving plate tectonics and continental drift.

The best seismological constraints on the depth of this convective circulation come from the seismic imaging of subducted slabs by residual-sphere techniques. In all deep-focus zones studied to date, the data favor slab penetration into the lower mantle to depths of at least 1000 km. The slab flux through the 650-km discontinuity is large, probably much greater than 100 km^3/yr. This conclusion appears to be inconsistent with the existence of major chemical discontinuity at 650-km separating an upper-mantle convecting system from a lower-mantle system. At larger horizontal scales, the downwellings associated with subduction zones account for much of the aspherical structure in the depth range 700–1700 km, suggesting that the plate-tectonic circulation extends throughout most of the mantle.

Lateral heterogeneity increases towards the base of the mantle. Recent tomographic mapping indicates that the aspherical structure of the CMB has an amplitude comparable to the Earth's outer layers. Although inconsistencies among the various seismological results inhibit detailed inferences, the magnitude and scale of CMB heterogeneity strongly suggests the existence of one or more CBLs whose large-scale accumulations on the surface of the core may in some respects be analogous to the continents.

We therefore envisage an Earth comprising four major dynamical parts: the two convecting layers of the mantle and core, and the two CBLs at the outer surface and core-mantle boundary. Although there has been little quantitative work on the dynamics of such a system, there can be little doubt that its behavior is very different from a planet lacking the CBLs.

6. References

Ahrens, T.J. and Y. Syono, Calculated mineral reactions in the Earth's mantle. *J. Geophys. Res.* **72**, 4181–4188, 1967.

Anderson, D.L., Recent evidence concerning the structure and composition of the Earth's mantle. *Phys. Chem. Earth* **6**, 1–131, 1966.

Anderson, D.L., Chemical plumes in the mantle. *Geol. Soc. Am. Bull.* **86**, 1593–1600, 1975.

Anderson, D.L., The 650 km mantle discontinuity. *Geophys. Res. Lett.* **3**, 347–349, 1976.

Anderson, D.L., The deep structure of continents. *J. Geophys. Res.* **84**, 7555–7560, 1979.

Anderson, D.L., A global chemical model for the evolution of the mantle. In R.J. O'Connell and W.S. Fyfe (eds.), *Evolution of the Earth, Geodynamics Series, Vol. 5*, Washington, D.C.: American Geophysical Union, pp. 6–18, 1981.

Anderson, D.L., A seismic equation of state II. Shear properties and the thermodynamics of the lower mantle. *Phys. Earth Planet. Int.* **45**, 307–323, 1987.

Anderson, D.L. and J.B. Minster, Seismic velocity, attenuation and rheology of the upper mantle. In C.J. Allegre (ed.), *Mechanismes et Prevision des Seismes, Hommage au Professeur J. Coulomb*, Paris: Editions du Centre National de la Recherche Scientifique, pp. 109–124. 1980.

Anderson, D.L. and C. Sammis, Partial melting in the upper mantle. *Phys. Earth Planet. Inter.* **3**, 41–50, 1970.

Anderson, D.L. and J. Regan, Upper mantle anisotropy and the oceanic lithosphere. *Geophys. Res. Lett.* **10**, 841–844, 1983.

Anderson, O.L., E. Schreiber, R.C. Liebermann, and N. Soga, Some elastic constants on minerals relevant to geophysics. *Rev. Geophys. Space Phys.* **6**, 491–524, 1968.

Andrews, D.J. and N.H. Sleep, Numerical modelling of tectonic flow behind island arcs. *Geophys. J. R. Astr. Soc.* **38**, 237–251, 1974.

Backus, G.E., Geographical interpretation of measurements of average phase velocities of surface waves over great circular and great semicircular paths. *Bull. Seismol. Soc. Am.* **54**, 571–610, 1964.

Barrell, J., The strength of the Earth's crust. *J. Geol.* **22**, 729–741, 1914.

Bassett, W.A. and L. Ming, Disproportionation of Fe_2SiO_4 to $2FeO+SiO_2$ at pressures up to 250 kbar and temperatures up to 3000°C. *Phys. Earth Planet. Inter.*, **6** 154–160, 1972.

Beaumont, C., On rheological zonation of the lithosphere during flexure. *Tectonophysics* **59**, 347–365, 1979.

Bell, P.M., T. Yagi, and H.K. Mao, Iron-magnesium distribution coefficients between spinel [$(Mg,Fe)_2SiO_4$], magnesiowüstite [(Mg,Fe)O], and perovskite, [$(Mg,Fe)SiO_3$]. *Carnegie Inst. Washington Year Book* **78**, 618–621, 1979.

Benioff, H., Seismic evidence for the fault origin of oceanic deeps. *Geol. Soc. Am. Bull.* **60**, 1837–1856, 1949.

Benioff, H., Orogenesis and deep crustal structure: Additional evidence from seismology. *Geol. Soc. Am. Bull.* **65**, 385–400, 1954.

Benioff, H., Source wave forms of three earthquakes. *Bull. Seismol. Soc. Am.* **53**, 893–903, 1963.

Bibee, L.D. and G.G. Shor, Compressional wave anisotropy in the crust and upper mantle. *Geophys. Res. Lett.* **3**, 639–642, 1976.

Billington, S., The morphology and tectonics of the subducted lithosphere in the Tonga-Fiji-Kermadec region from seismicity and focal mechanism solutions, Ph.D. dissertation, Cornell Univ., Ithica, New York, 1980.

Billington, S. and B.L. Isacks, Identification of fault planes associated with deep earthquakes. *Geophys. Res. Lett.* **2**, 63–66, 1975.

Birch, F., Elasticity and constitution of the Earth's interior. *J. Geophys. Res.* **57**, 227–286, 1952.

Birch, F., Composition of the Earth's mantle. *J. Geophys. Res.* **69**, 4377–4388, 1961.

Bird, P., Stress and temperature in subduction shear zones: Tonga and Mariana. *Geophys. J. R. Astr. Soc.* **55**, 411–434, 1978.

Bodine, J.H., M.S. Steckler, and A.B. Watts, Observations of flexure and the rheology of the oceanic lithosphere. *J. Geophys. Res.* **86**, 3695–3707, 1981.

Bonczar, L.J. and E.K. Graham, The pressure and temperature dependence of the elastic properties of polycrystal magnesiowüstite. *J. Geophys. Res.* **87**, 1061–1078, 1982.

Boyd, F.R., A pyroxene geotherm. *Geochim. Cosmochim. Acta* **37**, 2533–2546, 1973.

Boyd, F.R., J.J. Gurney, and S.H. Richardson, Evidence for a 150–200-km thick Archaean lithosphere from diamond inclusion thermobarometry. *Nature* **315**, 387–389, 1985.

Boyd, F.R., and R.H. McCallister, Densities of fertile and sterile garnet peridotites. *Geophys. Res. Lett.* **3**, 509–512, 1976.

Brune, J.N. and J. Dorman, Seismic waves and Earth structure in the Canadian shield. *Bull. Seismol. Soc. Am.* **53**, 167–210, 1963.

Bullen, K.E., An Earth model based on a compressibility-pressure hypothesis. *Mon, Mot. R. Astr. Soc., Geophys. Suppl.* **6**, 50–59, 1950.

Burdick, L.J., A comparison of the upper mantle structure beneath North America and Europe. *J. Geophys. Res.* **86**, 5926–5936, 1981.

Busse, F.H., Quadruple convection in the lower mantle? *Geophys. Res. Lett.* **10**, 285–288, 1983.

Cara, M., Regional variations of higher Rayleigh-mode phase velocities: A spatial filtering method. *Geophys. J. R. Astr. Soc.* **54**, 439–460, 1978.

Cara, M., Lateral variations of *S* velocity in the upper mantle from higher Rayleigh modes. *Geophys. J. R. Astr. Soc.* **57**, 649–670, 1979.

Cara, M., A. Nercessian, and G. Nolet, New inferences from higher mode data in western Europe and northern Eurasia. *Geophys. J. R. Astr. Soc.* **61**, 459–478, 1980.

Cardwell, R.K. and B.L. Isacks, Geometry of the subducted lithosphere beneath the Banda Sea in eastern Indonesia from seismicity and fault plane solutions. *J. Geophys. Res.* **83**, 2825–2838, 1978.

Cathles, L. M., III, *The Viscosity of the Earth's Mantle*. Princeton, N.J.: Princeton Univ. Press, 386 pp., 1975.

Cazenave, A., Thermal cooling of the oceanic lithosphere: New constraints from geoid height data, *Earth Planet. Sci. Lett.* **70**, 395–406, 1984.

Cazenave, A., B. Lago, and K. Dominh, Thermal parameters of the oceanic lithosphere estimated from geoid height data. *J. Geophys. Res.* **88**, 1105–1118, 1983.

Chapman, D.S. and H.N. Pollack, Regional geotherms and lithospheric thickness. *Geology* **5**, 265–268, 1977.

Chase, C., Tectonic history of the Fiji Plateau. *Geol. Soc. Amer. Bull.* **82**, 3087–3109, 1971.

Chinnery, M.A., Velocity anomalies in the lower mantle. *Phys. Earth Planet. Int.* **2**, 1–10, 1969.

Christensen, N.I. and S.M. Lundquist, Pyroxene orientation within the upper mantle. *Geol. Soc. Am. Bull.* **93**, 279–288, 1982.

Christensen, N.I. and M.H. Salisbury, Structure and composition of the lower oceanic crust. *Rev. Geophys.* **13**, 57–86, 1975.

Christensen, N.I. and M.H. Salisbury, Seismic anisotropy in the oceanic upper mantle: Evidence from the Bay of Islands ophiolite complex. *J. Geophys. Res.* **84**, 4601–4610, 1979.

Christensen, U.R. and D.A. Yuen, The interaction of a subducting lithospheric slab with a chemical or phase boundary. *J. Geophys. Res.* **89**, 4389–4402, 1984.

Clayton, R.W. and R.P. Comer, A tomographic analysis of mantle heterogeneities from body wave travel times (abstract). *Eos, Trans. AGU* **62**, 776, 1983.

Crampin, S., E.M. Chesnokov, and R.G. Hipkin, Seismic anisotropy—state of the art, II. *Geophys. J. R. Astr. Soc.* **16**, 1–16, 1984.

Creager, K.C., Geometry, velocity structure, and penetration depths of descending slabs in the western Pacific. Ph.D. dissertation, University of California, San Diego, 1984.

Creager, K.C. and T.H. Jordan, Slab penetration into the lower mantle. *J. Geophys. Res.* **89**, 3031–3049, 1984.

Creager, K.C. and T.H. Jordan, Slab penetration into the lower mantle beneath the Mariana and other Island Arcs of the Northwest Pacific. *J. Geophys. Res.* **91**, 3573–3589, 1986a.

Creager, K.C. and T.H. Jordan, Aspherical structure of the core-mantle boundary from *PKP* travel times. *Geophys. Res. Lett.* **13**, 1497–1500.

Crough, S.T., Hotspot epeirogeny. *Tectonophysics* **61**, 321–333, 1979.

Crough, S.T. and G.A. Thompson, Numerical and approximate solutions for lithospheric thickening and thinning. *Earth Planet. Sci. Lett.* **31**, 397–402, 1976.

Daly, R.A., *Strength and Structure of the Earth*. Englewood Cliffs, N.J.: Prentice-Hall, 434 pp., 1940.

Daly, S.F. and B. Parsons, Layered convection and its effects on gravity anomalies and surface topography (abstract). *Eos, Trans. AGU* **65**, 1092, 1984.

Davies, G.F., Thickness and thermal history of continental crust and root zones. *Earth Planet. Sci. Lett.* **44**, 231–238, 1979.

Davies, D. and D.P. McKenzie, Seismic travel-time residuals and plates. *Geophys. J. R. Astr. Soc.* **18**, 51–63, 1969.

Davis, E.E. and C.R.B. Lister, Fundamentals of ridge crest topography. *Earth Planet. Sci. Lett.* **21**, 405–413, 1974.

DePaolo, D.J. and G.J. Wasserburg, Inferences about magma sources and mantle structure from variations of $^{143}Nd/^{144}Nd$. *Geophys. Res. Lett.* **3**, 743–746, 1976.

Detrick, R.S., An analysis of geoid anomalies across the Mendocino fracture zone: implications for thermal models of the lithosphere. *J. Geophys. Res.* **86**, 11751–11762, 1981.

Doornbos, D.J., Present seismic evidence for a boundary layer at the base of the mantle. *J. Geophys. Res.* **88**, 3498–3505, 1983.

Doornbos, D.J., S. Spiliopoulos, and F.D. Stacey, Seismological properties of D″ and the structure of the thermal boundary layer. *Phys. Earth Planet. Inter.* **41**, 225–239, 1986.

Doyle, H.A. and A.L. Hales, An analysis of the travel times of *S* waves to North American stations in the distance range 28° to 82°. *Bull. Seismol. Soc. Am.* **57**, 761–771, 1967.

Duschenes, J.D. and S.C. Solomon, Shear wave travel time residuals from oceanic earthquakes and the evolution of oceanic lithosphere. *J. Geophys. Res.* **82**, 1985–2000, 1977.

Dziewonski, A.M., Upper mantle models from 'pure-path' dispersion data. *J. Geophys. Res.* **76**, 2587–2601, 1971.

Dziewonski, A.M., On regional differences in dispersion of mantle Rayleigh waves. *Geophys. J. R. Astr. Soc.* **22**, 289–325, 1971.

Dziewonski, A.M., Mapping the lower mantle: Determination of lateral heterogeneity in *P*-velocity up to degree and order 6. *J. Geophys. Res.* **89**, 5929–5952, 1984.

Dziewonski, A.M. and D.L. Anderson, Preliminary reference Earth model. *Phys. Earth Planet. Inter.* **25**, 297–356, 1981.

Dziewonski, A.M. and D.L. Anderson, Travel times and station corrections for *P*-waves at teleseismic distances. *J. Geophys. Res.* **88**, 3295–3314, 1983.

Dziewonski, A.M., B.H. Hager, and R.J. O'Connell, Large-scale heterogeneities in the lower mantle. *J. Geophys. Res.* **82**, 239–255, 1977.

Dziewonski, A.M. and J.M. Steim, Dispersion and attenuation of mantle waves through waveform inversion. *Geophys. J. R. Astr. Soc.* **70**, 503–527, 1982.

Dziewonski, A.M. and J.H. Woodhouse, Global images of the Earth's interior. *Science* **236**, 37–48, 1987.

Edmonds, A.R., *Angular Momentum in Quantum Mechanics*, Princeton, NJ: Princeton University Press, 146 pp., 1960.

Elsasser, W.M., Convection and stress propagation in the upper mantle. In W.K. Runcorn (ed.), *The Application of Modern Physics to the Earth and Planetary Interiors*, New York: Interscience, New York, 1969.

Elsasser, W.M., P. Olsen, and B.D. Marsh, The depth of mantle convection. *J. Geophys. Res.* **84**, 147–155, 1979.

Engdahl, E.R., Effects of plate structure and dilatancy on relative teleseismic *P*-wave residuals. *Geophys. Res. Lett.* **2**, 420–422, 1975.

Engdahl, E.R., N.H. Sleep, and M.-T. Lin, Plate effects in north Pacific subduction zones. *Tectonophysics* **37**, 95–116, 1977.

Finnerty, A.A. and F.R. Boyd, Evaluation of thermobarometers for garnet perioditites. *Geochim. Cosmochim. Acta* **48**, 15–27, 1984.

Fischer, K., K.C. Creager, and T.H. Jordan, Mapping the Tonga slab. *Eos, Trans. AGU* **67**, 316, 1986.

Fischer, K., T.H. Jordan, and K.C. Creager, Seismic constraints on the morphology of deep slabs, *J. Geophys. Res.* **93**, 4773–4783, 1988.

Fitch, T.J., Compressional velocity in source regions of deep earthquakes: An appliication of the master earthquake technique. *Earth Planet. Sci. Lett.* **26**, 156–166, 1975.

Fitch, T.J., In situ *P*-wave velocities in deep earthquake zones of the SW Pacific:

Evidence for a phase boundary between the upper and lower mantle. In M. Talwani and W.C. Pitman (eds.), *Island Arcs, Deep Sea Trenches and Back-Arc Basins, Maurice Ewing Ser., Vol. 1*, Washinton, D.C.: American Geophysical Union, pp. 123–136, 1977.

Forsyth, D.W., The early structural evolution and anisotropy of the oceanic upper mantle. *Geophys. J. R. Astr. Soc.* **43**, 103–162, 1975.

Forsyth, D.W., The evolution of the upper mantle beneath mid-ocean ridges. *Tectonophysics* **38**, 89–118, 1977.

Forte, A.M. and W.R. Peltier, Plate tectonics and aspherical Earth structure: The importance of poloidal-toroidal coupling. *J. Geophys. Res.* **92**, 3645–3679, 1987.

Frohlich, C. and M. Barazangi, A regional study of mantle velocity variations beneath eastern Australia and the southwestern Pacific using short-period recordings of *P, S, PcP, ScP* and *ScS* waves produced by Tonga deep earthquakes. *Phys. Earth Planet. Inter.* **21**, 1–14, 1980.

Fujita, K., E.R. Engdahl, and N.H. Sleep, Subduction zone calibration and teleseismic relocation of thrust zone events in the central Aleutian Islands. *Bull. Seismol. Soc. Am.* **71**, 1805–1828, 1981.

Garfunkel, Z., Growth, shrinking and long-term evolution of plates and their implications for the flow pattern in the mantle. *J. Geophys. Res.* **80**, 4425–4432, 1975.

Garfunkel, Z., C.A. Anderson, and G. Schubert, Mantle circulation and the lateral migration of subducted slabs. *J. Geophys. Res.* **91**, 7205–7224, 1986.

Gee, L.S., A.L. Lerner-Lam, and T.H. Jordan, Resolving power of higher-mode waveform inversion for Eurasian upper-mantle structure. *AFGL Technical Report 85-0206*, Air Force Geophysical Laboratory, Cambridge, pp. 11–36, 1985.

Giardini, D., Systematic analysis of deep seismicity: 200 centroid-moment tensor solutions for earthquakes between 1977 and 1980. *Geophys. J. R. Astr. Soc.* **77**, 883–914, 1984.

Giardini, D., X.-D. Li, and J.H. Woodhouse, Three dimensional structure of the earth from splitting in free oscillation spectra. *Nature* **325**, 405–411, 1987.

Giardini, D. and J.H. Woodhouse, Deep seismicity and modes of deformation in Tonga subduction zone. *Nature* **307**, 505–509, 1984.

Giardini, D. and J.H. Woodhouse, Evidence for a horizontal shear flow in the mantle beneath the Tonga Arc. *Nature* **319**, 551–555, 1986.

Gilbert, F., Excitation of normal modes of the Earth by earthquake sources. *Geophys. J. R. Astr. Soc.* **22**, 223–226, 1971.

Gilbert, F. and A.M. Dziewonski, An application of normal mode theory to the retrieval of structural parameters and source mechanisms from seismic spectra. *Phil. Trans. R. Soc. Lond., Ser. A* **278**, 187–269, 1975.

Given, J.W. and D.V. Helmberger, Upper mantle structure of northwestern Eurasia. *J. Geophys. Res.* **85**, 7183–7194, 1980.

Grand, S.P., Tomographic inversion for shear velocity beneath the North American plate. *J.Geophys. Res.* **92**, 14065–14090, 1987.

Grand, S.P. and D.V. Helmberger, Upper mantle shear structure of North America. *Geophys J. R. Astr. Soc.* **76**, 399–438, 1984a.

Grand, S.P. and D.V. Helmberger, Upper mantle shear structure beneath the northwest Atlantic Ocean. *J. Geophys. Res.* **89**, 11465–11475, 1984b.

Grand, S.P. and D.V. Helmberger, Uppermantle shear structure beneath Asia from multi-bounce *S* waves. *Phys. Earth Planet. Int.* **41**, 154–169, 1985.

Green, D.H. and R.C. Liebermann, Phase equilibria and elastic properties of a pyrolite model for the oceanic upper mantle. *Tectonophysics* **32**, 61–92, 1976.

Griggs, D.T., The sinking lithosphere and the focal mechanism of deep earthquakes. In E.C. Robertson (ed.), *The Nature of the Solid Earth*, New York: McGraw-Hill, pp. 361–384, 1972.

Gurney, J.J., J.W. Harris, and R.S. Rickard, Silicate and oxide inclusions in diamonds from the Finsch kimberlite pipe. In F.R. Boyd and H.O.A. Meyer (eds.) *Kimberlites, Diatremes, and Diamonds: Their Geology, Petrology, and Geochemistry*, Washington, D.C.: American Geophysical Union, pp. 1–15, 1979.

Gutenberg, B., H. Benioff, J.M. Burgers, and D. Griggs, Colloquium on plastic flow and deformation within the earth. *Trans. AGU* **32**, 497–543, 1951.

Gutenberg, B. and C.F. Richter, Depth and geographical distribution of deep-focus earthquakes. *Geol. Soc. Am. Bull.* **49**, 249–288, 1938.

Gutenberg, B. and C.F. Richter. *Seismicity of the Earth and Associated Phenomena*, Princeton, N.J.: Princeton Univ. Press, 310 pp., 1954.

Hager, B.H., Subducted slabs and the geoid: constraints on mantle rheology and flow. *J. Geophys. Res.* **86**, 6003–6015,1984.

Hager, B.H. and R.J. O'Connell, Subduction zone dip angles and flow driven by plate motion. *Tectonophysics* **50**, 111–133, 1978.

Hager, B.H. and R.J. O'Connell, A simple global model of plate dynamics and mantle convection. *J. Geophys. Res.* **86**, 4843–4867, 1981.

Hager, B.H., R.J. O'Connell, and A. Raefsky, Subduction, back-arc spreading and global mantle flow. *Tectonophysics* **99**, 165–189, 1983.

Hager, B.H., R.W. Clayton, M.A. Richards, R.P. Comer, and A.M. Dziewonski, Lower mantle heterogeneity, dynamic topography and the geoid. *Nature* **313**, 541–545, 1985.

Hales, A.L. and E. Herrin, Travel times of seismic waves. In E.C. Robertson (ed.), *The Nature of the Solid Earth*, New York: McGraw-Hill, pp. 172–215, 1972.

Harte, B., K.G. Cox, and J.J. Gurney, Petrography and geological history of upper mantle xenoliths from the Matsoku kimberlite pipe. *Phys. Chem. Earth* **9**, 477–506, 1975.

Hasebe, K., N. Fujii, and S. Uyeda, Thermal processes under island arcs. *Tectonophysics* **10**, 335–355, 1970.

Haxby, W.F. and D.L. Turcotte, On isostatic geoid anomalies. *J. Geophys. Res.* **83**, 5473–5478, 1978.

Haxby, W.F., D.L. Turcotte, and J.M. Bird, Thermal and mechanical evolution of the Michigan Basin. *Tectonophysics* **36**, 57–75, 1976.

Heestand, R.L. and S.T. Crough, The effect of hot spots on the oceanic age-depth relation. *J. Geophys. Res.* **86**, 6107–6114, 1981.

Herrin, E. *et al.*, 1968 seismological tables for *P* phases. *Bull. Seismol. Soc. Am.* **58**, 1196–1241, 1968.

Hess, H.H., Seismic anisotropy of the uppermost mantle under oceans. *Nature* **203**, 629–631, 1964.

Holmes, A., *Principles of Physical Geology*, 2nd ed., New York: Ronald, 1288 pp., 1965.

Honda, H., Earthquake mechanism and seismic waves. *J. Phys. Earth* **10**,1–97, 1962.

Honda, H. and A. Masatsuka, On the mechanism of the earthquakes and the stresses producing them in Japan and its vicinity. *Tohoku Univ., Sci. Rep., Ser. 5, Geophys.* **4**, 42–60, 1952.

Hubert, L.N., and C. Frohlich, The *P* velocity within the Tonga Benioff zone determined from traced rays and observations. *J. Geophys. Res.* **86**, 3771–3782, 1981.

Hunter, D.R. Crustal development in the Kaapvaal craton. *Precambr. Res.* **1**, 259–326, 1974.

Isacks, B. L., and M. Barazangi, Geometry of Benioff zones: Lateral segmentation and downwards bending of the subducted lithosphere. In M. Talwani and W.C. Pitman III (eds.), *Island Arcs, Deep Sea Trenches, and Back-Arc Basins, Maurice Ewing Ser., Vol. 1*, Washington, D.C.: American Geophysical Union, pp. 99–114, 1977.

Isacks, B. and P. Molnar, Mantle earthquake mechanisms and sinking of the lithosphere. *Nature* **223**, 1121–1124, 1969.

Isacks, B. and P. Molnar, Distribution of stress in the descending lithosphere from a global survey of focal-mechanism solutions of mantle earthquakes. *Rev. Geophys. Space Phys.* **9**, 103–174, 1971.

Isacks, B., J. Oliver, and L.R. Sykes, Seismology and the new global tectonics. *J. Geophys. Res.* **73**, 5855–5899, 1968.

Ito, E. and H. Yamada, Stability relations of silicate spinels, ilmenites, and perovskites. In S. Akimoto and M.H. Manghnani (eds.), *High Pressure Research in Geophysics*, Tokyo: Center Acad. Pub. Japan, pp. 405–419, 1982.

Jackson, I.N.S., R.C. Liebermann, and A.E. Ringwood, Disproportionation of spinels to mixed oxides: Significance of cation configuration and implications for the mantle. *Earth Planet. Sci. Lett.* **24**, 203–208, 1974.

Jacob, K.H., Global tectonic implications of anomalous seismic *P* travel times from the nuclear explosion Longshot. *J. Geophys. Res.* **77**, 2556–2573, 1972.

Jacobsen, S.B. and G.J. Wasserburg, The mean age of mantle and crustal reservoirs. *J. Geophys. Res.* **84**, 7411–7427, 1979.

Jarvis, G.T. and W.R. Peltier, Flattening of ocean bathymetry profiles due to radiogenic heating in a convecting mantle. *Nature* **285**, 649–651, 1980.

Jarvis, G.T. and W.R. Peltier, Lateral heterogeneity in the convecting mantle. *J. Geophys. Res.* **91**, 435–451, 1986.

Jeanloz, R., High pressure chemistry of the earth's mantle and core, this volume.

Jeanloz, R. and F.M. Richter, Convection, composition, and the thermal state of the lower mantle. *J. Geophys. Res.* **84**, 5497–5504, 1979.

Jeanloz, R. and A.B. Thompson, Phase transitions and mantle discontinuities. *Rev. Geophys. Space Phys.* **21**, 51–74, 1983.

Johnson, L.R., Array measurements of *P* velocities in the upper mantle. *J. Geophys. Res.* **72**, 6309–6325, 1967.

Johnson, L.R., Array measurements of *P* velocities in the lower mantle. *Bull. Seismol. Soc. Am.* **59**, 973–1008, 1969.

Jones, G.M., Thermal interaction of the core and the mantle and the long term behaviour of the geomagnetic field. *J. Geophys. Res.* **82**, 1703–1709, 1977.

Jordan, T.H., The continental tectosphere. *Rev. Geophys. Space Phys.* **13**, 1–12, 1975a.

Jordan, T.H., Lateral heterogeneity and mantle dynamics. *Nature* **257**, 745–750, 1975b.

Jordan, T.H., Lithospheric slab penetration into the lower mantle beneath the Sea of Okhotsk. *J. Geophys.* **43**, 473–496, 1977.

Jordan, T.H., Composition and development of the continental tectosphere. *Nature* **274**, 544–548, 1978a.

Jordan, T.H., A procedure for estimating lateral variations from low-frequency eigenspectra data. *Geophys. J. R. Astr. Soc.* **52**, 441–455, 1978b.

Jordan, T.H., Mineralogies, densities, and seismic velocities of garnet lherzolites and their geophysical implications. In F.R. Boyd and H.O.A. Meyer (eds.), *The Mantle Sample: Inclusions in Kimberlites and Other Volcanics*, Washington, D.C., American Geophysical Union, pp. 1–14, 1979a.

Jordan, T.H., Structural geology of the Earth's interior. *Proc. Natl. Acad. Sci. U.S.A.* **76**, 4192-4200, 1979b.

Jordan, T.H., The deep structure of the continents. *Sci. Am.* **240**, 92–107, 1979c.

Jordan, T.H., Global tectonic regionalization for seismological data analysis. *Bull. Seismol. Soc. Am.*, **71**, 1131–1141, 1981a.

Jordan, T.H., Continents as a chemical boundary layer. *Phil. Trans. R. Soc. Lond., Ser. A* **301**, 359–373, 1981b.

Jordan, T.H. and D.L. Anderson, Earth structure from free oscillations and travel times. *Geophys. J. R. Astr. Soc.* **36**, 411–459, 1974.

Jordan, T.H. and W.S. Lynn, A velocity anomaly in the lower mantle. *J. Geophys. Res.* **79**, 2679–2685, 1974.

Julian, B.R. and D. Gubbins, Three-dimensional seismic ray tracing. *J. Geophys.* **43**, 95–113, 1977.

Julian, B.R. and M.K. Sengupta, Seismic travel time evidence for lateral inhomogeneity in the deep mantle. *Nature* **242**, 443–447, 1973.

Kanamori, H., Velocity and *Q* of mantle waves. *Phys. Earth Planet. Int.* **2**, 259–275, 1970.

Kanamori, H. and F. Press, How thick is the lithosphere? *Nature* **226**, 330–331, 1970.

Katsumata, M. and L.R. Sykes, Seismicity and tectonics of the western Pacific: Izu-Mariana-Caroline and Ryukyu-Taiwan Regions. *J. Geophys. Res.* **74**, 5923–5948, 1969.

Kawakatsu, H., Can "pure path" models explain free oscillation data? *Geophys. Res. Lett.* **10**, 186–189, 1983.

Kay, R., N.J. Hubbard, and P.W. Gast. *J. Geophys. Res.* **75**, 1585–1613, 1970.

Keen, C.E. and D.L. Barrett, A measurement of seismic anisotropy in the northeast Pacific. *Can. J. Earth Sci.* **8**, 1056–1064, 1971.

King, D.W. and G. Calcagnile. *P*-wave velocities in the upper mantle beneath Fennoscandia and western Russia. *Geophys. J. R. Astr. Soc.* **46**, 407–432, 1976.

Knittle, E. and R. Jeanloz, High-pressure metallization of FeO and implications for the earth's core. *Geophys. Res. Lett.* **13**, 1541–1544, 1986.

Knopoff, L., Observation and inversion of surface wave dispersion. *Tectonophysics* **13**, 497–519, 1972.

Knopoff, L. and M.J. Randall, The compensated linear-vector dipole: a possible mechanism for deep earthquakes. *J. Geophys. Res.* **75**, 4957-4963, 1970.

Kono, Y. and M. Amano, Thickening model of the continental lithosphere. *Geophys. J. Roy. Astr. Soc.* **54**, 405–416, 1978.

Kramers, J.D., Lead, uranium, strontium, potassium, and rubidium in inclusion-bearing diamonds and mantle-derived xenoliths from Southern Africa. *Earth Planet. Sci. Lett.* **42**, 58–70, 1979.

Kusznir, N. and G. Karner, Dependence of the flexural rigidity of the continental lithosphere on rheology and temperature. *Nature* **316**, 138–142, 1985.

Lachenbruch, A.H. and J.H. Sass, Heat flow and energetics of the San Andreas Fault Zone. *J. Geophys. Res.* **85**, 6185–6222, 1980.

Langseth, M.G., X. LePichon, and M. Ewing, Crustal structure of mid-ocean ridges, 5; Heat flow through the Atlantic Ocean floor and convection currents. *J. Geophys. Res.* **71**, 5321–5355, 1966.

Lay, T., Localized velocity anomalies in the lower mantle. *Geophys. J. R. Astr. Soc.* **72**, 483–516, 1983.

Lay, T., Structure of the earth: Mantle and core. *Rev. Geophys. Space Phys.* **25**, 1161–1167, 1987.

Lay, T. and D.V. Helmberger, A shear velocity discontinuity in the lower mantle. *Geophys. Res. Lett.* **10**, 63–66, 1983a.

Lay, T. and D.V. Helmberger, A lower mantle *S* triplication and the shear velocity structure of D″. *Geophys. J. R. Astr. Soc.* **75**, 799–837, 1983b.

Leeds, A.R., Lithospheric thickness in the western Pacific. *Phys. Earth Planet. Inter.* **11**, 61–64, 1975.

Leeds, A.R., L. Knopoff, and E.G. Kausel, Variations of upper mantle structure under the Pacific Ocean. *Science* **186**, 141–143, 1974.

Lees, A.C., M.S.T. Bukowinski, and R. Jeanloz, Reflection properties of phase transition and compositional change models of the 650-km discontinuity. *J. Geophys. Res.* **88**, 8145–8159, 1983.

Leith, A. and J.A. Sharpe, Deep focus earthquakes and their geological significance. *J. Geol.* **44**, 877–917, 1936.

Lerner-Lam, A.L. and T.H. Jordan, Earth structure from fundamental and higher-mode waveform analysis. *Geophys. J. R. Astr. Soc.* **75**,759–797, 1983.

Lerner-Lam, A.L. and T.H. Jordan, How thick are the continents? *J. Geophys. Res.* **92**, 14007–14026, 1987.

Leveque, J.J. and M. Cara, Long-period Love wave overtone data in North America and the Pacific Ocean: new evidence for upper mantle anisotropy. *Phys. Earth Planet. Inter.* **33**, 164–179, 1983.

Liu, H.-P., D.L. Anderson, and H. Kanamori, Velocity dispersion due to anelasticity: implications for seismology and mantle composition. *Geophys. J. R. Astr. Soc.* **47**, 41–58, 1976.

Liu, L., The high-pressure phases of $MgSiO_3$. *Earth Planet. Sci. Lett.* **31**, 200–208, 1976a.

Liu, L., The post-spinel phase of forsterite. *Nature* **262**, 770–772, 1976b.

Liu, L., On the 650-km seismic discontinuity. *Earth Planet. Sci. Lett.* **42**, 202–208, 1979.

Loper, D.E. and F.D. Stacey, The dynamical and thermal structure of deep mantle plumes. *Phys. Earth Planet. Int.* **33**, 304–317, 1983.

MacDonald, G.J.F., The deep structure of continents. *Rev. Geophys.* **1**, 587–665, 1963.

Masters, G. and F. Gilbert, Structure of the inner core inferred from observations of its spheroidal shear modes. *Geophys. Res. Lett.* **8**, 569–571, 1981.

Masters, G., T.H. Jordan, P.G. Silver, and F. Gilbert, Aspherical earth structure from fundamental spheroidal-mode data. *Nature* **298**, 609–613, 1982.

Mathais, M., J.C. Siebert, and P.C. Rickwood, Some aspects of the mineralogy and petrology of ultramafic zenoliths in kimberlite. *Contr. Mineral. Petrol.* **26**, 75–123, 1970.

McKenzie, D.P., The viscosity of the lower mantle. *J. Geophys. Res.* **71**, 3995–4010, 1966.

McKenzie, D.P., Some remarks on heat flow and gravity anomalies. *J. Geophys. Res.* **72**, 6261–6273, 1967.

McKenzie, D.P., Speculations on the consequences and causes of plate motions. *Geophys. J. R. Astr. Soc.* **18**, 1–32, 1969.

McKenzie, D.P., Some remarks on the development of sedimentary basins. *Earth Planet. Sci. Lett.* **40**, 25–32, 1978.

McKenzie, D.P., The earth's mantle. *Sci. Am.* **249**, 67–78, 1983.

McKenzie, D.P. and R.L. Parker, The North Pacific: An example of tectonics on a sphere. *Nature* **216**, 1276–1280, 1967.

McKenzie, D.P. and F.M. Richter, Convection currents in the earth's mantle. *Sci. Am.* **235**, 72–89, 1976.

McKenzie, D.P. and F.M. Richter, Parameterized thermal convection in a layered region and the thermal history of the earth. *J. Geophys. Res.* **86**, 11667–11680, 1981.

McKenzie, D.P., J.M. Roberts, and N.O. Weiss, Convection in the earth's mantle: Towards a numerical simulation. *J. Fluid Mech.* **62**, 465–538, 1974.

McKenzie, D.P. and N. Weiss, Speculations on the thermal and tectonic history of the Earth. *Geophys. J. R. Astr. Soc.* **4**, 131–174, 1975.

McNutt, M.K., Lithospheric flexure and thermal anomalies. *J. Geophys. Res.* **89**, 11180–11194, 1984.

McNutt, M.K. and H.W. Menard, Constraints on yield strength in the oceanic lithosphere derived from observations of flexure. *Geophys. J. R. Astr. Soc.* **71**, 363–394, 1982.

Means, J.D. and T.H. Jordan, Slab penetration below 650 km: further evidence (abstract). *Eos, Trans. AGU* **61**, 1047, 1980.

Menzies, M. and V. Rama Murphy, Enriched mantle: Nd and Sr isotopes in diopsides from kimberlite nodules. *Nature* **283**, 634–636, 1980.

Minear, J.W. and M.N. Toksöz, Thermal regime of a downgoing slab and new global tectonics. *J. Geophys. Res.* **75**, 1397–1419, 1970.

Minster, J.B. and T.H. Jordan, Present-day plate motions. *J. Geophys. Res.* **83**, 5331–5354, 1978.

Minster, J.B. and T.H. Jordan, Present-day plate motions: A summary. In C.J. Allegre (ed.), *Mechanismes et Prevision des Seismes, Hommage au Professeur J. Coulomb*, Paris: Editions du Centre National de la Recherche Scientifique, pp. 109–124, 1980.

Mitchell, B.J. and G.K. Yu, Surface wave dispersion, regionalized velocity models and anisotropy of the Pacific crust and upper mantle. *Geophys. J. R. Astr. Soc.* **63**, 497–514, 1980.

Mitronovas, W. and B.L. Isacks, Seismic velocity anomalies in the upper mantle beneath the Tonga-Kermadec island arc. *J. Geophys. Res.* **76**, 7154–7180, 1971.

Molnar, P., D. Freedman, and J.S.F. Shih, Lengths of intermediate and deep seismic zones and temperatures in downgoing slabs of lithosphere. *Geophys. J. R. Astr. Soc.* **56**, 41–54, 1979.

Morelli, A. and A.M. Dziewonski, Topography of the core-mantle boundary and lateral homogeneity of the liquid core. *Nature* **325**, 678–683, 1987.

Morelli, A., A.M. Dziewonski, and J.H. Woodhouse, Anisotropy of the inner core inferred from *PKIKP* travel times. *Geophys. Res. Lett.* **13**, 1545–1548, 1986.

Morris, G.B., R.W. Raitt, and G.G. Shor, Velocity anisotropy and delay-time maps of the mantle near Hawaii. *J. Geophys. Res.* **74**, 4300–4316, 1969.

Muirhead, K.J. and A.L. Hales, Evidence for *P* wave velocity discontinuities at depths greater than 650 km in the mantle. *Phys. Earth Planet. Int.* **23**, 304–313, 1980.

Nakanishi, I. and D.L. Anderson, Measurements of mantle wave velocities and inversion for lateral heterogeneity and anisotropy, 1, Analysis of great circle velocities. *J. Geophys. Res.* **88**, 10267–10284, 1983.

Nakanishi, I. and D.L. Anderson, Measurements of mantle wave velocities and inversion for lateral heterogeneity and anisotropy, 2, Analysis by the single-station method. *Geophys. J. R. Astr. Soc.* **78**, 573–618, 1984.

Nataf, H.-C., I. Nakanishi, and D.L. Anderson, Anisotropy and shear-velocity heterogeneities in the upper mantle. *Geophys. Res. Lett.* **11**, 109–112, 1984.

Nataf, H.-C., I. Nakanishi, and D.L. Anderson, Measurements of mantle wave velocities and inversion for lateral heterogeneities and anisotropy, 3, Inversion. *J. Geophys. Res.* **91**, 7261–7307, 1986.

Navrotsky, A., Lower mantle phase transitions may generally have negative pressure-temperature slopes. *Geophys. Res. Lett.* **7**, 709–711, 1980.

Niazi, M. and D.L. Anderson, Upper mantle structure of western North America from apparent velocities of *P* waves. *J. Geophys. Res.* **70**, 4633–4640, 1965.
Nicolas, A. and J.F. Violette, Mantle flow at oceanic spreading centers: models derived from ophiolites. *Tectonophysics* **81**, 319–339, 1982.
Nicholayson, L.O., and R.J. Hart, The Vredefort radioelement profile extended to supracrustal strata at Carletonville, with implications for continental heat flow. *J. Geophys. Res.* **86**, 10653–10661, 1981.
Nixon, P.H., F.R. Boyd, and A.M. Boullier, The evidence of kimberlite and its inclusion on the constitution of the outer part of the earth. In P.H. Nixon (ed.) *Lesotho Kimberlites*, Lesotho: Lesotho National Development Corp., pp. 312–318, 1973.
Nolet, G., Higher Rayleigh modes in western Europe. *Geophys. Res. Lett.* **2**, 60–62, 1975.
Nolet, G., The upper mantle structure under western Europe inferred from the dispersion of Rayleigh modes. *J. Geophys.* **43**, 265–285, 1977.
Nolet, G. and J. Panza, Array analysis of seismic surface waves: limits and possibilities. *Pure Appl. Geophys.* **114**, 775–790, 1976.
O'Connell, R.J., On the scale of mantle convection. *Tectonophysics* **38**, 119–136, 1977.
O'Hara, M.J., Is there an Icelandic mantle plume? *Nature* **253**, 708–710, 1975.
Okal, E.A., The effect of intrinsic oceanic upper-mantle heterogeneity on regionalization of long-period Rayleigh-wave phase velocities. *Geophys. J. R. Astr. Soc.* **49**, 357–370, 1977.
Okal, E.A. and D.L. Anderson, A study of lateral inhomogeneities in the upper mantle by multiple *ScS* travel-time residuals. *Geophys. Res. Lett.* **2**, 313–316, 1975.
Oliver, J. and B. Isacks, Deep earthquake zones, anomalous structures in the upper mantle, and the lithosphere. *J. Geophys. Res.* **72**, 4259–4275, 1967.
O'Nions, R.K., N.M. Eversen, and P.J. Hamilton, Geochemical modeling of mantle differentiation and crustal growth. *J. Geophys. Res.* **84**, 6091–6101, 1979.
Oxburgh, E.R. and E.M. Parmentier, Compositional and density stratification in oceanic lithosphere—causes and consequences. *J. Geol. Soc.* **133**, 343–355, 1977.
Oxburgh, E.R. and E.M. Parmentier, Thermal process in the formation of continental lithosphere. *Phil. Trans. R. Soc. Lond., Ser. A* **288**, 415–429, 1978.
Parker, R.L., Inverse theory with grossly inadequate data. *Geophys. J. R. Astr. Soc.* **29**,123–138, 1972.
Parsons, B. and D.P. McKenzie, Mantle convection and the thermal structure of the plates. *J. Geophys. Res.* **83**, 4485–4496, 1978.
Parsons, B. and F.M. Richter, A relation between the driving force and geoid anomaly associated with mid-ocean ridges. *Earth Planet. Sci. Lett.* **51**, 445–450, 1980.
Parsons, B. and J.G. Sclater, An analysis of the variation of ocean floor bathymetry and heat flow with age. *J. Geophys. Res.* **82**, 803–827, 1977.
Pascal, G., J. Dubois, M. Barazangi, B.L. Isacks, and J. Oliver, Seismic velocity

anomalies beneath the New Hebrides Island Arc: Evidence for a detached slab in the upper mantle. *J. Geophys. Res.* **78**, 6998–7004, 1973.

Pascal, G., B.L. Isacks, M. Barazangi, and J. Dubois, Precise relocations of earthquakes and seismotectonics of the New Hebrides Island Arc. *J. Geophys. Res.* **83**, 4957–4973, 1978.

Peltier, W.R., Glacial isostatic adjustment—II. The inverse problem. *Geophys. J. R. Astr. Soc.* **46**, 669–706, 1976.

Peltier, W.R., Mantle convection and viscosity. In A. Dziewonski and E. Boschi (eds.), *"Physics of the Earth's Interior," Proceedings of the Enrico Fermi International School of Physics* (Course LXXVIII), New York: North Holland, pp. 362–427, 1980.

Poirier, J.P., On the kinetics of olivine-spinel transition. *Phys. Earth Planet. Inter.* **26**, 179–187, 1981.

Polyak, B.G. and Y.A. Smirnov, Relationship between terrestrial heat flow and the tectonics of continents. *Geotectonics* **4**, 205–213, 1968.

Poupinet, G., On the relation between *P*-wave travel time residuals and the age of continental plates. *Earth Planet. Sci. Lett.* **43**, 149–161, 1979.

Poupinet, G. and B. de Voogd, Some consequences of the differences in seismic properties in old and young continental plates for isostatic compensation. *Earth Planet. Sci. Lett.* **56**, 278–286, 1981.

Poupinet, G., R. Pillet, and A. Souriau, Possible heterogeneity of the earth's core deduced from *PKIKP* travel times. *Nature* **305**, 204–206, 1983.

Raitt, R.W., G.G. Shor, T.J.G. Francis, and G.B. Morris, Anisotropy of the Pacific upper mantle. *J. Geophys. Res.* **74**, 3095–3109, 1969.

Randall, M.J., Elastic multipole theory and seismic moment. *Bull. Seis. Soc. Am.* **61**, 1321–1326, 1971.

Regan, J. and D.L. Anderson, Anisotropic models of the upper mantle. *Phys. Earth Planet. Int.* **35**, 227–263, 1984.

Richards, M.A. and B.H. Hager, Geoid anomalies in a dynamic earth. *J. Geophys. Res.* **89**, 5987–6002, 1984.

Richardson, S.H., J.J. Gurney, A.J. Erlank, and J.W. Harris, Origins of diamonds in old enriched mantle. *Nature* **310**, 198–202, 1984.

Richter, F.M., Focal mechanisms and seismic energy release of deep and intermediate earthquakes in the Tonga-Kermadec region and their bearing on the depth extent of mantle flow. *J. Geophys. Res.* **84**, 6783–6795, 1979.

Richter, F.M., Models for the Archean thermal regime. *Earth Planet. Sci. Lett.* **73**, 350–360, 1985.

Richter, F.M. and C.E. Johnson, Stability of a chemically layered mantle. *J. Geophys. Res.* **79**, 1635–1639, 1974.

Richter, F.M. and D.P. McKenzie, On some consequences and possible causes of layered mantle convection. *J. Geophys. Res.* **79**, 6133–6142, 1981.

Richter, F.M., and B. Parsons, On the interaction of two scales of convection in the mantle. *J. Geophys. Res.* **80**, 2529–2541, 1975.

Ringwood, A.E., The chemical composition and origin of the earth. In P. Hurley (ed.), *Advances in Earth Composition*, Cambridge, Mass: M.I.T. Press, 1966.

Ringwood, A.E., Composition and evolution of the upper mantle. In P.J. Hart (ed.), *The Earth's Crust and Upper Mantle*, *Am. Geophys. Union Monograph* **13**, pp. 1–17, 1969.

Ringwood, A.E., *Composition and Petrology of the Earth's Mantle*, New York: McGraw-Hill, 619 pp., 1975.

Ringwood, A.E., Phase transformations and differentiation in subducted lithosphere: Implications for mantle dynamics, basalt petrogenesis, and crustal evolution. *J. Geol.* **90**, 611–643, 1982.

Ritzwoller, M., G. Masters, and F. Gilbert, Observations of anomalous splitting and their interpretation in terms of aspherical structure. *J. Geophys. Res.* **91**, 10203–10228, 1986.

Ruff, L., and D.L. Anderson, Core formation, evolution, and convection: a geophysical model. *Phys. Earth Planet. Inter.* **21**, 181–201, 1980.

Ruff, L., and D.V. Helmberger, The structure of the lowermost mantle determined by short-period *P*-wave amplitudes. *Geophys. J. R. Astr. Soc.* **68**, 95–119, 1982.

Salisbury, M.H., and N.J. Christensen, The seismic velocity structure of a traverse through the Bay of Islands ophiolite complex, Newfoundland, an exposure of oceanic crust and upper mantle. *J. Geophys. Res.* **82**, 805–817, 1978.

Sammis, C.G., J.C. Smith, G. Schubert, and D.A. Yuen, Viscosity-depth profile of the earth's mantle; effects of polymorphic phase transformations. *J. Geophys. Res.* **82**, 3741–3761, 1977.

Sandwell, D.T. and G. Schubert, Geoid height versus age for symmetric spreading ridges. *J. Geophys. Res.* **85**, 7235–7241, 1980.

Sandwell, D.T. and G. Schubert, Geoid height-age relation from SEASAT altimeter profiles across the Mendocino fracture zone. *J. Geophys. Res.* **87**, 3949–3958, 1982.

Sasatani, T., Source process of a large deep-focus earthquake of 1970 in the Sea of Okhotsk. *J. Phys. Earth* **24**, 27–42, 1976.

Sasatani, T., Source parameters and rupture mechanism of deep-focus earthquakes. *J. Fac. Sci. Hokkaido Univ., Ser. VII (Geophys.)* **6**, 301–384, 1980.

Satô, Y., Attenuation, dispersion, and the wave guide of the *G* wave. *Bull. Seismol. Soc. Am.* **48**, 443–459, 1958.

Schlue, J.W. and L. Knopoff, Shear-wave polarization anisotropy in the Pacific Basin. *Geophys. J. R. Astron. Soc.* **49**, 145–165, 1977.

Schlue, J.W. and L. Knopoff, Inversion of surface-wave phase velocities for an anisotropic structure. *Geophys. J. R. Astr. Soc.* **54,** 697–702, 1978.

Schubert, G., C. Froidevaux, and D.A. Yuen, Oceanic lithosphere and asthenosphere: thermal and mechanical structure. *J. Geophys. Res.* **81**, 3525–3540, 1976.

Schubert, G., D.A. Yuen, and D.L. Turcotte, Role of phase transitions in a dynamic mantle. *Geophys. J. R. Astr. Soc.* **42**, 705–735, 1975.

Sclater, J.G. and P.A.F. Christie, Continental stretching: An explanation of the post mid-Cretaceous subsidence of the central graben of the North Sea. *J. Geophys. Res.*

85, 3740–3750, 1980.

Sclater, J.G. and J. Francheteau, The implications of terrestrial heat flow observations on current tectonic and geochemical models of the crust and upper mantle of the earth. *Geophys. J. R. Astr. Soc.* **20**, 509–542, 1970.

Sclater, J.G., C. Jaupart, and D. Galson, The heat flow through oceanic and continental crust and the heat loss of the earth. *Rev. Geophys. Space Phys.* **18**, 269–311, 1980.

Sclater, J.G., L.A. Lawver, and B. Parsons, Comparisons of long wavelength residual elevation and free air gravity anomalies in the North Atlantic and possible implications for the thickness of the lithospheric plate. *J. Geophys. Res.* **80**, 1031–1052, 1975.

Sclater, J.G., B. Parsons, and C. Jaupart, Oceans and continents: Similarities and differences in the mechanisms of heat loss. *J. Geophys. Res.* **86**, 11535–11552, 1981.

Sengupta, M.K., and B.R. Julian, Radial variation of compressional and shear velocities in the Earth's lower mantle. *Geophys. J. R. Astron. Soc.* **54**, 185–219, 1978.

Sengupta, M.K., and M.M. Toksöz, Three dimensional model of seismic velocity variation in the Earth's mantle. *Geophys Res. Lett.* **3**, 84–86, 1976.

Sereno, T. and J. Orcutt, Synthesis of realistic oceanic P_n wave trains. *J. Geophys. Res.* **90**, 12755–12766, 1985.

Shearer, P. and J. Orcutt, Anisotropy in the oceanic lithosphere—theory and observations from the Ngendei seismic refraction experiment in the southwest Pacific. *Geophys. J. R. Astr. Soc* **80**, 493–526, 1985.

Shimamura, H., Anisotropy in the oceanic lithosphere of the northwestern Pacific Basin. *Geophys. J. R. Astr. Soc.* **76**, 253–260, 1984.

Shimamura, H. and T. Asada, Velocity anisotropy extending over the entire depth of the oceanic lithosphere. In T.W.C. Hilde and S. Uyeda (eds.), *Geodynamics of the Western Pacific-Indonesian Region, Geodynamics Series, Vol. 11*, Washington, D.C.: American Geophysical Union, pp. 121–125, 1983.

Shiono, K. and N. Sugi, Life of an oceanic plate: Cooling time and assimilation time. *Tectonophysics* **112**, 35–50, 1985.

Shor, G.G., R.W. Raitt, M. Henry, L.R. Bentley, and G.H. Sutton, Anisotropy and crustal structure in the Cocos plate. *Geofis. Int.* **13**, 337–362, 1973.

Sipkin, S.A. and T.H. Jordan, Lateral heterogeneity of the upper mantle determined from the travel times of *ScS, J. Geophys. Res.* **80**, 1474–1484, 1975.

Sipkin, S.A. and T.H. Jordan, Lateral heterogeneity of the upper mantle determined from the travel times of multiple *ScS, J. Geophys. Res.* **81**, 6307–6320, 1976.

Sipkin, S.A. and T.H. Jordan, Multiple *ScS* travel times in the western Pacific: Implications for mantle heterogeneity. *J. Geophys. Res.* **85**, 853–861, 1980.

Silver, P.G. and T.H. Jordan, Fundamental spheroidal mode observations of aspherical heterogeneity. *Geophys. J. R. Astr. Soc.* **64**, 605–634, 1981.

Sleep, N.H., Thermal effects of the formation of Atlantic continental margins by continental breakup. *Geophys. J. R. Astr. Soc.* **24**, 325–350, 1971.

Sleep, N.H., Teleseismic *P*-wave transmission through slabs. *Bull. Seismol. Soc. Am.* **63**, 1349–1373, 1973.

Sleep, N.H. and N.S. Snell, Thermal contraction and flexure of mid-continent and Atlantic marginal basins. *Geophys. J. R. Astr. Soc.* **45**, 125–154, 1976.

Solomon, S.C., Geophysical constraints on radial and lateral temperature variations in the upper mantle. *Amer. Mineral.* **61**, 788–803, 1976.

Sondergeld, C.H., B.L. Isacks, M. Barazangi, and S. Billington, A search for velocity anomalies near the deep portions of the inclined seismic zone of Tonga island arc. *Bull. Seismol. Soc. Am.* **67**, 537–541, 1977.

Souriau, A. and M. Souriau, Test of tectonic models by great circle Rayleigh waves. *Geophys. J. R. Astr. Soc.* **73**, 533–551, 1983.

Spencer, C.P. and E.R. Engdahl, A joint hypocentre location and velocity inversion technique applied to the Central Aleutians. *Geophys. J. R. Astr. Soc.* **72**, 399–415, 1983.

Spudlich, P. and J. Orcutt, Petrology and porosity of an oceanic crustal site: results from wave form modeling of seismic refraction data. *J. Geophys. Res.* **85**, 1409–1433, 1980.

Stacey, F.D. and D.E. Loper, The thermal boundary-layer interpretation of D″ and its role as a plume source. *Phys. Earth Planet. Inter.* **33**, 45–55, 1983.

Stark, P.B. and C. Frohlich, The depths of the deepest deep earthquakes. *J. Geophys. Res.* **90**, 1859–1869, 1985.

Stauder, W. and L. Mualchin, Fault motion in the larger earthquakes of the Kuril-Kamchatka Arc and of the Kuril-Hokkaido Corner. *J. Geophys. Res.* **81**, 297–308, 1976.

Steckler, M.S. and A.B. Watts, Subsidence of the Atlantic-type margin of New York. *Earth Planet. Sci. Lett.* **41**, 1–13, 1978.

Strelitz, R., The September 5, 1970, Sea of Okhotsk earthquake: A multiple event with evidence of triggering. *Geophys. Res. Lett.* **2**, 124–127, 1975.

Stevenson, D., Limits on lateral density and velocity variations in the earth's core. *Geophys J. R. Astr. Soc.* **88**, 311–319, 1987.

Suito, K., Phase relations of pure Mg_2SiO_4 up to 200 kilobars. In M.H. Manghnani and S. Akimoto (eds.), *High Pressure Research, Applications in Geophysics*, New York: Academic Press, pp. 255–266, 1977.

Sumino, Y., The elastic constants of Mn_2SiO_4, Fe_2SiO_4, and Co_2SiO_4, and the elastic properties of olivine group minerals at high temperature. *J. Phys. Earth* **27**, 209–238, 1979.

Sumino, Y., O.L. Anderson, and I. Suzuki, Temperature coefficients of elastic constants of single crystal MgO between 80° and 1300° K. *Phys. Chem. Minerals* **9**, 38–47, 1983.

Sumino, Y., O. Nishizawa, T. Goto, I. Ohno, and M. Ozima, Temperature variation of elastic constants of single-crystal forsterite between −190° and 400°C. *J. Phys. Earth* **25**, 377–392, 1977.

Sung, C.-M., Kinetics of the olivine → spinel transition under high pressure: Experimental results and geophysical implications. In K.D. Timmerhaus and M.S.

Barber (eds.), *High-Pressure Science and Technology, Vol. 2*, New York: Plenum Press, pp. 31–42, 1979.

Sung, C.-M. and R.G. Burns, Kinetics of the olivine → spinel transition: Implications to deep-focus earthquake genesis. *Earth Planet. Sci. Lett.* **32**, 165–170, 1976a.

Sung, C.-M. and R.G. Burns, Kinetics of high-pressure phase transformations: implications to the evolution of the olivine → spinel transition in the downgoing lithosphere and its consequences of the dynamics of the mantle. *Tectonophysics* **31**, 1–32, 1976b.

Sutton, G.H. and D.A. Walker, Oceanic mantle phases recorded on seismographs in the northwestern Pacific at distances between 7° and 40°. *Bull. Seismol. Soc. Am.* **62**, 631–655, 1972.

Suyehiro, K., and I.S. Sacks. *P*- and *S*-wave velocity anomalies associated with the subducting lithosphere determined from travel-time residuals in the Japan region. *Bull Seismol. Soc. Am.* **69**, 97–114, 1979.

Suzuki, I., O.L. Anderson, and Y. Sumino, Elastic properties of a single-crystal forsterite Mg_2SiO_4, up to 1200°K. *Phys. Chem. Minerals* **10**, 38–46, 1983.

Sykes, L.R., The seismicity and deep structure of island arcs. *J. Geophys. Res.* **71**, 2981–3006, 1966.

Tanimoto, T. and D.L. Anderson, Lateral heterogeneity and azimuthal anisotropy of the upper mantle: Love and Rayleigh 100–250 s. *J. Geophys. Res.* **90**, 1842–1858, 1985.

Toksöz, M.N. and D.L. Anderson, Phase velocities of long period surface waves and structure of the upper mantle. *J. Geophys. Res.* **71**, 1649–1658, 1966.

Toksöz, N., J.W. Minear, and B.R. Julian, Temperature field and geophysical effects of a downgoing slab. *J. Geophys. Res.* **76**, 1113–1138, 1971.

Toksöz, M.N., N.H. Sleep, and A.T. Smith, Evolution of the downgoing lithosphere and the mechanisms of deep focus earthquakes. *Geophys. J. R. Astr. Soc.* **35**, 285–310, 1973.

Trehu, A.M., Depth versus $(age)^{1/2}$: A perspective on mid-ocean rises. *Earth Planet. Sci. Lett.* **27**, 287–304, 1975.

Turcotte, D.L. and E.R. Oxburgh, Finite amplitude convective cells and continental drift. *J. Fluid Mech.* **28**, 29–42, 1967.

Turcotte, D.L. and E.R. Oxburgh, A fluid theory for the deep structure of dip-slip fault zones. *Phys. Earth Planet. Inter.* **1**, 381–386, 1968.

Turcotte, D.L. and G. Schubert, Structure of the olivine-spinel phase boundary in the descending lithosphere. *J. Geophys. Res.* **76**, 7980–7987, 1971.

Turcotte, D.L. and G. Schubert, Frictional heating of the descending lithosphere. *J. Geophys. Res.* **78**, 5876–5886, 1973.

Turner, H.H., On the arrival of earthquake waves at the antipodes, and on the measurement of the focal depth of an earthquake. *Mon. Not. R. Astr. Soc., Geophys. Suppl.* **1**, 1–13, 1922.

Turner, H.H., Deep focus. In *International Seismological Summary for 1927, January, February, March*,pp. 1–108, Oxford, 1930.

Utsu, T., Seismological evidence for anomalous structure of island arcs with special reference to the Japanese region. *Rev. Geophys. Space Phys.* **9**, 839–890, 1971.

Vassiliou, M.S., The state of stress in subducting slabs as revealed by earthquakes analysed by moment tensor inversion. *Earth Planet. Sci. Lett.* **69**, 195–202, 1984.

Vassiliou, M.S., B.H. Hager, and A. Raefsky, The distribution of earthquakes with depth and stress in subducting slabs. *J. Geodynamics* **1**, 11–28, 1984.

Veith, K.F., The relationship of island are seismicity to plate tectonics, Ph.D. dissertation, Southern Methodist Univ., Dallas, Tex., 1974.

Vening Meinesz, F.A., Indonesian Archipelago: A geophysical study. *Geol. Soc. Am. Bull.* **65**, 143–164, 1954.

Vink, G.E., W.J. Morgan, and W.-L. Zhao, Preferential rifting of continents: A source of displaced terranes. *J. Geophys. Res.* **89**, 10072–10076, 1984.

Vitorello, I. and H.N. Pollack, On the variation of continental heat flow with age and the thermal evolution of continents. *J. Geophys. Res.* **85**, 983–995, 1980.

Wadati, K., On shallow and deep earthquakes. *Geophys. Mag.* **1**, 161–202, 1928.

Wadati, K., On the activity of deep-focus earthquakes in the Japan Islands and neighborhoods. *Geophys. Mag.* **8**, 305–325, 1935.

Wadati, K. and K. Masuda, On the travel time of earthquake waves (Part V). *Geophys. Mag.* **7**, 269–290, 1933.

Walck, M.C., Teleseismic array analysis of upper mantle compressional velocity structure, Ph.D. Dissertation, Calif. Inst. of Technol., Pasadena, Calif., 1984a.

Walck, M.C., The *P*-wave upper mantle structure beneath an active spreading centre: The Gulf of California. *Geophys. J. R. Astr. Soc.* **76**, 697–723, 1984b.

Walker, D., High-frequency P_n and S_n phases recorded in the western Pacific. *J. Geophys. Res.* **82**, 3350–3360, 1977.

Watson, J.V.. *Phil. Trans. R. Soc. Lond., Ser. A.* **280**, 629–640, 1976.

Watts, A.B., J.H. Bodine, and M.S. Steckler, Observations of flexure and the state of stress in the oceanic lithosphere. *J. Geophys. Res.* **85**, 6369–6376, 1980.

Watts, A.B., J.R. Cochran, P. Patriat, and M. Doucoure, A bathymetry and altimetry profile across the Southwest Indian Ridge at 31°S latitude. *Earth Planet. Sci. Lett.* **73**, 129–139, 1985.

Weichert, D.H., *Earth Planet. Sci. Lett.* **17**, 181, 1972.

Weidner, D.J., A mineral physics test of a pyrolite mantle. *Geophys. Res. Lett.* **12**, 417–420, 1985.

Wickens, A.J. and G.G.R. Buchbinder, S-wave residuals in Canada. *Bull. Seismol. Soc. Am.* **70**, 809–822, 1980.

Windley, B.F., *The Evolving Continents*, Chichester, John Wiley & Sons, 385 pp., 1977.

Woodhouse, J.H. and F.A. Dahlen, The effect of a general aspherical perturbation on the free oscillations of the earth. *Geophys. J. R. Astr. Soc.* **53**, 335–354, 1978.

Woodhouse, J.H. and A.M. Dziewonski, Mapping the upper mantle: three-dimensional modeling of earth structure by inversion of seismic waveforms. *J. Geophys. Res.* **89**, 5953–5986, 1984.

Woodhouse, J.H., D. Giardini, and X.-D. Li, Evidence for inner core anisotropy from splitting in free oscillation spectra. *Geophys. Res. Lett.* **13**, 1549–1552, 1986.

Woodhouse, J.H. and T.P. Girnius, Surface waves and free oscillations in a regionalized earth model. *Geophys. J. R. Astr. Soc.* **68**, 653–673, 1982.

Wortel, R., Seismicity and rheology of subducted slabs. *Nature* **296**, 553–556, 1982.

Yagi, T., P.M. Bell, and H.K. Mao, Phase relations in the system MgO-FeO-SiO_2 between 150 and 700 kbar at 1000°C. *Carnegie Inst. Washington Year Book* **78**, 614–618, 1979.

Yoder, H.S., Jr.. *Generation of Basaltic Magma*, Washington, D.C.: National Academy of Sciences, 265 pp., 1976.

Yoshii, T., Regionality of group velocities of Rayleigh waves in the Pacific and thickening of the plate. *Earth Planet. Sci. Lett.* **25**, 305–312, 1975.

Yoshii, T., Cross sections of some geophysical data around the Japanese Islands. In T.W.C. Hilde and S. Uyeda (eds.), *Geodynamics of the western Pacific-Indonesian Region, Geodynamics Series, Vol. 11*, Washington, D.C., American Geophysical Union, pp. 343–354, 1983.

Yu, G.K. and B.J. Mitchell, Regionalized shear velocity models of the Pacific upper mantle from observed Love and Rayleigh wave dispersion. *Geophys. J. R. Astr. Soc.* **57**, 311–341, 1979.

Yuen, D.A. and W.R. Peltier, Mantle plumes and the thermal stability of the D″ layer. *Geophys. Res. Lett.* **7**, 625–628, 1980.

CHAPTER 4

High Pressure Chemistry of the Earth's Mantle and Core

Raymond Jeanloz

Department of Geology and Geophysics, University of California
Berkeley, California 94720

1. Introduction

High-pressure experiments are beginning to provide detailed and reliable information on the nature of the Earth's deep interior. For example, the single most abundant mineral of the interior, silicate perovskite, was only recently discovered as a result of experiments with the diamond-anvil cell (Liu and Ringwood, 1975; Liu, 1976). Although conditions existing in the mantle and core can be simulated in the laboratory, this is not the only reason that high-pressure experiments are important. Rather, by studying the response of minerals to a wide range of temperature and pressure, measuring physical properties and studying the nature of transformations or reactions that occur, it is possible to systematically understand the mineral properties that ultimately determine the evolution of the planetary interior.

That mineral chemistry and the evolution of the planet are interconnected is a major conceptual breakthrough of the past decade or two. Thus, any model of the composition or structure of the Earth's interior must be compatible with models of its thermal state and tectonic evolution. Constraints on the composition of regions in the mantle or core are derived largely from seismology, and these have an immediate bearing on the dynamics of the planetary interior. The interpretation of seismological data, however, is based on a knowledge of mineral properties which are experimentally determined at high pressures and temperatures.

The purpose of this article is to summarize the basic lines of evidence that are currently available for understanding the chemical nature of the mantle and core. In particular, I outline what is known of the crystal structures and physical properties of the minerals comprising these regions. Experimental rather than theoretical approaches are emphasized here because most of the important advances have come from laboratory studies. Nevertheless, making realistic theoretical models of the Earth's internal dynamics is a primary application of mineral properties that are measured.

2. Composition of the Earth's Interior: Overview

2.1. Petrology and Geophysics without Seismology

Compositionally, the Earth is divided into two regions: the metallic core, and the surrounding nonmetallic mantle consisting of oxide minerals (mainly silicates). This conclusion is based not so much on direct sampling of rocks from the Earth's interior, but mainly on geophysical lines of evidence. The trouble is that no samples are known to have come to the surface from depths greater than 150–200 km, the deepest that partial melting has been shown to occur in the mantle (Fig. 4.1). Thus, geophysical properties of the Earth's interior must be used to indirectly constrain the composition at depth.

Among the key observations is that the Earth's magnetic field is mainly generated in the core. Magnetohydrodynamics, which provides the only mechanism that has been identified for producing the geomagnetic field, requires the core to be electrically conducting. Consistent with this hypothesis is the fact that the outer core is liquid, as indicated by seismology (see

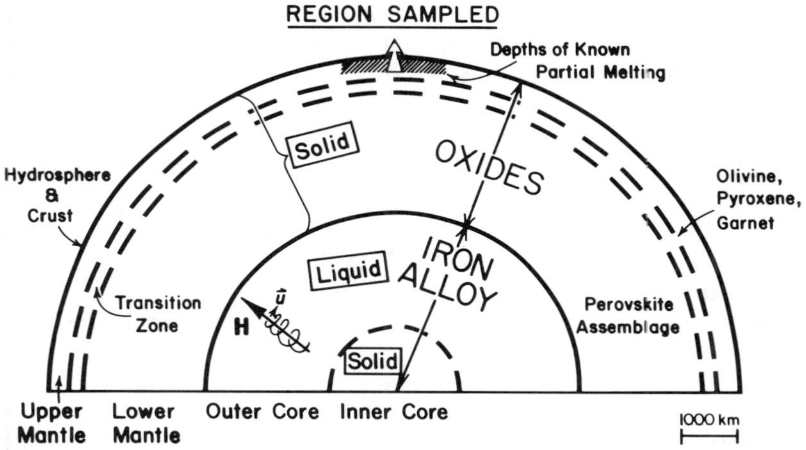

Figure 4.1. Schematic cross-section of the Earth illustrating the distinction between the oxide mantle and the iron-rich core. Based on seismological observations, the mantle and inner core are known to be essentially solid, whereas the outer core is fully molten. Cyclonic motions (u) of the metallic fluid in the outer core generate the main geomagnetic field (**H**). The greatest depths from which rock fragments are brought to the surface, and hence the greatest depth for which there is direct evidence of partial melting, is 150–200 km. Thus, only part of the upper mantle has been sampled; no samples have yet been identified with origins in the transition zone or lower mantle. At the conditions of these deeper regions of the mantle, the olivine, pyroxene and garnet of the upper mantle are transformed to a high-pressure assemblage dominated by the perovskite phase.

below), and hence motions in the core can be sufficiently vigorous to sustain the magnetic field (see Jacobs, 1975; Merrill and McElhinny, 1983).

Measured time variations of the Earth's magnetic field indicate that the mantle is not highly conducting (Banks, 1972; Anderssen *et al.*, 1979). Rather, the properties of the mantle are consistent with what would be expected for minerals that are known to exist in the uppermost mantle, such as olivine, pyroxene and garnet. That is, if one takes into account the effects of pressure and temperature on mineral properties it is found that the seismological properties of the whole mantle are consistent with its being made up of the oxides and silicates that are characteristic of the uppermost mantle.

Two types of sampling provide information on the mineral content of the mantle near the surface. On the one hand, xenoliths are brought up from depths as great as 150 to 200 km by kimberlites, and also by a limited number of basaltic magmas. These discrete rock or mineral fragments, typically millimeters to centimeters in dimension, are foreign to the fluid (kimberlite or basalt) that brings them up, and so can be thought of as samples that have been torn out of the mantle by the ascending magma (Yoder, 1976; Dawson, 1980).

Petrological studies on the origin of basaltic magmas provide a second, but indirect constraint on the minerals existing in the uppermost mantle. Given the extent of the mid-ocean ridges at which basaltic magma is continuously forming new crust, it is clear that basalt is by far the most voluminous melt to come from the Earth's interior. An understanding of the source rock which, when partially melted under the appropriate conditions of pressure and temperature, forms basalt should therefore provide a powerful insight into the average mineralogical composition of the upper mantle (Basaltic Volcanism Studies Project, 1981). This advantage of studying basalt genesis can be contrasted with the difficulty that arises in studying xenoliths, namely that they occur rarely and under what might be considered geologically anomalous conditions (e.g., the explosive, fluid-rich eruption that characterizes kimberlites). Although they are direct samples of the mantle, there is always a possibility that xenoliths represent anomalous regions within the mantle.

Despite the potential problems due to biased sampling (xenoliths) or ambiguous interpretations of indirect evidence (basalt petrogenesis), there is good agreement on the overall mineralogy and composition of the uppermost mantle based on petrological observations. The minerals olivine ($(Mg,Fe)_2 SiO_4$), pyroxene (typically two separate phases, clinopyroxene and orthopyroxene, of approximate composition $(Mg,Fe,Ca)SiO_3$) and garnet (approximately $(Mg,Fe,Ca)_3Al_2Si_3O_{12}$) are the most abundant in xenoliths brought up from depth. Also, the resulting rock type, which corresponds to garnet peridotite, does produce basalt upon partial melting (e.g., Ringwood, 1975; Yoder, 1976; Basaltic Volcanism Studies Project, 1981). The minerals are typically rich in the magnesian components (e.g., forsteritic olivine of composition $(Mg_{0.9}Fe_{0.1})_2SiO_4$ is characteristic), and the proportion of minerals constituting the upper mantle rock is approximately 60 percent olivine, 30 percent pyroxene and 10 percent garnet. Refinements beyond this level of discussion are arguable, for example whether more picritic (more Mg-rich) compositions are the most appropriate for basalt genesis. Nevertheless, the properties of the mineral assemblage outlined here has long been known to be consistent with the seismologically observed properties of the upper mantle (e.g., Fujisawa, 1968; Basaltic Volcanism Studies Project, 1981; Jeanloz and Thompson, 1983).

2.2. Seismology and Earth Structure

By far the most detailed information available on the properties of the lower mantle and core comes from seismology. Aside from the observations summarized in the preceding paragraphs, seismology provides the only direct information on the nature of the Earth's deep interior: seismic waves generated by earthquakes (or artificial sources) are the only sensitive probes that directly sample regions at all depths within the planet. The resulting determinations of elastic wave velocity and density as functions of depth define the basic structure of the Earth's interior (Fig. 4.2).

Earthquakes generate elastic waves travelling through and around the globe (body and surface waves), as well as free oscillations of the whole planet. Typically these range in frequency from mHz for the normal modes of free oscillation to 1 Hz or more for body waves, with surface waves at intermediate frequencies. Thus, models of the Earth's internal structure are based on combining measurements of the travel times for waves traversing the planet with determinations of the normal mode frequencies; integral properties, such as the known mass and moment of inertia of the planet are also included. This information is sufficient for deriving the longitudinal and transverse wave velocities as well as the density as functions of depth.

What should be emphasized, however, is that the inversion of the raw seismological data to yield a final model of the Earth's interior is not unique. The situation can be described by analogy with a least-squares problem in which there are more parameters to solve for than the number of observations available. In principle, defining a continuous distribution, such as density as a function of depth, requires an infinity of points and therefore can only be done with an infinite number of observations. Not only is the number of seismological observations finite, but the sampling of properties by seismic waves is highly nonuniform both laterally and in depth. For example, most earthquakes occur along crustal-plate boundaries and all recording stations are located at the Earth's surface. Thus, there is an inherent loss of resolution with depth, and the final model can not be uniquely defined.

What is done in this situation is to find a self-consistent Earth model that satisfies all of the observations as closely as possible (note that the observations, though numerous, are not necessarily independent). The importance of the starting model in such an inversion cannot be overemphasized, because the final solution is made to conform as closely as possible to the

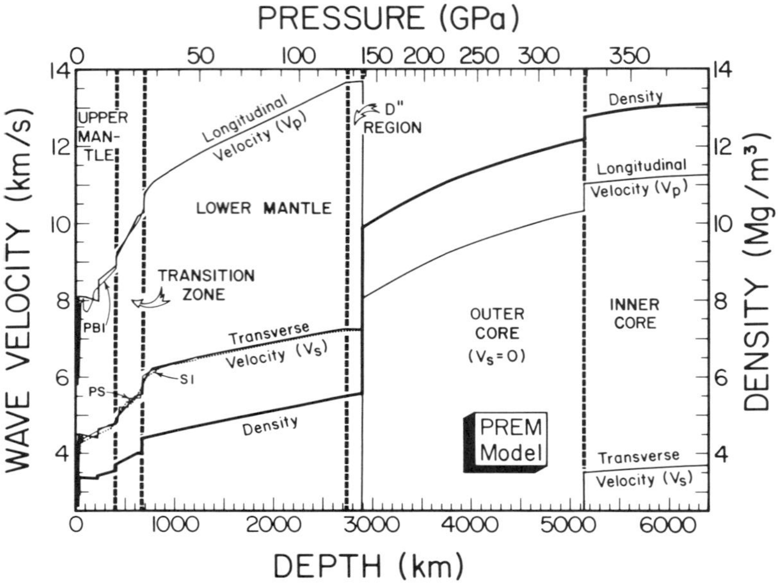

Figure 4.2. The present structure of the Earth's interior is given by the seismologically determined longitudinal velocity (V_p), transverse velocity (V_s) and density as a function of depth (the corresponding pressure at each depth is shown on the upper scale: see Table 4.1). This figure is based on the Preliminary Reference Earth Model of Dziewonski and Anderson (1981). For comparison, models of the elastic-wave velocity based on waveform analysis are also shown for the mantle (S1: Hart, 1975; PB1: Burdick and Powell, 1975; PS: Fukao *et al.*, 1982).

observed data while departing as little as possible from the initial model. Thus, an ever present danger is that what appears to be significant structure in the velocity or density profiles with depth may, in fact, be merely due to a bias that was built into the starting model. Nevertheless, the tradeoffs between resolution in depth and reliability of the final model can be objectively defined in this approach (Backus and Gilbert, 1970). In general, one can expect the spatial (depth) resolution to degrade with depth, but it is nonetheless quite good: on the order of a few hundred kilometers throughout the Earth.

Whole-Earth models, such as PREM (Fig. 4.2) represent an idealized planet which is, to first order, elastic, spherically symmetric (laterally homogeneous) and isotropic within each region. It should be noted, however, that much of the recent activity in seismology has centered on going beyond these idealized conditions. For example, anelasticity and one degree of anisotropy (but only near the surface) are taken into consideration in PREM (Dziewonksi and Anderson, 1981). Moreover, lateral heterogeneity is presently a topic of active research which is encouraging the use of tomographic back-projection techniques in seismology (e.g., Clayton, 1984; Anderson, 1984). This recent development is of considerable interest in allowing the properties at depth to be mapped laterally, thus giving new insights into mantle dynamics and its relation to the geological structures observed at the surface. Due both to the attenuation of seismic waves (this includes both scattering, and inherent attenuation or energy dissipation), as well as to the finite number and nonuniform distribution of seismic sources and receivers the tomographic problem for the Earth is still intrinsically nonunique (e.g., Deans, 1983). Nevertheless, an important direction of current research is to attempt to identify lateral variations in density and velocity at depth (Masters, *et al.*, 1982; Dziewonski, 1984; Woodhouse and Dziewonski, 1984).

Returning to the global model (Fig. 4.2), and treating each region as isotropic, the two wave velocities, longitudinal (or compressional, V_P) and transverse (or shear, V_S), are related to the adiabatic bulk modulus (K_S), the shear modulus (μ) and density (ρ) by:

$$V_P = \sqrt{\frac{K_s + \frac{4}{3}\mu}{\rho}}, \tag{4.1}$$

$$V_S = \sqrt{\frac{\mu}{\rho}}. \tag{4.2}$$

As the density is independently determined from V_P and V_S, this yields the bulk and shear moduli as functions of depth. Also, knowledge of the density with depth allows the pressure to be calculated throughout the Earth.

The pressure is essentially hydrostatic in the mantle and core because minerals are not strong enough at high temperature to support shear stresses (i.e., local stress variations) that are even remotely comparable in magnitude to the hydrostatic pressure. For example, long-term yield strengths are thought to be on the order of 1–100 MPa (10 bar–1 kbar), whereas comparable hydrostatic pressures are achieved within a few kilometers depth. This result has long been known (Birch, 1952), but it is even further reinforced by more recent findings that minerals are relatively weak at high pressures (Davies, 1982). Of course, it is these "negligible" nonhydrostatic stresses that drive convection in the mantle and core, so they are ultimately of great dynamic importance.

The hydrostatic pressure at depth z_1 is given by

$$P(z_1) = \int_0^{z_1} \rho g \, dz, \tag{4.3}$$

where the local acceleration of gravity (g) depends on radial distance (r) and hence depth ($z = 6371$ km $- r$):

$$g(r_1) = \frac{G}{r_1^2} \int_0^{r_1} 4\pi r^2 \rho \, dr, \tag{4.4}$$

with G being the gravitational constant. The density of a material is, in turn, thermodynamically related to the pressure by the bulk modulus:

$$K_S = \rho \left(\frac{\partial P}{\partial \rho} \right)_S \tag{4.5}$$

under adiabatic conditions.

If the density increases with depth due only to compression (i.e., if the composition and mineral phases remain uniform) and if the temperature variation with depth is adiabatic in that same region, the local change in density with depth (or pressure) should be given by the bulk modulus that is measured seismologically (equations 1, 2, and 5). Deviations from these conditions of adiabaticity and compositional homogeneity can be described by the inhomogeneity parameter (cf. Birch, 1952; Bullen, 1975)

$$\eta = \frac{\left(V_P^2 - \frac{4}{3} V_S^2 \right)}{\rho g} \frac{d\rho}{dr}. \tag{4.6}$$

From equations 4.1 to 4.5, it is evident that η is the ratio of the adiabatic bulk modulus, K_s, to the effective bulk modulus of the Earth at the same location: $K_E = \rho(dP/d\rho)$. The seismic velocities determine K_s, whereas K_E is constrained from the density-depth relation. Therefore, when the change in density with depth is equal to the change in density along an adiabat $\eta = 1$. That η is close to one for the lower mantle and outer core (e.g., Masters, 1979; Dziewonski and Anderson, 1981) is consistent with the idea that both of these regions are convecting vigorously, and hence the interiors of these regions are homogeneous and lie on an adiabat. Heinz and Jeanloz (1983) have noted, however, that the value of η is sensitive not only to inhomogeneity but also to anelastic effects and to the differences in elastic properties of the minerals coexisting in a given region.

The present structure of the Earth, as given by seismological data, is summarized in Table 4.1 (cf. Fig. 4.2). The most significant changes in velocity and density occur at the core-mantle boundary and near the crust; these are followed in importance by the discontinuities at 670 and 400 kilometers depth, which define the upper mantle, transition zone and lower mantle. The outer core is fluid, based on the observation that $V_S = 0$ in this region, whereas the inner core and the mantle are, within seismological resolution, solid (i.e., local partial melting that corresponds to small fractions of the global average are not precluded). Three observations indicate that the transverse-wave velocity vanishes in the outer core: horizontally polarized shear waves are not propagated from the mantle into the core (note that vertically polarized shear waves can generate compressional waves in the outer core), and both normal-mode frequencies and reflection coefficients for waves bouncing off the core-mantle interface differ significantly from what would be expected if the outer core were solid.

A comparison of the seismological profiles with high-pressure measurements (Fig. 4.3) confirms that the properties of the lower mantle are compatible with those of oxides and silicates at high pressures, whereas the properties of the core are consistent with those of iron alloys (e.g., iron-sulfur or iron-oxygen compositions). Iron is considered to be the main constituent of the core because it is the element with the largest cosmochemical abundance (by orders of magnitude) that satisfies the observed properties of the core, including the high (metallic) electrical conductivity and the pressure-density-velocity relation (see Stevenson, 1981). Hence the basic distinction between the silicate-oxide mantle (i.e., similar in gross composition to that

Table 4.1. Summary of Earth structure†

Depth range (km)	Region‡	Fraction of earth§ Mass	Atomic	Pressure‖ (GPa)	Temperature¶ (K)
0–400	Upper mantle	10.3%	12.8% (21)	0–13.4	1200–1700
400–670	Transition zone	7.5%	9.4% (21)	13.4–23.8	1400–2100
670–2890	Lower mantle	49.2%	61.4% (21)	23.8–135.8	2000–3600
2890–5150	Outer core	30.8%	15.0% (54)	135.8–328.9	3500–4800
5150–6370	Inner core	1.7%	0.8% (56)	328.9–363.9	4500–5500

†Based on PREM model (Dziewonski and Anderson, 1981).
‡Excluding crust, oceans, and atmosphere.
§Total mass of Earth is 5.973×10^{24} kg. Values in parentheses indicate assumed mean atomic weight in gm/mole of atoms = 1.66×10^{-27} kg/atom (probable uncertainty is 5 percent).
‖1 GPa = 10 kilobar, and 100 GPa = 1 Mbar ≅ 10^6 atmospheres pressure.
¶Range of estimates: horizontally averaged temperatures throughout each depth interval.

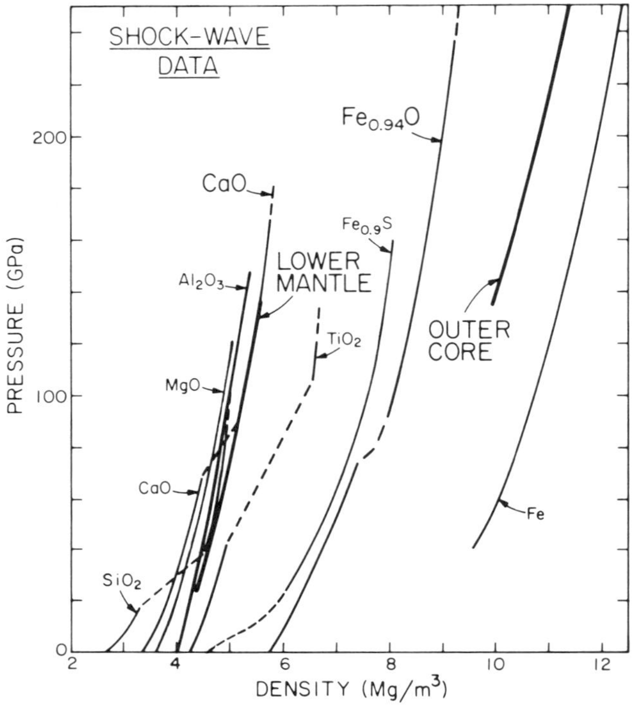

Figure 4.3. Shock-wave equations of state (Hugoniots) of simple-oxide and Fe phases are compared with the seismologically determined pressure-density profiles of the Earth's lower mantle and outer core. The dashed sections indicate regions of anomalous compression associated with phase transformations that are inferred to occur under shock-loading. Samples of both the oxide and sulfide of iron are nonstoichiometric. From Jeanloz (1981a), where the sources of data are given.

observed in samples from the uppermost mantle) and the iron-rich core is largely based on a comparison with the mineral properties measured at high pressures.

This distinction is reflected in the difference in estimated mean atomic weights for the two regions: ~21 and ~55 gm/mole of atoms for the mantle and core (Birch, 1961). It is evident, then, that although the core represents nearly one-third the mass of the Earth, it corresponds to less than 16 percent on an atomic basis (Table 4.1). In fact, the bulk of the planet is located in the lower mantle, and this motivates special interest in understanding the properties of minerals at the conditions of high pressure and temperature existing between 670 and 2900 km.

2.3. Effects of High Pressure and Temperature: Phase Transitions

There are two ways in which pressure and temperature affect mineral properties. On the one hand, density increases with increasing pressure and decreases with increasing temperature, as described by the bulk modulus (equation 4.5) and the coefficient of thermal expansion

$$\alpha = -\rho\left(\frac{\partial T}{\partial \rho}\right)_P \tag{4.7}$$

(α can be negative, but only at temperatures near 0 K). On the other hand, the much more profound effect of temperature and pressure is to transform minerals to new phases. Either solid–solid transformations (changes in crystal structure) or solid–liquid transformation (melting) can occur with increasing temperature; in general, only solid–solid transitions occur with increasing pressure. The result is that drastic changes in material properties occur at phase transitions. One of the primary objectives of high-pressure research, therefore, is to identify the nature of phase transitions and the resulting changes in properties, as well as to determine the influence of pressure on the properties of a single mineral phase.

Pressure-induced transitions differ fundamentally from temperature-induced transitions because the former are driven by the volume change on transformation (ΔV_{tr}, the high-pressure phase having the smaller volume), whereas the latter are driven by the entropy change on transformation (ΔS_{tr}, with the high-temperature phase having the larger entropy). In general, both factors can come into play, and the slope of the coexistence boundary of the phases in pressure-temperature space is given by the Clapeyron equation:

$$\frac{dP}{dT} = \frac{\Delta S_{tr}}{\Delta V_{tr}}. \tag{4.8}$$

In actuality, one or the other, ΔS_{tr} or ΔV_{tr}, often dominates so that most transitions can be thought of as either due to high pressure ($\Delta S_{tr} \sim 0$) or high temperature ($\Delta V_{tr} \sim 0$), as illustrated in Figure 4.4. This is useful also for considering the transitions at a microscopic level. Specifically, a volume decrease (pressure-induced transition) involves packing the atoms more tightly in the high-pressure phase. Entropy, however, is determined by the

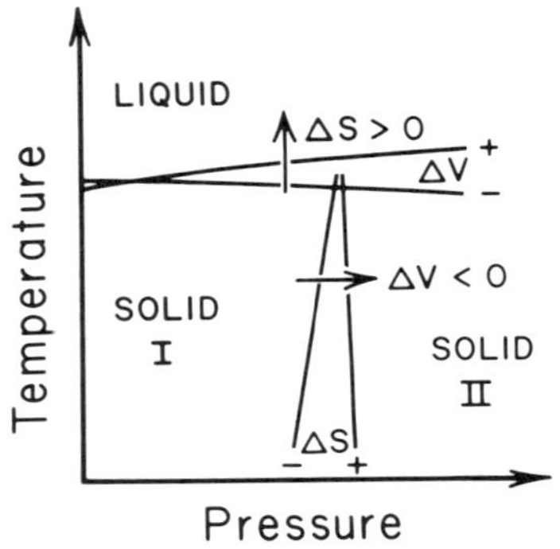

Figure 4.4. Pressure-temperature phase diagram illustrating the difference between pressure-induced phase transitions (driven by $-\Delta V$, as in the Solid I-Solid II transformation) and temperature-induced phase transitions (driven by ΔS, as in the Solid-Liquid or melting transition). Both positive and negative Clapeyron slopes are possible for the phase boundaries at either type of transition, depending on the signs (+ or −) of the volume and entropy charges (see equation 4.8).

vibrational frequencies (hence, interatomic bond strengths and distances) and also by the atomic packing configuration. For example, a major contribution to the entropy increase on melting is due to the fact that the melt is disordered relative to the crystalline order characterizing the solid.

To give an idea of the energetics involved in phase transitions, consider first a pressure-induced transformation. The internal energy change is

$$\Delta U_{tr} \sim -P_{tr}\Delta V_{tr} \tag{4.9}$$

(assuming $\Delta S_{tr} = 0$). For mantle silicates, transformations occur at pressures of the order of 10 GPa (10^2 kbar) and with volume decreases of several percent magnitude (corresponding to $\sim$0.5 cm^3/mol of atoms). Thus, energy changes of about 5–10 kJ/mol of atoms, equivalent to about 0.1 eV or 10^3 K temperature, are involved in high-pressure transformations. Similarly, the energy on melting,

$$\Delta U_{tr} \sim T_{tr}\Delta S_{tr} \tag{4.10}$$

(assuming $\Delta V_{tr} = 0$), is determined by the entropy change, which is roughly of the magnitude of the gas constant, R (8.314 J/K·mol of atoms). As melting temperatures are of order 10^3 K, the energy change is again found to be around 10 kJ/mol of atoms, or 0.1 eV.

For comparison, the energy changes on heating or compressing a mineral in a given phase are shown in Figure 4.5. Energy changes of several kJ/mol of atoms (tenths of eV) are achieved with either temperature changes of thousands of degrees or pressure changes of tens of GPa (hundreds of kilobars). Two conclusions can be drawn from this thermodynamic comparison. On the one hand, it illustrates how large an energy perturbation is

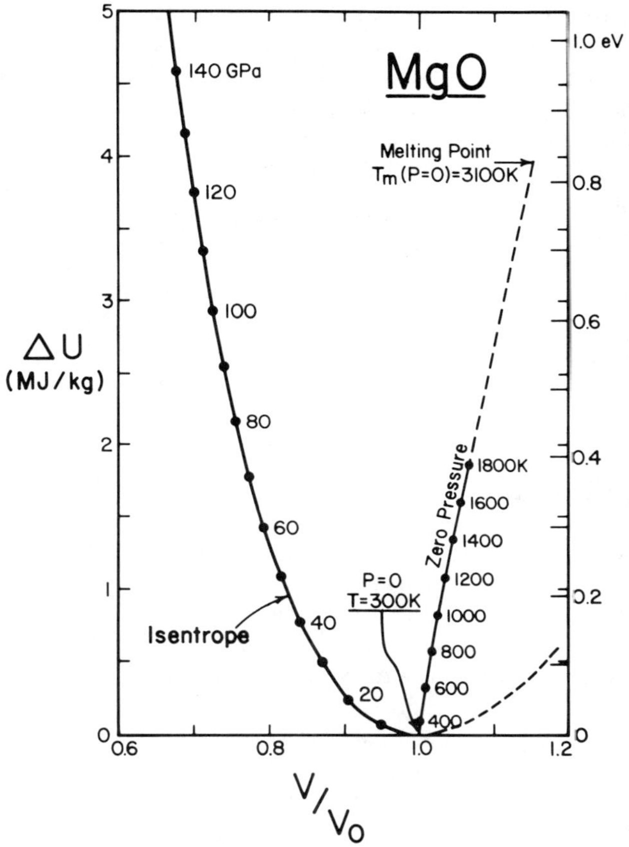

Figure 4.5. Change in the internal energy of MgO (periclase) as a function of volume, with the reference condition being at a temperature of 300 K and zero pressure. Two thermodynamic paths are illustrated: the isentrope shows the effect of pressure (marked at 10 GPa intervals), whereas the zero-pressure isobar shows the effect of temperature (marked at 200 K intervals) on the thermodynamic state of MgO. Internal energy is given in SI units (left scale) and in electron volts (right scale; 1 eV = 11605 K thermal energy). The figure is based on data given by Robie *et al.* (1978) and Jackson and Niesler (1982); a fourth-order Eulerian finite strain equation of state is used and the isobar is dashed where reliable data are not available. Note that the pressures and temperatures spanned in this figure are analogous to those found in the Earth's mantle.

associated with either high-pressure or high-temperature transformations (perturbations corresponding to large changes in pressure or temperature). On the other hand, it is clear that the ranges in pressure and temperature in the mantle and core can produce energy changes in a given mineral phase that are of comparable magnitude to those associated with phase transitions (Table 4.1). It is in this sense that the conditions of the Earth's deep interior represent a major perturbation to the thermodynamic state of minerals. The result is that minerals transform to phases that are not seen at the Earth's surface, except in the laboratory. Fortunately, techniques are available for producing in the laboratory pressures of hundreds of GPa and temperatures of thousands of degrees. Hence the conditions of the deepest interior can be reproduced and the response of minerals to the correspondingly large thermodynamic perturbation can be examined in detail.

3. High-Pressure Experiments

Dynamic and static experiments represent the two approaches used for achieving pressures of tens to hundreds of GPa (millions of atmospheres) in the laboratory (see Rice *et al.*, 1958; Ahrens, *et al.*, 1969; Bassett, 1979). Large, centimeter-sized samples can be compressed dynamically, but as the compression is achieved by a transient high-pressure wave (a shock wave), the experiments are of short duration ($\lesssim 10^{-6}$ s.). In contrast, elevated pressures can be maintained indefinitely in static experiments, but these involve squeezing extremely small ($\sim 10^{-6}$ cm^3) samples between anvils. Thus, both approaches are complementary in that they involve different limitations.

The contrast between static and dynamic techniques is even more significant from a thermodynamic point of view. High-pressure reactions typically occur under nonequilibrium conditions in shock-wave experiments, whereas equilibrium can be achieved in static experiments. Samples are heated during shock loading (in fact, the heating results from the irreversible nature of shock compression), and thus temperatures of the Earth's interior can readily be achieved at high dynamic pressures. In static experiments, however, high sample temperatures must be produced by independent means, and this can be quite difficult at elevated pressures. Finally, one of the most important features of shock-wave experiments is that the pressure and density of the sample are measured directly and unambiguously, whereas indirect determinations of pressure (relative to an independently calibrated pressure scale) characterize static experiments.

3.1. Shock-wave Experiments

High pressure is achieved in most shock-wave experiments by the impact of a projectile travelling at a velocity of several km/s into a stationary target which consists of the sample material. On impact, shock waves travel both into the target (forward) and into the projectile (backward), thus creating a highly compressed region (Fig. 4.6). The shock wave, being a large-amplitude stress wave, travels at a speed greater than the sound velocity in the uncompressed medium. Also, for a sufficiently high impact velocity, u_0, the shock pressure far exceeds the dynamic yield strength of the material. Therefore, in most experiments the sample behaves like a fluid under shock loading and simple conservation relations from hydrodynamics apply.

Consider a shock wave traversing a volume element of material with unit cross sectional area, as illustrated in Figure 4.6. In unit time, the shock front encompasses a volume equal to U_s, the shock-wave velocity. The initially stationary material ($u = 0$) is accelerated to a particle velocity u_p, so that the volume element is compressed to a size $U_s - u_p$. That is, the ratio of initial and final volumes is simply

$$\frac{V_0}{V_1} = \frac{U_s}{U_s - u_p}. \tag{4.11}$$

In terms of density ($\rho = 1/V$), this just expresses conservation of mass for the volume element:

$$\rho_0 U_s = \rho_1 (U_s - u_p). \tag{4.12}$$

The pressure change across the shock front, equal to the force exerted per unit area, is given by the mass of the volume element, $\rho_0 U_s$, times its acceleration (u_p in unit time):

$$P = \rho_0 U_s u_p, \tag{4.13}$$

assuming an initial pressure $P_0 = 0$

It is interesting to note that the internal energy change ($E - E_0$) is well characterized in this process, despite the fact that shock compression is irreversible. That is, the first law of thermodynamics stated as

$$dE = dW + dQ \tag{4.14}$$

is valid regardless of whether or not equilibrium is achieved (W and Q are the work and heat content, respectively, but $dQ < TdS$ in this case). As there is insufficient time for heat flow across the shock front, the compression is adiabatic and $dQ = 0$. The work is given by the compressional energy $-PdV$, however care must be taken because the material is not only compressed but

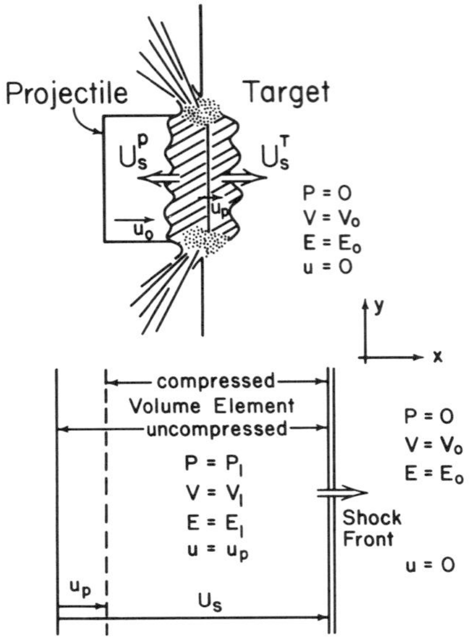

Figure 4.6. Illustration of a shock experiment. In two dimensions (top), the impact of a projectile travelling at velocity u_0 into a stationary target generates shock waves travelling forward into the target (at velocity U_S^T) and backwards into the projectile (shock velocity U_S^P). In the compressed region (shaded), the projectile is decelerated to a particle velocity u_P, which is the same velocity to which the target material is accelerated. Wherever the shocked region encounters vacuum, a rarefaction or decompressing is generated which tears the material apart (dotted region). The rarefaction front eats into the shocked region and creates the outward spray of material. In one dimension (bottom), a volume element of unit cross-sectional area is shown as it is encompassed by a shock wave travelling at velocity U_S. The material is initially at rest (velocity $u = 0$) and zero pressure, but is accelerated to velocity u_P and taken to pressure P_1. The initial volume (V_0) and energy (E_0) are decreased and increased, respectively, across the shock front.

also accelerated by the shock. That is, the kinetic energy of the volume element (mass of $\rho_0 U_s$ accelerated to velocity u_p),

$$KE = \tfrac{1}{2}\rho_0 U_s u_p^2\,, \qquad (4.15)$$

does not contribute to the internal energy. Recalling that the element is changed in volume by an amount u_p (Fig. 4.6), the energy change across the shock front is (using equations 4.13 and 4.14)

$$E - E_0 = -P\Delta V - KE = (\rho_0 U_s u_p)u_p - \tfrac{1}{2}\rho_0 U_s u_p^2 = \tfrac{1}{2}\rho_0 U_s u_p^2. \quad (4.16)$$

Fully half of the compressional energy change is "lost" from the internal (thermal) energy to kinetic energy.

Equations 4.12, 4.13 and 4.16, known as the Hugoniot relations, involve six variables, of which one (the initial sample density ρ_0) is readily measured prior to the experiment. Of the remaining five variables, the shock and particle velocities are usually measured during the experiment, so that the pressure, density (at high pressure) and energy change are uniquely defined. The shock velocity in the target is simply measured by observing the time required for the shock front to traverse the sample. Typically, samples are a few mm thick, so that with $U_s \sim 10$ km/s the corresponding travel time of about 10^{-7} s must be measured with nanosecond resolution.

The particle velocity, u_p, is derived from the impact velocity by way of impedance matching. For simplification, assume that the target and projectile are made of the same material. Then by symmetry $u_p = \tfrac{1}{2}u_0$, and the projectile is slowed down at the impact to the same velocity achieved by the target material (Fig. 4.6). In the case of unlike projectile and target materials the compression of the projectile must be accounted for, which is easily done if the projectile equation of state has been independently determined (i.e., the contrast in impedance, $\rho_0 U_s$, at the projectile-target interface must be accounted for).

Historically, shock-wave experiments have played an important role in high-pressure research because these provided, until recently, the only means of achieving pressures above a few tens of GPa. The fact that the pressure-volume-internal energy equation of state of materials can be directly measured, by way of the Hugoniot relations and the determination of U_s and u_p, has meant that shock experiments have continued to be of major importance. For example, virtually all calibrations of pressure scales used in ultrahigh pressure static experiments ultimately rely on equations of state measured in shock experiments. It should be noted, however, that one experiment is required for each P-V-E point defining a shock-compression curve (or Hugoniot). Thus, many experiments were required to define each of the Hugoniots illustrated in Figure 4.3.

For simple compounds and elements, which either undergo kinetically easy (e.g., displacive) transformations or do not transform, Hugoniot experiments have provided invaluable information on thermodynamic properties at high pressures. The application of shock experiments to structurally more complex substances, such as most minerals, is more controversial because these are characterized by kinetically difficult structural (solid–solid) transformations. As a result, the degree to which equilibrium is achieved under shock loading and the heterogeneous nature of the shocked state are areas of current study (e.g., Grady, 1980; Jeanloz, 1980; Watt and Ahrens, 1983); detailed studies of mineral high-pressure phases instead rely mainly on static experiments, at present.

Many of the new advances in shock experiments involve the development of new techniques that go beyond the Hugoniot measurements. For example, the high temperatures that are achieved during shock compression are now being measured directly. Such data, which are usually collected by radiometry, provide significant thermodynamic information on materials at high pressures and temperatures (e.g., Ahrens *et al.*, 1982). This can be illustrated by the fact that the temperature and Hugoniot energy can be combined to determine the specific heat ($C_V = (\partial E/\partial T)_V$). Furthermore, the change in entropy on melting causes a large change in temperature along the Hugoniot. This is important because, as noted above, the volume change on melting is small at high pressures (equation 4.10), so it has usually not been possible to determine when a sample melts under shock loading simply from Hugoniot equation-of-state measurements. With the new capability to measure temperature, however, the ~1000 K decrease in temperature associated with melting at constant volume ($T_m \Delta S_m / C_V$) can readily be observed. Melting temperatures at pressures of tens to hundreds of GPa are being determined in this way (Lyzenga *et al.*, 1983; Ahrens *et al.*, 1982). However, application of this technique is new in geophysics, so data are only available for a few minerals thus far.

Another important development has been the measurement of the sound velocity in shock-compressed samples. This is achieved by measuring the velocity of the rarefaction wave which decompresses the sample from the shocked state. As shown in Figure 4.6, a rarefaction is generated wherever the shock front emerges into vacuum. Surprisingly, as the rarefaction travels into the shock-compressed region (tearing apart the sample in the process) the decompression is essentially reversible (Jeanloz and Ahrens, 1979). Hence, when the shock wave in the projectile reaches the back end, a rarefaction is generated that travels through the projectile and target at the isentropic sound velocity, c. Because the rarefaction is carried along with the shocked material, its velocity is $c + u_p$. As $c + u_p > U_s$, the rarefaction catches up with the shock front if the sample is sufficiently thick (Fig. 4.7). By measuring the distance travelled by the shock through a target, before being attenuated, one directly measures the sound velocity at high pressures and temperatures (cf. Al'tshuler, 1967).

Normally the shock-compressed solid is unmelted, so the measured sound velocity is equivalent to the longitudinal velocity, V_p (equation 4.1). At

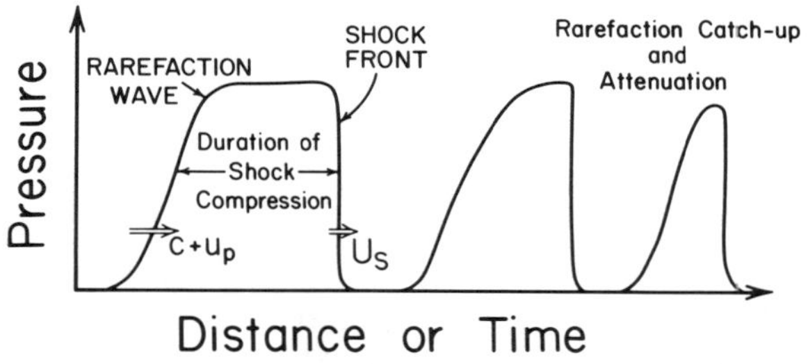

Figure 4.7. Summary of the rarefaction catch-up technique of measuring the sound velocity at high pressures and temperatures. In shock-impact experiments a rarefaction is generated when the backwards-travelling shock-wave emerges at the back of the projectile (see Figure 4.6). The rarefaction travels forward at the velocity $c + u_P$ relative to the shock front, where c is the longitudinal or bulk sound velocity in the shocked material. For a sufficiently thick sample, the rarefaction ultimately catches up with and attenuates the shock front.

sufficiently high shock pressures, however, Hugoniot temperatures can exceed the high-pressure melting point, so that the shear modulus vanishes and the velocity becomes the bulk sound speed

$$c = \sqrt{\frac{K_s}{\rho}}. \qquad (4.17)$$

The shock pressure at which the sound velocity drops to the bulk velocity (i.e., $\mu = 0$) defines the intersection of the Hugoniot with the melting curve. As shown below, recent data of this type for iron provide important constraints on the state of the Earth's core.

As more data become available the rarefaction velocity technique promises to increase substantially the ability to interpret the seismological observations on the Earth's deep interior. Specifically, Hugoniot sound velocity measurements can be used to determine the shear modulus of solids at high pressures and temperatures. The density and bulk modulus are determined from the Hugoniot equation of state (cf. equation 4.5), and when combined with the longitudinal velocity measured from the rarefaction wave this yields μ. Although limited to pressures below which melting occurs along the Hugoniot, this is nevertheless one of the only techniques currently known for measuring the shear modulus at pressures and temperatures up to hundreds of GPa and thousands of degrees. It has previously not been possible to compare the transverse wave velocity in the mantle and core with experimental data at the appropriate conditions of pressure and temperature.

3.2. Static Experiments

Two types of instrumentation are available for experiments at static pressures exceeding 10 GPa. These are large-volume presses and diamond-anvil cells, which can typically accommodate samples of 10^{-1} to 10^{2} mm^3 and 10^{-5} to 10^{-2} mm^3, respectively. There is a direct tradeoff between sample volume and maximum achievable pressure for either instrument; also, pressures above 100 GPa are readily produced in the diamond cell, but there are few convincing examples of large-volume presses reaching pressures above 30 GPa. Except for the synthesis of large quantities of new phases at elevated pressures, it is not obvious that large-volume presses provide any experimental advantages over diamond cells, particularly in light of the fact that it is not possible to observe samples directly *in situ*. Consequently, these large-volume instruments are not discussed any further here, but the interested reader is directed to Kumazawa and Endo's (1984) recent review.

The diamond cell is one of the most versatile instruments for high-pressure research, and the highest sustained pressures have been achieved with it (Mao and Bell, 1978; most recently, over 250 GPa peak pressure was reported by Goettel *et al.*, 1984). The fact that diamond is transparent to a broad range of electromagnetic radiation, i.e., that the sample can be directly observed through the diamond anvils while at high pressures (and temperatures), makes the diamond cell an exceptionally powerful tool.

The basic principle behind the diamond cell is that a sample is compressed between the points of two, gem-quality diamond crystals (Fig. 4.8). Because of the strength of diamond and the small dimension of the culets (the small faces between which the sample is compressed) pressures of tens of GPa are easily achieved. How the force is exerted on the diamonds is not critical: with steel and tungsten carbide anvils backing the diamonds, forces up to 10^3 kg can readily be produced on the table (the large face on the opposite side from the culet). This magnitude of force yields pressures on the order of 100 GPa for a typical culet area of 0.1 mm^2 (the ratio of the table to culet areas is of the order of 10^2). What is critical, however, is that the culet faces on either side of the sample be maintained in close alignment. The two faces are typically aligned to a parallelism better than 0.1μm across their width; retaining this parallelism at pressures of 100 GPa or more is a major experimental difficulty. The applied force at these pressures is sufficient to deform the whole

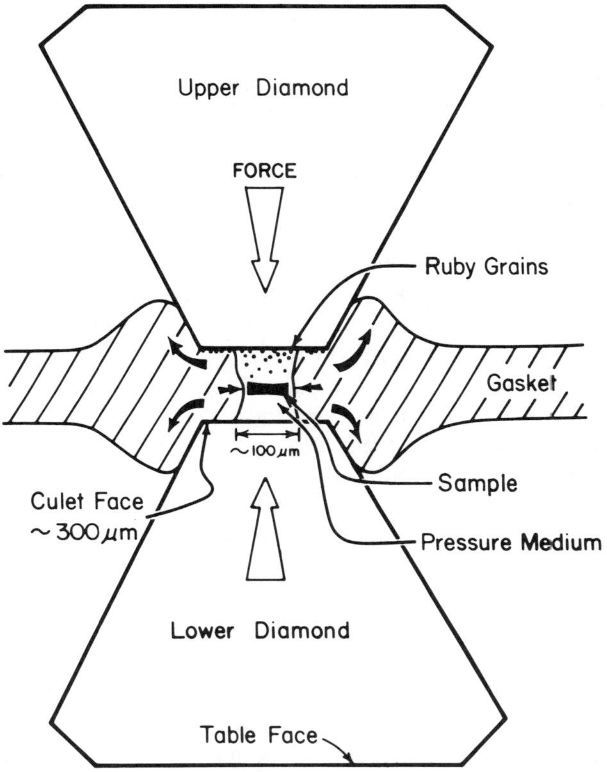

Figure 4.8. Schematic cross-section of the diamond-anvil high-pressure cell illustrating the way in which a sample is contained in a metal gasket while being squeezed between the points (culet faces) of two, single-crystal diamonds. The gasket pushes inward on the pressure medium surrounding the sample, and also oozes outward to provide support on the sides of the diamond. Ruby grains spread over the sample area are used for pressure calibration. Typical sample and culet dimensions are indicated.

diamond cell assembly, and therefore diamond alignment is not easily maintained.

A gasket, consisting of a metal foil with a hole for containing the sample, is used at high pressures. The metal squeezes around the ends of the diamonds to provide additional support at the culet end and to minimize pressure gradients across the sample. In the sample area (~10μm thick at high pressures and ranging between 20 and 300 μm in diameter), a pressure medium and calibrant are usually included along with the sample. Pressure media present a special problem because no fluid is known above about 10–14 GPa (Piermarini *et al.*, 1973; Munro *et al.*, 1979). Thus, ultrahigh pressure experiments are always carried out under nonhydrostatic conditions, although considerable effort has been put into devising ways of minimizing pressure variations (achieving "pseudohydrostatic" conditions) by use of soft solid media, specially prepared gasket assemblies or high-temperature annealing at elevated pressures (Bell and Mao, 1981; Knittle and Jeanloz, 1984; Heinz and Jeanloz, 1984).

Pressure can easily be measured at room temperature using the ruby fluorescence technique (Barnett *et al.*, 1973). Finely powdered ruby is spread across the top of the sample, and a light source (typically a laser) is focused onto the ruby through the upper diamond (Fig. 4.8). The light is used to excite the R_1 fluorescence line in the red end of the visible spectrum (the well known laser line of ruby), which shifts to lower energy (deeper red) with increasing pressure. The pressure shift has been calibrated to well over 100 GPa, mainly by way of shock-wave data (Mao *et al.*, 1979; Heinz and Jeanloz, 1984). One of the important features of the ruby-fluorescence calibration, however, is that the incoming light can be focused to excite ruby within an area of a few μm dimension. Therefore, pressure can be measured point-by-point across the sample, and pressure variations as small as a few percent can be resolved on a spatial scale of microns.

Not only can samples be examined *in situ* by spectroscopy, at wavelengths ranging from the ultraviolet to the far infrared, but crystal structure can be determined and lattice parameters can be measured by X-ray diffraction directly through the diamond anvils (or, more appropriately in this context, the diamond windows). Small sample sizes in the diamond cell ($\sim$1 μg is typical) either require access to a high-intensity X-ray source (rotating anode generator or synchrotron source) or else require long run times; weeks can be needed for one X-ray diffraction run on a conventional X-ray generator. Nevertheless, accuracy in determining unit cell parameters on the order of 1 part in 10^3 is achieved in ultrahigh pressure X-ray diffraction experiments.

Of particular importance for geochemical, petrological and geophysical applications is the fact that samples can be heated in the diamond cell. Both external heating, with a small furnace around the sample and diamonds, and laser heating have been successfully used to reproduce conditions of the Earth's interior (Fig. 4.9). The advantage to external heating is that it is possible to maintain steady, relatively uniform and well characterized temperature for hours to days (Sung, 1976; Ming *et al.*, 1983). Temperatures, however, are limited to 500–1000°C maximum because at higher tempera-

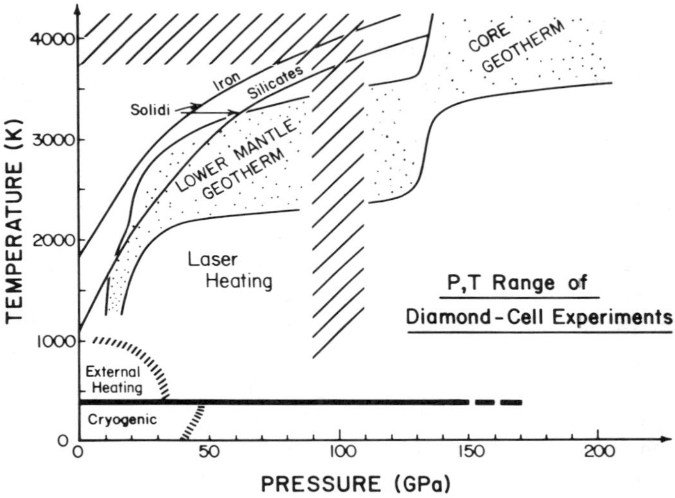

Figure 4.9. Pressure-temperature range that is currently accessible in diamond-cell experiments is compared with the range of estimated temperatures in the lower mantle and core (geotherms), as well as with current estimates of the melting curves of iron and mantle silicates. The highest pressures so far achieved with the diamond cell are at room temperature, but external heating and cooling (for cryogenic experiments) allow temperatures below about 1000 K to be reached. With laser heating, temperatures exceeding 2000–4000 K have been reached at pressures up to 100 GPa.

tures the whole body of the diamond cell, the metallic parts supporting the diamonds, become sufficiently hot that the alignment and the force on the diamonds can no longer be maintained.

Laser heating is used to achieve temperatures as high as 5000–6000 K (Jeanloz and Heinz, 1984). Thus, conditions of the Earth's deep interior can be achieved in the laboratory. The disadvantage, however, is that the laser beam is focused to a spot some 10–50 μm across in the sample area. Recall that the diamonds on either side of the ~10 μm thick sample are excellent thermal conductors. Due to the presence of these large heat sinks, as well as to the Gaussian intensity variation across the laser beam, sample temperatures are highly nonuniform.

Typically, a Nd:YAG continuous laser is used for heating because its radiation, at 1064 nm, is not absorbed by the diamonds but is readily absorbed by most silicates and oxides containing even small amounts of Fe

(Ming and Bassett, 1974). Alternatively, an inert absorber can be mixed in if the sample is otherwise transparent. As a result, mantle-like compositions can be directly studied without the diamond cell heating too rapidly: temperatures of several thousand degrees have been maintained for 30 minutes or more. The fact that temperatures are nonuniform with laser heating is a benefit in that the hot, reactive part of the sample is surrounded by its own cold equivalent rather than being in direct contact with the diamond. This self-buffering keeps the sample from reacting with the diamond anvils (such reaction usually damages the diamonds: e.g., Liu *et al.*, 1982).

Currently, much work is being done to measure the temperature and its variation across the sample during laser heating (e.g., by radiometry), and to use this information to determine melting and transformation temperatures at high pressures. To date, laser heating has been used nonquantitatively to overcome reaction kinetics in high-pressure mineral transformations, the reaction products being quenched in temperature and examined (mainly by X-ray diffraction) at high pressures but room temperature. The only *in situ* experiments at elevated pressures and temperatures have been under the better controlled conditions of external heating. Thus, one of the most significant advances from a geochemical or geophysical perspective would be the ability to use the laser heated diamond cell for quantitative phase equilibrium and disequilibrium (e.g., kinetic or diffusion) experiments.

Aside from the current efforts to better quantify and control the nonhydrostatic stresses and, during laser heating, the nonhomogeneous temperatures achieved in the diamond cell, there is also progress in devising new ways of observing the sample *in situ*. For example, viscosity measurements have been made on fluids up to ~10 GPa (Munro *et al.*, 1979). The technique involves measuring the velocity of a metal sphere falling across the sample area in the diamond cell. Similarly, optical techniques may allow creep deformation of solids to be studied at pressures of tens of GPa (Scott and Jeanloz, 1984). Radial distribution functions at high pressures (Munro *et al.*, 1982) giving the atomic packing of liquids, and electron microscopy of solids heated and deformed in the diamond cell (Lacam *et al.*, 1980; Madon *et al.*, 1980) represent important complementary observations for understanding the rheology of materials at conditions of the Earth's deep interior.

4. State of the Core

Evidence that the outer core is an iron-rich fluid surrounding a solid inner core (presumably of similar composition) comes from seismology and the magnetohydrodynamic requirement that the core be a good electrical conductor (e.g., Jacobs, 1975). Iron is thought to be the main constituent of the core because it is cosmochemically the most abundant element that satisfies the observed velocities and densities.

It has long been known, however, that the core is slightly less dense than would be expected if it were pure iron (Birch, 1952, 1972). Considerable effort has gone into identifying the estimated 10 percent of "contaminant" that is alloyed with the iron: sulfur and oxygen are the primary geochemical candidates at present (Brett, 1976, 1984; Stevenson, 1981). The nature of the alloying component must be known in order to estimate the amount that is present (i.e., the amount required to make up the density deficit of the core relative to iron). Also, models of the composition of the core play a role in estimating the temperature of the outer core. Thus, if the melting relations of iron or of an appropriate iron alloy were known at core pressures, then it would be possible to bracket the lowest temperature of the liquid outer core or the highest temperature of the solid inner core.

Much of the work to date has been based on the premise that the density change of iron on melting at high pressures is negligible (e.g., Birch, 1964; Jeanloz, 1979). As noted above, this idea is made plausible by the observation that melting is virtually impossible to observe along the pressure-volume equations of state of materials measured under shock loading. The resulting simplification is that the amount of contaminant in the core is estimated from a direct comparison of Hugoniot data for solid iron at high pressures with the seismologically observed density of the core (after a small thermal correction: Jeanloz, 1979). The melting point of iron at high pressures has then been estimated by extrapolating data collected at relatively low pressures (Higgins and Kennedy, 1971; Stacey, 1977). The melting temperature of iron should be an upper limit for the melting point of the core, assuming relatively uncomplicated thermodynamic mixing of the actual components present. The estimated melting point can then be corrected based on the amount and nature of alloying component that is present in the core.

This work has now been significantly extended by Brown and McQueen's (1982) determination of the melting point of iron at approximately 250 GPa. The major conclusions of their work are that the melting point of iron is higher than has previously been thought, and the volume change on melting is sufficiently large ($\sim$3 percent) to reduce significantly the amount of alloying component that has been estimated for the outer core. Thus, on both counts of the endmember melting for pure iron and the lesser amount of solute (i.e., lesser freezing-point depression), Brown and McQueen conclude that the temperature of the outer core is higher than has been expected from many of the previous estimates.

The results of Brown and McQueen (1982) are based on measurements of the rarefaction-wave velocity in shock experiments, as summarized above. They observed a drop in the sound velocity of iron at about 200 GPa and then another drop at 250 GPa to the bulk sound speed. Brown and McQueen's explanation is that the Hugoniot of iron crosses first the ε (hexagonal closest packed: hcp) phase to γ (face centered cubic or cubic closest packed: fcc) phase boundary at 200 GPa, and subsequently enters the liquid field of the phase diagram at 250 GPa. Hence, the main features of the phase diagram of iron, as illustrated in Figure 4.10, have now been experimentally constrained at high pressures.

From this result, Brown and McQueen (1982) estimate a melting point of 5000 (± 500) K for iron at 250 GPa (this is near the ε–γ–liquid triple point). The weakest part of the study is that the shock temperature must be calculated—it is not experimentally determined. Nevertheless, the temperature is well constrained by a full thermal equation of state that Brown and McQueen construct on the basis of: 1) the Hugoniot of iron; 2) determinations of the

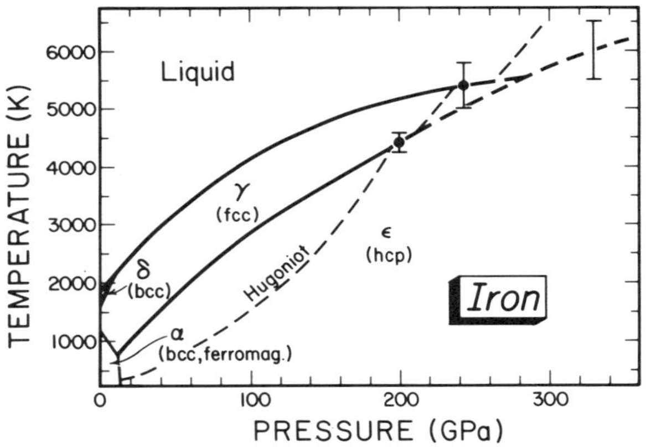

Figure 4.10. Phase diagram of iron based on Brown and McQueen's (1982) observed crossing of the Hugoniot (shown dashed) with the ε–γ and γ–liquid phase boundaries, at 200 and 250 GPa respectively (estimated uncertainty of calculated temperatures is shown by the error bars). Below 20 GPa the diagram is based on static data summarized by Liu (1975); the phase-boundary slopes measured in the static experiments require significant curvature for the ε–γ and γ–liquid boundaries in order that the shock-wave data be satisfied. The γ and ε phases are closest-packed (cubic and hexagonal, respectively), whereas the α and δ phases are in body-centered cubic (bcc) structures. Only the α phase is ferromagnetic.

Grüneisen parameter at high pressures and temperatures based on their shock-wave experiments; 3) the 300 K isotherm of ε-iron (Mao and Bell, 1979); and 4) a theoretically based estimate of the electronic contribution to the thermal properties of iron at high pressures (Bukowinski, 1977). To fit their data, Brown and McQueen find that the slope of the melting curve, dT_m/dP, is sufficiently high that iron must change in density by about 3 (± 2) percent on melting (see equation 4.8). The reason that this is not readily observed on the Hugoniot itself is largely due to the heat of fusion causing the sample to cool as it melts during the adiabatic (but not isentropic) process of shock loading (cf. Kormer *et al.*, 1968; Ahrens *et al.*, 1982). Effectively, the thermal contraction resulting from the heat of fusion masks the volume increase on melting. Note, however, that the curvature of the γ–liquid and ε–γ phase boundaries, evident in Figure 4.10, imply that the volume change across both of these transitions decreases with increasing pressure; ΔS increasingly becomes the driving force for the solid–solid as well as the melting transition (cf. Fig. 4.4).

From these results, Brown and McQueen (1982) find that the properties of the outer core are consistent with the presence of only 5–10 weight percent sulfur, rather than the 10–15 percent contamination that had previously been the best estimate. The nearly 50 percent change in the estimated amount of alloying of iron is due to the lower density of liquid iron than has previously been expected. Sulfur and oxygen are both prime candidates for the alloying element in the core at present; based on the work of Jeanloz and Ahrens (1980), a similar value would apply to the amount of oxygen (or a combination of S and O) in the outer core. Reanalyses of the data, including new Hugoniot measurements for iron sulfide, yield similar estimates for the composition of the core: ≃95 wt.% and ~90 wt.% Fe according to Spiliopoulous and Stacey (1984) and Brown *et al.* (1984), respectively.

Both the Fe-S and Fe-O systems are nonideal at low pressures due to the vastly different chemical properties of the endmember elements (metal vs. nonmetal). These systems may be more nearly ideal at high pressures, however, because both S and O are thought to be metallic at the conditions of the core (Jeanloz, 1983; Brown *et al.*, 1984). Thus, concerns that alloying in the core is complicated by nonideality are less compelling than previously thought (cf. Verhoogen, 1973). Quite possibly, S and O are merely the dominant species of a long list of contaminants in the core (Stevenson, 1981). Nevertheless, the amount of alloying components is sufficiently small that the state of the core can be adequately modelled without a more detailed knowledge of its composition.

Applying their new results, Brown and McQueen (1982) found that a temperature of 4000 (± 500) K at the top of the outer core satisfies all of the available data, with 7½ (± 2½) weight percent S inferred to be present. They estimate the melting point depression associated with 5–10 percent contaminant and assume that the average temperature gradient in the outer core is adiabatic, consistent with the occurrence of vigorous convection in this fluid region. If oxygen is the main contaminant in the core, temperatures are likely to be somewhat higher, (Ringwood, 1979; Jeanloz and Ahrens, 1980). Taking this into account, recent analyses are in agreement with temperatures in the outermost core being between 3800 and 5000 K (Brown *et al.*, 1984; Spiliopoulous and Stacey, 1984). According to the phase diagram of Figure 4.10, this range of temperatures is not very different from the melting point of iron at comparable pressures.

5. Mineral Phases of the Mantle

5.1. Crystal Chemistry

Starting with the basic garnet peridotite mineral assemblage of the upper mantle, static experiments have been used to delineate the corresponding high-pressure phases that would be produced at greater depth within the Earth. As summarized in Figure 4.11, the major mineral phases can be grouped into two categories: 1) the olivine phases, and 2) the pyroxene and garnet phases. At pressures beyond 20 GPa, however, these groups become indistinguishable.

The low-pressure mineral phases that occur in the crust and uppermost mantle can be usefully described in terms of their constituent molecular units, such as the silicon-oxygen tetrahedra and metal (e.g., Mg, Fe, Ca)-oxygen octahedra. Thus, the classical scheme outlined by Bragg and co-workers (1965) is readily applied for distinguishing the olivine, garnet and silicate spinel structures from the pyroxenes. The former are orthosilicates, characterized by separate SiO_4 tetrahedra, whereas the pyroxenes are chain silicated consisting of chains of linked SiO_4 tetrahedra. Between the tetrahedra reside the metal cations, often in octahedral sites with varying degrees of distortion. At the present level of discussion, it is not necessary to distinguish orthopyroxenes from clinopyroxenes; not only are the basic structures similar, but the physical properties with which we are concerned are nearly identical (Jeanloz and Thompson, 1983).

With increasing pressure the density of phases increases, as is thermodynamically required (see Fig. 4.4). The increase in density is achieved by a

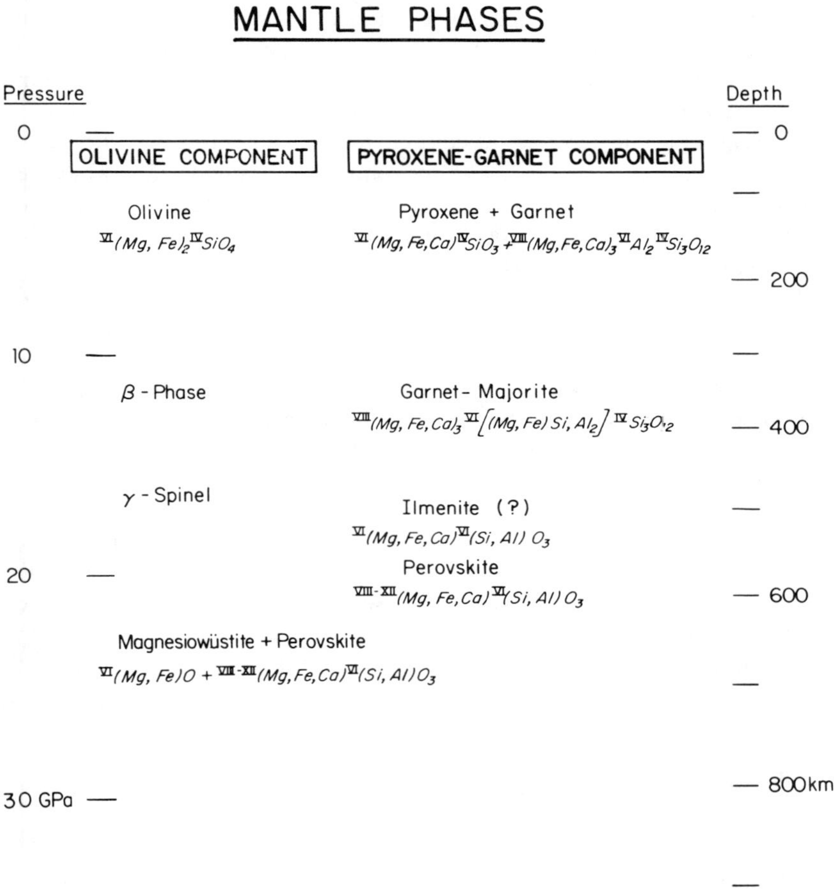

Figure 4.11. Summary of mineral phases corresponding to the olivine component and the pyroxene-garnet component of an upper-mantle assemblage as a function of pressure (or equivalent depth in the mantle). Crystal-chemical formulae indicating the coordination numbers of the cations (Roman numeral superscripts) are included under the corresponding names of the structure types to illustrate the increase in coordination with depth (the formulae of β-phase and γ-spinel are identical to the formula of olivine). The question mark indicates that the ilmenite structure may not be stable for the compositions occurring in the mantle (see Jeanloz and Thompson, 1983, from which this figure is taken).

tighter packing of the ions that can be described in terms of a decrease in the average volume of the phase per oxygen ion (Table 4.2). This increased packing efficiency of high-pressure mineral phases makes the molecular-unit based description less useful than for the low-pressure phases. Instead, it is more convenient to describe structures in terms of dominant sublattices that are essentially close packed, with the remaining ions populating available interstices (Bragg *et al.*, 1965). For example, the olivine structure is based on a hexagonal closest packed (hcp) sublattice of oxygen ions with Mg and Si located in the octahedral and tetrahedral interstices, respectively. In contrast, the two high-pressure structures of olivine, γ-spinel and β-phase, are based on a cubic closest packed (fcc) oxygen sublattice with Mg and Si in octahedral and tetrahedral sites. The β-phase and γ-spinel structures are closely related, except that the SiO_4 tetrahedra are linked in pairs in the β-phase. That is, β-phase is one of the relatively few known sorosilicate structures (see Akimoto *et al.*, 1976b).

For the mineral phases presented in Figure 4.11, the fcc sublattice is more stable at high pressures than the hcp sublattice (contrast this with the greater stability of the hcp over the fcc structure of iron at pressure: Fig. 4.10). Thus, olivine is replaced by the high-pressure phases β-phase, γ-spinel, perovskite and magnesiowüstite, all of which are based on fcc sublattices; similarly, the hcp sublattice of ilmentite is replaced by the fcc sublattice of perovskite with pressure. What should be noted, however, is that the fcc sublattice of

Table 4.2. Properties of mantle minerals

Mineral phase	Volume per oxygen ($Å^3 = 10^{-30}$ m^3)	Density (Mg/m^3)	Adiabatic bulk modulus (GPa)	Pressure derivative of bulk modulus
Olivine (ol)	18.1	3.22	129	5.2
Pyroxene (px)	17.3	3.21	108	7
Garnet-majorite (gt-maj)	15.7–15.8	3.56–3.51	174–220	4.5
β-phase (β)	16.8	3.47	174	4
γ-spinel (γ)	16.5	3.55	184	4.8
Ilmenite (ilm)	14.6	3.81	unknown	unknown
Magnesiowüstite (mw)	18.7	3.58	163	4.1
Perovskite (pv)	13.5	4.10	262	4

For simplicity, only values for the Mg-endmember at zero pressure and room temperature are given. Abbreviations for mineral phases are given in parentheses.
Sources of data are listed in Jeanloz and Thompson (1983), with additional data from Weidner *et al.* (1984).

perovskite is made up of both the oxygen and the metal ion (e.g., Mg). No longer is oxygen the sole constituent of the mineral sublattice: at sufficiently high pressure the cations can also be involved in defining the basic sublattice, and the molecular (polyhedral) clusters making up the structure become less important in determining the nature and properties of the phase.

The tighter packing of ions that is characteristic of high-pressure phases is achieved by reducing the average anion-anion separation, in many cases with an increase in coordination number around the cations (Fig. 4.11). As noted by Jeanloz (1982), however, increasing coordination across a phase transition can produce unexpected changes in thermodynamic properties because cation-anion (first neighbor) distances typically *increase*, in contrast to the tighter anion packing.

The most notable coordination change among silicates is the appearance of octahedrally (6-fold) coordinated silicon at high pressures, rather than the more familiar tetrahedrally (4-fold) coordinated Si of crustal minerals. The type example of this coordination increase is the case of the high-pressure stishovite (rutile-structured) phase of SiO_2, as opposed to the lower pressure phases such as α-quartz and coesite. Similarly, Si is 6-coordinated in the silicate perovskite, ilmenite and majorite (garnet-structured) phases.

Coordination around the metal cation (Mg, Fe, Ca) is octahedral in most minerals, but it is increased in the garnet and perovskite structures to 8- and 12-fold, respectively. Actually, $MgSiO_3$ occurs in an orthorhombically distorted version of the perovskite structure, so that the metal site is pseudoeightfold coordinated, with four oxygens lying at a greater distance (Yagi *et al.*, 1978). The high coordination of the metal sites in the garnet and perovskite structures may help to explain the remarkable stability of these minerals at high pressures. Garnet, which is unusually densely packed for a crustal mineral, is stable to 20 GPa or more (Table 4.2; qualitatively, garnet is already a high-pressure phase with high coordination). Perovskite is stable to over 70 GPa, implying that silicate perovskite is likely to occur throughout the whole lower mantle and is therefore the single most abundant mineral in the Earth.

Magnesiowüstite, however, presents an interesting contrast because the cation is only 6-fold coordinated, making this a low-pressure structure. This phase, however, is stable to extremely high pressures (70 GPa, or more), perhaps due to the fact that every octahedral site is occupied in the fcc oxygen sublattice (Jeanloz and Thompson, 1983; note the high density and poor packing, or large volume per oxygen ion, indicated in Table 4.2).

It is interesting to note that the availability of large (high-coordination) sites allow the garnet and (especially) perovskite structures to accommodate a wide variety of cations, including minor and relatively incompatible elements. Thus, partitioning of elements between solid phases or between solid and liquid is likely to be qualitatively different in the lower mantle than in the upper mantle and crust due to crystal chemical differences, as well as to the effects of pressure and temperature. Global geochemical models cannot be considered reliable until the nature of partitioning between high-pressure solid and liquid phases is better documented than at present.

5.2. Physical Properties

That the properties of the mantle are similar to those of oxides and silicates at high pressures has long been recognized (Birch, 1952). The more specific question than can now be addressed is: To what degree are the seismological properties of the mantle compatible with the garnet peridotite mineralogy of the uppermost mantle, or its high-pressure equivalents? This question cannot be fully answered because estimates of composition and temperature in the mantle are not uniquely determined, and not all of the necessary physical properties have been measured for the mineral phases of the mantle. Nevertheless, it is clear that the density and bulk modulus (or bulk sound speed: equation 4.17) of the upper mantle and transition zone are well matched by the properties of an olivine dominated mineralogy, with lesser pyroxene and garnet components.

A specific comparison for the olivine and pyroxene-garnet minerals, and their high-pressure phases, is shown in Figure 4.12. The zero-pressure changes in properties (e.g., density) reflect the effects of the pressure-induced phase transitions. Clearly, the 400 km discontinuity can be attributed to the appearance of β-phase in the olivine component, whereas the 670 km discontinuity seems to be associated with the appearance of the perovskite assemblage. When the effects of pressure on the physical properties are included, the conclusion that the upper mantle and transition zone properties are consistent with those of the proposed olivine-rich (garnet peridotite) composition is strengthened.

The effects of temperature are not shown in Figure 4.12 due to the paucity of experimental data on the appropriate temperature corrections. The depth of the 400 km discontinuity, however, is known to be in accord with the high-temperature transition pressure to the β-phase (Jeanloz and Thompson, 1983). Also, the 5½ percent difference in both density and bulk sound speed between the mantle and γ-spinel at the pressures of the transition zone can be readily explained in terms of thermal expansion associated with plausible temperatures at this depth ($\sim$ 2000 K; note the empirical observation that

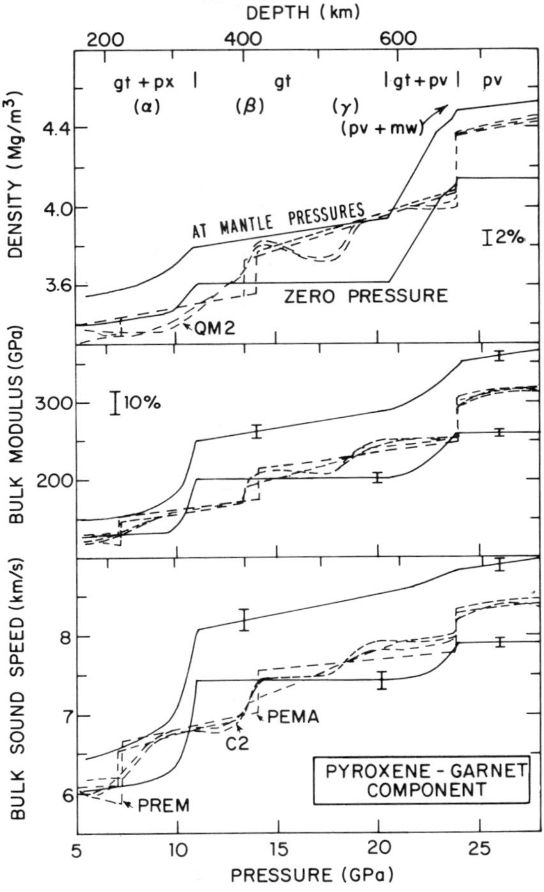

Figure 4.12. Effect of pressure on the physical properties (density, bulk modulus and bulk sound velocity) of the olivine and pyroxene-garnet components in the mantle (see Figure 4.11). Only room-temperature values of the properties are shown, both at zero-pressure and at the corresponding mantle pressures (based on data summarized by Jeanloz and Thompson, 1983, and including Weidner *et al.*'s, 1984, results). The olivine and pyroxene-garnet phases existing at each depth (or corresponding pressure) are indicated at the top (abbreviations given in Table 4.2; olivine phases coexisting with garnet-pyroxene phases are shown in parentheses). For comparison, mantle properties based on seismological whole-earth models are shown as dashed (PREM and PEMA are discussed by Dziewonski and Anderson, 1981, and Dziewonski *et al.*, 1975, whereas C2 and QM2 are discussed by Anderson and Hart, 1976, and Hart *et al.*, 1977).

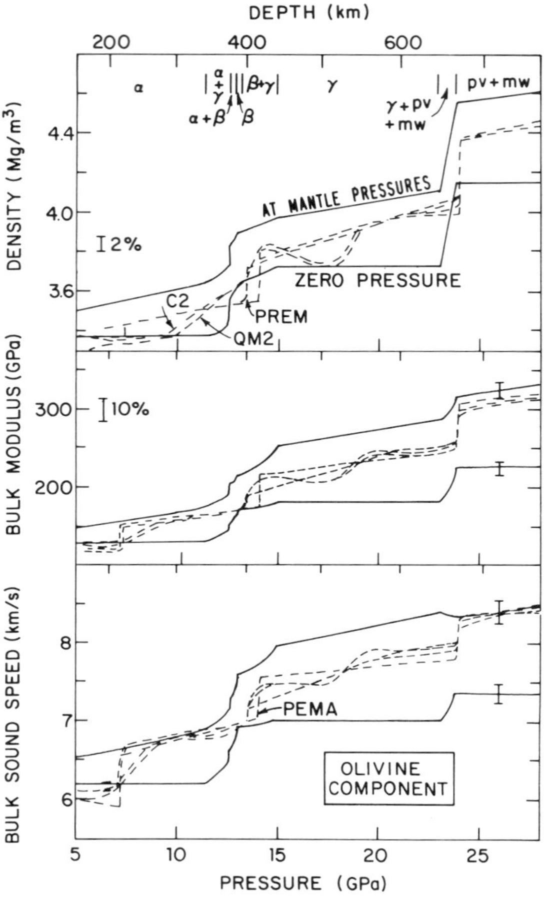

Figure 4.12. *Continued*

temperature is generally found to affect bulk sound velocity according to $(\partial \ln c)/(\partial \ln \rho) \cong 1$ to 2).

There is an important tradeoff here between composition and density; in contrast, the bulk modulus is relatively insensitive to composition. Specifically, an increase in density between 0.23 and 0.73 percent occurs for each percent increase in iron component of the mineral phases (these extreme values are for garnet and magnesiowüstite compositions, respectively; values for pyroxene and olivine are 0.31 to 0.44). Other major elements influence density less than iron, but the important conclusion is that an increase of one percent iron component affects density by as much as a decrease in temperature of about 200 to 500 degrees (assuming $\alpha \sim 1$ to 3×10^{-5} K^{-1}). In fact, the seismologically determined density is a major constraint on the average atomic weight, and hence the composition or iron content, of the upper mantle (e.g., Fujisawa, 1968).

Although the upper mantle and transition zone properties are compatible with those of an olivine-rich composition, this is not the case for the lower mantle. Instead, it is the pyroxene component that is more nearly compatible with the observed density and sound velocity below the transition zone. Magnesiowüstite, with its low-pressure structure and correspondingly low sound velocity, results in too small a predicted sound velocity for the olivine component at lower mantle pressures: a velocity decrease with depth is actually predicted at 670 km depth for an olivine-rich composition (Fig. 4.12).

6. Properties of the Lower Mantle at Zero Pressure and Nature of the 670 Km Discontinuity

One of the difficulties in comparing the measured density and velocity distributions of the lower mantle with the properties of model assemblages is that the effects of both pressure and temperature must be taken into account. One way to simplify such a comparison is to extrapolate the properties of the lower mantle to zero pressure. There is an important assumption in this approach, however: the region from which the extrapolation is made must be homogeneous. Excluding the top and bottom few hundred kilometers, the seismological properties of the lower mantle are consistent with its being homogeneous and adiabatic (e.g., Masters, 1979; Dziewonski and Anderson, 1981). Therefore, it is possible to extrapolate the density and velocity profiles along the adiabat to yield the properties of the lower mantle at zero pressure and high temperature (T_{0S}, which is the zero-pressure temperature of the lower mantle adiabat). These zero-pressure values of density and elastic properties can then be used to test the hypothesis that the whole mantle is uniform in composition.

The present analysis is based on the Eulerian finite-strain formalism developed by Birch (1947, 1952). This formalism is an extension of classical elasticity theory, and it is needed here because the compressions, or volume strains, achieved in the Earth are large. Elasticity theory, which is rigorously derived from thermodynamics, applies only to infinitesimally small strains (e.g., Wallace, 1972). For larger compressions, second-order terms in strain must be included in the thermodynamic development, and these terms lead to trouble because it is no longer possible to uniquely define the strain tensor. Physically, the ambiguity corresponds to there being no unique frame of reference for defining the amount of deformation.

The thermodynamic analysis rests on developing the strain energy density, the term $dW = -PdV$ in equation 4.14, as a series expansion in strain. Obviously, the quality and convergence of this expansion depends on the strain being appropriately defined. There is no fundamentally based definition of strain because a knowledge of crystal energy as a function of interatomic separation (or volume compression) is equivalent to knowing the bond strengths, which can only be theoretically derived from a complete quantum mechanical calculation. Therefore, a phenomenological approach is preferred.

As in fluid dynamics, two popular choices for reference frames, and hence strains, are Lagrangian (material) and Eulerian (spatial). They correspond to the following definitions of strain, which are positive with increasing compression (Davies, 1924; Birch, 1978):

$$f_{\text{Lagrangian}} = \frac{1}{2}\left[1 - \left(\frac{\rho}{\rho_0}\right)^{2/3}\right] \tag{4.18}$$

and

$$f_{\text{Eulerian}} = \frac{1}{2}\left[\left(\frac{\rho}{\rho_0}\right)^{2/3} - 1\right]. \tag{4.19}$$

The reason that powers of ⅔ appear is that the strain is based on the norm (length squared) of a line element within the deformed material (proportional to $V^{1/3} = \rho^{-1/3}$). Empirically, the Eulerian finite-strain formalism has proven to be remarkably successful in reproducing compression measurements to high pressures (e.g., Davies, 1973; Birch, 1977, 1978). For example, high-precision data collected on the pressure-calibration standard Au, as well as data on CsI and CsBr to volume compressions $V/V_0 \lesssim 0.5$, are in good agreement with Eulerian but not Lagrangian equations of state (Heinz and Jeanloz, 1984; Knittle and Jeanloz, 1984; Knittle *et al.*, 1985). Thus, equation 4.19 represents the best available definition of strain. With this definition, it is possible to expand the internal energy in strain and obtain the (adiabatic) pressure by differentiation. The result is:

$$P = 3K_{0S}f(1 + 2f)^{5/2}[1 - 2\xi f + 4\zeta f^2 + \ldots] , \tag{4.20}$$

with K_{0S} being the adiabatic bulk modulus at zero pressure, and the terms ξ and ζ involving first and second derivatives of the bulk modulus with pressure (Birch, 1978). Equation 4.20 includes terms up to fourth order in the expansion of energy in strain.

Equations 4.19 and 4.20 cannot be directly used to fit the pressure-density relations of regions within the Earth because the initial density, ρ_0, is not known. Jeanloz (1981b) showed how this difficulty can be circumvented if the pressure-density relation is considered in terms of the normalized pressure

$$S = \frac{P}{3}(1 + 2g)^{-5/2} \tag{4.21}$$

as a function of the effective strain

$$g = \frac{1}{2}\left[\left(\frac{\rho}{\rho^*}\right)^{2/3} - 1\right]. \tag{4.22}$$

with ρ^* being a reference density which can take on an arbitrary normalization value. The normalized pressure S is then a polynomial expansion in strain g: it is evident from Figure 4.13 that third- or fourth-order equations of state (quadratic or cubic polynomials) satisfy the seismological data for most of the regions of the Earth's interior.

The density and bulk modulus of the lower mantle extrapolated in this fashion are listed in Table 4.3, and the results are in accord with previous analyses (e.g., Davies and Dziewonski, 1975; Jackson, 1983). Separate estimates of K_{0S} and K'_{0S} are provided by extrapolating the pressure-density and velocity-density relations (the latter by way of a direct extension of Jeanloz's (1981b) approach). The consistency between these results, shown in the table, and the agreement between the third- and fourth-order fits gives an indication of the uncertainties in the values. At least a third-order fit is required to satisfy the velocity-depth profiles for the lower mantle, and it appears that the third-order fit is internally more consistent than the fourth order fit (this may well reflect the starting model in the seismological inversion more than the actual equation of state of the lower mantle). Thus, the initial density $\rho_0 = 4.00$ Mg/m^3 is determined to 0.2 percent, but the preferred value of $K_{0S} = 220\ (\pm\ 3)$ GPa may be off by 5 percent (e.g., $K_{0S} = 210 \pm 5$ GPa for the fourth-order fit).

In order to interpret these results, it is necessary to estimate the value of T_{0S}. The ratio of temperature at the top of the lower mantle to T_{0S} is well determined because the Grüneisen parameter

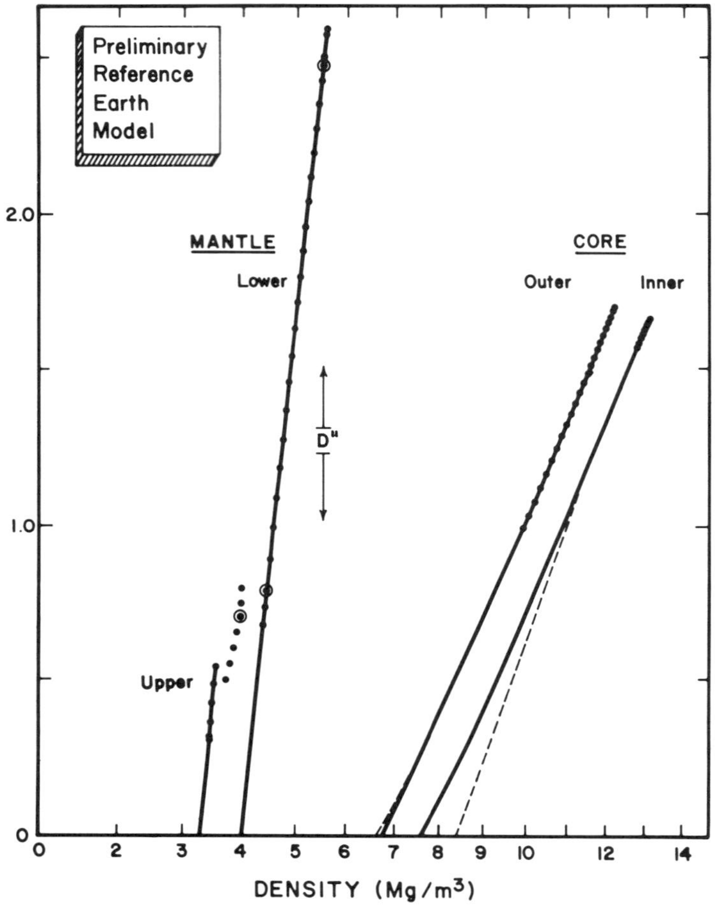

Figure 4.13. Finite-strain analysis of the pressure-density relations of regions within the Earth. The seismologically derived values from PREM (Dziewonski and Anderson, 1981) are shown as points (circled where overlapping). Third- and fourth-order equations of state appear as quadratic (dashed) and cubic (solid) curves in this plot of normalized pressure (S) as a function of strain (g: only the corresponding density is shown) (Jeanloz, 1981b). Note that whereas the zero-pressure densities of the outer core and (especially) the inner core are ill constrained, the zero-pressure density of the lower mantle is well determined (Table 4.3). Further details are given in the text.

Table 4.3. Lower mantle properties extrapolated to zero pressure and high temperature†

	Third order fit	Fourth order fit
Density, ρ_0 (Mg/m^3)	4.003	3.994
Bulk modulus, K_{0S} (GPa)	222.5‡	214.5‡
	216.5§	204.7§
$\left(\frac{\partial K_{0S}}{\partial P}\right)_S = K'_{0S}$	3.76‡	4.05‡
	3.80§	4.38§

†Preliminary Reference Earth Model (Dziewonski and Anderson, 1981) extrapolated to zero pressure and $T_{0S} \cong 1750$ K by way of Eulerian finite-strain equations of state (see text). Subscripts zero and S indicate zero pressure and adiabatic conditions respectively.
‡From density (pressure)-depth profile.
§From velocity-depth profile.

$$\gamma = \left(\frac{\partial \ln T}{\partial \ln \rho}\right)_S \tag{4.23}$$

falls within range $1 \lesssim \gamma \lesssim 2$ (Jeanloz and Thompson, 1983). Hence, the density distribution in the lower mantle and the value $\rho_0 = 4.00$ Mg/m^3 constrain this ratio to be

$$\frac{T(670\text{ km})}{T_{0s}} = 1.14\ (\pm\ 0.06). \tag{4.24}$$

As will be noted below, an estimate from the seismological data indicates that the zero-pressure $\gamma_0 = 1.4$ ($\pm$ 0.1). Assuming the mantle to be of uniform composition, a value of $T_{0S} = 1750$ ($\pm$ 80) K is derived from the petrologically determined temperature in the transition zone.

The temperature in the transition zone is best constrained by the measured phase equilibria among olivine and spinel polymorphs. The transformation of olivine and pyroxene components to a perovskite assemblage at higher pressures involves a small (possibly negative) Clapeyron slope which is not well constrained, and therefore is not used to estimate the temperature in the transition zone (Jeanloz and Thompson, 1983). In contrast, the pressure-temperature Clapeyron slope of the olivine + γ-spinel $\rightarrow$ olivine + β-phase reaction has been extensively investigated by Akimoto *et al.* (1976a). As the 400 km discontinuity is thought to result from this reaction, the temperature is relatively well determined near the top of the transition zone. Accounting for other possible transformations in the transition zone leads to an estimated value T(670 km) $\cong$ 2000 K, and hence the value of T_{0S} quoted above (Jeanloz and Richter, 1979; Jeanloz and Thompson, 1973). In reality, this is a minimum estimate because the lower mantle would be considerably hotter if a thermal boundary layer exists at 670 km depth due to a change of composition at this depth. Nevertheless, this value of T_{0S} may be used to test the hypothesis that the mantle is of uniform composition. The question then arises: Are the properties of the lower mantle compatible with those of the perovskite and magnesiowüstite high-pressure assemblage corresponding to an upper mantle composition?

First, the zero-pressure density of 4.00 Mg/m³ can be compared with the densities of high-pressure phases, as shown in Figure 4.14. For an upper mantle composition, a total thermal strain of 0.04 (± 0.01) is required to match the high temperature value of ρ_0. The thermal strain is defined by

$$\int_{290\,\mathrm{K}}^{T_{0s}} \alpha dT = \langle\alpha\rangle_0^{LM} (1750\ \mathrm{K} - 290\ \mathrm{K}), \qquad (4.25)$$

with brackets indicating an average over the interval from room temperature, 290 K, to $T_{0S} = 1750$ K. Hence, the mean coefficient of thermal expansion for the perovskite + magnesiowüstite assemblage must be

$$\langle\alpha\rangle_0^{LM} = 2.7\ (\pm\ 0.6) \times 10^{-5}\ \mathrm{K}^{-1} \qquad (4.26)$$

in order that a uniform-composition model be satisfied.

This value is low compared to what would be expected based on the following comparisons (Jeanloz and Thompson, 1983; Jackson, 1983). Given the measured value of $\langle\alpha\rangle_0^{mw} = 4.4 \times 10^{-5}\ \mathrm{K}^{-1}$ for magnesiowüstite, the coexisting perovskite must have a value of $\langle\alpha\rangle_0^{pv} \lesssim 2 \times 10^{-5}\ \mathrm{K}^{-1}$ in order that equation 4.26 be satisfied. In contrast, titanate and fluoride analog perovskites have $\langle\alpha\rangle_0^{pv} \gtrsim 3 \times 10^{-5}\ \mathrm{K}^{-1}$ above room temperature (Lytle, 1964; Naimon and Granato, 1973; Jones, 1979). Also, Jeanloz and Roufosse (1982) estimate that $\langle\alpha\rangle_0^{pv} \sim 3$ to $4 \times 10^{-5}\ \mathrm{K}^{-1}$ for $MgSiO_3$ perovskite based on a crystal chemical model, and a very recent measurement on $Mg_{0.9}Fe_{0.1}SiO_3$ perovskite confirms this estimate (Knittle *et al.*, 1986). Thus, a uniform mantle composition requires either that the coefficient of thermal expansion of silicate perovskite be anomalously low or that a much lower temperature be invoked for the lower mantle (e.g., T (670 km) $\lesssim$ 1650 K).

The latter alternative is highly unattractive because it results in calculated temperatures at the core-mantle boundary that are completely unacceptable in light of the core temperatures inferred from the shock-wave data on iron: $T_{CMB} \lesssim 2400$ K is implied rather than the required values $\gtrsim 3800$ K. Attempts have been made to reconcile estimates of low temperatures in the

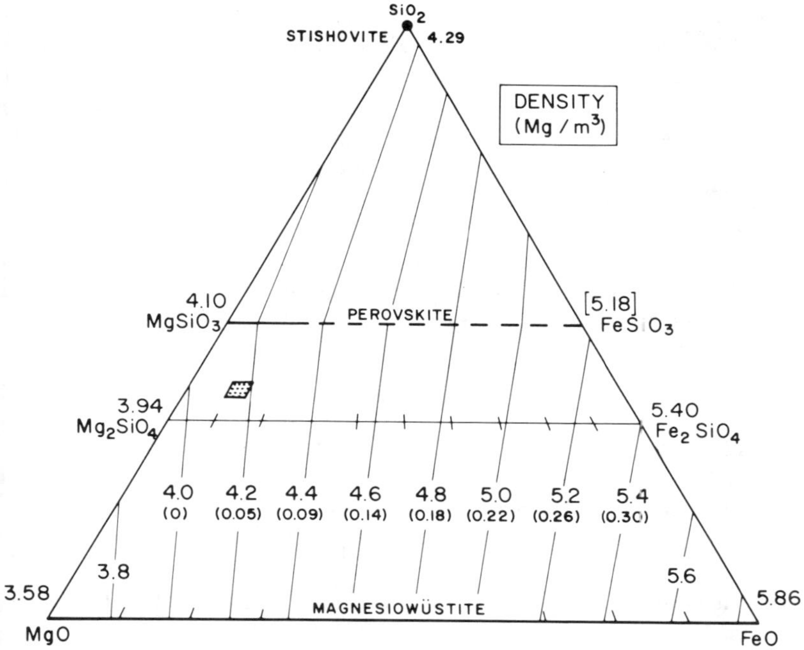

Figure 4.14. Measured densities of lower mantle phases at zero pressure and room temperature are shown contoured as a function of composition in the system MgO-FeO-SiO_2. As Ca and Al are expected to enter the perovskite structure, for a plausible range of lower mantle compositions, these components are projected onto $MgSiO_3$ because they do not measurably change the density of the perovskite phase (Jeanloz and Thompson, 1983, and references therein). The composition of the upper mantle is indicated by the stippled region, and values of total thermal strain required to match the lower mantle density at zero pressure are given in parentheses (cf. equation 4.25). Increasing either the Fe or Si content from the upper mantle value increases the value of thermal strain that is required.

lower mantle with much higher estimated temperatures at the top of the core by invoking the presence of a thermal boundary layer at the base of the mantle (e.g., Spiliopoulos and Stacey, 1984). Such a boundary layer with a high, conductive geothermal gradient is likely to be present because current estimates of elemental partitioning strongly suggest that substantial amounts of radioactive, heat-producing elements, such as uranium or potassium, can dissolve into the molten iron alloy of the core at high pressures (Stevenson,

1981; Jeanloz, 1983). The amount present is highly uncertain, but sufficient radioactive heat production is likely to be available to power the core dynamo and produce the geomagnetic field (cf. Verhoogen, 1980). Even so, models of the heat transfer across a thermal boundary layer inferred to be present at the base of the mantle result in a temperature offset near the core mantle boundary that is far smaller than the difference $\Delta T_{CMB} \gtrsim 3800–2400 \text{ K} = 1400 \text{ K}$ that is apparently required for uniform mantle composition (Jeanloz and Richter, 1979).

Similarly, it is not easy to conclude that $\langle\alpha\rangle_0^{pv}$ of $MgSiO_3$ perovskite is anomalously small if the seismological properties of the lower mantle are to be matched. From the definition (cf. equation 4.23)

$$\gamma = \frac{\alpha K_T V}{C_V} \tag{4.27}$$

and the known values of K_T, V and C_V (the latter being given by the Dulong-Petit value; see Jeanloz and Thompson, 1983), the Grüneisen parameter is related to the thermal expansion coefficient by

$$\gamma_0^{pv} = 0.51 \times 10^5 \ \alpha_0^{pv} \tag{4.28}$$

for silicate perovskite at zero pressure. Thus, if $\langle\alpha\rangle_0^{pv} \lesssim 2 \times 10^5 \text{ K}^{-1}$ is assumed for silicate perovskite, as required for uniform mantle composition, then the corresponding $\gamma_0^{pv} \lesssim 1.02$. Combining this with the value measured for magnesiowüstite results in a predicted high-temperature, zero-pressure Grüneisen parameter of 1.1 (± 0.1) for the assemblage of high-pressure phases existing in the lower mantle. The seismological (i.e., long-wavelength, acoustic) value for the lower mantle is, however,

$$\gamma_0^{LM} = 1.4 \ (\pm \ 0.1) \tag{4.29}$$

(e.g., Anderson and Sumino, 1980; Brown and Shankland, 1981). Hence, treating the lower mantle as a continuum (i.e., ignoring dispersive effects due to the atomic structure of minerals), the observed value of γ_0^{LM} (equation 4.29) is not consistent with $\gamma_0^{pv} \lesssim 1$ or $\langle\alpha\rangle_0^{pv} \lesssim 2 \times 10^{-5}$ for silicate perovskite. Rather, a value of

$$\gamma_0^{pv} = 1.5(\pm 0.1) \tag{4.30}$$

is in accord both with equation 4.29 and with the value of $\langle\alpha\rangle_0^{pv} \cong 3 \times 10^{-5}$ K^{-1} for $MgSiO_3$ perovskite (estimated via equation 4.28). Also, the work of Knittle, *et al.* (1986) now gives experimental support for these values. Of course, such large values of γ_0^{pv} and $\langle\alpha\rangle_0^{pv}$ are not consistent with a model of uniform mantle composition. That is, the lower mantle appears to be too dense for an upper mantle composition.

The bulk modulus of the lower mantle (Table 4.3) can similarly be compared with measured values for the magnesiowüstite + perovskite high-pressure assemblage of upper mantle composition. Subtracting out the contribution for magnesiowüstite, based on recent measurements of the elastic constants at high temperatures (Sumino *et al.*, 1983), the value of the bulk modulus required for coexisting perovskite is shown in Figure 4.15. The zero-pressure high-temperature value K_{0S}^{pv} is indicated as a function of composition, with T_{0S} = 1750 K being appropriate for uniform mantle composition. As will be evident, this temperature yields a conservative result

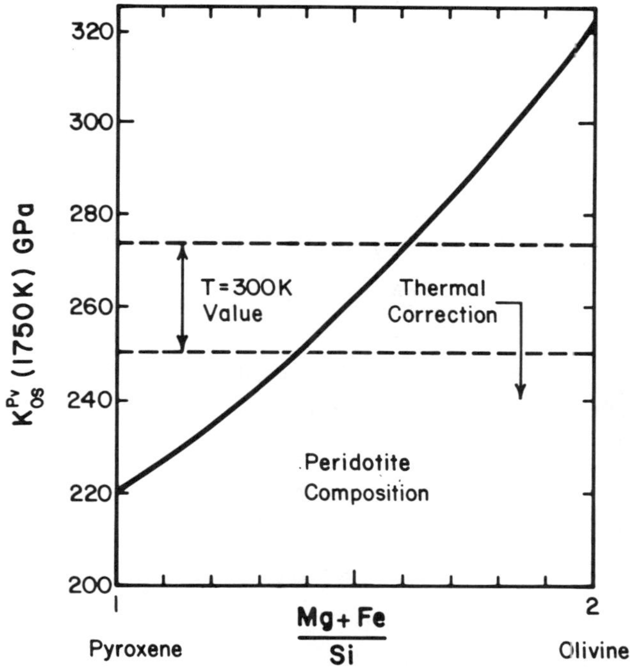

Figure 4.15. Zero-pressure bulk modulus of perovskite required to match the observed values for the mantle and for magnesiowüstite at high temperatures is shown as a function of the assumed composition of the lower mantle (expressed as the ratio of (Mg + Fe)/Si components). A temperature of 1750 K, in accord with uniform mantle composition, is assumed for this analysis. The bulk modulus of $MgSiO_3$ perovskite measured at 300 K lies between the dashed lines. According to the result shown, a peridotitic upper mantle composition is not acceptable for the lower mantle; the measured K_{0S} of perovskite (corrected to 1750 K) requires a greater pyroxene content below 670 km depth than is present in the upper mantle.

because raising T_{0S} increases the calculated value of K_{0S}^{pv} at each composition. The advantage of examining the bulk modulus is that, unlike density, it depends mainly on the olivine/pyroxene ratio and is insensitive to Fe-Mg content; in contrast, density is dependent on both of these factors (Anderson, 1976; Jackson *et al.*, 1978; Jeanloz and Thompson, 1983). Clearly, an upper mantle composition of (Mg + Fe)/Si $\cong$ 1.5 is incompatible with the measured bulk modulus of perovskite, because a peridotitic (olivine rich) composition would require K_{0S} at 1750 K to equal the value measured at room-temperature.

How large a temperature correction is required for the bulk modulus? This can be estimated from Birch's (1968) relation

$$K'_{0T} \cong \delta_{0T} \cong \delta_{0S} + \gamma_0, \tag{4.31}$$

(also, Anderson, 1967) with

$$\delta_0 \equiv \left(\frac{\partial \ln K_0}{\partial \ln \rho}\right)_P \tag{4.32}$$

and subscripts S and T denoting adiabatic and isothermal conditions. The static-compression data for $MgSiO_3$ perovskite are consistent with

$$\frac{dK_{0T}^{pv}}{dP} = K_{0T}^{pv\prime} = 4 \tag{4.33}$$

(Yagi *et al.*, 1982: a higher value of K'_{0T} would lower K_{0S}^{pv} from the value of 262 ± 12 GPa at 300 K which is shown in Fig. 4.15). Thus, from equations 4.30, 4.31 and 4.33

$$\delta_{0S}^{pv} \cong 2.5. \tag{4.34}$$

and the bulk modulus of silicate perovskite at 1750 K is estimated to be

$$K_{0S}^{pv} \cong 240 \text{ GPa}. \tag{4.35}$$

According to Figure 4.15, this result is only compatible with a pyroxene/olivine ratio of 2 or more for the lower mantle. Even considering the uncertainties in the measured value of K_{0T} for perovskite, a model with upper mantle composition apparently results in too high a compressibility for the lower mantle.

An independent line of evidence bearing on lower mantle composition lies in the nature of the 670 km seismological discontinuity. As pointed out by Richards (1972) and Lees *et al.*. (1983), the observed reflectivity of longitudinal waves at this depth is sufficiently high that the discontinuity must be extremely sharp. Lees *et al.* estimate that for any value of acoustic-impedance change consistent with the seismological data the 670 km discontinuity must be confined to less than a 7 km depth range (Fig. 4.16).

The only phase transition that has been observed in mantle assemblages at the pressure of the 670 km discontinuity is the breakdown of the olivine component (γ-spinel) to form perovskite and magnesiowüstite (Figs. 4.11 and 4.16; Jeanloz and Thompson, 1983). Hence, much of the change in properties between the transition zone and lower mantle is undoubtedly due to the occurrence of this transformation, but the issue is whether or not there is additionally a change in composition across the discontinuity. The equilibrium transformation is multivariant, and is therefore spread out over a

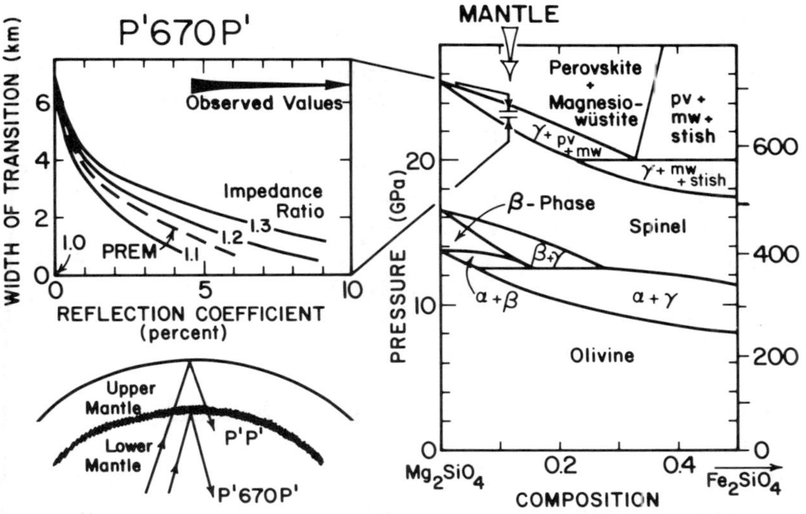

Figure 4.16. By comparing P′P′ and P′670P′ seismic phases it is possible to derive the reflection coefficient at the 670 km discontinuity. The wave-paths correspond to underside reflections from the surface and the 670 km discontinuity (lower left). A minimum reflection coefficient of 5 percent is observed at 670 km depth, and for seismologically acceptable impedance ratios at the 670 km discontinuity (~1.1 to 1.2; PREM is shown dashed), this implies a transition of only a few kilometers (upper left). In contrast, a much broader transition, of order 20–50 km, would be expected if the 670 km discontinuity were due solely to the equilibrium transformation of γ-spinel to a perovskite assemblage (right; the upper mantle composition is shown at the top). No other transformation is known for the major upper mantle minerals at these pressures (see Jeanloz and Thompson, 1983, and Lees *et al.*, 1983, for further details).

substantial depth interval if there is no change of composition. Based on the observed partitioning (Bell *et al*., 1979; Ito and Yamada, 1982), the perovskite-forming reaction alone leads to a width of 30–50 km for the 670 km discontinuity (Fig. 4.16). This result is clearly at variance with the seismological data. With a change of composition at 670 km depth, however, the velocity and density could readily change discontinuously.

Jeanloz and Thompson (1983) were concerned by a potentially serious ambiguity in the phase-equilibrium measurements. It appeared that the data on which the present discussion depends were collected at much lower temperatures (~1300 K) than are of interest for the lower mantle ($\gtrsim$2000 K). Thus, one could argue that the partitioning varies sufficiently with temperature or new reactions appear at mantle conditions, such that a sharp 670 km discontinuity could, after all, be explained in terms of transformations in an assemblage of constant composition. More recently, however, it has become evident that the optical pyrometry that has been used to determine temperature in the laser heated diamond-cell experiments systematically yields an underestimate: temperatures of 2000 K and above were probably achieved in the high-pressure experiments (Ito *et al*., 1984; Yagi, 1983, personal communication; Heinz and Jeanloz, 1984, unpublished work; Jeanloz and Heinz, 1985). Therefore, it seems that the pressures and temperatures of the lower mantle and transition zone have in fact been reproduced experimentally. The significant conclusion is that the available phase equilibrium data are in disagreement with the observed sharpness of the seismological discontinuity at 670 km depth.

7. Mantle Stratification, Recycling and Kinetics

Until recently, the preferred model has been that the mantle is homogeneous in composition. Accordingly, the geochemical nature of both the upper and lower mantle could be determined by examining xenoliths or by studying the petrogenesis of basalts (e.g., Ringwood, 1975; Yoder, 1979). In this homogeneous model, the seismological transition zone between 400 and 670 km depth corresponds only to progressive phase transitions of an upper mantle garnet peridotite assemblage. From the results of high pressure experiments, there is little doubt that most, if not all, of the changes in density and seismic velocity at these depths are due to phase transformations.

What is uncertain, however, is whether or not there is, in addition to the phase transitions, a change of composition at the 670 km seismological discontinuity that marks the boundary between the upper and lower mantle

(Fig. 4.17). The problem is that the presence of only a slight change in major-element composition prevents the lower mantle from being intermixed into the upper mantle. If the lower mantle is at least 3 percent denser than the upper mantle at equivalent conditions, then there is effectively a barrier which keeps these regions dynamically separate (Richter and Johnson, 1974; Richter and McKenzie, 1981; Christensen and Yuen, 1984).

The composition of the lower mantle is a major geochemical issue because this apparently homogeneous region contains the largest proportion of the Earth (Table 4.1). Thus, it is important to know whether material from the lower mantle is accessible to geochemical sampling at the surface, at least in being directly involved in near-surface petrological processes, or whether the lower mantle is essentially hidden beneath the upper mantle. In either case, convection is assumed to occur at all depths of the mantle, in accordance with the finding that there is no significant difference between the viscosity of the upper mantle and that of the lower mantle (Peltier, 1983). Hence, the main question becomes: Is there significant mass transfer between the lower and upper mantle, or are these regions separately convecting systems?

Aside from the compositional issues, there are important thermal consequences of stratification as well (Fig. 4.17). The lack of vertical mass transfer (i.e., advection of heat) through the top of the lower mantle makes the upper mantle act as an insulating layer. Thus, average temperatures in the lower mantle are significantly higher if the mantle is stratified, and a thermal boundary layer with a large, conductive gradient of temperature with depth would be expected near 670 km depth (Jeanloz and Richter, 1979; Richter and McKenzie, 1981). Furthermore, stratification of the mantle results in a much longer time scale for thermal evolution of the Earth ($\sim 10^{10}$ yr) than if the mantle is being uniformly stirred ($\sim 10^{9}$ yr): In the former case the planet is probably still evolving *away* from a global condition of thermal steady state (McKenzie and Richter, 1981).

A number of arguments have been put forth to suggest that the mantle is stratified, with the lower mantle differing in composition from the upper mantle (e.g., Anderson, 1976; Jeanloz and Richter, 1979; Richter, 1979; DePaolo and Wasserburg, 1976; Lees *et al.*, 1983). Among the classical arguments concerning whether or not the upper and lower mantle are intermixed is the interpretation that downgoing slabs cannot penetrate into the lower mantle (e.g., Richter, 1979). This conclusion is based on the observation that subduction-zone seismicity extends only to 650–700 km depth, and near this depth earthquake occurrences appear to be systematically more frequent and to have focal mechanisms that are consistent with the presence of down-dip compressive stresses. More recently, the modelling of geochem-

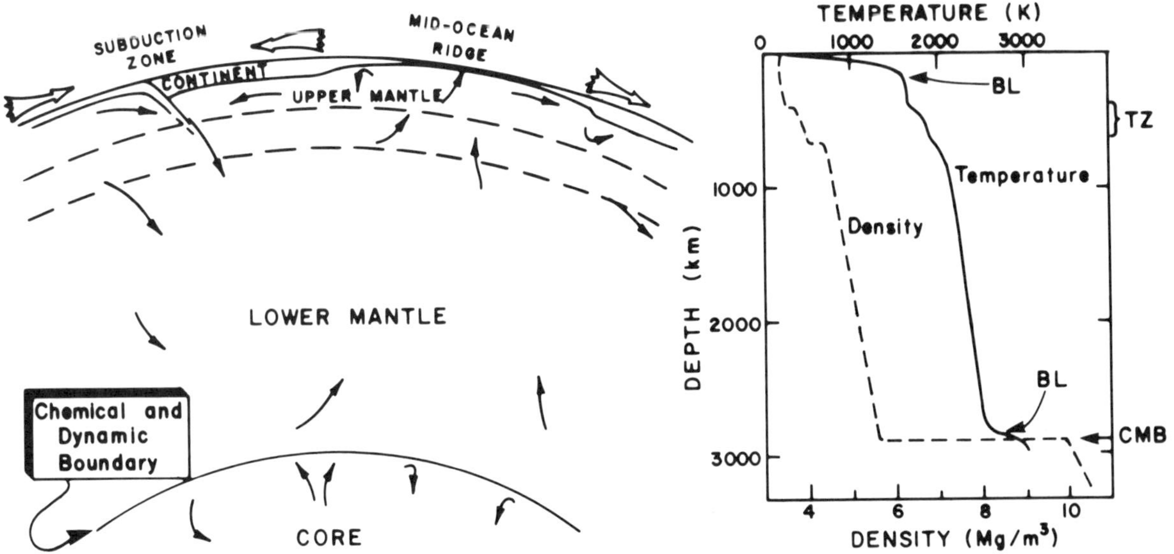

Figure 4.17. Two possible models of the mantle: uniform in composition and homogeneously mixed (upper panel) or stratified into separately convecting upper and lower regions (lower panel). Observed density and calculated temperature distributions as functions of depth are shown on the right. Stratification can occur due to a change in composition (chemical boundary) which prevents the lower mantle from being mixed into the upper mantle. As a result, thermal boundary layers (BL) occur at the base of the transition zone (TZ) and higher temperatures are reached at the core-mantle boundary (CMB) than in the case of uniform composition (from Jeanloz, 1981a).

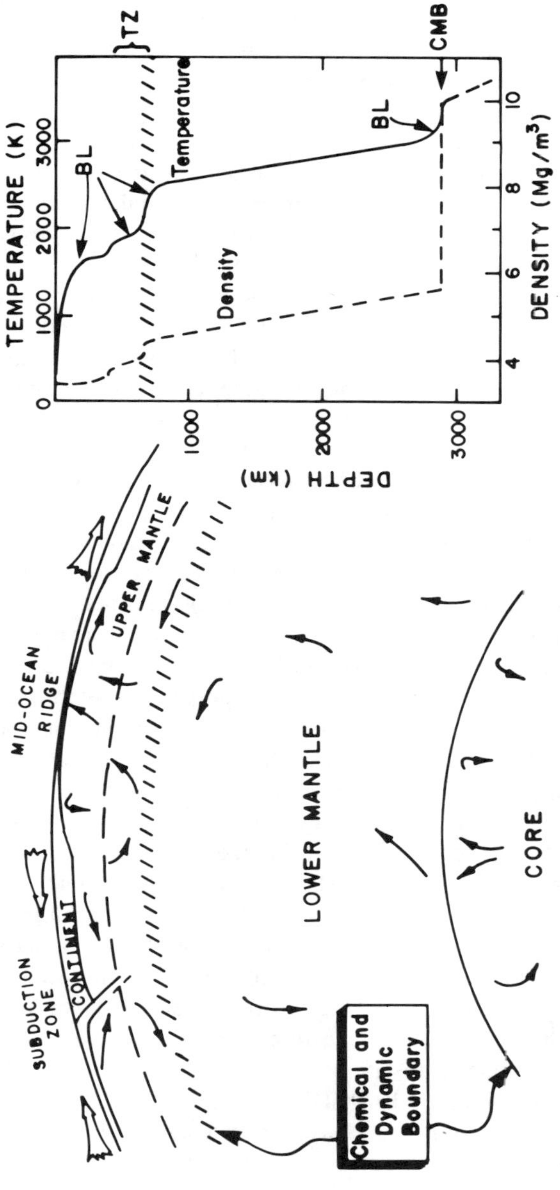

Figure 4.17. *Continued*

ical data (e.g., Sm-Nd, Rb-Sr, and He isotope systems) has also been used to argue that the upper and lower mantle correspond to distinct geochemical reservoirs that have been separate for billions of years (DePaolo and Wasserburg, 1976; O'Nions *et al.*, 1979; Jacobsen and Wasserburg, 1979; DePaolo, 1980; O'Nions and Oxburgh, 1983). The spatial locations of these depleted and undepleted reservoirs, respectively, cannot be unambiguously determined on the basis of the geochemical measurements alone. Nevertheless, the isotopic data suggest a ratio of volumes corresponding to the volumes of the upper and lower mantle, in agreement with what would be expected for a stratified mantle (see Davies, 1984, however).

These lines of argument are appealing in that they involve direct seismological or geochemical measurements on the Earth's mantle. The only difficulty lies in the validity or uniqueness of the ways in which the data are interpreted. In this paper, complementary lines of evidence are emphasized. The approach based on modelling the chemical and thermal state of the mantle and core has become useful only recently, as phase-equilibrium data at sufficiently high pressures have become available. In this case, the biggest uncertainty lies in the precision of the data, which may still not be sufficient to distinguish between stratified and homogeneous models of the mantle.

From the preceding discussion of mineral properties at high pressures it appears that a model of uniform mantle composition is not compatible with the data available from seismology. Rather, all indications are that a change in composition is required at 670 km depth and that the mantle is stratified. From the data considered, however, it is not possible to determine uniquely how large a difference in composition exists between the upper and lower mantle. If one attempts to find an acceptable lower mantle composition that differs *least* from that of the upper mantle, only a slight increase in silica and iron content are required. Accordingly, a lower mantle composition with pyroxene component between ⅔ and 1 (olivine component $\lesssim 1/3$) satisfies the bulk modulus data; an increase relative to the upper mantle of $\sim$3 to 5 percent in the molar ratio of Fe to Fe + Mg components is sufficient for the density (cf. Jeanloz and Thompson, 1983). The data do not provide sufficient resolution to distinguish changes between the upper and lower mantle of other major elements, such as Ca or Al. Nevertheless, these proposed compositions are consistent with the differences that could exist between the depleted and undepleted reservoirs of global geochemical models (e.g., Jacobsen and Wasserburg, 1979; O'Nions *et al.*, 1979). Also, a denser composition for the lower mantle, as proposed here, would be sufficient to stabilize the 670 km discontinuity as a barrier to convection.

Having recognized the evidence for mantle stratification, it is worth noting that the boundary between the upper and lower mantle is unlikely to be completely impenetrable. Indeed, it is some of the data supporting stratification, namely subduction-zone seismology and trace element geochemistry, that also indicate that this boundary is "leaky." Recent analyses of travel time anomalies associated with Benioff-zone earthquakes are especially noteworthy. These anomalies give a strong indication for a high-velocity region extending beyond the observed seismicity of at least some downgoing slabs, which are therefore inferred to pierce well into the lower mantle (to at least 1000 km depth: Jordan, 1977; Creager and Jordan, 1984). Similarly, the fact that specimens of undepleted mantle have been sampled and analyzed requires that some material has leaked out of the lower mantle. In both cases the mass flux between upper and lower mantle is not well determined, but could be very small. For example, a leakage across the 670 km boundary that amounts to no more than 10^{-3} to 10^{-2} of the vertical mass transfer within the convecting regions is probably consistent with both the seismological and geochemical data (see also, Christensen and Yuen, 1984). Although the suggestion that there is recycling between the upper and lower mantle is highly speculative, it is completely analogous to the recycling of crust into the upper mantle (e.g., DePaolo, 1980).

As pointed out by Ahrens and Schubert (1975) and Sung (1979), reaction kinetics can play an important geodynamic role in downgoing slabs, where temperatures are relatively low. To the extent that the formation of dense, high-pressure assemblages makes a contribution to the negative buoyancy forces pulling the slab downwards (cf. Schubert *et al.*, 1975), kinetics may become significant in determining whether or not a slab can penetrate into the lower mantle. If kinetics play a role in slab penetration, a delicate balance emerges between the requirement for low temperatures to keep the slab denser than the surrounding mantle (due to thermal expansion) and sufficiently high temperatures to overcome kinetic hindrances in the transformations to high pressure phases.

Based on high-pressure measurements of reaction kinetics, Sung (1979) found that the cold interior region of a slab moving at 10 cm/yr would probably not undergo the olivine $\rightarrow$ spinel transformation (a reasonable estimate of the activation volume was used: Ross *et al.*, 1979). Similarly, Ahrens and Schubert (1975) found that unless hydrous species are present, the gabbro-eclogite transition is unlikely to go to completion (at the shallower depths involved for this reaction, dehydration along the downgoing slab may, in fact, be important). Too few data currently exist on the kinetics of mineral reactions at high pressures to do more than speculate, but what measurements are available indicate that transformations are likely to occur unimpeded at the temperatures of most of the upper mantle ($T \gtrsim 1000$ to 1500°C). Within rapidly moving slabs, however, temperatures can be sufficiently low to preclude transformation, and reaction kinetics can therefore determine the location and rate of leakage into the lower mantle.

8. Summary

A relatively simple model adequately describes the composition of the mantle and core. The upper mantle and transition zone are viewed as consisting of an olivine-rich assemblage such as garnet peridotite (or compositionally equivalent rock), in accordance with most previous work. There is evidence suggesting that the lower mantle differs in composition from the upper mantle, probably in being somewhat richer in silica and iron. With such a change in composition, the coefficient of thermal expansion, Grüneisen parameter and bulk modulus predicted for the lower mantle come into good agreement with the available observations. Also, the resulting stratification is compatible with geochemical models of the mantle being made up of depleted and undepleted regions (upper and lower mantle, respectively). The core has long been known to be an iron alloy, but it is now recognized to be more nearly pure iron than was previously thought: a mere 5–10 weight percent of alloying component is required by the seismological and high-pressure data. This small amount of alloying, combined with new data on the melting of iron at 250 GPa, require temperatures in the outer core to be fairly high; values between 3800 and 5000 K are currently estimated for the outermost core, regardless of the detailed composition. In fact, iron may well be alloyed with a mixture of many components, of which oxygen or sulfur (or both in combination) are thought to be the most significant. A point to be emphasized, however, is that compositional models are intimately related to thermal and dynamic models of the Earth's interior. Evidence for stratification of the mantle, such as the P′670P′ reflectivity, is also evidence for a relatively high temperature in the lower mantle, consistent with the high estimated temperatures in the core (Figure 4.17). In addition, the inferred compositional difference that acts to keep the upper and lower mantle dynamically separate results in a much longer time constant for heat transfer out of the Earth's deep interior. If it is present this stratification should not be viewed statically, however, as the existing seismological and geochemical evidence also suggest that there may be a small amount of leakage or recycling between the upper and lower mantle. Thus, the detailed nature of the transition zone is likely to be of continuing interest in future dynamical studies of the Earth's mantle.

9. Acknowledgements

I thank O.L. Anderson, D. Heinz, E. Knittle, C. Meade, A. Rudy, H.R. Wenk, and G. Wolf for comments, and E. Ito and T. Yagi for providing unpublished material. This work was supported by the National Science Foundatin, the National Aeronautics and Space Administration, and the A.P. Sloan Foundation.

10. References

Ahrens, T.J., Anderson, D.L., and Ringwood, A.E., *Rev. Geophys.* **7**, 667 (1969).

Ahrens, T.J., Lyzenga, G.A., and Mitchell, A.C., in Akimoto, S. and Manghnani, M.H. (eds.), *High-Pressure Research in Geophysics*. Tokyo: Center Acad. Pub. Japan, 579 (1982).

Ahrens, T.J. and Schubert, G., *Rev. Geophys. Space Phys.* **13**, 383 (1975).

Akimoto, S., Akaogi, M., Kawada, K., and Nishizawa, O., in Sutton, G.H., Manghnani, M.H., and Moberly, R. (eds.), *The Geophysics of the Pacific Ocean Basin and Its Margin*. Washington, D.C.: Am. Geophys. Union, Washington, D.C., 399 (1976a).

Akimoto, S., Matsui, Y., and Syono, Y., in Strens, R.G.J. (ed.), *The Physics and Chemistry of Minerals and Rocks*. New York: Wiley, 327 (1976b).

Al'tshuler, L.V., *Sov. Phys. Uspekhi* **8**, 52, (1965).

Anderson, D.L., *Geophys. Res. Lett.* **3**, 347 (1976).

Anderson, D.L., *EOS* **65**, 147 (1984).

Anderson, D.L. and Hart, R.S., *J. Geophys. Res.* **81**, 1461 (1976).

Anderson, O.L., *J. Geophys. Res.* **72**, 3661 (1967).

Anderson, O.L. and Sumino, Y., *Phys. Earth Planet. Int.* **23**, 314 (1980).

Anderssen, R.S., DeVane, J.F., Gustafson, S.A., and Winch, D.E., *Phys. Earth Planet. Int.* **20**, p. 15 (1979).

Backus, G.E., and Gilbert, F., *Phil. Trans. Roy. Soc. London* **A266**, 123 (1970).

Banks, R.J., *J. Geomag. Geoelectr.* **24**, 337 (1972).

Barnett, J.D., Block, S., and Piermarini, G.J., *Rev. Sci. Instrum.* **44**, 1 (1973).

Basaltic Volcanism Studies Project, *Basaltic Volcanism on the Terrestrial Planets*. New York: Pergamon (1981).

Bassett, W.A., *Ann. Rev. Earth Planet. Sci.* **7**, 357 (1979).

Bell, P.M. and Mao, H.K., *Carnegie Inst. Washington Yearbook* **80**, 404 (1981).

Bell, P.M., Yagi, T., and Mao, H.K., *Carnegie Inst. Washington Yearbook* **78**, 618 (1979).

Birch, F., *Phys. Rev.* **71**, 809 (1947).

Birch, F., *J. Geophys. Res.* **57**, 227 (1952).
Birch, F., *Geophys. J. R. Astr. Soc.* **4**, 295 (1961).
Birch, F., *J. Geophys. Res.* **69**, 4377 (1964).
Birch, F., *J. Geophys. Res.* **73**, 817 (1968).
Birch, F., *Geophys. J. R. Astr. Soc.* **29**, 373 (1972).
Birch, F., *J. Phys. Chem. Solids* **38**, 175 (1977).
Birch, F., *J. Geophys. Res.* **83**, 1257 (1978).
Bragg, L., Claringbull, G.F., and Taylor, W.H., *Crystal Structures of Minerals*. Ithaca, N.Y.: Cornell University Press (1965).
Brett, R., *Rev. Geophys. Space Phys.* **14**, 375 (1976).
Brett, R., *Geochim. Cosmochim. Acta* **48**, 1183 (1984).
Brown, J.M., Ahrens, T.J., and Shampine, D.L., *J. Geophys. Res.* **89**, 6049 (1984).
Brown, J.M., and McQueen, R.G., in Akimoto, S. and Manghnani, M.H. (eds.), *High-Pressure Research in Geophysics*. Tokyo: Center Acad. Pub. Japan, 611 (1982).
Brown, J.M. and Shankland, T.J., *Geophys J. R. Astr. Soc.* **66**, 579 (1981).
Bukowinski, M.S.T., *Phys. Earth Planet Int.* **14**, 333 (1977).
Bullen, K.E., *The Earth's Density*. London: Chapman and Hall (1975).
Burdick, L.J. and Powell, C., *J. Geophys. Res.* **85**, 3845 (1980).
Christensen, U.R. and Yuen, D.A., *J. Geophys. Res.* **89**, 4389 (1984).
Clayton, R.W., *EOS* **65**, 236 (1984).
Creager, K.C. and Jordan, T.H., *J. Geophys. Res.* **89**, 3031 (1984).
Davies, G.F., *J. Phys. Chem. Solids* **34**, 1417 (1973).
Davies, G.F., *J. Phys. Chem. Solids* **35**, 1513 (1974).
Davies, G.F., *Geophys. Res. Lett.* **9**, 1267 (1982).
Davies, G.F., *J. Geophys. Res.* **89**, 6017 (1984).
Davies, G.F. and Dziewonski, A.M., *Phys. Earth Planet. Int.* **10**, 336 (1975).
Dawson, J.B., *Kimberlites and Their Xenoliths*. New York: Springer (1980).
Deans, S.R., *The Radon Transform and Some of Its Applications*. New York: Wiley (1983).
DePaolo, D.J., *Geochim. Cosmochim. Acta* **44**, 1185 (1980).
DePaolo, D.J. and Wasserburg, G.J., *Geophys. Res. Lett.* **3**, 743 (1976).
Dziewonski, A.M., *J. Geophys. Res.* **89**, 5929 (1984).
Dziewonski, A.M. and Anderson, D.L., *Phys. Earth Planet. Int.* **25**, 297 (1981).
Dziewonski, A.M., Hales, A.L., and Lapwood, E.R., *Phys. Earth Planet. Int.* **10**, 12 (1975).
Fujisawa, H., *J. Geophys. Res.* **73**, 3281 (1968).
Fukao, Y., Nagahashi, T., and Mori, S., in Akimoto, S. and Manghnani, M.H. (eds.), *High-Pressure Research in Geophysics*. Tokyo: Center Acad. Pub. Japan, 285 (1982).
Goettel, K.A., Mao, H.K., and Bell, P.M., *EOS* **65**, 283 (1984).
Grady, D.E., *J. Geophys. Res.* **85**, 913 (1980).
Hart, R.S., *J. Geophys. Res.* **80**, 4889 (1975).
Hart, R.S., Anderson, D.L., and Kanamori, H., *J. Geophys. Res.* **82**, 1647 (1977).

Heinz, D.L. and Jeanloz, R., *Nature* **301**, 138 (1983).
Heinz, D.L. and Jeanloz, R., *J. Appl. Phys.* **55**, 885 (1984).
Higgins, G. and Kennedy, G.C., *J. Geophys. Res.* **76**, 1870 (1971).
Ito, E., Takahashi, E., and Matsui, Y., *Earth Planet Sci. Lett.* **67**, 238 (1984).
Ito, E. and Yamada, H., in Akimoto, S., and Manghnani, M.H. (eds.). *High-Pressure Research in Geophysics*. Tokyo: Center Acad. Pub. Japan, 405 (1982).
Jackson, I. *Earth Planet Sci. Lett.* **62**, 91 (1983).
Jackson, I., Liebermann, R.C., and Ringwood, A.E., *Phys. Chem. Min.* **3**, 11 (1978).
Jackson, I. and Niesler, H., in Akimoto, S., and Manghnani, M.H. (eds.), *High-Pressure Research in Geophysics*, Tokyo: Center Acad. Pub. Japan, 579 (1982).
Jacobs, J.A., *The Earth's Core*. New York: Academic Press (1975).
Jacobsen, S.B. and G.J. Wasserburg, *J. Geophys. Res.* **84**, 7411 (1979).
Jeanloz, R., *J. Geophys. Res.* **84**, 6059 (1979).
Jeanloz, R., *J. Geophys. Res.* **85**, 3163 (1980).
Jeanloz, R., *Proc. Am. Chem. Soc. 17th State-of-the-Art Symposium*, Washington, D.C.: Am. Chem. Soc., 70 (1981a).
Jeanloz, R., *Geophys. Res. Lett.* **8**, 1219 (1981b).
Jeanloz, R., in Akimoto, S., and Manghnani, M.H. (eds.), *High-Pressure Research in Geophysics*. Tokyo: Center Acad. Pub. Japan, 479 (1982).
Jeanloz, R., *Scientific American* **249**, 56 (1983).
Jeanloz, R. and Ahrens, T.J., *J. Geophys. Res.* **84**, 7545 (1979).
Jeanloz, R. and Ahrens, T.J., *Geophys. J. R. Astr. Soc.* **62**, 505 (1980).
Jeanloz, R. and Heinz, D.L., *J. Physique* (Paris) **45**, (C8) 83 (1984).
Jeanloz, R. and Richter, F.M., *J. Geophys. Res.* **84**, 5497 (1979).
Jeanloz, R. and Roufosse, M., *J. Geophys. Res.* **87**, 10763 (1982).
Jeanloz, R. and Thompson, A.B., *Rev. Geophys. Space Phys.* **21**, 51 (1983).
Jones, L.E.A., *Phys. Chem. Minerals* **4**, 23 (1979).
Jordan, T.H. *Geophys. J.* **43**, 474 (1977).
Knittle, E. and Jeanloz, R., *Science* **223**, 53 (1984).
Knittle, E., Jeanloz, R., and Smith, G.L., *Nature* **319**, 214 (1986).
Knittle, E., Rudy, A., and Jeanloz, R., *Phys. Rev. B* **31**, 588 (1985).
Kormer, S.B. *Sov. Phys. Uspekhi* **11**, 229 (1968).
Kumazawa, M., and Endo, S., in Sunagawa, I, (ed.), *Materials Science of the Earth's Interior*. Tokyo: Terra Scientific Pub. Co., 587 (1984).
Lacam, A., Madon, M., and Poirier, J.P., *Nature* **288**, 155 (1980).
Lees, A.C., Bukowinski, M.S.T., and Jeanloz, R., *J. Geophys. Res.* **88**, 8145 (1983).
Liu, L.G., *Geophys. J. R. Astr. Soc.* **43**, 697 (1975).
Liu, L., *Phys. Earth Planet. Int.* **11**, 289 (1976).
Liu, L. and Ringwood, A.E., *Earth Planet. Sci. Lett.* **28**, 209 (1975).
Lyttle, F.W., *J. Appl. Phys.* **35**, 2212 (1964).
Lyzenga, G.A., Ahrens, T.J., and Mitchell, A.C., *J. Geophys. Res.* **88**, 2431 (1983).
Madon, M., Bell, P.M., Mao, H.K., and Poirier, J.P., *Geophys. Res. Lett.* **7**, 629 (1980).
Mao, H.K. and Bell, P.M., *Science* **200**, 1145 (1978).

Mao, H.K., and Bell, P.M., Dunn, K.J., Chrenko, R.J., and DeVries, R.C., *Rev. Sci. Instrum.* **50**, 1002 (1979).

Mao, H.-K. and Bell, P.M., *J. Geophys. Res.* **84**, 4533 (1979).

Masters, G., *Geophys. J. R. Astr. Soc.* **57**, 507 (1979).

Masters, G., Jordan, T.H., Silver, P.G., and Gilbert, F., *Nature* **298**, 609 (1982).

McKenzie, D. and Richter, F.M., *J. Geophys. Res.* **86**, 11667 (1981).

McQueen, R.G., Marsh, S.P., Taylor, J.W., Fritz, J.N., and Carter, W.J., in Kinslow, R. (ed.), *High Velocity Impact Phenomena*. New York: Academic Press, 294 (1970).

Merrill, R.T. and McElhinney, M.W., *The Earth's Magnetic Field*. New York: Academic Press (1983).

Ming, L.C. and Bassett, W.A., *Rev. Sci. Instrum.* **45**, 1115 (1974).

Ming, L.C., Manghnani, M.H., Balogh, J., Qadri, S.B., Skelton, E.F., and Jamieson, J.C., *J. Appl. Phys.* **54**, 4390 (1983).

Munro, R.G., Block, S., and Piermarini, G.J., *J. Appl. Phys.* **50**, 6779 (1979).

Munro, R.G., Block, S., Mauer, F.A., and Piermarini, G.J., *J. Appl. Phys.* **53**, 7080 (1982).

Naimon, E.R. and Granato, A.V., *Phys. Rev.* **137**, 2091 (1973).

O'Nions, R.K., Evensen, N.M., and Hamilton, P.J., *J. Geophys. Res.* **84**, 6091 (1979).

O'Nions, R.K. and Oxburgh, E.R., *Nature* **306**, 429 (1983).

Peltier, W.R., *Nature* **304**, 434 (1983).

Piermarini, G.J., Block, S., and Barnett, J.D., *J. Appl. Phys.* **44**, 5377 (1973).

Rice, M.H., McQueen, R.G., and Walsh, J.M., *Solid State Physics* **6**, 1 (1958).

Richards, P.G., *Z. Geophys.* **38**, 517 (1972).

Richter, F.M., *J. Geophys. Res.* **84**, 6783 (1979).

Richter, F.M. and Johnson, C.E., *J. Geophys. Res.* **79**, 1635 (1974).

Richter, F.M. and McKenzie, D.P., *J. Geophys. Res.* **861**, 6133 (1981).

Ringwood, A.E. *Composition and Petrology of the Earth's Mantle*. New York: McGraw-Hill (1975).

Ringwood, A.E., *Origin of the Earth and Moon*. New York: Springer (1979).

Robie, R.A., Hemingway, B.S., and Fisher, J.R., *U.S. Geol. Furv. Bull* **1452** (1978).

Ross, J.V., Ave'lallement, H.G., and Carter, N.L., *Science* **203**, 261 (1979).

Schubert, G., Yuen, D.A., and Turcotte, D.L., *Geophys. J.R. Astr. Soc.* **42**, 705 (1975).

Spiliopoulos, S. and Stacey, F.D., *J. Geodynamics* **1**, 61 (1984).

Stacey, F.D., *Phys. Earth. Planet. Int.* **15**, 341 (1977).

Stevenson, D.J., *Science* **214**, 611 (1981).

Sung, C.-M., *Rev. Sci. Instrum.* **47**, 1343 (1976).

Sung, C.-M., in Timmerhaus, K.D. and Barber, M.S. (eds.), *High-Pressure Science and Technology* **2**. New York: Plenum Press, 31 (1979).

Verhoogen, J., *Phys. Earth Planet. Int.* **7**, 47 (1973).

Verhoogen, J., *Energetics of the Earth*. Washington, D.C.: National Academy of Sciences (U.S.) (1980).

Wallace, D.C., *Thermodynamics of Crystals*. New York: Wiley (1972).

Watt, J.P. and Ahrens, T.J., *J. Geophys. Res.* **88**, 9500 (1983).
Weidner, J.D., Sawamoto, H., Sasaki, S., and Kumazawa, M., *J. Geophys. Res.* **89**, 7852 (1984).
Woodhouse, J.H. and Dziewonski, A.M., *J. Geophys. Res.* **89**, 5953 (1984).
Yagi, T., Mao, H.K., and Bell, P.M., *Phys. Chem. Minerals* **3**, 97 (1978).
Yagi, T., Mao, H.K., and Bell, P.M., *Adv. Phys. Geochem.* **2**, 317 (1982).
Yoder, H.S., Jr., *Generation of Basaltic Magmas*. Washington, D.C.: National Academy of Sciences (U.S.) (1976).

Note Added in Proof

This chapter was written in 1983, and updated in 1985 with the addition of three references. Recent developments pertaining to individual sections within the chapter are described in the following references. **2.1.** Jeanloz, R. and Morris, S., *Ann. Rev. Earth Planet. Sci.* **14**, 377 (1986); Li, X. and Jeanloz, R., *Geophys. Res. Lett.* **14**, 1075 (1987). **2.2.** Meade, C. and Jeanloz, R., *J. Geophys. Res.* **93**, 3261 (1988). **2.3.** Jeanloz, R., *Rev. in Mineralogy* **14**, 389 (1985); Jeanloz, R., *J. Geophys. Res.* **92**, 10352 (1987). **3.1.** Ahrens, T.J., *Methods of Experimental Physics* **24(A)**, 185 (1987); Hemley, R.J., Jephcoat, A.P., Mao, H.K., Ming, L.C., and Manghnani, M.H., *Nature* **334**, 52 (1988). **3.2.** Jayaraman, A., *Rev. Sci. Instrum.* **57**, 1013 (1986); Jephcoat, A.P., Mao, H.K., and Bell, P.M., in Ulmer, G.C. and Barnes, H.L. (eds.), *Hydrothermal Experimental Techniques*. New York: Wiley, 469 (1987); Hemley, R.J., Bell, P.M., and Mao, H.K., *Science* **237**, 605 (1987); Heinz, D.L. and Jeanloz, R., *J. Geophys. Res.* **92**, 11437 (1987). **4.** Knittle, E. and Jeanloz, R., *Geophys. Res. Lett.* **13**, 1541 (1986); Williams, Q., Jeanloz, R., Bass, J., Svendsen, B., and Ahrens, T.J., *Science* **236**, 181 (1987); Ahrens, T.J. and Jeanloz, R., *J. Geophys. Res.* **92**, 10363 (1987); Jeanloz, R. and Wenk, H.R., *Geophys. Res. Lett.* **15**, 72 (1988). **5.1.** Knittle, E. and Jeanloz, R. *Science* **235**, 668 (1987); Jackson, W.E., Knittle, E., Brown, G.E., Jr., and Jeanloz, R., *Geophys. Res. Lett.* **14**, 224 (1987). **5.2.** Weidner, D.J. and Ito, E., *Phys. Earth Planet. Int.* **40**, 65 (1985). **6.** Jeanloz, R., *Phys. Rev. B* **38**, 805 (1988); Jeanloz, R., and Knittle, E., *Adv. Physical Geochem.* **6**, 275 (1986); Jeanloz, R. and Rudy, A., *J. Geophys. Res.* **92**, 11433 (1987); Williams Q., Jeanloz, R., and McMillan, P., *J. Geophys. Res.* **92**, 8116 (1987). **7.** Williams, Q. and Jeanloz, R. *Science* **239**, 902 (1988); Jeanloz, R. and Knittle, E., *Phil. Trans. Roy. Soc. (London), Ser. A* (in press) (1988).

CHAPTER 5

Constraints on the Nature and Development of Chemical Heterogeneities in the Mantle

Stan Hart[†] and Alan Zindler[‡]

[†]Center for Geoalchemy, Department of Earth, Atmospheric and Planetary Sciences, Massachusetts Institute of Technology, Cambridge, Massachusetts 02139

[‡]Lamont-Doherty Geological Observatory and Department of Geological Sciences, Columbia University, Palisades, New York 10964

1. Introduction

The case for a chemically heterogeneous sub-oceanic mantle was firmly established by Gast *et al.* (1964), who showed that the Sr and Pb isotopic composition of basalts from Ascension and Gough Islands was measurably different. This finding was corroborated by Pb and Sr isotopic data for basalts from a variety of other oceanic islands (Tatsumoto, 1966a, 1966b). Evidence for a systematic difference between the mantle sources for oceanic islands and oceanic spreading ridges was first presented by Tatsumoto *et al.* (1965), and later firmly established as a general world-wide feature by Hedge and Peterman (1970), and Hart (1971). Since that time, geochemists have been grappling with the question of how these mantle heterogeneities developed, how they are spatially organized, and how the various components and reservoirs interact with each other. The evidence for a chemically heterogeneous mantle is unequivocal; the question of how these heterogeneities develop and where they are situated in the convecting mantle has proven to be unruly, and has occupied the attention of geochemists and geophysicists for more than a decade. It is our intent here to evaluate the various approaches which are relevant to this vexatious question of chemical mantle dynamics.

We first present the geochemical evidence, organized in terms of three principal boundary conditions: 1) initial chemical composition of the Earth; 2) scale lengths (sizes) inferred for mantle heterogeneities; and 3) correlation and interrelationships of various mantle isotopic tracers (Sr, Nd, Pb and He). We then scrutinize various models for the development of physical and chemical structure in the mantle, and the nature of mantle convection. Each of the various subsections of this paper is initiated with a short introduction.

2. Composition of Bulk Earth

A precise knowledge of the composition of the bulk silicate earth (BSE) can be used to constrain the dynamics and evolution of the crust-mantle system. For example, knowing the U content of BSE can be used to accurately estimate a value for the Urey ratio (ratio of total heat flow from the Earth, to the total internal radioactive heat production), which in turn may provide constraints on the nature of mantle convection (Richter, 1984; see Section

VI). In addition, the determination of how much mantle has been depleted in LIL elements by continental crust extraction (e.g., Jacobsen and Wasserburg, 1979; Goldstein *et al.*, 1982; Allegre *et al.*, 1983b, 1983c) depends heavily on knowledge of BSE Sr and Nd contents. In fact, our inability to specify BSE Sr and Nd contents is presently the "weak link" in deciding whether the whole mantle or just the upper mantle (~670 km) has been depleted. Obviously these considerations also impact the question of whole mantle versus layered mantle convection.

Several pioneering papers have attempted to estimate the composition of primitive upper mantle (PUM) by comparing the chemistry of upper mantle lherzolites with the chemistry of meteorites (Loubet *et al.*, 1975; Jagoutz *et al.*, 1979a). Loubet *et al.* (1975) showed that the REE concentrations in PUM were in the range of 2.5–3.1 times C1 chondrites. Jagoutz *et al.* (1979a) proposed that the intersection of lherzolite and chondrite arrays in Mg-Al-Si space could be used to determine a PUM composition. They concluded that the BSE was about 2.5×Cl chondrites for refractory incompatible elements such as Ca, Al and the REE. This approach appears to be very promising, and, in this section, we pursue it further to evaluate the underlying assumptions and look at possible reasons for the second-order scatter in the data. Our goal was to see if BSE could be determined to better than 10 percent, which is the level required for providing useful constraints on mantle models such as those mentioned above.

First, let us set down the assumptions involved in this approach:

1. The ultramafic samples accessible to geochemists cannot be proven to sample more than the upper ~150–200 km of the mantle, thus this approach will define only uppermost mantle chemistry. Extrapolation of this to BSE will involve additional assumptions regarding the mode of accretion of the Earth and subsequent intramantle chemical fractionations.
2. The Earth is presumed to have some chemical "kinship" with meteorites. In particular, the BSE is assumed to have similar ratios of those refractory (non-volatile) elements which can be shown to be constant throughout all classes of chondrites, and similar to solar ratios.
3. Various upper mantle ultramafic rocks are presumed to be "related" to each other by simple petrogenetic processes. In particular, since all known ultramafic samples appear to have undergone at least some modification in the light REE (LREE), and none have present day $^{143}Nd/^{144}Nd$ isotope ratios consistent with closed system growth for 4.55 b.y., some extrapolation will be required to project observed residual chemistries back to pre-depletion values. This projection requires application of some model for the process, and acceptance of any such model represents a

fundamental assumption to this approach. The usual model which is adopted, and which we will develop below, is that depletion of a "primitive" lherzolite occurs by extraction of a partial melt, leading to residual lherzolites with lower Ca, Al and REE, and higher Mg.

2.1. Methodology

We have assembled a data set comprising averages of 7 classes of chondrites (E, H, L, LL, C1, CM2 and CV&CO chondrites) and 33 individual lherzolites (Table 5.1: see Hart and Zindler, 1986, for details of data selection). For the chondrite REE, only analyses made by isotope dilution were used, and even these were selected only if the chondrite-normalized patterns were "smooth" (i.e., no anomalous features, such as zig-zag patterns, Yb anomalies, etc.). For the lherzolites (all but one of our samples are spinel lherzolites), we have not restricted ourselves to xenoliths as did Jagoutz *et al.* (1979a), but have included orogenic lherzolites as well. Our principle criteria was to select those samples which had the least-depleted LREE patterns (using both neutron activation and isotope dilution data sources), and the highest Ca and Al contents. Light-REE-enriched (metasomatized?) nodules were not included, though it is clear that some flat-pattern lherzolites may once have been "depleted" samples which have been metasomatized just enough to produce the nearly chondritic REE. In addition to this set of most "fertile" lherzolites, we included a variety of more depleted samples, to see if a depletion model could be constructed to fit observed ranges of depletion.

We have used only 3 REEs, though in principle any or all could be used. We chose Nd as a fairly incompatible REE, Sm as an intermediate REE, and Yb as a moderately compatible REE. Nd and Sm were also chosen because a variety of high precision data exists for these elements. Of the heavy REE, Yb was chosen as it has the best precision by neutron-activation techniques. Where modes and REE contents on separated minerals were available, we calculated "whole rock" REE data, to avoid the "labile" Nd which is common in whole rock lherzolites (e.g., Zindler *et al.*, 1983). All analyses were recalculated to an anhydrous basis. We should note here that our subsequent consideration of this lherzolite data set does not give equal weight to every sample; this is not a matter of subjective prejudice, but is based on detailed consideration of the samples and data which clearly show that some are more reliable than others (Hart and Zindler, 1986).

Figure 5.1a is a plot of our data on the Mg/Si–Al/Si plot used by Jagoutz *et al.* (1979a). The appearance of the plot is very similar to theirs, though only 8 samples are common to both diagrams. We have plotted the position of

Jagoutz *et al.*'s, (1979a) preferred composition of primitive upper mantle (JPUM), and drawn the terrestrial and meteorite arrays to intersect at this point. Note that the CM2 and CV-CO chondrites appear to be fractionated away from C1 and ordinary chondrites by a process which did not affect these latter chondrite classes (Larimer, 1979) and, following Jagoutz *et al.* (1979a), we will not consider them in our analysis. Two immediate conclusions are obvious from Figure 5.1a: 1) the primitive upper mantle (PUM hereafter) is chemically distinct from known chondrite classes, and, 2) Mg/Si and Al/Si of PUM may be defined to better than ± 5 percent in the context of the assumptions outlined earlier. It was our hope that similar plots involving Ca and the REE would equally well define Ca/Si and REE/Si for PUM. Figure 5.1b shows a Mg/Si–Ca/Si plot, and again, the terrestrial and chondrite arrays are well defined—however, a number of samples, including JPUM, fall to the right of the intersection, indicating the failure of one of the assumptions stated above.

Let us digress briefly here to discuss chemical fractionation in chondrites. Certain element ratios in chondrites do not appear to show any variation between classes; Table 5.2 compares chemical fractionations between chondrites for various element ratios. Concentrating on the chondrite classes exclusive of CM2 and CV-CO, it is apparent that ratios involving Nd, Sm, Ca and Al are constant to better than 6 percent, and part or all of this may be analytical. Ratios involving Yb show variations of 9–13 percent, clearly beyond analytical error. This is probably related to the relatively low condensation temperature (Tc) of Yb compared to the other REE (Boynton, 1975). Ratios involving Mg show variations of 23–39 percent, except for Yb/Mg which is only 13 percent. Mg also has a comparatively low Tc, relative to Ca and Al (Grossman and Larimer, 1974), and apparently follows Yb to some extent. Ratios involving Si are highly fractionated (factors of 1.6–1.7); again, Tc for Si is less than that for Mg (Grossman and Larimer, 1974). Thus, four ratios are of particular importance in estimating terrestrial chemistry from meteorite analogues: Sm/Nd and Ca/Al which are constant in *all* chondrite classes, and Nd/Ca and Nd/Al which are constant in *all* chondrite classes, exclusive of CM2 and CV-CO.

Returning now to the lherzolite data, the Ca/Al ratio ranges from 0.556–1.60, versus the constant value of 1.09 for meteorites (Ahrens and Von Michaelis, 1969; Ahrens *et al.*, 1969). Furthermore, all but one of the lherzolites lying to the right of the meteorite–terrestrial array intersection in Figure 5.1b, have Ca/Al ratios > 1.20. The terrestrial array is commonly viewed as a "melt-depletion" array (Loubet *et al.*, 1975; Suen, 1978; Frey *et al.*, 1985; O'Hara *et al.*, 1975). However, recognizing that all common

Table 5.1. Meteorite and lherzolite major and rare earth element concentrations

Sample I.D.	SiO_2	Al_2O_3	MgO	CaO	Nd, ppm	Sm, ppm	Yb, ppm	References
Meteorites								
av.E chondrite	35.27	1.47	18.33	1.178	0.422	0.1366	0.1597	1, 4, 7, 12
av.L chondrite	39.7	2.191	24.72	1.791	0.659	0.212	0.2309	1, 4, 6, 11, 12
av.LL chondrite	39.38	2.21	25.4	1.82	0.652	0.214	0.2403	2, 4, 6, 10–12
av.H chondrite	35.83	2.034	23.09	1.659	0.5934	0.1922	0.2077	1, 4–6
av.C1 chondrite	22.77	1.64	16.41	1.298	0.463	0.1519	0.1616	1, 3–5, 8, 9, 13
av.CM2 chondrite	27.1	2.117	19.02	1.684	0.6526	0.2129	0.2149	1, 4, 5, 8, 9, 11–13
av.CV+CO chondrite	33.9	3.213	24.19	2.567	1.028	0.3358	0.326	1, 4–6, 8, 9, 11, 13
Xenolith lherzolites								
D1, Dreiser W.	43.59	2.44	43.97	1.97	0.376	0.152	0.252	14, 16, 17
Kh1, Kilbourne H.	46.44	3.45	39.86	2.78	0.418	0.223	0.444	14, 15
Po1, Potrillo	45.39	3.52	39.12	2.84	0.479	0.227	0.384	14
Fr1, Landoz	45.96	4.02	38.39	3.25	1.09	0.362	0.407	14, 17
Ka168, Kapfenstein	44.02	3.732	39.87	3.082	0.768	0.274	0.375	14, 37
Sc1, San Carlos	45.37	4.01	37.61	3.24	1.18	0.425	0.404	14, 15
Pa15a, San Carlos	43.71	3.34	39.57	2.68	0.515	0.234	0.315	35, 36
Ib/6 Dreiser W.	43.61	2.09	43.91	1.69	0.476	0.274	0.191	18–20
Ib/5, Dreiser W.	42.91	2.16	43.64	1.74	0.13	0.0649	0.171	18, 19
Ib/K1, Dreiser W.	44.1	2.27	43.09	1.83	0.483	0.1996	0.307	16, 18, 19
D58, Dreiser W.	44.06	2.61	42.52	1.94	0.393	0.147	0.253	18, 19
Ib/8, Dreiser W.	45.87	4.05	37.41	3.27	1.134	0.372	0.45	16, 18–20
2728, Victoria	44.05	2	44.55	1.62	0.338	0.115	0.159	21
2642 Victoria	45.14	2.97	40.7	2.4	0.432	0.179	0.302	21
SL-7, Victoria	42.1	2.91	40.8	2.35	0.737	0.2	0.264	22
Le76-1L, N. Queensland	44.19	4.11	36.51	3.14	0.95	0.278	0.29	23
UM-14, San Quintin	45.26	2.86	41.3	2.16	0.37	0.147	0.26	24

Table 5.1. *Continued*

Sample I.D.	SiO_2	Al_2O_3	MgO	CaO	Nd, ppm	Sm, ppm	Yb, ppm	References
UM12, Chochise, Ariz.	45.62	3.33	39.42	2.69	0.416	0.192	0.273	24
5702 Velay	45.01	3.44	38.69	2.78	0.655	0.229	0.304	25
PHN1611, Lesotho	42.55	2.79	38.98	2.25	0.768	0.279	0.225	26–28
Orogenic Lherzolites								
Ronda R255	44.3	3.47	39.6	2.8	0.681	0.247	0.348	29
Ronda R717	44.7	3.87	38.28	3.13	0.877	0.308	0.4	29
Lanzo 705b	42.62	2.49	41.76	2.01	0.205	0.109	0.232	30
Freychinede 71-335	43.5	2.57	41.55	2.07	0.323	0.122	0.258	30
Rivara R1	43.02	2.6	41.29	2.1	0.216	0.112	0.236	30
Lanzo 775	43.08	3.02	40.66	2.44	0.457	0.192	0.285	30
Beni Bouch.M5-102	43.32	3.29	40.12	2.65	0.537	0.215	0.312	30
Alpe Arami AA	42.12	3.36	40.84	2.71	0.5	0.204	0.436	30
Beni Bouch.BB5	43.42	3.61	39.85	2.91	0.638	0.252	0.395	30
Liguria GE-1	45.1	3.17	39.39	2.21	0.582	0.225	0.319	31, 32
Liguria GE-3	45.19	3.28	39.58	2.55	0.479	0.255	0.363	31, 32
Liguria GE-11	44.38	3.9	39.35	2.43	0.411	0.196	0.287	31, 32
Lizard 90683	44.6	4.18	40.45	1.72	0.144	0.121	0.36	33, 34
Lizard 90686	45.12	4.96	37.97	3.1	0.73	0.315	0.44	33, 34

Note: Many of the lherzolite analyses have been modally corrected; see text.

References: 1: Von Michaelis *et al.*, 1969; 2: Mason, 1971; 3: Anders and Ebihara, 1982; 4: Evensen *et al.*, 1978; 5: Nakamura, 1974; 6: Masuda *et al.*, 1973; 7: Nakamura and Masuda, 1973; 8: Kallemeyn and Wasson, 1981; 9: Van Schmus and Hayes, 1974; 10: Keil, 1969a; 11: Jacobsen and Wasserburg, 1980; 12: Jacobsen and Wasserburg, 1984; 13: Knab, 1981; 14: Jagoutz *et al.*, 1979a; 15: Jagoutz *et al.*, 1980; 16: Stosch *et al.*, 1980; 17: Jagoutz *et al.*, 1979b; 18: Stosch and Seck, 1980; 19: Stosch, 1982; 20: Sachtleben and Seck, 1981; 21: Frey and Green, 1974; 22: Chen and Frey, personal communication; 23: Irving, 1980; 24: Basaltic Volcanism Study Project, 1981; 25: Chauvel and Jahn, 1984; 26: Shimizu, 1975; 27: Boyd and Nixon, 1973; 28: Nixon and Boyd, 1973a; 29: Frey *et al.*, 1985; 30: Loubet and Allegre, 1982; 31: Ernst and Piccardo, 1979; 32: Ottonello *et al.*, 1979; 33: Green, 1964; 34: Frey, 1969; 35: Frey and Prinz, 1978; 36: Zindler and Jagoutz, 1987; 37: Kurat *et al.*, 1980.

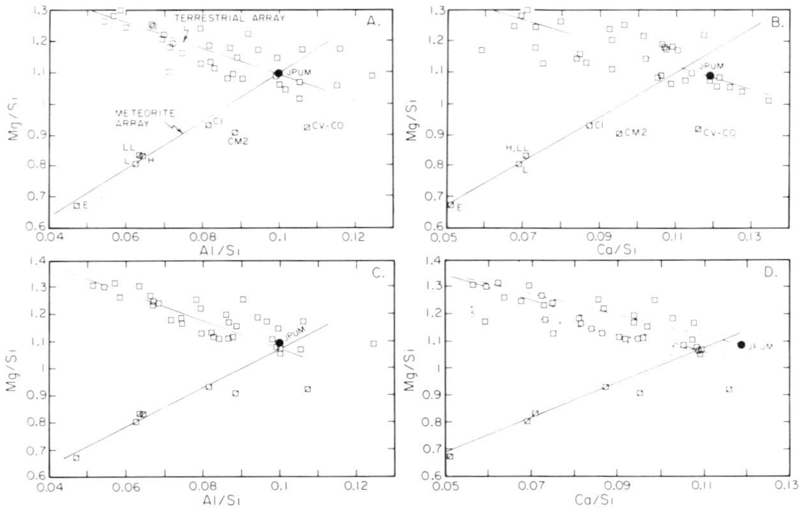

Figure 5.1. Mg/Si–Al/Si–Ca/Si relationships of meteorites and lherzolites. Panels a and b, data as published; panels c and d show lherzolites after subtraction of cpx to force Ca/Al = 1.09 (—chondrites); see text for full discussion and Table 5.1 for corrected data and references. Solid circle labelled "JPUM" is estimate of primitive mantle composition from Jagoutz *et al.* 1979a. Solid lines are eye-ball fits to the data (forced to go through "JPUM" in panels a and b). Dashed line in d is the best-fit terrestrial line from b, for reference. Solid line with the tic marks in d is trajectory of residual compositions resulting from melt removal; percent of melt removed is indicated by numbers (see text for description of this model). CM2, CV and CO chondrite data ignored in drawing best-fit line to meteorite array. Note that all concentration ratios are given as metal weight ratios.

mantle melts have approximately chondritic Ca/Al (average MORB—1.02; oceanic alkali basalts—0.91; all basalt types, worldwide—0.88; Manson, 1967; Melson *et al.*, 1976), it can easily be shown that these high Ca/Al ratios cannot be produced by melt extraction.

The simplest explanation of the high Ca/Al ratios is to postulate that these samples have high and unrepresentative clinopyroxene (cpx) contents, not related to any petrogenetic process but simply a function of local heterogeneity in mineral abundances, compounded by the fact that, in the field, geologists and geochemists will go for the "pretty green" cpx-rich rocks more often than the barren harzburgites. We have therefore "corrected" the

Table 5.2. Ranges of chemical fractionation in chondrites

		Observed range	
	Element ratio	All classes	Excluding CM2 and CV+CO
Least or not fractionated[†]	Sm/Nd	2%	2%
	Nd/Ca	12%	3%
	Ca/Al	4%	4%
	Nd/Al	13%	6%
Mildly fractionated	Yb/Ca	9%	9%
	Yb/Sm	20%	10%
	Yb/Al	10%	10%
	Yb/Mg	55%	13%
Moderately fractionated	Ca/Mg	65%	23%
	Nd/Mg	85%	23%
	Al/Mg	66%	25%
	Mg/Si	39%	39%
Highly fractionated	Yb/Si factor of:	2.1	1.6
	Ca/Si	2.3	1.7
	Nd/Si	2.5	1.7
	Al/Si	2.3	1.7

[†]May all be analytical.

lherzolite data set for excess cpx, by subtracting cpx until all Ca/Al ratios are $\leqslant 1.09$ (see Hart and Zindler, 1986, for details of this correction). Minor corrections to two other samples (Pa 15 and Ka 168), which did not require cpx corrections, have been made and are discussed in Hart and Zindler (1986). The final data set, after the corrections discussed above, is given in Table 5.1.

Some might argue that these various "corrections" to original data are ad-hoc and unwarranted; while perhaps ad-hoc they are certainly not unwarranted. Unless the upper and lower mantle are strongly fractionated in Ca/Al, any estimate for upper mantle composition which does not have Ca/Al close to chondritic appears unacceptable. Our approach above has been to simply assert that the modal abundances of minerals in lherzolites may not be representative of the local mantle region from which they were derived. In our view, it is plausible that all of the "scatter" in plots such as Figure 5.1 (and others to be discussed below) reflects this problem, and that the initial upper mantle composition was actually very homogeneous with respect to refractory element ratios. The alternative to this view, that substantial initial heterogeneities existed within the mantle, is of course possible. However, acceptance of this notion should be deferred until the homogeneous PUM model can be shown to be untenable, as a model involving substantial initial heterogeneity is largely one from which further progress will be severely limited.

In a four phase lherzolite, one could obviously "adjust" any or all of the four major mineral phases; we have chosen to adjust only a single mineral abundance in any given sample, as that is all that can be justified with the constraints available to us. These first order corrections have proven, in every case, to *reduce* the observed scatter in various data plots, even those which contain information which was not considered in determining the magnitudes of necessary adjustments. As an example, comparisons of corrected and uncorrected data are shown in Figures 5.1a and 5.1c, and 5.1b and 5.1d. For the corrected data, only one sample (a serpentinized, recrystallized lherzolite from the Lizard) is anomalously high in Al/Si (Fig. 5.1c), and all of the anomalously high Ca/Si samples have moved to lower Ca/Si ratios (Fig. 5.1d), so they are now at or to the left of the chondrite-terrestrial array intersection. There now tends to be a clustering near the intersection (somewhat circular, as there is some latitude in the position of the meteorite array, and we have chosen to draw it through the cluster). The fit to the terrestrial data is not a least-squares fit, because as mentioned above, we do not consider all the samples to be equally reliable. Based on *subjective* consideration of the data, we have placed the intersection (and its uncertainty) at Mg/Si $= 1.06 \pm 0.06$, Al/Si $= 0.100 \pm 0.003$ and Ca/Si $= 0.109 \pm 0.003$. Only our Ca/Si value differs significantly from that of Jagoutz *et al.* (1979a). The three lherzolites which fall closest to this intersection are all xenolith samples (Fr1, Sc1, Ib/8). R717, an orogenic lherzolite from Ronda, is the next closest.

Figure 5.2 is a similar plot, with Mg/Si versus Nd/Si. The terrestrial array appears somewhat fan-shaped, with a convergence on the meteorite array at Nd/Si $\sim 0.054 \pm 0.003$ (note that we will give REE/major element ratios in ppm/wt% throughout this discussion; true weight ratios can be derived simply by dividing by 10^4). Fewer of the samples have Nd/Si ratios near the intersection, simply reflecting the fact that Nd is depleted more rapidly than Ca or Al during partial melt extraction.

Since most of the lherzolites in the data set have Sm/Nd > chondrites (>0.325), it is clear that some depletion of LREEs is typical. In order to understand how the REE and major elements are coupled during the depletion process, we have formulated a simple depletion model, to gain at least a first-order understanding of what depletion trajectories will look like on various compositional plots. We will use a simple melt depletion model (O'Hara *et al.*, 1975; Loubet *et al.*, 1975; Suen, 1978) in which a small

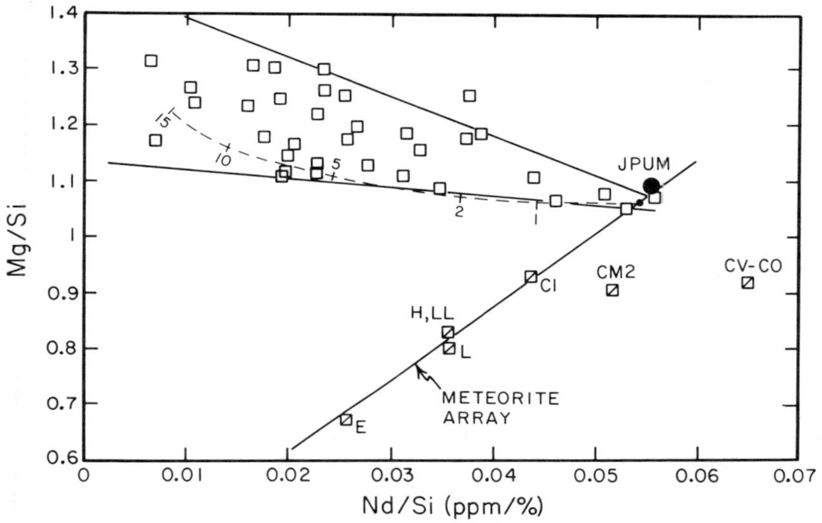

Figure 5.2. Mg/Si–Nd/Si relationships of meteorites and lherzolites (cpx corrected). Nd/Si ratios given as ppm/wt.% metal. Large solid circle (JPUM) is upper mantle estimate of Jagoutz *et al*. (1979a); small solid circle is our estimate for PUM derived from subsequent figures. Dashed line is trajectory of residual compositions following melt extraction, with percent melt extracted shown at tic marks.

degree of partial melt is extracted, batchwise, from a primitive mantle composition estimated on the basis of the relationships in Figures 5.1–5.7. This model has numerous free parameters and can hardly be certified as unique; we are principally interested in setting the free parameters so that the model provides a reasonable fit to the data. Our principal interest is in establishing the initial compositional trajectory for small degrees of melt extraction (2–3 percent), so that minor extrapolations can be made from our least depleted lherzolite data back to PUM. For this purpose, the model is very robust in terms of reasonable perturbations of input parameters.

For the major elements, we assume that the composition of the melt is invariant with the degree of melting F (probably a reasonable assumption for small values of F, considering that all of the lherzolites comprise 4-phase assemblages). We have recast our PUM compositional estimate into an initial mode of 72 percent ol+opx, 21 percent cpx, and 7 percent garnet. For all but a few samples, melting with residual garnet is absolutely required by Yb/Nd ratios. The major element composition of the derived liquid was taken as a

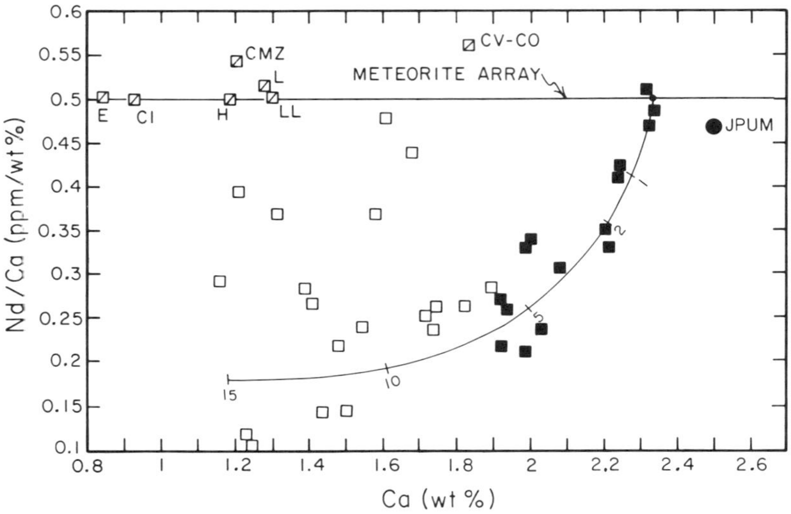

Figure 5.3. Nd/Ca versus Ca relationships for meteorites and lherzolites (cpx corrected). Discrepancy between our estimate for PUM (small solid circle) and that of Jagoutz *et al.* (1979a), solid circle marked JPUM, reflects the cpx correction we have applied to the lherzolite data. The fifteen most primitive lherzolites (Ca > 1.9 percent, Mg/Al < 14) are shown as solid squares; many of the low Ca lherzolites, mostly xenoliths, have anomalously high Nd/Ca, perhaps due to metasomatic LREE enrichment (Frey and Green, 1974). Note that many of the most metasomatized refractory xenoliths of Frey and Green (1974), not included in this data set, have Nd/Ca>chondrite. The intersection of the lherzolite array and the meteorite array at Nd/Ca (ppm/wt.%) = 0.500 is taken to define Ca and Nd in PUM (= 2.34 ± 0.02 wt.% Ca. and 1.17 ± 0.02 ppm Nd). Al in PUM then = 2.15 percent, from chondrite Ca/Al = 1.09. Solid curve is trajectory for residual lherzolite compositions derived from partial melt extraction model (see text); the model predicts 20 ppm Nd and Sm/Nd = 0.24 for the 2 percent melt, and 13.7 ppm and Nd, Sm/Nd = 0.26 for the 5 percent melt.

subjective blend of the various primary basalt compositions given in Chapter 1, Basaltic Volcanism Study Project (1981). This is essentially a MOR tholeiite, with Al adjusted down and SiO_2 adjusted up. As can be seen in Figure 5.1d, this composition provides a reasonable fit to the lherzolite data, at least for small F (<5 percent). For the REE, we have used the non-modal batch partial melting formulation of Allegre and Minster (1978). The mode of minerals entering the melt was derived approximately from the actual melt composition used; the calculations are relatively insensitive to this choice.

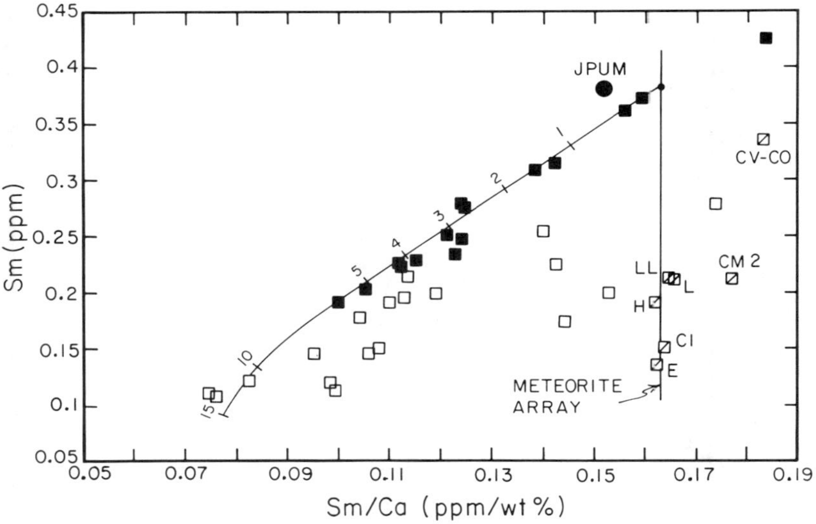

Figure 5.4. Sm versus Sm/Ca relationships for meteorites and lherzolites (cpx corrected). Best fit array, as defined by the 15 most primitive lherzolites (see Fig. 5.3 caption), intersects the meteorite array (Sm/Ca = 0.162) at Sm = 0.380 ± 0.006 ppm (small solid circle). Note that the Jagoutz *et al.* (1979a) estimate for (JPUM) falls neither on the terrestrial nor meteorite arrays, due to the cpx correction problem; note also that sample SC-1, the most primitive of Jagoutz *et al.*, falls well away from the intersection point and is therefore quite anomalous in Sm content. The Sm value derived from the intersection, coupled with the Nd from Figure 5.3, leads to Sm/Nd = 0.325 (exactly chondritic), providing an independent verification of these Sm and Nd estimates for PUM. As on the previous figures, the solid curve is the trajectory of residual compositions resulting from melt extraction, and is seen to be an excellent fit to the most primitive lherzolites. As on the Nd/Ca–Ca plot (Fig. 5.3), many of the more residual lherzolites (open squares), mostly xenoliths, have been apparently enriched in Sm relative to Ca, probably by metasomatic processes.

For REE partition coefficients, we chose set 5 of Frey *et al.* (1978); again, for small F, either set 1 or 5 would work equally well, but set 5 provided a better fit for large F (this set helps to avoid rapid depletion of Nd and Sm relative to Ca). A summary of the model parameters chosen is given in Table 5.3. Model depletion trajectories are shown in Figures 5.1–5.7, where it can be seen that the lherzolites which are least depleted (closest to PUM) have suffered < 0.5 percent melt extraction; typically, the average level of melt extraction for the more primitive lherzolites (black symbols) is on the order of

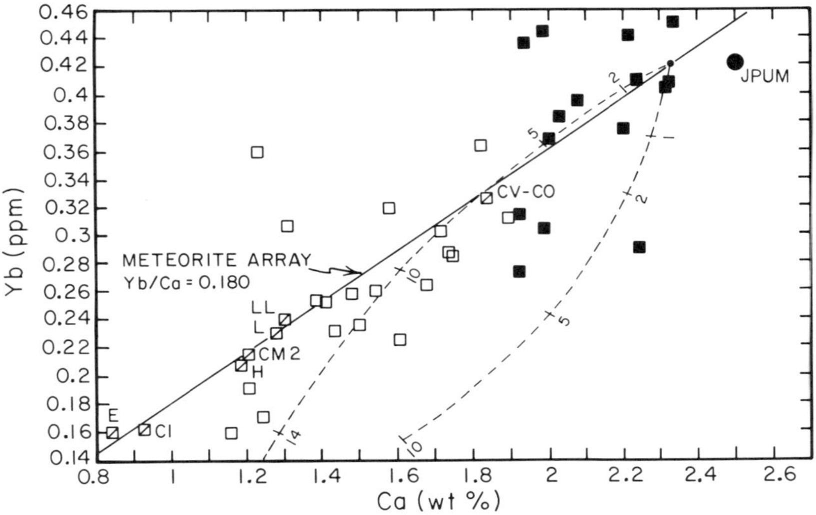

Figure 5.5. Yb–Ca relationships for meteorites (barred open squares) and lherzolites (open and filled squares). Unlike previous plots, all meteorite classes fit a single linear relationship here with Yb/Ca—0.180 (ppm/wt.%). The Yb estimate for PUM (= 0.420 ppm, small solid circle) is derived from the Ca value of 2.34 percent (Fig. 5.3) and meteorite Yb/Ca = 0.180. Two model curves are shown for the trajectory of residual compositions resulting from melt extraction; the upper one is for the same model used in previous figures (7 percent modal garnet in the source), whereas the lower one assumes no garnet in the source; at least one of the lherzolites (Le 76-1L) appears to be a residue from garnet-absent melting. The four lherzolites with Yb > PUM may indicate the need for a higher K_d for Yb between garnet and liquid than was used in the model (4.1), so that the bulk distribution coefficient can be >1.

2–3 percent. While this produces relatively insignificant depletions in major elements, it does significantly change REE concentrations.

We can now use all the above considerations in concert to develop a best estimate for PUM. From a large variety of compositional plots, Figures 5.1–5.7 were selected to illustrate the approach. Having forced all samples to Ca/Al ratios ≤1.09, we consider the next most invariant ratio in chondrites (Nd/Ca or Sm/Ca). Based on Figure 5.3, a plot of Nd/Ca vs. Ca (and a similar plot for Sm/Ca) we fix Ca in PUM at 2.34 ± 0.02 wt.% and Nd = 1.17 ± 0.02 ppm (using a Nd/Ca ratio of 0.500 in chondrites). Figure 5.4 can then be used to constrain Sm at 0.380 ± 0.006 ppm. Coupled with the Nd = 1.17, ppm, this gives Sm/Nd = 0.325, which is the chondritic value and thus provides a

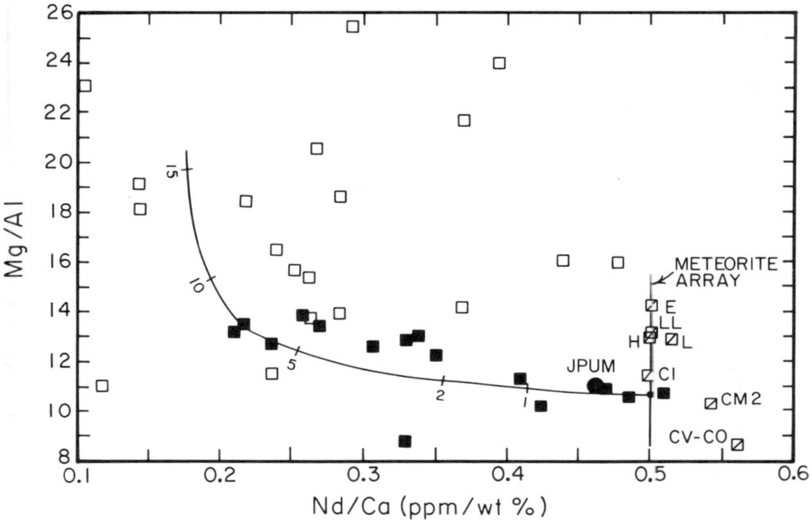

Figure 5.6. Mg/Al–Nd/Ca relationships for meteorites and lherzolites (cpx corrected). Intersection of terrestrial array and chondrite array (Nd/Ca—0.5) delimits the Mg/Al value of PUM as 10.6 ± 0.3, and consequently Mg = 22.8 percent (using Al = 2.15 percent from Fig. 5.3). Symbols and model curve, and explanation for high Nd/Ca in the most refractory (high Mg/Al) lherzolites, is as given in Figure 5.3.

good consistency check on these calculations. Note that one sample on Figure 5.4 (SC-1) appears to be anomalously high in Sm, perhaps suggesting some trapped melt in this lherzolite. Yb for PUM is not well constrained by Figure 5.5, but is derived as = 0.42 ± 0.02 ppm by using the average Yb/Ca ratio of chondrites (0.180). Finally, the Mg/Al ratio of PUM is determined as 10.6 ± 0.3 from Figure 5.6. Based on Ca/Al of 1.09, and Ca of 2.34 percent, we determine that Al = 2.15 ± 0.02 percent, and Mg = 22.8 ± 0.6 percent. Also, from Al/Si = 0.100 (Fig. 5.1c), we calculate Si = 21.5 ± 0.6 percent.

A summary of our estimate for PUM is given in Table 5.4, along with our admittedly subjective estimates of the uncertainties (which reflect only the application of the chosen data set, not the various assumptions concerning the development of the data set or its applicability to actual upper mantle). Other refractory element abundances in PUM may be estimated based on C1 abundances relative to Ca (Anders and Ebihara, 1982). Also given in Table 5.4 is the JPUM estimate derived by Jagoutz *et al*. (1979). The agreement for

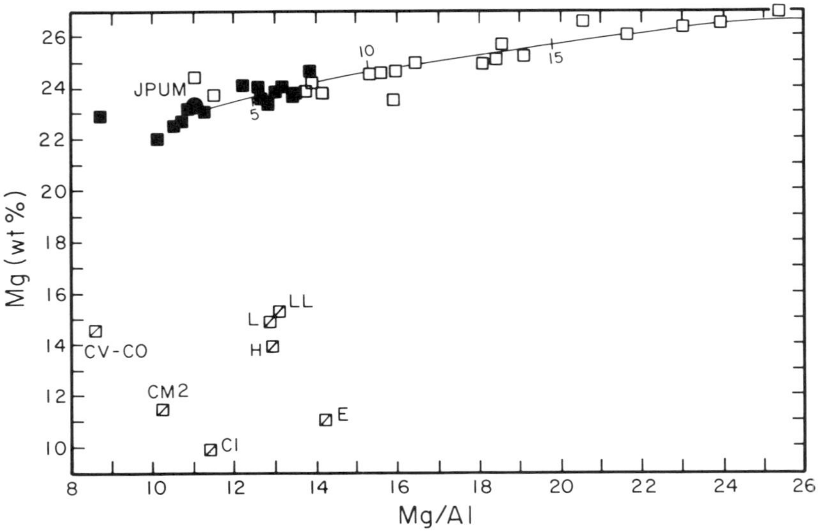

Figure 5.7. Mg–Mg/Al array for lherzolites, showing excellent fit of the melt extraction model (solid line) to the data. Chosen value for PUM (small solid circle) as derived in Figure 5.3 (Al) and Figure 5.6 (Mg). The sample with anomalously low Mg/Al (90686), which also appears Al enriched in Figure 5.1c and Figure 5.6, is a highly serpentinized lherzolite from the Lizard high temperature peridotite and is not given serious weight in any of these figures.

the REE is very good. The differences in major elements arise largely from our insistence that PUM must be chondritic in Ca/Al, thereby forcing a downward revision in the Jagoutz *et al.*, Ca estimate.

2.2. The Earth as a C1 Chondrite

The most popular Earth composition model today holds that the bulk earth is equivalent to devolatilized C1 chondrites (Ringwood, 1975, 1979; Wanke, 1981; Wanke *et al.*, 1984; Sun, 1982; Anderson and Bass, 1986). This model is attractive because C1 chondrites have overall compositions which are very similar to solar abundances (within 10 percent; Anders and Ebihara, 1982); that is, they are the least chemically fractionated (most primitive) material available in the solar system. Since the Earth is clearly not C1 in overall composition, the difficulty comes in specifying just which elements are under-represented in the Earth relative to C1s, either due to volatilization or

Table 5.3. Partial melt extraction model parameters

	Initial solid	Extracted melt	Partition coefficients		
			Nd	Sm	Yb
Ol + opx	72%	10%	0.0014	0.0022	0.008
Cpx	21%	45%	0.18	0.26	0.23
Garnet	7%	45%	0.019	0.107	4.1
SiO_2	46.02%	51.4%			
Al_2O_3	4.06%	13.2%			
MgO	37.8%	10.8%			
CaO	3.27%	12.5%			

Table 5.4. Primitive upper mantle (PUM) composition

	This paper	Jagoutz *et al.* (1979a)
Si	21.5 ± 0.6 wt.%	21.1%
Al	2.15 ± 0.02%	2.10%
Mg	22.8 ± 0.6%	23.1%
Ca	2.34 ± 0.02%	2.50%
Nd	1.17 ± 0.02 ppm	1.17 ppm[†]
Sm	0.380 ± 0.006 ppm	0.38 ppm
Yb	0.420 ± 0.02 ppm	0.42 ppm
SiO_2	46.0%	45.2
Al_2O_3	4.06%	3.97
MgO	37.8%	38.3
CaO	3.27%	3.50
Sr	19.6 ppm[‡]	28 ppm
U	20.8 ppb[§]	26 ppb

[†]Derived from Sm, and Sm/Nd chondrite = 0.325.
[‡]Derived using Sr data for meteorite classes E, H, L, LL, and C1; data from Allegre *et al.* (1983). This Sr data was coupled with Nd and Ca data for these meteorite classes from Table 5.1 to generate an average value for these 5 classes of meteorites of Sr/Nd = 16.7, and Sr/Ca = 8.40 (ppm/wt.%).
[§]Derived using U concentration data for E, H, L, LL, and C1 chondrites, from Chen and Tilton, 1976; Tilton, 1973; Manhes and Allegre, 1978; Tatsumoto, Knight, and Allegre, 1973; Tatsumoto, Unruh, and Desborough, 1976. These were coupled, where possible, with Nd from the same meteorite (references given in Table 5.1); otherwise from the same meteorite class (Table 5.1). The average U/Nd for the five meteorite classes so derived is 0.01780, with a maximum range between classes of 13 percent. The use of the average of all 5 classes was felt preferable to using only C1 data, as only one isotope dilution analysis exists for C1s. There is no obvious trend of U/Nd with fractionation indicators such as Al/Si for the 5 classes; inclusion of CM2, CV+CO meteorites, however, suggests a trend of decreasing U/Nd with increasing Al/Si. Use of the average U/Nd of .0178, coupled with Nd = 0.463 ppm in C1, gives U = 8.2 ppb which compares well with the value of 8.1 ppb for C1 from Anders and Ebihara (1982). Using U/Nd = 0.0178, and Nd in PUM = 1.17 ppm gives U in PUM = 0.0208 ppm.

non-accretion. In other words, are just the highly volatile elements missing (e.g., Pb, K, and various gasses), or does the volatile deficit extend even to elements such as Mg and Si, which are predicted to have moderately high condensation temperatures in the solar nebula? Our development of relatively precise concentration estimates for PUM should allow a better test of the C1 model than has heretofore been possible. It is immediately clear from Figures 5.1 and 5.6 that the Earth's upper mantle is depleted in Si and Mg (relative to Ca and Al) relative to C1s. To what extent can the missing Si and Mg be located in other parts of the Earth (lower mantle and core)?

To develop this test, we first need to recast C1 chondrite compositions into three components—a volatile component, a silicate (mantle) component, and a metal (core) component. This will allow us to compare absolute abundances of refractory elements in the silicate component with our estimate for PUM. Table 5.5 documents this calculation (complete details may be found in Hart and Zindler, 1986). First, the abundances of all the major silicate components of C1 chondrites, from Anders and Ebihara (1982), are recast as oxides. This is shown in column A, Table 5.5. The oxides total 69.79 percent, the remaining 30 percent are volatiles (H_2O, C, N, S) and organics (this result agrees with the summary of C1 compositions given by Keil, 1969). Because the Earth is clearly deficient in alkalies and phosphorus relative to C1s, part of this budget is removed (column B), leaving only enough for estimated mantle compositions (e.g., Ringwood, 1975). Now the major fractionation arises from splitting this C1 composition into silicate mantle and "metallic" core. We first do this by assuming the core is only Fe+Ni+Co metal; sufficient FeO is left in the silicate fraction to give Mg/Mg+Fe = 0.90 (atomic), a ratio typical of most mantle materials. We also leave a small amount of NiO and CoO in the mantle (to provide characteristic levels of these elements in typical mantle materials). What we obtain (column D) is a core which accounts for 27.6 wt. percent of the Earth, compared to the Earth's core mass of 32.5 percent (Dziewonski and Anderson, 1981). An adequate mass for the core cannot be achieved even by removing all of the Fe+Ni+Co from the silicate fraction.

This problem has been recognized by many authors (e.g., Ringwood, 1979; Anderson, 1983; Wanke, 1981); the solution to this problem is usually to add some other component (Si, S, O) to the core, both to settle the mass deficit and to lower the density to agree with geophysical constraints. If part of the volatile budget (HVE, MVE or O, column C) is used to bring the core mass up to 32.5 percent, the final core would contain 21 percent of this volatile component, as shown in column G. While there is easily enough of these elements present in C1s to accomplish this, the geophysical constraints

Table 5.5. C1 chondrite Earth model

	A	B	C	D	E	F	G
SiO_2	22.77	22.77	22.77	35.85	22.77	33.42	49.52
Al_2O_3	1.640	1.640	1.64	2.58	1.64	2.41	3.56
FeO	24.49	24.49	3.28	5.16	3.28	4.81	7.14
MgO	16.41	16.41	16.41	25.84	16.41	24.09	35.68
CaO	1.298	1.298	1.298	2.04	1.298	1.90	2.82
Na_2O	0.669	0.134	0.134	0.211	0.134	0.197	0.292
K_2O	0.067	0.013	0.013	0.020	0.013	0.019	0.028
Cr_2O_3	0.386	0.189	0.189	0.298	0.189	0.277	0.412
MnO	0.256	0.053	0.053	0.083	0.053	0.078	0.115
TiO_2	0.073	0.073	0.073	0.115	0.073	0.107	0.159
NiO	1.39	1.39	0.112	0.176	0.112	0.164	0.244
CoO	0.064	0.064	0.0052	0.008	0.0052	0.008	0.012
P_2O_5	0.274	0.008	0.008	0.0126	0.008	0.012	0.018
Silicate sum	69.79	68.532	45.985	72.392	45.985	67.50	100
HVE†	30.21	30.21	30.21	—	30.21	—	
MVE†	—	1.258	1.258	—	1.258	—	
Oxygen†	—	—	5.01	—	0.40	—	
	30.21	31.468	36.478	—	31.868	—	
Fe	—	—	16.487	25.955	16.487	24.199	74.44
Ni	—	—	1.004	1.581	1.004	1.474	4.53
Co	—	—	0.046	0.072	0.046	0.068	0.209
Oxygen†	—	—	—	—	4.61	6.766	20.81
Core sum			17.537	27.608	22.147	32.50	100
Total	100	100	100	100	100	100	

†HVE = highly volatile elements (H_2O, S, C, organics); MVE = moderately volatile elements (Na, K, P, Cr, Mn); oxygen is that obtained from FeO, NiO, CoO reduction.

A: All major components of C1 chondrites, from Anders and Ebihara (1982); B: Separation of MVE to leave MVE/Al ratios in silicate portion similar to Jagoutz *et al.* (1979) mantle; C: Separation of metallic core, leaving FeO, NiO, and CoO to Al ratios in silicate mantle of Jagoutz *et al.* (1979); D: Removal of HVE + MVE + oxygen from earth, renormalizing remained to 100 percent; E: Partial oxygen budget left in metallic core; F: Removal of HVE + MVE + part of oxygen from earth, renormalizing remainder to 100 percent; G: Silicate mantle and oxide core renormalized separately to 100 percent.

on the density of the core will not allow 21 percent of it to be made of light elements. In fact, Jeanloz (this volume) concludes that the core need only contain 5–10 percent of some light element. Anderson and Bass (1986) purport to be able to achieve an Earth-size core from C1s, with only some 12 percent of light element, but we are unable to reproduce their calculation.

Ringwood (1966) originally solved the core mass problem and the apparent Si deficiency in Earth's mantle in one stroke by postulating Si as the light element in the core (this position has since been recanted; Ringwood, 1979). Wanke (1981) advocates a C1 Earth model in which Si, Mn and Cr are all used to bring the core mass up to size, as well as to account for the deficit of these elements in the upper mantle. However, this model results in a somewhat oversize core (33.5 percent), and also does not solve the upper mantle Mg deficit (~8 percent, relative to Al or Ca in C1s); one simply cannot put enough Si and Mg in the core to solve the upper mantle deficits of these elements without either producing an oversize core or a core with too low a density (assuming PUM = BSE).

A further problem with the C1 model is that the abundances of Ca and Al in PUM (Table 5.4) are some 15 percent higher than those in the C1 silicate mantle (Table 5.5, column G). This discrepancy would be partly eliminated if Si (± Mg) were substituted for O as the light element in the core, though the Mg and Si deficits in the upper mantle cannot be totally accounted for in a core reservoir. Anderson's (1983) model circumvents these Ca, Al, Si and Mg mantle budget problems *vis à vis* C1 chondrites by arguing that the upper and lower mantles are different in composition (i.e., the compositions derived for PUM, Table 5.4, do not apply to the whole mantle). Wanke (1981) similarly is willing to accept small (~10 percent) chemical differences between the upper and lower mantle in order to reconcile the differences in Mg/Al between PUM and C1 chondrites.

While numerous geophysical arguments may be brought to bear on the question of compositional differences between upper and lower mantle (see Anderson and Bass, 1986, and Jeanloz, this volume) there is as yet no clear consensus on this issue (see Section VI). From a geochemical standpoint, one key observation made initially by Ringwood (1966), and reiterated by Wanke (1981), seems to us to be fundamentally important. The primitive upper mantle (PUM) has refractory element ratios very close to those in all chondrites (except CM2 and CV-CO). It is unlikely that ratios such as Nd/Ca, Sm/Nd and Ca/Al could persist at chondritic values if any significant degree of intramantle chemical differentiation has occurred, especially if it involves crystal-melt fractionation. Notice, for example, in Figure 5.3, that extraction of 1 percent melt from PUM decreases the Nd/Ca ratio in the residue by almost 20 percent; or Figure 5.1, where 5 percent melt extraction decreases the Ca/Si ratio by 15 percent, etc. This argument is simple yet potent. How are chondritic refractory element ratios preserved in any mantle model involving large chemical fractionations, especially models such as Anderson's (1983) which involves magma oceans, cumulate layers, and refractory residua?

There is admittedly a certain circularity in our argument here. PUM is derived by *assuming* that it must be chondritic in ratios such as Ca/Al, Nd/Ca, Sm/Nd. However, it is clear from Figures 5.1–5.7 that the upper mantle lherzolites which are intuitively most primitive (highest Ca, etc.) approach chondritic ratios very closely, and that all of the melt-depletion trends for these lherzolites also appear to precisely *terminate* right at chondritic refractory element ratios. Unless some means can be found of circumventing this argument, it appears to us to be compelling evidence against any significant post-accretional mega-scale, intramantle chemical differentiation, and, by inference, against any significant major element compositional differences between Earth's upper and lower mantle. We need to qualify this remark by saying that the mantle is clearly chemically heterogeneous—the isotopic evidence for this is irrefutable. However, this heterogeneity largely involves trace elements, and is related to various processes (melt extraction, metasomatism, oceanic plate reinjection, etc.) which significantly post-date accretion and possible initial differentiation.

We cannot, of course, rule out initial mantle chemical variations such as would result, for example, from heterogeneous accretion processes. Based on the evidence from meteorites, it is clear that substantial chemical variations can exist for some elements (e.g., Ca/Si) without accompanying variations in the refractory lithophile element ratios on which we have focussed. However, almost all current Earth models which involve chemical differences between upper and lower mantle (Anderson, 1983; Herzberg, 1983) relate these to intramantle differentiation events accompanying and post-dating accretion—it is this class of process for which our above argument seems entirely compelling.

Based on these arguments, we suggest that our PUM composition can be applied to the whole mantle. If this assumption is valid, then the overall Earth composition is not matched by C1 chondrites except in terms of highly refractory lithophile element ratios. Relative to Ca in C1s, the Earth is ~19 percent deficient in Si and 9 percent deficient in Mg (and of course strongly to grossly deficient in elements more volatile than Si). With respect to Mg/Ca and Si/Ca, etc., the Earth represents an extreme end member when compared to chondrites, with the ranking, in terms of increasing degrees of depletion, being: E, L ~H, C1, CM2, Earth.

We will now take the C1 model one step further. Recognizing that the Earth is grossly deficient in highly volatile elements, and apparently in Si and Mg as well (based on the primary assumption that the whole mantle is of PUM composition), we take our primary C1 composition (Table 5.5) and remove 19 percent SiO_2 and 8 percent MgO, so as to bring the Si and Mg ratios, relative to Ca and Al, into accord with PUM. This obviously has the effect of *raising* all the other oxide abundances, and should help the core mass, and Ca and Al deficiencies, discussed above. From this Si and Mg adjusted composition for C1s (Table 5.6, column B), we fractionate out Fe+Ni+Co for a core, again leaving a "mantle complement" of FeO, NiO and CoO behind. There is still only enough Fe+Ni+Co (as metal) to make 94 percent of Earth's core (column D). Leaving sufficient volatiles behind to bring the core up to full mass results in a core composed of 85.6 percent Fe, 5.4 percent Ni and Co and 9 percent light element (Table 5.6, column G). The fraction of light element is now similar to the light-element abundances

Table 5.6. LOSIMAG C1 Earth model

	B	C	D	E	F	G
SiO_2	18.58	18.58	31.96	18.58	31.02	45.96
Al_2O_3	1.640	1.64	2.82	1.64	2.74	4.06
FeO	24.49	3.05	5.25	3.05	5.09	7.54
MgO	15.27	15.27	26.26	15.27	25.50	37.78
CaO	1.298	1.298	2.23	1.298	2.17	3.21
Na_2O	0.134	0.134	0.230	0.134	0.224	0.332
K_2O	0.013	0.013	0.022	0.013	0.022	0.032
Cr_2O_3	0.189	0.189	0.325	0.189	0.316	0.468
MnO	0.053	0.053	0.091	0.053	0.088	0.130
TiO_2	0.073	0.073	0.126	0.073	0.122	0.181
NiO	1.39	0.112	0.193	0.112	0.187	0.277
CoO	0.064	0.0052	0.009	0.0052	0.009	0.013
P_2O_5	0.008	0.008	0.014	0.008	0.013	0.019
Silicate sum	63.202	40.425	69.530	40.425	67.50	100
HVE†	30.21	30.21	——	30.21	——	
MVE†	1.258	1.258	——	1.258	——	
SVE†	5.33	5.33	——	5.33	——	
Oxygen†	——	5.06	——	3.31	——	
	36.798	41.858	——	40.108	——	
Fe	——	16.665	28.664	16.665	27.825	85.62
Ni	——	1.004	1.727	1.004	1.676	5.16
Co	——	0.046	0.079	0.046	0.077	0.237
Oxygen†	——	——	——	1.75	2.922	8.99
Core Sum		17.715	30.470	19.465	32.50	100
Total	100	100	100	100	100	

†SVE = slightly volatile elements (Mg, Si); others defined as in Table 5.5.
B: Column B, Table 5.5, with enough Si and Mg removed to give Si/AL and Mg/Al ratios as in PUM, Table 5.4; C: As in column C, Table 5.5; D: As in column D, Table 5.5, with SVE also removed; E: As in column E, Table 5.5; F: As in column F, Table 5.5, with SVE also removed; G: As in column G, Table 5.5.

in the core predicted on the basis of geophysical constraints (e.g., Jeanloz, this volume).

Thus, by starting with a basic C1 composition, removing all or almost all of the highly volatile elements, and allowing small but significant deficiencies in Si (19 percent) and Mg (8 percent), we can construct an Earth which not only has a proper size core, with an appropriate composition, but also one which has the right *absolute* mantle abundances of the refractory lithophile elements. We will call this the LOSIMAG C1 Earth model. In other words, the Earth is not like any chondrite, but is composed of its own blend of accretion products, with a bias toward a higher proportion of the high-temperature refractory components (or a lower proportion of the volatile elements). C1 chondrites still remain a good guide to Earth's refractory element composition, with terrestrial mantle abundances equal to 2.47 times ave. C1 (as given by Anders and Ebihara, 1982). Our more direct derivation of mantle composition (PUM) is equivalent to an average enrichment over C1s of 2.51 times. This model has no obvious weaknesses until and unless the geophysical evidence for a chemically stratified mantle becomes unequivocal. While this would not rule out our preferred model, it would allow or confirm one of the principal requirements of the direct C1 model.

3. Scale Lengths for Mantle Heterogeneities

3.1. Kinetic Constraints on Maintaining Heterogeneities

Estimates for the scale length over which demonstrable isotopic heterogeneities exist range from mineralogical (centimeter) scale, to scale lengths comparable to mantle dimensions ($>$1000 km). The short end of this range can be evaluated by use of existing data on diffusion of cations in relevant mantle phases. Hofmann and Hart (1978) adopted this approach and concluded that mineralogical scale disequilibria of isotope ratios (trace elements) could only survive for the necessary billions of years in relatively cool ($<$1000° C) regions of the mantle where no partial melt existed (as for example in the upper levels of the subcontinental lithosphere). In regions of higher temperature, particularly in the presence of a melt phase (which will act as a short circuit for diffusion), homogenization of isotopes between phases of centimeter size will occur on time scales of $<10^5$ years.

A considerable quantity of new diffusion data has emerged since the study of Hofmann and Hart (1978), and it is pertinent to reconsider the arguments advanced therein. Diffusion data are now available for Fe, Mg, Co, Ni, Mn,

Ca, O, He and Si in olivine, Mg, Ca, Fe, Al, U, Sr, and Sm in diopside, Mg in garnet (pyrope), and Li, Na, K, Rb, Mn, Co, Ca, Sr, Ba, Gd, Eu, V, He, H_2O and Cs in basaltic melt. Of the likely mantle phases which might act as hosts for the elements whose isotope ratios are employed in mantle studies, only phases such as amphibole and phlogopite are virtually unstudied for diffusion systematics. Some constraints are available for these minerals, however, from geochronological studies in metamorphic terrains, which indicate that elements such as Ar, Sr and Nd diffuse relatively rapidly (relative to phases such as diopside). Thus amphibole and phlogopite are not likely to be rate-limiting phases with respect to mineralogical-scale isotopic equilibration in the mantle.

For the purposes of this discussion, it is not necessary to review all the above-mentioned diffusion data in detail. We note only that cation diffusion in olivine, diopside and basalt melt is observed to follow a compensation law (Hart, 1980; Sneeringer *et al.*, 1984), and this allows several brief generalizations to be made. Compensation, the existence of a well-defined relationship between the pre-exponential factor and the activation energy in the Arrhenius-type description of diffusion, implies a temperature at which individual element Arrhenius-lines intersect. For olivine, diopside and basalt melt, the crossover temperatures are 1360° C, 1300° C and 1370° C respectively; the diffusion coefficients at these crossover temperatures are $\sim 4 \times 10^{-11}$ cm^2/s, $3 \times 10^{-14} cm^2/s$ and 2×10^{-7} cm^2/s respectively. Note that the compensation effect for olivine and clinopyroxene involves not only various diffusing species, but variations due to pressure and crystallographic direction.

By virtue of the compensated nature of diffusion in olivine, diopside and basalt, we need not separately consider the differences in diffusion rate as a function of species, pressure or diffusion direction; at the temperature of particular interest to isotopic homogenization between mantle phases (1300–1400° C), diffusion in each of these phases may be characterized by a single diffusion coefficient. Hofmann and Hart (1978) used 10^{-13} cm^2/s for solid phases and 10^{-7} cm^2/s for melt phases; the more comprehensive set of diffusion data discussed above supports these values, and thus the conclusions of Hofmann and Hart (1978) remain valid. To reiterate these conclusions:

1. In the presence of a melt phase, mantle minerals will achieve isotopic equilibrium on time scales of 10^5 yrs (at 1350° C).
2. Maintenance of isotopic disequilibrium between minerals for a period of 1 b.y. requires a melt-free system where the host phases are not in contact (i.e., are separated by low-solubility barrier phases such as olivine), and where the temperature does not exceed 1100–1200° C.

3. In a mantle containing 5 percent basaltic partial melt at 1370° C, and for trace elements with bulk crystal/liquid partition coefficients <0.1, the maximum scale length ($\sim\sqrt{Dt}$) over which these elements can be diffusively transported is $\sim 500\sqrt{t}$ (in meters), where *t* is in billions of years (Hofmann and Magaritz, 1977). Thus, over the age of the Earth, diffusion will not eradicate isotopic heterogeneities which are developed on larger than a km scale even in the presence of a melt.

This latter calculation can be used to constrain the minimum dimension of a heterogeneity which can survive a melting event in a mantle segment with a *homogeneous* solidus temperature. For a mantle diapir rising with an ascent velocity of 5cm/y, it will take ~0.3 m.y. to generate 5 percent partial melt (for 180 cal/g heat of fusion, and 2°C/km difference between the melting curve and adiabat). If isotopic heterogeneities existed in this diapir prior to the onset of melting, they will be diffusively erased over scale lengths of some 10 meters. That is, isotopic differences between basaltic melts derived from a single diapiric melting event must be reflecting isotopic mantle heterogeneities on scales larger than 10 m, even assuming no convective mixing of the melt. Since most basaltic eruptions almost certainly integrate much larger volumes of mantle than this, we conclude that the minimum size of mantle heterogeneities which can be visualized by studying basalts is set by the volumes and dynamics of melt segregation and transport rather than by diffusion considerations.

In contrast, we can consider a diapir with relatively enriched zones which have lower solidus temperatures than the surrounding depleted matrix. At the onset of melting, the melt will be restricted to the enriched zones, and thus it is conceivable that very small-scale heterogeneities (meter-scale?) may be sampled (since an interconnected melt network is not present throughout the diapir). Thus, when heterogeneities are observed in basalts, we cannot be sure whether they document kilometer or meter scale heterogeneity in the source mantle.

3.2. Domain Sizes Inferred from Ultramafic Rocks

Scale lengths of mantle isotopic heterogeneities may be investigated by direct observations of mantle-derived rocks. Two types of material are accessible for this kind of study: 1) ultramafic xenoliths erupted in alkaline volcanic rocks and kimberlites; and, 2) tectonically-exposed ultramafic rocks, such as alpine ultramafics and high-temperature peridotites. Xenolith material, by virtue of its small size (<30 cm), is useful for studying only very small scale lengths. In contrast, high-temperature peridotites may have outcrop scale lengths of up to 35 km (e.g., Ronda massif, Spain).

Numerous early studies reported isotopic disequilibrium between minerals in ultramafic xenoliths (e.g., Stueber and Ikramuddin, 1974; Dasch and Green, 1975); it now appears that much of this observed heterogeneity was a result of alteration and analytical artifacts. The problem is that very few minerals in ultramafic xenoliths provide happy hosts for the elements which are useful as isotopic tracers (Sr, Nd and Pb). In spinel and garnet-facies peridotites, Sr is present at significant concentration levels only in clinopyroxene, and Nd is present at significant levels only in cpx and garnet. The levels of Sr and Nd in olivine, orthopyroxene, and spinel are usually not amenable to analysis when pure, inclusion-free mineral separates are obtained. Typically, as these minerals are purified to higher and higher degrees, the Sr and Nd contents decrease continuously until, finally, a pure mineral separate is obtained which contains only traces of the original Sr and Nd (Zindler and Jagoutz, 1988). Several recent studies have managed to circumvent the contamination problem and have concluded that no demonstrable inter-mineral isotopic disequilibria exists for Sr or Nd in spinel-facies peridotites (e.g., Jagoutz *et al.*, 1980; Zindler and Jagoutz, 1988).

The situation for garnet-facies peridotites would appear to be more promising, since both cpx and garnet contain appreciable Nd. However, garnet in these xenoliths is typically highly altered, and thus it is correspondingly more difficult to obtain pure separates. In an in-depth study of garnet-facies xenoliths from South African kimberlites, Richardson *et al.* (1985) found that Nd in cpx and garnet was not isotopically distinct, once very pure garnet separates were produced, except in peridotites where relatively recent metasomatism was evident.

Ultramafic xenoliths which contain hydrous (metasomatic?) phases often display Sr and Nd isotopic disequilibria between minerals such as phlogopite, amphibole, and cpx (Menzies and Murthy, 1980a, 1980b; Kramers *et al.*, 1983; Richardson *et al.*, 1985). However, it is not yet clear whether these metasomatic phases are long-term residents of mantle environments or whether they simply represent young events in some way related to the early stages of the volcanic event in which the xenoliths were erupted. We conclude from the xenolith studies that intermineral isotopic disequilibria is not a common long-lived feature of the upper mantle.

In contrast, larger-scale isotopic variations have been clearly documented in orogenic lherzolite massifs. In a pioneering study, Polve and Allegre (1980) analyzed diopside separates from various parts of the Lherz, Lanzo and Beni Bousera massifs, and observed ranges in $^{87}Sr/^{86}Sr$ of 0.7020–0.7032, 0.7020–0.7030 and 0.7020–0.7033, respectively. Significant variations were observed on scale lengths as small as tens of meters; the maximum variations were usually observed on scale lengths of 1–5 km. The numbers of samples analyzed from each massif ($n < 8$) does not allow one to firmly demonstrate any positive relationship between scale length and the amplitudes of the isotopic variations, though there is a consistent increase in isotopic variability with increasing sampling distance for both Lherz and Lanzo (Fig. 5.9). Results of a systematic scale length study of the Ronda massif (Reisberg and Zindler, 1987; see Fig. 5.31), using Sr and Nd isotopes in peridotite cpx separates, show significant isotopic variation on scale lengths of 10–20 meters (0.7031–0.7034 for Sr and 0.51255–0.51285 for Nd; see Fig. 5.8). Reisberg and Zindler (1987) have further shown that mean $^{143}Nd/^{144}Nd$ ratios for 5–10 km^2 terrains (as documented by Cr-diopside separates from river sediments) vary from 0.51325–0.51300 for terrains separated by less than 10 km. The total range in $^{143}Nd/^{144}Nd$ measured by Reisberg and Zindler (1985, 1987) over a scale length of ~30 km is on the order of 95 percent of the total variation observed in oceanic basalts (Fig. 5.9).

The massif of St. Paul's rocks, in the equatorial mid-Atlantic, is characterized by rock types ranging from dunites, to amphibole-bearing lherzolites, to hornblendites, with isotopic signatures similar to typical ocean island basalts (Fig. 5.31; Roden *et al.*, 1984). Surface exposures are limited in scale to a few hundred meters, but dredge sites on the flanks may reflect exposures on scales up to ~20 km. While very little Sr isotopic variation was found on any scale, Nd and Pb show variations which cover ~30 percent of the total range observed in oceanic basalts throughout the world (Fig. 5.9). Apparent Nd and Pb isotopic variations in surface outcrop are present on scales as small as tens of meters (Fig. 5.9). However, the small-scale sampling defines a crude isochron of ~160 m.y., which is argued to represent a metasomatic event of that age. That is, the observed small-scale length Nd and Pb variations may only date back to this time, and may not represent long-standing mantle heterogeneities. The larger-scale lengths (kms), represented by the dredge samples, apparently reflect much older heterogeneities, and cover about 30 percent of the total oceanic range for Pb and Nd. Thus, as with the orogenic lherzolite massifs discussed above, isotopic variations appear to exist on scales as small as tens of meters, with a strong indication of a positive relationship between the amplitude of the isotopic variation and the scale length (see Fig. 5.9).

Preliminary Nd and Sr data for Zabargad, an uplifted block of amphibole and spinel peridotite in the Red Sea, have been reported by Brueckner *et al.* (1988; see Fig. 5.31). $^{84}Sr/^{86}Sr$ and $^{143}Nd/^{144}Nd$ ratios, respectively, show ranges from 0.7022–0.7030, and from 0.51275–0.51339, over scale

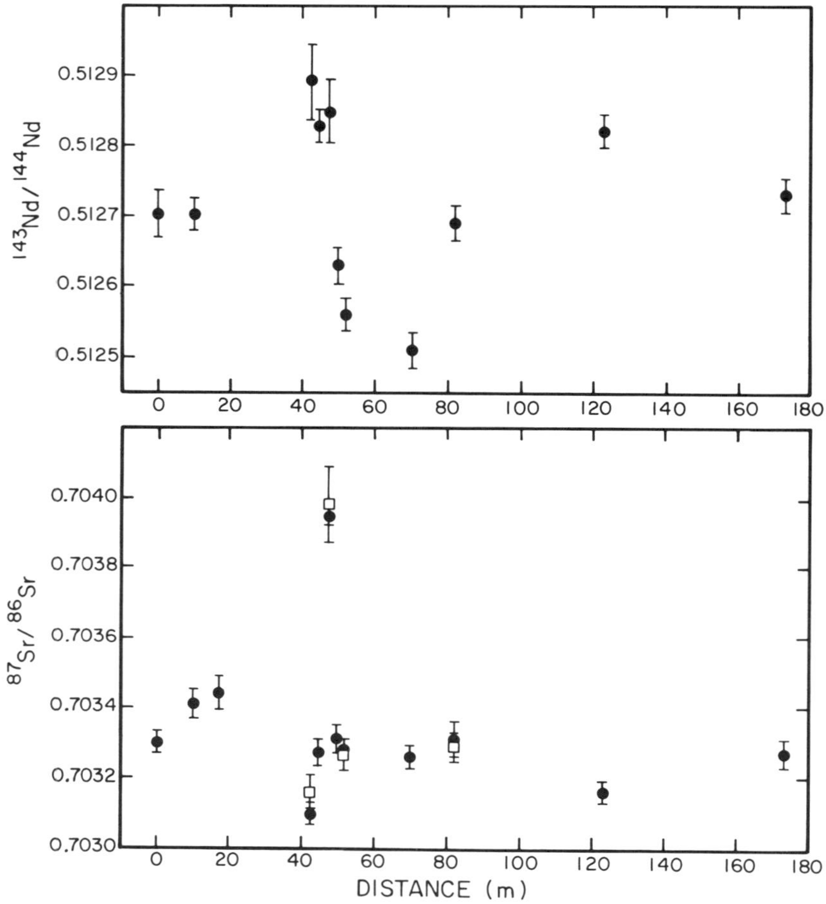

Figure 5.8. $^{87}Sr/^{86}Sr$ and $^{143}Nd/^{144}Nd$ versus distance (in meters) along a transect in the eastern portion of the Ronda ultramafic massif (Reisberg and Zindler, 1987). All samples are cpx separates from relatively fresh plagioclase peridotite (ol + opx + cpx + plag + spinel). Although the peridotites show no obvious indications of metasomatism, all samples are LREE-enriched, with Sm/Nd $<$ 0.3. Double error bars on the Sr diagram indicate individual analyses of different cpx separates from the same peridotite sample.

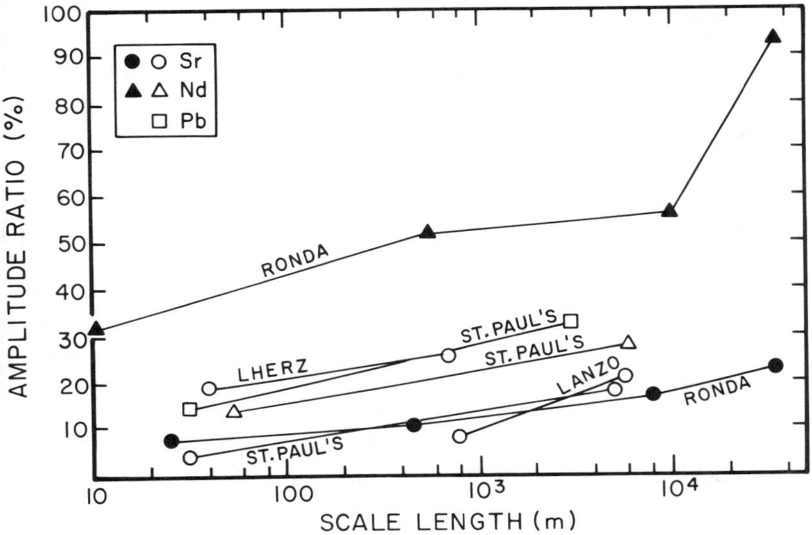

Figure 5.9. Isotopic heterogeneity versus scale length for ultramafic rocks from three orogenic massifs (Lherz, Ronda, and Lanzo) and one oceanic massif (St. Paul's Rocks). Isotopic variability is expressed as a percentage of the total isotopic range observed in all oceanic basalts (MORB and OIB, see discussion later in text). Two end points plotted represent the scale length of the maximum observed variation in each massif (right), and the largest variation observed at the smallest scale length (left). St. Paul's samples are whole rocks; Lherz, Ronda and Lanzo are diopsides. Data are from Polve and Allegre, 1980; Roden *et al.* 1984, and Reisberg and Zindler (1987).

lengths of <100 m. There is currently not enough data from Zabargad to evaluate possible relationships between the amplitude of the variations and the scale lengths.

3.3. Domain Sizes Inferred from Mantle-Derived Basalts

Inferring domain sizes of mantle heterogeneities based on observations of erupted basalts is a very complex issue, though this avenue is the principle source of the proof that mantle geochemical heterogeneities do in fact exist. The basic problem is one of trying to elucidate three-dimensional domain sizes from "planform" observations.

First, we do not truly know the degree of partial melting represented by any given basaltic melt. Estimates vary widely, from values <1 percent to values as high as 20–30 percent. This means that, even if the total volume of a given basalt extraction event could be accurately determined (including both erupted and non-erupted materials), the volume of the mantle domain which this melt sampled would not be specified to better than a factor of 10–20.

Secondly, we have very poor control on the depth of origin of basaltic magmas. Based on phase equilibria considerations, and trace element patterns, conventional wisdom has it that MORB melts are segregated at pressures of 10–25 kb (30–75 km depth), whereas alkali basalts are segregated at greater than 25 kb. In particular instances (e.g., kimberlites carrying diamonds and mantle xenoliths with interpretable P-T source conditions), depths of segregation of the order of 200 km may be certified. However, it must be admitted that, in most cases, the uncertainty in segregation depth is at least 50 km, and this sets the intrinsic limit of how well we can transform an isotopic planform into a three-dimensional map.

An even more fatal ambiguity exists in interpreting isotopic planforms. Basalt melting is presumed to occur in response to adiabatic decompression, thus melts are viewed as deriving from rising diapirs of mantle material. Even if the depth of magma segregation could be precisely determined, this would not delimit in any way the depth of diapir origin. Consider, as an extreme case, the plume model of Morgan (1971), as frequently applied to the generation of hot spots (oceanic islands). If mantle plumes originate near the core-mantle boundary, then basalts erupted from these plumes could be sampling (with only a relatively short time lag, allowing for plume transit time on the order of 10^8 years) the very lowest mantle. Furthermore, if the uprising plumes or blobs undergo any solid-state mixing during their ascent, the final basaltic melt product may represent some confused average of large vertical sections of mantle.

Thus, it is obvious that basalts from two volcanos on Hawaii, for example, separated by 10 km horizontal distance, need not be giving us interpretable information regarding isotopic mantle variations of 10 km scale lengths. Indeed, if this depth ambiguity did not exist in the isotopic signatures in basalts, the problem of delimiting mantle heterogeneities and convection kinematics would already be solved and this chapter would not be necessary! It may appear, then, that isotopic data is fundamentally ambiguous with respect to mantle convection models—so why is this section necessary? While the middle ground may remain murky, it appears to us that some useful constraints may nevertheless be placed using the extreme ends of the spectrum.

For example, significant isotopic heterogeneity has been observed (Zindler *et al.*, 1984) within the eruptives of a single small seamount (10 km base diameter, 42 km^3 volume); it seems unlikely that this variability can be ascribed to several major mantle plumes coming from disparate depths. Zindler *et al.*, (1984) used this data to convincingly argue for a mantle which is heterogeneous on a km or smaller scale but homogeneous on a 20–100 km scale.

At the other extreme (large scale lengths), there appear to be large areas of the Earth which, in planform, show coherent and characteristic isotopic signatures (Dupre and Allegre, 1983). One of these, termed the Dupal anomaly (Hart, 1984b) is more or less continuously present world wide between 0° and 45°S latitude. Figure 5.10 shows a planform of this anomaly, contoured using several different isotopic criteria. It is difficult to conceive of the isotopic delineation of such a mega-scale anomaly by melting of mantle which is anything but heterogeneous on a large (>1000 km) scale. Taken together with the aforementioned seamount data, we conclude that mantle heterogeneities must exist *at least* on both small and very large scales, as originally argued by Davies (1981). This seems like useful information, and encourages us to proceed with this section!

First, let us try to quantify the existing data as best we can. While geochemists have much of this information in random access memory, it has never been assembled in a way which is accessible to the uninitiated. As a start, we will simply establish a measure of the amplitude of isotopic variability, and then try to relate that to planform scale length. As an amplitude measure, we will take the extreme values of $^{87}Sr/^{86}Sr$, $^{143}Nd/^{144}Nd$ and $^{206}Pb/^{204}Pb$ measured on any oceanic basalt (both OIB and MORB). The isotopic "variability" of a given basalt suite (island, volcano, section of ridge, etc.) will then be expressed as a fraction of this total possible variation. We will call this AR, the amplitude ratio (expressed in percent of the total). As the extremes, we take Sr values of 0.70221 and .70694, Nd values of 0.512376 and 0.513365, and Pb values of 17.35 and 21.1 (see Fig. 5.28). The highest Nd and lowest Sr values are those measured on MORB glasses from the Kane Fracture Zone (Machado *et al.*, 1982); the lowest Pb is for a MORB sample from the Carlsberg ridge (Dupre and Allegre, 1983). The lowest Nd value is for a basalt drilled at site 525A on the Walvis Ridge (Richardson *et al.*, 1982); the highest Pb is for an alkali basalt from Tubuai Island (Vidal *et al.*, 1984); the highest Sr is for a basalt from Tahaa, Society Islands (White and Hofmann, 1982). The distances separating these sample localities for these isotopic extremes are 16,700 km (Pb), 7800 km (Nd) and 12,400 km (Sr). Other sample locality pairs could have been chosen which are separated by less than 3000 km for all 3 isotope systems, but which still show variations in excess of 90 percent of the maximum possible.

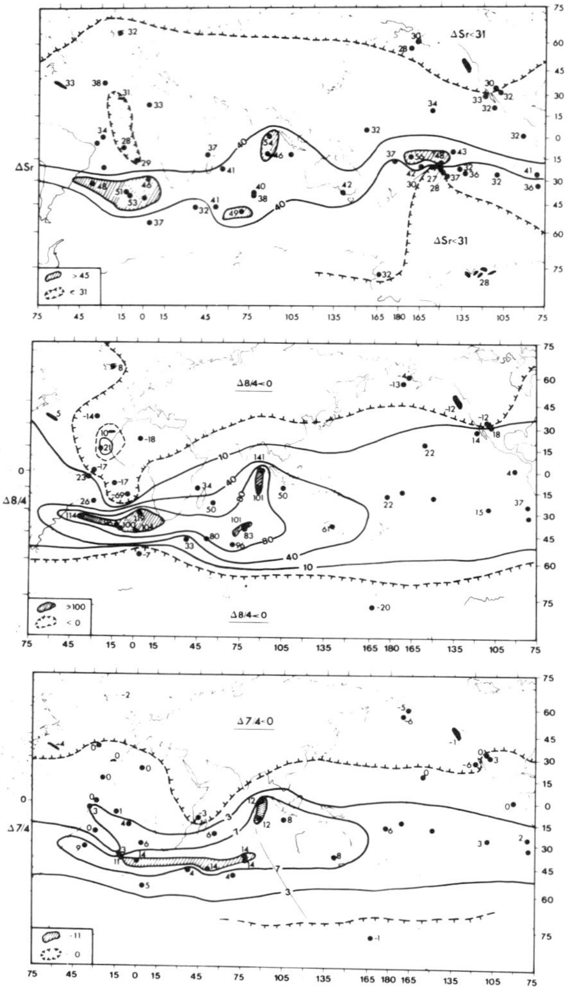

Figure 5.10. World maps showing the distribution of three isotopic criteria used to outline the very large-scale anomalous region which encircles the globe between the equator and 60°S. This region is termed the Dupal anomaly (Hart, 1984b). The three anomaly criteria are: ΔSr, upper figure (absolute $^{87/86}$Sr value, third and fourth significant figure only); Δ8/4, middle figure (vertical deviation in 208/204 from a northern hemisphere reference line (NHRL) on a $^{208/204}$Pb–$^{206/204}$Pb plot) and Δ7/4, lower figure (vertical deviation in 207/204 Pb relative to the NHRL on a $^{207/204}$Pb–$^{206/204}$Pb plot).

The amplitude ratios for a given locality are thus expressed as follows:

$$AR_{Sr} = [(\Delta^{87}Sr/^{86}Sr)/0.00473]100$$

$$AR_{Nd} = [(\Delta^{143}Nd/^{144}Nd)/0.000989]100$$

$$AR_{Pb} = [(\Delta^{206}Pb/^{204}Pb)/3.75]100$$

We have not attempted to assemble all existing isotopic data; rather, we have chosen areas for which a significant quantity of high quality data exists, and which tend to delimit the *maximum* amplitude ratios attainable for *minimum* scale length; a plot of this selected data set is shown in Figure 5.11. Also shown for comparison is the field of "first order" observations derived from ultramafic massifs (taken from Fig. 5.9). What is apparent from this plot is that the maximum attainable amplitude ratio increases with scale length, whereas relatively small amplitude ratios can be observed at all scale lengths. Thus, the figure provides a "wedge" of data, pinching out to insignificant amplitude ratios (~10 percent) for scale lengths less than ~1 km. The area of Figure 5.11 for scale lengths of greater than 500 km could be filled in by comparing various island pairs; we have limited our data set to either individual islands, related islands in a coherent island group or archipelago (Samoa, Societies, Galapagos, etc.), or contiguous segments of spreading ridge where large isotopic gradients exist.

The important feature of this plot is the *upper* bound of the wedge-shaped field. The fact that small amplitude ratios (<30 percent) can occur for essentially all scale lengths can be taken to mean simply that there are regions of the mantle which are relatively homogeneous on very large scale lengths (>1000 km). On the other hand, the upper bound convincingly shows that large amplitude ratios do not occur for small scale lengths, but occur only for large scale lengths. Note that the data from orogenic massifs suggests an upper bound which is shifted toward considerably smaller scale lengths. This difference suggests that the magmatic processes involved in basalt production are capable of considerable averaging on a 10 km scale. We have fit the upper bound of the data with two arbitrary functions, which are meant only as descriptions of the upper bound without implying any particular physical process. The upper bound so described shows that the maximum amplitude ratio increases slowly with scale length (for example, as the one-fifth power of SL) insofar as scale length varies over four orders of magnitude while maximum amplitude ratios increase from 0 to 100 percent.

This discussion suggests several important implications with respect to the nature of mantle heterogeneities. While heterogeneities of relatively small isotopic amplitude may exist on small scales (1–10 km), as argued by Zindler

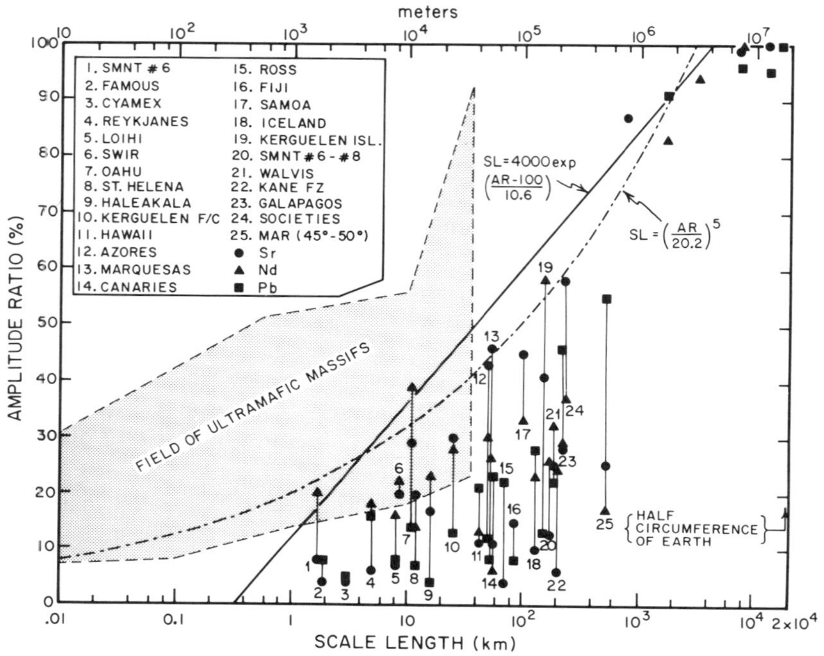

Figure 5.11. Amplitude ratio (AR) versus scale length (SL) for various volcanos, islands, island groups and ridge segments. Each vertical line is the observed range of AR observed for the three isotope systems. Points for Sr, Nd and Pb plotted at 100 percent AR are the normalizing values, given for Nd by Kane-Walvis (SL—7780 Km), for Sr by Kane-Societies (SL—12,400 Km), and for Pb by Mid-Indian Ocean ridge to Tubuai (16,700 Km). Other points plotted in upper right quadrant are for other various island pairs which provide large AR for minimal scale lengths (Nd: Walvis– Mid Atlantic ridge, 94 percent, 3030 Km; Sr: Tahaa–Juan de Fuca, 99 percent, 7100 Km: Tahaa–Tubuai, 87 percent, 770 Km; Pb: Walvis–Tubuai, 95 percent 13,500 Km; St. Helena–Mid Indian Ocean ridge, 96 percent, 7600 Km; Walvis–St. Helena, 91 percent, 1700 Km). Arbitrary exponential and power law curves are shown as possible upper bound fits to the data. Also shown is the outline of the field for ultramafic massifs, taken from Figure 5.9. Data references may be found in Zindler *et al.* (1982) and Hart (1984b).

et al. (1984), it does not appear possible to explain the whole observed range of oceanic mantle heterogeneity with small scale ($<$10 km) heterogeneities alone. To do this with the preferential melting models of Zindler *et al*. (1984) or Sleep (1984), would require that melting processes in certain broad regions of the oceanic mantle can selectively tap only the more enriched portions of small heterogeneities, whereas in other regions the melting process is unable to be so selective, thereby averaging in a larger fraction of more depleted matrix.

As it seems difficult to envision a mechanism for such large scale modulation of the melting process, one could argue that the small-scale heterogeneities (blobs) are themselves variable in composition, with large volumes of the mantle typified by blobs which are more enriched than those in other regions. For example, in this model, all of the mantle underlying the Dupal anomaly region (Fig. 5.10) would contain blobs of a more enriched character than those in mantle lying outside the Dupal region. Then the preferential melting models of Zindler *et al*. (1984) and Sleep (1984) could be applied with little difficulty. In essence, this model implies that the isotopic *contrast* between blobs and their depleted mantle matrix is systematically larger in Dupal mantle regions than elsewhere. The upper bound in Figure 5.11 would then be providing a measure of the degree of isotopic contrast between small scale blobs and their matrix, but only on short time scales. But how are the enriched blobs kept from dispersing away from the Dupal region over large time scales? In other words, the blobs themselves are isotopically variable on very large ($>$1000 km) scale lengths, while still being of modest size (1–10 km).

As a corollary of this, large isotopically anomalous regions (e.g., the Dupal region) should be defineable not only by their isotopically anomalous characteristics (as originally used to define the Dupal anomaly) but also by using an isotopic *contrast* parameter. It is true that three of the islands of largest scale length defining the upper bound of Figure 5.11 are part of the Dupal anomaly (Samoa, Kerguelen and the Society Islands); other islands defining the upper bound are smaller in scale length, and are external to the Dupal region (Azores, Oahu). In other words, if this model is correct, it should be possible to contour a world map in terms of isotopic contrast or variability, and produce a figure akin to those used to delineate the Dupal anomaly.

Alternative models deriving from the data of Figures 5.10 and 5.11 are equally possible, and perhaps more plausible. The simplest of these would view the oceanic mantle as isotopically heterogeneous on all scales (Davies, 1981), with the extent of heterogeneity increasing with scale length. In essence, this model would derive from Figure 5.11 as taken at face value. Because the isotopic variability (contrast) also seems to increase with scale length, this model would still require a means of generating both the large amplitude variations coupled with large isotopic contrasts. In the context of a plume model, as an example, this could be accomplished by allowing the lower mantle to carry the enriched isotopic signatures, with sub-Dupal lower mantle being substantially more "enriched" than other regions, and then allowing this material to mix with (be polluted by) the depleted upper mantle, which could be viewed as a relatively homogeneous reservoir worldwide. In this model, proposed by Dupre and Allegre (1983), the isotopic contrast would be that between the upper and lower mantles, with Dupal regions showing the largest contrast due to the more enriched nature of the lower mantle component.

Obviously, these two interpretations (models) for the isotopic characteristics depicted in Figures 5.10 and 5.11 lead to very different views of convective processes in the mantle. The first model would be consistent with whole-mantle convection, the second would require a two-layered-mantle with limited exchange between the layers over geologic time.

4. The Effects of Magma Degassing on Mantle Helium Signals

Measurements of $^3He/^4He$ ratios in oceanic basalts have established beyond any reasonable doubt that primordial gasses continue to escape from the Earth's mantle. $^3He/^4He$ ratios in oceanic basalts range from about 1.4×10^{-6} to 48.4×10^{-6} (Kurz *et al.*, 1983; Graham *et al.*, 1984, 1988), or from about 1 to 32 times the atmospheric value of 1.38×10^{-6} (Ra). Primordial values of $^3He/^4He$ are in the range $1.4–3 \times 10^{-4}$, or about 100–200 Ra (Craig and Lupton, 1978, 1981; Jeffrey and Anders, 1970). The relatively low $^3He/^4He$ ratios in the Earth's mantle are due to radiogenic accumulations of ^4He from U and Th decay over the course of Earth history. ^3He is dominantly primordial, although a minor amount is produced via tritium by (n,α) reactions on Li.

$^3He/^4He$ variations in oceanic basalts are displayed in histogram fashion in Figure 5.28 (along with Sr, Nd and Pb isotope ratios, for reference). In contrast to Sr, Nd and Pb, MORB $^3He/^4He$ values occupy a relatively small portion of the OIB array (1–32 Ra) and lie totally within that array, rather than defining one end of it, as is the case for the other isotope ratios. Basalts from Loihi Seamount, a young submarine volcano located on the southeast flank of Kilauea on the big island of Hawaii, span nearly the entire known range of oceanic $^3He/^4He$ values (4–32 × Ra; Kurz *et al.*, 1983) while displaying only minor variation in $^{87}Sr/^{86}Sr$, from a 0.70335 to 0.70370 (Staudigel *et al.*, 1984). Some of the lowest, or least primitive $^3He/^4He$ ratios found on oceanic islands come from Tristan da Cunha and Gough (~5 × Ra; Kurz *et al.*, 1982), known for their Sr and Nd isotope ratios which approximately coincide with many estimates for the bulk silicate earth (BSE). Recently, Graham *et al.*, (1984, 1988) have measured $^3He/^4He$ ratios of ~1 Ra in alkali basalts from several small seamounts adjacent to the East Pacific Rise at about 12° N.

Collectively, these observations have led to the widely held view that the U-Th-He system has been "decoupled" from the U-Th-Pd, Sm-Nd, Rb-Sr, and Lu-Hf systems; that is, the U-Th-He system, by virtue of the gaseous daughter product, has been subject to perturbations which have not affected the other systems. It is the nature, location and timing of these perturbations that are of interest to us if we are to accurately interpret the He isotope signal in oceanic basalts.

One of the most often proposed mechanisms for the decoupling of He has been that of very rapid solid-state diffusion, relative to the other ions of interest (e.g., O'Nions and Oxburgh, 1983; Allegre *et al.*, 1985). This mechanism has been suggested to account for the relatively high $^3He/^4He$ ratios and He concentrations in MORBs, which are unexpected in view of the long-term depleted nature of the MORB source. He diffusion data for olivine, however, suggest that this is not a viable mechanism (Hart, 1984a). As seen in Figure 5.12, the olivine data, taken at face value, preclude very extensive solid-state diffusional transport of He (>100 m), even over time scales similar to the age of the Earth. Thus, as for other ions of interest, convective transport of He within the mantle is orders of magnitude more efficient than solid-state diffusional transport.

Parent-daughter fractionation processes associated with magma generation and transport throughout Earth history are thought to be responsible for much, if not all, of the variability of Sr, Nd, Pb and Hf isotope ratios in the mantle. Variations in these isotope ratios may be produced via equilibrium fractionations of parent from daughter isotopes followed by variable rates of accumulation of the radiogenic daughter isotope over time. All the elements of interest are incompatible in likely mantle phases (k_ds < 1 to << 1) and large variations in parent-daughter ratios are probably produced only at very small degrees of partial fusion or very large degrees of fractional crystallization. Although He is expected to behave as a very incompatible element and partition strongly into the liquid during melting in the mantle, U and Th

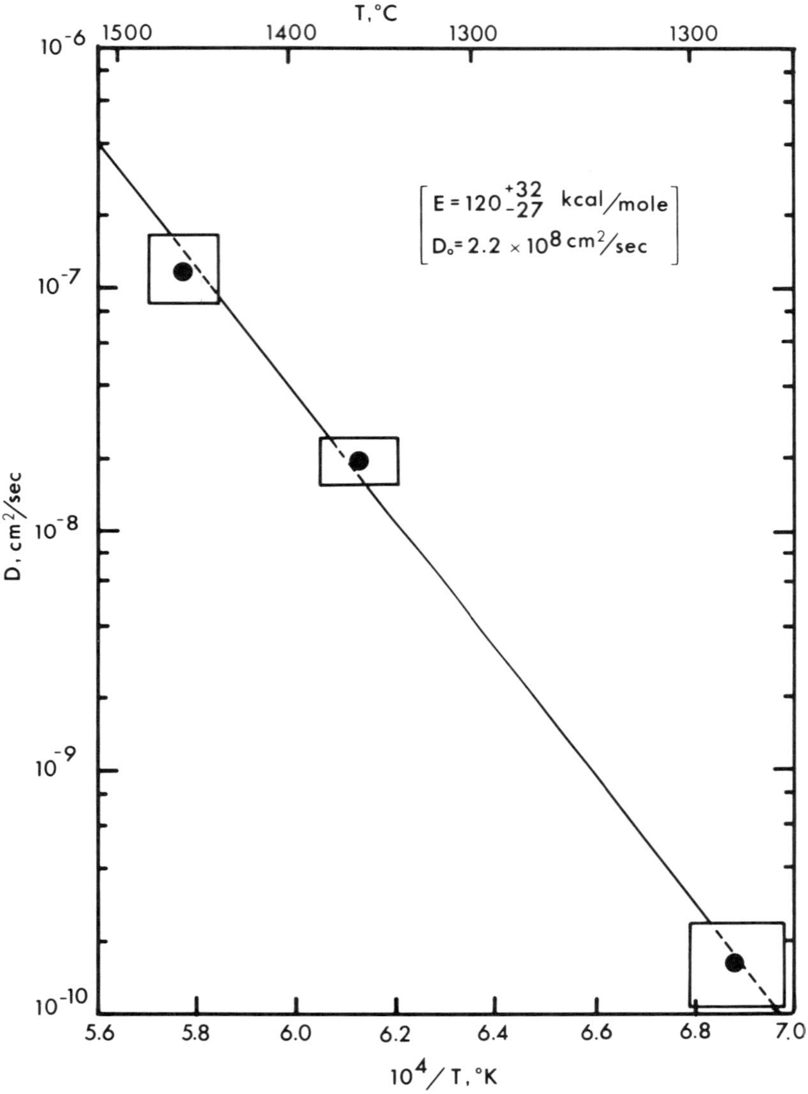

Figure 5.12. Arrhenius plot, diffusion coefficient versus reciprocal temperature, for He diffusion in olivine. Activation energy (E) and pre-exponential factor (D_0) given (from Hart, 1984a).

behave similarly and, thus, the U-Th-He system will probably be unremarkable relative to the other parent–daughter systems during mantle melting and crystallization processes which are governed by solid–liquid equilibria.

Interestingly, measured crystal–liquid partition coefficients for Ne, Ar, Kr and Xe (Hiyagon and Ozima, 1982) defy our expectations of incompatible behavior by having values surprisingly close to unity (0.1 to 0.6). Ozima and Podosek (1983) caution against blind acceptance of these values due to the considerable experimental difficulties involved in their measurement. Nevertheless, as they are the only such values which are available, they merit our consideration. By comparison of these values for the heavy rare gasses, we can estimate a k_d for He of $\sim$0.5 (cpx/liquid) and compare this to a recently determined cpx/liquid k_d for U of 0.0008 (Watson *et al.*, 1987) and so estimate changes in U/He during melting. Assuming a modal cpx content of 10 percent, and that cpx dominates the bulk partition coefficient for U and He, a 0.1 percent melt produced by fractional melting has U/He which is about five times the source value; at 5 percent melting, this value drops to $\sim$2. Compared to measured variations in U/He which range over 6 orders of magnitude, changes of this magnitude are relatively minor. We conclude, therefore, that, even if the partition coefficients measured by Hiyagon and Ozima (1982) are accurate, it is not likely that solid–liquid equilibria are responsible for producing all of the observed variations in U/He.

Another process for perturbing mantle He signals, suggested by Condomines *et al.* (1983), involves wall-rock assimilation in magma chambers. Basalts which are degassed during eruption and subsequently allowed to "age," thereby generating low $^3He/^4He$ ratios, could on later assimilation into a magma chamber, be expected to significantly alter the mantle He signal of the magma chamber.

4.1. Degassing of Magmas and Accumulation of Radiogenic ^4He

Unlike the processes discussed thus far, degassing of magma is expected to have a large effect on U-Th-He systematics without causing significant changes in the other parent–daughter systems. The ubiquity of degassing before volcanic eruptions is well known, as the chemical compositions of these gases are often monitored to assess hazard and the continuing evolution of a given system. The exsolution of gases during eruption is primarily responsible for the spectacular character of many eruptions. The important question in the present context is whether or not significant amounts of ^4He can accumulate subsequent to degassing and prior to sampling and analysis (see also comprehensive discussion in Kurz and Jenkins, 1981).

By and large, $^3He/^4He$ ratios are determined on "zero-age" rocks, or in fluid (or melt) inclusions in phenocrysts, so that *in situ* production of ^4He is not presumed to be significant. Even in rocks which are not "zero-age," the extent of the problem can be assessed by measuring U, Th, and He concentrations, and correcting $^3He/^4He$ ratios for *in situ* accumulation of ^4He. However, $^3He/^4He$ ratios measured in volcanic rocks are typically equated with mantle source values and thus it is implicitly assumed that ^4He accumulation during mantle or crustal magma storage is not important. We suggest that, in the face of abundant evidence for pre-eruptive degassing of magmas, this may be a dangerous presumption.

The change in $^3He/^4He$ with time in a given system can be assessed by defining a factor $f_{3/4}$ which represents the fractional change in $^3He/^4He$ with time. Because the radiogenic production ratio for $^3He/^4He$ is on the order of $1 \pm 0.05 \times 10^{-8}$ for lithologies of interest (O'Nions and Oxburgh, 1983), and the rocks we are considering have $^3He/^4He$ ratios in the range 1.4×10^{-6} to 4.5×10^{-5}, we will ignore the relatively minor radiogenic production of ^3He in the following calculations. Using the standard closed-system decay equation, we can calculate the total production of radiogenic ^4He as:

$$^4He^* = 8[^{238}U(e^{\lambda_8 t} - 1)] + 7[^{235}U(e^{\lambda_5 t} - 1)] + 6[^{232}Th(e^{\lambda_2 t} - 1)],$$

where $^4He^*$ is radiogenic ^4He, λ_8, λ_5, and λ_2 are decay constants for ^{238}U, ^{235}U, and ^{232}Th, respectively, t is in years, and the numerals 6, 7 and 8 represent the number of He atoms produced by each decay of the parent nuclide.

By assuming a $^{232}Th/^{238}U$ ratio of 3.5 and using the known ratio for $^{235}U/^{238}U$ of 1/137.88 (Steiger and Jager, 1977) we can write

$$^4He^* = {}^{238}U[8(e^{\lambda_8 t} - 1) + 7/137.88(e^{\lambda_5 t} - 1) + 21(e^{\lambda_2 t} - 1)].$$

Dividing by the total ^4He in the system (where $^4He_0$ is the inherited ^4He), we obtain

$$[^4He^*/(^4He^* + {}^4He_0)] = [^{238}U/(^4He^* + {}^4He_0)]\,[8(e^{\lambda_8 t} - 1) + 7/137.88(e^{\lambda_5 t} - 1) + 21(e^{\lambda_2 t} - 1)]. \quad (1)$$

We can now define $f_{3/4}$ so that

$$f_{3/4} = [^3He/(^4He^* + {}^4He_0)]/(^3He/^4He)_0 = [^4He/(^4He^* + {}^4He_0)] = 1-[^4He^*/(^4He^* + {}^4He_0)],$$

and, using (1),

$$f_{3/4} = 1 - (^{238}U/^4He)[8(e^{\lambda 8t} - 1) + 7/137.88(e^{\lambda 5t} - 1) + 21(e^{\lambda 2t} - 1)], \quad (2)$$

where $^4He = {}^4He^* + {}^4He_0$.

For magmatic systems where $t < 10^7$ years, we can make the approximation that $e^{\lambda t} - 1 \sim \lambda t$ and, using (2) and values for the decay constants from Steiger and Jager (1977), write that

$$f_{3/4} = 1 - [2.178 \times 10^{-13}(\mathrm{U}/^4\mathrm{He})t], \tag{3}$$

where U is now in ppm total U and ^{4}He is in std cc/g.

Using (3) we can now see that there will be essentially no change in the ^{3}He/^{4}He ratio of a system, that is $f_{3/4} \geqslant 0.95$, when the product $(\mathrm{U}/^4\mathrm{He})t \leqslant 2.3 \times 10^{11}$. U/^{4}He ratios (in many cases, estimated from K contents using K/U of 1.27×10^4 from Jochum *et al.*, 1983) in the range 5×10^5 to $>10^8$ are not uncommon in basalts from ocean islands and seamounts (Graham *et al.*, 1984, 1988; Kurz *et al.*, 1983; Condomines *et al.*, 1983; Kurz *et al.*, 1985;) and "zero age" ($f_{3/4} \geqslant 0.95$) in such cases may stipulate times since degassing of 2000 to 500,000 years. These relationships are shown in Figure 5.13 where U/^{4}He (ppm/std cc) is plotted versus time and contoured for values of $f_{3/4}$ ranging from 0.9999 to 0.0001. Thus, for a given present-day U/^{4}He ratio at the left side of the diagram, changes in ^{3}He/^{4}He back in time can be traced. For example, a rock with U/^{4}He-10^7 and ^{3}He/^{4}He-5 Ra today would have had a ^{3}He/^{4}He ratio of 30 Ra 383 thousand years ago (had it or the predecessor magma existed as a closed system over that time interval). Based on U/^{4}He ratios shown in Figure 5.13 for Hawaii, Iceland, and several small seamounts near the East Pacific Rise, it is clear that accumulation of radiogenic ^{4}He during magma transit and storage may result in perturbation of indigenous mantle He isotope signatures.

Closed system evolution curves for the MORB mantle and BSE are also shown in Figure 5.13. These curves, which suggest present day U/^{4}He of 630 and 310 in the MORB mantle and bulk Earth respectively, are based on initial and present-day ^{3}He/^{4}He ratios of 100 and 50×Ra, respectively, and the bulk Earth MORB and mantle U concentrations given in Table 5.7 (initial ^{3}He/^{4}He value from Jeffrey and Anders, 1970; present-day value arbitrarily chosen to be significantly higher than the highest measured terrestrial ^{3}He/^{4}He ratio—choice of 32 or 80 would not drastically alter the arguments presented here). The U/^{4}He ratio of 630 in the MORB mantle presumes outgassing at 4400 m.y. (e.g., Staudacher and Allegre, 1982; Allegre *et al.*, 1983a) and no loss of He relative to U since that time. This implies simply that magmas which outgas near the surface are not subsequently reassimilated into the MORB reservoir. If such reassimilation of degassed basalts does occur, the U/^{4}He ratio may have been less than the indicated value of $\sim 2 \times 10^4$ at the time of initial outgassing and/or somewhat higher than 630 today. The value of 630 may therefore be considered a minimum for the MORB mantle; a maximum of ~3000 is defined by the lowest U/^{4}He ratio measured in MORB (Kurz, 1982).

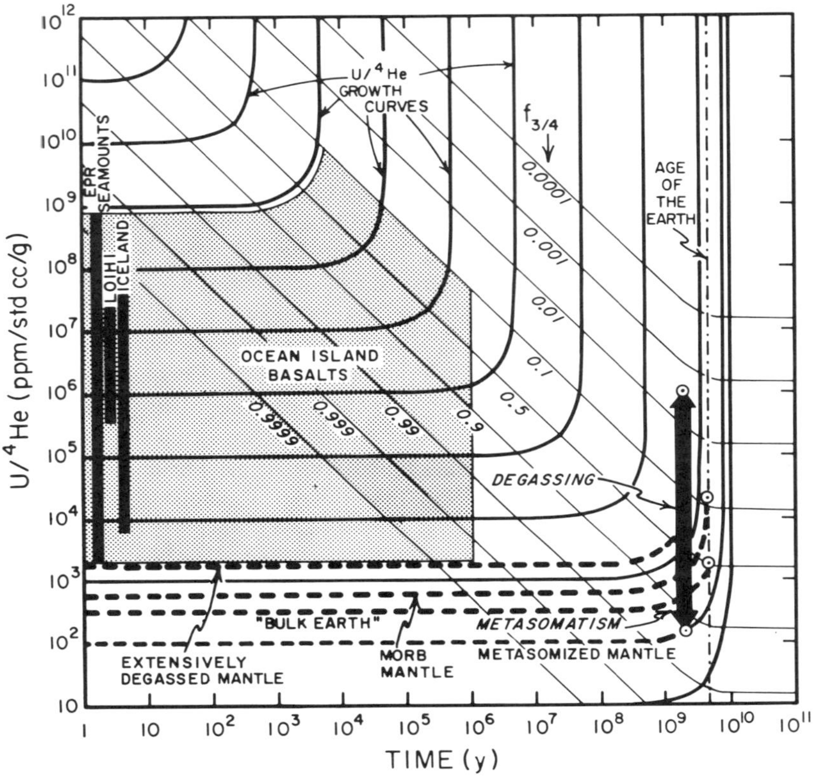

Figure 5.13. U/^{4}He (ppm/std cc) versus time (in years). Diagram is designed to aid in the assessment of the evolution of ^{3}He/^{4}He ratios in systems with a given U/^{4}He ratio. Heavy solid lines, labeled "U/^{4}He growth curves," document closed-system changes in U/^{4}He due to growth of radiogenic ^{4}He from U and Th decay. Contours are shown for values of $f_{3/4}$ ranging from 0.9999 to 0.0001 (thin solid lines with slopes of ~ -1); $f_{3/4} = (^3\text{He}/^4\text{He})_p/(^3\text{He}/^4\text{He})_t$, where $(^3\text{He}/^4\text{He})_p$ is the present-day He isotope ratio in a system which has been closed since time t. ^{232}Th/^{238}U is taken as 3.0; use of value of 3.5 or 4.0 would shift the $f_{3/4}$ contours slightly to the left, without qualitatively changing the appearance or implications of the diagram.

The inner two heavy dashed lines toward the base of the diagram indicate closed-system evolution paths for the MORB source reservoir and BSE over the past 4.4 and 4.55×10^9 y, respectively. U concentrations are from Table 5.7 and initial ^{3}He/^{4}He ratios are taken as 100 Ra, although use of 200 Ra would not significantly shift the positions of the paths, due to the log-log scale. Present-day ^{3}He/^{4}He ratios in the MORB source and BSE reservoirs are assumed to be 8 Ra and 50 Ra, respectively; again use of 30 Ra or 80 Ra for BSE would not significantly change the diagram. The resulting present-day $U/^4He$ ratios in these reservoirs are ~ 630 and ~ 310, for MORB and BSE, respectively. Note that these values are substantially lower than $U/^4He$ ratios measured in basalts from Iceland (Condomines *et al.*, 1983; Kurz *et al.*, 1985), Loihi (Kurz *et al.*, 1983), and several small seamounts adjacent to the East Pacific Rise at ~12°N (Graham *et al.*, 1984; 1988), suggesting large amounts of pre- or post-eruptive degassing.

The outer two heavy dashed lines show hypothetical closed-system evolution paths for extensively degassed and metasomatized mantle segments over the past 2×10^9 y. These mantle segments are presumed to have evolved from the MORB reservoir at this time. The metasomatized reservoir, considered a possible analog for EM I (see Section 5.2.5), has incorporated a He-rich, U-poor fluid component resulting in a factor of 10 reduction in its $U/^4He$ ratio. This has the effect of "freezing in" its 2 b.y. $^3He/^4He$ ratio, as indicated by the downward traverse across $f_{3/4}$ contours from ~0.70 to ~0.93. The extensively degassed reservoir is considered to be a possible analog for HIMU, originating as either a metasomatic residuum or a degassed magma (see Section 5.2.4). This increase in $U/^4He$ shifts the material above the $f_{3/4} = 0.01$ contour, leading to a present-day $^3He/^4He$ ratio less than the atmospheric value. Note that even this extensively degassed mantle material will have a relatively low $U/^4He$ ratio today, owing to the 2. b.y. aging period.

Heavy solid bars at the left side of the diagram show ranges for ocean island and seamount basalts from sources mentioned above (note that these bars are for reference and their positions have no time significance). The shaded field labeled "ocean island basalts" is drawn to include these bars, and is arbitrarily terminated at 10^6 y and $f_{3/4}$ = 0.1 (i.e., factor of ten closed-system reduction in $^3He/^4He$). This field documents the potential for $^3He/^4He$ reduction in real systems on time scales which may be appropriate for high-level magma evolution. "Worst case" scenarios involve the pre-eruptive loss of He from a magma (via CDL or EL models; see Sections 4.2 and 3) resulting in an increase in $U/^4He$ from ambient mantle values (heavy dashed lines) into the shaded field. Subsequent "aging" of the magma results in $^3He/^4He$ reductions, as indicated by $f_{3/4}$ contours. For example, a mantle melt which is degassed 2×10^5 y ago to a $U/^4He$ ratio of $\sim 2 \times 10^7$, will have suffered about a 50 percent reduction of its original mantle $^3He/^4He$ ratio by the present-day; this magma will then have a present-day $U/^4He$ ratio of 10^7. As another example, a magma which degasses 5×10^4 y ago to a $U/^4He$ ratio of $\sim 10^9$, will suffer a 90 percent reduction in $^3He/^4He$ by today, at which time it will have a $U/^4He$ ratio of $\sim 10^8$. The observed disparity between mantle and basalt $U/^4He$ ratios clearly documents the potential for $^3He/^4He$ reduction during high-level magma evolution.

Table 5.7. U-Pb budget for various terrestrial reservoirs

	U, ppm	Pb, ppm	μ	Nd, ppm	Sr, ppm	Sr/Nd	U/Nd	Pb/Nd
Cl[†]	0.0200	6.12	0.146	1.13	19.5	17.3	0.0177	5.42
PUM[‡]	0.0208	0.155	8.4	1.17	19.6	16.8	0.0178	0.132
MORB[§]	0.081	0.541	9.4	7.7	128	16.6	.0105	0.070
MORB mantle	0.0084	0.056	—	0.8	13.3	—	—	—
Upper crust[‖]	2.4 ± 0.5	18 ± 4	8.7 (6–14)	32.5	300	9.2	.073	0.55
Lower crust[‖]	0.7 ± 0.5	11 ± 3	3.8 (1–9)	32.5	300	9.2	.022	0.34
Total crust[¶]	1.2 ± 0.5	13.1 ± 4	5.7 (3–12)	32.5 ± 1.5	300 ± 30	9.2	.037	0.40
Mass in crust / Mass in PUM	37%	54%	—	18%	10%	—	—	—

[†]Element ratios are for Cl chondrites, Anders and Ebihara (1982); abundances from Cl chondrites multiplied by factor of 2.47.

[‡]From Table 5.4; Pb calculated for μ = 8.4 (see text).

[§]Average for 17 fresh MORB glasses and 4 leached whole rocks (Cohen and O'Nions, 1982a, 1982 b; Cohen *et al.*, 1980; Vidal and Clauer, 1981); average K for this suite is 1027 ppm. Twenty-two glasses from Jochum *et al.* (1983), average 0.075 ppm U.

[‖]Data sources: Allegre and Rousseau, 1984; Dupuy *et al.*, 1979; Fahrig and Eade, 1968; Goldstein *et al.*, 1984; Gray, 1977; Gray and Oversby, 1972; Heinrichs *et al.*, 1980; McLennan and Taylor, 1980; Michard *et al.*, 1985; Montgomery, 1977; Reilly and Shaw, 1967; Shaw, 1967; Shaw *et al.*, 1976; Taylor and McLennan, 1981; Taylor *et al.*, 1983; Tilton and Barreiro, 1980; White *et al.*, 1985.

[¶]Calculated at 30 percent upper crust–70 percent lower crust.

These $U/^4He$ values for the mantle are very low when compared with oceanic basalt values shown in Figure 5.13. If most or all of the degassing associated with the high $U/^4He$ ratios occurs during eruption, mantle source $^3He/^4He$ values can be accurately estimated (in conjunction with U and Th determinations). However, if a significant fraction of the gas loss occurs prior to eruption, estimated mantle $^3He/^4He$ ratios may be too low.

Extensive pre-eruptive degassing of magmas may result in correlation of $U/^4He$ ratios with factors which document increasing magma differentiation. Figure 5.14 is a plot of $U/^4He$ (ppm/std cc) versus Mg#. For many samples, U contents are estimated from K contents, taking advantage of the relative constancy of K/U in basalts (Jochum *et al.*, 1983; see figure caption for details). Data for Loihi are from Kurz *et al.* (1983), and for Iceland are from Kurz *et al.* (1985) and Condomines *et al.* (1983). Samples from Kurz *et al.* (1985) represent He extracted by crushing only and $U/^4He$ in these samples must be taken as a maximum. The relationships in Figure 5.14 document a tendency for cogenetic suites to display an inverse correlation of $U/^4He$ with Mg#. A number of factors, however, preclude a strictly quantitative analysis of this kind of diagram: 1) He abundances are not determined by the same method in all samples; 2) outgassing during eruption probably accounts for variable fractions of the total gas loss in different samples; 3) gas loss and heat loss, resulting in fractional crystallization, will likely occur at different relative rates in different physical settings; and 4) measured variations in $^{143}Nd/^{144}Nd$ and $^{87}Sr/^{86}Sr$ in several of the areas document the involvement of chemically distinct mantle source materials which interact in complex ways (Zindler *et al.*, 1979; Staudigel *et al.*, 1984; Condomines *et al.*, 1983; Kurz *et al.*, 1983). Even so, the convincing, and in some cases, striking negative correlations observed in Figure 5.14 support the contention that pre-eruptive degassing may be an important phenomenon.

The inverse correlation between age and $^3He/^4He$ for volcanos on the island of Hawaii (Kurz *et al.*, 1983) may also be a result of pre-eruptive degassing, with the older volcanos suffering more gas loss and correspondingly more *in situ* production of ^4He during magma chamber residence. It is interesting to note that, despite problems with measuring He concentrations, $^3He/^4He$ and ^4He data for Kilauea, Mauna Loa, and Hualalai, excluding Loihi, define an approximate 400,000 year U-He isochron. Since the Mauna Loa lava is historic, this again suggests an "aging" period in a magma chamber.

Another observation which supports the pre-eruptive degassing hypothesis is that MORB, derived from depleted and degassed mantle sources, have consistently lower $U/^4He$ ratios than basalts with hot-spot affinities. For

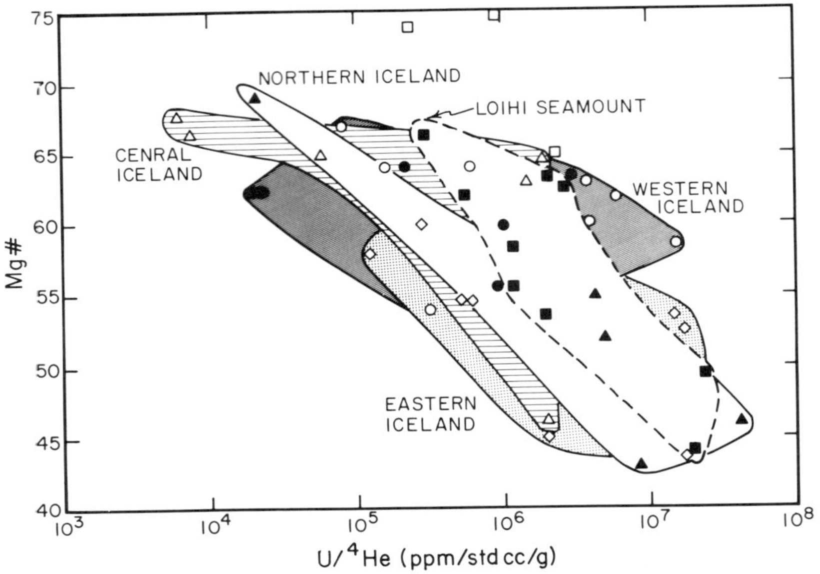

Figure 5.14. Mg# versus U/^{4}He (ppm/std cc) variation diagram for samples from Iceland (Condomines *et al.*, 1983; Kurz *et al.*, 1985) and Loihi (Kurz *et al.*, 1983). Mg# (atomic) calculated with Fe_2O_3/FeO—0.2. In many cases, U had to be estimated from K contents, taking advantage of the relative constancy of K/U in basalts (Jochum *et al.*, 1983). Data for Iceland are broken down into northern (closed triangles), central (open triangles), eastern (diamonds), and western (open and closed circles) regions (see Kurz *et al.*, 1985). The Kurz *et al.* data was by crushing only, and so may tend to underestimate He contents. Loihi data include closed and open squares; open squares are not included in Loihi field as they have clearly accumulated olivine, and therefore have anomously high Mg#s. Note that all fields, to greater and lesser degrees, display negative correlations between the two plotted parameters. We interpret this as strong evidence that degassing of magmas, resulting in increased U/^{4}He, proceeds in conjunction with crystal fractionation, which causes reduction of the Mg#.

example, Loihi seamount, which is characterized largely by ^{3}He/^{4}He ratios three to four times the MORB value (and which therefore derive from a lower U/^{4}He source than MORB), has U/^{4}He ratios which average more than a factor of 400 higher than the MORB value (2.6×10^6 relative to 6×10^3; Kurz *et al.*, 1983). We suggest that this apparent paradox may be resolved by considering differences between magma generation processes which operate at mid-ocean ridges and ocean islands. Mid-ocean ridges represent a high through-put environment where "primitive" basalts are produced by relatively large amounts of partial melting, resulting in low initial gas concentrations and gas-phase saturation at correspondingly late-stages. Thus, MORBs may suffer relatively minor gas loss prior to eruption. At ocean islands, relatively long staging times, particularly at older alkalic islands, and initially high volatile contents due to relatively small degrees of partial melting, will tend to increase the extent and importance of pre-eruptive degassing.

In the following sections (see also Zindler and Hart, 1986) we consider two-end-member models for pre-eruptive He loss: 1) a model of continuous diffusive loss (CDL) from a crystallizing magma chamber; and 2) a model of exhalative loss (EL) by supersaturation and exsolution of a carrier gas (CO_2) during magma ascent or crystallization. Controlling factors for the two processes are very different and often ill-constrained. We feel, however, that even the results of a semi-quantitative analysis clearly demonstrate the potential for $^3He/^4He$ change during at least some types of magmatism.

4.2. The Continuous Diffusion Loss Model

Diffusion measurements for He in basalt glass were reported by Kurz and Jenkins (1981), over the temperature range 125–400°C. Extrapolation to 1200° suggests a D^{He} of 7.5×10^{-5} cm^2/sec. If the Arrhenius relationship is not linear to such high temperatures, it is likely that we have underestimated D^{He}, so that the above value may be viewed as a minimum. For this value of D^{He}, diffusion from a spherical magma body 1 km in diameter will result, for example, in 99 percent He loss over about 4×10^5 years. Thus, diffusive loss of He clearly merits further consideration.

The CDL model assumes: 1) filling of a spherical magma chamber, followed by *in situ* production of ^4He (from U and Th), and diffusive loss of both the initial He and radiogenic He; 2) zero concentration of He at the surface of the chamber (finite wall rock permeability, allowing the venting of He to the sea and ultimately, the atmosphere); 3) non-convecting magma (certainly not true, but because convection will accelerate the rate of loss of He, the model provides an "upper bound" for $^3He/^4He$); and 4) equal rates of diffusion for ^3He and ^4He (again, if this condition is untrue, and ^3He diffuses faster than ^4He, the decrease of $^3He/^4He$ with time will be increased, so the model provides an upper bound on $^3He/^4He$ ratios at any given time). To describe this model, the closed-system decay equation must be modified to allow for continuous diffusive loss of initial ^3He and ^4He, and radiogenic

^{4}He. The equation for coupled production and loss of ^{4}He, considering only the ^{238}U terms, is

$$\frac{\partial \mathrm{He}}{\partial t} = D\left(\frac{\partial^2 \mathrm{He}}{\partial r^2} + \frac{2}{r}\frac{\partial \mathrm{He}}{\partial r}\right) + \lambda\, {}^{238}\mathrm{U}_0\, e^{-\lambda t}.$$

where D is the diffusion coefficient, r is the radius of the sphere, $^{238}U_0$ the initial concentration of ^{238}U, λ is the ^{238}U decay constant, and t is elapsed time. The solution to this equation can be written (Tilton, 1960) as

$$\left(\frac{^4\mathrm{He}}{^{238}\mathrm{U}}\right)_p = \frac{6P}{\pi^2}\sum_{n=1}^{\infty}\left(\frac{\lambda(e^{(\lambda - n^2B)t} - 1)}{n^2(\lambda - n^2B)}\right) = [\mathrm{PDS,238}], \tag{4}$$

where P is the number of ^{4}He atoms produced per decay of ^{238}U (= 8). Similar equations can be written for ^{235}U and ^{232}Th (with $P = 7$ and 6, respectively) and the λs are the respective decay constants. The equations for diffusive loss of the initial ^{3}He and ^{4}He are the same, with the deletion of the production term, and their solution can be written as

$$\left(\frac{^3\mathrm{He}_p}{^3\mathrm{He}_0}\right) = \left(\frac{^4\mathrm{He}_p}{^4\mathrm{He}_0}\right) = \frac{6}{\pi^2}\sum_{n=1}^{\infty}\left(\frac{e^{(-n^2Bt)}}{n^2}\right) = [\mathrm{DS}]. \tag{5}$$

where subscripts "0" and "p" stand for initial and present concentrations, and $B = \pi^2 D/r^2$. Finally, combining (4) and (5) into a form similar to the closed-system equation, we have

$$\left(\frac{^4\mathrm{He}}{^3\mathrm{He}}\right)_p = \left(\frac{^4\mathrm{He}}{^3\mathrm{He}}\right)_0 + \left(\frac{^{238}\mathrm{U}}{^3\mathrm{He}}\right)_0\left(\frac{[\mathrm{PDS,238}] + [\mathrm{PDS,235}] + [\mathrm{PDS,232}]}{[\mathrm{DS}]}\right), \tag{6}$$

where proper substitution is made in [PDS, 235] for ^{235}U (= ^{238}U/137.88) and [PDS, 232] carries ^{232}Th/^{238}U as a leading constant.

The general behavior of this model is illustrated in Figure 5.15, which shows the rapid loss of initial ^{3}He and ^{4}He, the build-up of radiogenic ^{4}He*, and a crossover of $^4He_0$ and ^{4}He* at $t \sim 2.3$ m.y. The total ^{4}He($^4He_0$ + ^{4}He*) becomes constant for $t > 4$ m.y., which represents a state of secular equilibrium between ^{4}He* loss and ^{4}He* production. The decrease of ^{3}He/^{4}He with time, also shown in Figure 5.15, is related to the fact that ^{4}He loss is being balanced by ^{4}He production while ^{3}He is simply lost. For the particular parameters chosen for Figure 5.15, the ^{3}He/^{4}He ratio drops by 10 percent in ~1.5 m.y., and by 50 percent in ~2.3 m.y.

Figure 5.16 shows the present day ^{3}He/^{4}He ratio of a magma chamber as a

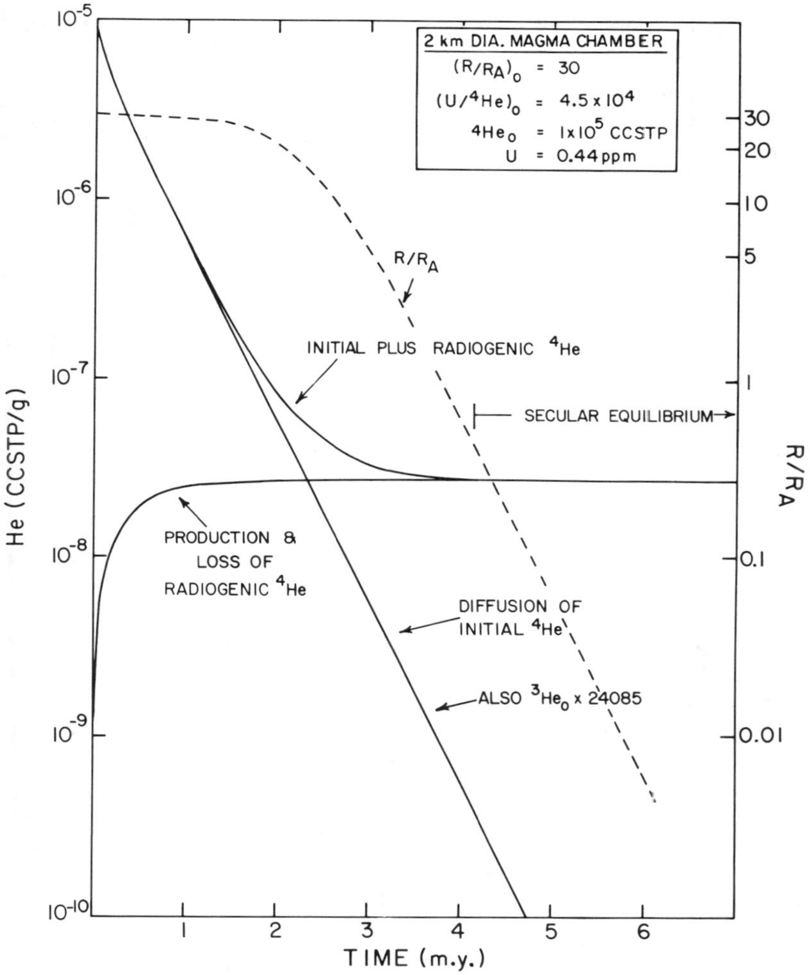

Figure 5.15. Variation of He concentrations and $^3He/^4He$ ratio (log scales) as a function of time, for CDL model. Model parameters are given on the figure; a He diffusion coefficient of 7.5×10^{-5} cm^2/s was used (see text). Curves show the diffusion loss of initial ^3He and ^4He (approximately log-linear), the *in situ* production and concurrent diffusional loss of radiogenic ^4He, and the sum of initial plus radiogenic ^4He. This latter quantity approaches steady state (secular equilibrium) for times >4 m.y. Also shown, as a dashed curve, is the evolution of $^3He/^4He$ ratios (right-side axis) as a function of time. For the chosen parameters, the $^3He/^4He$ ratio decreases very rapidly for times >2 m.y. Governing equations are derived in the text.

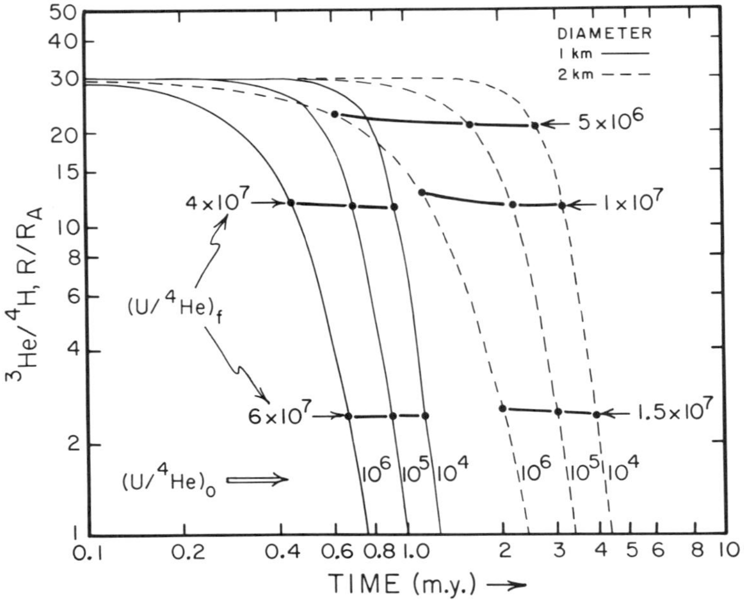

Figure 5.16. Variation of $^3He/^4He$ with time, for CDL model. Curves are shown for both 1 Km and 2 Km spherical magma chambers, using a He diffusion coefficient of 7.5×10^{-5} cm^2/sec, and Th/U = 3.5 (weight ratio). As an analogy for the Loihi seamount He system, we assumed a mantle melt with initial $^3He/^4He$ of 30× atmospheric. Numbers adjacent to lower parts of curves are various initial $U/^4He$ ratios, expressed as (ppm/std cc). Short semi-horizontal cross lines with numbers are the $U/^4He$ ratios at various times. Note that these ratios vary only slightly over large ranges of $^3He/^4He$, and, for significant perturbations of the $^3He/^4He$ ratio, are in the range 1×10^7–6×10^7. These values may be compared to ratios actually measured in various erupted basalts, shown in Figures 5.13 and 5.14.

function of magma chamber radius, time, and initial and final $U/^4He$ ratios. Note that while lower $(U/^4He)_0$ ratios require longer times to reach a given R/Ra, this is always reached at approximately the same value of $(U/^4He)_p$, for a given size magma chamber. Each curve is marked by a "knee," beyond which R/Ra decreases almost catastrophically with time. This "knee" occurs at shorter and shorter times as the size of the magma chamber decreases, and as initial $U/^4He$ increases.

How relevant is this model to real world processes? With regard to *scale*, the plumbing system under Hawaii has been delineated by Ryan *et al*. (1981), and while considerably more complex than a simple spherical body, the various conduits and chambers are all characterized by scale lengths of 1–3 km. The *duration* over which a given volcano is active has been best studied in the Hawaiian chain, where the duration of the shield-building stage may be as long as 1 m.y., and the total duration, including post-erosional eruption phase, is commonly a few million years (e.g., Shaw *et al.*, 1980). Hawaii is, to some extent a limiting case, since the duration of a given volcano is circumscribed by the surface propagation of the volcano away from the hot-spot source. Volcanos which are situated on ridges (e.g., Iceland), or on slow-moving plates, may persist for longer times.

It is clear, however, that a volcanic magma chamber cannot be a closed system for the whole lifetime of the volcano. Based on eruption rates of various volcanos (Crisp, 1984) which generally fall in the range of 10^{-4} to 10^{-1} km^3/year, it is obvious that magma residence times (defined as the time required to empty a magma chamber) will be on the order of 10^2–10^5 years (assuming the magma chamber is no larger than 4 km). While this consideration would seem to preclude any direct relevance of the above CDL model, it must also be noted that differentiated rocks are common in hot-spot volcanos, and thus a holding-tank for differentiation must exist. Furthermore, a given filling of the magma chamber must have a mean lifetime which is long enough for cooling and crystallization to occur.

Ultimately then, the evolutionary history of magma chambers depends on thermal budget considerations. By noting that thermal diffusivities are ~100 times higher than He diffusivities (at 1200°C), it is clear that it may be difficult to keep a magma chamber molten long enough for He diffusion to cause significant reduction in $^3He/^4He$. For example, a 2 km static magma chamber will cool below its solidus in ~20,000 years (For ΔH fusion = 150 cal/g, thermal diffusivity = 0.006 cm^2/sec and ΔT (magma-wall rock), = 150°C). The only conceivable way of avoiding such a short lifetime is to intrude, or refill, a magma chamber which has been in prior existence long enough to substantially pre-heat the environment so that ΔT between wall rock and magma approaches zero.

Figure 5.17 is a cartoon of magma chamber cooling with periodic replenishment, patterned after the thermal model results of Usselman and Hodge (1978). The time axis is arbitrary and depends on magma chamber size and eruption rates. During the preheating stage, radiogenic ^4He will accumulate slowly, ^3He will be maintained at about its initial level by the refillings, and $^3He/^4He$ will decrease only imperceptibly. The wall rock contact temperature will initially increase rapidly to ~1220°C and hover there; if the preheating stage lasts > 1 m.y., a steady state thermal gradient

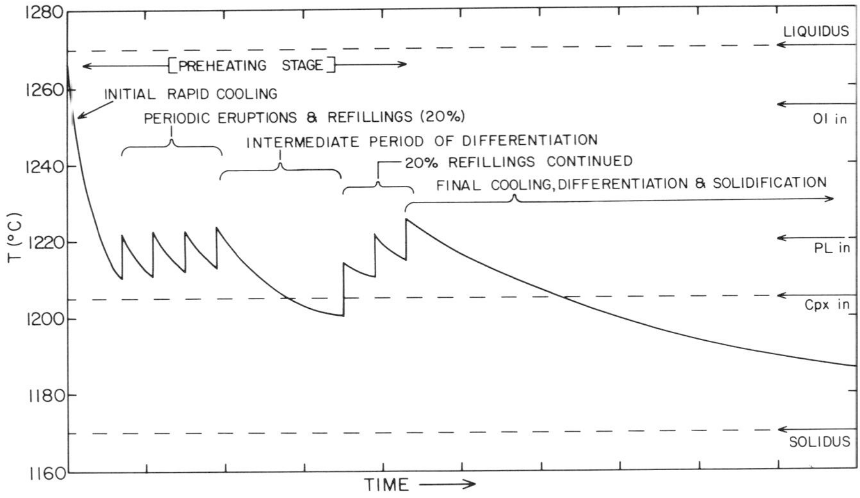

Figure 5.17. Cartoon of magma chamber cooling with periodic replenishment, patterned after the thermal model results of Usselman and Hodge (1978). Time axis is arbitrary and depends on magma chamber size and eruption rate. During the preheating stage, radiogenic ^4He will accumulate slowly, ^3He will be maintained at about its initial level by the refillings, and $^3He/^4He$ will decrease only imperceptibly. Wall-rock contact temperature will initially increase rapidly to ~1200C and hover there; if the preheating stage lasts > 1 m.y., a steady-state thermal gradient will be established between the surface of the magma chamber and the earth's surface temperature. During final cooling and differentiation, diffusion loss of $^3He/^4He$ will become significant, and a steady state will be established between ^4He loss and radiogenic ^4He production; this ultimately will lead to a catastrophic collapse of the $^3He/^4He$ ratio in the magma chamber and erupted magmas.

will be established between the surface of the magma chamber and the Earth's surface temperature. During final cooling and differentiation, diffusional loss of He will become significant, and a steady state will be established between ^4He loss and radiogenic ^4He production (see Fig. 5.15); this ultimately will lead to a catastrophic collapse of the $^3He/^4He$ ratio in the magma chamber and erupted magmas.

Figure 5.18 gives solidification times in millions of years as a function of ΔT for spherical magma chambers of various diameters (0.1–20 km) which have been refilled for the last time. Curves represent the time required by the CDL model to decrease $^3He/^4He$ by 10 percent as a function of various initial $U/^4He$ ratios (10^3–3×10^7). Clearly, the effect on $^3He/^4He$ is a sensitive

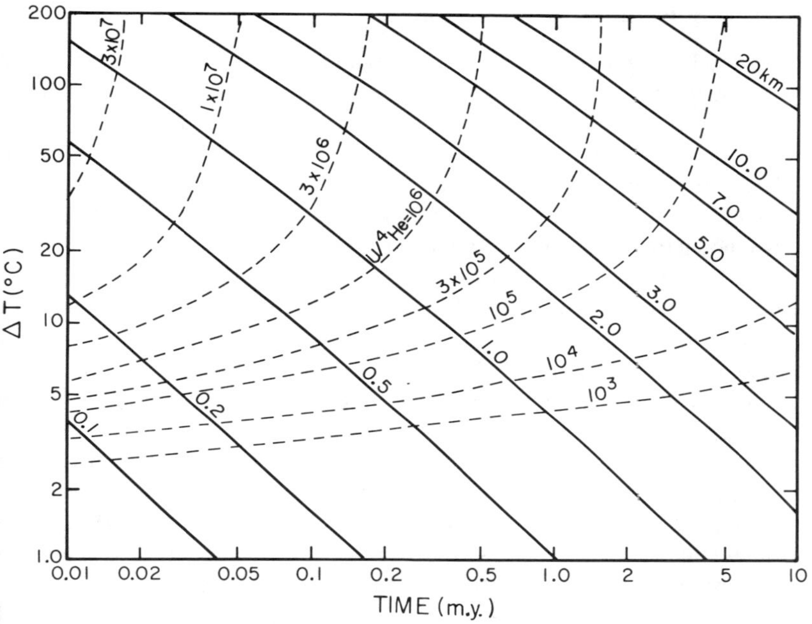

Figure 5.18. Solidification time, in millions of years, versus temperature contrast (ΔT°C) between a refilled spherical magma chamber and uniform wall rock temperature subsequent to refilling for the last time. Magma chambers of various diameters (in km) are shown as solid diagonal lines. Curves represent the time required by the CDL model to decrease the $^3He/^4He$ ratio by 10 percent, as a function of various initial $U/^4He$ ratios (ppm/std cc) ranging from 10^3 to 3×10^7, magma chamber diameters, and values of ΔT. Cooling models calculated for conductivity of 2.5×10^{-3} cal sec^{-1} cm^{-1} deg^{-1}, specific heat of 0.36 cal gm^{-1} °C^{-1} (basalt at 1200°C), density of 2.9 g/cm^3 (Birch *et al.*, 1942); latent heat taken as 110 cal/gram (Yoder, 1976). The non-convecting fixed melting point formulation of Flemings (1974) was used for the thermal model; the fact that basalt has a melting interval of ~100°C was accounted for to first order by using an "effective" latent head of 146 cal/g.

function of the initial $U/^4He$ ratio. For magmas which start at mantle values ($\leq 10^3$), changes in $^3He/^4He$ will likely not exceed 10 percent, even with a magma chamber life of several million years and $\Delta T = 10$°C. However, for magmas with high initial $U/^4He$ (perhaps resulting from degassing at depth, as discussed in the next section) changes in $^3He/^4He$ due to evolution in such a system may be dramatic.

4.3. The Exhalative Loss Model

This model assumes that He is lost from a magma by partitioning into, and subsequent loss of, a CO_2-rich gas phase, once CO_2 solubility is exceeded at a given pressure. Experimental work has shown that basaltic magmas may dissolve large quantities of CO_2 at depth. Mysen *et al.* (1975) determined CO_2 solubilities in andesite, tholeiite, and olivine nephelinite of about 2.0, 3.3 and 7.0 wt.%, respectively, at 30 kb and 1625°C. The solubilities were also shown to increase as a function of increasing pressure, temperature, and molar $CO_2/CO_2 + H_2O$ up to a value of about 0.6. Brey and Green (1977) showed that undersaturated olivine melilitite can be produced by melting of pyrolite at 27 kb and 1160°C with 7–8 wt.% H_2O and 6–7 wt.% CO_2 dissolved in the melt (under which conditions, the CO_2 saturation limit of the melt is not exceeded). In contrast, average CO_2 in MORB basalts is less than 400 ppm (Fine and Stolper, 1985/86). Greenland *et al.* (1985) have recently constructed a mass balance model for several volatile species in Kilauea magmas. They estimate that the magmas arrive at the summit magma chamber (2–5 km depth) laden with 0.32 wt.% of both CO_2 and H_2O. More than 90 percent of this CO_2 is lost by pre-eruptive degassing, ~6 percent is lost during eruption, and less than ~3 percent is retained in the magma (in contrast, H_2O solubility is not exceeded within the magma chamber and most of the H_2O is outgassed at eruption).

These observations suggest that pre-eruptive degassing of CO_2 from magmas is not only important, but dominates the CO_2 flux from the mantle to the atmosphere, with eruptive degassing accounting for a relatively minor component of this flux. During the exsolution of a CO_2-rich gas phase from a magma, He is expected to behave as an infinitely dilute trace element, and partition according to Henry's law between the silicate melt and gas or fluid phases. The Henry's law constant (K), for solubility of He in basaltic melt, has been determined experimentally by Jambon *et al.* (1986) to be $\sim 5 \times 10^{-4}$ std cc/(g atm) at 1300°C. This is in good agreement with the value determined by Kurz and Jenkins (1981) by measuring the equilibrium partitioning of He between basaltic melt and CO_2-rich gas in MORB glass vesicles. For modelling, it is more convenient to work with a distribution coefficient, K_D, in units of $(\text{moles/g})_m/(\text{moles/g})_g$, where the subscripts m and g stand for magma and gas respectively. If we assume an ideal vapor phase that is essentially pure CO_2, K_D and K are related as follows:

$$K_D = 1.989 \times 10^{-3}\ KP$$

(where P is the total pressure in bars). The importance of this relationship is that the rate of outgassing of He, for a given amount of CO_2 loss, is a sensitive function of the depth at which the outgassing occurs, particularly at lower pressures.

In order to predict the change in U/^{4}He in a system which has outgassed CO_2 prior to eruption, we need to know the amount of CO_2 which has been lost and the depth of outgassing. Figure 5.19 shows the relationship between the change in the U/^{4}He ratio and F, the wt.% of CO_2 which has been outgassed at various pressures. Unfortunately, the CO_2 budget of the mantle is ill-constrained so that primary magma CO_2 contents are difficult to

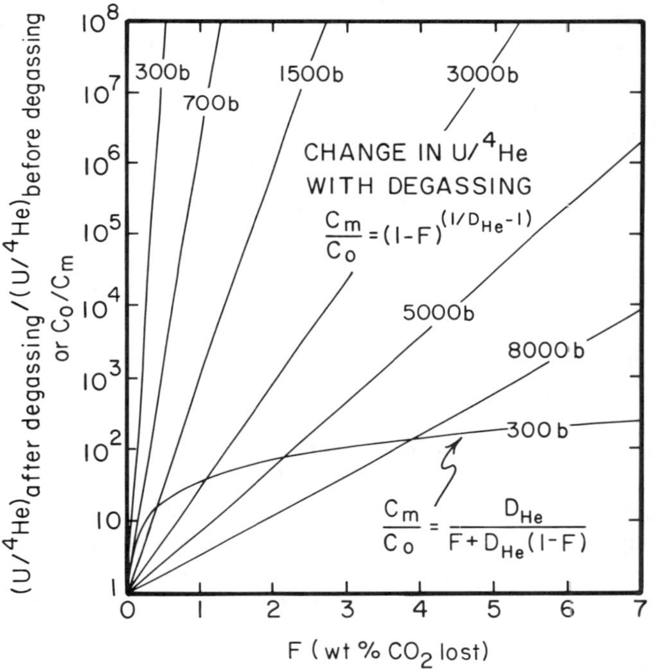

Figure 5.19. Changes in U/^{4}He are shown as a function of F, the wt.% CO_2 gas released from a magma. Curves for 300 b (bars), 700 b, and 1500 b, 3000 b, 5000 b and 8000 b are shown for a Rayleigh degassing model; 300 b curve for an equilibrium degassing model is also shown for reference ($C_m/C_o = D/[F + D(1 - F)]$). In these equations, the subscript m refers to the final magma and o refers to the magma prior to degassing. The Raleigh model essentially assumes that CO_2 is rapidly removed from the system as it exsolves; that is, it either escapes or is segregated at the top of a body of magma. Note that, for a given F or wt.% CO_2 released, He is lost much more efficiently at lower pressures.

estimate. We can, nevertheless, compare increases in $U/^4He$ in magmas relative to the mantle (see Fig. 5.13) with predicted CO_2 loss (Fig. 5.19) and measured CO_2 contents in magmas. MORBs, for example, have $U/^4He$ ratios on the order of 6×10^3, or about a factor of 10 higher than a model mantle source. If we assume that this outgassing occurs in the upper kilometer of the oceanic crust ($P \sim 300$ to 700 bars), we would predict the loss of something between 700 and 1600 ppm CO_2, compared to <400 ppm CO_2 retained in the erupted lava (Fine and Stolper, 1985/86). Basalts with the highest U/He ratios approach a factor of 10^6 enrichment over single stage mantle values (see Fig. 5.13); such enrichments may be produced by CO_2 losses of between 0.4 and 6.7 wt.% occurring at pressures between 300 bars and 5 kbars.

The potential for producing changes in $^3He/^4He$ in the context of the EL model is greater than for the CDL model because the He loss may occur over much shorter time spans and is not strongly coupled to heat loss. That is, in the CDL model, He diffusion from the magma chamber occurs as the magma cools and effective evolution times at high values of U/He will be limited. In the EL model, however, significant outgassing of a high-level magma may occur with very little associated cooling, allowing more time for evolution of the high U/He system. For example, with the CDL model and magma chamber diameter of 1 km, a magma with $(U/^4He)_0 = 5 \times 10^3$ and $(U/^4He)_f = 4.2 \times 10^6$, reduction of $^3He/^4He$ over 7×10^6 years is on the order of 7 percent (with initial $^3He/^4He$ of 30 Ra; see Fig. 5.16). By comparison, magma which attains a $U/^4He$ ratio of 4.2×10^6 by degassing will suffer more than 50 percent reduction of $^3He/^4He$ over 7×10^5 years, or a 6.7 percent reduction in less than 10^5 years (see Fig. 5.13).

The arguments presented here (and in Zindler and Hart, 1986) are necessarily qualitative, due to the lack of constraints on magma residence times, CO_2 budgets, and depths for degassing. We cannot, therefore, make statements regarding specific changes in $^3He/^4He$ ratios measured at individual localities. We can however, draw several general conclusions which may be used as guidelines for the interpretation of $^3He/^4He$ data until such time as a more quantitative treatment is feasible:

1. Low $U/^4He$ ratios in MORBs preclude any significant pre-eruptive reduction of $^3He/^4He$ ratios.
2. Low $U/^4He$ ratios in some OIBs ($U/^4He < 10^5$) are subject to a maximum of a 10 percent reduction of $^3He/^4He$ over time scales up to 5×10^6 years.
3. OIBs with $U/^4He$ between 10^6 and $\sim 5 \times 10^8$ may be subject to significant reduction of $^3He/^4He$ ($\geqslant$10 percent) with pre-eruptive aging over time periods ranging from 1×10^3 to 5×10^5 years.

4. The potential for pre-eruptive $^{3}He/^{4}He$ reduction may be considered to increase with decreasing magma throughput in the system, and with the degree of differentiation of the host magma.

5. Isotope Systematics in Oceanic Basalts

5.1. Background

In recent years, isotope geochemists have devoted considerable attention to investigating the relationships between various isotope systems in oceanic basalts, and their implications for mantle chemical structure. There are now seven decay schemes for which a significant body of analytical data exist for oceanic basalts: ^{87}Rb-^{87}Sr; ^{147}Sm-^{143}Nd; ^{238}U-^{206}Pb; ^{235}U-^{207}Pb; ^{232}Th-^{208}Pb; ^{176}Lu-^{176}Hf; and $(^{238}U+^{235}U+^{232}Th)$-$^{4}He$. Over the past decade, our understanding has evolved from thinking in terms of linear correlations and two-component mixing for two isotope systems (e.g., the Sr-Nd isotope correlation; DePaolo and Wasserburg, 1976a; O'Nions *et al.*, 1977; Allegre *et al.*, 1979), to thinking in terms of at least three components, which may define an approximate planar or two dimensional mixing array in three, four, or five-dimensional space (e.g., $^{87}Sr/^{86}Sr$, $^{143}Nd/^{144}Nd$, $^{206}Pb/^{204}Pb$, $^{207}Pb/^{204}Pb$, $^{208}Pb/^{204}Pb$; Zindler *et al.*, 1982). White (1985) and Allegre and Turcotte (1985) have recently proposed that there are five types of sources from which all oceanic basalts may be produced by variable mixing relationships.

The present status of oceanic isotopic data is summarized in a series of two-dimensional variation diagrams (Figs. 5.20–5.26). Hafnium isotopic data are not represented here because, in general, the overall high degree of correlation between Hf and Nd isotope ratios permits accurate inference of the behavior of $^{176}Hf/^{177}Hf$ based on $^{143}Nd/^{144}Nd$. Brief consideration of the data arrays in Figures 5.20–5.26 clearly demonstrates the inadequacy of simple two-component mixing models to explain isotopic variations in oceanic basalts (as originally pointed out by Sun and Hanson, 1975, based on Sr-Pb systematics). Even the Sr-Nd variation diagram (Fig. 5.20), until recently thought to represent a linear array (the "mantle array" of, for example, DePaolo and Wasserburg, 1977), requires a substantially more complex interpretation. This is an important, first-order conclusion based on all currently available data and is not subject to interpretation. Thus, we must consider that any model for the evolution of the mantle-crust system that presumes only two geochemically distinct components within the mantle is necessarily inaccurate.

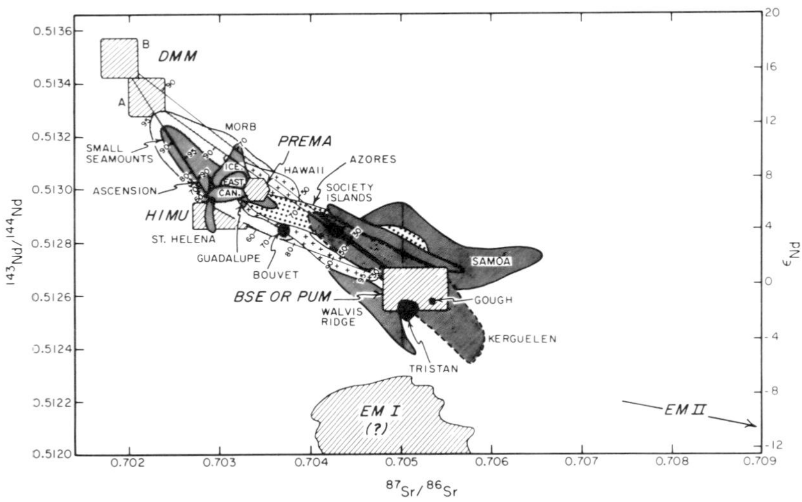

Figure 5.20. $^{143}Nd/^{144}Nd$ versus $^{87}Sr/^{86}Sr$ variation diagram for oceanic basalt suites (data sources may be found in Zindler *et al.*, 1982; Hart, 1984b, and White, 1985). Proposed mantle components are shown as cross-hatched boxes and blobs labelled DMM (a and b), HIMU, PREMA, BSE or PUM, EM I and EM II. Mixing lines between DMM (a and b), HIMU, and BSE or PUM, with tick marks indicating percentages of one end member, are constructed using the following source compositions for Sr and Nd, respectively, (in ppm): DMM (1)—6, 0.33; DMM (2)—12, 0.65; HIMU—120, 6.5; BSE—18.4, 1.0. The HIMU composition assumes a magmatic origin for this component; had we assumed it to be a high μ peridotite, with lower concentrations of Sr and Nd, but a similar Sr/Nd ratio, the shapes of the mixing curves would not change, only the positions of the tick marks (see discussion in Section 5.2.4). The two values given for DMM delimit the possible concentration range for this component, and result in two sets of tick marks on each mixing line which goes to DMM A or DMM B. Note that none of the mixing lines are highly curved, because of the similarity in Sr/Nd ratios in all components.

Until recently many geochemists felt that the apparent differences in the relative positions of various localities in the Sr-Nd and Pb-Pb variation diagrams (Figs. 5.20 and 5.23) documented a decoupling of the U-Th-Pb system from Rb-Sr and Sm-Nd. Zindler *et al.* (1982) proposed that there was a coherence in these variations which became apparent when they were viewed as different two-dimensional projections of a three-dimensional surface which is approximately planar. Delineation of this 3-D mantle plane is shown in Figure 5.27, where every data point represents the averaged

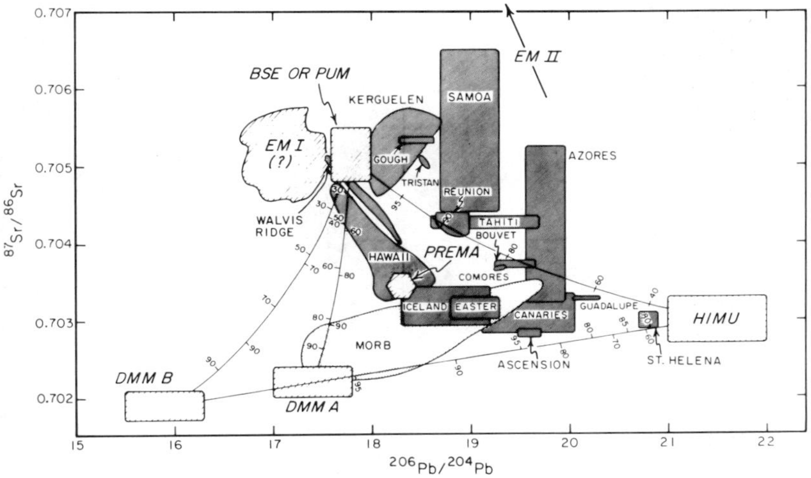

Figure 5.21. $^{87}Sr/^{86}Sr$ versus $^{206}Pb/^{204}Pb$ variation diagram for oceanic basalt suites. Data sources, components, and mixing lines are as in Figure 5.20. Pb concentrations are (in ppm): DMM (1), 0.02; DMM (2), 0.04; HIMU, 0.4; BSE, 0.135. Islands lying outside the DMM-HIMU-BSE mixing triangle document the existence of EM II, and are geographically situated within the Dupal anomaly. The Koolau Series of Oahu (Stille *et al.*, 1983) and the Walvis Ridge (Richardson *et al.*, 1982) document the existence of EM I.

isotopic values for an island or ridge system. It should be noted that this plane is defined by averaged data, and that many localities show intra-island isotopic variations which do not lie in the plane (e.g., Stille *et al.*, 1983; Staudigel *et al.*, 1984). There are also several localities for which data has recently been reported which, even when averaged, do not lie on the plane (e.g., Koolau series, Oahu, Stille *et al.*, 1983; Walvis Ridge, Richardson *et al.*, 1982; Samoa, White and Hofmann, 1982, and White, 1985). While for many localities, the mantle plane remains a useful representation, an accumulating body of evidence suggests that it, too, is an over simplification.

The mantle plane model is based on the premise the Sr, Nd and Pb isotopic variations in oceanic basalts may be explained as being due to mixing between three chemically distinct and independent components. These were presumed to be the MORB mantle, BSE, and a third component, similar to St. Helena in composition. As is seen clearly in Figure 5.20, data from the Walvis Ridge, the Society Islands, and Samoa suggest the existence of at least two distinct "enriched" components which plot at low values of

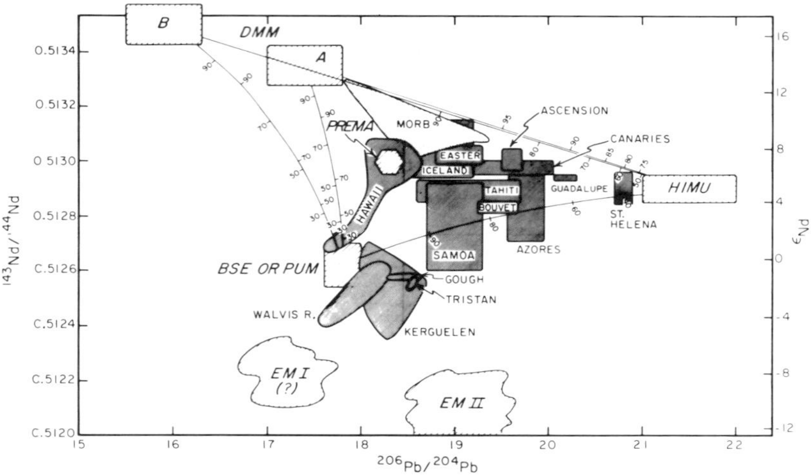

Figure 5.22. $^{144}Nd/^{144}Nd$ versus $^{206}Pb/^{204}Pb$ variation diagram for oceanic basalt suites. Data sources, components, mixing lines and other considerations are as in Figures 5.20 and 5.21.

$^{143}Nd/^{144}Nd$ in this diagram. Thus, it would seem that four or more distinct chemistries are required to explain the first-order isotopic variation among oceanic basalts as pointed out by White (1985). Furthermore, the significance of the proverbial "bulk earth" must come into question as it would now appear that, at least in terms of the heavy radiogenic isotopes, it has become somewhat superfluous (or perhaps non-existent) as an end-member composition in the mantle.

The interrelationships displayed by the various isotope ratios in Figures 5.20–5.26 delineate a number of important constraints regarding the definition of isotopically distinct mantle components:

1. The "depleted" isotopic character of N-type MORB (high $^{143}Nd/^{144}Nd$, low $^{87}Sr/^{86}Sr$ and $^{206}Pb/^{204}Pb$) requires the existence of a depleted end-member MORB-mantle component (DMM) which is clearly identifiable along ridge segments well removed from hot-spot volcanic activity.
2. The very high $^{206}Pb/^{204}Pb$ ratios observed at St. Helena and Tubuai (Fig. 5.23), coupled with low $^{87}Sr/^{86}Sr$ and intermediate $^{143}Nd/^{144}Nd$, suggest a mantle component which is markedly enriched in U and Th relative to Pb, without an associated increase in Rb/Sr; Zindler *et al.* (1982) suggested that this component, which we will call "HIMU," might represent

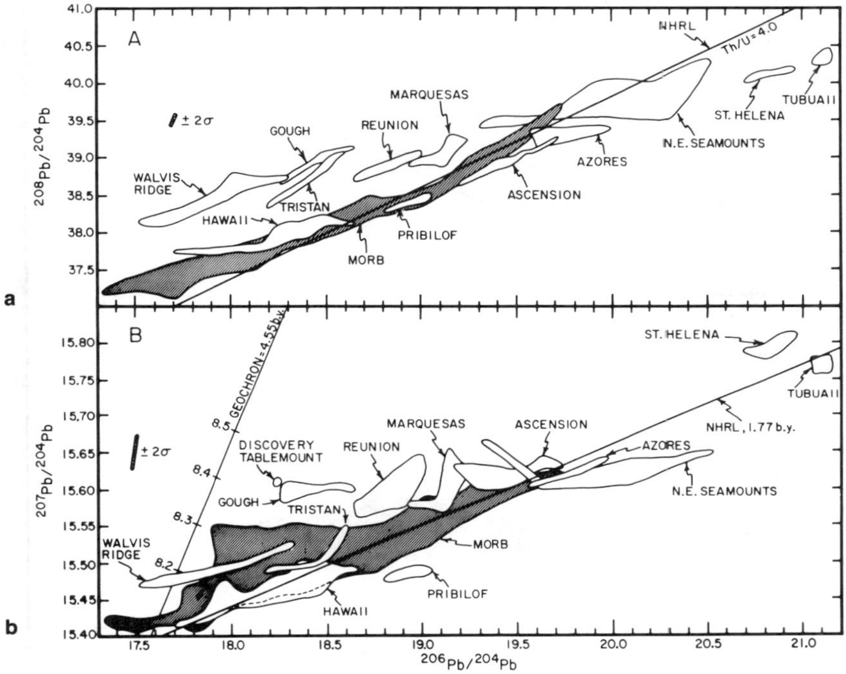

Figure 5.23. (a–d) 208/204 Pb–207/204 Pb–206/204 Pb isotope variation diagrams for a variety of oceanic basalts. A 4.55 b.y. geochron is shown, with single stage μ values (tick marks), along with the Northern Hemisphere reference line (NHRL) of Hart (1984b). This NHRL, when interpreted as a secondary isochron, has a slope of 1.77 b.y., and a second stage Th/U (wt. ratio) of 4.0. Islands lying significantly above the NHRL are part of the Dupal anomaly belt. Data sources may be found in Hart (1984b). Typical 2σ analytical precision is also shown (± 0.05 percent/amu). (e) In this panel, DMM, HIMU, BSE, EM I, and EM II component compositions are indicated on a $^{207}Pb/^{204}Pb$ versus $^{206}Pb/^{204}Pb$ diagram, with MORB and selected island fields from panels b and d shown for reference. Parameters and considerations for mixing lines are as in Figures 5.20 and 5.21.

recycled oceanic crust which has been stripped of alteration phases during subduction. The general appearance of mixing toward St. Helena in the Pb-Pb diagram (Fig. 5.23) suggests an "age" for the U/Pb enrichment in this component of ~1.5–2.0 b.y.

3. Trends for the Walvis Ridge, Kerguelen, and Samoa in Figure 5.21, which extend beyond assumed BSE values for $^{143}Nd/^{144}Nd$ and/or $^{87}Sr/^{86}Sr$,

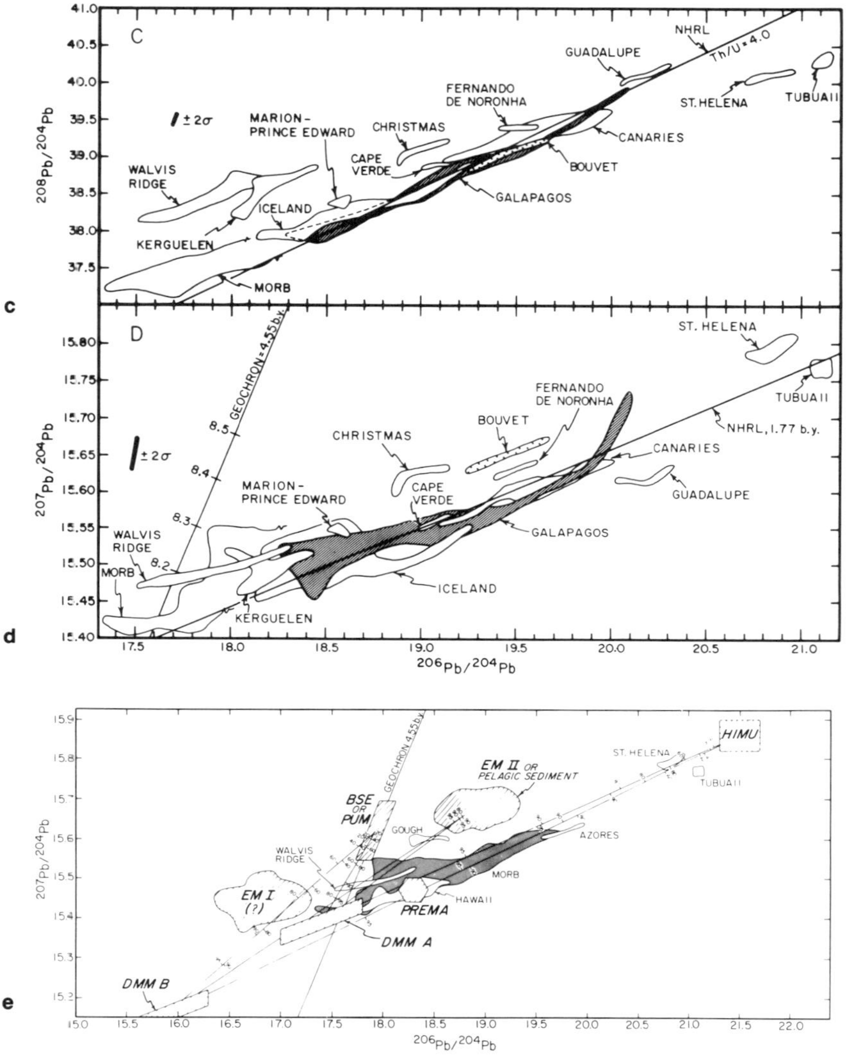

Figure 5.23. *Continued*

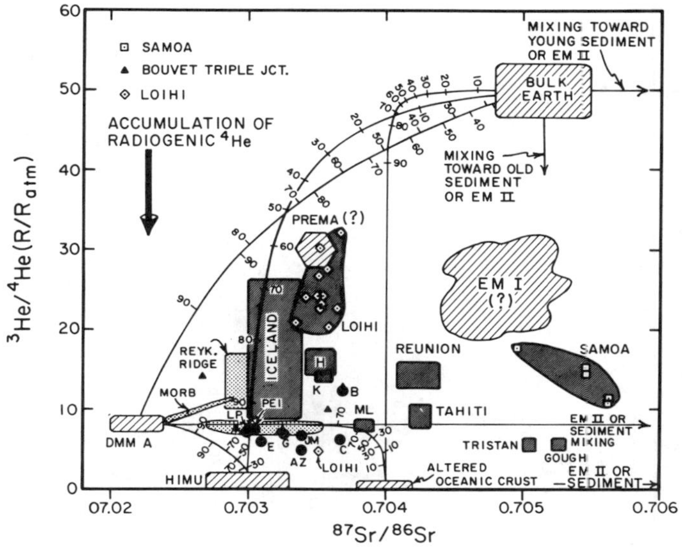

Figure 5.24. $^3He/^4He$ versus $^{87}Sr/^{86}Sr$ variation diagram of oceanic basalt suites. DMM, HIMU, BSE, PREMA, EM I and EM II components are shown as cross-hatched fields. Mixing lines and considerations are as in Figure 5.20; He concentrations for the various components are (std cc/g): DMM (1), 6.5×10^{-6}; DMM (2), 1.3×10^{-5}; HIMU, 2.7×10^{-5}; EM II (for "young" sediment or EM II case), 5.3×10^{-6}. Sr concentrations are given in Figure 5.20. Mixing lines are also shown to an altered crust component presumed to have a He concentration of 1.5×10^{-6} std cc/g. In general, we have assumed fairly low He concentrations in those components presumed to have low $^3He/^4He$ ratios. However, this may be highly variable as it is strongly dependent on the age of these components, due to the rapid growth of ^4He in materials with high $U/^4He$ ratios. This situation is qualitatively depicted by contrasting mixing trajectories shown for "young" and "old" EM II or sediment mixing. Due to the potential for pre- or post-eruptive accumulation of radiogenic ^4He (as indicated by the black arrow), $^3He/^4He$ ratios measured at ocean islands, particularly those in the waning stages of their volcanic cycles, must be considered as minimum values for mantle sources. The large number of components, together with the potential for radiogenic ^4He accumulation, precludes the unique definition of mixing scenarios for individual localities. Localities given by abbreviation are: LP, LaPalma, Canary Islands; PEI, Prince Edward Island; E, Easter Island; G, Galapagos; AZ, Azores; JM, Jan Mayen; B, Bouvet; C, Comores; H, Hualalai, Hawaii; K, Kilauea, Hawaii; ML, Mauna Loa, Hawaii. Data sources for $^3He/^4He$ shown in this diagram (as well as Figs. 5.25 and 5.26) are as follows: Poreda *et al.* (1980); Rison (1981); Kurz and Jenkins (1981); Rison and Craig (1982); Craig and Rison (1982); Kurz (1982); Kurz *et al.* (1982; 1983; 1985); Allegre *et al.* (1983a); Condomines *et al.* (1983).

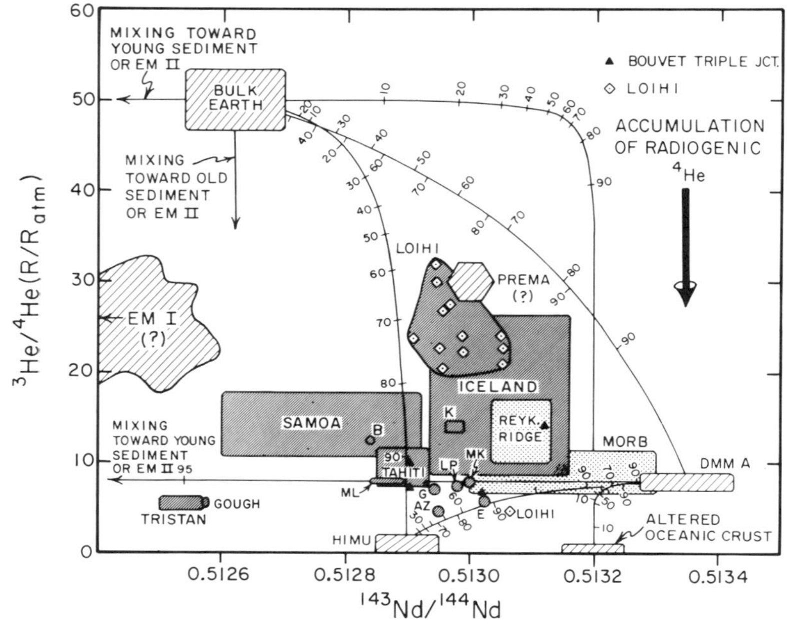

Figure 5.25. $^3He/^4He$ versus $^{143}Nd/^{144}Nd$ variation diagram for oceanic basalt suites. Components, considerations, mixing lines, locality abbreviations, and data sources are as in Figure 5.24. He and Nd concentrations for mixing lines are given in captions to Figures 5.24 and 5.20, respectively.

require the existence of at least two "enriched" mantle components (EM). These EMs may form by recycling of sediment, continental crust, or continental lithosphere into the mantle, or by intramantle metasomatism. Note that most of the islands requiring these EM components are those which delineate the Southern hemisphere Dupal anomaly (see Fig. 5.10). The relatively high $^{207}Pb/^{204}Pb$ values which characterize at least the high— $^{87}Sr/^{86}Sr$ EM (EM II) suggests that this component has an ancient lineage. The low $^{87}Sr/^{86}Sr$ EM (EM I) apparently contains B-type Pb (i.e., Pb with an isotopic composition which lies to the left of the geochron), as evidenced by Walvis ridge in Figure 5.23, and thus has an isotopic signature reminiscent of lower continental crust;

4. High $^3He/^4He$ ratios which are found at Iceland and Hawaii demand the involvement of relatively undegassed mantle, or a mantle segment that was enriched in volatiles during the Archean (see Fig. 5.13). Although

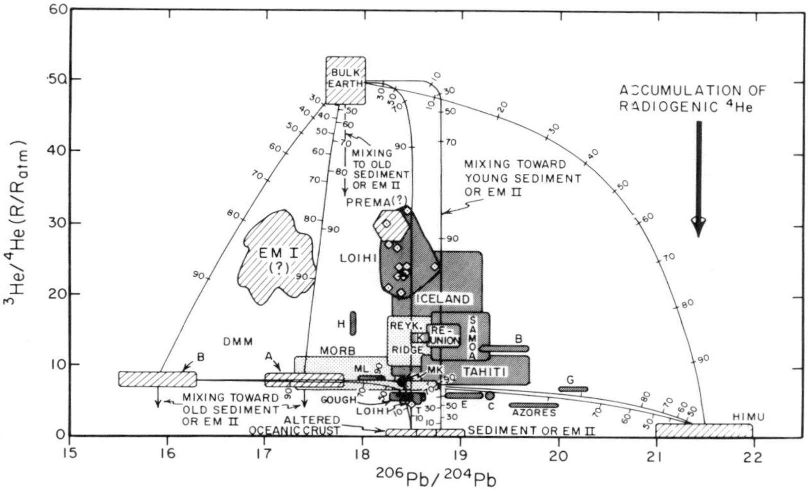

Figure 5.26. ^{3}He/^{4}He versus ^{206}Pb/^{204}Pb variation diagram for oceanic basalt suites. Components, considerations, mixing lines, locality abbreviations, and data sources are as in Figure 5.24. He and Pb concentrations for mixing lines are given in captions to Figures 5.24 and 5.23, respectively.

this component is often described as "primitive" or undifferentiated mantle, the Sr, Nd, and Pb character of Iceland and Hawaii (Figs. 5.24–5.26) are not consistent with such an interpretation; and

5. The high frequency of Nd and Sr isotopic compositions at about 0.5130 and 0.7033, respectively (Figs. 5.20 and 5.28), suggests the existence of a mantle component with this isotopic character (at least when scale lengths for melting volumes are considered), rather than calling on a constancy of mixing proportions between spatially distant components which are far removed from these isotopic values. It is notable that intraoceanic island arcs and the depleted end members of many continental basalt provinces (Fig. 5.29) also fall at this prevalent mantle composition (PREMA). The coincidence of PREMA with Iceland and Hawaii and, therefore, high ^{3}He/^{4}He ratios, suggests that at least portions of this reservoir are relatively undegassed.

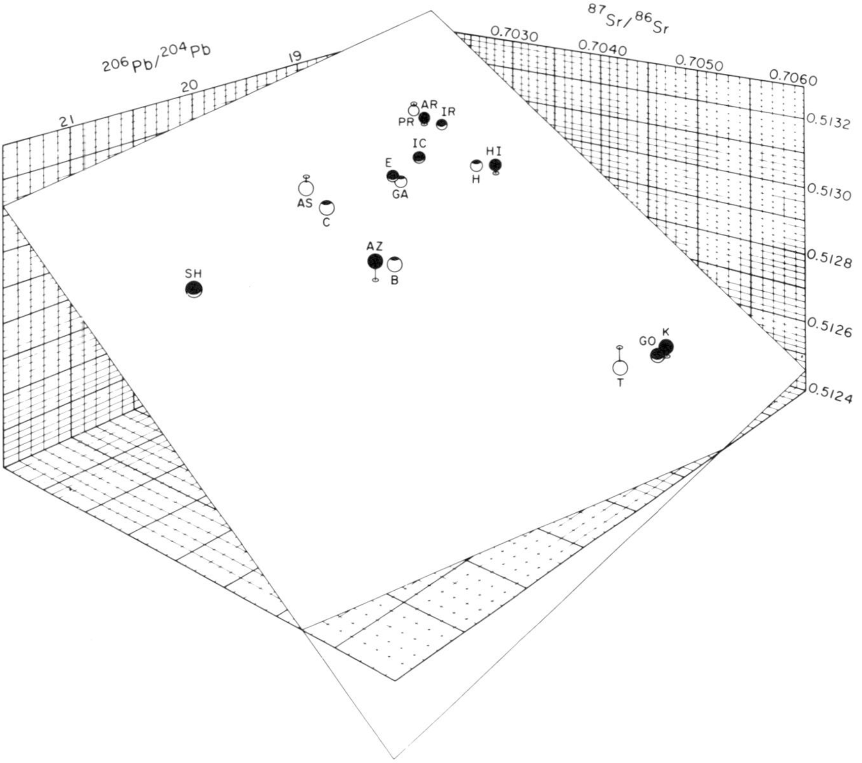

Figure 5.27. Three dimensional plot of average $^{206}Pb/^{204}Pb$, $^{143}Nd/^{144}Nd$, and $^{87}Sr/^{86}Sr$ for basalts from oceanic ridges and islands. The best-fit plane is shown. Closed and open symbols, whose size approximates the uncertainty of a single analysis, lie above and below the "mantle plane" respectively. Lines connecting data points to the plane are parallel to the $^{143}Nd/^{144}Nd$ axis and indicate those points which do not intersect the plane. Localities are: AR, Mid-Atlantic Ridge; PR, East Pacific Rise; IR, Indian Ocean ridges; C, Canary Islands, AZ, Azores; AS, Ascension, SH, St. Helena; GO, Gough; T, Tristan da Cunha; B, Bouvet; EA, Easter Island; HI, Hawaiian Islands; H, Island of Hawaii, GA, Galapagos; K, Kerguelen; IC, Iceland. Data sources are given in Zindler *et al.* (1982).

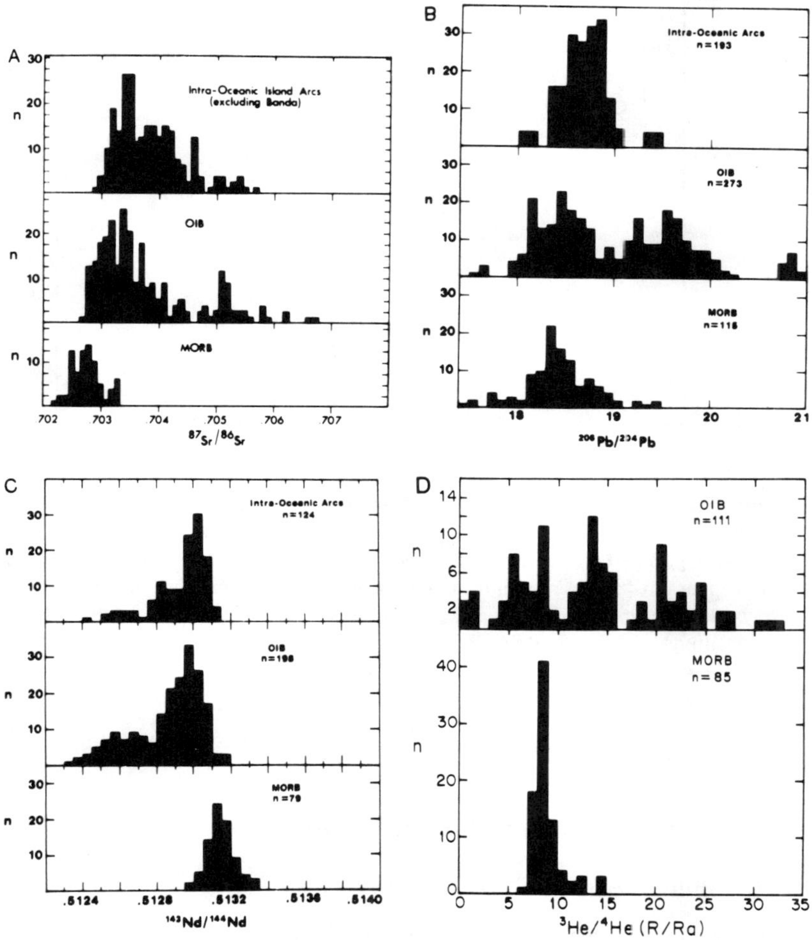

Figure 5.28. Histograms of $^{87}Sr/^{86}Sr$, $^{143}Nd/^{144}Nd$, $^{206}Pb/^{204}Pb$, and $^{3}He/^{4}He$ in MORB, oceanic island basalts and intra-oceanic arcs (adapted from Morris and Hart, 1983). Data sources for the He data in panel d are those given in Figure 5.24 plus Ozima and Zashu (1983).

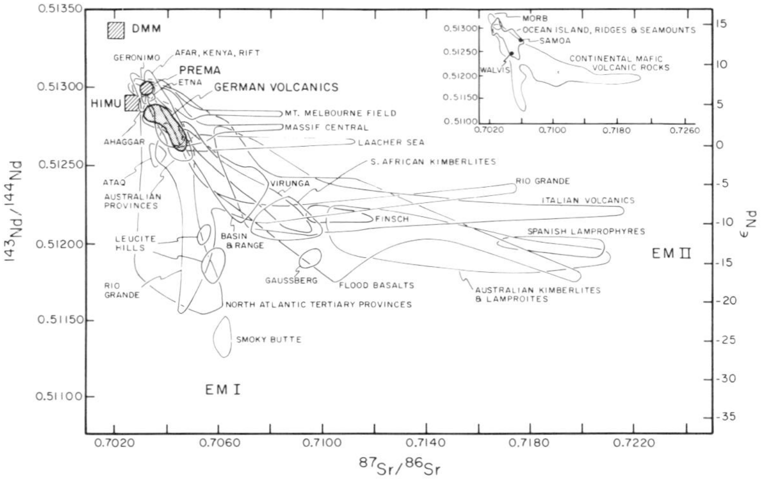

Figure 5.29. Sr and Nd isotopic variations for continental volcanic rocks (from Worner *et al.* [1986]), showing fields for individual provinces (except that all flood basalts are grouped into a single field). Note that the fields converge toward PREMA in the upper left-hand corner of the diagram. Individual "branches" of the combined arrays diverge toward EM I-type compositions at low $^{87}Sr/^{86}Sr$, and EM II-type compositions at high $^{87}Sr/^{86}Sr$. These mixing trends are similar to those observed in oceanic basalts (see inset), except that the continental basalts extend to much more extreme values.

5.2. Characterization of Mantle Components

5.2.1. Bulk Earth

The question of whether or not there exists some primitive, undifferentiated segment of the Earth's mantle is central to the development of a viable and comprehensive model for the geochemical evolution of the Earth. The concept of such a primitive mantle segment has become quite common in the modern literature although the terminology used to describe it has been varied. Essentially, the idea is that subsequent to core formation and associated degassing, the silicate earth achieved a fairly homogeneous state due to high heat flow and rapid convection. Those elements that do not enter the core or the atmosphere will then be represented in their initial, or primordial, abundance ratios in the silicate portion of the Earth. The elusive, primitive mantle segment (BSE or PUM) is then one which is supposed to have survived from this time as a closed system.

Documentation of the existence of such material today has been sought via two separate lines of reasoning: 1) that some oceanic and continental basalts are observed to have Nd isotope ratios similar to chondrites, and the presumed BSE (e.g., DePaolo and Wasserburg, 1976a,b; O'Nions *et al.*, 1977; Wasserburg and DePaolo, 1979; Jacobsen and Wasserburg, 1979; DePaolo, 1981); and, 2) that some oceanic and continental basalts have $^3He/^4He$ ratios which are significantly higher than the atmospheric value (e.g., Lupton and Craig, 1975; Kurz *et al.*, 1982; Kurz *et al.*, 1983). Although tacitly accepted throughout much of the literature, these two lines of reasoning do not represent unique interpretations of the available data, nor, in fact, are they altogether consistent with each other (as the highest observed $^3He/^4He$ ratios do not represent samples or locales with chondritic $^{143}Nd/^{144}Nd$, or single-stage Pb which lies on the geochron).

As fate would have it, the first series of high-precision measurements of $^{143}Nd/^{144}Nd$ in young basalts ranged from $\epsilon_{Nd} \sim 13$ in MORBs to $\epsilon_{Nd} \sim 0 \pm 2$ for continental flood basalts and the island of Tristan da Cuhna (Richard *et al.*, 1976; DePaolo and Wasserburg, 1976a, 1976b; and O'Nions *et al.*, 1977). The observed correlation of $^{143}Nd/^{144}Nd$ with $^{87}Sr/^{86}Sr$ gave rise to the "mantle array" and permitted estimation of an average $^{87}Sr/^{86}Sr$ values for BSE (0.7045–0.7050). Implicit in this line of reasoning, and explicit in many of the models which followed (e.g., Jacobsen and Wasserburg, 1979; DePaolo, 1980; O'Nions *et al.*, 1979; Allegre *et al.*, 1980), was the idea that isotopic variations in mantle-derived basalts result from mixing between depleted and undepleted (or primitive) mantle reservoirs. The data, however, which were taken to justify the notion of primitive undifferentiated mantle, would always have been equally well or better explained by mixing between a MORB-source component and an enriched component with $\epsilon_{Nd} < 0$ (Anderson, 1982). As more data have been acquired, the need for enriched mantle material has become more evident, and, consequently, compelling arguments for an undifferentiated mantle reservoir have been diminished. In addition, with the exception of a few diehards, the notion of a primitive source for continental basalts has been laid to rest (e.g., Carlson *et al.*, 1981; Carlson, 1984; Worner *et al.*, 1986; see Fig. 5.29).

Even when we explain away chondritic $^{143}Nd/^{144}Nd$ ratios as mixing phenomena, we must deal with the source for ^3He in the mantle. Irregardless of the problems concerning production of ^4He discussed in Section 4, the range of $^3He/^4He$ ratios measured in mantle-derived basalts clearly demonstrate that there are at least more and less "primitive," if not primordial, materials in the suboceanic mantle. However, as noted earlier, when we take $^3He/^4He$ ratios at face value, there is little evidence to suggest that high $^3He/^4He$ ratios occur in rocks that have ϵ_{Nd} values close to zero (see Fig. 5.25). Furthermore, high $^3He/^4He$ ratios need not indicate the existence of primitive mantle and could as well document intramantle metasomatism of a previously depleted mantle segment, resulting in a relative stabilization of high $^3He/^4He$ ratios some time in the past (see Fig. 5.13). If, as hypothesized earlier, changes in U/He are effected by degassing or mobilization of fluids, but not by separation of silicate liquid from residual solid, then it is conceivable that even residual, depleted mantle materials could have high $^3He/^4He$ ratios. Conservatively then, we can take high $^3He/^4He$ to be indicative of a time-integrated history of little or no degassing.

Based on the above discussion, we deem it reasonable to at least examine the consequences of a mantle which no longer contains any primitive undifferentiated component. Such a scenario could evolve in two different ways. First, enriched and depleted segments could evolve via continuous differentiation of a homogeneous mantle which had chondritic Sm/Nd subsequent to accretion. Or, second, borrowing from Armstrong (1981), we might consider that separation of the core did not result in homogenization of the silicate portion of the Earth. Certainly, with separation of the core and atmosphere, extensive differentiation of the silicate earth may occur in conjunction with a substantial upward flux of incompatible elements. This likely resulted in at least the transient production of a granitic or alkali basaltic crust. Whether or not the silicate portion of the Earth was ever homogenized subsequent to this time depends on how effective early mantle convection was with regard to destruction of these earliest differentiates. If they survived to be reworked and cratonized during the Archean, then even mantle which survives as a closed system for the past 4.4 b.y. may have a depleted Nd isotopic signature, due to the separation of early differentiates with low Sm/Nd ratios.

Thus, even if "primordial" mantle material with the composition of BSE or PUM has survived unfractionated since accretion or core formation, we contend that, to date, no geochemical observation demands this survival. However, neither do the data demand that no such "primordial" component survives. Accordingly, we shall consider both possibilities in our subsequent evaluation of geochemical mantle models.

5.2.2. MORB Mantle

The MORB source reservoir has long been characterized as being depleted in incompatible trace elements (e.g., Tatsumoto *et al.*, 1965; Kay *et al.*, 1970).

Relative to presumed bulk earth values, Rb is depleted relative to Sr by factors of 5–50, and Nd is depleted relative to Sm by 15–40 percent. This has led to the tacit assumption that LIL elements and basaltic constituents are strongly depleted in this mantle segment. Using the composition of PUM and the simple melting model discussed in Section 2.1, it is possible to approximate just how depleted the MORB source might be. The Sm/Nd ratio of N-type MORB is typically in the range 0.35–0.40; our partial melt extraction model will generate this degree of "depletion" in Sm/Nd with 1–13 percent melt extraction. This degree of melt extraction will only decrease Nd concentrations by 17–47 percent and Ca and Al contents by 2–9 percent. In other words, with respect to major elements, the MORB mantle is only marginally different from PUM, and for Nd, it still retains more than half of its original budget. Nd concentrations in MORBs range from ~4–10 ppm; allowing for ~50 percent high-level fractional crystallization for the more evolved MORBs (Hart and Staudigel, 1982) leads to an estimate of about 4–5 ppm Nd in primary MORB magmas. In conjunction with the above estimate of Nd in the MORB source of ~0.6 to 1 ppm, the degree of melting to produce MORB is constrained to be between ~10 and 23 percent. This result is in general agreement with the early suggestions of Gast (1968), Kay *et al.* (1970), and Schilling (1971, 1975).

The observation that MORBs which are well-removed from any hot-spot volcanism display unique "depleted" isotopic characteristics which are not observed elsewhere in oceanic or continental volcanic domains (see Figs. 5.20–5.26, and 5.28) documents the existence of a depleted mantle segment which constitutes an important volumetric fraction of the MORB source reservoir. We refer to this depleted segment as end-member depleted MORB mantle (DMM) and suggest that MORBs with more "enriched" isotopic signatures result from the inclusion of relatively enriched components (EMs or HIMU) in a DMM matrix.

This view of the MORB mantle, as containing relatively enriched components in a depleted matrix, is not a unique interpretation of the isotopic variability of MORBs. We could suggest (as pointed out by Allegre, 1982, and Allegre *et al.*, 1984) that MORBs with enriched or ocean island-type isotopic signatures (E-type MORB) result from melting of a homogeneous mantle segment which has just this signature. However, recent work and modelling of mid-ocean ridge isotopic variations (Cohen and O'Nions, 1982a; leRoex *et al.*, 1983; Allegre *et al.*, 1984), and studies of seamounts situated near the EPR (Batiza and Vanko, 1984); Zindler *et al.*, 1984; Graham *et al.*, 1984, 1988), all support the existence of small-scale heterogeneities within the MORB source mantle. In fact, this is one of the more notable successes of recent geochemical work on the mantle and we embrace it without reservation in our subsequent discussion of mantle chemical dynamics.

The question of whether the MORB source mantle is heterogeneous on a large or global scale (100–10,000 km) is a bit more difficult to approach. The scale of melting and rate of magma production at a given ridge segment will likely affect the chemical influence of the dispersed enriched components. This will be the case if solidii for the dispersed components are lower than that of DMM, and hence, their influence on the final product will be more dramatic at the early stages of melting. (Zindler *et al.*, 1984; Sleep, 1984). The chemical influence of the dispersed components will also be diminished by processing through large magma chambers which may be present at faster spreading ridges (Allegre *et al.*, 1984). Within this context, we might expect that some large-scale geochemical variations between MORBs may result from variable styles of magma processing at ridges with variable rates of magma throughput.

Recent work by Dupre and Allegre (1983) on MORBs from the Indian Ocean has shown that these rocks, even when far-removed from hot-spots, tend to be isotopically distinct from MORBs of other latitudes and oceans (Hart, 1984). Specifically, they tend toward higher values of $^{87}Sr/^{86}Sr$ and $\Delta 7/4$, (see Fig. 5.10 caption for definition of this parameter) and lower values of $^{143}Nd/^{144}Nd$, $^{206}Pb/^{204}Pb$, and $^{208}Pb/^{204}Pb$. In fact, several MORBs from the Carlsberg Ridge and the Southwest Indian Ridge have been found to lie to the left of the geochron in a $^{206}Pb/^{204}Pb$ vs. $^{207}Pb/^{204}Pb$ diagram (Dupre and Allegre, 1983; Fig. 5.23). For present purposes, the critical question with regard to these data is whether they document distinct differences in the isotopic character of the DMM component, the EM and HIMU components, or both, or whether they simply represent an increased abundance of EM components. While none of these explanations can be categorically excluded, the latter is consistent with the large amounts of subduction known to have characterized the southern hemisphere during the Phanerozoic (Anderson, 1982) and the hypothesized origins for EMs (to be discussed subsequently).

The MORB source mantle is generally assumed to be global in lateral extent and confined to the upper mantle (e.g., Morgan, 1971; Schilling, 1973; Sun and Hanson, 1975; Jacobsen and Wasserburg, 1979; Allegre, 1982) although other configurations have been proposed (e.g., Hofmann *et al.*, 1978; Anderson and Bass, 1986). This assumption is supported by the observation that ridge volcanism is generally characterized by a distinctive isotopic signature wherever it occurs. This has, however, been the subject of some debate, as recent geochemical mass balance models have shown that larger fractions of the mantle may have been processed and depleted to form the continental crust (Goldstein *et al.*, 1982; Allegre *et al.*, 1983b,c; Allegre *et al.*, 1986). With regard to the location of the MORB source reservoir within the mantle, we know that: 1) The MORB source mantle is present beneath all active ridges; 2) ocean island hot-spots, presumably derived from deeper mantle levels, as evidenced by the need for garnet in residual source materials and by the more or less stationary positions of hot-spots in an absolute frame of reference (Morgan, 1971), are derived from mantle sources which are chemically distinct from the MORB reservoir, although they may mix toward DMM (see Figs. 5.20–5.26 and 5.28); 3) the least contaminated representatives of continental basalt suites are isotopically similar to OIBs or enriched MORBs (see Fig. 5.29); and 4) intraoceanic island arc magmas (e.g., Marianas and New Britain, DePaolo and Wasserburg, 1977; DePaolo and Johnson, 1979; Aleutians, Goldstein *et al.*, 1981; McCulloch and Perfit, 1981; Morris and Hart, 1983; Goldstein *et al.*, 1988); are, again, isotopically similar to OIBs (see Fig. 5.28). Essentially, all volcanic materials known to have feeder systems which extend to deeper mantle levels ($\geq$100 km) than mid-ocean ridges tend to have OIB isotopic signatures. Taken at face value, this may mean that the MORB source mantle is restricted to the uppermost hundred or several hundred km of the upper mantle. Such a suggestion is perhaps reasonable if DMM comprises relatively low-density depleted peridotite.

We should note, in the context of the present discussion, that the only rocks at the surface of the Earth, other than normal MORBs, which are known to have DMM isotopic signatures are a subset of ultramafic xenoliths which are erupted in alkalic continental volcanics, and some parts of ultramafic peridotite massifs (see Figs. 5.30 and 5.31). This suggests that DMM is peridotitic, as opposed to eclogitic or piclogitic (the latter having been suggested by Anderson, 1983; Anderson and Bass, 1986), and documents the existence of at least some DMM beneath continents, particularly the relatively young mobile zones in which the xenolith localities and orogenic peridotites occur (e.g., Rio Grande Rift, Rhine Graben, East Africa, and Alpine zones of the western Mediterranean).

Recent work with rare gasses has shown that DMM is characterized by: 1) $^{3}He/^{4}He$ ratios of 8.0 ± 1.0 Ra (e.g., Kurz *et al.*, 1982); 2) $^{40}Ar/^{36}Ar$ ratios which are probably in excess of 25,000 (Allegre *et al.*, 1983a; Sarda *et al.*, 1985); and 3) occasionally observed excesses of ^{129}Xe, formed by the ancient decay of now-extinct ^{129}I (Staudacher *et al.*, 1982; Allegre *et al.*, 1983a). The xenon and argon data, taken together, strongly support a catastrophic degas-

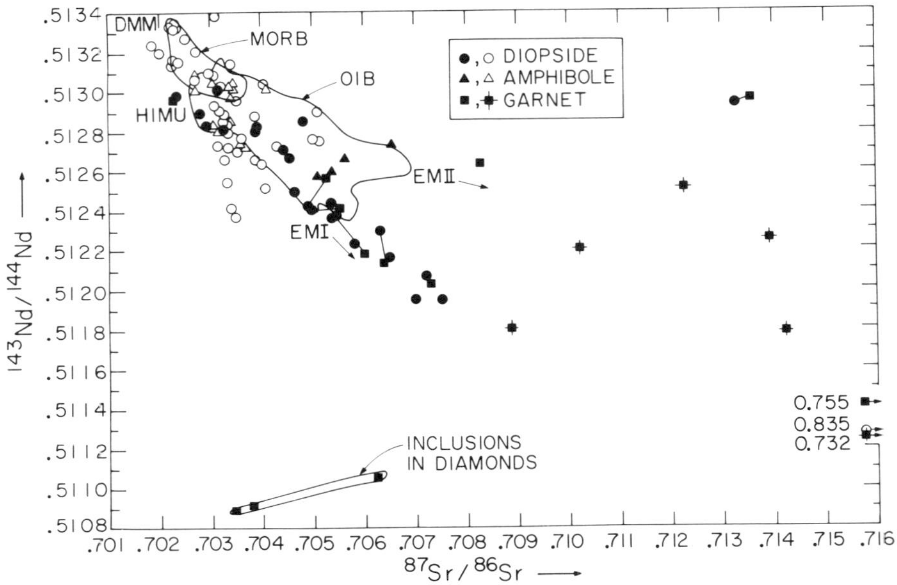

Figure 5.30. Nd-Sr isotope correlation plot, showing analyses of various minerals from mantle xenoliths and megacrysts. Localities are divided either as African kimberlites (solid symbols) or other localities, usually with alkali basalt hosts (open symbols). Xenolith data of Richardson *et al.* (1985) has first been age corrected back to 90 m.y., using mineral Sm/Nd, then corrected ahead to present using average bulk rock Sm/Nd ~ 0.10; solid lines connect coexisting garnet-cpx pairs. Other African kimberlite data is plotted as present day values, as insufficient data is available to make age corrections (though these will shift points only very slightly on this plot—generally <.00006 in 143/144). Data references: Basu and Tatsumoto (1979, 1980); Cohen *et al.* (1984); Jagoutz *et al.* (1980); Kramers *et al.* (1981, 1983); Menzies and Murthy (1980a,b); Menzies and Wass (1983); Richardson *et al.* (1984, 1985); Stosch *et al.* (1980); Zindler and Jagoutz (1988).

sing of proto-DMM during the first ~10^8 years of Earth history (Staudacher *et al.*, 1982; Sarda *et al.*, 1985). $^3He/^4He$ ratios of 8 Ra, although substantially lower than "planetary" values of ~100 to 200 Ra (Jeffrey and Anders, 1970; Craig and Lupton, 1978), are still markedly enriched in primordial ^3He compared to the atmosphere, and suggest that the early degassing was more efficient for Ar and Xe than for He; this is consistent with the enhanced solubility of the lighter noble gasses in silicate melts (e.g., Ozima and Podosek, 1983).

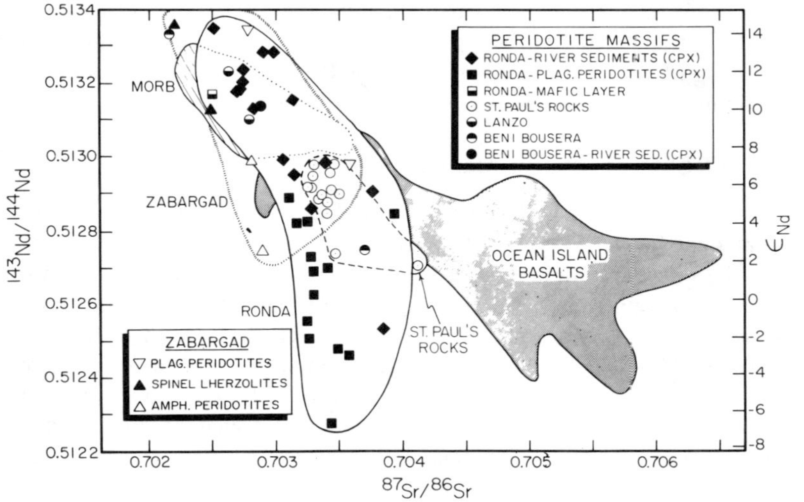

Figure 5.31. $^{87}Sr/^{86}Sr$ versus $^{143}Nd/^{144}Nd$ variation diagram for peridotite massifs. While all localities, with the exception of St. Paul's Rocks, contain some samples which fall within the depleted MORB field ($^{143}Nd/^{144}Nd > 0.5132$), extreme heterogeneity occurs within these massifs. The enriched portions of the Ronda, Beni Bousera, and Zabargad massifs tend to be EM I type, lending support to the contention of an intra-mantle metasomatic origin for this component. Scale lengths for the observed isotopic heterogeneities are discussed in Section 3.2, and shown in Figure 5.9. Data sources are: Ronda, Reisberg and Zindler (1987) and Zindler *et al.* (1983); St. Paul's Rocks, Roden *et al.* (1984); Zabargad, Brueckner *et al.* (1988); Lanzo and Beni Bousera, Richard and Allegre (1980).

When the He isotope data for MORB are compared to the other isotopic systems displayed in Figures 5.24–5.26 and 5.28, it is apparent that MORB $^3He/^4He$ ratios cover both a relatively small fraction of the total oceanic basalt range and, in contrast to all the other systems, MORBs fall *within* the oceanic range, rather than defining one end of it. The latter observation may relate to the rapid accumulation of radiogenic ^4He in degassed and/or recycled materials and the influence of such processes and/or materials at some ocean islands (see Section 5). The relatively restricted range of $^3He/^4He$ ratios in MORBs may document the enhanced mobility of He over the other elements of interest in the MORB source mantle. As discussed previously, there may be a tendency to reduce solid-state inter-component isotopic variability through magmatic processing. The relatively rapid diffusion of He in the presence of a small amount of melt (D 7 $\times$ 10^{-5} cm^2/sec at 1200°C, extrapolated from Kurz and Jenkins, 1981) may contribute to the smoothing of $^3He/^4He$ variations in the MORB reservoir during magmatic processing, if heterogeneities are on the order of 1 km or less.

5.2.3. PREMA (Prevalent Mantle)

Isotopic compositions of ocean islands and enriched MORBs, which are intermediate between DMM, BSE, HIMU, and EM components in Figures 5.20–5.23 have traditionally been interpreted as mixtures of these or other end-member components. We suggest, as an alternate view, the existence of a distinct mantle component with this isotopic character ("PREMA," for *PRE*valent *MA*ntle). The position of this component is shown in Figures 5.20–5.26, and 5.29, and is seen to coincide with: 1) fields for many ocean islands including Iceland and Hawaii; 2) the most depleted representatives of many continental basalt suites; and 3) fields for many intraoceanic island arcs (see Fig. 5.28). The coincidence of this component with Iceland and Hawaii, and their respective high $^3He/^4He$ ratios (Figs. 5.24–5.26), suggests that at least some portion of PREMA is relatively undegassed.

If PREMA represents a mixture of DMM, HIMU, EM, ±BSE (or PUM) on a scale of 100 m or less, then the process of sampling during melting may necessarily result in a fairly restricted range of compositions, as we have already discussed for the MORB mantle. Such a result could not be easily distinguished from a mantle segment with PREMA characteristics which is homogeneous down to a m or cm scale. On the other hand, if distinct components which comprise PREMA are separated by large distances in the mantle ($>$ 100 km) or reside within separate convective regimes, then the PREMA composition must result from a very reproducible mixing process which acts over large distances. We believe that the ramifications of both of these scenarios should be considered, and that the current geochemical data base is consistent with either of them.

If PREMA does exist as a distinct component, at scale lengths of $>$1 m, the implications for geochemical mass balance are significant. Because the PREMA signature is observed in so many different tectonic settings, it is likely that PREMA accounts for a volumetrically significant fraction of the mantle. In this context, PREMA would represent a major depleted mantle component which is, along with DMM, geochemically complementary to the continental crust. Because PREMA is not as depleted as DMM, simple mantle-crust mass balance will permit a larger fraction of the mantle to exist as PREMA than as DMM. There is, of course, an inverse relationship between permissible volumes for PREMA and DMM; as we hypothesize more PREMA, we can tolerate less DMM. If PREMA represents the major depleted mantle component, then it must comprise significantly more than 30 percent of the mantle, and must, therefore, be present below 670 km depth (see discussion in Section 5.5).

The origin of a PREMA component may be envisioned in several different scenarios. First, as discussed earlier in Section 2, we might imagine that significant differentiation of the silicate portion of the Earth occurred contemporaneously with core segregation (e.g., Armstrong, 1981). If something on the order of 50 percent of the present-day continental crust (or the trace element equivalent) were segregated at this time, then PREMA today might represent the most primitive remaining mantle, having essentially survived unscathed since the earliest days of Earth history. In this context, the high $^{3}He/^{4}He$ ratios at Iceland and Hawaii would not be surprising.

A variation of this scenario might suggest that, over the first part of Earth history ($\sim$1–2 $\times$ 10^9 years), the entire mantle mixed convectively resulting in a PREMA-type depletion of the whole mantle. Subsequent splitting into upper and lower mantle convective regimes would then preserve PREMA in the lower mantle today, while the upper mantle would continue to be depleted by crust extraction, evolving toward DMM.

Alternatively, PREMA may evolve via the quasi-continuous separation of a crustal component from the mantle over time. We hypothesize that the creation of new continental crust at volcanic arcs results in the complementary production of enriched and depleted mantle segments (EM, HIMU, and DMM). The EM, HIMU and DMM component may then remix into the convecting mantle with time constants of 1–2 $\times$ 10^9 years, so that at any given time, the mantle contains "quasi-steady-state" amounts of these materials. The ambient remixed mantle material will then represent PREMA. It would then follow that: 1) PREMA is the chemical complement of the continental crust; 2) PREMA is approximately the mean composition of the mantle; and 3) that DMM + EM I + EM II + HIMU = PREMA.

These scenarios allow significant latitude for the location of PREMA in the mantle today, providing that it does represent a distinct component. If PREMA is close to the average mantle composition and is complementary to the continental crust, then it may occupy much of the mantle and be overlain by a relatively thin layer dominated by low density DMM. Alternatively PREMA may reside totally within the lower mantle, with the upper mantle comprising a mixture of DMM, EMs and HIMU. Volcanic rocks from arcs, continents, and islands with PREMA signatures are generally thought to be derived from depths in excess of 100 km so that either of the above scenarios would be feasible.

Of the various isotope ratios under consideration, $^3He/^4He$ ratios are the most difficult to reconcile with the concept of PREMA. While Iceland, Hawaii, and oceanic arcs have PREMA signatures of Sr, Nd, and Pb ratios, high $^3He/^4He$ ratios at Iceland and Hawaii (~30 Ra) contrast strongly with $^3He/^4He$ ratios of ~6 Ra at island arcs (Craig and Rison, 1983). Although, as discussed in Section 4, $^3He/^4He$ ratios at arcs, where pre-eruptive differentiation of magmas is generally extensive, might be taken as minimum values, the relative constancy of $^3He/^4He$ ratios thus far measured at arcs argues for a less random process. It is conceivable that inclusion of a small subducted sediment component (with high U/He) in arc magmas might contribute to the reduction of $^3He/^4He$ ratios, but, again, this would likely account for very variable contributions and not result in a constant $^3He/^4He$ ratio for arcs. In any event, it may be that the PREMA component has a variable $^3He/^4He$ ratio or that, in fact, the helium data argue against the existence of PREMA.

5.2.4. HIMU (High U/Pb Mantle)

Isotopic evidence for the existence of the HIMU component derives largely from consideration of the Pb isotope systematics displayed in Figure 5.23, where oceanic basalts appear to mix from the depleted MORB zone near the geochron out toward the islands of St. Helena and Tubuaii. The extreme enrichments of ^{206}Pb and ^{208}Pb observed at these islands suggest an enrichment in U+Th relative to Pb on the order of 1.5–2.0 $\times$ 10^9 years ago (Tatsumoto, 1978). These enrichments are not paralleled by an increase in Rb/Sr and, thus, the islands are characterized by relatively low $^{87}Sr/^{86}Sr$ (~0.7029). This unradiogenic Sr which characterizes HIMU suggests the possibility that HIMU has evolved from DMM some time in the past.

Chase (1981) and Hofmann and White (1982) suggested that HIMU might evolve from subduction of ancient altered oceanic crust, which is significantly enriched in U by interaction with seawater. At face value, it would appear difficult to provide U enrichments by sea water alteration without introducing a high $^{87}Sr/^{86}Sr$ signature. However, if the oceanic crust underwent alteration a few billion years ago, when the sea water Sr isotope curve was much closer to the mantle Sr evolution curve, a significant sea water Sr imprint may be avoidable. Also, U (and Rb) enrichment by crust interaction with sea water proceeds more rapidly than Sr exchange with sea water (Hart, 1969; Hart *et al.*, 1974; Hart and Staudigel, 1982; Staudigel *et al.*, 1981). It is possible, then, that U/Pb ratios can be increased from typical unaltered MORB values (~9.4) to the value of ~16.5 necessary to generate HIMU $^{206}Pb/^{204}Pb$ ratios over 2 b.y. without significant concommitant increase of $^{87}Sr/^{86}Sr$. The increase in μ due to alteration will be accompanied by a comparable increase in Rb/Sr, which in 2 b.y. will lead to HIMU-type $^{87}Sr/^{86}Sr$ (~.7029), that is, only a slightly accelerated growth of radiogenic Sr compared to DMM evolution. To generate HIMU-type Nd (~.5129) in 2 b.y. from a DM evolution curve requires $^{147}Sm/^{144}Nd$ of ~0.22, which is close to the value for average erupted MORB. Neither Sm/Nd nor Nd isotopic ratios will be significantly perturbed by crust-sea water alteration processes. Thus the Hofmann and White (1982) model for HIMU is quantitatively consistent with what is known about the effect of sea water alteration of Rb-Sr, Sm-Nd, and U-Pb systems. It may be, however, that the fractionation of U from Th in the alteration cycle is sufficient to preclude the model. This aspect of the problem requires further consideration.

Zindler *et al.* (1982) proposed that HIMU might result from the subduction of ancient oceanic crust from which the alteration component had been removed by subduction related processes. This proposal was made largely to avoid the problem of Th/U fractionations during alteration. The viability of this model hinges on the likelihood of a primary magmatic increase in U/Pb and Th/Pb during the generation of MORB melts. Relatively low μ values in hand-picked MORB glasses (~9.4) however, now seem to suggest that significant enrichments of U relative to Pb do not occur during MORB magmatism. However, further processing of MORB through subduction zones may effect such enrichments, as evidenced by low μ values in island arc volcanics (e.g., Tatsumoto, 1978).

Allegre and co-workers (e.g., Allegre *et al.*, 1980; Allegre, 1982) and Vidal and Dosso (1978) have proposed that HIMU may result from extraction of Pb from some portion of the mantle to the core over a significant portion of Earth history (~15 percent of the core formed since 4.4×10^9 years). A lack of constraints on not only the time scales for core formation, but also the nature of the light element in the core (S, Si, O), and chalcophilic tendencies for Pb at current lower mantle pressures and temperatures, makes it difficult to evaluate this mechanism. For present purposes, we must allow that it remains a possibility (see further discussion in Section 5.4).

Another mechanism which must be considered as a plausible means by which to produce HIMU is the mobilization of Pb (and Rb?) in a metasomatic fluid and subsequent removal from some mantle segment. The behavior of Pb in the continental crust suggests its affinity for volatile-rich ore-bearing fluids and provides qualitative support for this hypothesis; there is no available data with which to evaluate such behavior in the mantle. Mantle metasomatism might thus result in production of residual degassed HIMU component and a complementary gas-rich low μ component elsewhere in the mantle (see discussion of EM I in next section).

Regardless of the mechanism by which HIMU is created, its ubiquitous presence in the interactive mantle is clearly documented by the mixing trends in Figure 5.23. We thus propose that HIMU is present as a dispersed component within a PREMA or DMM matrix. It is particularly important to note that, were it not for HIMU, we would see very little variation amongst islands and ridges in Figure 5.23. The main variation in Figure 5.20, for example, is between MORB and Kerguelen, which only differ by a small amount in the $^{207}Pb/^{204}Pb$ direction in Figure 5.23. Also, islands which approach the HIMU component in Figure 5.23 are characterized by relatively low $^{87}Sr/^{86}Sr$ and high $^{143}Nd/^{144}Nd$. These observations document the important differences between Figures 5.20 and 5.23 and were essentially those which led to the mantle-plane model (Zindler *et al.*, 1982).

Unfortunately, we know nothing of the He isotopic character of HIMU. However, our best guess, based on possible modes of production, is that HIMU has been relatively degassed (whether by removal of a fluid component or by degassing of a magma which is then recycled) and has subsequently evolved low $^{3}He/^{4}He$ ratios due to the consequent high U/He ratios (see Figs. 5.24–5.26).

5.2.5. Enriched Mantle Components

The isotopic character of islands which lie outside the DMM, HIMU, BSE mixing domains outlined in Figures 5.21 and 5.22 necessitate the existence of EM components, independent of whether or not we think a BSE or PUM component actually exists. The origin of EM components has been the subject of much speculation since their existence was first suggested (Anderson, 1982; Cohen and O'Nions, 1982b; Richardson *et al.*, 1982). Proposed modes of origin now include: 1) injection into the mantle, by subduction, of continentally derived sediment or continental crust (Cohen and O'Nions, 1982b), altered oceanic crust (Hofmann and White, 1982), islands, or seamounts; 2) delamination of subcontinental lithosphere (McKenzie and O'Nions, 1983); and 3) mantle metasomatism (e.g., Menzies and Murthy, 1980a,b). It may be significant that all of these mechanisms can operate in convergent margin tectonic settings.

EM components are characterized by low $^{143}Nd/^{144}Nd$, variable $^{87}Sr/^{86}Sr$ and high $^{207}Pb/^{206}Pb$ and $^{208}Pb/^{204}Pb$ (at a given value of $^{206}Pb/^{204}Pb$). We will refer to the low and high $^{87}Sr/^{86}Sr$ EMs as EM I and EM II, respectively. Comparison of Figures 5.20 and 5.29 reveals that the tendency for the continental data array to "fan out" at low $^{143}Nd/^{144}Nd$ is mimicked, to a lesser degree, by the oceanic data array (Worner *et al.*, 1986). This behavior in the continental suites has been attributed both to the existence of enriched subcontinental mantle components as well as the contamination of some mantle-derived magmas by upper and lower continental crustal materials, lying at high and low values of $^{87}Sr/^{86}Sr$, respectively (e.g., Brooks *et al.*, 1976; Carter *et al.*, 1978; Hawkesworth *et al.*, 1983; Worner *et al.*, 1986). The similarity between the continental and oceanic arrays suggests similar styles of enrichment for the subcontinental and suboceanic mantle (although enrichment of the continental mantle is clearly more extreme) as well as the possible involvement of the continental crust in the production of EMs.

A review of the isotopic character of ultramafic nodules shows that many of those which contain hydrous phases plot below the oceanic array (Fig. 5.30). If these materials represent metasomatized DMM, then the effect has been to decrease Sm/Nd substantially while increasing Rb/Sr only moderately (Zindler and Jagoutz, 1988). These nodules, as well as some portions of ultramafic massifs (Reisberg and Zindler, 1985, 1987; Brueckner *et al.*, 1988; Fig. 5.31), are displaced from the oceanic array toward measured compositions for lower continental crustal materials (e.g., Hamilton *et al.*, 1979; Stosch and Lugmair, 1984). Although data regarding the Pb isotopic character of ultramafic nodules or massifs is sparse (Kramers *et al.*, 1983; Cohen *et al.*, 1984), we suggested earlier that metasomatism might produce low U/Pb ratios and, with time, unradiogenic Pb, similar to that thought to characterize the lower continental crust. It is thus difficult to distinguish between recycled lower continental crust and metasomatized mantle as candidates for the EM I component.

EM II, characterized by high $^{87}Sr/^{86}Sr$ and low $^{143}Nd/^{144}Nd$, has a strong similarity with upper continental crust or continentally derived sediment (e.g., Hawkesworth and Vollmer, 1979; Carlson, 1984; Worner *et al.*, 1986). Subducted continental material has been strongly implicated in several island arc volcanic suites (e.g., Banda arc, Morris, 1984; Martinique, Davidson, 1983), and the isotopic signatures of these "contaminated" arcs bear a strong kinship with the EM II clan of OIB. (In contrast, there are no arc suites which mimic the EM I signature.) There are, as well, some ultramafic nodules, predominantly from South African Kimberlites, which display this EM II isotopic signature (Fig. 5.30). Menzies (1983) has proposed that the high $^{87}Sr/^{86}Sr$, or high Rb/Sr (EM II-type) character of these nodules is imprinted by incorporation of a hydrous metasomatizing agent, whereas EM I-type signatures are produced by CO_2-rich fluids. Experimentally determined solubilities for CO_2 and H_2O in mantle melts (e.g., Khitarov and Kadik, 1973) document the very high solubility of H_2O, relative to CO_2, and make it likely that mantle-derived metasomatic fluids are dominated by CO_2. EM II-type metasomatism may thus require a source of crustal water (perhaps in a subduction environment) and derive its isotopic character, indirectly, from the continental crust. EM II signatures may thus evolve via: 1) subduction of continental materials, which may subsequently imprint their isotopic signature on subjacent mantle; 2) direct high-level interaction with continental crust or sediment (particularly, but not exclusively relevant to continental basalts); or, 3) intra-mantle differentiation or metasomatism.

As noted earlier, the mechanisms proposed for production of EMs may all be associated with subduction processes. Clearly, mechanisms which involve the injection of near-surface materials into the mantle can only operate at convergent margins. Delamination or "peeling back" of metasomatized subcontinental lithosphere might be facilitated by subduction-related tectonic activity. Metasomatism of the mantle wedge may be initiated by dehydration or decarbonation of the down-going slab. The fact that EM I signatures are rare in arcs suggests that EM I is predominantly metasomatic in origin (possibly induced in arc settings, but requiring a subsequent fermentation period in the mantle). The fact that most intraoceanic arcs do not show an EM II signature (Kay, 1980; Morris and Hart, 1983; Goldstein *et al.*, 1988) suggests that, if EM II represents a continental imprint, it is a result of reinjection predominantly in those arcs which are consuming abundant terrigenous material, and not in the more typical intraoceanic arcs.

As noted earlier, the oceanic islands exhibiting strong EM II signatures are almost exclusively restricted to the southern hemisphere, and serve to delineate the Dupal anomaly. Whatever mechanism is envisioned to produce the EM components must obviously address this strong geographic localization. Based on the high $\Delta 7/4$ (Hart, 1984b) character of the EM components, they must have been isolated from other mantle components (DMM, PREMA, etc.) for a long period of time (>3 b.y.). Isolation in the continental crust or subcontinental lithosphere is one obvious solution, although we must then ask why the terrigenous-dominated subduction systems, which have been the vehicle for reinjection of this old isolated continental or subcontinental material, have been localized in southern latitudes? If the EMs are purely the result of intramantle processes (e.g., metasomatism), how has this EM reservoir been preserved for such a long period of time in the convecting mantle, and why is it currently localized in the sub-Dupal belt mantle? Although we have no simple answers for these questions, we will return to them in Section 6.

5.3. Mixing Systematics

Implicit in our discussion of component characterization is our contention that oceanic basalt isotope systematics reflect mixing. This interpretation is supported by studies which document mixing at individual localities (e.g., Langmuir *et al.*, 1978; Zindler *et al.*, 1979), as well as the various lines of reasoning which suggest relatively small-scale heterogeneities within magma source regions (see Section 5.2.2). As discussed in Section 5.1, early interpretations of basalt isotope systematics suggested two- or three-component mixing to explain the various arrays. Here, we identify a minimum of four distinct components (DMM, HIMU, EM I, EM II) and it is thus not possible to delineate unique "component recipes" for individual ocean islands or ridge segments. It is, however, useful to qualitatively discuss constraints on mixing which derive from simultaneous consideration of the two-dimensional variation diagrams (Figs. 5.20–5.26).

Our estimated component isotopic compositions are schematically depicted in Figures 5.20–5.22 and 5.23e. Mixing lines between DMM, HIMU and BSE are also shown for reference (element concentrations used for mixing calculations are given in the figure captions). Essentially, these mixing lines lie on the mantle plane (Fig. 5.27; Zindler *et al.*, 1982) and, as discussed earlier, those localities which lie outside the DMM-HIMU-BSE mixing domain (Kerguelen, Walvis, Tristan, Gough, Samoa, Azores, Reunion, Tahiti) demand the existence of the EM components. Having invoked the EM components, BSE or PUM is no longer necessary as a mixing end member, and all observed compositions can result from mixing between DMM, HIMU, EM I and EM II. If this mixing hypothesis is accurate, then the relative positions of individual localities, with respect to the components, should remain constant from one diagram to the next. Comparison of Figures 5.20–5.22 provides general support for this hypothesis, and the relationships in these figures relative to those in Figure 5.23 reveals that, after taking account of the relatively large analytical uncertainties in $^{207}Pb/^{204}Pb$, the Pb isotope systematics are also consistent with the mixing hypothesis. The islands mentioned above, which lie outside the DMM-HIMU-BSE mixing domain in the EM II direction, tend toward high $^{207}Pb/^{204}Pb$ ratios at a given value of $^{206}Pb/^{204}Pb$ (high $\Delta 7/4$Pb in the terminology of Hart, 1984b). Those islands which plot between DMM and HIMU at low $\Delta 7/4$ values also lie between DMM and HIMU in Figure 5.20. Islands which mix toward EM I in Figures 5.20–5.22 mix toward unradiogenic EM I with variable $\Delta 7/4$Pb in Figure 5.23.

The estimated composition of PREMA is also shown in Figures 5.20–5.23. In general, it is intermediate between DMM and HIMU but is displaced slightly toward EM II; this relationship to EM II is not as clear in Figure 5.23, but again, it may be obscured by the relatively large uncertainties in $^{207}Pb/^{204}Pb$ ratios. Insofar as PREMA does represent a distinct mantle component, and not simply a highly reproducible mixture, then PREMA may represent the depleted end-member OIB component, obviating the need for DMM in the OIB source reservoir. This is readily apparent from the position of PREMA in Figures 5.21 and 5.22, even though mixing lines to PREMA are not shown (i.e., they are not expected to be highly curved). In this context, only the isotopic composition of N-type MORB documents the existence of DMM.

Direct extrapolation of the mixing relationships discussed above to the He variation diagrams (Figs. 5.24–5.26) is not straightforward, due at least in part to the considerations discussed in Section 4. Intramantle decoupling of He and lithophile trace elements (e.g., during fluid phase metasomatism) may also contribute to this difficulty. We have, nevertheless, estimated $^3He/^4He$ ratios in the various components to facilitate direct comparison of Figures 5.24–5.26 with Figures 5.20–5.23.

While the $^3He/^4He$ ratio in DMM is tightly constrained by the narrow range of $^3He/^4He$ observed in MORBs (see Fig. 5.28), values for other components are ill-constrained. The BSE $^3He/^4He$ ratio is taken to be 50 Ra. This presumes that undifferentiated mantle should be richer in primordial ^3He than all other differentiated materials, and that it should be less than the primordial $^3He/^4He$ ratio of 100–200 Ra (Craig and Lupton, 1978). It is possible, however, that the highest terrestrial $^3He/^4He$ ratios occur in mantle segments which have been enriched in He (relative to U and Th) by metasomatic processes. In any case, whether we choose a value of 30 or 100 Ra for this component, the qualitative relationships shown in Figures 5.24–5.26 will not change significantly.

Assuming BSE U of 20.8 ppb (Table 5.4) and BSE $U/^4He$ of ~300 (Fig. 5.13), the BSE He concentration (for single-stage evolution) will be on the order of 7×10^{-5} std cc/g. A similar analysis for DMM, taking $U/^4He$ to be 600 (Fig. 5.13) and U to be 8.4 ppb (Table 5.7), yields a He concentration of $\sim 1.4 \times 10^{-5}$ std cc/g. Using the 9.6 enrichment factor for MORB over the MORB source, (Section 5.2.2 and Table 5.7) He concentrations on the order of 1.3×10^{-4} are predicted for MORBs. This is about a factor of four higher than the highest He concentration measured in a MORB glass (Kurz and Jenkins, 1981), and thus implies ~80 percent He loss from even the most He-rich MORB (equivalent to ~0.2 percent CO_2 loss; see Fig. 5.19). These considerations suggest that BSE will be relatively gas rich compared to DMM and that mixing lines between them in Figures 5.24–5.26 will be convex upward.

Given that $^3He/^4He$ ratios at ocean islands can only safely be interpreted as minimum values (Section 4), we cannot use inter-island relationships in Figures 5.24–5.26 to document component $^3He/^4He$ ratios. For the case of

HIMU, we lack even the first-order constraint which would derive from $^3He/^4He$ analyses of St. Helena or Tubuai lavas. Even so, our preferred modes of production of HIMU, involving either magmatism and degassing or loss of a metasomatic fluid within the mantle ("intramantle degassing"), are expected to lead to enrichments of U over He in HIMU source regions. Essentially, we know of no simple mechanism which can effect a strong increase in U over Pb, without also increasing U over He. A hypothetical evolution curve for proto-HIMU is shown in Figure 5.13 as a heavy dashed curve labeled "extensively degassed mantle." The production of this component is modeled as a ~2 b.y. degassing of previously (4.4 b.y.) degassed MORB mantle. The arrow labeled "degassing" takes the $U/^4He$ ratio up past the $f_{3/4} = 0.01$ curve (arbitrarily), setting the stage for more than two orders of magnitude reduction in $^3He/^4He$ over the subsequent 2 b.y. Accordingly, we propose that HIMU will have a $^3He/^4He$ ratio which is close to or less than the atmospheric value (see also Allegre and Turcotte, 1985).

He concentrations in HIMU, as a function of U concentrations, can be gleaned from Figure 5.13. Based on a present-day $U/^4He$ ratio of ~2000, He concentrations will range between 7.5×10^{-6} and 7.5×10^{-5} std.cc/g for U concentrations of 15 and 150 ppb, respectively. These U concentrations are appropriate for metasomatically depleted peridotite and degassed basalt respectively. The mixing lines from DMM and BSE to HIMU (Figs. 5.24–5.26) are based on the lower of these U concentrations; taking the higher one, the BSE-HIMU mixing line will be nearly straight and the DMM-HIMU mixing line will be slightly convex downward.

Because of its affinity with upper continental crust and/or sediment, EM II is presumed to have $^3He/^4He$ in the range of HIMU. EM I, on the other hand, which may represent ancient metasomatized mantle, is likely to be at least moderately enriched in ^3He (Rison and Craig, 1982). He concentrations in EM I will be strongly dependent on the age of the material. He concentrations, if produced as suggested by the "metasomatism" arrow in Figure 5.13, will be on the order of 2.5×10^{-4} std cc/g (assuming 20–50 ppb U in metasomatized peridotite).

PREMA is shown near the top of the Loihi field in Figures 5.24–5.26; this is because of the coincidence of high $^3He/^4He$ Hawaiian and Icelandic localities with the overall PREMA isotopic character. As mentioned in Section 5.2.3, the high $^3He/^4He$ ratio for PREMA is by no means certain. In fact, PREMA-like intraoceanic arcs with $^3He/^4He$ ratios of ~6 Ra (Craig and Rison, 1983) argue against such a hypothesis. The He isotopic character of PREMA is, then, somewhat problematic, as is the existence of PREMA itself.

Having placed the various components on the He variation diagrams (Figs. 5.24–5.26), we see that the possibilities for generating individual island compositions are essentially endless, particularly in conjunction with possible recent accumulation of radiogenic ^{4}He. While it is tempting to call on mixing between BSE and either HIMU or DMM, perhaps in conjunction with late-stage radiogenic He accumulation, to produce Hawaiian and Icelandic arrays, mixing between PREMA, HIMU, and the EMs will serve as well. EM II islands (e.g., Samoa, Reunion, and Tahiti), identified on the basis of Sr, Nd, and Pb systematics, may represent mixing between the Hawaii-Iceland array out toward EM II.

We propose the mixing relationships in Figures 5.24–5.26 as an interim working hypothesis. Given the large number of possible components, isotope systematics alone cannot rule out the possibility that mixing can account for all observed He isotope systematics. That is, the relationships in Figures 5.24–5.26 are broadly consistent with those depicted in Figures 5.20–5.23. However, the large range in ^{3}He/^{4}He observed at Iceland and Hawaii (note that no He isotopic ratios have been measured at Koolau, which accounts for the Hawaiian "tail" in Figures 5.20–5.23), where large numbers of samples have been analyzed, is not reconcilable with restricted ranges in Pb, Sr, and Nd isotope ratios observed for these locales. For example, at face value, Loihi is an end member on Figures 5.24–5.26 whereas it occupies an interior position on the various Sr-Nd-Pb diagrams (Figs. 5.20–5.23). If we try to consider only maximum values for given islands, then an extreme bias is introduced because so many islands are characterized by only one or two ^{3}He/^{4}He determinations. Our best guess is that the relationships in Figures 5.24–5.26 are significantly obscured by recent accumulation of radiogenic ^{4}He at some or all of the OIB localities, perhaps in conjunction with the paucity of data for many localities. A more quantitative understanding of He isotope systematics, as they relate to Pb, Nd and Sr, will become possible as more He data are obtained in conjunction with U and Th determinations, and as more quantitative models for degassing time-scales and magnitudes at specific localities are developed.

5.4. The Pb Paradox

The so-called "Pb-paradox" refers to the observation that Pb in ocean islands and MORBs is, almost without exception, more radiogenic than a 4.55 b.y. bulk earth "geochron" (Fig. 5.23). If the age of the Earth is 4.55 b.y. (= chondrite age), the bulk earth must lie somewhere on this geochron. Because Pb in oceanic basalts comprises most of what is known of the interactive portions of the mantle, we must look to some less obvious reservoir of unradiogenic Pb, which lies to the left of the geochron, to balance the oceanic mantle. The paradox arises from the fact that no such reservoir can be unambiguously identified, although several suggestions, including the lower continental crust (e.g., O'Nions *et al.*, 1979), DMM (Anderson, 1982), the core (Vidal and Dosso, 1978; Oversby and Ringwood, 1971; Allegre *et al.*, 1980; Vollmer, 1977; Allegre, 1982), and the lower mantle (Anderson, 1983), have appeared in the recent literature. Before going on to evaluate these various suggestions, we will summarize pertinent constraints on U and Pb budgets in the crust and mantle.

5.4.1. U and Pb Budgets

A variety of Pb data for various average crustal rock types and composites has been compiled and is given in Table 5.7. There is a fair degree of uniformity among the various data sources, with the maximum range being 8.5–24.5 ppm Pb for both upper and lower crustal rocks. There is a small but significant difference in Pb between upper and lower crustal rock types; this is largely related to the more basic character of lower crustal rocks, and not to Pb depletion during granulite facies metamorphism. Relative to a PUM silicate earth, the crust appears to contain an unusually large fraction of the silicate earth's Pb inventory (37–70 percent; Table 5.7). This Pb enrichment is in contrast to values derived for crustal Sr and Nd inventories of 10 percent and 18 percent respectively (Table 5.7). Based on our survey of available data, this strong crust versus mantle enrichment in Pb appears inescapable.

Extreme variations of U in lower crustal (granulite facies) rocks hinder an accurate assessment of the U contents of, and distribution within, the continental crust. Our estimates are given in Table 5.7 and are in reasonably good agreement with Taylor's (1982) estimates (upper crust = 2.5 ppm U, lower crust = 0.63 ppm U, total crust = 1.25 ppm U), though they have been compiled largely from independent data sources. Relative to a PUM silicate earth, ~22–52 percent of Earth's U is in the continental crust.

Our estimates for U and Pb in the continental crust lead to rather low μ values (in terms of conventional wisdom), particularly for the upper crust (mean $\mu = 8.7$; range = 6–14; Table 5.7). In contrast, most data which exist for upper crustal materials strongly indicate its radiogenic nature ($^{206}Pb/^{204}Pb$ ratios for pelagic and terrigenous sediments range from 18.5 to 19.1 and from 18.8 to 19.8, respectively; Sun, 1980; White *et al.*, 1985). For crust with a mean age of 2.0 b.y. (Goldstein *et al.*, 1984; Allegre and Rousseau, 1984), a range of μ from 10 to 14 is required to generate $^{206}Pb/^{204}Pb$ ranging from 18.5 to 19.8. While the estimated uncertainties in upper crustal U and

Pb (Table 5.7) encompass μ values as high as 14, this is barely sufficient to generate the most radiogenic Pbs observed in oceanic terrigenous sediments. In terms of the total crust estimate, the uncertainty in μ is large enough to tolerate a crustal reservoir lying either to the left or the right of the bulk earth geochron, though the best estimate will be well to the left of the geochron.

Insofar as U appears to be slightly more incompatible than Pb in most igneous processes (Watson *et al.*, 1987), it might appear contradictory to speak of a crustal μ which is lower than BSE, as the crust is the end result of a series of melting and crystallization processes. However, we must consider that arc processes are complex, and appear to produce numerous other element fractionation patterns which contravene conventional wisdom (i.e., high field strength element depletions, unusual enrichments of Ba relative to La, etc; Morris and Hart, 1983). Insofar as many arc volcanics have Pb isotope signatures which appear to demand a sediment Pb component (Morris, 1984; Davidson, 1983), and in view of the enrichment of Pb relative to U, Nd, and Sr in sediments, there may indeed be a major fractionation (enrichment) of Pb relative to other LIL elements in the arc factories which produce continental crust. This contention is supported by the relatively low μ values measured in many arc volcanics (e.g., Tatsumoto and Knight, 1969; Oversby and Ewart, 1972; Tatsumoto, 1978).

Also shown in Table 5.7 is the average Pb content (and U, Sr and Nd) of 17 fresh MORB glasses, compiled from the literature. While the various element ratios may approximate those of the MORB mantle itself, the concentrations have obviously been raised by partial melting and fractional crystallization processes. Approximations to the content of U and Pb in the MORB mantle itself can be derived by referring back to Section 5.2.2, where Nd in MORB mantle was estimated at 0.6–1.0 ppm. Working from U/Nd and Pb/Nd ratios in MORB (Table 5.7), this would place the composition of MORB mantle, for 0.8 ppm Nd, at 13.3 ppm Sr, 0.0084 ppm U and 0.056 ppm Pb. Thus, relative to PUM, the MORB mantle is depleted in Nd, Sr, U and Pb by factors of 0.68, 0.68, 0.40 and 0.36 respectively (i.e., U and Pb are behaving more incompatibly than Sr and Nd).

5.4.2. Discussion

The silicate portion of the Earth is clearly depleted in Pb relative to the LOSIMAG C1 chondrite model. This is an old observation, and has been attributed to removal of Pb to the core or to volatilization or non-condensation of Pb during Earth accretion (e.g., Murthy, 1975; Oversby and Ringwood, 1971). In the event that any of these processes were synchronous with accretion, and provided that accretion took place rapidly at an age given by chondrites (~4.55 b.y.), then the 4.55 b.y. geochron accurately specifies the locus of possible BSE Pb isotope compositions. The Earth is also depleted in a variety of other volatile elements, many of which are not expected to be chalcophilic (core compatible; e.g., Sun, 1982), and thus, there is no compelling reason, except for the Pb paradox itself, to assert that the depletion of Pb in the silicate earth is due to "core-pumping." However, it is clearly possible that the Pb depletion is a result of *both* non-accretion and core-pumping. In the event that some of the "missing" Pb was co-precipitated with the Fe-Ni core, and that core formation took a finite time to run to completion (>100 m.y.), then the bulk silicate earth is not required to lie on a 4.55 b.y. geochron. Various two-stage isochrons are shown on Figure 5.32 to illustrate this process. First-stage Pb evolution is presumed to occur from 4.55 b.y. to t, with a μ of 0.21 (40 percent higher than the μ of a LOSIMAG C1 Earth). Second-stage isochrons (for various μ's) represent growth from t to present. If the mean age of core formation was as large as 150 m.y. (4.40 b.y. isochron), then a significant fraction of oceanic basalt Pbs will lie to the left of this isochron, and the Pb paradox is no longer an obvious problem.

While the $^{206}Pb/^{204}Pb$ ratio of PREMA is not well constrained, it is probably ~ 18.2–18.5; insofar as the continental crust is the total complement to PREMA, and insofar as the crust contains ~ 50 percent of the silicate earth Pb budget (Table 5.7), then, relative to PREMA, the crust must lie on the opposite side, and equidistant from, the relevant bulk silicate earth isochron. If the average crustal μ is indeed as low as 5.7, as suggested earlier (Table 5.7), PREMA plus CRUST will lie on a 4.55 b.y. geochron, and little core-pumping of Pb is required. If the crustal μ is the upper limit allowed (12), then about 150 m.y. of core-pumping is required. At this point, the most that can be said is that the missing Pb needed to explain the Pb paradox may be accommodated either in the crust or the core, or both.

We may also examine the other solutions to the Pb paradox introduced earlier. Can the depleted MORB mantle be the reservoir which "balances" the other oceanic Pb components? Certainly some MORB have Pb isotope compositions lying to the left of the geochron (see Fig. 5.32), although the great majority lie to the right. Using the data from Table 5.7, the estimate for Pb in MORB mantle is 0.056 ppm, and for DMM, which represents the unradiogenic end-member MORB, this Pb content is an upper limit. Relative to PUM, the MORB mantle contains only ~11 percent of the silicate earth Pb budget (assuming the MORB reservoir is 30 percent of the mantle mass), and again, DMM likely contains much less. Clearly the continental crust dominates the Pb budget of the Earth, and it is unlikely that an unradiogenic DMM

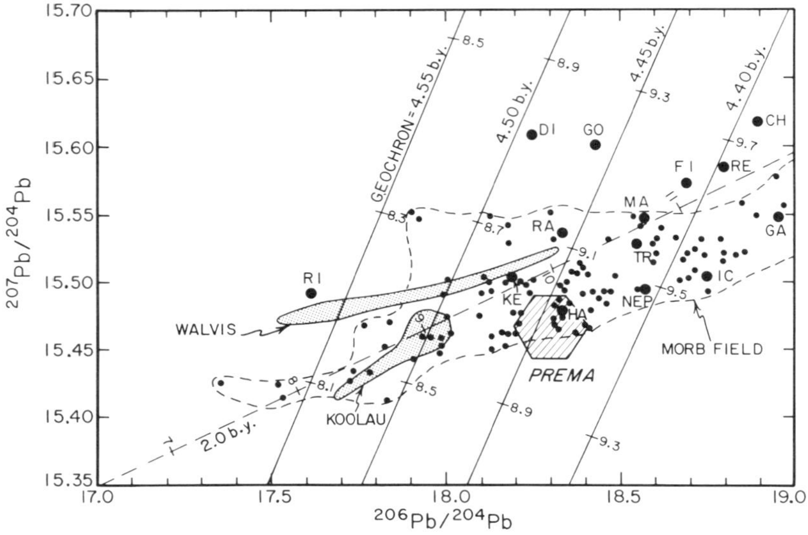

Figure 5.32. $^{207}Pb/^{204}Pb$–$^{206}Pb/^{204}Pb$ plot, showing MORB data (small dots), selected averaged oceanic islands (large dots), and the field for Koolau (Oahu) and Walvis Ridge basalts, in comparison with various terrestrial isochrons. The true age (t_0) of the earth (geochron) is taken to be 4.55 b.y.; the other isochrons represent possible core-mantle differentiation "ages." These isochrons are calculated for a first stage $\mu = 0.21$ from t_0 to t, and second stage from t to present. Various closed-system μ values for the second stage are indicated as cross-marks on the isochrons. Also shown for reference is a 2 b.y. secondary isochron (dashed), calculated with a first stage $\mu = 8.1$ (t_0 to 2 b.y.)—second stage μ values shown as cross-marks. This isochron may be used to estimate positions of various crust and mantle reservoirs having a ~2.0 b.y. mean age, as a function of second-stage-reservoir μ value. Shaded hexagon indicates possible position for PREMA. Data references given in Zindler *et al.* (1982), Hart (1984b) and Hamelin and Allegre (1985); Koolau data from Stille *et al.* (1983) and Hart (unpublished); primordial Pb parameters are from Tatsumoto *et al.* (1973). Oceanic islands: CH, Christmas; DI, Discovery; FI, Fiji; GA, Galapagos; GO, Gough; HA, Hawaii; IC, Iceland; KE, Kerguelen; MA, Marion; RA, Raratonga; RE, Reunion; RI, Rio Grande Rise; NEP, Northeast Pacific Seamounts; TR, Tristan da Cunha.

component can be effective in balancing the mantle segment represented by the oceanic Pb array, (presuming that much or all of the mantle, exclusive of the MORB reservoir, is characterized by radiogenic Pb similar to that observed in oceanic islands). Obviously one can postulate, on an ad hoc basis, that the lower mantle is unradiogenic (and also never sampled in erupted volcanics). For example, Anderson's (1983) model derives a lower mantle μ which is 25 percent lower than the μ of the upper mantle; however, his bulk silicate earth also has a very high μ (10.5) which would demand more than 150 m.y. of core-pumping of Pb.

It might appear to the reader that the Pb paradox is at least a "pink" herring, in view of the serious lack of constraints of terrestrial Pb budgets. While we admit the situation is murky at present, we feel it is an important issue insofar as it may be one of the few ways available to us to constrain the time interval of core formation if, in fact, Pb did partition into the core. In this context, however, it seems that a modern Pb budget is not a promising approach and that much better constraints are available by working with Pb isotope data from Archean rocks. Various work of this kind has been reported (Manhes *et al.*, 1979; Gancarz and Wasserburg, 1977; Tera, 1980); generally these give ages for the BSE (= core formation?) of 4.4–4.5 b.y., but generally with uncertainties which include the probable accretion age of 4.55 b.y.

5.5. Geochemical Constraints on the Constitution of the Mantle

The derivation of constraints on the relative volumes of geochemically distinct mantle components has received considerable attention in the recent literature (e.g., Jacobsen and Wasserburg, 1979; O'Nions *et al.*, 1979; Allegre *et al.*, 1980). Most of these models are based on only Sr and Nd systematics, and have assumed that the depleted MORB mantle is the sole complement of the enriched continental crust (the remainder of the mantle being BSE), and have thus not dealt with much of the geochemical complexity known to characterize the mantle. In addition, many modelers have chosen to utilize input parameters without considering the effects of substantial uncertainties known to be associated with them. Consequently, the conclusions based on these models (e.g., 32–45 percent of mantle depleted) should not be taken as evidence favoring layered mantle convection.

Goldstein *et al.* (1982) considered a more complex model, via inclusion of a HIMU component and U-Th-Pb systematics, and investigated the effects of parameter uncertainties using a Monte Carlo approach. They concluded that the mass fraction of the depleted mantle is ill-constrained by the present-day geochemical data base and could conceivably range from <20 percent to >90 percent of the mantle. They also concluded that the HIMU component could be geochemically similar to E-type MORB (Sun, 1980). Allegre *et al.* (1983b,c) used an inversion method to investigate the effects of input uncertainties and, even without the inclusion of a HIMU component, concluded that the error bounds on the estimate of depleted mantle mass were large and could encompass values ranging from 30 percent to 90 percent of the entire mantle. Even with clearly over-simplified models, it has not been possible to delineate definitive constraints on the mass fraction of depleted mantle.

One of the principal uncertainties in these mass balance models was the fact that initial (bulk earth) values for Sr and Nd concentration were poorly known. The refinement afforded by our PUM model should enable a much better constraint to be realized from these simple crust-mantle budget models. Allegre *et al.* (1983b,c) inverted data for two model variants, one where the depleted mantle was isotopically characterized by average MORB, and one which used the most depleted MORB (on the basis that most MORB are polluted by material from less depleted reservoirs). Using Nd $=$ 1.17 ppm for PUM, and 32.5 ppm for the continental crust (Hart and Zindler, 1986) the mean MORB model gives a depleted mantle mass of 65 $\pm$ 17 percent, the extreme MORB model gives a depleted mantle mass of 48 $\pm$ 8 percent. Neither result is at all consistent with the MORB mantle being confined above 670 km ($\sim$30 percent).

In this section, we derive constraints on the nature of mantle reservoirs by making simplifying assumptions which are consistent with the degree of chemical complexity known to characterize the mantle. It is appropriate here that we clarify our usage of the terms "reservoir" and "component." We have used component, in the preceding discussion, to refer to a geochemically distinct mantle segment that is homogeneous on scales sampled by the smallest effective melting volume. Obviously it may also be homogeneous on smaller or larger scales. Thus a component may comprise the entire upper mantle or be dispersed as meter- or even centimeter-scale veins or blobs. In contrast, we use the term reservoir to denote a portion of the mantle which is under discussion; it need not be chemically homogeneous nor physically contiguous. That is, we can speak of the MORB source reservoir, which is thought to be physically contiguous but chemically heterogeneous, comprising HIMU and EM components in a DMM matrix, or we can refer to an enriched reservoir which includes the sum of HIMU and EM components, and thus is neither contiguous nor homogeneous.

As a starting point, we consider the simplest form of crust-mantle mass balance where the mantle and crust constitute complementary reservoirs. In addition, we consider only the Sm-Nd system as, geochemically speaking, it constitutes the best-behaved parent-daughter pair. This approach does not deny that the mantle is very heterogeneous; it simply asks what the mean composition of the heterogeneous mantle is. The mass balance equation for this model may be rewritten (from Allegre *et al.*, 1983b):

$$\mu_m = (0.0064\,Nd_c\mu_c - \mu_t Nd_t)/(0.0064\,Nd_c - Nd_t),$$

where μ is $^{147}Sm/^{144}Nd$, Nd is concentration in ppm, and subscripts c, m and t refer to crust, depleted mantle, and bulk earth, respectively. The factor 0.0064 is simply the mass ratio of continental crust to total mantle plus crust (Turcotte and Schubert, 1982).

The selection of input parameters is given in the caption to Figure 5.33, with uncertainties which represent a subjective estimate of the maximum allowable range. The value for Nd_t is taken as that in PUM (see above). Fortunately, the three quantities which are least well known (μ_c, Nd_c, Nd_t) all become error-demagnified by about a factor of 10 during propagation through the above equation. Errors in μ_t are slightly magnified, but this value is also known very well. The calculated value for μ_m is 0.215 ± 1.1 percent; the error in this value was propagated by allowing the errors in the individual parameters to combine in the worst possible way. The maximum value for μ_m occurs when μ_t and Nd_c are maximized and μ_c and Nd_t are minimized (and vice versa); this infers a strong negative correlation between μ_c and Nd_c whereas in fact the available data suggest a very mild negative correlation at best. The calculated error bounds on μ_m are thus fairly conservative. Figure 5.33 also shows how μ_m varies as a function of Nd_c only. From simple mass balance, using the chosen parameters, ~18 percent of the Earth's Nd budget is stored in the crust (Table 5.7), and the Nd concentration of mean depleted mantle is 0.96 ppm (18 percent depleted from 1.17 ppm).

This model value of μ_m can be used to derive the $^{143}Nd/^{144}Nd$ of mean depleted mantle if we can estimate the mean age of the total crust-mantle system. As discussed by Allegre *et al.* (1983b,c), the continental crust clearly shows a large range in age; however as proven by Allegre *et al.* (1983b,c), the mean age of a system of variable age reservoirs is the same as the model age of an equivalent simple two-stage system. If continental crust extraction has occurred at a constant rate over the age of the Earth, the mean age of the system will be ~2.28 b.y.; if crust extraction was more rapid during the Archean when there was more heat available for differentiation, as argued for by Allegre and Rousseau (1984), then the mean age will be older.

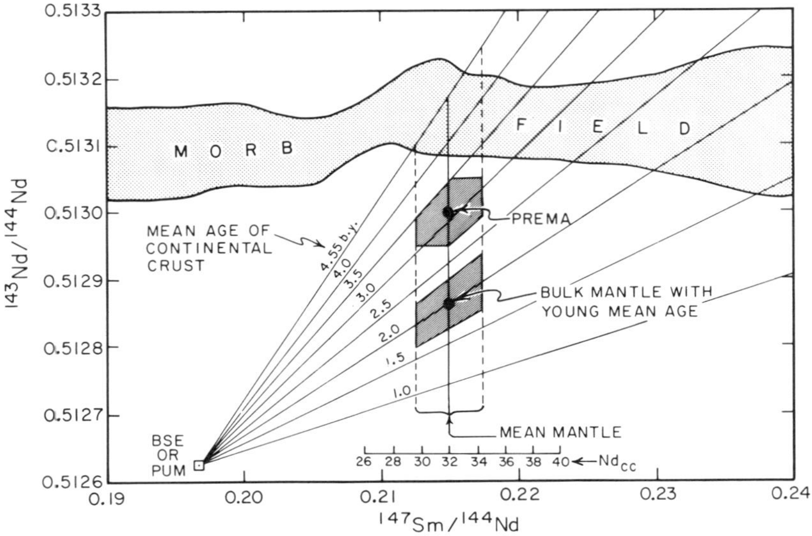

Figure 5.33. Sm/Nd isochron diagram showing various mean age isochrons, and the value for $^{147}Sm/^{144}Nd$ of the mean depleted mantle (= 0.2148 ± .0024) calculated from a bulk earth-continental crust depletion model. MORB field taken from literature values. Parameters and uncertainties chosen for model: primitive mantle (bulk earth) Nd = 1.17 ± 0.02 ppm (PUM value, see text), $^{147}Sm/^{144}Nd$ = 0.1967±.0003 (Jacobsen and Wasserburg, 1984); continental crust $^{147}Sm/^{144}Nd$ = 0.113 ± 0.003, Nd = 32.5 ± 1.5 ppm (Goldstein *et al.*, 1984; Taylor *et al.*, 1983; Allegre and Rousseau, 1984; Michard *et al.* 1985; Allegre *et al.*, 1983c). The effect of other choices for crustal Nd concentration may be judged from the scale labelled Nd_{cc}. Shaded polygons delimit solutions for two choices of mean age: 2.0 ± 0.3 b.y. and 3.1 ± 0.4 b.y. The lack of significant overlap between the mean depleted mantle field and that of MORB mantle simply means that the whole mantle is, on average, unlikely to be of MORB isotopic composition, unless the mean age of depletion is significantly greater than 3.5 b.y. See text for discussion of error propagation.

However, for either of these scenarios, significant recycling of average crustal material into the mantle, which is consistent with Sm-Nd systematics in modern and ancient sediments (Goldstein *et al.*, 1984; Allegre and Rousseau, 1984) will tend to push the crust toward younger mean ages (perhaps as young as 1.7 b.y.). If all or a significant fraction of the continental crust was extracted early, as in continental big-bang models (e.g., Patterson and Tatsumoto, 1964; Armstrong, 1968, 1981), the mean age might be as old as 3.8 b.y. All things considered, there is an extremely large range of mean ages which have been proposed as characterizing the continental crust.

We have chosen to consider the consequences of two "end-member" mean-age ranges for the continental crust: 2.0 ± 0.3 b.y. corresponding to mean mantle $^{143}Nd/^{144}Nd$ of $0.51286 \pm ^{8}_{6}$; and 3.1 ± 0.4 b.y. with mean mantle $^{143}Nd/^{144}Nd$ of 0.51300 ± 5 (Fig. 5.33). The latter estimate has the additional significance that it corresponds to the composition of PREMA. Based on the concentration of Nd and the $^{143}Nd/^{144}Nd$ ratio in the MORB reservoir (0.5–0.9 ppm; 0.51315, see Fig. 5.28), we can now investigate mass and compositional relationships between the MORB reservoir and the remainder of the mantle, which we refer to as the "complementary reservoir." This complementary reservoir is enriched relative to the MORB reservoir, though it may be depleted or enriched relative to BSE or PUM. Figure 5.34 shows the relationship between the relative masses of the MORB and complementary reservoirs and the $^{143}Nd/^{144}Nd$ of the complementary reservoir for the two mean mantle $^{143}Nd/^{144}Nd$ ratios estimated above. Figure 5.34 shows that as the MORB reservoir becomes larger, $^{143}Nd/^{144}Nd$ in the complementary reservoir decreases. Contours for Nd concentration in the complementary reservoir are also shown in Figure 5.34 and document a positive correlation between this parameter and the mass of the MORB mantle. Comparison of Figures 5.34a and b shows, predictably, that for a given MORB reservoir mass, the Nd isotopic composition of the complementary reservoir correlates positively with that of the mean mantle.

If the complementary reservoir is primitive (BSE or PUM), as has been assumed in most geochemical models, then the relationships in Figure 5.34 can be used to constrain the mass fraction of that PUM reservoir. This is obtained by ascertaining the mass of the complementary reservoir which corresponds to the intersection of the PUM $^{143}Nd/^{144}Nd$ value with the PUM (1.17 ppm) Nd contour. The resulting mass fractions for a PUM reservoir are ~30 and 48 wt.%, respectively for the models shown in Figure 5.34a and b. Taken at face value, this result suggests that if any primitive mantle volume survives today, it must constitute significantly less than the ~70 percent of the mantle which lies below the 670 km seismic discontinuity. This result lies within the permissible ranges calculated by Goldstein *et al.* (1982) and Allegre *et al.* (1983c); the higher precision and, therefore, enhanced significance of the current result is largely due to our improved understanding of Nd in PUM (see Section 2).

The current modelling does not rule out the existence of a PUM reservoir (or component) today, it simply constrains its size. Having established that such a reservoir cannot comprise the entire lower mantle (below 670 km), it

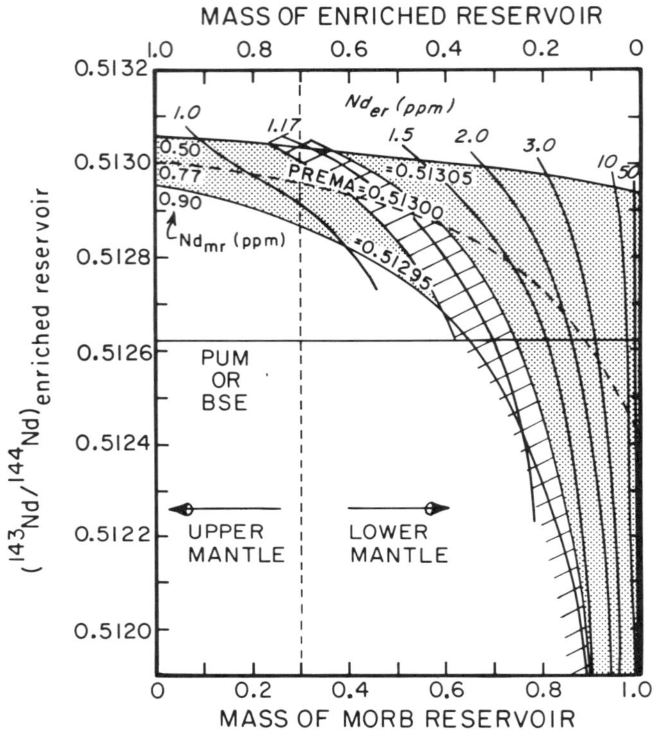

a

Figure 5.34. $^{143}Nd/^{144}Nd$ for the complementary enriched reservoir as a function of the mass of the MORB source reservoir (expressed as a mass fraction of the whole mantle). Panels a and b are for different mean mantle compositions corresponding to "old" and "young" mean ages for the continental crust, respectively (see Fig. 5.33). In Panel a, the mean mantle has the composition of PREMA. "Upper" and "lower" mantle volumes are defined by the position of the 670 km seismic discontinuity. The shaded areas correspond to the uncertainties indicated in Figure 5.33 for the two mean mantle compositions in conjunction with Nd concentrations in the MORB reservoir (Nd_{mr}) ranging from 0.5 to 0.9 ppm. Nd concentrations in the complementary enriched reservoir (Nd_{er}) are shown as contours across the shaded areas. $^{143}Nd/^{144}Nd$ in PUM or BSE is shown as a horizontal line across the diagrams. Approximate error bounds are shown for the $Nd_{er} = 1.17$ (= PUM) contour. The intersection of this error band with the PUM $^{143}Nd/^{144}Nd$ value defines the mass fraction of the complementary reservoir, if it comprises primitive, undifferentiated mantle. Resulting estimates for such a PUM reservoir are 26–38 percent and 42–53 percent of the mantle for models in Panels a and b, respectively. This suggests that if a primitive, undifferentiated mantle reservoir exists, it cannot constitute the entire lower mantle (~70 percent).

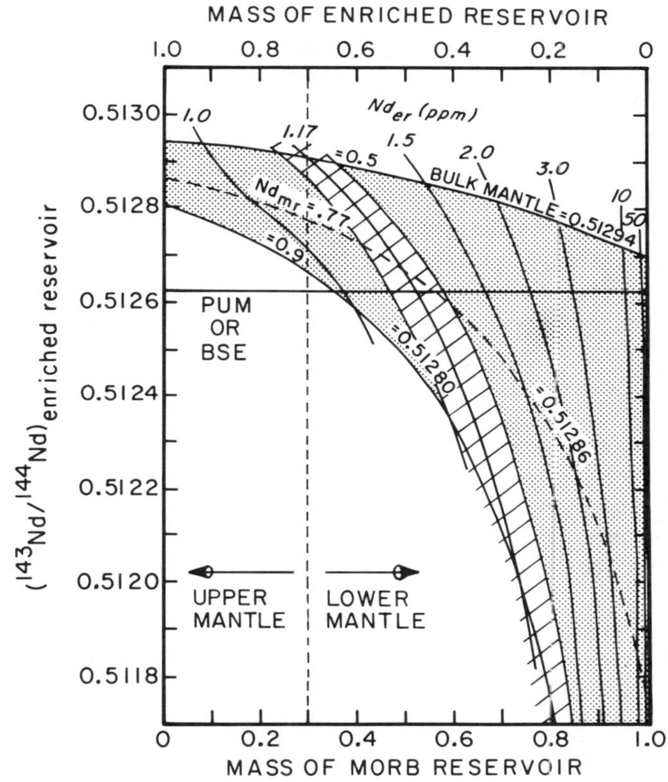

b

Figure 5.34. *Continued*

follows that the lower mantle must be at least somewhat differentiated (depleted). This implies a material flux into and/or out of the lower mantle at some time during Earth history and, thus, is not compatible with the oft-proposed persistent unbreachable chemical boundary layer at 670 km; such a boundary may exist today, but it cannot have existed continuously since the earliest Archean. Further, it seems likely that there is no convectively isolated reservoir for PUM today, and if PUM survives, it must exist as a dispersed component within a convecting medium. Hence, arguments which have cited the need for convective isolation of a PUM or BSE component (e.g., O'Nions and Oxburgh, 1983) need to be re-evaluated.

Earlier, we suggested that PREMA may be a physically distinct mantle component. If we now consider that the complementary reservoir is PREMA then it must be volumetrically dominant, with the MORB reservoir constituting less than ~10 percent of the mass of the mantle (Fig. 5.34). Alternatively, if the complementary reservoir comprises PREMA + HIMU + EM I + EM II ± PUM, then the relative masses of the MORB and complementary reservoirs are ill-constrained.

These mass balance considerations, together with component characterizations, document a mantle which is composed of DMM, HIMU, EM I and EM II components, and may also include PREMA and/or PUM. The MORB reservoir is volumetrically dominated by DMM (due to its highly depleted character) but contains various amounts of the enriched components, so that its mean composition is essentially that of average MORB (see Fig. 5.28). Several lines of reasoning suggest that the MORB reservoir resides in the uppermost mantle: 1) it is available for processing at all active spreading centers; 2) its depleted character likely results, at least in part, from the incorporation of recycled depleted harzburgite which will be somewhat less dense than surrounding "fertile" peridotite; and, 3) with the exception of normal ridge segments, the isotopic signature of depleted MORB is only observed in ultramafic nodule suites and in some portions of orogenic lherzolites (see Figs. 5.30 and 5.31), which are undoubtedly derived from the uppermost mantle.

This heterogeneous MORB reservoir is underlain, or grades downward into, mantle with an overall more enriched isotopic character. If the mean $^{143}Nd/^{144}Nd$ of this underlying reservoir is taken to be no lower than the most enriched OIB (~0.5123—Kerguelen), then the relationships in Figure 5.34 allow it to account for as little as 20 percent of the mantle, or, based on a mean isotopic character similar to the mean mantle composition, as much as ~90 percent of the mantle. The geochemical character of this relatively enriched underlying reservoir may be satisfied by any number of scenarios involving variable fractions of HIMU, EM I, EM II, PREMA, PUM, and DMM. The relatively enriched reservoir could contain dispersed enriched components in the DMM matrix, differing from the MORB reservoir only in the relative abundance or representation of the enriched components, or it could represent a convectively isolated mantle domain.

Three factors tend to preclude derivation of definitive geochemical constraints on the placement and physical relationships of the various geochemical components within the mantle: 1) most observed isotopic compositions represent mixtures between two or more components; 2) studies of small seamounts (Zindler *et al.*, 1984), orogenic lherzolites (Richard and Allegre, 1980; Polve and Allegre, 1980; Reisberg and Zindler, 1985, 1987), and ultramafic nodule suites (e.g., Stosch *et al.*, 1980; Menzies, 1983; Zindler and Jagoutz, 1988) document small-scale heterogeneities within the upper mantle; and 3) the number of components identified here, considered to be a minimum, precludes unique definition of mixing scenarios to explain isotope ratios observed in specific tectonic settings. As geochemical philosophers, unaware of the various and sundry geophysical constraints on the structure of the mantle, we can objectively make and defend the following statements:

1. The MORB source reservoir, which is situated in the uppermost mantle, contains small scale heterogeneities.
2. The MORB reservoir cannot constitute the whole mantle unless the mean age of the continental crust is on the order of 4.0 b.y.
3. The MORB source may either grade downward into mantle which is, on average, less depleted, or it may abruptly contact such mantle across a mid-mantle chemical boundary layer.
4. Mantle of PUM composition cannot constitute more than half of the mantle. Some depleted mantle must be present below the 670 km discontinuity.
5. The mantle below the MORB reservoir may represent a mixture similar in style to the MORB reservoir, but with less DMM component, or it may be highly stratified with homogeneous or heterogeneous layers.
6. The southern hemisphere Dupal anomaly (Hart, 1984b) documents mega-scale lateral heterogeneity within the mantle (Hart, 1984b; Dupre and Allegre, 1983).
7. Basalts with PREMA isotopic characteristics are found in a wide variety of oceanic, arc and continental tectonic settings. These basalts must represent melting of a PREMA reservoir, or a highly reproducible mixing between other components.
8. The constancy of $^{3}He/^{4}He$ ratios in MORBs likely results from a mixing between He in the various components which comprise the MORB reservoir. High $^{3}He/^{4}He$ components may include PUM, EM I, and/or PREMA; low $^{3}He/^{4}He$ components may include DMM, EM II, and HIMU. The small range of $^{3}He/^{4}He$ ratios in MORB, relative to the other isotope ratios of interest, documents an enhanced mobility of He in the upper mantle, due either to high diffusion rates in the presence of melt, or to migration of volatile-rich phases in the uppermost mantle.
9. The association of excess ^{129}Xe and high $^{40}Ar/^{36}Ar$ in some of the most depleted MORBs, (e.g., Allegre *et al.*, 1983a) reflects the early ($\geq$4.4 b.y.) and continuing degassing of at least some portions of the MORB reservoir (e.g., Staudacher and Allegre, 1982; R. Hart *et al.*, 1985; Sarda *et al.*, 1985).

10. Mechanisms which account for the production of EM I, EM II, and HIMU components operate in conjunction with magmatism or metasomatism and associated tectonic activity within the mantle. Some or all of the proposed mechanisms may be operative in subduction environments and, thus, some enriched mantle components (e.g., EM II) may incorporate or represent recycled crustal materials.

Together, these statements constitute a geochemist's view of the mantle and, without the benefit of geophysical constraints, say disappointingly little regarding the physical structure of the mantle. We offer these as hopefully objective constraints which can be used to build mantle models in the context of the various geophysical constraints which are being developed. In the next section, we consider the interplay of these geochemical constraints with the geophysical constraints, as seen through the eyes of a geochemist.

6. The Nature of Mantle Convection

Perhaps the most fundamental question in solid-earth geophysics today pertains to the nature of the ~670 km seismic discontinuity (e.g., Dziewonski *et al.*, 1975; Dziewonski and Anderson, 1981). Early workers (e.g., Anderson, 1967; Ringwood, 1975; Anderson, 1976) tended to attribute the seismic discontinuity to a phase change in a chemically homogeneous mantle. Recently, a number of compelling arguments have been made in support of a compositional boundary at 670 km. The existence of a chemical boundary layer at 670 km (with higher SiO_2 and/or FeO in the lower mantle) is expected to act as a barrier to convection and limit mixing between upper and lower mantle reservoirs (e.g., Richter and McKenzie, 1981; Olsen and Yuen, 1982), whereas convection across a phase boundary is likely permissible (e.g., Schubert and Turcotte, 1971). Thus, resolution of this question is critical to the development of realistic models for the chemical and thermal evolution of the Earth.

Both whole-mantle and layered-mantle camps have argued persuasively for their respective mantle models. Unfortunately, individual authors have not always addressed apparently conflicting arguments made on the basis of similar observations by other authors. The current state of the art is, therefore, difficult to assess, particularly for the naive but well-intentioned geochemist. Hence, in the following paragraphs, we present a summary of the present state of geophysical arguments which pertain to the nature of the 670 km discontinuity.

6.1. Underside Reflections

The high reflection coefficient inferred for the 670 km discontinuity on the basis of P′670P′ underside reflections (Richards, 1972) suggests that the transition in physical properties at this boundary occurs over 5 km or less. Through comparison with calculated reflection coefficients, based on equilibrium phase changes occurring in probable mantle mineral assemblages (Jeanloz and Thompson, 1983), Lees *et al.* (1983) have shown that observed reflectivities cannot be produced by phase transitions alone, and require an associated discontinuous change in chemical composition. However, Muirhead (1985) has summarized arguments, largely from the works of Cleary, which apparently permit an alternative interpretation, for much of the P′670P′ data, involving a core-mantle scattering phase. Richards (personal communication) acknowledges the viability of the Muirhead-Cleary hypothesis for some of the P′670P′ reflections, but asserts that in other cases, the evidence for a mid-mantle reflector is incontrovertible. It is noteworthy that the P′670P′ reflection is not observed at all localities (Richards, personal communication) and thus the reflection at 670 km may be laterally discontinuous.

The discontinuous nature of the 670 km discontinuity is supported by a comparison of two recent investigations of different subduction zones. Jordan (1977) and Creager and Jordan (1984) have compiled evidence for a high velocity path extending to depths of at least 1,000 km along an extension of the subducting slab in the Kuriles. These authors believe that this documents an aseismic portion of the slab itself which is extending into the lower mantle. In contrast, Giardini and Woodhouse (1984) have shown that seismicity along the Benioff Zone at Tonga documents the en echelon stacking of broken slab fragments, suggesting the impenetrable character of the 670 km discontinuity at this locality. We suggest that at the latter locality, true P′670P′ reflections might be very sharp whereas, at the former locality, they might be missing all together.

6.2. Convective Destruction of Heterogeneities

The need to preserve chemical heterogeneities within the mantle has led a number of convection dynamicists to investigate the fate of such heterogeneities in a convecting mantle (e.g., Richter and Ribe, 1979; Richter *et al.*, 1982; Olsen *et al.*, 1984; Hoffman and McKenzie, 1985). The results of these studies are in general accord; in interpreting these results, it is necessary to consider the scale of the initial heterogeneity and scale over which the resulting mixed mantle is sampled. Within this context, the results of the modelling can be summarized as follows: 1) large heterogeneities (similar in scale to the depth of the convecting layer) require very long time scales for dispersal, and regional (large wave length) variations will likely persist for the age of the Earth; 2) small-scale heterogeneities will be mixed over "regional" volumes over time scales significantly shorter than the age of the Earth; and 3) total destruction of even the smallest scale heterogeneities depends on chemical diffusion, and thus, the chemical heterogeneity observed in a series of melts will be proportional to the scale of sampling in the source region. It is important to note that in all of this modelling, heterogeneities were considered to be passive, that is, rheologically indistinguishable from the ambient mantle matrix. It is, therefore, somewhat difficult to generalize these results to the more realistic situation where the heterogeneities are rheologically dissimilar to the surrounding mantle (such as subducted eclogite in a peridotite matrix). In such cases, the survival of heterogeneities may be fostered by distinct rheologies, and, in fact, density differences may produce a heterogeneity gradient with depth (e.g., Davies, 1983). A further consequence of distinct chemical properties is a difference in solidus temperatures, which may facilitate sampling of even the smallest scale heterogeneities during melting (Zindler *et al.*, 1984; Sleep, 1984). All in all, we see nothing in the modelling which precludes the long term survival of heterogeneities within a convecting mantle domain, at least when very small scale lengths are considered to be important. We therefore contend that convective isolation is not necessary to preserve ancient mantle heterogeneity.

6.3. Physical Properties of Mantle Minerals

The elastic properties of mantle phases are being used to test the suitability of candidate bulk compositions to explain observed seismic velocities and inferred densities at various mantle levels (Lees *et al.*, 1983; Bass and Anderson, 1984). Results from both of these studies agree that above 400 km, both olivine- and pyroxene-rich assemblages ("pyrolite" and "piclogite," respectively) can satisfy seismic constraints. This may also be true for the lower mantle below 670 km, though Anderson (personal communication) and Knittle *et al.* (1985) now consider, respectively, that the relatively low velocity and density of olivine in the lower mantle (perovskite plus magnesiowustite) may preclude an olivine-rich (e.g., pyrolite) lower mantle. Within the transition region, however, there are a number of problems: 1) all of the phases investigated by Lees *et al.* (1983) (olivine, orthopyroxene, and garnet) are too fast to satisfy PREM or PEM Earth models (Dziewonski *et al.*, 1975; Dziewonski and Anderson, 1981), even when uncertainties in the calculations are propagated and evaluated (Lees *et al.*, 1983); and 2) all of the deeper phase transitions summarized by Jeanloz and Thompson (1983) and studied by Lees *et al.* (1983) occur at substantially shallower depths than can be made consistent with a 670 km seismic discontinuity. Bass and Anderson (1984) and Anderson (personal communication) have suggested that these apparent enigmas may be explained if Ca-rich pyroxene persists in its low-pressure manifestation through much of the transition region and if increased alumina in the garnet and pyroxene component pushes the perovskite transition to higher pressures. The contentions of Bass and Anderson (1984), taken at face value, suggest that the transition region is rich in clinopyroxene and, therefore, that piclogite is more appropriate than pyrolite as a bulk composition, and that eclogite cannot penetrate the 670 km discontinuity because of its retarded transformation to perovskite (Bass and Anderson, 1984; Anderson and Bass, 1986). On the other hand, Weidner (1985) has argued that a classic pyrolite composition is consistent with seismic data in and above the transition region. Because cosmochemical constraints are seriously violated by a piclogite transition region and lower mantle, Anderson and Bass (1986) propose that the 670 km discontinuity represents a fundamental chemical boundary across which the piclogite transition region is in contact with an orthopyroxene-rich lower mantle. They suggest that this chemical discontinuity was formed as a result of extensive melting of the Earth during accretion, and that the enstatite-rich lower mantle is essentially devoid of incompatible trace elements. This model is consistent with the physical properties inferred for lower mantle materials (e.g., Knittle *et al.*, 1985), tends to account for most of the MgO and some of the SiO_2 deficiency in the upper mantle discussed in Section 2, and would not be expected to affect most refractory lithophile element ratios in upper mantle materials. However, this model leads to absolute abundance levels of these elements in the upper mantle which are 2.5–3 times higher than the PUM value derived in Section 2. No upper mantle materials are as yet known which are both this enriched in refractory lithophile elements and which preserve chondritic abundance ratios of these elements. At the present time, the uncertainties associated with the basic physical properties data are significant and, as suggested by the transition region problems mentioned above, may be more important than is indicated by the experiments. Although a chemically layered mantle appears favored in current interpretations of the physical properties data, the data cannot preclude a more-or-less homogeneous composition (e.g., pyrolite or PUM) extending from the upper mantle, through the transition region, into the lower mantle.

An interesting consequence of Anderson's hypothesized sub-670 transformation of eclogite to perovskite, in conjunction with the density relationships shown by Lees *et al.* (1983) and Bass and Anderson (1984), is that the transition region would act as a "density trap" for subducted eclogite. Having undergone the eclogite transformation, subducted oceanic crust will drop through the upper mantle into the transition region where it becomes neutrally buoyant. Penetration into the lower mantle will be retarded by the alumina-induced late transformation to the perovskite structure (Anderson, personal communication). The production of eclogite through the generation and subduction of oceanic crust can account for something like 12 percent of the mantle, if the average rate of crust production has been about twice the present rate. The trapping of this eclogite in the transition region can satisfy the velocity constraints of the Bass and Anderson (1984) model without invoking fractional crystallization of an early magma ocean, as suggested by Anderson and Bass (1986).

6.4. Thermal Structure and Viscosity of the Mantle

Observations pertaining to postglacial rebound, non-tidal acceleration, true polar wander, and changes in the degree two component of the Earth's gravitational potential field all suggest that the viscosity of the mantle is constant, within about a factor of 5, across its entire thickness (Peltier and Andrews, 1976; Peltier, 1981; Peltier, 1983). This condition is most easily met in the context of whole-mantle convection models. In terms of conventional wisdom, a constant viscosity across a mid-mantle thermal boundary layer, with an associated temperature increase of 500–700° (Jeanloz and Richter, 1979), implies that the melting temperature (*Tm*) of the lower and upper mantle materials would have to be different by a comparable amount if, in fact, viscosity is proportional to homologous temperature (*Tm*/*T*; e.g., Schubert and Spohn, 1981; Spohn and Schubert, 1982). However, accepting a mantle with nearly constant viscosity, only *one* of the following conditions must hold: 1) there is no thermal boundary layer at 670 km; 2) the change in chemical properties associated with a chemical discontinuity at 670 km fortuitously results in a nearly constant homologous temperature across the boundary; or 3) the relationship between the homologous temperature and viscosity is not as simple as has been supposed. Although the contention that the mantle is isoviscous has been taken as support for whole-mantle convection (e.g., Peltier, 1981; Spohn and Schubert, 1982), this requirement is not absolute as there are clearly other alternatives.

Assuming a constant homologous temperature throughout the mantle,

Spohn and Schubert (1982) have further argued that the temperature increase across a mid-mantle thermal boundary layer predicts core-mantle temperatures significantly in excess of the 3200–3500 K estimated by Jeanloz and Richter (1979). Jeanloz (personal communication) has made preliminary measurements of the melting temperature of perovskite as a function of pressure which suggest little or no pressure dependance of *Tm*. In their calculation, Spohn and Schubert (1982) assumed a melting temperature gradient of 0.64°C/km for the lower mantle and, thus, the new data of Jeanloz may be qualitatively consistent with an isoviscous lower mantle and reasonable core-mantle temperatures, even in the presence of a thermal boundary layer 670 km. Essentially, this result predicts that the geotherm proposed by Jeanloz and Richter (1979) may be reconcilable with an isoviscous lower mantle.

6.5. Mantle Heat Flow and the Urey Ratio

The ratio of heat flow to heat production in the Earth (the so-called Urey ratio) can potentially constrain the nature of mantle convection by offering a measure of the efficiency of heat removal from the mantle (Richter, 1984). Using heat flow data from Sclater *et al.* 1980) and estimates for PUM U (20.8 ppb; Table 5.4) with Th/U–3.8 and K/U–1.27 $\times$ 10^4, the Urey ratio is approximately 2.1. However, eliminating that fraction of the total Earth heat flow (~15 percent) and heat production (~37 percent; equivalent to total crust U–1.20 ppm; see Table 5.7) thought to originate in the continental crust where thermal steady state is closely approached, we obtain a "reduced" Urey ratio for mantle heat flow to heat production of about 2.8. The conclusion from this is that mantle heat flow requires a significant component (~64 percent) of secular cooling. As discussed by Richter (1984), predictions from whole-mantle convection calculations suggest <20 percent secular cooling (given no significant heat source within, or cooling of, the core) due to the efficiency of heat removal in these models. Layered models, on the other hand, are capable of accounting for much larger percentages of secular cooling (~50 percent). Typical convection calculations, however, ignore the insulating effects of continental lithosphere, a consideration which will tend to increase the secular component available in whole- and layered-mantle models. Even so, taken at face value, these arguments suggest either a layered mantle or a significant heat source in the core, given that no more than about 5–10 percent of the present day heat flow can result from cooling of the core (Stacey, 1969). Speculation regarding the nature and magnitude of possible core heat sources is not fruitful at this time other than to say that K has long been considered a possibility (Hall and Murthy, 1971; Lewis, 1971), and that more than about one silicate-earth budget of K in the core would violate cosmochemical constraints related to K depletion during accretion of the Earth (Sun, 1982). One silicate earth budget of K in the core would account for 15 percent of the present total Earth heat flow (at steady state), or up to 60 percent of the total heat flow if the core heat is fossil production from a few billion years ago. The above analysis assumes that the heat flow estimates of Sclater *et al*. (1980) are representative of time-integrated values. If the rate of present-day volcanic activity at mid-ocean ridges, and associated heat flow, is higher than the time-integrated average, then the appropriate mantle Urey ratio would be smaller than that calculated here. Evaluation of the importance of this consideration is beyond the scope of this paper. Further, Richter, Peltier, and Turcotte (personal communication) all concur that, when realistic uncertainties in the convection calculations are considered, the possibility of an extensive secular component in whole-mantle convection models cannot be excluded.

O'Nions and Oxburgh (1983) have argued that the relationship between the oceanic ^4He flux (Craig *et al.*, 1975) and heat flow supports the contention of a depleted upper mantle which is convectively isolated from a primitive lower mantle. Implicit in their reasoning are the assumptions that ^4He and heat are removed from the upper mantle with the same efficiency (i.e., that the entire oceanic crustal section, including both extrusive and intrusive components, is outgassed completely) and that radiogenic ^4He has a very short residence time in the upper mantle. This is in marked contrast to their contention that heat is transported across the 670 km boundary layer much more efficiently than is He. In principle, the He flux and heat flow data could be used to calculate a "fractionation factor" for heat and He at the ridge.

He arguments aside, the conclusions of O'Nions and Oxburgh (1983) are essentially based on the inferred depletion of U in the MORB source reservoir (a fact easily documented by examination of U in fresh MORBs) in conjunction with convection arguments similar to those summarized by Richter (1984). In this context, we note that the U budget of O'Nions and Oxburgh (1983) ($\sim$1.5 ppm U in the crust and 20 ppb U in the BSE) has about 47 percent of the Earth's U concentrated in the continental crust, and thus precludes total convective isolation of a primitive lower mantle (since the mantle above 670 km constitutes only $\sim$30 percent of the total mantle).

6.6. Discussion

The geophysical arguments summarized above do not clearly support either the whole-mantle or layered-mantle convection hypotheses; at some level, all observations could be reconciled with either hypothesis, although some special begging would be required. Based on the geochemical arguments presented in the preceding section, we contend that, if the lower mantle is chemically distinct and convectively isolated from the upper mantle, it does not have a primitive or undifferentiated PUM (or BSE) composition. Thus, if there is an unbreachable chemical boundary layer at 670 km at present, it could not have existed for all of Earth history. At some time, this boundary must have been permeable, to permit differentiation (depletion) of the lower mantle. We also know that the MORB source reservoir, which probably comprises the uppermost mantle, cannot represent the mean composition of the mantle (see Fig. 5.35) and therefore, the present-day mantle must contain large-scale chemical gradients or discontinuities (though, as discussed earlier, these heterogeneities are largely documented by trace elements and isotopic ratios, and need not infer significant major element variations).

A compromise model which is compatible with many geochemical and geophysical observations might involve a chemical boundary layer at 670 km which is subject to periodic instability in time and/or space (similar to the model proposed by Herzberg, 1983 and Herzberg and Forsythe, 1983). Such a model offers a means to achieve some differentiation of the lower mantle, while providing for the current existence of a chemical boundary or gradient at 670 km. In fact, material exchange across the 670 km discontinuity may be necessary in order to effect the chemical differentiation between the upper and lower mantle which is necessary to stabilize a convective boundary layer. Heat and material flux across the boundary will serve to "buffer" the mean mantle temperature distribution near to an adiabat and thus satisfy the viscosity predictions of Peltier (1981) and thermal constraints discussed by Spohn and Schubert (1982). The high reflectivity of some of the P′670P′ reflections would then document stable portions of the boundary at a given time and place; in fact the geographically elusive nature of this phase (Richards, personal communication) might be nicely explained in the context of this compromise model. We stress that, with regard to geochemical budgets and mass balances, this compromise model has many of the same ramifications as whole-mantle convection.

7. Summary and Conclusions

We have explored three principal boundary conditions which relate to the nature and development of chemical heterogeneities in Earth's mantle: 1) initial bulk composition of the Earth; 2) scale lengths inferred for mantle heterogeneities, and 3) correlation and interrelationship of the isotopic tracers Sr, Nd, Pb and He. These boundary conditions are integrated with various geophysical arguments in order to evaluate possible models for the development of physical and chemical structure in the mantle. The major points of this synthesis are:

1. The composition of the primitive upper mantle (PUM) can be ascertained via comparison of mantle-derived lherzolites and chondrites, using a procedure modified after Jagoutz *et al.* (1979a). Resulting abundances of lithophile refractory elements in PUM are constrained to better than ± 5 percent.
2. The composition of the bulk silicate earth (BSE) can be inferred, with a few reasonable assumptions, to be similar to PUM. The BSE and core can be quantitatively derived from a C1 chondrite composition which is strongly deficient in the volatile elements and mildly deficient in the slightly volatile elements Mg and Si (8 and 19 percent, respectively). The absolute abundance levels of the highly refractory elements in the BSE is $2.51 \times$C1 chondrite.
3. Solid-state diffusion is not effective in the destruction of chemical heterogeneities larger than a few meters in size, even over time scales similar to the age of the Earth. Even in the presence of a small amount of basaltic melt, km-size heterogeneities can survive diffusive equilibration for billions of years. Diffusive equilibration of He is not drastically different than for other trace elements of interest; over 4.5 b.y. in the solid state, or in the presence of a melt, diffusive scale lengths for He are on the order of 1 km and 30 km, respectively.
4. The mantle is chemically heterogeneous on both very small ($<$10m) and very large ($>$1000 km) scales. There is a clear positive correlation between the amplitudes of isotopic variation and the maximum scale lengths of observation. A coherent belt of distinctive isotopic signature (the Dupal anomaly) can be traced world-wide between the equator and 60°S latitude.
5. The isotopic heterogeneities in the mantle can be characterized by 4 end-member and perhaps 1 or 2 intermediate components: 1) depleted MORB mantle (DMM); 2) high U/Pb mantle (HIMU); 3) two "enriched" mantle components (EM I and EM II); $\pm$4) a prevalent mantle composition

(PREMA) which is evidenced at many ocean islands, intraoceanic arcs, and in continental volcanic provinces; and, ±5) BSE. An undepleted BSE or PUM component can make up no more than ~60 percent of the total mantle, all uncertainties considered. The depleted MORB mantle can make up as much as 90 percent or as little as 10 percent of the total mantle, in trade-off with the relatively enriched mantle components. The whole mantle is, on average, depleted, with $^{143}Nd/^{144}Nd$ constrained to lie between 0.51280 and 0.51305 (for any reasonable continental crust mean age between 1.7 and 3.5 b.y.). If the prevalent mantle (PREMA) composition exists as a discrete component, as opposed to being a mixture of other components, it may constitute a very large portion of the mantle, and be tapped in a wide variety of tectonic settings (islands, arcs, continents).

6. The He isotope data cannot be simply interpreted in terms of the same mantle components as delineated by Sr, Nd and Pb isotopes. Solid-state diffusion of He is not a viable explanation for this decoupling. Episodic and/or continuous pre-eruptive degassing of He from magmas is shown to lead to high U/He ratios; consequent ingrowth of radiogenic ^4He may significantly perturb mantle He "signals." Oceanic islands with long lifetimes and low magma supply rates will be most susceptible to this process.
7. The Pb paradox is shown to be essentially non-paradoxical. Numerous low μ reservoirs are available to balance the ubiquitous high-μ oceanic basalt data.
8. The U content of the BSE is tightly constrained to be ~21 ppb. The fraction of this which now resides in the continental crust is less well constrained, but lies between 22 and 52 percent. The reduced mantle Urey ratio (crustal heat flow and heat production subtracted) derived from these values is ~2.8; this high value suggests a poor thermal efficiency for mantle convection or a significant component of heat deriving from the core.

Taken together, the above "boundary" conditions appear to us to be inconclusive with respect to the question of whole-mantle versus layered-mantle convection. All of the constraints can, with some special-pleading, be made consistent with either model. The 670 km discontinuity, however, is almost certainly not a boundary between a depleted upper mantle and an undifferentiated BSE lower mantle; on the other hand it could be a boundary between a depleted upper mantle and a less depleted lower mantle. The correspondence of the primitive upper mantle (PUM) composition, derived from ultramafic rocks, with the LOSIMAG C1 Earth model derived from C1 chondrites, suggests that there was no significant chemical layering of the early mantle (especially in the major elements which will control the density of mantle assemblages). Any present day chemical stratification which is significant enough to provide a density-block to convection must have developed since the early Archean by major intra-mantle differentiation, or by reinjection of fractionated lithospheric material. These processes would be required to have operated in such a way that the uppermost mantle (<150 km) has not been involved (otherwise the PUM composition would differ significantly from LOSIMAG C1, and the ^{129}Xe anomaly would be destroyed).

The strongest arguments for a convective boundary layer at 670 km appear to be geophysical; namely the underside reflections and the inability to match physical property changes in the transition region with isochemical phase changes. Even these evidences are not irrefutable, however, and it appears to us that a periodically unstable (in space and/or time) boundary layer model provides a nice compromise between the two end-member extremes into which convection models are usually cast. We propose that early Archean convection regimes may have been mantle-wide. During this time, a mean depletion of the whole mantle could have effected. At some intermediate time, intra-mantle differentiation and/or lithospheric reinjection occurred to an extent sufficient to occasionally or partially stabilize a boundary layer at 670 km, and the extent of this stability has increased with time to the present, though the boundary layer is even now not uniformly present around the world.

If an unequivocal answer is not yet in hand, where is it likely to come from? Certainly the continued integration of geophysical and geochemical studies will be a central theme. By their very nature, geophysical arguments cannot do more than say what is happening right now; geochemical arguments, on the other hand, supply the necessary historical or time-averaging power. Geochemical studies, however, are fundamentally weak in providing three-dimensional information, and thus need to be complemented with geophysical information. The two approaches are thus necessary and complementary to each other.

What are the most promising areas in these two disciplines? As outsiders, it would appear that tomographic mapping of the mantle, in conjunction with laboratory characterization of physical properties (density and velocity), particularly under P-T conditions relevant to the transition zone, are going to be very powerful tools. From a geochemical standpoint, further delineation (mapping) of the megascale isotopic anomalies (DUPAL, etc.) can be keyed into mantle tomographic studies. Understanding of small scale isotopic heterogeneities (ultramafic-massif and small-volcano scale) will provide the groundwork for convective modelling of production and destruction time scales for heterogeneities. Further analytical work on lherzolites, aimed at establishing the extent to which our derived PUM composition truly applies to all of the upper mantle, will strongly constrain the extent to which the mantle can be chemically stratified in major elements. Careful work to decipher mantle $^3He/^4He$ signals, for ultimate rectification with the heavy isotope tracer results, is imperative. A much broader geographic coverage of ^{129}Xe data is needed, as this is our only available tool for mapping heterogeneities which have preservation times in excess of 4 b.y.

While we have undoubtedly disappointed many of the mantle fraternity who want clean and simple solutions now, it does not seem overly optimistic to say that our science is indeed rumbling ahead on many interrelated fronts. The continued growth of the fairly young marriage between mantle geophysics and geochemistry clearly shows much promise for future advances in our understanding of the physical and chemical evolution of the Earth. Although it would now appear that the honeymoon is probably over, we have high hopes for a long, if contentious, union!

8. Acknowledgements

This work contains a myriad of old and new ideas which we have attempted to weave into a coherent self-consistent view of the Earth's mantle. We have endeavored to accurately cite appropriate literature references for old ideas, and we apologize in advance for where we have failed. The development of our new ideas has been immeasurably aided by conversations with many friends and colleagues. While we cannot hope to name them all, there are those whom we cannot fail to mention: Dave Graham, Don Anderson, Claude Allegre, Bill White, John Ludden, Jim Rubenstone, Charlie Langmuir, Ed Stolper, Jill Pasteris, Hubert Staudigel, Gerd Worner, Emil Jagoutz, Steve Goldstein, Laurie Reisberg, Ray Jeanloz, Frank Richter, Don Turcotte, and Dick Peltier. We would also like to thank Mark Kurz, Don Anderson, Ray Jeanloz, Claude Allegre, Fred Frey, Roger Hart, Dave Graham, and Gerd Worner for furnishing helpful preprints and unpublished data; Claude Allegre for his early role as a silent partner; Emil Jagoutz for inspiring "PUM"; and Dick Peltier for being patient, three year's worth. A.Z. would like to thank the participants in his recent mantle geochemistry seminar for acting as a sounding board for earlier manifestations of much of this material, and assure them that the learning process was always a two-way street. S.R.H. acknowledges the denizens of the 11th floor for abiding his absence during the writing of this manuscript and MIT for paying him to stay home for a year. This manuscript was skillfully produced by Michelle Lezie, who suffered through untold versions while thoroughly crash-testing Proofwriter version 2.09, and Joan Totton and Diane Watson-Mitchell, who provided timely help with the final staging. As usual, Patty Catanzaro did a superb job with the graphics, under no small amount of pressure from us. And finally, the work we present would not have been possible without the generous support of the National Science Foundation (grants EAR 82-11461, EAR 83-20066, and OCE 84-10615 to A.Z. and EAR 83-08809 to S.R.H.). (This work represents L.D.G.O. contribution #4360.)

7. References

Ahrens, L.H. and Von Michaelis, H. (1969) The composition of stony meteorites III. Some inter-element relationships. *Earth and Planet. Sci. Lett.* **5**, 395–400.

Ahrens, L.H., Von Michaelis, H., Erlank, A.J., and Willis, J.P. (1969) Fractionation of some abundant lithophile element ratios in chondrites. In: P.M. Millman (ed.) *Meteorite Research*. Amsterdam: D. Reidel, 166–173.

Allegre, C.J., (1982) Chemical geodynamics. *Tectonophysics* **81**, 109–132.

Allegre, C.J., Brevart, O., Dupre, B., and Minster, J.F. (1980) Isotopic and chemical effects produced in a continuously differentiating convecting earth mantle. *Philos. Trans. R. Soc. London Ser. A*. **297**, 447–477.

Allegre, C.J., Dupre, B., and Lewin, E. (1986) Thorium/uranium ratio of the earth. *Chem. Geo.* **56**, 219–227.

Allegre, C.J. and Minster, J.-F. (1978) Quantitative models of trace element behavior in magmatic processes. *Earth and Planet. Sci. Lett.* **38**, 1–25.

Allegre, C.J., Staudacher, T., Sarda, P. (1987) Rare gas systematics: Formation of the atmosphere, evolution and structure of the earth's mantle. *Earth and Planet. Sci. Lett.* **81**, 127–150.

Allegre, C.J., Staudacher, T., Sarda, P., and Kurz, M. (1983a) Constraints on evolution of Earth's mantle from rare gas systematics. *Nature* **303**, 762–766.

Allegre, C.J., Othman, D.B., Polve, M., and Richard, P. (1979) The Nd-Sr isotopic correlation in mantle materials and geodynamic consequences. *Physics of the Earth and Planet. Int.* **19**, 293–306.

Allegre, C.J., Hamelin, B., and Dupre, B. (1984) Statistical analysis of isotopic ratios in MORB: The mantle blob cluster model and the convective regime of the mantle. *Earth and Planet. Sci. Lett.* **71**, 71–84.

Allegre, C.J., Hart, S.R., and Minster, J.-F. (1983b) Chemical structure and evolution of the mantle and continents determined by inversion of Nd and Sr isotopic data, I. Theoretical methods, *Earth and Planet. Sci. Lett.* **66**, 177–190.

Allegre, C.J., Hart, S.R., and Minster, J.-F. (1983c) Chemical structure and evolution of the mantle and continents determined by inversion of Nd and Sr isotopic data, II. Numerical experiments and discussion. *Earth and Planet. Sci. Lett.* **66**, 191–213.

Allegre, C.J. and Rousseau, D. (1984) The growth of the continent through geological time studied by Nd isotope analysis of shales. *Earth and Planet. Sci. Lett.* **67**, 19–34.

Allegre, C.J. and Turcotte, D.L. (1985) Geodynamic mixing in the mesosphere boundary layer and the origin of oceanic islands. *Geophys. Res. Lett.* **12**, 207–210.

Anders, E. and Ebihara, M. (1982) Solar system abundances of the elements, solar system abundances of the elements. *Geochim. Cosmochim. Acta.* **46**, 2363–2380.

Anderson, D.L. (1967) Phase changes in the upper mantle. *Geophys. J. R. Astr. Soc.* **14**, 135–164.

Anderson, D.L. (1976) The 650 km mantle discontinuity. *Geophys. Res. Lett.* **5**, 347–349.

Anderson, D.L. (1982) Hotspots, polar wander, mesozoic convection and the geoid. *Nature* **297**, 391–393.

Anderson, D.L. (1982) Isotopic evolution of the mantle: The role of magma mixing. *Earth and Planet. Sci. Lett.* **57**, 1–12.

Anderson, D.L. (1983) Chemical composition of the mantle. Proc. Fourteenth Lunar and Planet Sci. Conf., *J. Geophys. Res.* **88**, B41–B52.

Anderson, D.L. and Bass, J.D. (1986) Transition region of the Earth's upper mantle. *Nature* **320**, 321–328.

Armstrong, R.L. (1968) A model for the evolution of strontium and lead isotopes in a dynamic Earth, *Rev. Geophys.,* **6**, 175-199.

Armstrong, R.L. (1981) Radiogenic isotopes: The case for crustal recycling on a near-steady-state no-continental growth Earth. *Philos. Tans. R. Soc. London, Ser. A* **301**, 443–472.

Basaltic Volcanism Study Project (1981) In: *Basaltic Volcanism on the Terrestrial Planets*. New York: Pergamon Press, 282–291.

Bass, J.D. and Anderson, D.L. (1984) Composition of the upper mantle: Geophysical tests of two petrological models. *Geophysical Res. Lett.* **11**, 237–240.

Basu, A.R. and Tatsumoto, M. (1979) Samarium-neodymium systematics in kimberlites and in the minerals of garnet lherzolite inclusions. *Science* **205**, 398–401.

Basu, A.R. and Tatsumoto, M. (1980) Nd-isotopes in selected mantle-derived rocks and minerals and their implications for mantle evolution. *Contr. Mineral. Petrol.* **75**, 43–54.

Batiza, R. and Vanko, D. (1984) Petrology of Young Pacific Seamounts. *J. of Geophys. Res.* **89**, 11235–11260.

Birch, F., Schairer, J.F., and Spicer, H.C. (1942) Handbook of Physical Constants. *Geo. Soc. of Amer. Spec.*, Paper 36.

Boyd, F.R. and Nixon, P.H. (1973) Structure of the upper mantle beneath Lesotho. *Carnegie Institution Yearbook* **72**, 431–445.

Boynton, W.V. (1975) Fractionation in the solar nebula: Condensation of yttrium and the rare earth elements. *Geochim. Cosmochim. Acta.* **3**, 569–584.

Brooks, C., James, D.E., and Hart, S.R. (1976) Ancient lithosphere: Its role in young continental volcanism. *Science* **193**, 1086–1094.

Brey, G. and Green, D.H. (1977) Systematic study of liquidus phase relations in olivine melilitite + H_2O + CO_2 at high pressures and petrogenesis of an olivine melilitite magma. *Contrib. Mineral. Petrol.* **61**, 141–162.

Brueckner, H.K., Zindler, A., Seyler, M., and Bonatti, E. (1988) Zabargad and the isotopic evolution of the sub-Red Sea mantle and crust. *Tectonophysics* **150**, 163–176.

Carlson, R.W. (1984) Isotopic constraints on Columbia River flood basalt genesis and

the nature of the subcontinental mantle. Geochim. *Cosmochim. Acta.* **48**, 2357–2372.
Carlson, R.W., Lugmair, G.W., and MacDougall, J.D. (1981) Crustal influence in the generation of continental flood basalts. *Nature* **289**, 160–162.
Carter, S.R., Evensen, N.M., Hamilton, P.J., and O'Nions, R.K. (1978) Continental volcanics derived from enriched and depleted source regions: Nd and Sr isotope evidence. *Earth and Planet. Sci. Lett.* **37**, 401–408.
Chase, C.G. (1981) Ocean island Pb: Two-stage histories and mantle evolution. *Earth and Planet. Sci. Lett.* **52**, 277–284.
Chauvel, C. and Jahn, B.-M. (1984) Nd-Sr isotope and REE geochemistry of alkali basalts from the Massif Central, France. *Geochim. Cosmochim. Acta.* **48**, 93–110.
Chen, C.-Y. and Frey, F.A. personal communication.
Chen, J.H. and Tilton, G.R. (1976) Isotopic lead investigations on the Allende carbonaceous chondrite. *Geochim. Cosmochim. Acta.* **40**, 635–643.
Cohen, R.S., Evensen, N.M., Hamilton, P.J., and O'Nions, R.K. (1980) U-Pb, Sm-Nd and Rb-Sr systematics of mid-ocean ridge basalt glasses. *Nature* **283**, 149–153.
Cohen, R.S. and O'Nions, R.K. (1982a) The lead, neodymium and strontium isotopic structure of ocean ridge basalts. *J. Petrol.* **23**, 299–324.
Cohen, R.S. and O'Nions, R.K. (1982b) Identification of recycled continental material in the mantle from Sr, Nd and Pb isotope investigations. *Earth and Planet. Sci. Lett.* **61**, 73–84.
Cohen, R.S., O'Nions, R.K., and Dawson, J.B. (1984) Isotope geochemistry of xenoliths from East Africa: Implications for development of mantle reservoirs and their interaction. *Earth and Planet. Sci. Lett.* **68**,209–220.
Condomines, M., Gronvold, K., Hooker, P.J., Muehlenbachs, K., O'Nions, R.K., Oskarsson, N., and Oxburgh, E.R. (1983) Helium, oxygen, strontium and neodymium isotopic relationships in Icelandic volcanics. *Earth and Planet. Sci. Lett.* **66**, 125–136.
Craig, H., Clarke, W.B., Beg. M.S. (1975) Excess ^{3}He in deep water on the East Pacific Rise. *Earth and Planet. Sci. Lett.* **26**, 125–132.
Craig, H. and Lupton, J.E. (1978) Helium isotope variations: Evidence for mantle plumes at Yellowstone, Kilauea and the Ethiopian Rift Valley. *EOS, Trans. Am. Geophys. Union* **59**, 1194.
Craig, H. and Lupton, J.E. (1981) Helium-3 and mantle volatiles in the ocean and the oceanic crust. In C. Emiliani (ed.), *The Sea* **7**, 391-428.
Craig, H. and Rison, W. (1982) Helium 3: Indian Ocean hotspots and the East African Rift. *EOS* **63**, 1144.
Craig, H. and Rison, W. (1983) Helium isotopes and mantle heterogeneity. *EOS* **64**, 348.
Creager, K.C. and Jordan, T.H. (1984) Slab penetration into the lower mantle. *J. Geophys. Res.* **89**, 3031–3050.
Crisp, J.A. (1984) Rates of magma emplacement and volcanic output. *J. Volcanol. and Geother. Res.* **20**, 177–211.
Dasch, E.J. and Green, D.H. (1975) Strontium isotope geochemistry of lherzolite

inclusions and host basaltic rocks, Victoria, Australia. *Amer. J. of Sci.* **275**, 461–469.
Davidson, Jon P. (1983) Lesser Antilles isotopic evidence of the role of subducted sediment in island arc magma genesis. *Nature* **306**, 253–256.
Davies, G.F. (1981) Earth's neodymium budget and structure and evolution of the mantle. *Nature* **290**, 208–213.
Davies, G.F. (1983) Viscosity structure of layered convecting mantle. *Nature* **301**, 592–594.
DePaolo, D.J. (1980) Crustal growth and mantle evolution: Inferences from models of element transport and Nd and Sr isotopes. *Geochim. Cosmochim. Acta.* **44**, 1185–1196.
DePaolo, D.J. (1981) Nd isotopic studies: Some new perspectives on Earth structure and evolution. *EOS* **62**, 137–140.
DePaolo, D.J. and Johnson, R.W. (1979) Magma genesis in the New Britain island arc: Constraints from Nd and Sr isotopes and trace-element patterns. *Contrib. Min. Petrol.* **70**, 367–380.
DePaolo, D.J. and Wasserburg, G.J. (1976a) Inferences about magma sources and mantle structure from variations of 143Nd/144Nd. *Geophys. Res. Lett.* **3**, 743–746.
DePaolo, D.J. and Wasserburg, G.J. (1976b) Nd isotopic variations and petrogenetic models. *Geophys. Res. Lett.* **3**, 249–252.
DePaolo, D.J. and Wasserburg, G.J. (1977) The sources of island arcs as indicated by Nd and Sr isotopic studies. *Geophys. Res. Lett.* **4**, 465–468.
Dupre, B. and Allegre, C.J. (1983) Pb-Sr isotope variation in Indian Ocean basalts and mixing phenomenoa. *Nature* **303**, 142–146.
Dupuy, C., Leyreloup, A., and Vernierest, J. (1979) The lower continental crust of the massif central (Bournac, France)—with special references to REE, U and Th composition, evolution, heat-flow production. In L.H. Aherns (ed.). *Origin and Distribution of the Elements, Physics and Chemistry of the Earth.* Vol. II, New York: Pergamon Press.
Dziewonski, A.M., Hales, A.L., and Lapwood, E.R. (1975) Parametrically simple earth models consistent with geophysical data. *Phys. Earth and Planet. Inter.* **10**, 12–48.
Dziewonski, A., Anderson, D.L. (1981) Preliminary reference Earth models. *Phys. Earth and Planet. Inter.* **25**, 297–356.
Ernst, W.G. and Piccardo, G.B. (1979) Petrogenesis of some Ligurian peridotites—I. Mineral and bulk-rock chemistry. *Geochim. Cosmochim. Acta.* **43**, 219–237.
Evensen, N.M., Hamilton, P.J., and O'Nions, R.K. (1978) Rare-earth abundances in chondritic meteorites. *Geochim. Cosmochim. Acta.* **42**, 1199–1212.
Fahrig, W.F. and Eade, K.E. (1968) The chemical evolution of the Canadian shield. *Canadian J. of Earth Sci.* **5**, 1247–1252.
Fine, G. and Stolper, E. (1986) Dissolved carbon dioxide in basaltic glasses: Concentrations and speciation. *Earth and Planet. Sci. Lett.*,**76**, 263–278.
Flemings, M.C. (1974) *Solidification Processing*. New York: McGraw-Hill.
Frey, F.A. (1969) Rare earth abundances in a high-temperature peridotite intrusion. *Geochim. Cosmochim. Acta.* **33**, 1429–1477.

Frey, F.A. and Prinz, M. (1978) Ultramafic inclusions from San Carlos, Arizona: Petrologic and geochemical data bearing on their petrogenesis. *Earth and Planet. Sci. Lett.* **38**, 129–176.

Frey, F.A., Suen, C.-Y.J., and Stockman, H.W. (1985) The Ronda high temperature peridotite: Geochemistry and petrogenesis. *Geochim. Cosmochim. Acta.* **19**, 463–513.

Frey, F.A. and Green, D.H. (1974) The mineralogy, geochemistry and origin of lherzolite inclusions in Victorian basanites. *Geochim. Cosmochim. Acta.* **38**, 1023–1050.

Frey, F.A., Green, D.H., and Roy, S.D. (1978) Integrated models of basalt petrogenesis: A study of quartz tholeiites to olivine melilitites from South Eastern Australia utilizing geochemical and experimental petrological data. *J Petrol.* **41**, 1283–1301.

Gancarz, A.J. and Wasserburg, G.J. (1977) Initial Pb of the Amitsoq gneiss, West Greenland, and implications for the age of the Earth. *Geochim. Cosmochim. Acta.* **41**, 1283–1301.

Gast, P.W. (1968) Trace element fractionation and the origin of tholeiitic and alkaline magma types. *Geochim. Cosmochim. Acta.* **32**, 1057–1086.

Gast P.W., Tilton, G.R., and Hedge, C. (1964) Isotopic composition of and strontium from Ascension and Gough Islands. *Science* **145**, 1181–1185.

Giardini, D., Woodhouse, J.H. (1984) Deep seismicity and modes of deformation in Tonga subduction zone. *Nature* **307**, 505–509.

Goldstein, S., Langmuir, C., Zindler, A., Weaver, S., Staudigel, H. (1981) Systematics of radiogenic isotope variations in Aleutian lavas: Implications for island arc petrogenesis. *EOS* **62**, 410.

Goldstein, S., Zindler, A., and Jagoutz, E. (1982) Evolution of the mantle-crust system: Quantitative geochemical modelling of Pb, Sr, and Nd isotopic systematics. *EOS* **63**, 460.

Goldstein, S.L., Rubenstone, J., Zindler, A., Langmuir, C.H., Staudigel, H., Weaver, S., and Kay, R.W. (1988) The geochemistry of Aleutian arc magmas and implications for petrogenesis of magmas at convergent plate margins, *Contr. Mineral. Petrol.*, in press.

Goldstein, S.L., O'Nions, R.K., and Hamilton, P.J. (1984) A Sm-Nd isotopic study of atmospheric dusts and particulates from major river systems. *Earth and Planet. Sci. Lett.* **70**, 221–236.

Graham, D.W., Zindler, A., Kurz, M.D., Jenkins, W.J., Batiza, R., and Staudigel, H. (1988) Helium, lead, strontium and neodymium isotope constraints on magma genesis and mantle heterogeneity beneath young Pacific seamounts. *Contrib. Miner. Petrol.*, in press.

Graham, D., Zindler, A., Kurz, M.D., and Batiza, R. (1985) He, Sr, Nd, and Pb isotopes in basaltic glasses from young Pacific seamounts. *Earth and Planet. Sci. Lett.*, in press.

Gray, C.M. (1977) The geochemistry of central Australian granulites in relation to the chemical and isotopic effects of granulite facies metamorphism. *Contrib. Min. Petrol.* **65**, 79–89.

Gray, C.M. and Oversby, V.M. (1972) The behaviour of lead isotopes during granulite facies metamorphism. *Geochim. Cosmochim. Acta.* **36**, 939–952.

Green, D.H. (1964) The petrogenesis of the high-temperature peridotite intrusion in the Lizard area. Cornwall, *J. of Petrol.* **5**, 134–188.

Greenland, L.P., Rose, W.I., and Stokes, J.B. (1985) An estimate of gas emissions and magmatic gas content from Kilauea Volcano. *Geochim. Cosmochim. Acta.* **49**, 125–129.

Grossman, L. and Larimer, J.W. (1974) Early chemical history of the solar system. *Rev. Geophys. Space Phys.* **12**, 71–101.

Hall, H.T. and Murthy, V. (1971) The early thermal history of the Earth. *Phys. Earth and Planet. Inter.* **2**, 19–29.

Hamelin, B. and Allegre, C.J. (1985) Large scale regional units in the depleted upper mantle: Revealed by an isotopic study of the Southwest Indian Ridge. *Nature* **315**, 196–199.

Hamilton, P.J., Evensen, N.M., O'Nions, R.K., and Tarney, J. (1979) Sm-Nd systematics of Lewisian gneisses: Implications for the origin of granulites. *Nature* **277**, 25–28.

Hart, R., Hogan, L., and Dymond, J. (1985) The closed system approximation for evolution of argon and helium in the mantle, crust, and atmosphere. *Isotope Geoscience* **52**, 45–73.

Hart, S.R. (1969) K, Rb, Cs contents and K/Rb, K/Cs ratios of fresh and altered submarine basalts. *Earth and Planet. Sci. Lett.* **6**, 295–303.

Hart, S.R. (1971) K, Rb, Sc, Sr and Ba contents and Sr isotope ratios of ocean floor basalts. *Phil. Trans. Roy. Soc. Land.* **A 26**, 573–587.

Hart, S.R. (1980) Diffusion compensation in natural silicates. *Geochim. Cosmochim. Acta.* **45**, 279–291.

Hart, S.R. (1984a) He diffusion in olivine. *Earth and Planet. Sci. Lett.* **70**, 297–302.

Hart, S.R. (1984b) A large-scale isotope anomaly in the Southern Hemisphere mantle. *Nature* **309**, 753–757.

Hart, S.R., Erlank, A.J., and Kable, J.D. (1974) Sea floor basalt alteration: Some chemical and Sr isotopic effects. *Contrib. Mineral. and Petrol.* **44**, 219–230.

Hart, S.R. and Staudigel, H. (1982) The control of alkalies and uranium in seawater by ocean crust alteration. *Earth and Planet. Sci. Lett.* **58**, 202–212.

Hart, S.R. and Zindler, A. (1986) In search of a bulk earth composition. *Chemical Geology* **57**, 247–267.

Hawkesworth, C.J. and Vollmer, R. (1979) Crustal contamination versus enriched mantle: $^{143}Nd/^{144}Nd$ and $^{87}Sr/^{86}Sr$ evidence from the Italian volcanics. *Contrib. Mineral. Petrol.* **69**, 151–169.

Hawkesworth, C.J., Erlank, A.J., Marsh, J.S., Menzies, M.A., and Van Calsteren, P. (1983) Evolution of the continental lithosphere: Evidence from volcanics and xenoliths in Southern Africa. In: C.J. Hawkesworth and M.J. Norry (eds.) *Continental Basalts and Mantle Xenoliths*. Cheshire, England: Shiva Publishing Ltd.

Hedge, C.E. and Peterman, Z.E. (1970) The strontium isotopic composition of basalts from the Gordo and Juan de Fuca Rises, Northeastern Pacific Ocean. *Contrib. Mineral. Petrol.* **27**, 114–120.

Heinrichs, H., Schulz-Dobrick, B., and Wedepohl, K.H. (1980) Terrestrial geochemistry of Cd, Bi, Tl, Pb, Zn and Rb. *Geochim. Cosmochim. Acta.* **44**, 1519–1533.

Herzberg, C.T. (1983) Chemical stratification in the silicate Earth. *Earth and Planet. Sci. Lett.* **67**, 249–260.

Herzberg, C.T. and Forsythe, R.D. (1983) Destabilization of a 650 km chemical boundary layer and its bearing on the evolution of the continental crust. *Physics of the Earth and Planet. Inter.* **32**, 352–360.

Hiyagon, H. and Ozima, M. (1982) Noble gas distribution between basalt melt and crystals. *Earth and Planet. Sci. Lett.* **58**, 255–264.

Hoffman, N.R.A. and McKenzie, D.P. (1985) The destruction of geochemical heterogeneities by differential fluid motions during mantle convection. *Geophys. J.R. Astron. Soc.* **82**, 163–206.

Hofmann, A.W and Hart, S.R. (1978) An assessment of local and regional isotopic equilibrium in the mantle. *Earth and Planet. Sci. Lett.* **38**, 44–62.

Hofmann, A.W. and Margaritz, M. (1977) Equilibrium and mixing in a partially molten mantle. In H.J.B. Dick (ed.) *Magma Genesis Bull. 96*. Protland, Ore.: Oregon Dept. of Geology & Mineral Industries.

Hofmann, A.W. and White, W.M. (1982) Mantle plumes from ancient oceanic crust. *Earth and Planet. Sci. Lett.* **57**, 421–436.

Hofmann, A.W., White W.M., and Whitford, D.J. (1978) Geochemical constraints on mantle models: The case for a layered mantle. *Carnegie Inst. Yearbook* **77**, 548–562.

Irving, A.J. (1980) Petrology and geochemistry of composite ultramafic xenoliths in alkalic basalt and implications for magmatic processes within the mantle. *Amer. J. Sci.* **280-A**, 389–426.

Jacobsen, S.B. and Wasserburg, G.J. (1980) Sm-Nd isotopic evolution of chondrites. *Earth and Planet. Sci. Lett.* **50**, 139–155.

Jacobsen, S.B. and Wasserburg, G.J. (1984) Sm-Nd isotopic evolution of chondrites and achondrites, II. *Earth and Planet. Sci. Lett.* **67**, 137–150.

Jacobsen, S.B. and Wasserburg, G.J. (1979) The mean age of mantle and crustal reservoirs. *J. Geophys. Res.* **84**, 7411–7427.

Jagoutz, E., Palme, H., Badenhausen, H., Blum, K., Cendales, M., Dreibus, G., Spettel, G., Lorenz, V., and Wanke, H. (1979a) The abundances of major, minor and trace elements in the Earth's mantle as derived from primitive ultramafic nodules. In: R.B. Merrill (ed.) *Proc. Lunar Planet, Sci. Conf., 10th*. New York: Pergamon Press, 2031–2050.

Jagoutz, E., Lorenze, V., and Wanke, H. (1979b) Major trace elements of Al-augite and Cr-diopsides from ultramafic nodules in European alkali basalts. In: F.B. Boyd and H.O.A. Meyer (eds.) *The Mantle Sample: Inclusions in Kimberlites and Other Volcanics*. Washington, D.C.: Am. Geophys. Union, 382–390.

Jagoutz, E., Carlson, R.W., and Lugmair, G.W. (1980) Equilibrated Nd-unequilibrated Sr isotopes in mantle xenoliths. *Nature* **296**, 708–710.

Jambon, A., Weber, H., and Braun, O. (1986) Solubility of He, Ne, Ar, Kr and Xe in a basalt melt in the range 1250–1600°C: Geochemical implications. *Geochim. Cosmochim. Acta.* **50**, 401–408.

Jeanloz, R. (1988) High pressure chemistry of the Earth's mantle and core, this volume.

Jeanloz, R. and Richter, F.M. (1979) Convection, composition, and the thermal state of the lower mantle. *J. Geophys. Res.* **84**, 5497–5504.

Jeanloz, R. and Thompson, A.B. (1983) Phase transitions and mantle discontinuities. *Rev. of Geo. and Space Phys.* **21**, 51–74.

Jeffrey, P.M. and Anders, E. (1970) Primordial noble gases in separated meteoritic minerals—I. *Geochim. Cosmochim. Acta.* **34**, 1175–1198

Jochum, K.P., Hofmann, A.W., Ito, E., Seufert, H.M., and White, W.M. (1983) K, U, and Th in mid-ocean ridge basalt glasses. The terrestrial K/U and K/Rb ratios and heat production in the depleted mantle. *Nature* **306**, 431–436.

Jordan, T.H. (1977) Lithospheric slab penetration into the lower mantle beneath the sea of Okhotsk. *J. Geophys.* **43**, 473–496.

Kallemeyn, G.W. and Wasson, J.T. (1981) The compositional classification of chondrites—I. The carbonaceous chondrite groups. *Geochim. Cosmochim. Acta.* **45**, 1217–1230.

Kay, R., Hubbard, H.J., and Gast, P.W. (1970) Chemical characteristics origin of oceanic ridge volcanic rocks. *J. Geophys. Res.* **75**, 1585–1613.

Kay, R.W. (1980) Volcanic arc magmas: Implications of a melting-mixing model for element recycling in the crust-upper mantle system. *J. Geol.* **88**, 497–522.

Keil, K. (1969) Meteorite composition. In Wedepohl, K.H. (ed.) *Handbook of Geochemistry*. Berlin: Springer Verlag, Vol. I, 78–115.

Khitarov, N.I. and Kadik, A.A. (1973) Water and carbon dioxide in magmatic melts and peculiarities of the melting process. *Contrib. Mineral. Petrol.* **41**, 205–215.

Knab, H.-J. (1981) The distribution of trace elements in carbonaceous chondrites. *Geochim. Cosmochim. Acta.* **45**, 1563–1572.

Knittle, E., Jeanloz, R., and Smith, G.L. (1985) The thermal expansion of silicate perovskite and stratification of the Earth's mantle. *Nature* **319**, 214–216.

Kramers, J.D., Smith, C.B., Lock, N.P., Harmon, R.S., and Boyd F.R. (1981) Can Kimberlites be generated from an ordinary mantle? *Nature* **291**, 53–56.

Kramers, J.D., Roddick, J.C.M., and Dawon, J.B. (1983) Trace element and isotopic studies on veined, metasomatic and "MARID" xenoliths from Bultfontein, South Africa. *Earth and Planet. Sci. Lett.* **65**, 90–106.

Kurat, G., Palme, H., Spettel, B., Baddenhausen, H., Hofmeister, H., Palme, C., and Wanke, H. (1980) Geochemistry of ultramafic xenoliths from Kapfenstein, Austria: Evidence for a variety of upper mantle processes. *Geochim. Cosmochim. Acta.* **44**, 45–60.

Kurz, M.D. (1982) Helium isotope geochemistry of oceanic volcanic rocks: Implications for mantle heterogeneity and degassing, Ph.D. Thesis, Mass. Inst. of Tech. and Woods Hole Oceanographic Inst., 281 pp.

Kurz, M.D. and Jenkins, W.J. (1981) The distribution of helium in oceanic basalt glasses. *Earth and Planet. Sci. Lett.* **53**, 41–54.

Kurz, M.D., Jenkins, W.J., Hart, S.R., and Clague, D. (1983) Helium isotopic variations in volcanic rocks from Loihi Seamount and the Island of Hawaii. *Earth and Planet. Sci. Lett.* **66**, 388–406.

Kurz, M.D., Jenkins, W.J., Schilling, J.G., and Hart, S.R. (1982) Helium isotopic systematics of ocean islands and mantle heterogeneity. *Nature* **297**, 43–47.

Kurz, M.D., Meyers, P.S., and Sigurdsson, H. (1985) Helium isotopic systematics within the neovolcanic zones of Iceland. *Earth and Planet. Sci. Lett.* **74**, 291–305.

Langmuir, C.H., Vocke, R.D., and Hanson, G.N. (1978) A general mixing equation with applications to icelandic basalts. *Earth and Planet. Sci. Lett.* **37**, 380–392.

Larimer, J.W. (1979) The condensation and fractionation of refractory lithophile elements. *Icarus* **40**, 446–454.

Lees, A.C., Bukowinski, M.S.T., and Jeanloz, R. (1983) Reflection properties of phase transition and compositional change models of the 670-km discontinuity. *J. Geophys. Res.* **88**, 8145–8159.

LeRoex, A.P., Dick, H.J.B., Erlank, A.J., Reid, A.M., Frey, F.A., and Hart, S.R. (1983) Geochemistry, mineralogy and petrogenesis of lavas erupted along the Southwest Indian Ridge between the Bouvet Triple Junction and 11 degrees east. *J. Petrol.* **24**, 267–318.

Lewis, J.S. (1971) Consequences of the presence of sulfur in the core of the Earth. *Earth and Planet. Sci. Lett.* **11**, 130–134.

Loubet, M. and Allegre, C.J. (1982) Trace elements in orogenic lherzolites reveal the complex history of the upper mantle. *Nature* **298**, 809–814.

Loubet, M., Shimizu, N., and Allegre C.J. (1975) Rare earth elements in Alpine peridotites. *Contrib. Miner. Petrol.* **53**, 1–12.

Lupton, J.E. and Craig, H. (1975) Excess ^{3}He in oceanic basalts: Evidence for terrestrial primordial helium. *Earth and Planet. Sci. Lett.* **26**, 133–139.

Machado, N., Ludden, J.N., Brooks, C., and Thompson, G. (1982) Fine-scale isotopic heterogeneity in the sub-Atlantic mantle. *Nature* **295**, 226–228.

McCulloch, M.T. and Perfit, M.R. (1981) ^{143}Nd/^{144}Nd, ^{87}Sr/^{86}Sr and trace element constraints on the petrogenesis of Aleutian island arc magmas. *Earth and Planet. Sci. Lett.* **56**, 167–179.

Manhes, G. and Allegre, C.J. (1978) Time differences as determined from the ratio of lead 207-206 in concordant meteorites. *Meteoritics* **13**, 543–548.

Manhes, G., Allegre, C.J., Dupre, B., and Hamelin, B. (1979) Lead-lead systematics, the "age of the Earth" and the chemical evolution of our planet in a new representation space. *Earth and Planet. Sci. Lett.* **44**, 91–104.

Manson, V. (1967) Geochemistry of basaltic rock: Major elements. In Hess, H.H. and A. Poldervaart (eds.) *Basalts*. New York: Interscience Publishers, 215–269.

Mason, B. (ed.) (1971) *Handbook of Elemental Abundances in Meteorites*. New York: Gordon and Breach.

Masuda, A., Nakamura, N., and Tanaka, T. (1973) Fine structures of mutually normalized rare-earth patterns of chondrites. *Geochim. Cosmochim. Acta.* **37**, 239–248.

McKenzie, D. and O'Nions, R.K. (1983) Mantle reservoirs and ocean island basalts. *Nature* **301**, 229–231.

McLennan, S.M. and Taylor S.R. (1980) Th and U in sedimentary rocks: Crustal evolution and sedimentary recycling. *Nature* **285**, 621–624.

Melson, W.G., Vallier, T.L., Wright, T.L., Byerly, G., and Nelen, J. (1976) Chemical diversity of abyssal volcanic glass erupted along Pacific, Atlantic, and Indian Ocean sea-floor spreading centers. The Geophysics of the Pacific Ocean Basin and its Margin. In: G.H. Sutton, M.H. Manghnani, R. Moberly (eds.) *Geophysical Monograph* **19**, 351–368.

Menzies, M. (1983) Mantle ultramafic xenoliths in alkaline magmas: Evidence for mantle heterogeneity modified by magmatic activity. In C.J. Hawkesworth and M.J. Norry (eds.) *Continental Basalts and Mantle Xenoliths*. Cheshire, Eng.: Shiva Publishing Ltd.

Menzies, M. and Murthy, V.R. (1980a) Enriched mantle: Nd and Sr isotopes in diopsides from kimberlite nodules. *Nature* **283**, 634–636.

Menzies, M. and Murthy, V.R. (1980b) Nd and Sr isotope geochemistry of hydrous mantle nodules and their host alkali basalts: Implications for local heterogeneities in meatsomatically veined mantle. *Earth and Planet. Sci. Lett.* **46**, 323–334.

Menzies, M.A. and Wass, S.Y. (1983) CO_2-and LREE-rich mantle below eastern Australia: A REE and isotopic study of alkaline magmas and apatite-rich mantle xenoliths from the Southern Highlands Province, Australia. *Earth and Planet. Sci. Lett.* **65**, 287-302.

Michard, A., Gurriet, P., Soudant, M., and Albarede, F. (1985) Nd isotopes in French Phanerozoic shales: External vs. internal aspects of crustal evolution. *Geochim. Cosmochim. Acta.* **49**, 601–610.

Montgomery, C.P.W. (1977) Uranium-lead isotopic investigation of the Archean Imataca complex, Guayana shield, Venezuela. Ph.D. Thesis, Mass. Inst. Tech. 261 pp.

Morgan, W.J. (1971) Convection plumes in the lower mantle. *Nature* **230**, 42–43.

Morris, J.D. (1984) Enriched geochemical signatures in Aleutian and Indonesian arc lavas: An isotopic and trace element investigation. Ph.D. Thesis, Mass. Inst. Tech.

Morris, J.D. and Hart, S.R. (1983) Isotopic and incompatible element constraints on the genesis of island arc volcanics from Cold Bay and Amak Island, Aleutians, and implications for mantle structure. *Geochim. Cosmochim. Acta.* **47**, 2015–2030.

Muirhead, K. (1985) Comments on "Reflection properties of phase transition and compositional change models of the 670 km discontinuity" by A.C. Lees, M.S.T. Bukowinski and R. Jeanloz. *J. Geophys. Res.* **90**, 2057–2059

Murthy, V.R. (1975) Composition of the core and the early chemical history of the Earth. In B.F. Windley (ed.) *The Early History of the Earth*. London: John Wiley & Sons.

Mysen, B.O., Arculus, R.J., and Eggler, D.H. (1975) Solubility of Carbon Dioxide in melts of andesite, tholeiite, and olivine nephelinite composition to 30 kbar pressure. *Contrib. Mineral. Petrol.* **53**, 227–239.

Nakamura, N. (1974) Determination of REE, Ba, Fe, Mg, Na and K in carbonaceous and ordinary chondrites. *Geochim. Cosmochim. Acta.* **38**, 757–775.

Nakamura, N. and Masuda A. (1973) Chrondrites with peculiar rare-earth patterns. *Earth and Planet. Sci. Lett.* **19**, 429–437.

Nixon, P.H. and Boyd, F.R. (1973) Petrogenesis of the granular and sheared ultrabasic nodule suite in kimberlites. In P.H. Nixon (ed.) *Lesotho Kimberlites.* Lesotho Nat. Devel. Corp., 48–56.

O'Hara, M.J., Saunders, M.J., and Mercey, E.L.P. (1975) Garnet peridotite, primary ultrabasic magma and eclogite; interpretation of upper mantle processes in kimberlite. In *Physics and Chemistry of the Earth*, 9, New York: Pergamon Press, 571–604.

Olson, P. and Yuen, D.A. (1982) Thermochemical plumes and mantle phase transitions. *J. Geophys. Res.* **87**, 3993–4002.

Olson, P., Yuen, D.A., and Balsiger, D. (1984) Mixing of passive heterogeneities by mantle convection. *J. Geophys. Res.* **89**, 425–436.

O'Nions, R.K., Evenson, N.M., and Hamilton, P.J. (1979) Geochemical modeling of mantle differentiation and crustal growth. *J. Geophys. Res.* **84**, 6091–6101.

O'Nions, R.K., Hamilton, P.J., and Evensen, N.M. (1977) Variations in $^{143}Nd/^{144}Nd$ and $^{87}Sr/^{86}Sr$ ratios in oceanic basalts. *Earth and Planet. Sci. Lett.* **34**, 13–22.

O'Nions, R.K. and Oxburgh, E.R. (1983) Heat and helium in the Earth. *Nature* **306**, 429.

Ottonello, G., Piccardo, G.B., and Ernst, W.G. (1979) Petrogenesis of some Ligurian peridotites—II. Rare earth element chemistry. *Geochim. Cosmochim. Acta.* **43**, 1273–1284.

Oversby, V.M. and Ewart, A. (1972) Lead isotopic compositions of Tonga Kermadec volcanics and their petrogenetic significance. *Cont. Min. Petrol.* **37**, 181–210.

Oversby, V.M. and Ringwood, A.E. (1971) Time of formation of the earth's core. *Nature* **234**, 463–465.

Ozima, M. and Podosek, F.A. (1983) *Noble Gas Geochemistry*, New York: Cambridge University Press.

Ozima, M. and Zashu, S. (1983) Noble gases in submarine pillow volcanic glasses. *Earth and Planet. Sci. Lett.* **62**, 24–40.

Patterson, C.C. and Tatsumoto, M. (1964) The significance of lead isotopes in detrital feldspar with respect to chemical differentiation within the earth's mantle. *Geochim. Cosmochim. Acta.* **28**, 1–28.

Peltier, W.R. (1981) Ice age geodynamics. *Ann. Rev. Earth and Planet. Sci.* **9**, 199–225.

Peltier, W.R. (1983) Constraint on deep mantle viscosity from Lageos acceleration data. *Nature* **304**, 434–436.

Peltier, W.R. and Andrews, J.T. (1976) Glacial isostatic adjustment. I. The forward problem. *Geophys. J. R. Astron. Soc.* **46**, 605–646.

Polve, M. and Allegre, C.J. (1980) Orogenic lherzolite complexes studied by ^{87}Rb-^{87}Sr: A clue to understand the mantle convection processes? *Earth and Planet. Sci. Lett.* **51**, 71–93.

Poreda, R. (1980) $^{3}He/^{4}He$ variations along the Reykjanes Ridge. *EOS* **61**, 1158.

Poreda, R., Craig H., and Schilling, J.G. (1980) $^3He/^4He$ variations along the Reykjanes Ridge. *EOS* **61**, p. 1158.

Reilly, G.A. and Shaw, D.M. (1967) An estimate of the composition of part of the Canadian shield in Northwestern Ontario. *Canadian J. Earth Sci.* **4**, 725–739.

Reisberg, L. and Zindler, A. (1985) The scale of mantle heterogeneity: Isotopic evidence from Ronda. *EOS* **66**, 413.

Reisberg, L. and Zindler A. (1987) Extreme isotopic variations in the upper mantle: Evidence from Ronda. *Earth and Planet. Sci. Lett.* **81**, 29–45.

Richard, P. and Allegre, C.J. (1980) Neodymium and strontium isotope study of ophiolite and orogenic lherzolite petrogenesis. *Earth and Planet. Sci. Lett.* **47**, 327–342.

Richard, P., Shimizu, N., Allegre, C.J. (1976) $^{143}Nd/^{144}Nd$, a natural tracer: An application to oceanic basalt. *Earth and Planet. Sci. Lett.* **31**, 269–278.

Richards, P.G. (1972) Seismic waves reflected from velocity gradient anomalies within the earth's upper mantle. *J. Geophys.* **38**, 517–527.

Richardson, S.H., Erlank, A.J., Duncan, A.R., and Reid, D.L. (1982) Correlated Nd, Sr, and Pb isotope variation in Walvis Ridge basalts and implications for the evolution of their mantle source. *Earth and Planet. Sci. Lett.* **59**, 327–342.

Richardson, S.H., Erlank, A.J., and Hart, S.R. (1985) Kimberlite-borne garnet peridotite xenoliths from old enriched subcontinental lithosphere. *Earth and Planet. Sci. Lett.* **75**, 116-128.

Richardson, S.H., Gurney, J.J., Erlank, A.J., and Harris, J.W. (1984) Origin of diamonds in old enriched mantle. *Nature* **310**, 198–202.

Richter, F.M. (1984) Regionalized models for the thermal evolution of the Earth. *Earth and Planet. Sci. Lett.* **68**, 471–484.

Richter, F.M. and McKenzie, D.P. (1981) On some consequences and possible causes of layered mantle convection. *J. Geophys. Res.* **86**, 6133–6142.

Richter, F.M. and Ribe, N.M. (1979) On the importance of advection in determining the local isotopic composition of the mantle. *Earth and Planet. Sci. Lett.* **43**, 212–222.

Richter, F.M., Daly, S.F., and Nataf, H.C. (1982) A parameterized model for the evolution of isotopic heterogeneities in a convecting system. *Earth and Planet. Sci. Lett.* **60**, 178–194.

Ringwood, A.E. (1966) The chemical composition and origin of the earth. In P.M. Hurley (ed.) *Advances in Earth Science*. Cambridge, Mass.: MIT Press, 287–356.

Ringwood, A.E. (1975) *Composition and Petrology of the Earth's Mantle*. New York: McGraw-Hill, 618 pp.

Ringwood, A.E. (1979) *Origin of the Earth and Moon*. New York: Springer-Verlag.

Rison, W. (1981) Loihi seamount: Mantle volatiles in the basalts. *EOS* **62**, 1083.

Rison, W. and Craig, H. (1982) Coming of age in Somoa. *EOS* **63**, 1144.

Roden, M.K., Hart, S.R., Frey, F.A., and Melson, W.G. (1984) Sr, Nd and Pb isotopic and REE geochemistry of St. Paul's Rocks: The metamorphic and metasomatic development of an alkali basalt mantle source. *Contrib. Min. Petrol.* **85**, 376–390.

Ryan, M.P., Koyanagi, R.Y., and Fiske, R.S. (1981) Modeling the three-dimensional structure of macroscopic magma transport systems: Application to Kilauea Volcano, Hawaii. *J. Geophys. Res.* **86**, 7111–7129.

Sachtleben, T. and Seck, H.A. (1981) Chemical control of Al-solubility in orthopyroxene and its implications of pyroxene geothermometry. *Contrib. Mineral. Petrol.* **78**, 157–165.

Sarda, P., Staudacher, T., and Allegre, C.J. (1985) $^{40}Ar/^{36}Ar$ in MORB glasses: Constraints on atmosphere and mantle evolution. *Earth and Planet. Sci. Lett.* **72**, 357–375.

Schilling, J.-G. (1971) Sea-floor evolution: Rare-earth evidence. *Phil. Trans. Roy. Soc. Lond.* **A268**, 663–706.

Schilling, J.G. (1973) Icelandic mantle plume: Geochemical evidence along the Reykjanes Ridge. *Nature* **242**, 565–571.

Schilling, J.-G. (1975) Rare-earth variations across "normal segments" of the Reykjanes Ridge, 60°-53° N, Mid-Atlantic Ridge, 29° S, and East Pacific Rise, 2°-19° S, and evidence on the composition of the underlying low-velocity layer. *J. Geophys. Res.* **80**, 1459–1473.

Schubert, G. and Spohn, T. (1981) Two-layer mantle convection and the depletion of radioactive elements in the lower mantle. *Geophys. Res. Lett.* **8**, 951–954.

Schubert, G.D. and Turcotte, D.L. (1971) Phase changes and mantle convection. *J. Geophys. Res.* **76**, 1424–1432.

Sclater, J.G., Jaupart, C., and Galson, D. (1980) The heat flow through oceanic and continental crust and the heat loss of the earth. *Rev. Geophys. Space Phys.* **18**, 269–311.

Shaw, D.M. (1967) U, Th, and K in the Canadian Precambrian shield and possible mantle compositions. *Geochim. Cosmochim. Acta.* **31**, 1111–1113.

Shaw, D.M., Dostal, J., and Keays, R.R. (1976) Additional estimates of continental surface Precambrian shield composition in Canada. *Geochim. Cosmochim. Acta.* **40**, 73–83.

Shaw, H.R., Jackson, E.D., and Baragar, K.E. (1980) Volcanic periodicity along the Hawaiian-Emperor chain. *Am. J. of Sci.* **280-A**, 667–708.

Shimizu, N. (1975) Rare Earth elements in garnets and clinopyroxenes from garnet lherzolite nodules in kimberlites. *Earth and Planet. Sci. Lett.* **25**, 26–32.

Sleep, N.H. (1984) Tapping of magmas from ubiquitous mantle heterogeneities: An alternative to mantle plumes? *J. Geophys. Res.* **89**, 10029–10041.

Sneeringer, M., Hart, S.R., and Shimizu, N. (1984) Strontium and samarium diffusion in diopside. *Geochim. Cosmochim. Acta.* **48**, 1589–1608.

Spohn, T. and Schubert, G. (1982) Modes of mantle convection and the removal of heat from the Earth's interior. *J. Geophys. Res.* **87**, 4682–4696.

Stacey, F.D. (1969) *Physics of the Earth*. New York: John Wiley & Sons, Inc.

Staudacher, T. and Allegre, C.J. (1982) Terrestrial xenology. *Earth and Planet. Sci. Lett.* **60**, 389–406

Staudigel, H., Hart, S.R., and Richardson, S.H. (1981) Alteration of the oceanic crust: Processes ad timing. *Earth and Planet. Sci. Lett.* **52**, 311-327.

Staudigel, H., Zindler, A., Hart, S.R., Leslie, T., Chen, C.-Y., and Clague, D. (1984) The isotope systematics of a juvenile intraplate volcano: Pb, Nd, and Sr isotope ratios of basalts from Loihi Seamount, Hawaii. *Earth and Planet. Sci. Lett.* **69**, 13–29.

Steiger, R.H. and Jager, E. (1977) Subcommission on geochronology: Convention on the use of decay constants in geo- and cosmochronology. *Earth and Planet. Sci. Lett.* **36**, 359–362.

Stille, P., Unruh, D.M., and Tatsumoto, M. (1983) Pb, Sr, Nd and Hf isotopic evidence of multiple sources for Oahu, Hawaii basalts. *Nature* **304**, 25–29.

Stosch, H.G. (1982) Rare earth element partitioning between minerals from anhydrous spinel peridotite xenoliths. *Geochim. Cosmochim. Acta.* **46**, 793–811.

Stosch, H.G., Carlson, R.W., and Lugmair, G.W. (1980) Episodic mantle differentiation: Nd and Sr isotopic evidence. *Earth and Planet. Sci. Lett.* **47**, 263–271.

Stosch, H.G. and Lugmair, G.W. (1984) Evolution of lower continental crust as recorded by granulite facies xenoliths from the Eifel/Germany. *EOS* **65**, 230.

Stosch, H.G. and Seck, H.A. (1980) Geochemistry and mineralogy of two spinel peridotite suites from Dreiserweiher, West Germany. *Geochim. Cosmochim. Acta.* **44**, 457–470.

Stueber, A.M. and Ikramuddin, M. (1974) Rubidium, strontium and the isotopic composition of strontium in ultramafic nodule minerals and host basalts. *Geochim. Cosmochim. Acta.* **38**, 207–216.

Suen, C.J. (1978) Geochemistry of peridotites and associated mafic rocks, Ronda ultramafic complex, Spain. Ph.D. Thesis, Mass. Inst. Tech., 305 pp.

Sun, S.-S. (1982) Chemical composition and origin of the Earth's primitive mantle. *Geochim. Cosmochim. Acta.* **46**, 179–192.

Sun, S.-S. (1980) Lead isotopic study of young volcanic rocks from mid-ocean ridges, ocean islands and island arcs. *Philos. Trans. R. Soc. London, Ser. A* **297**, 409–445.

Sun, S.-S. and Hanson, G.N. (1975) Evolution of the mantle: Geochemical evidence from alkali basalts. *Geology* **3**, 297–302.

Sun, S.-S. and Nesbitt, R.W. (1977) Chemical heterogeneity of the Archean mantle, composition of the Earth and mantle evolution. *Earth and Planet. Sci. Lett.* **32**, 429–448.

Tatsumoto, M. (1966a) Genetic relations of oceanic basalts as indicated by lead isotopes. *Science* **153**, 1094–1101.

Tatsumoto, M. (1966b) Isotopic composition of lead in volcanic rocks from Hawaii, Iwo Jima, and Japan. *J. Geophys. Res.* **71**, 1721–1733.

Tatsumoto, M. (1978) Isotopic composition of lead in oceanic basalt and its implication to mantle evolution. *Earth and Planet. Sci. Lett.* **38**, 63–87.

Tatsumoto, M., Hedge, C.E., and Engel, A.E.J. (1965) Potassium, rubidium, strontium, thorium, uranium, and the ratio of strontium-87 to strontium-86 in oceanic tholeiitic basalt. *Science* **150**, 886–888.

Tatsumoto, M. and Knight, R.J. (1969) Isotopic composition of lead in volcanic rocks from central Honshu—with regard to basalt genesis. *Geochemical J.* **3**, 53–86.

Tatsumoto, N., Knight, R.J., and Allegre, C. (1973) Time difference in the formation of meteorites as determined from the ratio of lead-207 to lead-206. *Science* **180**, 1278–1283.

Tatsumoto, N., Unruh, D.M., and Desborough, G.A. (1976) U-Th-Pd and Rb-Sr systematics of Allende and U-Th-Pb systematics of Orgueil. *Geochim. Cosmochim. Acta.* **40**, 617–634.

Taylor, S.R. (1982) Lunar and terrestrial crusts: A contrast in origin and evolution. *Physics of the Earth and Planet. Inter.* **29**, 233–241.

Taylor, S.R. and McLennan, S.M. (1981) The composition and evolution of the continental crust: Rare earth element evidence from sedimentary rocks. *Philos. Trans. R. Soc. London* **A-301**, 381–399.

Taylor, S.R., McLennan, S.M., and McCulloch, M.T. (1983) Geochemistry of loess, continental crustal composition and crustal model ages. *Geochim. Cosmochim. Acta.* **47**, 1897–1905.

Tera, F. (1980) Reassessment of the "age of the Earth" Carnegie Inst. *Wash. Year Book* **79**, 524–531.

Tilton, G.R. (1960) Volume diffusion as a mechanism for discordant lead ages. *J. Geophys. Res.* **65**, 2933–2945.

Tilton, G.R. (1973) Isotopic lead ages of chondritic meteorites. *Earth and Planet. Sci. Lett.* **19**, 321–329.

Tilton, G.R. and Barreiro, B.A. (1980) Origin of lead in Andean calc-alkaline lavas, Southern Peru. *Science* **210**, 1245-1247.

Turcotte, D.L. and Schubert, G. (1982) *Geodynamics*. New York: John Wiley, 435 pp.

Usselman, T.M. and Hodge, D.S. (1978) Thermal control of low-pressure fractionation processes. *J. Vol. and Geother. Res.* **4**, 265–281.

Van Schmus, W.R. and Hayes, J.M. (1974) Chemical and petrographic correlations among carbonaceous chondrites. *Geochim. Cosmochim. Acta.* **38**, 47–64.

Vidal, P. and Dosso, L. (1978) Core formation: Catastrophic or continuous? Sr and Pb isotope geochemistry constraints. *Geophys. Res. Lett.* **5**, 169–172.

Vidal, P. and Clauer, N. (1981) Pb and Sr isotopic systematics of some basalts and sulfides from the East Pacific Rise at 21°N (project RITA). *Earth and Planet. Sci. Lett.* **55**, 237–246.

Vidal, P., Chauvel, C., and Brousse, R. (1984) Large mantle heterogeneity beneath French Polynesia. *Nature* **307**, 536–538.

Vollmer, R. (1977) Terrestrial lead evolution and formation time of the Earth's core. *Nature* **270**, 144–147.

Von Michaelis, H., Ahrens, L.H., and Willis, J.P. (1969) The composition of stony meteorites II. The analytical data and an assessment of their quality. *Earth and Planet. Sci. Lett.* **5**, 387–394.

Wanke, H. (1981) Constitution of terrestrial planets. *Phil. Trans. R. Soc. Lond.* **A-303**, 287–302.

Wanke, H., Dreibus, G., and Jagoutz, E. (1984) Mantle chemistry and accretion history of the Earth. In A. Kroner *et al.* (eds.) *Archaean Geochemistry*. Berlin: Springer-Verlag, Berlin, 1–24.

Wasserburg, G.J. and DePaolo, D.J. (1979) Models of earth structure inferred from neodymium and strontium isotopic abundances. *Proc. Nat. Acad. Sci. USA* **76**, 3594–3598.

Watson, E.B., Othman, D.B., Luck, J.M., and Hofmann, A.W. (1987) Partitioning of U, Pb, Cs, Yb, Hf, Re and Os between chromian diopsidic pyroxene and haplobasaltic liquid. *Chemical Geology* **62**, 191–208.

Weidner, D.J. (1985) A mineral physics test of a pyrolite mantle. *Geophys. Res. Lett.* **12**, 417–420.

White, W.M. (1985) Sources of oceanic basalts: Radiogenic isotopic evidence. *Geology* **13**, 115–118.

White, W.M., Dupre, B., and Vidal, P. (1985) Isotope and trace element geochemistry of sediments from the Barbados Ridge-Demerara Plain region, Atlantic Ocean. *Geochim. Cosmochim. Acta.* **49**, 1875–1886.

White, W.M. and Hofmann, A.W. (1982) Sr and Nd isotope geochemistry of oceanic basalts and mantle evolution. *Nature* **296**, 821–825.

Worner, G., Zindler, A., Staudigel, H., and Schmincke, H.U. (1986) Sr, Nd and Pb isotope geochemistry of Tertiary and Quarternary alkaline volcanics from West Germany. *Earth and Planet. Sci. Lett.* **79**, 107–119.

Yoder, H.S., Jr. (1976) *Generation of Basaltic Magma*. Washington, D.C.: Nat. Acad. Sci.

Zindler, A. and Hart, S.R. (1986) Helium: Problematic primordial signals. *Earth and Planet. Sci. Lett.* **79**, 1–8.

Zindler, A., Hart, S.R., Frey, F.A., and Jakobsson, S.P. (1979) Nd and Sr isotope ratios and rare earth elements abundances in Reykjanes Peninsula basalts: Evidence for mantle heterogeneity beneath Iceland. *Earth and Planet. Sci. Lett.* **45**, 249–262.

Zindler, A., Jagoutz, E., and Goldstein, S. (1982) Nd, Sr and Pb isotopic systematics in a three-component mantle: A new perspective. *Nature* **58**, 519–523.

Zindler, A. and Jagoutz, E. (1988) Mantle cryptology. *Geochim. Cosmochim. Acta.* **52**, 319–333.

Zindler, A., Staudigel, H., and Batiza, R. (1984) Isotope and trace element geochemistry of young Pacific seamounts: Implications for the scale of upper mantle heterogeneity. *Earth and Planet. Sci. Lett.* **70**, 175–195.

Zindler, A., Staudigel, H., Hart, S.R., Endres, R., and Goldstein, S. (1983) Nd and Sr isotopic study of a mafic layer from Ronda ultramafic complex. *Nature* **304**, 226–230.

CHAPTER 6

Mantle Viscosity

W.R. Peltier

Department of Physics, University of Toronto
Toronto, Ontario, Canada, M5S 1A7

1. Introduction

The issue as to whether the planetary mantle is at all characterizable by a steady-state viscosity is a question which, until relatively recently, was considered a central issue of geophysical science. Sir Harold Jeffreys (e.g., 1972), for example, held the view that the mantle was incapable of any steady-state mode of material deformation and therefore that the convection hypothesis of continental drift was untenable. Although he believed that the transient anelastic relaxation processes embodied in "Lomnitz Law" were crucial in determining seismic Q, it did not seem to him that mantle material should be capable of the continuous deformation in response to a fixed applied shear stress which the thermal convection phenomenon clearly demands. Beginning in the early 1970s, however, laboratory data on the creep of the mineral Olivine ($[Mg,Fe]_2SiO_4$), which is the dominant constituent of the Earth's upper mantle (see Chapter 4), demonstrated that such material did exhibit a steady-state mode of material deformation at sufficiently high temperatures (e.g., Kohlstedt and Goetze, 1974; Durham and Goetze, 1977). These experiments, which were all performed on single crystals of Olivine, also established that the stress-strain relation in the steady-state regime was nonlinear, with strain rate depending upon the stress raised to a power near three. Although it remains a major issue in this field as to whether such non-Newtonian behavior persists in the polycrystalline material of which the mantle is actually composed, it must be considered well established at present that the Earth's solid mantle does indeed behave as a fluid when it is subjected to stresses of sufficient temporal duration. It must therefore be characterizable by a "steady-state" viscosity, although this viscosity is liable to be a strong function of both pressure and temperature, since the creep of a solid is a thermally activated process, and also of stress if grain boundary effects in the polycrystalline material of the mantle do not suffice to return the stress-strain relation to a Newtonian one. Although the experimentally demonstrated existence of a high temperature regime of steady deformation has established that the hypothesis of mantle convection is possible in principle, in order that this hypothesis be established as fact clearly requires that the effective viscosity of the mantle have a rather special value.

Even though it is evident that some latitude does exist within the context of the convection hypothesis for variations of mantle viscosity, it is also clear that such variations are severely restricted. This follows from the fact, which was very clearly established in Chapter 2 (see also Chapter 7), that the vigour of a convective circulation in a Newtonian fluid of essentially infinite Prandtl number (and in fact most physical characteristics of such a circulation) is determined entirely by the system Rayleigh number. For a spherical shell of fluid of thickness d this non-dimensional parameter may be written as:

$$\mathrm{Ra} = \frac{g\alpha d^3(F + Qd)}{\kappa K \nu}, \tag{1}$$

in which g is the assumed constant gravitational acceleration, α is the coefficient of thermal expansion, K is the thermal conductivity, and κ and ν are respectively the diffusivities for heat and momentum. Also in (1), $F = -K\, dT/dr\ (r = a)$ is the heat flux across the outer surface of the Earth and Q is the rate of heat generation per unit volume in the mantle. Now F is clearly measurable and known while Q is bounded. g is similarly known on the basis of the internal distribution of density, while α, K, and $\kappa = K/\rho C_p$ are measurable (and have been measured) in the laboratory. To understand the manner in which the validity of the convection hypothesis may be established by a measurement of the remaining parameter in (1), the kinematic viscosity ν, we have only to understand the manner in which the various characteristics of a convective circulation operating at a specific value of Ra depend upon this parameter.

At sufficiently high Rayleigh number the dependencies of these characteristics upon Ra become especially simple since the circulation is then controlled by the dominant balances of heat and vorticity respectively in the horizontal boundary layers and vertical thermal plumes which characterize the flow in this regime (see especially Chapters 2 and 7). For a plane layer circulation which is entirely forced by heating from below ($Q = 0$) the mean thermal boundary layer thickness, $\bar{\delta}$, the characteristic horizontal velocity $\bar{u}$, and the mean surface heat flow $\bar{q}$ are well approximated by the following simple power law relations which are functions of Ra alone:

$$\bar{\delta} = \beta_1(a)d(\mathrm{Ra}/R_c)^{s_1}; \tag{2a}$$

$$\bar{u} = \beta_2(a)(\kappa/d)(\mathrm{Ra}/R_c)^{s_2}; \tag{2b}$$

$$\bar{q} = \beta_3(a)(K\Delta T/d)(\mathrm{Ra}/R_c)^{s_3}, \tag{2c}$$

in which ΔT is the temperature difference impressed across the layer of thickness d. The beauty of equations (2a–c) is that each of the quantities on the left hand sides of these equations is a geophysical observable. $\bar{\delta}$ is known from the observed thickness of the lithosphere in which the viscosity is extremely high because the temperature in the outer thermal boundary layer is low, $\bar{u}$ is known from the observed relative velocities of the surface plates, while $\bar{q}$ is known on the basis of direct measurements of the surface heat flow in vertical bore holes. For steady state two dimensional circulations the vectors $\boldsymbol{\beta}$ and $\mathbf{s}$ are $\boldsymbol{\beta} = (0.300, 8.468, 2.30)$ and $\mathbf{s} = (-0.278, 0.645, 0.313)$ based upon the finite difference simulations of Jarvis and Peltier (1982) (see also Chapter 7) for flows of aspect ratio $a = 1$ which is the normal ratio (see Chapter 2).

If we combine (2b) and (2c) to eliminate the unknown temperature drop across the mantle ΔT ($F = K\Delta T/d$), the result may be solved for the depth of the convecting layer to give:

$$d_1 = \frac{\beta_1^{1/2}}{\beta_2}\left[\frac{C_p \nu}{R_c^{1/3}\bar{q}\alpha g}\right]^{1/2}\bar{u}. \tag{3a}$$

Now (2a) and (2b) may be similarly employed to eliminate ΔT and the result solved for a second estimate of the layer depth as:

$$d_2 = \frac{(\bar{\delta}^2/K)}{\beta_2\beta_1^2 R_c^{2/3}}\bar{u}. \tag{3b}$$

Following the analysis in Peltier (1980) with $\bar{q} = 80$ mW m^{-2}, $\bar{\delta} \simeq 100$ km, and $\bar{u} = 4$ cm yr^{-1} we obtain $d_1 \simeq 3 \times 10^3$ km and $d_2 \simeq 4 \times 10^3$ km (i.e., compatible with each other and with a radial length scale close to the thickness of the mantle ($d = 2.9 \times 10^3$ km). If we set (3a) = (3b) and solve for the viscosity ν, we then obtain

$$\nu = \bar{\delta}^4 \bar{q} \frac{(\alpha_g)}{(C_p K^2)} \frac{1}{\beta_1^4 \beta_3 R_c}. \tag{4}$$

Since $2 \times 10^{-5} \leqslant \alpha \leqslant 3.5 \times 10^{-5}\ {}^\circ\mathrm{K}^{-1}$, $10^{-6} \leqslant K \leqslant 2 \times 10^{-6}\ \mathrm{m}^2\ \mathrm{s}^{-1}$, and $C_p \simeq 1.25 \times 10^3\ \mathrm{J\ kg}^{-1}\ {}^\circ\mathrm{K}^{-1}$, substitution into (4) gives a molecular viscosity $\rho_0 \nu$ such that

$$\rho_0 \nu \simeq 10^{21}\ \mathrm{Pa\ s}. \tag{5}$$

This is the value which the molecular viscosity of the mantle must have if the convection hypothesis of continental drift is to be capable of reconciling the surface observables $\bar{\delta}$, $\bar{u}$, and $\bar{q}$. Clearly some variations on this value will be allowed since it has been inferred on the basis of a number of highly restrictive assumptions. In particular the inference was based on the observed properties of two-dimensional plane layer, steady state, aspect ratio 1, heated from below convection in a Newtonian fluid with constant viscosity. The combined effects of three dimensionality, sphericity, time dependence, internal heating, and variable viscosity, not to mention the possible influence of non-Newtonian behavior (see Chapter 8) will doubtless combine to introduce some allowed variations of the value inferred above on the basis of the simplest boundary layer considerations. Nevertheless, these considerations are so fundamental to the convection process that the viscosity of the Earth's mantle could not depart substantially from (5) without raising very severe doubts about the validity of the convection hypothesis of continental drift. In this sense an accurate inference of the viscosity of the mantle does indeed constitute a direct test of the validity of the convection hypothesis.

In the following sections of this chapter I will review the main avenues of inference which have enabled direct estimates of mantle viscosity to be obtained. Most of this work has involved analysis of various signatures of the planet's response to the continuous cycle of glaciation and deglaciation to which it has been subject throughout the Pleistocene epoch (last 2×10^6 years) of Earth history. These signatures consist of relative sea level observations in the time window from 18 kyr ago until the present, of observations of the "anomalies" in the surface gravitational acceleration (g) which are observed over the geographic regions which were once ice covered, and of observations of two rather subtle "anomalies" in the state of the Earth's rotation. The latter consist of the so-called non-tidal component of the acceleration of the planet's rate of axial rotation and the true secular drift (wander) of the rotation pole relative to the surface geography. The non-tidal rate of the acceleration of rotation was first observed on the basis of analyses of ancient eclipse data (e.g., Müller and Stephensen, 1975) but has been recently reconfirmed through the use of satellite laser ranging techniques (SLR) to measure the acceleration of the node of the orbit of the LAGEOS satellite (Yoder *et al.*, 1983, Rubincam 1984). Correspondingly, the currently ongoing drift of the rotation pole towards eastern Canada at a rate near 0.95 ± 0.15 degrees per 10^6 yrs was first determined on the basis of data from the photo-zenith-tube observatories of the International Latitude Service. This observation has also been recently reconfirmed on the basis of both SLR and very long baseline interferometry (VLBI) measurements (Carter *et al.*, 1986). The final ice age related phenomenon which I shall describe here that may be employed to constrain the viscosity of the planetary mantle is the cyclic oscillation of continental ice volume that has dominated the last 10^6a of Pleistocene climatic change. This variability is very strongly influenced by the isostatic adjustment process, so much so that the dominant 10^5a cyclic component of the ice volume signal inferred from $\delta^{18}O$ variations in deep sea sedimentary cores is apparently completely inexplicable in its absence. The period of this "ice-age cycle" is in fact a strong function of viscosity and so may be used to constrain its value if other parameters of the climate model can be assumed known.

As will be made clear on the basis of the constraints which each of these observations may be invoked to place upon the viscosity of the planetary mantle, each one of the dynamical phenomena mentioned above either individually require that the viscosity of the mantle equal the value demanded by the convection hypothesis stated in equation (5) or are fully compatible with this value. This will be construed as direct verification of the validity of the convection hypothesis of continental drift and sea floor spreading since it demonstrates, according to the brief synopsis of the implications of the boundary layer theory of thermal convection provided above, that surface geophysical observables such as lithospheric thickness, plate velocities, and surface heat flow are related just as they would be in a layer of vigorously convecting infinite Prandtl number Newtonian fluid in a laboratory apparatus. In establishing this remarkable fact the direct inference of the effective viscosity of the planetary mantle plays a crucial role.

2. The Viscoelastic Field Theory of Glacial Isostasy

In order to infer mantle viscosity on the basis of observations of various glaciation induced deformations of the planet, we clearly require a theory which may be employed to make accurate predictions of the response to a prescribed forcing for arbitrary radial viscosity profile. An appropriate linear viscoelastic field theory for use in this application has been developed over the past decade of research which began with that described in Peltier (1974), a paper which showed how one could construct the Green function for the impulse response of the viscoelastic planet to gravitational excitation by a point mass load. The mathematical ideas presented there have since been considerably extended to cover application to the prediction of each of the glaciation-deglaciation related effects mentioned in the last section. A brief review of the fundamental ingredients of this theory will be presented here, following which the remaining sections of the chapter will discuss the extensions of theory required for each of the specific predictions and will provide a summary of the constraints on mantle viscosity which each of the observations may be invoked to obtain. In the concluding section the results of each of these separate analyses will be integrated to provide a single "best" radial viscosity profile for the planet as a whole.

All of the detailed discussion in this chapter will be concerned with the deformation of spherically symmetric viscoelastic models of the planet by surface ice sheet loads. The reason why we believe such laterally homogeneous models should provide a reasonable first order description of the glaciation induced deformation is ultimately connected with the structure which we believe to be characteristic of the convective circulation responsible for the generation of the actual lateral heterogeneity which fills the mantle. At sufficiently high Rayleigh number that the boundary layer equations (2) may be assumed to apply, the temperature field in the convecting region is characterized by being essentially adiabatic and therefore laterally homogeneous throughout the cores of the individual cells (see Chapter 7) with all lateral variation focussed into the vertically oriented hot and cold plumes whose lateral dimensions are on the order of the boundary layer thickness $\bar{\delta}$. From (2a) this scale falls rapidly with increasing Ra and Ra is now such that $\bar{\delta}$ is near 100 km. This implies that over 90 percent of the volume of the mantle should be adiabatic and thus radially stratified. This inference must be qualified by the fact that the convective circulation is not entirely forced by heating from below as is assumed in (2) since internal heating is known to induce some weak lateral heterogeneity even in the cell cores. However, this heterogeneity is sufficiently modest (see Chapter 7) that it will be strongly dominated by that associated with the plumes. To the extent that the mantle convective circulation is of sufficiently high Rayleigh number to be well within the boundary layer regime, we may reasonably expect that a spherically symmetric approximation to the internal viscoelastic structure will constitute an acceptable first approximation.

If, furthermore, the deformations induced by surface ice sheet loading are sufficiently small as to constitute relatively insignificant modifications to the overall structure of the planet, then first order perturbation theory should provide a sufficiently accurate description of the dynamical process. Linearizing the equation of momentum balance and Poisson's equation for the gravitational potential in small deviations from a spherically symmetric hydrostatic equilibrium configuration then reduces these equations to the respective forms (e.g., Backus, 1967):

$$\boldsymbol{\nabla}\cdot\underline{\sigma} - \boldsymbol{\nabla}(\rho_0 g_0 \mathbf{u}\cdot\hat{e}_r) - \rho_0 \boldsymbol{\nabla}\phi_1 - g_0\rho_1\hat{e}_r = 0, \tag{6a}$$

$$\nabla^2\phi_1 = 4\pi G\,\rho_1, \tag{6b}$$

in which the density perturbation ρ_1 is obtained from the linearized continuity equation as:

$$\rho_1 = -\rho_0\boldsymbol{\nabla}\cdot\mathbf{u} - \mathbf{u}\cdot(\partial_r\rho_0)\hat{e}_r. \tag{7}$$

The background hydrostatic equilibrium configuration (ρ_0, p_0, ϕ_0) about which (6a) has been linearized is one which satisfies

$$\boldsymbol{\nabla}p_0 = -\rho_0 g_0\hat{e}_r \tag{8a}$$

$$\nabla^2\phi_0 = 4\pi G\rho_0 \tag{8b}$$

In (6a) and (6b) the gravitational potential perturbation ϕ_1 is in general the sum of two parts, ϕ_2 and ϕ_3, which are respectively the potential of any externally applied force field (the surface ice sheet and ocean loads) and that due to the internal redistribution of mass effected by the load induced deformation. Also in (6a) the field $\mathbf{u}$ is the displacement field while $\underline{\sigma}$ is the stress tensor which is to be determined from $\mathbf{u}$ via an appropriate constitutive relation involving the material properties (such as the viscosity). This equation has a number of further general characteristics which are worthy of mention here. First among these is the fact that the inertial force ($\rho_0 d^2\mathbf{u}/dt^2$) has been dropped from the r.h.s. in consequence of the extraordinarily long timescales which are characteristic of the phenomena of interest to us. Secondly, one should note that the second term on the l.h.s. is one which describes a body force determined by the gradient of the radially advected prestress p. As clearly demonstrated by Love (e.g., 1911) this term may be interpreted in the context of the Lagrangian theory for small deformations of an elastic solid by imagining that in the course of such deformation an individual element of material will transport its prestress with it. In virtually all elastic problems of geodynamic interest, however, (e.g., the theory of the elastic gravitational free oscillations) this effect turns out to be very modest indeed. In the viscoelastic theory of glacial isostasy, on the other hand, the incorporation of this term turns out to be the key to the successful development of a general model of the process since it is through this term that a buoyant restoring force is induced by the deflection of surfaces across which non-adiabatic variations of density occur, such as the outer surface of the Earth for example. Such deformation induced buoyancy is of course central to the physics underlying the glacial isostatic adjustment process (e.g., see Peltier, 1986b for detailed discussion of a simple model which illustrates this fundamental point).

Since equations (6) are linear we may understand each of the unknown fields which appear in them to be Laplace transformed from the time domain into the domain of the complex frequency (Laplace transform variable) s. In this domain the material constitutive relation connecting the elements of the stress tensor $\underline{\sigma}_{ij}$ to the elements of the strain tensor e_{ij} may be written as:

$$\underline{\sigma}_{ij} = \lambda(s)\mathsf{e}_{ii}\delta_{ij} + 2\mu(s)\mathsf{e}_{ij}, \tag{9}$$

if the material is linearly viscoelastic. For such models of mantle rheology the precise frequency dependence of the modulii $\lambda(s)$ and $\mu(s)$ determine the nature of the material. Figure 6.1 shows a series of one-dimensional spring and dashpot analogues which vary in complexity from the single spring representing the Hooke's Law elastic solid in series with a single dashpot which together describe the so-called Maxwell solid (plate a). The material described by this composite is such that its initial behavior in response to an applied shear stress is Hookean elastic whereas its final behavior is Newtonian viscous. It is the simplest model which is capable of describing the transformation of properties which is required to understand the ability of the Earth's mantle to deform as an elastic solid on short timescales that is demanded by seismology and its ability to deform as a viscous fluid on long timescales that is demanded by the thermal convection process. The Standard Linear Solid described by the analogue in plate b differs from the Maxwell solid in that it is incapable of steady state deformation. Rather, in response to a constant applied shear stress this material initially responds elastically, subsequently relaxes somewhat at a rate determined by the viscosity of the Kelvin-Voight element ν_2, but eventually enters a new steady state in which the shear modulus has been reduced from μ_1 but remains nonzero. The Burgers body solid represented by the one dimensional analogue in plate c is clearly a combination of a and b in which both transient anelastic stress relaxation (governed by ν_2) and steady state deformation (governed by ν_1) are possible. Plate d is intended to represent the generalized Burgers body in which the transient anelastic relaxation is governed either by a discrete or continuous spectrum of relaxation times.

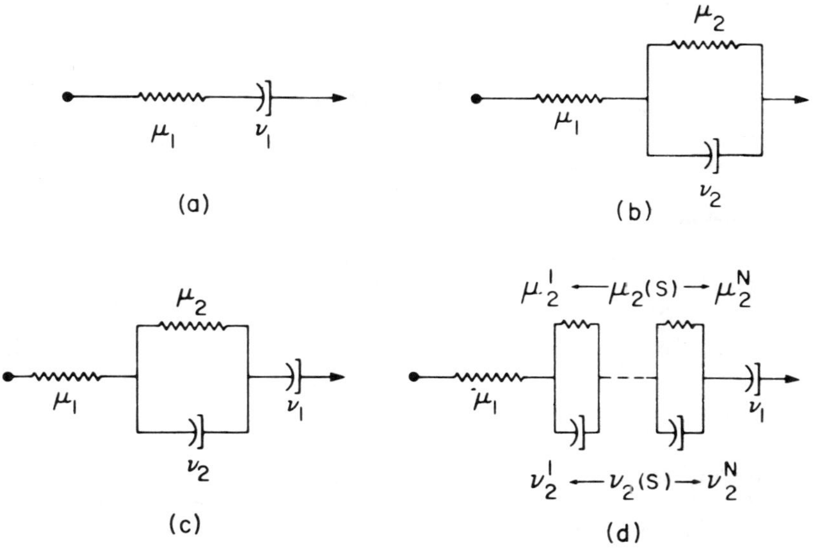

Figure 6.1. Spring and dashpot analogues of several common linear viscoelastic rheologies: a, Maxwell solid; b, Standard Linear Solid; c, Burger's body solid, d, the "generalized" Burger's body.

As discussed in detail in Peltier *et al.* (1986), the specific form of the generalized Burgers body that might be required to understand the full spectrum of geodynamic processes shown on the phenomenological diagram of Figure 6.2 is one for which the modulii $\lambda(s)$ and $\mu(s)$ are given by

$$\lambda(s) = K - \tfrac{2}{3}\mu(s), \tag{10a}$$

$$\mu(s) = \frac{\mu_1 s}{s + \mu_1/\nu_1}\left[\frac{(s + \mu_2/\nu_2)(s + \mu_1/\nu_1)}{(s + \mu_2/\nu_2)(s + \mu_1/\omega_1) + \mu_2 s/\nu_1}\right] \cdot \left[1 + \frac{2}{\Pi Q_m} \ln\left(\frac{s + T_2^{-1}}{s + T_1^{-1}}\right)\right]. \tag{10b}$$

In (10a) $K = \lambda(s) + \tfrac{2}{3}\mu(s) = \lambda + \tfrac{2}{3}\mu_1$ is the elastic, i.e., frequency independent, bulk modulus and thus the solid has no bulk dissipation. In (10b) the first term on the r.h.s. is the frequency dependent shear modulus of the Maxwell solid. The last term in (10b) describes the weak constant Q absorption band transient anelasticity that is required to understand the finite

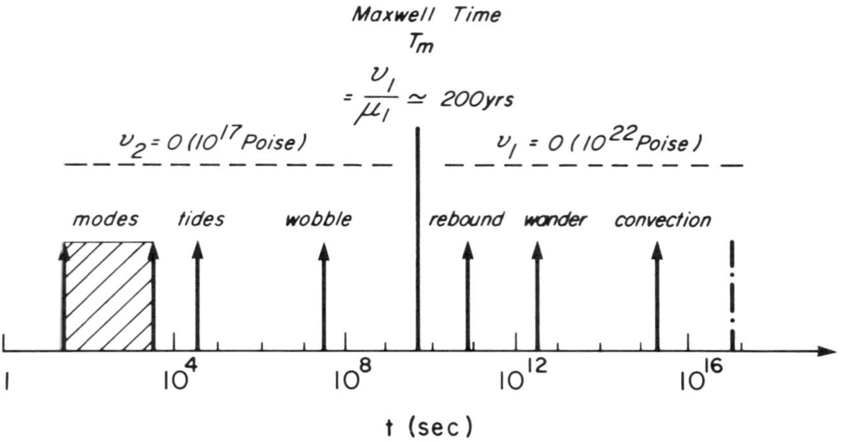

Figure 6.2. Characteristic timescales of various geodynamic process compared with the Maxwell time of the Earth's mantle. Phenomena with characteristic times shorter than the Maxwell time are governed by anelasticity, whereas those with longer characteristic time should be governed by steady state viscous deformation.

*Q*s of the modes of elastic gravitational free oscillation. The parameters which specify the strength of this component of the anelasticity must be near the values $Q_m = 250$, $T_1 = 10^{-2}$ s, and $T_2 = 10^4$ s of Minster (1978) in order to fit free oscillation *Q*s (Yuen and Peltier, 1982). With these parameter values this element of the anelasticity is responsible for a very small elastic defect $\Delta = (\mu_1 - \mu_r)/\mu_r = .03$ in which μ_1 is the initial elastic modulus and μ_r is the value towards which it is relaxed by the dissipation in the absorption band. The final term on the r.h.s. of (10b), the first factor in square brackets which contains the dependence on ν_2 and μ_2, embodies the contribution of an additional single isolated Debye peak in the anelastic relaxation spectrum to the total viscoelastic relaxation. It is a secondary peak whose existence was first postulated in Peltier (1985b) and Peltier *et al.* (1986) to account for a possible discrepancy between the estimates of mantle viscosity obtained by analysis of the isostatic adjustment process with the Maxwell analogue and that obtained by analysis of isostatic geoid anomalies forced by the lateral heterogeneity of mantle density inferred on the basis of seismic tomographic imaging (see Chapters 9 and 10). As demonstrated in Peltier (1985b) and Peltier *et al.* (1986), however, the form of this extra transient peak suggested by the rebound and geoid data is one for which $\mu_2/\mu_1 << 1$. In the limit $\mu_2 \to 0$ (10b) degenerates to

$$\mu(s) = \frac{\mu_1 s}{s + \mu_1/\nu_{\text{eff}}}\left[1 + \frac{2}{\Pi Q_m}\ln\left(\frac{s + T_2^{-1}}{s + T_1^{-1}}\right)\right]. \tag{10c}$$

in which

$$\nu_{\text{eff}} = \frac{\nu_1 \nu_2}{\nu_1 + \nu_2}$$

is the effective molecular viscosity of the material. Since the defect associated with the constant Q absorption band component of the rheology is so small, it is clear that postglacial rebound will then be well approximated by the following limiting form of the shear modulus,

$$\mu(s) = \frac{\mu_1 s}{s + \mu_l/\nu_{\text{eff}}}. \tag{10d}$$

Now it should not be inferred from this that transient relaxation is necessarily unimportant, only that for sufficiently small μ_2 the relaxation will proceed at a rate governed by ν_{eff} which may clearly differ considerably from the steady state viscosity ν_1 that determines the strain rate in the limit of very long time. For example, in a region of the Earth in which $\nu_1 >> \nu_2$ the effective viscosity ν_{eff} will equal the transient viscosity ν_2 rather than the steady state value. Peltier (1985b) and Peltier *et al.* (1986) have shown that for $\mu_2/\mu_1 <$ 0.1 the asymptotic condition (10d) effectively applies insofar as glacial rebound phenomenology is concerned. Since the limit $t \rightarrow \infty$ corresponds to the limit $s \rightarrow 0$ from the Tauberian theorems, it is clear by inspection of (10b) that

$$\lim_{s \to 0} \mu(s) = \frac{\mu_r}{\mu_1}\nu_1 s, \tag{11}$$

where the relaxed modulus $\mu_r = \mu_1\left[1 + \frac{2}{\Pi Q_m}\ln(T_1/T_c)\right] \approx \mu_1$ since $\Delta =$ 0.03. Therefore irrespective of the fact that (10d) will govern the transition from elastic to viscous behavior for very small but finite μ_2, the eventual steady state viscous behavior will *not* be governed by ν_{eff} but rather by the steady state viscosity ν_1. For most of the analyses of mantle viscosity structure to be described in this chapter, we will therefore employ the frequency dependent shear modulus given in (10d).

In order to solve the Laplace transformed field equations (6) for a model planet with constitutive relation determined by (9) and (10a, d) we employ the correspondence principle (e.g., Biot, 1954) which asserts that we may construct the Laplace transform of the solution of interest simply by solving an equivalent elastic problem, for a range of values of the complex frequency s and thus values of the viscoelastic modulii $\lambda(s)$ and $\mu(s)$, and then inverting the resulting spectrum to obtain the time domain response for the related viscoelastic problem. The equivalent elastic problem in this instance is clearly the surface loading problem for a radially stratified elastic sphere.

Since the deformation induced by an arbitrary surface load may be obtained from the principle of superposition, the field equations being linear, it may be expressed as a convolution of the actual load with an appropriate impulse response Green function. On the basis of symmetry considerations it will be clear that this function can depend only on the radial position r, the Laplace transform variable s, and the angular distance θ from the position of application of the point load. The fundamental solutions are therefore spheroidal and thus have the following vector harmonic decompositions:

$$\mathbf{u} = \sum_{l=0}^{\infty} \left(U_l(r, s) P_l(\cos\theta)\hat{e}_r + V_l(r, s) \frac{\partial P_l}{\partial \theta}(\cos\theta)\hat{e}_\theta \right); \tag{12a}$$

$$\phi_1 = \sum_{l=0}^{\infty} \phi_l(r, s) P_l(\cos\theta); \tag{12b}$$

$$\nabla\cdot\mathbf{u} = \sum_{l=0}^{\infty} \chi_l(r, s) P_l(\cos\theta). \tag{12c}$$

Substitution of expansions (12) into the field equations (6) reduces them to the following set of three simultaneous second order equations:

$$\ddot{\phi} + \frac{2}{r}\dot{\phi} - \frac{l(l+1)}{r}\phi = -4\Pi G(\pi_0 + \dot{\pi}_0 U); \tag{13a}$$

$$0 = -\rho_0\dot{\phi} + \rho_0 g_0 - \rho_0 d(g_0 U)/dr + d(\lambda + 2\mu\dot{U})/dr + (\mu/r^2)4r\dot{U} - 4U + l(l+1)(3V - U - r\dot{V}) \tag{13b}$$

$$0 = -\rho_0\phi - \rho_0 g_0 U + \lambda\chi + rd[\mu(\dot{V} - V/r + U/r)]/dr + [5U + 3r\dot{V} - V - 2l(l+1)V]/r, \tag{13c}$$

in which the dot denotes differentiation with respect to r and

$$\chi = \dot{U} + \frac{2}{r}U - \frac{l(l+1)}{r}V, \tag{13d}$$

$$\dot{g}_0 + \frac{2g_0}{4} = 4\Pi G\rho_0, \tag{13e}$$

in all of which it is understood that U, V, ϕ, χ, λ, and μ stand for $U_l(r, s)$, $V_l(r, s)$, $\phi_l(r, s)$, $\chi_l(r, s)$, $\lambda(r, s)$, and $\mu(r, s)$ respectively. Equations (13a–c) may be re-written as a set of six simultaneous first order equations by defining a solution six-vector **Y** as

$$\mathbf{Y} = (U_l,\ V_l,\ T_{rl},\ T_{\theta l},\ \phi_l,\ Q_l)^t, \tag{14}$$

in which

$$T_{rl} = \lambda_l + 2\mu U_l,$$

$$T_{\theta l} = \mu(\dot{V}_l - \frac{V_l}{r} + \mathbf{U}_l),$$

$$Q_l = \dot{\phi}_l + \frac{(l+1)}{r}\phi_l + 4\Pi G\rho_0 U_l.$$

In terms of **Y** the set (13a–c) is then equivalent to:

$$\frac{dY}{dr} = \mathsf{A}\mathbf{Y}, \tag{15}$$

in which the elements of the A matrix of coupling coefficients are:

$$a_{1i} = (-2\lambda/r\beta,\ l(l+1)\lambda/r\beta,\ 1/\beta,\ 0,\ 0,\ 0), \tag{16a}$$

$$a_{1i} = (-1/r,\ 1/r,\ 0,\ 1/\mu,\ 0,\ 0), \tag{16b}$$

$$a_{3i} = \left(\frac{4}{r}\left[\frac{\gamma}{r} - \rho_0 g_0\right],\ -\frac{l(l+1)}{r}\left[\frac{2\gamma}{r} - \rho_0 g_0\right],\right.$$
$$\left. -\frac{4\mu}{4\beta},\ \frac{l(l+1)}{r},\ -\frac{\rho_0(l+1)}{r},\ \rho_0\right), \tag{16c}$$

$$a_{4i} = \left(\frac{1}{r}\left[\rho_0 g_0 - \frac{2\gamma}{r}\right],\ -\frac{1}{r^2}\left[2\mu - l(l+1)(\gamma+\mu)\right],\right.$$
$$\left. -\frac{\lambda}{4\beta},\ \frac{-3}{r},\ \frac{\rho_0}{r},\ 0\right), \tag{16d}$$

$$a_{5i} = (-4G\rho_0,\ 0,\ 0,\ 0,\ -(l+1)/r,\ 1) \tag{16e}$$

$$a_{6i} = (-4\Pi G\rho_0(l+1)/r,\ 4\Pi G\rho_0 l(l+1)/r,\ 0,\ 0,\ 0,\ (l+1)/r) \tag{16f}$$

where

$$\beta = \lambda(r, s) + 2\mu(r,s),$$

$$\gamma = \mu(r, s)\cdot\frac{3\lambda(r, s) + 2\mu(r, s)}{\lambda(r, s) + \mu(r, s)}.$$

For impulsive point mass forcing (assuming a load of mass 1 kg) (15) is subject to the following s-independent boundary conditions at the Earth's surface $r = a$:

$$Y_3(a) = -g(2l+1)/4\Pi a^2 \tag{17a}$$

$$Y_4(a) = 0, \tag{17b}$$

$$Y_6(a) = (-g/m_e)(2l+1). \tag{17c}$$

The remaining three boundary conditions are those of regularity at $r = 0$. The system (15) may be solved using the parallel shooting algorithm discussed in Peltier (1974) which delivers a general solution in the form of a linear combination of three linearly independent solutions. If we call these linearly independent solutions $\mathbf{T}_1$, $\mathbf{T}_2$, $\mathbf{T}_3$, then in general:

$$\mathbf{Y}(r, s) = c_1(s)\mathbf{T}_1(r, s) + c_2(s)\mathbf{T}_2(r, s) + c_3(s)\mathbf{T}_3(r, s). \tag{18}$$

If we denote the boundary conditions (17) by $\mathbf{B}$ (a) then the coefficients $c_i(s)$ are determined by:

$$\mathbf{B} = \mathbf{Y}(a, s) = c_1(s)\mathbf{T}_1(a, s) + c_2(s)\mathbf{T}_2(a, s) + c_3(s)\mathbf{T}_3(a, s). \tag{19a}$$

If we further define a vector $\mathbf{C}$ as:

$$\mathbf{C} = [c_1(b), c_2(s), c_3(s)]^t, \tag{19b}$$

and denote by M the matrix whose column vectors are $\mathbf{T}_1(a, s)$, $\mathbf{T}_2(a, s)$, $\mathbf{T}_3(a, s)$, then

$$\mathsf{B} = \mathsf{M}\mathbf{C}, \tag{19c}$$

or

$$\mathsf{C} = \mathsf{M}^{-1}\mathbf{B}. \tag{19d}$$

Now since $\mathsf{M}^{-1} = \mathsf{M}^*/\det \mathsf{M}$ where M^* is the transpose matrix of cofactors M^*_{ij}, therefore

$$c_i = \sum_{j=1}^{3} \frac{M^*_{ij}(s)b_j}{\det \mathsf{M}}, \tag{20}$$

and thus the solution satisfying the boundary conditions is

$$Y(r, s) = \sum_{ij} \frac{M^*_{ij}b_j}{\det \mathsf{M}} T_i(r, s) = \frac{P(r, s)}{\det \mathsf{M}}, \tag{21}$$

where

$$P = \sum_{ij} M_{ij}^{*} b_j \mathbf{T}_i(r,\ s).$$

The solution (21) is now in a form which is useful for inversion back into the time domain. Suppose we define the "viscous part" of the solution as

$$\begin{aligned}\mathbf{Y}^{v}(r,s) &= \mathbf{Y}(r,s) - \mathbf{Y}^{E}(r,s) \\ &= \frac{\mathbf{P}^{v}(r,s)}{\det\mathsf{M}},\end{aligned} \tag{22}$$

where

$$\mathbf{Y}^{E}(r) = \lim_{s \to \infty} \mathbf{Y}(r,s).$$

The time domain form of (21) may now be written as

$$\mathbf{Y}(r,t) = \frac{1}{2\pi i}\int_L \frac{\mathbf{P}^{v}(r,s)}{\det\mathsf{M}} e^{st}\, ds + \mathbf{Y}^{E}\delta(t), \tag{23}$$

where L is the Bromwich path. If s_j are the zeros of the secular determinant detM, then (23) may be simply evaluated using Cauchy's Theorem as

$$\mathbf{Y}(r,t) = \sum_{k=1}^{N} \operatorname{Res}\left[\frac{\mathbf{P}^{v}(r,s)\, e^{st}}{\det\mathsf{M}\,(r,s)};\ s_k\right]. \tag{24}$$

Since $d\,(\det\mathsf{M})/ds \neq 0$ in general at $s = s_j$, it follows that the residues at the poles s_j are

$$\operatorname{Res}\left[\frac{\mathbf{P}^{v}(r,s)\, e^{st}}{\det\mathsf{M}\,(r,s)};\ s_j\right] = \frac{\mathbf{P}^{v}(r,s)\, e^{s_j t}}{\left.\dfrac{d}{ds}(\det\mathsf{M})\right|_{s=s_j}}, \tag{25}$$

and therefore that

$$\mathbf{Y}(r,t) = \sum_{k=1}^{N} \frac{\mathbf{P}^{v}(r,s_j)\, e^{s_j t}}{\left.\dfrac{d}{ds}(\det\mathsf{M})\right|_{s=s_j}} + \mathbf{Y}^{E}\delta(t). \tag{26}$$

If we define a "vector of residues" $\mathbf{R}_k$ as

$$\mathbf{R}_k = \frac{\mathbf{P}^{v}(r,s_k)}{\left.\dfrac{d}{ds}(\det\mathsf{M})\right|_{s=s_k}}, \tag{27}$$

then

$$\mathbf{Y}(r,t) = \mathbf{Y}^E\delta(t) + \sum_{j=1}^{N} \mathbf{R}_j e^{-\bar{s}_j t}, \tag{28}$$

where we have written $s_j = -\bar{s}_j$ since the N zeros of det**M** are all (usually) located on the negative real axis of the complex s-plane. The set of N zeroes of det**M** (a different set for each spherical harmonic degree l) clearly determines the inverse decay times of N normal modes of purely exponential viscous gravitational relaxation. These modes are the complex frequency equivalents of the purely oscillatory modes of elastic gravitational free oscillation which are supported by a perfectly elastic rheological model. The form of the exact normal mode inverse (28), as first presented in Peltier (1985a), is identical to the collocation inverse which I previously employed though in it the s_j are not now arbitrarily chosen. This exact solution has a considerable advantage over the approximate collocation solution in that it is stable and accurate for all time.

It is of some advantage to express the solution (28) in terms of so called surface load Love numbers, a triplet of dimensionless scalars in terms of which each individual signature of the isostatic adjustment process may be represented. These Love numbers are defined through the relations:

$$U_l(r,s) = \phi_{2,l}(r)\; h_l^L(r,s)/g_0(a) \tag{29a}$$

$$V_l(r,s) = \phi_{2,l}(r)\; l_l^L(r,s)/g_0(a) \tag{29b}$$

$$\phi_{3,l}(r,s) = -\phi_{2,l}(r)\; k_l^L(r,s) \tag{29c}$$

in which $g_0(a)$ is the surface gravitational acceleration and $\phi_{2,l}$ and $\phi_{3,l}$ are the spherical harmonic coefficients in the potential perturbation produced by the load and by the load induced deformation, respectively. From (28) it is clear that in the time domain these Love numbers have forms like, e.g., for h_l:

$$h_l(r,t) = h_l^E(r)\delta(t) + \sum_{j=1}^{N} r_j^l(r)\, e^{-\bar{s}_j^l t}. \tag{30}$$

Using (29) we may write the total potential perturbation as

$$\phi_{1,l} = -\phi_{2,l} + \phi_{3,l} = -\phi_{2,l}(1 + k_l), \tag{31}$$

where the coefficients in the expansion of the potential of the surface load $\phi_{2,l}$ are obtained from the definition

$$\phi_2(\mathbf{r}) = \int \frac{G_\gamma(r')d^2r'}{|\mathbf{r} - \mathbf{r}'|}, \tag{32}$$

in which integration is over the Earth's surface. If we substitute for $\gamma(\mathbf{r})$ from its Legendre expansion

$$\gamma = \sum_{l=0}^{\infty} \Gamma_l P_l \cos(\theta), \tag{33}$$

with

$$\Gamma_l = \frac{2l + 1}{4\Pi a^2},$$

and use the addition theorem for spherical harmonics we find

$$\phi_2(a, \theta) = \sum_{l}^{\infty} \frac{4\Pi aG}{2l + 1} \Gamma_l P_l(\cos\theta),$$

so that

$$\phi_{2,l}(a) = \frac{4\Pi aG\Gamma_l}{2l + 1} = \frac{ag_0(a)}{m_e}, \tag{34}$$

where m_e is the Earth's mass, a its radius, and $g_0(a)$ its surface gravitational acceleration as before.

With this point mass normalization and using (29) and (12) we may therefore proceed to construct Green functions for various surface signatures of the planet's response to gravitational excitation by a point mass load. Three examples of such Green functions are especially important, namely those for the radial displacement, the gravitational acceleration, and the perturbation to the gravitational potential field. Longman (1963) demonstrates that these have the following respective forms in terms of the Love numbers defined in (29):

$$u(\theta,t) = \frac{a}{m_e} \sum_{l=0}^{\infty} h_l(r,t) P_l(\cos\theta), \tag{35a}$$

$$\alpha(\theta,t) = \frac{g}{m_e} \sum_{l=0}^{\infty} [l + 2h_l(r,t) - (l + 1)k_l(a,t)] P_l(\cos\theta), \tag{35b}$$

$$\phi(\theta,t) = \frac{ag}{m_e} \sum_{l=0}^{\infty} [1 + k_l(r,t) - h_l(r,t)]\, P_l\,(\cos\theta). \tag{35c}$$

By convolution of these Green functions with any prescribed history of surface mass variation one may predict the associated signature of the planet's time dependent response to this forcing.

Before proceeding to employ this theoretical structure to predict any of the specific responses to the realistic glaciation deglaciation events for which data actually exists, it will prove useful to illustrate the various ingredients of the above described theory for a specific Earth model in order to fix ideas. To this end Figure 6.3 illustrates, in plate a, the elastic structure of a typical realistic Earth model which has been derived on the basis of analyses of elastic-gravitational free oscillations data. This structure, which is determined by $\rho_0(r)$, $\lambda(r)$, and $\mu(r)$, is that for model 1066B of Gilbert and Dziewonski (1975) and is actually described in terms of the compression wave velocity $V_p(r) = \sqrt{(\lambda + 2\mu)/\rho_0}$ and the shear wave velocity $V_s(r) = \sqrt{\mu/\rho_0}$ as well as the density $\rho_0(r)$. In terms of the Maxwell analogue the description of the viscoelastic structure is completed by specification of the

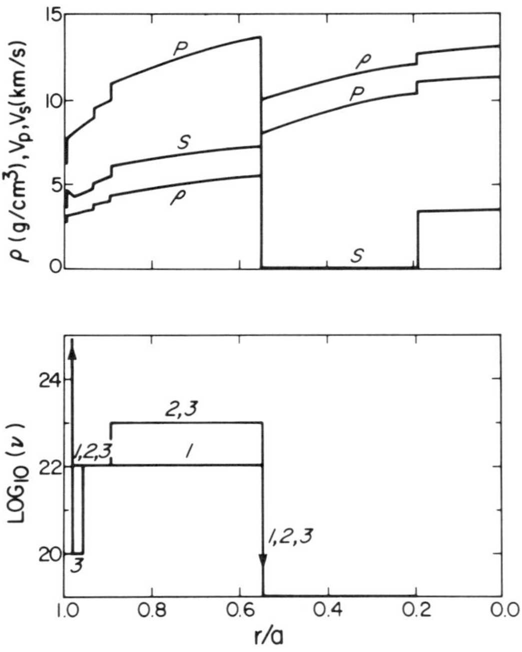

Figure 6.3. The upper plate shows the variation with radius of the parameters of the elastic earth model 1066B of Gilbert and Dziewonski (1975) in terms of density ρ and compression and shear wave velocities V_p and V_s. The lower plate illustrates three different simple models of the variation of steady state viscosity (labelled in Poise) with depth which is required to completely specify a Maxwell rheology for the interior.

viscosity profile $\nu(r)$ and three different variants of this are shown on plate b of the figure. Notable from the elastic structure of plate a are the two discontinuities of elastic parameters at 420 and 650 km depth which are now understood to be associated with solid–solid equilibrium thermodynamic phase transformations induced by the increase of pressure with depth. The shallowest of these is associated with the transformation of olivine to spinel while the deepest and strongest is due to the further transformation of spinel to a mixture of perovskite and magnesiowüstite (e.g., Chapter 4). As will already have become clear from the preceding chapters of this book, much of the continuing debate on the issue of convection in the Earth's mantle has become focussed upon the question of what effect these phase transitions might be expected to have on the style of the circulation (e.g., Chapters 3 and 8) and even upon the issue as to whether the deeper of the two transitions might be accompanied by a significant chemical discontinuity, i.e., a change of mean atomic weight (Chapter 4). As I will show in the following sections of the present chapter, these discontinuities may also play an important role in the phenomenology of postglacial rebound.

Figure 6.4 shows relaxation spectra for the three different viscoelastic Earth models of Figure 6.3, which are plots of $-\bar{s}_j$ as a function of spherical harmonic degree. The inverse decay times ($-\bar{s}_j$) on these plots have been non-dimensionalized with a characteristic time of 10^3 years so that where $\log_{10}(-s) = 0$ the relaxation time equals 10^3 years. Evident by inspection of these plates is the fact that the relaxation spectra are characterized by six different modal branches which are separately labelled MO, CO, LO, M1, M2 and C1, the latter of which is a weak numerical artifact and will not be discussed further. The first five modal branches are each supported by a particular feature in the radial viscoelastic structure. MO, the fundamental mantle mode according to the nomenclature of Peltier (1976), is supported by the buoyant restoring force across the free outer surface of the model, CO by the core-mantle interface, and LO by the discontinuity of viscosity across the lithosphere mantle boundary which has been assumed to be located at 120 km depth. The branches M1 and M2 respectively exist because it has been assumed that the density discontinuities across the 650 and 420 km horizons behave non-adiabatically on the rebound timescale. As I shall show, it appears that such behavior is required in order to understand the free air gravity anomalies which are found in association with regions that were once ice covered.

The mode labelling which I employ is justified by Figure 6.5 on which are plotted shear kernels for the five main modes plus two additional "transition" modes (labelled T1, T2) from the sequence of weak poles which exist near the

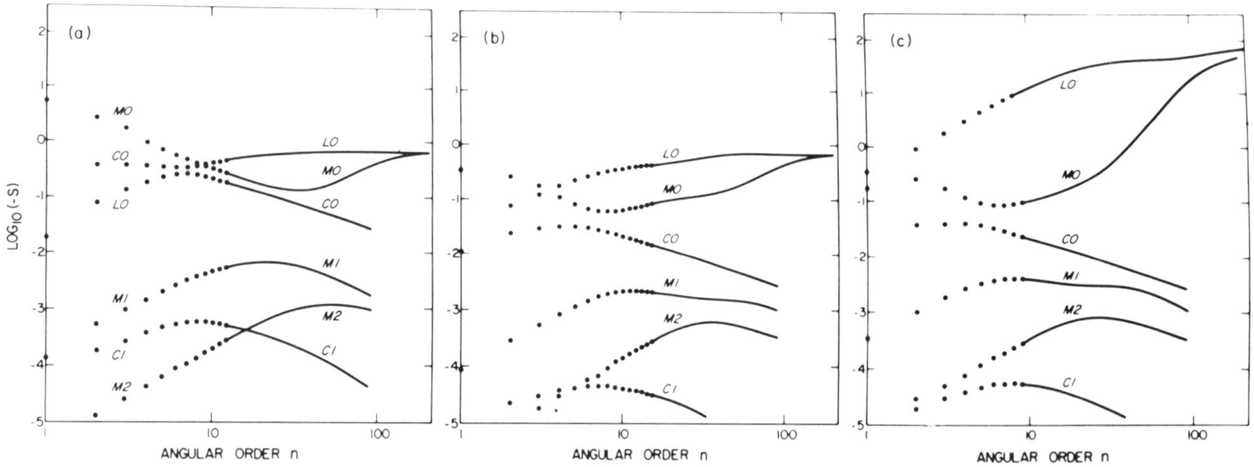

Figure 6.4. Relaxation diagram for models 1, 2, and 3 of Figure 6.3. The inverse relaxation times ($-s$) for each of the normal modes of viscous gravitational relaxation have been nondimensionalized with a characteristic time of 10^3 years. Model 1 (plate a) has a uniform viscosity mantle of 10^{21} Pa s, model 2 (plate b) has a high viscosity lower mantle with $\nu = 10^{22}$ Pa s, while model 3 is identical with 2 but has a low viscosity channel beneath the lithosphere in which $\nu = 10^{19}$ Pa s.

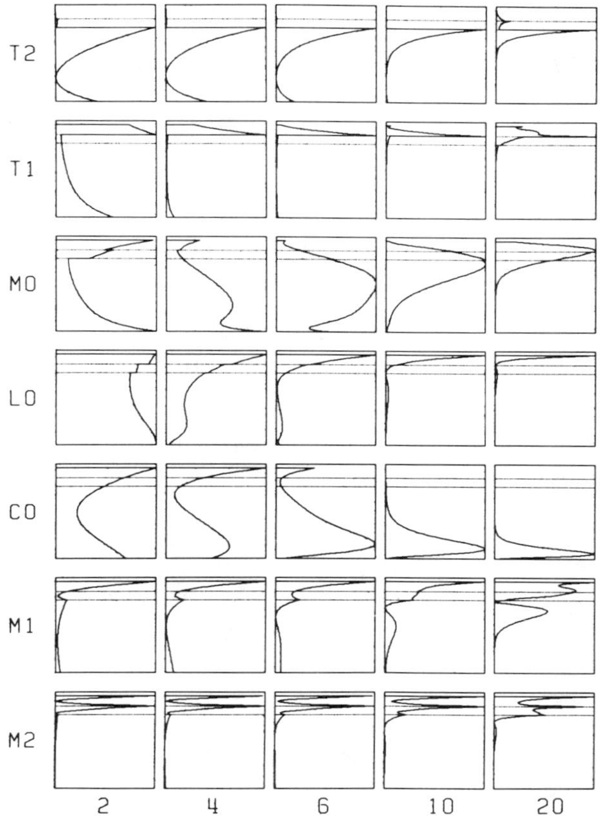

Figure 6.5. Shear kernels showing the variation of shear energy with depth for a number of modes along the MO, CO, LO, M1, and M2 branches in Figure 6.4a.

Maxwell time $T_m = \nu/\mu$ but which carry no significant component of the total response. The shear kernels illustrate the distribution of shear energy with depth in the mode and are determined from the s-domain shear energy distribution S as

$$S = 2\mu(s)\Delta_{ij}\Delta_{ij}, \tag{36}$$

where Δ_{ij} is the strain deviator,

$$\Delta_{ij} = \tfrac{1}{2}(\partial_i u_j + \partial_j \mu_i) - \tfrac{1}{3}\partial_k u_k \delta_{ij}. \tag{37}$$

The radial part of $\Delta_{ij}\Delta_{ij}$ is just

$$K_l = \tfrac{1}{3}(2\dot{U}l - F_l)^2 + r^{-2}l(l+1)\,[r\dot{V}_l - V_l + U_l]^2 + r^{-2}\,l(l-1)\,(l+1)\,(l+2)\,V_l^2, \qquad (38)$$

in which

$$F_l = r^{-1}[2U_l - l(l+1)\,V_l].$$

The viscoelastic extension of Rayleigh's Variational Principle (Peltier, 1976), then shows that the normalized shear kernel,

$$S_l(r) = \left[\frac{K_l[(\mu^2/\nu)/(s+\mu/\nu)^2]}{\int_0^a K_l[(\mu^2/\nu)/(s+\mu/\nu)^2]r^2dr}\right]_{s=s_j}, \qquad (39)$$

may be employed to calculate the shift in the inverse relaxation time of a free decay pole in the relaxation spectrum which is forced by a perturbation $\delta\,\nu(\mathrm{r})$ of the viscosity profile as:

$$\frac{\delta s}{s} = -\int_0^a S_l(r)r^2\frac{\delta\nu(r)}{\nu(r)}\,\mathrm{dr}. \qquad (40)$$

It is the function $S_l\,(r)$ which has been plotted in Figure 6.5. Inspection of the individual plates of this figure shows that the shear energy maximizes adjacent to the boundary in the model which is responsible for supporting the mode in question.

When the Love numbers are synthesized in terms of these normal modes by computing the residue vector according to (27) and used in expressions like (30), we eventually obtain time dependent structures such as those illustrated in Figure 6.6 which shows representative $h_l(t)$ Love numbers for viscoelastic model 1 of Figure 6.3, or rather the viscous part of the "Heaviside" form of these Love numbers which are obtained by convolving (30) with the unit step which gives:

$$h_l^H(t) = h_l^E + \sum_{j=1}^{N}\frac{r_j^l}{s_j^l}(1 - e^{-s_j^l t}) \qquad (41)$$

$$= h_l^E + h_l^{H,V}(t).$$

It is the parameters $h_l^{H,V}(t)$ which are illustrated in Figure 6.6 for several values of l. Adjacent to the right hand margin of this plate are shown the asymptotic limits towards which each of the $h_l^{H,V}(t)$ tend as $t \to \infty$.

An example of a Green function synthesized from these time dependent Love numbers is shown in Figure 6.7, again for model 1 of Figure 6.3 and using the Heaviside form of the Love numbers defined in (41). The figure

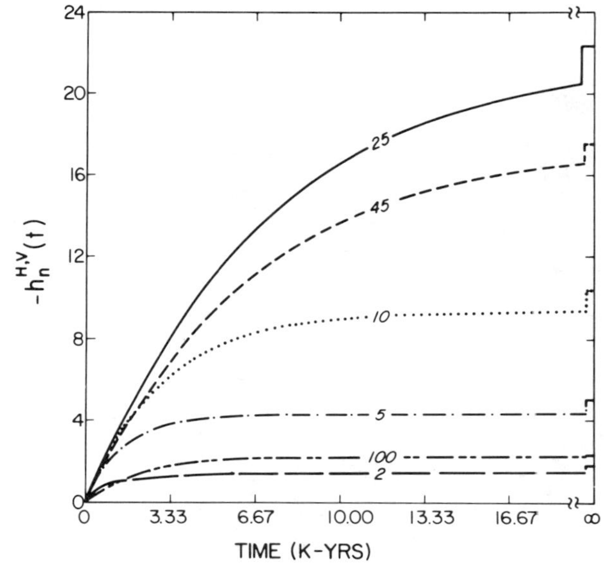

Figure 6.6. Decay spectra $h_n^{H,V}(t)$ for realistic earth model 1 of Figure 6.3. The spherical harmonic degree n is marked on each decay curve and time is measured in thousands of years. The asymptotic value which is obtained at infinite time is shown for each n adjacent to the right hand margin of the graph. Note that, even after 2×10^4 yr, substantial disequilibrium remains for most spherical harmonic degrees.

therefore actually shows the viscous part of the Green function for radial displacement (35a) convolved with a unit step. This function describes the manner in which the viscous component of the deformation builds under the point mass as time progresses, from a value of zero at $t = 0$ to an equilibrium level as $t \to \infty$. In the new equilibrium state which obtains in the limit of long time the buoyancy force induced by the deflection of the outer boundary just balances the weight of the load and we say that the loaded system is in a new state of isostatic equilibrium.

In the following sections of this chapter, I will describe the extensions of the above described field theory that have been required to enable its application to the understanding of each of the previously mentioned signatures of the isostatic adjustment process. The focus of each section will be on the constraint upon the mantle viscosity profile provided by each type of data.

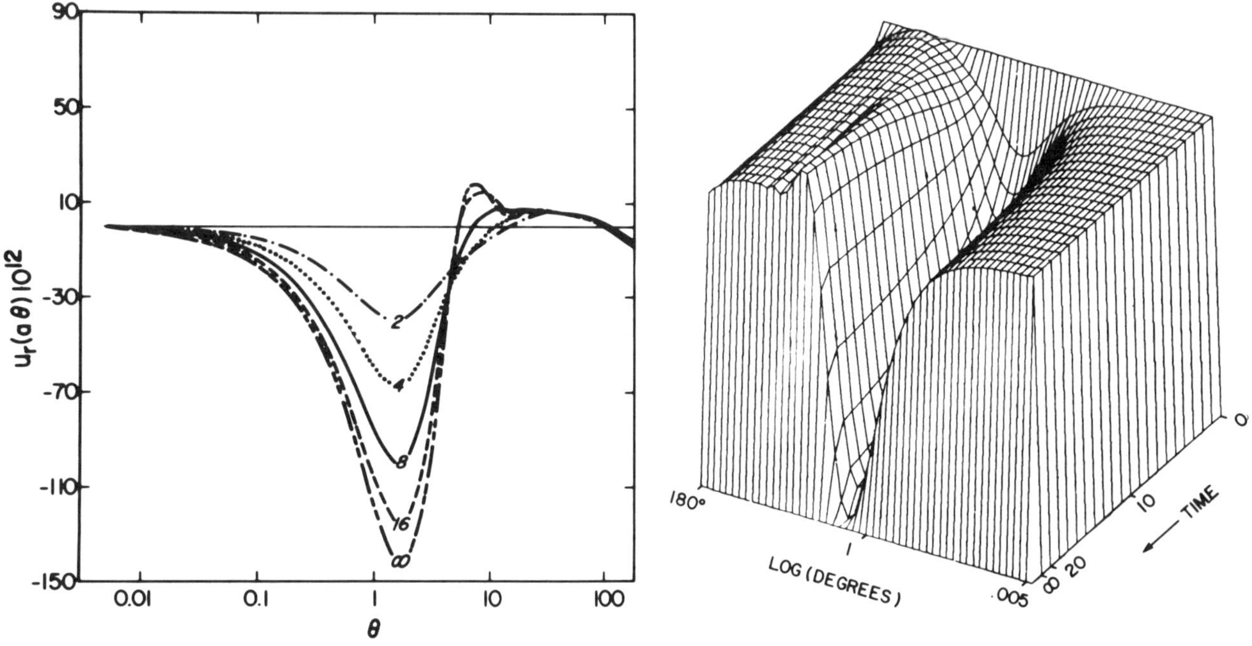

Figure 6.7. a, Shows several time slices through the viscous part of the Heaviside Green function for radial displacement for model 1 of Figure 6.3. The function has been normalized by multiplication with $a\theta$ to remove the geometric singularity for plotting purposes. The number on each curve is the time measured in thousands of years. Note the inward migration of the peripheral bulge as time proceeds for this uniform viscosity model. b, A three-dimensional view of the same Green function.

3. Mantle Viscosity from Postglacial Sea Level Variations

Because the last deglaciation event of the current ice age began only about 18,000 years ago and ended approximately 6000 years ago, the variations of sea level caused by the melting of glacial ice may be accurately timed using ^{14}C dating methods, either conventional counting of β-decays (e.g., Libby, 1952) from large specimens of interest, or using modern accelerator based techniques which are capable of delivering accurate dates on small samples of mg mass (e.g., Litherland, 1980). Figure 6.8 shows four examples of the relative sea level curves so derived which form the most important data base for studies of mantle viscosity using isostatic adjustment observations. The reason for the pre-eminent importance of such data in this work is that such curves constitute extensive, though incomplete, *histories* of the deglaciation induced deformation at different locations with respect to the major ice sheets. Obviously the nature of the response is a strong function of location.

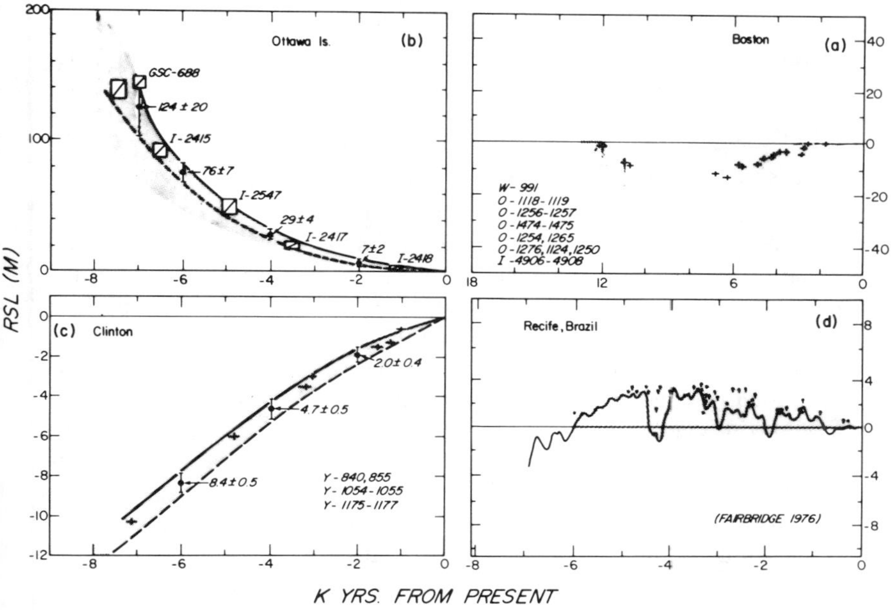

Figure 6.8. Illustrative relative sea level data for four sites: b, the Ottawa Islands in Hudson Bay; a, Boston near the southern margin of the Laurentide ice sheet at glacial maximum; c, Clinton Connecticut to the south of Boston on the U.S. east coast; and d, Recife, Brazil which is in the far field of the late Pleistocene ice masses.

At sites like the Ottawa Islands, which are located in Hudson Bay near the center of the ancient Laurentian ice sheet, the last 8000 years have been characterized by a monotonic emergence of the land out of the sea at a rate which appears to be an exponentially decreasing function of time. Near the edge of this ice sheet, however, at a place such as Boston, the relative sea level (rsl) history has been strikingly non-monotonic with an initial fall of sea level relative to the land later followed by rising sea levels. Further south along the U.S. east coast, at Clinton, Connecticut, the local history of relative sea level variation is one of monotonically rising level throughout the past 8000 years. The last plate on Figure 6.8d shows the rsl history from a site on the east coast of the continent of South America, Recife Brazil, within the same window of time. At each of these sites the observed variations of relative sea level can be understood in terms of the interplay between the deformation of the solid Earth induced by the melting ice and subsequent ocean loading, and the redistribution of water within the global ocean required to ensure that the ocean's surface, the geoid, remains an equipotential surface at all times within and subsequent to the deglaciation event. Such continuous records of deformation should be seen as distinct from almost all other signatures of the isostatic adjustment process, such as free air gravity anomalies and Earth rotation effects, which essentially constitute "snapshots" of some aspect of the isostatic adjustment process at one instant of time. It is because of the length of the deformation time series which they constitute that rsl data are able so powerfully to discriminate between competing models of the radial viscosity profile. However, because relative sea level data constitute records of the separation in time of two deforming surfaces, the surface of the solid Earth and the surface of the ocean, the theory for the prediction of rsl variations must be elaborated somewhat beyond the basic ingredients developed in the last section.

The "sea level equation" which embodies the full complexity of this prediction is based upon the Green function for the gravitational potential perturbation defined in equation (35c). The structure of this equation will be most clearly understood if we begin by supposing that all of the ice sheets which were on the surface at glacial maximum melted simultaneously. If $L(\theta,\phi)$ is used to denote the ice thickness removed from position (θ,ϕ) at $t = 0$ and $S(\theta,\phi)$ the amount of water added to the ocean at position (θ,ϕ) and time t, then we may compute the net change of potential at any position on the surface by convolution of the surface loads L and S with the Green function ϕ in (35c) to obtain:

$$\Phi(\theta,\phi,t) = \rho_I \phi^H \underset{I}{*} L + \rho_w \, \phi^H \underset{O}{*} S \tag{42}$$

where $\underset{I}{*}$ and $\underset{O}{*}$ indicate convolution over the ice and water respectively and ρ_i and ρ_w are the ice and water densities. From (35c) it is clear that (42) includes the change of potential due to the vertical displacement of the surface of the solid Earth since the Green function contains the Love number h_l. The change of potential (42) will clearly force an adjustment of the thickness of the sea water locally in the amount (Platzman, 1971; Farrell and Clark, 1976):

$$S = \frac{\phi(\theta,\phi,t)}{g} + C, \tag{43}$$

in which the constant C is fixed by the requirement that mass must be conserved. Now (43) is clearly a result of first order perturbation theory which is valid for sufficiently small changes of the local bathymetry S. It is important to note that S is, by construction, the local variation of sea level with respect to the deformed surface of the solid Earth and is thus precisely the quantity which is recorded in RSL data such as those shown in Figure 6.8. Substitution of (42) into (43) leads to the complete "sea level equation,"

$$S = r_i(\phi^H/g)\underset{I}{*}L + \rho_w\,(\phi^H/g)\underset{O}{*}S + C. \tag{44}$$

To obtain an explicit expression for C we note that the integral of $\rho_w S$ over the oceans must equal the instantaneous value of the total mass which has been lost by ice sheet disintegration at time t. Therefore

$$\begin{aligned}\langle\rho_w S\rangle_O &= \rho_w\langle\rho_I(\phi^H/g)\underset{I}{*}L + \rho_w(\phi^H/g)\underset{O}{*}S\rangle_O + \langle C\rangle_O\rho_w \\ &= -M_i(t).\end{aligned} \tag{45}$$

The minus sign on the r.h.s. of (45) is required as $M_i(t)$, the mass loss history for all ice sheets combined, is defined as negative for load removal. In (45) the symbol $\langle\ \rangle_O$ denotes integration over the oceans and since C is independent of position $\langle C\rangle_O = C(t)\,A_O$ where A_O is the assumed constant area of the oceans. Thus (45) gives

$$C(t) = -\frac{M_I(t)}{\rho_w A_O} - \frac{1}{A_O}\langle\rho_I(\phi^H/g)\underset{I}{*}L + \rho_w(\phi^H/g)\underset{O}{*}S\rangle_O. \tag{46}$$

With C given by (46), eq. (44) is an integral equation since S appears not only on the l.h.s. but also in the convolution integral on the r.h.s. Given the deglaciation history $L(\theta,\phi,t)$ and thus $M_i(t)$, and the potential perturbation Green function ϕ^H for a specific radially stratified viscoelastic model of the planet, we may invert this integral equation to find the history of relative sea level change $S(\theta,\phi,t)$ at any point on the Earth's surface. The numerical methods which have been developed to solve the sea level equation were first described in Peltier *et al*. (1978). More recent discussions will be found in Peltier (1982), Wu and Peltier (1983), and Peltier (1986).

The above summary should serve to make clear the basic structure of the isostatic adjustment problem. Two inputs to the theoretical model are required before it can be employed to make a prediction. The first of these is a model of the planet's radial viscoelastic structure, while the second is a model of ice sheet disintegration. Neither of these functionals of the model is perfectly known *a priori* and we are obliged to proceed iteratively to refine our knowledge of them. Only because we are in a position to construct a good first guess to the deglaciation history is it possible to begin the process of iterative refinement with some hope that the process will converge. In what follows, I will attempt to illustrate the manner in which this process of iteration has been made to work and will summarize what has been learned from it.

The ice sheet disintegration model with which work on this problem was begun, was the model called ICE-1 tabulated in Peltier and Andrews (1976). This model included a representation of only the main northern hemisphere ice masses of Laurentia, Fennoscandia, and Greenland. Both of the former ice sheets of course completely disappeared at the end of the Wurm-Wisconsin period while the latter, although it was diminished somewhat in the last glacial-interglacial transition, is still essentially intact. Wu and Peltier (1983) published a somewhat refined version of this melting history called ICE-2, the most important modification in which, however, was the addition of a melting chronology for the Antarctic continent. One of the most important new advances which have followed from the process of iterative refinement of the isostatic adjustment model has been the recent demonstration (Peltier, 1987) that the timing of Antarctic deglaciation in the ICE-2 chronology of Wu and Peltier (1983) was far too early to reconcile rsl data from the southern oceans. I will demonstrate this point below following a summary of the implications of solutions of the sea level equation concerning mantle viscosity stratification.

Figure 6.9 shows three time slices through the northern hemisphere ice sheet components of the ICE-2 disintegration history of Wu and Peltier (1983). Substituting the $L(\theta,\phi,t)$ of this model into eq. (44), and employing ϕ^H for viscoelastic model 1 of Figure 6.3 to define the structure of the solid Earth, results in a solution $S(\theta,\phi,t)$ which is shown contoured in metres in Figure 6.10 for four times since deglaciation commenced 18,000 years ago. For purposes of these relative sea level calculations which included no

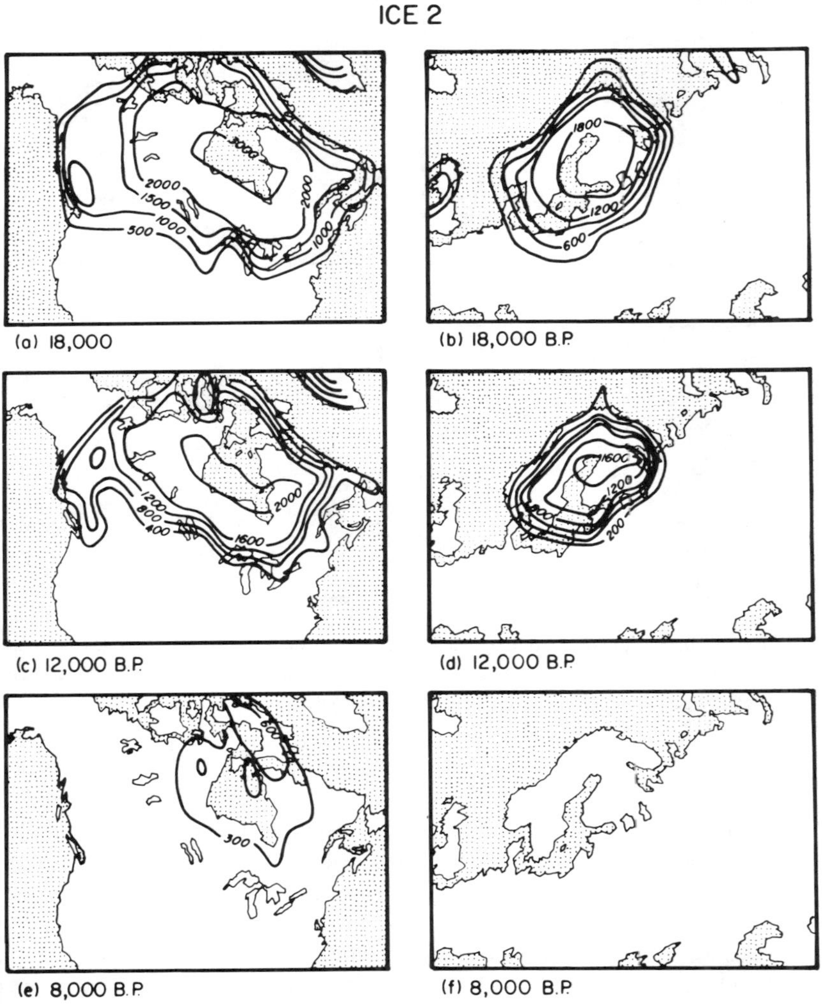

Figure 6.9. Three time slices through the ICE-2 deglaciation chronology of Wu and Peltier (1983) for both Laurentia (a, c, e) and Fennoscandia (b, d, f).

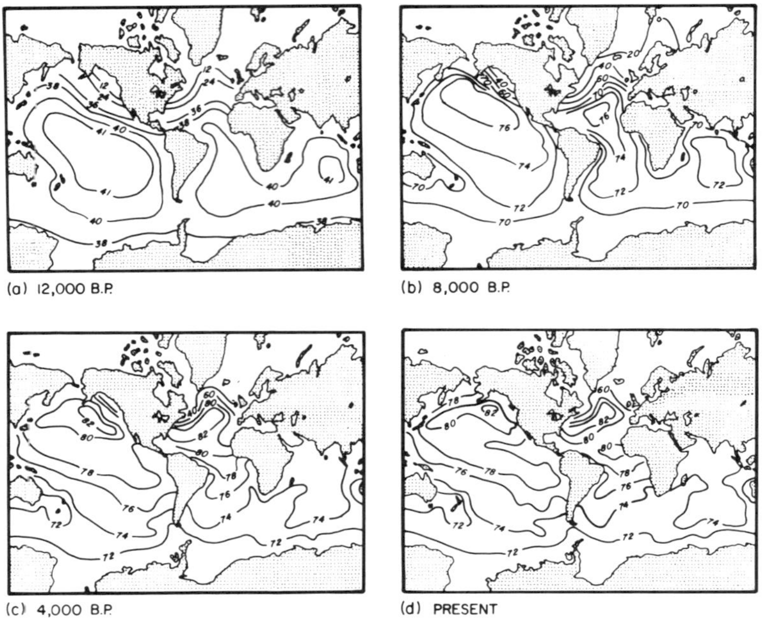

Figure 6.10. Predicted change in sea level (meters) since the onset of deglaciation at 18 kyr. The deglaciation chronology is ICE-2 and the earth model has uniform 10^{21}Pa s mantle viscosity and 120 km lithospheric thickness.

Antarctic component of the melting history, it should be understood that the system is assumed to have been in isostatic equilibrium at glacial maximum prior to the onset of melting. Wu and Peltier (1983) have shown that this assumption does not effect the r.s.l. predictions in any significant way, although, as I will show in the following sections, it does have a major effect on free air gravity anomaly and polar wander predictions.

In order to compare the theoretical predictions with observations we simply pierce the stack of output maps, such as those of the sample shown in Figure 6.10, at any point on the Earth's surface at which we have a ^{14}C record of relative sea level change. All calculations which have been performed with this theory to date have been based upon a time step between successive output times of 1000 years. Figures 6.11–6.14 (from Peltier, 1986) show a sequence of comparisons of predicted and observed relative sea level histories for twelve sites on the North American continent which are arranged to illustrate the ability of such data to discriminate the contrast in viscosity

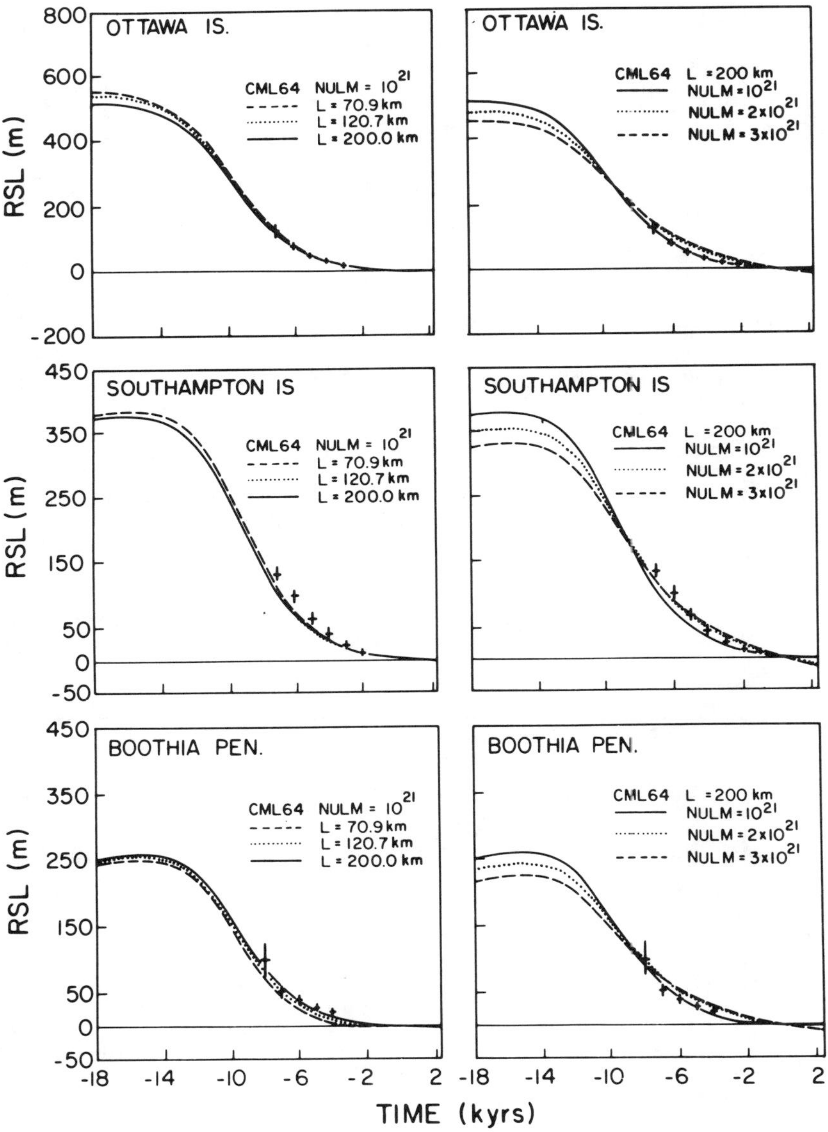

Figure 6.11. Comparisons between observed and predicted relative sea level curves at three sites which were under the central Laurentide ice dome. The elastic basic state is CML64.

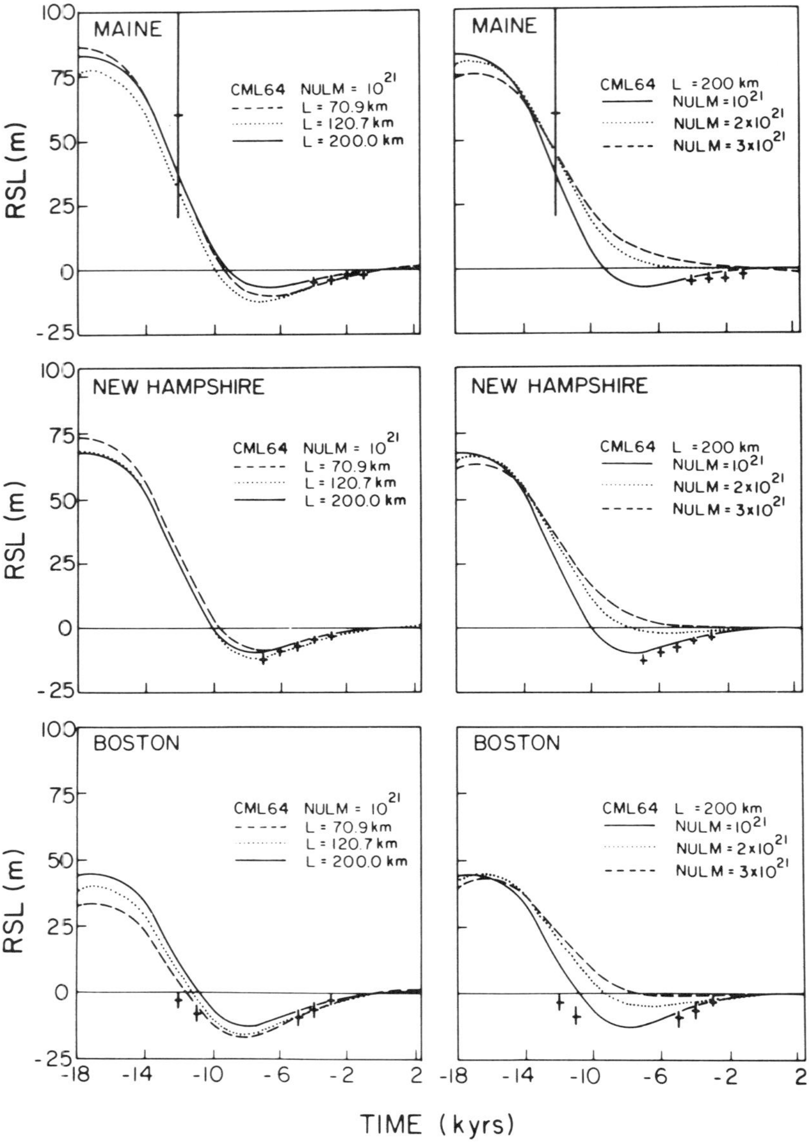

Figure 6.12. Same as Figure 6.11 but for three sites near the southern edge of the Laurentian ice sheet on the U.S. east coast.

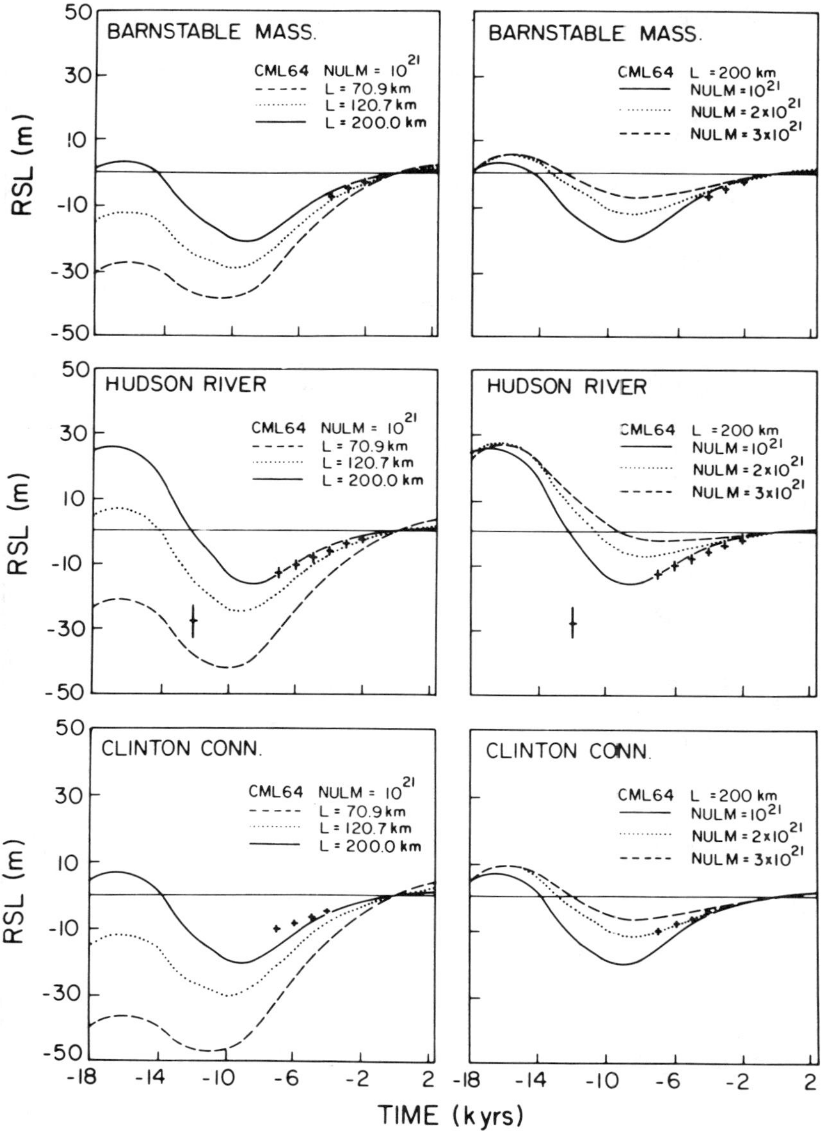

Figure 6.13. Same as Figure 6.11 but for three sites near the crest of the glacial forebulge on the U.S. east coast.

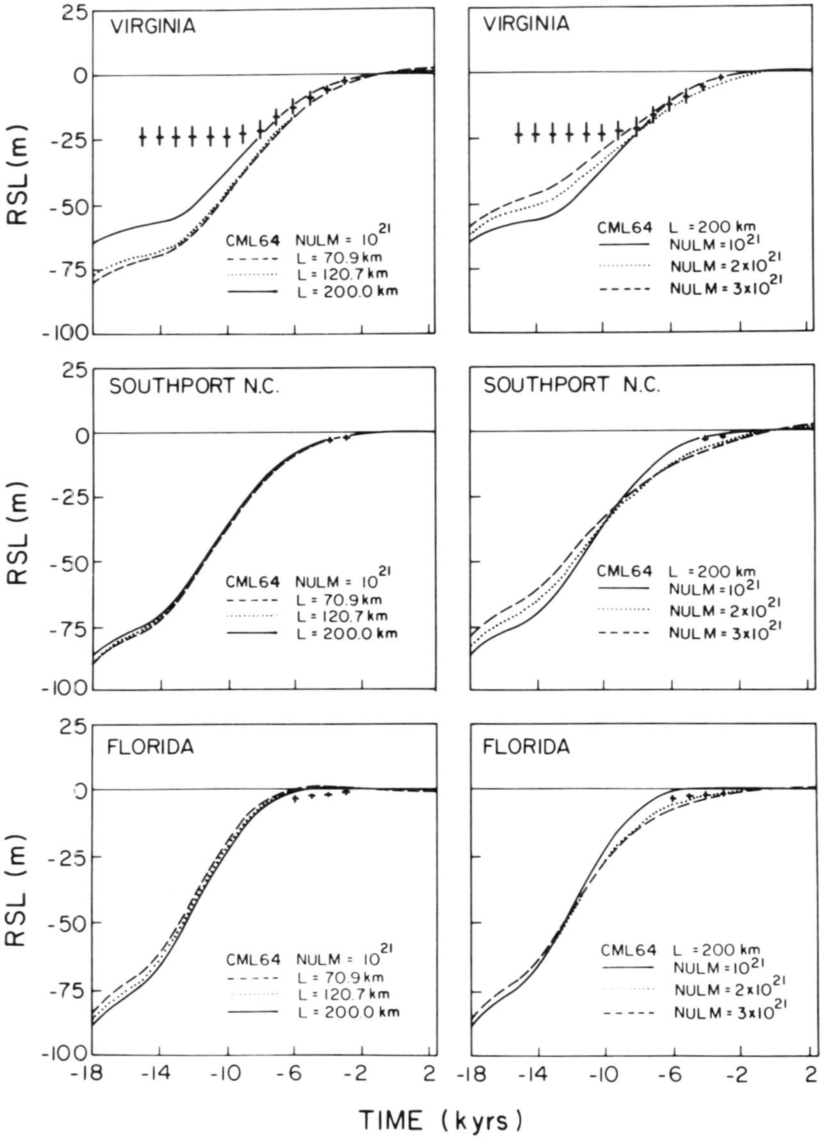

Figure 6.14. Same as Figure 6.11 but for three sites on the southern flank of the forebulge along the U.S. east coast.

between the upper and lower mantle and the thickness of the lithosphere in the vicinity of the Laurentian ice sheet. In each case the upper mantle viscosity is fixed to the 10^{21} Pa s value of Haskell (1935, 1936, 1937). In the left hand column of plates on each figure the lower mantle viscosity is fixed to the same value of 10^{21} Pa s (NULM) and computations are shown for three values of lithospheric thickness L (70.9 km, 120.7 km, and 200.0 km). In the right hand column of plates the lithospheric thickness is fixed to $L = 200$ km and computations are shown for three values of the lower mantle viscosity (NULM = 10^{21} Pa s, 2×10^{21} Pa s, and 3×10^{21} Pa s). In each case the elastic basic state employed in the calculations is that of model CML64 (Peltier, 1986) which is a multiple homogeneous layer approximation to model 1066B which differs from that seismically realistic model mainly in the amount of buoyancy which is assumed to be operative across the transition zone (which is reduced in the CML64 model). Inspection of these comparisons illustrates a number of fundamental points.

For sites near the center of the ice sheet (Fig. 6.11) the rsl response is clearly insensitive to lithospheric thickness as should be expected on physical grounds. At such sites the response is dominated by such long horizontal wavelengths that the relatively thin lithosphere is essentially transparent. These data are much more sensitive discriminants of the lower mantle viscosity, however, as even a factor of three increase above the upper mantle value causes a sufficiently sharp increase in the predicted present day emergence rate as to be excluded by the data at most sites. Since the rsl data are characterizable by a simple exponential history at these interior locations the present day emergence rate (slope), and the height above present sea level of any one temporal horizon, suffice to completely characterize the rsl history. Figure 6.15 (from Peltier *et al.*, 1986) demonstrates the manner in which these two data from each site combine to completely exclude models in which the viscosity of the lower mantle is significantly in excess of the viscosity of the upper mantle. These calculations employ a circular disk load approximation to the Laurentian ice sheet and approximate the rsl history by the variation of radial displacement, an approximation which Wu and Peltier (1983) have shown to be accurate at near field sites for times subsequent to deglaciation. Only models with very modest contrast in viscosity are capable of simultaneously fitting the height of the 7000 year beach and the present day rate of emergence. As evident from Figure 6.11, even modest increases of lower mantle viscosity (to values $\gtrsim 3 \times 10^{21}$ Pa s) cause the predicted present day rate of emergence to greatly exceed the observed rate which is near 1 cm yr^{-1}, while further increases of the lower mantle viscosity to remedy this misfit results in such sharp reductions in the predicted height of the 7 kyr beach that these models are completely excluded by the data.

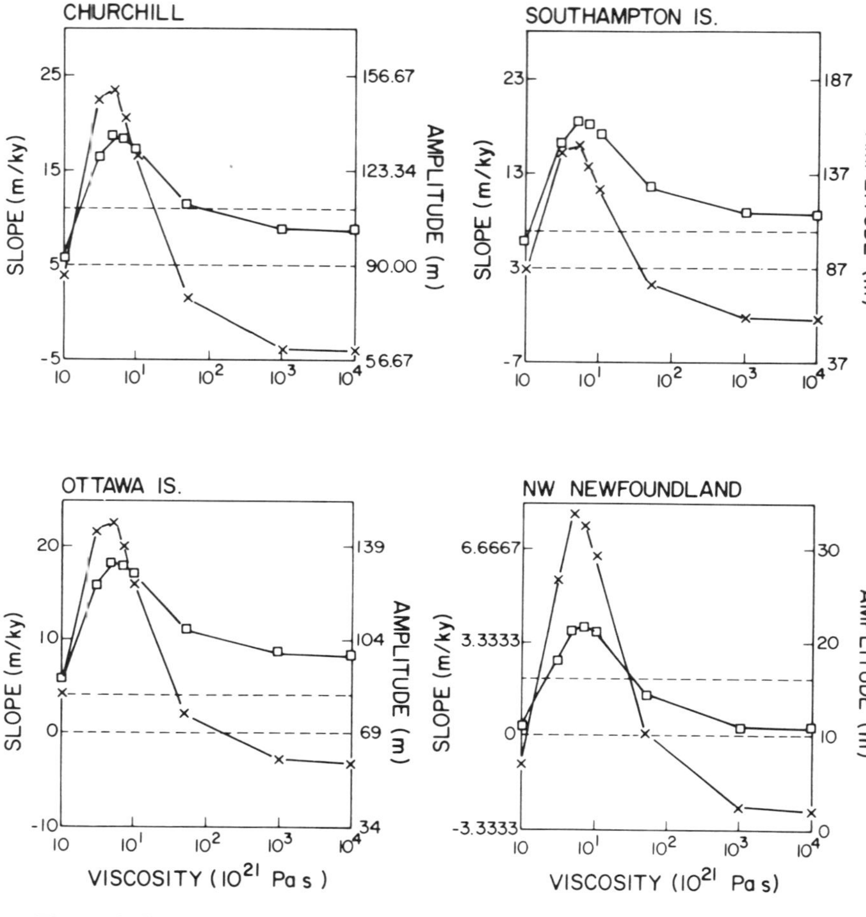

Figure 6.15. Comparisons of the predicted height of the 7 kyr beach and the present day emergence rate with the observations at four sites within the margin of the Laurentide ice sheet. The predictions are shown as a function of lower mantle viscosity in the Maxwell parameterization and the elastic basic state is CML64. Note that only the constant viscosity model provides a satisfactory fit to both these data at all sites. The upper mantle viscosity is fixed to the Haskell value of 10^{21}Pa s for all calculations. Amplitude (x), slope (□).

This conclusion is reinforced by rsl data from sites at locations which are successively further removed from the ice sheet centre. For example in Figure 6.12 we show comparisons between theory and observation at three sites located near the margin of the Laurentian ice sheet. At all such locations the observations are characterized by non-monotonic rsl histories with initial emergence of the land followed by submergence which is continuing today. Inspection of the individual plates shows that although the response at such sites is still relatively insensitive to changes of lithospheric thickness, it continues to be extremely sensitive to lower mantle viscosity, so much so that an increase of this parameter by a factor of three to 3×10^{21} Pa s is sufficient to completely eliminate the non-monotonic rsl signature which is the most prominent feature of the data. Although Figure 6.12 shows that even a viscosity increase by a factor of 2 is ruled out by the data in the context of the CML64 elastic parameterization, Peltier (1986) shows that an increase by this amount *is* accepted when the CML64 basic state is replaced by the seismically realistic model 1066B.

Moving further south along the U.S. east coast Figure 6.13 shows comparisons between theory and observation at three sites located near the crest of the glacial forebulge, the collapse of which following deglaciation is responsible for the rising sea levels which characterize this ring shaped region surrounding the glaciated area (e.g., see Figure 12 of Peltier, 1986). As first discussed in Peltier (1984) the rsl histories at such sites are extremely sensitive to variations of lithospheric thickness and Figure 6.13 shows that in the context of the CML64 elastic parameterization the preferred lithospheric thickness L is near 200 km. As shown in Peltier (1986) this increases to a value near 250 km when the elastic structure is represented by model 1066B. At these sites the response is also sensitive to lower mantle viscosity with increases in lower mantle viscosity trading-off against increases of lithospheric thickness. Most such data are quite content with the uniform viscosity model with relatively high lithospheric thickness.

As one moves even further south along the U.S. east coast and onto the southern flank of the forebulge to sites such as Florida, which is shown on Figure 6.14, one finds a further feature of the data which bound the lower mantle viscosity from below as effectively as data from edge sites bound the lower mantle viscosity from above. Inspection of the comparisons for Florida shows that at this site, and others similarly located, the uniform viscosity model predicts that raised beaches should be observed whereas the observations show that no such features exist. Also evident by inspection of these comparisons however, is the fact that a very small increase of lower mantle viscosity to a value near 2×10^{21} Pa s broadens the forebulge sufficiently to remove this feature. The conclusion then, based upon the totality of analyses such as those shown explicitly in Figures 6.11–6.14, is that with the upper mantle viscosity fixed to a value near 10^{21} Pa s the lower mantle viscosity is somewhat higher but very modestly so and near 2×10^{21} Pa s. The relative sea level data therefore very strongly support the convection hypothesis of continental drift in that they show that the Earth's mantle has rather precisely the viscosity derived in eq. (5) which is required if the boundary layer form of this hypothesis is to reconcile the surface observables.

In concluding this section illustrating the nature of the constraints on the radial viscoelastic structure which may be derived through detailed application of the sea level equation (44) I would like to return briefly to the point made previously that this equation requires an iterative solution since it embodies two unknown functionals: the radial viscoelastic structure and the deglaciation history. All of the rsl comparisons described above were based upon the deglaciation history ICE-2 of Wu and Peltier (1983) and the comparisons were restricted to northern hemisphere sites at which the Antarctic component of this melting history was of minor importance. At sea level sites in the southern ocean, however, this is no longer the case and having settled upon a best radial viscoelastic structure on the basis of northern hemisphere data, we are now in a position to refine the southern hemisphere melting chronology by invoking the data from the southern oceans.

Figure 6.16 shows three different versions of the Antarctic melting chronology, described in terms of equivalent eustatic water rise, for which illustrative calculations will be shown. The Northern hemisphere melting chronology tabulated in Wu and Peltier (1983) is the solid curve while the three dashed curves represent spatially identical Antarctic melting chronologies but delayed either by 5000 yrs or 7000 years from that of ICE-2 which is earliest melting of the three approximations. The original chronology was based upon the ice sheet reconstruction in Denton and Hughes (1981) which clearly has the deglaciation of West Antarctica *leading* the deglaciation of North America and western Europe. Figure 6.17 (from Peltier, 1987) shows a sequence of sixteen comparisons between observed and predicted relative sea level data from sites in the far field of the Northern hemisphere ice sheets which include several at which the ^{14}C time series extend back to 10–12 kyr bp into the time in which deglaciation was still taking place. At such locations as New Zealand (and many others described in Peltier, 1987) the data clearly require that the Antarctic ice sheet must have disintegrated contemporaneously with the ice masses of the northern hemisphere. As I will discuss briefly in Section 6 below, this scenario is also assumed in the model

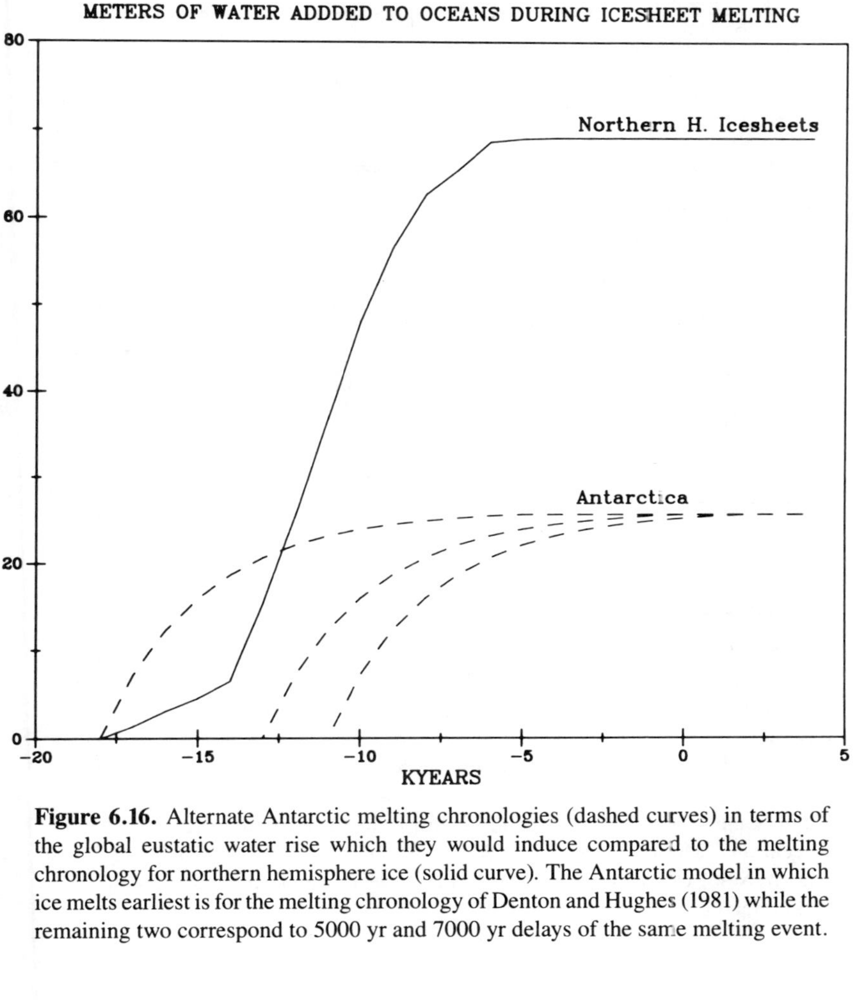

Figure 6.16. Alternate Antarctic melting chronologies (dashed curves) in terms of the global eustatic water rise which they would induce compared to the melting chronology for northern hemisphere ice (solid curve). The Antarctic model in which ice melts earliest is for the melting chronology of Denton and Hughes (1981) while the remaining two correspond to 5000 yr and 7000 yr delays of the same melting event.

of the ice age cycle which is discussed there. Its verification through analysis of southern hemisphere relative sea level histories is an important biproduct of the iterative application of the sea level equation (44) to the inference of both the radial viscoelastic structure of the planet and the deglaciation history through the last glacial–interglacial transition. In the remaining sections of this chapter, I will review the extent to which other signatures of the isostatic adjustment process may be employed to confirm the validity of

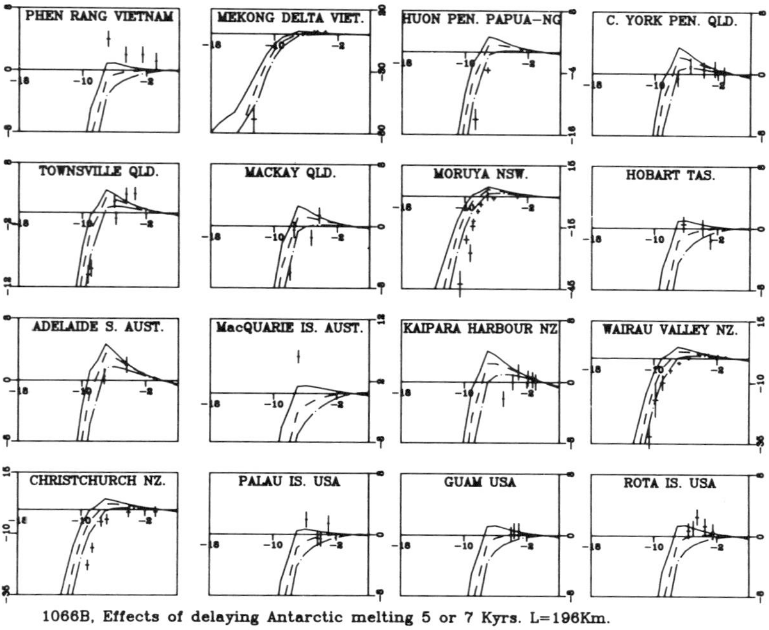

Figure 6.17. Observed (crosses) and predicted relative sea level histories for 16 sites in the southern hemisphere. The solid, dashed, and dash-dotted curves are the predicted rsl histories based upon the three variants of the Antarctic melting chronology shown on Figure 6.16 and respectively correspond to the earliest, intermediate, and latest choices for the onset of the disintegration of Antarctic ice. Note that all sites which have ^{14}C records extending sufficiently into the past (Mekong Delta, Vietnam; Huon Penninsula of Papua New Guinea; Townsville, Queensland, Australia; Mackay Queensland; Moruya, New South Wales; Wairu Valley, New Zealand; and Christchurch, New Zealand) are all best fit by the model in which Antarctic ice melts latest (7 kyr delay).

the radial viscosity profile which has been inferred on the basis of these relative sea level data.

4. Mantle Viscosity from Free Air Gravity Anomalies

Free air gravity anomaly maps for both Canada (Walcott, 1970) and northwestern Europe (Balling, 1980) are shown in Figure 6.18. Comparison of

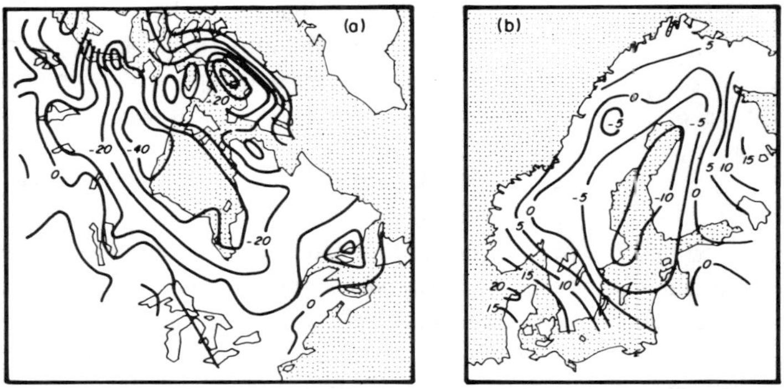

Figure 6.18. a, Free air gravity anomaly map for northeastern Canada (adapted from Walcott). b, Mean residual gravity anomaly map for Fennoscandia (adopted from Balling).

these maps with those of ice sheet topography at glacial maximum, shown previously in Figure 6.9, reveals that both free air anomalies are well correlated with the local ice thickness data and should therefore be attributable to the degree of isostatic disequilibrium which has yet to be compensated by vertical crustal motion in the Hudson Bay and Gulf of Bothnia regions. The interpretation of these data in terms of the theory of glacial isostasy, however, has been extremely controversial. Jeffreys (e.g., 1972), for example, based his objection to the notion that steady state creep in the mantle was possible, in part on the fact that the constant density and viscosity Newtonian viscous model of isostatic adjustment of Haskell (1935, 1936, 1937) was incapable of simultaneously reconciling relative sea level records in Fennoscandia (which Haskell had shown required an upper mantle viscosity near 10^{21} Pa s) and the negative free air gravity anomaly there, which at 15–20 mgal amplitude was far too large to be accounted for by this model. Walcott (e.g., 1970, 1980) has argued that the large observed free air anomalies over both Canada and Europe require a very much higher viscosity in the lower mantle than 10^{21} Pa s in order to reconcile them. However, we saw in the last section that North American rsl data rule out a mantle viscosity profile which has any substantial increase in viscosity from the upper to the lower mantle. In the same connection it is interesting to realize that both O'Connell (1971) and Cathles (1975) were obliged to argue that there were no significant free air gravity anomalies over either Laurentia or Fennoscandia in order to support their respective arguments for the uniform viscosity model. As I will show in what follows, the modern theory of glacial isostatic adjustment reviewed in Section 2 has led to a reconciliation of these two important geophysical observables by demonstrating the important role played by the radial *elastic* component of the viscoelastic structure in determining the magnitude of the free air gravity anomaly associated with the isostatic adjustment process.

This may be seen most easily through a brief consideration of the form which the theoretical prediction of the free air anomaly takes in the new theory. It will suffice for this purpose to restrict consideration to the free air gravity anomalies induced by disk load approximations to the actual deglaciation events over Canada and northwestern Europe. We predict the free air anomaly simply by convolving the appropriate Green fraction (35b) over the glaciation history $L(\theta,t)$. Clearly (35b) may be expanded as:

$$\alpha(\theta,t) = \alpha^E(u)\delta(t) + \sum_{l=0}^{\infty}\sum_{j=1}^{M} p_j^l e^{-s_j^{-l}t} P_l(\cos\theta), \tag{47}$$

in which

$$\alpha^E(\theta) = \frac{g}{m_e}\sum_{l=0}^{\infty} [l + 2h_l^E - (l + 1)k_l^E] P_l(\cos\theta), \tag{48}$$

and

$$p_j^l = 2r_j^l - (l+1)\, q_j^l, \tag{49}$$

since

$$h_l = h_l^E\delta(t) + \sum_{j=1}^{M} r_j^l e^{-s_j^{-l}t}, \tag{50a}$$

and

$$k_l = k_l^E\delta(t) + \sum_{j=1}^{M} q_j^l e^{-s_j^{-l}t}, \tag{50b}$$

from (28). The free air gravity anomaly $\Delta g(\theta,t)$ is then just:

$$\Delta g(\theta,t) = \int_{-\infty}^{t} \alpha(t - t') * L(t')dt', \tag{51}$$

where the * in eq. (51) denotes spatial convolution over the loaded surface and the spatial dependence of α and L have been suppressed. Suppose we assume that the time of the last glacial maximum determines the origin of time $t = 0$ and that at that time the load thickness is $h_0(\theta)$, whereas for $t < 0$ the prehistory of loading is $L_p(\theta,t)$ and for $t > 0$ the thickness history is represented by $h_0 + h(\theta,t)$, i.e.

$$L(\theta,t) = L_p(\theta,t) \qquad t < 0 \\ = h_0 + h(\theta,t) \quad t \geqslant 0, \tag{52}$$

with $L_p(\theta,0) = h_0(\theta)$ and $h(\theta,0) \equiv 0$. For $t > 0$ we may therefore expand (51) as:

$$\Delta g(\theta,t) = \int_{-\infty}^{0} \alpha(t - t')*L_p(t')dt' + \int_0^t \alpha(t-t')*h_0 dt' \\ + \int_0^t \alpha(t-t')*h(t')dt. \tag{53}$$

In Peltier (1982) it is shown that (53) may be written as:

$$\Delta g(\theta,t) = \Delta\tilde{g}(\theta,t) - h_0*E(t) \tag{54}$$

in which $\Delta\tilde{g}(\theta,t)$ is the free air anomaly which would be observed if the load had been in isostatic equilibrium at $t = 0$ while the second term $h_0*E(t)$, which has the form of a convolution of the load at $t = 0$ with a "correction" or "error" Green function, determines the contribution to the anomaly due to the specific degree of disequilibrium which existed at $t = 0$ due to the specific loading prehistory $L_p(\theta,t)$ which occurred prior to the model glacial maximum. For the free air anomaly, unlike the relative sea level data discussed in the last section, the influence of initial isostatic disequilibrium is rather important.

Since the degree of initial disequilibrium is important to the accurate determination of free air gravity anomalies we require some means of estimating it and this clearly requires a suitable representation of the actual history of loading and unloading to which the Earth's surface has been subjected. In fact reasonably accurate data on the history of continental ice volume fluctuations throughout the Pleistocene period (last 2.5×10^6 years of Earth history) is available in the form of $\delta^{18}O$ data from deep sea sedimentary cores. Since the evaporation of sea water fractionates mass in the sense that the vapour is inevitably enriched in $H_2{}^{16}O$ relative to $H_2{}^{18}O$ compared to standard oceanic abundances, it is clear that at a time in the past in which large continental ice sheets existed the contemporaneous ocean would have been anomalously heavy isotopically. Thus a down core measurement of $\delta^{18}O$, which simply represents the ratio, $(^{18}O/^{16}O)$, of concentrations relative to a suitable standard (the PDB standard is usually employed), is directly interpretable in terms of ice volume variations. Based primarily on the work of Shackleton (e.g., 1967 and thereafter) which has led to a calibration of these data which includes a correction for temperature changes, it is generally accepted that these data demand a peak-to-peak ice volume variation such as to imply a global fall and rise of sea level during a single glaciation-deglaciation episode of 110–130 m.

Figure 6.19a shows the SPECMAP record of Imbrie *et al*. (1984) which is the best available globally representative $\delta^{18}O$ time series. The record is actually a composite generated by isotopic information from five open ocean cores from which obvious defects of the coring process where eliminated. Events impacting the cores whose ages were well known through other evidence (isotopic events radiocarbon dated at 18 KBP and 21 KBP, an isotopic boundary at 127 KBP, the Brunhes-Matuyama magnetic boundary near 730 KBP, and the Jamarillo magnetic event at 900 KBP) were used as control points to map the depth in each core into time. The ages of events between the points were assigned by correlating them with the astronomical fluctuations assumed to be responsible for the observed ice volume variations, and allowing for a frequency dependent phase lag. The resulting individual time series were then adjusted to have zero mean and unit standard deviation, following which they were stacked and smoothed with a nine point time filter. The final result, shown in Figure 6.19a, with its power spectrum in Figure 6.19b, demonstrates that during the last 5×10^5 years of the Pleistocene period the $\delta^{18}O$ time series, and thus continental ice volume variability, has been dominated by an oscillation of period very close to 10^5 years. Each cycle of this long timescale oscillation is characterized by a relatively slow glaciation phase and a relatively fast deglaciation phase or termination, the former lasting on average 9×10^4 years and the latter 10^4 years. This is the basis upon which we may approximate the history of loading and unloading by the simple sawtooth waveform shown in Figure 6.20 where it is compared not to the SPECMAP stack but rather to one specific $\delta^{18}O$ record. Because there is no evidence in the surface geology that the Pleistocene ice sheets responsible for inducing the observed variations of $\delta^{18}O$ ever occupied positions that were significantly different from those at the last glacial maximum it seems reasonable to assume as a first approximation that the single 10^5 year cyclic sawtooth history shown on Figure 6.20 was common to each of the major ice sheets whose appearance and disappearance has characterized the last 2.5×10^6 years of Earth history. This is the assumption which we shall make in evaluating equation (53) for the free air gravity anomalies associated with the regions which were once ice covered.

Figure 6.21 shows an initial comparison between the theoretically predicted and observed peak free air gravity anomalies for Laurentia and Fennoscandia respectively in plates a and b. On each plate the observed peak free air anomaly is shown as the hatched region and theoretical predictions are shown as a function of lower mantle viscosity with the upper mantle value

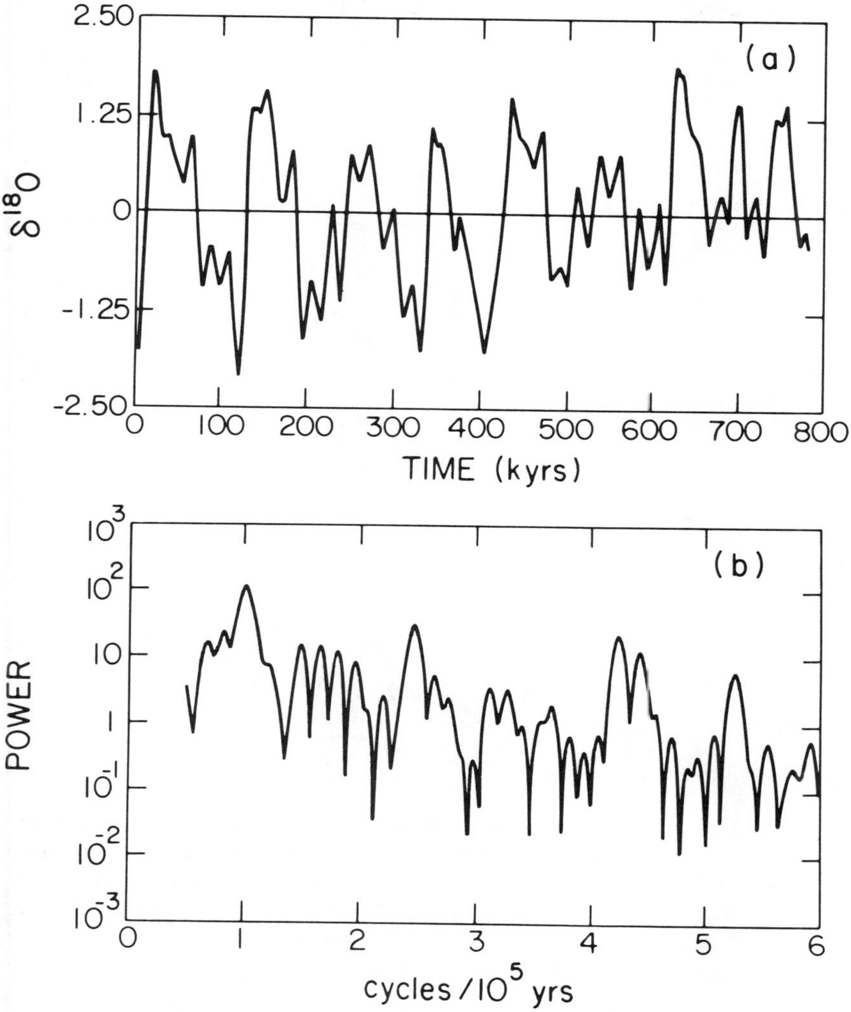

Figure 6.19. a, The SPECMAP $\delta^{18}O$ time series of Imbrie *et al.* (1984). b, The power spectrum of the SPECMAP time series. Note the dominance of the spectral "line" corresponding to an oscillation of period 10^5 years.

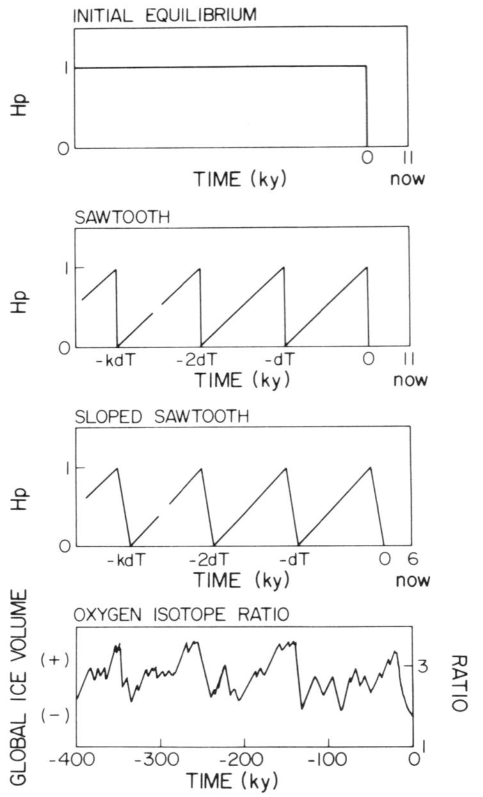

Figure 6.20. a–c, from top to bottom, three increasingly realistic approximations to the history of ice sheet loading and unloading employed to assess the effect on the theoretical free air gravity anomaly of initial isostatic disequilibrium. d, The final plate shows Oxygen isotope ratio as a function of time from a typical deep sea sedimentary core (from Hays *et al.*, 1976) that constrains the actual history of ice sheet loading and unloading.

fixed to the Haskell number of 10^{21} Pa s. Three different versions of the theoretical predictions are shown for each region: the dash dotted curve is computed on the basis of the assumption that isostatic equilibrium prevailed prior to the onset of deglaciation at 18 KBP, the dashed curve includes the influence on the prediction of the particular degree of isostatic disequilibrium due to an assumed loading prehistory consisting of 30 cycles of the sawtooth form shown in Figure 6.20 under the assumption that the radius of

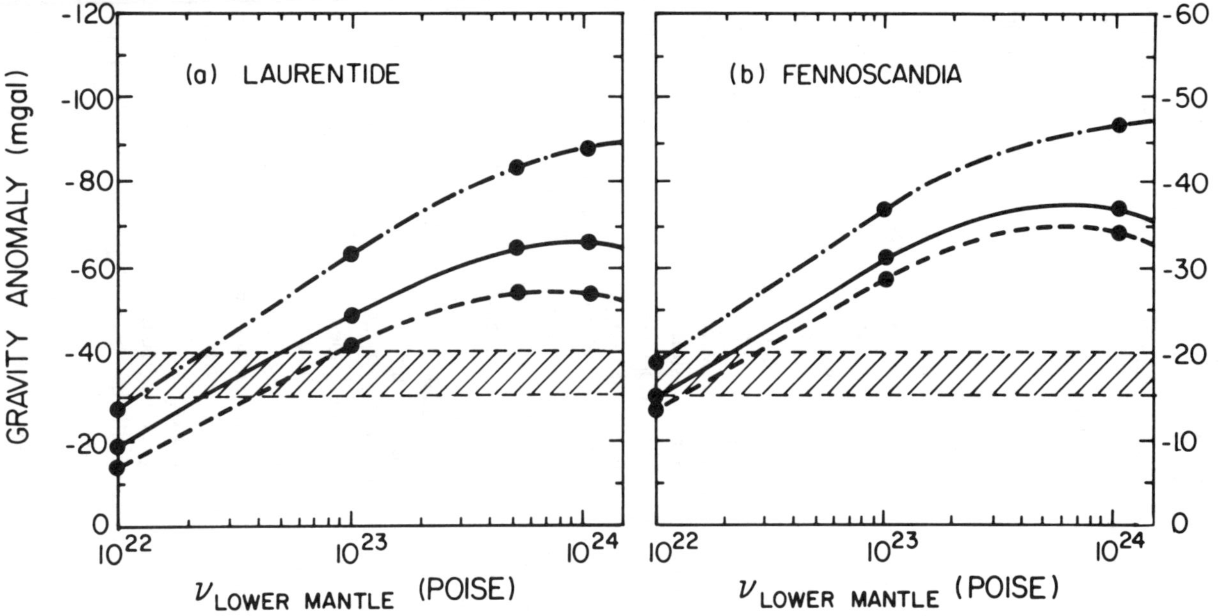

Figure 6.21. Predicted peak present day free air gravity anomalies for both Laurentia and Fennoscandia as a function of lower mantle viscosity with the upper mantle value fixed at 10^{21}Pa s. Calculations are based on parabolic disk load approximations to each ice mass. The horizontal hatched region is the observed range of the anomaly over these deglaciated areas shown on Figure 6.18. The dash-dotted lines are predictions based upon the assumption of initial isostatic equilibrium. The dashed lines show the influence of initial disequilibrium based upon the model glaciation history shown in Figure 6.20 based on $\delta^{18}O$ data, under the assumption that the ice sheet radius does not change while the volume fluctuates. The solid curve should be the most accurate since it also includes the influence of the variation of ice sheet radius with time.

each ice cap remains constant while its volume fluctuates, and finally the solid curve shows predictions based upon the same assumed prehistory but with allowance for the variation of ice sheet radius that in fact accompanies the volume fluctuation. All of these calculations employ an elastic radial structure which is identical to that of model 1066B of Gilbert and Dziewonski (1975). Inspection of these predictions, the most accurate of which is that shown as the solid line, shows that the lower mantle viscosity preferred by these data is 2×10^{21} Pa s $\leqslant \nu_{LM} \leqslant 5 \times 10^{21}$ Pa s according to the data from Laurentia whereas it is 10^{21} Pa s $\leqslant \nu_{LM} \leqslant 2 \times 10^{21}$ Pa s according to the data from Fennoscandia. The small region of overlap near $\nu_{LM} = 2 \times 10^{21}$ Pa s is the same value of the viscosity of the lower mantle as was shown in the last section to be required by the relative sea level data. In this sense the new theory of glacial isostasy elaborated in the preceding sections of this chapter fully resolves the previous controversy concerning the inability of Haskell type models (such as the spherical incarnation elaborated by Cathles, 1975) to simultaneously reconcile both relative sea level and free air gravity data. To appreciate how it is that the new theory has managed to effect this reconciliation, however, requires that we isolate the property of the models which is responsible.

Precisely which aspect of the new theory is responsible for the successful reconciliation of the requirements of these two types of geophysical observation in terms of a single profile of mantle viscosity is established in Figure 6.22. In this figure we compare predicted peak present day free air gravity anomalies as a function of lower mantle viscosity (with the upper mantle value fixed at 10^{21} Pa s) for a sequence of models which differ from one another only in the elastic components of their radial structures. These models include a series of four models which are based upon multiple homogeneous layer approximations to the mantle of 1066B and two seismically realistic models including the "PREM" of Dziewonski and Anderson (1981) and well as 1066B. As described in the figure caption, these models constitute a sequence in which the amount of internal mantle buoyancy is the main distinguishing feature of each model. Inspection of the figure shows that as the amount of internal mantle buoyancy is increased, the upper mantle–lower mantle viscosity contrast which is required to fit the data decreases. For the seismically realistic models in which all of the radial variation of density is treated as non-adiabatic the viscosity of the lower mantle which is required to reconcile the observed free air anomalies is the same viscosity required by the sea level data as previously discussed. Clearly then it is internal mantle buoyancy which allows the model to reconcile both RSL and free air gravity data with the same profile of mantle viscosity.

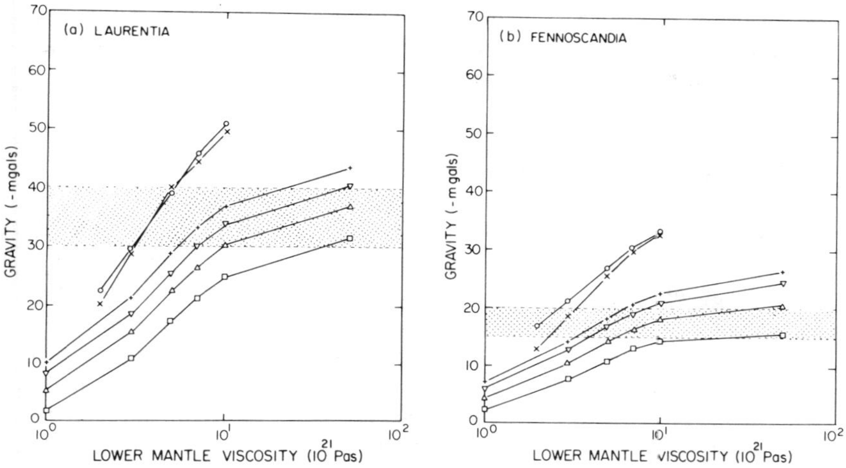

Figure 6.22. Comparisons of predicted and observed peak free air gravity anomalies for Laurentia and Fennoscandia as a function of lower mantle viscosity with the upper mantle value fixed to 10^{21} Pa s. Computations are shown at each site for a set of five different earth models which differ from one another only with respect to the amount of buoyancy which is contained in their mantles. These are respectively: □, for a model with 6.2 percent increase of density at 670 km depth; ∇ for a model with 6.2 percent and 3.8 percent density increases at 670 km and 420 km depth respectively; + for a model with a 12.4 percent increase of density at 670 km depth; × for 1066B; O for PREM.

As demonstrated in Section 2 of this chapter, the way in which internal mantle buoyancy enters the theory is through the extra modes of viscous gravitational relaxation which it supports, mainly the M1 and M2 modes associated with the 650 km and 420 km seismic discontinuities, respectively. Of these two discontinuities, that at 650 km is obviously the more important. It might be thought that the requirement that this interface induce a buoyant restoring force when it is deflected argues strongly that the boundary itself could not be an equilibrium phase boundary but would have to be in some substantial part chemical, which would in turn require that the mantle convective circulation be of the layered form (see Chapter 4). However, as discussed in Peltier (1985c) based upon calculations of O'Connell (1971) and Mareschal and Gangi (1977) in another context, to the extent that the equilibrium phase boundaries which are actually responsible for these discontinuities are nearly univariant (see Chapter 4), they would behave as if they were chemical discontinuities on the timescales of postglacial rebound since latent heat will then be removed during transformation only on a rather long thermal diffusion timescale (see also Chapter 8).

5. Mantle Viscosity from Earth Rotation Observations

The above analyses clearly establish that both of the "local" signatures of the isostatic adjustment process which can be invoked to constrain mantle viscosity are fit by the same radial profile of this parameter and that this radial profile appears to be independent of geographic location. This is precisely what we expect to be the case based upon the idea, discussed in section 1, that the mantle is filled by a convective circulation that is well within the boundary layer regime so that the lateral variations of viscosity through the thermal plumes are confined to a fraction of the mantle volume which is small compared to its total volume. The utility of the Earth rotation observations which will be discussed in the present section is that, being sensitive only to the behavior of the spherical harmonic degree 2 components of the deformation spectrum, they provide on optimal measure of the planetary average of the upper mantle–lower mantle viscosity contrast. To the extent that this planetary average contrast agrees with that deduced above on the basis of geographically local observations of relative sea level and free air gravity we will have strong further confirmation of the view of the geometric structure of the circulation which follows from the high Rayleigh number scaling relations discussed in the introduction to this chapter. There are two different rotational observables which may be brought to bear on this question, these being the so-called non-tidal component of the acceleration of planetary rotation and the true wander of the pole with respect to the surface geography. I will discuss the interpretation of these two observations in turn in the following subsections.

5.1 The Non-Tidal Acceleration of Planetary Rotation

The ongoing secular deceleration of the Earth's rate of axial rotation due to friction associated with the lunar semidiurnal tide in the oceans is a well understood phenomenon (e.g., Lambeck, 1980). The loss of spin angular momentum of the Earth due to this effect is compensated by the increase of orbital angular momentum of the moon through a continuous increase of the Earth–Moon distance which is directly observable by laser ranging to the moon. If it is assumed that the torque on the Earth due to lunar tidal friction has been constant through time, specifically over the past few thousand years of Earth history, then one may reconstruct the relative geometric positions of the Earth and Moon with respect to the Sun at any time in the past and, in particular, predict when and where total eclipses of the Sun and Moon should have been observed on Earth back into antiquity. Detailed examination of the astronomical records of early civilizations, particularly the Babylonian, first led to the remarkable conclusion that such predictions were subject to a successively larger systematic error the further back into the past that the prediction was made and to the conclusion that some non-tidal acceleration of rotation related to internal dynamical processes was at work on the Earth which was acting to somewhat reduce the tidal deceleration which was also known to be active (Currot, 1966; Müller, 1975; Müller and Stephenson, 1975). Lambeck's (1977) review settled on a value of $(0.69 \pm 0.3) \times 10^{-10}$ yr^{-1} for the non-tidal contribution to the ratio $\dot{\omega}_3/\Omega$ where $\dot{\omega}_3$ is the time derivative of the perturbation to the basic axial rate of rotation Ω. This observation was rather stunningly confirmed recently by Yoder *et al.* (1983) and Rubincam (1984) through analyses of laser ranging data for the LAGEOS satellite. Their independent analysis of 5½ years of such data established rather convincingly that there exists a residual acceleration in the node of the satellite's orbit in the amount -14.7 ± 1.3 m arc s yr^{-2} (according to Yoder *et al.*). This acceleration is predominantly attributable to a secular decrease of the degree two axial component of the Earth's gravitational potential field, J_2, at the rate $\dot{J}_2 = -3.5 \pm 0.3 \times 10^{-11}\ yr^{-1}$. Since $\dot{J}_2 = (-3C/2\ m_e\ a^2)\cdot(\dot{\omega}_3/\Omega)$, this implies a non-tidal acceleration which is very close to that inferred from the eclipse data. Rubincam's (1984) independent estimate of $\dot{J}_2$ gave the value $-2.6 \pm 0.6 \times 10^{-11}\ yr^{-1}$ which, although close to that of Yoder et al., is significantly lower. Since it is not yet known why these two reductions of the same data give discrepant estimates of $\dot{J}_2$, I shall discuss the slight differences of their separate implications below.

The first suggestion that the non-tidal acceleration of rotation could be a forced response to the glaciation–deglaciation process appears to be that by Dicke (1966, 1969) who also recognized that it could be employed to infer the viscosity of the mantle. O'Connell (1971) attempted to use this observation to show that the viscosity of the lower mantle was close to the upper mantle value but his argument is incorrect (as discussed in the last section) as it relied upon the assumption that there was no significant negative free air gravity anomaly correlated in position with the location of the ancient Laurentide ice sheet. That a satellite observation of $\dot{J}_2$ could be used to confirm the inference of the non-tidal acceleration from ancient eclipse data appears to have been first suggested in Peltier (1982). Detailed subsequent discussions using the published results for $\dot{J}_2$ from LAGEOS (Peltier 1983, Peltier and Wu 1983, Wu and Peltier 1984, Peltier 1985a) demonstrated a basic consistency of the viscosity structure obtained from RSL and free air gravity observations with this additional datum. More recent analyses of the observation have further confirmed this basic conclusion but have demonstrated in the process that account must be taken of the additional rotational forcing associated with present day melting of the small ice sheets and glaciers of the world which is required by modern tide gauge recordings and that a more accurate model of the late Pleistocene ice sheets than that employed in the original papers cited above is also necessary.

Both of the rotational observables to be discussed in this section are predicted as solutions of the following Euler equations which express the conservation of angular momentum in a system subject to no external torque:

$$\frac{d}{dt}[J_{ij}\,\boldsymbol{\omega}_j] + \varepsilon_{ijk}\,\boldsymbol{\omega}_j\,J_{kl}\,\boldsymbol{\omega}_l = 0 \tag{55}$$

in which J_{ij} is the (time dependent) moment of inertia tensor of the planet, $\boldsymbol{\omega}$ is the angular velocity vector, and ε_{ijk} is the Levi-Cevita (alternating) tensor. For sufficiently small perturbations away from an initial state of steady axial rotation with angular velocity $\boldsymbol{\Omega}$, (1) may be linearized with the following perturbation expansion:

$$\boldsymbol{\omega}_i = \boldsymbol{\Omega}\,(\delta_{i3} + m_i), \tag{56}$$

$$J_{ij} = I_{ij} \qquad i \neq \mathrm{j}, \tag{57a}$$

$$J_{11} = A + I_{11}, \tag{57b}$$

$$J_{22} = A + I_{22}, \tag{57c}$$

$$J_{33} = C + I_{33}. \tag{57d}$$

Substituting (56) and (57) into (55) and dropping terms of second and higher order in the perturbations m_i and I_{ij}, decouples (55) into the following two independent equations:

$$\frac{i}{\sigma_r}\mathbf{m} + \mathbf{m} = \boldsymbol{\psi}, \tag{58}$$

$$\dot{m}_3 = \psi_3. \tag{59}$$

In (58) $\sigma_r = \Omega(C - A)/(A)$ is the Chandler wobble frequency of the rigid Earth and the vectors $\mathbf{m}$ and $\boldsymbol{\psi}$ are such that

$$\mathbf{m} = m_1 + im_2 \tag{60}$$

$$\boldsymbol{\psi} = \psi_1 + i\psi_2 \tag{61}$$

with

$$\psi_1 = \frac{I_{13}}{(C - A)} + \frac{\dot{I}_{23}}{\Omega(C - A)}, \tag{62a}$$

$$\psi_2 = \frac{I_{23}}{(C - A)} + \frac{\dot{I}_{13}}{\Omega(C - A)}, \tag{62b}$$

$$\psi_3 = -\frac{I_{33}}{C}. \tag{62c}$$

Equation (58) describes the forced polar motion which we will discuss in the next subsection while (59) describes the forced variation of the axial rate of rotation. Focussing first on (59) it is clear that to solve it we must be able to compute the perturbation of the axial component of inertia I_{33} which is produced by glacial loading and unloading.

In the domain of the Laplace transform variable s, the solution of (59) is simply:

$$m_3(s) = -\frac{I_{33}(s)}{C}. \tag{63}$$

For an arbitrary radially stratified viscoelastic Earth model Peltier (1982) shows that I_{33} (s) may be written as:

$$I_{33}(s) = [1 + k_2^L(s)]f(s)I_{33}^R \tag{64}$$

in which $k_2^L(s)$ is the Laplace transform domain form of the surface load Love number of degreee 2, $f(s)$ is the Laplace transform of the loading history $f(t)$ introduced in the last section, and I_{33}^R is the perturbation of the axial component of inertia which would be produced by the changing surface load if the Earth were rigid. As discussed in detail in Wu and Peltier (1984), I_{33}^R has a particularly simple form if we assume that during a glaciation–deglaciation cycle we may approximate the active ice sheets by circular disk loads of fixed radii α_i whose centroids are located at colatitudes θ_i and which accrete ice from and uniformly discharge their meltwater to, ocean basins of realistic shape. Under these assumptions, I_{33}^R may be expressed as:

$$I_{33}^R = \sum_{i=1}^{M} m_i a^2 \left[\frac{2}{15}\frac{a_{20}}{a_{00}} - \frac{1}{3}\cos\alpha_i(1 + \cos\alpha_i)P_2^0(\cos\theta_i)\right]. \tag{65}$$

In (65) the number $a_{20}/a_{00} = -0.1925$ is a ratio of spherical harmonic coefficients in the expansion of the "ocean function" which describes the shape of the ocean basins [e.g., Lambeck, 1980]. The M_i are the masses of the M ice sheets at glacial maximum and a is the Earth's radius. The values of the parameters required to determine the I_{ij}^R for each of the four main Pleistocene ice sheets are listed in Table 6.1. Besides information on the Laurentian, Fennoscandian, and Antarctic ice masses, mentioned previously, the table also provides information on the Barents Sea ice sheet for reasons which will become clear momentarily.

The influence of radial viscoelastic Earth structure upon the predicted acceleration of rotation (63) is entirely contained in the Love number $k_2^L(s)$. As shown by Peltier (1976), this Love number may be written in the form of a normal mode expansion as in section 2:

$$k_2{}^L(s) = k_2^{LE} + \sum_{j=1}^{N} \frac{r_j}{s + s_j}. \tag{66}$$

If we define an "isostatic factor" l_s as

$$l_s = \lim_{s \to 0} [1 + k_2^L(s)], \tag{67}$$

then, since

$$\lim_{s \to 0} k_2^L(s) = k_2^{LE} + \sum_j \frac{r_j}{s_j}, \tag{68}$$

we may rewrite (66) as

$$k_2^L(s) = (l_s - 1) - s\sum_{j=1}^{N} \frac{(r_j/s_j)}{(s + s_j)}. \tag{69}$$

Table 6.1a. Ice sheet parameters

	Laurentia	Fennoscandia	Antarctic	Barents Sea
Mass M_i (kg × 10^{19})	2	0.56	0.70	0.60
Radius α_i (deg)	15	9.5	20.0	6.0
Colatitude θ_i (deg)	30	25.5	180.	15.0
East longitude ϕ_i (deg)	270	25.0	—	50.0
Eustatic sea level rise (m)	55.4	15.5	19.4	16.6

Table 6.1b. Inertia perturbations I_{i3}^R with realistic oceans

	I_{13}^R (kg m^2)	I_{23}^R (kg m^2)	I_{33}^R (kg m^2)
Laurentia only	1.203×10^{31}	3.491×10^{32}	-3.420×10^{32}
Fennoscandia only	-7.504×10^{31}	-3.227×10^{31}	-1.130×10^{32}
Antarctica only	4.211×10^{30}	5.370×10^{30}	-1.799×10^{32}
Barents Sea only	-3.521×10^{31}	-4.165×10^{31}	-1.511×10^{32}

Substituting from (69) for $1 + k_2^L(s)$ into (64) and from (64) into (63), Laplace inversion of $m_3(s)$ yields the formal solution for the increment of axial angular velocity as

$$m_3(t) = -\frac{I_{33}^R}{C}[D_1 f(t) + \sum_j r_j(f*e^{-s_j t})], \qquad (70)$$

in which

$$D_1 = l_s - \sum_j \frac{r_j}{s_j} = 1 + k_2^{LE}.$$

In (70) the asterisk denotes convolution. The predicted acceleration of rotation is obtained from (70) by direct time differentiation as

$$\dot{m}_3(t) = \frac{\dot{\omega}_s}{\Omega} = -\frac{I_{33}^R}{C}[D_1 \dot{f}(t) + \sum_j r_j \frac{d}{dt}(f*e^{-s_j t})]. \qquad (71)$$

This is the solution upon which all further discussion in this subsection will be based. For the glaciation history $f(t)$ the initial calculations with this theory employed the same saw-tooth approximation to the individual glacial cycles that was used for the free air gravity analyses of the last section.

These analyses delivered the result shown on Figure 6.23 which shows the model prediction of present day $\dot{\omega}_3/\Omega$ (or equivalently $\dot{J}_2$) as a function of lower mantle viscosity with the upper mantle value held fixed to the 10^{21} Pa s preferred by the RSL data. The elastic basic state used for these calculations was model 1066B, the lithospheric thickness was fixed to the value of 120.7 km, and seven 10^5-year cycles of glaciation and deglaciation were included in the history of loading and unloading. The Yoder *et al.* (1983) value of $\dot{\omega}_3/\Omega$ is shown on the figure as the upper hatched region which extends one standard error on either side of the experimentally determined value cited above and the Rubincam (1984) number is similarly denoted below. The curves marked L, $L+F$, and $L+F+A$ respectively show the theoretical results for calculations which include only the Laurentide ice sheet, the Laurentian and Fennoscandian ice complexes, and finally all three of the main land based complexes including that of West Antarctica. Inspection of these results demonstrates that as each additional ice mass is added to the rotational forcing the two lower mantle viscosities allowed by the data diverge until, with all three ice sheets included, the observation may be fit either by a model with lower mantle viscosity near 3×10^{21} Pa s (if the Yoder *et al.* value were correct, lower if the Rubincam number is preferred) or near 10^{23} Pa s. Clearly the latter root is ruled out by the relative sea level data discussed previously while the former is essentially identical with the lower mantle viscosity required by these previously discussed observations.

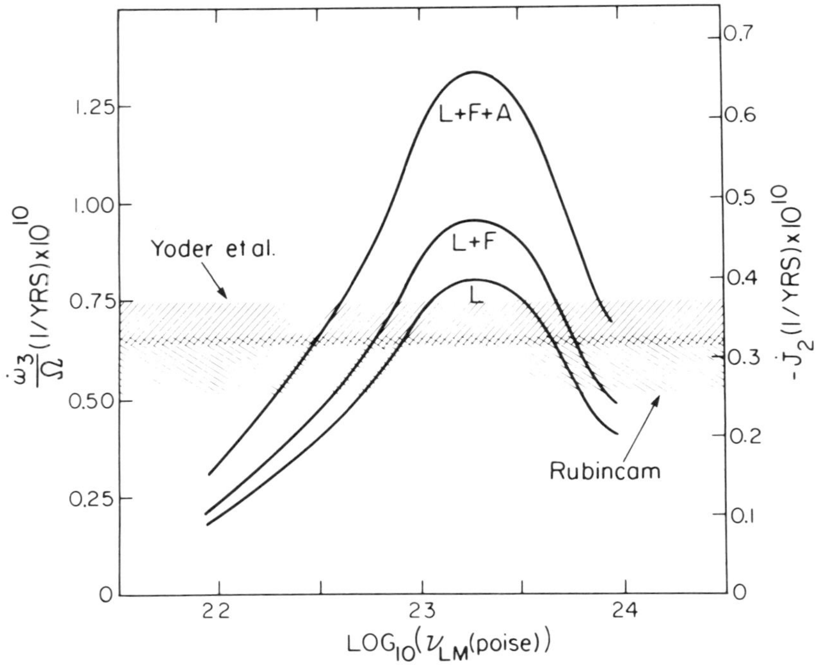

Figure 6.23. Predicted nontidal acceleration as a function of lower mantle viscosity based upon the use of 1066B fully non-adiabatic elastic structures. The three curves are for a loading model including only the Laurentide ice sheet (L), the Laurentian plus Fennoscandian ice sheets (L+F), and also including the Antarctic ice sheet (L+F+A). The hatched regions show the observed nontidal acceleration (and $\dot{J}_2$) inferences based upon LAGEOS laser ranging observations.

Very recently new information has become available which has, superficially at least, suggested that the conclusion implicit in the above described analyses may not be warranted. Clearly if there were other known sources of rotational forcing acting on the planet than those due to late Pleistocene glacial melting this would require readjustment of the parameters of the preferred rheological model which could make it less compatible with the relative sea level observations than demonstrated above. In fact, Mark Meier (1984) has recently provided a rather convincing argument that such additional forcing is presently active and in particular argued that it is associated with the ongoing melting of the numerous small ice sheets and glaciers of the world whose locations are shown in Figure 6.24. That this scenario, which I shall refer to as "Meier's hypothesis," is fully plausible has been demonstrated on the basis of recent analyses of secular sea level trends based upon tide gauge recordings of particularly long duration (e.g., Barnett, 1982). The global average of the secular rates of sea level rise based on such data suggests that global sea level has risen by 10–15 cm in the past 100 years (Gornitz *et al.*, 1980; Hansen *et al.*, 1981) and a great deal of attention has recently been drawn to this result because of its importance in the area of planetary climatology. It is thought that this present day eustatic rise could be due to the climate amelioration expected due to the increasing atmospheric load of CO_2 and other "greenhouse" gases which has been ongoing throughout the last century due to the increased burning of fossil fuels which has accompanied global industrialization. Although much more of the tide gauge data on which those ideas are based have come from sites, like those on the east coast of the United States of America, which are very strongly influenced by the ongoing process of glacial isostatic adjustment, Peltier (1986) has recently shown that when these data are corrected for this effect using the theory previously discussed in Section 3 a residual does remain which *is* in the range 1–1.5 mm/yr. As argued by Gornitz *et al.* (1982), whose analysis was accepted by Meier (1984), only about .5 mm yr^{-1} of this secular sea level rise appears to require an increase in ocean mass to explain it, the reminder being attributable to the steric effect of thermal expansion.

When the 31 contemporaneously melting small ice sheets and glaciers are added to the rotational forcing function (65) with the strengths assigned to them by Meier (1984) which are listed in Table 6.2, it is found that the action of this enhanced forcing has a very significant effect upon the predicted non-tidal acceleration of rotation that is such as to sharply reduce it from the previous predictions. That this should be the case is clear on the basis of the fact that the small ice sheets and glaciers are for the most part located in high latitudes so that, as they melt back, matter is transferred to a relatively more equatorward location which increases the axial component of the inertia tensor and thus decreases the rate of rotation. The magnitude of this effect is illustrated in Figure 6.25 which compares the original prediction (computed on the basis of an assumed 5 cycles of loading and unloading) with the three different versions of the additional forcing according to Meier's hypothesis described in the inset text. Plates a and b of this figure compare the results as a function of the elastic basic state employed in the model, the former illustrating results for model CML64 and the latter the results for 1066B. Clearly the dependence of these predictions on the elastic parameterization is very weak, in sharp contrast to the results for the free air gravity anomaly

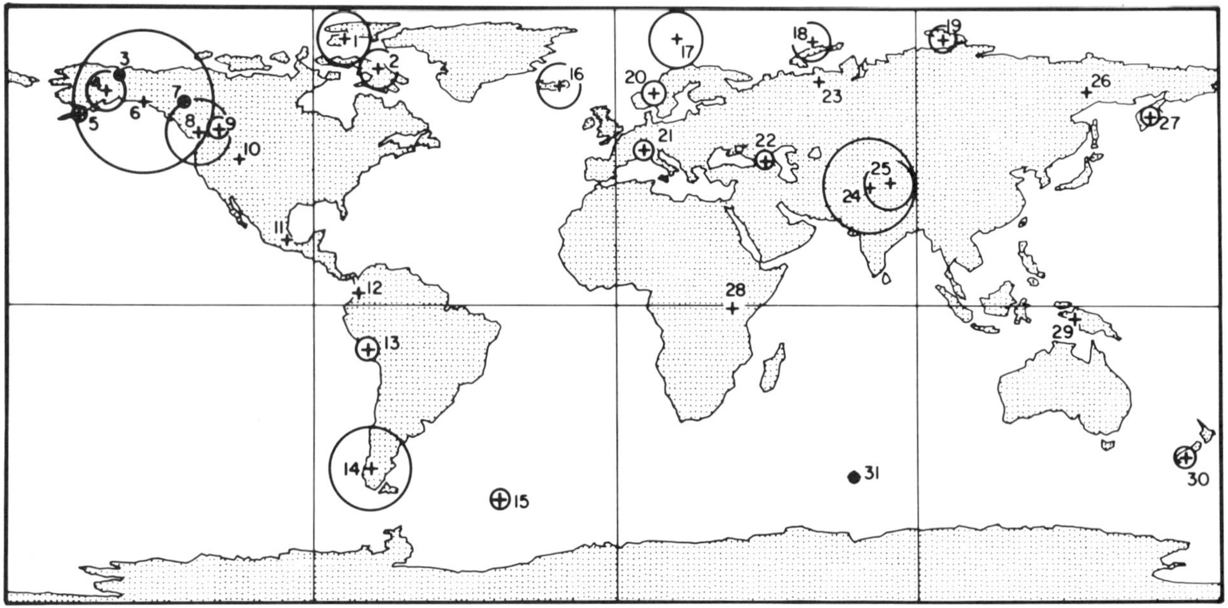

Figure 6.24. The locations of the 31 small ice sheets and glaciers which are currently contributing to rising sea levels according to Meier (1984). The relative importance of the sources, whose masses are listed in Table 6.2, are indicated by the radius of the circle centered on each geographic location.

Table 6.2.

Glacial region	Rad	Colat	Elong	Case I† Mass × 10^{14} kg	I^R_{13}	I^R_{23} × 10^{28} kg m²	I^R_{33}	Water mm/yr	Case II† Mass × 10^{14} kg	I^R_{13}	I^R_{23} × 10^{28} kg m²	I^R_{33}	Water mm/yr
1 Canadian Arctic-North	.0	11.0	279.	6.900	.04	−.57	1.84	0.019					
2 Baffin and Bylot Islands, Labrador	.0	20.0	289.	3.910	.14	−.51	.91	.011					
3 Brooks Range, Kigluaik Mountains	.0	22.0	213.	.138	−.02	−.01	.03	.000	.140	−.02	−.01	.03	.000
4 Alaska Range; Talkeetna, Kilbuk Mountains	.0	27.0	209.	3.680	−.55	−.32	.73	0.010	8.195	−1.23	−.72	1.62	.023
5 Aleutian Is., Alaska Peninsula	.0	34.0	201.	.506	−.09	−.04	.08	.001					
6 Coastal Mountains, Kenai Peninsula to 55N	.0	30.0	220.	51.750	−7.28	−6.24	9.29	.143	35.360	−4.97	−4.27	6.35	.098
7 MacKenzie, Selwin, Rockies N of 55N	.0	30.0	232.	.207	−.02	−.03	.04	.001					
8 Coast Mountains S of 55N, Cascades, Olympics	.0	39.0	236.	10.350	−1.21	−1.78	1.24	0.029	9.940	−1.16	−1.71	1.19	.028
9 Rockies S of 55N, Selkirks (Canada), N. Rocky Mtns.	.0	38.0	242.	1.196	−.12	−.22	.15	.003	.880	−.09	−.16	.11	.002
10 Middle & S. Rockies, Sierra Nevada	.0	47.0	248.	.046	.00	−.01	.00	.000	0.34	.00	−.01	.00	.000
11 Mexico	.0	71.0	262.	.002	.00	.00	.00	.000					
12 Andes (10N–0N)	.0	86.0	283.	.012	.00	.00	.00	.000					
13 Andes (0S–30S)	.0	103.0	286.	1.173	−.04	.09	−.12	.003					
14 Andes (30S–55S)	.0	138.0	287.	17.710	−1.15	3.28	1.76	.049					
15 S. Atlantic Islands	.0	147.0	325.	.943	−.15	.09	.15	.003					
16 Iceland, Jan Mayen	.0	25.0	343.	4.830	.69	−.26	1.01	.013	7.119	1.02	−.38	1.48	.020

(continued)

Table 6.2. *Continued*

Glacial region	Rad	Colat	Elong	Case I[†] Mass $\times 10^{14}$ kg	I^R_{13}	I^R_{23} $\times 10^{28}$ kg m^2	I^R_{33}	Water mm/yr	Case II[†] Mass $\times 10^{14}$ kg	I^R_{13}	I^R_{23} $\times 10^{28}$ kg m^2	I^R_{33}	Water mm/yr
17 Svalbard	.0	11.0	18.	7.820	.52	.12	2.08	.022					
18 Franz Josef Is., Novaya Zemlya	.0	12.0	59.	3.450	.13	.22	.91	.010	4.901	.18	.31	1.29	.014
19 Severnaya Zemlya, other Soviet Arctic islands	.0	11.0	98	1.702	−.03	.12	.45	.005					
20 Norway, Sweden	.0	27.0	11	1.541	.24	.04	.30	.004	.651	.10	.02	.13	.002
21 Alps, Pyrenees	.0	44.0	8.	.851	.17	.02	.07	.002	1.1218	.24	.03	.10	.003
22 Caucasus, Turkey, Iran	.0	47.0	45.	.805	.11	.11	.05	.002	.420	.06	.06	.03	.001
23 Ural Mountains, Byrranga, Putorana	.0	24.0	61.	.046	.00	.01	.01	.000	.035	.00	.00	.01	.000
24 Hindukush, Pamir, Alai, Karakorum, Himalaya, . . .	.0	55.0	76.	23.690	.95	4.20	.21	.066					
25 Tien Shan, Kun Lun, Altai, Qilian Shan, . . .	.0	54.0	82.	6.210	.13	1.14	.10	.017	.642	.01	.12	.01	.002
26 Kodar, Orulgan, Chersky, Sun-Kho, Koryasky	.0	27.0	141.	.092	−.01	.01	.02	.000					
27 Kamchatka	.0	34.0	160.	.759	−.14	.04	.12	.002					
28 Africa	.0	91.0	35.	.002	.00	.00	.00	.000					
29 West Irian	.0	94.0	137.	.009	.00	.00	.00	.000					
30 New Zealand	.0	135.0	170.	1.104	.21	−.05	.09	.003					
31 Kerguelen, Heard Islands	.0	141.0	71.	.207	−.01	−.04	.02	.001					
Totals				151.6	−7.5	−.6	21.5	.42	136.3	−5.6	.0	20.0	.38

[†]Case I: Mass loss per year = 0.23 aG; Case II: Mass loss per year = $\bar{b}$G when available, else = 0.23 aG.

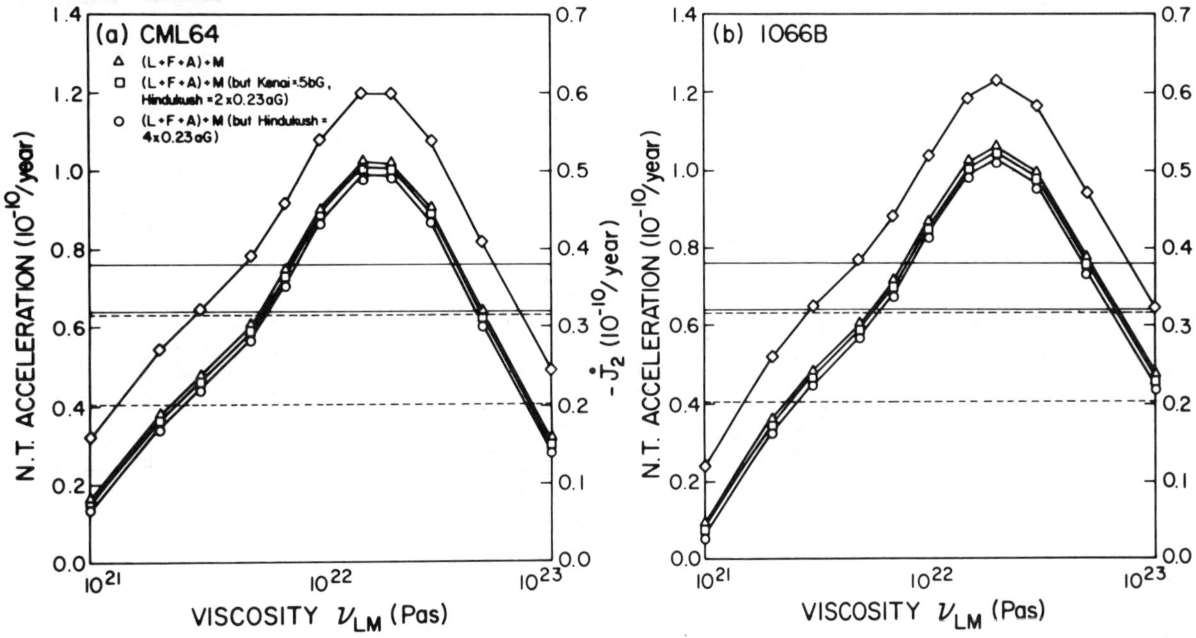

Figure 6.25. Theoretical predictions of the non-tidal acceleration of rotation as a function of lower mantle viscosity with the upper mantle value fixed to 10^{21} Pa s. Plates a and b display the influence on the theoretical predictions of the radial elastic structure assumed for the earth model. On each plate the curve marked with the diamonds shows the predictions for model L+F+A of Figure 6.24. The remaining three curves illustrate the effect on the prediction of three different versions of Meier's hypothesis described in the legend. On each plate the horizontal solid and dashed lines illustrate the $\dot{J}_2$ observations of Yoder *et al*. (1983) and Rubincam (1984) respectively.

discussed previously. The thin horizontal solid and dashed lines on both plates respectively denote the standard error bars on the LAGEOS $\dot{J}_2$ observation according to Yoder *et al*. (1983) and Rubincam (1984). The important point to note concerning these comparisons is that, even if the latter observation were the more accurate, the effect of the action of Meier's sources upon the Earth's rotation is such as to require the model which fits the data to possess a much *higher* lower mantle viscosity ν_{LM} (again with the upper mantle value held fixed to 10^{21} Pa s) than is the case in the absence of such additional rotational excitation. This increase by a factor of 3–5 is sufficient to be excluded by the previously discussed relative sea level data.

A clue as to what might be the problem with the above analysis is provided by the data in the previously discussed Table 6.1. In particular the last line of Table 6.1a provides a best estimate of the eustatic water rise delivered by each of the Laurentian, Fennoscandian, and Antarctic ice sheets employed in the previous rotational analyses. Summing these three separate contributions we get a total eustatic rise of only 90.3 m. This is considerably below the *a priori* constraint which is available from $\delta^{18}O$ data and from altitude observations of dated corral terraces which together appear to require a total eustatic variation across the last glacial–interglacial transition of about 130 m (Chapell and Shackleton 1986). As discussed in greater detail in Peltier (1987c) the most probable explanation for the majority of this discrepancy is that no allowance in these calculations has been made for the presence of the Barents Sea and Kara Sea ice complex whose existence at glacial maximum is strongly suggested by the patterns of land uplift in this geographic region to the north and east of Fennoscandia. The last column of Table 6.1 provides an estimate of the magnitude of the eustatic variation (16.6 m) which was likely delivered by the disintegration of this marine ice complex. When this additional late Pleistocene ice mass is inserted into the rotational forcing the modification which its presence makes to the viscoelastic model which is required to fit the data is illustrated on Figure 6.26. On this figure (which is in other respects completedly analogous to Figure 6.25) the curve marked L+F+A is for a model which includes neither Meier's sources nor a Barents Sea ice mass, that marked L+F+A+B includes the influence of the Barents Sea ice sheet but not Meier's sources, whereas that marked L+F+A+B+M includes the effect of all five sources of excitation. Clearly the effect of the Barents Sea ice sheet is to almost exactly annihilate the net influence of Meier's sources on the predicted non-tidal acceleration and thereby to allow the data to be fit once more by a model with the same modest viscosity contrast between upper and lower mantle as is required by the relative sea level data. The addition of further sources of late Pleistocene meltwater, such

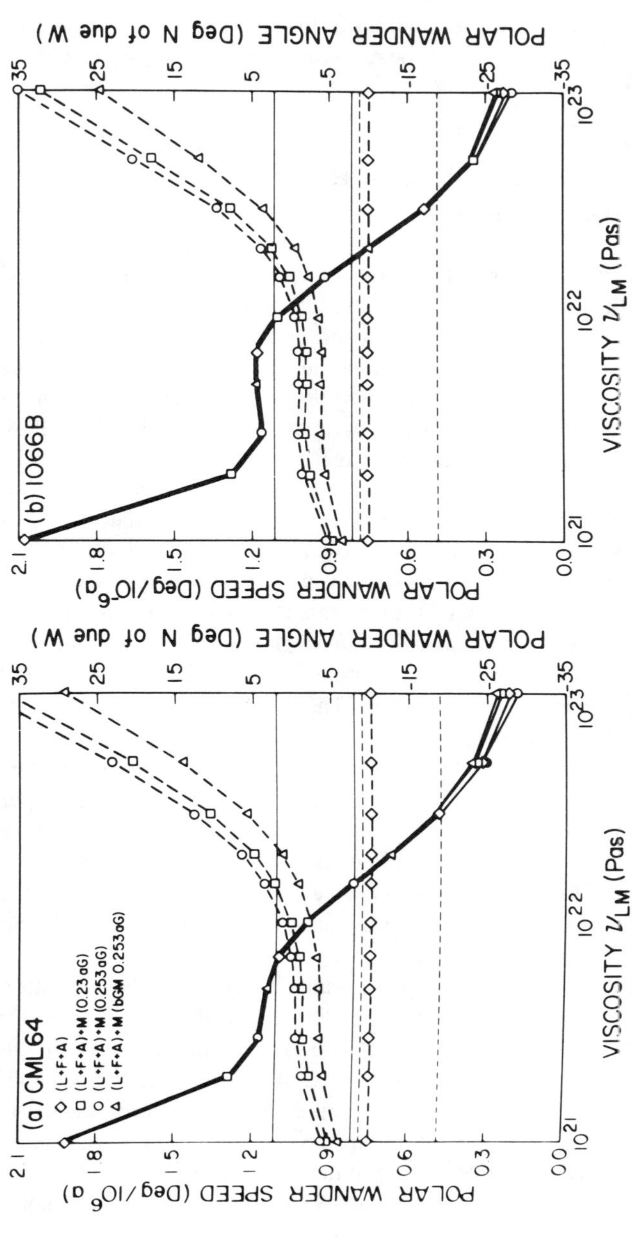

Figure 6.26. Theoretical predictions of the present day polar wander speed (solid curves) and direction (dashed curves) for the same sequence of models employed in Figure 25. Note that every version of the calculation which includes Meier's sources (L+F+A)+M predicts a present day direction of polar wander which does not match the direction observed in the ILS data, the standard error bounds on which are shown as the horizontal dashed lines. The solid horizontal lines show the corresponding error bounds on polar wander speed.

as might be required to further rectify the remaining eustatic misfit, will simply shift the viscosity profile required to fit the data even more sharply back toward the almost uniform profile required by the previously discussed relative sea level data.

6. The Wander of the Pole with Respect to the Surface Geography

The second rotational observable which we can include as an additional constraint upon the internal viscoelastic structure is one which, like the free air gravity anomaly measurements discussed previously, has engendered a good deal of controversy. Figure 6.27 shows the photo-zenith tube data of the ILS (International Latitude Service) which depict the wander of the north pole of rotation relative to the surface geography over the period from 1900 AD to about 1975. The polar motion is presented as decomposed into x and y components relative to the coordinate system shown on the inset polar projection. Clearly the data are dominated by a strong oscillatory signal which even visual inspection will reveal to consist of an oscillation with a period near 1 year which is modulated at a period near seven years. The dominant signal is accounted for by the interference at the 14 month Chandler wobble and the 12 month annual wobble as is well known. Of greatest interest to us here is the slow secular drift of the pole upon which the more evident oscillatory fluctuations are superimposed. The direction of this secular drift is shown by the arrow on the inset polar projection which points towards eastern Canada. The rate of drift in this direction, as determined by analysis of these observations (e.g., Dickman, 1977), is 0.95 $\pm$ 0.15 degrees per million years. On the basis of the orientation of this direction of drift relative to the centroids of the Laurentian and Fennoscandian ice sheets one should not be too surprised to learn that it also is an effect due at least in part to the melting of these ice complexes which began about 18,000 years ago and ended approximately 6000 years ago.

The theoretical solution for the polar wander forced by the glaciation deglaciation cycle follows from equation (58). Analysis in Peltier (1982) following Munk and McDonald (1960) shows that the elements of the excitation function Ψ, ψ_1 and ψ_2 defined in (62a,b), have contributions both due to the changes in the elements of the moment of inertia tensor due to the load induced deformation ($\boldsymbol{\psi}^L$) and due to the centrifugal force associated with the changing rotation ($\boldsymbol{\psi}^{\mathrm{ROT}}$). Both of these effects must be included in the calculation to achieve first order accuracy in the prediction of the polar

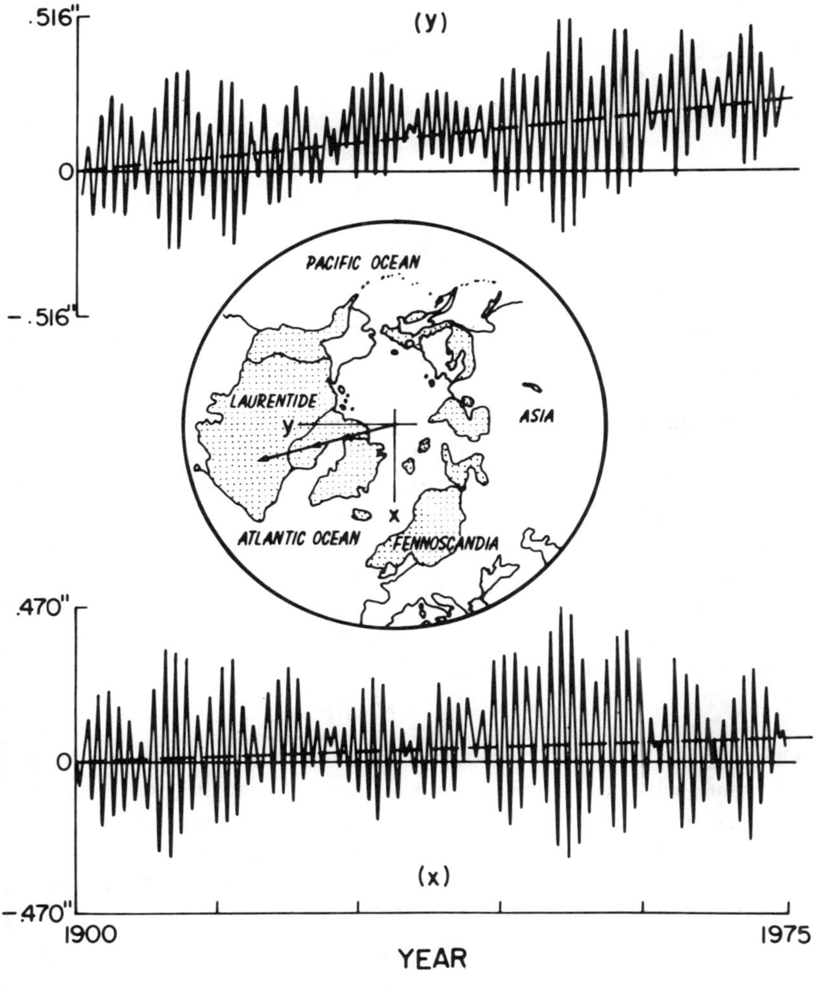

Figure 6.27. International Latitude Service (ILS) data showing the location of the rotation pole relative to the CIO (Conventional International Origin) as a function of time since AD 1900.

wander speed and direction. An application of MacCullagh's formula (e.g., Munk and McDonald, 1960) shows that:

$$I_{13}^{\mathrm{ROT}} \simeq \frac{k_2^T}{k_f} * m_1(C - A), \tag{72a}$$

$$I_{23}^{\mathrm{ROT}} \simeq \frac{k_2^T}{k_f} * m_2(C - A), \tag{72b}$$

in which the asterisk denotes temporal convolution, k_2^T is the "tidal" Love number of degree two, and $k_f = \lim_{t\to\infty} k_2^T$ is its infinite time limit. The tidal Love number as an element of the solution to the same sixth order system of ordinary differential equations (15) whose integration determines the surface load Love number $k_2{}^L$defined in (29). The difference concerns only the surface boundary conditions which the solution is constructed to satisfy. For the surface load Love numbers the boundary conditions are given in equations (17) which differ from those appropriate for the determination of the tidal Love number only in that in the latter case the normal stress at the surface Y_3 (a) is now identically zero, the system being forced only by an externally derived gravitational perturbation.

The contribution to the excitation function due to the direct effect of the load and its induced deformation $\boldsymbol{\Psi}^L$ is given in (61), in which the perturbations of inertia I_{12} and I_{23} are given by:

$$I_{ij}^L(s) = [1 + k_2^L(s)]f(s)I_{ij}^R, \tag{73}$$

where $f(s)$ is the Laplace transform of the temporal history of loading and unloading, which is assumed common to every point on the surface, while I_{ij}^R is the inertia perturbation which would be caused by this temporal variability if the earth were rigid. Thus the surface load Love number k_2^L carries the influence on the inertia perturbations of the internal redistribution of mass effected by the deformation. Using

$$\boldsymbol{\Psi} = \boldsymbol{\Psi}^{\mathrm{ROT}} + \boldsymbol{\Psi}^L \tag{74}$$

and the definition of the $\boldsymbol{\Psi}$s in terms of the I_{ij} in (62) the Laplace transform domain solution of (58) is

$$\mathbf{m}(s) = \frac{-i\sigma_r \boldsymbol{\Psi}^L(s)}{[s - i\sigma_r(1 - k_2^T(s)/k_f)]}, \tag{75}$$

with $\boldsymbol{\Psi}^L = \boldsymbol{\Psi}_1^L + i\Psi_2^L$ and

$$\Psi_1^L(s) = \frac{I_{13}(3)}{(C - A)} + \frac{[sI_{23}(s) - I_{23}(\mathrm{t} \equiv 0)]}{\Omega(C - A)}, \tag{76a}$$

$$\Psi_2^L(s) = \frac{I_{23}(s)}{(C - A)} - \frac{[sI_{13}(s) - I_{13}(t \equiv 0)]}{\Omega(C - A)}, \tag{76b}$$

Assuming the system to be unperturbed for $t \leqslant 0$ we may write

$$\Psi^L(s) = [1 + k_2^L(s)]f(s)[(\Psi_{13}^R + i\Psi_{23}^R) + \frac{s}{\Omega}(\Psi_{23}^R - i\Psi_{13}^R)], \tag{77}$$

in which

$$\Psi_{13}^R = \frac{I_{13}^R}{(C - A)} = \frac{\Omega}{A\sigma_r} I_{13}^R, \tag{78a}$$

$$\Psi_{23}^R = \frac{I_{23}^R}{(C - A)} = \frac{\Omega}{A\sigma_r} I_{23}^R. \tag{78b}$$

Expanding both k_2^L and k_2^T in terms of normal modes following Peltier (1976) we may write them as:

$$k_2^L(s) = k_2^{LE} + \sum_{j=1}^{N} \frac{r_j}{s + s_j}, \tag{79a}$$

$$k_2^T(s) = k_2^{TE} + \sum_{j=1} \frac{t_j}{s + s_j}. \tag{79b}$$

Using the notation $k_2^T(0) = k_2^T\ (s = 0)$ and $1 = 1 + k_2^L(0)$ these may be re-written as:

$$k_2^L(s) = (-1 + 1_s) - s\sum_{j=1}^{N} \frac{r_j/s_j}{(s + s_j)}, \tag{80a}$$

$$k_2^T(s) = k_2^T(0) - s\sum_{j=1}^{N} \frac{t_j/s_j}{(s + s_j)}, \tag{80b}$$

Substitution of these forms into the solution (75) and making use of the evident relation $|\sigma_0/\Omega| << 1$ reduces the Laplace transform domain solution for the forced polar motion to:

$$\mathbf{m}(s) = \frac{\Psi^L(s)}{\dfrac{s}{k_f} \displaystyle\sum_{j=1}^{N} \frac{(t_j/s_j)}{(s + s_j)}}, \tag{81}$$

with

$$\mathbf{\Psi}^L(s) = \frac{\Omega}{A\sigma_r}[1 + k_2^L(s)]f(s)[I_{13}^R + iI_{23}^R]. \tag{82}$$

The formal inversion of (81) into the time domain was described in Peltier (1982). The final form of the solution obtained there is

$$m_j = \frac{\Omega}{A\sigma_0} I_{j3}^R \left[D_1 \dot{f}(t) + D_2 f(t) + \sum_{i=1}^{N-1} E_i \frac{d}{dt}(f * e^{-\lambda_i t}) \right], \tag{83}$$

in which the constants D_1, D_2, and E_i are

$$D_1 = l_s - \sum_{j=1}^{N} \frac{r_j}{s_j} = 1 + k_2^{LE}, \tag{84a}$$

$$D_2 = -l_s q(0) \prod_{i=1}^{N-1} \lambda_i, \tag{84b}$$

$$E_i = \left[\frac{l_s q(-\lambda_i)}{\lambda_i} + \sum_{i=1}^{N} \frac{r_j R_j(-\lambda_i)}{s_j} \right] \Big/ \prod_{k \neq i}^{N-1} (\lambda_k - \lambda_i), \tag{84c}$$

where $R_j(s)$ and $q(s)$ are the following respective degree $N - 1$ and degree N polynomials:

$$R_j(s) = \prod_{i=1}^{N-1} (s + \lambda_i) - \prod_{i \neq j} (s + s_i), \tag{85a}$$

$$q(s) = s \prod_{i=1}^{N-1} (s + \lambda_i) - \prod_{i=1}^{N} (s + s_i), \tag{85b}$$

and λ_i are the roots of the degree $N - 1$ polynomial Q_{n-1} defined as

$$Q_{N-1}(s) = \sum_j g_j \left[\prod_{i \neq j} (s + s_i) \right] = \prod_{i=1}^{N-1} (s + \lambda_I), \tag{86}$$

where $g_j = (t_j/s_j) \Big/ \sum_{i=1}^{N} (t_i/s_i)$ is the relative strength of the jth mode in the decay spectrum of the tidal Love number of degree 2.

In order to evaluate the formal solution (83) we require the I_{j3}^R and $f(t)$ appropriate to a specific model of the glaciation-deglaciation process. With the same circular disk load representations of the Pleistocene ice sheets as were employed in the discussion of the non-tidal acceleration in the last subsection, I_{j3}^R are

$$I_{13}^R = \sum_{i=1}^{M} M_i a_2 \left[\frac{1}{6} \cos\alpha_i (1 + \cos\alpha_i) P_2(\cos\theta_1') \cos\phi_i' - \frac{a_{21}}{5a_{00}} \right], \tag{87a}$$

$$I_{23}^R = \sum_{i=1}^{M} M_i a_2 \left[\frac{1}{6} \cos\alpha_i (1 + \cos\alpha_i) P_2(\cos\theta_1') \sin\phi_i' - \frac{b_{21}}{5a_{00}} \right], \tag{87b}$$

where the circular ice caps have been assumed to interact with geographically realistic oceans described by the ocean function (e.g. Lambeck 1980) of which b_{21} and a_{00} are usual spherical harmonic coefficients. All of the computations which have been performed using (83) have employed the same history function $f(t)$ employed in the last subsection which consists of a 10^5 year periodic sawtooth approximation to the $\delta^{18}O$ data from deep sea sedimentary cores.

Examples of a sequence of specific polar wander speed predictions made using equations (83) are provided in Figure 6.28. For the purpose of these initial calculations each ice sheet was assumed to have undergone seven 10^5-year cycles of loading and unloading and the observed polar wander speed, to which the prediction of the model is to be compared, is denoted on the figure by the cross centered on the time of 6 kyr which is measured relative to a zero of time assumed to coincide with the end of the last model deglaciation event. For purposes of these calculations only the Laurentian, Fennoscandian, and Antarctic ice complexes have been taken to contribute to the rotational forcing and the parameters used to specify the I_{ij}^R in (87) are those listed previously in Table 6.1. On Figure 6.28 the observed polar wander speed is compared with the predictions of six models, each of which has a lithosphere of thickness $L = 120$ km but which otherwise differ from one another considerably in the nature of their internal viscoelastic stratifications. The models are all members of a sequence denoted LNF in which N denotes the number of internal radial density discontinuities in the mantle. Thus LOF (no core) is a model which not only has no mantle discontinuities but also has no core; it is therefore a model which is viscoelastically homogeneous throughout except for the presence of the lithosphere. Model LOF has a high density core and a density jump across the core mantle boundary equal to that in 1066B. L1F has a non-adiabatic density jump across the 670 km depth horizon while model L2F also includes the density jump across the 420 km seismic discontinuity.

Inspection of the comparisons on Figure 6.28 show that the predicted present day polar wander speed for the homogeneous model (LOF (no core)) is almost identically zero in accord with the previous analysis of the forced polar wander problem by Munk and Revelle (1952). It was apparently not understood by Munk and Revelle that this null result was an artifact of their having employed a homogeneous model in their analysis since they concluded that the observation of a non-zero value for the present day polar wander speed *required* some ongoing melting of ice from either Greenland and/or Antarctica. That this conclusion is unjustified is clearly demonstrated by the additional results on Figure 6.28 for radially stratified models. As first

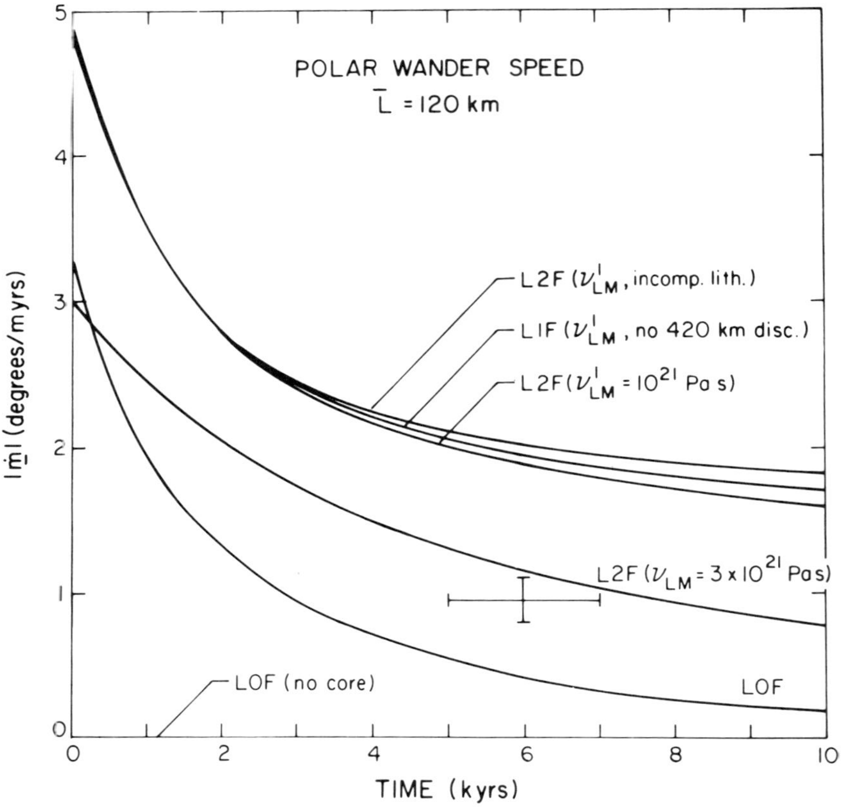

Figure 6.28. Predicted polar wander speed as a function of time for a series of multiple homogeneous layer elastic models. The zero of time is taken to be coincident with the end of the final deglaciation event of a glaciation history consisting of seven complete cycles. The polar wander speed implied by the data in Figure 6.27 is shown as the cross at a time (t = 6 kyr) such as to provide a reasonable fit of the final deglaciation event to that of the ICE-2 chronology of Wu and Peltier (1983).

pointed out in Peltier (1982), the null result for the homogeneous model is a consequence of the fact that for this model, whose viscoelastic relaxation spectrum consists of only a single decay time, the rotational forcing due to the load $\boldsymbol{\Psi}^L$, and that due to the changing rotation $\boldsymbol{\Psi}^{\mathrm{ROT}}$, exactly annihilate one another at any time during which the surface is not subject to a changing load. When a more realistic radially stratified model of the planet is employed then the symmetry which causes this cancellation in the one mode model is broken and the planets' current rotational state includes a non-zero memory of previously active rotational forcing. This memory is embodied in the third term of (83) which contains the convolution integral. Even if $f \equiv 0$, which it is at a time like the present of apparent hiatus in a previously continuous 10^5-year cycle of glaciation and deglaciation, $\dot{m}_j$ need not vanish and that it does not is explicitly established for stratified models in Figure 6.28.

The addition of a density jump across the CMB (model (LOF)) results in a marked increase in the present day speed prediction but this is still to slow, by a factor of 2, to fit the data when the viscosity of the mantle is held fixed to the Haskell value of 10^{21} Pa s throughout its volume. Adding a 670 km discontinuity to the uniform viscosity model, however, sharply increases the speed prediction by more than a factor of 3, to a level which is now in excess of the observation. Adding the second discontinuity at 420 km depth has no substantial further effect. The excess speed prediction in the uniform viscosity model with internal mantle buoyancy is nicely rectified, as illustrated by the prediction for the model L2F($\nu_{LM} = 3 \times 10^{21}$ Pa s), simply by increasing the viscosity of the lower mantle by the same factor of about 3 which was previously shown to be required by relative sea level, free air gravity, and non tidal acceleration of rotation observations. That this model also reconciles the observed direction of polar wander, which from Figure 6.27 is directed along the 76° ± 5° West meridean, is demonstrated on Figure 6.29 which shows the predicted direction for the same models L, L+F, and L+F+A as were previously employed in the analysis of the nontidal acceleration datum. Although these analyses do therefore establish that the same global model which fits this other rotational observable is compatible with the ILS inferred polar wander, further analysis in Wu and Peltier (1984) has demonstrated that the prediction is also sufficiently sensitive to lithospheric thickness that it cannot serve as a particularly powerful independent constraint upon the upper mantle–lower mantle viscosity contrast.

A further potential difficulty should be apparent on the basis of our previous remarks concerning the probable action in the system of a significant ongoing loss of mass from the small ice sheets and glaciers of the world. In the last subsection it was established that the rotational effect of the melting of Meier's (1984) sources was exactly annihilated by the late Pleistocene melting of the substantial ice cap which most probably existed over the Barents Sea. We should clearly be very concerned as to the influence which these sources might have on the polar wander datum. Their effect is quantitatively assessed in Figures 6.30 and 6.31 which should be taken as

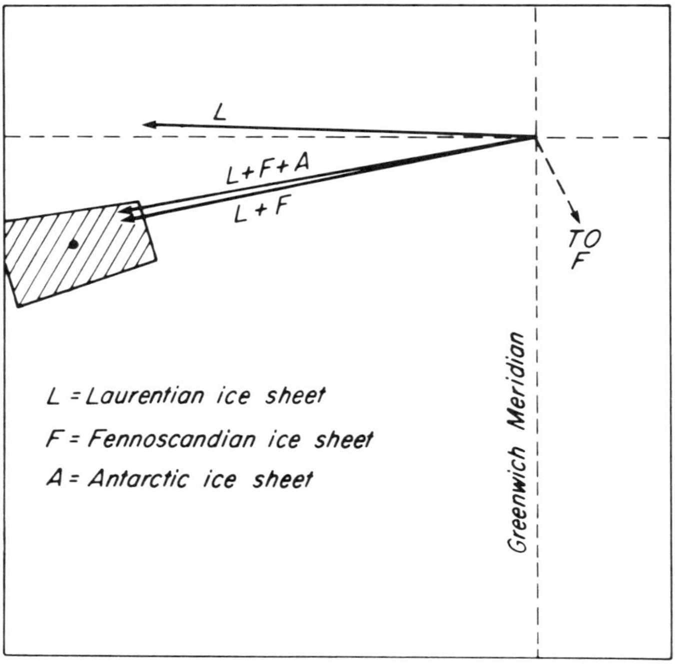

Figure 6.29. Variation of the direction and length of the polar wander velocity vector with the number of ice sheets employed in the loading history. The earth model has lithospheric thickness $\bar{L}$ = 120 km, the upper mantle viscosity equals 10^{21}Pa s, and the lower mantle viscosity equals 3×10^{21}Pa s.

parallel to the previously described Figures 6.25 and 6.26. As in those previous calculations Meier's sources are introduced into the theoretical prediction as 31 point sources of deglaciation with the strengths tabulated in Table 6.2, which combine to deliver a present day eustatic rise of about 0.45 mm per year in accord with the tide gauge observations of secular sea level rise appropriately corrected for the thermal expansion of the oceans. As in Figures 6.25 and 6.26 the polar wander speed and direction predictions (shown as the solid and dashed lines on Figures 6.30 and 6.31, respectively) are shown for two different versions of the elastic basic state parameterization (CML64 and 1066B) as a function of the lower mantle viscosity ν_{LM} with the upper mantle value fixed to 10^{21} Pa s.

Inspection of Figure 6.30 shows that although model L+F+A (result denoted by the diamonds) very nicely fits the observed polar wander direc-

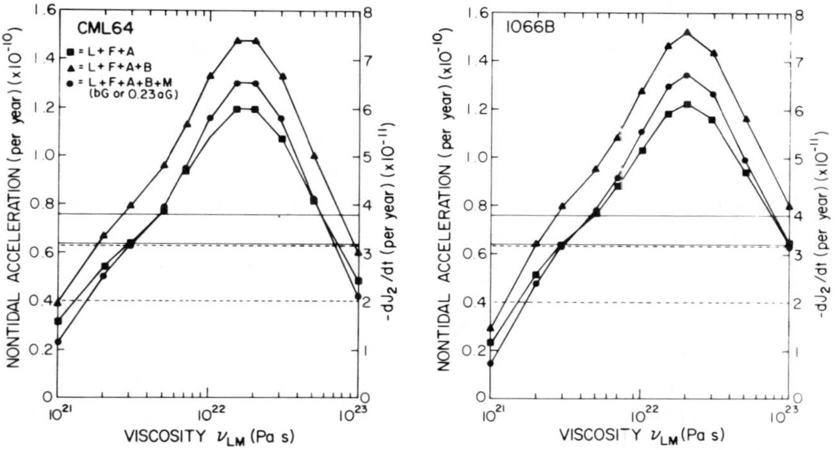

Figure 6.30. Same as Figure 6.25 for the non-tidal acceleration of rotation but including the extra rotational forcing associated with a late Pleistocene Barents Sea ice sheet.

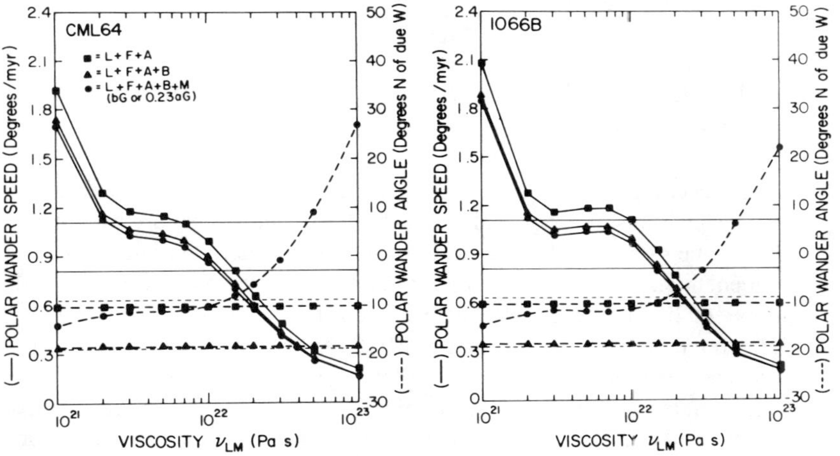

Figure 6.31. Same as Figure 6.26 for polar wander speed and direction but including the extra rotational forcing associated with a late Pleistocene Barents Sea ice sheet.

tion, when Meier's 31 sources are added to the rotational forcing the predicted direction swings markedly to the west and well outside the error bars on the observed direction. It should be noted, as above, that although the speed prediction, which is clearly rather insensitive to the presence of absence of Meier's sources, fits the observed datum only for rather high viscosity contrast, this may be rectified simply by thinning the lithosphere. The predicted direction of polar wander is especially useful in this analysis because it, unlike the speed, is relatively independent of the radial viscoelastic structure. The analyses presented in Figure 6.31 illustrate the impact upon this prediction of the presence of the same Barents Sea ice sheet as was previously shown to be required to allow the non-tidal acceleration of rotation datum to be fit by the same weak viscosity contrast model required by the relative sea level data. These analyses establish that the action of the Barents Sea ice sheet has the same salutory effect upon the predicted polar wander direction as it had on the nontidal acceleration. Its effect annihilates the misfit to the observed direction which would otherwise exist due to the action of Meier's sources. It also clearly allows the observed speed to be fit with a thicker lithosphere than would otherwise be required. It is important therefore that the independent evidence for the existence of the Barents Sea ice sheet discussed previously is so compelling. We also need this additional ice sheet in the system to improve agreement with the estimate of the total eustatic sea level rise during the last glacial–interglacial transition which is provided by sea sediment $\delta^{18}O$ measurements and corral reef observations (Chapell and Shackleton 1986).

7. Mantle Viscosity Constraints from the Nature of the Ice Age Cycle

In the preceding sections of this chapter I have demonstrated that all of the known signatures of the glacial isostatic adjustment process are well reconciled by the same radial viscoelastic structure. This structure has a viscous component such that with the upper mantle viscosity fixed to the Haskell number of 10^{21} Pas, the lower mantle viscosity is required to be somewhat higher but modestly so and near 2×10^{21} Pa s. The elastic component of the structure also plays a crucial role in allowing all of the data to be reconciled by this single viscosity profile. In particular both the free air gravity anomalies observed over present day centers of post-glacial rebound and the wander of the rotation pole with respect to the surface geography agree in their requirement that the mantle must contain significant internal buoyancy when it is deformed on the timescales of post glacial rebound if the appropriate constitutive relation is the Maxwell relation (10d). As argued in Peltier (1985) and Peltier *et al.* (1986), the existence of some transient component in the relaxation spectrum might alleviate this requirement somewhat but there does not appear to be any particularly compelling argument for the necessity of any such further complexity in the constitutive relation (see Chapters 9 and 10 for alternative views of the requirements of isostatic geoid anomalies in this regard). In this penultimate section of the present chapter I will summarize recent analyses of one additional large scale glacial isostatic adjustment related phenomenon which may be invoked to further constrain the value of the effective viscosity of the planetary mantle. This phenomenon concerns the variation of continental ice volume through the Pleistocene epoch that has been inferred through analyses of $\delta^{18}O$ data from deep sea sedimentary cores, the same data which were invoked in Section 4 of this chapter to constrain the ice volume history for use in the free air gravity and earth–rotation computations. As pointed out in the context of that discussion these data demonstrate (Fig. 6.19) that the ice volume variations have been dominated by an oscillation with characteristic period 10^5 years which typically explains over 65 percent of the variance in the data. My purpose in the present section is to demonstrate that this period is determined in large part by the viscosity of the mantle through the agency of the glacial isostatic adjustment process that apparently plays a crucial role in the physics underlying the volume oscillations.

Hays *et al.* (1976), who first published power spectra of the $\delta^{18}O$ time series extracted from deep sea cores, argued rather persuasively that the structure of their spectra could be taken as establishing the essential validity of the astronomical theory of ice ages as this had been elaborated by Milankovitch (1941), who, following Croll, Köppen and others, suggested that the ice volume oscillations were forced by the small variations of summertime seasonal solar insolation at the earths surface produced by time variations of the parameters (obliquity, eccentricity, precession) that control the geometry of the earths' orbit around the sun. The data in Figure 6.32 may be taken as providing strong support for this contention. Plate a of this figure shows the variation of summertime (caloric) seasonal insolation for 65° North Latitude over the past several hundred thousand years of earth history and plate b its power spectrum. This should be compared to the previously shown Figure 6.19 on which the SPECMAP $\delta^{18}O$ observations were presented. Clearly the astronomical forcing shown in Figure 6.32 has no power at the period of 10^5 years whereas the dominant cryospheric oscillation has precisely this period. In spite of this discrepancy at the dominant 10^5 year

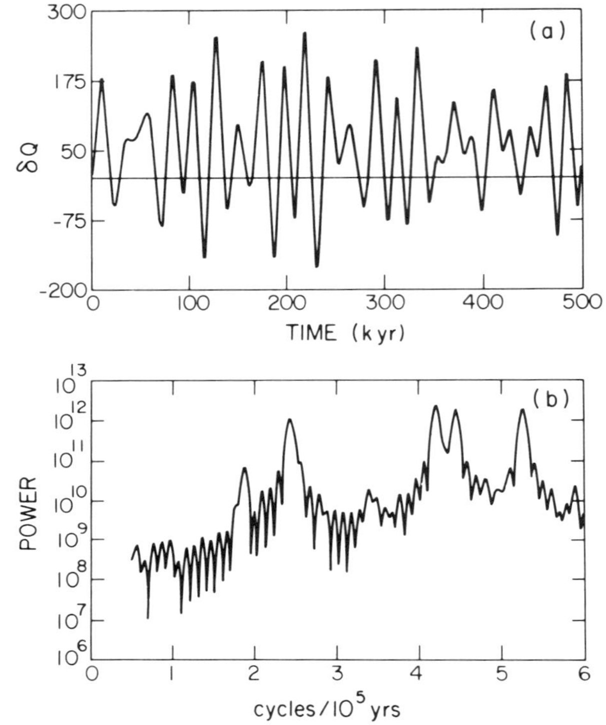

Figure 6.32. a, Summertime seasonal insolation anomaly for 65°N latitude computed from the formulae in Berger (1978). b, Power spectrum of the summertime seasonal insolation anomaly.

period it is nevertheless obvious that ice volume does respond very sensitively to the astronomical forcing since all of the spectral lines which are present in the astronomical input are also present in the output. Since the input summertime seasonal insolation anomaly ΔQ_s (e.g., see Berger, 1978) has the mathematical form

$$\Delta Q_s = \Delta R \Delta \varepsilon + m \Delta [e \sin \tilde{\omega}], \tag{88}$$

with

$$m = \frac{2TS \cos \phi}{\pi^2 (1 - e^2)^{1/2}},$$

in which T is the duration of the tropical year, S the solar constant, and R_s is the total insolation received at the top of the atmosphere over the caloric summer, and where ε is the orbital obliquity with period 41 kyrs, e is the orbital eccentricity with periods 100 and 400 kyrs and $\tilde{\omega}$ is the longitude of the perehelion with respect to the moving vernal equinox with period near 22 kyrs, it is clear that the only effect of the eccentricity variation is to modulate the amplitude of the precessional effect. It is this modulation which causes the splitting of the single precessional line into the triplet of lines in the precessional band. Because the multiple peaks in this band very closely duplicate the corresponding peaks in the SPECMAP series, it would appear that the climate system is indeed responding, through the variation of continental ice volume, to the summertime seasonal insolation anomaly as hypothesized by Köppen. However, the dominant response at a period of 10^5 years clearly presents an enigma. If the climate system were linear, this component of the response should not be present at all. In the remainder of this section I will summarize the results of recent analyses which have culminated in the development of a nonlinear model which appears to explain the reason for the appearance of the 10^5 year cycle in the $\delta^{18}O$ record of Pleistocene climatic change.

A schematic illustration of this new climate model is shown in Figure 6.33. The model contains two active and explicitly modeled ingredients, namely an ice sheet and the earth beneath it. The ice sheet is assumed to consist of a zonal ring of ice surrounding a polar ocean. Expansions and contractions of the ice sheet are assumed to occur in response to the small variations of summertime seasonal insolation associated with the above mentioned variations of $\tilde{\omega}$, ε, and e. This astronomical forcing is introduced into the model using a climate point-climate surface scheme of the sort first proposed by Weertman (1976) in his steady state analysis of the factors determining the size of ice age ice sheets. In the absence of surface ice the climate point is defined as the southern boundary of the perennial snow field. Fluctuations δQ_s are assumed to cause this climate point to migrate in the north–south direction by a distance $\delta x = \alpha \delta Q_s/(dQ/dx)$ where dQ/dx is the present day north-south insolation gradient and α is a meteorological amplification factor associated principally with ice albedo feedback. Energy balance analyses by North *et al.* (1983) suggest $1 < \alpha < 3$ with a preferred value near 2. In the presence of an ice sheet, initially produced by the buildup of snow north of the climate point, after it has made an excursion onto land, the boundary separating the accumulation area from the ablation area (see Fig. 6.33) is located considerably to the south of the climate point. This is essentially because temperature decreases with increasing atmospheric

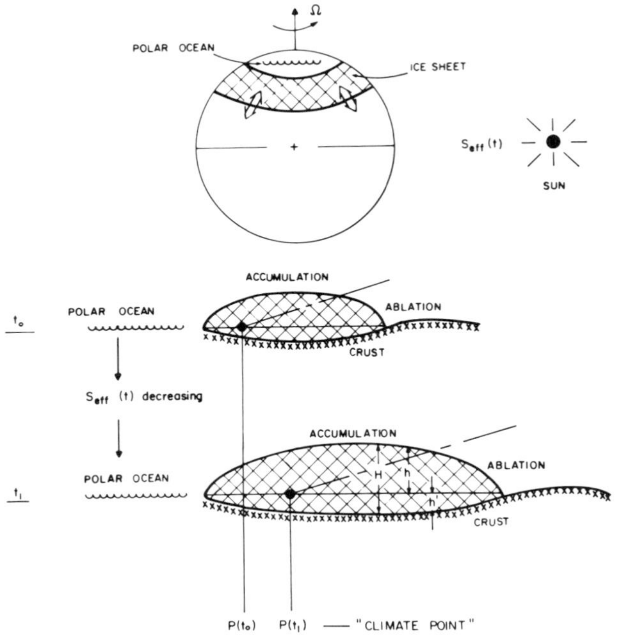

Figure 6.33. Schematic diagram illustrating the ingredients of the new paleo climate model.

height so that an increase of elevation of the ice sheet pushes the critical isotherm separating accumulation from ablation further south. In the model it is assumed that the slope of the climate surface is identical with the slope of the 0°C isotherm in a standard atmosphere, that is $s = .0067$ km/rad.

All of the meteorological and hydrological processes which determine the precipitation and melting fields are assumed to be parameterizable in the form:

$$A(\theta,t) = \max\{a[1-bh(\theta,t)],0\}, \text{ above the climate surface}$$
$$= \min\{a'[1-bh(\theta,t)],0\}, \text{ below the climate surface.} \quad (89)$$

The field $h(\theta,t)$ in (89) is the instantaneous elevation of the ice sheet above sea level (see Fig. 6.33), a (> 0) and a' (< 0) are respectively the sea level accumulation and ablation rates, and b is an extremely important parameter which describes the rate of decrease of accumulation and ablation rates with increasing topographic height. This describes the so-called "elevation desert" effect. The parameter b which controls its strength is estimated on the basis of the observed rate of decrease of the saturation mixing ratio of water in air as a function of height in a standard atmosphere. Because of this dependence of A upon the ice sheet topography h, it is quite clear that any process which contributes to changes of the elevation of the ice sheet above sea level will have important dynamical consequences. In this model it is hypothesized that the only one of such effects which is important is the effect of glacial isostatic adjustment. That this effect should be important must be clear on the basis of the simple observation that the deflections of the Earth's surface caused by this process are such as to allow the buoyant restoring force thereby induced to cancel the weight of the load. Thus the equilibrium deflections are approximately equal to one-third the maximum ice thickness, where this fraction is determined as the ratio of the density of ice to that of rock. For ice sheets 3–4 km thick the deflections are therefore of order 1 km which from (2) given b is sufficient to induce significant variations of A. In Figure 6.33 the deflection of the Earth's surface is labelled $h'(\theta,t)$.

The mathematical structure of the dynamical model which links the evolutionary fields $h(\theta,t)$ and $h'(\theta,t)$ is based upon simple conversation laws and was originally derived in Peltier (1982). The two dimensional ice dynamics is governed by the vertically integrated continuity equation for ice mass which has the form:

$$\frac{\partial H}{\partial t} = -\nabla\cdot\mathbf{U} + A(\theta,t), \tag{90}$$

where $H = h + h'$ is the ice thickness and the ice flux $\mathbf{U}$ is determined by the Glen flow law as:

$$\mathbf{U} = \frac{\tilde{c}}{5}\left[\frac{\rho g}{a}\frac{\partial h}{\partial\theta}\right]^3 H^5\boldsymbol{\theta}, \tag{91}$$

in which $\tilde{c}$ is an empirical constant (Birchfield and Weertman, 1981), ρ is the density of ice, g the gravitational acceleration, a the Earth's radius, and $\boldsymbol{\theta}$ a unit vector in the direction of increasing colatitude. Equation (91) may clearly be rewritten in the form of a nonlinear diffusion equation as

$$\frac{\partial h}{\partial t} = \frac{1}{\alpha\sin\theta}\frac{\partial}{\partial\theta}\left[\sin\theta\ K(\theta,t)\frac{\partial h}{\partial h\theta}\right] - \frac{\partial h'}{\partial t} + A(\theta,t), \tag{92}$$

in which the diffusion coefficient $K(\theta,t)$ is given by:

$$K = \frac{\hat{c}}{5}\left[\frac{\rho g}{a}\right]^3\left[\frac{\partial h}{\partial\theta}\right]^2 H^5(\theta). \tag{93}$$

As discussed in detail in Peltier (1982), a separate evolution equation for the field $h'(\theta,t)$ may be obtained from the theory of glacial isostasy which delivers a prediction of $h'(\theta,t)$ in the form of a convolution of the ice thickness $H(\theta,t)$ with the viscoelastic Green function for radial displacement u given previously in (35). Explicitly this is:

$$h'(\theta,t) = \iiint d\Omega'\, dt'\, \rho H(\theta',t')u(\theta/\theta',t/t'). \qquad (94)$$

Theory reviewed in Section 2 of this chapter shows that u may be represented in the form of the normal mode expansion,

$$u(\theta/\theta',t/t') = \frac{a}{m_e}\sum_{l=0}^{\infty} q_l(t-t')P_l(\cos\gamma), \qquad (95)$$

where the surface load Love numbers q_l have the Dirichlet expansions,

$$q_l(t) = \sum_{j=1}^{M} r_j^{lt} + q_l^E\delta(t), \qquad (96)$$

as in (30) where we are now using q rather than h to represent the radial displacement Love number since h in the present section is used to denote the ice sheet topography with respect to sea level.

All of the results which have been obtained to date with this model have been extracted on the basis of the assumption that an adequate approximation to the isostatic adjustment process could be derived with $M = 1$ and the complete neglect of the elastic component of the response determined by q_l^E. Under these assumptions Peltier (1982) showed that (92) and (94) reduce to the following double infinity of coupled evolution equations for the individual spherical harmonic contituents of the fields $h(\theta,t)$ and $h'(\theta,t)$:

$$\frac{\partial h_l}{\partial t} = B_{lmn}K_m h_n - C_l r^l h_l + (s^l - C_l r^l)h'_l + A_l, \qquad (97a)$$

$$\frac{\partial h'_l}{\partial t} = C_l r^l h_l - (s^l - C_l r^l)h'_1. \qquad (97b)$$

In (97) the numbers h_l and h'_l are the Legendre coefficients in the spherical harmonic expansions of the h and h' fields, $C_l = 3\rho/(2l+1)\rho_e$ where ρ_e is the average density of that part of the mantle which is involved in the isostatic adjustment process, and the B_{lmn} are spectral interaction coefficients defined as:

$$B_{lmn} = -\frac{(2l+1)}{2a}\int_{-1}^{+1}(1-x^2)\frac{\partial P_l}{\partial x}P_m\frac{\partial P_n}{\partial x}dx, \qquad (98)$$

where P_l is the Legendre polynomial of degree l and x is its argument. When the expansions for h and h' are truncated at a finite number of terms, 48 for the solutions to be described below with l restricted to even values, the resulting system of equations can be integrated forward from prescribed initial conditions to produce a synthetic history of ice volume fluctuations which may be compared directly to the $\delta^{18}O$ data. The main results which have been obtained to date from such analyses will be summarized briefly here.

Figure 6.34 shows the result obtained from an initial integration of the system when the astronomical forcing is approximated by a single sinusoidal oscillation with a period of 20,000 years which is intended to mimic the precessional component of the actual insolation changes. Plates a, b, and c of this figure (from Peltier and Hyde, 1984) respectively show the synthetic ice volume history, its power spectrum, and the cross-correlation vector between

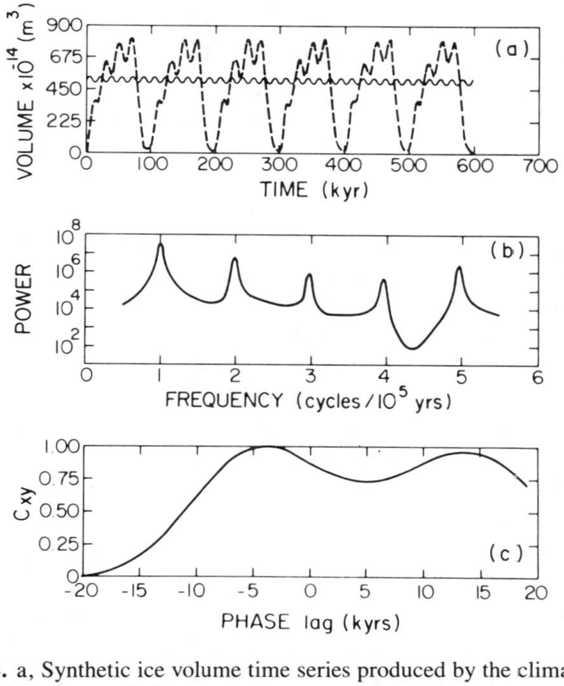

Figure 6.34. a, Synthetic ice volume time series produced by the climate model in response to 20 kyr harmonic forcing. b, Power spectrum of the synthetic in a. c, Cross correlation vector between the 20 kyr harmonic input and the synthetic ice volume output in a.

the input 20,000 year sinusoidal forcing and the output ice volume record. The former shows that the response of the model is dominated by an oscillation of period 10^5 years and therefore establishes that the nonlinear dynamical system is fully capable of reconciling the existence of the strong spectral line at this period in the SPECMAP record even when there is precisely zero forcing at this period. For simple harmonic forcing Hyde and Peltier (1985) have provided an exhaustive analysis of the conditions under which the model will continue to deliver this result. Their calculations established that one of the most important parameters in the model is that labelled b in equation (89) which controls the strength of the feedback between accumulation rate and ice sheet topographic height. Of equal importance, as we will shortly see, is the viscosity of the mantle which governs the rate of relaxation under the weight of the ice and thus the timescale on which glacial isostasy may contribute to the mass balance by changing ice sheet topographic height.

The pulse shape of the 10^5 year oscillation which this climate model supports is strongly suggestive of a "relaxation oscillation" of the sort which occurs in a simple dc circuit consisting of a resistor and a capacitor in series with a source of emf when a discharge tube is connected across the capacitor and set to fire when the voltage approaches its saturation value. The nonlinear physics in the climate model which is analogous to the discharge tube is discussed in detail in Hyde and Peltier (1985) where it is shown that the collapse mechanism associated with a model "termination" onsets when a sufficiently large ice sheet begins to melt back into the low warm Earth depression created by the isostatic adjustment process. This process causes a sharp increase in slope of the southern face of the ice sheet and thus from (91) a dramatic increase of ice flux which is so marked that the entire ice mass is exhausted into the depression on a timescale which is near one-half the precessional cycle (i.e., 10,000 years).

That this oscillation is not an artifact of the approximation of the astronomical forcing by a simple sinusoidal oscillation is demonstrated in Figure 6.35 which shows the synthetic ice volume time series delivered by the model when it is subject to realistic astronomical forcing (which is shown above the synthetic ice volume record in plate (7) for phase reference). The forcing employed for this calculation was that appropriate to 65°N latitude. Plate b of this figure shows the power spectrum of the response which demonstrates that all five of the lines present in the insolation forcing are indeed present in the output which nevertheless remains dominated by the relaxation oscillation with period near 10^5 years. Plate c shows an overlay of the synthetic ice volume prediction with the SPECMAP record, demonstrat-

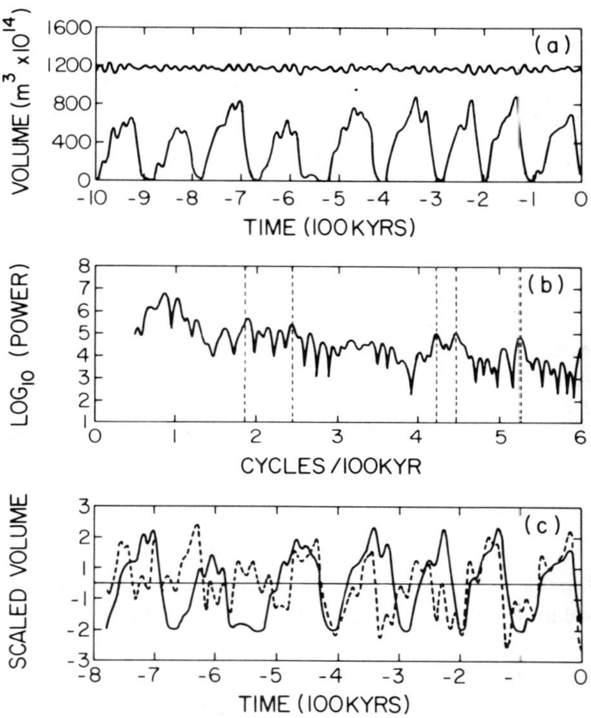

Figure 6.35. a, Synthetic ice volume time series produced by the climate model in response to realistic Milankovitch forcing consisting of the summertime seasonal insolation anomaly for 65°N latitude. b, Power spectrum of the synthetic which should be compared to the power spectrum of the SPECMAP record shown in Figure 6.4 (a). c, Overlay of the synthetic ice volume record (solid) with SPECMAP (dashed).

ing that in spite of its simplicity the model does deliver a rather successful simulation of the actual $\delta^{18}O$ record. A detailed account of such realistic experiments will be found in Hyde and Peltier (1987).

For present purposes the main interest in the phenomenon concerns the extent to which it may be invoked to further constrain the value of mantle viscosity. To this end Figure 6.36 provides explicit evidence of the crucial role which the timescale of glacial isostatic adjustment plays in the 10^5 year oscillation by comparing the ice volume histories delivered by models with infinite viscosity ($\nu = \infty$) with those delivered when $\nu = 10^{21}$Pa s for both

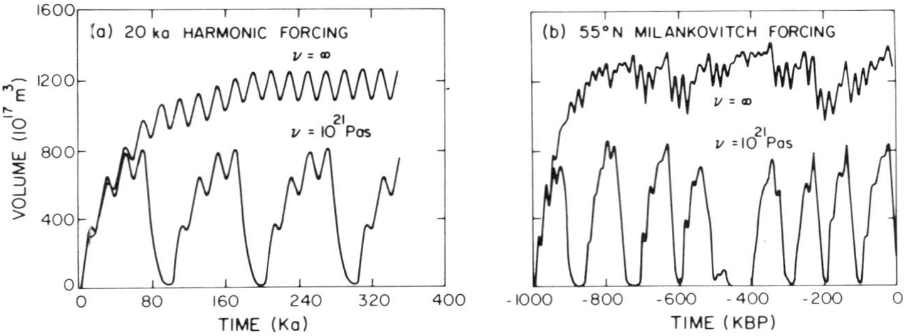

Figure 6.36. a, Compares the synthetic ice volume history predicted by a harmonically forced model in which the isostatic adjustment process is governed by a realistic viscosity of 10^{21}Pa s with that for the same model when isostatic adjustment is suppressed by taking the mantle viscosity to be infinite. b, Same as plate a but for a realistic Milankovitch experiment.

harmonic and astronomically realistic forcing (plates a and b, respectively). and then fluctuates around this glaciated state in a way which mimics the applied insolation variations. Deglaciation never occurs. Figure 6.37 demonstrates the sensitivity of the period of the relaxation oscillation to small changes in the viscosity of the mantle for a sequence of harmonically forced experiments. These data establish the extreme variations of the period of the relaxation oscillation which are caused by changes of the viscosity of the mantle by even a factor of 2 away from the value of 10^{21} Pa s which were shown to be required by all of the previously discussed signatures of the isostatic adjustment process. Lowering the viscosity decreases the predicted period of the ice age cycle whereas increasing the viscosity increases the period. This final example should suffice to establish the rather extensive set of geophysical observations which can be brought to bear on the issue of mantle viscosity and to demonstrate the consistency of the results which are obtained by applying them to infer its value.

8. Conclusions

The radial profile of this parameter which is required to fit the totality of the observational constraints described above is most tightly constrained by relative sea level data in the age range 0–18 kyr. These data require the existence of a lithosphere near the Earth's surface which has essentially

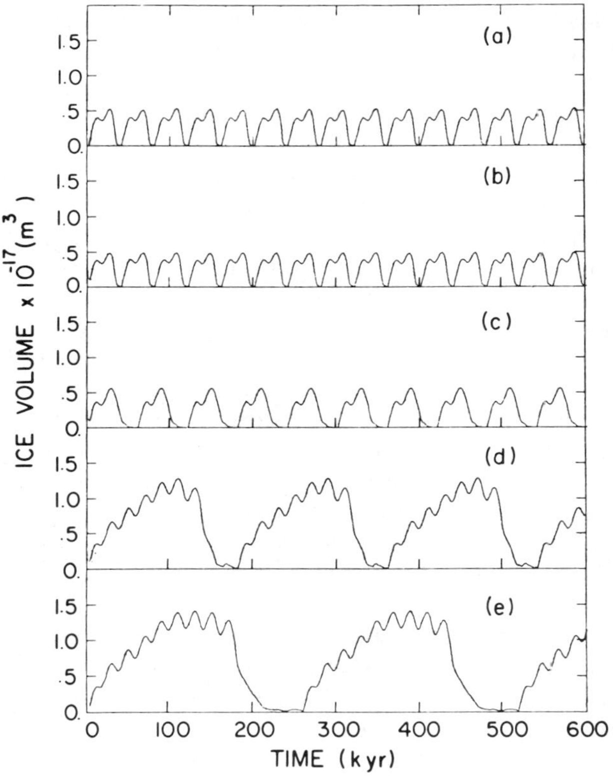

Figure 6.37. A sequence of ice volume histories predicted by the model when it is subject to 20 kyr simple harmonic forcing for different choices of the viscosity which governs the isostatic adjustment process. ν_{ref} is the viscosity required by the geophysical evidence discussed previously, i.e., 10^{21}Pa s.

infinite viscosity and thickness varying from a few tens of kilometers near mid-oceanic ridges to somewhat greater than 200 km for continental cratons. Beneath the lithosphere the upper mantle is required to have a viscosity near the 10^{21} Pa s value which is demanded by the boundary layer theory for convection at high Rayleigh number when this is fit to plate tectonic observables. In the lower mantle beneath the depth of the 670 km seismic discontinuity the viscosity is required by the sea level data to be somewhat higher but very modestly so and near a value of 2×10^{21}Pa s. Free air gravity anomaly observations agree with this requirement of ^{14}C controlled RSL histories only if the density discontinuities at 420 and 670 km depth which bracket the transition zone are assumed to behave non-adiabatically on the timescale of postglacial rebound. Even if these discontinuities are equilibrium phase transitions, as might be assumed on the basis of the high pressure diamond anvil and shock wave data discussed in Chapter 4, we cannot rule out the possibility of such behavior because, to the extent that the transitions are close to univariant the timescale for migration of the phase boundary in response to the pressure variations associated with the changing surface load will occur on the timescale of thermal diffusion which is long compared to the period of the ice age cycle. The rotation data fully concur on the viscosity profile determined from RSL histories. The non-tidal acceleration of rotation datum is especially important because it is relatively insensitive to the presence or absence of plausible internal buoyancy in the mantle and to lithospheric thickness. Even when expected modern sources of ice sheet melting are fully accounted for these data require an upper mantle-lower mantle viscosity contrast which is very close to that inferred from RSL observations. The ILS observation of the present day rate and direction of polar wander further corroborates the validity of this viscosity profile in the context of a Maxwell parameterization of the viscoelastic structure. This datum also appears to require the presence of internal buoyancy in the interior and so complements the original argument for the importance of this effect which was based upon free air gravity data. Finally, in the last section of this chapter we showed that not only is the viscosity of the mantle crucial to understanding the dynamic response of the planet to the last deglaciation event, it is also crucial to understanding why the Earth has been subjected to the 10^5 year cycle of glaciation and deglaciation which has characterized the Pleistocene period of earth history.

9. References

Backus, G.E., 1967. Converting vector and tensor equations to scalar equations in spherical co-ordinates, *Geophys. J. Roy. Astr. Soc.* **13**, 71–103.

Balling, N., 1980. The Land uplift in Fennoscandia, gravity field anomalies and isostasy. In N.-A. Mörner (ed.) *Earth Rheology, Isostasy, and Eustasy*. New York: Wiley, pp. 297–321.

Barnett, T.P., 1983. Recent changes in sea level and their possible causes, *Clim. Change* **5**, 15–38.

Berger, A.L., 1978. Long term variations of daily insolation and Quaternary climatic changes, *J. Atmos. Sci.* **35**, 2363–2367.

Birchfield, G.E., Weertman, J., and Lunde, A.T., 1981. A paleoclimatic model of Northern Hemisphere ice sheets, *Quat. Res. (N.Y.)* **15**, 126–142.

Carter, W.E., Robertson, D.S., Pyle, T.E., and Diamante, J., 1986. The application of geodetic radiointerferometric surveying to the monitoring of sea level, *Geophys. J. Roy. Astr. Soc.* **87**, 3–13.

Cathles, L.M., 1975. *The Viscosity of the Earth's Mantle*. Princeton, N.J.: Princeton University Press.

Chappell, J. and Shackleton, N.J., 1986. Oxygen isotopes and sea level, *Nature* **324**, 137–140.

Currot, D.R. 1966. Earth deceleration from ancient solar eclipses, *Astron. J.* **71**, 264–269.

Denton, G.H. and Huges, T.J., 1981. *The Last Great Ice Sheets*. New York: John Wiley and Sons.

Dicke, R.H., 1966. The secular acceleration of the earth's rotation and cosmology. In B.G. Marsden and A.G.W. Cameron (eds.)*The Earth-Moon System*. New York: Plenum Press, pp. 98–163.

Dicke, R.H., 1969. Average acceleration of the earth's rotation and the viscosity of the deep mantle, *JGR, J. Geophys. Res.* **74**, 5895–5902.

Dickman, S.R., 1977. Secular trend of the earth's rotation pole: Consideration of motion of the latitude observatories. *Geophys. J. Roy. Astron. Soc.* **57**, 41–50.

Durham, W.B. and Goetze, C. 1977. Plastic flow of oriented single crystals of olivine. 1. Mechanical data. *JGR, J. Geophys. Res.* **82**, 5737–5754.

Dziewonski, A.M. and Anderson, D.L., 1981. Preliminary reference earth model, *Phys. Earth Planet. Int.* **25**, 297–356.

Farrell, W.E. and Clark, J.A., 1976. On postglacial sea level, *Geophys. J. Roy. Astr. Soc.* **46**, 647–667.

Gilbert, R. and Dziewonski, A.M., 1975. An application of normal mode theory to the retrieval of structural parameters and source mechanisms from seismic spectra, *Philos. Trans. R. Soc., London, Ser. A* **278**, 187–269.

Gornitz, V., Lebedeff, L., and Hansen, J., Global sea level trend in the past century, *Science* **215**, 1611–1614.

Hansen, J.D., Johnson, D., Lacis, A., Lebedeff, S., Lee, P., Reid, D., and Russell, G., 1981. Climate impact of increasing atmospheric carbon dioxide, *Science* **213**, 957–966.

Haskell, N.A., 1935. The motion of a viscous fluid under a surface load, 1. *Physics* (N.Y.) **6**, 265–269.

Haskell, N.A., 1936. The motion of a viscous fluid under a surface load, 2. *Physics* (N.Y.) **7**, 56–61.

Haskell, N.A., 1937. The viscosity of the asthenosphere, *Am. J. Sci.* **33**, 22–28.

Hays, J.D., Imbrie, J., and Shackleton, N. J., 1976. Variations in the earth's orbit: Pacemaker of the ice ages, *Science* **194**, 1121–1132.

Hyde, W.T. and Peltier, W.R., 1985. Sensitivity experiments with a model of the ice age cycle: The response to harmonic forcing, *J. Atmos. Sci.* **42**, 2170–2188.

Hyde, W.T. and Peltier, W.R., 1987. Sensitivity experiments with a model of the ice age cycle: The response to Milankovitch forcing, *J. Atmos. Sci.,* in press.

Imbrie, J., Shackleton, N.J., Pisias, N.G., Morely, J.J., Prell, W.L., Martinson, D.G., Hays, J.D., McIntyre, A., and Mix, A.C., 1984. The orbital theory of Pleistocene climate: Support from a revised chronology of the marine $\delta^{18}O$ record. In A. Berger, J. Imbrie, J. Hays, G. Kukla, and B. Saltzman (eds.),*Milankovitch and Climate*. Dordrecht: D. Reidel.

Jarvis, G.T. and Peltier, W.R., 1982. Mantle convection as a boundary layer phenomenon. *Geophys. J. Roy. Astron. Soc.* **68**, 389–427.

Jeffreys, H., 1972. Creep in the earth and planets, *Tectonophysics* **13**, 569–581.

Kohlstedt, D.L. and Goetze, C. 1974. Low stress and high temperature creep in olivine single crystals. *JGR, J. Geophys. Res.* **79**, 2045–2051.

Lambeck, K., 1977. Tidal dissipation in the oceans: Astronomical, geophysical and oceanographic consequences, *Philos. Trans. R. Soc. London Ser. A* **287**, 545–594.

Lambeck, J., 1980. *The Earth's Variable Rotation: Geophysical Causes and Consequences*. New York, London: Cambridge University Press.

Libby, W.F., 1952. *Radiocarbon Dating*. Chicago: University of Chicago Press.

Litherland, A. E., 1980. Ultrasensitive mass spectrometry with accelerators, *Annu. Rev. Nucl. Part. Sci.* **30**, 437–473.

Longman, I. M., 1963. A Greens' function for determining the deformation of the earth under surface mass loads. 2. Computations and numerical results, *J. Geophys. Res.* **68**, 485–496.

Love, A.E.H., 1911. *Some Problems of Geodynamics*. New York: Dover.

Mareschal, J-C. and Gangi, A.F., 1977. Equilibrium position of phase boundary under horizontally varying surface loads, *Geophys. J. Roy. Astr. Soc.* **49**, 757–772.

Meier, Mark F., 1984. Contribution of small glaciers to global sea level, *Science* **226**, 1418–1421.

Milankovitch, M., 1941. Canon of Insolation and the Ice Age Problem, (translated from the German), U.S. Department of Commerce, Israel Program for Scientific Translations (distr. by Davey, Hartford, Connecticut, 1969).

Minster, J.B., 1978. Transient and impulse responses of a one-dimensional linearly attenuating medium. II. A parametric study. *Geophys. J. Roy. Astr. Soc.* **52**, 503–524.

Müller, P.M., 1975. An analysis of the ancient astronomical observations with the implications for geophysics and cosmology, Ph.D. Thesis, University of Newcastle-upon Tyne.

Müller, P.M. and Stephenson, F.R., 1975. The acceleration of the earth and moon from early observations. In G.D. Rosenburg and S.K. Runcorn (eds.), *Growth Rhythms and History of the Earth's Rotation*. New York: Wiley, pp. 459–534.

Munk, W.H. and MacDonald, G.F., 1960. *The Rotation of the Earth*. London, New York: Cambridge University Press.

Munk, W.H. and Revelle, R., 1952. On the geophysical interpretation of irregularities in the rotation of the earth, *Mon. Not. R. Astr. Soc., Geophys. Supp.* **6**, 331–347.

North, G.R., Mengel, J.G., and Short, D.A., 1983. Simple energy balance model resolving the seasons and the continents: Applications to the astronomical theory of the ice ages, *J. Geophys. Res.* **88**, 6576–6586.

O'Connell, R.J., 1971. Pleistocene glaciation and the viscosity of the lower mantle, *Geophys. J. Roy. Astr. Soc.* **23**, 299–327.

Peltier, W.R., 1974. The impulse response of a Maxwell earth. *Rev. Geophys. Space Phys.* **12**, 649–669.

Peltier, W.R., 1976. Glacial isostatic adjustment. II. The inverse problem, *Geophys. J. Roy. Astr. Soc.* **46**, 669–706.

Peltier, W.R., 1980. Mantle convection and viscosity. In A.M. Dziewonski and E. Boschi (eds.), *Physics of the Earth's Interior*. Amsterdam: North Holland Publ., pp. 362–431.

Peltier, W.R., 1987a. Lithospheric thickness, Antarctic deglaciation history, and ocean basin discretization effects in a global model of post glacial sea level changes, *Quat. Res.*, in press.

Peltier, W.R. and Andrews, J.T., 1976. Glacial isostatic adjustment. I. The forward problem, *Geophys. J. Roy. Astr. Soc.* **46**, 605–646.

Peltier, W.R., Farrell, W.E. and Clark, J.A., 1978. Glacial isostasy and relative sea level: A global finite element model, *Tectonophysics* **50**, 81–110.

Peltier, W.R., 1982. Dynamics of the ice age earth, *Advances in Geophysics* **24**, 1–146.

Peltier, W.R., 1983. Constraint on deep mantle viscosity from LAGEOS acceleration data, *Nature* **304**, 434–436.

Peltier, W.R., 1984. The thickness of the continental lithosphere, *J. Geophys. Res.* **89**, 11303–11316.

Peltier, W.R. and Wu, P., 1983. Continental lithospheric thickness and deglaciation induced true polar wonder, *Geophys. Res. Lett.* **10**, 181–184.

Peltier, W.R., 1985a. The LAGEOS constraint on deep mantle viscosity: Results from a new normal mode method for the inversion of viscoelastic relaxation spectra, *J. Geophys. Res.* **90**, 9411–9421.

Peltier, W.R., 1985b. New constraints on transient lower mantle rheology and internal mantle buoyancy from glacial rebound data, *Nature* **318**, 614–617.

Peltier, W.R., 1985c. Mantle convection and viscoelasticity, *Ann. Rev. Fluid Mech.* **17**, 561–608.

Peltier, W.R., 1986a. Deglaciation induced vertical motion of the North American continent and transient lower mantle rheology, *J. Geophys. Res.* **91**, 9099–9123.

Peltier, W.R., 1986b. Slow changes in the earths' shape and gravitational field: Constraints on the glaciation history and internal viscoelastic stratification, in *Space Geodesy and Geodynamics*, Anderson, A. J., and Casanave, A., eds. pp. 75–109.

Peltier, W.R., Drummond, R.A., and Tushingham, A. M., 1986. Postglacial rebound and transient lower mantle rheology, *Geophys. J. Roy. Astr. Soc.* **87**, 79–116.

Peltier, W.R. and Hyde, W.T., 1984. A model of the ice age cycle. In A. Berger, J. Imbrie, J. Hays, G. Kukla, and B. Saltzman (eds.), *Milankovitch and Climate*. Dordrecht: D. Reidel.

Platzman, G.W., 1971. Ocean tides and related waves. In W.H. Reid (ed.), *Mathematical Problems in the Geophysical Sciences*. Providence, R.I.: AMS, pp. 239–291.

Rubincam, D.P., 1984. Postglacial rebound observed by LAGEOS and the effective viscosity of the lower mantle, *J. Geophys. Res.* **89**, 1077–1087.

Shackleton, N.J., 1967. Oxygen isotope analyses and Pleistocene temperatures re-addressed, *Nature* (London) **215**, 15–17.

Walcott, R.I., 1970. Isostatic response to the loading at crust in Canada, *Can. J. Earth Sci.* **7**, 716–727.

Walcott, R.I., 1980. Rheological models and observational data of glacio-isostatic rebound. In N.-A. Mörner (ed.), *Earth Rheology, Isostasy, and Eustasy*. New York: Wiley, pp. 3–10.

Weertman, J., 1976. Milankovitch solar radiation variation and ice age ice sheet size, *Nature* (London) **261**, 17–20.

Wu, P. and Peltier, W.R., 1983. Glacial isostatic adjustment and the free air gravity anomaly as a constraint on deep mantle viscosity, *Geophys. J. Roy. Astr. Soc.* **74**, 377–449.

Wu, P. and Peltier, W.R., 1984. Pleistocene deglaciation and the earth's rotation: A new analysis, *Geophys. J. Roy. Astr. Soc.* **76**, 753–791.

Wu, P. and Peltier, W.R., 1984. Pleistocene deglaciation and the earth's rotation: A new analysis, *Geophys. J. Roy. Astr. Soc.* **76**, 753–791.

Yoder, C.F., Williams, J.G., Dickey, J.O. Schultz, B.E., Eanes, R.J., and Tapley, B.D., 1983. Secular variation of earth's gravitational harmonic J_2 coefficient from Lageos and nontidal acceleration of earth rotation, *Nature* **303**, 757–762.

Yuen, D.A. and Peltier, W.R., 1982. Normal modes of the viscoelastic earth, *Geophys. J. Roy. Astr. Soc.* 495–526.

CHAPTER 7

Convection Models and Geophysical Observations

G.T. Jarvis[†] and W.R. Peltier[‡]

[†]York University
North York, Ontario, Canada

[‡]University of Toronto
Toronto, Ontario, Canada

To gain an understanding of the thermal and dynamic state of the Earth's mantle we must combine the basic physics of convection, outlined in Chapter 2, with a knowledge of the geophysical and geochemical structure of the mantle as reviewed in Chapters 3 through 6. This is most successfully accomplished by studying idealized models of mantle convection, using the geophysical observations to provide the constraints which successful models must satisfy. In this chapter we shall discuss recent and current models of mantle convection, their implications for mantle dynamics and their limitations.

1. The Philosophy of Convection Modelling

There is no uniqueness principle for models of mantle convection; different models incorporating different physical processes may each satisfy the observational constraints. Thus even "successful" models may be ambiguous and we should not expect that they will reproduce mantle dynamics in detail. Moreover all models incorporate simplifying conditions or assumptions which prevent them from replicating the complex geodynamic behavior of the Earth's interior. These simplifications are necessary to render the model studies feasible; but they are not completely undesirable. It is only through the study of simple models that we gain a basic understanding of the physical processes involved. If all of the complexities of the Earth's mantle were included in one convection model, the model might simulate the Earth's behavior but would lead to little new understanding.

The basic philosophy of the modelling approach is to isolate the various processes in idealized models and assess their relative importance. Once these are understood, more complex models are developed to study the interactions of the individual effects. We do not attempt to simulate the mantle but rather to understand the governing physics. Thus we draw the following distinction between the terms simulation and model: a simulation is designed to reproduce a particular behavior, whereas a model is designed to reveal the processes responsible for that behavior. Each new modelling study adds to our cumulative understanding of the physics of convection in the mantle. From these models we can extract general implications which may be related to large scale surface features of the planet.

2. Application of Models to the Earth

Laboratory and numerical convection studies are specific examples of the physics described by the general hydrodynamic equations discussed in Chapter 2. These equations include a large number of thermodynamic parameters whose values, in general, change with changing pressure and/or temperature. Under the conditions existing in the Earth's mantle some of these parameters are believed to vary by more than 100 percent. However, the mathematical structure of the equations governing natural convection is extremely complicated when all of these variations are included. In practice a number of simplifying assumptions is made to render the mathematics tractable to either analytic or numerical solution. These assumptions are generally chosen in accord with laboratory conditions so as to allow direct comparisons between laboratory observations and mathematical predictions. Successful comparisons establish the validity of the mathematical solutions and confirm that the simplifying assumptions are appropriate under laboratory conditions. It seems unlikely however that they should also be appropriate to the Earth's mantle where length scales are greater by six to seven orders of magnitude. Nevertheless, following the philosophy outlined in Section 1 we shall initially discuss models of the mantle which include these simplifications and subsequently examine the importance of relaxing various assumptions.

2.1. The Boussinesq Approximation

In 1916 Lord Rayleigh wrote a short paper (Rayleigh, 1916) which laid down the mathematical foundation for all subsequent studies of convection in viscous fluids. The purpose of this paper was to account for the first careful laboratory observations of convection in a layer of viscous fluid which was heated from below (Bénard, 1901). In his theoretical treatment Rayleigh made a number of simplifying assumptions which he attributed to Boussinesq (1903). Although Oberbeck (1879, 1880) had previously made similar assumptions in a study of a gas in a spherical shell, it was Boussinesq who argued that the resulting approximate mathematical formulation could be used for a wide variety of problems in fluid dynamics. This formulation has since become known as the Boussinesq approximation, and fluids for which this approximation can be justified are often referred to as "Boussinesq Fluids." In its strongest form the Boussinesq approximation includes the following assumptions:

1. Density is independent of pressure and depends only weakly on temperature; and

2. The thermodynamic parameters g (gravitational acceleration), α (coefficient of thermal expansion), K (thermal conductivity), C_p (specific heat at constant pressure) and η (dynamic viscosity) are all independent of temperature and pressure (i.e., constant).

Less restrictive forms of the Boussinesq approximation allow for variations in one or more of the thermodynamic parameters listed under (2) but maintain assumption (1) concerning density. Thus the principal assumption is that density depends (weakly) on temperature but not on pressure. This condition is formalized by the linearized equation of state which is generally adopted for Boussinesq fluids, namely

$$\rho = \rho_r(1 - \alpha\Delta T), \tag{1}$$

where ρ is density, ρ_r is a reference density and ΔT is the difference between the prevailing temperature and the reference temperature, T_r for which $\rho = \rho_r$. ($\Delta T = T - T_r$, where T is temperature in degrees K.)

It is clear from equation (1) that the assumption of a *weak* dependence of density upon temperature requires that the maximum value of $\alpha\Delta T$ be small compared to one (i.e., $\alpha\Delta T^* << 1$, where ΔT^* is the maximum value of ΔT). When this condition obtains, it can be shown rigorously that the role of density variations due to changing temperature are negligible in both the continuity and heat flow equations, but dominant in the equation of motion (Mihaljan, 1962; Malkus, 1964). Thus curiously, although thermal expansion provides the buoyancy force which drives the flow it does not cause significant density variations, and the fluid may otherwise be treated as incompressible.

The Boussinesq approximation is most appropriate for shallow layers of liquid (as in laboratory experiments) where hydrostatic compression is not important. However for deep layers, or compressible fluids, assumption (1) of the Boussinesq approximation is clearly invalid. Thus in addition to the requirement of small values of $\alpha\Delta T$, the Boussinesq approximation is only appropriate when the depth of the convecting layer, d, is small compared to the scale height H_T over which significant density variations occur (Spiegel and Veronis, 1960). For a hydrostatically compressed liquid the scale height is

$$H_T = \frac{C_p}{g\alpha}. \tag{2}$$

Thus the Boussinesq approximation can be justified formally, through scale analysis, provided that the convecting system satisfies the two conditions

$$\alpha \Delta T^* << 1 \tag{3}$$

and

$$d/H_T << 1. \tag{4}$$

Under these conditions the hydrodynamic equations describing the conservation of mass, momentum and thermal energy become:

$$\nabla \cdot \mathbf{v} = 0 \tag{5}$$

$$\rho_r \frac{D\mathbf{v}}{Dt} = -\nabla P + \rho g + \nabla \cdot \tau, \tag{6}$$

$$\rho_r C_p \frac{DT}{Dt} = K\nabla^2 T + H, \tag{7}$$

where $\mathbf{v}$ is velocity, t is time, D/Dt is the Langrangian time derivative (i.e., $D/Dt = [\partial/\partial t + \mathbf{v}\cdot\nabla]$), P is pressure, g is the gravitational acceleration, τ is the deviatoric stress tensor and H is the rate of volumetric heating due to heat sources within the fluid (e.g., the spontaneous decay of radiogenic elements within the mantle). If the constitutive relationship between deviatoric stress and strain rate is Newtonian, and the fluid is incompressible and isotropic, τ may be expressed in terms of v (using tensor notation) as

$$\tau_{ij} = \eta\left(\frac{\partial v_i}{\partial x_j} + \frac{\partial v_j}{\partial x_i}\right), \tag{8}$$

where v_i is the ith component of $\mathbf{v}$ and x_i is the ith spatial (Cartesian) coordinate. This definition of τ has been used in the vast majority of mantle convection studies. When substituted into equation (6) it leads to the simple form for the momentum equation

$$\rho_r \frac{D\mathbf{v}}{Dt} = -\nabla P + \rho \mathbf{g} + \eta\nabla^2\mathbf{v}. \tag{6$'$}$$

2.2. Dimensionless Variables, Equations and Parameters

The hydrodynamic equations are frequently expressed in terms of dimensionless variables by scaling the four basic units (length, time, mass and temperature) by values which are characteristic of the problem under consideration. This enables order of magnitude estimates of the various terms in the equations and facilitates numerical calculations. Solutions of the resulting dimensionless equations depend only on dimensionless combinations of the physical variables and may thus be scaled to different problems. For example, convection in a 1 cm layer of oil in the laboratory and convection in a 700 km layer in the Earth's mantle are both governed by the same dimensionless equations.

For the simple case of a plane layer of depth d with an imposed temperature difference across the layer of ΔT^*, characteristic values may be chosen as follows: length, d; time, $\gamma = d/V$ where V is a characteristic velocity; mass, $\rho_r d^3$; and temperature, ΔT^*. Assuming a balance between buoyancy and viscous forces, an expression for V can be derived from the momentum equation as $V = g\alpha\Delta T^* d^2/\nu$ where $\nu = \eta/\rho_r$ is the kinematic viscosity. All of the variables in the hydrodynamic equations can be written in terms of these characteristic values and dimensionless variables as

$$(x, y, z) = d(x', y', z'); \qquad t = (d/V)t'; \qquad \mathbf{v} = V\mathbf{v}';$$
$$\rho = \rho_r\rho'; \qquad P = \rho_r g d P'; \qquad T = \Delta T^* T' + T_r, \tag{9}$$

where the primed variables are dimensionless. Substitution of the expressions (9) into equations (5), (6)′ and (7) yields the following set of dimensionless equations:

$$\nabla\cdot\mathbf{v}' = 0 \tag{10}$$

$$\frac{R_B}{Pr}\frac{D\mathbf{v}'}{Dt'} = -\frac{\nabla P'}{\alpha\Delta T^*} - \frac{\rho'}{\alpha\Delta T^*}\mathbf{e}_z + \nabla^2\mathbf{v}', \tag{11}$$

$$\frac{DT'}{Dt'} = \frac{1}{R_B}\nabla^2 T' + \epsilon', \tag{12}$$

where

$$R_B = \frac{g\alpha(\Delta T^*)d^3}{\kappa\nu}, \tag{13}$$

$$Pr = \frac{\nu}{\kappa}, \tag{14}$$

and $\kappa = K/\rho_r C_p$ is the thermal diffusivity. In (12) ϵ' represents the dimensionless rate of change of temperature due to internal heat sources; it is related to H as

$$\epsilon' = \frac{Hd}{\rho_r C_p \Delta T^* V} = \frac{H\nu}{\rho_r C_p g\alpha(\Delta T^*)^2 d}. \tag{15}$$

The term R_B^{-1} in (12) represents a dimensionless thermal diffusivity. From (9) and equation (1) it may be shown that $\rho' = (1 - \alpha\Delta T^* T')$.

Equations (10), (11) and (12) depend on three dimensionless parameters:

the Rayleigh number, R_B; the Prandtl number, Pr; and the internal heating rate, ϵ'. All convecting layers which have the same values of these three parameters (and the same boundary conditions) will have the same dimensionless solutions for T', $\mathbf{v}'$, and P'. These solutions may be scaled to particular cases using the relationships listed in (9).

The Prandtl number only enters the dimensionless equations in the factor R_B/Pr which multiplies the inertial term, $D\mathbf{v}'/Dt'$, in the momentum equation (11). This indicates that for infinitely large Prandtl numbers the inertial forces are negligible compared to the pressure, buoyancy and viscous forces and hence equation (11) may be written as

$$\mathbf{0} = -\nabla P' - \rho'\mathbf{e}_z + \alpha\Delta T^*\nabla^2\mathbf{v}'. \tag{16}$$

This equation is a considerable simplification of (11) since the non-linear term $\mathbf{v}'\cdot\nabla\mathbf{v}'$, representing the advection of momentum, has been eliminated. Since the inertial term in equation (11) is of order (R_B/Pr), equation (16) becomes a valid approximation to (11) when $R_B/\text{Pr} << 1$.

Prandtl numbers for typical fluids used in laboratory studies of convection (at 25°C) are listed in Table 7.1. Since R_B must be greater than about 10^3 for convection to occur, Table 7.1 indicates that the condition $R_B/\text{Pr} << 1$ is only roughly met for the high viscosity oils. However, in the Earth's mantle $\nu \approx 10^{18}$ m^2/sec (Peltier, 1974) and $\kappa \approx 10^{-6}$ m^2/sec (Schatz and Simmons, 1972) so that the Prandtl number has a value of $\text{Pr} \approx 10^{24}$, essentially infinite. For reasonable values of g, α and d equal to 10 m/sec^2, 3×10^{-5} °C^{-1} and 3×10^6 m respectively, the Rayleigh number has a value of $R_B \approx 10^4\Delta T^*$ and hence $R_B/\text{Pr} \approx 10^{-20}\Delta T^*$ which easily satisfies the condition $R_B/\text{Pr} << 1$ for any conceivable value of ΔT^*. The appropriate form of the momentum equation for the mantle is therefore equation (16). This "infinite Prandtl number approximation" has the effect of eliminating the Prandtl number from the convection equations.

The set of dimensionless equations employed in modelling the temperature and velocity fields in the mantle are (10), (12) and (16). These depend on two dimensionless parameters R_B and ϵ'. In the absence of internal heat sources (the case studied by Bénard, 1900, 1901), $\epsilon' = 0$ and the form of the convection depends on the single parameter R_B, where the subscript B refers

Table 7.1. Prandtl numbers of laboratory fluids

	Mercury (1)	Air (2)	Water (2)	Silicon oils (2)
Prandtl number	0.024	0.071	6.7	57–8500

Sources: 1, Rossby, 1969; 2, Krishnamurti, 1970a.

to the Bénard configuration. When $\epsilon' > 0$ it is often useful to construct two alternate parameters. These are μ, the ratio of the rate of internal heating to the rate of heat flow across the upper surface of the convecting layer, and R_R, a Rayleigh number defined in terms of the heat flux at the upper surface. If the heat flux into the base of the layer is F, then the heat flux at the upper surface is $E = F + Hd$, so that

$$\mu = Hd/E \tag{17}$$

and

$$R_R = \frac{g\alpha d^4 E}{\rho_r C_p \kappa^2 \nu}. \tag{18}$$

The subscript R on the Rayleigh number defined by equation (18) refers to the convection configuration studied by Roberts (1967) in which $\mu > 0$ and the heat flux, rather than the temperature, was specified at the lower boundary.

2.3. Rayleigh Numbers for Mantle Convection

The value of the Rayleigh number appropriate to convection in the Earth's mantle depends strongly upon d, the depth extent of the convective circulation, and upon values of the thermodynamic parameters which are characteristic of this depth extent. Unfortunately the depth of the mantle circulation which involves plate creation and destruction is not known and is currently a matter of intense debate. The two main hypotheses are: (1) that convection is confined to the upper 700 km of the mantle (e.g., Turcotte and Oxburgh, 1967, 1972; Richter, 1973; McKenzie *et al.*, 1974; McKenzie and Weiss, 1975, 1980; Parmentier and Turcotte, 1978; Richter and McKenzie, 1978; O'Nions *et al.*, 1979; Jacobsen and Wasserburg, 1979; McKenzie and O'Nions, 1983), and (2) that convection extends throughout the mantle from the Earth's surface to the core–mantle boundary (e.g., Hess, 1962; Peltier, 1972; Davies, 1977; O'Connell, 1977; Sharpe and Peltier, 1978, 1979; Elsasser *et al.*, 1979; Olson and Corcos, 1980; Peltier, 1980; Olson, 1981; Peltier and Jarvis, 1982; Allegre *et al.*, 1983). Geophysical and geochemical data are not yet sufficiently discriminating to rule out completely either hypothesis. The discussions below will reflect this impasse; we shall not argue in favor of either hypothesis, but rather will consider the implications of scaling simple numerical models to both cases.

Associated with this uncertainty in the depth of mantle convection is an ambiguity in the value of the Rayleigh number for this circulation. Values of the physical parameters which we adopt as characteristic of upper- and whole-mantle convection are listed in Table 7.2. Here we have equated E in (18) with the mean oceanic heat flow given by Sclater *et al.* (1980). Due to a considerable uncertainty in estimates of K, and hence of κ, in the lower mantle our estimates of these parameters may be in error by as much as a factor of two.

The values listed in Table 7.2 suffice to define the (Roberts) Rayleigh number R_R for either an upper mantle (0–670 km) or a whole mantle (0–2900 km) circulation. For the upper mantle we have from equation (18) $R_R = 1.38 \times 10^6$, and for the whole mantle $R_R = 6.04 \times 10^7$. Estimates of the appropriate (Bénard) Rayleigh number R_B can similarly be made once an estimate of ΔT^*, the temperature difference across the convecting zone in excess of the adiabatic temperature difference, is obtained. This is most readily accomplished for the whole mantle model since the lithosphere is normally considered to be a part of the convecting system and the temperature difference across the lithosphere represents approximately 50 percent of the non-adiabatic temperature variation across the convection zone. Thus an appropriate value of ΔT^* may be 2400°C and hence from equation (13) we have for the whole mantle, $R_B = 7.47 \times 10^6$. For a layered mantle Jeanloz and Richter (1979) have estimated a temperature difference across the upper 670 km as 1950°C, of which about 200°C could be due to adiabatic temperature variation, implying that $\Delta T^* \approx 1750$°C for the upper mantle. Using this value of ΔT^* in (13) we have for the upper mantle, $R_B \approx 2.55 \times 10^5$. However most of this ΔT^* occurs across the lithosphere, whereas models of upper mantle convection usually do not include the lithosphere in the convective circulation. For these models $\Delta T^* \approx 600$°C and $d = 600$ km, in

Table 7.2. Physical parameters for mantle convection models

Upper mantle		Whole mantle	
$d = 6.7 \times 10^5$ m (670 km)		$d = 2.90 \times 10^6$ m (2900 km)	
$g = 9.8$ m s^{-2}		$g = 9.8$ m s^{-2}	
$\alpha = 2 \times 10^{-5}$ K^{-1}	(1)	$\alpha = 1.4 \times 10^{-5}$ K^{-1}	(1)
$\kappa = 1.5 \times 10^{-6}$ m^2 s^{-1}	(2)	$\kappa = 2.5 \times 10^{-6}$ m^2 s^{-1}	(6)
$\nu = 2.7 \times 10^{17}$ m^2 s^{-1}	(3)	$\nu = 4.3 \times 10^{17}$ m^2 s^{-1}	(7)
$\rho_r = 3.7 \times 10^3$ kg m^{-3}	(4)	$\rho_r = 4.7 \times 10^3$ kg m^{-3}	(4)
$\eta = 1.0 \times 10^{21}$ kg m^{-1} s^{-1} (10^{22} Poise)	(3)	$\eta = 2 \times 10^{21}$ kg m^{-1} s^{-1} (2×10^{22} Poise)	(7)
$C_p = 1.26 \times 10^3$ J kg^{-1} K^{-1}	(1)	$C_p = 1.26 \times 10^3$ J kg^{-1} K^{-1}	(1)
$E = 99$ mW m^{-2}	(5)	$E = 99$ mW m^{-2}	(5)
$K = 6.66$ W m^{-1} K^{-1}	(2)	$K = 14.8$ W m^{-1} K^{-1}	(6)

Sources: 1, Stacey, 1977; 2, McKenzie *et al.*, 1974; 3, Peltier, 1974; 4, Dziewonski *et al.*, 1975; 5, Sclater *et al.*, 1980; 6, Jarvis and Peltier, 1982; 7, Wu and Peltier, 1983.

which case $R_B \approx 6.27 \times 10^4$. The various estimates of R_R and R_B for upper and whole mantle convection are compared in Table 7.3. This wide range of possible values for the Rayleigh number complicates studies of mantle convection and diffuses the efforts of Earth scientists studying the dynamics of the Earth's interior, since all possibilities must be examined with equal care. Once the issue of the depth extent of the mantle convective circulation is resolved, efforts may be concentrated on the relevant models and our understanding of geodynamics should advance rapidly.

In the following sections of this chapter we shall initially discuss the simplest convection models, employing the strong form of the Boussinesq approximation, and subsequently examine more complex models in which some of the initial restrictions are relaxed. The initial discussions will be concerned with incompressible fluids in a plane layer with a Newtonian rheology. From even these simple models we can draw inferences about the nature of the mantle flow. The more complex models in some cases serve to justify the simplifications and in others to reveal important additional physics.

3. Simple Boussinesq Models

Both laboratory and numerical experiments are most readily conducted on incompressible viscous fluids. In laboratory experiments the layer depth d and range of temperatures ΔT^* are invariably so small that the conditions $\alpha\Delta T^* << 1$ and $d/H_T << 1$ are easily satisfied, and hence the convecting layer may be treated as a Boussinesq fluid. In numerical experiments there are no *a priori* restrictions on the size of d or ΔT^*. However, the assumptions that $\alpha\Delta T^* << 1$ and $d/H_T << 1$ result in a considerable simplification of the mathematical structure of the convection equations. This simplification is often necessary in order to obtain a reasonably accurate solution to the equations.

Laboratory and numerical convection models complement each other in that each provides a check on the other. At high Rayleigh numbers analytic

Table 7.3. Approximate mantle Rayleigh numbers

	Upper mantle	Whole mantle
R_R	1.38×10^6	6.04×10^7
R_B	6.27×10^4–2.55×10^5	7.47×10^6

solutions to the non-linear convection equations are not available. It is therefore essential to compare mathematical predictions of quantities which can also be measured experimentally with laboratory observations. If such comparisons are favorable one has confidence that the mathematical formulation of the problem is correct and hence that predictions of quantities that cannot be measured experimentally may also be correct.

3.1. Laboratory Experiments

The first careful laboratory experiments on convection in a horizontal layer of viscous fluid were those by Bénard (1900, 1901). Bénard was struck by the fact that after a brief initial transient stage a remarkably uniform set of steady convection cells became established. The cell pattern was consistently reproducible and occurred with the same aspect ratio (cell width: layer depth) for different layer depths. Although it is now known that many of the details which Bénard observed were the result of surface tension at the exposed upper surface (Block, 1956; Pearson, 1958; Neild, 1964; Scanlon and Segal, 1967; Koschmeider, 1967), the interest generated by his experiments led theoreticians to develop the theory of convection in viscous fluids. In particular Rayleigh (1916) and Jeffreys (1926, 1928) developed the mathematical formulation from which it is possible to predict both the critical value of the Rayleigh number, R_c, which must be exceeded before convection occurs and the corresponding horizontal wavenumber, k_c, of the resulting flow pattern.

Critical Rayleigh Numbers and Aspect Ratios

The aspect ratio, $\bar{a}$, of individual cells is related to the wavenumber through the relations

$$\bar{a} = \frac{\lambda}{2d} = \frac{2\pi/k}{2d} = \frac{\pi}{kd}, \tag{19}$$

where λ is the horizontal wavelength of the periodic flow pattern and k is the corresponding wavenumber. For convection between two free boundaries (where shear stresses vanish at the boundaries) the critical values of R_c and k_c are predicted analytically to be $R_c = {}^{27}\!/\!_4\pi^4 = 657.51$ and $k_c = \pi/\sqrt{2}d = 2.22/d$; the corresponding value of the aspect ratio is, according to (19), $\bar{a} = \sqrt{2} = 1.41$. At wavenumbers different from k_c neutral stability occurs for values of R_c given by

$$R_c = \frac{(\pi^2 + k^2d^2)^3}{k^2d^2}, \tag{20}$$

or, in terms of aspect ratio

$$\mathrm{R_c} = \bar{a}^2\pi^4\left(1 + \frac{1}{\bar{a}^2}\right)^3. \tag{21}$$

For convection between rigid horizontal boundaries (where horizontal velocities vanish) the critical conditions for the onset of convection are determined numerically as $\mathrm{R_c} = 1707.76$ and $k_c = 3.1117$, with a corresponding aspect ratio $\bar{a} = 1.01$.

Although the case of convection between free boundaries allows an analytic solution it is a difficult configuration to model in the laboratory. Most laboratory experiments are conducted on fluids between rigid boundaries; and these have verified the numerically derived predictions (Schmidt and Milverton, 1935; Silveston, 1958). The precise experimental results of Silveston (1958), for example, demonstrated that the onset of convective heat transport consistently occurred at $\mathrm{R_c} = 1700 \pm 51$ for each of the five different fluids he employed. This close agreement between theory and experiment for the case of rigid boundaries gives us confidence that the theoretical predictions for the simpler case of free boundaries are also correct.

Bénard's original experiments were intermediate in nature; the lower boundary was rigid while the upper boundary was free. For this case numerical solutions yield $\mathrm{R_c} = 1100.65$, $k_c = 2.682$ and $\bar{a} = 1.171$ (Chandrasekhar, 1961). Bénard (1901) noted that in his steady cells $\bar{a}$ had a value between 1.3 and 1.9. Given that Bénard's experiments were influenced by the effects of surface tension the discrepancy here between theory and experiment is surprisingly small.

The two major conclusions derived from these early experiments are (1) that convection will occur in a fluid layer whenever the value of the Rayleigh number exceeds a critical value of order 10^3, and (2) that in a uniform incompressible layer of convecting fluid the aspect ratio $\bar{a}$ will have a value close to unity regardless of the nature of the mechanical boundary conditions at the upper and lower surfaces. Transferring these results to the geophysical context, we can infer that (1) for the high values of $\mathrm{R_B}$ estimated above (see Section 2) the mantle must be actively convecting and (2) if the mantle can be treated as a uniform incompressible fluid we can expect that the horizontal and vertical extents of the fluid motion are comparable. Thus for whole mantle convection we expect horizontal motions to occur on a length scale of 3000–4000 km (comparable to plate dimensions), while for upper mantle convection below the lithosphere we expect horizontal motions to occur on a length scale of 600–700 km (much smaller than plate dimensions).

Stability of Rolls

One of the striking features of Bénard's experiments was that once the flow pattern was established it remained very steady for long periods of time. This result is, however, due to the low Rayleigh numbers in Bénard's experiments. In later experimental studies between two rigid boundaries it has been noted that as the Rayleigh number is increased, there is a series of discrete transitions in heat transfer properties of the layer (Schmidt and Saunders, 1938; Malkus, 1954; Krishnamurti, 1970a,b, 1973) and that these are accompanied by discrete changes in the flow pattern (Krishnamurti, 1970b, 1973; Busse and Whitehead, 1971; Whitehead and Parsons, 1978). In a series of careful experiments on fluids with Prandtl numbers ranging from 2.5×10^{-2} to 8.5×10^{3} Krishnamurti (1970a,b, 1973) has shown that initially steady two-dimensional flows at low Rayleigh numbers will change with increasing R_B first to three-dimensional flow and ultimately into completely disordered time-dependent three-dimensional flow.

The geophysical implications of these results are that if convection in the mantle (for which $Pr \approx \infty$) occurs between rigid boundaries and $R_B > 34\,R_c$, mantle convection occurs in a continually changing three-dimensional pattern. The case of convection between rigid boundaries is most relevant to models of convection in the upper mantle below the rigid lithosphere and above a barrier at a depth of 670 km. For such a model $R_B = 6.27 \times 10^4 = 37R_c$ (see Table 7.3) and hence time-dependent flow is likely. Since the Rayleigh number for the whole mantle is much larger we likewise expect that whole mantle convection would be time-dependent. However whole-mantle convection would occur between "free" (i.e., zero shear stress) boundaries so that neither Krishnamurti's regime diagram nor Busse's stability predictions are directly relevant. In fact theoretical investigations by Straus (1972) and by Skilbeck and McKenzie (1979) of the stability of two-dimensional rolls between free boundaries found that rolls were stable in a limited range of wavenumbers up to $R_B = 1.83 \times 10^5 = 278R_c$, the highest value considered. Because of the difficulty of constructing laboratory models of convection between two free boundaries, these predictions have not been verified experimentally.

Nusselt Numbers

The fact that the Earth's mantle overturns convectively is clearly important in terms of accounting for the large scale horizontal motion of the planetary surface. Of equal importance in terms of the thermal history of the planet is the fact that heat is transported across the mantle much more efficiently by convection than by conduction. A measure of this efficiency is provided by the Nusselt number, Nu, which is the dimensionless ratio of the rate of heat transport across a convecting layer to that which would exist in the absence of convection. At sufficiently high values of R_B the Nusselt number is related to the Rayleigh number by a power law of the form

$$\mathrm{Nu} = \beta \left(\frac{R_B}{R_c} \right)^S, \tag{22}$$

where the coefficient β and exponent S are, in general, functions of the aspect ratio of the convection cells. The values of β and S also depend on whether the bounding surfaces are free or rigid. Empirical determinations of β and S have been attempted by several experimentalists. In a careful series of experiments with water (Pr = 6.7) Rossby (1969) found $\beta = 1.222$ and $S = 0.30$ while Chu and Goldstein (1973) found $\beta = 1.449$ and $S = 0.278$. In higher Prandtl number silicon oils (Pr = 200) Rossby (1969) found $\beta = 1.490$ and $S = 0.281$. Sommerscales and Gazda (1969) obtained a similar result, $\beta = 1.55$ and $S = 0.282$ at high Prandtl number.

The Nusselt number–Rayleigh number power law relation is an important tool which is used in modelling the thermal history of the Earth and other planetary bodies. For a given Rayleigh number, it allows one to predict the rate of heat loss from a convecting layer without the necessity of solving the full set of hydrodynamic equations. The parameterization of heat flow in terms of Rayleigh number results in such a substantial simplification that it is possible to model entire planetary thermal histories. This approach, now commonly referred to as "parameterized convection" (Sharpe and Peltier, 1979), has been followed recently in a variety of forms by several research groups (e.g., Tozer, 1967, 1972, 1974; McKenzie and Weiss, 1975, 1980; Sharpe and Peltier, 1978, 1979; Kaula, 1979; Cassen *et al.*, 1979; Stevenson and Turner, 1979; Turcotte *et al.*, 1979; Daly, 1980; Davies, 1980; Peltier and Jarvis, 1982; Richter, 1984). This subject will be discussed in more detail in subsequent chapters.

Applying equation (22) to the case of upper mantle convection between rigid boundaries (for which $R_B/R_c \approx 37$; see Table 7.3) and using the values of $\beta = 1.55$ and $S = 0.282$ (Sommerscales and Gazda, 1969) we find Nu = 4.9. Similarly, for the case of whole-mantle convection between free boundaries (for which $R_B/R_c = 11{,}361$; see Table 7.3), using the numerically derived values of $\beta = 2.24$ and $S = 0.318$ (Jarvis, 1984) we find Nu = 43.6. Thus models of whole mantle convection transfer heat to the surface almost ten times more efficiently than some models of upper mantle convection. Clearly this difference is relevant to the cooling of the planetary interior. This wide range of possible Nusselt numbers in the mantle emphasizes the need to resolve the issue of the depth extent of mantle convection and the related issue of the dynamic role of the lithosphere.

Although the parameterization of the Nusselt number in terms of Rayleigh number is useful in a wide variety of applications, the numerical values of β and S have only been determined for very simple experimental configurations. The effects of variable physical parameters, spherical geometry and partial internal heating on such parameterizations are not well understood. Consequently, application of equation (22) to the Earth's mantle involves the implicit assumption that such effects are not important; as yet this assumption has not been justified. Indeed recent studies by Christensen (1984a, 1984b, 1985) on variable viscosity fluids at high Rayleigh numbers suggest that this assumption is not valid.

General

At low Rayleigh numbers, laboratory experiments on uniform fluids with no internal heat sources produce steady symmetric rolls (or bimodal cells) of uniform size and spatial periodicity. However at high Rayleigh numbers this uniform system becomes highly time-dependent producing transient asymmetric cells of various sizes and shapes. Whitehead and Parsons (1978) noted that under such conditions convective velocities could vary by a factor of four between cells of different sizes. These characteristics of high Rayleigh number flow are in accord with the geophysical observations that plate sizes and speeds are highly variable and the directions of plate motions change on time-scales shorter than the recycling time of the lithosphere (as evidenced, for example, by the change in strike of the Hawaii-Emperor seamount chain). Thus the general character of plate motions at the planetary surface may be understood, in terms of convective motions in the mantle below, on the basis of even the simplest laboratory models.

There is, however, a number of limitations to laboratory models which restrict their usefulness to predictions of gross thermal properties such as the Nu–R_B relationship. A major limitation in terms of the whole-mantle convection model is the difficulty of performing laboratory experiments between two free boundaries. Consequently laboratory information on rates of heat transfer and cell stability is not available for this model. Similarly it is difficult in laboratory experiments to obtain pointwise information such as velocity gradients or the separate contributions of conduction and advection to the total local heat flow. For reasons such as these, laboratory models must be complemented by numerical solutions of the hydrodynamic equations. These solutions predict at each point in the fluid the values of temperature and velocity from which other quantities of interest can be computed. This is particularly useful for time-dependent flows since a "snapshot" of the temperature and velocity fields is provided at each time step. Similar analysis of time-dependent flows is not feasible in laboratory experiments.

3.2. Numerical Experiments

The motivation for conducting numerical experiments is the same as for laboratory experiments: to determine how a given fluid responds to various heating configurations, boundary conditions and variations of physical properties. In numerical experiments computers are used in the place of laboratory tanks. The disadvantages of numerical experiments are in general related to limited availability of computer resources. These include problems of resolving small scale effects and the restriction of most high Rayleigh number numerical studies to two dimensions. The latter restriction results in the suppression of all three dimensional instabilities. As a result, steady numerical solutions are often found under conditions where time-dependent flow would normally occur in the laboratory. In spite of their limitations, numerical experiments have provided major contributions to our understanding of convection in the Earth's mantle and its geophysical implications.

Early Developments

In one of the earliest attempts to solve the full non-linear equations, Malkus and Veronis (1958) used a perturbation method to expand the non-linear equations as a sequence of linear equations whose solutions followed from those of the linear problem for the onset of convection. Extending this expansion technique to investigate the stability of the possible steady solutions, Schluter *et al.* (1965) found that two dimensional rolls were the only stable stationary solutions at Rayleigh numbers close to critical. Inspired by this result Veronis (1966) used a numerical expansion of the dependent variables in terms of eigenfunctions of the linear stability problem to investigate large amplitude two-dimensional convection. Veronis' solutions, valid for $R_B \leqslant 30R_c$, covered a range of Prandtl numbers from 0.01 to 100. All of Veronis' solutions settled to a steady state and he speculated that this was likely due to the imposed two dimensionality of the solutions.

The work of Schluter *et al.* (1965) was extended by Busse (1967) to examine the stability of two dimensional rolls at larger Rayleigh numbers. In order to isolate the gravitational cause of instability and to simplify the mathematical problem, Busse considered infinite Prandtl number fluids (between rigid boundaries). He found that there was a closed region in the

Rayleigh number–wavenumber plane within which steady two dimensional flow is stable with respect to any three dimensional disturbance.

Mathematical descriptions of the various instabilities responsible for limiting the region of stability of rolls in the Rayleigh number–wavenumber plane were given by Busse (1967, 1971) and Clever and Busse (1974). They considered infinitesimal perturbations of arbitrary two-dimensional rolls aligned parallel to the y-direction. The horizontal dependencies of the perturbations and the initial rolls were of the form $e^{i(cx+by)}$ and e^{inkx} respectively, where c and b are horizontal wavenumbers of the perturbation, k is the horizontal wavenumber of the initial rolls and n is an integer. For given values of Rayleigh number, k and b, Busse (1967) found that the two-dimensional rolls were most unstable to disturbances for which $c = 0$. This is of particular relevance to two-dimensional numerical models since the only disturbances possible in these models have $b = 0$ and $c \neq 0$. These are two-dimensional disturbances aligned parallel to the original rolls. At high Prandtl numbers the stability boundary defined by these (Eckhaus type) instabilities lies outside those for the three-dimensional instabilities (Busse, 1971; Clever and Busse, 1974). Consequently the numerical solutions are artificially stabilized with respect to the latter disturbances. Furthermore, the horizontal periodicity imposed on most mathematical models eliminates all Eckhaus disturbances with $c < 2\pi/\lambda$ where λ is the wavelength of the periodicity. Thus the simplifications of numerical models all tend to stabilize the flow.

The direct numerical integration of the Boussinesq equations, using finite difference methods, provides an alternative to the expansion technique for studying finite amplitude convection. Pioneering work in this field was reported by Fromm (1964, 1965) and Deardorff (1964, 1965). However, limited computer resources restricted these early investigations to the consideration of only a few values of R_B and Pr. Subsequently Moore and Weiss (1973) described an extensive investigation of two dimensional convection between free boundaries for $R_c \leqslant R_B \leqslant 10^3 R_c$ and $0.01 \leqslant \Pr \leqslant \infty$. Moore and Weiss (1973) indicated that a reliable, if somewhat cautious, criterion by which a solution may be considered accurate is that at least three grid intervals span every thermal boundary layer. In order to satisfy this criterion for the largest estimates of the Rayleigh number appropriate to mantle convection, very fine numerical meshes are required. For the relatively low Rayleigh numbers appropriate to upper-mantle convection, coarser grids may be used.

Some of the earliest numerical studies of convection in the upper mantle using finite difference grids with up to 48 vertical grid intervals were described by Richter (1973), Turcotte *et al.* (1973) and McKenzie *et al.*

(1974). Recently whole-mantle convection has been modelled by Jarvis and Peltier (1980, 1981, 1982) using 96 × 96 numerical grids and by Jarvis and Campbell (1983) and Jarvis (1984) using 200 × 200 numerical grids. The geophysical implications of these models of both upper- and whole-mantle convection are discussed below.

Stream Function and Vorticity

Most numerical experiments have been concerned with "two-dimensional" convection cells. These take the form of infinitely long parallel rolls such that temperature and velocity are invariant along the direction parallel to the axis of the rolls. As such, the temperature and velocity fields on all planes perpendicular to the axis of the rolls are identical and hence the mathematical problem is reduced to that of solving the hydrodynamic equations for temperature and velocity on one such plane. The sole reason for our preoccupation with this type of convection is that, in this special case, the relevant equations become simple enough that numerical solutions are feasible with commonly available computer resources.

For two-dimensional convection between free boundaries the mathematical formulation of the convection equations is simplified by the introduction of the dimensionless stream function $\boldsymbol{\Psi}'$ and the vorticity $\boldsymbol{\omega}'$ which are related to the dimensionless velocity $\mathbf{v}'$ as

$$\nabla \times \boldsymbol{\Psi}' = \mathbf{v}', \tag{23}$$

$$\nabla \times \mathbf{v}' = \boldsymbol{\omega}'. \tag{24}$$

For convection in the x-z plane (in Cartesian coordinates) velocity has the form $\mathbf{v}' = (u', 0, w')$ where u' and w' are the horizontal and vertical components of $\mathbf{v}'$. Hence from inspection of equations (23) and (24) we can deduce that $\boldsymbol{\Psi}'$ and $\boldsymbol{\omega}'$ have y-components only, hereafter denoted ψ' and ω' respectively, and that ψ', ω', and $\mathbf{v}'$ are related as

$$\mathbf{v}' = \left(\frac{-\partial\psi'}{\partial x'}, 0, \frac{\partial\psi'}{\partial z'}\right), \tag{25}$$

and

$$\nabla^2\psi' = -\omega'. \tag{26}$$

The expression for $\mathbf{v}'$ in terms of ψ' automatically satisfies the continuity equation, $\nabla\cdot\mathbf{v}' = 0$, for the conservation of mass.

The dimensionless momentum equation (16) may be re-expressed in terms of vorticity by applying the curl operator ($\nabla\times$) to both sides of the equation.

This operation eliminates the pressure term $\nabla P'$ from the equations since $\nabla \times \nabla P' = \mathbf{O}$. The new form of the momentum equation becomes

$$\nabla^2 \omega' = \frac{\partial T'}{\partial x'}, \tag{27}$$

where the equation of state has been used to substitute for ρ'. Finally, expanding the Lagrangian derivative of T' in the temperature equation (12) and substituting for $\mathbf{v}'$ from (25), the temperature equation becomes

$$\frac{\partial T'}{\partial t'} = \frac{\partial \psi'}{\partial z'} \frac{\partial T'}{\partial x'} - \frac{\partial \psi'}{\partial x'} \frac{\partial T'}{\partial z'} + \frac{1}{\mathrm{R_B}} \nabla^2 T' + \epsilon'. \tag{28}$$

Equations (26), (27) and (28) form a closed system of equations for the three unknowns ψ', ω' and T'. Since only (28) includes a time derivative, the temporal evolution of these three quantities is due entirely to changes in temperature. This system of equations is solved cyclically as an initial and boundary value problem. Because it is an initial value problem the solution prevailing at any time is a function of the assumed starting values of T'.

Two-Dimensional Fields

Examples of numerical solutions of the temperature and stream function fields in steady two dimensional convection rolls are shown in Figures 7.1 and 7.2, respectively. Contours of T' and ψ' are shown for various Rayleigh numbers covering a range of five orders of magnitude. These solutions were obtained for convection between free boundaries with constant temperatures at the upper and lower surfaces. Each set of contours in these figures corresponds to a pair of counter rotating cells with a cold sinking plume at the center and hot rising plumes at the outer edges. (Since the outer edges are planes of mirror symmetry only half of each hot plume is seen in these diagrams.)

Considering the temperature solutions initially, Figure 7.1a demonstrates that at a Rayleigh number just marginally greater than $\mathrm{R_c}$, fluid motion produces only small distortions of the isotherms from horizontal (the pure conduction case). At $\mathrm{R_B} = 10\mathrm{R_c}$ (Fig. 7.1b) velocities are sufficiently large to advect the cold isotherms downwards and the hot isotherms upwards, generating a broad gravitationally stable central region in which warm fluid lies above cold. As the Rayleigh number is continually increased in passing through frames c to f of Figure 7.1, isotherms are swept clear of the interior and become concentrated into narrow boundary layers about the edges of each roll. This results in a central isothermal region in each cell. It is evident

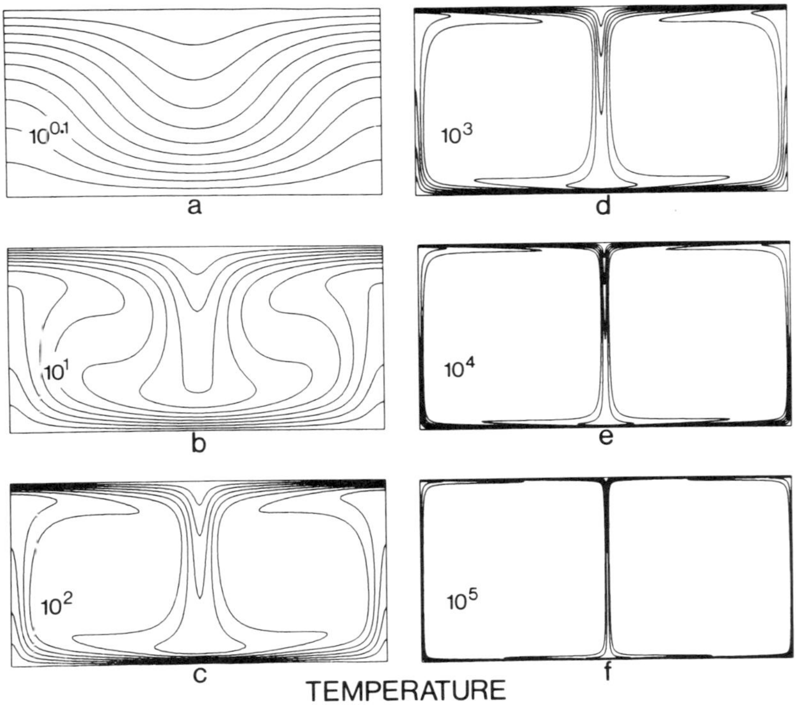

Figure 7.1. Steady two dimensional temperature fields for pairs of counterrotating convection rolls. Contours of temperature are plotted with a constant interval of $\Delta T/11$ in each frame. Each frame is labelled with the corresponding value of R_B/R_c. Solutions were obtained on the following numerical meshes: a and b, 24×24; c, 80×80; d, e, and f, 200×200.

from Figure 7.1 that the thickness of the boundary layers decreases with increasing Rayleigh number. At the highest Rayleigh number shown here the thickness of the horizontal boundary layers is only a few percent of the layer depth. It may also be seen from inspection of Figure 7.1 that the horizontal thermal boundary layers are significantly thinner than the vertical boundary layers (i.e., plumes). Since the stability of the horizontal boundary layers determines the nature of the flow, the numerical grids upon which these solutions are obtained must be sufficiently fine to resolve these boundary layers. Grids with up to 200 vertical intervals were used for the solutions shown in Figure 7.1.

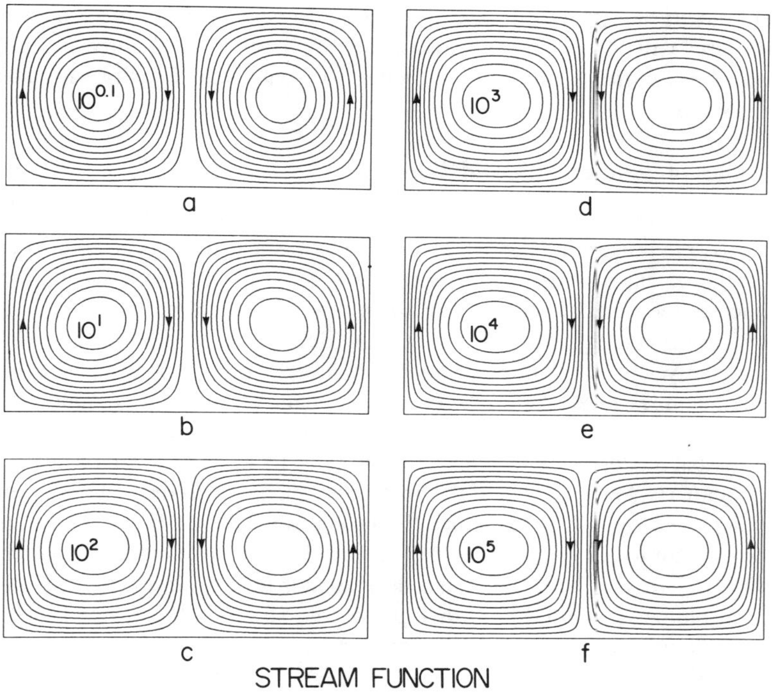

Figure 7.2. Steady two dimensional stream function fields, corresponding to the temperature fields shown in Fig. 7.1, for pairs of counterrotating convection rolls. Contours of stream function are plotted with a constant interval of $\psi_{max}/11$, where ψ_{max} is the maximum value of ψ for each solution. Each frame is labeled with the corresponding value of the ratio R_B/R_c.

Figure 7.2 displays contours of the stream function fields corresponding to T' shown in Figure 7.1. Since solving equations (27) and (26) for ψ' is essentially a fourth order smoothing of $\partial T'/\partial x'$, ψ' is a very smooth field and accordingly it masks most of the short wavelength variations of T'. At large values of R_B, the only signature in the contours of ψ' of the extreme concentrations of $\partial T'/\partial x'$ is a decrease in the spacing of the streamlines (i.e., contours of constant ψ') near the plumes relative to that near the horizontal boundaries. Since velocity in the direction of the streamlines is inversely proportional to the distance between adjacent streamlines, the closer spacing at vertical plumes indicates a higher vertical velocity at plumes than horizontal velocity at upper and lower surfaces. Moreover since this asymmetry increases with R_B, the mean vertical velocity in plumes evidently increases more rapidly with R_B than does the mean horizontal velocity at free horizontal surfaces.

Profiles of Temperature and Velocity

Although the contour plots of T', shown as a function of R_B in Figure 7.1, are effective in illustrating the overall tendency of isotherms to become concentrated into narrow boundary layers as R_B increases, this concentration conceals the detailed structure of the temperature field at high R_B. Thus, particularly at high Rayleigh numbers, details of the temperature field (especially within the thermal boundary layers) are better revealed by selected temperature profiles across the domain of solution. Examples of horizontal and vertical profiles of temperature are shown in Figure 7.3 for a numerical model of steady convection at $R_B = 10^4 R_c$ as obtained on a 200 × 200 numerical mesh. Each of the profiles is labelled with the column or row number of the mesh along which the profile was taken. The set of profiles corresponds to one roll rotating in the clockwise sense (i.e., the left-hand roll in Figs. 7.1e and 7.2e); column 0 corresponds to the mid-plane of the hot rising plume, column 200 to that of the cold sinking plume, row 0 to the bottom boundary and row 200 to the top boundary.

Figure 7.3a displays a series of vertical profiles of temperature across the convecting layer. Away from the plumes (columns 0 and 200) the profiles are characterized by an isothermal interior and steep temperature gradients in the thermal boundary layers adjacent to the upper and lower surfaces. Immediately below the cold upper boundary layer and above the hot lower boundary layer are zones of temperature "overshoot" in which temperatures are respectively higher and lower than the constant temperature of the central region. The profile at the left in Figure 7.3a (along the mid-plane of the hot plume) shows that as hot material is swept upwards its temperature decreases little until it approaches very close to the upper boundary. The thickness of the thermal boundary layer is least above the rising hot plume where vertical advection is dominant. The rising plume material has higher than average temperatures as it spreads horizontally below the upper surface and thus accounts for the temperature overshoot above that of the isothermal core. This material loses heat through diffusion both to the cold upper surface and into the underlying central core. Consequently the thickness of the temperature overshoot zone increases towards the cold descending plume while the amount of temperature overshoot decreases.

The horizontal temperature profiles displayed in Figure 7.3b reveal the

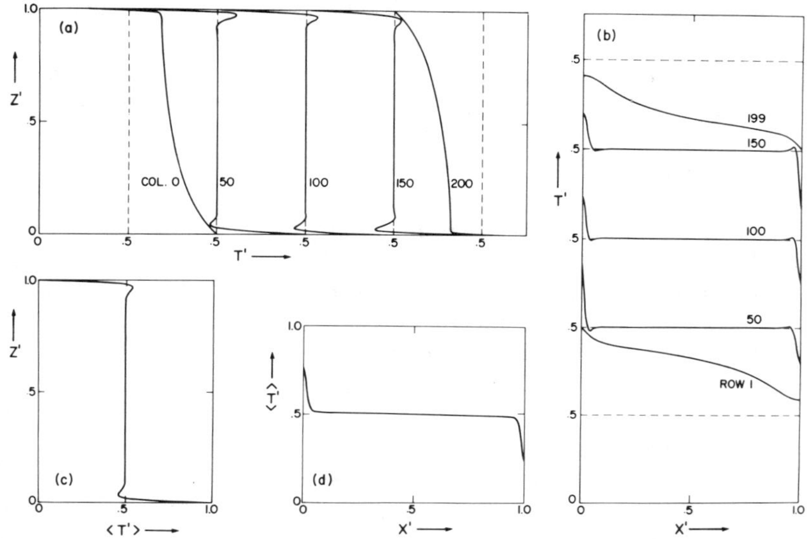

Figure 7.3. Vertical and horizontal profiles of dimensionless temperature across a convection cell for which $R_B = 10^4 R_c$, obtained on a 200 × 200 finite difference grid. a, Vertical profiles along columns of the numerical grid as indicated. Column 0 corresponds to the center of the rising plume (at the left of the box) while column 200 corresponds to the center of the descending plume (at the right of the box). The broken vertical lines indicate the mean temperature of the box of fluid. b, Horizontal profiles along rows of the numerical grid as indicated. Row 1 is one grid-plane above the (constant temperature) bottom of the box. Row 199 is one grid-plane below the (constant temperature) top of the box. The broken horizontal lines indicate the mean temperature of the box of fluid. c, Mean vertical profile. d, Mean horizontal profile.

narrow hot rising plume at the left and cold sinking plume at the right. The rise in temperature adjacent to the sinking plume and the dip in temperature next to the rising plume are due respectively to the entrainment of hot material (from the rising plume) by the cold sinking jet, and of cold material (from the sinking jet) by the hot plume.

The mean vertical and mean horizontal variations of temperatures are plotted in Figure 7.3c, d. (The symbols $\langle\ \rangle$ and $\hat{\check{\ }}$ denote the horizontal and vertical average, respectively, of the enclosed quantity.) The mean vertical temperature profile (Fig. 7.3c) is similar to the central profile shown in Figure 7.3a except that the overshoot zones are broader and the magnitude of the overshoot is reduced. The mean horizontal temperature profile (Fig.

7.3d) shows a similar broadening of the thermal boundary layers relative to the central profile of 7.3b. In addition there is a distinct horizontal gradient of temperature across the central zone of the mean profile. This is due to the contribution of temperature variations within the horizontal boundary layers (such as shown for rows 1 and 199 in Fig. 7.3b) to the mean horizontal variation of temperature. Such a horizontal gradient is *not* characteristic of the interior of the convecting layer.

In Figure 7.4 mean vertical temperature profiles are plotted for a series of models with Rayleigh numbers ranging from $R_B = R_c$ to $R_B = 10^4 R_c$. As R_B increases from the critical value, the mean temperature profiles initially acquire a reversed temperature gradient in the central region. This central gradient continues to increase in magnitude and vertical extent until $R_B \approx 20R_c$. At $R_B \geqslant 50R_c$ the central zone of reversed temperature gradient is split by an isothermal region, the vertical extent of which increases with R_B. In other words for $R_B \geqslant 50R_c$ the upper and lower boundary layer structures are isolated by the isothermal core. From inspection of Figure 7.4 it is evident that the reversed central temperature gradient at $R_B \leqslant 20R_c$ is due to a merging of the upper and lower boundary layer structures, and in particular of the lateral spreading hot and cold plume material. At such low Rayleigh numbers the flow velocities are not sufficient to produce isolated thermal boundary layers.

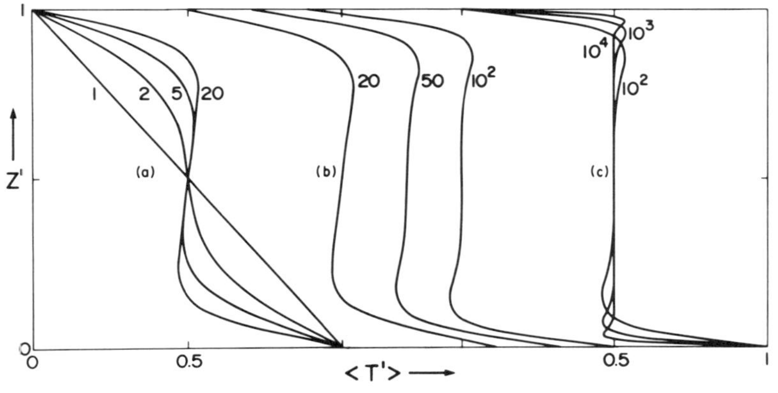

Figure 7.4. Vertical profiles of horizontally averaged dimensionless temperature $\langle T' \rangle$ for several values of R_B. Each profile is labelled with the corresponding value of R_B/R_c. a, Low Rayleigh number regime. b, Transition to boundary layer regime. c, Boundary layer regime.

Of particular interest is the striking similarity of the dimensionless temperature profile in the boundary layer regime. The primary effect of increasing R_B is to increase the vertical extent of the isothermal zone at the expense of the thickness of the boundary layers. The relative overshoot of the boundary layer temperatures beyond that of the isothermal core is not significantly altered. Richter and McKenzie (1981) and Jarvis and Peltier (1982) describe how this similarity may be used to predict the mean temperature profile at any Rayleigh number within the boundary layer regime.

The Parameterization of Mean Quantities

In section 3.1 the Nusselt number–Rayleigh number power law relationship was introduced as a means of predicting the overall heat flux across a convecting layer, as a function of the ratio R_B/R_c. The boundary layer theory introduced by Turcotte and Oxburgh (1967) predicts that similar power law relationships exist between R_B/R_c and the boundary layer thickness and mean horizontal and vertical velocities. This prediction is borne out by plotting the logarithm of the latter quantities as a function of the logarithm of R_B/R_c. Power law relationships appear linear on such a graph. Figure 7.5 demonstrates that linear relationships exist between the logarithms of R_B/R_c and each of the Nusselt number, Nu; boundary layer thickness, δ'; mean horizontal velocity at the upper surface, $\bar{u}'$; and the mean vertical velocity in a rising plume $\bar{w}'$. Here the numerically obtained values of the latter four quantities are plotted for a number of models with different values of R_B/R_c ranging from 1.26 to 10^4. Straight lines have been fit (in the least squares sense) through the individual points for which $R_B/R_c \geqslant 100$. At high values of R_B/R_c the data points all lie very nearly on the straight lines. At low values of R_B/R_c the data points depart from the straight lines. The transition into the power-law regime as R_B increases from its critical value, R_c, occurs at approximately $R_B = 50R_c$. This is the minimum Rayleigh number noted previously for which the mean temperature field adopts its characteristic boundary layer form (see Fig. 7.4). Thus the power law relationships are only valid in the boundary layer regime.

The slopes of the linear curves in Figure 7.5 are equal to the exponents in the power law relations, while the left-hand intercepts of these curves (at $R_B/R_c = 1$) are equal to the coefficients of proportionality. Thus from Figure 7.5 we find

$$\mathrm{Nu} = 2.24\left(\frac{R_B}{R_c}\right)^{0.318}, \tag{29}$$

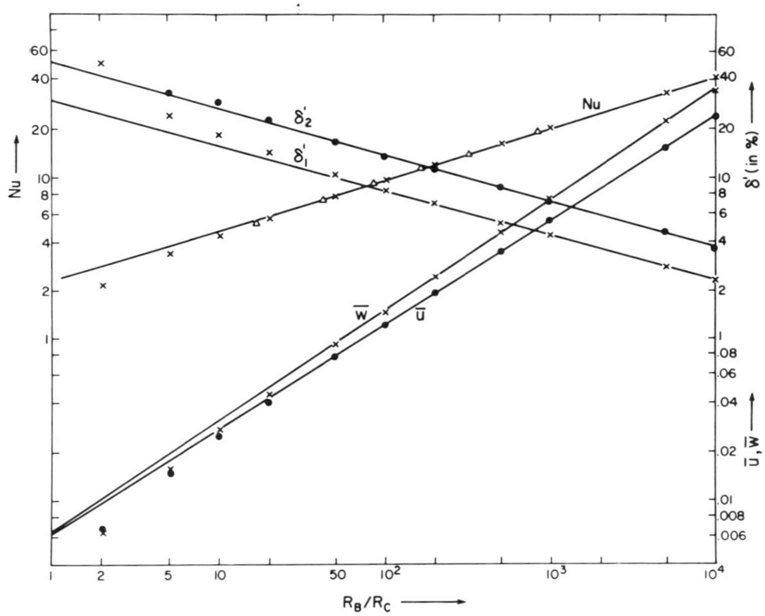

Figure 7.5. Nusselt number Nu, dimensionless boundary layer thickness δ', mean (dimensionless) horizontal velocity at upper surface $\bar{u}$, and mean (dimensionless) vertical velocity in center of rising plume $\bar{w}$, all plotted against the ratio R_B/R_c. All variables are plotted on logarithmic scales. Points indicated by solid circles or Xs at different values of R_B/R_c correspond to different numerical solutions. Straight lines are least-squares best fits to points in the range $R_B/R_c \geqslant 100$. The lines labelled δ'_1 and δ'_2 refer to the boundary layer thickness as defined by the depth at which the mean temperature is first reached, and the depth of the peak overshoot of temperature above the mean, respectively. The triangles plotted along the Nu line indicate values of Nu computed by Moore and Weiss (1973). Velocities may be re-dimensionalized by multiplying by 3.71 cm/yr for a whole mantle circulation or by 9.24 cm/yr for an upper mantle circulation.

$$\delta'_1 = 0.300\left(\frac{R_B}{R_c}\right)^{-0.278}, \tag{30}$$

$$\delta'_2 = 0.518\left(\frac{R_B}{R_c}\right)^{-0.285}, \tag{31}$$

$$\bar{u}' = 8.468\left(\frac{R_B}{R_c}\right)^{0.645}, \tag{32}$$

$$\overline{w}' = 8.671\left(\frac{R_B}{R_c}\right)^{0.684}, \tag{33}$$

where dimensional values of boundary layer thickness and velocity are obtained by multiplying the appropriate dimensionless expressions above by d and (in this instance only) κ/d, respectively. These numerically determined relations are compared to the predictions of boundary layer theory in Table 7.4 below.

Our value of 0.318 for the exponent in the Nusselt number–Rayleigh number relation agrees with the predictions of boundary layer theory (blt) to within 6 percent but, significantly, is less than the blt value of one third. This is in accord with the experimental results discussed previously for the case of rigid boundaries (see Section 3.1). Similarly the magnitudes of the exponents in expressions (30) and (31) for boundary layer thickness are less than the blt prediction of one third. They are also less than the exponent in (29) for the Nusselt number. Thus as R_B/R_c increases, the boundary layer thickness is predicted to decrease less rapidly than the Nusselt number increases.

The exponents of 0.645 and 0.684 for $\overline{u}'$ and $\overline{w}'$ are both within 3 percent of the blt prediction of a value of two-thirds for both quantities. However our numerically determined exponent for $\overline{w}'$ is greater than two thirds while that for $\overline{u}'$ is less than two-thirds. This quantifies the statement made earlier (in the discussion of the streamlines plotted in Fig. 7.2) that the vertical velocities in plumes increase more rapidly with R_B than do the horizontal velocities at the upper and lower surfaces.

The success of boundary layer theory in predicting the correct power law dependencies of the various quantities upon R_B/R_c is well demonstrated by Figure 7.5 and Table 7.4. Moreover although Turcotte and Oxburgh's (1967) original boundary layer theory could not account for the overshoot of the boundary layer temperature profiles beyond the temperature of the isothermal core, recent blt analyses by Roberts (1979) and Olson and Corcos (1980)

Table 7.4. Comparison of numerical and theoretical power law relations

	Coefficient (β_i)		Exponent (S_i)	
	Numerical	Theoretical	Numerical	Theoretical)
$Nu = \beta_1(R_B/R_c)^{S_1}$	2.24	0.93	0.318	0.333
$\delta_1 = \beta_2 d(R_B/R_c)^{S_2}$	0.300	0.465	−0.278	−0.333
$\delta_2 = \beta_3 d(R_B/R_c)^{S_3}$	0.518	0.465	−0.285	−0.333
$\overline{u} = \beta_4(\kappa/d)(R_B/R_c)^{S_4}$	8.468	10.81	0.645	0.666
$\overline{w} = \beta_5(\kappa/d)(R_B/R_c)^{S_5}$	8.671	18.98	0.684	0.666

do predict this feature. However, for convection between free boundaries, boundary layer theory does not appear to determine the constant of proportionality very accurately. Roberts (1979) compared his blt predictions of the coefficient in the Nu-R_B relationship with values derived from the Galerkin analysis of Straus (1972) and from the finite-difference solutions of Moore and Weiss (1973). For rolls with an aspect ratio of one, Roberts' value disagrees with the numerically derived values by approximately 40 percent while the mean values obtained numerically agree to within 1 percent of each other. The blt coefficient also shows a much stronger dependence on aspect ratio than do those obtained from the exact numerical solutions. It would therefore appear that for the case of convection between free boundaries the absolute value of the Nusselt number, and hence of the temperatures within the boundary layers, cannot be accurately determined by boundary layer theory. However, for convection between rigid boundaries Mitrovica and Jarvis (1987) have recently found that numerically derived Nusselt numbers agree, to within 6 percent, with the blt predictions of Roberts (1979). Thus it appears that boundary layer theory is most limited in its application to the case of free boundaries. Mitrovica and Jarvis (1987) suggested that this limitation is due to the neglect of significant contributions to the local energetics of heat transfer within the thermal boundary layers of free boundary convection.

Boundary Layer Energetics

Horizontal Boundary Layers

A basic assumption made in most boundary layer theories of convection is that the horizontal component of velocity, u, is a constant function of the horizontal coordinate within the horizontal boundary layers. In Figure 7.6 $u(x)$ at the upper surface is plotted as a function of R_B/R_c. Relatively constant velocities, at high Rayleigh numbers, are confined to a central zone covering about 50 percent of the cell. Although the extent of this zone increases with R_B the dependence is very weak. Since the maximum value of R_B/R_c appropriate to mantle convection is about 10^4, plate-like surface motions (i.e., $\partial u/\partial x = 0$)could not be produced in a constant viscosity mantle.

The importance of $\partial u/\partial x \neq 0$ can be appreciated by considering the steady-state temperature equation. In the absence of internal heat sources

$$u\frac{\partial T}{\partial x} + w\frac{\partial T}{\partial z} = \kappa\frac{\partial^2 T}{\partial x^2} + \kappa\frac{\partial^2 T}{\partial z^2}. \tag{34}$$

At the upper surface $w = 0$, and from continuity $\partial w/\partial z = -\partial u/\partial x$; hence if $\partial u/\partial x = 0$ in the boundary layer, then $w = 0$ and the second term in (34),

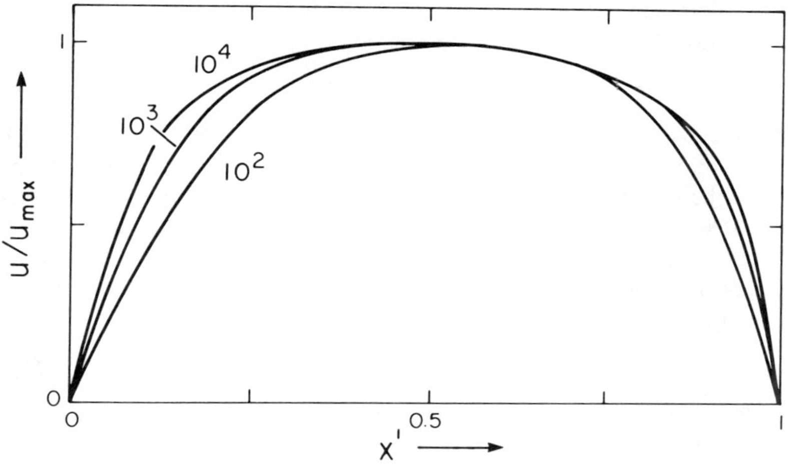

Figure 7.6. Horizontal velocity at the upper surface of model convection cells for various values of R_B/R_c (as labelled) expressed as a ratio of their respective maximum values u_{max}.

which represents the rate of temperature change due to vertical advection, vanishes. With the additional assumption that $\partial^2T/\partial x^2 << \partial^2T/\partial z^2$ in horizontal boundary layers, the equation governing heat transfer in the thermal boundary layer becomes

$$u\frac{\partial T}{\partial x} = \kappa\frac{\partial^2 T}{\partial z^2}. \tag{35}$$

However, when $\partial u/\partial x \neq 0$, $w \neq 0$, and although $u >> w$ we also have $\partial T/\partial x << \partial T/\partial z$ and, therefore, the neglect of vertical advection cannot be justified.

Jarvis and Peltier (1982) examined the relative importance of conduction and advection in the horizontal and vertical directions by integrating (34) over volumes δV with square cross-section of dimension $2h$ (where h is the mesh spacing of the finite-difference grid) centered on grid-points within the thermal boundary layer.

The net contribution to the local energy flux balance due to horizontal advection, vertical advection, horizontal conduction and vertical conduction were denoted ΔAD_H, ΔAD_V, $\Delta COND_H$ and $\Delta COND_V$ respectively. Each contribution represents the net energy flux out of a volume element due to one mode and direction of heat transport. Away from the plumes it is generally assumed that horizontal advection balances vertical diffusion (as in equation (35)) or, in the above notation

$$\Delta AD_H = -\Delta COND_V, \tag{36}$$

and that ΔAD_V and $\Delta COND_H$ are negligible. A plot of the variations of each contribution is shown in Figure 7.7 along a row of elements one grid plane below the upper surface of a model with $R_B = 10^3 R_c$. Figure 7.7 confirms that $\Delta COND_H \approx 0$. However, throughout much of the central zone vertical advection (ΔAD_V) is comparable to or greater than vertical conduction. Except in the near vicinity of the midpoint of the cell, where $w \approx 0$, the correct energetic balance is

$$\Delta AD_H + \Delta AD_V = -\Delta COND_V. \tag{37}$$

The overall effect at the surface is to produce a surface heat flow which is somewhat greater than predicted by boundary layer theory to the left of center and less than predicted to the right.

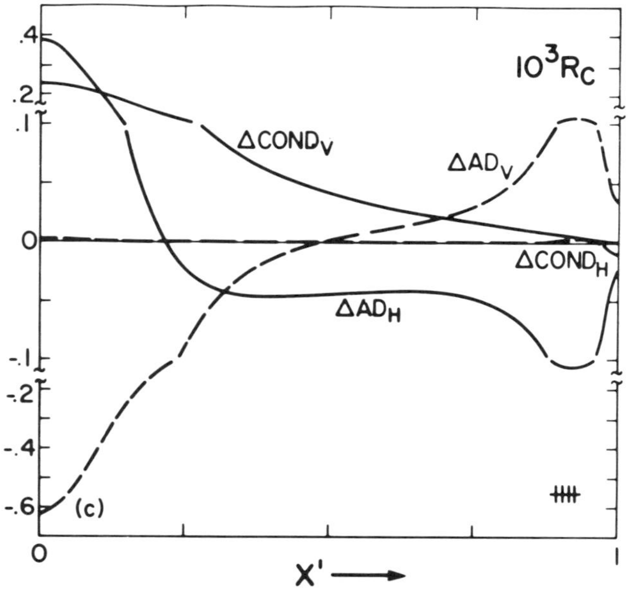

Figure 7.7. Net horizontal and vertical, advective and conductive fluxes outwards across the bounding surface, S, surrounding square volume elements one grid plane below the upper surface.

In Figure 7.8 we have plotted the surface heat flow F_u, computed at the upper surface of a convection model with $R_B = 5 \times 10^3 R_c$ and compared this with a curve which is proportional to $x^{-1/2}$ and constrained to fit the computed curve at the point indicated by the small arrow. A Rayleigh number $R_B = 5 \times 10^3\ R_c$ is close to our estimate of the maximum value appropriate to convection in the Earth's mantle (see Table 7.3). As anticipated from the discussion of Figure 7.7 the computed heat flow falls below the reference $x^{-1/2}$ curve to the right of center and above it to the left of center. The discrepancy to the left of center is mitigated by the fact that the reference curve approaches infinity as $x \rightarrow 0$, whereas the computed flux remains finite, thus necessitating a second cross-over of the two curves. Away from the edges, the overall trend of the computed curve does not deviate substantially from an $x^{-1/2}$ behavior. Thus boundary layer theory provides a reasonable prediction of the power law dependence of F_u on x, particularly to the left of center. (However, as noted above, the actual magnitude of F_u is accurately predicted by blt only for the case of convection between rigid boundaries.)

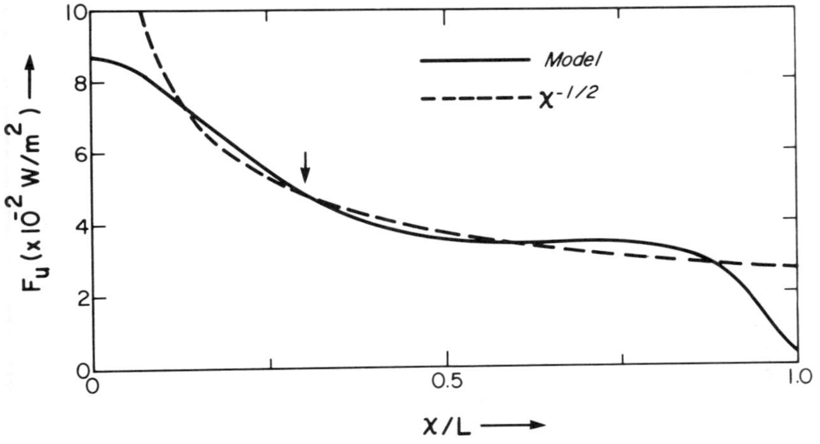

Figure 7.8. Variation of surface heat flux across the upper surface of a convection model with $R_B = 5 \times 10^3\ R_c$ (solid line) compared to a reference curve proportional to $x^{-1/2}$ (broken line) which is constrained to have the same value as the model prediction at the point indicated by the small arrow.

Vertical Plumes

The thermal structure and energetics of rising vertical plumes are briefly examined here. The characteristic shape of a plume, as defined by the location of the mean temperature isotherm, is shown in Figure 7.9 for a sequence of models at different Rayleigh numbers. For each plume the vertical arrow indicates the direction of the flow and the center line of the plume. The tail of each arrow indicates the height at which the plume is narrowest. The horizontal component of velocity, converging towards the plume axis in the lower half of the cell, constrains the plume to be relatively narrow in this region. Moving upwards from the base of the arrow, the plume broadens diffusively as horizontal velocities initially decrease and then change sign. In the uppermost regions diverging horizontal flow sweeps the isotherms horizontally. The horizontal bar on each plume axis (above the arrow) indicates the depth of the thermal boundary layer above the diverging flow.

An examination of the local energy balance within the plume, in the manner described previously for the horizontal boundary layers, indicates that the energetic balance may be expressed as

$$\Delta AD_V + \Delta AD_H \approx 0; \tag{38}$$

that is, vertical advection is balanced primarily by horizontal advection. Horizontal conduction plays a very minor role in the central 15–20 percent of

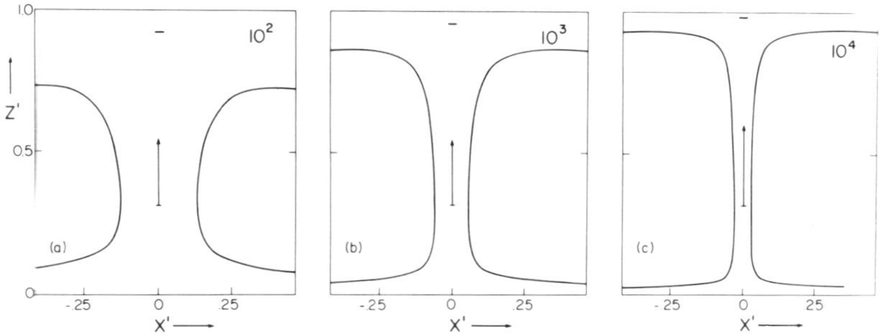

Figure 7.9. Cross sectional shapes of two dimensional plumes (as defined by the isotherm of the mean temperature of the convecting layer) for models at various values of R_B/R_c, as indicated on the figures. Plumes are narrowest at depths indicated by the tail of the central arrows. The depths of the upper thermal boundary layers are indicated by the horizontal bars above the arrows.

the plume and is negligible elsewhere (Jarvis and Peltier, 1982). This result invalidates the assumption that vertical advection is balanced by horizontal conduction, which was employed in the boundary layer theory of Turcotte and Oxburgh (1967) and Oxburgh and Turcotte (1978), and also by Turcotte and Schubert (1982).

3.3. Effects of Internal Heating

In this section we discuss the effects of partial internal heating on the thermal and flow structure of a convecting layer. This is of particular relevance to the geophysical problem of mantle convection since radiogenic elements are believed to be distributed throughout the mantle and are commonly assumed to be important in driving the mantle circulation. Since the actual distribution of the radiogenic elements in the mantle is not known, we will consider only the simple case in which the heat sources are uniformly distributed.

The proportional amount of internal heating in the mantle is described by the ratio μ, defined in equation (17), which represents the total rate of heat generation within the layer expressed as a fraction of the total heat flux across the upper surface of the layer. Thus when $\mu = 0$, heating is entirely from below and when $\mu = 1$, heating is entirely from within. In order to prescribe μ, we must apply a (known) constant heat flux boundary condition at the lower surface and use the Rayleigh number R_R defined in equation (18), to characterize the model solutions.

The major effect of adding internal heat sources into a convection model is to introduce asymmetry into both the stream function and temperature fields. This asymmetry results from the asymmetric modes of heating and cooling the layer as a whole. When internal heat sources are present, less heat is conducted into the bottom of the layer than is conducted out of the top; some of the heat originates throughout the volume of the layer. Consequently, the upper horizontal boundary layer and the cold plumes originating in it are more prominent than the lower thermal boundary layer and the hot plumes rising from it.

The nature of the asymmetry produced by partial internal heating is illustrated in Figure 7.10 in which numerically derived two dimensional fields of T and ψ are juxtaposed for values of $\mu = 0.0, 0.1, 0.2, 0.3$ and 0.6, and $R_R = 5 \times 10^6$. For $\mu \leqslant 0.3$ steady flow solutions are obtained with a unit aspect ratio. As μ increases above zero a relative decrease in prominence of the lower thermal boundary layer is evident in the temperature field and the center of circulation, as revealed by the ψ field, shifts towards the downwelling plumes. For a value of $R_R = 5 \times 10^6$, the flow becomes time dependent

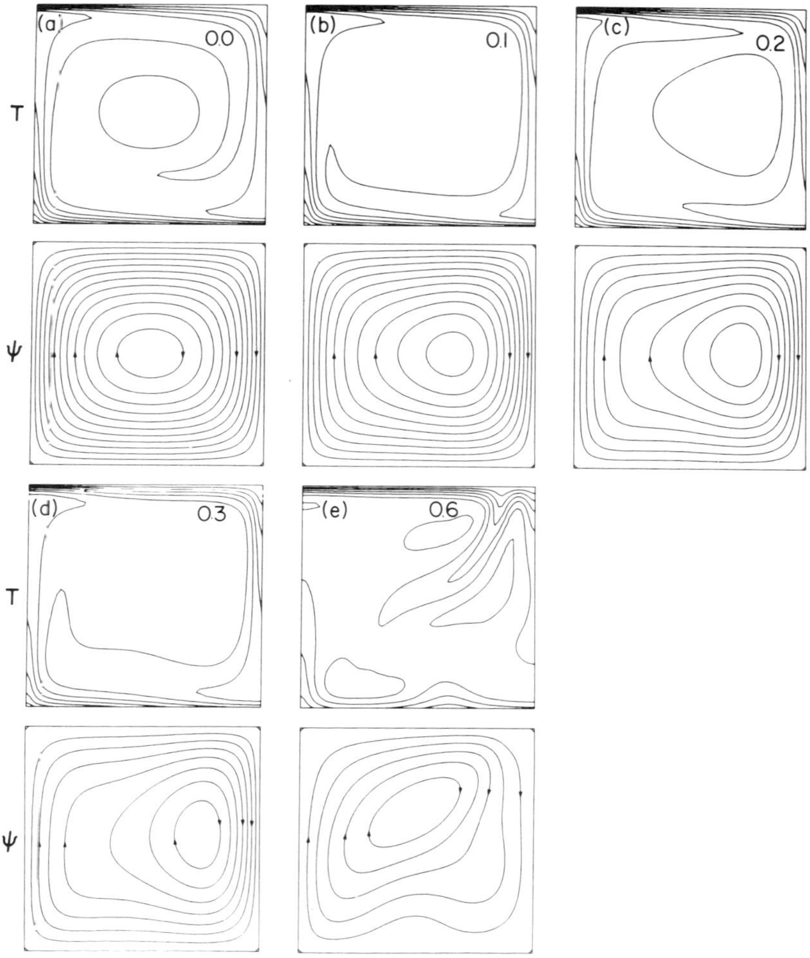

Figure 7.10. Contours of temperature, T, and stream function, ψ, for models with $R_R = 5 \times 10^6$ and, a, $\mu = 0.0$; b, $\mu = 0.1$; c, $\mu = 0.2$; d, $\mu = 0.3$; and, e, $\mu = 0.6$.

when μ is sufficiently large; Figure 7.10e illustrates an instantaneous view of the numerical solution for such a case.

In Figure 7.11a we have plotted horizontal temperature profiles at mid-depth across the steady convection models shown in Figure 7.10 for μ increasing in increments of 0.1 from 0.0 to 0.3. The profiles have been offset vertically for the sake of clarity. The dashed line represents the same reference temperature on each profile. As μ increases, the symmetry between hot and cold temperature fluctuations (associated with hot and cold plumes) is lost. The interior temperature increases towards the cold plume so that the temperature contrast between the cold plume and adjacent material is larger than that between the hot plume and adjacent material.

In Figure 7.11b we have plotted vertical temperature profiles half-way across the temperature fields for the same cases as in Figure 7.11a. In this series of profiles it is the symmetry between the upper and lower thermal boundary layers which is lost as μ increases. Moreover, as μ increases the distribution of temperatures in the interior deviates from the isothermal distribution which is characteristic of the case when $\mu = 0$, and the temperature overshoot below the upper thermal boundary layer becomes less prominent.

As illustrated in Figure 7.10, the higher the proportion of internal heating (i.e., μ) the greater the asymmetry becomes. Although this qualitative aspect of the results may have been anticipated, it is perhaps less obvious that for a given fractional amount of internal heating, the degree of asymmetry will

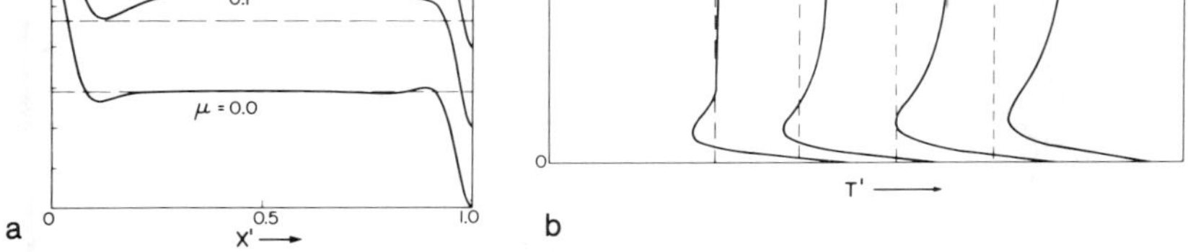

Figure 7.11. a, Horizontal variation of temperature at mid-depth for four model solutions each with $R_R = 5 \times 10^6$ but with different values of μ as indicated on each profile. b, Vertical variation of temperatures at horizontal mid-points of cells for the same models shown in a.

increase rapidly with increasing Rayleigh number. This effect is illustrated in Figure 7.12, wherein we have plotted contours of T and ψ (similar to Figure 7.10) for a fixed value of $\mu = 0.2$ and three different values of $R_R = 6 \times 10^5$, 5×10^6 and 1.25×10^7. Although R_R only varies by a factor of 20 from frames (a) to (c) in Figure 7.12, the increased asymmetry in the solutions is as great as that seen in Figure 7.10 (for constant R_R and variable μ). In the geophysical context Figure 7.12 indicates that for a given value of μ, greater asymmetry could be expected in a whole mantle circulation than in an upper mantle circulation. However, as discussed below, at the large Rayleigh numbers appropriate to mantle convection the simple form of asymmetry illustrated in Figure 7.12 gives way to a more complex time dependent flow structure (even at low values of μ).

At sufficiently low values of R_R it is possible to achieve steady two dimensional flow solutions for entirely internally heated models (i.e., $\mu = 1$). Figure 7.13 presents an example of the typical thermal and dynamic structure for a steady flow solution at $R_R = 6 \times 10^5$ and $\mu = 1.0$. This solution is characterized by the lack of lower thermal boundary layer and upwelling plume, a pronounced upper thermal boundary layer and downwelling plume and a center of circulation shifted substantially toward the downwelling stream. The close spacing of the descending streamlines and wide spacing of

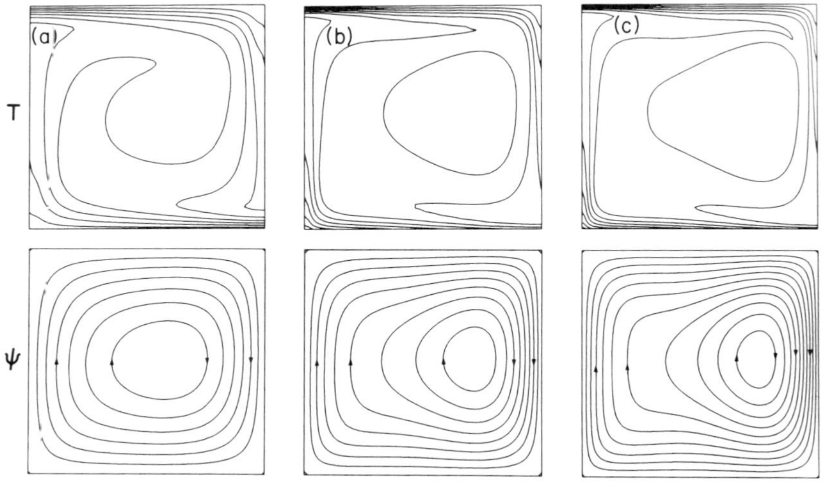

Figure 7.12. Contours of T and ψ for three models each with $\mu = 0.2$ but with different values of R_R: a, $R_R = 6 \times 10^5$; b, $R_R = 5 \times 10^6$; and c, $R_R = 1.25 \times 10^7$.

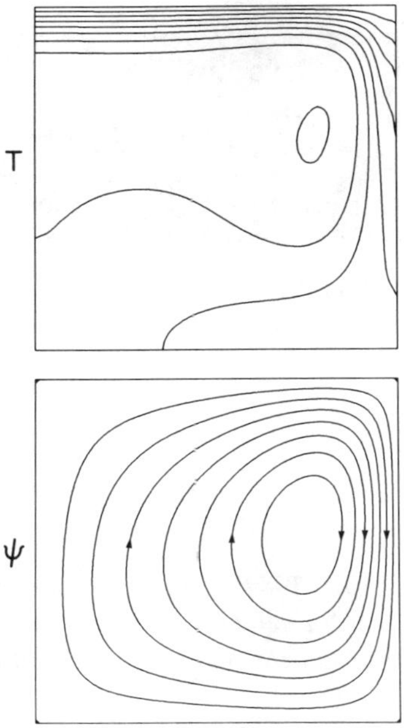

Figure 7.13. Steady model solution for internally heated convection at low Rayleigh number: $R_R = 6 \times 10^5$, $\mu = 1.00$.

upwelling streamlines indicates a rapid narrow descending stream and a broad region of slow upwelling.

Patterns of circulation similar to that shown in Figure 7.13 have also been observed in laboratory experiments of convection in internally heated fluid layers. Figure 7.14 presents an example of a time exposed "streak photograph" obtained by De la Cruz (1973) of flow patterns in an electrolytically heated fluid. The narrow downwelling and broad upwelling regions are readily apparent. This agreement in the general form of the circulation as determined experimentally and numerically is encouraging since it implies that the numerical solutions are describing the physics of internally heated convection correctly. However, the parameter range in which steady internally heated flows occur is quite limited. For example, laboratory experiments on internally heated fluids at high Rayleigh numbers yield time dependent, disordered, flows (Carrigan, 1982). Similarly, McKenzie *et al.* (1974) found from numerical modelling that internally heated flow solutions, for aspect ratio one Boussinesq fluids, were unstable for $R_R > 1.4 \times 10^6$. At the same Rayleigh number, however, steady flow solutions were obtained with $\mu = 0.0$ and $\mu = 0.5$. At a larger value of $R_R = 2.5 \times 10^7$, Jarvis and

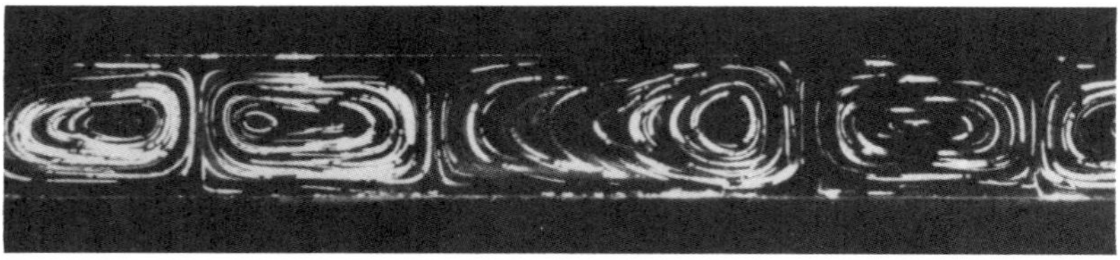

Figure 7.14. Streak photograph of internally heated convective flow pattern in the vertical plane, for the case $\mu = 1$.

Peltier (1982) found that aspect ratio one model solutions were time dependent for values of μ as low as 0.1, and Jarvis (1984) subsequently reported time dependence for models with $\mu = 0.0$ at Rayleigh numbers of $R_R = 1.5 \times 10^9$.

A time sequence of instantaneous views of the time dependent temperature and stream function fields of the latter model solution is reproduced here as Figure 7.15. In the first frame (i.e., the left frame) of the temperature field, the head of a mature hot plume (labelled A) has risen almost to the surface while its source region (labelled B) has been swept into the major upwelling at the left-hand vertical boundary. In the next two temperature frames of Figure 7.15, the source region B rises rapidly at the left, while a new cold plume develops from the upper thermal boundary layer.

This discussion indicates that time dependence likely occurs at large Rayleigh numbers for all values of μ. Furthermore, we expect that time dependence would occur in the Earth even at low Rayleigh numbers. The steady model solutions discussed above are all artificially stabilized by virtue of their two dimensionality, assumed incompressibility (Jarvis and McKenzie, 1980; section 4 below) and imposed unit aspect ratio. Whereas McKenzie *et al.* (1974) obtained steady internally heated convection solutions with $\mu = 0.0$ and 0.5 at $R_R = 1.4 \times 10^6$ in models with unit aspect ratios, the same models yielded time-dependent flows when the aspect ratios were increased to a value of 2.0. Nevertheless, employing steady model solutions does facilitate the comparisons between models obtained with partial internal heating and those, discussed above, which are heated entirely from below.

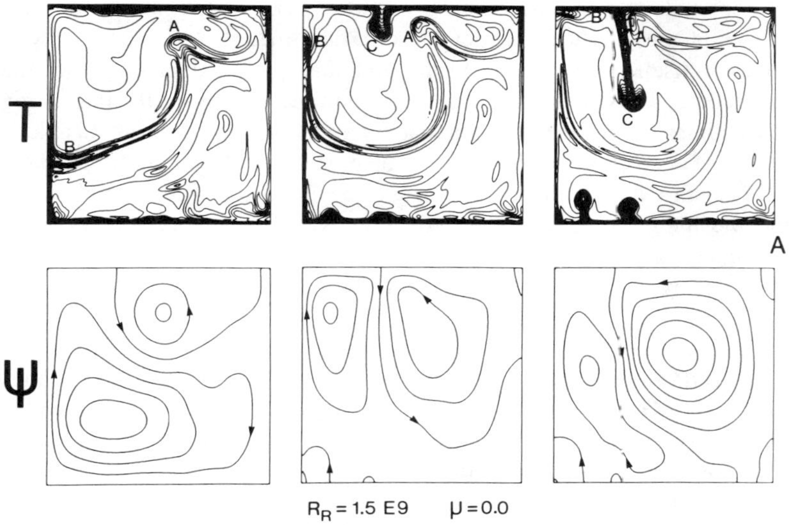

Figure 7.15. Time dependent whole mantle convection with $R_R = 1.5 \times 10^9$ and $\mu = 0.0$. Contours of T and ψ are shown at three successive times. Times relative to the left-hand frame are: 0.00 Ma, 6.96 Ma, and 13.5 Ma. Contour intervals are the same for each frame: dT, 31.3°C, $d\psi$, 6.08×10^{-5} m^{2}s^{-1}.

4. Compressibility and Viscous Dissipation

Before attempting to draw geophysical implications from the simple Boussinesq models discussed above, it is necessary to establish: (1) the conditions under which the Boussinesq approximation would be valid for convection in the Earth's mantle; (2) whether or not these conditions are met in the mantle; and (3) what departures from the predicted model results could be expected if these conditions are not met. As discussed in section 2.1 above, the Boussinesq approximation can be justified, through scale analysis, provided $\alpha\Delta T^* << 1$ and $d/H_T << 1$ (where ΔT^* is the maximum temperature variation in the layer of depth d, and H_T is the density scale height—see equation 2). However, in the geophysical context Jarvis and McKenzie (1980) have estimated that the appropriate value of d/H_T is either

0.6 for whole mantle convection, or 0.3 for upper mantle convection. Since neither of these values satisfies the criterion $d/H_T << 1$, scale analysis cannot in itself justify the use of the Boussinesq approximation for mantle convection. Thus although valid for a number of physically interesting convecting systems, the Boussinesq approximation may not be applicable to studies of deep convection in, for example, stellar interiors, the atmosphere and planetary mantles in general. In such deep layers d is comparable to H_T and the influence of viscous dissipation, variable adiabatic temperature gradients and density gradients (effects of order d/H_T) cannot be neglected. The basic hydrodynamic equations must be solved retaining all the relevant non Boussinesq terms.

The Boussinesq condition on d/H_T was relaxed by Ogura and Phillips (1962) and Gough (1969) in their one-parameter expansions for deep gaseous layers. These expansions lead to the anelastic equations for a gas, provided the relative potential temperature fluctuation (Ogura and Phillips) or the square of the Mach number M (ratio of convective velocity to sound velocity) (Gough) is small. In the anelastic approximation $\partial\rho/\partial t = 0$ and consequently acoustic modes do not exist. In the limit of shallow layers the anelastic-gas equations reduce to the Boussinesq equations.

In this section we shall consider an approximate set of equations, introduced by Jarvis (1978) and Jarvis and McKenzie (1980), for an infinite Prandtl number Newtonian fluid in which it is assumed that $d = O(H_T)$ and that $\alpha T << 1$, where T is the absolute temperature. In gases $\alpha \approx 1/T$ and hence αT is O(1). The condition that αT be small thus restricts the application of these equations to liquids and creeping solids for which α is generally small and only weakly dependent on T. For the mantle it can be shown that $M^2 << 1$ (Jarvis and McKenzie, 1980) and accordingly elastic waves can be ignored. Jarvis and McKenzie referred to their set of approximate equations as the *anelastic-liquid* equations. These have the virtue of being simpler than the anelastic-gas equations and facilitate the extension of studies of finite amplitude convection in infinite Prandtl number fluids into the compressible domain in a mathematically consistent manner.

Turcotte *et al.* (1974) included the non-Boussinesq effects of dissipative heating and adiabatic gradients in their study of incompressible steady state convection. The effect of increasing d/H_T in their study was to reduce the vertical heat transport. This was attributed to the corresponding increase of the adiabatic gradient. However, when all of the non-Boussinesq effects of order d/H_T were retained Jarvis and McKenzie (1980) found that the variation of the adiabatic gradient has the dominant effect on the solutions for $d/H_T \leqslant 1.0$ but that density gradients become important when $d/H_T > 1.0$.

The energetics of convection in compressible liquids were studied by Hewitt *et al.* (1975) and Backus (1975). These authors concluded that the ratio of the overall rate of viscous heating, Φ, to the total heat flux emerging from the system, F_u, is bounded above by the expression $(T_{max} - T_{min})/T_{min}$, where T_{max} and T_{min} are the maximum and minimum temperatures in the system. Hewitt *et al.* (1975) showed further that the ratio Φ/F_u is independent of Rayleigh number and proportional to d/H_T. Hence for $T_{max} > 2T_{min}$, or $d/H_T > 1$, Φ may be greater than F_u. Although Φ/F_u may be regarded as the efficiency of conversion of heat into mechanical energy within the convecting layer (Malkus, 1973), the heat generated by dissipation remains within the system and Φ/F_u is therefore not constrained to be less than unity. Hewitt *et al.* (1975) derived the following relation for a vigorously convecting system in which most of the heat is carried by convection:

$$E = \frac{\Phi}{F_u} = \frac{d}{H_T}\left[1 - \frac{r}{2}\right], \tag{39}$$

where r (in this section only) denotes the ratio of the rate of internal heating to the total heat flux across the upper surface. When heating is entirely from below, $E = d/H_T$. Equation (39) was verified by Hewitt *et al.* in the Boussinesq limit of $d/H_T = 0.117$ by numerical calculations for a Newtonian fluid with infinite Prandtl number. For the case $r = 0$, this equation was further verified by Jarvis and McKenzie (1980) in terms of their numerical models of convection in compressible liquids for which $d = O(H_T)$. In particular, these authors provided an example of convection in which $\Phi > F_u$.

Although it may seem curious (or even contradictory) that the overall rate of internal dissipative heating in a convecting layer may be greater than the heat flux emerging from the layer, this result can be understood in terms of the simple energetic considerations discussed below.

4.1. Convection Energetics

The general form of the pointwise differential hydrodynamic equations describing the conservation of mass, momentum and energy may be written as

$$\frac{\partial \rho}{\partial t} + \boldsymbol{\nabla}\cdot(\rho \mathbf{v}) = 0, \tag{40}$$

$$\rho\left(\frac{\partial \mathbf{v}}{\partial t} + \mathbf{v}\cdot\boldsymbol{\nabla}\mathbf{v}\right) = -\boldsymbol{\nabla}P + \rho\mathbf{g} + \frac{\partial \tau_{ij}}{\partial x_j}, \tag{41}$$

$$\rho C_p\left\{\frac{\partial T}{\partial t} - \frac{\alpha T}{\rho C_p}\frac{\partial P}{\partial t} + \mathbf{v}\cdot[\boldsymbol{\nabla} T - (\boldsymbol{\nabla} T)_s]\right\}$$

$$= \boldsymbol{\nabla}\cdot(K\boldsymbol{\nabla} T) + H + \tau_{ij}\frac{\partial v_i}{\partial x_j}, . \qquad (42)$$

Under the Boussinesq approximation these equations simplify to those of equations (5) through (7), given earlier. Summation over repeated indices is implied in the terms involving the stress tensor. The term $(\boldsymbol{\nabla} T)_s$ in equation (42) refers to the adiabatic temperature gradient and is defined as

$$(\boldsymbol{\nabla} T)_s = \frac{\alpha T}{\rho C_p}\boldsymbol{\nabla} P . \qquad (43)$$

An equivalent form of (42), which will be useful in the discussion below, is

$$\frac{D\epsilon}{Dt} + P\boldsymbol{\nabla}\cdot\mathbf{v} = \boldsymbol{\nabla}\cdot(K\boldsymbol{\nabla} T) + H + \tau_{ij}\frac{\partial v_i}{\partial x_j}, \qquad (44)$$

where ϵ is the internal energy per unit mass.

In a convecting system energy is continuously transformed from one form to another. Heat conducted into the bottom of a layer of liquid increases the internal energy, thermal expansion creates potential energy, the resulting motions convert potential into kinetic energy, viscous forces dissipate kinetic energy as heat and finally heat is conducted out the top of the layer. Figure 7.16 illustrates this energy transfer cycle schematically. Equation (42), or (44), expresses the local rate of change of thermal energy per unit volume. A similar expression for the local rate of change of mechanical (i.e., kinetic and potential) energy per unit volume may be derived by forming the scalar product of $\mathbf{v}$ with the momentum equation (41). After some manipulation the resulting expression is

$$\frac{\partial(\frac{1}{2}\rho v^2)}{\partial t} + \frac{\partial(-\rho\xi)}{\partial t} = \frac{1}{2}\frac{v^2\partial\rho}{\partial t} - \rho v_j\frac{\partial(\frac{1}{2}v^2)}{\partial x_j}$$

$$- v_i\frac{\partial P}{\partial x_i} + \frac{\partial(\rho v_j\xi)}{\partial x_j} - \frac{\rho\partial\xi}{\partial t} + \frac{\partial(v_i\tau_{ij})}{\partial x_j} - \tau_{ij}\frac{\partial v_i}{\partial x_j}, \qquad (45)$$

where $v^2 = \mathbf{v}\cdot\mathbf{v}$ and ξ is the gravitational potential such that $\mathbf{g} = \boldsymbol{\nabla}\xi$. (If $\mathbf{g}$ is assumed to have a constant value and to be directed vertically downwards, then $\xi = -gz + c$, where c is a constant.) The processes contributing to the *overall* rates of energy transfer are determined by evaluating the integrals of the differential equations over the volume of the convecting system.

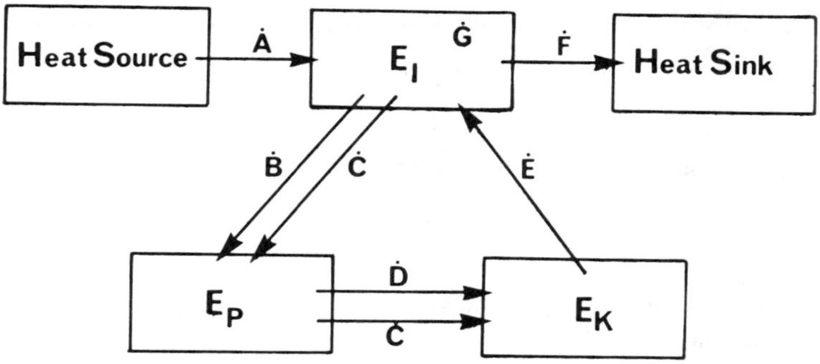

Figure 7.16. Energy transfer cycle: Internal energy, E_I, is increased at a rate $\dot{A}$ by heat conducted into the system and $\dot{G}$ by internal heat sources. Internal energy is converted into potential energy, E_p, at a rate $(\dot{B} + \dot{C})$, potential energy is converted into kinetic energy, E_k, at a rate $(\dot{D} + \dot{C})$ and kinetic energy is converted into internal energy at rate $\dot{E}$. Finally internal energy is reduced by conduction of heat out of the system at a rate $\dot{F}$.

The total potential energy, E_p, kinetic energy, E_k, and internal energy, E_I, are defined as

$$E_p = \int_v -\rho\xi \, dV, \tag{46}$$

$$E_k = \int_v \tfrac{1}{2}\rho v^2 \, dV, \tag{47}$$

and

$$E_I = \int_v \rho\epsilon \, dV. \tag{48}$$

The overall rates of change are therefore

$$\dot{E}_p = \int_v -\frac{\partial(\rho\xi)}{\partial t} dV = -\frac{d}{dt}\int_v \rho\xi \, dV, \tag{49}$$

$$\dot{E}_k = \int_v \frac{\partial(\tfrac{1}{2}\rho v^2)}{\partial t} dV = \frac{d}{dt}\int_v \tfrac{1}{2}\rho v^2 \, dV, \tag{50}$$

$$\dot{E}_I = \int_v \frac{\partial(\rho\epsilon)}{\partial t} dV = \frac{d}{dt}\int_v \rho\epsilon \, dV, \tag{51}$$

where d/dt is used to represent the time derivative, since the volume integral is a function of time only.

With some manipulation of the above equations it can be shown that

$$E_p = \int_v P_H \boldsymbol{\nabla}\cdot\mathbf{v}\, dV + \int_v \rho_1 g w\, dV, \tag{52}$$

where P_H is hydrostatic pressure and $\rho_1 = \rho - \rho_H$ is the deviation of density from a hydrostatic density distribution;

$$\dot{E}_k = \int_v P_1 \boldsymbol{\nabla}\cdot\mathbf{v}\, dV - \int_v \rho_1 g w\, dV - \int_v \tau_{ij}\frac{\partial v_i}{\partial x_j} dV, \tag{53}$$

where $P_1 = P - P_H$ is the deviation of P from hydrostatic and w is the vertical component of fluid velocity;

$$\dot{E}_I = -\int_v P\boldsymbol{\nabla}\cdot\mathbf{v}\, dV + \int_s K\boldsymbol{\nabla}T\cdot\mathbf{n}\, dS + \int_v H\, dV + \int_v \tau_{ij}\frac{\partial v_i}{\partial x_j} dV, \tag{54}$$

where S is the bounding surface of V and $\mathbf{n}$ is a unit outward pointing normal on S; and

$$\dot{E}_{\text{total}} = (\dot{E}_p + \dot{E}_k + \dot{E}_I) = \int_s K\boldsymbol{\nabla}T\cdot\mathbf{n}\, dS + \int_v H\, dV, \tag{55}$$

(e.g., Jarvis, 1978). We can now identify the labels of Figure 7.16 as follows:

$$(\dot{A} - \dot{F}) = \int_s K\boldsymbol{\nabla}T\cdot\mathbf{n}\, dS \tag{56}$$

represents the net inward flux of heat across the bounding surface, S;

$$\dot{B} = \int_v P_H\boldsymbol{\nabla}\cdot\mathbf{v}\, dV \tag{57}$$

is the rate of work done by the fluid expanding against the hydrostatic pressure;

$$\dot{C} = \int_v P_1\boldsymbol{\nabla}\cdot\mathbf{v}\, dV \tag{58}$$

is the rate of work done by the fluid expanding against the non-hydrostatic pressure;

$$\dot{D} = -\int_v \rho_1 g w\, dV \tag{59}$$

is the rate at which buoyancy forces (with respect to a hydrostatic reference field) do work;

$$\dot{E} = \int_v \tau_{ij}\frac{\partial v_i}{\partial x_j} dV \tag{60}$$

is the rate of viscous dissipation of mechanical energy; and

$$\dot{G} = \int_v H\, dV \tag{61}$$

is the rate of total internal heating.

In terms of these labels we have

$$\dot{E}_{\text{total}} = (\dot{A} - \dot{F}) + \dot{G}, \tag{62}$$

$$\dot{E}_p = (\dot{B} + \dot{C}) - (\dot{D} + \dot{C}), \tag{63}$$

$$\dot{E}_k = \dot{C} + \dot{D} - \dot{E}, \tag{64}$$

$$\dot{E}_I = (\dot{A} - \dot{F}) + \dot{G} - (\dot{B} + \dot{C}) + \dot{E}, \tag{65}$$

and

$$\dot{E}_m = (\dot{B} + \dot{C}) - \dot{E}, \tag{66}$$

where $\dot{E}_m = (\dot{E}_p + \dot{E}_k)$ is the overall rate of change of mechanical energy. Equation (62) states that changes in the total energy of the system are due entirely to differences in the rate at which the heat is conducted into and out of the system, $(\dot{A} - \dot{F})$, and to the generation of thermal energy by internal heat sources (such as phase changes, for example). In particular the heating produced by viscous dissipation of mechanical energy $(\dot{E})$ does not contribute to the total energy balance.

In a steady state equations (65) and (66) yield

$$\dot{F} = \dot{A} + \dot{G}, \tag{67}$$

and

$$\dot{E} = (\dot{B} + \dot{C}) = (\dot{D} + \dot{C}). \tag{68}$$

These decoupled equations do not relate the rates of energy transfer in the compression cycle to the heat flux across the system (e.g., $\dot{E}/\dot{F}$). Thus $\dot{E}$ may be arbitrarily large relative to $\dot{F}$ without violating any basic conservation laws. It is this decoupling of the dissipative heating from the global energy balance that enables the ratio Φ/F_u to be greater than one. A constraint can be placed on the ratio of Φ/F_u, however, by equating the left hand sides of (42) and (44) and integrating over V in steady state. The net result is equation (39) derived originally by Hewitt *et al.* (1975).

4.2. Anelastic Liquid Equations

In a Newtonian fluid the deviatoric stress tensor τ_{ij} is related to the velocity field through the dynamic viscosity η and "second" or "bulk" viscosity η_2 as

$$\tau_{ij} = \eta\left(\frac{\partial v_i}{\partial x_j} + \frac{\partial v_j}{\partial x_i} - \tfrac{2}{3}\delta_{ij}\frac{\partial v_k}{\partial x_k}\right) + \nu_2\delta_{ij}\frac{\partial v_k}{\partial x_k}. \tag{69}$$

(For incompressible fluids $\partial v_k/\partial x_k = 0$ and equation (69) reduces to (8).) The bulk viscosity is associated with rapid changes in volume and depends on the relative rates of volume change and molecular relaxation processes (Landau and Lifshitz, 1959). Since volume changes in mantle convection occur over times greater than 10^7 years, η_2 is generally assumed to be unimportant and the last term in (69) may be ignored. In addition Jarvis and McKenzie (1980) also assumed that g, C_p, α, K, η, and Γ (Gruneisen's parameter) were constant parameters. The latter is defined as

$$\Gamma = \frac{\alpha K_s}{\rho C_p} = \frac{\alpha K_T}{\rho C_v}, \tag{70}$$

where K_s and K_T are respectively the isentropic and isothermal moduli of bulk compressibility, and C_v is the specific heat at constant volume.

Jarvis and McKenzie (1980) considered a reference state with hydrostatic pressure, P_H, and an adiabatic temperature distribution, T_s. Provided that fluid motions cause only small departures from this reference state, the equation of state may be written as a Taylor expansion of $\rho(T, P)$ about the reference state density distribution, ρ_r, as

$$\rho(T, P) = \rho_r[1 - \alpha(T - T_s) + K_T^{-1}(P - P_H)]. \tag{71}$$

The reference state density ρ_r is given by the Adam–Williamson relation as

$$\frac{1}{\rho_r}\frac{\partial \rho_r}{\partial z} = -\frac{1}{H_T\Gamma}, \tag{72}$$

or

$$\rho_r(z) = \rho_0 \exp\left(\frac{d - z}{H_T\Gamma}\right), \tag{73}$$

where $\rho_0 = \rho_r(z = d)$ is the value of ρ_r at the upper surface. The equations in this section may be non-dimensionalized as described in Section 2.2 above, with the additional scaling for non-hydrostatic pressure, $P_1 = (P - P_H)$, given by

$$P_1 = \alpha T_1 \rho_0 g d P_1', \tag{74}$$

where T_1 is a characteristic temperature of the system. In terms of dimensionless variables the equation of state (71) becomes

$$\rho = \rho_r\left\{1 - \mu(T - T_s) + \mu\frac{D}{\rho_r\Gamma}[1 + \mu\Gamma T_0 \exp((1 - z)D)]P_1\right\}, \quad (75)$$

where $\mu = \alpha T_1$, $D = d/H_T$ and (70) has been used to express K_T in terms of Γ. Thus to order μ, $\rho = \rho_r$ and (as in the Boussinesq approximation) for $\mu << 1$ slight deviations in ρ from the reference density may be ignored, except in the body force term of the momentum equation where the temperature and pressure dependence of ρ will be seen to provide the driving force of the flow. Since in the geophysical case $\alpha \approx 10^{-5}$ °K^{-1} for mantle type materials, the condition $\mu << 1$ is satisfied for any reasonable value of T_1 in the mantle.

To order μ, the continuity equation becomes $\nabla \cdot \rho_r \mathbf{v} = 0$, or

$$\nabla \cdot \left[\mathbf{v} \exp\left(-\frac{zD}{\Gamma}\right)\right] = 0. \quad (76)$$

The solenoidal nature of equation (76) enables a two-dimensional representation in terms of a dimensionless stream function ψ, such that

$$\mathbf{v} = (u, 0, w) = \left(-e^{zD/\Gamma}\frac{\partial\psi}{\partial z}, 0, e^{zD/\Gamma}\frac{\partial\psi}{\partial z}\right). \quad (77)$$

Furthermore the stream function and velocity components are related to the y-component of vorticity, ω, to order μ, as

$$\nabla^2\psi + \frac{D}{\Gamma}\frac{\partial\psi}{\partial z} = -\omega e^{-zD/\Gamma}, \quad (78)$$

and

$$\nabla^2\omega = e^{(1-z)D/\Gamma}\frac{\partial T}{\partial x} - \frac{D}{\Gamma}\left(\nabla^2 u + \frac{D}{3\Gamma}\frac{\partial w}{\partial x}\right). \quad (79)$$

The dimensionless thermal energy equation, to order μ, takes the form

$$e^{(1-z)D/\Gamma}\frac{\partial T}{\partial t} = -\nabla \cdot [Te^{(1-z)D/\Gamma}\mathbf{v}] - De^{(1-z)D/\Gamma}(T + T_0)w + \kappa_0\nabla^2 T + \epsilon_0$$

$$De^{2zD/\Gamma}\left[\left(\frac{2\partial^2\psi}{\partial x\partial z} + \frac{D}{\Gamma}\frac{\partial\psi}{\partial x}\right)^2 + \left(\frac{\partial^2\psi}{\partial x^2} - \frac{\partial^2\psi}{\partial z^2} - \frac{D}{\Gamma}\frac{\partial\psi}{\partial z}\right)^2 + \frac{1}{3}\left(\frac{D}{\Gamma}\frac{\partial\psi}{\partial x}\right)^2\right]. \quad (80)$$

The coupled system of dimensionless equations (77) through (80) completely determines the velocity and temperature fields as functions of space and time for any given set of initial and boundary conditions. These equations apply for $\mu << 1$ and $D = O(1)$. In the limit $D << 1$ this set of anelastic-liquid equations reduces to the simpler Boussinesq set outlined earlier.

4.3. Finite Amplitude Compressible Convection

Solutions of the anelastic liquid equations can be characterized by a dimensionless parameter, analogous to the classical Rayleigh number, defined as

$$R^0 = \frac{g\alpha d^4}{\kappa_0{}^2 \nu_0}\left(\frac{\bar{\rho}}{\rho_0}\right)^2 f_0, \tag{81}$$

where κ_0, ν_0 and ρ_0 are surface values of κ, ν and ρ, $\bar{\rho}$ is the mean density of the layer, and f_0 is the mean thermometric heat flux across the layer (i.e., $f_0 = -\kappa_0\, \partial T/\partial z$). (In the absence of internal heating f_0 may be specified as a boundary condition at the lower boundary.) R^0 has the same form as the Rayleigh number for incompressible fluids, and accounts for the density variation across the layer. However, it does not give a true indication of the stability of the layer when compared to the usual values of the critical Rayleigh number. Consequently critical values of R^0 must be computed separately for each degree of compressibility; that is for each value of D ($= d/H_T$).

The critical points $[R^0_c(D), k_c(D)]$ as computed by Jarvis and McKenzie (1980) are tabulated in Table 7.5. These data illustrate a general trend of increasing critical wavenumber with increasing D. Jarvis and McKenzie attributed this effect to the shallower depths of unstable regions at larger values of D, the horizontal wavelength being reduced proportionately.

In two-dimensional numerical solutions, the features predicted by linear theory are observed at low values of $R^0/R^0_c(D)$ but are eliminated as R^0 is increased. Moreover, the linear conductive temperature profile in the interior is replaced by an adiabatic central temperature distribution.

Figure 7.17 illustrates the effects of increasing the value of D on suites of

Table 7.5. Critical Values of R_0 and k

d/HT	R^0_c	k_c
0.00	384.7	1.76
0.117	9109.2	1.79
0.25	48,751	2.90
0.50	260,116	4.46
1.00	1,897,970	6.92

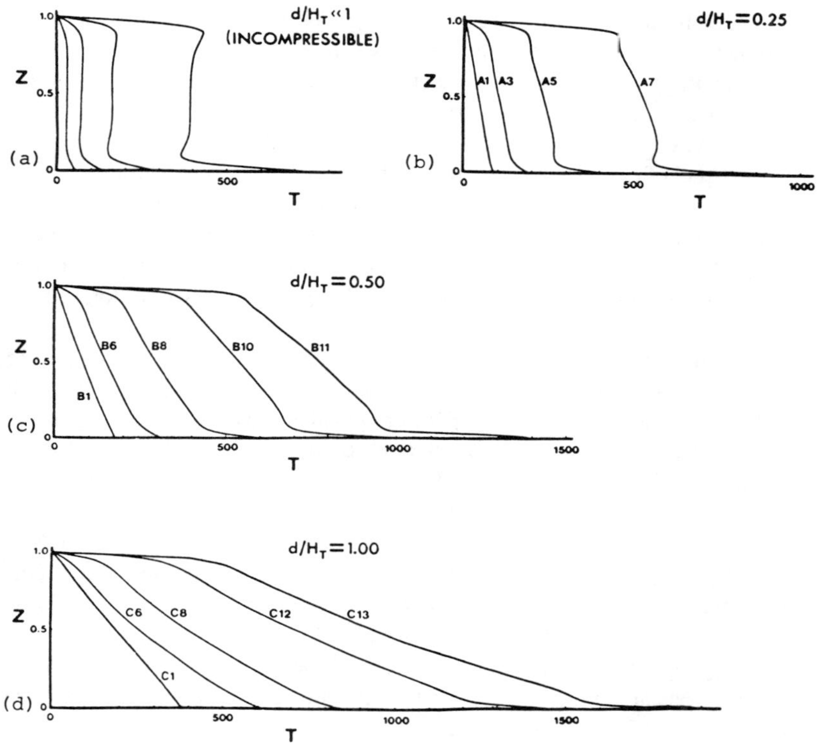

Figure 7.17. Vertical profiles of horizontally averaged temperature. a, Boussinesq solutions from McKenzie *et al.* (1974). b, c and d, Compressible solutions with d/H_T = 0.25, 0.50 and 1.00, respectively. Labels on the compressible profiles indicate specific models of Jarvis and McKenzie (1980). Temperature and height are in dimensionless units.

mean vertical temperature profiles for model solutions at various Rayleigh numbers. It is clear that the approximately isothermal central regions of the incompressible models (Fig. 7.17a) are replaced by approximately adiabatic central cores in the compressible models. The thin conductive thermal boundary layers at the upper surface vary little while those at the bottom become less pronounced as D increases. This behavior occurs because more heat is *conducted* down the steeper adiabatic gradient near the bottom of the layer where *convection* is (accordingly) less vigorous, an effect which becomes more important as D increases. Indeed if D is increased sufficiently, for a fixed bottom flux F, first the lower regions and then the entire layer are stabilized as the amount of heat which is capable of being conducted down the adiabatic gradient exceeds the actual heat flux F. For example, penetrative convection occurs for $D = 1.00$ and R^0 between 2.0×10^6 and 4.5×10^6; that is between curves C1 and C6 in Figure 7.17d.

If a parcel of descending (ascending) fluid is heated (cooled) adiabatically, its entropy remains constant. Thus calculation of the horizontally averaged value of entropy as a function of depth can be used to determine the degree to which the central zone approximates a region of adiabatic variation and to better distinguish the gravitationally unstable boundary layers, notably the lower.

Vertical profiles of the horizontally averaged value of entropy relative to the upper surface, ΔS, are displayed for several models in Figure 7.18. The similarity of these curves to temperature profiles of Boussinesq fluids illustrates the parallel roles of entropy in compressible fluids and temperature in incompressible fluids. The central regions of the compressible models are isentropic, analogous to the isothermal regions of Boussinesq fluids. This behavior also occurs in compressible gases (Graham, 1975). Interior zones of positive entropy gradient (corresponding to sub-adiabatic temperature gradients) analogous to the positive temperature gradients in Boussinesq fluids can also be seen on the profiles of Figure 7.18. The gravitationally unstable boundary layers are clearly demarcated in these profiles. In addition *stable* lower regions are evident on the profiles labelled B1, C1 and C3; these have no counterpart in Boussinesq fluids.

Figures 7.19 and 7.20 show examples of the two dimensional flow solutions obtained for $D = 1.00$ at two different Rayleigh numbers. In each figure contours are shown for temperature T, vorticity ω, stream function ψ, entropy (relative to the upper surface) S, the rate of viscous dissipation (dissipative heating) ϕ $(= \tau_{ij}\partial v_i/\partial x_j)$, and the rate of working due to expansion and contraction W $(= P\nabla\cdot\mathbf{v})$. As with the vertical profiles of Figure 7.18, the entropy contours are the compressible analogs of Boussinesq isotherms. They delineate regions of non-adiabatic temperature variations. Figure 7.19 reproduces Jarvis and McKenzie's (1980) model C1 which convects weakly with a Rayleigh number just slightly above critical; $R^0/R^0_c = 1.07$. The entropy contours indicate the region of buoyant convection as an isentropic layer overlying a gravitationally stable zone. The zero contours of ψ reveal three pairs of countercells in the stable zone which are driven by viscous coupling to the upper pair of convection cells. Since this is a steady state solution, Equations (54) and (55) imply that the volume integrals of ϕ and W should be equal. Numerical integration of the ϕ and W fields confirms this

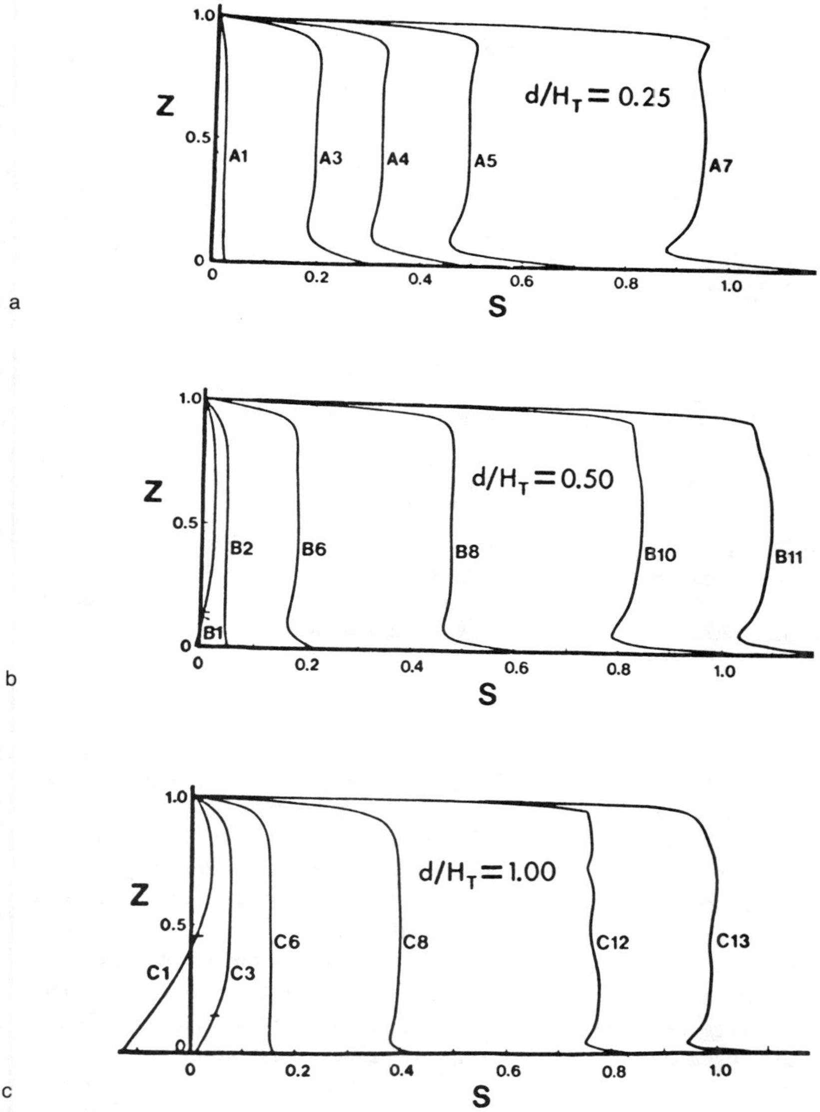

Figure 7.18. Vertical profiles of horizontally averaged entropy (measured relative to the value at the upper surface) for compressible model solutions for: a, $d/H_T = 0.25$; b, $d/H_T = 0.50$; and c, $d/H_T = 1.00$. The Rayleigh number increases from left to right in each set of profiles. Entropy and height are in dimensionless units.

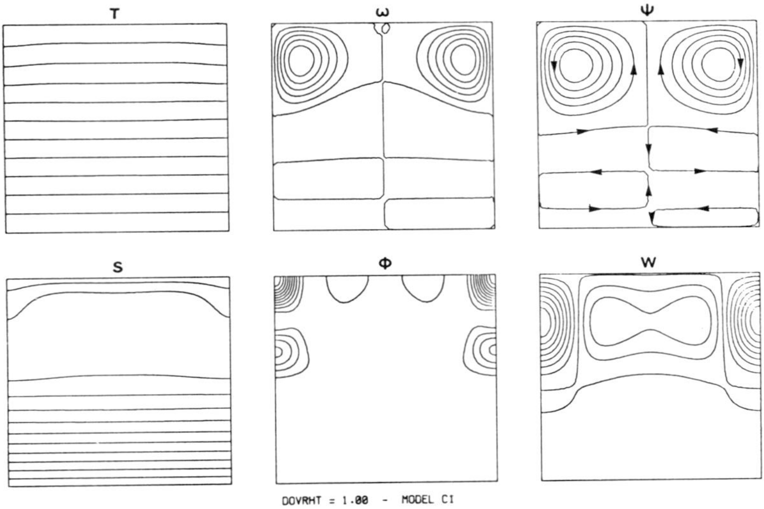

Figure 7.19. Steady numerical solution of Jarvis and McKenzie's (1980) compressible convection model C1, with d/H_T = 1.00 and R^0/R^0_c = 1.07. Contours of temperature T, vorticity ω, stream function ψ, entropy S, rate of viscous dissipation of mechanical energy ϕ, and rate of working in adiabatic volume changes, W.

equality to within 2.5 percent despite the disparate distribution of contours of the two fields. This numerical agreement is to be expected in such steady state models since it is predicted theoretically. However, even in the time dependent models discussed below the equality of the volume integrals of ϕ and W is preserved although it has not been demonstrated theoretically that this result should persist in time dependent flows.

As the Rayleigh number is increased above that for marginal convection the main convection cells penetrate deeper into the stable bottom layer and the flow becomes time dependent at relatively low Rayleigh numbers. Figure 7.20 shows a typical view of the time dependent solution of a model (C10) with $R^0/R^0_c = 15$. At this instant two rolls occupy the solution domain. The extremely concave side of the left-hand roll is a result of intensive frictional heating (ϕ) adjacent to the descending plume. The temperature contours (T) show a marked rise in temperature next to the descending plumes which is most intense at the left. This generates vorticity of the opposite sense to the main circulation of the left-hand cell. Concave streamlines result. The same

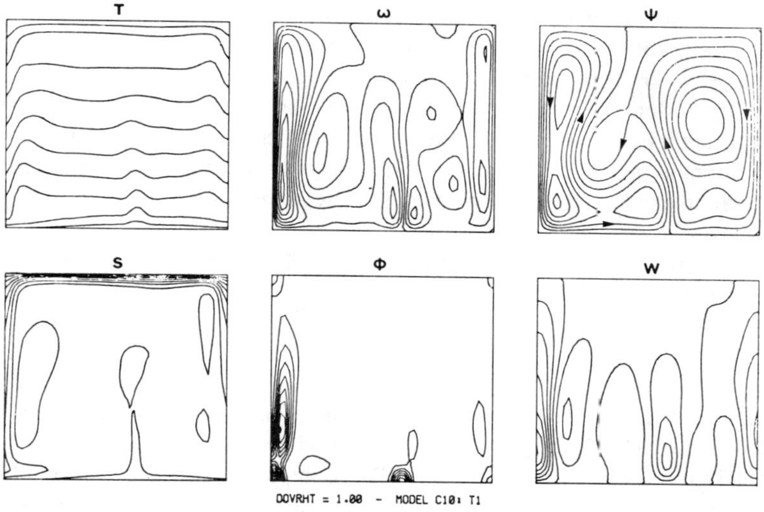

Figure 7.20. An example of a time dependent compressible convection solution from Jarvis and McKenzie's (1980) model C10, with $d/H_T = 1.00$ and $R^0/R^0_c = 14.9$.

effect occurs, though less intensely, in the right-hand cell. The rising plume near the middle of the box although masked in the temperature field by the overall adiabatic rise of temperature with depth, is well revealed by contours of the entropy field (S).

The development of time dependent flow at relatively low Rayleigh numbers is a striking result of increasing the compressibility factor d/H_T (or D). The simplifications of the two dimensional numerical models employed by Jarvis and McKenzie (1980) all tend to stabilize the flow. Nevertheless, time dependent solutions occur for values of R^0/R^0_c as low as 1.5 when $d/H_T = 1.0$, and 6.3 when $d/H_T = 0.5$. Models with $d/H_T = 0.25$ behave more like incompressible solutions with steady flow persisting up to $R^0/R^0_c = 107$. Although the value of R^0 at which time dependence first occurs is probably an upper bound to the range of Rayleigh numbers for steady flow, it would appear that the Rayleigh number range for stable two dimensional rolls diminishes with increasing values of d/H_T.

The importance of viscous heating clearly increases with d/H_T. In the series A models (with $d/H_T = 0.25$), shear heating is insignificant, whereas in the vigorously convecting models of series C ($d/H_T = 1.00$), localized dissipative heating near descending plumes is of sufficient intensity to modify both temperature and flow solutions. This is particularly true when an instability of the upper boundary layer is swept downwards with the flow. The amount of viscous dissipation thus changes with time and contributes to the complexity of the time dependent flow solutions of the compressible models.

A quantitative measure of the importance of viscous dissipation is given by equation (39) which takes the form

$$E = \frac{\Phi}{F_u} \approx \frac{d}{H_T} \tag{82}$$

when heating is entirely from below. This relationship was proposed by Hewitt *et al.* (1975) for vigorously convecting systems in which most of the heat is transported by convection. The compressible convection models enable a check on the validity of this relationship. Jarvis and McKenzie (1980) computed the ratio of the global viscous dissipation rate Φ to the heat flux out of the upper surface F_u at each time step of their numerical solutions. For the time dependent solutions running averages of the ratio Φ/F_u were constant for periods long compared to typical periods of oscillation. Figure 7.21 summarizes the results for five different values of d/H_T and a range of values of R^0/R^0_c. Of particular interest is the value of E (i.e., Φ/F_u) $= 1.30$ for model $D1$; this model provides the first known physical example of a convecting system in which $E > 1.00$. Each solid circle in Figure 7.21 indicates the value of E as computed from one model. The horizontal lines labelled with values of d/H_T represent the values of E predicted by equation (82). Figure 7.21 provides strong evidence that equation (82) is valid for $d \leqslant 1.50$. The value of d/H_T appears to be an upper limit which is approached asymptotically by the vigorously convecting systems. Values of E for the low Rayleigh number models fall well below the upper limit because a significant proportion of the total heat flux is *conducted* along the adiabatic gradients, violating the assumption implicit in Hewitt *et al.*'s (1975) theory that convection carries the total heat flux across the layer. This assumption was required in order to equate the mean convective flux F at any depth to the total surface flux F_u. Since it is actually F which may be equated to d/H_T, Jarvis (1978) suggested that equation (82) be generalized to allow for $F < F_u$ by introducing a constant of proportionality $p = F/F_u$ and noting that

$$\frac{\Phi}{F} = \frac{\Phi}{pFu} = \frac{d}{H_T}. \tag{83}$$

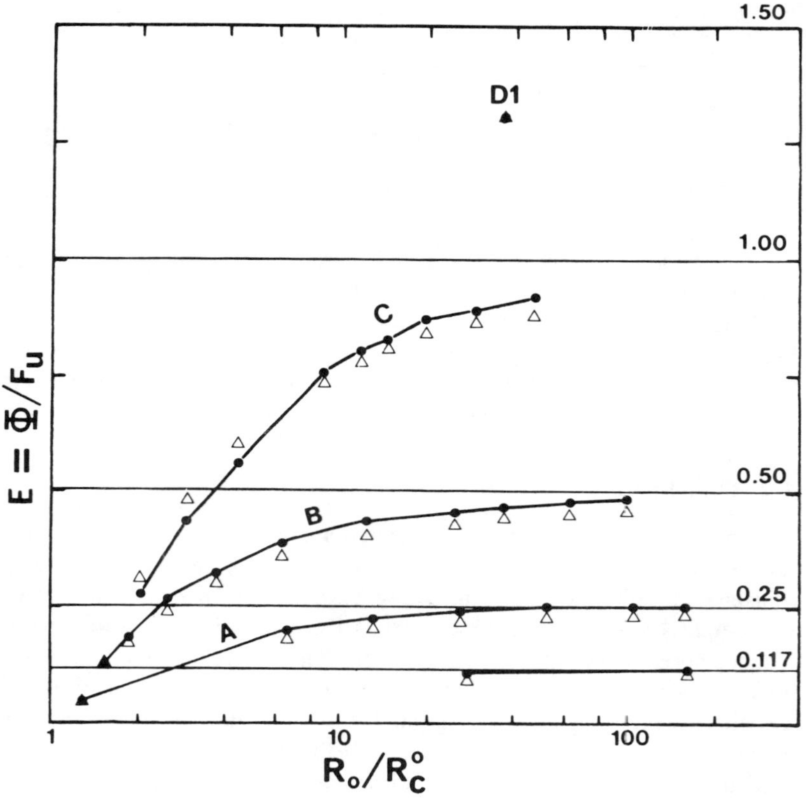

Figure 7.21. Efficiency E ($= \Phi/F_u$) as a function of R_0/R_c°. Each solid circle represents the value of E computed from one numerical model. The lines labelled A, B and C connect points from models with the same value of d/H_T. The horizontal lines labelled with values of E indicate the theoretically predicted limit for each series at high values of R_0/R_c°. Open triangles represent values of E' as computed from equation (85) based on the Nusselt number.

The generalized version of (82) is therefore

$$E' = \frac{\Phi}{F_u} = p\frac{d}{H_T}, \tag{84}$$

or, estimating p from the Nusselt number N as $p = 1 - N^{-1}$, we have

$$E' = (1 - N^{-1})\frac{d}{H_T}. \tag{85}$$

At low Rayleigh numbers $N \to 1$ and therefore $E' \to 0$, while at high Rayleigh numbers $N \to \infty$ and $E' \to d/H_T$ (as in equation (82)).

Values of E' predicted from the model Nusselt numbers are plotted on Figure 7.21 as small open triangles. The close agreement at all values of R^0/R^0_c of the computed ratio Φ/F_u and the value of E' based on the Nusselt number, substantiate the relationships (83)–(85) for $d = 0(H_T)$.

4.4. Conclusions

Cold plumes descending from the gravitationally unstable upper boundary layer dominate the compressible flow solutions. At high values of R^0/R^0_c weaker rising plumes also occur. However, due to the decrease in density with height, rising plumes expand and thus dissipate their buoyancy (through reduced horizontal temperature gradients) whereas descending plumes contract and concentrate their buoyancy. This asymmetry in buoyancy forces associated with rising and sinking plumes is believed to be responsible for much of the observed time dependence of the flow.

In the model solutions, viscous dissipation is concentrated in narrow vertical zones adjacent to sinking sheets of cold material; there is no evidence of horizontal planes of dissipation. In these constant viscosity, free-boundary solutions localized kinks in the isotherms, and consequently odd-shaped convection rolls, appear to be the prime effect of viscous dissipation in deep layers.

The influence of density stratification is primarily geometric for $d/H_T \leq 1$ and does not have any significant stabilizing influence on the layer. However the linear analysis by Jarvis and McKenzie (1980) demonstrated that for $d/H_T = 3.0$ the density variation introduces a high degree of stability. This result suggests that the conclusions which are relevant to the Earth's mantle (for which $d/H_T < 1.0$) may not be valid in deeper layers. Of the non-Boussinesq features discussed above, the variation of the adiabatic gradient has the dominant effect on the character of the flow. Viscous dissipation and density stratification appear to affect only details of the two dimensional flow structure, while the decrease in magnitude of the adiabatic gradient with height acts as an internal source of buoyancy. This effect will, itself, be minimized if the adiabatic gradient is approximately constant in the Earth as is often assumed. Thus if $d/H_T \lessapprox 0.6$ and $(\nabla T)_s$ is approximately constant, non-Boussinesq effects are likely to be of secondary importance to the overall character of the flow and it would seem reasonable to initially investigate other complications such as variable viscosity and spherical geometry in terms of the simpler Boussinesq models.

5. Geophysical Implications of Simple Convection Models

In order to draw geophysical implications from simple Boussinesq models of convection, some justification is required for treating the Earth's mantle as a Boussinesq fluid. As described at the outset of Section 4, above, mantle depths are too great relative to the density variation scale height to satisfy Spiegel and Veronis' (1960) criterion that $d/H_T << 1$ in Boussinesq fluids. However, the numerical solutions of compressible convection, described briefly in Section 4, do appear to provide the required justification. These calculations show that for Earth-like values of d/H_T (≈ 0.5) the main influence of compressibility and viscous dissipation combined is that due to the variation of the adiabatic gradient. Since the magnitude of the adiabatic gradient is proportional to the product αT and α was assumed constant in Section 4, the adiabatic gradient was proportional to T and its magnitude therefore increased significantly with depth. However, in the Earth's interior α is known to decrease significantly with depth due to increasing pressure (Stacey, 1977); this effect counteracts that of increasing T on the product αT, and thus minimizes variations of the adiabatic gradient within the Earth. Assuming that the adiabatic gradient in the mantle is reasonably constant, the properties of the circulation should be reasonably well represented by Boussinesq models of convection. In the application of Boussinesq results to the Earth, however, it must be remembered that all references to temperature differences and heat flux are to values in excess of those associated with a constant adiabatic gradient.

A major limitation to drawing inferences about mantle convection from simple Boussinesq models arises from the fact that many of the complexities of the mantle (such as its curved geometry, variable composition and the temperature and pressure dependence of its viscosity) are not included in the simple models. It is nevertheless instructive to examine the predictions of the simple models. These will provide a basis of understanding and a reference standard against which models with added complexity may be compared.

5.1. Gravity

In one of the earliest comprehensive studies of mantle convection McKenzie *et al.* (1974) predicted regional variations of gravity, horizontal velocity, surface topography and heat flow at the upper surface of unit aspect ratio convection rolls in a Boussinesq fluid. These predictions pertained to convection in the upper 700 km of the mantle. The major geophysical result was the demonstration that a positive gravity anomaly occurs above hot upwelling streams in the layer, and a negative anomaly occurs above downwellings. This result was explained in terms of the two main contributions to the gravity anomaly: (1) the upward (downward) deformation of the surface near a rising (sinking) current, and (2) the lower (higher) than average density of the hot (cold) rising (sinking) material. These effects act in opposite senses and are of similar order of magnitude; in constant viscosity models the effect of surface deformation invariably outweighs that of density variations at depth. In a more extensive investigation of this problem McKenzie (1977) found that in variable viscosity calculations the sense of the surface deformation could be reversed so that negative gravity anomalies could be associated with upwelling currents and positive with downwelling currents. However, this result has been questioned recently by Jian and Parmentier (1985) who find positive gravity anomalies over upwellings and negative over downwellings, even with large variations of viscosity. Jian and Parmentier (1985) also suggest that small scale gravity anomalies (with wavelengths of ≈200 km) in the central Pacific (Haxby and Weissel, 1983) may be caused by small convection rolls of approximately 100 km depth immediately below the oceanic lithosphere. These have total gravity anomalies of 20–30 mgal. The upper mantle models of McKenzie *et al.* (1974) predicted gravity anomalies of ≈400 mgal between upwellings and downwellings. Similar calculations for the whole mantle models of Jarvis and Peltier (1982) also predict a range of approximately 400 mgal.

5.2. Heat Flow and Bathymetry

Models Without Internal Heating

Heat Flux

In their boundary layer theory of mantle convection, Turcotte and Oxburgh (1967) considered only the major temperature difference across the boundary layer and ignored the overshoot of the temperature profile beyond the mean value of the central region. This theory predicts surface heat flow $F_u \propto x^{-1/2}$.

In Figure 7.22a we compare the surface heat flow F_u computed at the upper surface of a convection model (with $R_B = 5 \times 10^3 R_c$) with a curve which is proportional to $x^{-1/2}$. As noted in the discussion of Figure 7.8 above, at such high Rayleigh numbers the power law dependence of F_u on x is in reasonable agreement with the predictions of boundary layer theory. This agreement is of geophysical interest since the same boundary layer approximations have been incorporated into the plate models of the lithosphere (McKenzie, 1967; Sclater and Francheteau, 1970; Davis and Lister, 1974) which have proven

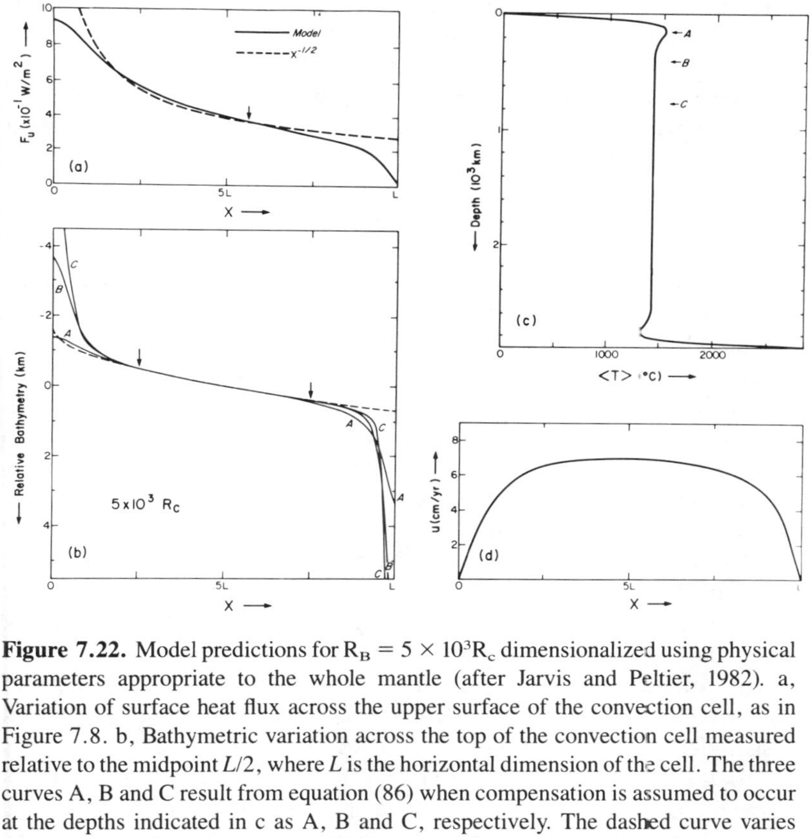

Figure 7.22. Model predictions for $R_B = 5 \times 10^3 R_c$ dimensionalized using physical parameters appropriate to the whole mantle (after Jarvis and Peltier, 1982). a, Variation of surface heat flux across the upper surface of the convection cell, as in Figure 7.8. b, Bathymetric variation across the top of the convection cell measured relative to the midpoint $L/2$, where L is the horizontal dimension of the cell. The three curves A, B and C result from equation (86) when compensation is assumed to occur at the depths indicated in c as A, B and C, respectively. The dashed curve varies linearly with $x^{1/2}$ and is constrained to fit curve B at the points indicated by the small arrows. c, Mean vertical temperature profile. Depths A, B and C indicate the values of z_1 (Equation 86) used in b. d, Variation of horizontal surface velocity across the top of the convection cell.

quite successful in matching regional variations of both oceanic heat flux and bathymetry. Figure 7.22a, for a Rayleigh number appropriate to whole mantle convection, is an example of a convection model which predicts a similar regional variation of heat flux. However, at the lower Rayleigh numbers appropriate to upper mantle convection, the surface heat flux does not follow an $x^{-1/2}$ variation (McKenzie *et al.*, 1974). Moreover the lateral scale of the variation ($\approx$600 km) is much smaller than average plate scales.

Topography

In terms of boundary layer theory the surface topography above a convection cell may be determined by integrating the temperature structure in the thermal boundary layer and assuming isostatic compensation below. The resulting surface topography varies as $x^{1/2}$. If we assume the upper surface of the convection cell to be overlain with sea water, the increase in the water depth $D(x)$, relative to that at $x = 0$, is predicted to be

$$D_1(x) = \frac{\alpha\rho_m}{\rho_m - \rho_w}\int_0^{z_1} [T(0, z^*) - T(\mathrm{x}, \mathrm{z}^*)]dz^* \qquad (86)$$

(Oxburgh and Turcotte, 1978) where ρ_m is the mean mantle density, ρ_w is the density of sea water, z^* is the depth measured downwards and z_1 is the assumed depth of isostatic compensation. The bathymetry profiles shown in Figure 7.22b were computed using (86) for three different values of z_1. The solid curves labelled A, B and C result when compensation is assumed to occur at the depths indicated in Figure 7.22c by the arrows labelled A, B and C respectively. When plotted relative to the depth at the mid-point the three curves overlie each other in the central region suggesting that compensation has occurred by depth A, at the base of the thermal boundary layer. The dashed line in Figure 7.22b varies linearly with $x^{1/2}$ (corresponding to the boundary layer theory prediction) and is constrained to fit curve B at the points indicated by the two arrows. Away from the vertical boundaries the model and theoretical curves are in excellent agreement.

The expression for $D_1(x)$ given by (86) breaks down close to the edges of the convection roll due to the explicit assumption of isostacy and the implicit assumption in the derivation of (86) that $T(0, z^*) \approx T(x, z^*)$ for $z^* > z_1$ (below the thermal boundary layer) for all x. In fact, at all depths in the hot rising plume $T > T_c$, where T_c is the temperature of the isothermal central core of the convection cell, and in the cold sinking plume $T < T_c$. Hence, the deeper we choose z_1 the greater the computed density deficit above the rising plume and density excess above the sinking plume will be. Nevertheless, by depth A just below the stagnation point of the rising plume, the accumulated density deficit is not sufficient to cause significant departure from the reference $x^{1/2}$ curve. It is noteworthy that depth A corresponds most closely to the depth of compensation assumed for plate models of the lithosphere. The results shown in Figure 7.22b are of geophysical interest since oceanic bathymetry is observed to vary linearly with the square root of distance away from mid-ocean ridges out to distances corresponding to half the age of the oldest sea floor (Parsons and Sclater, 1977). At older sea floor ages, bathymetry increases more slowly than the $x^{1/2}$ variation illustrated in Figure 7.22b. This "flattening" of the bathymetry variation relative to an $x^{1/2}$ variation is accounted for by plate models of the lithosphere (Parsons and Sclater, 1977) but not by boundary layer theory or simple Bénard convection models. Nevertheless the fact that both of the latter predict the initial variation of bathymetry correctly suggests that the plate-like nature of the lithosphere could be due to some relatively minor differences between convection in the mantle and the simple models considered in Figure 7.22.

Application of equation (86) is restricted to high Rayleigh number convection cells for which the boundary layer approximations can be justified. An alternative expression for $D(x)$, which is free of this restriction, can be derived from a balance of the dynamic forces due to the fluid motion at the upper surface (McKenzie *et al.*, 1974; McKenzie, 1977). Assuming mechanical equilibrium at the upper surface, the normal stress must vanish. For small perturbations, ξ, of the upper surface from horizontal it may be shown that

$$\xi(x) = \frac{\nu}{g}\int_0^x \left.\frac{\partial\omega}{\partial z}\right|_{z=d} dx - \left.\frac{2\nu}{g}\frac{\partial^2\psi}{\partial x\partial z}\right|_{z=d} + C, \tag{87}$$

or

$$\xi(x) = \xi_1 + \xi_2 + C, \tag{88}$$

where ω is vorticity, C is an arbitrary constant, ξ_1 represents the first term on the right hand side of (87) and ξ_2 the second. If the convection cell is overlain with water of density ρ_w, the increase in water depth across the cell is therefore

$$D_2(x) = \frac{\rho}{\rho - \rho_w}[\xi(0) - \xi(x)]. \tag{89}$$

$D_2(x)$ is derived from a consideration of the dynamics of fluid flow, whereas $D_1(x)$ derives from a consideration of the resulting thermal structure. Unlike equation (86), equation (89) is exact at all x, for any Rayleigh number. In particular, it does not break down in the vicinity of the plumes and does not require any assumptions about isostacy and the corresponding depth of compensation.

The spatial variation of ξ and of the separate contributions to ξ (given on the right side of (88)) across the upper surface of a convection cell are shown in Figure 7.23a and b for $R_B = 50R_c$ and $R_B = 10^4R_c$, respectively. ξ_1 is the contribution due to non-hydrostatic pressures while ξ_2 is that due to deflection of the streamlines. At low Rayleigh numbers ξ_2 is smaller than, but comparable to, ξ_1 across most of the upper surface. However, at high Rayleigh numbers, non-negligible values of ξ_2 are confined to the vicinity of the plumes; ξ_2 is essentially zero across the central portion of the cell (Fig. 7.23b). Consequently, at high Rayleigh numbers $\xi \approx \xi_1$, except near the plumes. In the limit $\xi = \xi_1$, equation (89) can be shown (within the context of boundary layer theory) to take a form similar to (86), namely

$$D_2(x) = D_1(x) + D_3(x) + \xi_{20}, \tag{90}$$

where ξ_{20} represents $(\rho/(\rho - \rho_w)\ \xi_2(x = 0)$, a constant, and

$$D_3(x) = \frac{\rho}{\rho - \rho_w}[\xi_1(0) - \xi_1(x)]|_{z=d-z_1}. \tag{91}$$

$D_3(x)$ represents the departure from horizontal of the level of constant vertical stress at a mean depth of z_1 below the upper surface. The assumptions implicit in boundary layer theory imply that $D_3(x) = 0$ (Jarvis and

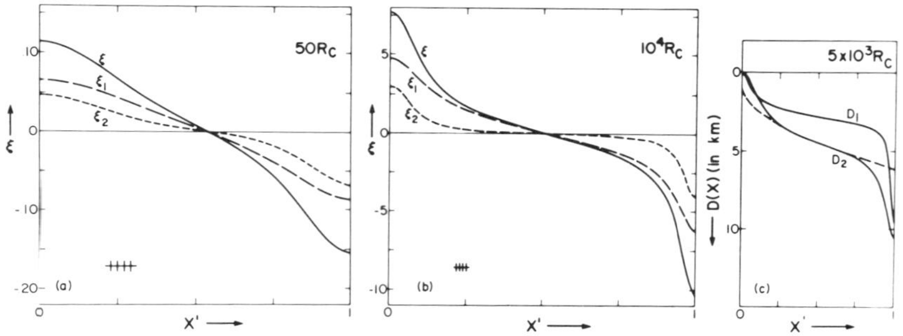

Figure 7.23. Topography ξ (in km) and components ξ_1 and ξ_2 (see Equation 88) produced at the upper surface of convection cells when: a, $R_B = 50R_c$; and b, $R_B = 10^4R_c$. c, Comparison of bathymetric variation $D_1(x)$ predicted by (86) and $D_2(x)$ computed from (89) when $R_B = 5 \times 10^3R_c$. The dashed line varies linearly with $x^{1/2}$. Dimensional values were obtained using values of physical parameters appropriate to whole mantle convection. For upper mantle convection the ordinate scales should be multiplied by 2.06 in a and b, and by 2.25 in c.

Peltier, 1982), and therefore (according to (90)) that the variation of $D_1(x)$ equals that of $D_2(x)$ in the central region. Similarly, in plate models of the lithosphere $D_3(x)$ is assumed to vanish; that is compensation is assumed to exist at some horizontal plane surface. However, a comparison of the profiles of $D_1(x)$ and $D_2(x)$ for a model with $R_B = 5 \times 10^3 R_c$ (Fig. 7.23c) shows that although $D_2(x)$ also varies linearly with $x^{1/2}$ (away from the vertical boundaries) this variation is not the same as that of $D_1(x)$. In fact Jarvis and Peltier (1982) argue that the variation due to $D_3(x)$ is comparable to that of $D_1(x)$. This result may have important geophysical implications. For if the thermal structure of oceanic plates were the consequence of whole mantle convection, then approximately 50 percent of the $x^{1/2}$ variation of oceanic bathymetry would be attributable to $D_3(x)$, the variation in depth of compensation, and only $\approx$50 percent of the observed bathymetric variation may be due to $D_1(x)$ (i.e., thermal contraction of the plate). In all plate models, to date, it has been assumed that the entire bathymetric variation is due to thermal contraction, and on the basis of this assumption various thermal parameters (e.g., the coefficient of thermal expansion, thickness or basal temperature) have been inferred for the lithosphere (Parsons and Sclater, 1977). Clearly the validity of these inferences rests upon the validity of the assumption that $D_3(x)$ is negligible for oceanic plates.

Models With Internal Heating

Heat Flow and Topography

The change in thermal structure of the boundary layers due to the asymmetric flow which occurs when heating is partially from within is sufficient to produce major departures from the power law variations of F_u, the surface heat flow, and of $D(x)$, the bathymetry, discussed above. As the fractional amount of internal heating, μ, increases from 0.0 to 1.0, F_u loses its $x^{-1/2}$ dependence and the bathymetry increases with x more slowly than the power law reference curve. This is a direct consequence of the additional heat advected into the base of the boundary layer for $\mu > 0$. We define P, the percentage "flattening" of the bathymetry profiles, as the maximum difference between the model bathymetry curve and the reference curve, expressed as a percentage of the value on the reference curve. As might be expected, for a fixed value of R_R, P is an increasing function of μ. A perhaps less intuitive result is that for a fixed value of μ, P is an increasing function of R_R. In Figure 7.24 a sequence of bathymetry and surface heat flow profiles is shown for three values of R_R but the same value of $\mu = 0.20$. The increasing departure of the model bathymetry curve (solid) from the refer-

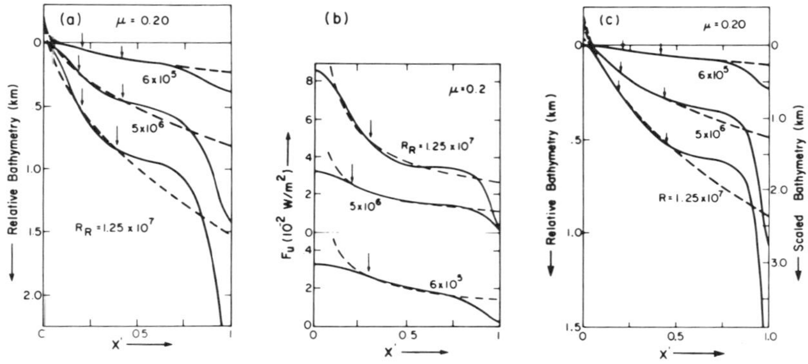

Figure 7.24. Horizontal variations of a, bathymetry $D_2(x)$; b, surface heat flux F_u; and c, bathymetry $D_1(x)$, across model convection cells with $\mu = 0.20$ and a range of R_R as indicated next to the individual curves. Solid curves are the model predictions while dashed curves are the appropriate reference curves fit to the model curves at the points indicated by the arrows. Dimensionalization for whole mantle convection.

ence curve (dashed) is apparent in Figure 7.24a. In contrast Figure 7.24b indicates that departures of the model heat flow curves from the reference curves are small and not very sensitive to the changing value of R_R. Thus at $R_R = 1.25 \times 10^7$ and $\mu = 0.20$, while the bathymetry varies as $x^{1/2}$ to the left of center and then flattens by an amount $P \approx 20$ percent to the right of center, the heat flow departs little from an $x^{-1/2}$ behavior. At lower values of R_R and higher μ the same value of P can be obtained but F_u departs considerably from an $x^{-1/2}$ power law curve (Jarvis and Peltier, 1982).

In the Earth it is generally assumed that μ must be greater than zero due to the presence of heat producing radioactive elements within the convecting region. The model results discussed above show that significant departures from the predictions of heated from below boundary layer theory do occur for convection cells which are partially heated from within. Consequently, we shall proceed to consider the geophysical implications of $\mu > 0$ first in the context of upper mantle convection, and then in the context of whole mantle convection.

5.3. Upper Mantle Convection

Models of convection in the upper mantle may be divided into two basic categories, which we shall refer to as "large scale" and "small scale." These are illustrated in Figure 7.25a and b, respectively. In the large scale model it

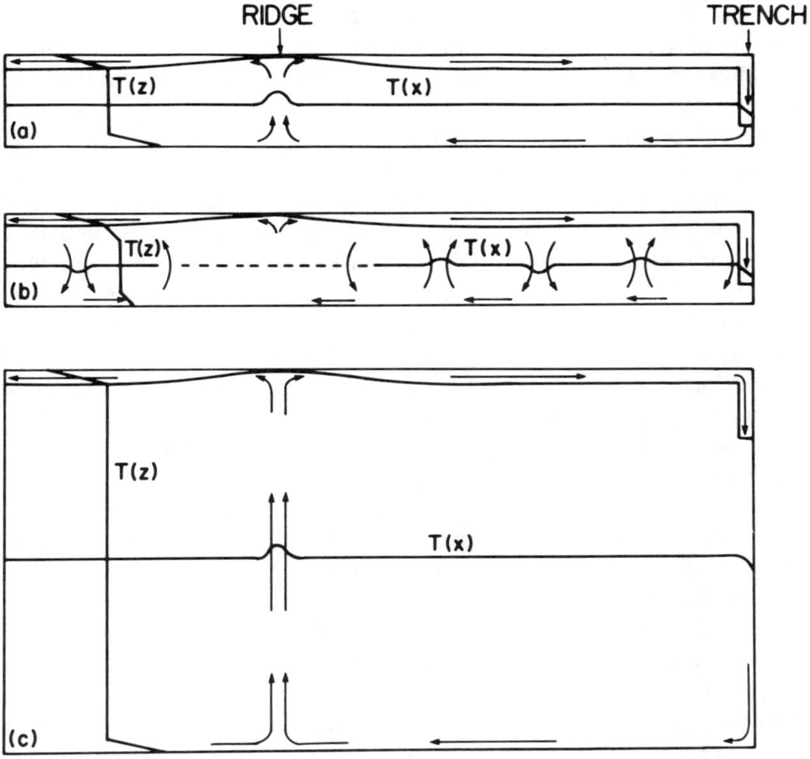

Figure 7.25. A comparison of three models of mantle convective circulation involving the lithospheric plates. a, The lithosphere forms the thermal boundary layer of an upper mantle convective circulation. b, The lithosphere moves above a smaller-scale convective circulation. c, The lithosphere forms the thermal boundary layer of a mantle-wide convective circulation.

is assumed that the convective circulation takes the form of large aspect ratio Bénard-like convection cells with a hot rising plume beneath oceanic ridges and a corresponding cold descending plume associated with the deep oceanic trenches (e.g., Turcotte and Oxburgh, 1967, 1972). In this model it is assumed that the circulation is restricted to the region above the seismic discontinuity at 670 km depth and that the oceanic lithosphere is to be associated with the thermal boundary layer itself. Below the lithosphere is an extremely elongate isothermal core. In order to achieve the large aspect ratios required to account for surface plate dimensions, it is assumed that the rigidity of the lithosphere (due to the temperature dependence of viscosity) stabilizes the cells with respect to aspect ratio one disturbances. It is also conceivable, in this picture, that the scale of the upper mantle flow associated with plate creation and destruction might be governed by that of a distinct circulation in the underlying lower mantle (Busse, 1981).

In the small scale model illustrated in Figure 7.25b, it is assumed that convection is occurring beneath a rigid lithosphere and that this circulation has a horizontal scale comparable to the depth of the upper mantle (Richter, 1973b; McKenzie *et al.*, 1974; Richter and Parsons, 1975). The many small cells maintain an almost isothermal region below the lithosphere except for the hot and cold plumes associated with the small-scale circulation. In this model, oceanic ridges are zones of plate divergence and are not in general associated with rising plumes of the small scale circulation. As the plates move apart, magma is tapped from the underlying convecting zone. A return flow from trench to ridge is superimposed on the small scale flow as required by mass conservation. (This would be weak if the lithosphere were mechanically decoupled from the mantle below by a thin low viscosity channel (Richter and McKenzie, 1978).) In this model the driving mechanism for the observed plate motions is not associated with the small-scale convection. Rather it is assumed that buoyancy forces localized at ridge crests and ocean trenches maintain the plate motions (Richter and McKenzie, 1978).

The two versions of upper mantle convection may be compared from the point of view of their predictions regarding surface heat flow and ocean floor bathymetry. In the large scale upper mantle model, if $\mu = 0$ surface heat flux is entirely due to cooling of the initially hot plume material as it drifts horizontally away from the ridge. There is virtually no heat delivered to the base of the thermal boundary layer and thus cooling continues across the length of the plate with heat flow decreasing as $x^{-1/2}$ and bathymetry increasing as $x^{1/2}$. In this model the rheological lithosphere and the thermal boundary layer of the convective circulation are coincident. Qualitatively our previous discussion of numerical results suggests that if $\mu > 0$ additional heat will be advected to the base of the lithosphere and thus reduce the rate of thermal contraction. Quantitative predictions of the magnitude and distribution of this effect are not possible, however, because of the wide disparity between the high aspect ratios required in this model of upper mantle convection and the unit aspect ratio cells of the numerical models.

In the small scale upper mantle model, if $\mu = 0$ the surface heat flux is due to cooling of the intruding magma as it moves away from the ridge crest, plus the background heat flux due to the small scale convection below the lithosphere. Although the upper thermal boundary layer of each small cell cools approximately as predicted by boundary layer theory, the large number of cells and the relatively small variation of heat flux across each produces an approximately uniform heat flux into the base of the lithosphere. Cooling of the lithosphere (and the consequent contraction) continues until the surface heat flux due to cooling of the horizontally moving lithosphere equals that delivered to the base of the lithosphere by the small scale convection. There, even with $\mu = 0$ this model predicts that surface topography will flatten away from an initial $x^{1/2}$ variation (Parsons and McKenzie, 1978). With $\mu > 0$ there would be asymmetries induced in the small scale convection cells but this would produce only relatively short wavelength effects. In this model surface heat flow and topography are therefore relatively insensitive to the degree of internal heating.

In both large and small scale upper mantle models, the implied ocean floor bathymetry initially varies as $x^{1/2}$. With an appropriate choice of physical parameters, this topographic variation can be accounted for by thermal contraction if one assumes isostatic compensation at the base of the lithosphere (Parsons and Sclater, 1977; Oxburgh and Turcotte, 1978). However, the numerical models have demonstrated that compensation does not occur at a constant depth in a dynamic flow. For the large scale upper mantle model this may produce a significant contribution to the surface topography. Attempts to fit the observed topography with the thermal contraction model could therefore lead to erroneous estimates of the relevant physical parameters of the lithosphere. Although our numerical solutions show that this error could be as large as 100 percent for aspect ratio one cells, we cannot extrapolate this result to the large scale upper mantle model which requires a much larger aspect ratio. Nevertheless this effect may be another source of error to be considered when the large scale model of upper mantle convection is employed. For the small scale upper mantle model, variations in the depth of compensation occur with a much shorter wavelength than that corresponding to the overall variation of topography. No systematic variation occurs over the length of the plate and hence the assumption of a constant depth of compensation is more appropriate.

5.4. Whole Mantle Convection

A whole mantle alternative to the upper mantle convection models is illustrated schematically in Figure 7.25c. In this model the lithosphere is identified with the thermal boundary layer of a large scale convective circulation which extends to the core-mantle boundary. Hot plumes rise beneath ocean ridges and cold plumes sink beneath ocean trenches. As in the large scale upper mantle model, the rigidity of the lithosphere is assumed to provide stability with respect to smaller scale circulation, although in the whole mantle model the largest aspect ratio required to account for surface motions is only about three. Because the lateral scale of coherent surface motion is in general comparable to the depth of the convection zone, the whole mantle model more readily accounts for the observed scale of plate motions at the Earth's surface. Moreover Figure 7.22 indicates that the magnitude of surface velocities predicted by whole mantle models ($\sim$ 5 cm/yr) is in close agreement with observed plate velocities. One difficulty in applying numerical results for convection in a plane layer to the whole mantle model arises from the neglect of curvature in the numerical calculations. In applying numerical results to this model it is implicitly assumed that convection in a rectangular domain approximates that in a curved wedge of the mantle. For an aspect ratio of one this assumption does not seem unreasonable. For aspect ratios as large as three this assumption is clearly suspect, although the few numerical solutions obtained in spherical coordinates (e.g., Hsui *et al.*, 1972; Young, 1974; Zebib *et al.*, 1980) show few qualitative differences from those in Cartesian coordinates. The main effects appear to be those of an increasing heat flux with depth, to account for the decreasing surface area of concentric spherical shells. This results in a larger temperature difference across the lower thermal boundary layer than that across the upper, since the boundary layer thicknesses are found to remain approximately equal.

In the whole mantle model an adiabatic core rotates passively. When $\mu = 0$ heat flows across the mantle primarily through the vertical plumes at the edges of the convection cells. There is no (superadiabatic) background heat flux from the mantle away from the plumes and consequently surface heat flux and topography vary as $x^{-1/2}$ and $x^{1/2}$ respectively. When $\mu > 0$ the numerical results discussed above suggest that, at low values of μ, the departure of the surface heat flow variation from that when $\mu = 0$ is relatively small. In contrast, even small amounts of internal heating produce a sufficient change in the thermal boundary layer structure to cause significant "flattening" of the topographic variation compared to the $x^{1/2}$ variation found when $\mu = 0$. The initial $x^{1/2}$ variation of the cooling lithosphere can only be partially accounted for by thermal contraction; there is also a contribution from hydrodynamic pressures beneath the lithosphere. Thus, if the whole mantle convection model were correct, fitting an isostatically compensated thermal contraction model to ocean bathymetry would lead to erroneous estimates of physical parameters of the lithosphere.

The geophysical data most relevant to this discussion are those for the variation of ocean floor bathymetry (Parsons and Sclater, 1977). The bathymetry of both the North Pacific and North Atlantic varies initially as $t^{1/2}$ where t is the age of the sea floor, or as $x^{1/2}$ on each (constant velocity) plate. For ocean floor older than about 80–90 Ma (the approximate mid-point of both ocean basins) the bathymetry systematically flattens with respect to an $x^{1/2}$ behavior. The percentage flattening in both cases is about 18 percent. The small scale upper mantle model appears to be able to account qualitatively for this behavior regardless of the value of μ. In the context of the whole mantle model the asymmetric flow structure developed when $\mu > 0$ may also account for (or at least contribute to) this behavior. In particular at Rayleigh numbers appropriate to the whole mantle model, a percentage flattening of $P \approx 18$ percent is obtained in constant viscosity fluids for $\mu \approx 0.20$ (Jarvis and Peltier, 1982). The main difficulty with this interpretation is that it does not account for the similarity of the depth-age curves for the Atlantic and Pacific oceans which occurs in spite of a presumably different large scale flow structure beneath the two. The amount of flattening achieved with relatively low values of μ and the qualitative similarity of the model and ocean floor bathymetry nevertheless suggest that internal heating may play a crucial role.

The small scale model of upper mantle convection appears capable of explaining the observed bathymetry, independent of the value of μ. It is therefore only in the context of the whole mantle model that topography constrains the value of μ to be small. Although this constraint depends upon the validity of the whole mantle model it is nevertheless interesting to compare the value of $\mu = 0.20$ to independent estimates.

Compared to the view of some authors (e.g., Oxburgh and Turcotte, 1978; Turcotte *et al.*, 1979) that the convective circulation is most probably driven entirely by internal heating, $\mu = 0.20$ is an extremely low estimate. With μ as low as 0.2 the circulation is, in contrast, driven primarily by heating from below. In the context of the whole mantle model this would imply that there is substantial heat flow across the core-mantle boundary. This implication is consistent with the thermal history calculations of Sharpe and Peltier (1978, 1979) which suggest that the heat capacity of the core is sufficiently high that the large heat transports associated with high Rayleigh number convection can be tolerated for times on the order of the age of the Earth without leading to freezing of the Earth's core. In these models the assumed mantle-wide convective circulation is driven entirely by the cooling of the Earth in bulk. Although this cannot be literally true their calculations established that this energy source for the circulation could well be of predominant importance.

Independent geochemical evidence has recently been forthcoming which suggests that an appropriate value of μ for the whole mantle may be as low as 0.3 (O'Nions *et al.*, 1978, 1979; Ringwood, 1979).

Parameterized Convection

The plausibility of a low value of μ in a mantle-wide convection model has important implications for planetary thermal history calculations. Tozer (1965, 1967), McKenzie and Weiss (1975), Sharpe and Peltier (1978, 1979) and Schubert *et al.* (1979), for example, all assumed $\mu \approx 0$ as a first approximation and made use of boundary layer scaling relationships appropriate for Bénard convection in a form similar to equations (29)–(33). In order to apply these scaling relations to the Earth's mantle, some *a priori* assumptions must be made. This is because the computed power law indices and coefficients apply strictly to two dimensional, steady state flows in constant viscosity fluids. Their use, therefore, involves the implicit assumption that departures from these conditions will not affect the end result. As yet these assumptions have not been justified. Indeed, recent calculations by Parmentier and Morgan (1982) and by Christensen (1984b, 1985) suggest that the parameterization of Nusselt number in terms of Rayleigh number may be very different in variable viscosity fluids than in constant viscosity fluids.

5.5. Time Dependent Convection

Most theoretical studies of mantle convection, including those described above, deal with steady state, single cell flow fields. However, convection on any scale in the Earth's mantle is likely to be highly time dependent because of (1) the high value of the Rayleigh number appropriate to the mantle, and (2) the variability of physical conditions (McKenzie and Richter, 1976; Jarvis, 1984). Indeed, most geological features at the Earth's surface have been produced by transient aspects of time dependent flow in the mantle. Examples of time dependence in plate scale convective motions include: the migration and/or jumping of ridges and trenches from one location to another; changes in the direction of motion of plates as recorded by magnetic anomalies and the strikes of ocean island chains; rifting of continents; opening and closing of major ocean basins; and the termination of subduction by triple point migration to produce transform faults.

The steady models have been useful in leading to an understanding of the basic physical processes operating within a convecting layer. But it is important to realize that such flows are deficient as models of planetary convection. Models of time dependent convection are characterized by transient plumes of hot and cold material breaking away from the corresponding boundary layers and traversing the vertical extent of the layer. The interior of the layer through which these plumes travel remains essentially isothermal in the models (or adiabatic in the Earth). A sample time sequence from a non steady flow model was shown previously in Figure 7.15. Such non steady convective flow models predict significant fluctuations in surface velocities and heat flow. This variability is illustrated in Figure 7.26 where dimensionless values of the surface heat flux, F'_s; total kinetic energy, E'_k; maximum horizontal velocity, U'_{max}; and maximum vertical velocity, W'_{max}, are each plotted as a function of time for the model shown in Figure 7.15. Arrows on the time axis indicate times corresponding to the different frames in Figure 7.15. In the relatively short interval shown in Figure 7.26, the surface heat flow fluctuates between dimensionless values of 33.2 and

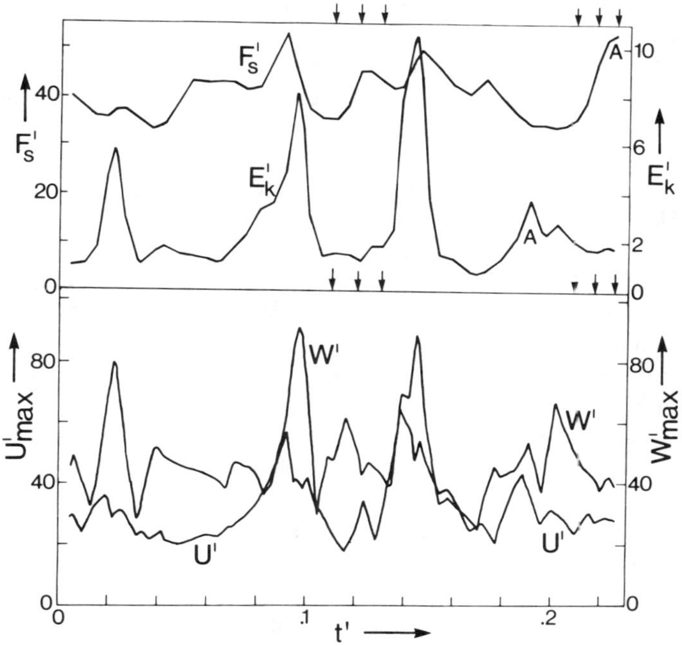

Figure 7.26. Time sequence of the dimensionless surface heat flux, F'_s; total kinetic energy, E'_k; maximum horizontal velocity, U'_{max}; and maximum vertical velocity W'_{max}; for a model with $R_R = 1.5 \times 10^9$ and $\mu = 0.0$ (after Jarvis, 1984).

53.1, with a mean value of ≈40. This range of variability corresponds to ≈50 percent of the mean value. The kinetic energy (which may be thought of as a proxy for the square of the mean convective velocity) varies in a pulsating manner from a base level of ≈100 up to a short lived peak value of 1050. The maximum horizontal and vertical velocities, U'_{max} and W'_{max}, vary (in a manner similar to E'_k) by a factor of ≈3 from base level to peak value. The time series for U'_{max} and W'_{max} have more structure than that for E'_k since these represent local rather than integral fluctuations. If the motion of the surface plates on Earth is related to mantle convective velocities, then Figure 7.26 suggests that episodic surface motions are a natural consequence of time dependent flow in the mantle.

Comparison of the four curves plotted in Figure 7.26 reveals little phase correlation between the various parameters of the circulation. The best correlation is between U'_{max} and W'_{max} although at times U'_{max} is a minimum while W'_{max} is a maximum. Similarly, F'_s may be at a local minimum while E'_k is at a maximum, and vice versa. These asynchronous responses occur because the time dependence results from the erratic rising and sinking of plumes. W'_{max}, and hence E'_k, is a maximum as a plume (or plumes) leaves its boundary layer of origin. However, at this stage a rising plume would not yet have reached the upper surface to produce the anomalously high (near surface) temperatures responsible for the increase in F'_s. By the time a hot rising plume reaches the upper surface its vertical velocity is significantly reduced. The resultant increase in F'_s may therefore be accompanied by a decrease in E'_k (unless another plume is concurrently breaking away from either of the boundary layers).

The possibility of a higher than average value of F'_s coexisting with a lower than average value of E'_k (and hence of U'_{max} and W'_{max}) is a significant departure from the predictions of steady state convection models. For steady convection cells, F'_s, E'_k, U'_{max}, and W'_{max} all increase together according to well defined power law dependencies on the appropriate Rayleigh number (e.g., equations 29–33; Fig. 7.5). These relationships have been used by Sharpe and Peltier (1979), Elsasser *et al.* (1979) and Peltier (1980) to estimate the depth of the mantle convective circulation. They are also fundamental to parameterized thermal history models of planetary interiors. The possibility of a breakdown of these relationships in the case of time dependent convection raises concerns about the validity of these applications. This is particularly so for inferences based upon present day (i.e., instantaneous) measurements of plate speed and heat flow. Thermal history calculations may not be affected as seriously by these considerations, since they deal with average values of heat flux and velocity on time scales which are long compared to the convective overturn time. However, all thermal history models of the Earth are pinned to one instantaneous datum—the present day total heat flux from the Earth's interior. If convection in the mantle is highly time dependent, the present day heat flow may not coincide with the appropriate mean value of heat flux over the past few convective overturns. Figure 7.26 indicates that in model calculations an instantaneous value of F'_s may deviate from the mean by as much as 25 percent. (Actual deviations may be larger than this for models which include internal heat sources.) One mitigating factor in the case of the Earth is that heat loss from the interior occurs due to the convective recycling of several different surface plates. Thus the total heat loss averages the effects of plates in different dynamic states and may not deviate from the temporal mean as seriously as Figure 7.26 suggests. Future models of time dependent convection in spherical shells or wide boxes may help to resolve this issue. Meanwhile it is important to consider this extra uncertainty when interpreting the predictions of thermal history models.

We have stressed that the observed present day plate velocities and global heat flow are instantaneous "snapshots" of highly variable time dependent surface dynamics. Figures 7.15 and 7.26 indicate that no one snapshot of such flow solutions is likely to be "characteristic" of the flow. Rather, statistical analyses of various time series data such as those plotted in Figure 7.26 will be required to establish characteristic features of these flows. (For example: means, standard deviations from the means, and typical frequencies of the fluctuations could be determined as functions of the Rayleigh number and degree of internal heating.) Time dependent flow features very similar to those of Figures 7.15 and 7.26 above have been found in studies of convection at high Rayleigh numbers between concentric cylindrical shells by Travis and Brandow (1984). In other recent studies of time dependent aspects of mantle convection Richter *et al*. (1982), Olson *et al*. (1984a,b) and Davies (1984) have examined mixing rates of chemical heterogeneities, Mitrovica and Jarvis (1985) have studied topographic effects of transient subduction adjacent to continental margins and Boss and Sacks (1986) have investigated time dependent layered mantle convection.

6. Lateral Heterogeneity Due to Convection

Traditionally, interest in the thermal structure of the Earth's mantle has centered on the *radial* variation of temperature. Until quite recently this has continued to be true despite a large number of laboratory and numerical models of convection in two and three dimensions, throughout the past two decades. The reason for this is that until recently, global seismology was only capable of determining the mean radial variation of seismic velocities; consequently our view of the planetary interior was an azimuthally averaged one. With no reliable information on lateral variations of mantle structure at depth, model implications for deep lateral variations were unconstrained and therefore of little value.

However, the recent growth of global geophysical data sets (for such fields as heat flow, gravity, chemistry of igneous rocks, and seismic surface and body wave velocities) has led to a heightened interest in the implications and origins of the lateral heterogeneity in the Earth's mantle which they seem to imply. As these data continue to be refined, it may ultimately be possible to use them to constrain models of mantle flow on the basis of the characteristic lateral heterogeneity implied by the different flow models. This possibility now provides the geophysical motivation for studying, and documenting, the characteristic lateral heterogeneity found in various convection models. In this section we will examine the characteristic signatures of lateral temperature variations predicted by simple convection models and discuss possible interpretations of seismic tomographic inversions on the basis of these signatures.

Seismology is the only geophysical tool capable of probing the deep interior on a local basis. As seismic data from global networks increase in quantity and quality, the newly developing field of global tomography may eventually reveal the detailed lateral structure of the mantle (Dziewonski, 1984). To date, tomographic studies have identified large scale lateral heterogeneities in seismic velocities within both the upper and lower mantle (Dziewonski *et al.*, 1977; Masters *et al.*, 1982; Anderson, 1984; Dziewonski, 1984; Nakanishi and Anderson, 1984a,b; Woodhouse and Dziewonski, 1984; Hager *et al.*, 1985; Clayton and Comer, 1987; Comer and Clayton, 1987). One cause of lateral variations in seismic velocities is lateral variations in temperature. Although crystal anisotropy and lateral variations of mineralogy would also produce heterogeneity of seismic velocities, these influences would likely be confined to the upper 400 km of the mantle, above the olivine-spinel phase change. Moreover, even at shallow depths, regions of fast and slow velocities occur where they are expected on the basis of plate tectonics (i.e., fast below subduction zones and slow below mid-ocean ridges). Accordingly, Anderson and Dziewonski (1984) concluded that "mapping seismic velocity anomalies is an accurate way of locating hot and cold regions and hence of locating ascending and descending currents in the mantle." If this is correct, then seismic tomography may be beginning to provide an image of the temperature field deep within the mantle. By comparing this tomographic image with the predictions of numerical models one may hope to obtain additional constraints on the possible modes of flow.

However, a major difficulty in making such comparisons arises from the fact that due to the uneven distribution and limited amount of seismic data currently available, only the lowest order harmonic representation of the tomographic image is available. Dziewonski (1984), for example, considers a spherical harmonic expansion of the heterogeneity, assuming smooth radial variations, and extracts the expansion coefficients from his data up to the sixth angular order and fourth Legendre function in radius. Thus only the longest wavelength features are revealed in the expansion. Using an iterative method of back projection of travel time residuals along ray paths, Clayton and Comer (1987) obtain a measure of heterogeneity on a finer scale in which the mantle is discretized into blocks 100 km deep and ~500 km square. Although this approach provides relatively fine resolution vertically, the lateral resolution is insufficient to detect individual vertical plumes. Moreover, Hager *et al.* (1985) point out that the inversions by Dziewonski (1984) and by Clayton and Comer (1987) only agree for angular orders of two and three.

Thus the tomographic image of the thermal structure of the deep mantle, which we would like to compare with model predictions, is a highly filtered smooth image. In contrast, the thermal structure predicted by numerical models of whole mantle convection (see Fig. 7.1) is characterized by narrow plumes and boundary layers containing extreme temperature variations—that is, it is composed of both low and high wavenumber harmonics.

Clearly, direct comparisons between observations and convection models are not as yet warranted. It is, nevertheless, instructive to examine the lateral variations of temperature predicted by models of mantle convection. Moreover, since the recent tomographic studies by Dziewonski (1984) and Woodhouse and Dziewonski (1984) have been conducted in terms of spherical harmonic analyses, it seems appropriate to subject the predictions of convection models to similar harmonic analyses. Thus by analyzing the spectral signature of various model solutions it may be possible to isolate qualitative similarities between model predictions and observations.

6.1. Horizontal Temperature Fluctuations

The nature of the lateral temperature variations found in steady convection models is illustrated by the contours of the two dimensional temperature fields shown in Figure 7.1 for a series of models at various values of the Rayleigh number, R_B. It is clear from Figure 7.1 that lateral temperature variations are primarily due to the hot upwelling, and cold downwelling streams of fluid. And, as the Rayleigh number increases the hot and cold plumes thin so that the major lateral temperature fluctuations are confined to ever narrower regions. This feature is emphasized in Figure 7.27 in which horizontal temperature profiles, at mid-depth, across convection cells at different Rayleigh numbers have been superimposed. In the case of a high Rayleigh number the heterogeneity is quite sharp and we can anticipate that the power spectrum will be rich at large wavenumbers relative to that for a low Rayleigh number case. The symmetry of the hot and cold temperature fluctuations shown here is due to the fact that the midpoints of these profiles coincide with the centers of the convection cells themselves. At any other depth this symmetry in the horizontal profiles is lost.

A noteworthy feature of Figure 7.27 is the fact that each of the profiles, regardless of its corresponding value of R_B, varies from a dimensionless temperature of 0.5 in the center of the profile to approximately 0.75 in the hot plume and to approximately 0.25 in the cold plume. In other words, the

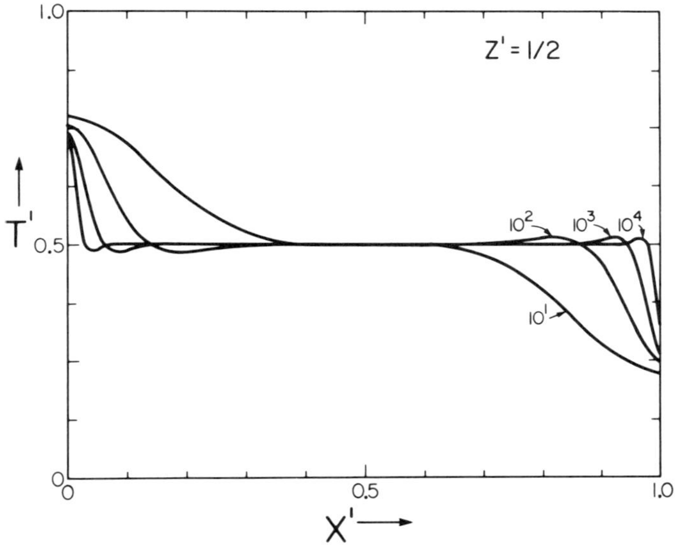

Figure 7.27. Horizontal profiles of dimensionless temperature across steady convection rolls at mid-depth. Each profile is labelled with the corresponding value of the ratio R_B/R_c. Temperatures may be re-dimensionalized by multiplying by the appropriate value of ΔT at each Rayleigh number.

temperatures at the centers of the hot and cold plumes differ from each other by $\Delta T/2$. This fact allows us to estimate the magnitude of the horizontal variations in temperature once the appropriate vertical temperature difference, ΔT, is known. For the whole mantle, for example, the oceanic lithosphere is normally considered to be a part of the convecting system (Peltier, 1980) and the temperature difference across the lithosphere represents approximately 50 percent of the non-adiabatic temperature variation across the convecting layer. Thus an approximate value of ΔT may be 2500°C (which when added to a 900°C adiabatic variation across the mantle implies a temperature of 3400°C at the core-mantle boundary). For Bénard convection the horizontal variation in temperature would be $\Delta T/2 = 1250$°C. This inferred temperature variation for whole mantle convection cannot be taken literally since it was derived from an over-simplified model of mantle convection. However, it does indicate the magnitude of the temperature fluctuation which may occur.

Both the symmetry and the magnitude of the hot and cold temperature variations change with height in the convecting layer. These changes are illustrated for the case of $R_B = 10^4 R_c$ in Figure 7.3b, where horizontal temperature profiles are plotted across a convection cell at five different levels. The middle curve in Figure 7.3b is the same mid-depth profile as that labelled 10^4 in Figure 7.27. Moving upwards, the increasing asymmetry about the midpoint of the temperature profiles becomes evident. The region of relatively high temperatures associated with the rising plume is broader than that at mid depth and has a lower maximum temperature while the region of relatively low temperatures associated with the sinking plume is narrower and the minimum temperature is lower. The uppermost profile in Figure 7.3b is taken at a level which is one grid level, Δz, below the upper surface of the numerical mesh upon which the temperature field is defined. This profile, well within the thermal boundary layer, differs significantly from the deeper interior profiles. Temperature varies continuously across this profile from a maximum at the left to a minimum at the right. There are no sharp fluctuations associated with the hot and cold plumes below, although there is a noticeable asymmetry between the left- and right-hand edges of the profile. The effect of the rising plume at the left is a broad zone of slightly increased temperature while that of the sinking plume at the right is a narrow zone of sharply decreased temperature. It will be seen, below, that the power spectra are sensitive to the changing character of these profiles with height.

Temperature profiles for the lower half of the convecting layer are very similar to those in the upper half. Due to the 180° rotational symmetry of steady Bénard convection cells about their centers, the shape of the hot rising plume at any distance from the bottom is identical to the shape of the cold sinking plume at the same distance from the top. The hot and cold plumes are most intense when closest to their boundary layer of origin. Thus at the upper surface, for example, the effects of the concentrated cold sinking plumes will be much more prominent than those of the diffuse hot rising plume. Since many geophysical signatures of the convection process are influenced primarily by the near surface thermal structure (McKenzie *et al.*, 1974; Jarvis and Peltier, 1982; Parsons and Daly, 1983), the effects of the sinking plumes dominate those of the rising plumes. Thus if oceanic plates were the upper lids of whole mantle Bénard-like convection cells, broad mid-ocean swells (ridges) above rising mantle plumes and narrow oceanic trenches above sinking plumes would be a natural consequence of the asymmetry seen in the near surface temperature profile of Figure 7.3b.

Bénard convection is clearly an oversimplified model of convection in the mantle. One obvious deficiency is the lack of internal heat sources which are believed to exist within the mantle due to radioactive heating. A second deficiency is the lack of a component of forcing due to secular cooling of the mantle in bulk (Sharpe and Peltier, 1978, 1979) which may also act, qualitatively, like a volume heat source (Krishnamurti, 1968a, 1968b). When internal heat sources are included in convection models the asymmetry between the hot and cold plumes is enhanced considerably (De La Cruz, 1973; McKenzie *et al.*, 1974). Nevertheless, as illustrated in Figure 7.3b, this asymmetry occurs to a lesser extent even in the absence of internal heating.

The horizontal profiles obtained from the two dimensional temperature fields run from $x = 0$ to $x = \lambda/2$, where λ is the wavelength of the periodic model fields. In order to obtain the corresponding Fourier amplitude spectra each profile is extended to $x = \lambda$ with its mirror image about the point $x = \lambda/2$ such that $T(x) = T(\lambda - x)$. The fast Fourier transform, $\overline{T}(k)$, is then obtained by applying the following algorithm to the extended profile

$$\overline{T}(k) = \sum_{n=0}^{N-1} T(n\Delta x) \exp\left[\frac{in2dk}{N}\right], \tag{92}$$

where N is the number of grid intervals in the extended profile, $\Delta x = 2d/N$ is the grid interval and k is the wavenumber. The power at each wavenumber is then given by the product $\overline{T}(k)\overline{T}^*(k)$ where $\overline{T}^*$ is the complex conjugate of $\overline{T}$. (Because the extended profiles have even symmetry the spectral amplitudes are real and $\overline{T} = \overline{T}^*$.) For convenience of presentation each power spectrum shown below is normalized by expressing the power at every wavenumber as a fraction of the maximum value of $\overline{T}\overline{T}^*$ in the spectrum.

Power spectra for the symmetric horizontal temperature profiles at mid-depth in convecting layers are shown in Figure 7.28 for seven Rayleigh numbers in the range $10^{0.1} \leq R_B/R_c \leq 3 \cdot 10^4$. The wavenumber is measured in units of π/d. Since the maximum wavelength of any feature in the horizontal profiles is $\lambda_{max} = 2d$, the minimum wavenumber is $k_{min} = 2\pi/\lambda_{max} = \pi/d$. Conversely, the minimum wavelength feature which can be resolved on a half-profile with NX grid intervals is $\lambda_{min} = 2d/NX$, and hence the Nyquist wavenumber is $k_{max} = NX(\pi/d)$. In the power spectra discussed in this section, power is plotted as a function of a dimensionless wavenumber k' which lies in the range $1 \leq k' \leq NX$. It may be dimensionalized for a particular application by multiplying by π/d.

The most prominent feature in the spectra shown in Figure 7.28 is the increasing power at high wavenumbers as the Rayleigh number increases. This is a direct consequence of the thinning of the vertical plumes and sharpening of the temperature fluctuations as the Rayleigh number increases (see Figs. 7.3b and 7.27). At the lowest value of R_B/R_c temperature varies

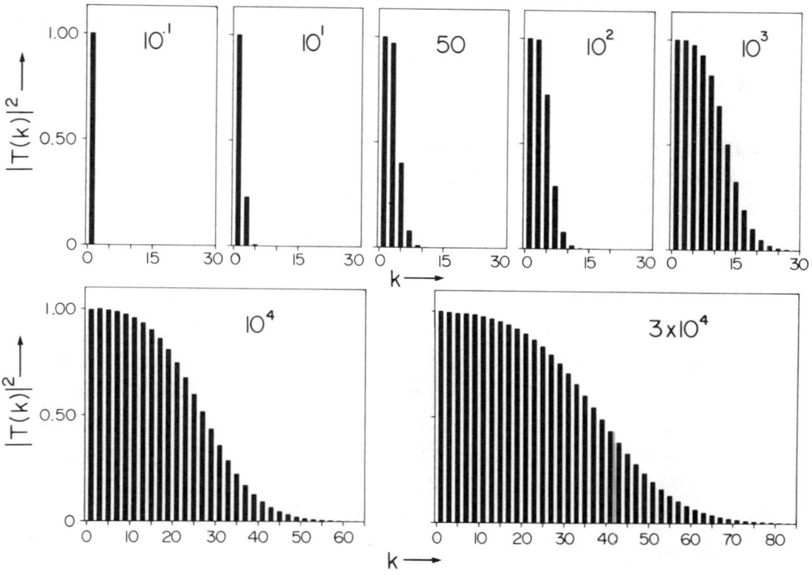

Figure 7.28. Power spectra $|T(k)|^2$ of the lateral variation of temperature at mid-depth in steadily convecting two dimensional Bénard rolls for different values of the Rayleigh number. Labels indicate the ratio R_B/R_c. Power has been normalized relative to the maximum value in each spectrum. Wave numbers, k, are indicated as multiples of the fundamental harmonic for which $k = \pi/d$.

sinusoidally across the horizontal mid-plane and hence all of the power resides in the lowest wavenumber harmonic. As the vigor of convection increases, the wavelength of the individual temperature fluctuations become smaller and higher wavenumber harmonics are involved. However, since the entire pattern of alternating hot and cold fluctuations repeats with a wavelength of $\lambda = 2d$, significant power remains in the fundamental mode. In the limiting case as $R_B \rightarrow \infty$, it may be shown analytically that the Fourier transform is real and the power spectrum consists of a sequence of Dirac delta functions of equal amplitude centered on odd integral wavenumbers, $k' = 2n+1$ (Jarvis and Peltier, 1986). The spectra of Figure 7.28 are consistent with this limiting prediction. Comparing cases representative of upper- and whole-mantle convection, at a value of $R_B = 10^2\ R_c$ (upper mantle) most of the power is confined to the range $k' \leqslant 9$, whereas for $R_B = 10^4 R_c$ (whole mantle) significant power exists in the range $k' \leqslant 45$.

The absence of power at even wavenumbers in Figure 7.28 is due to the symmetry of the temperature profiles about their mid points. The lack of symmetry of the hot and cold temperature variations at levels other than $z' = \frac{1}{2}$ was indicated in Figure 7.3 for the case of $R_B = 10^4\ R_c$. Power spectra corresponding to these profiles are shown in Figure 7.29. The spectrum labelled $z' = \frac{1}{2}$ in Figure 7.29 is the same as that labelled 10^4 in Figure 7.28. There are two major changes in the spectral signature with increasing height (or depth) in the layer. The first is the loss of power in the higher wavenumber modes; whereas power is distributed over more than 40 harmonics at $z' = \frac{1}{2}$, in the boundary layer (i.e., $z' = 1 - \Delta z$) almost all of the power is concentrated in the lowest wavenumber harmonic. The second change is the introduction of power into the even wavenumber harmonics, necessary to produce an asymmetry between the hot and cold temperature fluctuations. (Because of the symmetry of temperature fluctuations about the center of the convection cell, power spectra at heights $z' = \frac{1}{4}$, $z' = \frac{1}{8}$ and $z' = \Delta z$ are identical to those at $z' = \frac{3}{4}$, $z' = 7/8$ and $z' = 1 - \Delta z$, respectively.) The same basic features, although somewhat less pronounced, are also apparent in the spectra for the lower Rayleigh number case appropriate to upper mantle convection (Jarvis and Peltier, 1986).

Although the spectral envelopes in Figure 7.29 gradually contract towards low wavenumbers with increasing distance from the midplane, the major contrast in spectral signature is between the interior, where power is spread over several harmonics, and the boundary layers where it is concentrated at $k' = 1$. This fundamental distinction may have important geophysical implications. Any thermally induced lateral heterogeneity will show the same distinction in terms of the distribution of power amongst the various har-

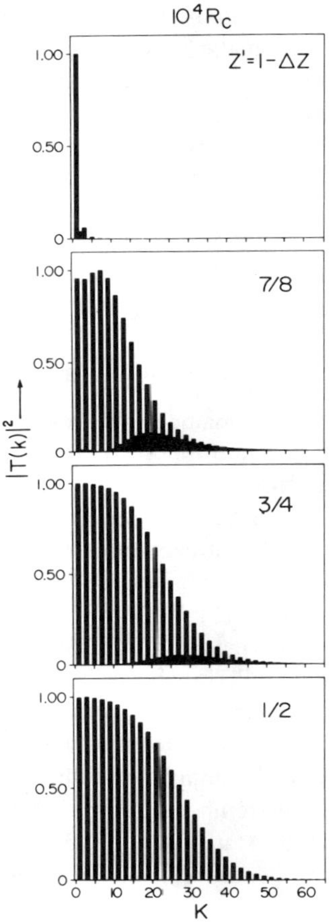

Figure 7.29. Power spectra $|T(k)|^2$ of the lateral variation of temperature as a function of dimensionless height z′ for the case $R_B = 10^4R_c$. Scales as in Figure 7.28. (Spatial variations for the same model solution were shown in Figure 7.3b.)

monics. For example, lateral variations of seismic velocities in the mantle could be due to lateral temperature fluctuations. Masters *et al.* (1982) have reported a large scale lateral heterogeneity in seismic shear wave velocities which they suggest is located in the seismic transition region in the depth range 420 km–670 km. Spectral analysis of the heterogeneity indicates that power is concentrated in the lowest harmonic in a manner very similar to that shown in Figure 7.29 at $z' = 1 - \Delta z$. This spectral signature suggests that the heterogeneity, if thermally induced, arises in a horizontal thermal boundary layer and not in the interior of a large scale convection cell. The location of the boundary layer is, unfortunately, not well determined. A more serious ambiguity, however, is that the lateral heterogeneity in the transition region could also be due to the differential vertical displacement of the phase boundaries which are located at average depths of 420 km and 670 km. Phase boundaries represent a non-thermal source of lateral variation in seismic velocities which has not been included in the simple models discussed above. Consequently a thermal interpretation of the observations by Masters *et al.* should not be applied until the implications of the phase changes have been fully investigated and understood.

Effects of Internal Heating

In the remainder of this subsection we shall briefly indicate how the above results are modified by the inclusion of internal heat sources distributed uniformly throughout the convecting layer. The major effect of adding internal heat sources into a convection model is to introduce asymmetry into both the stream function and temperature fields. This asymmetry results from the asymmetric modes of heating and cooling in the convecting layer as a whole (McKenzie *et al.*, 1974).

The nature of the asymmetry in the temperature field is illustrated in Figure 7.11a, above, by the juxtapositioning of horizontal temperature profiles taken at mid-depth across steady convection models with $R_R = 5 \times 10^6$ and $\mu = 0.0, 0.1, 0.2$ and 0.3. As μ increases, the symmetry between hot and cold temperature fluctuations is lost. The power spectra corresponding to the temperature profiles shown in Figure 7.11a are displayed in Figure 7.30. At $\mu = 0.0$, the spectrum is very similar to those shown in Figure 7.28 for Bénard convection, with power at odd integral wavenumbers. As μ increases, power in the fundamental ($k' = 1$) harmonic decreases substantially and significant power enters the low even wavenumber harmonics. In addition there is a loss of power at large k'. As in the case of Bénard profiles, away from the mid-depth power in the even harmonics can again be attributed to the increasing asymmetry between the hot and cold plumes with increasing μ. However, substantial loss of power from the $k' = 1$ harmonic is not observed in Bénard models and appears to be a distinctive feature of the internally heated models. The loss of power at large k' as μ increases is most likely due to the elimination of the constant temperature region when $\mu > 0$. Despite the non-monotonic interior spectra shown in Figure 7.30, the major distinction between the spectra of lateral heterogeneity within the thermal boundary layers and in the deep interior of the layer, is the same as that noted above for Bénard convection. That is, the interior spectra are rich in higher

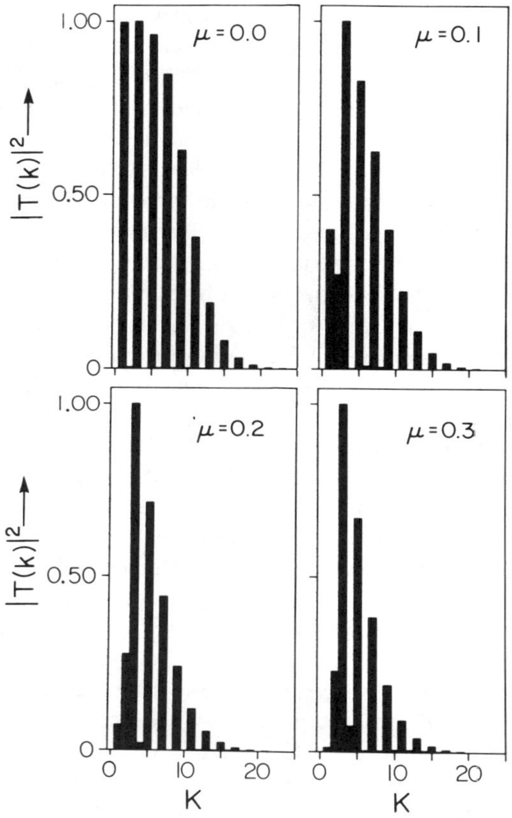

Figure 7.30. Power spectra $|T(k)|^2$ corresponding to the temperatures shown in Figure 7.11a for $R_R = 5 \times 10^6$ and values of μ from 0.0 to 0.3 as indicated on each spectrum.

wavenumber harmonics while the boundary layer spectra are dominated by the single fundamental harmonic with $k' = 1$ (Jarvis and Peltier, 1986).

The preservation of this distinctive form of the power spectrum of lateral heterogeneity within thermal boundary layers, regardless of the degree of internal heating, may find an important geophysical application in terms of locating thermal boundary layers within the mantle. This is because the newly developing field of seismic tomography is beginning to identify the long wavelength spectral components of lateral heterogeneity in the mantle

(Dziewonski *et al.*, 1977; Masters *et al.*, 1982; Anderson, 1984; Dziewonski, 1984; Nakanishi and Anderson, 1984a,b; Woodhouse and Dziewonski, 1984; Hager *et al.*, 1985; Clayton and Comer, 1987; Comer and Clayton, 1987). From Figure 7.29 it is clear that only the longest wavelength components are required to capture most of the power in the lateral heterogeneity in boundary layers. In contrast, outside of the boundary layers tomographic inversions do not have the required resolution to identify the short wavelength lateral temperature variations associated with narrow plumes.

As tomographic data improve it may eventually be possible to "prospect" for thermal boundary layers in the mantle by examining the spectral components of the lateral heterogeneity inferred at various levels in the mantle. Unfortunately, from a diagnostic viewpoint, the red spectrum so typical of thermal boundary layers is not unique to such boundary layers. Jarvis and Peltier (1986) discuss a counterexample in which heating is entirely from within the convecting layer (i.e., $\mu = 1$). In this case there is no thermal boundary layer at the lower boundary. Nevertheless the power spectrum of lateral temperature variation at the base of the layer is dominated by the single $k' = 1$ harmonic. In this case temperature varies smoothly along the base of the layer, being coolest below the sinking plumes and warmest mid-way between sinking streams. Thus the *shape* of the power spectrum is the same as that for a well defined thermal boundary layer which is being heated from below.

Clearly the shape of the spectrum of lateral heterogeneity is not sufficient on its own to indicate either the presence or absence of a thermal boundary layer within a convecting layer. However, an examination of the amounts of power in the spectra at different levels in the convecting layer may, when considered together with the shape of the various spectra, enable the detection of internal thermal boundary layers. Each of the spectra shown above has been normalized with respect to the maximum power found at any wavenumber in that spectrum. The value of this maximum power varies considerably from one level in the layer to another, and from one model to another.

Figure 7.31, for example, shows the variation of the magnitude of the $k' = 1$ spectral amplitude with height in a convecting layer which is heated from below at a Rayleigh number of $R_B = 10^4 R_c$. The boundary layers adjacent to the upper and lower surfaces are clearly demarcated by narrow regions of significantly larger magnitudes. In this example the spectral amplitudes are approximately one order of magnitude larger in the boundary layers than in the interior; and the power is therefore approximately two orders of magnitude greater in the boundary layers. In contrast, for the counterexample discussed by Jarvis and Peltier (1986) in which a steadily convecting layer is

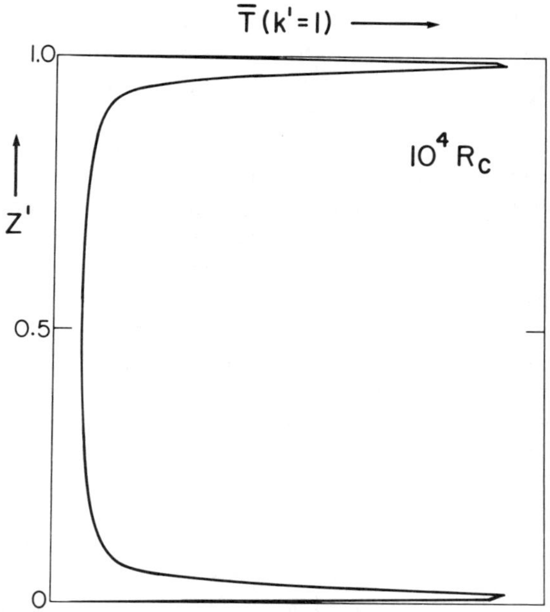

Figure 7.31. Vertical variation of the amplitude of the fundamental harmonic, $T(k = \pi/d)$, for a model solution of Bénard convection with $R_B = 10^4 R_c$.

heated entirely from within, the maximum power in the interior is actually greater than that at either the top or bottom of the layer.

It would therefore appear that there are two characteristics of the vertical variation of the spectrum of lateral heterogeneity which, taken together, may be diagnostic of the presence of a thermal boundary layer: (1) the spectra should change from one dominated by the fundamental harmonic at one depth to a more nearly white spectrum at higher levels, and (2) the maximum power in the "single harmonic" spectrum should be much larger than that in the "white" spectra found at higher levels.

6.2. Long Wavelength Spectral Components of Model Temperature Fields—Implications for Seismic Tomography

Seismic tomographic images of the mantle contain only the longest wavelength components of the actual lateral heterogeneity. In contrast model temperature fields for a convecting mantle contain substantial power at short wavelengths at all depths except within horizontal thermal boundary layers (Fig. 7.29). Because it is, in general, difficult to visualize the long wavelength component of a rapidly varying spatial field, direct comparisons of predicted thermal structures with the highly smoothed seismic observations are not particularly useful. However, by applying spectral analysis techniques to the two dimensional model temperature fields the long wavelength components of the model fields can be compared with the known full temperature solutions (Jarvis, 1985). Such comparisons contribute to the development of a theoretical basis for the inference of the unknown full temperature field in the mantle from the long wavelength component obtained by tomography. The extent to which tomographic images are "out of focus" will be illustrated here with two numerical examples. The approach taken is to Fourier analyze the two dimensional temperature fields predicted by numerical models of convection, eliminate the high wavenumber spectral amplitudes and then re-synthesize the remaining low wavenumber components of the temperature field.

To use fast Fourier transform techniques the range of each temperature field is extended by generating mirror images of the field about the planes $x = \lambda/2$ and $z = d$, such that $T(x, z) = T(\lambda - x, 2d - z)$. The horizontal extension simply completes one full wavelength of the horizontally periodic solution. The vertical extension, however, has no physical meaning; it is strictly a mathematical device used to satisfy the requirement of a periodic variation for the fast Fourier transform algorithm. The two dimensional FFT (fast Fourier transform) of the temperature field on the extended grid (consisting of $N \times N$ mesh intervals) was obtained by first applying the one dimensional FFT discussed above (equation 92) across each row of the numerical grid, and then applying the same algorithm to the vertical variation of each spectral amplitude of the horizontal transforms. The wavenumbers in the horizontal (x) and vertical (z) directions are denoted k_x and k_z, respectively, and have non-dimensional forms $k_x' = k_x d/\pi$ and $k_z' = k_z d/\pi$.

Because the extended temperature fields have even symmetry in both x- and z-directions, the spectral amplitudes $\overline{T}(k_x, k_z)$ are real and the inverse transform is obtained by re-applying equation (92) to $\overline{T}/N$ as

$$T(n\Delta x,\ z) = \frac{1}{N}\sum_{k'_x=0}^{K} \overline{T}(k_x,\ z)\ \exp\left[\frac{in2dk_x}{N}\right]. \tag{93}$$

To recover the complete field $T(x, z)$, the summation limit K in equation (93) must be $K = N-1$. However, if only a subset of low wavenumber components is to be included in the re-synthesized field, the value of K is set to a lower value. Such incomplete fields are denoted $T^*_{kx\max,kz\max}$ where the pair of subscripts indicates the maximum wavenumbers in the x- and z-directions (e.g., $T^*_{3,6}$ includes all harmonics with $k_x' \leqslant 3$ and $k_z' \leqslant 6$). For kxmax = kzmax, the horizontal thermal boundary layers and the vertical plumes become equally apparent as the number of terms retained in the Fourier expansion of T^* increases. While the vertical plumes contain large lateral variations of temperature, the horizontal boundary layers do not. Rather they are defined primarily by a large overall vertical variation of temperature, upon which relatively small lateral variations are superimposed. The mean vertical variation is contained in the *DC* component of the Fourier series expansion (i.e., the $k_x' = 0$ term in equation (93). Whereas this mean reference value (upon which the lateral variation is superimposed) is known at each height in the convection model solutions, it is not in general known in seismology. Tomographic results are normally presented as lateral variations about an azimuthal average (e.g., Masters *et al.*, 1982; Dziewonski, 1984; Hager *et al.*, 1985). Such a presentation neglects the major signature of horizontal thermal boundary layers—the vertical variation. Consequently the tomographic image of a horizontal boundary layer would appear less pronounced than would that of a vertical plume, at the same level of resolution. This is an important consideration in the interpretation of lateral variations of seismic velocity as an indication of temperature variations in the mantle. The DC component of lateral variation can be eliminated from the truncated expansions, T^*, by setting the first term in the summation in equation (93) equal to zero (i.e., $\overline{T}(k_x = 0, z) = 0$). The resulting field is denoted δT^* (where the symbol δ indicates deviations about the mean) and has a horizontal mean value of zero at every level. In this respect δT^* is a closer analogue to the long wavelength tomographic images than is T^*.

A sequence of six incomplete temperature fields, $\delta T^*_{kx\max,kz\max}$, for each of which kxmax = kzmax = $k$max, is shown in Figure 7.32 for the case of $R_B = 10^2 R_c$. The values of kxmax and kzmax are indicated on each frame. The model solution was obtained on an 80 × 80 numerical mesh, so that frame f represents the complete expansion except for the DC component of lateral variation at each height. The full temperature field for this model is shown in Figure 7.1c. The contour interval is the same in each frame of Figure 7.32, and hence the increasing number of contours, with increasing $k$max, indicates the increasing proportion of the total temperature fluctuation that is recovered as additional harmonics are added. For $k\max \leqslant 7$ the main features apparent in the truncated expansions are the anomalously hot and cold patches of fluid in the source regions of the hot and cold vertical plumes. At larger values of kmax the vertical plumes become apparent, but owing to

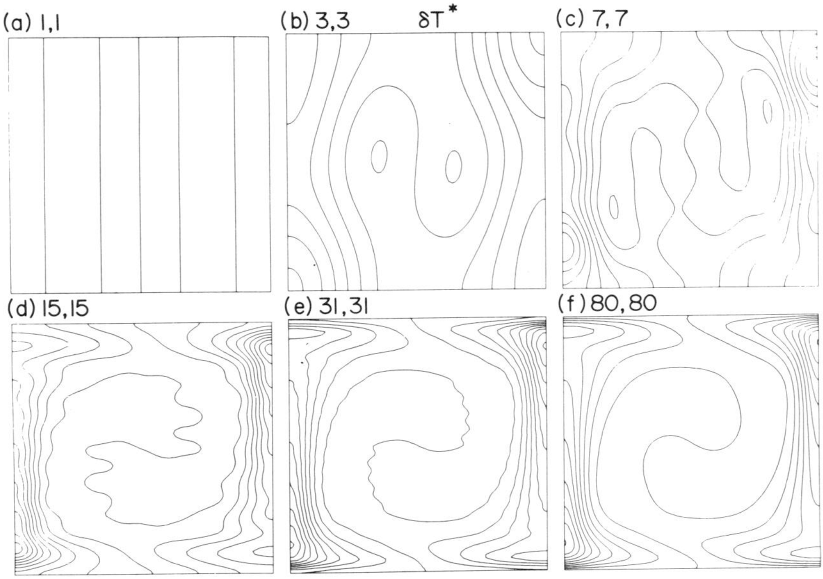

Figure 7.32. Truncated two dimensional Fourier expansions of a model temperature solution with the DC component of lateral variations set equal to zero at all heights. The temperature solution is from a model of Bénard convection with $R_B = 100\ R_c$, appropriate to upper mantle convection—see Figure 7.1c. The truncated fields are denoted: a, $\delta T^*_{1,1}$; b, $\delta T^*_{3,3}$; c, $\delta T^*_{7,7}$; d, $\delta T^*_{15,15}$; e, $\delta T^*_{31,31}$; and f, $\delta T^*_{80,80}$, where the subscripts indicate the number of harmonics (in the x- and z-directions) retained in the series expansions.

the elimination of the $k_x = 0$ terms the horizontal thermal boundary layers are largely obscured. The main signature of the boundary layers, at high values of $k$max, is a near-surface horizontal extension of the isotherms associated with the vertical plumes.

As illustrated by Figure 7.32, tomographic analyses inherently defocus boundary layers to a greater extent than plumes. However, in practice this aspect of tomography is somewhat compensated for by the fact that tomographic *data* usually have better resolution in the vertical (radial) than in the horizontal (azimuthal) direction. For example, Dziewonski's (1984) tomographic image of the lower 2200 km of the mantle contains four radial harmonics and six angular harmonics. This provides a relatively high vertical resolution of ~500 km, as compared to a lateral resolution of ~2340 km at mid-depth in the mantle. Similarly, Clayton and Comer's (1987) tomographic data have a resolution of 100 km vertically compared to 5° horizontally, or ~430 km at mid-depth in the mantle. Similar resolution to that obtained in these tomographic studies is obtained in a square box whole-mantle convection model by the truncated expansions $\delta T^*_{1,6}$ and $\delta T^*_{7,29}$.

To illustrate the effects of differential horizontal and vertical resolution, and a higher Rayleigh number, a sample temperature field from a highly time dependent whole mantle convection model is subjected to the same analysis as above (Fig. 7.33). The nature of the time dependence and the boundary layer instabilities which produce it are discussed in detail by Jarvis (1984). Here only a single, instantaneous "snapshot" of the continually evolving temperature field is considered. Isotherms of the model temperature at this instant are shown in Figure 7.33a; in addition to the hot upwelling at the left-hand, and cold downwelling at the right-hand edges of the solution domain, a new cold plume is in the process of sinking from an intermediate location on the upper boundary and a new hot plume is rising from the lower boundary.

Frames b–f in Figure 7.33 show isotherms of the same temperature field after its two dimensional Fourier series expansion has been truncated at various maximum values (*kx*max and *kz*max) of the wavenumbers k'_x and k'_z, and the DC component of lateral variations has been set equal to zero. The values of *kx*max and *kz*max for these series expansions are indicated at the top of each frame in Figure 7.33b–f. Frame b differs from a only in that the DC component has been removed from the lateral variation at each height. (This expansion has not been truncated at a limiting wavenumber.) Frame c is analogous to Figure 7.32c, but at this Rayleigh number there is even less evidence of the actual boundary layer and plume structure which characterizes the total field (Fig. 7.33a). With seven harmonics in each direction, the mushroomed heads of the transient plumes are partially recovered. However, the trailing plume material is too narrowly confined to be detected at this level of resolution.

Frames d and e show the expansions $\delta T^*_{1,6}$ and $\delta T^*_{7,29}$ with truncation at the levels thought to approximate the resolution of Dziewonski's (1984) and Clayton and Comer's (1987) tomographic images. Both plumes and boundary layers are poorly represented in 7.33d, while the horizontal boundary layers but not the vertical plumes are evident in 7.33e. In addition the relatively broad heads of the transient plumes are detected in Figure 7.33e, but appear to be detached from the upper and lower surfaces. Frames d and e illustrate the degree to which the enhanced vertical resolution can offset the inherent bias against resolving horizontal boundary layers.

The final frame in Figure 7.33 is included to indicate the level of recovery

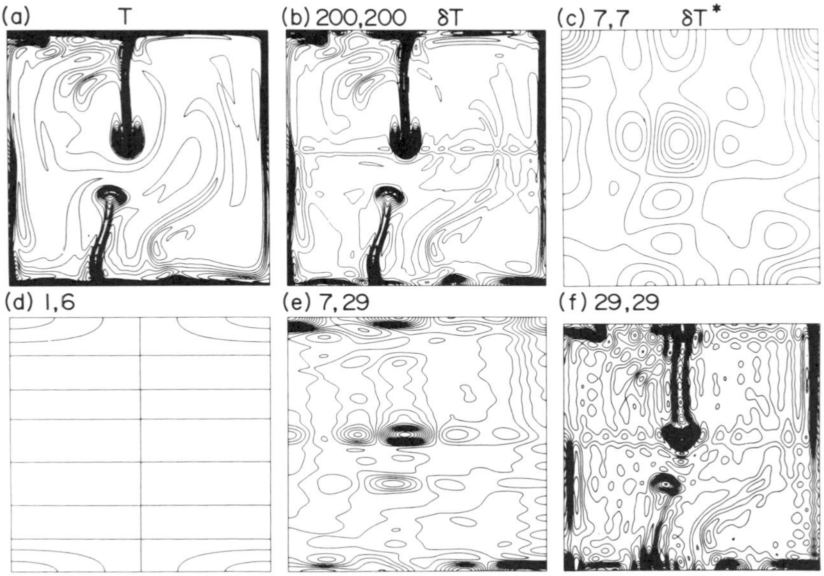

Figure 7.33. Isotherms at one instant in a time dependent whole mantle convection model with $R_R = 1.5 \times 10^9$ and $\mu = 0.20$. Contour interval is 45°C in each frame. The complete field is shown in a. Frame b is the total field less the DC component, denoted δT. Truncated two dimensional Fourier expansions, with the DC component removed, δT^*, of the solution shown in a are shown in c–f. The truncated fields are denoted: c, $\delta T^*_{7,7}$; d, $\delta T^*_{1,6}$; e, $\delta T^*_{7,29}$; and f, $\delta T^*_{29,29}$, where the subscripts indicate the number of harmonics (in the x- and z-directions) retained in the series expansions.

of the actual temperature field which could be achieved if it were possible (using Clayton and Comer's technique, for example) to obtain a resolution of 100 km in both azimuthal and radial directions. This expansion, which includes 29 harmonics in each direction, succeeds in recovering the major features of the thermal field. However, because the recovered image shown in Figure 7.33f is still somewhat "out of focus," the width of both plumes and boundary layers is significantly over-estimated. Since the actual widths of plumes and boundary layers are diagnostic of the Rayleigh number of the flow (e.g., Turcotte and Oxburgh, 1967; Sharpe and Peltier, 1979) and thus indirectly of the depth extent of convection (Elsasser *et al.*, 1979), considerable caution will have to be exercised in interpreting similar "blurred" images of the mantle's temperature field if these are ever obtained.

6.3. Conclusions

Spectral decomposition of the lateral temperature variations predicted by numerical convection models appears to be a promising technique for the interpretation of the results of seismic tomography. In particular it allows the long wavelength component of model solutions to be compared to the long wavelength component of mantle heterogeneity as determined by tomography. Although this approach has, as yet, only been applied to very simple models, potentially useful geophysical inferences may be drawn. The possibility of searching the tomographic data for evidence of thermal boundary layers within the mantle was raised in Section 6.1. In Section 6.2 it has been argued that the signature of mantle convection to be expected in the tomographically reconstructed image of mantle heterogeneity should consist of patches of anomalously hot and cold fluid (i.e., low and high seismic velocities) at depths corresponding to thermal boundary layers. Away from thermal boundary layers the current seismic data cannot be expected to reveal significant lateral variations, nor vertical continuity of features, although they may well exist. Thus the data are not yet of sufficient quality to auger preferentially for either layered or whole mantle convection. The spectral approach may also be employed to assess the relative recovery of different signatures of convection in the mantle. For example, for the simple Bénard whole mantle model discussed above ($R_B = 10^4 R_c$), Jarvis (1985) found that approximately 50 percent of the total range of topographic variation (at the surface of the cell) is contained in the lowest order harmonics, while only about 10 percent of the total range of lateral variation of temperature is contained in the same harmonics.

These initial results from the application of spectral analysis to simple convection models suggest that the spectral approach may prove to be an invaluable tool in the interpretation of tomographic data in the future.This should be particularly true when applied to more complex numerical models which more closely approximate conditions in the mantle.

7. Implications of Near Surface Model Geotherms

The temperature solutions provided by the numerical models of mantle convection predict the general character of the geotherm (vertical temperature profile) at different positions across the top of a convection cell. Although the models discussed in this chapter are undoubtedly too simple to account for the details of crustal temperature variations, it may be expected that certain qualitative features would prevail even in more complex models.

Two such features, the implications of which we shall examine here, are (1) the steep temperature gradient characterizing the upper thermal boundary layer and (2) the temperature overshoot beyond the mean temperature of the layer which occurs at the base of the thermal boundary layer, immediately above the uniform temperature cell interior. Because the viscosity of the fluids in the simple models described above was assumed to be spatially constant these two near surface characteristics of the geotherm had no direct mechanical effect on the flow dynamics. This is not the case for the Earth, however; the viscosities of mantle materials are known to depend exponentially on temperature (and pressure) so that spatial variations of temperature must be accompanied by corresponding spatial variations in mantle viscosity (Kolstedt and Goetze, 1974). Specifically, the viscosity of the cold upper portion of the thermal boundary layer must be higher than that of the uniform temperature cell interior, while the viscosity of the temperature overshoot region must be lower than the mantle below. (This naturally assumes that the inclusion of a temperature dependent viscosity in the governing hydrodynamic equations would not invalidate the prediction of a temperature overshoot below a cold upper thermal boundary layer.)

7.1. Lithospheric Rigidity

Despite the substantial variation of viscosity across the upper thermal boundary layer, it can be argued that non-hydrostatic effects such as melting and fracturing, acting at the sites of ocean ridges and trenches, may be of sufficient importance in regulating plate motions that the high viscosity of the cold lithosphere itself would exert little active influence on the flow below. In other words the rigid plates may be carried passively as the upper lids of deep seated convection cells (e.g., Peltier, 1980). Within this context the most obvious mechanical effect of the rigidity of the plates is the integration of tractions along the base of each plate to produce a uniform horizontal velocity. In contrast, Figure 7.6 indicates that in constant viscosity fluids convection is characterized by pronounced variations in horizontal velocity at the upper surface, even at very high Rayleigh numbers. This variation, $\partial u/\partial x$, results in vertical advection of heat within the thermal boundary layers and causes the regional heat flow variation to differ from the $x^{-1/2}$ behavior predicted by cooling half-space/plate models. A temperature dependent viscosity acts to minimize horizontal variations in surface velocities and hence to reduce the influence of vertical advection upon the boundary layer temperature field.

A simple means of investigating the effects of suppressing $\partial u/\partial x$ in the horizontal boundary layers is to impose a constant horizontal velocity on the upper surface as a boundary condition. Viscous coupling to the underlying fluid will then suppress $\partial u / \partial x$ close to the surface. Imposing a constant surface velocity produces a shear stress on the upper boundary; hence vorticity along the upper surface, which is proportional to shear stress, no longer vanishes. Rather it must be obtained iteratively from the ψ field (Richter, 1973a; Lux *et al.*, 1979). With this approach the effects of a rigid lithosphere are mimicked in the context of a constant viscosity model.

For illustrative purposes we consider two partially internally heated convection models, both with $R_R = 5 \times 10^6$ and $\mu = 0.20$. The models differ only in that one has free-slip boundary conditions imposed, while the other has a constant surface velocity, u_0, imposed. The value of u_0 is taken to be the mean surface velocity of the corresponding model with free boundaries. This value is determined by the overall balance between buoyancy and viscous forces and, since the net mass flux along the upper surface remains unchanged, results in little distortion of the overall circulation (e.g., Lux *et al.*, 1979). It is also consistent with the condition that the integral of the surface shear stress over the half cell is identically zero. If this integral constraint were not satisfied the flow would be externally forced or inhibited.

In Figure 7.34 we compare the temperature, vorticity and stream function fields which result when the mechanical boundary condition at the surface is either (a) no shear stress, or (b) constant velocity with no net shear stress over the half-cell. The vorticity field is the most diagnostic. The constant surface velocity u_0 in solution b is greater than the surface velocity occurring near the plumes in a and less than that in the central region of solution a. Thus the mechanical boundary condition in b produces an increase in vorticity near the vertical boundaries of the cell (drawing the contours of ω towards the corners) and a reversal of vorticity, or drag, in the central region. The net effect on the streamlines is to square off the flow in the upper regions—that is, streamlines are more nearly parallel to the upper surface. However, the overall flow pattern of solution b is very similar to that of a.

Except near the vertical boundaries, the thermal field is not very sensitive to the change in the mechanical boundary condition. With a constant surface velocity, vertical advection is enhanced above the plumes (due to the increased horizontal surface velocities) and suppressed throughout the central region. The net effect on the variation of surface heat flow, shown in Figure 7.35, is that it now more closely approximates the $x^{-1/2}$ variation predicted by boundary layer theory and observed on the seafloor.

Figure 7.35 also compares the predicted topographic variation for the cases with: A, no shear stress; and B, constant surface velocity. In spite of

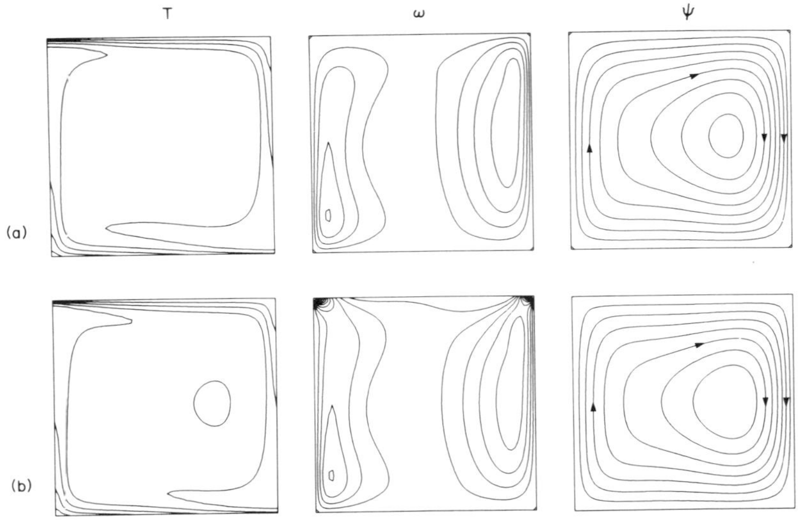

Figure 7.34. Contours of temperature T, vorticity ω, and stream function ψ, for model solutions with $R_R = 5 \times 10^6$, and $\mu = 0.20$. a, Solution with free-slip boundary condition at the upper surface. b, Solution with constant velocity boundary condition at the upper surface. Contour levels of each field are the same in a and b.

finite shear stresses on the upper surface in case B, the topography deduced from the normal stresses is very similar in both cases (except near the plumes). The reason for this similarity is that in both cases topographic variations are determined by variations in $\partial\omega/\partial z$ at the upper surface, which in turn is approximately equal to the sum of $\partial\omega/\partial z$ at the base of the themal boundary layer and the vertical integral of $\partial T/\partial x$ over the depth of the thermal boundary layer (Jarvis and Peltier, 1981). Thus the similarity of the temperature fields within the upper thermal boundary layers and of the overall flow structures below, combine to produce similar topographic expressions at the surface. The reversed topographic gradients close to the vertical boundaries in solution B can be understood in terms of horizontal diffusion of vorticity (Jarvis and Peltier, 1982). This is a local effect resulting from unrealistically high shear stresses generated near the plumes by the imposed constant surface velocity. On the ocean floor brittle fracture and partial melting in the vicinity of mid-ocean ridges preclude the possibility of such large stresses.

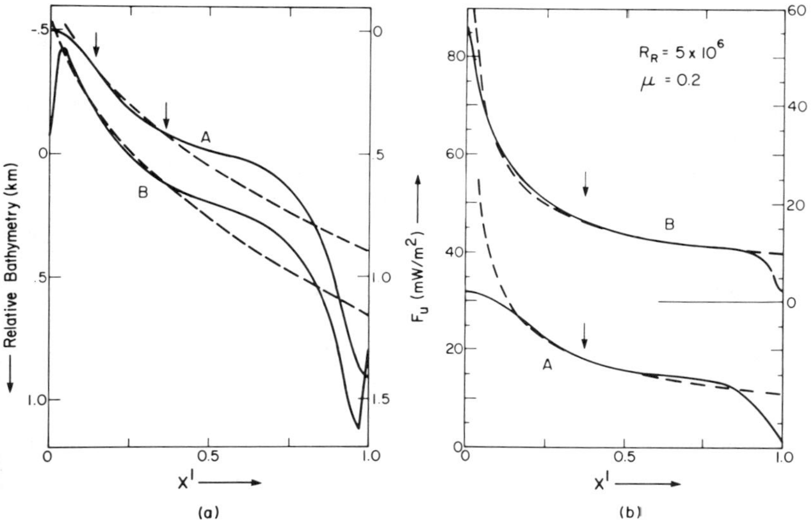

Figure 7.35. Comparison of a, relative bathymetry $D_2(x)$ and b, surface heat flux F_s, as computed from the model solutions with free-slip upper boundary conditions (curves labelled A) and constant velocity upper boundary condition (curves labelled B) (see Fig. 7.34). In a the left ordinate scale refers to curve B while the right scale refers to curve A. Dimensionalization is appropriate to whole mantle convection (after Jarvis and Peltier, 1981).

Thus the major effect of the lithosphere in this example is to adjust the distribution of surface heat flux so as to agree more closely with plate models and cooling half-space models, despite the presence of internal heat sources and the asymmetric circulation below the surface. The variation in surface topography appears relatively insensitive to the presence of the plates.

Another effect commonly attributed to the rigidity of the plates is the large horizontal extent of the implied underlying mantle convection cells. Thermal instabilities within the thermal boundary layer at the Earth's surface (i.e., the lithosphere) are suppressed due to its extremely high viscosity. In contrast boundary layers at depth, being hotter than the bulk of the mantle, will be destabilized by their lower viscosity. Thus in terms of the relative degrees of stability of a cold boundary layer at the surface and a hot boundary layer at depth, we can qualitatively understand the existence of small scale upwellings in the form of hot spots but the lack of similar small scale sinking plumes. This issue may also be studied by simulating the rigid effects of the lithosphere in terms of mechanical boundary conditions at the upper surface. Lux *et al.* (1979), for example, studied the influence of imposed surface velocities on the aspect ratio of steady convection cells. At sufficiently high surface velocities, very wide aspect ratios were obtained. However, in the extreme cases the forcing produced by the imposed boundary condition greatly exceeded the total buoyancy force available in the convecting layer. Consequently, the entire flow fields were predominantly viscously driven (Stokes flow) and thermal instabilities at the lower boundary were inhibited. A similar result was obtained by Parmentier and Turcotte (1978) by imposing plate-like velocities (and variable temperatures) on the surface of an upper mantle model.

It is not yet clear whether, with sufficient heating (i.e., buoyancy force) and low enough (constant) surface velocities, it would be possible to generate a model with a thermally unstable lower boundary layer and a stable upper boundary layer. Although such a result may be beyond the simple constant viscosity models, it would nevertheless be an interesting exercise to attempt to achieve it.

7.2. Low Viscosity and Partial Melt Zones

In a narrow zone below the rigid lithosphere the temperature profiles predicted by simple convection models, in which there is significant heating from below, typically overshoot beyond the mean temperature of the layer (Fig. 7.3). As temperatures in this region approach the local solidus temperature the viscosity would drop rapidly. The relationship between viscosity and solidus temperature may be expressed simply in terms of the homologous temperature T_{sol}/T (where T_{sol} is the solidus temperature and T is temperature, both measured in °K) as

$$\nu = \nu_0 \exp \frac{cT_{sol}}{T}, \tag{94}$$

where ν_0 is a constant with dimensions of viscosity and c is a dimensionless constant with a value of ≈ 30 (Weertman, 1970). If temperatures were sufficiently high to exceed the local solidus temperature, a partial melt would be produced and a further significant reduction in mean viscosity of this region would occur. Thus although the effects of melting and variable viscosity are not included in the simple models discussed above, the model prediction of a temperature overshoot zone has geophysical implications with regard to these effects.

In the context of whole mantle convection the temperature overshoot below the cold thermal boundary layer may be a few hundred degrees, sufficient to produce partial melting at depths of ≈100 km. This possibility was investigated by Campbell and Jarvis (1984) by comparing model geotherms with the mantle solidus. Zones in which the local geotherm exceeded the solidus were interpreted as zones of partial melting, although the model geotherms in these zones would no longer be appropriate since some of the thermal energy would be stored as latent heat in the melt (Oxburgh and Turcotte, 1968).

In order to employ the model temperature fields in such a comparison a uniform adiabatic variation of temperature with depth was added to the model fields. This procedure approximates the increase in temperature with depth owing to hydrostatic compression within the mantle. Although not valid for compressible fluids in general, this procedure is justified for the physical conditions within the mantle (Jarvis and McKenzie, 1980; Section 4). The result of adding an adiabatic variation of 0.3°C/km to the Bénard whole mantle model, used previously, for which $R_B = 10^4 R_c$ is illustrated in Figure 7.36. The thermal boundary layers and temperature overshoots are still clearly identifiable, while the isothermal central region has become a zone of adiabatic temperature variation. The total temperature variation across the mantle has now increased by almost 900°C. In Figure 7.36b the mantle solidus as estimated by Kennedy and Higgins (1972) has been superimposed. The geotherm approaches the solidus most closely in the zone of temperature overshoot at the base of the upper thermal boundary layer. The temperature field shown in Figure 7.36 is a highly idealized model of the present-day mantle. It is employed here to illustrate a principle, not to make precise quantitative predictions. The cold upper boundary layer with a maximum thickness of ≈120 km represents the lithosphere. Figure 7.36b indicates that no partial melting is predicted below the lithosphere at a point halfway across the cell. However, closer to the upwelling plume temperatures in the boundary layer are higher and a zone of partial melting centered on the plume axis is predicted by the model. In Figure 7.37 a series of vertical temperature profiles is plotted at successively greater distances from the rising plume axis. Close to the axis the temperature overshoot intersects the solidus curve but at greater distances it falls below the solidus at all depths.

If the mantle temperatures were higher in the past, close to an upwelling plume the geotherm would exceed the solidus by a greater amount (suggesting more melt and a higher degree of partial melting) and the zone of partial melting would extend further from the upwelling axis towards the sinking plumes. Thus as the mantle slowly cooled an initially extensive zone of partial melting below the lithosphere would have retreated towards the

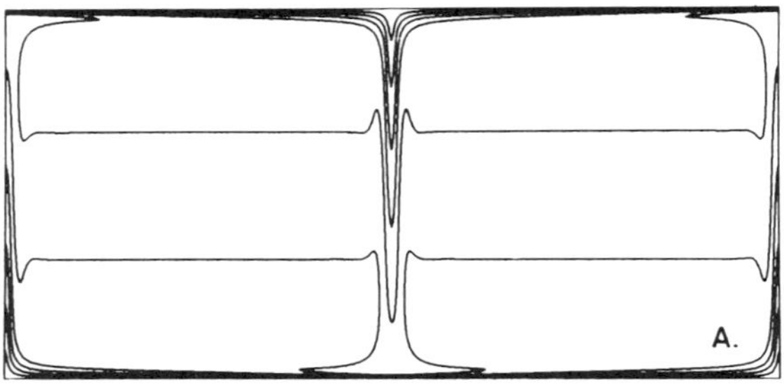

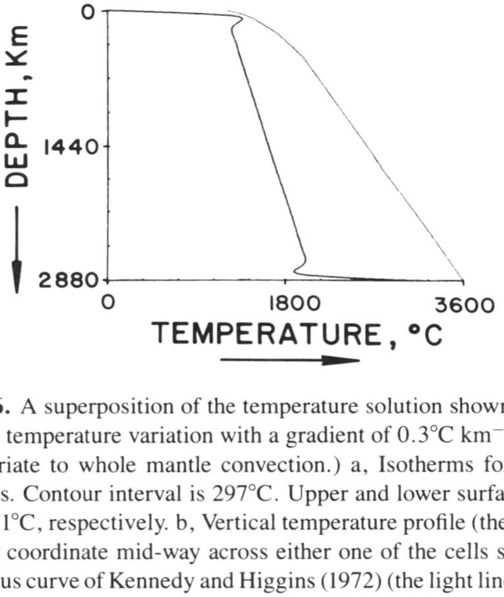

Figure 7.36. A superposition of the temperature solution shown in Figure 7.1e and an adiabatic temperature variation with a gradient of 0.3°C km^{-1}. (Dimensionalization appropriate to whole mantle convection.) a, Isotherms for a pair of counter-rotating cells. Contour interval is 297°C. Upper and lower surface temperatures are 0°C and 3271°C, respectively. b, Vertical temperature profile (the heavy solid line) at a horizontal coordinate mid-way across either one of the cells shown in a, plus the mantle solidus curve of Kennedy and Higgins (1972) (the light line). Solidus temperatures at the upper and lower surfaces are 1200°C and 3600°C, respectively.

centers of major upwelling. This retreat of the melt zone is illustrated in Figure 7.38. Here contours of the temperature *excess* above the solidus are plotted for several models of the mantle, each convecting under different conditions of temperature and viscosity. In each frame of Figure 7.38 the zero contour encloses areas of partial melt. The contours of temperature excess shown in frames b, c and d are for pairs of steady convection cells with a common descending limb at the center. Frame 7.38c for example corre-

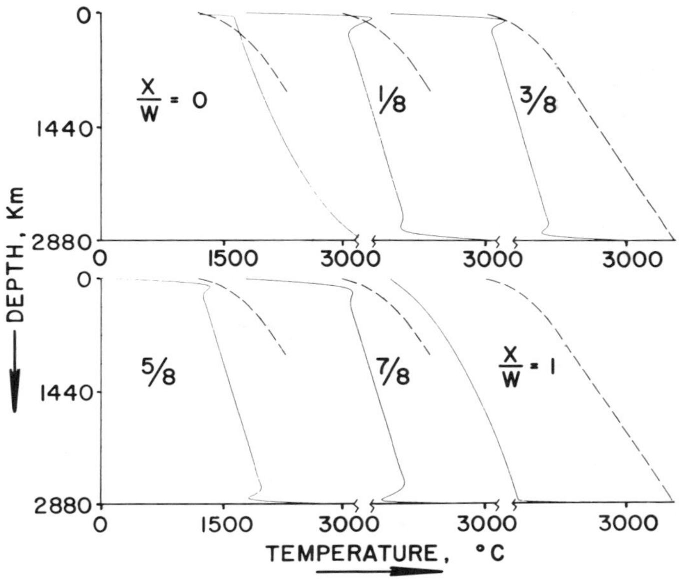

Figure 7.37. Vertical temperature profiles (solid curves) and superimposed mantle solidus profiles (dashed) at increasing horizontal distance, X, from the center of the upwelling plumes shown in Figure 7.36a. The fractional distance X/W across the cell width W is indicated next to each profile. $X/W = 0$ corresponds to the central plane of the hot rising plume, while $X/W = 1$ corresponds to the central plane of the cold sinking limb of the convection cell.

sponds to the temperature solution shown in Figure 7.36a. Figure 7.38a shows contours for a model of disordered, time-dependent convection. Each frame in Figure 7.38 is labelled with the maximum temperature in the overshoot zone at a horizontal coordinate half way across either of the two convection cells. The depth at which this temperature occurs and the Rayleigh number of each convection model are also indicated. The juxtaposition of these labels for the models of Figure 7.38 illustrates the extreme sensitivity of mantle convection to temperatures in the mantle. For example, although temperatures in the melt zones in frames b and c differ by only 7 percent, the depth at which they occur differs by more than 100 percent and the Rayleigh numbers differ by an order of magnitude. This sensitivity is due to the exponential dependence of the viscosity of mantle materials on temperature. From equation (94) it is clear that if the viscosity is known to

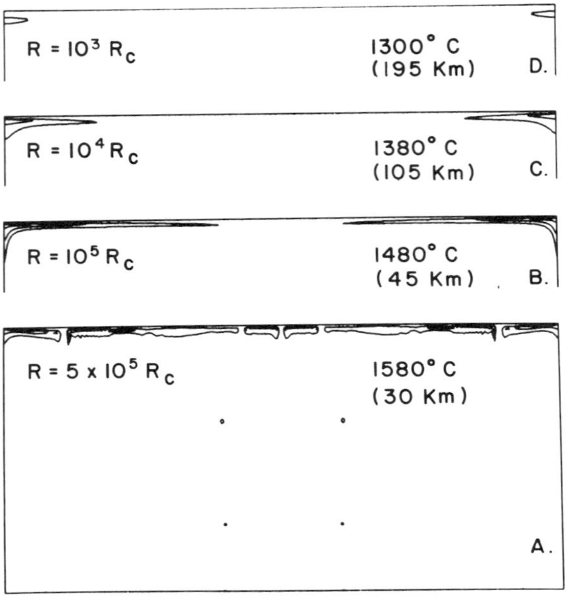

Figure 7.38. Contours of the temperature-excess above the solidus (i.e, positive values of $T - T_{\rm sol}$) for convection models at four different values of the Rayleigh number. Contour interval is 200°C with a minimum contour value of 0°C. Labels on each frame indicate the Rayleigh number, R_B; the maximum temperature in the overshoot zone at a horizontal coordinate half-way across one of the pair of convection cells; and the depth below the upper surface (in km) at which this maximum temperature occurs. Only the upper portions of frames b, c and d are shown since there are no contours at greater depths. Each of the model solutions shown was obtained on a 200 × 200 finite difference grid.

have a value of ν_1 at temperature T_1, then at a new temperature T_2 the viscosity ν_2 is given as

$$\nu_2 = \nu_1 \exp\left(cT_{\rm sol}\frac{T_1 - T_2}{T_1 T_2}\right). \tag{95}$$

Similarly, the new Rayleigh number R_2 is

$$\mathrm{R}_2 = \mathrm{R}_1\frac{\Delta T_2}{\Delta T_1}\exp\left(cT_{\rm sol}\frac{T_2 - T_1}{T_1 T_2}\right). \tag{96}$$

where R_1 is the Rayleigh number defined with $\nu = \nu_1$ and $\Delta T = \Delta T_1$. Equation (96) illustrates explicitly the exponential dependence of the mantle Rayleigh number on temperature.

The thickness of the upper boundary layer, δ, and hence the depth to the maximum thermal overshoot (indicated on Fig. 7.38) varies approximately as the inverse cube root of the Rayleigh number (e.g., Turcotte and Oxburgh, 1967). Thus increasing the Rayleigh number by a factor of 10, decreases δ by approximately a factor of 2.15. In particular, raising mantle temperatures by 200°C from Figure 7.38c to Figure 7.38a causes the Rayleigh number to increase from 10^4R_c to 5×10^5R_c and a typical depth to the thermal overshoot zone to be reduced from 105 km to 30 km.

Owing to the relatively higher mantle temperatures and shallower temperature overshoot zone, an almost continuous shallow partial-melt zone is predicted at $R_B \geqslant 5 \times 10^5 R_c$ (Fig. 7.38a). The only breaks in the lateral continuity of the melt zone occur where unstable portions of the cold upper boundary layer sink into the interior as narrow cold jets. Since these are transient events which only interrupt the melt zone temporarily, a melt zone exists below all points on the surface most of the time. Convection at such high Rayleigh numbers in the Earth's mantle may not be conducive to crustal stabilization, although the Earth's lithosphere may be more stable than the model's upper boundary layer due to the influence of temperature on viscosity. Indeed, Campbell and Jarvis (1984) suggested that the model conditions of Figure 7.38a correspond to the mantle prior to the earliest stabilization of continental crust (i.e., pre-Archean).

Campbell and Jarvis (1984) considered the model for Figure 7.38b, with $R_B = 10^5R_c$, to be representative of conditions in the Archean. This model is a steady state solution for which the melt zone extends across approximately 75 percent of the convection cell, having retracted away from the sinking jets. The stability of this model is artificial in that it results from the symmetry of the mathematical boundary conditions and the restriction to two dimensions. It is extremely unlikely that an actual three dimensional convecting system would remain steady at this high Rayleigh number (McKenzie and Richter, 1976).

The models shown in Figures 7.38c and d are also artificially stabilized steady state solutions. We expect that convection in the mantle will also be time dependent at $R_B = 10^4R_c$ and 10^3R_c, although boundary layer instabilities will be less disruptive and less frequent as the Rayleigh number decreases. Moreover, instabilities at the upper surface may be completely suppressed by the higher viscosity of the cold thermal boundary layer (Yuen *et al.*, 1981). As the mantle cools, Figures 7.38b–d indicate that the melt zones continue to retract horizontally away from the sites of subduction and become restricted to the vicinity of spreading centers (e.g., mid-ocean ridges). Campbell and Jarvis (1984) suggested that the model of Figure 7.38c best represents present conditions in the mantle. The temperature of 1380°C in the overshoot zone below the lithosphere agrees well with Parson and Sclater's (1977) estimate of 1350°C ± 275°C at a depth of 125 km ± 10 km, based on analyses of sea floor topography. Figure 7.38d would therefore represent conditions in the mantle at some future time when plate thicknesses have doubled.

Although the models used to generate the melt zones of Figure 7.38 are highly idealized, a general trend of melt zones retracting away from sinking regions and becoming confined to the vicinity of upwelling plumes, as the Rayleigh number decreases, may also apply to the case of a cooling mantle. There is, of course, considerable uncertainty in any attempt to attach a time scale to this process.

7.3. Discussion

An important feature of the numerical convection model is that at high Rayleigh numbers it predicts a partial melt zone in the upper mantle of global dimensions which begins to shrink during the early Archean (Fig. 7.38a, b). A consequence of mantle cooling is that the degree of partial melting produced by hot rising plumes will gradually decrease. Since the density of mafic-ultramafic rocks decreases with decreasing partial melting, the density of the resulting crust would decrease with time. Thus Archean oceanic crust likely had a higher average MgO content than modern oceanic crust, perhaps comparable to that of high MgO komatiites. If so, then Archean oceanic crust likely consisted primarily of high MgO komatiite, and its differentiation products, with lesser amounts of basalt formed by lower degrees of partial melting. The high density of komatiite relative to basalt may account for the fact that no Archean oceanic crust has been preserved.

Numerical models of the type discussed in this chapter treat the mantle as a solid and ignore the possibility that the upper mantle will develop a molten phase. Thus differences in the thermal properties of liquids and solids, the latent heat of fusion and the cooling effect of magma loss (volcanism) from the upper mantle are not considered. The last of these factors may be especially important since volcanism is a form of convection and as such is an efficient mechanism for removing heat from the upper mantle. This process will obviously be most effective where the degree of partial melting is greatest; that is, where the contours are highest in Figure 7.38. As a result, temperatures within the melt zone will be buffered by magma loss and will not approach the maximum contours predicted by Figure 7.38. Campbell and Jarvis (1984) suggested that the region of the upper mantle outlined by the super-solidus contours of Figure 7.38 represents a broad region of incipient melting which includes numerous pockets of magma which have not yet escaped to the surface.

The prediction of partial melt zones close to the upper surface depends on the existence of a hot plume rising from a thermal boundary layer at depth. The models shown in Figure 7.38 were heated entirely from below, a heating configuration which maximizes the strength of the upwelling plumes. If, however, some of the surface heat flux were generated by internal heating (owing to radioactive decay) the hot lower boundary layer and rising plume would be less pronounced (McKenzie *et al.*, 1974). Consequently, the temperature overshoot below the upper boundary layer would be less pronounced and the degree of partial melting reduced. Since radioactive elements are likely to exist within the Earth's mantle, the models of Figure 7.38 (e.g., Fig. 7.36) overestimate the role of the hot lower boundary layer in this respect. On the other hand, the effects of curvature of the planetary interior, which cause the lower boundary layer to be more prominent than the upper (Zebib *et al.*, 1980), have not been included in this simple model. Thus the models of Figure 7.38 underestimate the prominence of the lower boundary layer in this respect. Although these two effects tend to cancel we do not expect that the simple model simulates the actual thermal structure of the boundary layer at the core-mantle boundary. Rather it is employed here to illustrate general principles. As long as the mantle is not entirely heated from within, both upper and lower thermal boundary layers will exist and the basic character of the temperature field is thus approximated by the models without internal heating.

7.4. Speculation on the Origin of Komatiites

The general concept of a cooling planetary interior is corroborated by the temporal distribution of the MgO content in magnesium rich magmas. On the basis of high pressure melting experiments the MgO content can be related to the liquidus temperature of the erupted lava.

The maximum MgO content of modern mid ocean ridge basalts (MORB) is estimated to be 18 percent (O'Hara, 1968; Clarke, 1979; Elthon, 1979). This implies a liquidus temperature of 1350°C–1400°C (Elthon and Scarfe, 1980). Similarly, the maximum MgO content in Proterozoic rocks is 19 percent (Francis *et al.*, 1981) and hence a similar liquidus temperature is inferred. In contrast, Archean komatiites have MgO contents as high as 32 percent, which implies a liquidus temperature in excess of 1650°C (Bickle *et al.*, 1977). Thus the Archean mantle must have been hotter than the present mantle and was capable of producing magmas with a liquidus temperature 250°C–300°C above the maximum temperature of modern magmas. On the basis of this fact it might be concluded that the mantle was 250°C–300°C hotter during the Archean than at present (Green, 1975). In fact Sleep (1979) argues on the basis of a liquidus temperature of 1270°C for primitive MORB (Bender *et al.*, 1978; Usselman and Hodge, 1978) that the early Archean mantle may have been almost 400°C hotter than at present.

The inference of a relatively hot Archean mantle from the MgO content of komatiites is contradicted by estimates of Archean geothermal gradients. These are based on the pressure-temperature conditions recorded in Archean metamorphic assemblages and indicate that geothermal gradients in Archean continents were only slightly higher than those in modern continents (Wells, 1976; Bickle, 1978; Burke and Kidd, 1978). From this, one might conclude that mantle temperatures were only slightly higher in the Archean than today.

Jarvis and Campbell (1983) attempted to reconcile this contradictory evidence which appears to require both a hot Archean mantle to produce the komatiites and a relatively cool Archean mantle to account for the geothermal gradients. On the basis of Figure 7.38, they suggested that the mean temperature of the Archean mantle was approximately 100°C higher than at present. The labels on Figure 7.38 indicate that such a model would have an upper boundary layer thickness approximately one half of that at present, and therefore a mean heat flux of about twice the present value. Although this mean value is higher than that inferred from Archean geotherms, lower than average near surface temperature gradients exist closer to downwelling jets and it may be that Archean cratons formed in association with such downwellings (Jarvis and Campbell, 1983; Campbell and Jarvis, 1984). Thus doubling the mean heat flux out of the mantle need not be inconsistent with the inferred Archean geotherms.

If the mantle were only 100°C hotter during the Archean than at present, the source regions for Archean komatiites must have been at least 200°C hotter than the bulk of the mantle. The only region in a whole mantle convection model capable of satisfying this condition is the lower thermal boundary layer at the core-mantle boundary. Here temperatures are several hundred degrees higher than in the overlying mantle (see Fig. 7.36, for example). Accordingly, Jarvis and Campbell (1983) suggested that Archean komatiites could be produced, in a relatively cool Archean mantle, from hot plumes rising from the core-mantle boundary. As parcels of hot material break away from the lower boundary layer they rise and cool adiabatically, but because the mantle solidus temperature decreases more rapidly with height than does the adiabat, the hot parcels of mantle material eventually cross the solidus and begin to melt. This process is illustrated in Figure 7.39 in which a vertical temperature profile is plotted through a model solution of time dependent convection. The profile passes through two hot rising parcels of material which have broken away from the lower boundary layer. Although the temperature profile (the heavy solid line) is irregular, there is a general adiabatic variation with height upon which spikes of higher temperature are superimposed indicating the presence of the hot rising parcels. As the hot parcels continue to rise, the peaks of the temperature spikes will follow the adiabatic gradient (dashed lines) becoming increasingly supersolidus and arriving at the surface with temperatures several hundred degrees higher than magmas originating elsewhere in the mantle. At lower positions the temperature spikes are equally prominent but are entirely subsolidus and therefore would not be apparent on contour plots of the temperature excess

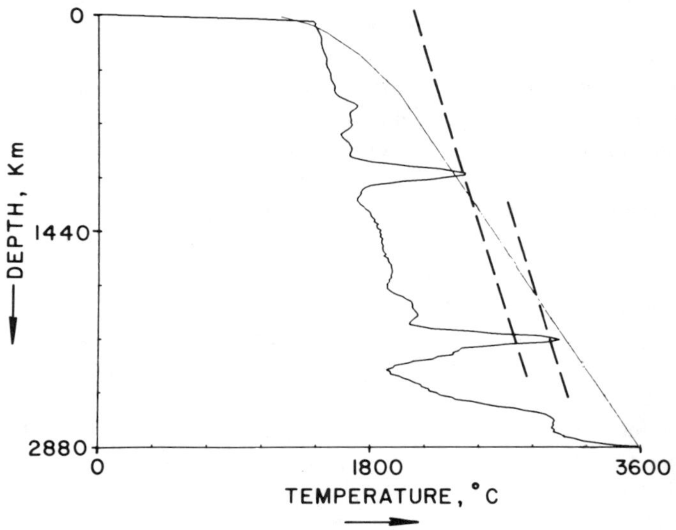

Figure 7.39. A vertical temperature profile (the heavy solid curve) through the temperature field of Figure 7.38a at a horizontal coordinate $X/W = 0.785$ passing through hot rising parcels of mantle material. The mantle solidus curve (the light solid curve) is also shown for comparison. Dashed lines show adiabatic temperature variations with height.

above the solidus (such as those of Fig. 7.38). The convection model from which this vertical profile is taken is the same pre-Archean model from which Figure 7.38a was generated. The isolated patches of partial melting seen at depths of 1000–2000 km in Figure 7.38a, have originated as illustrated in Figure 7.39. Very high degrees of partial melting could be expected before these hot parcels reach the surface.

The pre-Archean model was used in this example since it provides a ready example of time dependent flow, and suffices to illustrate the principle, whereas the Archean model (with $R_B = 10^5 R_c$) is artificially steady. However, similar rising plumes could be expected during the Archean but with temperatures at least 100°C lower than those shown in Figure 7.39. These plumes would have to rise further before crossing the solidus—perhaps to depths of 450 km (Jarvis and Campbell, 1983). Since this is still considerably deeper than the source region of modern MORB, a higher degree of partial melting will occur as these hot parcels of Archean mantle continue to rise. This could account for the high MgO content of Archean komatiites. (The lack of high MgO content komatiites after Archean times does not imply a lack of rising plumes or hot spots in the present mantle. Rather, plumes rising in the cooler present mantle must rise to shallower depths before crossing the solidus.)

This mechanism for the formation of komatiites does not require the very high Archean geothermal gradients which are predicted if the source region of Archean komatiites is assumed to be the same as for MORB, but 300°C hotter (e.g., Green, 1975). An Archean mantle with a temperature 300°C above that of the modern mantle is inconsistent with pressure–temperature estimates for Archean high grade metamorphic rocks which, as mentioned above, suggest that Archean continental geothermal gradients were not appreciably greater than modern gradients.

Although the above suggestion for the production of komatiites from a relatively cool mantle is based on a numerical model of whole mantle convection, a similar qualitative explanation can be made in terms of convection confined to the upper mantle (0–700 km depth). The essential factor is the existence of a lower thermal boundary layer where temperatures are much hotter than average.

The above discussion is highly speculative. Nevertheless, it serves to illustrate how an appreciation of the thermal regime to be expected during vigorous mantle convection can lead to new interpretations of unexplained, or seemingly contradictory, observations. We cannot expect simple models to explain the observations in detail; however, if they help us to ask new questions about the data they will have served their purpose.

8. References

Allegre, C.J., Hart, S.H., and Minster, J.-F., 1983. Chemical structure and evolution of the mantle and continents determined by inversion of Nd and Sr isotopic data. Part 2. Numerical experiments and discussion. *Earth Planet. Sci. Lett.* **66**, 177–213.

Anderson, D.L., 1984. Surface wave tomography. *EOS, Trans. Am. Geophys. Union* **65**, 147–148.

Anderson, D.L. and Dziewonski, A.M., 1984. Seismic tomography. *Scientific American* **251**, 60–68.

Backus, G.E., 1975. Gross thermodynamics of heat engines in the deep interior of the Earth. *Proc. Natl. Acad. Sci. U.S.A.* **72**, 1555–1558.

Bénard, M.H., 1900. Les tourbillons cellulaires dans une nappe liquide. *Revue générale des Sciences pures et appliquées* **11**, 1261.

Bénard, M.H., 1901. Les tourbillons cellulaires dans une nappe liquide transportant de la chaleur par convection en régime permanent. *Annales de Chimie et de Physique*, 7e série, Tome XXIII, 62–144.

Bickle, M.J., 1978. Heat loss from the Earth: a constraint on Archean tectonics from the relation between geothermal gradients and the rate of plate production. *Earth Planet. Sci. Lett.* **40**, 301–315.

Bickle, M.J., Ford, C.E., and Nisbet, E.G., 1977. The petrogenesis of peridotitic komatiites: evidence from high-pressure melting experiments. *Earth Planet. Sci. Let.* **37**, 97–106.

Bender, J.F., Hodges, F.N., and Bence, A.E., 1978. Petrogenesis of basalts from the project FAMOUS area: Experimental study from 0 to 15 kbar. *Earth Planet. Sci. Lett.* **41**, 277–302

Block, M.J., 1956. Surface tension as the cause of Bénard cells and surface deformation in a liquid film. *Nature* **178**, 650–651.

Boss, A.P. and Sacks, I.S., 1986. High spatial resolution models of time-dependent layered mantle convection. *Geophys. J. R. Astr. Soc.* **87**, 241–264.

Boussinesq, J., 1903. *Théorie analytique de la Chaleur mise en Harmonie avec la Thermodynamique et avec la Théorie mécanique de la Lumière*, Tome II, Paris: Gauthier-Villars, pp. 157–176.

Burke, K. and Kidd, W.S.F., 1978. Were Archean continental geothermal gradients much steeper than those of today? *Nature* **272**, 240–241.

Busse, F.H., 1967. On the stability of two-dimensional convection in a layer heated from below. *J. Math. Phys.* **46**, 140–150.

Busse, F.H., 1971. Stability regions of cellular fluid flow. Proceedings of the IUTAM Symposium, Herrenalb, 1969. In H. Leipholz (ed.), *Instability of Continuous Systems*, Berlin, Heidelberg, New York: Springer-Verlag, pp. 41–47.

Busse, F.H., 1972. The oscillatory instability of convection rolls in a low Prandtl number fluid. *J. Fluid Mech.* **52**, 97–112.

Busse, F.H., 1981. On the aspect ratios of two-layered mantle convection. *Phys. Earth Planet. Int.* **24**, 320–324.

Busse, F.H., 1982. Multiple solutions for convection in a two component fluid. *Geophys. Res. Lett.* **9**, 519–522.

Busse, F.H. and Whitehead, J.A., 1971. Instabilities of convection rolls in a high Prandtl number fluid. *J. Fluid Mech.* **47**, 305–320.

Campbell, I.H. and Jarvis, G.T., 1984. Mantle convection and early crustal evolution. *Precambrian Res.* **26**, 15–56.

Carrigan, C.R., 1982. Multiple-scale convection in the Earth's mantle: a three-dimensional study. *Science* **215**, 965–967.

Cassen, P., Reynolds, T., Graziani, F., Summers, A., NcNellis, J., and Blalock, L., 1979. Convection and lunar thermal history. *Phys. Earth Planet. Int.* **19**, 183–196.

Chandrasekhar, S., 1961. *Hydrodynamic and Hydromagnetic Stability*. Oxford, Clarendon Press, 652.

Christensen, U., 1984a. Convection with pressure- and temperature-dependent non-Newtonian rheology. *Geophys. J. R. Astr. Soc.* **77**, 343–384.

Christensen, U., 1984b. Heat transport by variable viscosity convection and implications for the Earth's thermal evolution. *Phys. Earth Planet. Int.* **35**, 264–282.

Christensen, U., 1985. Heat transport by variable viscosity convection II: Pressure influence, non-Newtonian rheology and decaying heat sources. *Phys. Earth Planet. Int.* **37**, 183–205.

Chu, T.Y. and Goldstein, R.J., 1973. Turbulent convection in a horizontal layer of water. *J. Fluid Mech.* **60**, 141–150.

Clarke, D.B., 1970. Tertiary basalts from Baffin Bay: possible primary magma from the mantle. *Cont. Mineral. Petrol.* **25**, 203–224.

Clayton, R.W. and Comer, R.P., 1987. Reconstruction of mantle heterogeneity by iterative back projection of travel times, II. Results for P-waves. To be submitted to: *J. Geophys. Res.*

Clever, R.M. and Busse, F.H., 1974. Transition to time-dependent convection. *J. Fluid Mech.* **65**, 625–645.

Comer, R.P. and Clayton, R.W., 1987. Reconstruction of mantle heterogeneity by iterative back projection of travel times, I. Theory and reliability. To be submitted to: *J. Geophys. Res.*

Daly, S.F., 1980. Convection with decaying heat sources: constant viscosity. *Geophys. J. R. Astr. Soc.* **61**, 519–547.

Davies, G.F., 1977. Whole-mantle convection and plate tectonics. *Geophys. J. R. Astr. Soc.* **49**, 459–486

Davies, G.F., 1980. Thermal histories of convective Earth models and constraints on radiogenic heat production in the Earth. *J. Geophys. Res.* **85**, 2517–2530.

Davies, G.F., 1984. Geophysical and isotopic constraints on mantle convection: an interim synthesis. *J. Geophys. Res.* **89**, 6017–6040.

Davis, E.E. and Lister, C.R.B., 1974. Fundamentals of ridge crest topography. *Earth Planet. Sci. Lett.* **21**, 405–413.

Deardorff, J.W., 1964. A numerical study of two-dimensional parallel-plate convection. *J. Atmos. Sci.* **21**, 419–438.

Deardorff, J.W., 1965. A numerical study of pseudo three-dimensional parallel-plate convection. *J. Atmos. Sci.* **22**, 419–435.

De La Cruz Reyna, S., 1973. Convection with internal heat generation. M.Sc. Thesis, University of Toronto, Toronto, Ontario, Canada, pp. 139.

Dziewonski, A.M., 1984. Mapping the lower mantle: determination of lateral heterogeneity in P velocity up to degree and order 6. *J. Geophys. Res.* **89**, 5929–5952.

Dziewonski, A.M., Hager, B.H., and O'Connell, R.J., 1977. Large scale heterogeneities in the lower mantle. *J. Geophys. Res.* **82**, 239–235.

Elsasser, W.M., Olson, P., and Marsh, B.D., 1979. The depth of mantle convection. *J. Geophys. Res.* **84**, 147–155. Correction: **84**, 613.

Elthon, D., 1979. High magnesia liquids as the parental magma for ocean floor basalts. *Nature* **278**, 514–518.

Elthon, D. and Scarfe, C.M., 1980. High-pressure phase equilibria of a high-magnesia basalt: Implications for the origin of mid-ocean basalts. Carnegie Institute, Washington, Yearbook 79, 277–281.

Francis, D.M., Hynes, A.J., Ludden, J.N., and Badard, J., 1981. Crystal fractionation and partial melting in the petrogenesis of a Proterozoic high-MgO volcanic suite, Ungava, Quebec. *Cont. Mineral. Petrol.* **78**, 27–36.

Fromm, J.E., 1964. The time dependent flow of an incompressible viscous fluid. *Meth. Comp. Phys.* **3**, 345–382.

Fromm, J.E., 1965. Numerical solutions of the nonlinear equations for a heated fluid layer. *Phys. Fluids* **8**, 1757–1769.

Gough, D.O., 1969. The anelastic approximation for thermal convection. *J. Atmos. Sci.* **26**, 448–456.

Graham, E., 1975. Numerical simulation of two-dimensional compressible convection. *J. Fluid Mech.* **70**, 689–703.

Green, D.H., 1975. Genesis of Archean peridotitic magmas and constraints on Archean geothermal gradients and tectonics. *Geology* **3**, 15–18.

Hager, B.H., Clayton, R.W., Richards, M.A., Comer, R.P., and Dziewonski, A.M., 1985. Lower mantle heterogeneity, dynamic topography and the geoid. *Nature* **313**, 541–545.

Haxby, W.F., and Weissel, J.K., 1983. Evidence for small-scale mantle convection from SEASAT altimeter data. *EOS, Transactions of the American Geophysical Union* **64**, 838.

Hess, H.H., 1962. History of the ocean basins. In A.E.J. Engel, H.L. James and B.F. Leonard (eds.), *Petrologic Studies: A Volume in Honor Of A.F. Buddington*, Geological Society of America, pp. 599–620.

Hewitt, J.M., McKenzie, D.P., and Weiss, N.O., 1975. Dissipative heating in convective flows. *J. Fluid Mech.* **68**, 721–738.

Hsui, A.T., Turcotte, D.L., and Torrance, K.E., 1972. Finite-amplitude thermal convection within a self-gravitating fluid sphere. *Geophys. Fluid Dynam.* **3**, 35–44.

Jacobsen, S.B. and Wasserburg, G.J., 1979. The mean age of mantle and crustal reservoirs. *J. Geophys. Res.* **84**, 7411–7427.

Jarvis, G.T., 1978. Convection in infinite Prandtl number compressible fluids. Ph.D. thesis, University of Cambridge, Cambridge, England.

Jarvis, G.T., 1984. Time dependent convection in the Earth's mantle. *Phys. Earth Planet. Int.* **36**, 305–327.

Jarvis, G.T., 1985. The long wavelength component of mantle convection. *Phys. Earth Planet. Int.* **40**, 24–42.

Jarvis, G.T. and Campbell, I.H., 1983. Archean komatiites and geotherms: solution to an apparent contradiction. *Geophys. Res. Lett.* **10**, 1133–1136.

Jarvis, G.T. and McKenzie, D.P., 1980. Convection in a compressible fluid with infinite Prandtl number. *J. Fluid Mech.* **96**, 515–583.

Jarvis, G.T. and Peltier, W.R., 1980. Oceanic bathymetry profiles flattened by radiogenic heating in a convecting mantle. *Nature* **285**, 649–651.

Jarvis, G.T. and Peltier, W.R., 1981. Effects of lithospheric rigidity on ocean floor bathymetry and heat flow. *Geophys. Res. Lett.* **8**, 857–860.

Jarvis, G.T. and Peltier, W.R., 1982. Mantle convection as a boundary layer phenomenon. *Geophys. J. R. Astr. Soc.* **68**, 389–427.

Jarvis, G.T. and Campbell, I.H., 1983. Archean komatiites and geotherms: solution to an apparent contradiction. *Geophys. Res. Lett.* **10**, 1133–1136.

Jarvis, G.T. and Peltier, W.R., 1986. Lateral heterogeneity in the convecting mantle. *J. Geophys. Res.* **91**, 435–451.

Jeanloz, R. and Richter, F.M., 1979. Convection, composition, and the thermal state of the lower mantle. *J. Geophys. Res.* **84**, 5497–5504.

Jeffreys, H., 1926. The stability of a layer of fluid heated below. *Philosophical Magazine, Series VII* **2**, 833–844.

Jeffreys, H., 1928. Some cases of instability in fluid motion. *Proc. Royal Soc., Series A* **118**, 195–208.

Jian, L. and Parmentier, E.M., 1985. Surface topography due to convection in a variable viscosity fluid: application to short wavelength gravity anomalies in the central Pacific Ocean. *Geophys. Res. Lett.* **12**, 357–360.

Kaula, W.M., 1979. Thermal evolution of Earth and Moon growing by planetesimal impacts. *J. Geophys. Res.* **84**, 999–1008.

Kennedy, G.C. and Higgins, G.H., 1972. Melting temperatures in the Earth's mantle. *Tectonophysics* **13**, 221–232.

Kolstedt, D.L. and Goetze, C., 1974. Low stress and high temperature creep in olivine single crystals. *J. Geophys. Res.* **79**, 2045–2051.

Koschmieder, E.L., 1967. On convection under an air surface. *J. Fluid Mech.* **30**, 9–15.

Krishnamurti, R., 1968a. Finite amplitude convection with changing mean temperature. Part 1. Theory. *J. Fluid Mech.* **33**, 445–455.

Krishnamurti, R., 1968b. Finite amplitude convection with changing mean temperature. Part 2. An experimental test of the theory. *J. Fluid Mech.* **33**, 457–463.

Krishnamurti, R., 1970a. On the transition to turbulent convection. Part 1. The transition from two- to three-dimensional flow. *J. Fluid Mech.* **42**, 295–307.

Krishnamurti, R., 1970b. On the transition to turbulent convection. Part 2. The transition to time-dependent flow. *J. Fluid Mech.* **42**, 309–320.

Krishnamurti, R., 1973. Some further studies on the transition to turbulent convection. *J. Fluid Mech.* **60**, 285–303.

Landau, L.D. and Lifshitz, E.M., 1959. *Fluid Mechanics*, Oxford: Pergamon Press.

Lux, R.A., Davies, G.F., and Thomas, J.H., 1979. Moving lithospheric plates and mantle convection. *Geophys. J. R. Astr. Soc.* **57**, 209–228.

Malkus, W.V.R., 1954. Discrete transitions in turbulent convection. *Proc. Royal Soc. London, Series A* **225**, 185–195.

Malkus, W.V.R., 1964. Boussinesq equations and convection energetics. *Geophys. Fluid Dynam.*, Woods Hole Oceanographic Institute, Ref. No. 64-46.

Malkus, W.V.R., 1973. Convection at the melting point: a thermal history of the Earth's core. *Geophys. Fluid Dynam.* **4**, 267–278.

Malkus, W.V.R. and Veronis, G., 1958. Finite amplitude cellular convection. *J. Fluid Mech.* **4**, 225–260.

Masters, G., Jordan, T.H., Silver, P.G., and Gilbert, F., 1982. Aspherical Earth structure from fundamental spheroidal-mode data. *Nature* **298**, 609–613.

McKenzie, D.P., 1967. Some remarks on heat flow and gravity anomalies. *J. Geophys. Res.* **72**, 6261–6273.

McKenzie, D.P., 1977. Surface deformation, gravity anomalies and convection. *Geophys. J. R. Astr. Soc.* **48**, 211–238.

McKenzie, D.P. and O'Nions. R.K., 1983. Mantle reservoirs and ocean island basalts. *Nature* **301**, 229–231.

McKenzie, D.P. and Richter, F.M., 1976. Convection currents in the Earth's mantle. *Scientific American* **235**, 72–89.

McKenzie, D.P. and Weiss, N.O., 1975. Speculations on the thermal and tectonic history of the Earth. *Geophys. J. R. Astr. Soc.* **42**, 131–174.

McKenzie, D.P. and Weiss, N.O., 1980. The thermal history of the Earth. In D.W. Strangway (ed.), *The Continental Crust And Its Mineral Deposits*, Geological Association of Canada, Special Paper 20, pp. 575–590.

McKenzie, D.P., Roberts, J.M., and Weiss, N.O., 1974. Convection in the Earth's mantle: towards a numerical simulation. *J. Fluid Mech.* **62**, 465–538.

Mihaljan, J.M., 1962. A rigorous exposition of the Boussinesq approximations applicable to a thin layer of fluid. *Astrophys. J.* **136**, 1126–1133.

Mitrovica, J.X., and Jarvis, G.T., 1985. Surface deflections due to transient subduction in a convecting mantle. *Tectonophysics* **120**, 211–237.

Mitrovica, J.X. and Jarvis, G.T., 1987. A numerical study of thermal convection between rigid horizontal boundaries. *Geophys. Astrophys. Fluid Mech.* **38**, 193–224.

Moore, D.R. and Weiss, N.O., 1973. Two-dimensional Rayleigh-Bénard convection. *J. Fluid Mech.* **58**, 289–312.

Nakanishi, I. and Anderson, D.L., 1984a. Aspherical heterogeneity of the mantle from phase velocities of mantle waves. *Nature* **307**, 117–121.

Nakanishi, I. and Anderson, D.L., 1984b. Measurements of mantle velocities and inversion for lateral heterogeneity and anisotropy, II, Analysis by the single-station method. *Geophys. J. R. Astr. Soc.* **78**, 573–618.

Nield, D.A., 1964. Surface tension and buoyancy effects in cellular convection. *J.*

Fluid Mech. **19**, 341–352.

Oberbeck, A., 1879. Uber die Warmeleitung der Flussigkeiten bei Berucksichtigung der Stromungen infolge von Temperaturdifferenzen. *Annalen der Physik und Chemie, Neue Folge* **7**, 271–292.

Oberbeck, A., 1880. Stromungen von Flussigkeiten infolge ungleicher temperature innerhalb derselben. *Annalen der Physik und Chemie Neue Folge* **11**, 489–495.

O'Connell, R.J., 1977. On the scale of mantle convection. *Tectonophysics* **38**, 119–136.

Ogura, Y. and Phillips, N.A., 1962. Scale analysis of deep and shallow convection in the atmosphere. *J. Atmos. Sci.* **19**, 173–179.

O'Hara, M.J., 1968. Are ocean floor basalts primary magmas? *Nature* **200**, 683–686.

Olson, P., 1981. Mantle convection with spherical effects. *J. Geophys. Res.* **68**, 4881–4890.

Olson, P. and Corcos, G.M., 1980. A boundary layer model for mantle convection with surface plates. *Geophys. J. R. Astr. Soc.* **62**, 195–219.

Olson, P., Yuen, D.A., and Belsiger, D., 1984a. Mixing of passive heterogeneities by mantle convection. *J. Geophys. Res.* **89**, 425–436.

Olson, P, Yuen, D.A., and Belsiger, D., 1984b. Convective mixing and the fine structure of mantle heterogeneity. *Phys. Earth Planet. Int.* **36**, 291–304.

O'Nions, R.K., Evensen, N.M., Hamilton, P. J., and Carter, S.R., 1978. Melting of the mantle past and present: isotope and trace element evidence. *Phil. Trans. Royal Soc. London, Series A* **258**, 547–559.

O'Nions, R.K., Carter, S.R., Evensen, N.M., and Hamilton, P.J., 1979. Geochemical and cosmochemical applications of Nd isotope analysis. *Ann. Rev. Earth Planet. Sci.* **7**, 11–38.

Oxburgh, E.R. and Turcotte, D.L., 1968. Mid-ocean ridges and geotherm distribution during mantle convection. *J. Geophys. Res.* **73**, 2643–2661.

Oxburgh, E.R. and Turcotte, D.L., 1978. Mechanisms of continental drift. *Rep. Prog. Phys.* **41**, 1249–1312.

Parmentier, E.M. and Morgan, J. III, 1982. Thermal convection in non-Newtonian fluids: volumetric heating and boundary layer scaling. *J. Geophys. Res.* **87**, 7757–7762.

Parmentier, E.M. and Turcotte, D.L., 1978. Two-dimensional mantle flow beneath a rigid accreting lithosphere. *Phys.Earth Planet. Int.* **17**, 281–289.

Parsons, B. and Daly, S., 1983. The relationship between surface topography, gravity anomalies and the temperature structure of convection. *J. Geophys. Res.* **88**, 1129–1144.

Parsons, B. and McKenzie, D.P., 1978. Mantle convection and the thermal structure of the plates. *J. Geophys. Res.* **83**, 4485–4496.

Parsons, B. and Sclater, J.G., 1977. An analysis of the variation of ocean floor bathymetry and heat flow with age. *J. Geophys. Res.* **82**, 803–827.

Pearson, J.R.A., 1958. On convection cells induced by surface tension. *J. Fluid Mech.* **4**, 489–500.

Pekeris, C.L., 1935. Thermal convection in the interior of the Earth. *Month. Not. Royal Astr. Soc., Geophys. Supp.* **3**, 343–367.

Peltier, W.R., 1972. Penetrative convection in the planetary mantle. *Geophys. Fluid Dynam.* **5**, 47–88.

Peltier, W.R., 1974. The impulse response of a Maxwell Earth. *Rev. Geophys. Space Phy.* **12**, 649.

Peltier, W.R., 1980. Mantle convection and viscosity. In A. Dziewonski, and E. Boschi (eds.), *Physics of the Earth's Interior, Proceedings of the Enrico Fermi International School of Physics* (Course LXXVIII), New York: North Holland.

Peltier, W.R. and Jarvis, G.T., 1982. Whole mantle convection and the thermal evolution of the Earth. *Phys. Earth Planet. Int.* **29**, 281–304.

Rayleigh, Lord, 1916. On convective currents in a horizontal layer of fluid, when the higher temperature is on the under side. *Philosophical Magazine, Series VI* **32**, 529–546.

Richter, F.M., 1973a. Dynamic models for sea floor spreading. *Rev. Geophys. Space Phys.* **11**, 223–287.

Richter, F.M., 1973b. Convection and the large-scale circulation of the mantle. *J. Geophys. Res.* **78**, 8735–8745.

Richter, F.M., 1984. Regionalized models for the thermal evolution of the Earth. *Earth Planet. Sci. Lett.* **68**, 471–484.

Richter, F.M. and McKenzie, D.P., 1978. Simple plate models of mantle convection. *J. Geophys.* **44**, 441–471.

Richter, F.M. and McKenzie, D.P., 1981. Parameterizations for the horizontally averaged temperature of infinite Prandtl number convection. *J. Geophys. Res.* **86**, 1738–1744.

Richter, F.M., and Parsons, B., 1975. On the interaction of two scales of convection in the mantle. *J. Geophys. Res.* **80**, 2529–2541.

Richter, F.M., Daly, S.F., and Nataf, H.-C., 1982. A parameterized model for the evolution of isotopic heterogeneities in a convecting system. *Earth Planet. Sci. Lett.* **60**, 178–194.

Ringwood, A.E., 1979. *Origin of the Earth and Moon*, New York: Springer-Verlag, pp. 295.

Roberts, G.O., 1979. Fast viscous Bénard convection. *Geophys. Astrophys. Fluid Dyn.* **12**, 235–272.

Roberts, P.H., 1967. Convection in horizontal layers with internal heat generation. Theory. *J. Fluid Mech.* **30**, 33–49.

Rossby, H.T., 1969. A study of Bénard convection with and without rotation. *J. Fluid Mech.* **36**, 309–335.

Scanlon, J.W. and Segel, L.A., 1967. Finite amplitude cellular convection induced by surface tension. *J. Fluid Mech.* **30**, 149–162.

Schatz, J.F. and Simmons, G., 1972. Thermal conductivities of Earth materials at high temperatures. *J. Geophys. Res.* **77**, 6966.

Schlüter, A., Lortz, D., and Busse, F., 1965. On the stability of steady finite

amplitude convection. *J. Fluid Mech.* **23**, 129–144.
Schmidt, R.J. and Milverton, S.W., 1935. On the instability of a fluid when heated from below. *Proc. Royal Soc., Series A* **152**, 586–594.
Schmidt, R.J. and Saunders, O.A., 1938. On the motion of a fluid heated from below. *Proc. Royal Soc., Series A* **165**, 216–228.
Schubert, G., Cassen, P., and Young, R.E., 1979. Subsolidus convective cooling histories of terrestrial planets. *Icarus* **38**, 192–211.
Sclater, J.G. and Francheteau, J., 1970. The implications of terrestrial heat flow observations on current tectonic and geochemical models of the crust and upper mantle of the Earth. *Geophysical J. R. Astr. Soc.* **20**, 509–542.
Sclater, J.G., Jaupart, C., and Galson, D., 1980. The heat flow through oceanic and continental crust and the heat loss of the Earth. *Rev. Geophys. Space Phys.* **18**, 269–311.
Sharpe, H.N. and Peltier, W.R., 1978. Parameterized mantle convection and the Earth's thermal history. *Geophys. Res. Lett.* **5**, 737–740.
Sharpe, H.N. and Peltier, W.R., 1979. A thermal history model for the Earth with parameterized convection. *Geophysical J. R. Astr. Soc.* **59**, 171–205.
Silveston, P.L., 1958. Warmedurchgang in waagerechten Flussigkeitsschichten, Part 1. *Forsch. Ing. Wes.* **24**, 29–32, and p. 59–69.
Skilbeck, J.N. and McKenzie, D.P., 1979. An approximate method for determining the stability of two-scale flow in the mantle. *Pageoph* **117**, 958–987.
Sleep, N.H., 1979. Thermal history and degassing of the Earth: Some simple calculations. *J. Geol.* **87**, 671–686.
Sommerscales, E.F.C. and Gazda, I.W., 1969. Thermal convection in high Prandtl number liquids at high Rayleigh numbers. *Int. J. Heat Mass Trans.* **12**, 1491–1511.
Spiegel, E.A. and Veronis, G., 1960. On the Boussinesq approximation for a compressible fluid. *Astrophys. J.* **131**, 442–447.
Stacey, F.D., 1977. A thermal model of the Earth. *Phys. Earth Planet. Int.* **15**, 341–348.
Stevenson, D.J. and Turner, J.S., 1979. Fluid models of mantle convection. In M.W. McElhinny (ed.), *The Earth, Its Origin, Evolution and Structure*, New York: J. Wiley, pp. 227–263.
Straus, J.M., 1972. Finite amplitude doubly diffusive convection. *J. Fluid Mech.* **56**, 353–374.
Tozer, D.C., 1965. Heat transfer and convection currents. *Philosoph. Trans. R. Soc., Series A* **258**, 252–275.
Tozer, D.C., 1967. Towards a theory of thermal convection in the mantle. In: T.F. Gaskell (ed.), *The Earth's Mantle*, New York: Academic Press, pp. 325–353.
Tozer, D.C., 1972. The present thermal state of the terrestrial planets. *Phys. Earth Planet. Int.* **6**, 182–197.
Tozer, D.C., 1974. The internal evolution of planetary-sized objects. *The Moon* **9**, 176–182.
Travis, B.J. and Brandow, B., 1984. A numerical study of mantle convection with radiochemical transport. *Terra Cognita* **4**, 224.

Turcotte, D.L. and Oxburgh, E.R., 1967. Finite amplitude convective cells and continental drift. *J. Fluid Mech.* **28**, 29–42.

Turcotte, D.L. and Oxburgh, E.R., 1972. Mantle convection and the new global tectonics. *Ann. Rev. Fluid Mechan.* **4**, 33–68.

Turcotte, D.L. and Schubert, G., 1982. *Geodynamics, Applications of Continuum Mechanics to Geological Problems*, New York: J. Wiley & Sons, pp. 450.

Turcotte, D.L., Cooke, F.A., and Willeman, R.J., 1979. Parameterized convection within the moon and the terrestrial planets. *Proceedings of the Tenth Lunar and Planetary Science Conference, Houston, Texas, March 19–23, 1979. Volume 3: Planetary Interiors and Surfaces*, New York: Pergamon Press, (*Geochim. Cosmochim. Acta*, Supp. 11), 2375–2392.

Turcotte, D.L., Hsui, A.T., Torrance, K.H., and Schubert, G., 1974. Influence of viscous dissipation on Bénard convection. *J. Fluid Mech.* **64**, 369–374.

Turcotte, D.L., Torrance, K.E., and Hsui, A.T., 1973. Convection in the Earth's mantle. *Meth. Comp. Phys.* **13**, 431–454.

Usselman, T.M., and Hodge, D.S., 1978. Thermal control of low pressure fractionation processes. *J. Volcanol. Geotherm. Res.* **4**, 265–281.

Veronis, G., 1966. Large-amplitude Bénard convection. *J. Fluid Mech.* **26**, 49–68.

Weertman, J., 1970. The creep strength of the Earth's mantle. *Rev. Geophys. Space Phys.* **8**, 145–168.

Wells, P.R.A., 1976. Late Archean metamorphism in the Buksefjorden region, S.W. Greenland. *Cont. Mineral. Petrol.* **56**, 229.

Whitehead, J.A. and Parsons, B., 1978. Observations of convection at Rayleigh numbers up to 760,000 in a fluid with a large Prandtl number. *Geophys.Astrophys. Fluid Dyn.* **9**, 201–217.

Woodhouse, J.H., and Dziewonski, A.M., 1984. Mapping the upper mantle: three dimensional modeling of Earth structure by inversion of seismic waveforms. *J. Geophys. Res.* **89**, 5953–5986.

Young, R.E., 1974. Finite-amplitude thermal convection in a spherical shell. *J. Fluid Mech.* **63**, 695–721.

Yuen, D.A., Peltier, W.R., and Schubert, G., 1981. On the existence of a second scale of convection in the upper mantle. *Geophys. J. R. Astr. Soc.* **65**, 171–190.

Zebib, A., Schubert, G., and Straus, J., 1980. Infinite Prandtl number thermal convection in a spherical shell. *J. Fluid Mech.* **97**, 257–277.

CHAPTER 8

Mantle Rheology, Constitution, and Convection

Ulrich R. Christensen

Institut für Geowissenschaften, Universität Mainz
and Max-Planck-Institut für Chemie, Mainz, Federal Republic of Germany

1. Introduction

With the success of the plate tectonic theory it was also recognized that the heat engine which drives the plates by transforming thermal energy into mechanical work must be a process of thermal convection in the Earth's mantle. Most of the geodynamical activity, the opening and closing of oceans, the formation of mountain belts, etc., depends directly or indirectly on mantle convection. Besides the mechanical aspect there is the thermal one: mantle convection dominates the radial heat transport and determines the thermal structure of the mantle. Relatively thin thermal boundary layers with strong temperature gradients surround the nearly adiabatic core of the convection cell. Convection controls the overall heat loss from the Earth and therefore the thermal evolution of our cooling planet. Finally the shear motion of the convective flow is potentially able to mix chemical heterogeneities in the mantle.

Concerning the theoretical treatment, mantle convection differs in some points from convection in laboratory fluids, in technical applications, or from convection in the atmosphere or hydrosphere. Due to the slow motion and high viscosity, inertial forces do not play a role in mantle convection—the Prandtl number is virtually infinite—which simplifies its treatment. On the other hand the rheology of mantle rocks is considerably more complex than usually encountered in convective problems. Its dependence on temperature and pressure may introduce viscosity differences of many orders of magnitude. Furthermore it is likely that the rheology is non-Newtonian, i.e., the strain rate varies with some power of the stress rather than linearly. Another important complication is that the mantle cannot be considered as a simple one-phase fluid. Solid–solid phase transitions occur in the depth range of 300–700 km and partial melting occurs beneath mid-ocean ridges, which will affect the dynamics. It is disputed whether the mantle is chemically uniform or radially layered. The preferred candidate for a chemical boundary is the seismic discontinuity at 670 km depth. Chemical layering both requires and potentially causes the convection to be organized in distinct layers rather than being mantlewide. Even if there is no gross layering of the mantle it seems inevitable that chemical heterogeneities of various scales must exist in its volume.

Despite these complications, mantle convection has often been studied in quite simple models, using plane layers of fluid with uniform viscosity heated from below (Rayleigh–Bénard convection). To some extent the assumption of isoviscosity appears justified because the analysis of postglacial rebound favors relatively uniform viscosity in the entire mantle (Cathles, 1975; Peltier, 1976). The models based upon simple Rayleigh–Bénard convection are able to predict some gross features correctly, such as average plate velocity, the magnitude of the heat flow and of gravity and topography anomalies. However, although mantle convection is accepted as a concept, many of its features are either unresolved or poorly understood. A major issue of debate is whether the whole mantle turns over or if convection is separated into layers. Another question concerns the nature of the oceanic lithosphere, which is now regarded as the upper boundary layer of the convective circulation rather than as an appendage to an underlying cell. But why does it move as a quasi-rigid unit? If this is due to a much higher viscosity compared to the asthenosphere, then why does it move at all? Is the convective circulation shaped by the geometry of the surface plates or vice versa? A further point of uncertainty concerns the concept of mantle plumes (Morgan, 1971), axisymmetric hot upwelling currents causing hot spot volcanism. Can their existence be established beyond doubt and where do they come from? Are there upwelling currents from deep within the mantle below normal oceanic ridges?

In order to shed further light on the structure of mantle convection, clearly more observational evidence from different fields of geoscience is required. On the other hand the theoretical understanding of the convection process regarding the complications of mantle rheology and constitution needs also improvement. The results of such theoretical studies can be used in two different ways. On the one hand, they could help to elucidate the possible structure and properties of mantle convection. On the other hand, many parameters are quite uncertain, and "inverse modelling" may help to put constraints on them.

The computer is an especially useful tool to study the influence of those features of mantle convection which are not readily simulated in laboratory convection, such as phase transitions, the effects of pressure-dependence of viscosity or power-law creep, free-slip boundaries, sphericity, or strong fluid compressibility. A disadvantage of numerical studies is that for reasons of economy one is normally restricted to two-dimensional, steady-state, single-cell models. Each of these restrictions can be relaxed in principle, but at costs which range from high to outrageous.

In this chapter I will review some of the numerical work which I have performed during the last few years. The philosophy underlying most of the case studies is to isolate a small number of possible complications and to determine their effect on the properties of convection and their mutual influence. With the exception of some models described in Sections 6 and 9 the intention is not to simulate mantle convection directly, but to learn something about the basic physical processes in the context of simple fluid dynamical models. Given the large number of possible complications, and the great uncertainty as to the quantitative values of mantle parameters, this seems a sensible way to procede.

In Section 2 a brief outline of the governing equations and of the finite element method which I have employed to solve them will be provided, and in Section 3 some experimental results concerning the rheology of mantle rock are reviewed. Numerical results on convection with exponentially temperature-dependent viscosity are discussed in Section 4. Quite surprisingly it was found that the heat transport through the convecting layer becomes nearly independent of the viscosity in the interior of this layer. This may have important consequences for the theory of the thermal evolution of the Earth (Section 5). The Archaean mantle would have been significantly hotter than the present mantle, but neither heat flow nor plate velocities would have been much greater than today. These results contrast with previous estimates based on the idea of parameterized convection. Section 6 deals with additional pressure influence on the rheology, and with the lower boundary layer of convection. The D″-layer above the core–mantle boundary is a candidate for this boundary layer, and problems regarding its instability and the ejection of plumes from it are looked into. In Section 7 a third feature is added to the pressure- and temperature-dependence of rheology: power-law creep. Its major effect is to reduce spatial viscosity contrasts which are build up by the p,T dependence. Furthermore it was found that stress-dependent rheology may lead to a platelike behavior of the upper thermal boundary layer. The state of the argument concerning a possible layering of mantle composition and mantle convection is discussed in Section 8, with special emphasis put on the mineralogical and dynamical role of the 670-km discontinuity as a phase or chemical boundary. It has been speculated from time to time that an endothermic phase boundary might induce layered convection, but only recently has it been possible to demonstrate this in numerical experiments (Section 9). However, the existence of layered convection would require a somewhat extreme p,T slope of the phase boundary. The question of layered versus whole-mantle convection is closely linked to the question of the fate of subducting lithospheric slabs at the depth were all seismic activity ceases—i.e. near the 670 km discontinuity. Convection models including the simulation of a subducting slab interacting with a chemical and/or phase boundary are also discussed in Section 9. In the concluding section I have tried to formulate some urgent problems which need to be solved for a completion of the puzzle of mantle dynamics.

2. Governing Equations and Numerical Analysis

Variable viscosity convection is governed by an equation of motion, an equation of energy transport, and a law relating the viscosity to temperature, pressure and possibly stress. Inertial forces can be safely neglected and usually the Boussinesq-approximation is applied, i.e., the fluid density is taken as constant (incompressible flow) except in computation of the buoyancy force. The Boussinesq approximation is well justified for laboratory convection but over the depth-range of the mantle the compressibility is significant. Although non-Boussinesq effects need not be small (Jarvis and McKenzie, 1980), they are usually neglected for mathematical simplicity. The Boussinesq equations of convection are extensively discussed in the literature (e.g., Oxburgh and Turcotte, 1978; McKenzie, *et al.*, 1974). For two-dimensional convection the use of the stream function ψ instead of the "primitive" variables velocity $\vec{u} = (u, 0, w) = (\psi_z, 0, -\psi_x)$ and pressure p can be advantageous. To facilitate the numerical treatment of the partial differential equations they are usually expressed in nondimensional form. With the nondimensionalization scheme given in Table 1 in Christensen (1984a) the equations of motion and of conservation of energy for a two-dimensional variable viscosity flow are, respectively,

$$(\partial_{zz} - \partial_{xx})\{\eta(\psi_{zz} - \psi_{xx})\} + 4\partial_{xz}(\eta\psi_{xz}) = \mathrm{Ra}_0 \frac{\partial T}{\partial x}, \tag{2.1}$$

$$\frac{DT}{Dt} = \frac{\partial T}{\partial t} + u\frac{\partial T}{\partial x} + w\frac{\partial T}{\partial z} = \nabla^2 T + Q_{\mathrm{int}}. \tag{2.2}$$

D/Dt is the substantial derivative, T the temperature and Q_{int} the rate of internal heating. The computational domain is $0 \leqq x \leqq l$ and $0 \leqq z \leqq 1$ where l is the aspect ratio of the cell. The Rayleigh number Ra_0 is given by

$$\mathrm{Ra}_0 = \frac{\alpha g \rho \Delta T h^3}{\kappa\eta(T = 0,\ z = 0)}, \tag{2.3}$$

where α is the coefficient of thermal expansion, g the gravitational acceleration ρ the density, ΔT the temperature contrast across the convecting layer of height h, and κ the thermal diffusivity. The reference viscosity $\eta_0 = \eta(T = 0,\ z = 0)$ is taken at the temperature of the top surface and if there is a pressure or depth-dependence of viscosity also at the pressure on this surface (i.e., zero). Modifications of eqs. 1 and 2 which are necessary to treat chemical heterogeneities, phase boundaries or non-Boussinesq effects will be discussed in the respective sections. Eq. 3 refers to the Rayleigh–Bénard case of a fixed applied temperature contrast; if instead the rate of bottom or internal heating is fixed, slightly different definitions are appropriate (e.g., Jarvis and Peltier, 1982; Christensen, 1985a, section 4).

Besides definition (2.3) of the "surface Rayleigh number" Ra_0, two more definitions are in use for variable viscosity convection. Booker (1976) proposed to use the viscosity value at the mean of top and bottom temperatures, which is $T = 0.5$ with the scaling employed above. If the viscosity is also pressure- or depth-dependent then we consistently take the average pressure or depth ($z = 0.5$):

$$\mathrm{Ra}_{1/2} = \frac{\alpha g \rho \Delta T h^3}{\kappa\eta(T = 0.5,\ z = 0.5)}. \tag{2.4}$$

A further option is to calculate the viscosity with the real average temperature of the convecting layer, which is of course not known *a priori*, and we obtain the "internal Rayleigh number"

$$\mathrm{Ra}_T = \frac{\alpha g \rho \Delta T h^3}{\kappa \eta(\langle T \rangle, z = 0.5)}. \tag{2.5}$$

For completeness we note the "bottom Rayleigh number"

$$\mathrm{Ra}_1 = \frac{\alpha g \rho \Delta T h^3}{\kappa \eta(T = 1, z = 1)}. \tag{2.6}$$

The parabolic–elliptic system 2.1 and 2.2 needs an initial temperature field and boundary conditions for ψ and T to be specified. The conditions for the lateral boundaries result from the assumption of reflective continuation with identical cells on both sides, i.e., $\partial T/\partial x = 0$ and $\partial^2\psi/\partial x^2 = 0$. The mechanical boundary conditions for top and bottom are either rigid (no slip, $\partial\psi/\partial z = 0$) or stress-free (free slip, $\partial^2\psi/\partial z^2 = 0$). Most of the models in this article use free-slip boundaries which is the natural condition for both the Earth's surface and the core-mantle boundary.

Usually finite difference techniques have been applied to solve the equations of convection. More recently also finite element methods have been successfully used (Schubert and Anderson, 1985; Hansen and Ebel, 1984). For the models discussed in this article a nonstandard finite element approach with B-splines was used. The equation of motion 2.1 is solved via the Ritz–Rayleigh variational principle and the elliptic part of the energy equation 2.2 by an upwind weighted residual approach. A more detailed account on the method is given in Christensen (1984a). In many cases only steady state solutions are of interest. It is then more economical to drop the term $\partial T/\partial t$ in eq. 2.2 *ab initio* and to iterate between eqs. 2.1 and 2.2 rather than to solve the whole time-dependent problem. To keep the iteration stable an underrelaxation of the temperature is required. This procedure leads to a steady state solution within about one-hundredth to one-tenth of the computer time needed to reach it by solving the time-dependent problem. However, sometimes the steady state solution is numerically constrained, i.e., a time-dependent disturbance would grow rather than decay (cf. Christensen, 1984a, Section 5.2).

When the temporal evolution is of interest, a predictor-corrector scheme must be applied to the parabolic equation 2.2 to ensure second-order convergence in time. The numerical method was tested thoroughly by (1) reproducing published convection models, (2) convergence tests, (3) cross-checks with other finite element codes, (4) internal consistency checks (such as checking the equality of mechanical power generation and dissipated heat) The numerical grid consists of typically 16–32 elements in the z-direction, and 16–40 elements in x-direction depending on the aspect ratio. An important feature is grid refinement in the horizontal and vertical boundary layers.

As far as the accuracy of steady state solutions is concerned, the code is much superior to equidistant-grid finite difference methods. Due to the use of spline functions it is also superior to standard finite element techniques, running faster by a factor of about five to achieve the same level of accuracy (Quareni *et al.*, 1985). Global quantities, like the Nusselt number, are determined with an error of about 1 percent even with a coarse grid of 16×16 elements for effective Rayleigh numbers of $0(10^6)$ and viscosity variations of $0(10^3)$. Even very large viscosity contrasts up to 10^6 do not cause numerical problems. However, the advantage of this method compared to others is reduced when time-dependent problems are to be solved.

3. Mantle Rheology

The rheological behavior of mantle minerals has been extensively studied both experimentally and theoretically. Reviews were published by Stocker and Ashby (1973), Weertmann and Weertmann, (1975); Ashby and Verrall, (1977); Yuen, Sabadini and Boschi, (1983). At high temperature the migration of point defects or of dislocations leads to steady state creep. Whatever the exact mechanism of migration is, the migration rate will be controlled by a thermally activated process. Due to the multiplication of dislocations with increasing stress the strain rate of dislocation creep will increase stronger than linearly with stress (non-Newtonian behavior; Weertman, 1970). Theoretically dislocation creep must be superseded at low stress by a Newtonian creep mechanism due to the diffusion of vacancies either through the bulk grains (Herring-Nabarro-creep) or along grain boundaries (Cobble creep). While the dislocation creep is independent of the grain size d, the rate of Herring-Nabarro creep changes as d^{-2} and that of Cobble creep as d^{-3}. The Newtonian flow mechanisms have not yet been conclusively identified experimentally, possibly because of the difficulty of making creep experiments at very low stress and hence strain rates. The level of transition stress between Newtonian and non-Newtonian creep is thus uncertain, although estimates usually range between 0.01 to 10 MPa (0.1 to 100 bar) depending on the assumed grain size. However, an additional complication is the possible adjustment of the average grain size to the level of stress (s) by recrystallization or subgrain formation. Post (1977), for example, found empirically $d \propto s^{-1.5}$. Through this mechanism and the grain-size dependence of Herring-Nabarro or Cobble creep, even these basically Newtonian creep mechanisms may lead to an effectively non-Newtonian flow law. In connection with the structural phase transitions in the Earth's mantle, mechanisms of transitional superplasticity (reaction-enhanced ductility) have been envisaged (Sammis and Dein, 1974). During the phase change the material may become much softer. A different mechanism of "structural superplasticity" was proposed by Rubie (1984), based on the fine-grained texture of the product of the phase change reaction. With a grain-size dependent flow mechanisms this would also lead to a reduction of the effective viscosity. Superplasticity might influence the dynamical role of a mantle phase boundary.

Experimental work has been carried out on single crystal forsterite and olivine, on polycrystalline aggregates and on natural dunite and peridotite. The results of the experiments are fitted by the partly empirical and partly theoretical law

$$\dot{e} = As^n \exp\left(-\frac{E^* + pV^*}{RT}\right), \tag{3.1}$$

where $\dot{e}$ is the strain rate, s the stress, E^* an activation energy, V^* an activation volume, R the gas constant, T the absolute temperature. A and the stress exponent n are empirical constants.

For brevity the activation enthalpy,

$$H^* = E^* + pV^*, \tag{3.2}$$

can be introduced. The stress exponent n is usually reported to be in the range of 2–5 with a significant concentration of values near $n = 3$ (Kohlstedt and Goetze, 1974; Durham and Goetze, 1977; Post, 1977; Chopra and Patterson, 1981). In some studies a change of n to values close to one at stresses below 5–10 MPa has been reported (Berckhemer *et al.*, 1979; Relandeau, 1981). This agrees roughly with theoretical estimates of the transition stress towards Newtonian flow. Stress estimates for mantle convection range from 0(1 MPa) in the low-viscosity zone to 0(100 MPa), i.e., the kilobar range, in subducting lithosphere. Thus possibly at least a part of the convective circulation is in the non-Newtonian power-law regime. Depending on the exact size of the transition stress, or due to the mechanism of stress adjustment of the grain size, mantle convection might be governed entirely by non-Newtonian rheology. The fact that postglacial rebound can be interpreted consistently in terms of a Newtonian mantle (Peltier, 1976; Cathles, 1975) is not a conclusive argument against power-law creep in the mantle. The stress produced by the ice load is possibly only a minor increment to the larger stress-level produced by mantle convection. In this case the postglacial isostatic adjustment process would "see" a quasilinear rheology with the effective viscosity controlled by the convective stress.

Experimental data also establish a strong temperature dependence of the creep mechanism. The activation energy E^* is found in the range of 350–670 kJ/mol (80–160 kcal/mol). 525 kJ/mol (125 kcal/mol) is the usually assumed value for "dry" olivine (Kohlstedt and Goetze, 1974). As a rule of thumb, this means that a temperature change of 100 degrees in the mantle will change the strain rate by an order of magnitude for a given stress. There are indications that the creep process is also strongly influenced by minor chemical differences, like oxygen fugacity or water content (Post, 1977; Jaoul *et al.*, 1980; Justice *et al.*, 1982). Experimental determinations of the activation volume V^* are difficult, and until recently only theoretical estimates based on the volume of O^{--}-ions (11 cm^3/mol) were available. Meanwhile experimental determinations have led to values in the range of 11–18 cm^3 (Ross *et al.*, 1979; Kohlstedt *et al.*, 1980; Karato and Ogawa, 1982). It has been proposed that the pressure effect on the creep rate is the same as on the melting temperature T_m (Weertmann, 1970), which leads to the formulation

$$\dot{e} \propto s^n \exp\left(-a\frac{T_m}{T}\right), \tag{3.3}$$

where a is a numerical constant of the order 30–45. The pressure dependence obtained from eq. 3 seems to be in agreement with the cited values for V^*. Along with the self-compression and due to phase changes, the activation volume must decrease with pressure. The value of V^* in the lowermost mantle has been estimated to be one quarter to one half of the near-surface value (O'Connell, 1977; Sammis *et al.*, 1977; Karato, 1981; Liebermann and Poirier, 1984). The pressure effect in the interior of the mantle will be partly impaired by the adiabatic rise of temperature in the core of convection cells. However, even when subtracting the adiabatic temperature effect an activation volume of 10 cm^3/mol implies a viscosity increase of one order of magnitude for every 150 km depth under constant stress. More reliable values for the rheological parameters E^* and V^* deep in the mantle would be needed to understand the radial viscosity structure of the mantle.

The analysis of postglacial rebound and more recently of data on deglaciation-induced true polar wander and nontidal acceleration of the Earth's rotation have been used to constrain the radial viscosity-distribution of the mantle (for a review, see Peltier (1985)). For the upper mantle 10^{21} Pa s (10^{22} Poise) has become the accepted standard value. A low-viscosity channel of less than 10^{20} Pa s and about 100 km thickness may be present below the lithosphere (Cathles, 1975). The viscosity in the lower mantle is found to be the same as in the upper mantle or at most moderately higher. This would require a fortuitous compensation of the effect of the adiabatic gradient and of the pressure influence on the viscosity. Such compensation appears only marginally compatible with the estimates from solid state physics and requires V^* to be at the low end and α and thus the adiabatic gradient at the high end of estimates. On the other hand, the strong positive geoid anomalies over subduction zones would favor an increase of viscosity from upper to lower mantle by a factor of at least 30 (Hager, 1984). It is noteworthy, that the rotational data also allow for a solution with high viscosity in the lower mantle, although the beach level emergence data do not (Peltier, 1985).

To summarize, the following points on mantle rheology should be considered when modelling mantle convection:

1. The solid state creep process of mantle minerals is certainly strongly temperature dependent. The appropriate activation energy in the upper mantle is known within 25 percent.
2. The creep process is also pressure dependent, but the exact magnitude of the pressure dependence is rather uncertain, especially regarding the lower mantle.
3. It is possible, that the mantle in general deforms according to a power-law with a stress exponent of the order of three, and it is very likely that the deformation is non-Newtonian at least in high-stress regions.
4. Large uncertainties remain concerning the influence of minor chemical differences (water content, oxygen fugacity), the occurance of reaction-enhanced ductility, the rheology of lower mantle phases, but also concerning the behavior at the low temperature ($T < 0.5T_m$) of the lithosphere.

4. Convection with Temperature-Dependent Viscosity

In the light of the strong temperature dependence of the mantle rheology it seems surprising that relatively few attempts have been made towards a thorough evaluation of its influence on convection. Although in a large number of numerical model studies the temperature dependence has been included, often along with a number of additional effects (e.g., Torrance and Turcotte, 1971; Houston and DeBremaecker, 1976; Parmentier and Turcotte, 1978; Rabinowicz *et al.*, 1980; Boss and Sacks, 1982), its influence has not been assessed in a systematic manner and isolated from other effects. More instructive in this respect were laboratory experiments with fluids of strongly temperature-dependent viscosity, like glycerine or glucose syrups (Booker, 1976; Richter *et al.*, 1983). The experiments demonstrate how the thermal structure of the convection cells becomes asymmetric, with a large temperature jump in the upper boundary layer and a small one in the lower boundary layer. At high enough viscosity contrasts a thick stagnant layer is formed near the top boundary. Numerical studies (e.g., McKenzie, 1977) confirm that the flow velocity decreases at the upper and sinking boundary of the cell and increases at the bottom and in the rising current. A chief goal of these experiments was to determine the Nusselt number–Rayleigh number relationship for variable viscosity convection. The Nusselt number Nu measures the ratio of the heat transported by convection to the heat carried by conduction alone. For constant viscosity convection at Rayleigh numbers well above critical a relation of the form

$$\mathrm{Nu} = a\mathrm{Ra}^{b}, \tag{4.1}$$

with the exponent b in the order of one quarter to one third is well documented. This relation was confirmed in the variable viscosity experiments with b in the same range as for constant viscosity if the Rayleigh number is defined with the viscosity at the mean of top and bottom temperature (i.e., $\mathrm{Ra}_{1/2}$, eq. 2.4). Here a is a function of the viscosity ratio η_0/η_1 between top and bottom boundary. $a(\eta_0/\eta_1)$ varies by 15 percent over the range $1 \leqq \eta_0/\eta_1 \leqq 10^5$, and this variation is readily explained by a corresponding change of the critical Rayleigh number (Booker and Stengel, 1978). The aspect ratio of the convection cells was found to decrease with increasing viscosity ratio (Nataf and Richter, 1982). However, it is uncertain to what extent these results for convection between rigid boundaries may be applicable to the plate-tectonic mode of mantle convection with its free upper boundary. For example, numerical results by Daly (1980) suggest that the aspect ratio of variable viscosity convection with free boundaries may be much larger than 1, which facilitates the movement of the stiff surface layer and increases the Nusselt number.

Recently I have performed a detailed numerical study on the properties of convection with temperature-dependent viscosity and free boundaries (Christensen, 1984b). In the remainder of this section these results will be discussed. A simple exponential temperature dependence was assumed of the form

$$\dot{\eta} = \eta_0 \exp(-\Theta T). \tag{4.2}$$

As a reminder, the temperature T is scaled to be zero on top and one at the bottom of the layer, thus the viscosity contrast is $\eta_0/\eta_1 = e^{\Theta}$. The convective state is determined by three parameters: Ra_0 (eq. 2.3), Θ, and l (aspect ratio). The variation of flow pattern and isotherms, when $\mathrm{Ra}_{1/2} = \mathrm{Ra}_0 e^{\Theta/2}$ is kept constant and Θ is increased, can be seen in Figure 8.1. As mentioned previously, the flow becomes more sluggish in the upper and faster in the bottom and rising boundary layer. The average temperature of the cell rises

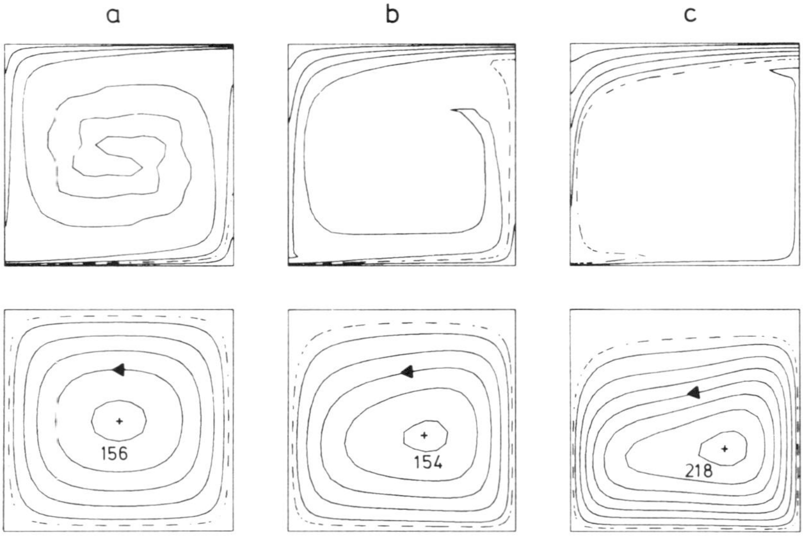

Figure 8.1. Convection cells with increasingly temperature dependent viscosity. $Ra_{1/2}$ = 316,230 in all cases, the ratio of top to bottom viscosity is 1 (constant viscosity) in a; 250 in b; and 4000 in c. Isotherms in the upper diagrams are drawn at 0.1, 0.3, 0.5, 0.7, 0.8 (broken), and 0.9 of the bottom temperature; stream-lines in the lower diagrams are drawn every 30 non-dimensional units (broken: 15).

with increasing Θ well above the mean of top and bottom temperature, exceeding 0.8 for viscosity ratios over 10^4. Concerning the aspect ratio dependence it is found that in the range $250 \leqq e^{\Theta} \leqq 16000$

1. The maximum of Nu(l) occurs at large l between 1.7–4 when the "surface Rayleigh number" Ra_0 is less then 3000, but close to 1 when Ra_0 is larger (Fig. 8.2a).
2. In this parameter range the mean velocity at the top boundary increases strongly with l while the variation of the average convective velocity measured by the stream function maximum is more moderate (Fig. 8.2b).
3. The average internal temperature decreases with l, i.e. the asymmetry between upper and lower boundary layer decreases (Fig. 8.2c).

Because the maximum Nusselt number occurs at aspect ratios near 1 at high Rayleigh number a more detailed study of the heat transport properties has been done for convection in a square box. This seems a sensible choice although a natural convecting system does not necessarily choose the aspect ratio which maximes the heat flow.

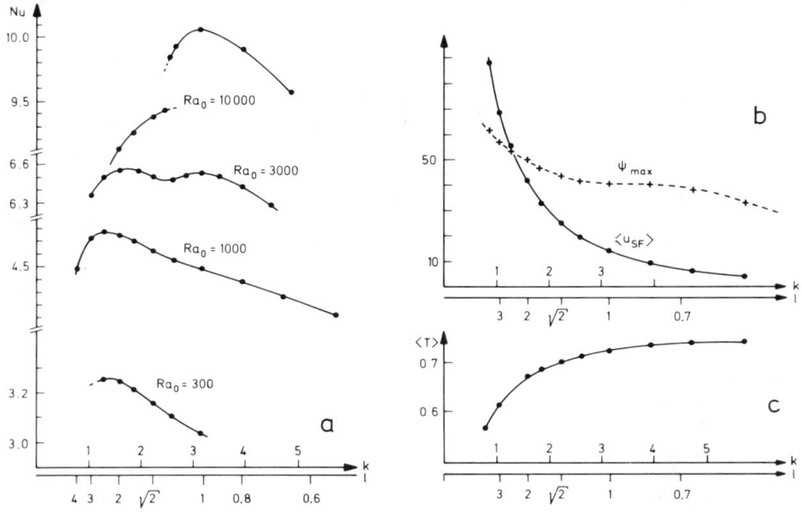

Figure 8.2. Dependence of various properties on the wavenumber k or the aspect ratio l of convection cells with temperature-dependent viscosity. The ratio of top to bottom viscosity is 1000 in all cases. a, Nusselt number for various values of the "surface Rayleigh number" Ra_0. $Ra_{1/2}$ is approximately $32Ra_0$. b, Stream function maximum ψ_{max} and average surface velocity $\langle u_{sf} \rangle$ for $Ra_0 = 1000$, $Ra_{1/2} = 31623$. c, Average temperature $\langle T \rangle$ of the convection cell for $Ra_0 = 1000$, $Ra_{1/2} = 31623$.

For constant viscosity convection the following Nusselt–Rayleigh number dependence was found

$$Nu = 0.2697\ Ra^{0.3185} \tag{4.3}$$

in good agreement with results by Jarvis and Peltier (1982) and Schubert and Anderson (1985). For $\Theta > 0$ the variation of Nu with $Ra_{1/2}$ is similar as in the constant viscosity case, in accord with the experimental results with rigid boundaries (Booker, 1976; Richter *et al.*, 1983). However, two differences are observed (Fig. 8.3): for a given $Ra_{1/2}$ the Nusselt number deviates from its constant viscosity value by a factor up to two when Θ is high, whereas with rigid boundaries the difference was less than 15 percent. Secondly, the slopes in the Nu-$Ra_{1/2}$-diagram differ slightly from the constant viscosity slope, and they are for a given Θ not particularly constant with changing Rayleigh number. The exponent b in eq. 4.1 is replaced by the logarithmic slope

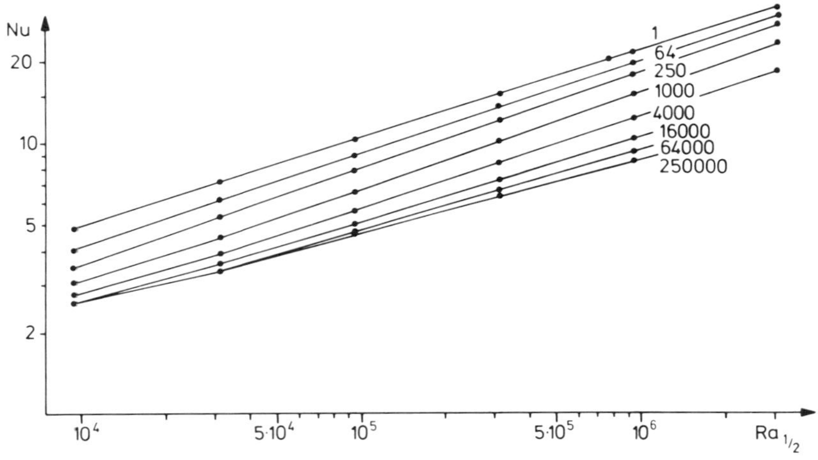

Figure 8.3. Nusselt–Rayleigh number relation for temperature-dependent viscosity, aspect ratio 1, free boundaries. The Rayleigh number is calculated with the viscosity at the mean of top and bottom temperature ($Ra_{1/2}$). The viscosity ratio $\eta_0/\eta_1 = e^{\Theta}$ is indicated for each curve.

$$b_1 = \left\{\frac{\partial \ln(\mathrm{Nu})}{\partial \ln(\mathrm{Ra}_{1/2})}\right\}_{\Theta=\mathrm{const}} \tag{4.4}$$

In the investigated parameter range b_1 is between 0.23 and 0.36. This variability with changing $Ra_{1/2}$ is obviously due to a gradual change of the convective style. At sufficiently high viscosity contrast, the upper boundary layer participates hardly in the circulation when the Rayleigh number is low, it behaves rather more like an insulating lid. In this regime the Nusselt number is almost identical whether the upper boundary condition is rigid or free (Fig. 8 in Christensen, 1984b). With increasing Rayleigh number (Θ kept constant) the upper boundary layer participates to an increasing extent, and thus the Nu-$Ra_{1/2}$-curve achieves a steeper slope. However, the transition from the insulating lid to the free boundary layer regime is a very gradual one and spread over a Rayleigh number range of many orders of magnitude.

Empirical Nusselt–Rayleigh number relations of the form of eq. 4.1 have been used to "parameterize" the convective heat transport. Such a parameterization may be useful to analyze the evolution of a slowly cooling convecting system, such as a magma chamber (e.g., Carrigan, 1984) or the whole Earth. In this way the necessity at solving a very complex three-dimensional and time-dependent convection problem is bypassed. The problem with the

Nusselt–Rayleigh number relation is that for variable viscosity convection no longer a single parameter—the Rayleigh number—suffices to describe the convective state, but at least one further parameter is needed. There is no unique choice for such parameters, but a number of possible combination exits, such as (Ra_0, Θ), ($Ra_{1/2}$, Θ), (Ra_0, Ra_1), ($Ra_{1/2}$, Ra_0), (Ra_T, Ra_0) . . . (recall eqs. 2.3 to 2.6 for the different definitions). Therefore no unique Nusselt–Rayleigh number relationship exists either. Obviously it must be specified how exactly the Rayleigh number is defined, and which second parameter we wish to keep fixed while varying the Rayleigh number. In the preceding discussion of Figure 8.3, $Ra_{1/2}$ (definition 2.4) was taken, and $\eta_0/\eta_1 = e^{\Theta}$ was fixed. However, to parameterize a cooling system, Ra_T (eq. 2.5) is a more useful definition (Nataf and Richter, 1982). Furthermore, if one envisages a situation in which the surface temperature of the system remains constant, while the interior and bottom temperatures drop, both Ra_T and η_0/η_1 will decrease. On the other hand, the surface viscosity η_0 remains unchanged and so, approximately, does Ra_0. The surface Rayleigh number Ra_0 depends on ΔT and thus decreases when the system cools, but this variation will be minute compared with the variation of Ra_T when the viscosity is strongly temperature-dependent. In Figure 8.4 the Nusselt number is shown as function of Ra_T for different fixed values of Ra_0. The slopes of these curves

$$b_2 = \left\{\frac{\partial \ln(\mathrm{Nu})}{\partial \ln(\mathrm{Ra}_T)}\right\}_{\mathrm{Ra}_0=\mathrm{const}} \tag{4.5}$$

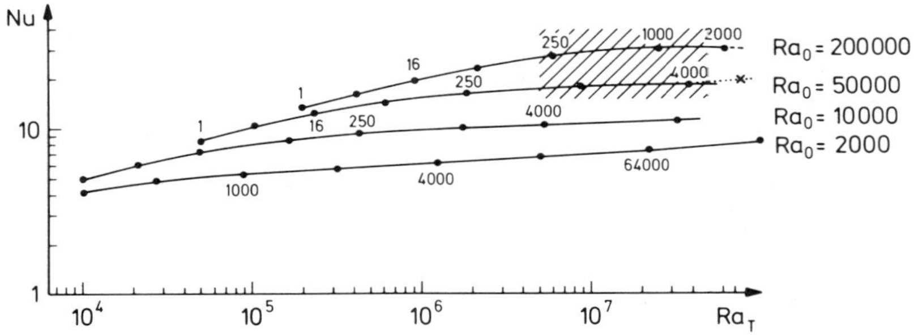

Figure 8.4. Data as in Figure 8.3, but plotting the Nusselt number in terms of the "internal Rayleigh number" Ra_T (Eq. 2.5) and connecting points of equal "surface Rayleigh number" Ra_0. The small numbers indicate the viscosity ratio. The hatched region is the estimated location of present whole mantle convection in the Nu-Ra_T-space.

start at low viscosity contrasts and low Rayleigh number near 0.3. However, when the viscosity contrast exceeds 250 the slope b_2 has dropped to 0.1 and at higher viscosity ratios b_2 may come very close to zero. At Rayleigh numbers Ra_T exceeding about 5×10^7 no steady state solutions can be obtained, rather time-dependent boundary layer instabilities occur. They cause a slight increase of b_2; however, the slope remains below 0.1 (Christensen, 1984b). Like the Nusselt number the average surface velocity becomes weakly dependent on Ra_T when Ra_0 is fixed (Fig. 8.5). On the other hand, the velocity in the interior of the convection cell increases with Ra_T. It seems that in this regime the velocity in the upper boundary layer is controlled by the surface viscosity and not by the viscosity in the bulk of the convecting region. Apparently the surface boundary becomes the bottleneck for the heat transport. Because its velocity no longer depends on Ra_T, the Nusselt number is likewise decoupled from Ra_T. This requires that the surface layer is *not* an insulating lid and that its movement and the related advection of heat is still dominant for the heat transport. Clearly, Nu-Ra_T-slopes close to zero can only be found in a transitional regime between the end members of a fully participating upper boundary layer and the case where the lid movement is insignificant for the heat transport. Increasing Ra_T further, b_2 should increase again after passing through a minimum. This increase is only perceptible in the case $Ra_0 = 2000$ in Figure 8.4 (see also Christensen, 1984b, Figure 14). The data suggest that the transitional regime is in fact important over a Rayleigh number range of many orders of magnitude.

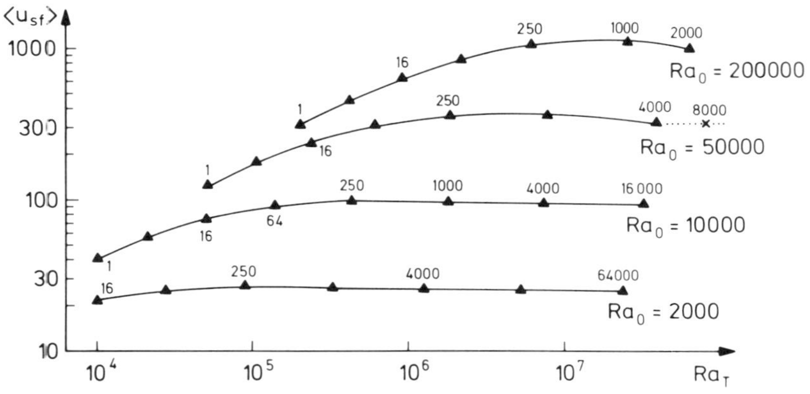

Figure 8.5. The average surface velocity $\langle u_{sf} \rangle$ plotted as function of Ra_T for fixed values of Ra_0. Small numbers indicate the viscosity ratio. Free boundaries, aspect ratio 1.

For the weak dependence of Nu on Ra_T the free slip condition at the upper boundary and the movement of the lid is important. This would refer to the plate-tectonic mode of mantle convection. In addition small scale secondary convection underneath the lithosphere has been considered (Parsons and McKenzie, 1978). In such a situation the dependence of the heat flow on the Rayleigh number may be stronger (Fleitout and Yuen, 1984). The main effect is here the thinning of the conductive lid on raising Ra_T.

The weak dependence of Nu on Ra_T with a moving upper boundary layer was also found in less idealized model cases (Christensen, 1985a), including for example additional pressure dependence of the viscosity or non-Newtonian rheology. Two time-dependent model studies with decreasing rate of bottom or internal heating are also worth mentioning. In all the cases under consideration the logarithmic slope becomes small, typically in the range 0.0–0.10, when the variability of the viscosity due to temperature is strong enough.

5. Parameterized Convection Models for the Thermal Evolution of the Earth

The thermal evolution of convecting planetary bodies has been studied in simplified mathematical models using a parameterization of the convective heat transport (e.g. McKenzie and Weiss, 1975; Sharpe and Peltier, 1979; Schubert *et al.*, 1980; Davies, 1980; Turcotte, 1980; Stacey, 1981). The key for this parameterization is the Nusselt–Rayleigh number relationship. The consequences of the weak dependence of Nu on Ra_T, as discussed in the previous section, for thermal evolution models of the Earth have been described in Christensen (1985b). The temporal change of the mean temperature $T^{\dagger}$ of a cooling planet is obtained from the principle of conservation of energy

$$C\frac{dT}{dt} = H(t) - Q(T), \tag{5.1}$$

where C is the heat capacity, H the time-dependent rate of heating due to radioactive decay and Q the surface heat loss. The temporal variation of H is determined by the relative abundance of the three elements uranium, thorium and potassium, which is reasonably well constrained for the Earth—

†In this section T means always the averaged temperature.

U:Th:K = 1:3.6:12600 (Jochum *et al.*, 1983). The absolute present value of H remains at disposition in the model calculations. The present day ratio H/Q is called the Urey ratio. An important supposition is that the heat loss Q depends only on the average temperature T of the system. The functional dependence $Q(T)$ is obtained by specifying a viscosity law $\eta(T)$ and a Nu-Ra_T relationship. Concerning the latter, most workers have taken an exponent b on the order of ¼–⅓ in the usual relation derived for constant viscosity convection (eq. 4.1). Booker's (1976) experimental results were often cited as argument that the spatial variability of η in the convection cells does not change the Nu-Ra_T relationship. This is quite possibly a misinterpretation for three reasons:

1. Booker's data are for convection with rigid boundaries, whereas three quarters of the heat from the mantle is released through sea-floor spreading and cooling of the oceanic plate (Sclater *et al.*, 1980), i.e., by a mode of convection with free surface.
2. Booker used the definition 2.4 for the Rayleigh number, i.e., $\mathrm{Ra}_{1/2}$, whereas the use of equation 5.1 demands a relation between Nu and Ra_T.
3. In Booker's parameterization the viscosity ratio η_0/η_1 is the fixed parameter and the slope b_1 (eq. 4.4) is obtained. However, while the interior temperature of the Earth has most possibly decreased, the surface temperature remained essentially constant during geological time. Thus the logarithmic slope b_2 (eq. 4.4) with fixed Ra_0 is perhaps more appropriate to obtain the proper exponent in the Nu-Ra relationship.

In Figure 8.4 the range in the Nu-Ra_T space in which the Earth is presumably found today in case of whole mantle convection is shaded. In this range b_2 is found between 0.0 and 0.10. The average surface velocity of the cells is somewhat low in this range, 0.3–1.0 cm/a when scaled to the Earth. These low values are possibly due to the neglect of the third parameter l, which was kept at unity in the calculations. The aspect ratio for whole mantle convection is in the range 2–3 on average, and in Figure 8.2b it is demonstrated that the surface velocity may strongly increase with l. The (effective) surface viscosity η_0, which is kept constant in the numerical models, is certainly an elusive quantity for the Earth. Extrapolating the high temperature creep law of mantle minerals down to 300 K leads to prohibitive high viscosity values. The fact that the plates move indicates that this cannot be the true effective viscosity. Rather processes like low-temperature plasticity, elastic bending, brittle deformation, and the presence of melt at ocean ridges, must govern the deformation of the lithosphere and make it softer. Whether the concept of constant effective surface viscosity conforms with these complications is highly uncertain, but at any rate it seems worthwhile to follow up this line and see what sort of thermal history has to be expected.

Assuming a constant exponent b in the relation between Nu and Ra_T, we obtain from eq. 5.1

$$C\frac{dT}{dt} = H(t) - a\frac{T^{1+b}}{\{\eta(T)\}^b}. \tag{5.2}$$

The coefficient a comprises a number of constants and is best obtained by adjusting the present heat loss from the mantle—about 3.7×10^{13}W—to the assumed present mantle temperature. This temperature was chosen as 1623K in Christensen (1985b) which is meant to be the mean mantle temperature extrapolated along an adiabat to zero pressure (potential temperature). For the temperature dependence of η an Arrhenius law (eq. 3.1) with an activation energy of 525 kJ/mol was taken. The ordinary differential equation 5.2 is integrated backwards in time, trying various degrees of present radioactive heating (or various Urey ratios). Unrealistic thermal histories were rejected. I considered those as unrealistic which are at 4.5 billion years age colder than presently, or which are at 3.8 billion years ago—the time where the geological record starts—more than 400 K warmer than nowadays. A rather high retention of accretional energy and simultaneous core formation are now favored ideas about the Earth's birth and imply a rather hot start. On the other hand it appears unlikely that the Earth was ever very much hotter than presently during the time which is geologically documented. A catastrophic degree of melting in the upper mantle would be the consequence for which we do not have evidence. The exact choice of bounds on the very early temperature hardly affects the evolution history later than 3 Ga B.P. when the exponent b is larger than 0.15, but has some effect with smaller b.

The Urey ratios which are demanded for the evolution models, in order to conform with the assumed bounds on the temperature, generally increase with increasing b. For $b > 0.2$ the Urey ratio is typically in the range 0.7–0.9, whereas for $0.0 \leqq b \leqq 0.1$ it is in the range of 0.3–0.6 (cf. also Schubert *et al.*, 1980). If indeed the low values of b from variable viscosity convection apply, perhaps less than half of the present heat loss of the Earth is replenished by radioactive heating and the main source is the cooling of the Earth. This is in agreement with geochemical models which assume the relative abundance of refractory elements, such as Ca, Al, U, Th, in the Earth to be the same as in Cl carbonaceous chondrites; whereas moderately volatile elements as potassium are depleted by a factor of 6 (Wänke, 1981). Depending on the Ca- or Al-abundance, the following mean concentrations have been estimated in the silicate shell of the Earth: U—20 ppb, Th—75 ppb, K—200 ppm (Taylor, 1979; O'Nions *et al.*, 1981; Jochum *et al.*, 1984). Making allowance for the crustal share of the radioactive elements, the Urey ratio of the mantle is estimated to be 0.4 ± 0.15 (Christensen, 1985b), in good agreement with the requirement of evolution models with low *b*.

In Figure 8.6 the variation of the mean temperature, mean heat flow, and mean plate velocity is displayed for evolution models with various exponents *b*. The surface velocity is assumed to vary according to

$$u \propto \mathrm{Ra}_T^{2b}, \tag{5.3}$$

which is reasonably well in accord with the numerical experiments, although in the cases where b_2 was less than 0.1 it is probably more an upper bound. Let us consider the "prediction" of the different models concerning the Archaean. With the traditional assumption $b = 0.3$ the heat flow was 2.5–4 times higher than now, and the plates were moving 5–15 times faster. With a low b on the other hand, say 0.05, the heat flow is only 50 percent larger and the plates were less than twice as fast. Concerning the mantle temperature the change becomes larger for a low exponent. With $b = 0.05$ the mantle is 250–

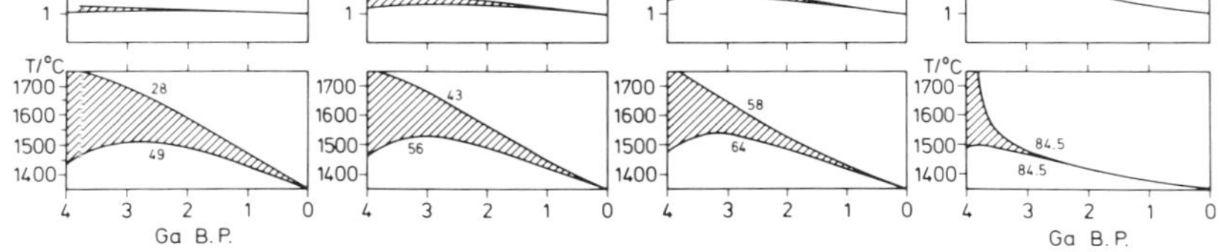

Figure 8.6. Thermal evolution models with different exponents *b* (β) in the Nusselt-Rayleigh number relation. The range of solutions is shown which satisfy the boundary conditions on the initial temperature (see text). In each column, from bottom to top, the variation of the average mantle temperature extrapolated to zero pressure, of the mean surface heat flow *Q*, and of the average plate velocity $\bar{u}$ are displayed. The present temperature is taken to be 1350°C and present heat flow and plate velocity are set to unity, proper dimensional values would be 80 mW m^{-2} and 3–4 cm/a respectively. In the temperature diagrams also the limiting Urey ratios in percent are indicated.

350 K hotter at 3 Ga B.P., whereas with $b = 0.3$ the difference is only 100–150 K.

A number of independent data and estimates to test the different "predictions" are summarized in Christensen (1985b). The frequent occurence of komatiites during the Archaean, very magnesia-rich basaltic rocks indicating a very high melting temperature, suggests a temperature difference between Archaean and present mantle of the order 250–300 K (Green, 1981). This agrees better with the low-b evolution models. England and Bickle (1984) studied the p,T conditions indicated by the metamorphism in Archaean terrains and concluded that the continental heat flow could not have been much higher than today. The result also supports the model with low b, but it is of limited value because the oceanic and not the continental heat flow dominates the global heat loss of the mantle. Finally, paleomagnetic data have been used to estimate minimum drift rates of continental blocks since the early Proterozoic (Ullrich and van der Voo, 1982) and even for the Archaean (Kröner *et al.*, 1984). These studies do not support large changes in the average level of plate velocity since the Archaean, rather they are compatible with essentially unchanged drift velocities.

Certainly the data used to test the different evolution models are burdened with large uncertainties and interpretational assumptions. Nevertheless, it is encouraging that data from different fields—geology, paleomagnetism, and geochemistry—consistently favor the evolution models with a low exponent b, which were originally stimulated from fluid-dynamical results. It seems that on the one hand some of the popular ideas about the Archaean must be revised, and that on the other hand variable viscosity may shape the properties of mantle convection so decisively that results of constant viscosity models should be considered with suspicion.

6. Pressure-Dependence, Lower Thermal Boundary Layer, and Mantle Plumes

While the strong decrease of mantle viscosity with rising temperature is undisputed, its variation with pressure—hence depth—is much less clear as outlined in Section 3. The pendulum swung from the assumption of an overwhelming increase with depth in the sixties, based on the interpretation of a "fossil nonhydrostatic equatorial bulge" of the Earth (McKenzie, 1966) to the assumption of an essentially isoviscous mantle in the seventies, based on the analysis of postglacial rebound (Cathles, 1975; Peltier, 1976). Lately, there has been again support for at least a moderate increase of viscosity with depth (Hager, 1984), based on the dynamical interpretation of geoid anomalies.

In order to study the influence of additional pressure-(depth-)dependence of the viscosity on the properties of convection, the viscosity law eq. 4. is changed to

$$\eta = \eta_0 \exp(-\Theta T + \zeta z), \tag{6.1}$$

where the z-coordinate is zero on top and one at the bottom of the convecting layer. The viscosity variation due to pressure alone, i.e., e^{ζ}, was taken either 64 or 1000. Such models are compared with models of the same Ra_0, l, and the same viscosity contrast across the layer $\eta_0/\eta_1 = e^{\Theta-\zeta}$. Thus Θ is also increased on making ζ larger, in order to balance the pressure effect across the layer. The values of $Ra_{1/2}$ are identical in such model series. Results are described in detail in Christensen (1985a). The main effects of the pressure-dependence are:

1. The mean temperature $\langle T \rangle$ of the convecting layer drops. With strong temperature dependence alone $\langle T \rangle$ is much above the mean of top and bottom temperature, but adding pressure influence it may be well below 0.5 (Fig. 8.7). In order to get $\langle T \rangle \geqq 0.5$ it was found necessary that $\Theta \geqq 2\zeta$ (as a rule of thumb). Accordingly the asymmetry between top and bottom boundary layer may be reversed, with the larger temperature jump in the bottom boundary layer.
2. The viscosity depth distribution has now a distinct "asthenosphere" (low-viscosity channel), and high viscosity in the lower part of the cell overlying a sharp drop of viscosity in the lower boundary layer.
3. The ratio of velocity in the "return flow" to the velocity in the upper boundary layer becomes small, the return flow extends over a larger depth range, and there is some channelling of the flow into the lower boundary layer (Fig. 8.7).
4. The aspect ratio which maximes the heat transport decreases below 1 (Christensen, 1985a).
5. As long as Θ is sufficiently larger than ζ, the Nu-Ra_T relationship is still characterized by low slopes b_2 (with Ra_0 fixed).

Some simple numerical experiments by Gurnis and Davies (1986) with pressure-dependence alone (i.e., $\Theta = 0$) corroborate points (1) and (3). However, it appears surprising that the pressure influence is still dominant in shaping the thermal and flow structure even when the temperature influence is stronger, i.e., when $\Theta > \zeta$, or $\eta_0/\eta_1 >> 1$.

The features of convection with pressure dependent viscosity support

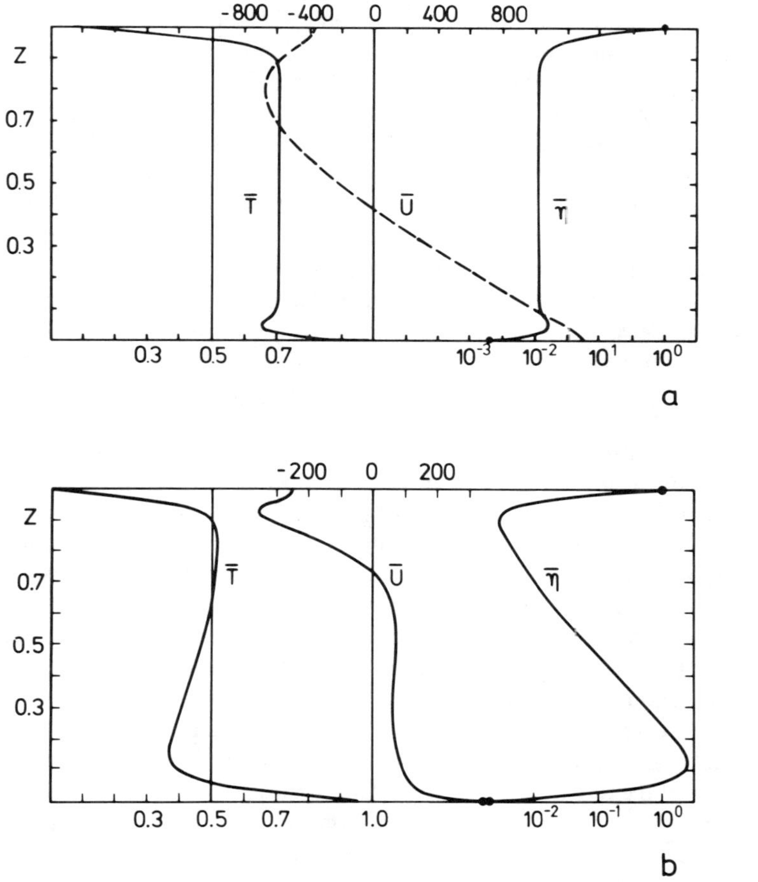

Figure 8.7. Depth profiles of horizontally averaged temperature $\bar{T}$, average horizontal velocity $\bar{u}$, and the average of the logarithm of the viscosity $\bar{\eta}$. Aspect ratio 1, free boundaries, $Ra_0 = 50{,}000$, $Ra_{1/2} = 1.1 \times 10^6$, and $\eta_0/\eta_1 = e^{\Theta-\zeta} = 500$ in both cases. a, No pressure dependence of viscosity ($\zeta = 0$); the Nusselt number is 17.6, $\psi_{max} = 366.6$, $\langle T \rangle = 0.689$. b, With pressure dependence, $e^{\zeta} = 1000$, $e^{\Theta} = 500{,}000$; $Nu = 11.1$, $\psi_{max} = 66.5$; $\langle T \rangle = 0.461$.

some ideas about the mantle circulation. The preferred explanation for hot spot volcanism is the concept of mantle plumes, narrow pipes of hot rising flow coming from the lower mantle (Morgan, 1971). The relative motion between hot spots seems to be clearly slower than the plate motion (e.g., Molnar and Francheteau, 1975). This requires that the velocity of the general circulation in the lower mantle must be less than the plate velocity, which could be achieved by the pressure effect on the viscosity, but which is hardly to be expected in an isoviscous core of a convection cell. Slow motion in the lower mantle may also be important for the survival of chemical heterogeneities which are observed on many length scales (e.g., Allègre *et al.*, 1980). They would be mixed efficiently by the shear motion of shallow isoviscous convection, but they may possibly survive on the billion-year time scale with whole mantle convection and sluggish movement in the lower mantle (Davies, 1984). However, the numerical data indicate that quite a substantial increase of viscosity with depth is required to slow down the velocity in the lower parts of the cell. In the example of Figure 8.7b a viscosity contrast of $0(10^2)$ is needed to establish a velocity difference of a factor of 5. In one-dimensional return-flow models Jacoby (1978) also found a relatively weak dependence of the local velocity on the local viscosity. A statistical analysis of the velocity-viscosity correlation in variable viscosity convection suggests that this is in fact a general rule (Christensen, 1985c).

A further point where pressure-dependent rheology may shape mantle convection concerns the lower thermal boundary layer. The D″-layer above the core–mantle boundary, which has probably anomalously low seismic gradients, has been identified as a thermal boundary layer with a temperature jump of the order of 1000 K (e.g., Doornboos and Mondt, 1979; Stacey and Loper, 1983), although the seismic evidence is not unequivocal (Mula and Müller, 1980; Doornboos, 1983). At least it seems undisputed that the lowermost mantle exhibits strong seismic lateral heterogeneity on both large and small scales Dziewonski, 1984; Sacks *et al.*, 1979). A possible problem of the whole-mantle convection scenario was pointed out by Jeanloz and Richter (1979). There is a gap of the order 500–1000 K between the temperature near the base of the mantle—estimated from the upper mantle temperature extrapolated along an adiabat—and the temperature of the outer core from estimates of the minimum melting temperature of possible iron alloys. The gap might be closed by the thermal boundary layer between core and mantle. However, Jeanloz and Richter estimated the possible ΔT to be only 300 K based on the properties of isoviscous convection. Also Fowler (1983) argued for a rather small temperature difference across any hot boundary layer because of the strong decrease of viscosity with rising temperature which would tend to destabilize it. Jeanloz and Richter proposed that additional thermal boundary layers either between upper and lower mantle (two-layer convection) or at a compositional different D″-layer itself could close the remaining temperature gap. However, the numerical results confirm that the pressure-dependence can help to stabilize a large ΔT in the low boundary layer, despite the strong viscosity drop, which is more than 3 orders of magnitude in the example of Figure 8.7. In this example ΔT is larger across the bottom layer than across the top layer.

Whether the pressure-dependence of rheology can sufficiently stabilize a boundary layer with a ΔT of the order of 500–1000 K and a related viscosity contrast of 4 to 10 orders of magnitude has been studied in Christensen (1984c). The term "stabilization" is always meant in a relative sense, at some time and somewhere such a boundary layer must become unstable. As far as the core–mantle boundary concerned, the requirement is that this does not happen so frequently and so violently, that too much heat is extracted out of the core. An upper limit for the average heat flow from the core is certainly the total heat loss of the Earth. An interesting suggestion is that the lower boundary layer instability is independent of the large scale circulation of the lower mantle and leads to the formation of mantle plumes. Previously this process has either been studied as incipient instability of the boundary layer by means of linear theory (Yuen and Peltier, 1980) or as a fully established steady state plume (Parmentier *et al.*, 1975; Loper and Stacey, 1983). In Christensen (1984c) the instability of a thermal boundary layer is pursued into the stage where a mature thermal diapir ascends. The models are two-dimensional and may thus reflect the behavior of axisymmetric plumes only in an approximate sense. Initially an isothermal stagnant lower mantle of "background viscosity" η_∞ is assumed, with a temperature jump of 1000 K at the core-mantle boundary. An "Arrhenius type" of temperature dependence is taken for the viscosity, leading to a viscosity contrast between 0 and 4.5 orders of magnitude across the boundary layer. The viscosity variation induced by a ΔT of 1000 K would probably be even higher, but it might be damped by non-Newtonian rheology (cf. Section 7). First the thermal boundary layer grows by conduction. On reaching a critical thickness depending on η_∞ and on the viscosity variation inside the layer, it becomes unstable, ejecting a thermal diapir or plume (Fig. 8.8a). An interesting phenomenon is found when the viscosity variation exceeds 10^3. The instability then has a small spatial scale as predicted by linear theory (Yuen and Peltier, 1980), and small scale convection confined to the boundary layer precedes the formation of a diapir (Fig. 8.8b). The small scale convection might explain lateral heterogeneity on the scale of 150 km in D″ found by Sacks *et al.* (1979).

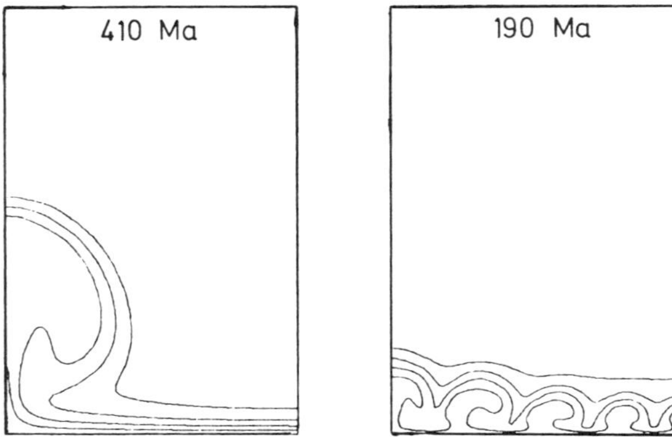

Figure 8.8. Left: Rise of a thermal diapir out of an unstable hot boundary layer with a "background viscosity" $\eta_\infty = 10^{23}$ Pa s and a temperature induced viscosity contrast of 100, ΔT is 1000 K. Isotherms are drawn in steps of 200 K. The height and width of the model box are 1200 km and 800 km respectively. 410 million years passed since the thermal boundary layer started growing. Right: Small-scale convection confined to the thermal boundary layer for $\eta_\infty = 10^{23}$ Pa s and a viscosity contrast of 10^4. The small-scale cells are destroyed with the growth of a diapir out of the boundary layer.

However, only a single cell develops into the plume stage, when this has happened the large scale flow of the ascending diapir sweeps away all the other little cells.

Comparing the average heat flow from below during the combined stage of boundary layer growth and of diapir ascent with the likely heat flux from the core or even with the maximum conceivable flux, it turns out that the lower mantle viscosity η_∞ must be clearly larger than 10^{22} Pa s (10^{23} Poise) and more likely of the order of 10^{24} Poise to prevent too vigorous instability. Thus we need a significant depth or pressure-dependence of viscosity in a scenario of whole mantle convection with a substantial lower boundary layer, in order to avoid outrageous rates of core cooling.

Another interpretation for the D″-layer is in terms of a chemical difference to the bulk of the mantle. If the density difference is large enough, it could stabilize D″ which would act as a thermal buffer between core and mantle. Plumes might then rise from above the D″-layer (Fig. 8.9a) due to the heat which is conducted (or convected) through it. The rise of such diapirs may introduce a considerable thickness variations on large length-scales in the

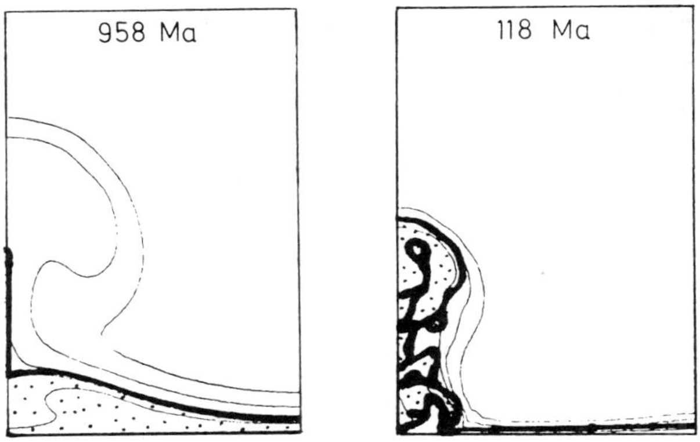

Figure 8.9. Models with a compositionally dense layer of 100 km average thickness at the bottom of the mantle. $\Delta T = 1000$ K, $\alpha = 2 \times 10^{-5}$ K^{-1}. a, Density contrast 3 percent $\eta_\infty = 10^{23}$ Pa s, and temperature-induced viscosity variation of 10^3. A thermal diapir rises from above the dense layer. b, Density contrast 1 percent, $\eta_\infty = 10^{22}$ Pa s, and viscosity variation 10^4. The dense material is incorporated into the rising diapir. Some mixing with overlying light material occured in the preceding stage of small-scale convection.

underlying layer which could explain the long wavelength heterogeneity at the base of the mantle (Dziewonski, 1984). If the chemical density contrast is smaller than the potential thermal density difference (i.e., $\rho_0\alpha\Delta T$) even a compositionally different layer may become unstable and be incorporated into the rising diapir (Fig. 8.9b). The compositional difference in D″ might be continuously renewed by the segregation of subducted oceanic crust—transformed to dense eclogite—as proposed by Hofmann and White (1982). An interesting feature in the model of Figure 8.9b is the mixing of heavy and light material by the small scale convection which precedes the plume formation, and which leads to a diapir which is chemically heterogeneous on a small scale. Such heterogeneity may explain the large variation of isotopic anomalies even in a single hot spot (White, 1985).

The numerical studies discussed in this section show that pressure-dependence of viscosity can play an important role in whole mantle convection. Further efforts in different fields, e.g., analysis of glacial isostatic and rotational dynamics, solid state creep experiments and theory, or dynamical interpretation of geoid anomalies, should be done to obtain further evidence about the radial viscosity structure of the mantle.

7. Comparison of Convection with Newtonian or Power-Law Rheology

The rheological law eq. 3.1, which is used to fit data for creep und uniaxial stress, is usually generalized to an isotropic law of the form

$$\dot{e}_{ij} = As^{n-1} \exp\left(-\frac{H^*}{RT}\right) s_{ij}, \tag{7.1}$$

where $\dot{e}i_j$ and s_{ij} are the tensor components of strain rate and deviatoric stress, and s is the second invariant of the stress tensor. Because 7.1 is isotropic, an effective viscosity can be defined

$$\eta = \frac{s_{ij}}{\dot{e}_{ij}} = \frac{A^{-1}}{s^{n-1}} \exp\left(\frac{H^*}{RT}\right), \tag{7.2}$$

which depends on the stress, temperature, and on the pressure which is implicit in the activation enthalpy H^*. Rather than as a function of stress, the effective viscosity may be expressed as a function of strain rate $\dot{e}$ (second invariant). The formulation is obtained by replacing $s^{n-1} = (\eta\dot{e})^{n-1}$ in the denominator of eq. 7.2 and by solving for η:

$$\eta = \frac{A^{-1/n}}{\dot{e}^{(n-1)/n}} \exp\left(\frac{1}{n}\frac{H^*}{RT}\right). \tag{7.3}$$

Note the factor of $1/n$ in the argument of the exponential, the explicit dependence of the viscosity on temperature and pressure is much weaker with the latter formulation. If we hold a piece of material under constant stress and change temperature or pressure, the change in effective viscosity will be documented by a changing strain rate with a magnitude obtained from the familiar Arrhenius term in eq. 7.1. However, if we keep the strain rate constant and change the temperature or pressure, the change of viscosity is now documented by the change of the required stress. Because of the power-law dependence between $\dot{e}$ and s the relative stress change will now be smaller than the relative change of $\dot{e}$ in the previous situation, which explains the weaker p,T dependence in the formulation where η is a function of strain rate. The significance of this difference has been noted before, and used for example by Karato (1981) in a discussion of the depth dependence of mantle viscosity. He argued in favor of a constant strain rate in the mantle, leading to a much weaker increase in viscosity with depth in the mantle than the assumption of constant stress. However, *a priori* it is not clear if any of the assumptions of constancy of $\dot{e}$ or s make sense in variable viscosity convection cells. This has to be determined empirically and as we shall see, often the truth lies inbetween.

The first systematic comparison of Newtonian and power-law convection has been done by Parmentier *et al.* (1976) and Parmentier and Morgan (1982). They found for 2-D stationary convection in square boxes with rigid boundaries that power-law convection with $n = 3$ is very similar to Newtonian convection at a properly defined effective Rayleigh number. A simultaneous pressure- or temperature-dependence was not considered. These results are often quoted as argument that the possible nonlinearity of mantle rheology is no point to worry about, and that the temperature- and pressure-dependence of viscosity is more important than the stress-dependence. However, later numerical work by Cserepes (1982) and Christensen (1983, 1984a) has demonstrated that the different factors which influence the viscosity cannot be separated, but interact. In the following text some numerical results are reviewed which are described in more detail in Christensen (1984a).

In Figure 8.10a a convection model with Newtonian viscosity published by Torrance and Turcotte (1971) was reproduced. High activation energy and activation volume impose strong viscosity differences up to 6 orders of magnitude, but the viscosities at the top and bottom boundary are equal. For the model in Figure 8.10b the same activation energy and volume are used and the stress exponent n is set to three (the parameter A in eqs. 7.1–7.3 is properly adjusted). The flow pattern and isotherms are quite different from those in Figure 8.10a, and the same applies to the heat transfer properties. Noteworthy is that the maximum viscosity differences inside the cell dropped from formerly six to now three orders of magnitude, which causes a more even distribution of the flow in the cell. It turns out that the stress level is high in regions with high viscosity and low at places with low viscosity in both models. When switching from Newtonian to power-law rheology the stress-dependence reduces the effective viscosity in high stress regions and increases it in low-stress regions, damping in this way the viscosity contrasts imposed by the p,T dependence. This dissimilarity between Newtonian and power-law convection was found over a wide range of parameters (Rayleigh numbers, activation energy and volume, aspect ratio, mode of heating or boundary conditions).

Despite these differences, a Newtonian analogue could be found to each non-Newtonian convection model. This was done by reducing the activation enthalpy, i.e., both E^* and V^*, as it is suggested by the previous considerations about the different p,T influence if the viscosity laws are expressed as function of stress or of strain rate (eqs. 7.2, 7.3). The proper reduction factor was found to be 0.5 in some cases and as low as 0.3 in others. In Figure 8.10c the Newtonian analogue to 8.10b with $H^*_{\text{eff}} = 0.5\ H^*$ is displayed, the similarity is striking. The original model in Figure 8.10a can be considered

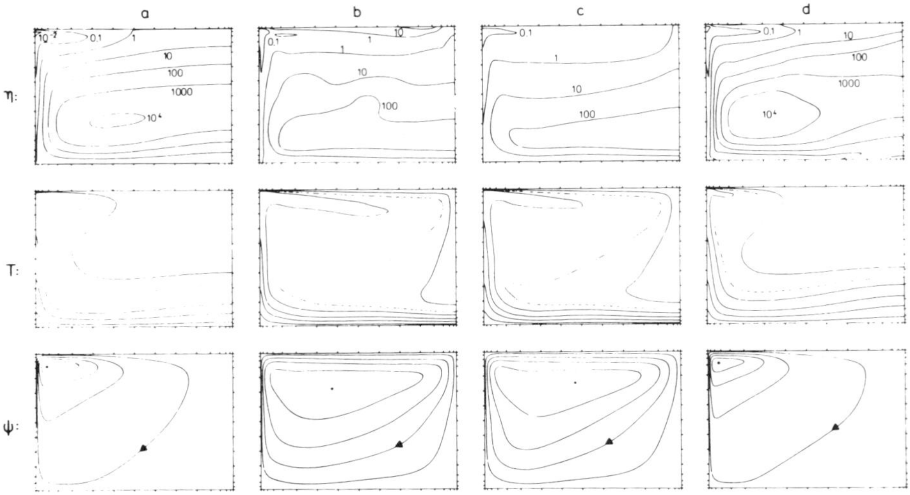

Figure 8.10. Convection with either Newtonian ($n = 1$) or power-law rheology ($n = 3$), free slip, aspect ratio 1.4. Pressure and temperature influence on the rheology cancel over the entire depth of the layer. The stream function is plotted at ⅛, ⅜, ⅝, and ⅞ of its maximum value, the temperature at 0.1, 0.2 (broken), 0.3, 0.5, 0.7, and 0.9 of the bottom temperature. a, $n = 1$, the rheological law is $\eta = 10^{-16} \exp((E^* + zV^*)/(2.088 + T))$ with z and T being 0 on top and 1 at the bottom, and $E^* = 76.912$, $V^* = 36.912$, $Ra_0 = 10^6$, Nu = 4.33, $\psi_{max} = 33.3$. b, $n = 3$, same rheological depth- and temperature-dependence as in a. Nu = 8.59, $\psi_{max} = 62.2$. c, $n = 1$, Newtonian analogue to b. $E^* = 38.456$, $V^* = 18.456$, Nu = 8.56, $\psi_{max} = 60.0$. d, $n = 3$, $E^* = 153.824$, $V^* = 73.824$. Model a is the Newtonian analogue to this case, Nu = 4.66, $\psi_{max} = 35.1$.

as the Newtonian analogue to a non-Newtonian model with twice the activation enthalpy (Fig. 8.10d), the similarity is again very high.

To return to the question whether constancy of stress or of strain rate inside a convection cell is more sensible: in the cases displayed in Figure 8.10 an intermediate choice seems appropriate. If we wish to express the effective viscosity as a function of the product of $\dot{e}$ and s, which is the rate of local viscous dissipation D, we arrive at

$$\eta(p, T, D) = \frac{A^{-2/(n+1)}}{D^{(n-1)/(n+1)}} \exp\left(\frac{2}{n+1} \frac{H^*}{RT}\right). \tag{7.4}$$

For $n = 3$ the factor in the argument of the exponential in eq. 7.4 is 0.5, which agrees with the empirical reduction factor for H^* in the Newtonian analogue.

In Christensen (1984a) also Newtonian models are described where stagnant layers are formed either on top or in the lower parts of the convection cell. When n was raised to three the viscosity contrasts dropped sufficiently to activate the formerly stagnant regions. The same effect was achieved by keeping $n = 1$ but reducing the activation enthalpy to obtain the Newtonian analogue. In the cases shown in Fig. 8.11a–c the reduction factor has to be 0.3 to get optimal agreement between the non-Newtonian case and the Newtonian analogue.

If the mantle rheology is in fact non-Newtonian, the application of these results to mantle convection is immediately obvious. The use of standard values of E^* (e.g., 525 kJ/mol = 125 kcal/mol) and V^* (say, 11 cm^3/mol) connected with a neglect of the stress dependence overestimates the pressure and temperature influence. Effective values of E^* and V^* which are only 30–50 percent of the nominal values should be used. This is interesting concern-

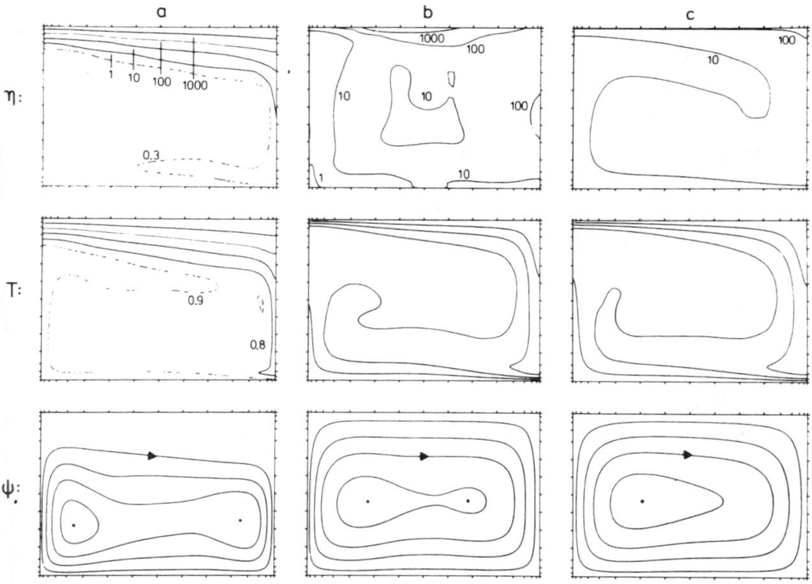

Figure 8.11. Convection with strongly temperature-dependent viscosity for Newtonian and power-law rheology. a, $n = 1$, rheological law as in Figure 8.10a with $E^* = 76.912$, $V^* = 1.302$, $\mathrm{Ra}_0 = 1000$. The Nusselt number is 3.88, $\psi_{max} = 111.4$. Convection takes place under a stagnant lid, $\langle u_{sf} \rangle = 6$. b, $n = 3$, same temperature/depth dependence as in a. Nu = 7.77, $\psi_{max} = 58.5$, $\langle u_{sf} \rangle = 142$. c, $n = 1$, Newtonian analogue to b with 0.3 times the activation energy. $E^* = 23.074$, $V^* = 0.391$, Nu = 7.63, $\psi_{max} = 56.4$, $\langle u_{sf} \rangle = 154$.

ing the rheology of the lower mantle. Taking an average $V^* = 6\ \mathrm{cm^3/mol}$ for the lower mantle, $E^* = 525$ kJ/mol, an upper mantle temperature of 1673 K, and an adiabatic gradient of 0.3 K/km, the viscosity increase from the upper to the lowermost mantle is of the order of 10^8 according to eq. 7.2 for $n = 1$. As outlined in the previous section a viscosity increase by two or even three orders of magnitude may be reasonable and welcome to explain some features of whole mantle convection, but eight orders of magnitude is outrageous and would prevent convection in the lower mantle. Of course, this estimate is fraught with great uncertainties, and the dilemma could be avoided if V^* would be lower, or the adiabatic gradient higher. On the other hand, with power-law convection the resulting viscosity contrast would be much smaller even with the quoted values, on the order of 10^4 or less. Whatever the appropriate size of V^* is, with stress-dependent rheology the mantle would appear closer to an isoviscous state than with Newtonian viscosity.

Another application of the damping influence of stress dependence was anticipated in the previous section. When a ΔT of 1000 K across a hot boundary layer (e.g., D″-layer) and proper values for H^* are assumed, a viscosity contrast of 6–10 orders of magnitude results with Newtonian rheology (Christensen, 1984c, Table 1). With power-law rheology it would be much smaller and of such as size as it was assumed in the models of the previous section. Thus a non-Newtonian flow lay may contribute substantially to the stability of a hot boundary layer. That this reduction of viscosity contrasts, or of the activation enthalpy in the Newtonian analogue, really applies to a thin hot lower boundary layer is for example demonstrated in Figure 14 in Christensen (1985a). In this non-Newtonian model the large ΔT across the bottom boundary layer would imply a viscosity contrast of 3×10^6 if the flow law were Newtonian, whereas it is actually only 600.

While the agreement between power-law convection and its reduced-H^* analogue is generally good, some remarkable differences are observed. When the free slip condition is applied to the upper surface and the temperature dependence of the viscosity is strong, the variability of the surface viscosity with length becomes clearly less for $n = 3$ than for the Newtonian analogue (Fig. 8.12). If we proceed to $n = 5$ the velocity in the specific model becomes virtually constant over most of the length of the surface, with the exception of narrow marginal regions where it must go to zero. The reason for this plate-like behavior is the concentration of stresses at the margins of the upper boundary layer. This leads to high effective viscosity in the main part of the "plate" and weak margins (see Fig. 8.11b). Such a viscosity distribution has been used previously as an *ad hoc* assumption to model

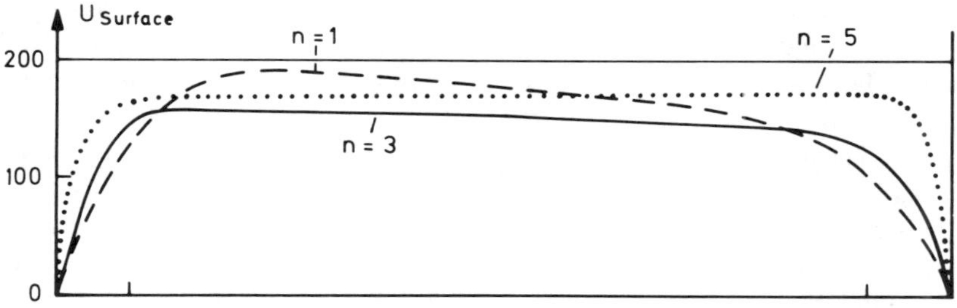

Figure 8.12. Variation of surface velocity with length for convection models with strongly temperature-dependent power-law rheology and their Newtonian analogue. The aspect ratio of the cells is 2.8. Rheological law as in Figure 8.10a with $E^* =$ 23.074, 76.912, and 153.824 for $n = 1, 3, 5$ respectively; V^* is insignificant. Excluding the marginal regions as indicated by tick marks, the standard deviation of the velocity from the mean is 12.3 percent, 5.4 percent, and 1.5 percent for n from 1 to 5.

convection including a moving lithosphere plate (Kopitzke, 1979; Schmeling and Jacoby, 1982). It is in accord with estimates of the effective plate viscosity in subduction zones of the order of 10^{22} Pa s, (DeBremaecker, 1978; Melosh, 1978) being thus much less than in the plate interior (Walcott, 1970). Apparently power-law rheology can contribute to the answer as to why the upper boundary layer of the large-scale convective circulation behaves as a plate with little internal deformation. This does not mean that a third-power law for high temperature creep as derived experimentally for mantle minerals can explain the rheological behavior of the entire lithosphere. Below 600°C the strain rate will be insufficient for plate bending in subduction regions even with the highest reasonable stress. Elastic deformation in a depth range of about 10–40 km and brittle failure in the uppermost plate have to be considered (e.g., McAdoo *et al.*, 1978). The elastic contribution may be the most important one for explaining the platelike behavior, and numerical models of viscoelastic convection are required. Nevertheless, the power-law creep regime in the lower half of an oceanic plate is certainly also of importance. Furthermore purely viscous convection models with $n \geqq 3$ are at least a good means to simulate plate-like movement of the upper boundary layer, and they are much easier to handle than models comprising visco-elasticity.

Another interesting difference between Newtonian and power-law convection is revealed when studying the time-dependent behavior. The spin-up from the conductive state was compared for two models with $n = 1$ (constant viscosity) and $n = 3$ (no p,T dependence) at an effective Rayleigh number of 10^4. The Newtonian model settles rapidly into steady state after a stage of exponential growth, an overshoot, and a few decaying oscillations (Fig. 8.13). The model with $n = 3$, on the other hand, exhibited a number of sharp peaklike bursts of flow, separated by longer intervals of virtual quiescence. This can be readily explained by the strain-rate dependence of the effective viscosity (eq. 7.3). In the unstable initial temperature distribution a lot of potential energy is stored. When the instability grows and the strain rate increases the effective viscosity will drop. This amplifies the instability in a catastrophic way. After half an overturn the thermal structure is inverted and almost stable. As soon as the strain rate starts to decrease, the effective viscosity will rise and bring the vigorous motion to a rapid end. Thereafter it takes some time to establish sufficiently thick boundary layers by conduction to get the cycle started once more. The amplitude of the spikes in Figure 8.13 decreases with time and the system will finally settle into a steady state, although after much longer time than the Newtonian model. More complex models at higher Rayleigh number and including the p,T dependence were also tested concerning their time dependence. They were not started from

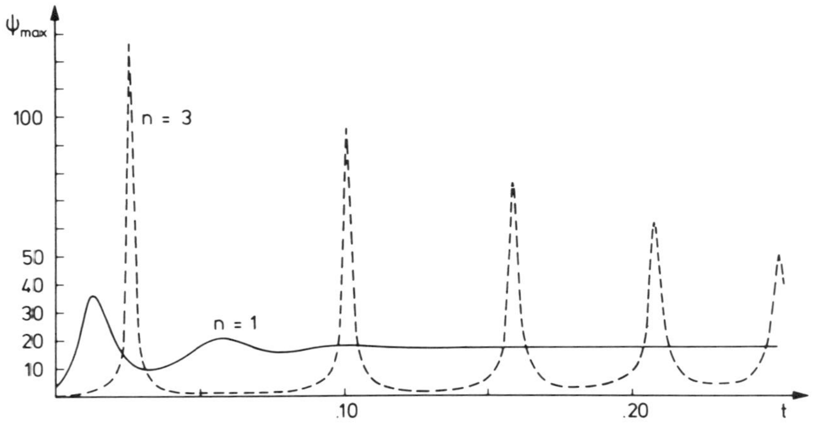

Figure 8.13. Time dependent spin-up (stream function maximum versus time) from a conductive state with superimposed harmonic disturbance. The aspect ratio is 1, there is no temperature or depth depndence of the rheology, free slip boundaries. The (effective) Rayleigh number is 10^4, the steady state values for the Nusselt number are 4.88 ($n = 1$) and 4.91 ($n = 3$), and for ψ_{max} 17.9 and 17.4 respectively.

such highly unstable state as a conductive temperature profile is, and the results are less dramatic. Nevertheless, the oscillations which were found in some non-Newtonian cases are strongly enhanced compared to the analogue Newtonian model (Christensen, 1984a).

In order to define a Rayleigh number for non-Newtonian convection a reference stress (or strain rate) must be specified in addition to a reference temperature and pressure to calculate η in the formula for the Rayleigh number. In Christensen (1985a, Section 3) the rheological law

$$\eta = \eta_0 s^{-2} \exp(-\Theta T) \tag{7.5}$$

has been used, and $Ra^s_{1/2}$ was defined with the effective viscosity at the mean of top and bottom temperature and at the scaling stress $s_0 = \eta_0 h^2/\kappa$. In Figure 8.14 the Nusselt number is plotted against $Ra^s_{1/2}$ for various Θ. The slope of these curves is of the order 0.52–0.55 as long as Θ is not too large. At the highest value of $e^\Theta = 10^7$ which was under consideration a sharp transition is found. At low Rayleigh number the Nusselt number is almost identical if the surface boundary condition is rigid or free. Thus an immobile lid is formed even with the free-slip condition. However, the activation of the lid at a certain critical $Ra^s_{1/2}$ is a very sudden one. On increasing the Rayleigh number in the critical range by 13 percent it was found that the Nusselt number increases by 50 percent and the average surface velocity even

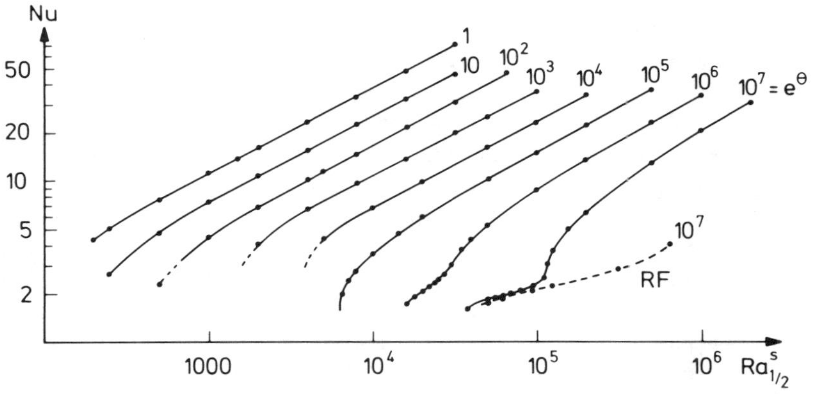

Figure 8.14. Nusselt number of non-Newtonian convection ($n = 3$) as a function of $Ra^s_{1/2}$ and temperature dependence according to Eq. 7.5. The aspect ratio is 1, and the boundary condition is free slip except for the broken line where the upper boundary is rigid and the lower free (RF).

sixfold! In contrast, the transition from the stagnant-lid to the active-lid regime was found to be very smooth with Newtonian rheology.

As outlined in Sections 4 and 5 the parameterization of Nu in terms of Ra_T with the surface temperature kept constant is more appropriate to study the evolution of a cooling planetary body. It leads to very low slopes b_2 (eq. 4.5) when the viscosity contrast is high enough. For a power-law rheology described by eq. 7.5 a very similar result is obtained (Christensen, 1985a). At low Θ the slope is of the order of 0.18 and it drops steadily to values close to zero on increasing Ra_T^s, which is defined in analogue to eq. 2.5. One of our major concerns in Section 5 (thermal evolution of the Earth) was the applicability of the Nu-Ra_T-relationship, derived from a simply temperature-dependent rheology, to the plate-tectonic style of mantle convection. We have seen that we can achieve nearly platelike behavior with power-law rheology. One more point is that the viscosity at the surface is no longer constant, even with fixed temperature, because of the stress-dependence. Regarding non-Newtonian rheology, the parameterized convection models with low b_2 no longer require the existence of an elusive constant surface viscosity for the Earth. Altogether then, we can feel somewhat more at ease about the question as to whether the numerical Nusselt–Rayleigh number relation for variable viscosity convection is appropriate for the Earth.

8. The 670-km Discontinuity

The seismic discontinuity at about 670 km depth in the mantle plays a key role in the discussion of the radial structure of mantle convection. Do upper and lower mantle form separately convecting layers, as proposed by McKenzie *et al.* (1974) and Richter and McKenzie (1981), or do convection currents cross the entire mantle, as held by Davies (1977) or Peltier and Jarvis (1982) among others? Observational evidence on the structure of mantle flow is best obtained in regions of lithospheric subduction. Does the subducting plate penetrate the lower mantle or is it deflected sidewards at the 670 km discontinuity? Traditionally the complete cessation of earthquakes at 700 km depth is taken as argument against penetration, augmented by the observations that the down-dip stress in deeper parts of the subducted plate is always compressive (Isacks and Molnar, 1971), and that the seismic energy release rises from a minimum around 400 km to a maximum close above the seismic cutoff (Richter, 1979). All these observations are taken as evidence for a barrier against slab penetration at 600–700 km. However, the state of stress could also be explained by a moderate increase of viscosity with depth, and the cessation of seismic activity below 700 km by either sufficient heating of the slab or by phase transitions causing the rheology to pass from the brittle to the ductile regime. There is also seismic evidence in favor of whole mantle convection: rays originating from deep earthquakes in some slabs, travelling in the direction of geometrical extension of the slab downwards, have a slightly positive travel-time anomaly, i.e., arrive earlier than expected from a radially symmetric earth model (Creager and Jordan, 1984). The simplest explanation is that the cold high velocity slab extends for at least several hundred kilometers into the lower mantle. In other subduction zones, however, the focal mechanism of deep earthquakes favors the interpretation that the slab fails to enter the lower mantle (Giardini and Woodhouse, 1984). A problem with layered convection is the double thermal boundary layer at the interface of the convecting system. It should leave its fingerprints in the variation of seismic velocities with depth, the viscosity-depth distribution, or the variation of the seismic Q-factor with depth. Spohn and Schubert (1982) even argue that it should lead to melting in the upper parts of the lower mantle. So far no evidence for an internal thermal boundary layer was found, but it cannot be excluded that the ΔT is so small that any possible effect lies below the threshold of resolution. Finally the strong depletion in lithophile elements of that part of the mantle which produces mid-ocean ridge basalts can be explained in mass-balance calculations when only one third of the mantle contributed to form the continental crust (e.g., Jacobsen and Wasserburg, 1979). This might be the upper mantle isolated from the primordeal lower mantle. However, the mass balance calculations are subject to large uncertainties and the geochemical data cannot determine the location of depleted and undepleted "reservoirs." An interpretation consistent with whole mantle convection is the assumption of randomly scattered lumps of enriched material floating around in a generally depleted matrix (Davies, 1984).

Obviously the mechanism which might enforce layered convection must be connected to the physical nature of the 670-km discontinuity. Seismic velocities and density increase there by some 6–11 percent. The other major discontinuity at 400 km depth is explained by the transformation of olivine into the β-spinel structure. The decomposition of spinel into a mixture of perovskite $(Mg,Fe)SiO_3$ and periclase (magnesiowüstite) $(Mg,Fe)O$ occurs probably at the right temperature and pressure and is accompanied by the right density change to explain the 670-km discontinuity (e.g., Jeanloz and Thompson, 1983). There is both experimental (Ito and Yamada, 1982) and theoretical (Navrotsky, 1980) evidence that the transition is endothermic. Thus it has a negative Clapeyron slope (pressure-temperature slope) γ:

$$\gamma = \frac{dp}{dT} = \frac{Q_L \rho_0^2}{T_a \Delta\rho}. \tag{8.1}$$

Q_L is the latent heat, $\Delta\rho$ the density difference between the phases, ρ_0 the mean density, and T_a the absolute temperature. Tentatively a value of $\gamma = -2$ MPa/K (-20 bar/K) was proposed by Ito and Yamada for the spinel $\rightarrow$ perovskite + periclase transition. Generally a phase boundary is not considered as a barrier for convection. Two effects must be regarded (Schubert *et al.*, 1975): the release of latent heat in a convection current crossing the boundary changes the temperature which affects the driving buoyancy forces. The second effect appears more important. According to the sign of the Clapeyron slope the phase boundary is depressed or elevated in warm rising or cold sinking currents. Due to the high density contrast between the two phases strong buoyancy forces are connected with the distorted regions. For an exothermic transition ($\gamma > 0$) this effect helps to drive convection, whereas in the endothermic case it is a hindrance. The crucial question is how negative must the Clapeyron slope be to arrest convection completely. Previous studies suggested that a mildly endothermic transition will not be sufficient (Richter, 1973; Christensen, 1982; Olson and Yuen, 1982).

The interpretation of the 670-km discontinuity in terms of an isochemical phase change is challenged by the observation of strong seismic reflections from that depth, which require the discontinuity to be no more that 5 km wide (Lees *et al.*, 1983). A divariant phase change boundary in the binary Mg-Fe-system should be spread over typically 20–30 km. The sharpness is therefore better explained by a compositional change. Two kinds of chemical difference are usually diagnosed: a higher Fe:Mg-ratio or a higher Si:Mg-ratio in the lower mantle compared to the upper mantle. The problem with iron-enrichment is that it increases the density but reduces the seismic velocities. Therefore only a moderate increase in iron could be superimposed to what is essentially a phase change boundary, leading to a compositional density difference of at most 2–4 percent (Jeanloz and Thompson, 1983). A higher density contrast can be achieved if the mantle is more silica-rich. The equation of state parameters α and K_S (incompressibility) of perovskite are somewhat uncertain. Therefore a lower mantle of either pure pyroxene ($MgSiO_3$) stoichiometry or of upper mantle composition would be compatible with seismic velocities and density of the lower mantle (Jackson, 1983; Jeanloz and Thompson, 1983). Liu (1979) has proposed that the transition from the ilmenite to the perovskite structure in the pyroxene system occurs at pressures well above the 670 km discontinuity, whereas the corresponding transition in the olivine system (spinel $\rightarrow$ perovskite + periclase) occurs well below 700 km (Fig. 8.15a). In his model the 670-km discontinuity marks the

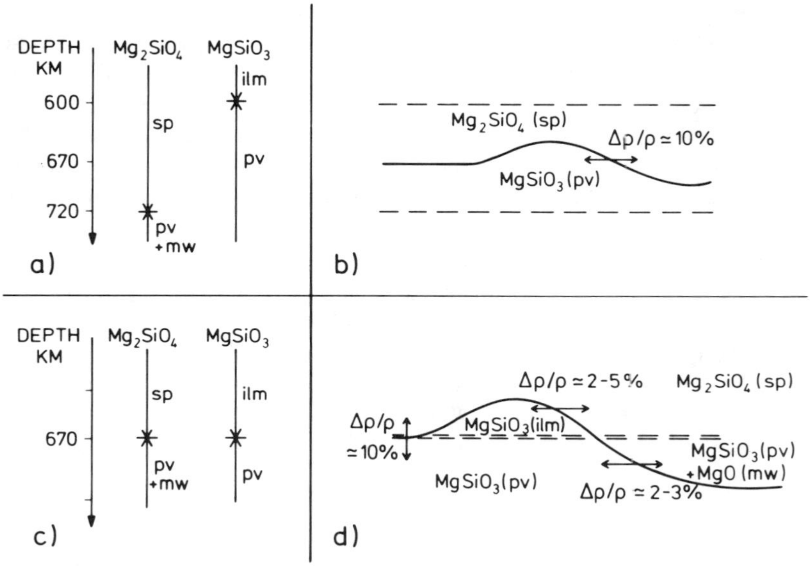

Figure 8.15. Schematical diagram of possible phase relations and associated dynamics of a chemical boundary at the 670-km discontinuity. Ideally pure olivine (Mg_2SiO_4) in the upper and pure pyroxene ($MgSiO_3$) stoichiometry in the lower mantle are assumed. a, Phase transitions according to Liu (1979). Sp stands for spinel, pv for perovskite, mw for magnesiowüstite, and ilm for ilmenite. b, During a small deflection of the chemical boundary (solid line) the phase assemblages of upper and lower mantle remain unchanged because of the gap in the transition pressures. c, Phase transitions according to Ito and Yamada (1982). d, During distortion the elevated lower mantle material transforms to ilmenite and depressed upper mantle material to perovskite and periclase.

boundary between ideally olivine stoichiometry in the upper and pure pryoxene stoichiometry in the lower mantle. Due to the gap in the transition pressures the large density contrast of the order of 10 percent is maintained even when the chemical boundary becomes deflected by some tens of kilometers (Fig. 8.15b). However, contrary to Liu's assumption the experimental data by Ito and Yamada (1982) suggest that both transitions fall close together and close to the depth of the discontinuity. In Figure 8.15c the end-member of a virtual coincidence of the transition pressures is assumed. In this case the elevated or depressed material will change phase when the chemical boundary is deflected. Although the vertical density contrast is still of the order of 10 percent, the lateral difference between the elevated/depressed regions and their environment is only of the order of 3 percent (Fig. 8.15d). This latter value is the dynamically active density difference which determines the magnitude of restoring buoyancy. It might thus be much smaller than the global density jump between upper and lower mantle.

It is obvious that chemical layers in the mantle can only survive when the convection currents do not break through them. On the other hand, how much compositional density difference is needed to keep separately convecting layers stable? An estimate based on linear stability analysis (Richter and Johnson, 1974; Sammis, 1976) predicted that already a $\Delta\rho$ of only 0.1 percent would be sufficient, but that does not hold for finite amplitude convection. Laboratory experiments by Richter and McKenzie (1981) confirmed a rule which seems intuitively appealing: the compositional density contrast $\Delta\rho$ must be larger than the potential thermal density contrast $\rho_0\alpha\Delta T$. In other words, the "boundary Rayleigh number,"

$$\mathrm{Rb} = \frac{\Delta\rho g h^3}{\kappa\eta}, \tag{8.2}$$

must exceed the thermal Rayleigh number to keep separately convecting layers stable (this is meant as a rule of thumb). The numerical experiments described in Section 6 with a compositionally different layer at the base of the mantle conform with this rule. The ΔT of mantle convection including subducting lithospheric slabs is of the order of 1000 K and Rb and Ra are approximately equal when the chemical density difference $\Delta\rho/\rho$ is 3 percent. This is within the range of estimates for a possible chemical contribution to the density difference at 670 km. Because the thermal density difference is not necessarily much smaller than a possible chemical density jump, the dynamical behavior of such a system can perhaps not be determined by a simple rule of thumb, and detailed modelling seems warranted. In the next section such models are discussed, assuming the 670-km discontinuity to be either a pure (endothermic) phase boundary, a compositional boundary, or a mixture of both.

9. Convection with Phase or Chemical Boundaries

When a compositional or phase boundary is present in a convecting system, deflections of these boundaries cause buoyancy forces in addition to the buoyancy due to thermal expansion. The equation of motion (2.1) must be supplemented by this buoyancy term and reads:

$$\hat{L}\psi = \mathrm{Ra}\frac{\partial T}{\partial x} - \mathrm{Rb}\frac{\partial \Gamma}{\partial x}, \tag{9.1}$$

where $\hat{L}$ is the linear differential operator from eq. 2.1, Rb is defined by eq. 8.2, and Γ is the so-called phase function indicating the relative fraction of the denser phase. Γ varies between 0 and 1, step-like in the case of a chemical boundary or an univariant phase change, or gradually for a divariant phase transition. When Γ stands for a difference in composition it is simply advected with the flow. In case of a phase change Γ depends on the p,T conditions. For simplicity we assume the pressure to be hydrostatic and the reaction kinetics fast enough, i.e., the readjustment of Γ to a change in pressure or temperature is instantaneous for all practical purposes. Transition pressure and transition temperature are linked through the Clapeyron slope, and Γ will actually depend only on a single parameter, which we call the "excess pressure,"

$$\pi = P - \gamma T. \tag{9.2}$$

For a divariant phase boundary $\Gamma(\pi)$ is taken as a tanh-function. The exact functional dependence appears insignificant as long as the transition width is much less than the thickness of the whole convecting layer.

The energy eq. 2.2 remains unchanged in case of a compositional boundary. In the case of a phase boundary one has to worry about the release of latent heat. In eq. 2.2 and 9.1 the Boussinesq approximation is applied, i.e., the density is taken as constant, except where buoyancy forces are to be determined. This implies that also the thermal effects of adiabatic compression and frictional heating are neglected. They both scale with the non-dimensional dissipation number,

$$Di = \frac{\alpha g h}{c_p}, \tag{9.3}$$

which is set to zero in the Boussinesq limit. It turns out, that also the latent heat effects of a phase boundary scale with Di and can thus be ignored in the Boussinesq limit (Christensen and Yuen, 1985). On the other hand it is not certain if the non-Boussinesq effects play a significant role in mantle convection or not. Fully compressible convection is difficult to treat (Jarvis and McKenzie, 1980). An intermediate choice is the "extended Boussinesq approximation," where the latent heat effect along with the adiabatic gradient and frictional heat are kept by taking a finite Di, but setting the incompressibility K_S (or the Grüneisen parameter) to infinity. The non-dimensional energy equation then reads

$$\frac{DT}{Dt} + Di\,wT - \bar{\gamma}\frac{\text{Rb}}{\text{Ra}}Di\,T\frac{D\Gamma}{Dt} = \nabla^2 T + Q + \frac{Di}{\text{Ra}}s_{ij}\frac{\partial u_i}{\partial x_j}. \quad (9.4)$$

T means here the absolute temperature, and $\bar{\gamma}$ is the Clapeyron slope non-dimensionalized according to $\bar{\gamma} = \gamma\Delta T/\rho gh$. The terms on the left hand side stand for change/advection of temperature, adiabatic heating, latent heat (in this sequence), and on the right hand side: diffusion, internal and frictional heat.

In Figure 8.16a a sequence of models with divariant phase boundary at mid-depth is displayed. The viscosity is constant, the Rayleigh number is 5×10^4, $Di = 0.0$, and the influence of the phase transition can be measured by the phase parameter P,

$$P = \bar{\gamma}\frac{\text{Rb}}{\text{Ra}} = \frac{\gamma}{\rho^2}\frac{\Delta\rho}{\alpha gh}. \quad (9.5)$$

An exothermic reaction ($P > 0$) enhances convection slightly, and a mildly endothermic reaction impedes it. If γ or P becomes negative enough, layered convection is indeed enforced (Fig. 8.16d+e). Especially in Figure 8.16d it can be seen that the separation into layers is not a perfect but a leaky one. This leakage has been studied by introducing passive markers randomly into the lower part of a stratified convecting system, separated by a phase change boundary (Christensen and Yuen, 1984). After a couple of overturns a significant number of them had passed into the upper layer. The leakage conforms with geochemical models of layered mantle convection: the lower primordeal reservoir has to be tapped at least occasionally to account for the isotopic signature of hot spot volcanism.

An important result is the decrease of the critical phase parameter P_{crit}, the point where breakdown into separate layers occurs, with increasing thermal Rayleigh number. This transition from single- to double-layer convection is marked in Figure 8.17 as sudden decrease of the Nusselt number and of the material flux across the boundary. In the range $10^5 \leqq \text{Ra} \leqq 2\times10^6$ P_{crit} decreases approximately with $\text{Ra}^{-0.2}$. In the model studies of Figures 8.16 and 8.17 the dissipation number and thus the latent heat effects were set to zero. However, taking a finite Di has no apparent influence on the point of breakdown into separate layers, although otherwise it can change the properties of convection quite profoundly. The location of the sudden decrease of the Nusselt number with decreasing P, marking the occurance of double layer convection, is unaffected by the size of Di (Fig. 8.18). Likewise it has little effect on P_{crit} if the thickness of the divariant transition zone is varied within reasonable limits, or if the phase change is taken to be univariant. No significant change in P_{crit} is observed if the transition depth is changed

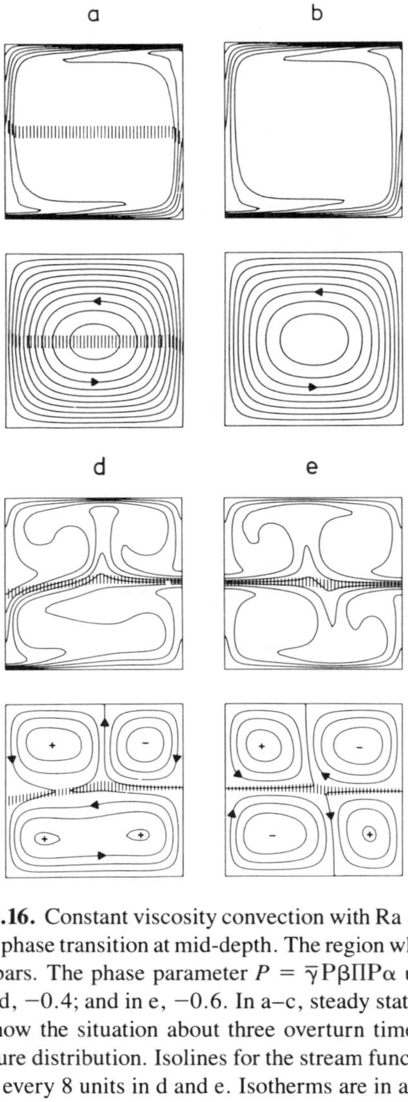

Figure 8.16. Constant viscosity convection with Ra $= 4 \times 10^5$, $Di = 0.0$ including a divariant phase transition at mid-depth. The region where $0.2 < \Gamma < 0.8$ is marked by vertical bars. The phase parameter $P = \overline{\gamma}$PβΠPα ισ ιν ακ +0.3; in b, 0.0; in c, −0.3; in d, −0.4; and in e, −0.6. In a–c, steady state results are displayed, while d) and e) show the situation about three overturn times after start from a single-cell temperature distribution. Isolines for the stream function are drawn every 20 units in a–c, and every 8 units in d and e. Isotherms are in a–c 0.05 (+0.10) 0.95, and in d and e 0.1 (+0.1) 0.9.

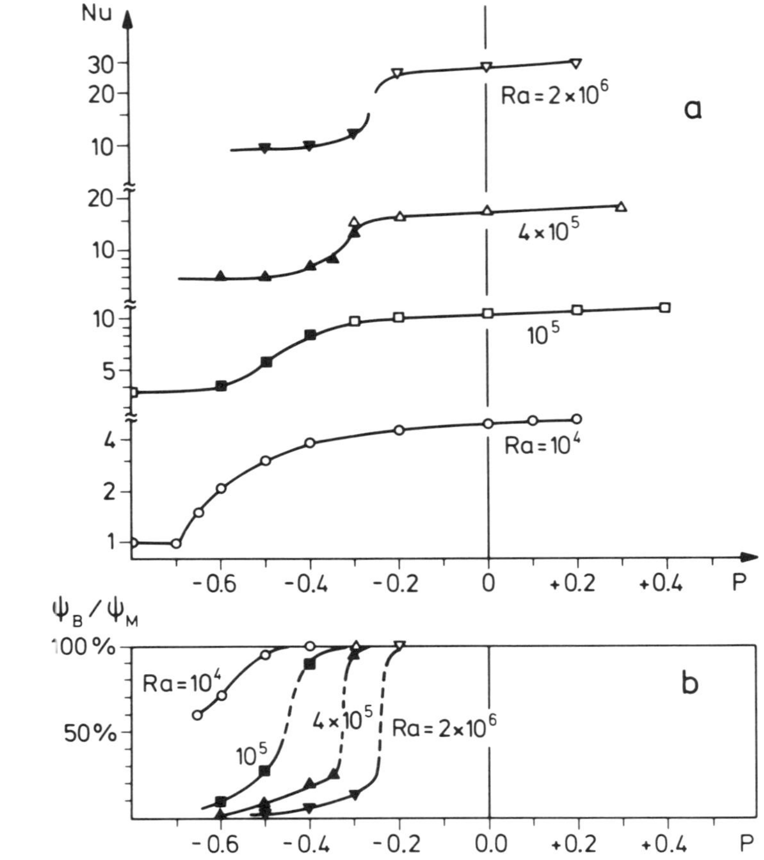

Figure 8.17. a, Nusselt number for different values of the Rayleigh number as a function of the phase parameter P of a divariant transition at mid-depth of a constant viscosity cell. Free slip, aspect ratio 1, $Di = 0.0$. Full symbols indicate the mean from time-dependent calculations, open symbols are steady state results. The step in Nu-vs.-P at negative P marks the transition from double to single layer canovection. For Ra $= 10^4$ double layer convection is not possible (the Rayleigh number for each sublayer is below critical), the flow is choked completely at very negative P. b, "Leaking rate", i.e., the ratio of the stream function maximum on the phase boundary (ψ_B) to the absolute maximum in the cell (ψ_M).

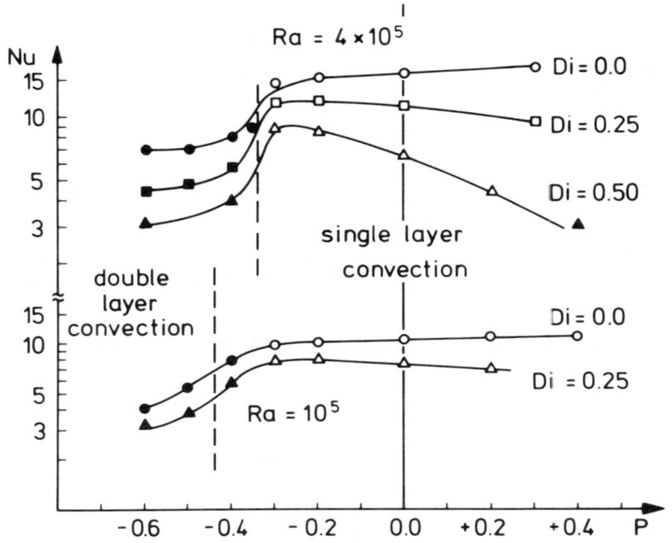

Figure 8.18. Nusselt number as a function of the phase parameter P for different dissipation number Di at Ra = 10^5 and 4×10^5. Open symbols are steady state results, full symbols the average from time-dependent runs. For whole mantle convection Di is estimated to be of the order of 0.5.

towards an asymmetric layering with the layer thickness ratio of 3:7 (Christensen and Yuen, 1985). We also tried to assess the influence of a possible superplastic rheology in the transition zone by reducing the viscosity in the mixed phase region by a factor of up to 1000. Superplasticity favors layered convection, however, the reduction of P_{crit} is not dramatic, amounting to about 30 percent.

At a Rayleigh number appropriate for the whole mantle, say 10^7, the likely range for P_{crit} is -0.15 to -0.30, taking into account uncertainties about the effects of variable viscosity, superplasticity, and so on. With typical mantle parameters in eq. 9.5, namely $\Delta\rho/\rho = 6 \ldots 12$ percent and $\alpha = 1.5 \ldots 2.5 \times 10^{-5}$ K^{-1}, this translates into a critical Clapeyron slope in the range of -4 to -8 MPa/K (-40 to -80 bar/K). That the spinal → perovskite + periclase transition has a Clapeyron slope of -8 MPa/K seems out of question. -4 MPa/K can probably not be excluded, although the prefered range is approximately -1 to -3 MPa/K. If the average thermal expansivity in the mantle is fairly low and superplasticity in the transition zone in fact occurs, the critical

Clapeyron slope for layered convection may be as low as −4 MPa/K. Layered convection induced by a pure phase boundary at 670 km thus appears marginally feasible.

Aside from the simple constant viscosity models, we have studied the interaction of a subducting lithospheric slab with a chemical or phase boundary in a rheologically more complex model (Christensen and Yuen, 1984). As outlined in Section 7, plate-like behavior can be achieved with temperature-dependent power-law rheology. To make things simple, we took a step-like rheology with high lithospheric viscosity of 10^{24} Pa s at temperatures below 1000°C and low asthenospheric viscosity of 10^{21} Pa s above 1000°C. This basic Newtonian rheology is superceded by a fifth power law dependence at a transition stress of 35 MPa (350 bar). This is not meant as the transition from Herring–Nabarro to dislocation creep, but rather it is a heuristic way to simulate the rheological behavior of the lithosphere. We started our calculation with a thermal lithosphere of about 100 km thickness with an "embryonic slab" attached at one side in order to get subduction started. Let us first consider the case of a purely compositional boundary at 650 km depth. The boundary is represented by a chain of markers, which is advected with the local flow. Restoring buoyancy forces act when the marker chain is distorted. Our initial temperature distribution had an internal boundary layer with a ΔT of 400 K.

In Figure 8.19 a sequence of snapshots is shown for a case where the compositional density contrast is high. The plate velocity first increases with increasing length of the subducting slab. When it hits the chemical boundary the slab bends and is deflected sidewards, the plate velocity decreases and two-layer convection persists. However, the chemical boundary is depressed by the load of the slab, the depression amounts to 70 km, 130 km, and 230 km for density contrasts $\Delta\rho/\rho$ of 9 percent, 6 percent, and 4.5 percent, respectively. In principal the depression must be detectable by seismology, for example by studying travel time anomalies of deep focus earthquakes. One would expect late arrival times for those rays which are radiated into the geometrical down-dip direction of the slab, because they have a longer path through the upper mantle material with its intrinsic low velocity. However, the actual data, as far as available, show the opposite anomaly (Creager and Jordan, 1984).

When the chemical density difference is decreased to 3 percent, or Ra/Rb is increased to 2, a different behavior is observed. Since the total $\Delta\rho/\rho$ at 670 km is certainly larger than 3 percent, one can think of the rest as being made up by a phase change, perhaps similar to the situation sketched in Figure 8.15c, d. When the Clapeyron slope of the phase change part is close to zero it would not affect the dynamics. The slab now intrudes the lower mantle (Fig. 8.20), though with decreasing velocity, and comes to a rest 700 km below the equilibrium depth of the boundary. The subducting material appears to accumulate in a kind of thickening root in the lower mantle. The final fate of this root is unclear, probably it will rise again after a long period of re-heating. When $\Delta\rho/\rho$ was lowered to 1.5 percent (Ra/Rb = 4) the slab descended rapidly to the bottom of the model system, crept along the bottom boundary, and then started to eject rising chemical diapirs. An intrinsic mixing of upper and lower mantle material after a few cycles of deep subduction and reascent would be expected in this case.

For comparison, we considered also constant viscosity convection with a chemical boundary. Approximately constant viscosity might be appropriate for a secondary mode of convection underneath the plates (e.g., Parsons and McKenzie, 1978). At a Rayleigh number ratio of Ra/Rb = 1.6 we found time-dependent fluctuations of the boundary of about 100 km amplitude. At the same Ra/Rb with a slab being present the penetration depth is more than 300 km. Apparently the mechanical strength of the dense subducting slab plays an important role. Due to the strength the whole weight of the heavy cold column is available to depress the chemical boundary.

The case of an endothermic phase boundary was also considered in models containing a subducting slab. Latent heat was neglected as a non-Boussinesq effect and the transition was taken as univariant and represented by the marker chain. The results closely agree with the estimates from the constant viscosity models. With $\Delta\rho/\rho = 9$ percent and $\gamma = -4.5$ MPa/K the slab penetrates the lower mantle, whereas with a Clapeyron slope of -6 MPa/K it is deflected at the phase boundary and two-layer convection is established. When the 670-km discontinuity is modelled as a combination of an endothermic phase boundary and a superimposed compositional difference, with a density contribution $\Delta\rho_c$, a more moderate Clapeyron slope is found to enforce layered convection, e.g. $\gamma = -3$ MPa/K when $\Delta\rho_c/\rho = 3$ percent. There is an interesting difference between the cases where the discontinuity is taken as a compositional boundary, and those where it is predominantly a phase boundary with negative γ. In the first case there is a gradual increase of penetration depth with decreasing $\Delta\rho$. In the second case there is a sudden switch from unlimited penetration of the slab to deflection at the boundary with a maximum depression of about 100 km.

In the domain diagram Figure 8.21 the results of the convection models containing a subducting slab are summarized. When the whole density contrast at the 670 km discontinuity is due to a change in composition, as in the model of Liu (1979), layered convection in a strict sense must be

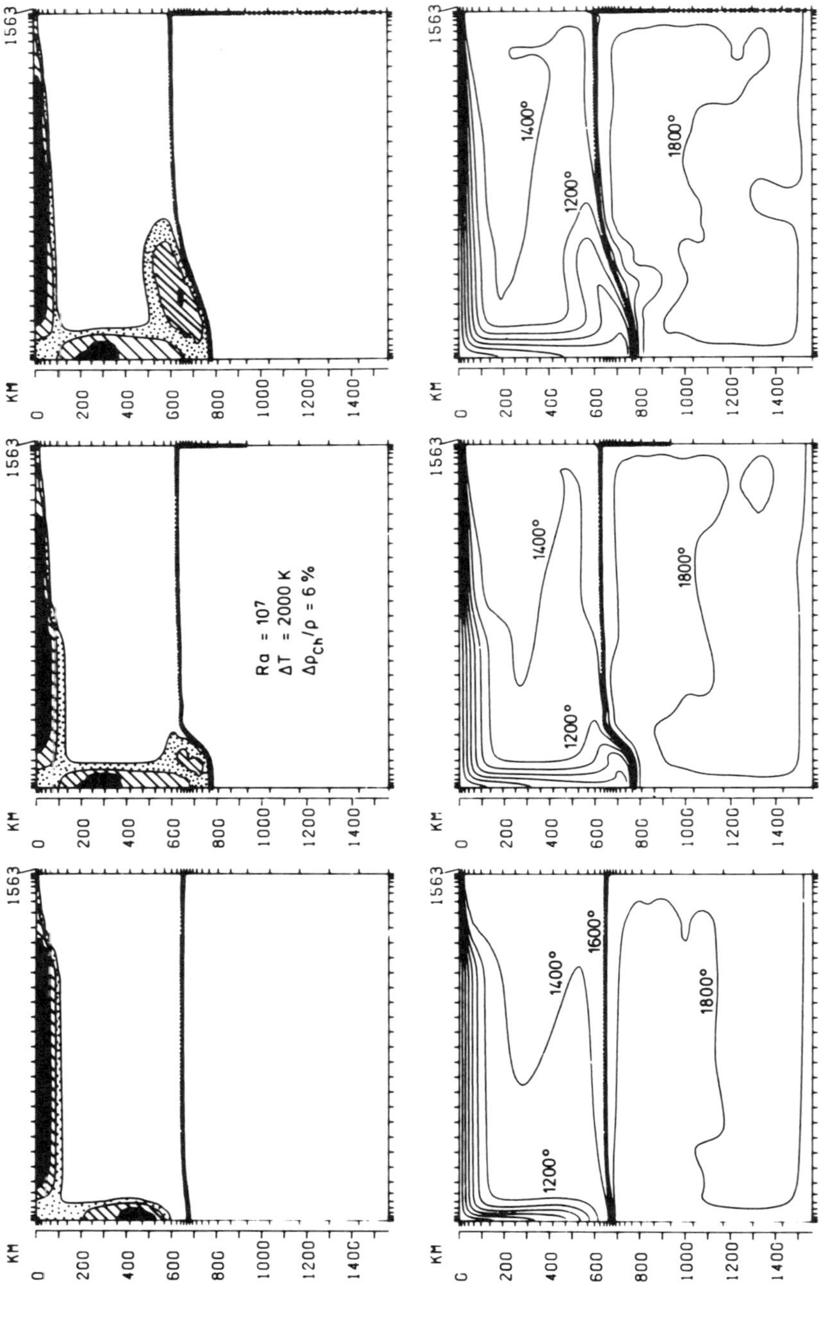

KM
0
200
400
600
800
1000
1200
1400
1563
Ra = 10^7
ΔT = 2000 K
Δρ_ch/ρ = 6 %
1200°
1400°
1600°
1800°

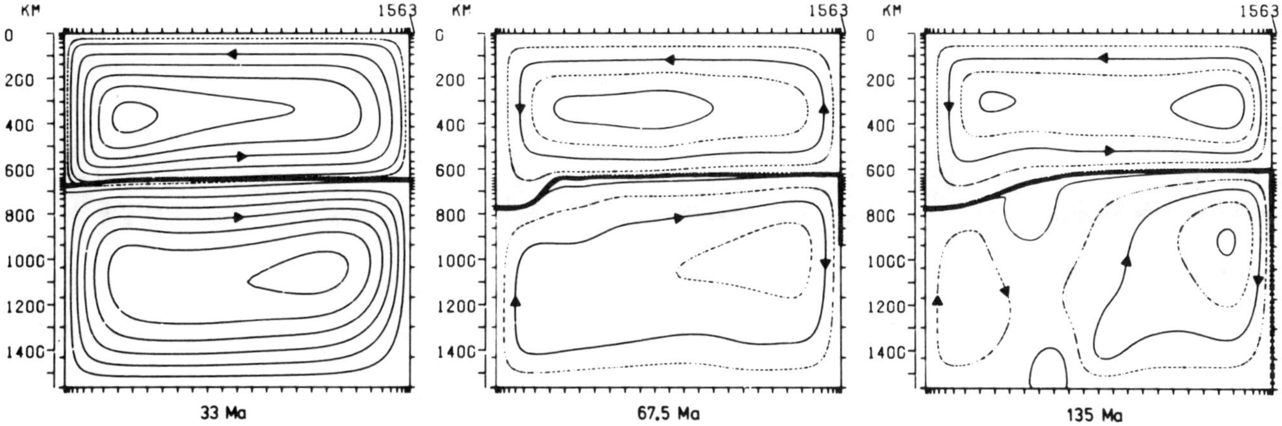

Figure 8.19. Snapshots of a convection model including a subducting lithospheric slab and a chemical boundary at 650 km depth. In each column, from top to bottom, the viscosity, isotherms in 200° intervals, and stream-lines are shown. Viscosity ranges from $<3 \times 10^{21}$ Pa s in the white region to $>3 \times 10^{23}$ Pa s in the black regions. The compositional density contrast is 6 percent.

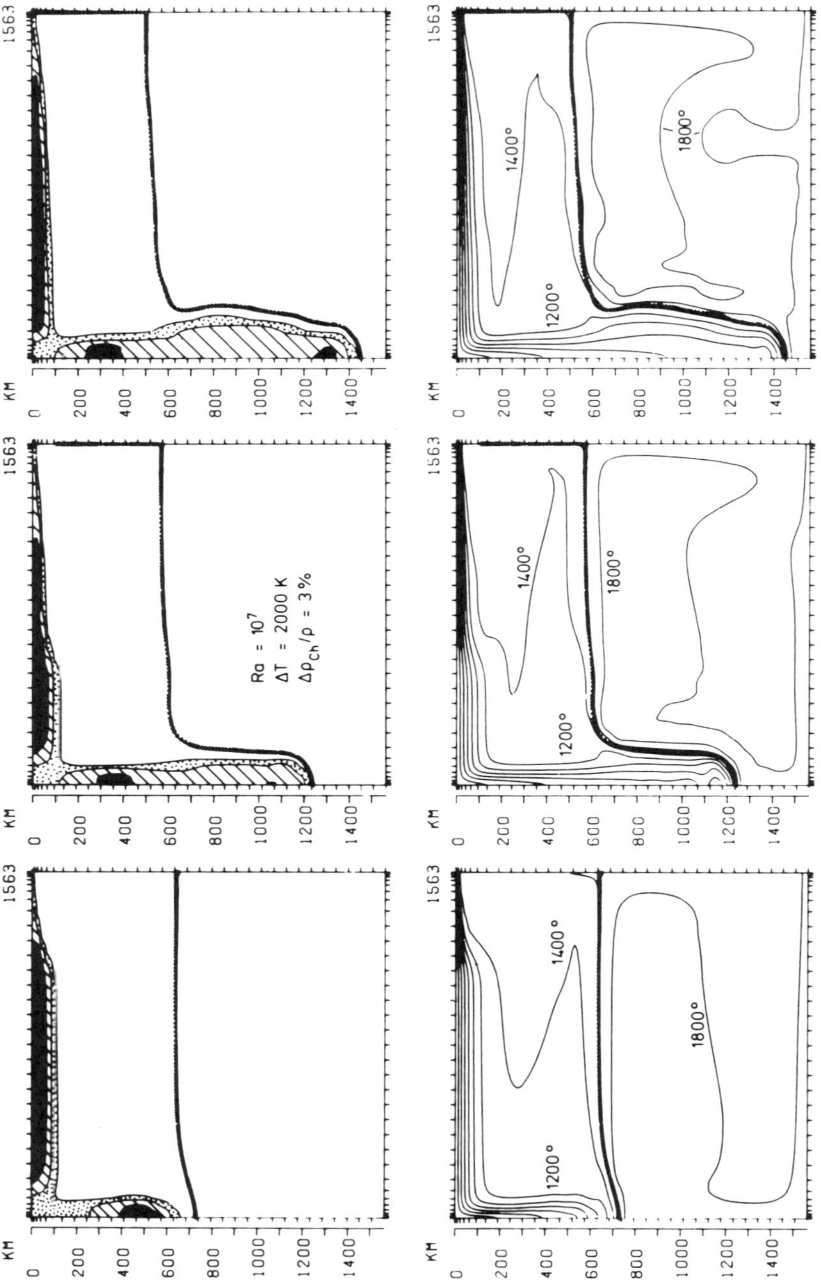

Ra = 10^7
ΔT = 2000 K
Δρ_ch/ρ = 3%
1200°
1400°
1800°
KM
0
200
400
600
800
1000
1200
1400
1563

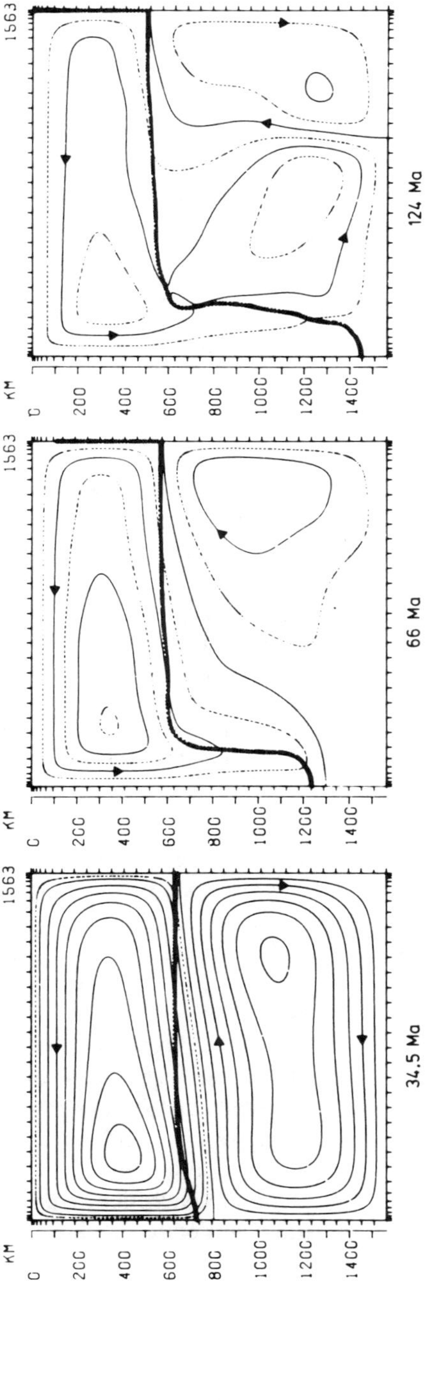

Figure 8.20. As in the previous figure, but with a chemical density difference of 3 percent. Note that stream-lines crossing the boundary indicate that the boundary itself moves, not a material exchange across the boundary.

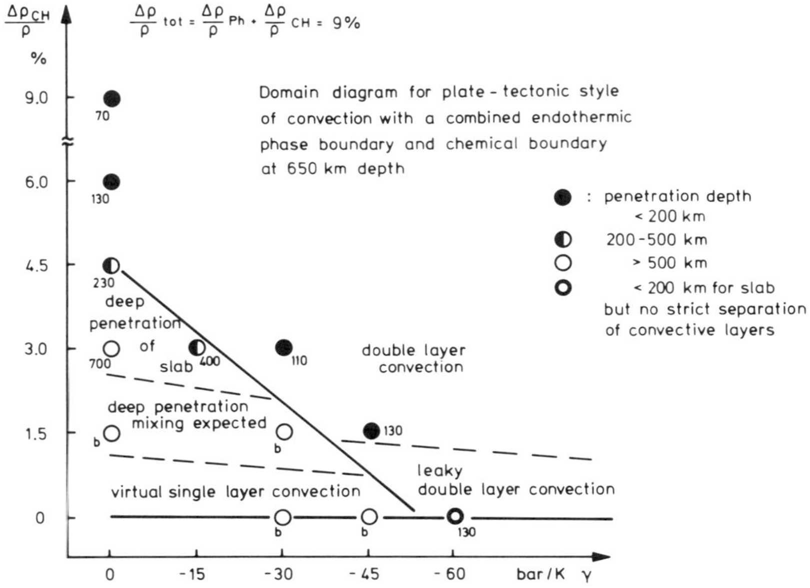

Figure 9.21. Domain diagram for a plate tectonic style of mantle convection with a combined phase and chemical boundary at 650 km depth. Model results are arranged according to the compositional part of the density contrast (the total being 9 percent) and the Clapeyron slope of the phase transition part. Numbers attached to the symbols indicate the maximum penetration of the slab below the mean depth of the boundary in kilometers, "b" means penetration to the bottom of the mantle.

expected. When the possible chemical contribution is small, on the order of a few percent, the size of the Clapeyron slope of the phase transition part becomes important. If γ is below -3 MPa/K layered convection appears likely. When γ is not low enough, slabs would plunge deeply into the lower mantle, and when $\Delta\rho_c/\rho$ is less than about 2 percent one can probably expect homogenization of the system within relatively short time. Even when the general circulation is layered, some mass flux across the boundary is expected. We found this numerically for the case of a phase boundary, and Olson (1984) demonstrated experimentally that viscous entrainment leads to the same effect in the case of a compositional difference.

The results of our model calculations do not give definite preference to either the whole mantle or the layered convection hypothesis. With the appropriate assumptions on the parameters of the 670-km discontinuity both are dynamically feasible. However, our numerical experiments have quantified the premises of layered convection more clearly than previous work, and determined the degree of depression of the boundary in subduction regions. This will be useful in collecting more definite evidence to decide the outstanding issue. The evidence must come from the combined efforts of seismology and high-pressure mineralogy: What are the elastic and thermodynamic properties of the perovskite phases in the lower mantle, what are the exact transition pressures and Clapeyron slopes in the MgO-FeO-SiO_2—system between 20 and 30 GPa? Is there seismic evidence for strong lateral heterogeneity of the 670-km discontinuity, or for a double thermal boundary layer? Can the existence of aseismic slabs below 700 km be substantiated, and can the sharpness of the discontinuity be determined more accurately? It would not be surprising, if our final picture of the dynamic mantle is more complex than both end members of whole mantle convection or strictly layered convection suggest.

10. Outlook

Our knowledge of mantle convection will only improve with the combination of a better theoretical understanding of the convection process, and with increased observations and inferences about the Earth's structure from the different fields of geosciences. Concerning the latter point, the following approaches appear to be especially promising: The interpretation of the earth's long-wavelength gravity field in terms of convection, the determination of seismic lateral heterogeneity and anisotropy ("seismic tomography"), studies of the fine structure of seismic discontinuities (670-km discontinuity, core–mantle boundary), experimental data pertaining to the resolution of the mantle constitution (phase transitions, equation-of-state data), and isotope geochemistry as means of studying the chemical heterogeneity of the mantle. The theoretical understanding of mantle convection is difficult because it requires the solution of complex nonlinear mathematical problems, involving a large number of parameters of potential importance. A most useful approach seems to be to attack this problem by careful and systematic numerical experiments, isolating a small number of parameters and studying their mutual influence, in order to understand the basic physics. In my opinion the two most rewarding fields for further studies are convection with complex rheology, and convection involving chemical heterogeneity. This puts the numerical study of three-dimensional convection off to the third place.

A peculiarity of mantle convection is the platelike movement of the upper thermal boundary layer, which might shape the character of the whole circulation. Until now this has either been neglected altogether in convection models, or a plate has been introduced by ad hoc means, as by imposing spatially constant surface velocity (e.g., Olson and Corcos, 1980) or weak zones at the margins of a high-viscosity boundary layer. In Section 7 evidence was presented that non-Newtonian temperature-dependent rheology is a possible cause for the plate tectonic style of mantle convection. However, it is unlikely that a simple power-law describes the rheology correctly, and for example viscoelasticity may play a decisive role. The active role of oceanic lithospheric plates may be important for the heat loss from the Earth and thus its thermal evolution (Sections 4 and 5), for mixing of mantle heterogeneities (Gurnis and Davies, 1986b), or for the dynamical behavior of a stratified convecting mantle (Section 9). Thus more efforts to understand the influence of the complex rheological behavior of the lithosphere on mantle convection are badly needed. A further point where the rheology may play a key role is in determining the size of the temperature jump across an internal boundary layer in a two-layer mantle. Reliable estimates how big ΔT must be, or how small it can be, would help the search for observational evidence of such a boundary layer.

The study of mantle convection has received new impetus by the data on isotopic heterogeneity and inferred geochemical models of the mantle. The conditions for the persistance of a gross chemical layering have been determined in Section 9. However, even if there is no gross layering of the mantle, substantial chemical heterogeneities must exist. They are continuously reproduced by magma segregation at ocean ridges and subsequent subduction of the oceanic lithosphere, which is layered into crust, depleted mantle, and fertile mantle. To date most numerical studies regarding this issue considered the dispersal of passive, i.e., not compositionally buoyant, heterogeneities (e.g., Richter *et al.*, 1982; Olson *et al.*, 1984). However, in the subducting plate density differences due to composition may be larger than those due to thermal expansion. Thus the numerical study of thermochemical convection appears necessary, where both sources of buoyancy are considered. First steps have been made with the calculations described in Sections 6 and 9 of this review. A next step would be to consider randomly scattered lumps or continuously injected chemical heterogeneities. Do they mix or do they segregate under their own buoyancy? The pressure-dependence of viscosity, controlling the convective velocity in the lower mantle (Section 6) may play a role for the billion-year survival of heterogeneities. A preferential segregation of the heavy former oceanic crust downwards, which is afterwards incorporated into mantle plumes (Hofmann and White, 1982), and of light depleted peridotite upwards, which contributes then mainly to mid-ocean-ridge volcanism, may explain the isotopic features of both. However, the dynamical feasibility of such a model remains to be demonstrated.

11. References

Allègre, C.J., Brévart, O., Dupré, B., Minster, J.-F., 1980. Isotopic and chemical effects produced in a cotinuously differentiating convecting earth mantle, *Phil. Trans. R. Soc. London, A* **297**, 447–477.

Ashby, M.F., Verall, R.A., 1977. Micromechanisms of flow and fracture, and their relevance to the rheology of the upper mantle, *Phil. Trans. R. Soc. London, A* **288**, 59–95.

Berckhemer, H., Auer, F., Drisler, J., 1979. High temperature elasticity and anelasticity of mantle peridotite, *Phys. Earth Planet. Inter.* **20**, 48–59.

Booker, J.R., 1976. Thermal convection with strongly temperature dependent viscosity, *J. Fluid Mech.* **76**, 741–754.

Booker, J.R., Stengel, K.C., 1978. Further thoughts on convective heat transport in a variable viscosity fluid, *J. Fluid Mech.* **86**, 289–291.

Boss, A.P., Sacks, I.S., 1982. Time-dependent numerical models of mantle convection, *Carnegie Inst. Wash. Yearbook*, **81**, 543–550.

Carrigan, C.R., 1984. Time and temperature dependent convection models of cooling reservoirs: Application to volcanic skills, *Geophys. Res. Lett.* **11**, 693–696.

Cathles, L. M., 1975. *The Viscosity of the Earth's Mantle*. Princeton, N.J.: Princeton University Press, 386 pp.

Chopra, P.N., Paterson, M.., 1981. The experimental deformation of dunite, *Tectonophys.* **78**, 453–473.

Christensen, U., 1982. Phase boundaries in finite amplitude mantle convection, *Geophys. J. R. Astr. Soc.* **68**, 487–497.

Christensen, U., 1983. Convection in a varible viscosity fluid: Newtonian versus power-law rheology, *Earth Planet. Sci. Lett.* **64**, 153–162.

Christensen, U., 1984a. Convection with pressure- and temperature-dependent non-Newtonian rheology, *Geophys. J. R. Astr. Soc.* **77**, 343–384.

Christensen, U., 1984b. Heat transfer by variable viscosity convection and implications for the Earth's thermal evolution, *Phys. Earth Planet. Inter.* **35**, 264–282.

Christensen, U., 1984c. Instability of a hot boundary layer and initiation of thermochemical plumes, *Ann. Geophysicae* **2**, 311–320.

Christensen, U., 1985a. Heat transfer by variable viscosity convention II: Pressure influence, non-Newtonian rheology, and decaying heat sources, *Phys. Earth Planet. Inter.* **37**, 183–205.

Christensen, U., 1985b. Thermal evolution models for the Earth, *J. Geophys. Res.* **90**, 2995–3008.

Christensen, U., 1985c. Viscosity-velocity correlation in convection cells with non-uniform viscosity, *J. Geophys.* **57**, 72–75.
Christensen, U., Yuen, D.A., 1984. The interaction of a subducting lithospheric slab with a chemical or phase boundary, *J. Geophys. Res.* **89**, 4389–4402.
Christensen, U., Yuen, D.A., 1985. Layered convection induced by phase transitions, *J. Geophys. Res.* **90**, 10291–10300.
Creager, K.C., Jordan, T.H., 1984. Slab penetration into the lower mantle, *J. Geophys. Res.* **89**, 3031–3050.
Cserepes, L., 1982. Numerical studies of non-Newtonian mantle convection, *Phys. Earth Planet. Inter.* **30**, 49–61.
Daly, S. F., 1980. The vagaries of variable viscosity convection, *Geophys. Res. Lett.* **7**, 841–844.
Davies, G.F., 1977. Whole mantle convection and plate tectonics, *Geophys. J. R. Astr. Soc.* **49**, 459–486.
Davies, G.F., 1980. Thermal histories of convective earth models and constraints on radiogenic heat production in the earth, *J. Geophys. Res.* **85**, 2517–2530.
Davies, G. F., 1984. Geophysical and isotopic constraints on mantle convection: An interim synthesis, *J. Geophys. Res.* **89**, 6017–6040.
DeBremaecker, J.-Cl., 1977. Is the oceanic lithosphere elastic or viscous, *J. Geophys. Res.* **82**, 2001–2004.
Doornbos, D.J., 1983. Present seismic evidence for a boundary layer at the base of the mantle, *J. Geophys. Res.* **88**, 3498–3505.
Doornbos, D.J., Mondt, J.C., 1979. P and S waves refracted around the core and the velocity structure at the base of the mantle, *Geophys. J. R. Astr. Soc.* **57**, 381–396.
Durham, W.B., Goetze, C., 1977. Plastic flow of oriented single crystals of olivine: Mechanical data, *J. Geophys. Res.* **82**, 5737–5753.
Dziewonski, A.M., 1984. Mapping the lower mantle: Determination of lateral heterogeneity in P velocity up to degree and order 6, *J. Geophys. Res.* **89**, 5929–5952.
England, P., Bickle, M., 1984. Continental thermal and tectonic regimes during the Archaean, *J. Geol.* **92**, 353–367.
Fleitout, L., Yuen, D.A., 1984. Steady state, secondary convection beneath lithospheric plates with temperature and pressure dependent viscosity, *J. Geophys. Res.* **89**, 9227–9244.
Fowler, A.C., 1983. On the thermal state of the Earth's mantle, *J. Geophys.* **53**, 42–51.
Giardini, D., Woodhouse, J.H., 1984. Deep seismicity and modes of deformation in Tonga subduction zone, *Nature* **307**, 505–509.
Green, D.H., 1981. Petrogenesis of Archaean ultramafic magmas and implications for Archaean tectonics. In A. Kröner (ed.), *Precambrian Plate Tectonics*. Amsterdam: Elsevier, pp. 469–489.
Gurnis, M., Davies, G.F., 1986a. Numerical study of high Rayleigh number convection in a medium with depth-dependent viscosity, *Geophys. J. R. Astro. Soc* **85**, 523–541.
Gurnis, M., Davies, G.F., 1986b. Mixing in numerical models of mantle convection incorporating plate kinematics, *J. Geophys. Res.* **91**, 6375–6395.

Hager, B.H., 1984. Subducted slabs and the geoid: Constraints on mantle rheology and flow, *J. Geophys. Res.* **89**, 6003–6016.

Hansen, U., Ebel, A., 1984. Experiments with a numerical model related to mantle convection: Boundary layer behavior of small- and large scale flows, *Phys. Earth Planet. Inter.* **36**, 374–390.

Houston, M.H., DeBremaecker, J.-Cl., 1976. Numerical models of convection in the upper mantle, *J. Geophys. Res.* **80**, 742–751.

Isacks, B., Molnar, P., 1971. Distribution of stresses in the descending lithosphere from a global survey of focal mechanism solutions of mantle earthquakes, *Rev. Geophys. Space Phys.* **9**, 103–174.

Ito, E., Yamada, H., 1982. Stability relations of silicate spinels, ilmenites, and perovskites. In S. Akimoto, M.H. Manghani (eds.), *High Pressure Research in Geophysics*, Tokyo: Center of Academic Publishing, pp. 405–419.

Jackson, I., 1983. Some geophysical constraints on the chemical composition of the lower mantle, *Earth Planet. Sci. Lett.* **62**, 91–103.

Jacobsen, S.B., Wasserburg, G.J., 1979. The mean age of the mantle and crustal reservoirs, *J. Geophys. Res.* **84**, 7411–7427.

Jacoby, W.R., 1978. One-dimensional modelling of mantle flow, *Pure Appl. Geophys.* **116**, 1231–1249.

Jaoul, O., Froidevaux, C., Durham, W.B., Michaut, M., 1980. Oxygen self-diffusion in forsterite: Implications for the high-temperature creep mechanism, *Earth Planet. Sci. Lett.* **47**, 391–397.

Jarvis, G.T., McKenzie, D.P., 1980. Convection in a compressible fluid with infinite Prandtl number, *J. Fluid Mech.* **96**, 515–583.

Jarvis, G.T., Peltier, W.R., 1982. Mantle convection as a boundary layer phenomenon, *Geophys. J. R. Astr. Soc.* **68**, 385–424.

Jeanloz, R., Richter, F.M., 1979. Convection, composition, and the thermal state of the lower mantle, *J. Geophys. Res.* **84**, 5497–5504.

Jeanloz, R., Thompson, A.B., 1983. Phase transitions and mantle discontinuities, *Rev. Geophys. Space Phys.* **21**, 51–74.

Jochum, K.P., Hofmann, A.W., Ito, E., Seufert, H.M., White, W.M., 1983. K, U, and Th in MORB glasses and heat production, K/U and K/Rb in the mantle, *Nature* **306**, 431–436.

Justice, M.G., Graham, E.K., Tressler, R.E., Isong, I.S.T., 1982. The effect of water on high temperature deformation of olivine, *Geophys. Res. Lett.* **9**, 1005–1008.

Karato, S., 1981. Rheology of the lower mantle, *Phys. Earth Planet. Inter.* **24**, 1–14.

Karato, S., Ogawa, M., 1982. High pressure recovery of olivine: Implications for creep mechanisms and creep activation volume, *Phys. Earth Planet. Inter.* **28**, 102–117.

Kohlstedt, D.L., Goetze, C., 1974. Low stress high temperature creep of olivine single crystals, *J. Geophys. Res.* **79**, 2045–2051.

Kohlstedt, D.L., Nichols, P.K., Hornack, P., 1980. The effect of pressure on the rate of dislocation recovery in olivine, *J. Geophys. Res.* **85**, 3122–3130.

Kopitzke, U., 1979. Finite element convection models: Comparison of shallow and deep mantle convection and temperatures in the mantle, *J. Geophys.* **46**, 97–121.

Kröner, A., Layer, P.W., McWilliams, M.O., 1984. Archaean paleomagnetism: Evidence for continental drift and existence of a dipolar magnetic field since ca. 3.5 billion years ago (abstract), *Terra Cognita* **4**, 78.

Lees, A.C., Bukowinski, M.S.T., Jeanloz, R., 1983. Reflection properties of phase transitions and compositional change models of the 650-km discontinuity, *J. Geophys. Res.* **88**, 8145–8159.

Liu, L.G., 1979. On the 650-km discontinuity, *Earth Planet. Sci. Lett.* **42**, 202–208.

Loper, D.E., 1983. The dynamical and thermal structure of deep mantle plumes, *Phys. Earth Planet. Inter.* **33**, 304–317.

McAdoo, D.C., Caldwell, J.G., Turcotte, D.L., 1978. On the elastic perfectly plasting bending of the lithosphere under generalized loading with application to the Kuril trench, *Geophys. J. R. Astr. Soc.* **54**, 11–26.

McKenzie, D.P., 1966. The viscosity of the lower mantle, *J. Geophys. Res.* **71**, 3995–4010.

McKenzie, D.P., 1977. Surface deformation, gravity anomalies and convection, *Geophys. J. R. Astr. Soc.* **48**, 211–238.

McKenzie, D.P., Roberts, J.M., Weiss, N.O., 1974. Convection in the Earth's mantle: Towards a numerical simulation, *J. Fluid Mech.* **62**, 465–538.

McKenzie, D.P., Weiss, N.O., 1975. Speculations on the thermal and tectonic history of the Earth, *Geophys. J. R. Astr. Soc.* **42**, 131–174.

Melosh, H.J., 1978. Dynamic support of the outer rise, *Geophys. Res. Lett.* **5**, 321–324.

Molnar, P., Francheteau, J., 1975. The relative motion of 'hot spots' in the Atlantic and Indian oceans during the Cenozoic, *Geophys. J. R. Astr. Soc.* **43**, 763–774.

Morgan, W.J., 1971. Convection plumes in the lower mantle, *Nature* **230**, 42–43.

Mula, A.H.G., Müller, G., 1980. Ray parameters of diffracted long-period P and S waves at the base of the mantle, *Pure Appl. Geophys.* **118**, 1270–1290.

Nataf, H.C., Richter, F.M., 1982. Convection experiments in fluids with highly temperature-dependent viscosity and the thermal evolution of the planets, *Phys. Earth Planet. Inter.* **29**, 320–329.

Navrotsky, A., 1980. Lower mantle phase transitions may generally have negative pressure-temperature slopes, *Geophys. Res. Lett.* **7**, 709–711.

O'Connell, R.J., 1977. On the scale of mantle convection, *Tectonophys.* **38**, 119–136.

Olson, P., 1984. An experimental approach to convection in a two-layered mantle, *J. Geophys. Res.* **89**, 11293–11302.

Olson, P., Corcos, G.M., 1980. A boundary layer model for mantle convection with surface plates, *Geophys. J. R. Astr. Soc.* **62**, 195–219.

Olson, P., Yuen, D.A., 1982. Thermochemical plumes and mantle phase transitions, *J. Geophys. Res.* **87**, 3993–4002.

Olson, P., Yuen, D.A., Balsiger, D., 1984. Mixing of passive heterogeneities by mantle convection, *J. Geophys. Res.* **89**, 425–436.

O'Nions, R.K., Carter, S.R., Evensen, N.M., Hamilton, P.J., 1981. Upper mantle geochemistry. In C. Emiliani (ed.) *The Oceanic Lithosphere, Vol. 7, The Sea*, New York: J. Wiley and Sons, pp. 49–72.

Oxburgh, E.R., Turcotte, D.L., 1978. The mechanism for continental drift, *Rep. Progress Phys.* **41**, 1249–1312.

Parmentier, E.M., Turcotte, D.L., Torrance, K.E., 1975. Numerical experiments on the structure of mantle plumes, *J. Geophys. Res.* **80**, 4417–4424.

Parmentier, E.M., Turcotte, D.L., Torrance, K.E., 1976. Studies of finite amplitude non-Newtonian thermal convection with application to convection in the earth's mantle, *J. Geophys. Res.* **81**, 1839–1846.

Parmentier, E. M., Turcotte, D. L., 1978. Two-dimensional mantle flow beneath a rigid accreting lithosphere, *Phys. Earth Planet. Inter.* **17**, 281–289.

Parmentier, E.M., Morgan, J., 1982. Thermal convection in non-Newtonian fluids: Volumetric heating and boundary layer scaling, *J. Geophys. Res.* **87**, 7757–7762.

Parsons, B., McKenzie, D.P., 1978. Mantle convection and the thermal structure of the plates, *J. Geophys. Res.* **83**, 4485–4496.

Peltier, W.R., 1976. Glacial isostatic adjustment II: The inverse problem, *Geophys. J. R. Astr. Soc.* **46**, 664–706.

Peltier, W.R., 1985. Mantle convection and viscoelasticity, *Ann. Rev. Fluid Mech.* **17**, 561–608.

Peltier, W.R., Jarvis, G.T., 1982. Whole mantle convection and the thermal evolution of the Earth, *Phys. Earth Planet. Inter.* **29**, 281–304.

Poirier, J.P., Liebermann, R.C., 1984. On the activation volume for creep and its variation with depth in the Earth's lower mantle, *Phys. Earth Planet. Inter.* **35**, 283–293.

Post, R.L., 1977. High temperature creep of Mt. Burnet dunite, *Tectonophys.* **42**, 75–110.

Quareni, F., Yuen, D.A., Sewell, G., Christensen, U., 1985. High Rayleigh number convection with strongly variable viscosity: A comparison between mean-field and two-dimensional solutions, *J. Geophys. Res.* **90**, 12633–12644.

Rabinowicz, M., Lago, B., Froidevaux, C., 1980. Thermal transfer between the continental asthenosphere and the subducting lithosphere: Its effect on subcontinental convection, *J. Geophys. Res.* **85**, 1839–1853.

Relandeau, C., 1981. High temperature creep of forsterite polycrystalline aggregates, *Geophys. Res. Lett.* **8**, 733–736.

Richter, F.M., 1973. Finite amplitude convection through a phase boundary, *Geophys. J. R. Astr. Soc.* **35**, 265–276.

Richter, F.M., 1979. Focal mechanisms and seismic energy release of deep and intermediate earthquakes in the Tonga-Kermadec region and their bearing on the depth-extent of mantle flow, *J. Geophys. Res.* **84**, 6783–6795.

Richter, F.M., Johnson, C.E., 1974. Stability of a chemically layered mantle, *J. Geophys. Res.* **79**, 1635–1639.

Richter, F.M., McKenzie, D.P., 1981. On some consequences and possible causes of layered mantle convection, *J. Geophys. Res.* **86**, 6133–6142.

Richter, F.M., Daly, S.F., Nataf, H.C., 1982. A parameterized model for the evolution of isotopic heterogeneities in a convecting system, *Earth Planet. Sci. Lett.* **60**, 178–194.

Richter, F.M., Nataf, H.C., Daly, S.F., 1983. Heat transfer and horizontally averaged temperature of convection with large viscosity variations, *J. Fluid Mech.* **129**, 173–192.

Ross, J.V., Ave'Lallemant, H.G., Carter, N.L., 1979. Activation volume for creep in the upper mantle, *Science* **203**, 261–263.

Rubie, D.C., 1984. The olivine → spinel transformation and the rheology of subducting lithosphere, *Nature* **308**, 505–508.

Sacks, I.S., Snoke, J.A., Beach, L., 1979. Lateral heterogeneity at the base of the mantle revealed by observations of amplitudes of PKP phases, *Geophys. J. R. Astr. Soc.* **59**, 379–387.

Sammis, C.G., 1976. On the effect of polymorphic phase boundaries on vertical and horizontal motions in the earth's mantle, *Tectonophys.* **35**, 169–182.

Sammis, C. G., Dein, J. L., 1974. On the possibility of transformational superplasticity in the Earth's mantle, *J. Geophys. Res.* **79**, 2961–2965.

Sammis, C.G., Smith, J.C., Schubert, G., Yuen, D.A., 1977. Viscosity-depth profile of the earth's mantle: Effects of polymorphic phase transitions, *J. Geophys. Res.* **82**, 3747–3761.

Schubert, G., Yuen, D.A., Turcotte, D.L., 1975. Role of phase transitions in a dynamic mantle, *Geophys. J. R. Astr. Soc.* **42**, 705–735.

Schubert, G., Stevenson, D., Cassen, P., 1980. Whole planet cooling and the radiogenic heat source content of the earth and moon, *J. Geophys. Res.* **85**, 2531–2538.

Schubert, G., Anderson, C.A., 1985. Finite element calculations of very high Rayleigh number thermal convection, *Geophys. J. R. Astr. Soc.* **80**, 575–601.

Sclater, J.G., Jaupart, C., Galson, D., 1980. The heat flow through the oceanic and continental crust and the heat loss of the earth, *Rev. Geophys. Space Phys.* **18**, 269–311.

Sharpe, H.N., Peltier, W.R., 1979. A thermal history model of the earth based on parameterized convection, *Geophys. J. R. Astr. Soc.* **59**, 171–205.

Spohn, T., Schubert, G., 1982. Modes of mantle convection and the removal of heat from the Earth's interior, *J. Geophys. Res.* **87**, 4682–4696.

Stocker, R.L., Ashby, M.F., 1973. On the rheology of the upper mantle, *Rev. Geophys. Space Phys.* **11**, 391–426.

Taylor, S., 1979. Lunar and terrestrial uranium abundances: Implications for the fission hypothesis, in: *Proc. Lunar Planet. Sci. Conf. 10th, Geochim. Cosmochim. Acta*, Suppl 11, 2017–2030.

Torrance, K.E., Turcotte, D.L., 1971. Thermal convection with large viscosity variations, *J. Fluid Mech.* **47**, 113–125.

Turcotte, D.L., 1980. On the thermal evolution of the earth, *Earth Planet. Sci. Lett.* **48**, 53–58.

Ullrich, W., van der Voo, R., 1982. Minimum continental velocities with respect to the pole since the Archaean, *Tectonophys.* **74**, 17–27.

Walcott, R. I., 1970. Flexural rigidity, thickness and viscosity of the lithosphere, *J. Geophys. Res.* **75**, 3941–3954.

Wänke, H., 1981. Constitution of the terrestrial planets, *Phil. Trans. R. Soc. London A* **303**, 287–302.
Weertman, J., 1970. The creep strength of the earth's mantle, *Rev. Geophys. Space Phys.* **8**, 145–168.
Weertman, J., Weertman, J.R., 1975. High temperature creep of rock and mantle viscosity, *Ann. Rev. Earth Planet. Sci.* **3**, 293–316.
White, W.M., 1985. The sources of oceanic basalts: The isotopic evidence, *Geology* **13**, 115–118.
Yuen, D.A. Peltier, W.R., 1980. Mantle plumes and the thermal instability of the D″-layer, *Geophys. Res. Lett.* **7**, 625–628.
Yuen, D.A., Sabadini, R., Boschi, E., 1983. Mantle rheology from a geodynamical standpoint, *Riv. Nuovo Cim.* **5**, 1–35.

CHAPTER 9

Constraints on the Structure of Mantle Convection Using Seismic Observations, Flow Models, and the Geoid

Bradford H. Hager and Robert W. Clayton

Seismological Laboratory, California Institute of Technology
Pasadena, California 91125, U.S.A.

1. Introduction

The establishment of the theory of plate tectonics in the late 1960s has left little doubt that the mantle is convecting. The plates themselves form the cold upper thermal boundary layer of the mantle convection system; the cooling of oceanic plates as they move away from midoceanic ridges provides the mechanism by which the Earth loses most of its heat (e.g., Sclater *et al.*, 1980; O'Connell and Hager, 1980). The mantle is in turn cooled by the cold slabs that plunge into Earth's interior at subduction zones.

Although plate tectonics implies that convective motions in the mantle are the dominant mechanism for heat transport, and we can measure the surface motions associated with it, we are remarkably ignorant of even the gross features of the interior flow field associated with this mantle circulation. Only at subduction zones, where seismicity presumably marks the particle trajectories of the cold descending boundary layer, do we have direct evidence for the interior flow pattern and state of stress. Most of what is understood, or thought to be understood, about convection in the Earth's interior is based on comparison of simplified models to observations taken at the surface.

Examples of these models of mantle convection are given in the other chapters of this book, as well as in the general geophysical literature. These include studies of convection in media with uniform rheology (Busse, this volume; Jarvis and Peltier, this volume), interpretation of travel time anomalies from deep earthquakes in terms of simple thermal models of subducted slabs (Jordan *et al.*, this volume), interpretation of geochemical anomalies in terms of models of the distribution of mantle heterogeneities (Hart and Zindler, this volume), and interpretation of changes in the Earth's shape and rotational parameters in terms of models of mantle rheology (Peltier, this volume).

In order to be useful, models must be simple enough to understand, and yet contain enough of the essential physics to be applicable. The line between oversimplification and overwhelming complexity is a fine one, and its positioning is a matter of subjective judgement, particularly when some observations have a fairly small signal to noise ratio. The ultimate test of a particular model is whether it can satisfy, within their uncertainties, the observations. If it cannot it must be rejected, although unfortunately, the converse is not true. The more types of observations a model can satisfy, however, the more likely it is to be correct.

The main emphasis in this paper is on the long-wavelength nonhydrostatic geoid, shown in Figure 9.1, which provides a fundamental observational constraint on models of mantle convection. This figure shows the geoid for degree and order 2–9 determined by satellite geodesy, (GEM-L2, Lerch *et al.*, 1983) with the hydrostatic oblateness resulting from the spinning of the Earth removed (Nakiboglu, 1982). The geoid is dominated by polar lows (the most prominent one over the Ross Sea south of New Zealand is −120m), and equatorial highs (the major one over New Guinea is +100 m). Note that this representation of the geoid is different from that commonly shown in textbooks, with the major low just south of India. This is because Figure 9.1 has the hydrostatic oblateness, rather than the total oblateness removed. The nonhydrostatic oblateness is of tectonic significance and provides a useful signal; there is no reason to ignore it—it should be treated the same as any other component of the field.

When viewed in the context of mantle convection and plate tectonics, the nonhydrostatic geoid presents some apparent paradoxes. One might expect

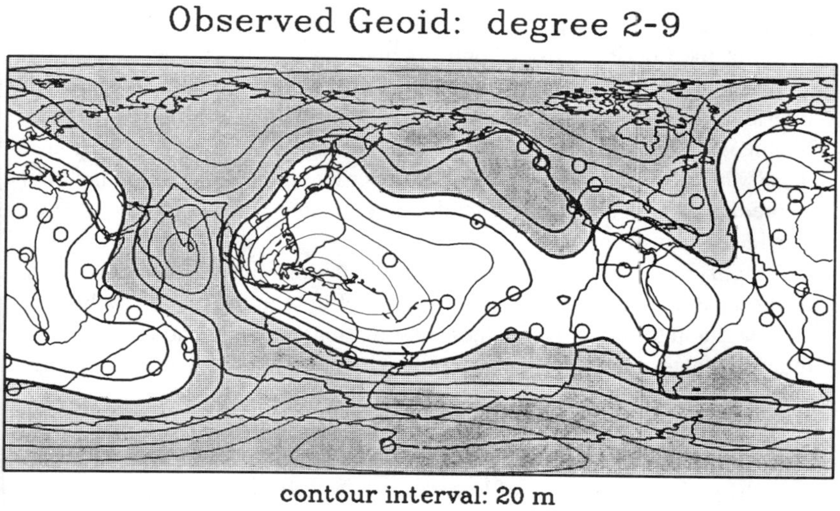

Figure 9.1. The observed long-wavelength geoid (Lerch *et al.*, 1983) referred to the hydrostatic figure of the Earth (f = 1/299.63), with plate boundaries and hotspots indicated. The contour interval is 20 m and geoid lows are shaded. Cylindrical equidistant projection.

that, since plate tectonics is the surface manifestation of mantle convection, the geoid would be dominated by features related to plate boundaries. Although subduction zones are always associated with relative highs of the long-wavelength geoid (e.g., Kaula, 1972; Chase, 1979; Crough and Jurdy, 1980; Hager, 1984), there is no very obvious association of long-wavelength geoid features with ridges. In the wavelength range shown in Figure 9.1, the slow-spreading Mid-Atlantic and Atlantic-Indian ridges are mainly associated with relative geoid highs, while the faster spreading South Indian Ridge, Albatross Cordillera, and East Pacific Rise are associated with relative geoid lows of variable amplitude. Quantitative correlations show that the rather weak correlation of plate velocities with the geoid (Hager and O'Connell, 1978; Kaula, 1980; Peltier and Forte, 1984) is due to the strong correlation of the convergence rate at subduction zones with the geoid (Hager, 1982). The spreading velocity at ridges, however, has no significant correlation (i.e., at greater than the 95 percent confidence level) with the geoid (Richards and Hager, 1988).

Furthermore, the large geoid highs over Africa and the central Pacific are associated with the major hotspot provinces there (Chase, 1979; Crough and Jurdy, 1980). It seems surprising that geoid highs are associated both with high density, cold subducting slabs and low density hotspots. This apparent paradox has up to now prevented quantitative interpretation of the geoid in terms of the interior density contrasts associated with mantle convection.

We describe in this chapter an approach to modeling the state of the Earth's interior that combines fluid mechanical models of slow, creeping flow in the mantle with models derived from seismic tomography of the distribution of density heterogeneity in the lower and upper mantle. These models are used to predict the long wavelength geoid; comparison with the observed geoid provides a test of the models.

Simplifying assumptions are made in calculating both the mantle structure using seismic tomography and the models of flow in the mantle. While we cannot guarantee *a priori* that these assumptions will not lead us astray, the independent test provided by the geoid gives us confidence that we are nearly on the right track.

As we show later, most of the variations in the Earth's longest-wavelength geoid are ultimately caused by density contrasts in the lower mantle. Mapping the global distribution of lower mantle heterogeneity using travel time anomalies for direct P wave arrivals has been the subject of several recent studies (Dziewonski *et al.*, 1977; Dziewonski, 1984; Clayton and Comer, 1983, 1984). In the next section we describe a particular algorithm, SIRT, that we are presently using for body wave tomography and present the results of an inversion for lower mantle structure (Clayton and Comer, 1984). We discuss the reliability of these results and their relation to what would be expected from the simple models of mantle convection presented in other chapters of this book. We also calculate the geoid anomalies predicted from a density field inferred from the model of mantle heterogeneity, assuming a static Earth.

The geoid predicted assuming a nonconvecting mantle matches the observed geoid reasonably well in pattern, but it is of opposite sign and larger in amplitude. This points out the need to consider the flow that would be induced over geologic time by the density variations in the mantle. This flow causes dynamic topography both at the Earth's surface and at the core–mantle boundary. The mass anomalies associated with this dynamic topography contribute to the geoid a signal comparable in magnitude and opposite in sign to the observed geoid and can lead to a reversal in sign of the total geoid anomaly. Thus accurate calculation of the geoid associated with a given distribution of density anomalies in a dynamic planet requires consideration of the effects of dynamically maintained topography. This dynamic topography is difficult to observe directly, but it can be calculated using fluid mechanical models.

In the fourth part of this chapter we develop the formalism for calculating dynamic topography and geoid anomalies in a viscous planet. We then calculate the predicted geoids for a number of simple Earth models and compare them to the observed geoid. We also include in these models upper-mantle density heterogeneities inferred from a thermal model of subducted slabs and a model of upper mantle heterogeneity based on surface wave tomography (Tanimoto, 1986). We find that we can explain ~90 percent of the variance in the long wavelength geoid using these models.

In the final sections we discuss the implications, as well as the limitations of these models. We end the chapter by suggesting future work that might prove productive in improving our understanding of mantle convection.

2. Travel Time Tomography of the Lower Mantle

The determination of lateral velocity variations is based on the travel times of seismic waves that propagate through the Earth's interior. The inversion of the travel time data is done with a technique that is now popularly known as tomography. This technique was pioneered in medical X-ray imaging (Gordon *et al.*, 1970; Jaffe, 1982), and in this section we show its adaptation and application to the seismic travel time problem.

The use of travel times to determine lateral variations in velocity has been investigated by several researchers. Aki *et al.* (1977) used a generalized inverse approach to determine the local structure beneath a seismic array. Dines and Lytle (1979) proposed an iterative inversion method for cross-borehole studies. Fawcett and Clayton (1984) have analyzed the reconstruction of velocity anomalies in the presence of an unknown velocity gradient. In Humphreys *et al.* (1984), an iterative Fourier method was used to map velocity anomalies in the mantle beneath Southern California. Hearn and Clayton (1986a, 1986b) combined the statics estimation problem and velocity inversion in an iterative procedure to determine the lateral variations in P_g and P_n velocities in Southern California. The tomographic algorithm presented in this paper has also been used on a small scale study of the Coso geothermal region (Walck and Clayton, 1987).

The lateral variations in lower mantle velocities have also been investigated by Dziewonski *et al.* (1977), Sengupta and Toksoz (1976), and Sengupta et al. (1981). These studies used a generalized inverse approach with a limited number of blocks spanning the mantle. More recently, Dziewonski (1984) has given an estimation of the lower-order spherical harmonic variations in mantle velocity. This study will be compared to the one presented here.

In this section, a simple, iterative inversion scheme is presented for determining lateral variations in seismic velocity from the observations of anomalous travel times. In the first part, a brief description of the method is given. In the next three parts, the data and necessary corrections are presented. The results, an error analysis, and a discussion conclude the section.

Tomographic Imaging

The travel time of a ray connecting a source at x and a receiver at x' is given by the deceptively simple equation

$$t(x, x') = \int_x^{x'} s(l)dl\,(x, x', s), \qquad (2.1)$$

where s is the slowness function (inverse of velocity), and l is the location of the ray path, which is a function of the slowness s and the ray path. This equation is nonlinear in s, since it appears as the integrand as well as in the ray path function.

For small variations in slowness, equation (2.1) can be linearized to

$$\Delta t(x, x') = \int_x^{x'} \Delta s(l)dl\,(x, x', s_0) + 0(\Sigma\Delta s^2), \qquad (2.2)$$

where Δs is the slowness perturbation, and s_0 is the reference slowness model. The travel time deviations (Δt) are the observed travel times minus those predicted by the reference slowness model. The linearization in equation (2.2) is discussed in Aki *et al.* (1977) and Fawcett and Clayton (1984). It is valid for $\Delta s/s << 1$, and for rays not near caustics. The basic assumption of equation (2.2) is that to second-order, velocity variations affect only the travel time and not the ray path. Near caustics, however, small velocity variations will cause large departures in the raypath, and the linearization may break down. Note that this equation lends itself to an iterative solution for the nonlinear ray path effects. However, its convergence properties are not generally known. For slowness variations that are less than 10 percent, numerical tests indicate that the nonlinear effects are negligible.

To apply equation (2.2) it is necessary to cast Δs as a parameterized model, and there are a number of ways of doing this. The simplest is to divide the region of interest (the mantle in the case of this chapter) into a number of blocks. Another approach would be to expand the Δs function as a harmonic series (spherical harmonics for the Earth), and solve for the coefficients of the expansion. With both procedures the number of parameters has to be decided *a priori*. Too few parameters will lead to under resolution of the image, and too many will lead to excessive model variance due to poor coverage by the rays. The situation is further complicated by the non-uniform coverage provided by the rays. One advantage of a block parameterization is that poorly sampled regions of the model are more easily dealt with than with parameterizations that use global basis functions. One disadvantage is that substantially more parameters are usually required.

In the study presented in this paper, a block parameterization was used, with the number of blocks (that is the block size) chosen by trial and error. The blocks were chosen as the smallest size that still retains a reasonable average number of ray intersections (hits) per block. The image-resolution/model-variance trade-off is then set by *a posteriori* filtering of the image. An alternative approach is provided by Dziewonski (1984) who used a spherical harmonic expansion in latitude and longitude, and a Legendre function expansion in radius. The two methods are compared at the end of this section.

With a block discretization of the Δs function, equation (2.2) can be cast as a linear system,

$$\Delta t_i = \sum_j l_{ij} \Sigma \Delta s_j, \tag{2.3}$$

where Δt_i is the travel time deviation of the ith ray, Δs_j is the slowness perturbation of the jth block and l_{ij} is the ray segment length of the ith ray in the jth block. For convenience, $l_{ij} \equiv 0$ for blocks not crossed by a particular ray.

For most tomographic problems, equation (2.3) is an extremely sparse system, since a single ray usually intersects no more than 1–10 percent of the blocks. Also, in most problems the number of observations is large (very large in some cases). This makes direct least squares inversions of equation (2.3) impractical, and instead, an iterative approach is used.

One of the first methods developed in medical imaging for inverting equation (2.3) is the SIRT (Simultaneous Iterative Reconstruction Technique) (Jaffe, 1982). This method is similar to Jacobi iteration for sparse matrix solutions. To derive this method, equation (2.3) is rewritten as the matrix system

$$\mathbf{L}\Delta\mathbf{s} = \Delta\mathbf{t}, \tag{2.4}$$

where $\mathbf{L} = \{l_{ij}\}$, $\Delta\mathbf{s} = \{\Delta s_j\}$, and $\Delta\mathbf{t} = \{\Delta t_i\}$. Equation (2.4) is then converted to a normal system by premultiplying by $\mathbf{L}^T$.

$$\mathbf{L}^T\mathbf{L}\Delta\mathbf{s} = \mathbf{L}^T\Delta\mathbf{t}. \tag{2.5}$$

To form an iterative set of equations, a matrix $\mathbf{D}$, whose inverse is well-defined, and can be easily computed, is added to equation (2.5)

$$\mathbf{D}\Delta\mathbf{s} = \mathbf{L}^T\Delta\mathbf{t} + (\mathbf{D} - \mathbf{L}^T\mathbf{L})\Delta\mathbf{s}. \tag{2.6}$$

Dividing by $\mathbf{D}$, and defining $\Delta\mathbf{s}^{(n)}$ and $\Delta\mathbf{s}^{(n-1)}$ to be the solution estimates at the nth and $(n-1)$th iterations, respectively, we have

$$\Delta\mathbf{s}^{(n)} = \mathbf{D}^{-1}\mathbf{L}^T\Delta t + (\mathbf{I} - \mathbf{D}^{-1}\mathbf{L}^T\mathbf{L})\Delta\mathbf{s}^{(n-1)}. \tag{2.7}$$

Equation (2.7) is a bootstrap procedure for determining $\Delta\mathbf{s}^{(n)}$ from $\Delta\mathbf{s}^{(n-1)}$. It can be placed in a more convenient form by rearranging the terms

$$\Delta\mathbf{s}^{(n)} = \Delta\mathbf{s}^{(n-1)} + \mathbf{D}^{-1}\mathbf{L}^T(\Delta\mathbf{t} - \Delta\mathbf{t}^{(n-1)}), \tag{2.8}$$

where $\Delta\mathbf{t}^{(n-1)}$ is the solution of the forward problem $\Delta\mathbf{t}^{(n-1)} = \mathbf{L}\Delta\mathbf{s}^{(n-1)}$.

At this point, the matrix $\mathbf{D}$ is unspecified, but it is clear from equation (2.7) that $\mathbf{D}$ should approximate $\mathbf{L}^T\mathbf{L}$. If $\mathbf{D}$ is chosen to be the diagonal of $\mathbf{L}^T\mathbf{L}$, then we have the SIRT reconstruction. That is, we set

$$\mathbf{D} = \{d_{ii}\} = \Big\{\sum_k l_{ik}{}^2\Big\}, \tag{2.9}$$

which allows equation (2.8) to be written as the algorithm

$$n_k = \sum_i l_{ik}\Delta t_i, \tag{2.10a}$$

$$d_k = \sum_i l_{ik}^2, \tag{2.10b}$$

and

$$s_k^{(n)} = s_k^{(n-1)} + \frac{n_k}{\mu + d_k}, \tag{2.10c}$$

where μ is a damping parameter. The numerator n_k, and denominator d_k, are calculated in one sequential pass through the data. After n_k and d_k have been determined, the entire slowness model is updated simultaneously using equation (2.10.c). The damping parameter, μ, is used to attenuate the solution in blocks that are poorly sampled. Note that d_k need only be calculated once.

For a single ray, with no damping, (2.10c) attributes the entire travel time anomaly to each block traversed by the ray, tending to overestimate the amplitude of the slowness anomaly. It was found through experimentation that a similar algorithm (termed modified SIRT) that conserves travel time anomalies along rays (in the undamped case), and that results in smaller variations in poorly sampled regions, converged more rapidly when applied to seismic data:

$$n_k = \sum_i l_{ik} \Delta t_i / L_i, \tag{2.11a}$$

$$d_k = \sum_i l_{ik}, \tag{2.11b}$$

and

$$s_k^{(n)} = s_k^{(n-1)} + \frac{n_k}{\mu + d_k}. \tag{2.11c}$$

Here L_i is the total length of the ith ray. It is proved in Comer and Clayton (1989) that this algorithm is absolutely convergent to the generalized inverse (Aki and Richards, 1980) of equation (2.4) for $\mu > 0$. We typically carried out 50 iterations of equation (2.11).

If in equation (2.8), we had chosen $\mathbf{D}$ to be either the upper or lower triangular portion of $\mathbf{L}^T\mathbf{L}$, then the inversion scheme would be the equivalent of the Algebraic Reconstruction Technique (ART) (Herman *et al.*, 1973) that is used in medical X-ray tomography. The ART can be implemented in the same fashion as SIRT, but its convergence depends on the ordering of the data. Usually, the observations are ordered in a highly structured manner, with each event followed by all the data associated with that event. It is time consuming to cycle through the data in any order other than forward, sequentially. For this reason it was decided to use the modified SIRT, even though ART could potentially converge more rapidly. The SIRT also appears to be much more easily adaptable to parallel computers than ART or other techniques.

The Observations

The data used in this study are the P-wave travel time residuals as reported by the International Seismic Centre (ISC) for the decade of 1970. The raw data were winnowed using several overlapping criteria.

1. The shortest distance between source and receiver was set at 25°. This avoids the travel time mispicks associated with the 670 km seismic discontinuity which produces a travel time triplication in the 18°–25° range. This also prevents the tomographic imaging from being applied to rays near caustics.
2. The largest source-receiver distance was set at 95° to avoid P-waves diffracted by the core.
3. A lower bound on event magnitude of $m_b = 4.3$ was used to eliminate most poorly recorded earthquakes.
4. Each station used in this study was required to record at least 50 sources, and each source had to be recorded by at least 50 stations. The restriction was applied to avoid poorly located sources, and to make the statics estimates (discussed below) robust.
5. Any pick having a travel time residual greater than 5 sec was assumed to be erratic and was removed.

The application of these criteria left some 25,000 earthquakes, 1260 stations, and 1.7×10^6 picks in the data set. The reference velocity model used in the study is the Jeffreys–Bullen (JB) model (Jeffreys, 1936). The residuals relative to the JB model are shown in Figure 9.2, as a function of distance; the residuals in the 50–75° range are on average negative, while those beyond 70° are positive. This indicates that the JB model is too fast in the mid-mantle and too slow in the lowest part of the mantle. In Figure 9.3 the distribution of all the travel time residuals is shown. The overall variance of the observations is 1.3 seconds.

The travel time residuals as reported by the ISC are corrected for the ellipticity of the Earth using the correction formula of Bullen (1937). These corrections can differ by as much as ±¼ sec from the more accurate ellipticity corrections given by Dziewonski and Gilbert (1976). Logically, and most efficiently, the more accurate corrections should be used at this point. However, in this study the new corrections were made after the fact by

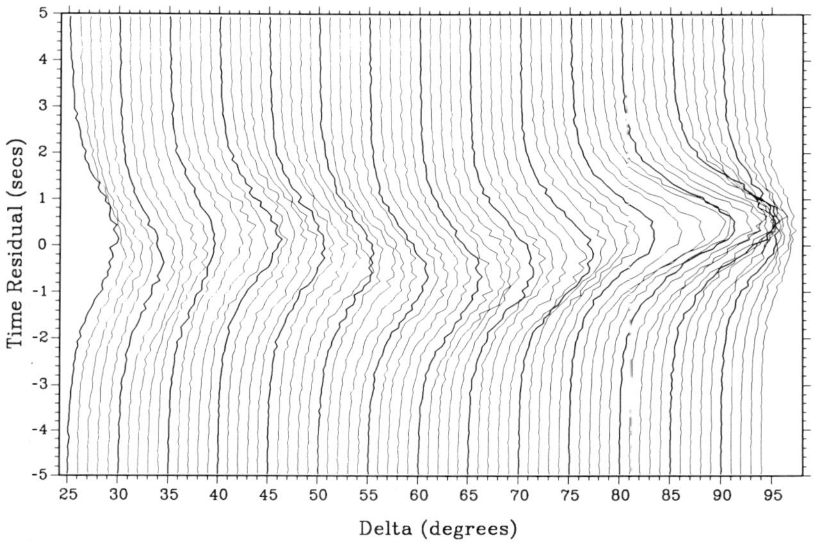

Figure 9.2. Travel time residuals versus distance. The JB travel time residuals are shown for the distance range 25° to 95°. The distributions for each one degree interval are shown, plotted side by side.

separately inverting the difference between the Bullen and the Dziewonski-Gilbert corrections and adding the result to the solution. Since the overall system is linear this is equivalent to but less efficient than correcting the travel times *a priori*.

Part of the scatter in the residuals is almost certainly due to picking errors (wrong phase identified, not picking the initial onset, and picks associated with the wrong earthquake). Another source of error is earthquake location. Since the travel times themselves are used to locate the event, biases in the distribution of stations and in the reference velocity model can affect the locations, and consequently, the residuals. Statics corrections, which are discussed in the next section, help remove this source of error. The source static is, in fact, a first-order correction to the event origin time.

The restriction that only data beyond 25° are used leads to poor ray coverage in the upper mantle. This poor coverage will in turn cause upper mantle anomalies to be spread into the lower mantle. To help prevent this, source and receiver statics are removed from the data before the inversion. This technique is commonly used in exploration seismology.

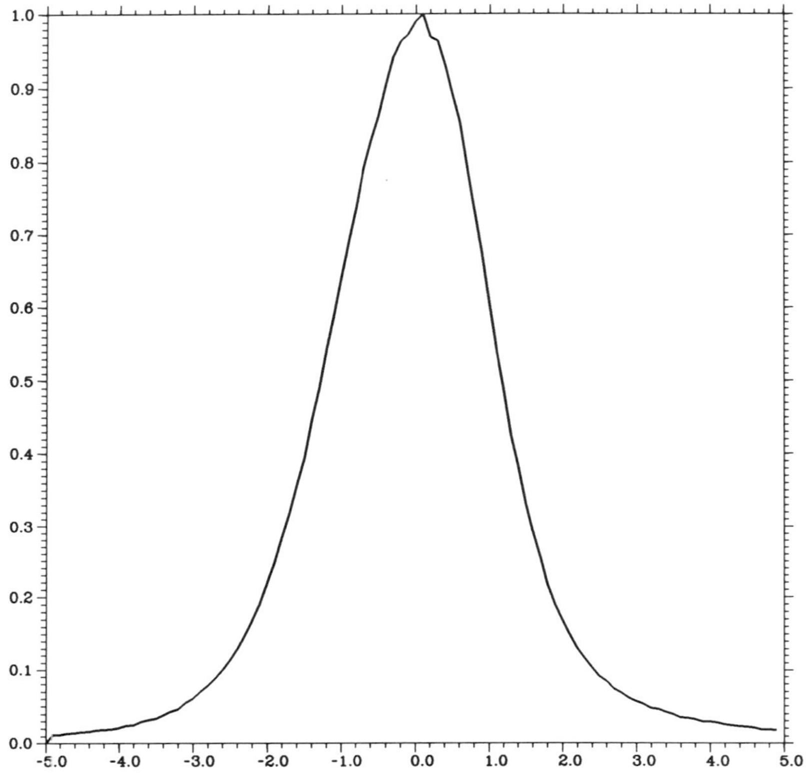

Figure 9.3. Histogram of JB travel time residuals.

Source and Receiver Statics

The misestimation of the origin time of an earthquake will add a bias to the residuals of all stations reporting that event. Depth errors, and near source velocity inhomogeneities, will have similar effects. To minimize this problem, a source static, which is simply the average of all of its residuals, is removed. Similarly, velocity anomalies near a receiver will cause a bias in the residuals that that particular station reports. A station static (the average of all residuals reported by the station) is also removed. In this case, however, since the station is presumably correctly located (a questionable assumption in some cases), the statics can generally be attributed to velocity variations in the crust and upper mantle.

The source and receiver statics are calculated iteratively. If the residuals are thought of as a matrix, with one column for each source and one row for each receiver, then calculating the statics amounts to removing the principle row and column eigenvectors from the matrix. This is done iteratively by first estimating and removing the average row vector (source statics), followed by estimating and removing the average column vector (station statics). For the ISC data, this took 5 iterations to converge, and had the net effect of reducing the variance of the travel time residuals by 20 percent.

The source statics for the ISC P-wave data set are very small (<0.1 sec in RMS amplitude). This simply reflects the fact that the source location and origin times are determined by a least-squares method that seeks to minimize the residuals at each station. The receiver statics, however, range up to ± 2.0 sec in size. The station statics for the entire world are shown in Figure 9.4. This plot also serves to show the station distribution used in this study. More detailed plots of the statics for North America, Europe, and Japan are shown in Figures 9.5, 9.6, and 9.7, respectively. In overall pattern, the statics match generally accepted crust/upper mantle tectonic features. For example, the mid-continental platform area of the USA shows up as fast station arrivals, in contrast with the slower arrivals that are seen in the tectonic regions of the western U.S.A. In Europe, the deep roots of the Alps and Calabrean subduction zone appear as fast residuals, along with the Scandinavian Shield. In Japan, stations over the subducted slab have faster arrivals than the nonslab stations.

One issue on which this study differs from that of Dziewonski (1984) is in the complexity of the static terms. In this study, only isotropic corrections are used. In Dziewonski (1984), isotropic and first-order anisotropic terms, reported in Dziewonski and Anderson (1983) are used. With perfectly uniform ray coverage, neither correction would be necessary. For limited and usually nonisotropic ray coverage, it is debatable which one should be used. In this study, only the isotropic corrections are used because their effect on the velocity inversion is more straightforward to understand. The static solution will generally remove upper mantle variations (primarily subducted slabs) from the inversion. Anisotropic corrections may also remove lower mantle variations in poorly sampled regions.

Composite Rays

The irregular distribution of the sources and receivers used in this study causes some regions of the mantle to be sampled by many redundant rays. While these contribute to the statistical robustness of the inversion, they do

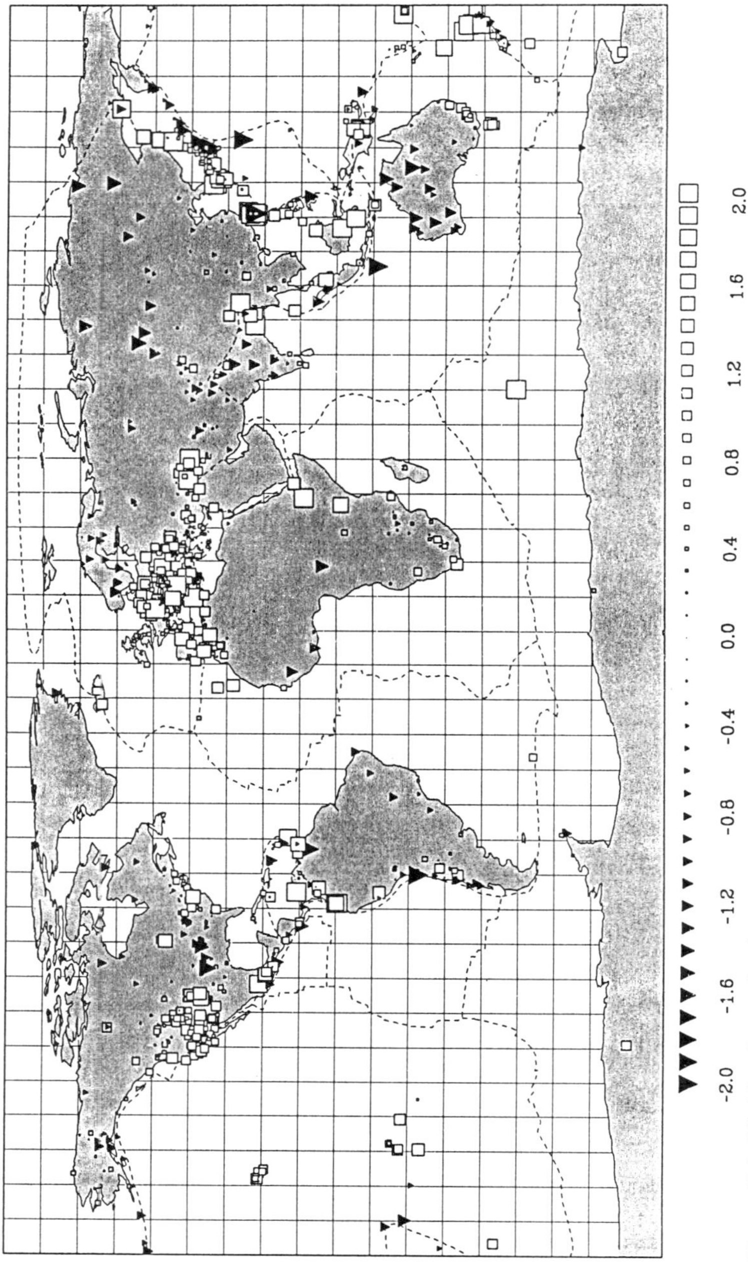

Figure 9.4. World station statics, in seconds. Positive statics mean slow stations. This figure also shows the distribution of stations used in this study.

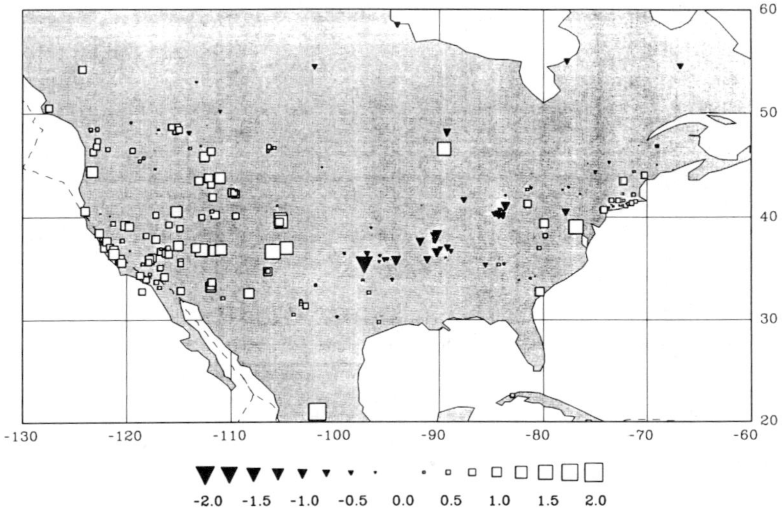

Figure 9.5. Station statics for North America, in seconds. The slow statics (positive) in the west correspond to the tectonically active region, while the fast statics (negative) in the mid-continental region correspond to the shield.

not add to its resolution. In terms of generalized inverse theory (Aki and Richards, 1980), these redundant rays form part of the observational null space.

To remove this type of redundancy, a set of composite or average rays are formed. To do this the Earth's surface is divided into 1676 approximately equal area cells that are 5° by 5° at the equator. The distribution of cells is shown in Figure 9.8. The travel time residuals of all rays that travel from sources within the *i*th cell to receivers in the *j*th cell are averaged together to form a single composite ray for the *i*-*j*th cell combination. Note that this step is done after the statics are removed to avoid intracell aliasing. The composite rays are formed separately for 11 source depth ranges spanning from 0 km to 700 km. The cell size used here is the same as the one used in the slowness inversion presented in the next section.

The formation of composite rays reduces the number of rays to 110,000 and reduces the relative travel time variance by a further 10 percent. A more profound effect is that it reduces the computer time from 2 CPU-days/iteration to 20 CPU-mins/iteration on a VAX 11-780 minicomputer. This allows much more flexibility in exploring the resolution and error aspects of the model.

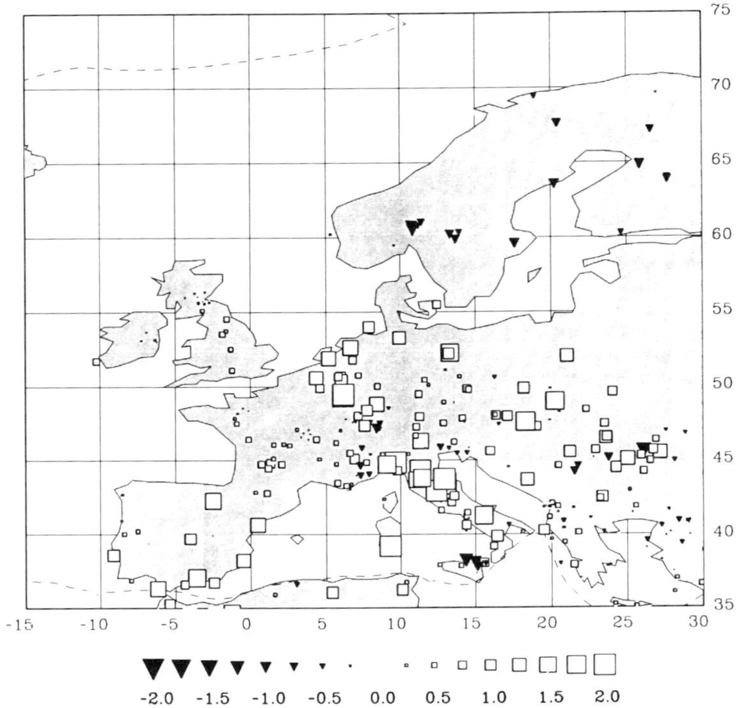

Figure 9.6. Station statics for Europe. The bulk of the European stations have positive statics, except for a few stations associated with the Scandinavian shield, the deep root under the Alps, and the region of deep seismicity beneath Sicily.

Results for the Lower Mantle

After the raw travel time residuals have been calculated from the JB model, the static corrections done, and the composite rays formed, the tomographic inversion technique is then applied. In this study, the modified SIRT algorithm given by equation (2.11) is used. The ray paths were determined by interpolation from a set of basis rays as described in Comer (1984).

The slowness model was divided into 48,604 blocks equally distributed amongst 29 shells that span the entire mantle from the core–mantle boundary (CMB) to the Earth's surface. Each shell is 100 km thick, and has its blocks distributed as shown in Figure 9.8. The approximate size of the blocks is 550 km × 550 km × 100 km at the Earth's surface, and is 300 km × 100 km at the CMB. The large aspect ratio was chosen in order to have sufficiently fine

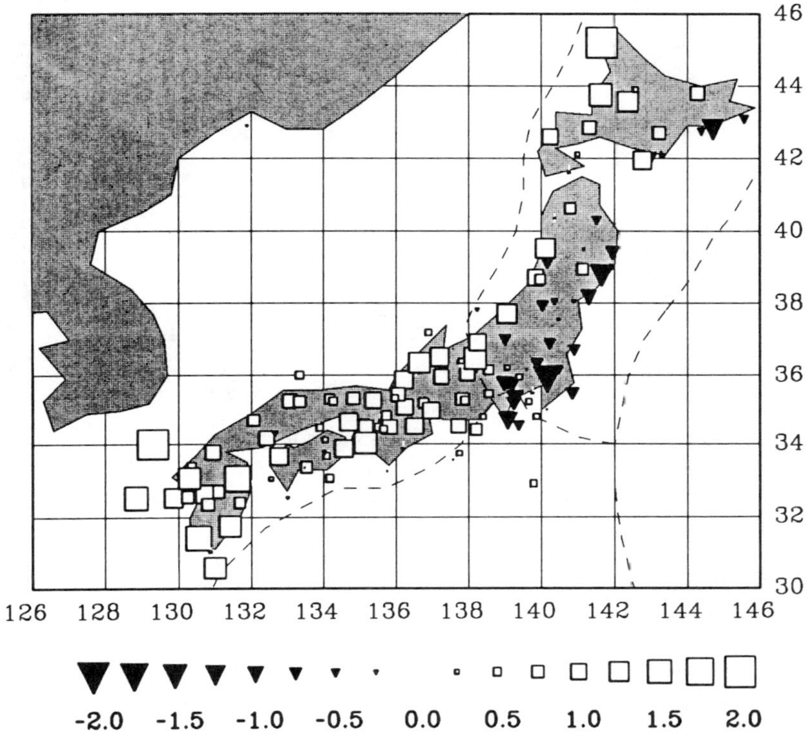

Figure 9.7. Station statics for Japan. The negative statics mark the fast slab in the upper mantle, while the rest of Japanese stations have positive statics.

radial spacing to avoid problems resulting from errors in the reference JB model (Figure 9.2) but still keep the blocks large enough to have a reasonable number of hits.

The coverage provided by the ISC data is surprisingly good in the mid-mantle range. Almost all blocks in the 800 km to 2800 km depth range are hit by 10 rays or more. In the upper mantle, the coverage is poor in some regions, particularly in the Pacific, because the rays are nearly vertical as they leave the sources or arrive at the receivers. Near the CMB, the coverage is marginal in the southern hemisphere, and particularly poor at the South Pole. Displays of the hit counts (number of hits per block) are given in Comer and Clayton (1989).

The solution was iterated using equation (2.11) until the relative changes

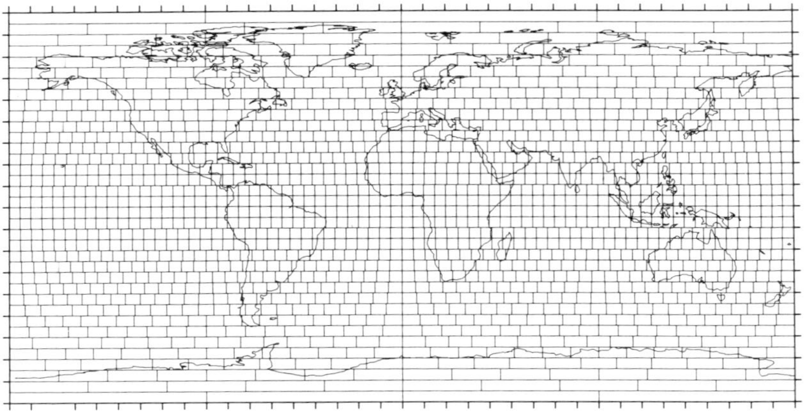

Figure 9.8. The surface distribution of blocks used in the composite ray averaging and the tomographic inversion is shown. In all, 1676 blocks are used to cover the surface in an approximately equal-area manner. Cylindrical equidistant projection.

between $\Delta\mathbf{s}^{(n)}$ and $\Delta\mathbf{s}^{(n-1)}$ became small (less than 10^{-3}). This required approximately 50 iterations. The variance of the travel-time residuals was further reduced by 36 percent by the inversion. Variance reduction is discussed further below.

The results of the inversion show distinct trends with depth. The average slowness perturbation in each radial shell is shown in Figure 9.9. The slowness perturbations are generally negative (faster velocities) in the mid-mantle and positive (slower velocities) in the lowest part of the mantle. This trend is almost certainly due to inadequacies of the reference slowness model (the JB model) itself. The erratic behavior near the surface reflects the poor sampling in the upper mantle. The result can be thought of as an overall correction (or update) that needs to be removed from the reference model. The correction is small in absolute terms (~0.1 percent).

Ideally, the reference model correction should be subtracted from the JB model, and the entire inversion process (including statics and composite rays) repeated. This will have the effect of slightly altering the raypaths, and reducing the distance bias in the residuals shown in Figure 9.2. If this process were iterated the result should be a better globally-averaged velocity model, and a set of slowness perturbations that have zero mean in radial shells. This step has not as yet been done. Instead, the shell means have been simply removed from the solution. Tests have shown that errors due to the slight mislocations of the raypaths only lead to slight mislocation of the anomalies

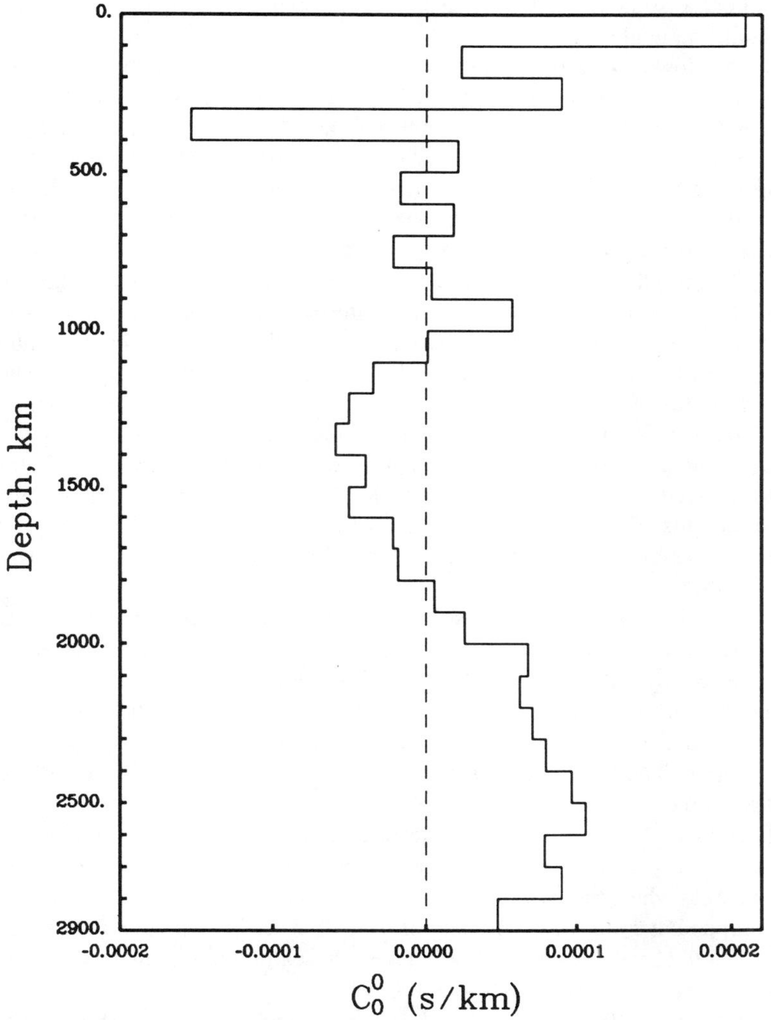

Figure 9.9. Layer averages of the model. The average slowness for each layer in the inversion is shown. This function is a correction to the J-B velocity model. The layer averages are removed from all of the following displays of the inversion.

and cause some reduction in resolution. At long wavelengths, these errors should be negligible.

The slowness perturbations (after the layer averages are moved) have a relative maximum amplitude of ± 1 percent. The largest anomalies occur near the CMB with the maximum anomaly beneath the southern tip of Africa. The small size of the perturbation justifies the assumptions that went into linearization of the travel time equation [equations (2.1) and (2.3)].

Displaying the results of a spherical 3D inversion is problematic. No single plot format seems to suit all needs. In this paper, we restrict ourselves to displaying a few representative cross-sections in a halftone display. This type of plot is useful for presenting spatial patterns, but not for measuring values. In all cases but one, the results have been filtered prior to plotting to remove the rapid variations. The rapid variations cannot by adequately displayed with halftones, and they may be substantially contaminated by noise.

To filter the solution, it was expanded to degree and order 12 in spherical harmonics, and then truncated to degree and order 6. The radial direction was filtered with a symmetric 7-layer moving average operator with triangular weights. The filter was shortened near the surface and the CMB. The filtered results are comparable to the solution of Dziewonski (1984) in resolution.

Figures 9.10 through 9.13 show four great-circle cross-sections. Figure 9.10 is unfiltered to show the nature of the raw solution, while the remaining three are filtered as described above. The map displayed above each slice shows its location; the Earth is sliced along the great circle in the center of the map and the resulting annulus split at the antipode and displayed as a rectangle. The line about a quarter of the way down is the 670 km seismic discontinuity. The top of each slice is the Earth's surface, and the bottom is the CMB. There is a 12 to 1 vertical exaggeration in the plots which significantly distorts dipping features. The uniformly shaded regions represent areas where there is no inversion due to a lack of ray coverage. The scale is clipped at a slowness anomaly of 5×10^{-4} sec/km, about a 0.5 percent variation. Using Creager and Jordan's (1986) scaling, this could be caused by a temperature variation of about 100°C.

In Figure 9.10, a slice centered on the Icelandic hot spot is shown. This cross-section is displayed without filtering to show the potential resolution of the inversion. The visual impression of the inversion is dominated by rapid variations in slowness. Some of these features can be matched to surface features, such as the juxtaposition of the slow region in the upper mantle beneath the Icelandic hotspot and the fast region beneath Greenland, or the fast stable platform beneath North America and slow, tectonically active

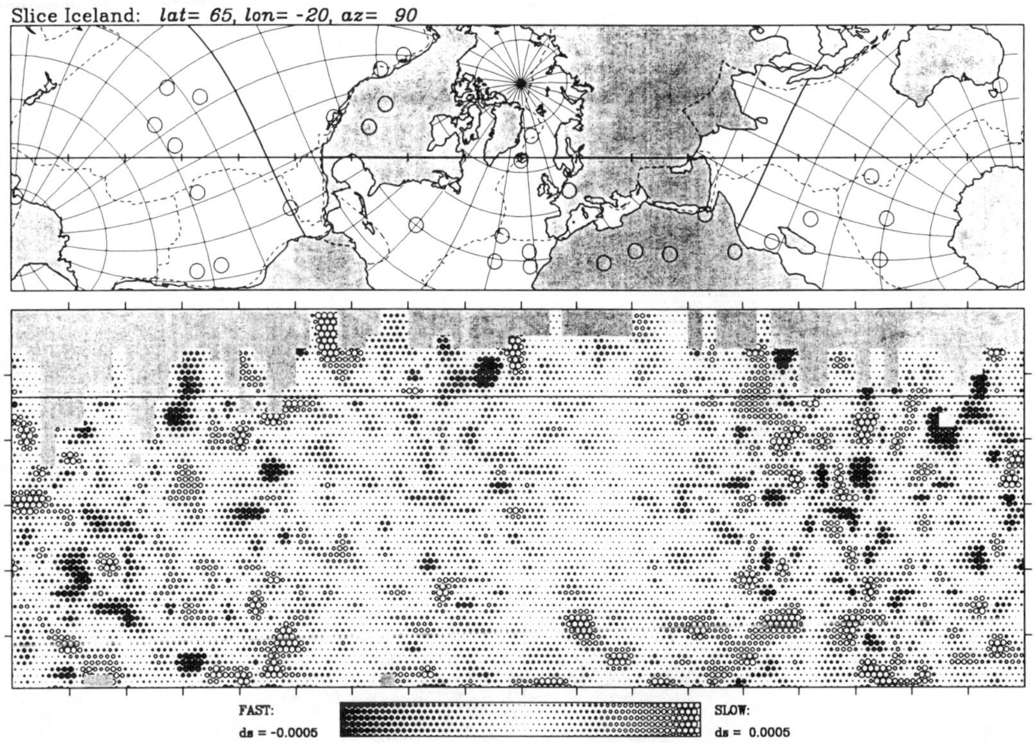

Figure 9.10. Unsmoothed results for a great circle slice passing through Iceland are shown. The numbers at the top show the longitude and coordinates of the center of the slice, and the azimuth of the great circle. The map in the top panel is a projection of the world along this slice. The open circles denote hot spots, and dashed lines mark plate boundaries. The bottom panel shows the results of the inversion from the surface to the core-mantle boundary. The dark circles indicate fast regions, while open circles indicate slow regions.

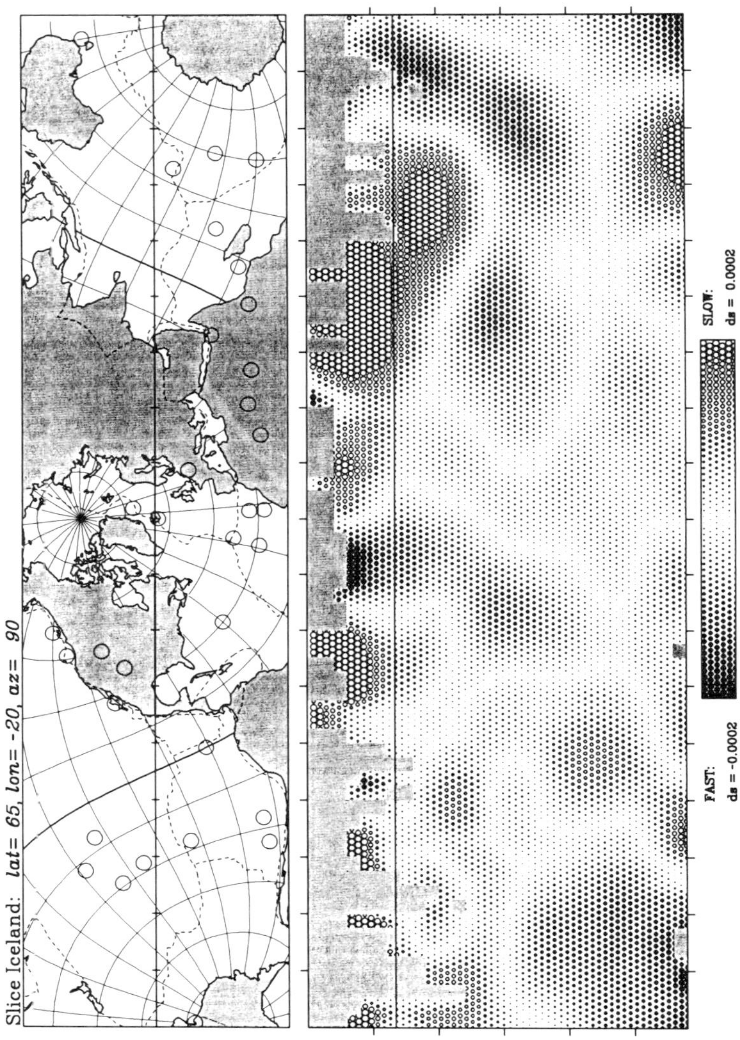

Figure 9.11. Smoothed cross-section through the Icelandic hotspot. The horizontal filtering is low-pass up to harmonic degree 6, while the radial filtering is a simple weighted average of 7 adjacent layers. The format of the figure is described in Figure 9.10.

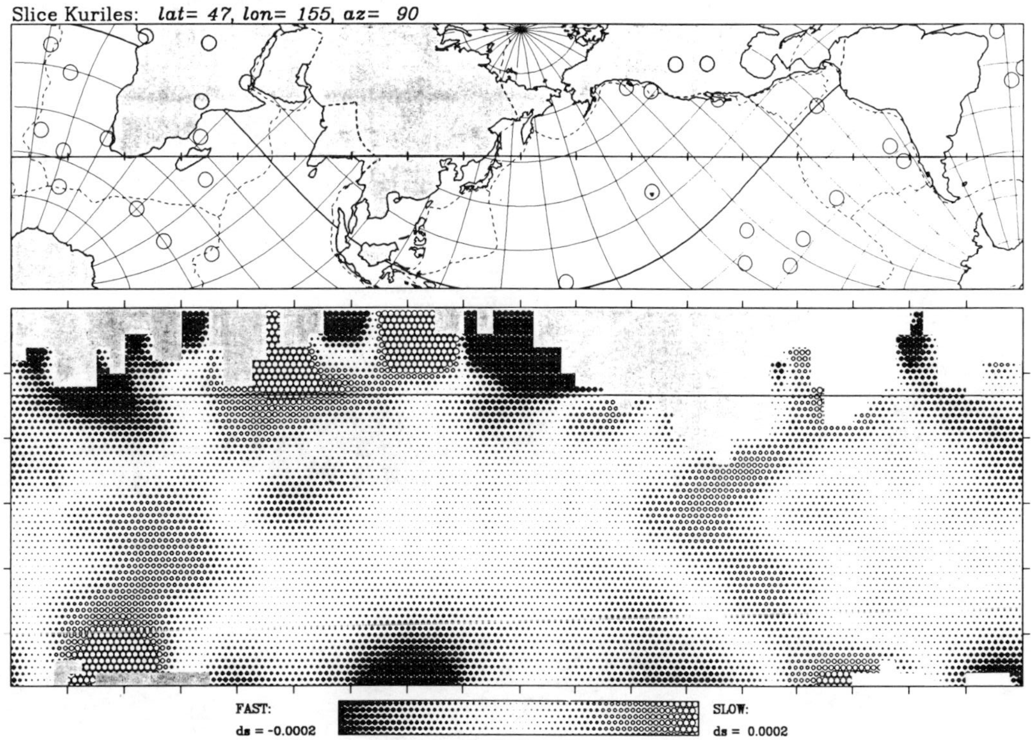

Figure 9.12. Smoothed cross-section through the Kuriles. The filtering is described in Figure 9.11, and the format of the figure is described in Figure 9.10.

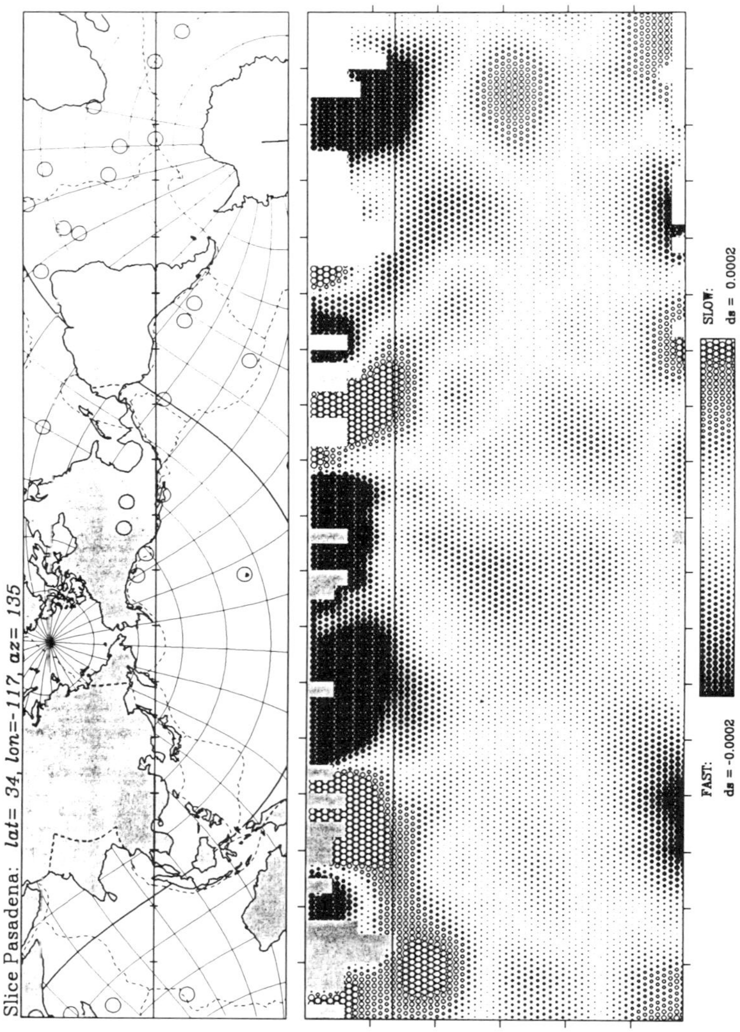

Figure 9.13. Smoothed cross-section slice passing through Pasadena. The filtering is described in Figure 9.11, and the format of the figure is described in Figure 9.10.

Central America. Most, however, have no immediately identifiable association and they may be primarily noise. The long wavelength components of the solution are masked by the rapid variations.

In Figure 9.11, a filtered cross-aection through the Icelandic hot spot is shown. The scale is now clipped at 2×10^{-4} sec/km. Some of the small-scale features shown in Figure 9.10 have coalesced to form smoother, larger scale features. Others have disappeared altogether. In particular, the sharp, discrete, low velocity "plume" beneath Iceland in Figure 9.10 has disappeared. The region beneath Iceland in Figure 9.11 is a zone of transition from fast to slow. Prominent lower mantle features in this figure include slow velocities beneath the mid-Indian Ridge, high velocities beneath the region between Australia and Antarctica, and slow velocities throughout the lower mantle beneath the East Pacific Rise. Note that the filtering process will combine out of plane features with the in plane anomalies. An example of this is the spreading out of the slow Basin and Range anomaly into the central U.S.A.

In Figure 9.12, a slice centered on the Kurile region is shown. The largest anomaly in the section is at the bottom of the mantle beneath southern Africa. There is a fast region at the CMB beneath China. Fast velocities in the upper part of the lower mantle are associated with the subduction zone in the Kurile region, which has been discussed by Jordan (1977), Creager and Jordan (1984), and Jordan *et al.* (this volume). Slow regions are evident in the upper mantle beneath the Carlsberg Rise and beneath China. The zone beneath the Carlsberg Rise appears to have some connection to the slow region at the CMB beneath Africa. Also the East Pacific Rise appears connected to a slow feature at the CMB, although the path is quite sinuous and the dip greatly distorted due to the 12:1 vertical exaggeration. Another feature is the fast region extending from the upper mantle beneath the South American subduction zone to a depth of 1500 km beneath the mid-Atlantic ridge. It is speculated that this is evidence for the subducted slab extending back to the opening of the Atlantic. Poor coverage beneath the central Pacific and near the CMB is evident.

The final cross-section, shown in Figure 9.13, cuts through some of the major subduction zones. In the center of the plot, upper mantle zones of faster velocities are coincident with the major north Pacific subduction zones and seem to extend into the lower mantle. The subduction zones west of Sumatra and S. America are also evident in the upper mantle. The slow zones in the upper mantle are associated with the Kerguelen hot spot, the China tectonic region, and the Galapagos hotspot. Again, coverage of the CMB in the southern hemisphere is relatively poor.

One of the most striking characteristics of these cross-sections in the context of mantle convection is their general lack of radial continuity. Coherent fast or slow regions have generally shallow dips (the 12 to 1 vertical exaggeration makes them appear much steeper than they are); fast features often lie above or below slow ones. The patterns are different than those expected from experience with steady, aspect ratio 1 convection with uniform viscosity.

In Figures 9.14 through 9.17, four maps at depths of 1000 km through 2500 km are shown. Although some features, such as the generally fast polar regions and slow equatorial regions, are fairly continuous with depth, there is substantial radial variation in velocity structure. The slice at 1,000 km depth, Figure 9.14, has major low velocity features in the central Pacific (although the coverage there is weak), beneath the East Pacific Rise, beneath the central Atlantic, from Afar through the central Indian Ocean, and southeast of New Zealand. High amplitude fast anomalies are prominent northwest of Kamchatka, southwest of Australia, and south of Africa and South America. Most hotspots are underlain by slow mantle at this depth. This map is fairly similar to the one presented by Dziewonski (1984), although there are notable differences, particularly in the southern hemisphere, such as beneath the East Pacific Rise and southwest of Australia.

At 1500 km depth, Figure 9.15, the low velocity regions beneath the central Pacific, the central Atlantic, and east Africa are still visible. So are the fast regions south of South America, southwest of Australia, and northwest of the Kuriles. There are some important changes at this depth, however. A prominent low velocity feature now occurs south of Africa and a high velocity anomaly is seen beneath India. A region of high velocity is present beneath North America. Again, most hotspots overlie slow mantle. There are many similarities between this map and that shown by Dziewonski (1984). There are a few prominent differences, such as beneath Hawaii (a region of poor coverage) and the eastern Pacific near the tip of South America and beneath New Zealand.

There are noticeable changes in the pattern at 2000 km depth, Figure 9.16, but in general it is similar to the pattern at 1500 km depth. The most pronounced feature is the low velocity anomaly beneath eastern Africa. Beneath North America and Siberia, the velocities are still high. At this depth, the correlation with Dziewonski's (1984) structure is not as good as it is at shallower depth.

At 2500 km depth, Figure 9.17, the map is dominated by the region of low velocity beneath southern Africa. This is the largest amplitude anomaly in the lower mantle. Other prominent anomalies are the slow region beneath Tonga and the fast region beneath Eurasia. This map is in general fairly

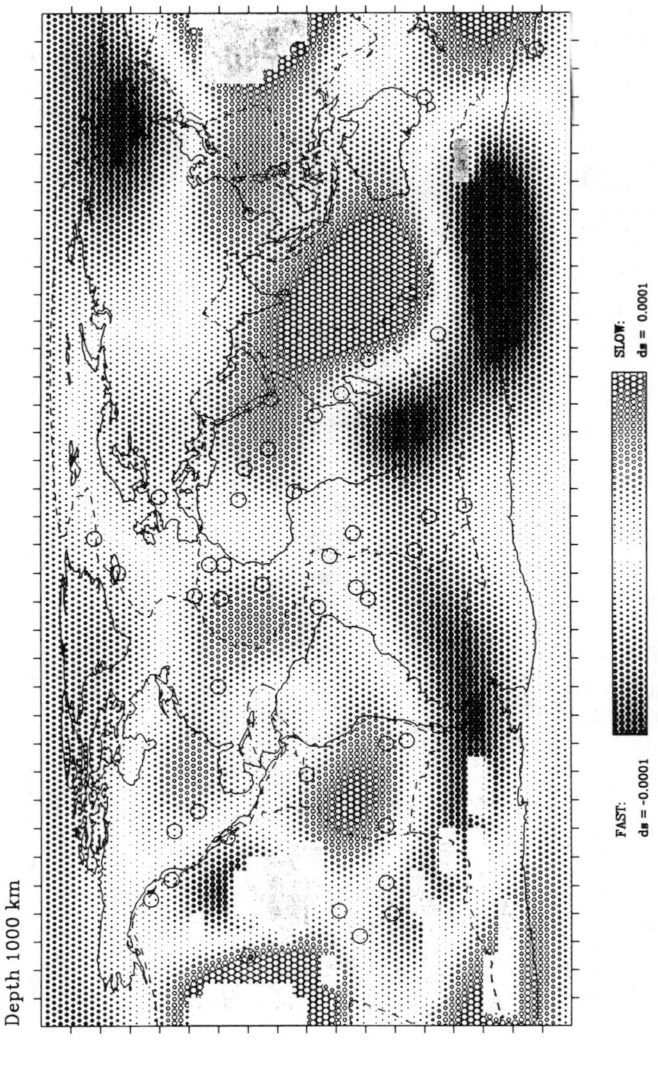

Figure 9.14. Smoothed depth-section at a depth of 1000 km. A world map is superimposed for reference. The filtering is described in Figure 9.11.

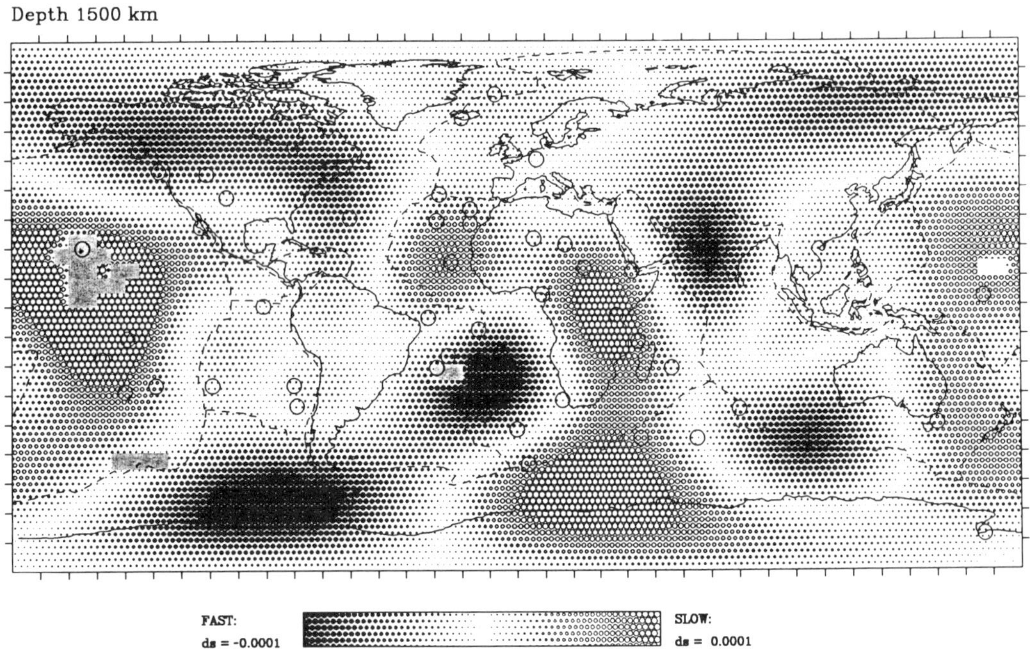

Figure 9.15. Smoothed depth-section at a depth of 1500 km. A world map is superimposed for reference. The filtering is described in Figure 9.11.

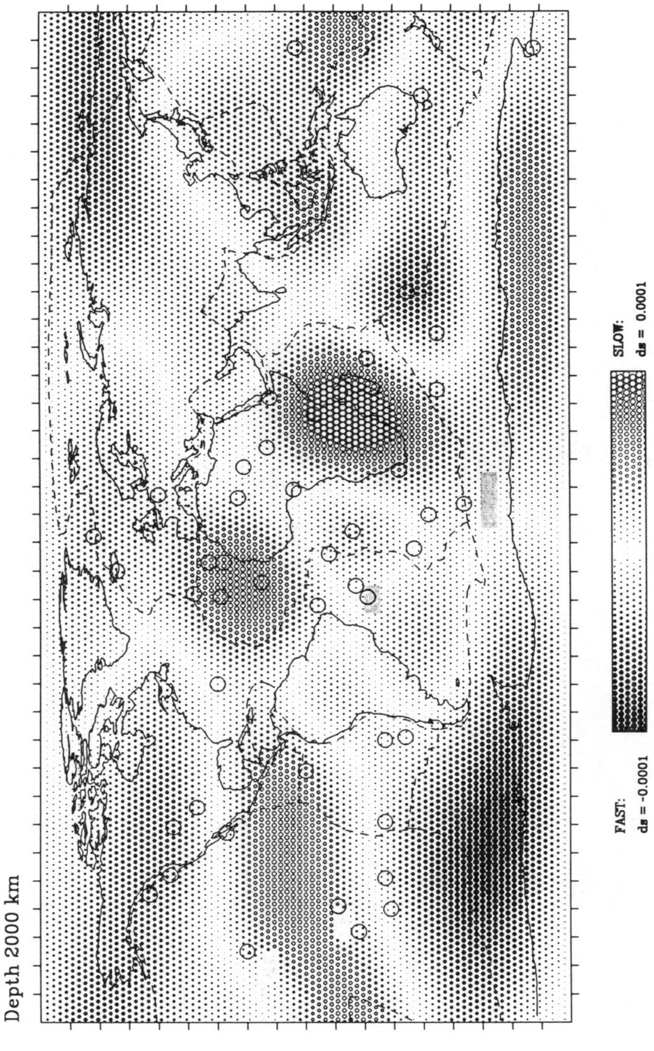

Figure 9.16. Smoothed depth-section at a depth of 2000 km. A world map is superimposed for reference. The filtering is described in Figure 9.11.

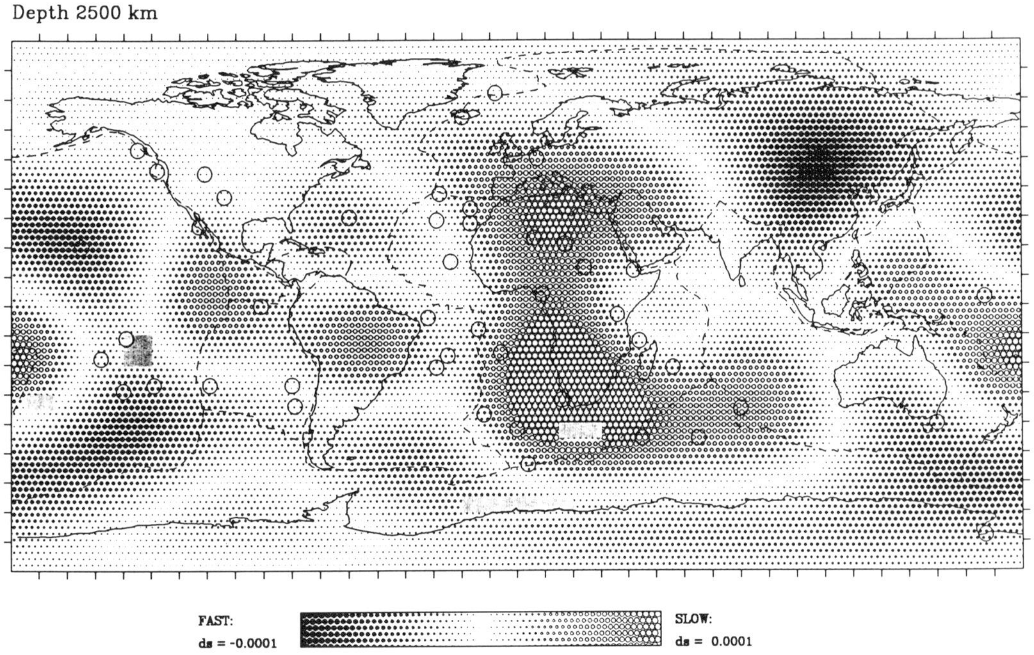

Figure 9.17. Smoothed depth-section at a depth of 2500 km. A world map is superimposed for reference. The filtering is described in Figure 9.11.

similar to the one published by Dziewonski (1984), but is different in the southeast Pacific and under South America.

One interesting attribute of the inversion is the power spectrum of the variations in each radial shell. If convection is dominated by narrow plumes and slabs, then it is expected that the spectrum should be white at these long wavelengths. On the other hand, noise in the model might also produce a white spectrum. In Figure 9.18, the power spectrum for several layers of the raw inversion is displayed. The spectrum appears to be white for all layers up to degree 12. Only the layer at 2700 km depth gives some indication of a reduction in power at higher harmonic degree.

Resolution and Variance of the Inversion

The iterative tomographic inversion presented in the previous sections was used in this study because of the large number of model parameters and observations. One disadvantage of this approach, compared to a generalized inverse method, is the lack of formal error and resolution estimates.

The resolution issue is really a moot point since the full resolution matrix for a model with 50,000 parameters would have 2.5×10^9 elements, which would be impossible to display in its entirety. The model error estimates are also problematic since they are directly proportional to the estimated errors and correlations in the data. These are difficult, if not impossible, to estimate, and are generally based on arguments of how accurately an arrival can be (theoretically) picked. In reality, erratic and systematic errors probably dominate this type of error.

An alternate approach is to sample the resolution matrix at a few select points. This can be done by performing the inversion with synthetic data calculated from an impulse model, in which only one parameter is fixed, and all others are set to zero. The raypaths used in the synthetic calculation are the same ones present in the real data set. The resolution of the inversion process is then determined by the degree to which the single impulse parameter is resolved or spread into other parameters. In practice, several impulses can usually be tested at once. An example of this is shown in Figure 9.19. This plot has not been filtered so the true potential resolution can be seen. The results show that all eight impulses have been reconstructed, and are only spread over a few blocks adjacent to the impulse points. Even the point in the Pacific near a region of marginal ray coverage is well imaged, although it has a slightly diminished amplitude.

The resolution calculations can also be performed with other types of test patterns. We have directly tested, for example, the resolution associated with

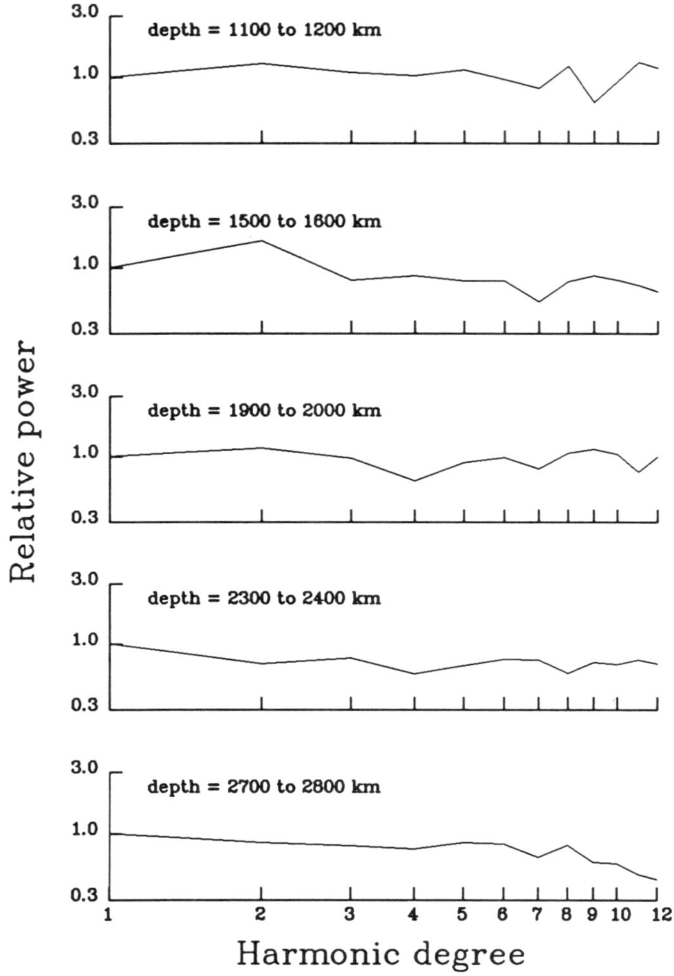

Figure 9.18. The spectrum of the slowness inversion. The power spectrum in the inversion is shown for several depths. At each depth, power is normalized relative to that at degree 1. The model is essentially white at all depths.

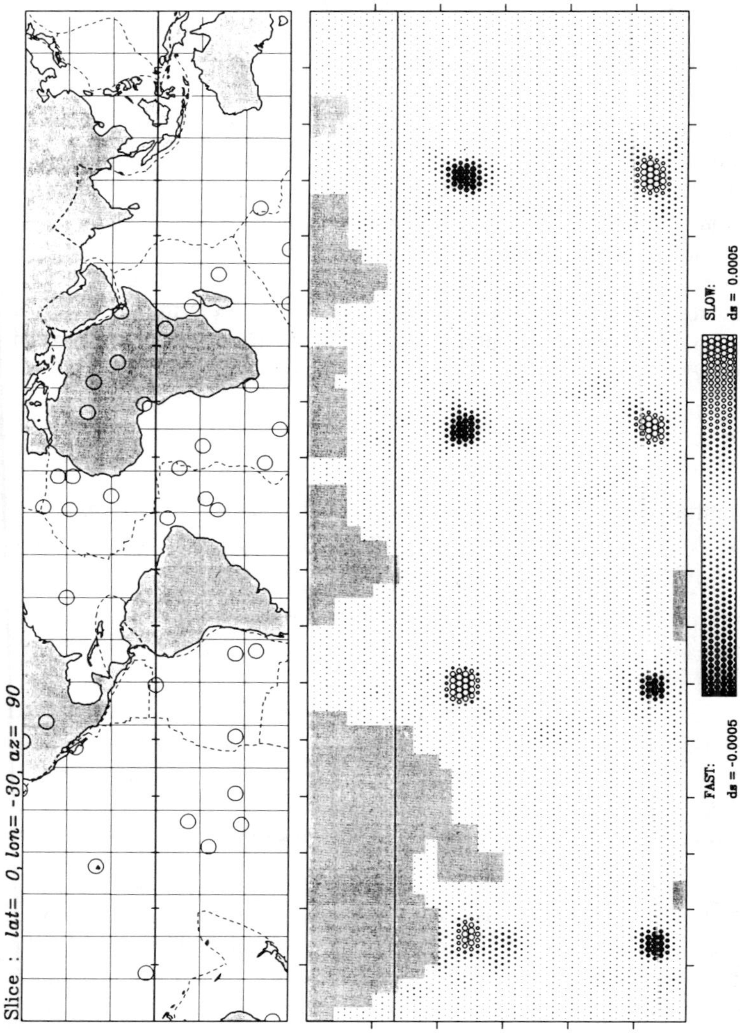

Figure 9.19. Synthetic cross-section slice. An unfiltered slice along the equatorial great circle is shown in the same format as Figure 9.10. The input model contained eight spots (impulse points) in this projection. All eight are reconstructed, although the ones near areas of poor coverage are diminished in amplitude.

reconstructing velocity variations distributed as low-order harmonics. The resolution in this case is excellent (Comer and Clayton, 1989).

To test the effect of data errors on the model, a synthetic inversion is again used. This time, however, the $\{\Delta t_i\}$ associated with each ray is replaced with a random number. These numbers are drawn from a distribution that best fits the one shown in Figure 9.3. This is a worst case analysis, because it assumes that all of the observations are random noise.

For a typical noise inversion, the variance of the model determined by an inversion of noise is one fifth of the variance of the model determined from the real data. The variance reduction of the noise residuals is 15 percent, which contrasts with the 36 percent achieved with the real data. These tests give us confidence that the inversion results obtained from the ISC data are more than just noise. In theory, if many noise simulations were done, a formal error bound for each model element could be determined. This, of course, would be dependent upon one's opinion of what the nature and size of the data errors are.

Discussion

The essential question concerning the lower mantle velocity variations is, *are they real features of the Earth?*. It is difficult of course, to give a definitive answer to this question. The arguments on the positive side are:

1. The error and resolution studies presented in the previous section indicate that the inversion is at least not all noise, and the anomalies can potentially be well resolved.
2. The overall variance of the winnowed ISC travel time residuals is reduced by 56 percent. This is the compounded combination of relative reductions of 20 percent for statics, 10 percent for composite rays, and 36 percent for the tomographic inversion.
3. The station statics solution appears to fit, in a general sense, the surficial tectonic settings of the stations.

On the negative side of this question, there is always the possibility of *backdoor* correlations affecting the results. An example of this is the correlation of station gain with tectonic region. Tectonically active regions have higher microearthquake noise levels, which causes the station amplification factors to be set lower than stations in more stable shield regions. This, in turn, may cause these stations to pick arrivals systematically later than the shield stations. This bias would generate an artificial correlation of travel time residuals with tectonic region. Fortunately, for this study, this type of systematic error will be taken out with the station statics. However, there may be other correlation errors that we are not aware of. Another perhaps negative feature is that the solution seems to violate the preconception that thermal anomalies in a convecting mantle should be, for the most part, coherent radially.

The comparisons with the results given by Dziewonski (1984) show that the two inversions loosely agree at the longest wavelengths, but disagree at short wavelengths (Hager *et al.*, 1985). The source of this discrepancy is not known at this time. This may be due to the differences in the data set used, or in the use of global basis functions (spherical harmonics in azimuth with Legendre polynomials in radius) to span the model. Some of the largest lower mantle anomalies determined by Dziewonski correspond to "no coverage areas" of the inversion results presented in this study.

The desire to seek corroborating evidence on the validity of the proposed lower mantle velocity variations led to a comparison with the Earth's gravity field (or geoid). The results for a nonconvecting Earth, presented in the next section are negative. However, the inclusion of dynamic effects presented in Sections 4 and 5 indicate that the long-wavelength seismic velocity variations in the lower mantle predict a substantial portion of the long-wavelength geoid anomalies.

3. Static Calculation of Geoid Anomalies

The P-wave velocity anomalies inferred from seismic tomography are plausibly interpreted as a consequence of convection, with slow regions representing hot, buoyant upwellings and fast regions representing cold downwellings. Such an interpretation should be subjected to any independent tests possible. One straightforward test is to compute the geoid anomalies given from the density contrasts inferred from the slowness anomalies and compare them to the observed geoid. Although, in principle, the relation between P-wave velocity anomalies and density anomalies could be complicated by effects such as dislocation relaxation, melting, or changes in composition, for simplicity we assume they are linearly related, with the proportionality constant independent of depth.

Basic Equations

The geoid represents a surface of constant gravitational potential V, which is related to the density through Poisson's equation.

$$\nabla^2 V = -4\pi\gamma\rho, \tag{3.1}$$

where γ is the universal gravitational constant, and ρ is the density. The potential δV from a density anomaly $\delta\rho$ also satisfies 3.1. The geoid anomaly δN (to first order), is found by dividing the anomalous potential δV at the surface by the normal derivative of the potential, which is just the value of surface gravity $g(a)$:

$$\delta N = \frac{\delta V(a)}{g(a)}, \tag{3.2}$$

where a is the radius of the Earth. In obtaining (3.2), the hydrostatic solution to (3.1) is subtracted out, leaving only the effect of density perturbations. We choose the negative sign in Poisson's equation such that positive density anomalies result in positive potentials. For this sign convention, gravitational acceleration is given by $\mathbf{g} = -\nabla V$.

In spherical coordinates, (3.1) becomes

$$\frac{1}{r^2}\frac{\partial}{\partial r}\left(r^2\frac{\partial \delta V}{\partial r}\right) + \frac{1}{r^2 \sin\theta}\frac{\partial}{\partial\theta}\left(\sin\theta\frac{\partial\delta V}{\partial\theta}\right) + \frac{1}{r^2 \sin\theta}\frac{\partial^2 \delta V}{\partial\phi^2} = -4\pi\gamma\delta\rho. \tag{3.3}$$

Although the slowness, and hence the density perturbation model, is discretized in terms of values within blocks, (3.3) is most conveniently solved for a (nearly) spherical Earth using spherical harmonics.

$Y_{lmi}(\theta,\phi)$ is defined as a surface spherical harmonic of degree l and order m normalized to have root mean square amplitude of unity over the sphere:

$$Y_{lmi}(\theta, \phi) = P_{lm}(\cos\theta)\begin{cases}\cos m\phi, & i = 1\\ \sin m\phi, & i = 2,\end{cases} \tag{3.4}$$

where θ is colatitude and ϕ is longitude. Also its derivatives are written as

$$Y^{\theta}_{lmi} = \frac{\partial Y_{lmi}}{\partial\theta} \quad \text{and} \quad Y^{\phi}_{lmi} = \frac{1}{\sin\theta}\frac{\partial Y_{lmi}}{\partial\phi}, \tag{3.5}$$

δV and $\delta\rho$ are expressed in terms of an infinite series of radial functions multiplied by surface spherical harmonics:

$$\delta V(r,\theta,\phi) = \delta V^{lmi}(r) Y_{lmi}(\theta,\phi), \tag{3.6}$$

$$\delta\rho(r,\theta,\phi) = \delta\rho_{lmi}(r) Y_{lmi}(\theta,\phi), \tag{3.7}$$

where sums over repeated indices ($l = 1, \infty$; $m = 0, l$; $i = 1, 2$) are implicit. Then, using the identity

$$Y_{lmi}^{\theta\theta} + \cot\theta\, Y_{lmi}^{\theta} + Y_{lmi}^{\phi\phi} = -LY_{lmi}, \tag{3.8}$$

with $L = l(l + 1)$, and substituting into (3.3 gives)

$$\left(-\frac{L}{r^2}\delta V^{lmi} + \frac{2}{r}\frac{d\delta V^{lmi}}{dr} + \frac{d^2\delta V^{lmi}}{dr^2}\right) Y_{lmi} = -4\pi\gamma\delta\rho^{lmi} Y_{lmi}. \tag{3.9}$$

(Here and in the following, functional dependences on r, θ, and ϕ are usually taken to be implicit.) A particular (l, m, i), e.g., (p, q, s), can be obtained by multipling (3.9) by Y_{pqs}, and integrating over the surface of the sphere. The orthonormality property of surface spherical harmonics will then give,

$$-\frac{L}{r^2}\delta V^{pqs} + \frac{2}{r}\frac{d\delta V^{pqs}}{dr} + \frac{d^2\delta V^{pqs}}{dr^2} = -4\pi\gamma d\rho^{pqs}. \tag{3.10}$$

The homogeneous part of this second order differential equation has solutions of the form r^l and $r^{-(l+1)}$. The first is bounded at the origin, while the second form is bounded at infinity. The particular form of the solution is found by considering a density perturbation of thickness dr at a specific radius, say $r = b$. Matching inner and outer solutions for potential at $r = b$ gives

$$\delta V^{lmi}(r) = \frac{4\pi\gamma b\delta\rho^{lmi}(b)dr}{2l + l}\begin{cases}(r/b)^l & r < b \\ (b/r)^{l+1} & r \geqslant b\end{cases} \tag{3.11}$$

The total potential at any radius is determined by integrating (3.11) throughout the volume where density contrasts are present.

If $\delta\rho(r,\theta,\phi)$ is known exactly everywhere, then $\delta\rho^{lmi}(r)$ is determined by the integration

$$\delta\rho^{lmi}(r) = \frac{1}{4\pi}\int_{\theta=0}^{\pi}\int_{\phi=0}^{2\pi}\delta\rho(r,\theta,\phi)Y_{lmi}(\theta,\phi)\sin\theta d\theta d\phi, \tag{3.12}$$

Application to Lower Mantle Tomography Results

Unfortunately, as we have discussed, there is uncertainty in our estimate of $\delta s(r,\theta,\phi)$ and substantial regions where there is no ray coverage, and hence no estimate. Thus there is great uncertainty in our estimates of $\delta\rho^{lmi}(r)$ for some shells. Using (3.12) to determine these coefficients, and integrating only over those regions where there is ray coverage, would be equivalent to assuming $\delta\rho(r,\theta,\phi) = 0$ in regions of no coverage, a poor assumption. Instead, we evaluate $\delta\rho^{lmi}(r)$ using a least squares technique and discard the results for those layers that are poorly determined due to lack of coverage. We have greatest confidence in our results for depths between 800 and 2800 km;

at other depths coverage is poor. For reasons discussed later, the results seem best determined at the longest wavelengths.

The geoid calculated from the density variations inferred from the velocity model derived from seismic tomography is shown in Figure 9.20. The tomography results for spherical harmonic degrees 2–6, in the depth range 800–2800 km are used, with a velocity perturbation to density perturbation scaling of 2.1 (km/sec)/(g/cm^3), a value discussed in Section 5. There are striking similarities to the observed geoid, Figure 9.1. For example, both are dominated by prominent features of one sign at the poles and opposite sign over Africa and the central Pacific. The formal correlations between the two geoids at degrees 2–6 are −0.86, −0.54, −0.04, +0.62, and −0.02, respectively. These correlations are significant at the 97, 85, and 92 percent confidence levels for degrees 2, 3, and 5, respectively. Correlations at degrees 4 and 6 are insignificant. Although the calculated patterns, dominated by degrees 2 and 3, are similar to the observed geoid, they are of opposite sign and the geoid calculated from the tomographic results is about a factor of 4 larger in peak to peak amplitude. Similar results were given in

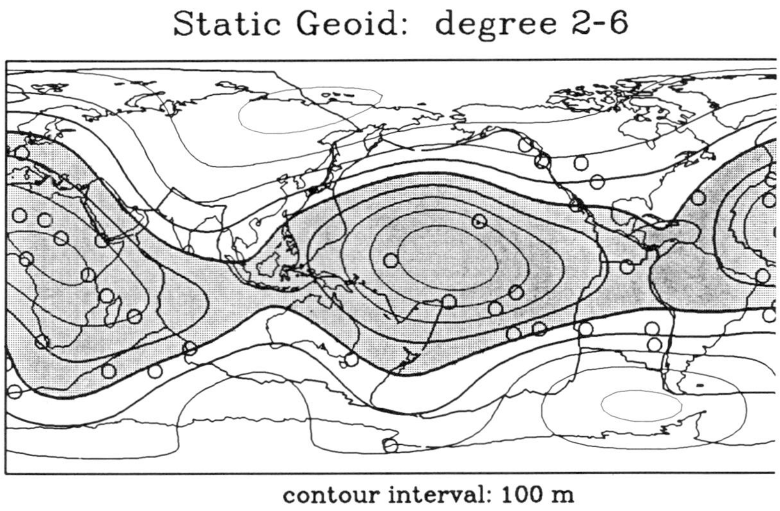

Figure 9.20. Static model geoid calculated from the tomographic inversion of lower mantle seismic heterogeneity assuming that velocity and density perturbations are directly proportional as expected for thermal convection. The contour interval is 100 m and geoid lows are shaded.

the earlier studies by Dziewonski *et al.* (1977) and Dziewonski (1984), although their correlations were not as good.

One interpretation of the predominant anticorrelation between the predicted and the observed geoid is that the sign and magnitude of the velocity to density perturbation ratio is incorrect. For example, as discussed in some detail by Dziewonski *et al.* (1977), seismic velocity and density perturbations resulting from variations in chemical composition can lead to a negative value for this ratio. We feel that such an interpretation is unlikely in a dynamic, convecting planet, as it would require long-lived sources and sinks of chemical heterogeneity at the top and bottom of the mantle, analogous to the equivalent sources and sinks of heat in a convecting mantle. Such a system of sources and sinks of chemical variation, if operative over the lifetime of the Earth, would probably have led to far more profound chemical differentiation than now exists.

4. Fluid Mechanical Models of the Geoid

A more likely reason that the geoid calculated from the density perturbations inferred from seismic tomography is essentially anticorrelated with the observed long-wavelength geoid is that we have up to this point in the calculation failed to take into account the dynamic surface deformation that is caused by the flow induced by these density contrasts. As pointed out by Busse (this volume) in his introductory section, the fact that the upper surface of the Earth has no significant tractions acting on it requires that convection within the mantle lead to a dynamically-maintained deformation of this surface away from its equilibrium position. Similarly, the CMB and any discontinuities in chemical composition that might exist in the interior of the mantle must be deformed. These boundary deflections lead to mass anomalies that have a substantial effect on the geoid, a point first discussed by Pekeris (1935), and later addressed by, e.g., Runcorn (1964, 1967), Morgan (1965a, 1965b), McKenzie (1977), Parsons and Daly (1983), Ricard *et al.*, (1984), and Richards and Hager (1984).

To determine these effects quantitatively, it is necessary to solve the equations of motion, the continuity equation, and Poisson's equation for gravitational potential subject to an assumed constituitive law. Since the Earth is to a first approximation spherical, it is convenient to make use of spherical harmonics in solving these equations. Analysis is further simplified if it is assumed that flow in the mantle is incompressible. This is a fairly crude approximation—material at the top of the mantle would be compressed by about a factor of $5/3$ in sinking to the bottom of the mantle.

However, as explained by Busse, the fact that the hydrodynamic stresses involved in mantle convection are much smaller than the hydrostatic pressure changes makes the assumption of incompressible flow a reasonable first approximation.

We also assume, for the sake of analytic tractability, that the viscosity of mantle rocks is Newtonian and a function of depth only. The actual details of mantle rheology are not known, but theory exists for extrapolating laboratory measurements of the rheology of mantle materials to conditions applicable to the mantle (e.g., Ashby and Verrall, 1977). At ambient conditions, a decrease of temperature by about 100 K is thought to lead to an increase in viscosity by about a factor of 10. An increase in stress by a factor of 3 may lead to a decrease in effective viscosity by about an order of magnitude; to the extent that stress is concentrated in regions which have higher viscosities due to lower temperatures, the effects of stress and temperature on effective viscosity tend to cancel (e.g., Christensen, 1984).

The effects of pressure on effective viscosity are less well constrained, but in the uppermost mantle, the effect of pressure on viscosity probably results in an increase in viscosity by about a factor of 10 for each 85 km increase in depth. This effect is mitigated to some extent by the increase of temperature with depth along an adiabat, leading to a probable net increase with depth of about a factor of 10 for each 150 km increase in depth. A decrease in activation energy as a function of depth will limit this increase in viscosity (O'Connell, 1977; Sammis *et al.*, 1977). Still, an increase in viscosity across the mantle by a factor of 100–1000 is expected on the basis of solid state physics (Sammis *et al.*, 1977).

Our mathematical technique allows us to address radial variations of effective viscosity, but not the effects of azimuthal variations in viscosity. Our interpretation of mantle heterogeneity inferred from seismic tomography in terms of temperature variations ($\sim\pm15$ K at long wavelengths, $\sim\pm100$ K at shorter wavelengths) suggests that lateral variations in viscosity may be important, but not so large as to be overwhelming. The effects of ignoring them will be discussed in more detail later.

Basic Equations

To solve for the flow, dynamic surface deformation and gravitational potential, the equations of continuity, the equations of motion, and Poisson's equation are reduced to a set of coupled ordinary differential equations in radius. In spherical coordinates, the incompressible equation of continuity can be written (Landau and Lifschitz, p. 52)

$$\frac{\partial v_r}{\partial r} + \frac{1}{r}\frac{\partial v_\theta}{\partial \theta} + \frac{1}{r \sin\theta}\frac{\partial v_\phi}{\partial \phi} + 2\frac{v_r}{r} = \frac{\cot\theta v_\theta}{r} = 0. \tag{4.1}$$

The stress tensor for a Newtonian fluid in spherical coordinates is (e.g., Landau and Lifschitz, p. 52):

$$\tau_{rr} = -p + 2\eta\frac{\partial v_r}{\partial r}, \tag{4.2}$$

$$\tau_{\theta\theta} = -p + 2\eta\left(\frac{1}{r}\frac{\partial v_\theta}{\partial \theta} + \frac{v_r}{r}\right), \tag{4.3}$$

$$\tau_{\phi\phi} = -p + 2\eta\left(\frac{1}{r \sin\theta}\frac{\partial v_\phi}{\partial \phi} + \frac{v_r}{r} + \frac{\cot\theta v_\theta}{r}\right), \tag{4.4}$$

$$\tau_{r\theta} = \eta\left(\frac{1}{r}\frac{\partial v_r}{\partial \theta} + \frac{\partial v_\theta}{\partial r} - \frac{v_\theta}{r}\right), \tag{4.5}$$

$$\tau_{\phi\theta} = \eta\left(\frac{1}{r \sin\theta}\frac{\partial v_\theta}{\partial \phi} + \frac{\partial v_\phi}{\partial \theta} - \frac{\cot\theta v_\phi}{r}\right), \tag{4.6}$$

$$\tau_{r\phi} = \eta\left(\frac{\partial v_\phi}{\partial r} + \frac{1}{r \sin\theta}\frac{\partial v_r}{\partial \phi} - \frac{v_\phi}{r}\right). \tag{4.7}$$

From here on it is assumed that a hydrostatic reference stress and potential have been removed. Stresses and potentials refer to departures from that reference state.

The equations of motion in a spherical shell with uniform viscosity (e.g., Landau and Lifschitz, p. 52), including self-gravitation, become

$$0 = \rho\frac{\partial \delta V}{\partial r} - \delta\rho g - \frac{\partial p}{\partial r} + \eta\left[\frac{1}{r}\frac{\partial^2(rv_r)}{\partial r^2} + \frac{1}{r^2}\frac{\partial^2 v_r}{\partial \theta} + \frac{1}{r^2 \sin^2\theta}\frac{\partial^2 v_r}{\partial \phi^2}\right.$$
$$\left. + \frac{\cot\theta}{r^2}\frac{\partial v_r}{\partial \theta} - \frac{2v_r}{r^2} - \frac{2}{r^2}\frac{\partial v_\theta}{\partial \theta} - \frac{2\cot\theta}{r^2}v_\theta - \frac{2}{r^2 \sin\theta}\frac{\partial v_\phi}{\partial \phi}\right], \tag{4.8}$$

$$0 = \frac{\rho}{r}\frac{\partial \delta V}{\partial \theta} - \frac{1}{r}\frac{\partial p}{\partial \theta} + \eta\left[\frac{1}{r}\frac{\partial^2(rv_\theta)}{\partial r^2} + \frac{1}{r^2}\frac{\partial^2 v_\theta}{\partial \theta^2} + \frac{1}{r^2 \sin^2\theta}\frac{\partial^2 v_\theta}{\partial \phi^2}\right.$$
$$\left. + \frac{\cot\theta}{r^2}\frac{\partial v_\theta}{\partial \theta} - \frac{2\cos\theta}{r^2 \sin^2\theta}\frac{\partial v_\phi}{\partial \phi} + \frac{2}{r^2}\frac{\partial v_r}{\partial \theta} - \frac{v_\theta}{r^2 \sin^2\theta}\right], \tag{4.9}$$

$$0 = \frac{\rho}{r \sin\theta}\frac{\partial \delta V}{\partial \phi} - \frac{1}{r \sin\theta}\frac{\partial p}{\partial \phi} + \eta\left[\frac{1}{r}\frac{\partial^2(rv_\phi)}{\partial r^2} + \frac{1}{r^2}\frac{\partial^2 v_\phi}{\partial \theta^2} + \frac{1}{r^2 \sin^2\theta}\frac{\partial^2 v_\phi}{\partial \phi^2}\right.$$
$$\left. + \frac{\cot\theta}{r^2}\frac{\partial v_\phi}{\partial \theta} - \frac{v_\phi}{r^2 \sin^2\theta} + \frac{2\cos\theta}{r^2 \sin^2\theta}\frac{\partial v_\theta}{\partial \phi} + \frac{2}{r^2 \sin\theta}\frac{\partial v_r}{\partial \phi}\right]. \tag{4.10}$$

As was the case with gravity, it is convenient to solve these equations by expressing all azimuthal variations in terms of spherical harmonics. This method is similar to that originally used for describing the Earth's seismic free oscillations (Alterman *et al.*, 1959), and its viscous response to Pleistocene deglaciation (Takeuchi and Hasegawa, 1965, O'Connell, 1971). It has been used for mantle flow problems by, e.g., Kaula (1975), and Hager and O'Connell (1978, 1979, 1981), and to calculate the geoid response by Richards and Hager (1984) and Ricard *et al.*, (1984).

The components of velocity in the radial, southerly, and easterly directions can be expressed as infinite series of functions of radius times vector spherical harmonics:

$$v_r(r,\theta,\phi) = y_1^{lmi}(r)\, Y_{lmi}(\theta,\phi), \tag{4.11}$$

$$v_\theta(r,\theta,\phi) = y_2^{lmi}(r)\, Y_{lmi}^{\theta}(\theta,\phi) + y_9^{lmi}(r)\, Y_{lmi}^{\phi}(\theta,\phi), \tag{4.12}$$

$$v_\phi(r,\theta,\phi) = y_2^{lmi}(r)\, Y_{lmi}^{\phi}(\theta,\phi) - y_9^{lmi}(r)\, Y_{lmi}^{\theta}(\theta,\phi), \tag{4.13}$$

Again, sums over repeated indices ($l = 1, \infty$; $m = 0, l$; $i = 1, 2$) are implicit.

Similarly, the nonhydrostatic stress components can be written

$$\tau_{rr}(r,\theta,\phi) = y_3^{lmi}(r)\, Y_{lmi}(\theta,\phi), \tag{4.14}$$

$$\tau_{r\theta}(r,\theta,\phi) = y_4^{lmi}(r)\, Y_{l\ mi}^{\ \theta}(\theta,\phi) + y_{10}^{lmi}\ (r)\, Y_{lmi}^{\phi}(\theta,\phi), \tag{4.15}$$

$$\tau_{r\phi}(r,\theta,\phi) = y_4^{lmi}(r)\, Y_{l\ mi}^{\ \phi}(\theta,\phi) - y_{10}^{lmi}\ (r)\, Y_{lmi}^{\theta}(\theta,\phi), \tag{4.16}$$

Finally, the nonequilibrium gravitational potential, its radial gradient (perturbed gravity), the nonhydrostatic pressure and density perturbations are expressed as

$$\delta V(r,\theta,\phi) = y_5^{lmi}(r)\, Y_{lmi}(\theta,\phi), \tag{4.17}$$

$$\frac{\partial \delta V}{\partial r}(r,\theta,\phi) = y_6^{lmi}(r)\, Y_{lmi}(\theta,\phi), \tag{4.18}$$

$$\delta p(r,\theta,\phi) = y_7^{lmi}(r)\, Y_{lmi}(\theta,\phi), \tag{4.19}$$

$$\delta \rho(r,\theta,\phi) = y_8^{lmi}(r)\, Y_{lmi}(\theta,\phi). \tag{4.20}$$

By selectively substituting (4.11)–(4.20) into (3.3) and (4.1)–(4.10), making use of the identify (3.8), picking out individual indices l, m, and i by surface integration, and eliminating the nonhydrostatic pressure through

$$y_7^{lmi} = -\frac{4\eta}{r} y_1^{lmi} + \frac{2\eta L}{r} y_2^{lmi} - y_3^{lmi}, \tag{4.21}$$

a set of 8 coupled ordinary differential equations in radius for $y_1^{lmi} - y_8^{lmi}$ result for each l, m, and i (for more details see, e.g., Kaula, 1975; Hager, 1978).

Both poloidal and toroidal fields are necessary to express an arbitrary velocity or stress. Coefficients $y_1^{lmi}(r)$ through $y_8^{lmi}(r)$ are associated with poloidal fields, while $y_9^{lmi}(r)$ and $y_{10}^{lmi}(r)$ are associated with toroidal fields. Coupling between poloidal and toroidal components of the flow only occurs through lateral variations in viscosity, which we ignore in our analysis. As a cautionary note, however, we point out that the power in the spectra of observed plate motions is nearly as large for toroidal fields (representing transform faults) as for poloidal fields (spreading and subduction) (Hager and O'Connell, 1978; Kaula, 1980; Peltier and Forte, 1984). This coupling may well influence the geoid (Forte and Peltier, 1987; Ricard, 1986), although the physics of such a process has yet to be worked out.

The equations may be simplified by defining the variable $\nu = \ln(r/a)$, where a is the radius of the Earth, taking η_0 as a reference viscosity and ρ_0 as a reference density, and for convenience, dropping the superscripts and radial dependences, henceforth taken as implicit. Setting

$$u_1 = y_1, \tag{4.22}$$

$$u_2 = y_2, \tag{4.23}$$

$$u_3 = ry_3/\eta_0 + \rho r y_5/\eta_0, \tag{4.24}$$

$$u_4 = ry_4/\eta_0, \tag{4.25}$$

$$v_1 = \rho_0 r y_5/\eta_0, \tag{4.26}$$

$$v_2 = \rho_0 r^2 y_6/\eta_0, \tag{4.27}$$

$$\omega_1 = y_9, \tag{4.28}$$

$$\omega_2 = ry_{10}\eta_0, \tag{4.29}$$

and assuming constant viscosity η within a shell, yields

$$\frac{d\mathbf{u}}{d\nu} = \mathbf{Au} + \mathbf{a}, \tag{4.30}$$

$$\frac{d\mathbf{v}}{d\nu} = \mathbf{Bv} + \mathbf{b}, \tag{4.31}$$

$$\frac{d\mathbf{w}}{d\nu} = \mathbf{Cw}, \tag{4.32}$$

where

$$\mathbf{A} = \begin{bmatrix} -2 & L & 0 & 0 \\ -1 & 1 & 0 & 1/\eta^* \\ 12\eta^* & -6L\eta^* & 1 & L \\ -6\eta^* & 2(2L-1)\eta^* & -1 & -2 \end{bmatrix}. \tag{4.33}$$

$$\mathbf{B} = \begin{bmatrix} 1 & 1 \\ L & 0 \end{bmatrix}, \tag{4.34}$$

$$\mathbf{C} = \begin{bmatrix} 1 & 1/\eta^* \\ (L-2)\eta^* & -2 \end{bmatrix}, \tag{4.35}$$

$$\mathbf{a} = (0,\ 0,\ r^2 g(r) y_8/\eta_0,\ 0)^T, \tag{4.36}$$

$$\mathbf{b} = [0,\ -4\pi r^3 \gamma \rho_0 y_8/\eta_0]^T, \tag{4.37}$$

and $\eta^* = \eta/\eta_0$.

These equations are similar to those used (correctly) but misprinted in papers by Hager and O'Connell (1978, 1979, 1981), Richards and Hager (1984), and Richards (1986), except for changes resulting from including the potential term y_5 in the definition of u_3, equation (4.24), which allows partial decoupling of the equations for velocities and stress, $u_1 - u_4$, from those for potential, v_1 and v_2 (O'Connell *et al.*, 1984).

The solution of (4.30) for a shell in which **A** is constant is

$$\begin{aligned} \mathbf{u}(\nu) &= \exp[\mathbf{A}(\nu - \nu_0)]\mathbf{u}(\nu_0) + \int_{\nu_0}^{\nu} \exp[\mathbf{A}(\nu - \epsilon)]\mathbf{a}(\epsilon)d\epsilon \\ &= \mathbf{P}_A(\nu,\nu_0)\mathbf{u}(\nu_0) + \int_{\nu_0}^{\nu} \mathbf{P}_A(\nu,\epsilon)\mathbf{a}(\epsilon)d\epsilon. \end{aligned} \tag{4.38}$$

Here $\mathbf{u}(\nu_0)$ is the starting vector at $\nu = \nu_0$ and $\mathbf{P}_A(\nu,\nu_0)$ is the propagator matrix that propagates this solution to ν (Gantmacher, 1960, vol. II). For a series of shells, the solution is propagated through the innermost shell using (4.38), contributions to **u** from boundary effects between shells are added, as discussed in more detail below, and (4.38) is applied again for the new medium.

The analysis is simplified if $\delta\rho^{lmi}(r)$ is approximated as a series of J discrete sheets or surface mass anomalies σ_j^{lmi} rather than as continuous (volumetric) density contrasts, e.g.,

$$\sigma_j^{lmi} = \sum_{r_j-\epsilon}^{r_j+\epsilon} \delta\rho^{lmi}(r)dr. \tag{4.39}$$

Then (4.38) becomes

$$\mathbf{u}(\nu) = \mathbf{P}_A(\nu,\nu_0)\mathbf{u}(\nu_0) + \sum_{j=1}^{i} \mathbf{P}(\nu,\nu_j)\mathbf{a}_j, \tag{4.40}$$

with

$$\mathbf{a}_j = [0,\ 0,\ r_j g(r_j)\sigma_j/\eta_0,\ 0]^T. \tag{4.41}$$

A similar expression holds for potential, with

$$\mathbf{b}_j = [0,\ -4\pi r_j^2 \gamma \rho_0 \sigma_j/\eta_0]^T. \tag{4.42}$$

Propagator Matrices

The form of the propagator for a matrix is given analytically in terms of its n eigenvalues λ_i. For a matrix $\mathbf{A}$ with no repeated eigenvalues, the propagator $\mathbf{P}_A$ is given by (Gantmacher, 1960, vol. II):

$$\mathbf{P}_A = \sum_{i=1} \exp[\lambda_i(\nu - \nu_0)] \prod_{s \neq i} (\lambda_s \mathbf{I} - \mathbf{A})/(\lambda_s - \lambda_i), \tag{4.43}$$

where $\mathbf{I}$ is the identity matrix. The eigenvalues of $\mathbf{A}$ are $(l + 1)$, $(l - 1)$, $-l$, and $-(l + 2)$. $\mathbf{P}_A$ is evaluated using (4.43).

The eigenvalues of $\mathbf{B}$ are $(l + 1)$ and $-l$. Substituting gives the relatively simple propagator for the potential:

$$\mathbf{P}_B = \frac{1}{(2l + 1)} \begin{bmatrix} (l + 1)R^{l-1} + lR^{-l} & R^{l+1} - R^{-l} \\ l(l + 1)(R^{l+1} - R^{-l}) & lR^{l+1} + (l + 1)R^{-1} \end{bmatrix}, \tag{4.44}$$

where $R = (r/r_0)$. This propagator simply satisfies (3.3), Poisson's equation for gravitational potential.

The eigenvalues of $\mathbf{C}$ are l and $-(l + 1)$. The toroidal propagator then can be written in the simple form:

$$\mathbf{P}_C = \frac{1}{(2l + 1)} \begin{bmatrix} (l + 2)R^{l} + (l - 1)\mathrm{R}^{-(l+1)} & (R^{l} - R^{-(l+1)})/\eta^* \\ (L - 2)\eta^*(R^{l} - R^{-(l+1)}) & (l + 1)R^{l} + lR^{-(l+1)} \end{bmatrix}. \tag{4.45}$$

Boundary Conditions

At any boundary within the mantle, continuity and force balance require that the primitive variables v_r, v_θ, v_ϕ, the total normal stress τ_{rr}, radial shear stresses $\tau_{r\theta}$, and $\tau_{r\phi}$, and δV are continuous. There are two types of boundaries that we consider in our models: boundaries between layers of the same intrinsic density but different viscosity and boundaries between layers of different intrinsic densities due to differing compositions, as well as with (perhaps) different viscosities. For the first type of boundary, which we use to simulate the effects of depth on viscosity, the vectors **u** and **v** are continuous. Consequently, only $\mathbf{P}_A$ changes.

The second type boundary is one in which the intrinsic density changes owing to a change in chemical composition. If the change in density is large enough, such a "chemical" boundary will lead to a stratification of the flow into separate layers. At such a boundary, the above primitive quantities are still all continuous, and in addition, at steady state, $v_r = 0$. There is, however, a jump in u_3 caused by the change in density in the second term.

The surface of the Earth and the CMB represent extreme cases of chemical discontinuities. Because the atmosphere and outer core have viscosities negligible compared to that of the mantle, shear tractions at the surface and at the CMB are negligible. In addition, $\delta\tau_{rr}$ at the Earth's surface vanishes.

Loading of the Earth by an external gravitational potential (e.g., tidal loading), by an external load (e.g., an ice cap), or by an internal load (e.g., the density contrasts associated with mantle convection) results in deformation of the surface, the CMB, and any other chemical boundaries that might exist, away from their spherical hydrostatic reference radii. (The hydrostatic effects due to centrifugal force from Earth rotation have been assumed removed.) The physical boundary conditions of continuity of velocity, tractions and potential apply on the actual (deformed) boundaries. The mathematical technique using propagator matrices is applicable only between surfaces of constant radius. Thus it is necessary to relate the physical boundary conditions on the deformed boundaries to mathematical conditions on surfaces of constant radius.

This is accomplished by making use of a first order Taylor's series expansion. If b is the mean radius of a boundary deformed by an amount δb, and if δb is much smaller than the wavelength of some y^{lmi} or the thickness of an adjacent layer, than the first order expansion

$$y^{lmi}(b) = y^{lmi}(b + \delta b) - \frac{\partial y^{lmi}}{\partial r}\delta b \tag{4.46}$$

is sufficiently accurate.

The derivatives of stresses associated with flow are of the order of the product of the spatial wave number times the stresses ($\approx l/a\,\tau_{ij}$). The radial derivative of the total radial normal stress has a contribution due to the lithostatic pressure gradient, ρg, as well as a contribution from flow. For the magnitudes of the stresses involved and the wavelength range of interest, $\rho g >> \tau_{ij} l/a$. Thus the only first order correction to the flow variables is for the radial stress:

$$\delta\tau_{rr}(b) = \delta\tau_{rr}(b + \delta b) - \rho g \delta b. \tag{4.47}$$

The gravitational effects of the deformed boundaries must also be considered. Upwarping of a boundary at radius b by δb^{lmi} introduces an excess effective surface mass $\Delta\rho\delta b^{lmi}$, where $\Delta\rho$ is the difference in density between the lower and upper layers. Since Poisson's equation is linear, the gravitational effects of this mass anomaly alone may be considered independently. The perturbed potentials above and below the surface mass are given by the solutions to Laplace's equation,

$$\delta V^{lmi}(r) = \begin{cases} \delta V^{lmi}(b)(b/r)^{l+1} & r > b \\ \delta V^{lmi}(b)(r/b)^{l} & r < b \end{cases}. \tag{4.48}$$

Using the standard trick of applying Green's theorem to Poisson's equation in the vicinity of b and integrating over a "pill box" centered about b leads to

$$\left. \frac{\partial \delta V^{lmi}(b)}{\partial r} \right|_{b-}^{b+} = -4\pi\gamma\sigma^{lmi}, \tag{4.49}$$

and

$$\delta V^{lmi}(b) = \frac{4\pi\gamma}{2l + 1}\sigma^{lmi}. \tag{4.50}$$

From (4.49), there is an effective jump condition in gravity at the spherical reference surface $r = b$.

In terms of $\mathbf{u}$ and $\mathbf{v}$, these boundary conditions lead to the conditions

$$\left. \mathbf{u} \right|_{b-}^{b+} = [0, 0, \Delta\rho b(g\delta b - \delta V)/\eta_0, 0]^T, \tag{4.51}$$

$$\left. \mathbf{v} \right|_{b-}^{b+} = [0, -4\pi\gamma b^2 \rho_0 \sigma_b/\eta_0]^T. \tag{4.52}$$

Within the fluid core, the perturbed normal stress at the reference core radius, $r = c$, is $-\rho_c \delta V(c)$. For a free-slip boundary condition at the CMB, the starting vector $\mathbf{u}(c)$ is given by

$$\mathbf{u}(c^+) = [0, u_2(c), u_3(c^+), 0]^T. \tag{4.53}$$

Just above the Earth's surface, the density and normal stress vanish, so $u_3(a^+) = 0$. Applying (4.51), just within the mantle,

$$\mathbf{u}(a^-) = [0, u_2(a), u_3(a^-), 0]^T, \tag{4.54}$$

with $u_3(a^-) = -\rho_\mu\alpha[g(a)\delta a - \delta V_a]/\eta_0$, where ρ_m is the density of the mantle and δa is the deformation of the upper surface of the Earth.

The differentiation of (4.48) just outside the Earth's surface gives

$$\frac{\partial \delta V^{lmi}}{\partial r}(a^+) = -\frac{(l+1)}{a}\delta V^{lmi}(a^+). \tag{4.55}$$

Using (4.52), just below the Earth's surface,

$$\mathbf{v}(\mathrm{a}^-) = [v_1(a^-), -(l+1)v_1(a^-) + 4\pi\gamma a^2\rho_0\rho_m\delta a/\eta_0]^T. \tag{4.56}$$

Similarly, just inside the CMB,

$$\frac{\partial \delta V^{lmi}}{\partial r}(c^-) = \frac{l}{c}\delta V^{lmi}(c^-). \tag{4.57}$$

Then, just above the CMB,

$$\mathbf{v}(\mathrm{c}^+) = [v_1(c^+), lv_1(c^+) - 4\pi\gamma c^2\rho_0\Delta\rho_{CMB}\delta c/\eta_0]^T. \tag{4.58}$$

Finally, it is necessary to relate the deformation of a given boundary to the flow and stress fields. From (4.47), the deformation of a boundary δb is related to the nonhydrostatic normal stress $\delta\tau_{rr}$ by

$$\delta b = -\ \Delta\rho g(b)\tau_{rr}(b). \tag{4.59}$$

Applying (4.40), (4.41), (4.42), (4.51), (4.52), (4.53), (4.54), (4.56), and (4.58) leads to six equations in six unknowns, which can be solved to obtain, e.g., the starting vectors $\mathbf{u}(c)$ and $\mathbf{v}(c)$. Then (4.40) and the related equation for potential can be applied to propagate the solution to any other radius.

For the purpose of this chapter, we are most interested in the values of the potential at the surface and the deflections of the surface, of the core-mantle boundary and of any interior chemical boundaries that might exist. Rather than focus on the values for a particular distribution of density contrasts, we solve for normalized response functions. Since the problem is linear, the total response is obtained by convolving these normalized response functions with a particular distribution of density contrasts.

Because the propagator matrices and constants in the **a** and **b** vectors depend on degree l, but not on order m or on i, the response functions for a given harmonic degree are degenerate, with no m or i dependence. (For a model Earth with azimuthal variations in effective viscosity, this degeneracy breaks down and the problem becomes much more complicated.) $G^l(r)$ is defined to be the normalized gravitational response function or kernel for a density contrast $\delta\rho^{lmi}(r)$. This kernel, or Green function, contains the effects of both the internal density contrast itself and the effects of all boundary deformations induced by the resulting flow. Then the potential at the surface for some distribution of internal density contrasts is given by

$$\delta V^{lmi}(a) = \frac{2\pi\gamma a}{2l + 1}\int_c^a G^l(r)\delta\rho^{lmi}(r)dr. \tag{4.60}$$

With this normalization, for a static Earth, with no dynamic boundary deformation, $G^l(r) = (r/a)^{(l+2)}$.

A normalized displacement kernel, or Green function, which is the ratio of the mass displaced at a given boundary to the mass driving the flow, can also be defined. Then the deformation of the surface, the CMB, and, say, a chemical boundary at $r = d$ are given by

$$\begin{aligned}
\delta a^{lmi} &= \frac{1}{\Delta\rho_a}\int_c^a A^l(r)\delta\rho^{lmi}(r)dr, \\
\delta c^{lmi} &= \frac{1}{\Delta\rho_{CMB}}\int_c^a C^l(r)\delta\rho^{lmi}(r)dr, \\
\delta d^{lmi} &= \frac{1}{\Delta\rho_d}\int_c^a D^l(r)\delta\rho^{lmi}(r)dr,
\end{aligned} \tag{4.61}$$

Displacements are inversely proportional to the density contrast across a given boundary.

Kernels for Earth Models

We present here kernels for a series of models of increasing complexity and, perhaps, realism, to illustrate how changes in mantle structure can affect the geoid calculated for a given distribution of density contrasts. These models, summarized in Table 9.1, can have up to 5 viscous layers. For each viscosity model, we calculate two flow models: model W, representing whole-mantle

Table 9.1. Relative viscosities as a function of depth for models used

	Depth range (km)				
Model	0–100	100–400	400–670	670–2800	2800–2900
1	1.0	1.0	1.0	1.0	1.0
2	10.	1.0	1.0	1.0	1.0
3	10.	0.032	1.0	1.0	1.0
4	10.	0.032	1.0	10.	10.
4a	1.0	0.032	1.0	10.	10.
5	10.	0.032	1.0	10.	0.01

flow, and model C, representing flow in a chemically stratified mantle, with separate flow systems above and below 670 km depth. We start with a simple model with uniform viscosity and add layers until we end up with a model that provides geoid responses that seem appropriate for the Earth.

Note that the inverse of η_0, the reference viscosity, enters linearly in the propagator matrix $\mathbf{P}_A$, as well as in $\mathbf{u}_3$, $\mathbf{u}_4$, $\mathbf{a}_3$ and $\mathbf{a}_4$, but not in $\mathbf{u}_1$ and $\mathbf{u}_2$. The form of the equations shows that for a given distribution of density contrasts, while the flow velocity is inversely proportional to the reference viscosity, the stresses and potential are independent of it. Thus the only sensitivity we have to the absolute level of mantle viscosity in models with flow driven by interior density contrasts is through flow velocity, and not potential. (For kinematic flows, models in which surface velocities are prescribed as boundary conditions (Hager and O'Connell, 1978, 1979, 1981; Ricard, 1986), stress does depend on viscosity.) We discuss these points later in our estimates of advective heat flux. For this section, we choose η_0 to be the (unspecified) viscosity of the lower part of the upper mantle, the transition zone.

The simplest model is model W1, representing mantle-wide convection in a uniform viscosity mantle. Normalized potential kernels $G^l(r)$ for $l = 2, 4$, and 8 are shown along with boundary deflection kernels $A^l(r)$ and $C^l(r)$, as well as the sum of $A^l(r)$ and $C^l(r)$.

Since the total geoid response $G^l(r)$ is made up from the gravitational effects of the deformed boundaries, as well as that of the mass itself, we first discuss the boundary deformations. For all degrees, and for driving density contrasts approaching the compositional boundaries at the surface or at the CMB, the total mass displaced at boundaries by the flow approaches the mass of the driving density contrast—an example of dynamic isostasy. For this case, the geoid response goes to zero.

Also for all degrees, the deformation of the upper surface decreases and the deformation of the CMB increases as the depth of the load increases. Loads right at the upper surface have no effect on the CMB and vice versa. The redistribution of deformation from the upper surface to the CMB as the load depth increases is not symmetric—the crossover point occurs nearer the upper surface than the CMB. This is a geometric effect resulting from the constriction of flow at smaller radii, resulting in relatively higher stresses due to greater strain rates at depth than near the surface.

A perhaps surprising result is that for the lower degree harmonics, the driving mass is overcompensated—the sum of the masses displaced at the upper and lower surfaces is larger than the mass driving the flow. This overcompensation is the result of two effects. One is that self-gravitation amplifies all boundary deformations. The second effect is a more obscure hydrodynamic one. Large aspect ratio flow generates substantial conical, or radially directed, shear tractions that act in the same direction as the body forces from the internal density contrasts. Force balance on a given column of the mantle then requires the sum of the normal tractions at the top and bottom of the mantle to exceed the force due to internal buoyancy alone.

This effect is reversed at short wavelengths, and radial shear tractions in the mantle tend to support the density contrast. For this reason, and also because the effects of self-gravitation are less important, loads with half wavelength less than the depth of the mantle tend to be somewhat undercompensated. For all wavelengths shown, the mass displaced by surface deformations is of the same order as that of the driving density contrasts.

The gravitational potential kernels in Figure 9.21 represent the effects of both the driving density contrasts and the deformed boundaries. For this model they are always negative. The gravitational effects of boundary deformation overwhelm the signal from the driving density contrasts. At all wavelengths, the main effect is due to the deformed upper surface having more influence on the geoid (at the surface) than does the driving density contrast. Gravitational potential decays more rapidly with distance than does the viscous stress causing surface deformation. The effect of distance is augmented by the overcompensation due to self-gravitation and viscous stresses.

These kernels show that the dynamic surface deformation caused by flow in the mantle can reverse the sign of the geoid anomaly associated with a given density field. Such a mechanism would explain the anticorrelation between the observed geoid and that predicted from the lower mantle density structure inferred from seismic tomography. But before making a detailed comparison, we first investigate the effects that possible variations in viscosity with depth might have on surface deformation and the geoid. We do this by introducing successively more detail into our rheological model and seeing how the relevant kernels are affected.

Probably the major rheological stratification in the Earth is the contrast between the relatively stiff lithosphere and the relatively weak asthenosphere beneath it. The lithosphere comprises the cold upper thermal boundary layer of the mantle convection system. Because it is cold, it is relatively strong. In plate interiors, the effective viscosity of approximately the outer 30 km is effectively infinite, and the lithosphere behaves as an elastic plate over geologic times (e.g., Watts, 1978).

Although locally the lithosphere behaves elastically, viewed on a global basis it does deform, as is demonstrated clearly by the relative motions

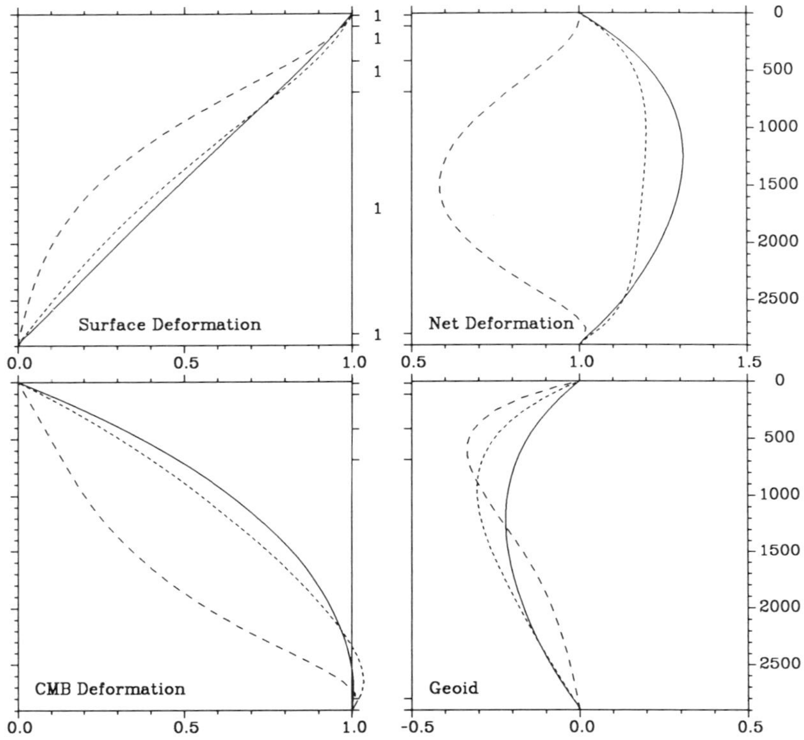

Figure 9.21. Normalized kernels for model W1, showing the effects of a surface mass σ placed at a given depth in the mantle for $l = 2, 4$, and 8. The mass displaced by surface deformation, by deformation of the CMB, and the total of these two are normalized by dividing by $(-\sigma)$. A value of unity implies perfect dynamic compensation. The geoid kernel, which includes the effects of the dynamic topography as well as σ, is normalized by the potential due to a mass σ at the surface of the Earth. The solid line is for degree 2, short dashes for degree 4, and long dashes for degree 8.

between plates. Hager and O'Connell (1981) have given an extensive discussion of how to incorporate the lithosphere into models of mantle flow. Lithosphere deformation is concentrated in the fairly narrow regions making up the plate boundaries. Presumably it is weaker in these regions due to through going faults (e.g., at subduction zones) or anomalously low temperatures (e.g., at ridges) or both. Unfortunately, our mathematical technique, which limits us to spherically symmetric viscosity structures, does not allow us to consider these obvious heterogeneities, and we must assign an effect viscosity to the lithosphere as a whole. (We will discuss possible effects of this parameterization later.) We can operationally assign an effective viscosity at a given harmonic degree as a stress divided by a strain rate (e.g., Hager and O'Connell, 1981; Kaula, 1980).

For the sake of illustration, model W2 has a 100 km thick "lithosphere" with a viscosity a factor of 10 larger than the reference viscosity of the mantle. (If this reference viscosity is 10^{22} p, the lithosphere would have a viscosity of 10^{23} p; comparable to that suggested by Kaula (1980)). Kernels for this model are shown in Figure 9.22. The high viscosity lid generally results in more surface deformation and less deformation of the CMB than was the case for the uniform viscosity model W1. The cross-over point where

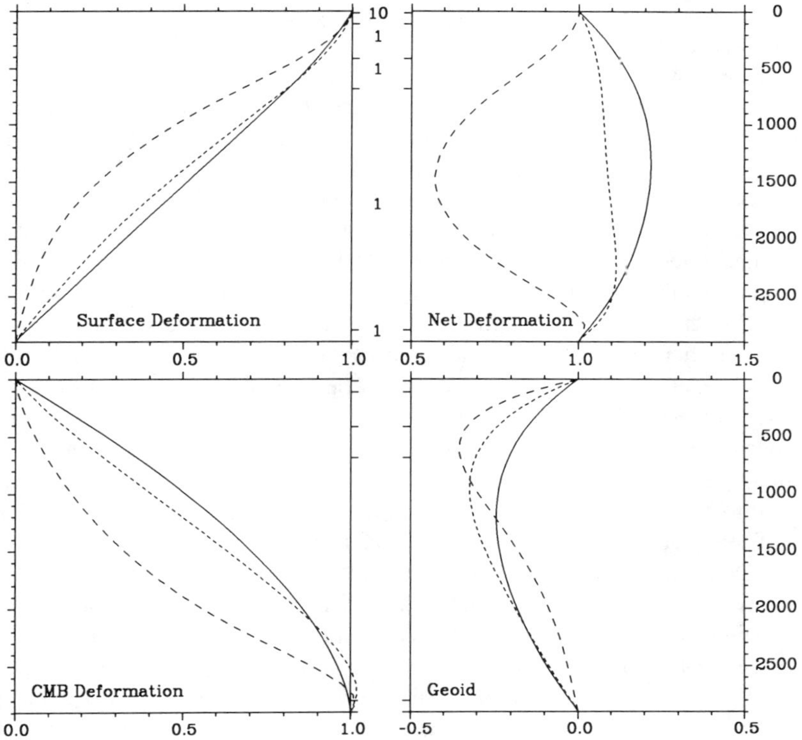

Figure 9.22. As in Figure 9.21, but for model W2, which differs from model W1 by the addition of a 100 km thick high viscosity lid at the surface.

an equal portion of the load is "supported" by deformation of the top surface and the CMB is pushed deeper into the mantle, as the retardation of flow by the high viscosity lid competes against the geometric effect of decreasing radius near the CMB. Interestingly, the total mass displaced by surface deformation decreases.

The net result is that the geoid kernels for all degrees become more negative than for the uniform viscosity example. This can be seen more clearly in Figure 9.27, where the kernels for all W type models are compared degree by degree.

The next level of complication is to introduce a global low viscosity channel, or asthenosphere, into our model. In model W3, a layer with a viscosity 1½ orders of magnitude lower than that of the reference viscosity is placed in the upper mantle between the depths of 100 and 400 km. (The bottom is placed at 400 km depth to coincide with the phase change there. Phase changes are expected to lead to viscosity jumps (Sammis *et al.*, 1977).) Kernels for this model are shown in Figure 9.23, and geoid kernels are compared with those in other models in Figure 9.27.

The effect of the addition of the low viscosity channel to the model is, in general, to decrease coupling of the flow to the surface and to increase coupling to the CMB. Crossover points for equal distribution of loads to the surface and the CMB become shallower than for model W2, and at degrees 4 and 8, shallower than for model W1. The addition of the low viscosity channel effectively cancels the effect of the high viscosity lid, giving a geoid response quite close to that of W2 for long wavelengths. For shorter wavelengths, the effects of this fine structure can begin to be discerned.

In addition to this effective cancellation of the effects of adjacent high and low viscosity regions, there is a trade-off between the thickness of the low viscosity channel and its viscosity contrast. A thinner channel with a proportionately lower viscosity would give kernels similar to those shown here, although subtle differences remain for shorter wavelength loads in the vicinity of the channel.

The effects of pressure on viscosity lead us to expect that, assuming that the deep mantle geotherm lies essentially along an adiabat, viscosity should vary continuously with depth in the deep mantle, increasing, decreasing, or remaining constant, depending upon the value of the adiabatic gradient. We also expect discrete increases to occur at phase transitions (Sammis *et al.*, 1977). As a first approximation to this effect, in model W4 the lower mantle viscosity is increased relative to that in the transition zone by a factor of ten. The resulting kernels, shown in Figure 9.24, show the effects of such an increase in viscosity. The increased viscosity of the lower mantle causes

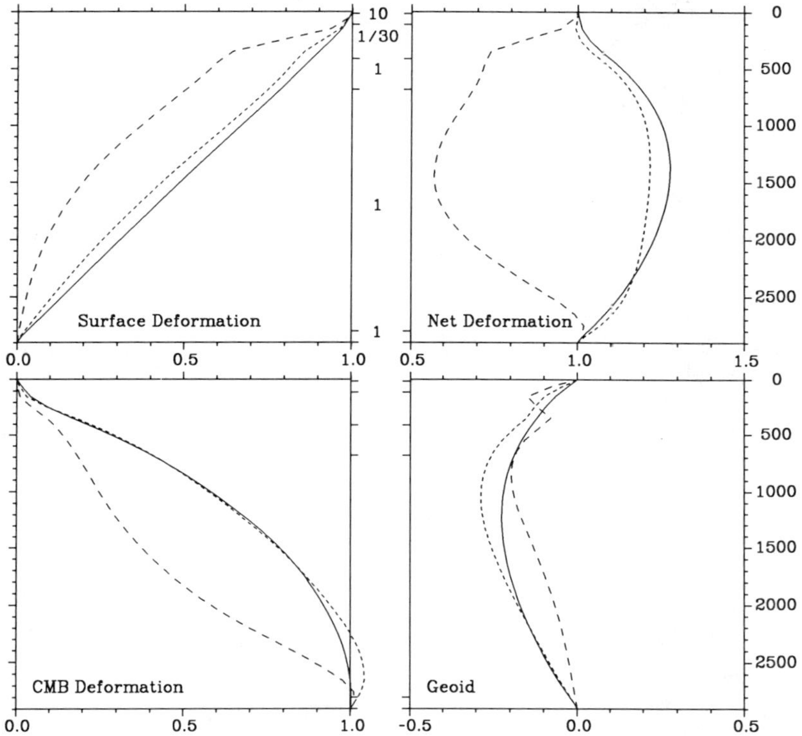

Figure 9.23. As in Figure 9.22, but for model W3, which differs from model W2 by the addition of a low viscosity asthenosphere in the depth range 100–400 km.

substantial shifts in the displacement kernels. Now the crossover point where equal mass displacement occurs for the surface and the CMB shifts upward and is shallower than 500 km for all degrees. Deformation of the top surface is dramatically reduced, leading to geoid kernels that are more positive than for the previous models. Except within the high viscosity lid, the geoid kernels are positive within the upper mantle. Kernels for shorter wavelengths remain positive to greater depths than for longer wavelengths. This is the first of these models to show that the sign of the geoid response can change as a function of radius. Perhaps more important is the variation in form of the kernels for different harmonic degrees—if the distribution of density contrasts can be determined reliably as a function of depth for a range of

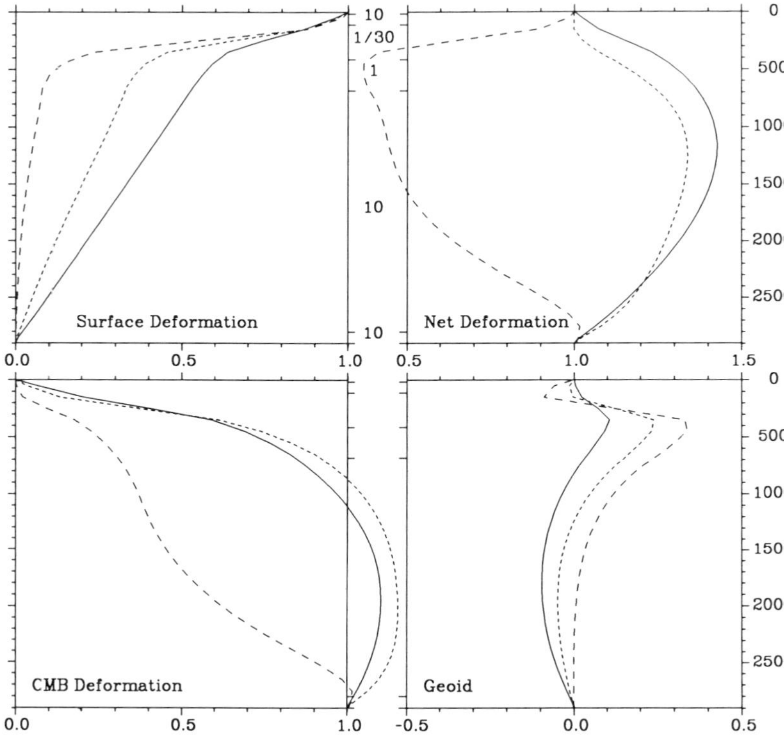

Figure 9.24. As in Figure 9.23, but for model W4, which differs from model W2 in having a higher viscosity below 670 km depth.

spherical harmonic degrees, fairly tight constraints can be placed on the general radial variation of mantle viscosity.

Parameterization of the lithosphere in terms of a single effective viscosity is one of the most questionable of the assumptions required by our modeling technique, so it is important to examine the sensitivity of the models to this parameter. The sensitivity of these kernels to the effective viscosity of the upper "lithospheric" layer is tested in model W4a (Figure 9.25), which has a high viscosity lid a factor of 10 less viscous than that in W4. Relative to W4, W4a has less deformation of the upper surface, more deformation of the CMB, shallower cross-over points of deformation kernels, and therefore more positive geoid kernels. The effects are most pronounced at longest

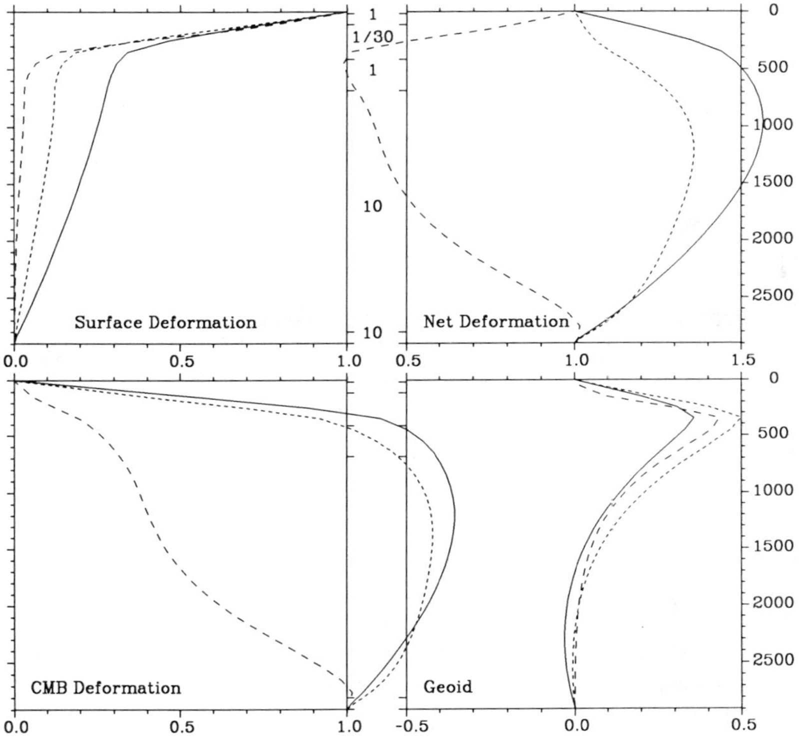

Figure 9.25. As in Figure 9.24, but for model W4a, which differs from model W4 by having a weaker "lithosphere."

wavelengths; at degree 2, the geoid kernel becomes positive in the upper two thirds of the mantle.

If substantial heat is coming from the core, we expect a hot thermal boundary layer to exist at the base of the mantle (e.g., Jeanloz and Richter, 1979). Such a thermal boundary layer is expected to have a decrease in viscosity associated with it. In model W5, the effects of a low viscosity layer at the base of the mantle are examined by including a 100 km thick layer with a viscosity a factor of 100 lower than that of the lower mantle. Kernels for this model are shown in Figure 9.26, with the geoid kernels for this model compared to those of the other models in Figure 9.27. Including a low viscosity zone at the base of the lower mantle decreases the coupling of

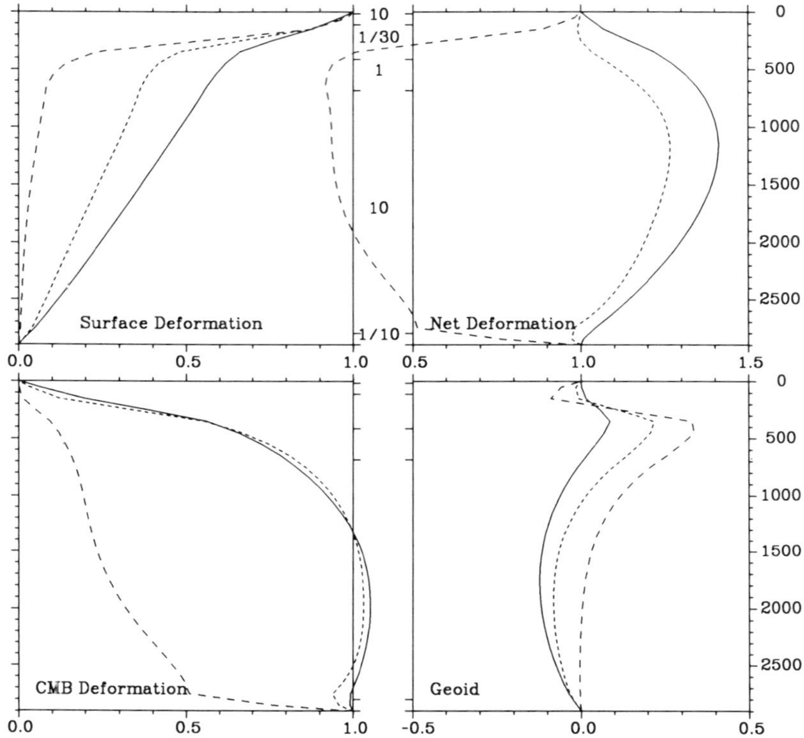

Figure 9.26. As in Figure 9.24, but for model W5, which differs from model W4 by having a 100 km thick low viscosity layer just above the CMB.

deformation to the CMB and increases the coupling to the surface. This makes the geoid kernels more negative. The effect on boundary deformation is much more pronounced for density contrasts in the lower mantle than it is for those in the upper mantle and increases dramatically at shorter wavelengths. Although the relative effect on CMB deformation is largest at shorter wavelengths, the absolute effect on the total geoid kernels is more pronounced for longer wavelengths, since due to the $(r/a)^{l+2}$ effect, kernels for shorter wavelength (higher l) density contrasts in the lower mantle are generally small.

Geoid kernels $G^l(r)$ for the whole-mantle flow models discussed are compared at individual degrees in Figure 9.27, which illustrates some of the trade-offs involved in the flow models. For all degrees, model W2 generally

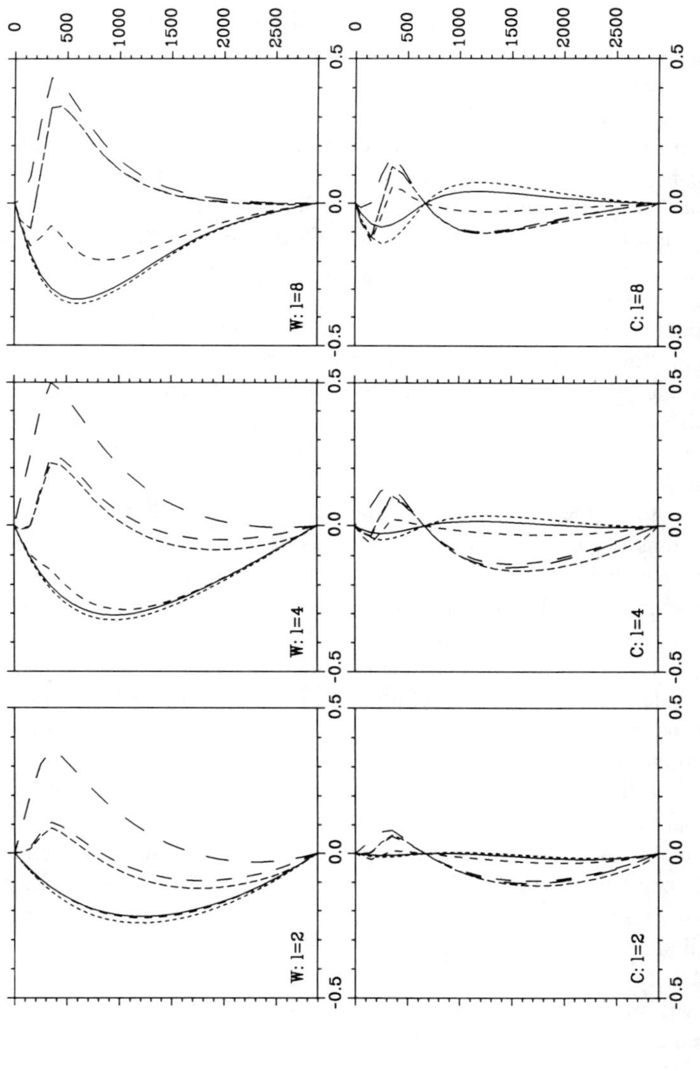

Figure 9.27. The upper panels show $G^l(r)$ as a function of radius for W (whole-mantle flow) models at degrees 2, 4, and 8. The lower panels are for chemically stratified flow models for the same degrees.

has the most negative kernels due to its strong coupling to the upper surface caused by its high viscosity lid, lack of asthenosphere, and its relatively low viscosity lower mantle. (Very near the CMB, the low viscosity layer near the CMB in model W5 allows even more negative kernels, although their amplitude is too small to be seen clearly in this figure.) Model W4a has the most positive kernels for all depths because of its weak "lithosphere," its asthenospheric layer, and its high viscosity lower mantle. For long wavelengths (aspect ratio greater than unity), models W1 and W3 have similar kernels; the effects of the high viscosity lid and low viscosity asthenosphere nearly cancel. At shorter wavelength (degree 8), the kernels become more sensitive to radial structure. At long wavelengths (degrees 2–4) the low viscosity layer near the CMB in model W5 counteracts the effect of the higher viscosity lower mantle in W4. At shorter wavelengths (degree 8), the kernels are not sensitive to the viscosity structure at the base of the mantle.

It is possible that the mantle is chemically stratified with a compositional barrier to flow at 670 km. The effects of chemically stratification are illustrated for the same viscosity models used above. (Chemical stratification might also complicate the viscosity structure, e.g., by causing a low viscosity channel just above the 670 km discontinuity and a high viscosity lid just below it due to a double thermal boundary layer there.)

Kernels for model C1, a model with uniform viscosity but a chemically induced density contrast providing stable stratification at a depth of 670 km, are shown in Figure 9.28. Again, normalized potential kernels $G^l(r)$ are shown, while $D^l(r)$, the normalized mass displaced at the 670 km discontinuity is given, in addition to boundary deformation kernels $A^l(r)$, $C^l(r)$. The sum of $A^l(r)$, $D^l(r)$, and $C^l(r)$, the total mass displaced is also shown.

For all degrees, the surface deformation kernel $A^l(r)$ is unity at the top surface and decreases nearly linearly to zero at a depth of 670 km, where the load is supported completely by deformation of the 670 km discontinuity. For loads in the lower mantle, shear coupling between the lower mantle and upper mantle leads to, for example, uplift of the upper surface over dense sinkers in the lower mantle, so the surface deformation kernel $A^l(r)$ actually becomes negative. A similar effect is shown for the CMB deformation kernel $C^l(r)$, which is negative for loads in the upper mantle, goes to zero at the 670 discontinuity, then increases to unity as a load in the lower mantle approaches the CMB. $D^l(r)$, representing the normalized dynamic deformation of the 670 km discontinuity, is always positive, i.e., the boundary is always displaced downward by dense sinkers. For loads just below the 670 discontinuity, $D^l(r)$ exceeds unity for the same reasons discussed for $A^l(r)$ for model W1—self-gravitation and large conical shear tractions. The total mass dis-

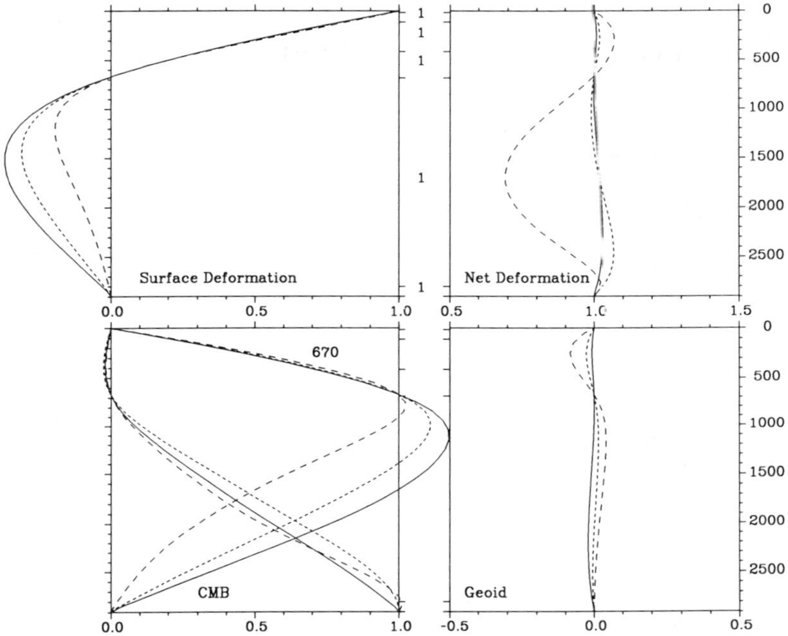

Figure 9.28. a, As in Figure 9.21, but for model C1 with stable chemical stratification at a depth of 670 km. The normalized deformation of the 670 km discontinuity is also shown. b, As in Figure 9.21, with the additional kernel $D^l(\tau)$, the normalized mass displacements at the 670 km (assumed) chemical boundary, are also plotted. $D^l(\tau)$ are zero at the surface and the CMB and reach unity at 670 km depth.

placed by boundary deformation is of the order of the mass of the load, and dynamic isostasy holds especially well at long wavelengths.

The geoid kernels $G^l(r)$ in the upper mantle are all negative and much smaller in amplitude than for the uniform model W1. Kernels have larger amplitude as the harmonic degree increases; for degree 2, the geoid kernel is almost zero. The $G^l(r)$ in the lower mantle are again small in amplitude and are of variable sign, due primarily to the effects of dynamic deformation caused by shear coupled flow.

Including a high viscosity lithosphere, as shown by model C2 in Figure 9.29, leads to stronger coupling of the flow to the upper surface, especially for loads in the upper mantle, and enhanced deformation of the upper surface with respect to the 670 km discontinuity and the CMB. This enhanced

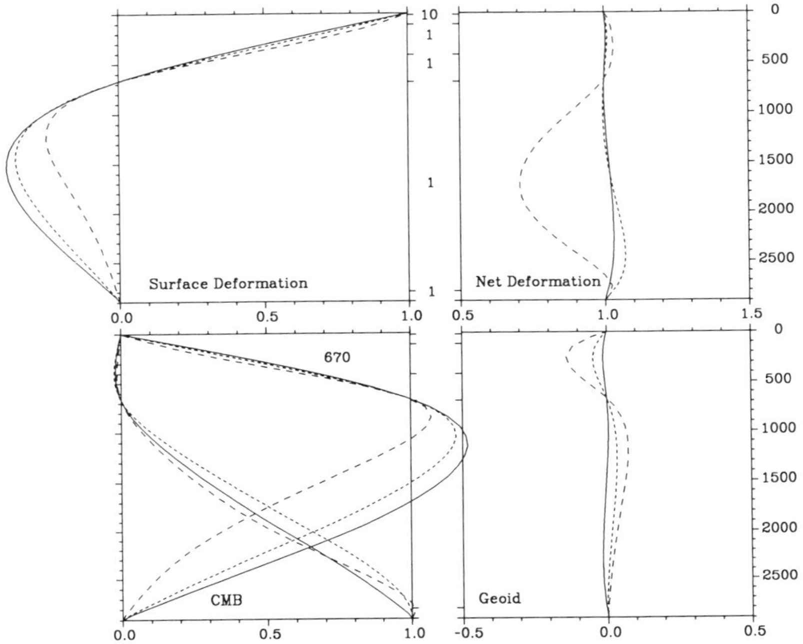

Figure 9.29. As in Figure 9.28, but for chemically stratified model C2, which has a high viscosity lid near the surface.

deformation of the upper boundary leads to more negative $G^l(r)$ for loads in the upper mantle and more positive $G^l(r)$ for loads in the lower mantle. Again, $G^l(r)$ have amplitudes that increase with l.

Adding a low viscosity asthenosphere, model C3 (Figure 9.30), has some interesting effects. Loads above this layer are more strongly coupled to the upper boundary, leading to more negative $G^l(r)$ for $r \lesssim a$. They are also more strongly coupled to the CMB. The decrease in viscosity above the 670 km discontinuity leads to larger $D^l(r)$ for upper mantle loads and smaller $D^l(r)$ for loads in the lower mantle. The net effect is to give more positive kernels for upper mantle loads and more negative kernels for lower mantle loads.

The $G^l(r)$ for model C4 (Figure 9.31), which has an increased viscosity in the lower mantle, are not entirely intuitively obvious. As expected, loads in the upper mantle are transferred more to the 670 km discontinuity and the CMB, resulting in more positive $G^l(r)$, especially below the asthenospheric layer. But despite the higher viscosity in the lower mantle, the $G^l(r)$ are more

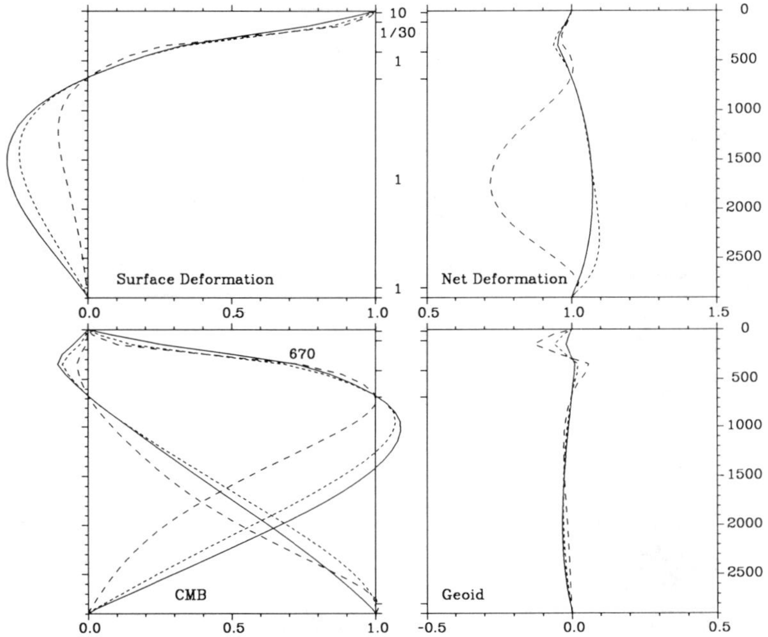

Figure 9.30. As in Figure 9.29 but for chemically stratified model C3, which has a low viscosity asthenosphere.

negative there. This is because the increased viscosity of the lower mantle leads to less shear coupling to the upper mantle for lower mantle loads. (Recall that this shear coupling tends to make $G^l(r)$, more positive for $r < d$.) Consequently, for loads in the lower mantle, Model C4 is quite similar to Model W1.

The effect of decreasing the viscosity of the "lithosphere" by a factor of 10, Model C4a, shown in Figure 9.32, is to decrease surface deformation in favor of deformation of the 670 km boundary and the CMB. The effects on the geoid kernels are quite small, however, except for loads in the uppermost mantle.

The addition of a low viscosity layer just above the CMB, model C5, Figure 9.33, has almost no effect for loads in the upper mantle, but a more pronounced effect for loads in the lower mantle. There, the deformation of the 670 km boundary is somewhat increased and the deformation of the CMB substantially decreased, leading to more negative kernels for all harmonics.

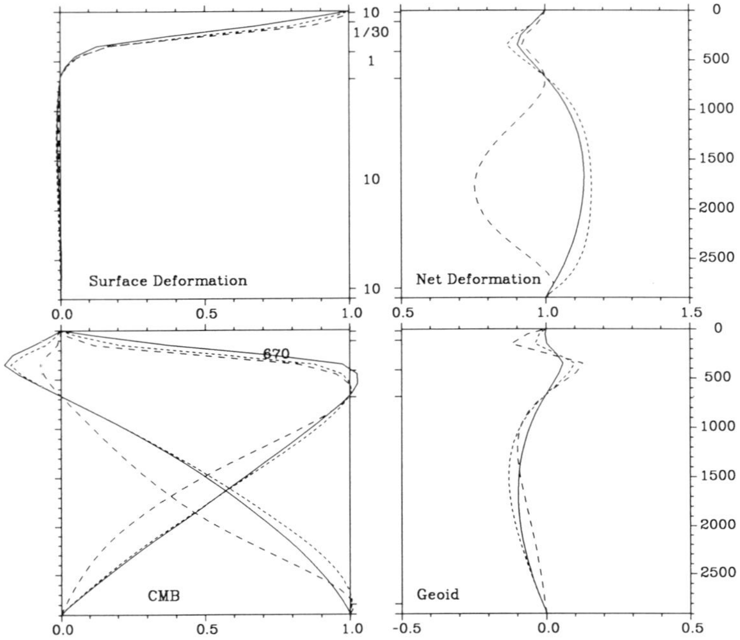

Figure 9.31. As in Figure 9.30, but for chemically stratified mantle model C4, which has a higher viscosity lower mantle.

Geoid kernels for degrees 2, 4, and 8 for the chemically stratified set of models are compared in Figure 9.27. The major factor influencing the general shape of these kernels is the average viscosity contrast between the upper mantle and the lower mantle. If the lower mantle has a greater viscosity than the upper mantle, geoid kernels for upper mantle loads are positive and for lower mantle loads are negative. Furthermore, the more the upper mantle viscosity increases with depth, the more positive the upper mantle kernels are.

For loads in the (shallow) upper mantle, the shorter the wavelength of the load, the larger is the geoid kernel. This illustrates nicely the effects of the ratio of depth to wavelength in determining the amplitude of geoid anomalies in a fluid layer. Interior density contrasts and the resulting dynamic compensation provides a mass quadrapole. The smaller the arm length of the quadrople (relative to the wavelength), the smaller the resulting gravitational effect.

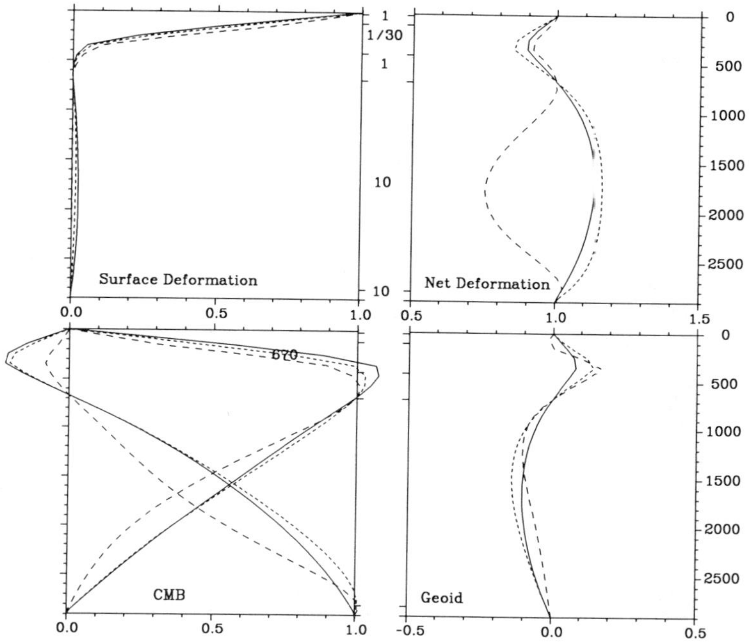

Figure 9.32. As in Figure 9.31, but for chemically stratified mantle models C4a, which has a less viscous near surface layer.

Based on the models shown here, as well as many other flow models that we have calculated but do not show, we can make several generalizations. Most important is that geoid kernels are sensitive to the radial structure of the mantle, being influenced by both changes of viscosity and chemical composition with depth. Density contrasts near boundaries between layers of different composition (including the surface and CMB) are locally compensated and have little effect on the geoid in a dynamic Earth; at long wavelengths, density contrasts in the interior of convecting regions have the most influence on the geoid. (Just the opposite is true for the effects of density contrasts on dynamic surface deformation.) Shorter wavelength geoid kernels are more sensitive to rapid variations in mantle structure than are longer wavelengths, which tend to average over greater depths. In order to get good resolution of radial viscosity structure, a broad spectrum of loads is required.

Increasing viscosity with depth tends to make geoid kernels more posi-

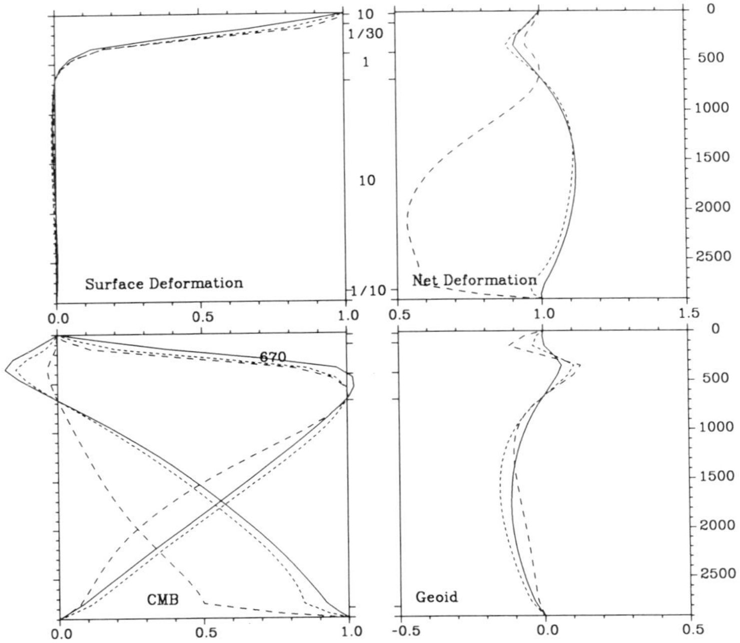

Figure 9.33. As in Figure 9.31, but for chemically stratified mantle model C5, which has a low viscosity layer just above the CMB.

tive, while decreasing viscosity with depth, such as going from a high viscosity lithosphere to a lower viscosity planetary interior, makes kernels more negative. Once a low viscosity asthenosphere is included in a model, the value of the effective viscosity of the lithosphere becomes especially important in determining the form of the geoid kernels.

Finally, bearing in mind likely radial variations in viscosity structure based on the competing effects of pressure and the geotherm with depth, i.e., a high viscosity lid underlain by a low viscosity asthenosphere, with viscosity then increasing somewhat with depth through the mantle, it is straightforward to obtain geoid kernels that are positive for loads in the upper mantle, yet negative for loads in the lower mantle. As we see in the next section, this allows us to resolve the apparent paradox of geoid highs associated both with high density subducted slabs in the upper mantle and with low velocity, presumably low density, material in the lower mantle.

5. Dynamic Calculations of the Geoid

Three sets of density contrasts identified as contributing to the net geoid are considered here. These are lower mantle density variations inferred from body wave tomography, upper mantle variations inferred from surface wave tomography, and density variations in subducted slabs. For each of these the pattern is given, but the magnitude of the density contrasts is not known *a priori*. We assume that the magnitude of the density contrast for each of these is given by a proportionality constant. In this section, all three constants of proportionality are determined in a combined inversion to predict the non-hydrostatic variations in the long wavelength geoid. For simplicity they are discussed one at a time; the details of the joint inversion are left for a later section.

Lower Mantle Structure and the Geoid

We now have developed the mechanism for calculating dynamic topography and geoid anomalies, given the distribution of heterogeneity within a dynamic Earth and a model of the distribution of viscosity with depth in the mantle. The geoid calculated by convolving the $G^l(r)$ for model W4 with the density heterogeneity inferred for the lower mantle from seismic tomography is shown in Figure 9.34. For reasons explained in Section 3, (4.60) is evaluated only over the depth range 800–2800 km. Velocity and density perturbations are taken to be proportional, with the proportionally constant, $\delta v_p/\delta\rho$ = (2.1 km/sec)/(g/cm^3) discussed later.

There is a striking similarity between the predicted geoid and the observed geoid (Figure 9.1). Both are dominated by degree 2 polar lows, with highs over Africa and the Pacific. Both encompass most of the hotspots within geoid highs. The major qualitative disagreements are that the geoid high over the Pacific in the model is shifted to the east relative to the observed geoid and the observed high over the South American subduction zone is missing in the model. The longer wavelength features, which contribute most of the power to the geoid, are better fit than the shorter wavelength features.

The major geophysical signatures in the predicted geoid are the geoid highs associated with large scale regions of low velocities in the lower mantle. These low velocity features at great depth are closely associated with hotspot provinces at the Earth's surface (Hager *et al.*, 1985; Richards and Hager, 1988; Richards *et al.*, 1986), suggesting mantle-wide convection or thermal coupling between the lower mantle and the surface. Statistically, the correlation coefficients between the observed and the model geoid at degrees

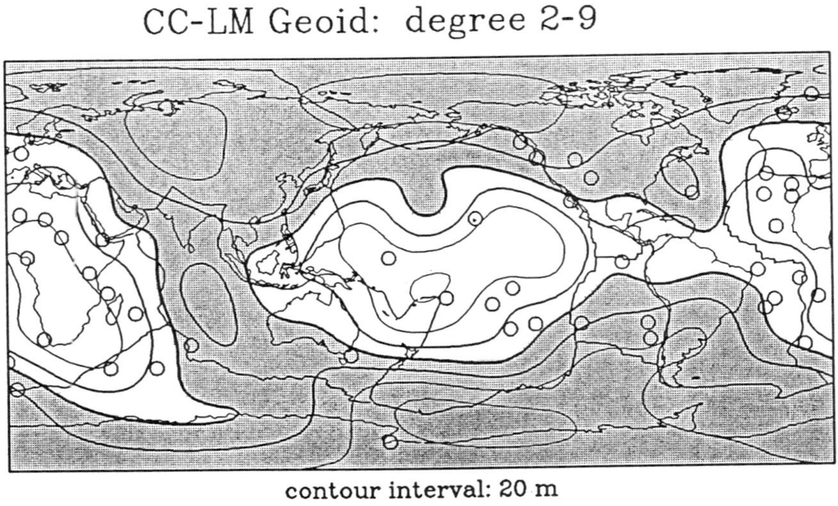

Figure 9.34. The geoid predicted using the dynamic geoid kernels for model W4 with the lower mantle (LM) tomographic model (CC) described in Section 2. Geoid lows are shaded.

2–6 are 0.82, 0.83, 0.63, 0.52, and −0.02, respectively. These correlations are significant at the 95, 99, 95, and 92 percent confidence levels for degrees 2–5, respectively. In contrast to the static case, the correlations are all positive and highly significant for degrees 2–5. The constant of proportionality relating velocity and density perturbations is, within the standard errors, independent of degree.

We have chosen to present the model geoid predicted using viscosity model W4, in part because a model like W4 incorporates some of our geophysical prejudice concerning mantle structure (e.g., a "lithosphere" and an asthenosphere) and in part because this type of viscosity structure will allow us to match the geoid signature, to be discussed shortly, associated with subducted slabs and broad scale upper mantle structure inferred from surface wave tomography. For long-wavelength lower mantle structure alone, there is much nonuniqueness in the kernels, and this complicated a viscosity model is not justified by the lower mantle results alone. Previously we used a two layer model that fit nearly as well (Hager *et al.*, 1985).

The mismatches between the predicted and the actual geoid are highlighted in Figure 9.35, a residual geoid produced by subtracting Figure 9.34 from Figure 9.1. The residual geoid is dominated by geoid highs associated

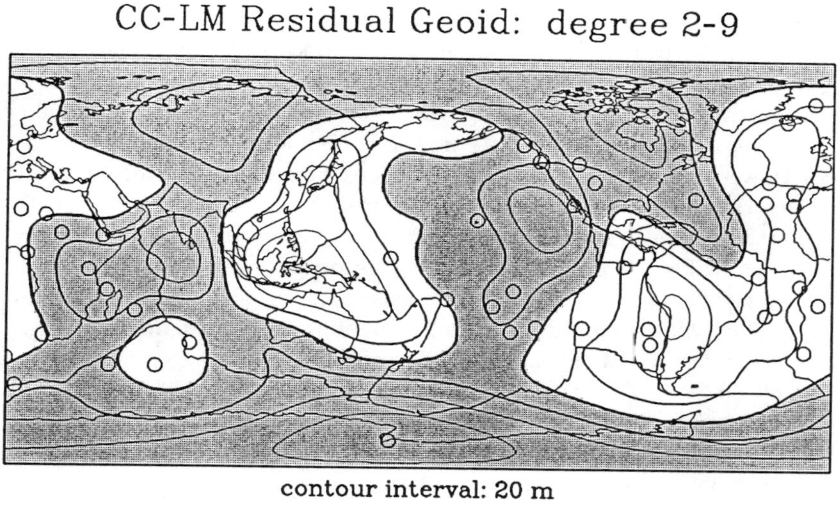

Figure 9.35. The residual geoid obtained by subtracting the model geoid shown in Figure 9.34 from the observed geoid, Figure 9.1. The contour interval is 20 m and geoid lows are shaded.

with subduction zones. This is perhaps expected given our earlier discussion about the association of observed geoid anomalies with both hotspots and subduction zones. Removing the signature of broadscale geoid highs associated with slow, presumably hot, lower mantle (and surface hotspots) leaves mainly the geoid highs associated with subduction zones.

As was the case with the observed geoid, removing the effects of lower mantle structure gives a long-wavelength residual geoid with no apparent association with midoceanic ridges. Ridges appear to be, in general, passive features with no significant geoid anomalies suggestive of a deep thermal structure associated with them. (There may be, of course, local exceptions such as the Atlantic hotspot provinces.) The lack of a geoid low associated with the rapid divergence at the East Pacific Rise can place a useful upper bound on mantle viscosity (Hager and O'Connell, 1979; Ricard, 1986), a point we will return to later.

Subducted Slabs and the Geoid

According to plate tectonic theory, cold, dense lithospheric slabs sink into the upper mantle beneath island arcs at subduction zones. The foci of deep earthquakes mark the present positions of the cold, central cores of the slabs, providing us with the only noncontroversial tracers of mantle convection currents presently available. Earthquakes do not occur deeper than the 670 km seismic discontinuity, leading some to conclude that subducted slabs do not sink any further and that the mantle is chemically stratified.

However, many geophysical observations suggest that convection does extend throughout the mantle and that the lack of seismicity deeper than 670 km reflects a change of material properties or other conditions necessary for seismic faulting (e.g., O'Connell, 1977). Among the observations we find more compelling are the travel time anomalies for deep earthquakes discussed by Jordan *et al.* (this volume), suggesting slab penetration to at least 1400 km depth, and the match of observed subduction zones dips to those predicted for whole mantle flow models by Hager and O'Connell (1978) and Hager *et al.* (1983). As we will see, the lower mantle residual geoid, Figure 9.35, can also be explained in a straightforward way by a model that allows whole-mantle convection (Hager, 1984).

If we knew the distribution of density contrasts associated with subducting slabs, we could calculate the effect on the geoid predicted for a given mantle model by using equation (4.60). Unfortunately, while the distribution of deep seismicity constrains to some extent the location of subducted slabs, it tells us little about their density distribution. Thus in calculating the geoid anomaly associated with subducting slabs, the density model, as well as the viscosity distribution must be assumed.

The mass anomaly associated with old lithosphere just before subduction is fairly well constrained to be about $8 \times 10^5 \mathrm{g/cm^2}$ by the 3.5 km subsidence due to thermal contraction of a plate with age as it moves from its ridge to its abyssal plain; for a plate thickness of 125 km (e.g., Parsons and Sclater, 1977) this corresponds to an average density contrast of 0.064 $\mathrm{g/cm^3}$. For subduction zones where seismicity extends to greater than 500 km depth, we assume that this mass anomaly is conserved, since thermal conduction is very slow and only redistributes mass anomalies. We assume that the slab neither lengthens nor extends, but that its mass anomaly per unit area normal to the plane of the slab remains constant, while the slab is rotated to an average dip of 60°. We also assume that subduction zones in which seismicity is confined to shallower than 300 km depth have a factor of two smaller mass anomaly associated with them, either due to younger lithosphere being subducted or to slower subduction. This parameterization is motivated by the concept that the depth to which slabs are seismically active is probably thermally controlled, and is observed to be proportional to the product of subduction velocity and age at the time of subduction (Molnar *et al.*, 1979; Davies,

1980). Thus mass anomaly and depth of the deepest seismicity should be related.

Although the seismicity cuts off at a given depth, possibly due to warming up of the slab, the mass anomaly associated with subducted slabs probably does not (warming the slab cools the surrounding mantle), so we must choose a lower limit to the integral in (4.60). In order to avoid double counting, we integrate the effects of subducted slabs only through the outer 800 km of the mantle, assuming that any deeper heterogeneities associated with subducted slabs are accounted for by the lower mantle tomography.

Clearly, in order to obtain positive geoid anomalies for the positive density contrasts associated with subducted slabs in the upper mantle, the geoid kernels must be positive there. Our best fit to the geoid, shown in Figure 9.36, is for model W4a. The geoid highs around the Pacific are reproduced quite well, although the residual geoid high over the Atlantic hotspot province is, of course, not explained by slab effects.

The correlation of subducted slabs and the observed geoid is best at degrees 4–9 (Hager, 1984), although it is still positive at degrees 2 and 3. As can be seen from Figure 9.24, the geoid kernel for $l = 2$ for model W4, the one we used for the lower mantle structure is positive but rather small in the

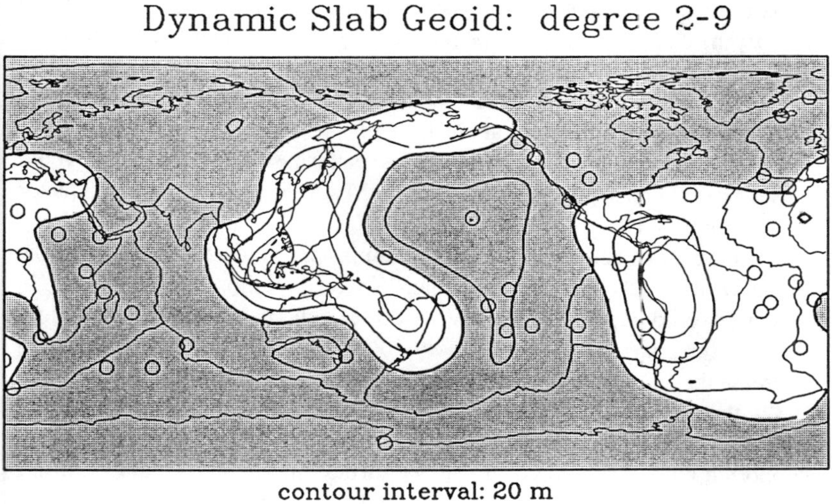

Figure 9.36. A dynamic geoid predicted for mantle wide flow, W4a and the slab density contrast model discussed in the text. The contour interval is 20 m and geoid lows are shaded.

upper mantle, while the geoid kernel $l = 2$ for model W4a is somewhat larger. For this reason, model W4 does not fit the lower mantle residual geoid with its prominent degree 2 anomaly quite as well as model W4a, with its weaker lithosphere, does.

The lithosphere at subduction zones is cut by through going faults, so it seems reasonable to expect it to have a lower effective viscosity (stress/strain rate) than the global average. Finite element calculations show that the effect of a weakened lithosphere near subduction zones perturbs geoid kernels for subducted slabs in a way similar to a globally weaker lithosphere (Richards, 1986; Richards *et al.*, 1988). Based on these calculations, we view model W4a as a reasonable approximation to a regionalized Earth model appropriate for subduction zones. Alternatively, a model with a lithospheric viscosity structure intermediate between W4 and W4a provides a somewhat larger geoid anomaly for slabs at $l = 2$ than model W4 without too seriously degrading the fit for the lower mantle model.

Up to here, we have shown only the results for models with mantle wide flow. Chemically stratified models such as C4a also have the positive upper mantle kernels and negative lower kernel models needed to fit the observed geoid using lower mantle tomography and subducted slabs. The slab geoid calculated for the density model described above (but confined to above the 670 km seismic discontinuity, consistent with chemical stratification at 670 km depth) using the kernels for model C4a is not shown because it is so similar in pattern to Figure 9.36. Model C4a has the largest upper mantle kernels of the models discussed above, yet the peak amplitude of the calculated slab geoid is only about 12 m, far less than the amplitude of the lower mantle residual geoid. Although the pattern is correct, either the geoid kernels or the density contrast used (or both) are too small, by about a factor of 5.

We have not been successful in obtaining a chemically stratified flow model with kernels larger by more than 25 percent of those for model C4a. In fact, we would expect that the low viscosity zone that would result from a thermal boundary layer just above 670 km in a stratified mantle would decrease the amplitudes of upper mantle kernels. As can be seen from Figure 9.32, the fraction of the load supported by deformation of the 670 km chemical boundary in these models is already quite large compared to that supported by deformation of the upper surface and it is difficult to transfer even more of the load to the 670 km boundary. The fundamental problem in obtaining large kernels for wavelengths so long compared to the thickness of the layer is that the dynamic compensation occurs so close to the load (the appropriate scale distance being the wavelength) that the net gravitational effect of the mass dipole is very small. If chemically stratified mantle convection is to be reconciled with the amplitude as well as the pattern of geoid anomalies associated with subduction zones, we must appeal to substantially larger mass anomalies than we have used in our slab model. We return to this point in the discussion.

Subtracting the model W4a dynamic slab geoid in Figure 9.36 from the lower mantle residual geoid in Figure 9.35, we obtain the "lower mantle and slab" residual geoid shown in Figure 9.37. The amplitude of this residual geoid is now substantially lower than that of the observed geoid (note the smaller contour interval), but it still has what appears to be "geophysical signal" associated with it. Shields tend to be in geoid lows, as do regions covered by Pleistocene deglaciation. The Atlantic hotspots, as well as the prominent Hawaiian and Kerguelen hotspots are in residual geoid highs.

Upper Mantle Structure and the Geoid

In parallel with studies of lower mantle structure using body wave tomography, there has recently been spectacular progress in addressing upper mantle structure using free oscillations (e.g., Masters *et al.*, 1982) and surface wave

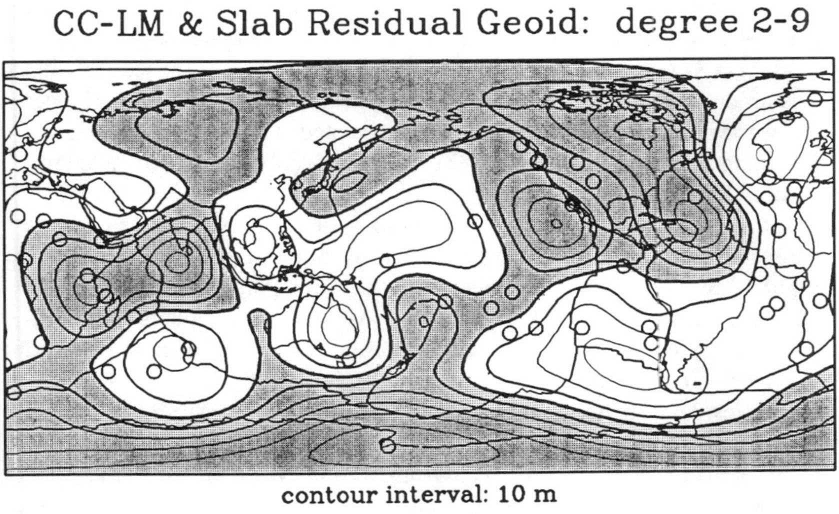

Figure 9.37. The lower mantle and slab residual geoid obtained by subtracting the slab model geoid in Figure 9.36 from the lower mantle residual geoid in Figure 9.35. The contour interval is now 10 m and geoid lows are shaded.

tomography (e.g., Nataf *et al.*, 1984; 1986; Nakanishi and Anderson, 1984; Woodhouse and Dziewonski, 1984; Tanimoto and Anderson, 1984; Tanimoto, 1986). Although at the longest wavelengths upper mantle geoid kernels tend to be small, sufficiently large density contrasts in the upper mantle might make a large enough contribution to the geoid to help explain some of the residual in Figure 9.37.

A basic problem in interpreting structure inferred from surface wave tomography in terms of the geoid is that the two quantities have opposite ranges of sensitivity with respect to depth. Geoid anomalies in a dynamic Earth are almost completely insensitive to near-surface heterogeneity and are most sensitive to heterogeneity at depth, while surface waves are most sensitive to near surface heterogeneity and have progressively less resolution with depth (e.g., Tanimoto, 1986). Geoid anomalies predicted from models of Earth structure derived from surface wave tomography are most sensitive in the region where inversions for mantle structure have the most variations among models.

Perhaps for these reasons, while all the upper mantle seismic models referred to above have striking similarities, we have had the most success in matching features of the geoid using only the longest wavelengths of Tanimoto's (1986) model, a model which does not exhibit the large rapid radial variations of some of the other models. The dynamic model of the geoid at degrees 2–4 predicted for this model is shown in Figure 9.38. For this model, an S-wave velocity to density variation ratio of 24 (km/sec)/(g/cm^3) was used. The main features in this model geoid are the degree 2 high over the western Pacific and the Atlantic, associated with fast material in the transition zone (Masters *et al.*, 1982), and lows over North America, Antarctica, and the Carlsberg Ridge.

If this dynamic model geoid is added to those predicted for lower mantle anomalies and subducted slabs, the total dynamic model geoid shown in Figure 9.39 is obtained. The match between this predicted geoid and the observed geoid, Figure 9.1, is excellent. All the major features are reproduced. In particular, the geoid highs over both subduction zones and hotspot provinces are apparent.

The residual obtained when the effects of interior density contrasts inferred from tomographic studies of upper and lower mantle structure and from subducted slabs are subtracted from the observed nonhydrostatic geoid is shown in Figure 9.40. The amplitude of the residual geoid is about a third that of the observed geoid, leading to a variance reduction of almost 90 percent.

While most of the geoid has been explained, many of the features of this residual geoid are probably associated with specific geologic phenomena,

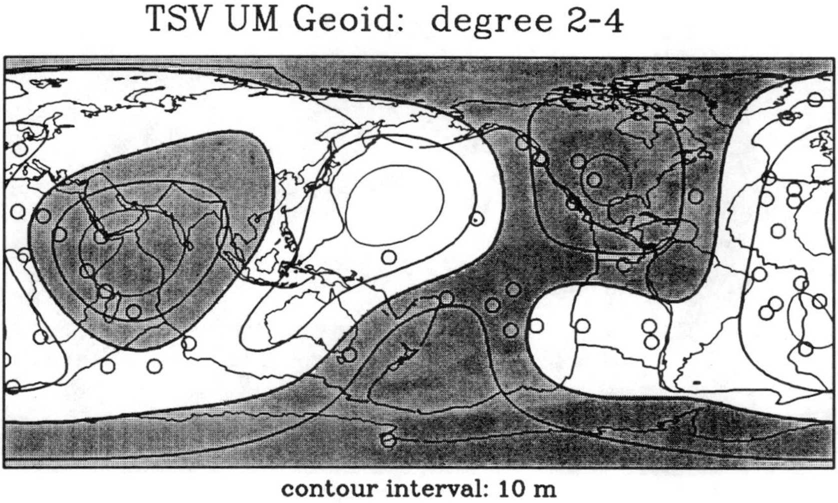

Figure 9.38. A dynamic geoid predicted using density contrasts inferred from Tanimoto's (1986) upper mantle (T-UM) structure derived from a tomographic inversion of surface wave data and geoid kernels for a model intermediate between W4 and W4a. The contour interval is 10 m and geoid lows are shaded.

rather than being just random noise. For example, most of the more prominent hotspot provinces remain associated with geoid highs. Hudson Bay and West Antarctica, loci of fairly recent and major deglaciation, remain as residual lows; including the gravitational effects of incomplete postglacial rebound allows over 90 percent of the geoid variance to be explained. Tectonically active areas such as the Alpine-Himalayan mountain belt, the Andes, Southeast Asia, and the Basin and Range remain as residual geoid highs.

Discussion of Dynamic Geoid Models

We have presented a group of dynamic models of the geoid arranged in a particular order. Questions of the uniqueness and resolution of the models naturally arise. For example, does the final predicted geoid depend on the order in which the effects of various density models are calculated? How well constrained are the velocity perturbation to density perturbation ratios used? How well constrained are the details of the viscosity distribution, or the presence or absence of chemical stratification?

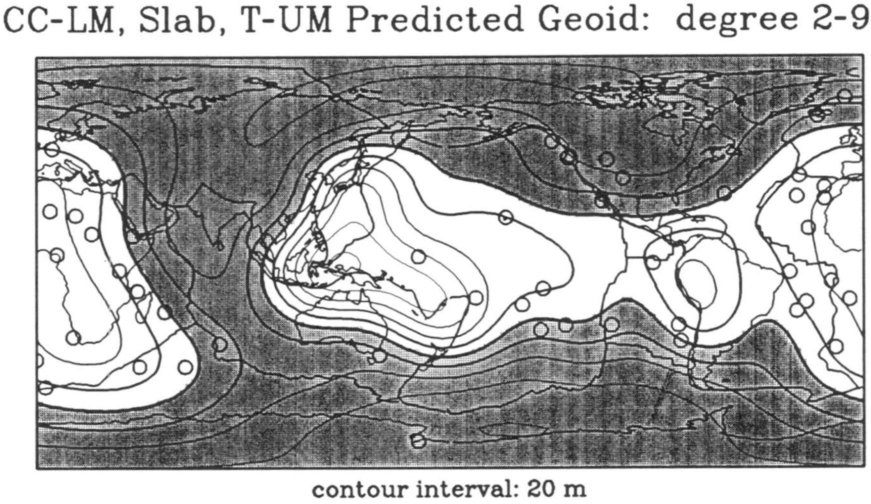

Figure 9.39. The total dynamic geoid, obtained by adding the contributions from Figures 9.34 and 9.38 for degree 2–4 and 9.36 for degree 2–9. The contour interval is 20m and geoid lows are shaded.

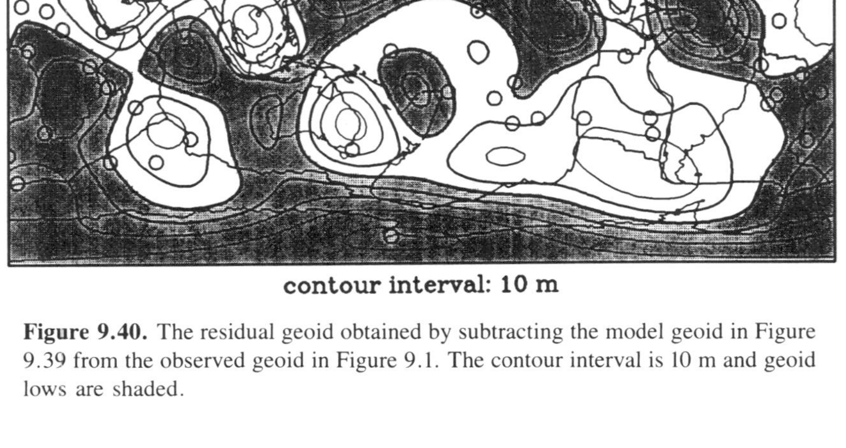

Figure 9.40. The residual geoid obtained by subtracting the model geoid in Figure 9.39 from the observed geoid in Figure 9.1. The contour interval is 10 m and geoid lows are shaded.

These questions are difficult to answer precisely, given our lack of formal error estimates for the tomographic inversions, our uncertainty in slab density contrasts, and the possibility that unphysical assumptions such as spherically symmetric viscosity distributions or incompressible flow contaminate the models. We have made an effort to experiment with a broad range of models in order to answer some of these questions, however.

The geoid models presented above were actually determined by solving the following set of linear equations for $2 \leqslant l \leqslant 9$; $m \leqslant l$; $i = 1{,}2$:

$$Rs^{lmi} + Pf_pV_p^{lmi} + Sf_sV_s^{lmi} = g^{lmi}. \tag{5.1}$$

The equations were solved for R, the density contrast of mature subducted slabs, (assumed distributed over a thickness of 125 km and dipping at 60°), P, the density perturbation to velocity perturbation ratio for lower mantle compressional wave velocity anomalies, and S, the density perturbation to velocity perturbation ratio for upper mantle shear wave velocity anomalies. In this equation, s^{lmi} is the geopotential coefficient obtained by integrating (4.60) using the slab model convolved with the geoid kernel for a particular Earth model, V_p^{lmi} is a similar integral involving the convolution of the expansion of lower mantle velocity heterogeneity with the geoid kernel for a particular (and perhaps different) Earth model, V_s^{lmi} represents the convolution of upper mantle velocity heterogeneity with a model geoid kernel, and g^{lmi} is a spherical harmonic coefficient in the observed geoid. We have found more success in using the tomographic results to explain the geoid at long wavelengths than at short wavelengths, so we include the weighting functions f_p and f_s. These weights are set equal to unity for $l \leqslant l_{\max}$ and to zero for $l > l_{\max}$, where $l_{\max}$ is not necessarily the same for f_p and f_s.

The model geoids we showed were, in fact, not computed sequentially in the order presented, but all together using (5.1) with $l_{\max} = 4$ for both the lower mantle and the upper mantle. (These values were determined by trial and error, with explaining variance in the geoid the selection criterion.) The values of R, P, and S, along with the covariance matrix for this model is given in Table 9.2 in column WO, for "optimal" whole mantle model. The geoid contributions from these three sources are not totally orthogonal and there is some trade off among the parameters. However, this trade-off is not severe and our conclusions are not very sensitive to whether all parameters are inverted for simultaneously, or whether the effects are estimated sequentially. In fact, this study was originally done by first characterizing the slab signal (Hager, 1984), then by determining the lower mantle structure by a fit to the slab residual geoid (Hager *et al.*, 1985). The results of these earlier studies are very similar to those presented here.

Further insights into questions of resolution and uniqueness is given by

Table 9.2. Parameters for dynamic flow models

Model	WO	W4	C4a
R (g/cm^3)	0.064 ± 0.006	0.16 ± 0.02	0.36 ± 0.09
P (g/cm^3)/(km/sec)	0.48 ± 0.02	0.49 ± 0.03	0.35 ± 0.03
S (g/cm^3)/(km/sec)	0.041 ± 0.005	0.11 ± 0.02	0.15 ± 0.02
C_{RP} (10^{-4})	−0.94	−0.48	−1.4
C_{RS} (10^{-5})	−0.50	0.13	−7.1
C_{PS} (10^{-4})	−0.20	−0.65	−0.60
Percent variance explained	89	84	82

R = slab density contrast for slab dipping at 60°; P = lower mantle density variation/P wave velocity variation; S = upper mantle density/S wave velocity variation. Model WO is the optimal model, discussed in the text. Models W4 and C4a use the kernels for models bearing the same name. C_{RP}, C_{RS}, and C_{PS} are the off-diagonal elements of the covariance matrix. Also given is the percent of the variance of the observed geoid explained by each model.

presenting the solutions for two other models we have looked at, also shown in Table 9.2. Model W4 used geoid kernels for model W4 throughout. Model C4a used geoid kernels for model C4a. These models do not differ greatly in their ability to explain the observed geoid. We prefer the "optimal" model, with its weaker "lithosphere" for subducted slabs, not only because it explains more of the variance in the observed geoid, but also because it fits our geophysical prejudice that the lithosphere is indeed weaker at subduction zones. Others might prefer a spherically symmetric model such as W4 or C4a, which explain 84 and 82 percent of the variance of the observed geoid, respectively, on the basis of mathematical consistency.

Although for a given flow model the uncertainties in R, P, and S are fairly small, there is a wide range in the values of these parameters, depending upon the flow model. The variation in parameter values among the models is far larger than their formal uncertainty in the context of a given model. As in many other geophysical models, formal error estimates are basically meaningless; the most important uncertainties arise in the parameterization of the model.

The most stable parameter is P, the ratio of density perturbation to lower mantle P wave velocity perturbation, which has a value between ⅓ and ½ (g/cm^3)/(km/sec). (A value of ¼ was obtained in our previous model (Hager *et al.*, 1985).) These values are reasonable based on laboratory measurements on upper mantle minerals, although few direct measurements on lower mantle minerals are available. The value of this ratio calculated from changes in density and P wave velocity from the top to the bottom of the lower mantle in standard Earth model PREM (Dziewonski and Anderson, 1981) is 0.4. Using Craeger and Jordan's (1986) inferred ratio of velocity to temperature changes of -0.4 m/s/K, a density of 5.0 g/cm^3, and a thermal expansion coefficient of 3×10^{-5}/K also gives a value of 0.4. The close coincidence of these values of P with the ones derived in our models gives us further confidence in both the tomography and the geoid modeling.

The range in R, the inferred slab density contrast, assumed distributed over a 125 km thick lithosphere, is quite large—R varies by over a factor of 5. The value of 0.064 g/cm^3 for our preferred model is just the value of the density contrast between cold slabs and hot mantle that is inferred from the variation of oceanic bathymetry with age (e.g., Hager and O'Connell, 1981). Within the uncertainties of slab dip, and extension or compression of the slab as it sinks through the upper mantle, the agreement is probably fortuitous. Elevation of the olivine-spinel phase boundary could increase the average density contrast between subducted slabs and the surrounding mantle by up to a factor of 2 (Schubert *et al.*, 1975) so on this basis we cannot reject model W4. It is conceivable that transformation of the slab to the ilmenite phase might increase this density contrast still further (D.L. Anderson, personal communication), so we cannot reject, on the basis of the geoid alone, even the huge density contrast of 0.41 g/cm^3 given by model C4a. Further uncertainty, discussed below, results from the tradeoff with possible geoid anomalies associated with the kinematic return flow from trenches to ridges.

A wide range of parameter values also occurs for S, the ratio of density perturbation to shear wave velocity perturbation in the upper mantle. This value ranges from about 1⁄10 that for P for our optimal model to about ½ that for P for model C4a.

If all upper mantle density and shear wave velocity heterogeneity were due to temperature variations, a reasonable value of S, assuming a thermal expansion coefficient of 3×10^{-5}/K would be in the vicinity of 0.2 (g/cm^3)/(km/sec). (Grand, 1986), on the order of that for models W4 or C4a. However, there are large variations in shear wave velocity observed in the upper mantle that are almost certainly not primarily of thermal origin. For example, Grand and Helmberger (1986) have found variations in shear wave velocity of over 10 percent between stable and tectonically active North America at depths of 100–200 km. Such a large variation would require over 1,000 K temperature variation and a density contrast of over 3 percent if it were to be explained in terms of temperature alone. There is no evidence for such large temperature or density contrasts. The most likely explanation of these large variations in shear velocity is variations in composition. The low value of S determined for preferred model WO is consistent with upper mantle shear wave velocity heterogeneities being primarily chemical, rather than thermal in origin.

6. Geophysical Implications of the Flow Models

We have derived mantle models with flow driven by interior density contrasts that predict, with reasonable success, the observed long-wavelength geoid. While computing the model geoids, we must compute dynamic surface topography, and also can easily calculate interior flow velocities. In this section, we present the dynamic topography and interior flow calculated for our preferred density and viscosity models. In our presentation to this point we have shown results separately for our models of lower mantle heterogeneity, slab density contrasts, and long-wavelength upper mantle structure. In light of the uncertainties for the density scaling just discussed, we continue to discuss the three contributions separately, as well as all together for our preferred model.

Dynamic Topography

The key to calculating geoid anomalies caused by density contrasts within a dynamic planet is the inclusion of the effects of dynamically maintained topography at the surface, CMB, and any interior chemical discontinuities that might exist. As is shown in Section 4, topography kernels for a given compositional model (e.g., whole mantle flow or chemically stratified flow) depend much less on the viscosity structure than do the geoid kernels. Even the comparison among topography kernels for different compositional structures is relatively straightforward. For this reason, we show only the topography for our preferred geoid model, WO, but discuss the topography generated for some other models as well.

The dynamic topography at the Earth's surface for degrees 1–6, supported by the density contrasts in the lower mantle inferred from P wave tomography, is shown in Figure 9.41. In generating this spherical harmonic expansion of dynamic topography, a density contrast at the surface of 2.3 g/cm^3, appropriate for regions beneath the oceans, was assumed worldwide. This leads to an over estimate of dynamic topography by about 50 percent over continental areas. The topography is quite similar in pattern to the geoid predicted for this model, Figure 9.34, with polar lows, and highs over Africa and the central Pacific. For lower mantle density contrasts in our preferred viscosity model, the gravitational effect of the deformed upper surface dominates. This is why the model geoid and dynamic surface topography are similar in both pattern and sign. Both the topography and geoid are referenced to the center of mass of the Earth. In this reference frame, although there is no degree 1 term in the potential, there is a small degree 1 term in the

LM Predicted Surface Topography: degree 1-6

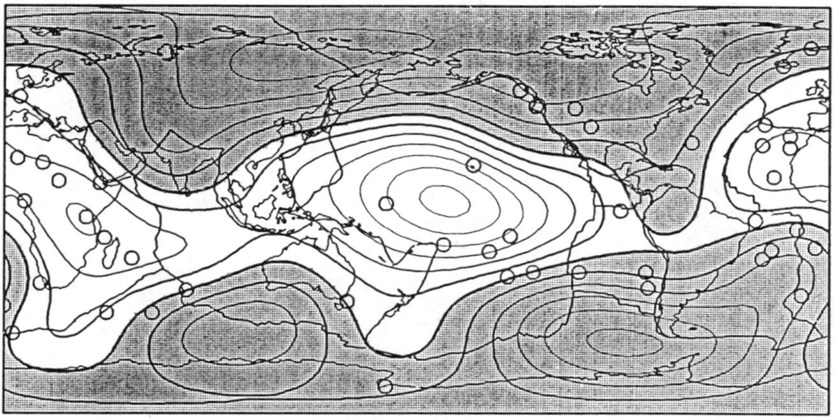

contour interval: 200 m

Figure 9.41. Dynamic topography at the surface of the Earth caused by lower mantle density contrasts calculated using the kernels for model W4. A density jump of 2.3 g/cm^3, appropriate for suboceanic areas, is assumed globally. Depressions are shaded.

dynamic topography, indicating an offset between the center of mass and the center of figure of the Earth.

There is very little difference in the density perturbation to P-wave velocity perturbation ratio inferred for model W4, another model we discussed earlier that does not fit the geoid quite as well as model WO, but that employs the same spherically symmetric viscosity structure for all density contrasts. Thus the surface deformation predicted from lower mantle density contrasts for that model is quite similar to that shown here. In contrast, for chemically stratified models, the surface topography kernels for lower mantle density contrasts are very small, so little dynamic topography is generated at the surface.

The dynamic surface topography predicted for subducted slabs at degrees 1–9 for model WO, which uses the kernels for viscosity model W4a for slabs, is shown in Figure 9.42. As expected, the dynamic topography is dominated by depressions of several hundred meters in subduction zone regions. This is consistent with back arc basins in general being deeper than oceanic crust of equivalent age (e.g., Davies, 1981). For long wavelength loads in the upper

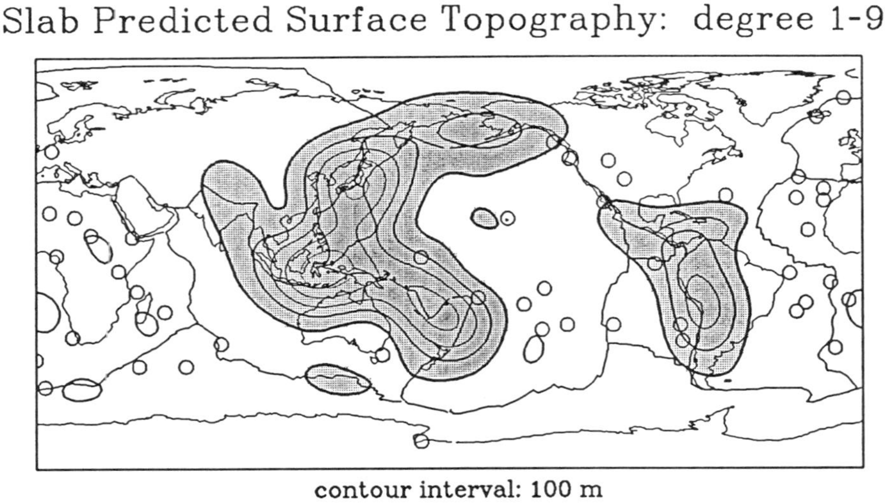

Figure 9.42. Dynamic topography at the Earth's surface maintained by density contrasts in subducting slabs, calculated using viscosity model W4a.

mantle, for our preferred viscosity model, have dynamic topography and geoid anomalies of opposite sign.

For model W4, the surface topography kernels in the upper mantle are about a factor of two larger than those for model W4a, while the estimate of average slab density contrast is larger by about a factor of 2.4. The dynamic surface topography driven by slabs in model W4 is similar in pattern, but about a factor of 5 larger than that shown in Figure 9.42.

For model C4a, the topography kernels are a bit smaller for upper mantle density contrasts than for model W4a, while the inferred mass anomaly is a factor of 5.6 larger. Again, the topography predicted from slabs is about a factor of 5 larger than that shown, but similar in pattern.

The surface topography for our preferred viscosity model, driven by the density contrasts inferred from Tanimoto's (1986) upper mantle structure at degrees 1–6, is shown in Figure 9.43. The dynamic topography is generally of opposite sign than the predicted geoid. It is dominated by uplifts over North America, New Zealand, and the Carlsberg Ridge and depressions over the Mid-Atlantic Ridge and western Pacific. The topography for model W4 is similar in pattern and about a factor of 4 larger, as is the dynamic topography for model C4a.

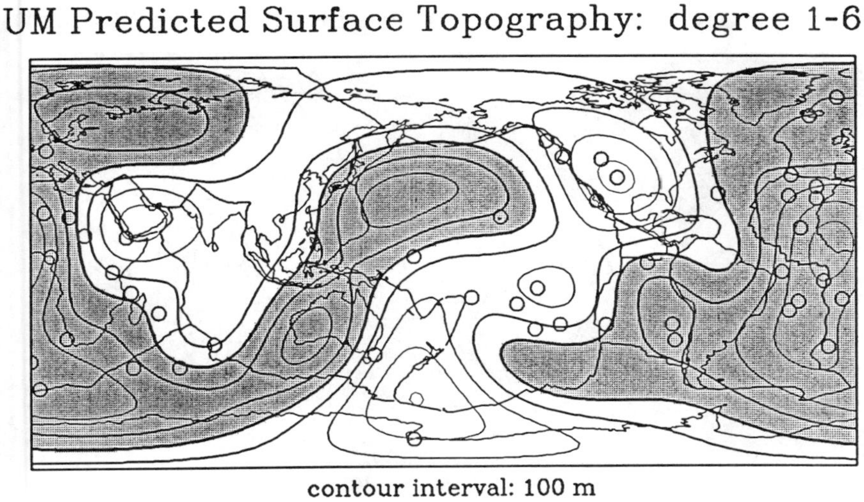

Figure 9.43. Dynamic surface topography supported by upper mantle density contrasts. This model was calculated assuming a viscosity structure intermediate between W4 and W4a.

The total dynamic surface topography for our preferred model WO, obtained by summing the topography shown in Figures 9.41–9.43, is shown in Figure 9.44. It is dominated by the long wavelength contribution from the lower mantle, with elevated topography over Africa and the central Pacific and depressions at the north pole and south of Australia.

Given the strong dependence of dynamic topography on the Earth model just presented (assuming the models are also constrained to fit the geoid), it seems clear that comparison of observed dynamic topography to that calculated for a model could provide additional strong constraints on the Earth's rheological and compositional structure. Unfortunately, much of the actual variation in surface topography is caused by variations in crust and lithospheric structure, rather than being dynamically generated by deeper density contrasts outside the surface boundary layer of mantle convection. Comparing the predicted dynamic topography to the total observed surface topography is not very useful. It is necessary first to remove effects associated with variations in crustal and lithospheric structure.

Crough (1983) attempted such an analysis for portions of the world's ocean basins where sufficient data exist to correct for effects such as sediment

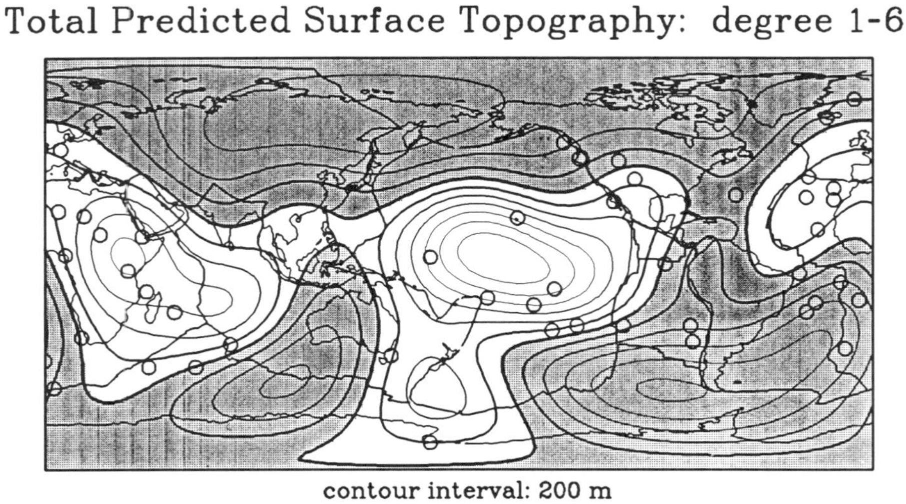

Figure 9.44. Total dynamic surface topography obtained by adding the contributions in Figures 9.41–9.43.

loading. While his coverage is spotty, and while it is always dangerous to compare points on a map with a long-wavelength harmonic expansion, we find it encouraging that Crough found residual bathymetric highs in the central Pacific, over the Canaries, and southeast of Africa. His major low is coincident with that in Figure 9.44 southwest of Australia, and is nearly identical in amplitude.

Further comparison can be made for continental shields, where erosion over geologic time is expected to have led to a fairly constant "continental freeboard." Africa is now highstanding relative to the North American and Siberian shields (e.g., Harrison *et al.*, 1981), consistent with our model prediction.

It is interesting to note that the dynamic topography predicted from the upper mantle density structure inferred from surface wave tomography is usually opposite in sign from that observed. North America is high, Africa, the Canaries, and the western Pacific are low in the upper mantle model. This again suggests that the density contrasts associated with the fairly large upper mantle seismic heterogeneity must be fairly small, perhaps of compositional rather than thermal origin.

The predicted dynamic topography is nearly independent of the present day locations of continents and oceans. Some ocean basins are elevated, while others are depressed, and similarly for continents. Presumably, as continents and ocean basins move with respect to each other, they also move with respect to the dynamic topography generated by mantle convection. Such motion would result in variations of sea level on both local and global scales.

Relative variations in global eustatic sea level can be recognized by comparing the percentage of continental areas flooded as a function of time. Absolute calibration of eustatic sea level variations is more difficult. Vail *et al.* (1978) estimate changes in eustatic sea level of over 400 m in the past 600 Myr, with about 250 m variation over the past 200 Myr. Harrison *et al.* (1983) have constructed relative sea level curves for the continents for the past 180 Myr. Their estimate of eustatic sea level variation over this time is about 150 m, while their estimated variation among continents reaches 300 m. These variations are determined by averaging over continent-wide areas, and local variations would be larger. These variations are of the order of magnitude to be explained by our models, although if Harrison *et al.*'s (1983) values are used, the actual variations may be a bit smaller than those predicted by our models. Given the uncertainty in the absolute calibration of eustatic sea level curves, we find the qualitative agreement encouraging.

The dynamic topography generated at the CMB is also important in determining the total geoid anomaly from a given internal density contrast (Richards and Hager, 1984). The nonhydrostatic core topography generated by each component of density contrasts considered for our preferred geoid model is shown in Figures 9.45–9.47. A density contrast of 4.5 g/cm^3 across the CMB is assumed. The total predicted core topography, shown in Figure 9.48, is generated by summing these contributions. In general, the patterns are quite similar to those of the predicted surface topography and of the geoid. Although the density contrast at the CMB is almost twice that between mantle and seawater, the topography of the CMB is about twice that calculated for the surface. This large deformation of the CMB is partly the result of constriction of flow by spherical geometry, and partly the result of increasing viscosity with depth.

At present, dynamic topography of the CMB is even more poorly constrained than is dynamic surface topography. However, there are some estimates available. The sun and moon exert periodic tidal torques on the Earth, resulting in forced nutation. The responses of the core and mantle to these torques are different, and the core and mantle are mechanically coupled by the ellipticity of the CMB. The amplitude of the observed forced nutation depends on the resonance frequency of this coupling, which in turn depends upon the ellipticity of the CMB, among other parameters. Gwinn *et al.* (1986) have recently suggested, based on an analysis of nutation data, that

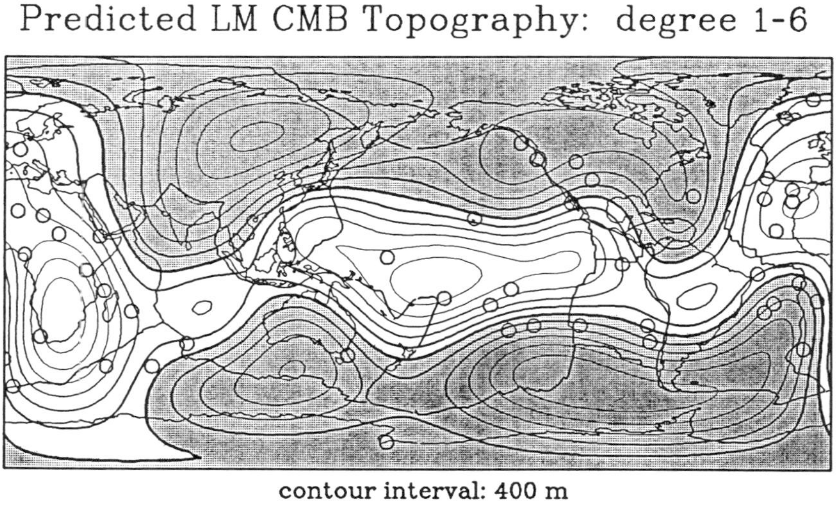

Figure 9.45. Predicted topography of the CMB caused by the lower mantle density contrasts inferred from the seismic tomography. Flow model W4 is used. Downwarpings are shaded.

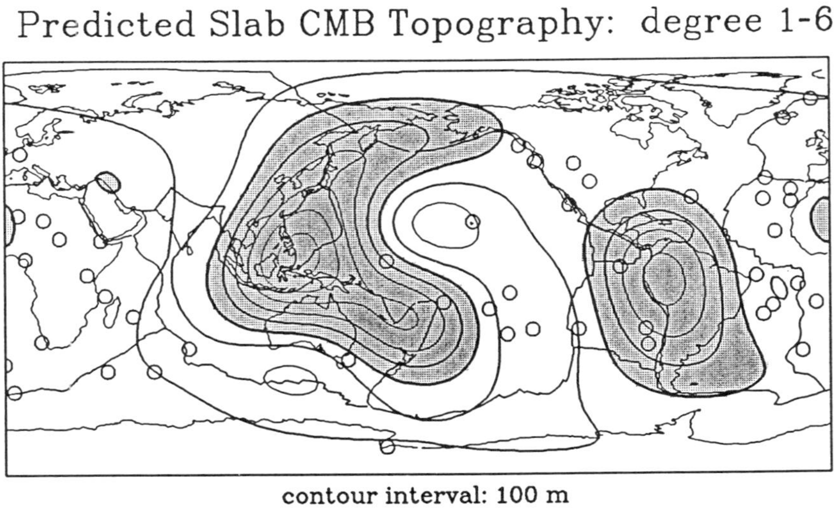

Figure 9.46. Topography of the CMB predicted using flow model W4a for our model of slab density contrasts. Downwarpings are shaded.

Predicted UM CMB Topography: degree 1-6

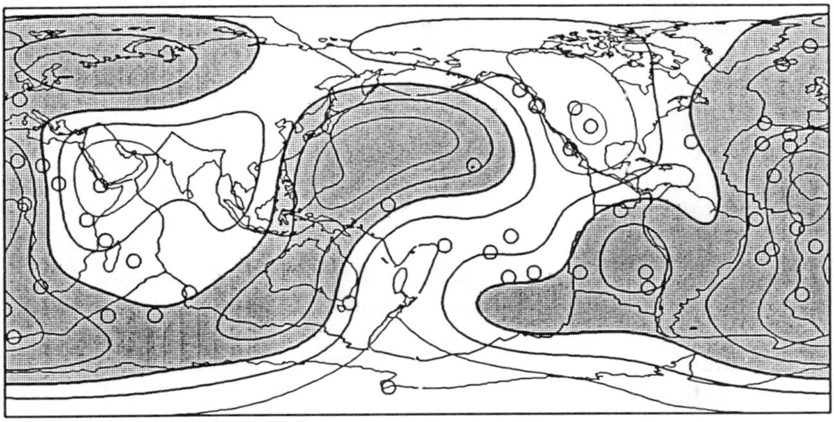

contour interval: 100 m

Figure 9.47. Predicted topography of the CMB caused by upper mantle density contrasts inferred from surface wave tomography. A flow model with a lid viscosity intermediate between that in model W4 and W4a was used. Downwarpings are shaded.

Total Predicted CMB Topography: degree 1-6

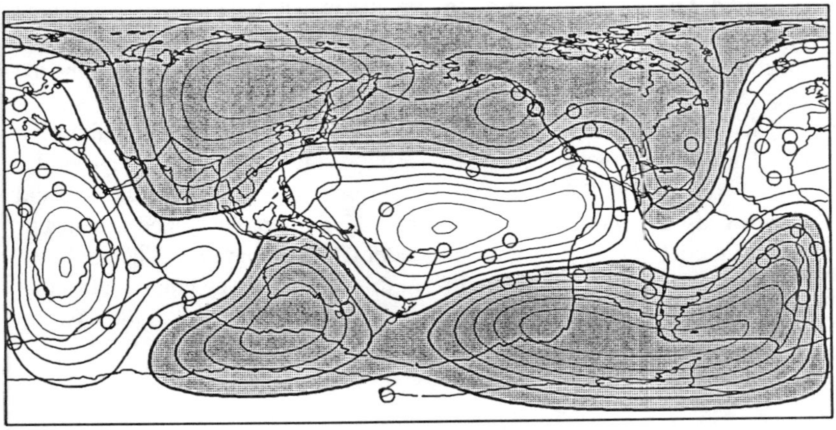

contour interval: 400 m

Figure 9.48. The total predicted CMB topography for our preferred Earth model, obtained by summing the contributions in Figures 9.45–9.47.

the excess ellipticity of the CMB is about 500 m. The excess ellipticity of the CMB shown in Figure 9.48 is 2 km, a factor of 4 larger.

Certain normal modes of the Earth that are sensitive to core radius show anomalous splitting (Ritzwoller and Masters, 1984), perhaps caused by topography of the CMB (Park and Gilbert, 1984). Preliminary seismological models based on these observations suggest about 20 km of topography (Giardini *et al.*, 1987).

Body wave studies (e.g., Morelli and Dziewonski, 1986) also are suggestive of substantial topography at the CMB, but in a sense generally opposite to that predicted by our dynamic models and with a pattern that is different than that predicted by free oscillation studies. Results are in a state of flux, but if current progress continues, it should be possible to test our models using seismology in the near future.

Constraints on the topography of the CMB would be very useful in understanding the dynamics of D″, the layer just above the CMB. If D″ has a very low viscosity as the result of being a hot thermal boundary layer, kernels for deformation of the CMB would be decreased substantially (e.g., Figure 9.26), as may be suggested by the nutation observations. If D″ is a chemically distinct layer, the sign of the kernels could change, as would be implied by some of the preliminary seismological models mentioned above.

Hide (see review, 1986) has suggested that some features in the Earth's magnetic field are caused by interaction of flow in the fluid core with kilometer size bumps at the CMB. He assumed in his analysis that the bumps at the CMB were responsible for most of the long-wavelength variations of the geoid. Our models support the hypothesis that dynamically maintained topography at the CMB is correlated with the geoid. If features in the magnetic field can indeed be related quantitatively to CMB topography, the magnetic field might be useful in testing these models.

Under certain assumptions, the secular variation of the magnetic field can be related to the flow velocity field near the surface of the fluid core (e.g., Le Mouel *et al.*, 1985). The torque resulting from coupling between the core and mantle caused by flow interacting with CMB topography can be calculated, given a model of core flow and a model of CMB topography. The predicted change in length of day resulting from this torque can be compared to that observed, providing a test of the combined models (e.g., Hide, 1986). We are in the process of undertaking such a comparison.

Bounds on Mantle Viscosity

All of the results of flow models that we have presented to this point are independent of the absolute value chosen for the reference mantle viscosity;

they depend only on the relative viscosity distribution and compositional structure. So long as the reference viscosity is large enough that the Reynolds number of the flow is negligible, for a given density distribution driving flow, the stresses, boundary deformation, and geoid anomaly depend only on relative viscosity variations. This can be seen from the structure of equation 4.30, where the reference viscosity cancels out of the terms relating stress to buoyancy. It is also apparent that flow velocity is inversely proportional to the reference viscosity.

The relative viscosity profile we have used to fit the geoid differs substantially from the approximately uniform viscosity commonly deduced from studies of postglacial rebound (e.g., Cathles, 1975; Peltier, 1981, this volume). It has been suggested that the viscosity sampled by postglacial rebound may be different than that governing quasi steady solid state convection, perhaps as the result of transient creep (e.g., Weertman, 1978; Peltier, 1985; Sabadini *et al.*, 1985). It is possible, with some simple assumptions, to use our flow models to place an independent bound on the mantle viscosity associated with convection.

We have hypothesized that the velocity anomalies in the tomographic models of mantle structure and the density anomalies we infer from them are primarily the result of temperature anomalies in the convecting mantle. This assumed convection would transport heat by advection, with the advective heat transport proportional to the mantle flow velocity. If the mantle is in a quasi steady state, this advective heat transport must not exceed that at the Earth's surface (O'Connell and Hager, 1984), a value of about 40 Tw (O'Connell and Hager, 1980).

The total advective heat flow Q through any surface S is given by

$$Q = \int_S \rho c_p v_r T dS, \tag{6.1}$$

where c_p is the heat capacity at constant pressure. Expanding v_r and T into surface spherical harmonics, integrating 6.1 over the area of a sphere at radius r, and applying the orthonormality properties of surface spherical harmonics yields:

$$Q = 4\pi r^2 \rho c_p \sum_{l=1}^{\infty} \sum_{m=0}^{l} \sum_{i=1}^{z} v_r^{lmi} T^{lmi}, \tag{6.2}$$

where v_r^{lmi} and T^{lmi} are the coefficients in the expansion of radial flow velocity and temperature. (There is no degree zero term in the sum because v_r^{0mi} is identically zero.) We have estimates for only a finite number of spherical harmonic degrees, so cannot evaluate the infinite sum. Making the reasonable assumption that v_r^{lmi} and T^{lmi} are predominantly of the same sign, i.e., positive temperature anomalies excite upward flow, truncating 6.2 at any harmonic degree will lead to a lower bound to the actual advective heat flux.

We assume that P, the proportionality constant relating density variation to lower mantle P-wave velocity variations does not depend on depth, and also that α, the thermal expansion coefficient, is independent of depth. Then

$$\rho^{\lambda\mu\iota} = \Pi\Omega_{\pi}^{\lambda\mu\iota}, \tag{6.3}$$

and

$$\delta T^{lmi} = \rho^{lmi}/(\rho\alpha) = PV_p^{lmi}/(\rho\alpha) \tag{6.4}$$

Each term in (6.2) is, then, proportional to $(PV_p^{lmi})^2/\alpha$.

O'Connell and Hager (1984) used this analysis and an earlier version of a lower mantle dynamic geoid model to conclude that the lower mantle viscosity is at least 10^{23} p in order not to violate the surface heat flow. For the remainder of this paper, we choose 3.2×10^{22} p as the normalizing viscosity in the transition zone, which gives a lower mantle viscosity of 3.2×10^{23} p, half an order of magnitude greater than their lower bound. The advected heat flux calculated using (6.3), (6.4) with $\alpha = 3 \times 10^{-5}$/deg and truncating (6.2) at degree and order 6 is plotted as a function of depth in Figure 9.49.

The radial flow velocity is determined by a weighted depth average of the model density perturbations, with density contrasts in regions near the boundaries given less weighting because of the requirement that radial flow velocity vanish there. The increase with height above the CMB might be the result of substantial internal heating within the mantle. It might also indicate that either α or P varies with depth.

The advected flux for degrees 1–6 reaches a maximum of 12 Tw, about one third of the surface heat flux. Truncating 6.2 at degree 6 probably underestimates the total flux by a significant amount. For example, using our expansion of slab density contrasts as a known function that can be expanded to much higher degree and order, truncating at degree 6 underestimates total flux by a factor of 3. Assuming that this factor of 3 is also applicable to lower mantle density contrasts, the normalizing viscosity we use leads to about the expected advected heat flux for degrees 1–6. If heat transport in the lower mantle is dominated by low viscosity plumes (e.g., Richards *et al.*, 1988), or if α decreases substantially with depth as expected on theoretical grounds (D. L. Anderson, personal communication, 1986), an even higher viscosity would be required. Schubert *et al.* (1978) have previously pointed out that convection in a 10^{22} p mantle is expected to be sufficiently vigorous to have frozen the core by now.

It is also possible to use the geoid in conjunction with spherical flow models to set an upper bound to the absolute mantle viscosity. The motion of

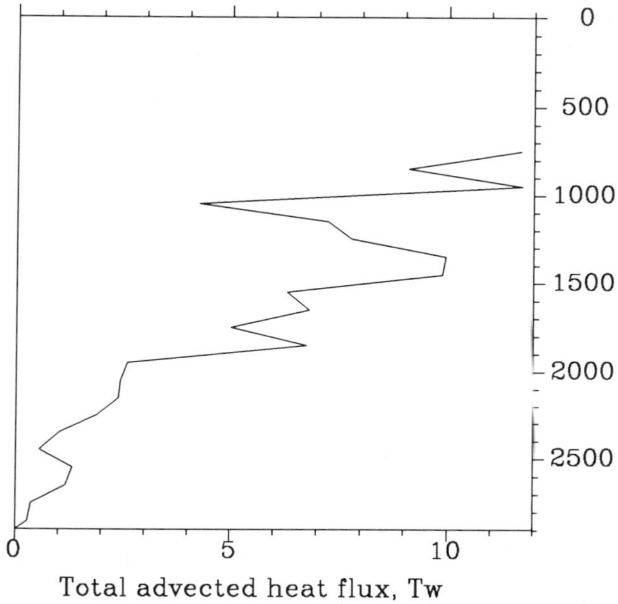

Figure 9.49. Advected heat flux as a function of depth in the lower mantle associated with the density contrasts inferred from lower mantle tomography including contributions for degrees 1–6.

the plates at the surface sets up a global return flow from ridges to trenches (e.g., Hager and O'Connell, 1978, 1979; Richter and McKenzie, 1978). This large-scale return flow has high pressures at convergence zones and low pressures at ridges associated with it. These pressure differences also lead to dynamic surface topography that results in geoid anomalies. An upper bound to mantle viscosity is obtained by requiring that these geoid anomalies (or equivalent gravity anomalies) not violate the observed geoid (Richter and McKenzie, 1978; Hager and O'Connell, 1979; Ricard, 1986).

Kinematic flow models can be calculated by solving 4.30 and 4.32 with **a** set to zero and with the spherical harmonic coefficients of the observed plate velocities applied as boundary conditions (e.g., Hager and O'Connell, 1978, 1979). For spherically symmetric viscosity profiles, toroidal flow does not couple into the geoid, so only 4.30 is of interest for the geoid problems, given our rheological assumptions.

Using 3.2×10^{22}p as the reference viscosity in the transition zone for our

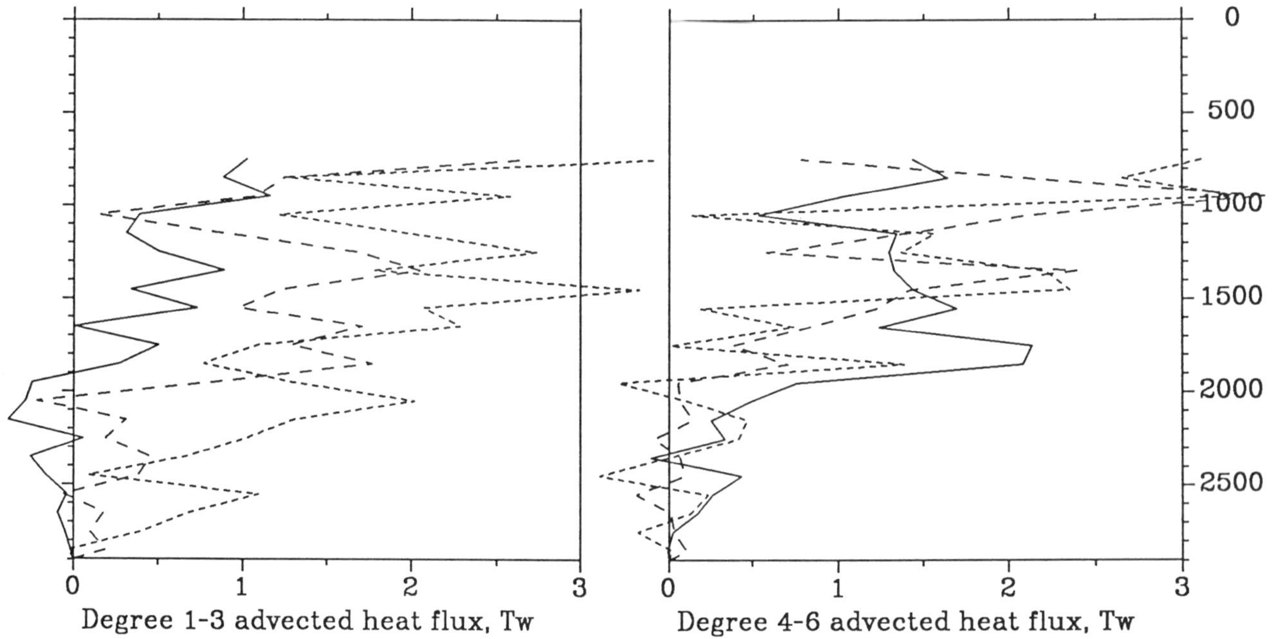

Figure 9.50. Individual degree contributions to the advected heat flux for degrees 1–6 plotted as a function of depth. Solid lines are for degrees 1 and 4; short dashes for degrees 2 and 5; and long dashes for degrees 3 and 6. The contribution for degree 1 has been multiplied by 5 for clarity in plotting.

preferred mantle model leads to geoid highs over convergence zones of ~60 m at degrees 2–10, with a low of −80 m over the East Pacific Rise. Although the amplitudes of the geoid highs are acceptable, the low over the East Pacific rise is not.

Ricard (1986) has pointed out, however, that lateral variations in viscosity, such as the low viscosity regions probably associated with ridges, could have a substantial effect on the geoid associated with kinematic flow. For the particular model he calculated, the geoid low over the East Pacific Rise was nearly eliminated and the geoid highs over subduction zones were reduced by a factor of two. Taking into account the possible effects of lateral variations in viscosity, our absolute viscosity calibrated on the basis of heat flux is probably near, but not over, the limits imposed by the amplitude of the geoid associated with the kinematic flow. (We return to possible tradeoffs with kinematic flow in the discussion section.)

Returning to the discussion of heat flux, it is instructive to plot (6.2) as a function of depth. The advected flux at a given depth depends upon the correlation of the inferred density anomalies at that depth with the weighted average of the density anomalies given by the radial flow velocity. Rapid variations in the advected flux curves are diagnostic of rapid variations in the seismic velocities as a function depth.

The contribution at degree 1 is small and changes sign midway through the lower mantle. This is the result of the rather curious phenomenon that the upper and lower parts of the degree 1 lower mantle tomographic model are strongly anticorrelated.

At degree 2, the tomographic model shows its greatest coherence as a function of depth. Still there are fairly rapid variations of about 50 percent superimposed on the relatively smooth background.

The local advected heat flux shows progressively more variation as higher degree harmonics are considered. At degree 6, there is essentially no correlation between the average radial flow velocity and the average inferred temperature anomalies in the lower half of the lower mantle. This may be the result of an increasingly noisy tomographic inversion as a function of harmonic degree. It might also be a real feature of Earth structure.

Interior Flow Patterns

Once a reference viscosity is chosen, equations (4.38) can be evaluated at any depth to give a flow velocity field. The flow at a depth of 670 km driven by the density contrasts inferred from our lower mantle tomographic model and our preferred viscosity model is shown in Figure 9.51. Upwellings are shown by boxes, downwellings by crosses, and horizontal velocities are

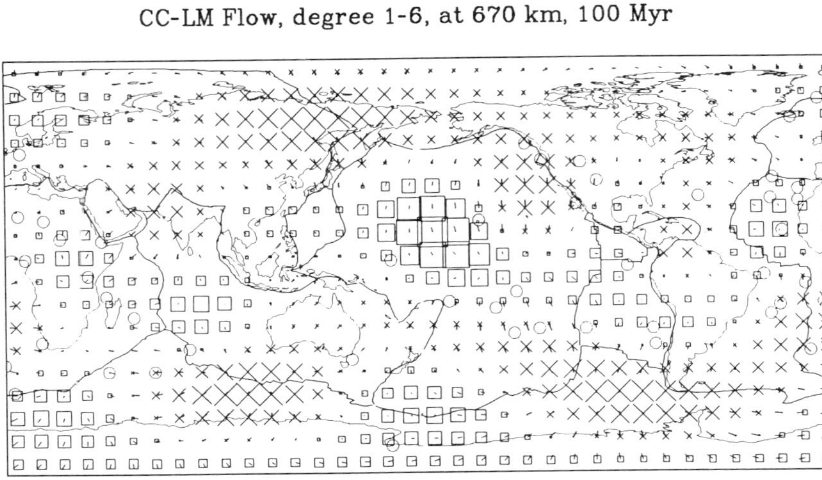

Figure 9.51. Flow velocities at 670 km depth driven by the densities inferred from lower mantle tomography. The crosses at the base of each vector represent downwelling, while the boxes indicate upwelling. The size of the symbols and vectors give the displacement that would occur by extrapolating the flow velocities for 100 Myr; touching symbols represent velocities of 1 cm/yr.

indicated by vectors originating at the centers of the symbols. At this depth the velocities are primarily radial. The pattern of flow in general matches the predicted geoid or surface topography, with upwellings underlying geoid and topographic highs. However, the flow has relatively more power at the higher harmonics than does the geoid or topography. Density contrasts at half-wavelengths in excess of the thickness of the mantle are more efficient in generating dynamic topography and less efficient at generating flow than those of scale lengths comparable to or less than the depth of the layer. Most, but not all hotspots are in regions of upwelling.

Maximum radial velocities at these long wavelengths are of the order of 1 cm/year, giving a transit time through the lower mantle of several hundred million years, a factor of 10 less than plate tectonic rates. Such a slow flow would tend to preserve heterogeneities in an unmixed state for longer times than are typical of more rapidly convecting systems.

Flow driven by subducting slabs, also at a depth of 670 km, is shown in Figure 9.52. This flow is dominated by downwelling at subduction zones, as expected, with a maximum rate of about ½ cm/yr. Part of the apparent slow-down, as compared to typical subduction rates of order 10 cm/yr, is the result

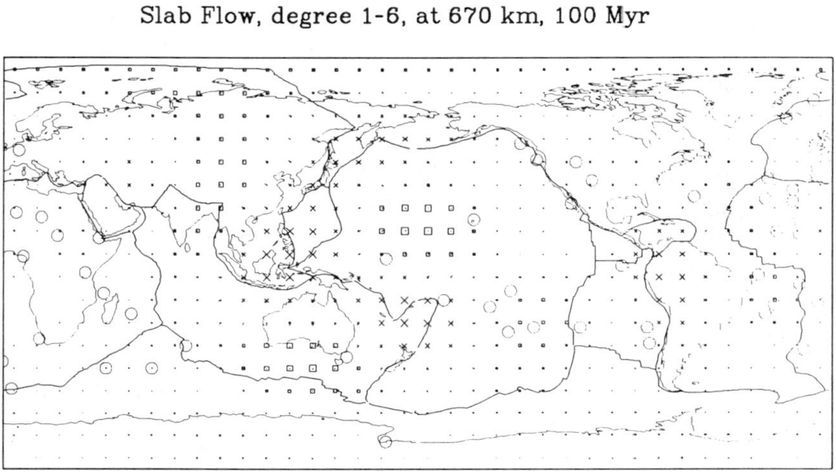

Figure 9.52. Flow velocities at 670 km depth driven by the density contrasts in the preferred slab model. Symbols are described in Figure 9.50.

of truncation of the flow field at degree 6. Numerical experiments carried out to higher degree and order indicate that truncation at degree 6 underestimates the slab velocity by about a factor of 5 locally. Still, our lower mantle viscosity of 3×10^{23} p slows down slabs sufficiently that they should undergo severe compressive strain and shortening.

7. Discussion and Summary

Using density contrasts inferred from global inversions given by seismic tomography and a simple density model of subducting slabs as input into fluid dynamical models of mantle flow, it is possible to explain almost 90 percent of the variance in the observed geoid. This is a reasonable first cut at the problem, but there are many improvements that remain to be made.

The lower mantle P-wave tomography suffers from noisy data, uneven sampling, and gaps in the coverage, notably in the uppermost and lowermost mantle. Incorporating more data from a longer timespan of observation would help to reduce the effects of noise in the data, as well as fill in some gaps in the coverage.

The addition of more phases (e.g., PP, PcP, PKP, PKIKP) also can help to fill in gaps in coverage. Unfortunately, adding these phases brings additional problems, as they are harder to pick than the first P arrival, so these data sets are noisier.

Grand (1986) has demonstrated that S and multiple S phases can also be used successfully for mantle tomography. His results, which span the region from upper to middle mantle, illustrate that it is possible to obtain coverage in this important region. A joint S and P wave inversion might be valuable, although it would require additional parameterization relating P and S wave velocity variations.

Improved data coverage and inclusion of more overtones should also improve the resolution of surface wave tomography. Additional station coverage would be valuable for these studies.

The tomographic inversions used in this paper are global. In some regions, localized inversions of higher quality are available (e.g., Humphreys *et al.*, 1984; Grand, 1986). Constraining these "known" regions in global inversions might provide better global solutions.

There are a number of technical issues that might also allow improved resolution. One example would be to use station corrections that are a function of event size and distance to counteract the gain problems mentioned in Section 2. Different parameterization, e.g., using blocks of more uniform aspect ratio, might prove useful.

For the geoid models, including additional geophysical constraints other than the global seismological models used here would be very useful. In particular, observations of dynamic topography at the surface, at the CMB, and at the 670 km seismic discontinuity would provide powerful constraints on models of mantle dynamics. Dynamic topography kernels are largest where geoid kernels are smallest, and topography provides a complementary constraint to the geoid.

As discussed in the previous section, the effects of crustal and lithospheric thickness variations obscure the dynamic topography due to mantle convection. It seems ironic that the dynamic topography of Venus appears to be better known than that of Earth (e.g., Kiefer *et al.*, 1986). It should be possible, although tedious, to separate out the effects of known crustal thickness variations and lithospheric age to provide estimates of dynamic topography for most of the Earth's surface.

Increased attention is now being paid to determining the shape of the CMB. P-wave tomography using PKP, PcP, and PKIKP (e.g., Morelli and Dziewonski, 1986) and free-oscillation studies (e.g., Giardini *et al.*, 1987) now provide conflicting results; both fields of research are fairly new, however, and rapid progress may be expected. Studies of forced nutation (e.g., Gwinn *et al.*, 1986) and changes in length of day caused by the interaction of flow in the fluid core and topography of the CMB (e.g., Hide, 1986) also show promise.

Determining the topography of the CMB is particularly important in constraining the structure and dynamics of the D″ region. The kernels for dynamic deflection of the CMB are much more sensitive to the viscosity of this region than are geoid kernels, as was demonstrated in Section 4. Although we did not show kernels for models in which D″ is not part of the convecting mantle, the kernels for CMB topography for these models are quite different from those in which D″ is only a thermal boundary layer.

The residual geoid (Figure 9.40) has anomalies with amplitudes of ~30 m. The geoid anomalies associated with known near-surface heterogeneities such as crustal thickness variations and variations in oceanic lithospheric thickness begin to become important at this level of detail (Chase and McNutt, 1982; Hager, 1983). These effects should be taken into account in future analyses.

Incomplete rebound following Pleistocene deglaciation is another geophysical phenomenon that contributes geoid anomalies not caused by mantle convection (e.g., Peltier, this volume; Peltier and Wu, 1982). A small part of the residual geoid shown in Figure 9.40 is probably the result of incomplete rebound. Including a simple model of this effect, as well as the effects of crust and lithosphere allows us to account for over 91 percent of the variance in the geoid.

There is additional geophysical signature in the residual geoid, Figure 9.40, that suggests that improvements are needed in the geoid modeling. In particular, residual geoid highs are associated with hotspot groups in the central Indian Ocean, the North Atlantic, the southeast Pacific, and the Basin and Range. This correlation is particularly strong at spherical harmonic degree 6, where there is a peak in the spectrum of the hotspot expansion. Richards *et al*. (1988) have discussed this problem extensively. One possible explanation is that these geoid highs come from lower mantle heterogeneity not well imaged by tomography. Another is that the lateral variations in viscosity associated with hotspots result in negative kernels for hot, low density upper mantle density contrasts, as opposed to the positive kernels appropriate for subducted slabs.

This points out what probably is the most problematic assumption used in the flow models—that the viscosity profile is spherically symmetric. Clearly, the temperature and stress variations associated with mantle convection will lead to azimuthal variations in viscosity that will violate this assumption.

Richards (1986) has carried out an extensive investigation of the effects of lateral variations in viscosity on the generation of geoid anomalies in a convecting planet. He used finite element calculations to avoid the constraints imposed by our analytic method in order to investigate more realistic viscosities. He showed that at half wavelengths long compared to the depth of the convecting layer, the lateral variations in viscosity expected in the interior of a convecting mantle are less important than the radial variations that we have been addressing here. At shorter wavelengths, the effects become more important.

Based on Richards' (1986) work, we feel that our conclusions about the source of long-wavelength geoid anomalies and our model of mantle structure are in general correct, although some details may be poorly resolved. In particular, there is a tradeoff among thickness of the asthenosphere, viscosity of the asthenosphere, and effective strength of the lithosphere that is not well resolved. The locations of significant jumps in viscosity are based on *a priori* assumptions justified by solid state theory (e.g., Sammis *et al.*, 1977), not on the demands of fitting the geoid.

The most significant unresolved issue is how best to parameterize the effective viscosity of the lithosphere. Because the lithosphere is broken into plates separated by faults, describing it in terms of simple fluid mechanical models is problematic. The "regionalized model" approach used here seems plausible, although other approaches, such as applying plate motions as kinematic boundary conditions (Hager and O'Connell, 1978, 1979, 1981; Ricard, 1986) provide alternative and complementary approaches. As discussed in Section 6, there is some additional uncertainty in geoid modeling resulting from the geoid anomalies produced by the kinematic return flow.

Both these approaches, while valuable, are somewhat *ad hoc*. What seems to be required are dynamic models of geoid anomalies with more realistic lithospheric rheology (e.g., Christensen, 1984; Schmeling and Jacoby, 1981). While much could be learned from simple two-dimensional models, ultimately the full three-dimensional problem must be attacked. This will require substantial computing resources, perhaps provided by concurrent processors.

Other oversimplifications that should be addressed in future work are the assumptions that P and S, the ratio of compressional and shear wave velocity perturbations to density contrasts, are independent of depth. There is obviously a tradeoff between the viscosity structure and the density anomaly resulting from a given seismic velocity anomaly. The variation of α with depth is also important. Further experimental and theoretical work is required.

An additional uncertainty in the viscosity structure is the neglect of compressibility in our flow models. Self-compression helps to counter the increase in flow stress with depth caused by sphericity. Neglecting this effect probably leads to an underestimate of the viscosity increase with depth.

Despite these caveats, we feel that the broad conclusions in this chapter are essentially correct; we feel justified in drawing some conclusions about mantle dynamics on the basis of these models. The model of mantle convection developed here, based on observations from seismic tomography and calculations of dynamic models of the geoid, differs in some important ways from models based on simple convection experiments. While there is radial coherence of the inferred temperature anomalies at some very long wavelengths, as exemplified by the correlation of surface hotspot distributions, slow, presumably hot, lower mantle, and geoid highs at degree 2, cross-sections in the spatial domain show substantial radial variability. Also, the anticorrelation of degree 1 velocity anomalies between the top and bottom of the lower mantle is notable.

Some of these counterintuitive features may well be the result of poor ray coverage and noisy data, although the discretization of the tomographic inverse used should minimize these artifacts. However, the independent test provided by the fit to the observed geoid suggests that the inversion contains substantial geophysical signal.

The viscosity model that we have found that best satisfies the geoid may help to explain some of the differences between the tomographic model of lower mantle structure and that expected on the basis of simple convection calculations. This model is motivated by the prejudice that the effective viscosity of the mantle decreases as a result of the strong temperature gradient from a strong "lithospheric" lid to a low viscosity asthenosphere, then increases across the transition zone due to the dominating effects of activation volume and phase changes along an adiabat (e.g., Sammis *et al.*, 1977). With the simple four layer parameterization used here, an increase in effective viscosity from the asthenosphere to the lower mantle of a factor of 300 is indicated by the strong correlation of geoid highs with subducted slabs at degrees 4–9. While the geoid generated by internal density contrasts depends only upon the relative viscosity distribution, an absolute value of $\sim 3 \times 10^{23}$p for the lower mantle is suggested by requiring that the advected heat flux associated with lower mantle density contrasts not exceed the observed heat flux.

Such a viscosity profile should have a significant effect on the convective circulation of the mantle. It would provide a partial barrier to penetration of subducted slabs into the lower mantle, in accord with compressive focal mechanisms in deep subducted slabs (Isacks and Molnar, 1971; Richter, 1979; Giardini and Woodhouse, 1984). Vassiliou *et al.* (1984) showed that this type of viscosity structure can also explain the rate of seismicity in subducted slabs as a function of depth if the number of earthquakes occurring in a given time depends on the deviatric stress magnitude.

Strain rates would be much lower in the lower mantle than in the upper mantle. This might allow the simultaneous occurence of well mixed reservoirs in the upper mantle beneath oceanic ridges and heterogeneities isolated for eons in the lower mantle, as required by geochemical observations (e.g., Hart and Zindler, this volume).

There is other geophysical evidence supporting the concept of slow flow velocities and long residence times in the lower mantle. For example, Grand (1986) used S-wave tomography to derive a more detailed structure of seismic velocity anomalies beneath the North American plate. One prominent feature in his structure is a high velocity anomaly extending to at least 1800 km depth, where resolution runs out. This feature lies directly beneath the previous location of the subduction zone along the west coast of North America before the breakup of Pangea; it may represent the fossil Farallon plate. If this interpretation is correct, it indicates a residence time of over 100 Myr in the lower mantle, consistent with a viscosity there of $>10^{23}$p.

Grand's (1986) interpretation of the high velocity anomaly in the lower mantle beneath eastern North America as an ancient slab that has not yet sunk to the core mantle boundary is consistent with Chase and Sprowl's (1983) interpretation of the generation of the long wavelength geoid anomalies associated with the antipodal hotspot provinces of Africa and the central Pacific. In their view, the Pangean supercontinent had subduction localized around its edges, resulting in cooling of the lower mantle there in contrast to the buildup of heat under Pangea and its antipode (see also Anderson, 1982). The contrast in temperature between this belt of cold mantle and the relatively warmer remainder of the mantle leads to the observed long wavelength geoid when dynamically maintained topography is included. The Chase and Sprowl hypothesis is consistent with the lower mantle tomographic model and suggests a high lower mantle viscosity in order to maintain heterogeneities for the over 100 Myr since Pangea broke up.

If subducted slabs take 100's of Myr to sink through the lower mantle, the residence time in the lower mantle is comparable to the timescale for migration of subduction zones over plate dimensions. The return of cold slabs to the lower mantle is the major way in which this part of the Earth is cooled. The thermal structure of the lower mantle may be closer to that of ribbon candy than of more familiar convection experiments.

The discussion above assumes a mantle with uniform composition. However, it is also possible to explain much of the observed geoid using a layered model. Only a rather special distribution of density contrasts will work—subducted slabs must be more dense than is usually supposed, and the large-scale convection pattern in the upper and lower mantles must be thermally coupled. In such a scenario, substantial dynamic topography must be maintained at the 670 km seismic discontinuity. Hager and Raefsky (1981) calculated dynamic topography of 25–150 km for slab mass anomalies of 8 $\times$ 10^5 g/cm^2. But in order to match the observed amplitudes of geoid anomalies over subduction zones, five times as much mass, resulting in five times as much dynamic deflection is required. This topography should be resolvable using seismology.

In summary, using density contrasts inferred from seismic tomography and a simple thermal density model of subducting slabs as input into fluid dynamical models of mantle flow, it is possible to explain 90 percent of the variance in the observed geoid. The model assumes a spherically symmetric viscosity profile, which is perhaps its most dangerous assumption, given the sensitivity of the kernels to the effective viscosity of the lithosphere. Further work addressing the problem of lateral variation in lithospheric properties is called for.

Despite this difficulty, we feel that the conclusions that most of the geoid is accounted for in terms of known density heterogeneities, and that the mantle has a substantial viscosity increase with depth, are robust. Dynamic implications of such a viscosity structure should be pursued.

8. Acknowledgements

The research reported in this chapter was supported by NASA contract NAG5-315 and NSF contract EAR-8351371 and by a Sloan Foundation Fellowship to Bradford H. Hager. The graphics were made possible in part by a grant from the W.M. Keck Foundation. We would like to specifically acknowledge the valuable contributions of Robert P. Comer and Mark A. Richards, who played central roles in the research reviewed here. Encouragement and valuable reviews of the manuscript were provided by Richard J. O'Connell and Walter Kiefer. Contribution number 4342, Division of Geological and Planetary Sciences, California Institute of Technology, Pasadena, CA 91125.

An updated summary of further progress in this area in the over two years since this paper was completed can be found in Hager and Richards (1989).

9. References

Aki, K. and W.H.K. Lee, Determination of three-dimensional velocity anomalies under a seismic array using first P arrival times from local earthquakes, 1. A homogeneous initial model. *J. Geophys. Res.* **81**, 4381–4399, 1976.

Aki, K. and P.G. Richards. *Quantitative Seismology, Vol. 2*, San Francisco: W.H. Freeman, 1980.

Aki, K., A. Christoffersson, and E.S. Husebye, Determination of the three-dimensional seismic structure of the lithosphere. *J. Geophys. Res.* **82**, 277–296, 1977.

Alterman, Z., H. Jarosch, and C.L. Pekeris, Oscillations of the earth. *Proc. Roy. Soc., A,* **252**, 80–95, 1959.

Anderson, D.L., Hotspots, polar wander, Mesozoic convection and the geoid. *Nature* **297**, 391–393, 1982.

Ashby, M.F. and R.A. Verrall, Micromechanisms of flow and fracture, and their relevance to the rheology of the mantle. *Phil. Trans., R. Soc. Lond. A.* **288**, 59–95, 1977.

Bullen, K.E., The ellipticity correction to travel times of P and S earthquake waves. *Mon. Not. R. Astr. Soc., Geophys. Suppl.* **4**, 143–157, 1937b.

Chase, C.G., Subduction, the geoid, and lower mantle convection. *Nature* **282**, 464–468, 1979.

Chase, C.G. and M.K. McNutt, The geoid: Effect of compensated topography and uncompensated oceanic trenches. *Geophys. Res. Lett.* **9**, 29–32, 1982.

Chase, C.G. and D.R. Sprowl, The modern geoid and ancient plate boundaries. *Earth Planet. Sci. Lett.* **62**, 314–320, 1983.

Christensen, Y., Convection with pressure- and temperature-dependent non-Newtonian rheology. *Geophys. J. R. astr. Soc.* **77**, 343–384, 1984.

Clayton, R.W. and R.P. Comer, A tomographic analysis of mantle heterogeneities from body wave travel times. *EOS Trans. AGU* **64**, 776, 1983.

Clayton, R.W. and R.P. Comer, A tomographic analysis of mantle heterogeneities. *Terra Cognita* **4**, 282–283, 1984.

Comer, R.P. and R.W. Clayton, Reconstruction of mantle heterogeneity by iterative back projection of travel times: 1. Theory and Reliability, to be submitted to *J. Geophys. Res.*, 1989.

Comer, R.P., Rapid seismic ray tracing in a spherically symmetric earth via interpolation of rays. *Bull. Seismo. Soc. Am.* **74**, 479–492, 1984.

Creager, K.C. and T.H. Jordan, Slab penetration into the lower mantle. *J. Geophys. Res.* **89**, 3031–3049, 1984.

Creager, K.C. and T.H. Jordan, Slab penetration into the lower mantle beneath the Mariana and other island arcs in the Northwest Pacific. *J. Geophys. Res.* **91**, 3573–3589, 1986.

Crough, S.T., Hotspot swells. *Ann. Rev. Earth Planet. Sci.* **11**, 165–193, 1983.

Crough, S.T. and D.M. Jurdy, Subducted lithosphere, hotspots and the geoid. *Earth Planet. Sci. Lett.* **48**, 15–22, 1980.

Davies, G.F., Mechanics of subducted lithosphere. *J. Geophys. Res.* **85**, 6304–6318, 1980.

Davies, G.F., Regional compensation of subducted lithosphere: effects on geoid, gravity, and topography from a preliminary model. *Earth Planet. Sci. Lett.* **54**, 431–441, 1981.

Dines, K.A. and R.J. Lytle, Computerized geophysical tomography. *Proc. IEEE* **67**, 1065–1073, 1979.

Dziewonski, A.M., Mapping the lower mantle: Determination of lateral heterogeneity in P-velocity up to degree and order 6. *J. Geophys. Res.* **89**, 5929–5952, 1984.

Dziewonski, A.M. and D.L. Anderson, Preliminary reference earth model (PREM). *Phys. Earth Planet. Int.* **25**, 297–356, 1981.

Dziewonski, A.M. and D.L. Anderson, Travel times and station corrections for P waves at teleseismic distances. *J. Geophys. Res.* **88**, 3295–3314, 1983.

Dziewonski, A.M. and F. Gilbert, The effects of small, aspherical perturbations on travel times and a re-examination of the corrections for ellipticity. *Geophys. J. R. Astr. Soc.* **44**, 7–17, 1976.

Dziewonski, A.M., B.H. Hager, and R.J. O'Connell, Large-scale heterogeneities in the lower mantle. *J. Geophys. Res.* **82**, 239–255, 1977.

Fawcett, J.A. and R.W. Clayton, Tomographic reconstruction of velocity anomalies. *Bull. Seismo. Soc. Am.* **74**, 2201–2219, 1984.

Forte, A.M. and W.R. Peltier, Plate tectonics and aspherical earth structure: The importance of poloidal-toroidal coupling. *J. Geophys. Res.* **92**, 3645–3679, 1987.

Gantmacher, F.R., *The Theory of Matrices*, vols. 1 and 2, translated from Russian by K.A. Hirsch, New York: Chelsea, 1960.

Giardini, D. and J. Woodhouse, Deep seismicity and modes of deformation in Tonga subduction zone. *Nature* **307**, 505–509, 1984.

Giardini, D., X-D. Li, and J.H. Woodhouse, Three dimensional structure of the earth from splitting in free oscillation spectra. *Nature* **325**, 405–411, 1987.

Gordon, R., R. Bender, and G.T. Herman, Algebraic reconstruction techniques (ART) for three-dimensional electron microscopy and X-ray photography. *J. Theor. Biol.* **29**, 471–481, 1970.

Grand, S.P., Shear velocity structure of the mantle beneath the North American Plate, Ph.D. thesis, California Institute of Technology, 228 pp., 1986.

Grand, S.P. and D.V. Helmberger, Upper mantle shear structure of North America. *Geophys. J. R. Astr. Soc.* **76**, 399–438, 1984.

Gwinn, C.R., T.A. Herring, and I.I. Shapiro, Geodesy by radio interferometry: Studies of the force nutations of the Earth. 2. Interpretation. *J. Geophys. Res.* **91**, 4755–4765, 1986.

Hager, B.H., Subduction velocity and the geoid: an empirical relation. *EOS, Trans. AGU* **62**, 906, 1982.

Hager, B.H., Global isostatic geoid anomalies for plate and boundary layer models of the lithosphere. *Earth Planet. Sci. Lett.* **63**, 97–109, 1983.

Hager, B.H., Subducted slabs and the geoid; constraints on mantle rheology and flow. *J. Geophys. Res.* **89**, 6003–6015, 1984.

Hager, B.H. and R.J. O'Connell, Subduction zone dip and flow driven by plate motions. *Tectonophysics* **50**, 111–133, 1978.

Hager, B.H. and R.J. O'Connell, Kinematic models of large-scale mantle flow. *J. Geophys. Res.* **84**, 1031–1048, 1979.

Hager, B.H. and R.J. O'Connell, A simple global model of plate dynamics and mantle convection. *J. Geophys. Res.* **86**, 4843–4867, 1981.

Hager, B.H. and A. Raefsky, Deformation of seismic discontinuities and the scale of mantle convection. *EOS, Trans. AGU* **62**, 1074, 1981.

Hager, B.H. and M.A. Richards, Long-wavelength variations in Earth's geoid: Physical models and dynamic implications. *Phil Trans. Roy. Soc. Lond., A*, in press, 1989.

Hager, B.H., R.J. O'Connell, and A. Raefsky, Subduction, back-arc spreading and global mantle flow. *Tectonophysics* **99**, 165–189, 1983.

Hager, B.H., R.W. Clayton, M.A. Richards, R.P. Comer, and A.M. Dziewonski, Lower mantle heterogeneity, dynamic topography and the geoid. *Nature* **313**, 541–545, 1985.

Harrison, C.G.A., K.J. Miskell, G.W. Brass, E.S. Saltzman, and J.L. Sloan II, Continental hypsography. *Tectonics* **2**, 357–377, 1983.

Harrison, C.G.A., G.W. Brass, E. Saltzman, J. Sloan II, J. Southam, and J.M. Whitman, Sea level variations, global sedimentation rates, and the hypsographic curve. *Earth Planet. Sci. Lett.* **54**, 1–16, 1981.

Hearn, T.M. and R.W. Clayton, Lateral velocity variations in southern California: 1 Results for the upper crust from Pg-waves. *Bull. Seismo. Soc. Am.* **76**, 495–509, 1986a.

Hearn, T.M. and R.W. Clayton, Lateral velocity variations in southern California: 2 Results for the lower crust from Pn-waves. *Bull. Seismo. Soc. Am.* **76**, 511–520, 1986b.

Herman, G.T., A. Lent, and S.W. Rowland, ART: Mathematics and Applications. *J. Theor. Biol.* **42**, 1–32, 1973.

Hide, R., Presidential address: The earth's differential rotation. *Q. Jl. Roy. Astr. Soc.* **27**, 30–20, 1986.

Humphreys, E., R.W. Clayton, and B.H. Hager, A tomographic image of mantle structure beneath southern California. *Geophys. Res. Lett.*, v. 11, 625–627, 1984.

Isacks, B. and P. Molnar, Distribution of stresses in the descending lithosphere from a global survey of focal-mechanism solutions of mantle earthquakes. *Rev. Geophys. Space Phys.* **9**, 103–174, 1971.

Jaffe, C.C., Medical imaging. *Am. Sci.* **70**, 576–585, 1982.

Jeffreys, H., On travel times in seismology. *Bur. Centr. Seismol. Trav. Sci.* **14**, 3–86, 1936.

Jeanloz, R. and F.M. Richter, Convection, composition and the thermal state of the lower mantle. *J. Geophys. Res.* **84**, 5497–5504, 1979.

Jordan, T.H., Lithospheric slab penetration into the lower mantle beneath the Sea of Okhotsk. *J. Geophys.* **43**, 473–496, 1977.

Kaula, W.M., Global gravity and tectonics. In E.C. Robinson (ed.) *The Nature of the Solid Earth*, New York: McGraw-Hill, pp. 386–405, 1972.

Kaula, W.M., Product-sum conversion of spherical harmonies with application to thermal convection. *J. Geophys. Res.* **80**, 225–231, 1975.

Kaula, W.M., Material properties for mantle convection consistent with observed surface fields. *J. Geophys. Res.* **85**, 7031–7044, 1980.

Kiefer, W.S., M.A. Richards, B.H. Hager, and B.G. Bills, A dynamic model of Venus's gravity field. *Geophys. Res. Lett.* **13**, 14–17, 1986.

Landau, L.D. and E.M. Lifshitz. *Fluid Mechanics*, New York: Pergamon, 536 pp., 1959.

Le Mouel, J.L., C. Gire, and T. Madden, Motions at the core surface in the geostrophic approximation. *Phys. Earth Planet. Int.* **39**, 270–287, 1985.

Lerch, F.J., S.M. Klosko, and G.B. Patch, A refined gravity model from LAGEOS (GEM-L2). *NASA Tech. Memo* **84986**, 1983.

Masters, G., T.H. Jordan, P.G. Silver, and F. Gilbert, Aspherical earth structure from fundamental spheroidal-mode date. *Nature* **298**, 609–613, 1982.

McKenzie, D., Surface deformation, gravity anomalies and convection. *Geophys. J. R. Astr. Soc.* **48**, 211–238, 1977.

Molnar, P., D. Freedman, and J.S.F. Shih, Lengths of intermediate and deep seismic zones and temperatures in downgoing slabs of lithosphere. *Geophys. J. R. Astr. Soc.* **56**, 41–54, 1979.

Morgan, W.J., Gravity anomalies and convection currents. 1. A sphere and cylinder sinking beneath the surface of a viscous fluid. *J. Geophys. Res.* **70**, 6175–6187, 1965a.

Morgan, W.J., Gravity anomalies and convection currents. 2. The Puerto Rico trench and the mid-Atlantic rise. *J. Geophys. Res.* **70**, 6189–6204, 1965b.

Morelli, A. and A.M. Dziewonski, 3D structure of the earth. *Terra Cognita* **6**, 308–309, 1986.

Nakanishi, I. and D.L. Anderson, Measurements of mantle wave velocities and inversion for lateral heterogeneity and anisotropy. II: Analysis by the single-station method. *Geophys. J. R. Astr. Soc.* **78**, 573–617, 1984.

Nakiboglu, S.M., Hydrostatic theory of the earth and its mechanical implications. *Phys. Earth Planet. Int.* **28**, 302–311, 1982.

Nataf, H.-C., I. Nakanishi, and D.L. Anderson, Anisotropy and shear-velocity heterogeneities in the upper mantle. *Geophys. Res. Lett.* **11**, 109–112, 1984.

Nataf, H.-C., I. Nakanishi, and D.L. Anderson, Measurements of mantle wave velocities and inversion for lateral heterogeneities and anisotropy part III: Inversion. *J. Geophys. Res.* **91**, 7261–7307, 1986.

O'Connell, R.J., Pleistocene glaciation and the viscosity of the lower mantle. *Geophys. J. R. Astr. Soc.* **23**, 299–327, 1971.

O'Connell, R.J., On the scale of mantle convection. *Tectonophysics* **38**, 119–136, 1977.

O'Connell, R.J. and B.H. Hager, On the thermal state of the earth. In A. Dziewonski and E. Boschi (eds.) *Physics of the Earth's Interior*, Enrico Fermi International School of Physics, Course 78, Soc. Italiana di Fisica, Bologna, 270–317, 1980.

O'Connell, R.J. and B.H. Hager, Velocity anomalies, convection, heat transport, and viscosity of the lower mantle. *EOS, Trans. AGU* **65**, 1093, 1984.

O'Connell, R.J., B.H. Hager, and M.A. Richards, Analytic models of deformation and flow of a viscous, self-gravitating mantle. *Terra Cognita* **4**, 261, 1984.

Park, J. and F. Gilbert, Computed splitting of multiplets by aspherical models of the structure of the mantle. *EOS, Trans. AGU* **65**, 1002–1003, 1984.

Parsons, B. and J.G. Sclater, An analysis of the variations of ocean floor bathymetry and heat flow with age. *J. Geophys. Res.* **82**, 803–827, 1977.

Parsons, B. and S. Daly, The relationship between surface topography, gravity anomalies, and the temperature structure of convection. *J. Geophys. Res.* **88**, 1129–1144, 1983.

Pekeris, C.L., Thermal convection in the interior of the earth. *Mon. Not. R. Astr. Soc., Geophys. Suppl.* **3**, 343–367, 1935.

Peltier, W.R. and P. Wu, Mantle phase transitions and the free air gravity anomalies over Fennoscandia and Laurentia. *Geophys. Res. Lett.* **9**, 731–734, 1982.

Peltier, W.R., New constraints on transient lower mantle rheology and internal mantle buoyancy from glacial rebound data. *Nature* **318**, 611–614, 1985.

Peltier, W.R. and A. Forte, The gravitational signature of plate tectonics. *Terra Cognita* **4**, 251, 1984.

Ricard, Y., Dynamique de la Terre et observations gravimetriques, D.S.N. Thesis, Universite de Paris-Sud, Orsay, France, 1986.

Ricard, Y., L. Fleitout, and C. Froidevaux, Geoid heights and lithospheric stresses for a dynamic earth. *Annales Geophysicae* **2**, 267–286, 1984.

Richards, M.A., Dynamical models for the Earth's geoid, Ph.D. thesis, California Institute of Technology, Pasadena, California, 91125, 1986.

Richards, M.A. and B.H. Hager, Geoid anomalies in a dynamic Earth. *J. Geophys. Res.* **89**, 5987–6002, 1984.

Richards, M.A. and B.H. Hager, The Earth's geoid and the large-scale structure of mantle convection. In S.K. Runcorn (ed.) *The Physics of Planets*, John Wiley and Sons., Ltd., 247–272, 1988.

Richards, M.A., B.H. Hager, and N. Sleep, Dynamically supported geoid highs over hotspots: Observations and theory. *J. Geophys. Res.* **93**, 7690–7708, 1988.

Richter, F.M., Focal mechanisms and seismic energy release of deep and intermediate earthquakes in the Tonga-Kermadec region and their bearing on the depth extent of mantle flow. *J. Geophys. Res.* **84**, 6783–6795, 1979.

Richter, F.M. and D.P. McKenzie, Simple plate models of mantle convection. *J. Geophys.* **44**, 441–471, 1978.

Ritzwoller, M. and G. Masters, The fine structure of resolvably split multiplets. *EOS, Trans. AGU* **65**, 1003, 1984.

Runcorn, S. K., Satellite gravity measurements and a laminar viscous flow model of the earth's mantle. *J. Geophys. Res.* **69**, 4389–4394, 1964.

Runcorn, S.K., Flow in the mantle inferred from the low degree harmonics of the geopotential. *Geophys. J. R. Astr. Soc.* **42**, 375–384, 1967.

Sabadini, R., D.A. Yuen, and P. Gasperini, The effects of transient rheology on the interpretation of lower mantle viscosity. *Geophys. Res. Lett.* **12**, 361–364, 1985.

Sammis, C. G., J.C. Smith, G. Schubert, and D.A. Yuen, Viscosity-depth profile of the earth's mantle: effects of polymorphic phase transitions. *J. Geophys. Res.* **82**, 3747–3761, 1977.

Schmeling, H. and W.R. Jacoby, On modelling the lithosphere in mantle convection. *J. Geophys.* **50**, 89–100, 1981.

Schubert, G., P. Cassen, and R.E. Young, Core cooling by subsolidus mantle convection. *Phys. Earth Planet. Int.* **20**, 194–208, 1979.

Schubert, G., D.A. Yuen, and D.L. Turcotte, Role of phase transitions in a dynamic mantle. *Geophys. J. R. Astr. Soc.* **42**, 705–735, 1975.

Sclater, J.G., C. Jaupart, and D. Galson, The heat flow through oceanic and continental crust and the heat loss of the earth. *Rev. Geophys. Space Phys.* **18**, 269–311, 1980.

Sengupta, M.K. and M.N. Toksoz, Three dimensional model of seismic velocity variation in the earth's mantle. *Geophys. Res. Lett.* **3**, 84–86, 1976.

Sengupta, M.K., R.E. Hassell, and R.W. Ward. Three-dimensional seismic velocity structure of the Earth's mantle using body wave travel times from intra-plate and deep-focus earthquakes. *J. Geophys. Res.* **86**, 3913–3934, 1981.

Takeuchi, H. and Y. Hasegawa, Viscosity distribution within the earth. *Geophys. J. R. Astr. Soc.* **9**, 503–508, 1965.

Tanimoto, T., The Backus-Gilbert approach to three-dimensional structure in the upper mantle: II. SH and SV velocity. *Geophys. J. R. Astr. Soc.* **84**, 49–69, 1986.

Tanimoto, T. and D.L. Anderson, Mapping convection in the mantle. *Geophys. Res. Lett.* **4**, 287–290, 1984.

Vail, P.R., R.M. Mitchum, and S. Thompson, Seismic stratigraphy and global changes of sea level, 4. Global cycles of relative changes of sea level. In C.E. Payton (ed.) *Seismic Stratigraphy—Applications to Hydrocarbon Exploration*, Am. Assoc. Pet. Geol., Mem. 26, 83–97, 1978.

Vassiliou, M.S., B.H. Hager, and A. Raefsky, The distribution of earthquakes with depth and stress in subducting slabs. *J. Geodyn.* **1**, 11–28, 1984.

Walck, M.C. and R.W. Clayton, P-wave velocity variations in the Coso Region, California, derived from local earthquake travel times. *J. Geophys. Res. 92*, 393–405, 1987.

Watts, A.B., An analysis of isostasy in the worlds oceans 1. Hawaiian-Emperor Seamount Chain. *J. Geophys. Res.* **83**, 5989–6004, 1978.

Woodhouse, J.H. and A.M. Dziewonski, 1984, Maping the upper mantle: Three dimensional modelling of Earth structure by inversion of seismic waveforms. *J. Geophys. Res.* **89**, 5953–5986.

CHAPTER 10

The Radial Structure of the Mantle General Circulation

W.R. Peltier,[†] G.T. Jarvis,[‡] A.M. Forte,[†] and L.P. Solheim[†]

[†]Department of Physics, University of Toronto
Toronto, Ontario, Canada M5S 1A7
[‡]Department of Earth and Atmospheric Science, York University
North York, Ontario, Canada, M3J 1P3

1. Introduction

A fundamental issue of mantle dynamics that remains outstanding today concerns the radial structure of the thermally forced flow that is responsible for sustaining the motion of the surface plates against the influence of viscous dissipation. Although this issue might be considered esoteric, it is nevertheless a crucial one insofar as a number of important questions in the earth and planetary sciences are concerned. For example, if the mantle convective circulation were layered in the manner illustrated in Figure 10.1b, then the rate of planetary cooling would be very much diminished from the rate which would obtain if the circulation were of the whole mantle style illustrated in Figure 10.1a. Clearly, therefore, we shall be in no position to understand the thermal history of planet Earth (as discussed in the final chapter of this book)

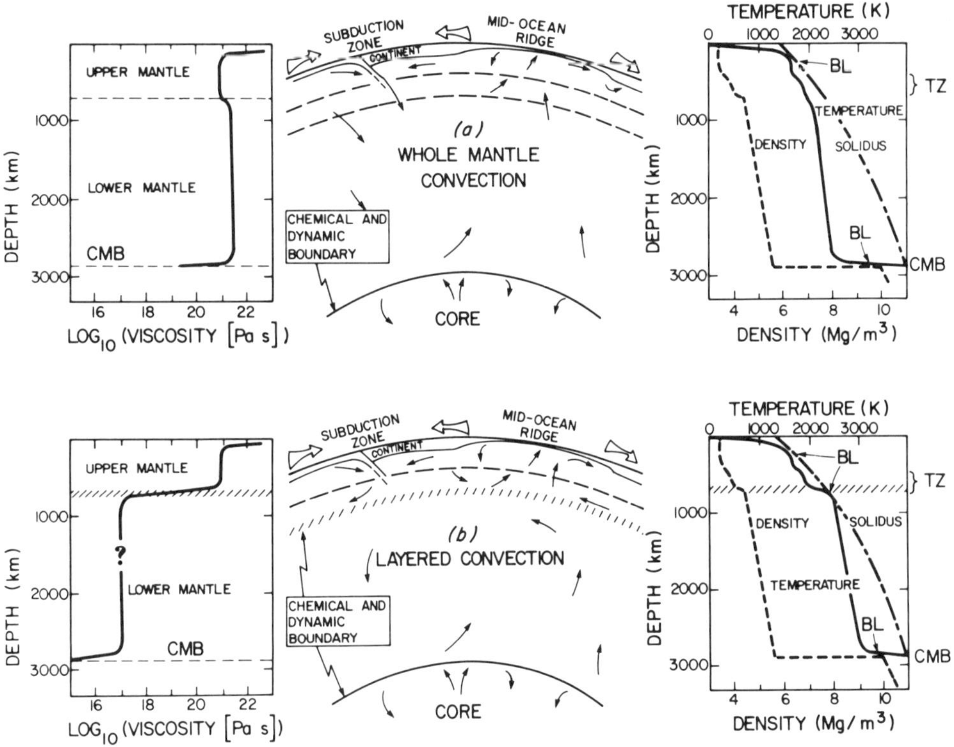

Figure 10.1. Schematic diagrams of the whole-mantle and layered convection models with associated viscosity, density, and temperature profiles. Note the question mark adjacent to the viscosity profile for the layered convection model. We would expect the large viscosity drop only if the activation parameters for the creep process did not change across this horizon.

until we can settle the issue of which of these models of reality is closer to the truth. Of course this issue of the rate at which the Earth is being cooled through the action of the convective circulation in its mantle also has far reaching consequences on the dynamo theory for the maintenance of the Earth's main dipole magnetic field. If the rate of core cooling induced by the mantle convective circulation is sufficiently high, then this source of convective stirring of the Earth's outer liquid iron core might provide a sufficient source of energy to maintain the field against dissipation, and exotic effects associated with the freezing of the inner core might play a much less significant role in the evolution of the planet than is currently believed (e.g., see the discussion by Stevenson in the following chapter). Our purpose in the present chapter will be to review and to extend the arguments that have been presented in the past in support of either one or the other of the two competing views of the mantle convection process that are represented by plates a and b of Figure 10.1.

Characteristic of the whole mantle model illustrated in Figure 10.1a is the fact that the entire mantle is presumed to be mechanically well mixed throughout its volume and to be characterized by horizontally averaged temperature and density fields which vary adiabatically with depth except in thin (of the order of 10^2 km thick) boundary layers adjacent to the Earth's surface and the core–mantle boundary (CMB). At the Earth's surface the oceanic lithosphere is presumed to be very tightly coupled to the motions in the underlying mantle and to be coincident with the cold upper thermal boundary layer of the convective circulation. In this model the continents are seen to constitute a combined chemical and thermal boundary layer which may be somewhat thicker than the oldest oceanic lithosphere, particularly beneath continental cratons.

This "whole mantle" picture of the mantle general circulation differs from that of the layered model in several fundamental ways as illustrated in Figure 10.1b. In the layered model, mass transfer across the seismic discontinuity at 670 km depth is assumed not to occur. Consequently, although the upper and lower mantles are individually well mixed by the thermal convection process they may have distinct chemical compositions. Because heat can be transferred across the boundary at 670 km depth only by diffusion, the layered mantle model has a thermal boundary layer there across which the temperature must increase by at least 500 K (as in Figure 10.1b) and more likely by 1000 K if the upper mantle is depleted of radioactivity relative to the lower as recent geochemical data (see Chapter 6) have sometimes been taken to suggest. Although the lower mantle of the layered model is also made adiabatic by its convective circulation, this adiabat must at all lower mantle depths be at a temperature which in from 500 to 1000 K hotter than that of the whole mantle model. In consequence of this, for a fixed known temperature at the CMB the layered model must have a weaker thermal boundary layer at this interface which implies a smaller heat flux from the core into the mantle and thus a slower rate of secular cooler of the Earth as a whole. This was discussed in some detail in Peltier and Jarvis (1982). There are also differences between the upper mantle of the layered model and that in the whole mantle alternative. Foremost among these, according to most advocates of this picture, is the notion that there exists a separate scale of convection which is uncorrelated with the motion of the plates and which may be responsible for the radial transport of a significant fraction of the heat flowing across the upper mantle.

As mentioned previously, the issue as to which of these two "models" of the circulation is closer to the truth has a number of rather profound consequences, a fact which explains the extent to which attention has come to be focussed upon this question during the course of the last decade. In Table 10.1 we list for future reference the main observations which have been invoked by various authors to favour one or the other of the two distinct views represented on Figure 10.1. Of this list of 16 entries, three (6, 11, and 14) have been invoked as strongly supporting the layered model, whereas a different three (1, 9, and 15) have been taken to imply the whole mantle alternative. In

Table 10.1. Constraints on models of the mantle convection process

1. Mean horizontal plate scale $l \simeq 4000$ km
2. Mean plate speed $u \simeq 4$ cm/y
3. Mean surface heat flow $q \simeq 80$ mW/m
4. Mean plate thickness $\delta \simeq 100$ km
5. Mean mantle viscosity $\mu \simeq 10^{21}$ Pa s
6. Seismicity ceases below the 670 km seismic discontinuity
7. Heat flow and topography $\alpha 1/\sqrt{\text{AGE}}$ of the ocean floor
8. Volcanism observed in plate interiors—"hot spots"
9. Viscosity increases only modestly from 10^{21} Pa s to $\sim 2 \times 10^{21}$ Pa s from the upper to the lower mantle
10. Boundary layer nature of the D″ layer above the CMB
11. Deep seismic focal mechanisms compressive
12. Bathymetry deviates from $(\text{AGE})^{-1/2}$ for AGE $\gtrsim 7 \times 10^7$ yr
13. Seismic LVZ exists only under young oceanic lithosphere
14. Rb-Sr and Nd-Sm systematics constrain models of mantle mixing and crustal growth
15. High pressure chemistry of the olivine binary constrains the nature of the 400 km and 650 km discontinuities
16. *A priori* geochemical constraints on the amount of radioactivity in the mantle.

fact it was only with the discovery of point 9 (Peltier, 1974, 1976; Cathles, 1975) that the whole mantle model became tenable at all.

As ideas of mantle convection have developed over the past decade, leading to the impasse illustrated on Figure 10.1, discussion has usually been focussed upon the layered model or some variant of it. The reason for this has to do with early geophysical understanding of the variation of mantle viscosity as a function of depth, which has been reviewed in detail in Chapter 6. As discussed there, modern work on this problem essentially began with McConnell's (1968) analysis of Sauramo's (1958) shoreline data from Fennoscandia. McConnell's results included an estimate of the thickness of the lithosphere in the Baltic Shield ($\sim$120 km) and a measurement of the viscosity of the upper mantle which he found to be close to $\sim 1 \times 10^{21}$Pa s, in agreement with original work of Haskell (1937). In order to retrieve information on the viscosity of the mantle beneath the transition region, McConnell was obliged to invoke information beyond that contained in the shoreline observations themselves. What he did was to assume, following Munk and MacDonald (1960), that the earth possessed a real nonhydrostatic equatorial bulge, i.e., a polar flattening in excess of that which would be in equilibrium with the current rotation rate. He further assumed that the excess bulge was produced by Pleistocene glaciation (see Chapter 6). It then followed from the continuing existence of the bulge that the relaxation time of the second degree harmonic must be extremely long. On this basis McKenzie (1966, 1967, 1968) and McConnell (1968) were led to construct models in which the viscosity beneath the transition zone was in excess of 10^{25} Pa s. This argument for high lower viscosity was completely undermined by Goldreich and Toomre (1969) who pointed out that the nonhydrostatic equatorial bulge did not in fact exist! It had been inferred from a spherical harmonic expansion which was improperly biased towards the second degree harmonic.

In spite of the fact that McKenzie's original argument in favor of extremely high viscosity in the Earth's lower mantle was shown to be incorrect, this did not lead to a rejection of the notion itself. In the model of mantle convection advocated by McKenzie *et al.* (1974) and discussed by Oxburgh and Turcotte (1978) it was assumed that thermal convection was confined to the upper mantle alone. Also, in McKenzie and Weiss (1975) it was argued that either the upper mantle convects alone or that the upper and lower mantles were filled by distinct circulations (the layered model of Figure 10.1b). The argument given by McKenzie and Weiss for the existence of a separate circulation in the upper mantle was again that there existed a large increase of viscosity across the 670 km seismic discontinuity, although no direct observational evidence was then available which might be invoked to support the assumption. They speculated, however, that the large increase of viscosity which they required was caused by an increase of the creep activation energy of about 2 ev. Observation 6 of Table 10.1 was invoked to support the argument that the upper mantle flow could not penetrate the 670 km horizon.

In Peltier (1972), an analysis in terms of linear stability theory was produced which demonstrated that the magnitude of the viscosity contrast across 670 km depth which would be required to focus a distinct circulation into the upper mantle was near three orders of magnitude. Peltier (1976) argued on the basis of this result and the then new postglacial rebound analyses (Peltier 1974; Cathles 1975; Peltier and Andrews 1976) that "if the radial variation of mantle viscosity is not 'extreme' then the radial mixing length for thermal convection in this system is liable to be on the order of the thickness of the mantle itself." This argument in favor of the whole mantle style of convection was later echoed by Davies (1977), O'Connell (1977), and Schubert (1979).

That the observed viscosity profile of the mantle was shown using glacial isostatic adjustment data to be rather uniform (see Chapter 6) did not completely win the argument for the whole mantle convection scenario was due to the fact that separation into distinct upper and lower mantle circulations could be effected by means other than a marked increase of viscosity. It has often been argued (e.g., see D.L. Anderson, 1981, for a recent review) that the mantle is chemically stratified (see also Chapter 4). If the seismic discontinuity at 670 km depth were a chemical boundary as shown in Figure 10.1b then the stabilizing effect of a change in mean atomic weight across this horizon could easily prevent mass transport across the boundary (e.g., Richter and Johnson 1974). This idea received very strong support from recent studies of mantle differentiation and crustal growth using the Sm-Nd and Rb-Sr isotopic systems (Wasserburg and DePaulo, 1979; Onions et al., 1979; point 15 of Table 10.1). These data were initially interpreted as requiring the existence of two distinct chemical reservoirs in the mantle and mass balance arguments were employed to argue that the two reservoirs should be associated with the upper and lower mantles. As discussed in detail in Chapter 5 of this book this interpretation has not stood the test of time as the idea of the existence of a simple "mantle array" upon which it was based, with OIB and MORB representing the separate pristine and depleted reservoirs, has not held up in the face of accumulating observational evidence (as discussed in Chapter 5). These data apparently cannot be employed to argue strongly that any specific volume of the mantle has been depleted to form continental crust and therefore they probably have nothing which is profound to say concerning the radial structure of the mantle general circulation.

The layered model is also vulnerable to criticism because it appears to require the validity of notions which are themselves extremely contentious. On the question of chemical stratification, for example, it has been demonstrated (e.g., Mao *et al.*, 1979) through direct static high pressure experiments on the olivine binary with the diamond anvil apparatus (see Chapter 4) that the seismic discontinuity at 670 km depth is explicable as a phase boundary. The observed change in density across the phase boundary, furthermore, appears sufficient to explain the requirements of the seismic data (e.g., Mao *et al.*, 1979). It therefore seems that there is very little room for any substantial density change due to a change of mean atomic weight and therefore that the mantle may have no appreciable chemical stratification. This model has long been advocated by Ringwood (e.g., 1979). Anderson (e.g., 1981) has therefore suggested that the 670 km discontinuity may be in part a phase boundary and in part a chemical boundary. This would appear rather *ad hoc*. The only evidence that a simple equilibrium phase boundary might not provide an adequate explanation of the 670 km discontinuity is that the boundary is required to be rather sharp in order to explain the good seismic reflections which are obtained from it, and somewhat sharper than it has been observed to be in the laboratory measurements (Jeanloz, 1981 and Chapter 4).

If both the 420 km and the 670 km mantle discontinuities are to be interpreted as equilibrium phase transitions then whole mantle convection would appear to be inevitable unless there were physical effects associated with the phase boundaries themselves which might be expected to enforce a layered style of mixing onto the convective circulation. In fact, as we will show in the following section, previous analyses of the influence of phase transitions upon convection (see Chapter 8 for a review) appear to have missed an important aspect of their behavior in consequence of certain inappropriate assumptions concerning scaling. The original analyses of such effects at infinitesimal amplitude by Busse and Schubert (1971), and Schubert *et al.* (1975) concluded that the olivine-spinel transition was strongly destabilizing to convection while the spinel-post spinel transition could be moderately stabilizing. These analyses were encouraging because it was known that the cold downgoing slab continued through the 420 km discontinuity (in the sense that the seismicity did). The fact that seismicity ceases below a depth of 670 km and that deep focal mechanisms are observed to be compressive (items 6 and 11 of Table 10.1) were also taken as being in accord with a stabilization of the flow by the deeper horizon. As we will show in the next section, however, when the small amplitude problem is properly scaled it turns out that the olivine-spinel transition is strongly stabilizing to convection whereas the spinel-post spinel transition is destabilizing.

The main body of the present chapter will be found in Section 3 (following the analysis of the influence of phase transitions in Section 2) where we shall employ the recent results from seismic tomographic imaging to demonstrate that the observed motions of the surface plates are in rather striking accord with expectations based upon the simplest possible theory of the mantle convection process. We believe that these analyses represent the first direct test of the validity of the convection hypothesis itself. By extending the analyses of Hager and others (see Chapter 9) of the predictability of the planet's gravitational field, based upon the depth dependent lateral heterogeneity of density delivered by seismic tomography, to include a prediction of surface plate velocities, we are able to produce a theory that may be employed to discriminate sharply between the competing whole mantle and layered models of the mantle convection process. Our conclusions based upon these analyses are summarized in Section 4.

2. Mantle Phase Transitions and Layered Convection

In the Boussinesq approximation, the problem of the onset of thermal convection in a spherical shell of fluid is embodied in the solution of the following linearized system of coupled partial differential equations which have been nondimensionalized using the scheme of Peltier (1972):

$$\nabla \cdot \mathbf{u} = 0, \tag{1}$$

$$\frac{\mathrm{Ra}}{\mathrm{Pr}} \frac{\partial u}{\partial t} = \theta \hat{r} - \frac{1}{\delta} \nabla \Pi + \nabla^2 \mathbf{u}, \tag{2}$$

$$\frac{\partial \theta}{\partial t} + (\mathbf{u} \cdot \nabla) \theta = \frac{1}{\mathrm{Ra}} \nabla^2 \theta, \tag{3}$$

in which Ra and Pr are the usual Rayleigh and Prandtl numbers, $\delta = \alpha \Delta t$ is the nondimensional thermal expansivity and $\mathbf{u}$, θ and Π are the nondimensional velocity, temperature, and reduced pressure perturbations. Analyzing this system by expanding the fluctuations into normal modes and invoking the principle of exchange of stabilities, the stability problem is reduced to the solution of the sixth order differential system

$$\frac{dY}{dr} = \mathsf{A}\mathbf{Y}, \tag{4}$$

in which the matrix of coupling coefficients A is

$$\mathsf{A} = \begin{bmatrix} 0 & 1 & 0 & 0 & 0 & 0 \\ 0 & 0 & 1 & 0 & 0 & 0 \\ 0 & 0 & 0 & 1 & 0 & 0 \\ \dfrac{L(2-L)}{r^4} & 0 & \dfrac{2L}{r^2} & \dfrac{-4}{r} & \dfrac{Lg(r)}{r} & 0 \\ 0 & 0 & 0 & 0 & 0 & 1 \\ \dfrac{\mathrm{Ra}dT_B}{rd_1} & 0 & 0 & 0 & \dfrac{L}{r^2} & \dfrac{-2}{r} \end{bmatrix}, \tag{5}$$

with $L = l(l+1)$ and $g(r)$ the nondimensional gravitational acceleration through the shell. The solution 6-vector $\mathbf{Y} = (W_{lm}, W'_{lm}, W''_{lm}, W'''_{lm}, \theta_{lm}, \theta'_{lm})$ in which W_{lm} are the coefficients in the spherical harmonic expansion of the vertical component of velocity and θ_{lm} are those for the temperature perturbation. The prime here indicates differentiation with respect to nondimensional radius. The fact that A in (5) is independent of m indicates that the individual eigenfunctions of degree l are $(2l + 1)$-fold degenerate.

As first demonstrated by Busse and Schubert (1971) and further elaborated by Schubert and Turcotte (1971), Peltier (1972), and Schubert *et al.* (1975) in connection with plane layer geometry, the effect of phase boundaries on the stability of the layer may be incorporated by imposing the following boundary conditions of discontinuity at the undisplaced position of the phase boundary:

$$\theta_2 - \theta_2 = \frac{1}{r} R_Q w(r), \tag{6a}$$

$$\frac{\Pi_2}{\delta} - \frac{\Pi_1}{\delta} = SW(r), \tag{6b}$$

where region 1(2) is the region of the light (heavy) phase of density $\rho_1(\rho_2)$. The nondimensional parameters R_Q and S are defined as follows:

$$R_Q = \frac{g\sigma g^3 Q}{K\nu C_p}, \tag{7a}$$

$$S = \frac{\Delta\rho/\rho}{d\alpha\left(\dfrac{g\rho}{\gamma} + \dfrac{dT}{dr}\right)}, \tag{7b}$$

in which γ is the Clapeyron slope of the phase boundary and α the coefficient of thermal expansion. Values for these parameters have been estimated in Table 10.2 and may be compared to previous estimates by Schubert *et al.* (1975) shown on Table 10.3. The new estimates of R_Q are clearly several

Table 10.2. Physical properties of the mantle

$2 \times 10^{-5} \leqslant \alpha \leqslant 3.5 \times 10^{-5}$/K	
$10^{-6} \leqslant \kappa \leqslant 2 \times 10^{-6}$ m²/s	
$10^{21} \leqslant \mu \leqslant 5 \times 10^{21}$ Pa s	
$C_p \simeq 1.25 \times 10^3$ J/kgK	
$\rho \simeq 4500$ km/m³	
Olivine-spinel phase change	Spinel-perovskite phase change
$0.046 \leqslant \Delta\rho/\rho \leqslant 0.05$	$\Delta\rho/\rho \simeq 0.11$
$1800 \leqslant T \leqslant 1900$ K	$2300 \leqslant T \leqslant 2400$ K
$-0.3 \times 10^{-3} \leqslant \beta \leqslant -1.3 \times 10^{-3}$ K/m	$-0.3 \times 10^{-3} \leqslant \beta \leqslant -1.3 \times 10^{-3}$ K/m
$3 \times 10^6 \leqslant \gamma \leqslant 6.2 \times 10^6$ Pa/K	$-3 \times 10^6 \leqslant \gamma \leqslant -1 \times 10^6$ Pa/K
$0.03 \leqslant S \leqslant 0.10$	$-0.08 \leqslant S \leqslant -0.03$
$7.3 \times 10^5 \leqslant R_Q \leqslant 4.0 \times 10^6$	$-1.6 \times 10^5 \leqslant R_Q \leqslant 8.6 \times 10^5$
($d \simeq 2.9 \times 10^6$ m)	($d \simeq 2.9 \times 10^6$ m)

Table 10.3. Comparison of control variables for phase change dynamics

	Schubert *et al.* (1975)	Spherical model
Olivine-spinel	$5 \leqslant R_Q \leqslant 6000$	$7.3 \times 10^5 \leqslant R_Q \leqslant 4 \times 10^6$
	$0.5 \leqslant S \leqslant 18$	$0.03 \leqslant S \leqslant 0.10$
Spinel-perovskite	$-700 \leqslant R_Q \leqslant -0.04$	$8.6 \times 10^5 \leqslant R_Q \leqslant 1.6 \times 10^5$
	$-0.4 \leqslant S \leqslant -0.1$	$-0.08 \leqslant S \leqslant -0.03$

orders of magnitude higher than those of Schubert *et al.*, whereas our values of S are similarly one or two orders of magnitude lower. There are two reasons for these differences. Firstly, the radial length scale d has been taken equal to the mantle thickness (the only natural length scale in the problem), and, secondly, the value of μ has been taken equal to that inferred on the basis of the glacial rebound data discussed in Chapter 6. This new scaling has a profound effect because it ensures that the importance of the phase boundaries (in linear theory) will be governed by the latent heat release (as measured by R_Q) and only marginally effected by the pressure difference arising from the density difference between the phases (as measured by S). For a spinel-post spinel phase change that has negative Clapeyron slope, the latent heat release effect is strongly destabilizing while for the olivine-spinel transition it is equally strongly stabilizing.

Illustrative results from such stability analyses demonstrate the profound effect which the new scaling has upon the radial structure of the convective circulation which is expected at the onset of instability. For example, Figure

10.2 shows the critical Rayleigh number for the spherical shell (minimum on the neutral curve of Ra vs. spherical harmonic degree l) subject to heated from below, stress free, boundary conditions as a function of R_Q for two different values of S. These calculations were performed with a single olivine-spinel type exothermic transition in the center of the shell (i.e., at midmantle depth rather than in the correct physical location). In the scaling of Schubert *et al.* (1975) we would be in the lower left hand corner of this figure at small R_Q and S where the critical wavenumber (l) is 3 and the layer is *less* stable than it would be in the absence of the phase transition since the critical Rayleigh number is reduced by the effect of S. In the new scaling, however, we are in the upper right hand corner where R_Q is large and the

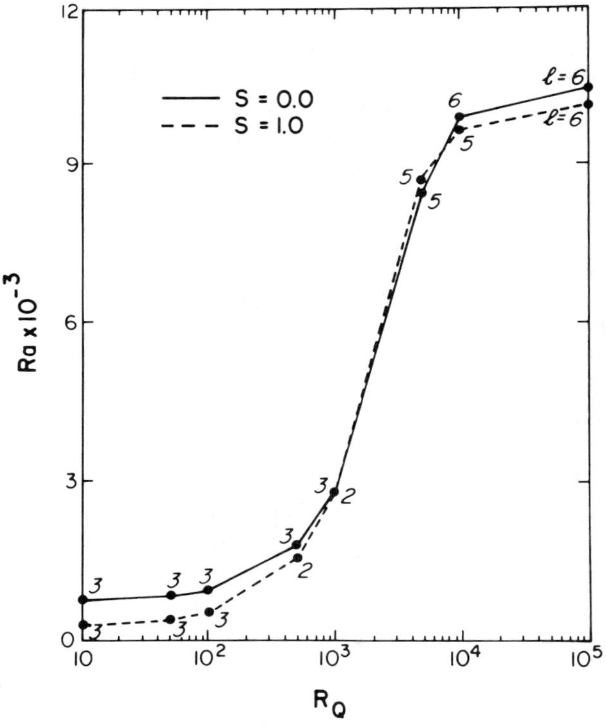

Figure 10.2. Critical Rayleigh number as a function of R_Q for two values of S. The phase change is of olivine-spinel type and is located in the center of the spherical shell. The numbers adjacent to the curves denote the critical spherical harmonic degree at the onset of instability.

critical wave number is now 6. This dramatic shift to shorter horizontal wavelength which is induced by increasing R_Q signals the transition of the circulation from one of whole mantle style to one which is layered, a fact which is clearly illustrated on Figure 10.3 which shows the radial eigenfunctions for W and θ for 3 points on the stability diagram of Figure 10.3 All of these results are for the case $S = 0$ and plates a, b, and c are, respectively, for $R_Q = 10, 10^3$, and 10^5. For the latter case the radial eigenfunction for W has a single node between the Earth's surface and the CMB and thus the circulation at the onset of convection would consist of two layers with the interface between them located at the phase boundary. Clearly then the influence of the olivine-spinel transition will be quite strongly stabilizing and will promote the occurence of layered convection rather than, as concluded in Schubert *et al.* (1975), being destabilizing. The reason for the opposite conclusion that was reached in these previous analyses has to do with an improper scaling which led to a sharp underestimate of the influence of latent heat release (R_Q) relative to the influence of the buoyancy force developed because of the density difference between the phases (S).

Because of this previous problem in scaling, conclusions concerning the influence of the spinel-post spinel transition were also questionable. Because this transition is endothermic, the influence of the latent heat of transition is destabilizing whereas the influence of the buoyancy force is stabilizing. The remarkable effect which such a phase transition may have on the structure of the convective circulation at the onset of instability is illustrated in Figure 10.4, plate a of which shows the variation of the critical Rayleigh number for the onset of convection in the shell as a function of $-R_Q$ with $S = 0$ for the case in which the transition is again placed in the center of the shell and therefore not at the correct geophysical depth. This figure demonstrates that for $R_Q \lessapprox -200$ the critical Rayleigh number actually becomes negative, indicating that the shell would be convectively unstable even if the background temperature gradient were such that the basic state density field were stably stratified with heavier fluid underlying light fluid at all depths. As this destabilizing transition occurs, it is accompanied by a marked shift in the critical wavenumber to higher values. At small negative R_Q the critical wavenumber is $l = 3$ (shown adjacent to the stability curve on plate a) whereas at large wavenumber it increases to $l = 8$ for $R_Q = -10^4$. It should be noted from Table 10.2 that even this large negative value of R_Q is considerably smaller than that which should be characteristic of the spinel-post spinel phase transition. Plates b and c of Figure 10.4 show radial eigenfunctions for the points $R_Q = -300$ ($l = 3$) and $R_Q = -10^4$ ($l = 8$), respectively, inspection of which shows that the flows are not caused to assume a layered form by the presence of the phase transition, even when the wavenumber becomes quite

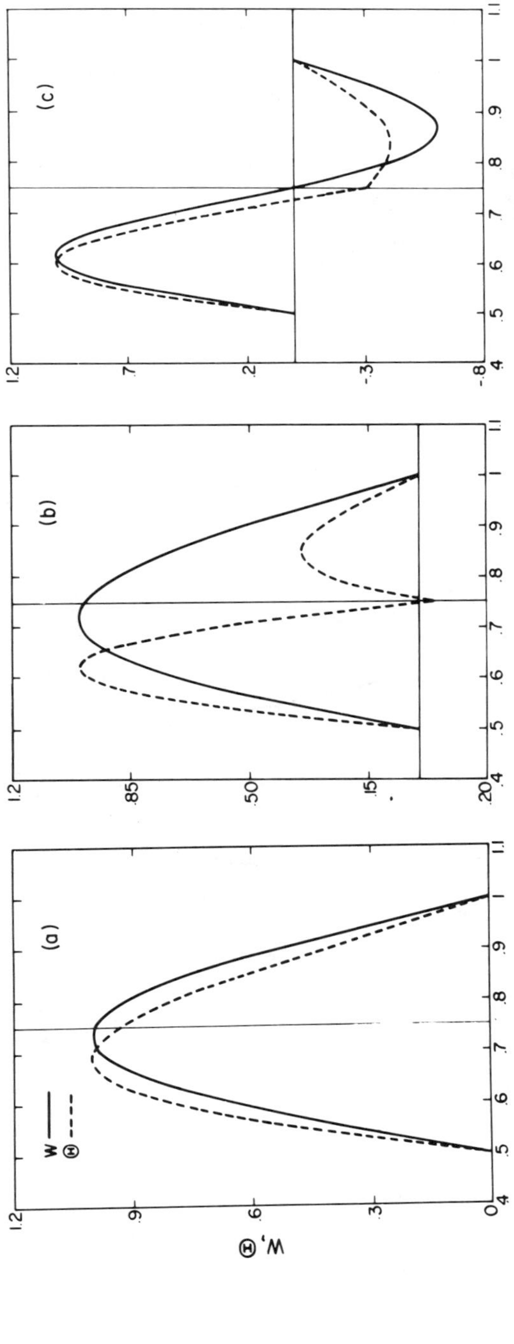

Figure 10.3. The eigenfunctions of the most unstable modes for the physical model corresponding to the data on Figure 10.2. All eigenfunctions are for the case $S = 0$. The values of R_Q and l are: a, (10, 3); b, $(10^3, 3)$; c, $(10^5, 6)$.

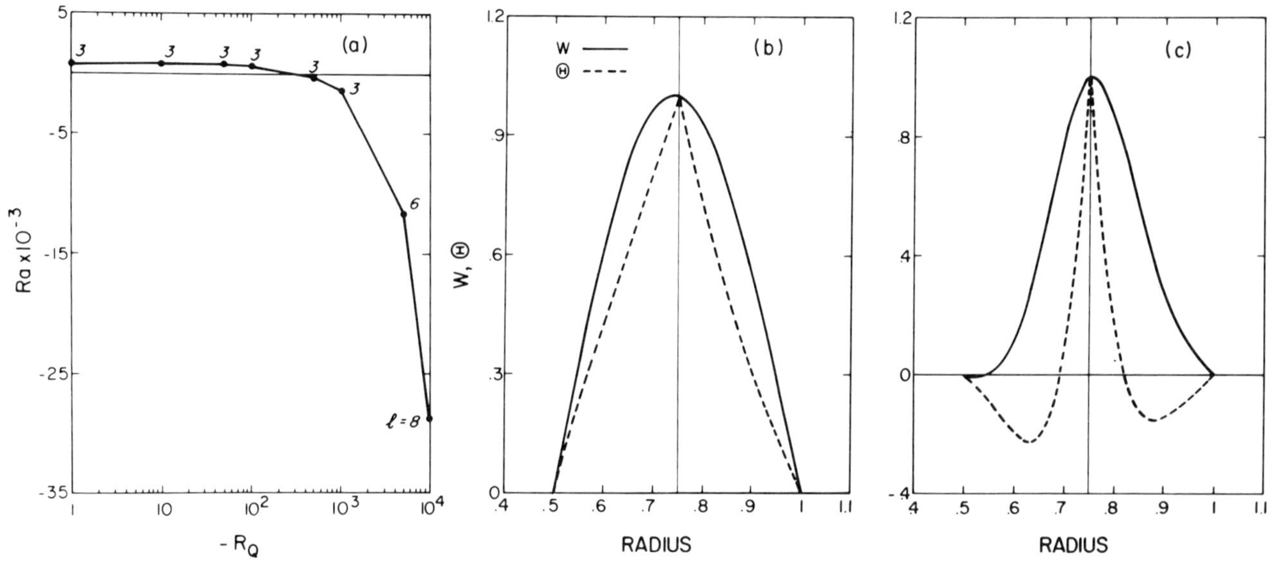

Figure 10.4. a, Critical Rayleigh number Ra as a function of $-R_Q$ for the case of a single endothermic phase transition at the center of the spherical shell with $S = 0$. Eigenfunctions are for b, $R_Q = 300, l = 3$; and c, $R_Q = -10^4$, $l = 8$.

large, as no radial node appears in the vertical velocity eigenfunction except perhaps for that very close to the CMB in the $R_Q = -10^4$ case.

The above described results, which are the first obtained for spherical geometry (see Peltier 1985), clearly reverse the conclusions previously reached by Schubert *et al.* (1975) concerning the influence of phase transitions and demonstrate that such boundaries could, of themselves, have a profound influence upon the radial structure of the mantle general circulation. One boundary, the olivine-spinel, is strongly stabilizing and tends to induce a layering of the circulation, whereas the second, the spinel-perovskite boundary, is strongly destabilizing and tends to promote a whole mantle circulation. Of course these predictions are valid only for convection of infinitesimal amplitude, and we must therefore be concerned as to what the influence would be of finite amplitude effects at Rayleigh numbers which are 10^4 times the critical value at onset. In order to carefully assess such effects we must perform fully nonlinear time dependent simulations.

Nevertheless the issue of what linear theory has to say concerning the competition between the respective stabilizing and destabilizing effects of the shallower and deeper phase boundaries is of some interest. Although there is some technical difficulty in solving the differential eigensystem (4) with both phase boundaries present *and* in their correct physical locations, because the system becomes extremely stiff, we have nevertheless managed to solve it using the Gram-Schmidt re-orthonormalization methods of Scott and Watts (1977) which are embodied in the software package SUPORE. It will suffice for present purposes to show a few results for this realistic geometry and these are in Figure 10.5. All of the calculations on this figure are for an exothermic (olivine-spinel) transition with $R_Q = 4 \times 10^6$ and $S = 0$ which optimizes the tendency toward a layered circulation. Plate a then shows the variation of the critical Rayleigh number for the shell as a function of $-R_Q$ for the endothermic (spinel-perovskite) transition which has its S fixed at -0.1 which is also maximally stabilizing. For sufficiently large $-R_Q$, plate a shows that the critical Rayleigh number is again reduced to negative values ($-R_Q \gtrsim 3 \times 10^4$) and these values are less (from Table 10.2) than that appropriate for the boundary at 670 km depth in the Earth. Plates b and c of Figure 10.5 show radial eigenfunctions for w and θ for the cases $-R_Q = 0$ ($l = 3$) and $-R_Q = 10^4$ ($l = 5$), respectively. Inspection reveals the presence of a single node in the radial velocity eigenfunction at the depth of the olivine-spinel boundary; however, the strength of the penetrative cell which develops between this depth and the surface is weak and would almost certainly be entrained in the dominant large scale flow by finite amplitude effects at large positive values of the system Rayleigh number. On this basis it seems clear that phase transitions alone could not be expected to induce a radially layered

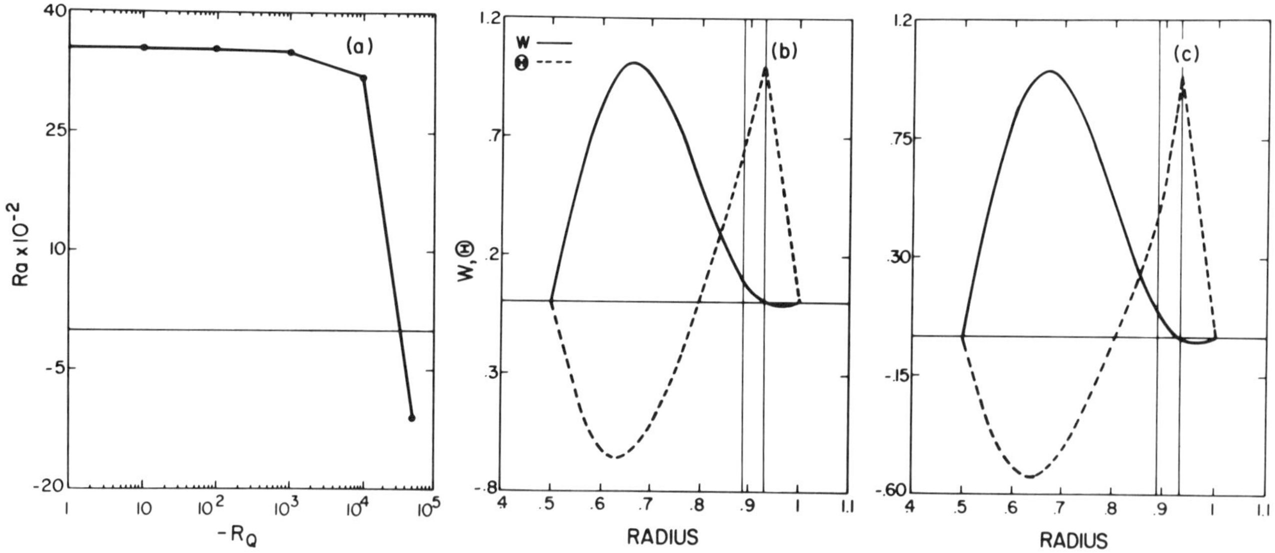

Figure 10.5. Results of the stability analysis for the case in which two phase transitions (one exothermic and one endothermic) in accord with the observed depths of the spinel-post spinel and olivine-spinel horizons. The deeper endothermic transition has S fixed at $S = -0.10$. Plate a shows Ra vs. $-R_Q$ as the value of $|R_Q|$ of the endothermic transition is increased. Eigenfunctions for b, $R_Q = 0$, $l = 3$; c, $R_Q = -10^4$, $l = 5$.

mode of circulation even though past analyses which also lead to this conclusion were not correctly performed. Since we have no *a priori* reason to expect a layered circulation in the absence of chemical stratification and since there does not appear to be any *a priori* argument on the basis of which chemical stratification is required we are therefore lead to attempt to infer the presence or absence of a layering of the circulation directly. Only in the past few years has this possibility become even marginally viable.

3. The Radial Structure of the Circulation from Seismic Tomographic Imaging

The earliest attempts to employ the results of global seismic tomographic imaging to address questions concerning aspects of global tectonics were those by Hager (1984) and Richards and Hager (1984) who focussed their attention exclusively upon the possibility that the largest scale features of the planet's gravitational field might be explained in terms of the tomographically inferred lateral heterogeneity of density. This work has been very thoroughly reviewed and extended in the previous chapter of this book. Here we will briefly review and further develop the recent work of Forte and Peltier (1987a, 1987b) which was concerned with an analysis of the extent to which similar models of the lateral density heterogeneity might be employed to predict the observed surface plate velocities as characterized, for example, by those for the Minster and Jordan (1978) model shown on Figure 10.6.

3.1 A 2-Scalar Representation of the Surface Kinematics

Peltier (1985a) demonstrated the utility of a decomposition of this vector field into two complimentary scalar fields, these being the horizontal divergence and the radial component of the vorticity. The fundamental importance of this decomposition follows from the fact that any solenoidal vector field $\mathbf{u}$, which is by definition such that $\nabla \cdot \mathbf{u} = 0$, may be expanded as

$$\mathbf{u} = \mathbf{T} + \mathbf{S}, \tag{8}$$

in which $\mathbf{T} = \Lambda\psi$ is the toroidal field, $\mathbf{S} = \nabla \times \Lambda\phi$ is the poloidal field, and the vector operator $\Lambda = \mathbf{r} \times \nabla$ (e.g., Backus, 1958). Since the assumption $\nabla \cdot \mathbf{u} = 0$ is obviously valid for rigid surface plates which are constrained to move with constant angular velocity, (8) is an expansion which may be employed to decompose the maps of Figure 6. Upon expanding ψ and ϕ in terms of spherical harmonics one may write:

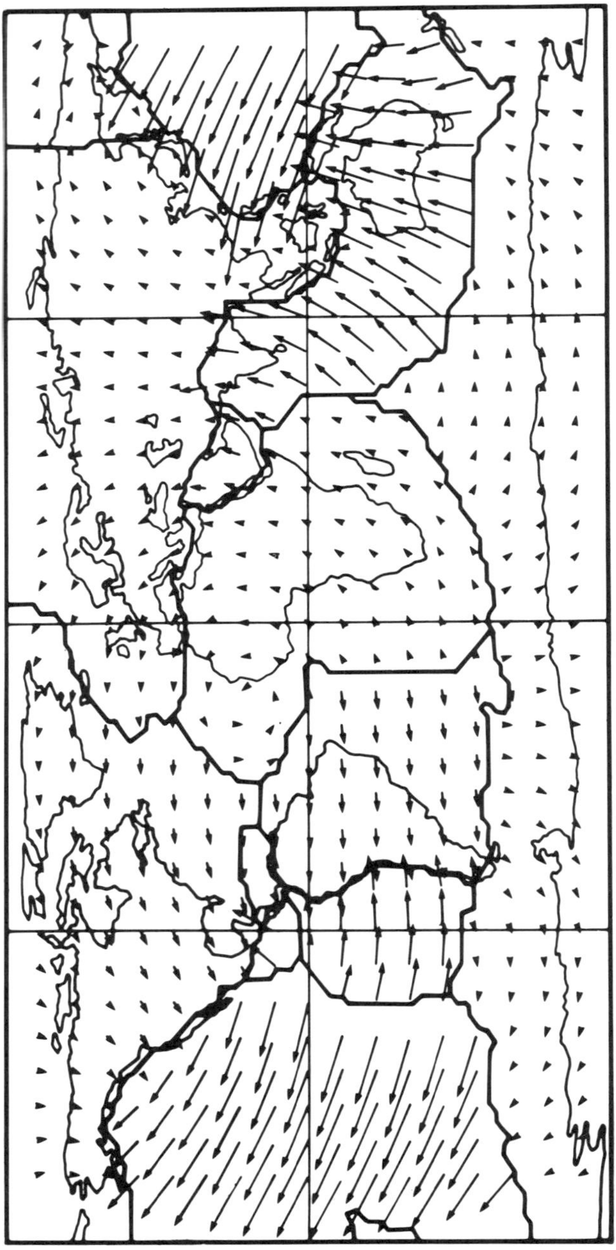

VELOCITY SCALE: ⟶ = 10 cm/Yr

Figure 10.6. Plate boundaries and velocities in the "hot-spot frame" from the analysis in Minster and Jordan (1978).

$$\mathbf{T} = \sum_{l,m} T_l^M \Lambda Y_l^M, \tag{9a}$$

and

$$\mathbf{S} = \sum_{l,m} S_l^m \hat{r} \times \Lambda Y_l^m, \tag{9b}$$

in which $T_l^m = \psi_l^m$ $(r = a)$ and $S_l^m = d\phi_l^m/dr$ $(r = a)$. In terms of the north-south $(u_\theta(\theta,\phi))$ and east-west $(u_\phi(\theta,\phi))$ components of the surface velocity field the T_l^m and S_l^m are obtained from:

$$T_l^{m(\substack{c\\s})} = \frac{1}{4\Pi l(l + 1)} \iint \left[\frac{u\theta}{\sin\theta} \frac{\partial Y_l^{m(\substack{c\\s})}}{\partial\phi} - u\phi \frac{\partial Y_l^{m(\substack{c\\s})}}{\partial\theta} \right] d\phi d\cos\theta, \tag{10a}$$

$$S_l^{m(\substack{c\\s})} = \frac{1}{4\Pi l(l + 1)} \iint \left[u_\theta \frac{\partial Y_l^{m(\substack{c\\s})}}{\partial\theta} + \frac{u\theta}{\sin\theta} \frac{\partial Y_l^{m(\substack{c\\s})}}{\partial\phi} \right] d\phi d\cos\theta, \tag{10b}$$

Insofar as the characterization of surface plate tectonics is concerned the utility of this decomposition follows from the fact that the horizontal divergence of the surface plate velocities is entirely determined by the poloidal scalar whereas the radial vorticity is entirely determined by the toroidal scalar. The relations are simply

$$(\nabla_H \cdot \mathbf{u})_l^m = -\frac{l(l + 1)}{a\sqrt{2 - \delta_{mo}}} S_l^m, \tag{11a}$$

$$\hat{\mathbf{r}} \cdot (\nabla_H \times \mathbf{u})_l^m = \frac{l(l + 1)}{a\sqrt{2 - \delta_{mo}}} T_l^m. \tag{11b}$$

When the Minster and Jordan (1978) model is employed to characterize the surface velocity field and the horizontal divergence and radial vorticity are computed using (10) and (11) then one obtains the maps of these scalar fields shown in Figure 10.7, respectively in plates (a) and (b). Inspection of these plates clearly demonstrates the manner in which the horizontal divergence picks out plate boundary segments which are oceanic ridges (positive divergence) or trenches (negative divergence) whereas the radial vorticity picks out plate boundary segments which are transform faults such as the San Andreas fault which is aligned with the west coast of the North American continent. Since these are the only types of plate boundary that exist in nature it is clear that the two scalar fields provide a complete characterization of surface plate tectonics. Given this scalar characterization we are in a position to investigate the connections which might exist between the surface kinematics and other geophysical data which should be related to the surface motions if the convection hypothesis of continental drift is correct. Before proceeding to describe the results of such analyses, however, we show in Figure 10.8 degree variance spectra for both the scalars S and T and for the

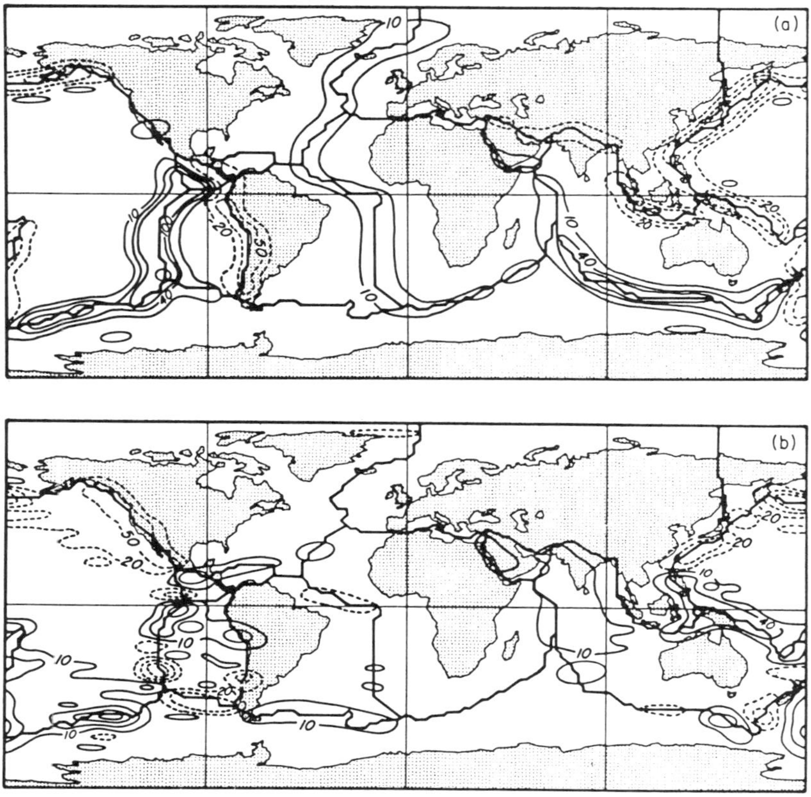

Figure 10.7. a, Surface divergence up to degree and order 32. The contour interval is 30×10^{-9} rad/yr and the individual contour levels are in units of 10^{-9} rad/yr. The dashed contour lines indicate negative and solid indicate positive divergence. b, Radial vorticity to degree and order 32. The contour interval and units are as in plate a. The dashed contour lines indicate negative vorticity (i.e., counter clockwise circulation) and one of the strongest such centers in San Andreas fault which parallels the west coast of the North American continent.

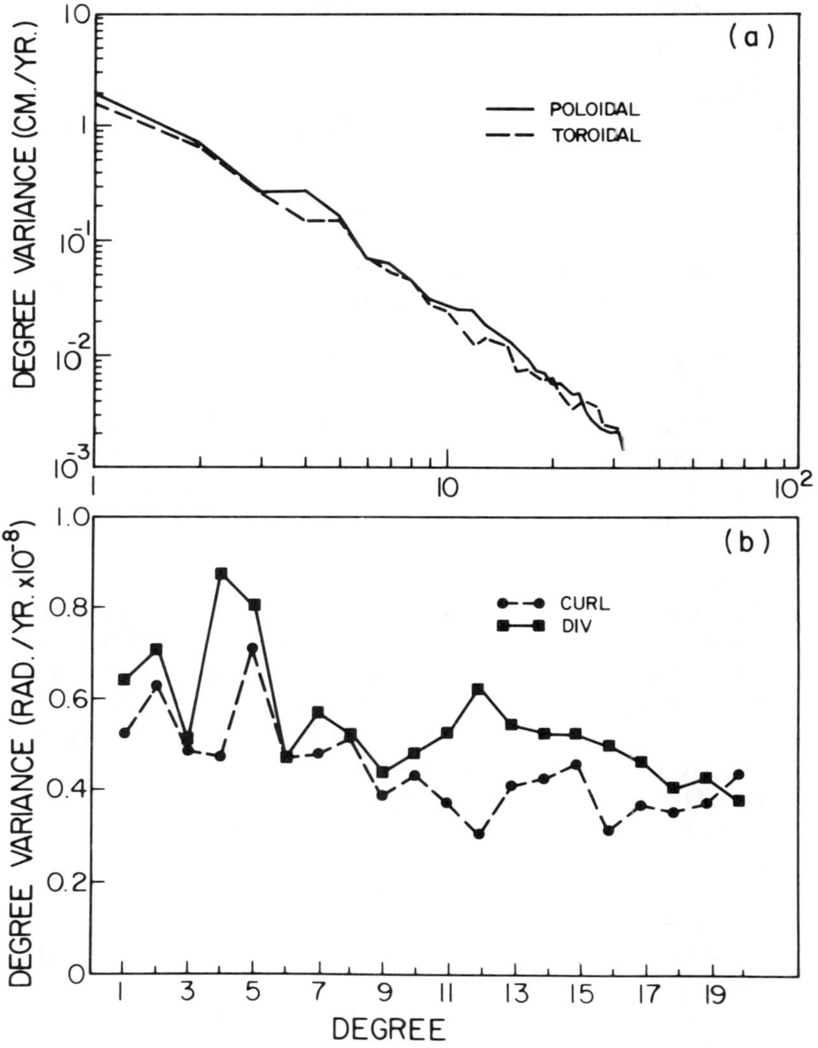

Figure 10.8. a, Degree variances of the poloidal and toroidal generating scalars of the surface velocity field. b, Degree variances of the horizontal divergence and radial vorticity fields derived from the surface plate velocities of Figure 10.6.

"process related" fields which are derivative of them, namely $\nabla_H\cdot\mathbf{u}$ and $\hat{r}\cdot(\boldsymbol{\nabla}_H\times\mathbf{u})$. These degree variance spectra as defined as, e.g., for the poloidal and toroidal scalars,

$$\sigma_l^2 \text{ (poloidal)} = \sum_{m=-l}^{+l} S_l^m S_l^{M*}, \tag{12a}$$

$$\sigma_l^2 \text{ (toroidal)} = \sum_{m=-l}^{+l} T_l^m T_l^{M*}, \tag{12b}$$

The interest in these particular spectra, from a physical point of view, arises because the mean kinetic energy of the surface motions is given by

$$\frac{1}{4\pi}\int_0^{2\pi}\int_{-1}^{+1} \mathbf{u}\cdot\mathbf{u}\, d\phi\, d\cos\phi = \sum_l l(l+1)[\sigma_l^2 \text{ (pol)} + \sigma_l^2 \text{ (tor)}]. \tag{13}$$

Therefore, the fact which is evident from Figure 10.8a, that the degree variances σ_l^2 (pol) and σ_l^2 (tor) are essentially identical implies that there is an equipartition of kinetic energy between the poloidal and toroidal components of the field that is maintained at least up to degree and order 32. Since, as we will discuss in detail below, internal mantle buoyancy could force only a poloidal flow if the Earth had spherically symmetric properties, the observed strength of the toroidal component of the flow demonstrates that aspherical effects are crucial. Although it is common to suppose that the most important such effect is due to the pressure and temperature dependence of viscosity, it is more probable that chemical heterogeneity plays a much more important role in forcing the flow of energy from the poloidal to the toroidal component which is required to understand the equipartition evident in Figure 10.8a (Peltier 1985). The degree variance spectra shown for vorticity and divergence in Figure 8b reveal much more structure than is evident in Figure 10.8a, a consequence of the fact that these "process related" fields are essentially second derivatives of the generating scalars (e.g., see (11)). Most evident in Figure 10.8b is the presence of the well defined spectral peak at spherical harmonic degrees 4 and 5 that is determined by the mean scale of the surface plates.

Figure 10.9 presents degree correlations between the two scalar fields which are required to characterize the surface kinematics and two other scalar fields which we expect to be equally intimately related to the mantle convection process, namely the gravitational field as represented by the GEM 10B field of geoid heights (Lerch *et al.* 1979) filtered by removal of the dynamically inferred flattening of Nakiboglu (1982) and the global variation of surface heat flow as represented by the global spherical harmonic model of Chapman and Pollack (1980). The correlation coefficients ρ_l are computed according to the definition

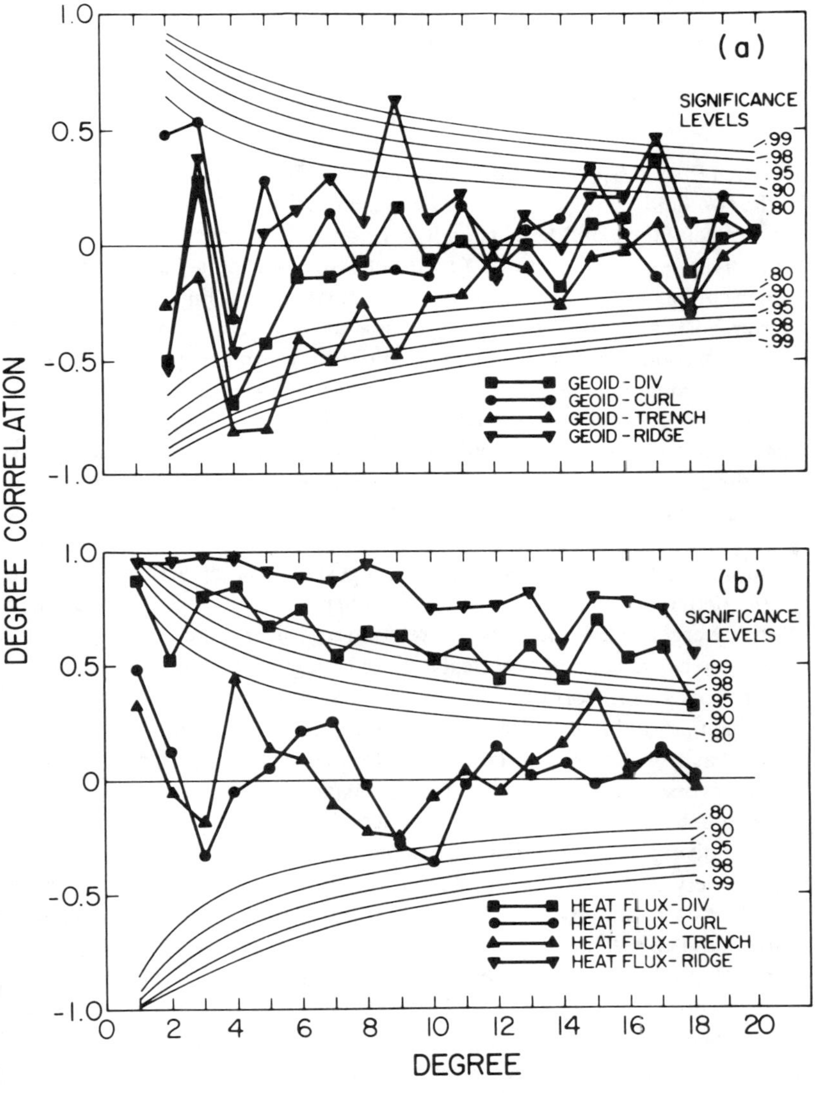

Figure 10.9. a, Degree correlations of the horizontal divergence (div), radial vorticity (curl), trench signal, and ridge signal with the corrected GEM 10B nonhydrostatic geoid of Lerch *et al*. (1979). b, Same as a but for the global heat flow model of Chapman and Pollack (1980).

$$\rho_l = \frac{\sum_{\mu=-l}^{l} F_l^m G_l^{m*}}{\left[\sum_{m=-l}^{l} F_l^m F_l^{m*}\right]\left[\sum_{m=-l}^{l} G_l^m G_l^{m*}\right]}, \qquad (14)$$

in which F_l^m and G_l^m are the coefficients in the spherical harmonic expansions of the two fields being correlated. The results shown on Figure 10.9a and b, respectively, for geoid heights and heat flow, include correlations not only with divergence and vorticity (curl) but also with the separate ridge and trench components of the divergence field. Inspection of Figure 10.9a, which is from Forte and Peltier (1987), shows that the best geoid correlation is with the trench component of the divergence field with strong negative correlation existing throughout the range $4 \leq l \leq 9$. Hager (1984) (see also the Chapter 9 of this book) also discovered a good correlation through this same range of degrees between the actual geoid and a geoid predicted by a model including only the density heterogeneity associated with downgoing slabs. Our similar result demonstrates that Hager's correlation was simply a consequence of his having put the downgoing slabs in the right places, i.e., in association with regions of surface convergence. The strongest correlation evident on Figure 10.9b for the surface heat flow, on the other hand, is with the ridge component of the divergence field. Here the correlation coefficients are essentially unity across the entire band of spherical harmonic degrees that have been investigated. This superb correlation should not be too surprising, however, since the Chapman and Pollack heat flow map has ridge crest heat flows assigned on the basis of the local surface divergence because they assume heat flow varies inversely with the square root of seafloor age.

In order to reinforce the goodness of fit of the geoid height field and that for surface divergence throughout the range $4 \leq l \leq 9$, we show these fields respectively in plates a and b of Figure 10.10. Inspection demonstrates an excellent correlation between geoid highs and trenches, but there is also a rather good correlation between geoid lows and ridges which is absent only in the North Atlantic ocean. Also shown on Figure 10.10 in plates c and d, respectively, is the map composed of the degree 2 constituents of the surface divergence field and the map of the lateral density heterogeneity in the transition zone of the mantle between the depths of 420 km and 670 km that was first inferred on the basis of analyses of anomalous splitting of the fundamental spheroidal modes of elastic gravitational free oscillation by Masters *et al*. (1982). The excellent correlation evident between these two fields that was first noted in Peltier (1985) was the result which motivated the extensive recent analyses by Forte and Peltier (1987) to test the hypothesis that surface plate velocities may be directly predictable on the basis of results

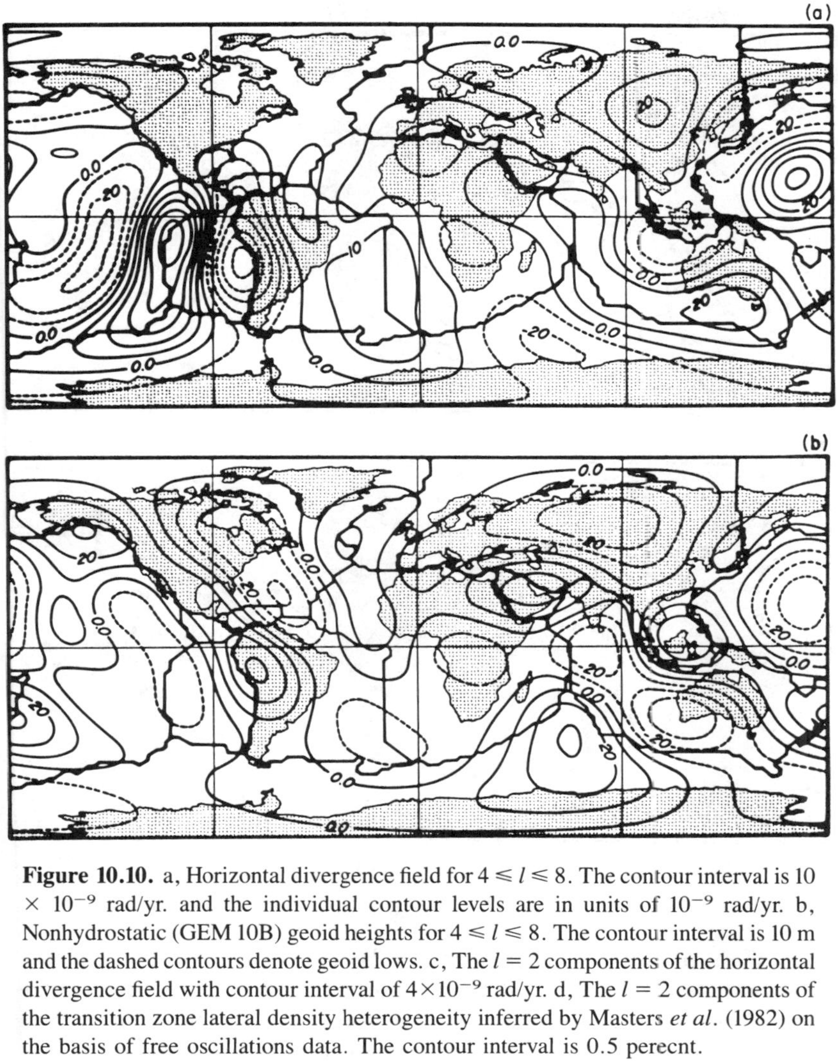

Figure 10.10. a, Horizontal divergence field for $4 \leqslant l \leqslant 8$. The contour interval is 10×10^{-9} rad/yr. and the individual contour levels are in units of 10^{-9} rad/yr. b, Nonhydrostatic (GEM 10B) geoid heights for $4 \leqslant l \leqslant 8$. The contour interval is 10 m and the dashed contours denote geoid lows. c, The $l = 2$ components of the horizontal divergence field with contour interval of 4×10^{-9} rad/yr. d, The $l = 2$ components of the transition zone lateral density heterogeneity inferred by Masters *et al.* (1982) on the basis of free oscillations data. The contour interval is 0.5 perecnt.

from seismic tomographic imaging. Close inspection of plates c and d of Figure 10.10 demonstrates that the two strongest centers of surface convergence in the quadrapolar pattern are located above similar regions in the transition zone in which the density is anomalously high, just as would be expected if the observed internal density heterogeneity were actually driving

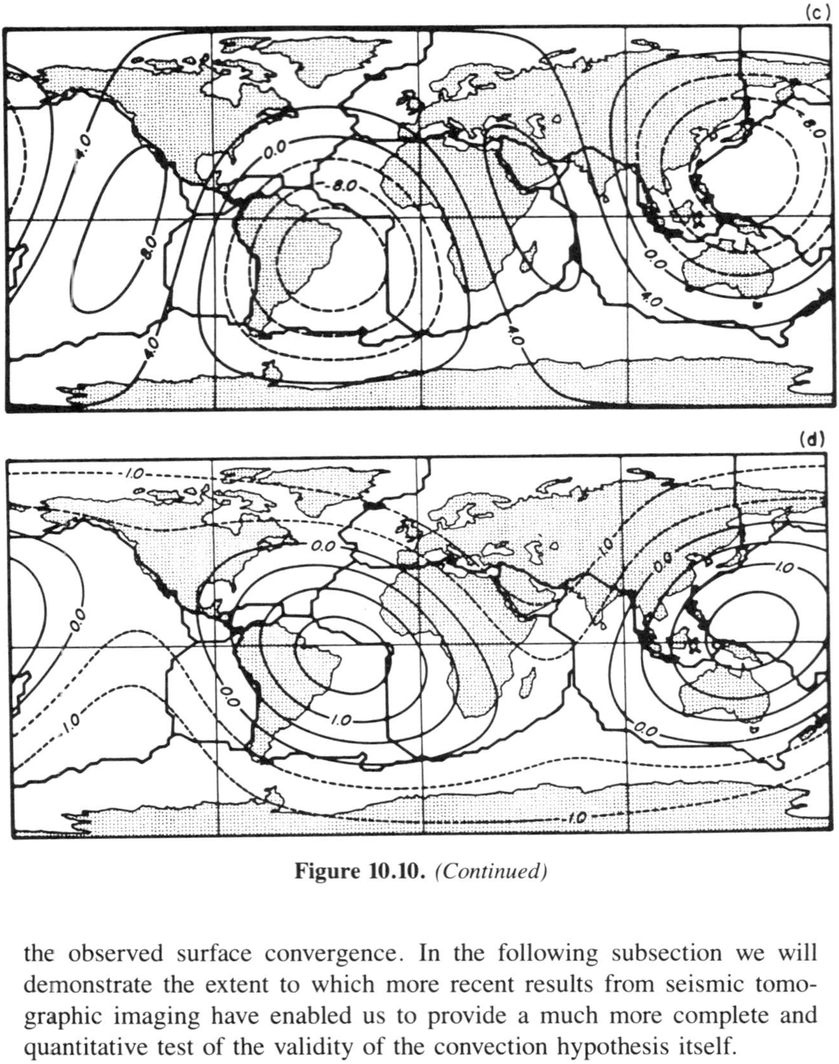

Figure 10.10. *(Continued)*

the observed surface convergence. In the following subsection we will demonstrate the extent to which more recent results from seismic tomographic imaging have enabled us to provide a much more complete and quantitative test of the validity of the convection hypothesis itself.

3.2 A Direct Test of the Convection Hypothesis of Continental Drift and Seafloor Spreading

Following the early success of Masters *et al.* (1982) in the use of free oscillations data to infer the longest wavelength features of the lateral heterogeneity of the mantle as a function of depth, Dziewonski (1984) described the application of seismic body wave data to the problem of retrieving the depth dependent lateral heterogeneity of the mantle to degree and order 6 for the lower mantle and Woodhouse and Dziewonski (1984) provided a parallel discussion of the application of surface wave data to recover upper mantle heterogeneity to degree and order 8. In this subsection of the present chapter we will review the recent analyses of Forte and Peltier (1987) which have directly employed these results from seismology (alternate versions of which have been obtained by Robert Clayton and are described in the Chapter 9 of this book) to determine the extent to which they may be employed to predict surface plate kinematics.

The manner in which this test is performed follows from the equations of continuity and momentum balance which we may write in the form

$$\nabla \cdot \mathbf{u} = 0, \tag{15}$$

$$\rho_O \mathbf{g}_1 + \rho_1 \mathbf{g}_O - \mathbf{v} p_1 + \eta \nabla^2 u = 0, \tag{16}$$

in which η is the molecular viscosity, ρ_0 is the background density field, and ρ_1 and p_1 are the density and pressure perturbation respectively. Assuming a background gravitational acceleration $g_0 = -g_0 \mathbf{r}/r$ with g_0 constant, and a perturbation of acceleration $g_1 = \nabla \phi_1$ with the potential perturbation determined by the density perturbation through the Poisson equation $\nabla^2 \phi_1 = -4\pi G \rho_1$ with G the universal gravitational constant, (16) may be simply solved for $\mathbf{u}$ given ρ_1. The method of solution is to use the poloidal-toroidal decomposition (8) which we will rewrite here in the form

$$u = \nabla \times \Lambda p + \Lambda q, \tag{17}$$

with p and q now used to denote the poloidal and toroidal scalars. Using the above representations for $\mathbf{g}_0$ and $\mathbf{g}_1$ and substituting (17) into (16) followed by application of the operator ∇x yields

$$\Lambda \frac{\rho_1 g_O}{r} + \eta \nabla \times \Lambda \nabla^2 q - \eta \Lambda \nabla^4 p = 0. \tag{18}$$

Applying the operator

$$\frac{1}{4\pi} \int_s ds \Lambda Y_l^m \tag{19}$$

to (18) and making use of the property that $\Lambda \cdot \Lambda = \wedge^2$ is such that $\wedge^2 Y_l^m = -l(l+1) Y_l^m$ then reduces (18) following application of (19) to the form:

$$D_l^2 P_l^m(r) = \frac{g_O}{\eta} \frac{(\rho_1)_l^m(r)}{r}, \tag{20}$$

in which

$$D_l = \frac{D^2}{dr^2} + \frac{2}{r}\frac{d}{dr} - \frac{l(l+1)}{r^2},$$

and P_l^m and $(\rho_1)_l^m$ are the depth dependent spherical harmonic coefficients of the poloidal scalar and the lateral heterogeneity of density respectively. Equation (20) demonstrates that when the properties of the Earth model are assumed spherically symmetric the internal buoyancy in the mantle which operates through the density heterogeneity ρ_1 drives only a poloidal flow. The toroidal scalar q does not appear in (20). The fact demonstrated in the last subsection that the surface velocity field has half of its kinetic energy partitioned to the toroidal component of the flow obviously means that a pure poloidal theory such as that embodied in (20) will not be able to account for this characteristic of tectonic plate velocities. As we will see, however, it does provide an adequate first order reconciliation of the poloidal component of the surface velocity field, that characterized by the pattern of surface divergence shown in Figure 10.7a.

In following the method of analysis outlined above it should be clear that the work described here is a rather sharp departure from the explanatory framework which is conventionally employed to address the question as to "what drives the plates." Most previous analyses in the geophysical literature have sought to explain the observed motions of the surface plates following Forsyth and Uyeda (1975) who attempt to decompose the driving forces into "ridge push" and "trench pull" and to lump the resisting force into one associated with viscous friction between the plate and the underlying mantle. In using (20) to predict plate velocities we are clearly ascribing the driving force to the deep lateral heterogeneity of density associated with the mantle convection process. This forcing is clearly not one which can be separated conveniently into "ridge push" and "trench pull" components except insofar as these separate elements of the surface divergence field have distinct hot and cold plumes vertically oriented beneath them which provide the required lateral density heterogeneity. As we will see, the density heterogeneity which is most efficient at driving surface divergence is *not* that nearest the surface but is rather that located from mid- to upper mantle depths, that is, well removed from the plates themselves.

In order to solve (20) to make a prediction of the divergent component of the surface plate velocities we may employ a Green function technique which, as shown in detail by Forte and Peltier (1987), delivers predictions of surface divergence and geoid height in the respective forms:

$$(\nabla_H \cdot \mathbf{u})_l^m \ (r = a) = \frac{g_0}{\eta_U} \int_b^a S_l(r', \eta_L/\eta_U, d)(\rho_1)_l^m(r')dr', \tag{21}$$

$$(\text{Ge})_l^m \ (r = a) = \frac{3}{(2l + 1)\rho} \int_b^a G_l(r', \eta_L/\eta_U, \delta)(\pi_1)_l^m(r')dr', \tag{22}$$

which are written here for the specific case in which the region from the Earth's surface to the core-mantle boundary is divided into two layers separated by an interface at depth d with the upper layer characterized by viscosity η_U and the lower layer characterized by viscosity η_L. Examples of the surface divergence and geoid kernels for layered models with interface at the specific depth of 1200 km depth are shown in Figure 10.11 for a number of different upper mantle-lower mantle viscosity contrasts. In equations (21) and (22) we have emphasized the fundamentally different way in which surface divergence and geoid anomaly depend upon mantle viscosity by employing kernels S_l and G_l which depend solely upon the ratio η_L/η_U. As shown in (22), geoid anomalies depend solely upon this ratio whereas surface divergence depends upon the absolute value of the viscosity through the pre-integral factor $1/\eta_U$. The two fields therefore provide highly complimentary constraints upon the radial viscosity structure and therefore a useful means of inferring the global viscosity profile required in order to allow the mantle convection hypothesis to best reconcile surface plate tectonic observations.

Forte and Peltier (1987) have implemented the above described theoretical structure using the previously mentioned models of the lateral heterogeneity of the mantle developed by Dziewonski (1984) and Woodhouse and Dziewonski (1984), the main results of which are illustrated in Figures 10.12 and 10.13. These figures respectively show the correlation between the predicted and observed GEM10B geoid and the degree variances of the predicted and observed surface divergence field for two layer models with the layer boundary at various depths and for various lower mantle-upper mantle viscosity ratios. Based upon the geoid data alone the preferred solution of Hager (1984) with the depth of the interface at 670 km has $\eta_L/\eta_U \simeq 18$ (see Chapter 9 of this book) while the solution preferred by Forte and Peltier (1987) is that which has the interface at 1200 km depth and is characterized by the lower ratio $\eta_L/\eta_U \simeq 8$. Forte and Peltier show that the latter model is superior to the former on quantitative grounds. When the surface divergence is added as an additional constraint (with allowance for the power in the toroidal component of the velocity field at each spherical harmonic degree) their analysis (see Figure 10.13) delivers absolute estimates of the upper and lower mantle viscosities of $(2 \pm .5) \times 10^{21}$ Pa s and $(23 \mp 4) \times 10^{21}$ Pa s with the upper and lower signs appropriate for layer boundaries at 1200 and 670

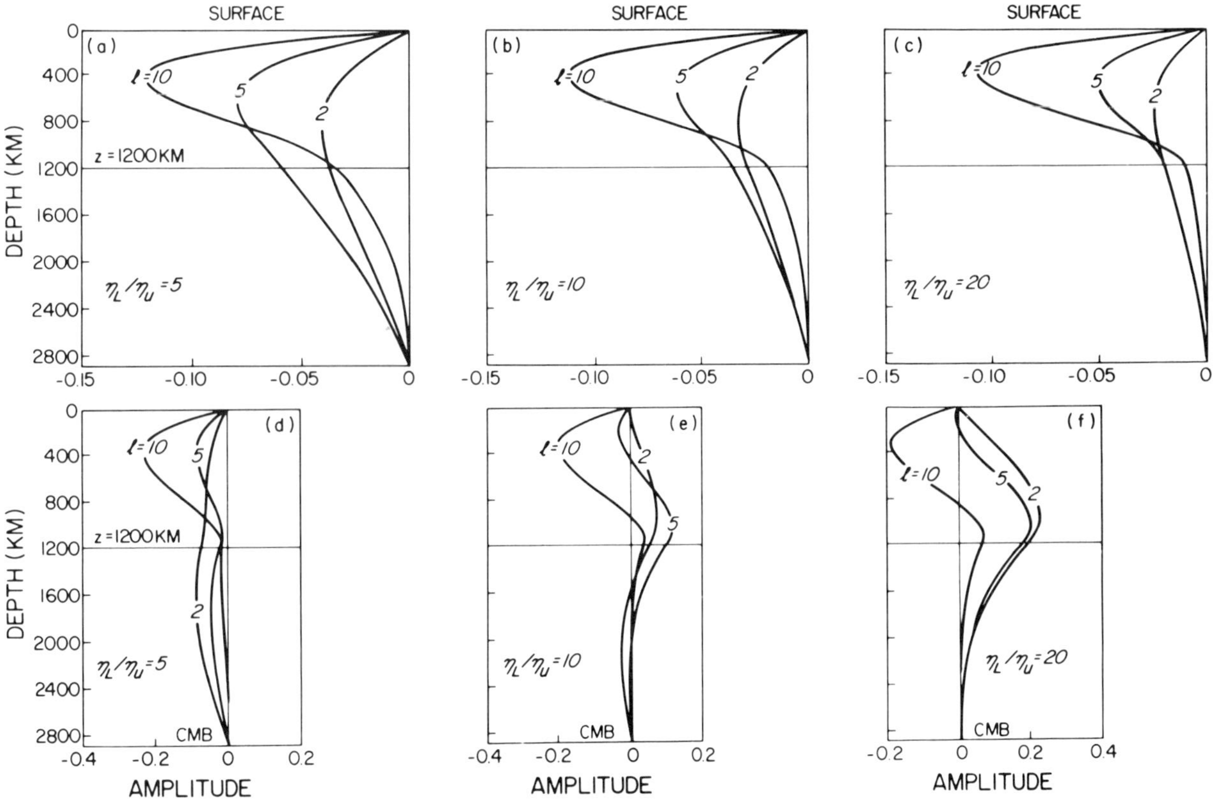

Figure 10.11. Radial kernels for a two layered mantle with free slip boundary conditions at the Earth's surface and the CMB and a viscosity discontinuity at depth $d = 1200$ km. a–c, horizontal divergence kernels for $\eta_L/\eta_u = 5$, 10, and 20, respectively; d–e, geoid kernels for $\eta_L/\eta_u = 5$, 10, and 20.

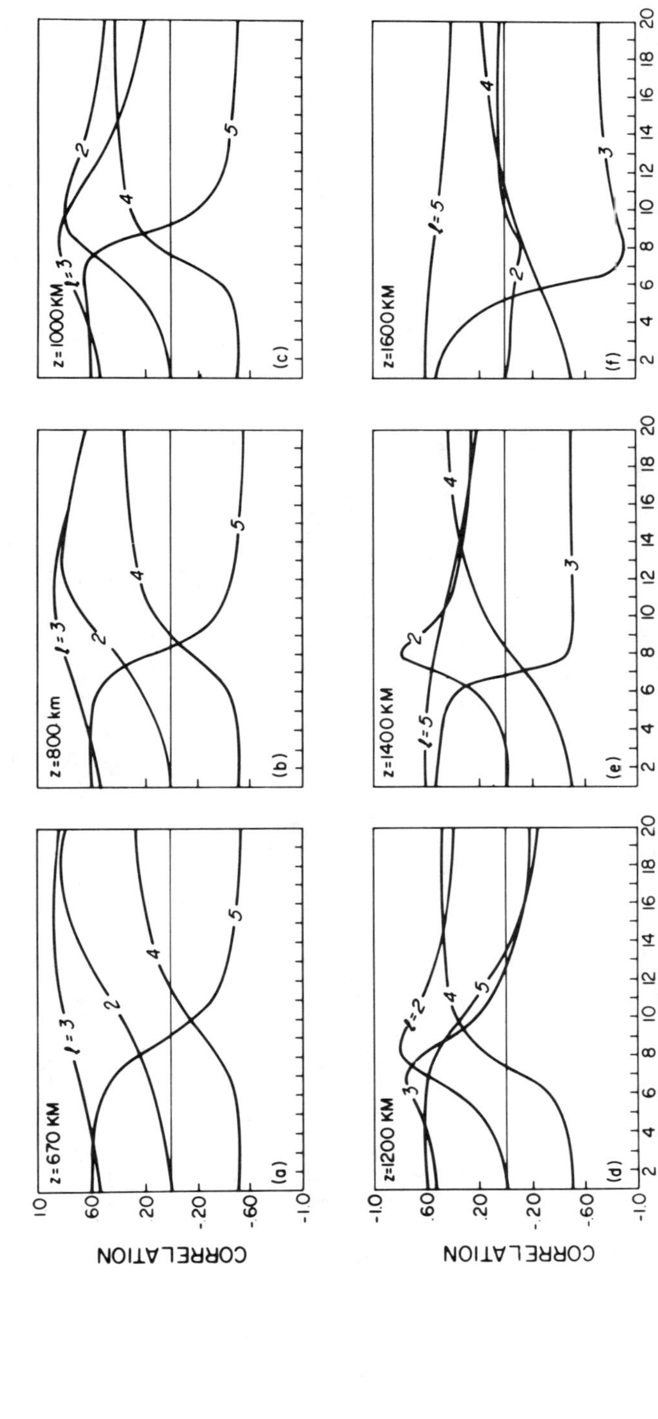

Figure 10.12. Degree correlations of the corrected nonhydrostatic GEM 10B geoid with the predicted geoid for a two layer mantle as a function of the viscosity contrast and depth of the viscosity jump.

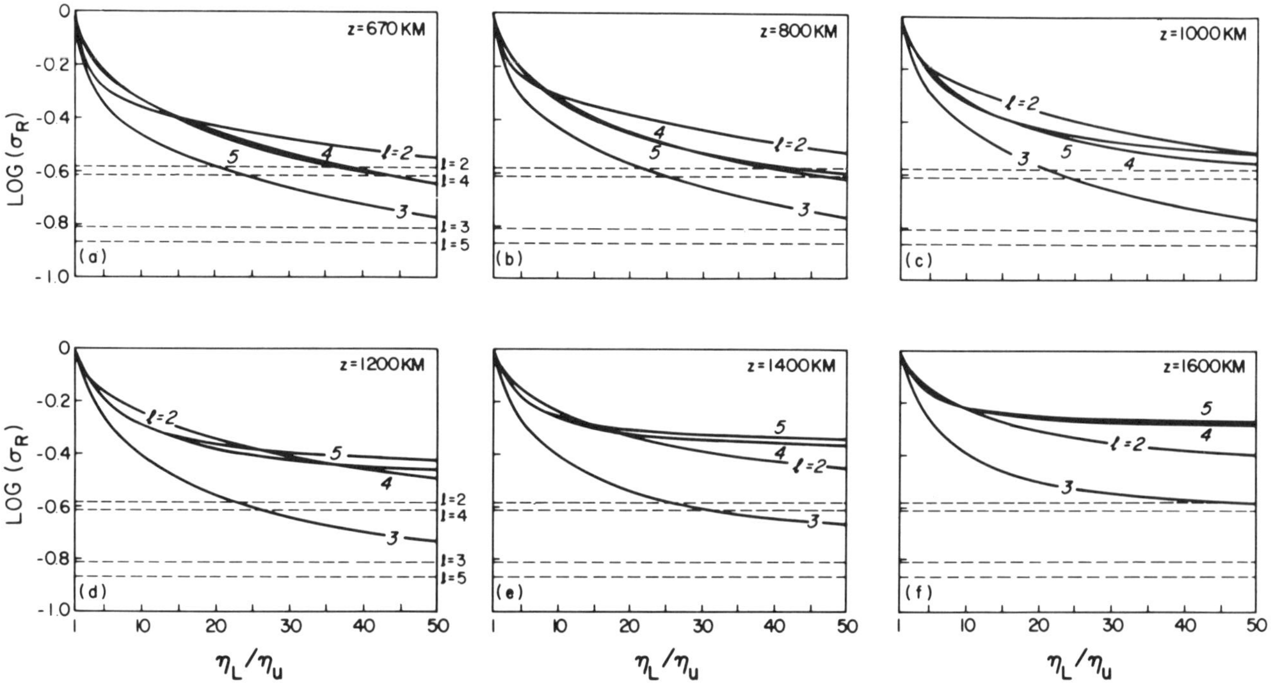

Figure 10.13. Degree variances of the predicted surface divergence field for a two layer mantle as measured by log (σ_R), see Forte and Peltier (1987) for details, as a function of viscosity contrast and depth of viscosity jump. The dashed horizontal lines are the degree variances of the observed surface divergence field as measured by log (σ_O).

km depth respectively. Although the upper mantle viscosity so inferred is within a factor of 2 of that found through the analysis of postglacial rebound data in Chapter 6, the lower mantle viscosity is a full order of magnitude higher than that compatible with glacial isostasy and is cause for some concern. Peltier (1985) and Peltier *et al.* (1986) suggested that this discrepancy might be rectified by invoking the notion of transient rheology wherein short timescale processes such as glacial isostasy sense a lower value of the creep resistance than the longer timescale process of mantle convection. This is in accord with laboratory experiments on the transient creep of olivine by Post (1977) and Karato *et al.* (1986), at least insofar as the viscosity discrepancy in the upper mantle is concerned. In fact it is only the upper mantle discrepancy that should be considered as plausibly realistic for, as shown by Forte and Peltier (1987), the upper mantle-lower mantle viscosity contrast inferred from the geoid height data is liable to be strongly biased to high values by virtue of the failure of the model to reconcile the toroidal component of the surface plate velocities.

Figure 10.14 compares the observed and predicted maps of the surface divergence constructed from all components in the range of spherical harmonic degrees $2 \leqslant l \leqslant 5$ of the best fitting 2 layered model with internal interface at 1200 km depth and $\eta_L/\eta_u = 8$. Inspection of the observed and predicted maps demonstrates that, although the gross features of the observed pattern are very well fit by the theoretical prediction, there are certain characteristic misfits which are worthy of note. The main discrepancy is in fact one which is simply explicable and that is associated with the fact that most old continental cratons are predicted to be actively subducting (Australia, North America, Eurasia) whereas these continental blocks are not observed to be doing any such thing! The reason for this systematic error is partly attributable to the fact that we have made no allowance for the chemical difference between continents and the surrounding ocean basins, and therefore no allowance for the positive buoyancy of the continents which is attributable to this chemical contrast. More important than this defect, however, is the fact that the upper mantle model of the lateral heterogeneity of Woodhouse and Dziewonski does not resolve the negative buoyancy associated with downgoing slabs. Aside from this rather easily correctable systematic error, however, our first order prediction of the divergent component of the surface plate velocities is in quite good accord with the observed locations of surface ridges and trenches if not with their relative strengths. On this basis we are obliged to conclude that the surface plates are simply doing "what they are told to do" by the buoyancy forces associated with the lateral heterogeneity of mantle density that has been revealed though the application

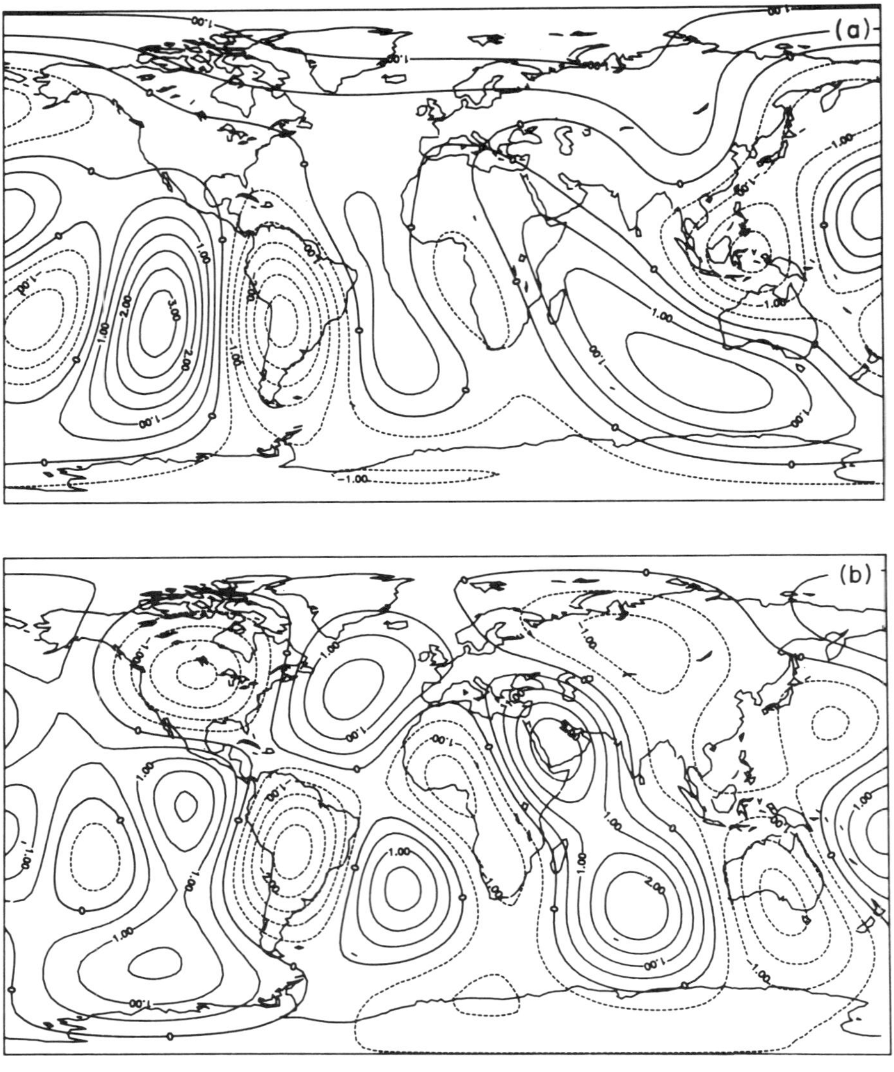

Figure 10.14. The normalized observed horizontal divergence of the surface plate velocities synthesized from harmonics in the range $2 \leqslant l \leqslant 5$. The total variance of the field shown is exactly unity. The contour interval is 0.5 (dimensionless) units. (b) The normalized horizontal divergence predicted using the divergence kernel for a two layered mantle with a factor of 8 viscosity increase at a depth of 1200 km. The total variance of this field, synthesized from harmonics in the range $2 \leqslant l \leqslant 5$ is again exactly unity and the contour interval is the same as in plate (a).

of seismic tomographic imaging. The forces which drive the plates are simply these buoyancy forces.

This mode of interpretation also has a great deal more to offer from the point of view of discriminating between competing models of the radial structure of the general circulation which is the main topic of the present chapter. It should be clear that the simultaneous inversion of geoid height and plate motion data discussed above to deliver radial profiles of mantle viscosity were performed on the basis of the *assumption* that the circulation in the mantle was of the whole mantle form. That is, the boundary conditions imposed at the interface separating the upper and lower mantle were taken to be conditions of continuity on the normal and tangential components of velocity and on tangential stress. If the circulation were of the layered type sketched in Figure 1b then the normal velocity at the internal layer boundary would be required to vanish. Although it is an algebraically more complicated task to derive analytic forms for the geoid and divergence kernels in this case, Forte and Peltier (1987b) have performed this analysis and representative results from the inversion of the geoid and divergence data for the mantle viscosity profile using these layered mantle kernels are presented on Figure 10.15 where they are compared to the previously described results obtained using the whole mantle kernels and to the complimentary result obtained on the basis of the inversion of glacial isostatic adjustment data described in Chapter 6. Inspection of this figure shows that there is a striking difference between the mantle viscosity profile required to fit the convection constraints when the inversion is performed on the basis of the layered convection hypothesis. This difference is such that upper mantle viscosity is required to be more than an order of magnitude lower in order, principally, to fit the rate of flow of material at the surface if the circulation has a layered style. This is very easily understandable on the simplest physical grounds. With an impermeable boundary at 670 km depth the aspect ratio (ratio of horizontal to vertical scales) of the convection cells driven by the internal density heterogeneity is obliged to become very large indeed. In order for the observed heterogeneity to drive the flow to acceptable rate in these circumstances then requires a sharply reduced viscosity.

As discussed in greater detail than we are able to provide here in Forte and Peltier (1987b), the three differently inferred viscosity profiles shown on Figure 10.15 allow for the construction of a new argument which is germane to the issue of whole mantle as opposed to layered convection. This argument, very simply stated, is as follows. The viscosity profile inferred on the basis of the whole mantle scenario is everywhere somewhat stiffer than the viscosity profile inferred on the basis of isostatic adjustment observations whereas that inferred on the basis of the layered hypothesis is very considera-

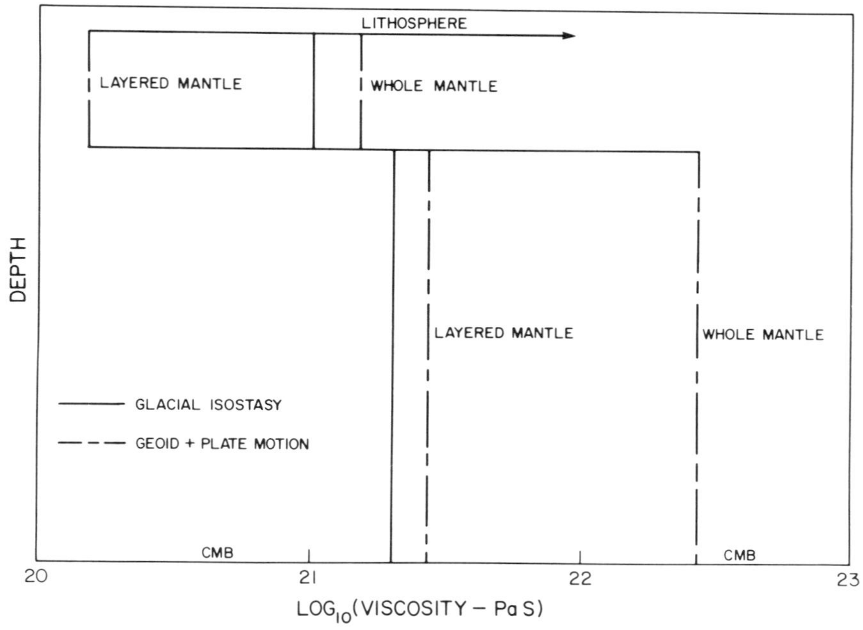

Figure 10.15. Mantle viscosity profiles deduced on the basis of the inversion of glacial isostatic adjustment observations (center) and on the basis of the inversion of geoid height and plate kinematic observations using the results of seismic tomographic imaging to prescribe the internal buoyancy (leftmost profile based on the layered mantle hypothesis and rightmost based upon the whole mantle hypothesis).

bly softer in the upper mantle than that inferred on the basis of isostatic adjustment considerations. The nub of the new argument which has been developed at length in Forte and Pelter (1987b) is that the previously cited experimental observations on the transient creep of polycrystalline olivine of Post (1977) and more recently Karato *et al.* (1986) are compatible only with the scenario in which the viscosity profile on the convection timescale is *stiffer* than that which governs the shorter timescale postglacial rebound process. Such transient creep as is observed in the laboratory is inevitably of the *strain hardening* variety. In the context of the simple two layer models of the viscosity stratification only the whole mantle version of the convection hypothesis is compatible with these experimental observations and the layered hypothesis would have to be rejected. Unfortunately this conclusion is not robust against the incorporation of a plausible low viscosity zone beneath the lithosphere and so the argument, although suggestive, is not definitive.

4. Lateral Heterogeneity and Chaotic Dynamics in Spherical Whole Mantle Convection

Although the results from seismic tomographic imaging described in the last section have provided an initial means whereby it has become possible to provide a direct test of the validity of the convection hypothesis, it should be "clear" that the image which these results have provided is at best a rather "blurred" one and that the scales of the structures which have been revealed in the Earth's deep interior may bear little relation to those which are characteristic of the lateral heterogeneity which is actually present. Put differently, the preliminary images which have become available, including those obtained by Clayton and described in the previous chapter of this book, are really low pass filtered representations of the actual structures. In order to better understand the scales which should actually be characteristic of the field of lateral density heterogeneity we must resort to numerical simulations of the convective circulation such as have been previously described in Chapters 7 and 8 of this book using models formulated in Cartesian geometry. Here we will elaborate somewhat further on such calculations by illustrating some of the effects to be expected in realistic spherical geometry. This is the present frontier of feasibility in three dimensional simulations given the capacity of existing computer systems such as the Cray X-MP. Here we will confine our attention to describing some characteristics of spherical convection in the axi-symmetric approximation which is the two dimensional spherical equivalent of the Cartesian plane layer simulations discussed previously in Chapters 7 and 8.

In the axi-symmetric approximation we may continue to employ a stream function—vorticity formulation of the infinite Prandtl number dynamical problem. With ω denoting the azimuthal vorticity and ψ the streamfunction, the nonlinear flows then satisfy

$$\nabla^2\omega = g(r)\frac{\partial T}{\partial \theta}, \tag{23}$$

$$\omega = -\frac{1}{r\sin\theta}\left[\frac{\partial^2\psi}{\partial r^2} + \frac{1}{r^2}\frac{\partial^2\psi}{\partial \theta^2} - \frac{\cot\theta}{r^2}\frac{\partial\psi}{\partial\theta}\right], \tag{24}$$

which are elliptic equations for ω given T and for ψ given ω. Knowing ψ, and thus the velocity field in merideonal section, the temperature field may be updated using the energy equation, either in the form

$$\frac{\partial T}{\partial t} = -\mathbf{u}\cdot\nabla T + \frac{1}{\mathrm{Ra}_B}\nabla^2 T, \tag{25}$$

if the flow is entirely forced by heating from below in which case Ra_B is the usual Rayleigh number of Bénard, or, alternatively,

$$\frac{\partial T}{\partial t} = -\mathbf{u}\cdot\nabla T + \frac{1}{\mathrm{Ra}_B}(\nabla^2 T + \mu), \tag{26}$$

if the flow is partly heated from within and partly from below through imposition of a constant heat flux bottom boundary condition. In the latter case the Robert's Rayleigh number (see Chapter 7) is defined as

$$\mathrm{Ra}_R = \frac{g\alpha d^4}{k\kappa\nu}\left[\frac{F}{1-\mu}\right], \tag{27}$$

where F is the constant heat flux across the bottom boundary and μ is the following dimensionless measure of internal relative to total heating,

$$\mu = \frac{Qd}{F + Qd}, \tag{28}$$

with Q the rate of internal heating per unit volume and d the thickness of the shell between the core-mantle boundary and the Earth's surface.

The system of equations (23), (24), and either (25) or (26) constitutes a closed model for the θ, r, and t dependent fields ω, ψ, and T. Since the model includes only a single prognostic equation it is straightforward to implement numerically. We simply begin with an initial guess for the laterally heterogeneous temperature field. For a Rayleigh number which is only slightly supercritical, a reasonable first guess would be $T = T_c + T'$ where $T_c(r)$ is the conduction profile and T' is a perturbation with structure determined by that of the fastest growing mode of linear stability theory and amplitude fixed arbitrarily. Given this first guess, the elliptic equation (23) is inverted to get ω using an appropriate numerical scheme. That which we continue to employ makes use of the package written by John C. Adams of the U.S. National Center for Atmospheric Research. It is based upon a direct method which first discretizes the operator, incorporating boundary conditions in the resulting block tridiagonal matrix equation ($\mathsf{b}\underline{\phi} = \mathbf{F}u$). Following this step an LU decomposition is done on the B matrix to give $\mathsf{B} = \mathsf{LU}$. A forward sweep to solve $\mathsf{L}\underline{\phi}' = \mathsf{F}$ followed by a backward sweep to solve $\mathsf{U}\underline{\phi} = \underline{\phi}'$ then yields the solution $\underline{\phi}$. Of course if the same operator is used on subsequent calls (i.e., same equation but different forcing F) then the discretization and decomposition steps may be done once and for all and simply used at each subsequent time step in the marching scheme. Given ω from (23) determined in this way we then insert this ω into (24) which is a second elliptic equation for the streamfunction ψ. This is inverted using exactly the same software that is employed for (23). Given ψ we may then determine the north south and vertical components of the velocity field which are required in the advection term on the right hand side of either (25) or (26). This term is represented by the second order accurate conservative flux form originally derived by Arakawa (1966). The time stepping in the model which uses (25) or (26) to update the temperature field is based on leapfrog centred time differencing, the intrinsic splitting in which is obviated by restarting and appropriate averaging every $n\Delta t$. Once we have obtained a new T field in this way we feed it back into (23) and continue the integration sequence either until we have obtained a steady state solution or until a sufficiently long integration has been performed to enable us to statistically characterize the nature of the unsteadiness.

Here we shall provide only a few illustrations of the results which have been obtained with the model described above in order to illustrate the nature of the lateral heterogeneity which we expect should actually exist in the Earth's mantle and the manner in which this is conditioned by spherical geometry. Figure 10.16 presents results of one particular characterization (from Solheim and Peltier 1988) of the spherical axially symmetric flows in (Ra, μ) space which is particularly revealing of the dominant effects of the variation of these model parameters. This characterization is in terms of the depth variation of the horizontal average of the temperature field and thus one in which the lateral heterogeneity does not appear explicitly. The four frames of the figure respectively show mean temperature profiles for $\mu = 0, 0.25, 0.50$, and 1.0. On each frame these temperature profiles are shown for several values of the Rayleigh number including the critical value for the onset of convection. It should be noted that the analyses done for the $\mu = 0$ cases employed constant temperature lower boundary conditions and therefore the Rayleigh number for these cases is that of Bénard, namely Ra_B. The main effect of spherical geometry is evident for the cases with $\mu = 0$ which are the spherical analogues of the classical Bénard problem for the plane layer which was described at great length in Chapter 7. In the plane layer case the upper and lower thermal boundary layers for flows with $\mu = 0$ are of identical strength. Evident by inspection of Figure 10.17, however, is the fact that comparable flows in the sphere are characterized as possessing strongly asymmetric boundary layers. Although the boundary layer thicknesses are identical (or nearly so) the temperature drop across the bottom boundary layer is about four times that across the top boundary layer. This boundary layer asymmetry is simply understandable in terms of the difference in surface area of the CMB and the Earth's surface, which is (in the model), precisely equal to the same factor of 4. As we will see, this symmetry

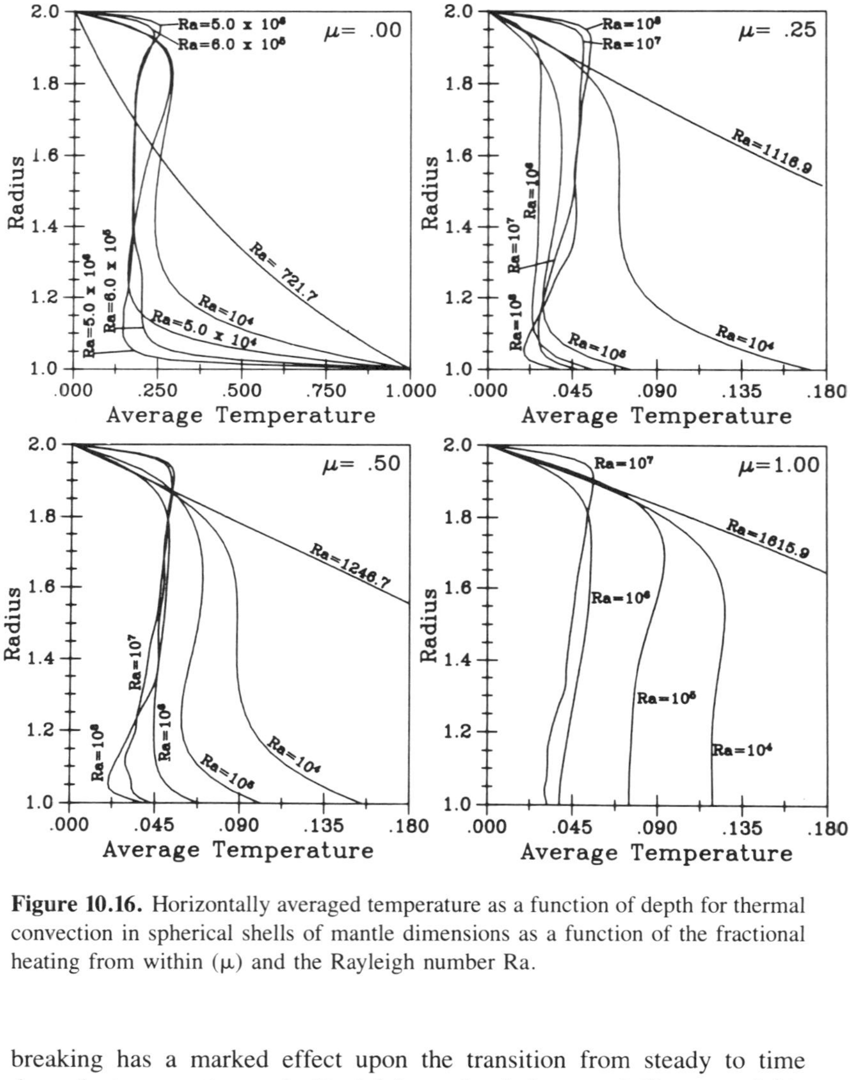

Figure 10.16. Horizontally averaged temperature as a function of depth for thermal convection in spherical shells of mantle dimensions as a function of the fractional heating from within (μ) and the Rayleigh number Ra.

breaking has a marked effect upon the transition from steady to time dependent convection as the Rayleigh number is increased. Because of this effect of geometry, the lower boundary layer of the convective circulation remains pronounced for much larger values of μ than would be the case in a Cartesian model and this may be an important point to keep in mind in connection with the interpretation of the D″ layer at the base of the mantle in terms of models of thermal convection with partial internal heating.

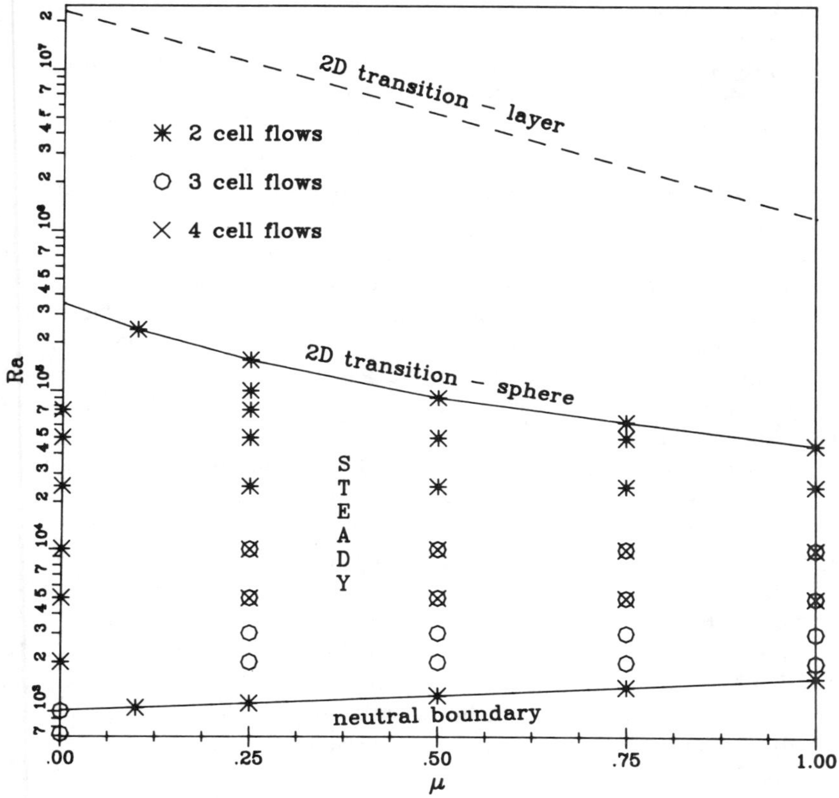

Figure 10.17. Regime diagram of the Ra-μ plane for axisymmetric spherical convection. Above the neutral stability boundary the steady flows pass from 3 cell to 3 or 4 cell to 2 cell in character as the Rayleigh number increases at finite $\mu \leqslant 1.0$. Above the curve marked 2D transition-sphere the convective circulation is time dependent. The curve marked 2D transition-layer denotes the same transition as it occurs in Cartesian geometry. The point is that spherical geometry is destabilizing in that it, of itself, sharply reduces the Rayleigh number required for the onset of time dependent flow.

Figure 10.17 illustrates one of the most important additional consequences of the spherical geometry in the form of curves on the Ra–μ plane denoting the transition from steady-state to strongly time-dependent convection as Ra is increased for fixed μ in either Cartesian or spherical axi-symmetric geometry. The curve for Cartesian geometry is from Jarvis and Peltier (1982) whereas that for spherical axi-symmetric geometry is from Solheim and Peltier (1988). Clearly, in spherical geometry strong time dependence onsets at a value of Ra for fixed μ which is more than an order of magnitude lower than in the Cartesian case. In this sense spherical geometry strongly enhances the tendency of the convective circulation to exhibit strong time dependence and thus a dynamical structure which is completely chaotic. This "thermal chaos," or thermal turbulence, is of course not accompanied by any mechanical turbulence since inertial forces are completely irrelevant in the high Prandtl number viscously dominated flows which are characteristic of the planetary interior.

Precisely how this chaos onsets is described in more detail by the sequence of time series of kinetic energy, mean interior temperature, and Nusselt number which obtain for flows on either side of the boundary drawn on Figure 10.17. These time series are shown on Figure 10.18. The flows all have $\mu = 0.25$, which is close to that suggested as an appropriate value for mantle convection by Jarvis and Peltier (1982) based on sea floor topography and heat flow observations (see Chapter 7), and are for Rayleigh numbers of (1.5, 1.6, 2.0, 2.2, 2.5) $\times$ 10^5 which therefore differ by less than a factor of two. On the "2-D stable" side of the boundary at $Ra_B = 1.5 \times 10^5$ the flow is steady. At $Ra_B = 1.6 \times 10^5$ the time dependence onsets as a perfectly harmonic variability. At $Ra = 2.0 \times 10^5$ the time series are still dominated by this same spectral line but the amplitude of the original harmonic is now markedly increased. At $Ra = 2.2 \times 10^5$ this dominant frequency is now modulated by a lower frequency component while at $Ra = 2.5 \times 10^5$ the time series have assumed a highly chaotic appearance with no discernable dominant periodicities characteristic of the temporal variability. The power spectra of these time series are shown on Figure 10.19 and display the properties characteristic of a simple Hopf bifurcation in which a steady solution loses stability to a periodic solution which in turn loses stability to an aperiodic solution whose power spectrum eventually becomes continuous. Since, as discussed in Chapter 7, the Rayleigh number of the mantle is roughly two orders of magnitude higher than that above which the transition from 2-D steady to 2-D unsteady solutions occurs it is clear that the convective flow in the Earth's mantle must be highly unsteady and thermally chaotic with the positions of ridges and trenches migrating through the convecting layer on timescales which are not much longer than the turnover times for individual

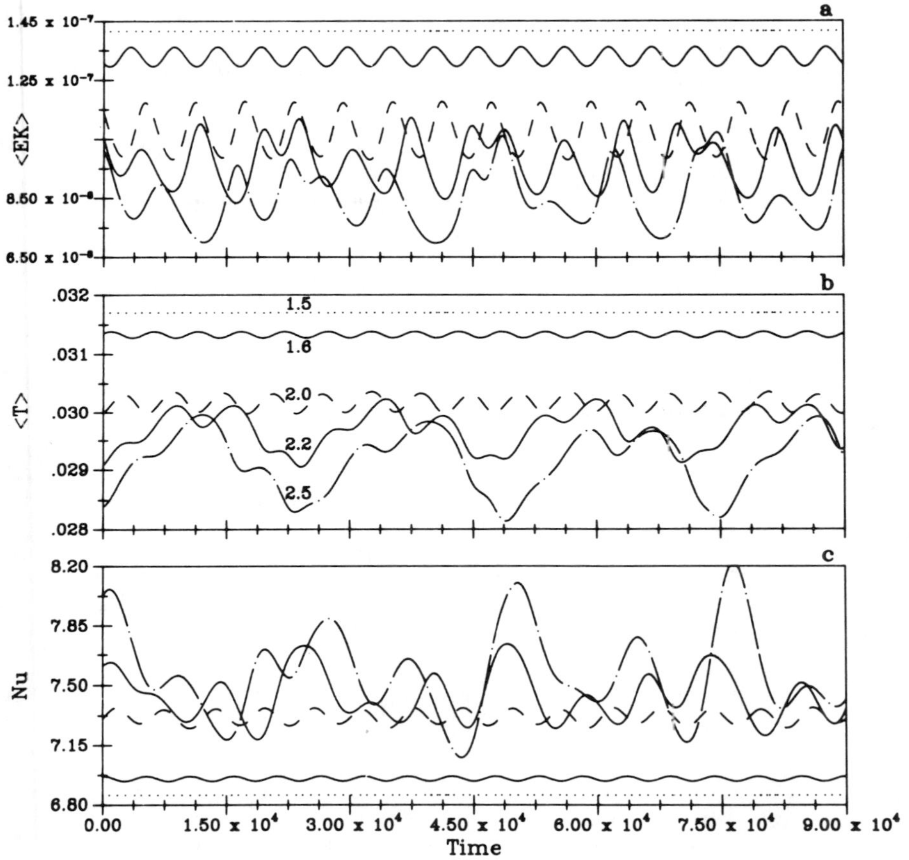

Figure 10.18. Time series for the variation of a, kinetic energy; b, mean temperature; and c, Nusselt number for a series of flows with $\mu = 0.25$ and Rayleigh number varying from the steady side of the transition on Figure 10.17 to the unsteady side. The Rayleigh numbers for which results are explicitly shown are (1.5, 1.6, 2.0, 2.2, 2.5) $\times 10^5$. Note the expected coincidence between times of low mean temperature and high heat flow (Nusselt number) as one would expect.

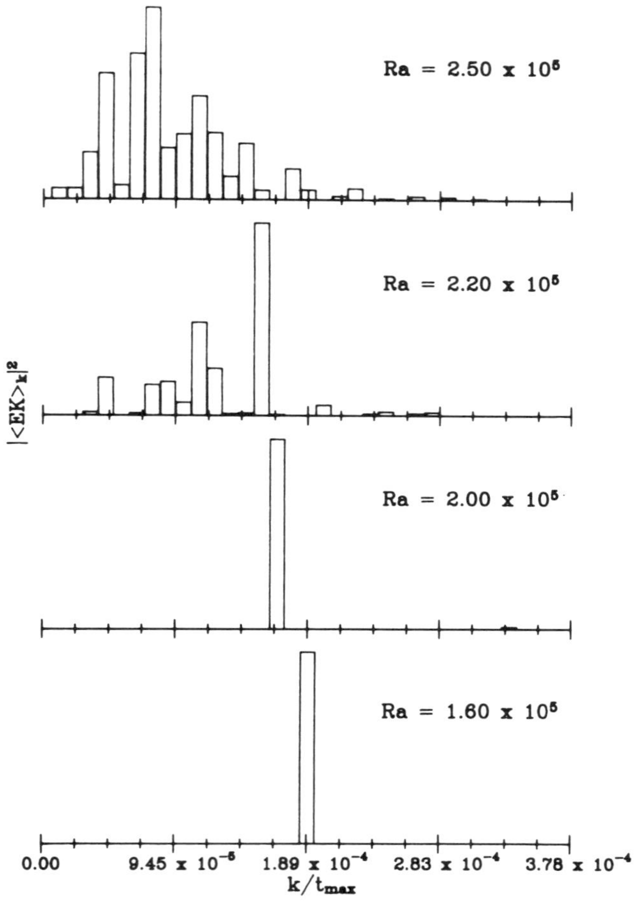

Figure 10.19. Power spectra of the time series shown on Figure 10.18 for kinetic energy and Nusselt number.

convection cells. Since ridges and trenches are *observed* to propagate rather than to remain stationary we may therefore understand these observations simply as a consequence of the chaotic nature of the thermal circulations which are characteristic of high Rayleigh number conditions in the sphere.

In this high Rayleigh number regime in which the Earth actually operates, the lateral heterogeneity of density will of course also be highly variable in time. At any particular time, however, the lateral scale of such heterogeneity will be on the same scale as the thickness of the horizontal thermal boundary

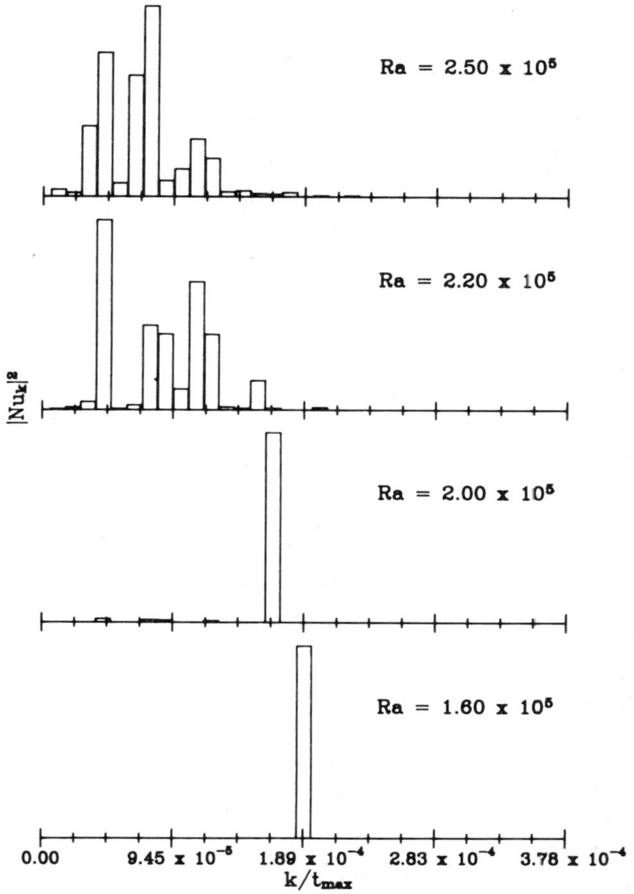

Figure 10.19. *(Continued)*

layers. Figures 10.20 and 10.21 respectively show solutions to the system (23), (24), and (26) for $\mu = 0.25$ at a Rayleigh number Ra_R of 1.5×10^5 which is near the transition from steady to unsteady flows and at $Ra_R = 2 \times 10^7$ which is near the Rayleigh number which we expect to be characteristic of the Earth. Solutions are shown for vorticity ω, streamfunction ψ, and temperature T. At the lower Rayleigh number the flow is clearly dominated by a degree 2, two cell pattern with upwelling at the poles and downwelling at the equator (it should be understood, of course, since rotation effects are

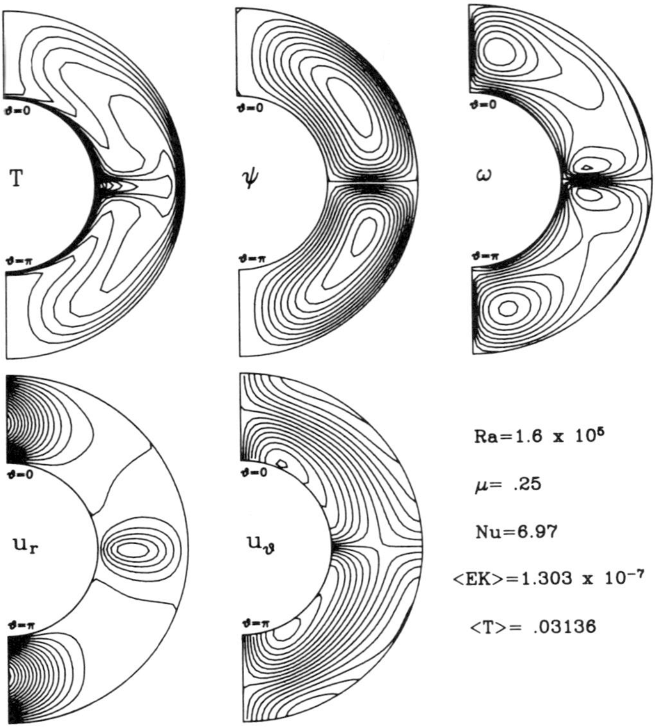

Figure 10.20. Example of a flow in the sphere at $\mu = 0.25$ and $\mathrm{Ra} = 1.6 \times 10^5$, i.e., just after the onset of the harmonic time dependence which is precursory to the transition to strongly aperiodic time dependent flow. The solution is presented in terms of temperature T, streamfunction ψ, and vorticity ω as well as the fields of radial (u_r) and azimuthal (u_θ) velocity.

unimportant in these circulations, that the poles of the model need have no relation to the geographic poles of the planet). At the most realistic higher Rayleigh number the flow is much more irregular and the characteristic horizontal scale of the lateral heterogeneities in the interior significally reduced to a value near 100 km that is equal to the thickness of the horizontal boundary layers out of whose instability the hot and cold rising and sinking plumes arise. Since this characteristic horizontal scale is more than an order of magnitude less than the shortest scale that is resolved by the existing seismic data that have been employed in seismic tomographic imaging it should be quite clear that the latter techniques cannot be expect to "see" the

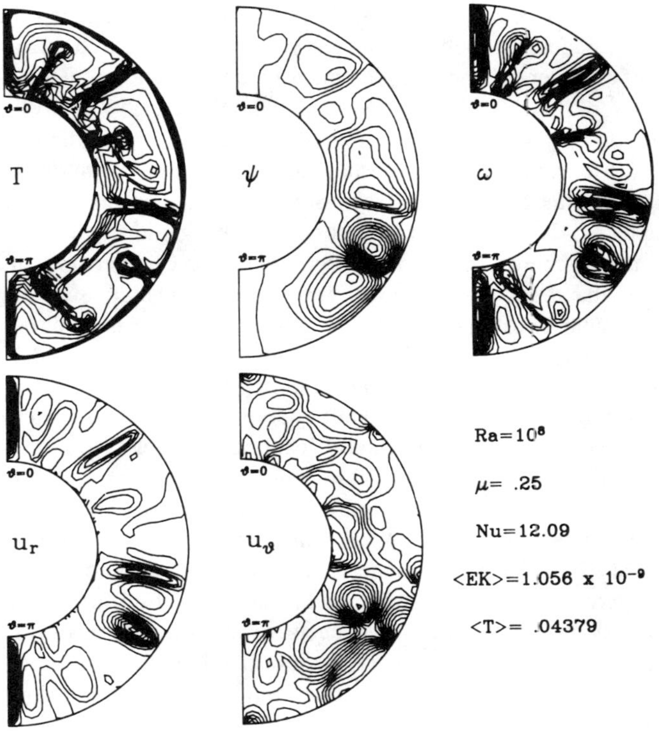

Figure 10.21. As in Figure 10.20 but for $\mu = 0.25$ and $\text{Ra} = 10^8$ which in near the Rayleigh number appropriate to the Earth's mantle.

thermal plumes which, according to the discussion in the last subsection, are actually responsible for driving the surface plates. In employing these data we must therefore be very careful not to forget that the deep mantle heterogeneity which actually exists in the Earth will doubtless have a form that bears little relation at all to those long wavelength features of it which have actually been recovered through seismic analysis.

5. Conclusions

In this chapter we have attempted to review, extend, and interrelate several of the ideas which have been developed in earlier chapters of this book, especially those which have some bearing on the radial structure of the mantle general circulation. Although the main issue concerning the whole mantle-layered mantle debate cannot as yet be regarded as settled in absolute favour of either one of these end member models, it does seem to us that the onus of proof has shifted rather markedly onto those who support the notion that the convective circulation is layered. We now know that the viscosity of the mantle does not increase substantially across the 670 km discontinuity, principally on the basis of the glacial isostatic adjustment analyses discussed in Chapter 6. Although recent analyses of isostatic geoid anomalies by Hager (1984) and Richards and Hager (1984) (e.g., see Chapter 9 of this book) suggested that the convection timescale viscosity may increase strongly with depth, the analysis of Forte and Peltier (1987) has shown that this conclusion is liable to be reversed when proper account is taken of the poloidal-toroidal coupling which is required to understand surface velocity spectra.

Furthermore, as discussed in detail by Jordan and colleagues in Chapter 3 of this book, when one does careful local analyses of seismic travel time residuals in the vicinity of downgoing slabs such as exist in the Northwest Pacific, then one concludes that the cold slab continues to exist on the downward extension of its upper mantle trajectory *through* the seismic horizon at 670 km depth. This clearly provides direct support for the whole mantle convection scenario. Also, and as discussed by Hart and Zindler in Chapter 5 of this book, previous geochemical analyses of the Nd-Sm and Rb-Sr isotopic systems are not nearly so definitive in their requirement for the existence of some large volume pristine chemical reservoir as the previously cited early analyses of these data were interpreted as demanding. A main difficulty with these early analyses appears to be that they seriously underestimated the volume of LIL enriched material at the surface by failing to recognize the existence of substantial enriched material in the region beneath the continental crust which seems to be required to understand the totality of the geochemical evidence. There is also the problem of the lack of strong support from high pressure experiments on representative mantle materials (discussed by Jeanloz in Chapter 4 of this book) for the interpretation of either the 420 or the 670 km discontinuities as other than equilibrium phase boundaries. Diamond anvil data seem to strongly support the interpretation of these interfaces as phase boundaries and if they are, then analysis by Christensen in Chapter 8 of this book, modified according to the discussion in Section 2 of this chapter, rather strongly suggest that the convective circulation would be of the whole mantle variety. In this regard an interesting recent exchange is that between Anderson (1987) on the one hand and Weidner (1985) and Bina and Wood (1987) on the other in which Anderson tion of the 420 km discontinuity as purely an equilibrium phase transformation.

There being no unequivocal argument supportive of the layered mantle hypothesis, and much evidence against its plausibility, we rather strongly suggest that the whole mantle alternative be adopted as a working hypothesis until such time as some telling argument is provided that this hypothesis should be abandoned. If the whole mantle scenario is correct then the structure of the lateral heterogeneities in the interior should be similar to that of the high Rayleigh number flows discussed in Section 4 of this chapter. These flows are "thermally chaotic" and there is a great deal that remains to be learned through the three dimensional numerical simulation of these highly nonlinear time dependent circulations. Our search for improved understanding of this dynamical system will likely tax the capacities of the most advanced supercomputers that will be available for the foreseeable future. However, we do foresee that the near future will witness marked further progress in understanding the high degree of physical complexity which is characteristic of the mantle convection process which governs the thermal evolution of the Earth. The latter problem forms the subject matter of the final chapter of this book.

6. References

Anderson, D.L., 1981. Hotspots, basalts and the evolution of the mantle. *Science* **213**, 82–89.

Anderson, D.L., 1987. The 400 km mantle discontinuity. *EOS* **68**, 409.

Arakawa, A., 1966. Computational design for long term integration of the equations of motion: Two dimensional incompressible flow. Part 1. *J. Computat. Phys.* **1**, 119–143.

Backus, G., 1958. A class of self sustaining dissipative spherical dynamos. *Ann. Phys.* **4**, 372–447.

Bina, C.R. and B.J. Wood, 1987. The 400-km seismic discontinuity and the proportion of olivine in the Earth's upper mantle. *Nature* **324**, 449–451.

Busse, F.H. and G. Schubert, 1971. Convection in a fluid with two phases. *J. Fluid Mech.* **46**, 801–812.

Cathles, L.M., 1975. *The Viscosity of the Earth's Mantle*, Princeton, NJ: Princeton University Press, 386 pp.

Davies, G.F., 1977. Whole mantle convection and plate tectonics. *Geophys. J. R. Astr. Soc.* **49**, 459–486.

Dziewonski, A.M., 1984. Mapping the lower mantle: Determination of lateral heterogeneity in P velocity up to degree and order 6. *J. Geophys. Res.* **89**, 5929–5952.

Forsyth, D.W. and S. Uyeda, 1975. *Geophys. J. R. Astr. Soc.* **43**, 163–

Forte, A..M. and W..R. Peltier, 1987a. Plate tectonics and aspherical earth structure: The importance of poloidal-toroidal coupling. *J. Geophys. Res.* **92**, 3645–3679.

Forte, A..M. and W.R. Peltier, 1987b. Mantle convection and seismic tomography: Rheological constraints on the layered convection scenario. *J. Geophys. Res.*, submitted.

Goldreich, P. and A. Toomre, 1969. Some remarks on polar wandering. *J. Geophys. Res.* **74**, 2555–2567.

Hager, B.H., 1984. Subducted slabs and the geoid: Constraints on mantle rheology and flow. *J. Geophys. Res.* **89**, 6003–6015.

Haskell, N.A., 1937. The viscosity of the asthenosphere. *Am. J. Sci.* **33**, 22–28.

Jarvis, G.T. and W.R. Peltier, 1982. Mantle convection as a boundary layer phenomenon. *Geophys. J. R. Astr. Soc.* **68**, 389–427.

Jeanloz, R., 1981. Phase transitions and mantle discontinuities. *Rev. Geophys. Space Phys.* **21**, 51–74.

Karato, S.I., M.S. Paterson, and J.D. Fitz Gerald, 1986. *J. Geophys. Res.*,

Lerch, F.S., S.M. Klosko, R.E. Laubscher, and C..A. Wagner, 1979. Gravity model improvement using GEOS 3 (GEM 9 and GEM 10). *J. Geophys. Res.* **84**, 3897–3916.

Masters, G., T.H. Jordan, P.G. Silver, and F. Gilbert. 1982, Aspherical earth structure from fundamental spheroidal-mode data. *Nature* **298**, 609–613.

Mao, H.K., P.M. Bell, and T. Yagi, 1979. Iron magnesium fractionation model for the earth. *Carnegie Inst. Wash. Yearbook* **78**, 621–625.

McConnell, R.K., 1968. Viscosity of the mantle from relaxation time spectra of isostatic adjustment. *J. Geophys. Res.* **73**, 7089–7105.

McKenzie, D.P., 1966. The viscosity of the lower mantle. *J. Geophys. Res.* **71**, 3995–4010.

McKenzie, D.P., 1967. The viscosity of the mantle. *Geophys. J. R. Astr. Soc.* **14**, 297–305.

McKenzie, D.P., 1968. The geophysical importance of high temperature creep. In R.A. Phinney (ed.) *The History of the Earth's Crust*. Princeton, NJ: Princeton University Press.

McKenzie, D.P. and N.O. Weiss, 1975. Speculations on the thermal and tectonic history of the earth. *Geophys. J. R. Astr. Soc.* **42**, 131–174.

McKenzie, D.P., J.M. Roberts, and N.D. Weiss, 1974. Convection in the earth's mantle: Towards a numerical simulation. *J. Fluid Mech.* **62**, 465–538.

Minster, J.B. and T.H. Jordan, 1978. Present day plate motions. *J. Geophys. Res.* **83**, 5331–5354.

Munk, W.H. and G.J.F. MacDonald, 1960. *The Rotation of the Earth*. Cambridge: Cambridge University Press.

Nakiboglu, S.M., 1982. Hydrostatic theory of the earth and its mechanical implications. *Phys. Earth Planet. Inter.* **28**, 302–311.

O'Connell, R.J. 1977. On the scale of mantle convection. *Tectonophysics* **38**, 119–136.

O'Nions, R.K., N.M. Evensen, and P.J. Hamilton, 1979. Geochemical modeling of

mantle differentiation and crustal growth. *J. Geophys. Res.* **84**, 6091–6101.
Oxburgh, E.R. and D.L. Turcotte, 1978. Mechanisms of continental drift. *Rep. Prog. Phys.* **41**, 1249–1312.
Peltier, W.R., 1972. Penetrative convection in the planetary mantle. *Geophys. Fluid Dyn.* **5**, 47–88.
Peltier, W.R., 1974. The impulse response of a Maxwell earth. *Rev. Geophys. Space Phys.* **12**, 649–669.
Peltier, W.R., 1976. Glacial isostatic adjustment II: The inverse problem. *Geophys. J. R. Astr. Soc.* **46**, 669–706.
Peltier, W.R., 1985. Mantle convection and viscoelasticity. *Ann. Rev. Fluid Mech.* **17**, 561–608.
Peltier, W.R. and J.T. Andrews, 1976. Glacial isostatic adjustment I: The forward problem. *Geophys. J. R. Astr. Soc.* **46**, 605–646.
Peltier, W.R. and G.T. Jarvis, 1982. Whole mantle convection and the thermal evolution of the earth. *Phys. Earth Planet. Int.* **29**, 281–304.
Post, R.L., 1977. High temperature creep of Mount Burnet dunite. *Tectonophysics* **42**, 75–110.
Richards, M.A. and B.H. Hager, 1984. Geoid anomalies in a dynamic earth. *J. Geophys. Res.* **89**, 5987–6002.
Richter, F.M. and C. Johnson, 1974. Stability of a chemically layered mantle. *J. Geophys. Res.* **79**, 1635–1637.
Ringwood, A.E., 1979. *Origin of the Earth and Moon*. New York: Springer Verlag, 295 pp.
Sauramo, M., 1955. Land uplift with hinge lines in Fennoscandia. *Ann. Acad. Sci. Fenn. Ser. A III* **44**, 1–25.
Schubert, G., 1979. Subsolidus convection in the mantles of the tenestrial planets. *Annu. Rev. Earth Planet. Sci.* **7**, 298–342.
Schubert, G. and D.L. Turcotte, 1971. Phase changes and mantle convection. *J. Geophys. Res.* **76**, 1424–1432.
Schubert, G., D.A. Yuen, and D.L. Turcotte, 1975. Role of phase transitions in a dynamic mantle. *Geophys. J. R. Astron. Soc.* **42**, 705–735.
Scott, M.R. and H.A. Watts, 1977. A computer code for two point boundary value problems via orthonormalization. *Siam J. Numer. Anal.* **14**, 40–68.
Solheim, L.P. and W.R. Peltier, 1987. Axisymmetric convection in a spherical shell at high Rayleigh number: The transition to time dependent flow and spectra of the lateral heterogeneity. *Geophys. J. R. Astr. Soc.*, submitted.
Wasserburg, G.J. and D.J. DePaulo, 1979. Models of earth structure inferred from neodymium and strontium isotopic abundances. *Proc. Nat. Acad. Sci. U.S.A.* **76**, 3594–3598.
Weidner, D.J., 1985. A mineral physics test of a pyrolite mantle. *Geophys. Res. Lett.* **12**, 417–420.
Woodhouse, J.H. and A.M. Dziewonski, 1984. Mapping the upper mantle: Three dimensional model of earth structure by inversion of seismic waveforms. *J. Geophys. Res.* **89**, 5953–5986.

CHAPTER 11

Formation and Early Evolution of the Earth

David J. Stevenson

Division of Geological and Planetary Sciences, California Institute of Technology
Pasadena, California 91125

Contribution number 4223 of the Division of Geological and Planetary Sciences, California Institute of Technology.

1. Introduction

We live on a planet where the evidence of our origins is obliterated or subtly hidden. Plate tectonics has reworked the surface of the globe, internal dynamics and differentiation have confused the chemical clues, and biological processes have modified our volatile reservoirs. A detailed reconstruction of the Earth's formation and early evolution is inconceivable except as a form of scientific mythology in which the trappings of respectable methodology lend credence to some plausible fiction consistent with data. The role of creation scenarios lies in identifying likely processes and general principles for Earth and nearby planets. Chemical data and comparative planetology must play especially important roles in discerning the relative merits of competing hypotheses, but it is premature to champion a single paradigm for Earth formation.

We may not know all the answers but we can at least attempt to ask the right questions. The following list of ten main questions are somewhat arbitrary subdivisions of the issues posed by Earth origin and early evolution; but they provide a focus and organization for the material in this chapter.

1. What is the environment of Earth formation? Did the Earth form in the presence of a gas phase (the hydrogen-rich solar nebula) or did the accumulation take place in the absence of gas?
2. What is the inventory of bodies from which the Earth formed? All plausible models assume Earth formation by the accumulation of smaller bodies (rather than, say, fissioning the Earth from a larger body). But the smaller bodies could be dust or droplets or planetesimals (e.g., asteroid-sized bodies) or even Mars-sized bodies. What was the size spectrum of these bodies, and what range of compositions did they have?
3. What is the timescale over which the Earth formed? This has important implications for the thermal state of the primordial Earth. If the timescale is $\lesssim 10^6$ years, then the surface of the proto-Earth is unavoidably hot. If the timescale is nearer 10^8 years, then the surface might be cold (depending on the presence of an atmosphere).
4. When and how did the primary differentiation occur? Did the Earth's core accrete directly or form by internal differentiation? What was the timing of this event? Did the Earth's hydrosphere and atmosphere form by outgassing or by late addition of volatiles, or both?

5. What were the thermal and compositional structures of the primordial Earth interior? To what extent was the early Earth molten? What implications does this have for mantle differentiation?
6. What were the thermal and compositional structures of the primordial Earth atmosphere? Did the proto-Earth have a dense, blanketing atmosphere? Were there oceans at this stage? How did this atmosphere evolve?
7. What were the early internal dynamics and thermal evolution? Did the early Earth have vigorous mantle convection and extensive volcanism? How rapidly did the internal temperature change?
8. What is the origin of continents? Was sialic crust created well before 3.8 b.yr.bp; if so, what happened to it? At what point was light crust partially stabilized against recycling?
9. What can we infer about the early Earth by looking at the other planets and the Moon? Did other planets have similar early histories? Does the Moon tell us about processes on the early Earth? Was the Moon derived from the Earth?
10. What memory has the Earth retained of early events? What aspects of modern processes are affected by Earth formation and early evolution?

A review of existing work on Earth origin shows a remarkable lack of consensus on the answers to most of these questions. Part of this divergence of opinion stems from an inadequate appreciation among some workers of the physical processes or the chemical constraints, or even both. It is common practice for workers in one area of expertise to dismiss or attach little weight to the constraints of another field . . . or, even worse, to choose the particular theoretical framework or data set that suits their prejudices. Inevitably, this review focuses more one the physical constraints because the author understands those better; but an effort is made to reduce the confusion of diverse opinion by distinguishing those model results which are "reliable" (in the sense that they are relatively insensitive to model assumptions) from results which are "suspect" (sensitive to model assumptions). Nevertheless, there must be some commonly agreed principles that form the basis of discussion, since the analysis would otherwise be too diffuse and discursive. The most important principle adopted here is that the solar system was derived from a solar nebula; a discoid assembly of gas and dust. This is described and developed in Section 2. Although other concepts have been proposed (see Dermott, 1978), the solar nebula stands out as the concept best supported by astrophysical observation and theory. This review also adopts the rather conventional view that electromagnetic processes (e.g., hydromagnetic torques, plasma processes) play a minor or negligible role in the issues posed by the ten questions above. The work of Alfven and Arrhenius (1976), in particular, has attracted few advocates because it offers no prospect of explaining planetary compositions. Other preferences, such as the important role of planetesimals in planetary formation, emerge and are justified as the review proceeds.

In Section 2, the solar nebula is described and discussed, with an emphasis on dynamics, timescales, and temperatures. Condensation and coagulation of particulates are analyzed and the onset of gravitational instabilities is described. We are left, at the end of this section, with a disk of small bodies, possibly still immersed in a gas phase. Sections 3 and 4 deal with the accumulation of Earth from these bodies, approached from the Safronov (gas-free) and Hayashi (gas-rich) viewpoints, respectively. In each case, the growth of bodies and the evolution of their velocity distributions are discussed. Accretion timescales and the temperature and internal structure of the proto-Earth are considered in detail. Comparison with different sized bodies is also made. Section 5 deals with the Hadean Earth (the Earth prior to the known rock record), focusing on questions of thermal state, heat loss, outgassing, progressive differentiation, and continent stabilization. Section 6 deals with the post-Hadean Earth, including consideration of convective self-regulation, and the relationship of the present Earth to the primordial state. Section 7 concludes the tentative answers to the ten questions above.

2. The Solar Nebula

Modern ideas of solar system formation have been greatly influenced by our improved understanding of the interstellar medium and nearby stars, including the existence and nature of dense, molecular hydrogen clouds (see *Protostars and Planets*, T. Gehrels (ed.), 1978; and *Protostars and Planets II*, Black and Matthews (eds.), 1985), recent work on T Tauri (Hanson *et al.*, 1983), IRAS observations (Aumann *et al.*, 1984), and ground based observations of nearby stars (Smith and Terrile, 1984; McCarthy *et al.*, 1985). These observations favor a solar system origin involving a gaseous disk formed from gravitational collapse of a rotating, interstellar cloud of dust and gas, with the disk having far greater extent than the central condensed body because of the angular momentum budget. Gaseous disk models are in the spirit of the early (18th century) models of Kant and Laplace. Although the detailed dynamics of the collapse and disk formation (interplay of rotation and magnetic fields, role of triggering events such as supernova-induced shock waves or spiral density waves) remain controversial, the availability of appropriate conditions for collapse is not seriously in doubt.

The mass distribution (or, equivalently, total angular momentum) of the forming disk is very uncertain, especially since there is no reason to suppose that the central concentration of mass is equal to the mass of the Sun, at least in the earliest stages. This uncertainty is the primary reason for the range of models encountered in the literature. The models developed by Cameron (1978, 1983a) are generally characterized as "high mass" models because the total mass is roughly twice the solar value. However, only about 0.1 $M_\odot$ (1 $M_\odot \equiv$ one solar mass) is in the part of the disk exterior to the orbit of Mercury and interior to the present orbit of Neptune. In this respect, his models of the disk are only about twice as massive as the so-called "minimal mass" nebula models (best represented in the work of Hayashi and collaborators; see Hayashi, 1981; Hayashi *et al.*, 1985). These operate on the pragmatic (but probably unreasonable) principle that what you can see now tells you what you had then. "Minimal mass" simply means taking an inventory of the total amount of involatile ("rock") constituents in the present planetary system and calculating the mass of the reservoir of assumed cosmic abundance from which this material was derived. The dominant contribution (and uncertainty) in this reconstitution is the amount of heavy material at the centers of the giant planets (apparently $\sim$10 $M_\oplus$ per planet, where 1 $M_\oplus \equiv$ 1 Earth mass, but with an uncertain ice/rock ratio; see Stevenson, 1982). The biggest uncertainty, however, lies in the treatment of disk dynamics during and after accretion.

2.1. Accretion Disk Dynamics

Most current models are based on, or related to, the now-classic work of Lynden-Bell and Pringle (1974), some aspects of which were present in the earlier work of Shakura and Sunyaev (1973). The essential idea is a disk in which radial redistribution of mass and angular momentum are achieved by "viscous" stresses $\rho\nu R(d\Omega/dR)$, where ρ is the mass per unit volume, ν is the effective kinematic viscosity and $\Omega(R)$ is the angular velocity at distance R from the central axis. In this disk (schematically illustrated in Fig. 11.1), radial mass flux and angular momentum redistribution occur because of radial dependence of $g \equiv 2\pi R^3\nu\sigma d\Omega/dR$, the viscous couple on material between R and $R + dR$. Here, σ is the surface density of the disk (mass per unit area).

There is a characteristic timescale $\tau_{\text{visc}} \sim R^2/\nu$ for the redistribution of a large fraction of the disk mass. For $R \sim 10^{15}$ cm, characteristic of our solar system, this time exceeds the age of the solar system if $\nu \lesssim 10^{13}$ cm^2/s. For comparison, the molecular kinematic viscosity of hydrogen at typical primordial solar nebula conditions is $\sim 10^5$ cm^2/s. Clearly, significant evolution

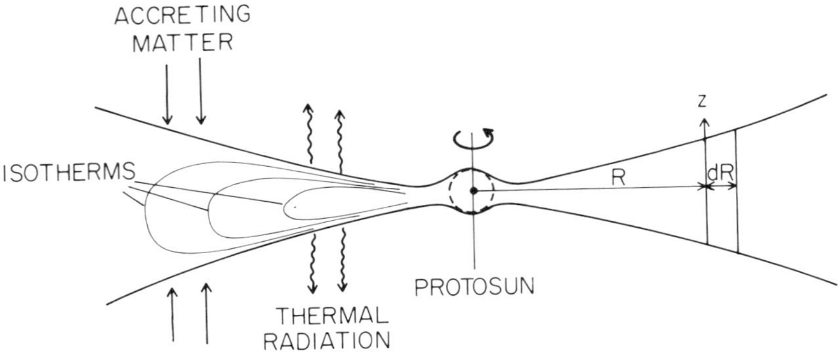

Figure 11.1. Geometry of the protosolar nebula accretion disk. The flared shape is determined by optical depth unity and is remarkably similar to the observed shape of a disk discovered around β Pictoris, a nearby star (Smith and Terrile, 1984).

of viscous accretion disks in a timescale $\lesssim 10^9$ years requires an effective viscosity caused by macroscopic motions, e.g. turbulence. This is perhaps the most important unsolved problem of disk dynamics. A popular choice is $\nu = \alpha c H$ where c is the sound speed, H is the scale height (essentially the disk thickness) and α is a fudge parameter, sometimes chosen to be as large as ~ 0.2 (e.g., Cameron, 1983a) but other times as low as $\sim 10^{-2}$ (Lin and Bodenheimer, 1982). If $\alpha = 0.1$, then $\tau_{\text{visc}} \sim 10^5$ yrs. Choices of α are usually either *ad hoc* or based on unverified assertions about the existence and nature of turbulence. The nearest to a predictive theory for α is the work of Canuto *et al.* (1984), which predicts $\alpha \sim 10^{-2}$ and is based on the expectation that the disk is thermally convective in the vertical direction (meaning the direction perpendicular to the plane of the disk). A necessary (but not sufficient) condition is that the disk is optically thick. However, more work is needed before the effective viscosity can be said to be even qualitatively understood.

The energetics of the disk can be appreciated by considering a vertical cylinder of unit cross section. If both top and bottom surfaces radiate into free space at an effective ("photopsheric") temperature T_e, then the energy loss is $2\sigma_{\text{SB}} T_e^4$, where σ_{SB} is the Stefan-Boltzmann constant. If this is balanced locally by viscous dissipation, then

$$\sigma \nu R^2 \left(\frac{d\Omega}{dR} \right)^2 \simeq 2\sigma_{\text{SB}} T_e^4. \tag{1}$$

In the present context, viscous dissipation simply means the local cascade of eddy motions into smaller eddies, eventually balanced by molecular viscosity. In a Kolmogorov cascade, this process depends only on the large scale characteristics of the flow. It is important to realize that the photospheric temperature is not necessarily the temperature of the disk interior because the disk may be optically thick. This means that the concept of a single temperature at a radial position R may be meaningless, an important point since models of Earth formation often assume some characteristic temperature at Earth orbit.

The thickness of the disk is readily calculated, at least for the simple case where the disk is approximately isothermal and the dominant contribution to the vertical component of gravity is provided by the central mass (the proto-Sun) of mass M_c. In this case, hydrostatic equilibrium implies

$$\frac{dp}{dz} \cong -\rho \frac{z}{R} \cdot \frac{GM_c}{R^2}, \tag{2}$$

where p is the pressure and ρ is the density.

Application of the ideal gas equation of state $p = (\rho/\mu)kT$, where μ is the molecular weight, then gives

$$\rho = \rho_c \exp\left(-\frac{z^2}{H^2}\right),$$

$$H \equiv \left(\frac{2kT}{\mu} \cdot \frac{R^3}{GM_c}\right)^{1/2}. \tag{3}$$

As a simple application of the above ideas, substitute equation (3) into $\nu \approx cH$ and then solve equation (1). The startlingly simple result is $T_e \sim 10^6 \rho^{2/5}$, where ρ is in g/cm^3 and molecular hydrogen is assumed. This result is independent of mass! If we further suppose that $\sigma \propto R^{-1}$, roughly compatible with the observed mass distribution of planets, then application of the above equations yields $H \propto R^{1.1}$, $\rho \propto R^{-2.1}$, and $T_e \propto R^{-0.83}$. (The choice $\sigma \propto R^{-2}$ is also acceptable and yields 0.92, −2.9, and −1.17 for the latter three exponents.) The latter result, in particular, enables one to see how the mass distribution in the disk is related to the temperature distribution. As an example, a nominal solar nebula model can be constructed by requiring that $T \sim 140$ K at 5 AU from the Sun, in order that water ice condense at the orbit of Jupiter. For $\sigma \propto R^{-1}$ and $\alpha = 0.1$, one then finds $T_e \simeq 700\,(1\text{ AU}/R)\,K$, $H \simeq 7 \times 10^{11}\,(R/1\text{ AU})^{1.1}$ cm, $\rho \simeq 8 \times 10^{-9}\,(1\text{ AU}/R)^{2.1}$ g/cm^3, and $\sigma \simeq 1.3 \times 10^4\,(1\text{ AU}/R)$ g/cm^2. The disk mass is $(0.03\,M_\odot)(R_{outer}/1\text{ AU})$, where R_{outer} is the outer radius of the disk.

2.2. Disk Instabilities

Details of nebula disk models change (the "evolution time" of Cameron's modeling effort is one or two years) but the enduring feature which distinguishes high mass from low mass models is the existence and nature of gravitational instabilities, a crucial issue for understanding the formation of planets and satellites. The theory of these instabilities was developed by numerous people (Edgeworth, 1949; Lin and Shu, 1966; Safronov, 1969; Goldreich and Ward, 1973) and only the basic principles are outlined here. Consider a patch of the accretion disk with a characteristic linear dimension (in the plane of the disk) of $l \lesssim H << R$. We can identify three characteristic timescales: an orbital timescale $\tau_{\text{rot}} \sim \Omega^{-1}$ associated with the effects of the Coriolis force and differential rotation, a sound wave travel-time $\tau_{\text{sound}} \sim l/c$, and a characteristic gravitational timescale $\tau_{\text{grav}} \sim \sqrt{l/\sigma G}$, associated with the "free-fall" collapse of the patch. (Notice that all these timescales are derivable purely from dimensional analysis.) Crudely speaking, gravitational collapse occurs provided τ_{grav} is less than either of the other timescales. If $\tau_{\text{grav}} > \tau_{\text{rot}}$ then subelements of the patch separate because of the stabilizing effect of the orbital motion before they can collapse towards each other by self-gravitation. If $\tau_{\text{rot}} > \tau_{\text{sound}}$ then the collapse can be prevented by a pressure change. In general, collapse will be possible for all length scales which satisfy $\tau_{\text{grav}} < \tau_{\text{rot}}$ (i.e., $l \lesssim \sigma G/\Omega^2$) *and* $\tau_{\text{grav}} < \tau_{\text{sound}}$ (i.e. $l \gtrsim c^2/\sigma G$). Clearly, the lowest surface density σ for which both inequalities can be satisfied is $\sigma_{\text{crit}} \sim c\Omega/G$ and the corresponding instability wavelength is $l \sim c/\Omega$. In a gas, c will be the actual sound speed; in a particulate disk (e.g., Saturn's rings) c will be essentially the random component of the velocity of the particles, and σ will be the surface density of solid matter only. In a mixture of gas and particles there is no completely general result.

2.3. Giant Gaseous Protoplanets

A gaseous disk with $T_e \sim 500$ K at $R = 1$ AU typically requires $\sigma \gtrsim 10^5$ g/cm^2 for gravitational instability. Cameron's disk models satisfy the (exact) criterion $\sigma > c\Omega/\pi G$ early in the evolution (elapsed time ~10^4 years, typically), leading him to propose that the primordial solar nebula fragmented into a large number of giant, gaseous protoplanets (Cameron, 1978; Cameron *et al.*, 1982). The existence of the instability is not well established, even for the high surface densities favored by Cameron, because the instability criterion is only marginally satisfied (i.e., by a factor ~2) and the thin disk approximation breaks down.

Possibly the greatest difficulty with giant, gaseous protoplanets is that it is very difficult to see how to evolve from them to the present solar system. The difficulties exist for both the giant planets and the terrestrial planets. Although the gaseous protoplanets may evolve in such a way as to cause rain-out of iron and silicates to the center (Slattery *et al.*, 1980), this is insufficient to explain the ten or so Earth mass "rock" cores of the present giant planets, unless additional (*ad hoc*) assumptions are made about feeding material from the solar nebula into the protoplanet or the size distribution of protoplanets. Although it is possible to eliminate some or all of a gaseous envelope at a later stage (by evaporation or tidal stripping or high early solar UV), it is also difficult to reconcile the terrestrial planetary properties with the properties of gaseous protoplanets. For example, these models seem to imply an almost purely iron core, incompatible with the core of the Earth which contains substantial amounts of lighter elements (Jacobs, 1975). It has to be acknowledged, however, that insufficient work has been done on the implications of the model yet that we can reach a conclusion on its capabilities. We choose to ignore this model in the remainder of this review for the pragmatic reason that its implications for the terrestrial planets have never been described or modeled adequately. To the extent that the evolution for the giant, gaseous protoplanet scenario eventually merges with other models discussed below, the neglect is only partial.

2.4. Condensation, Sedimentation, Coagulation

The starting material for the solar system is a cloud of gas and interstellar dust. Some, but not all, of the dust is evaporated by the high temperatures produced in the formation and evolution of the accretion disk. In the outermost part of the solar system, a lot of interstellar dust may survive. This may explain, for example, the very high D/H observed in parts of some meteorites (Yang and Epstein, 1983). At any radial position in the disk, there will be a complicated thermal history, but the long term trend will be a cooling one, leading to an additional condensation of material from the gas phase. It is possible to predict which materials condense and the order in which they condense, provided one assumes equilibrium chemistry (Grossman and Larimer, 1974). For our purposes, it suffices to note some very general points which are common to most models. First, free iron condenses at a very similar temperature to major silicate phases. Consequently, it is difficult and probably unreasonable to explain iron-rich bodies (e.g., Mercury) or iron-poor bodies (Moon) solely on the basis of the temperature in their regions of formation (Ringwood, 1979). Second, water ice condenses at around 160 K at the low pressures encountered in the solar nebula and this appears to demarcate approximately the terrestrial planet zone from the giant planet zone. By virtue of its abundance, water ice is the most important condensate, probably exceeding in mass the combined total of all the other condensates in the solar system. The third point about condensation scenarios is that they should not be taken too seriously! The above two properties are relatively reliable statements about condensation whereas detailed implications of the models must be treated with suspicion.

The general picture is an accreting disk in which the temperature first increases and then subsequently decreases (either from a Lagrangian or from an Eulerian point of view). "Meteorological" effects (upwelling and downwelling convective motions) are superimposed on these long term "climatic" trends. In the cooling phase, nucleation of solid condensates occurs from the gas phase. This nucleation can be either heterogeneous (on the surfaces of preexisting refractory seed nuclei) or homogeneous. Growth of these condensation centers to micron-sized grains is by diffusion alone and rapid ($\lesssim 10^3$ years). Subsequent growth can be slower but is aided by the relative motion between gas and grains or because of the possible sticking of grains during collisions. Condensation is likely to begin near the disk photosphere, where the temperature is lowest, and cause the formation of a cloud of micron-sized grains. Unlike the gas, which is supported by pressure gradients, the dust grains settle toward the midplane of the disk. Since the molecular mean free path is greater than the grain size, the drag exerted on the grains is ballistic rather than viscous. In the grain frame of reference, molecules impacting from upwind tend to be going faster on average than molecules from downwind by $\sim v$, where v is the grain settling velocity. The drag force is then $\sim \pi rmncv$, where r is the grain radius, n is the gas number density, m is the gas molecular weight, and c is the sound speed. Since the vertical (downward) component of gravity is $\sim z\Omega^2$, an estimate of the sedimentation time is then

$$\tau_{\text{sed}} \sim (3 \times 10^5 \text{ years})\left(\frac{1\ \mu}{r}\right). \tag{4}$$

A more detailed analysis (Weidenschilling, 1980; Nakagawa *et al.*, 1981) suggests short sedimentation times ($\sim 10^3$ years) because of the growth of large (up to 1 cm sized) particles by sticking. The situation is complicated and not yet well understood because of the probable presence of turbulence.

If turbulence is weak or absent then the concentration of solid particles toward the disk midplane leads inevitably to gravitational instabilities. This can be appreciated by applying $\sigma_{\text{crit}} = c\Omega/\pi G$ and realizing that the near-midplane effective sound speed c is decreasing as sedimentation proceeds. If the gas-solid medium is thought of as a pseudogas in which most of the density ρ arises from the solid particles but most of the pressure p arises from the gas, then $c^2 \sim p/\rho << p/\rho_{gas}$. The precise form of the instability is likely to be more complicated than the above crude analysis of c suggests (see Sekiya, 1983) but the result, independent of details, is likely to be the clumping of solid grains into planetesimals $\sim$ few kilometers in radius. The estimated planetesimal mass $\sim\sigma_{solid}^3 G^2/\Omega^4$ can be estimated by dimensional analysis alone or by finding the mass within a patch of area $\sim\pi(\pi/k)^2$ where $k \sim \pi/l$ is the critical wavevector of the instability.

If substantial turbulence is present, gravitational instabilities may not occur but planetesimals $\sim$1 kilometer in radius may still be possible by sticking of particles (Weidenschilling, 1980). The only difficult situation may be strong turbulence where collisional disruption could exceed coagulation.

At this point, between 10^5 and 10^6 years have elapsed since the initiation (collapse of an interstellar cloud) and the system consists of a thin disk of planetesimals in closely spaced, Keplerian orbits, immersed in a relatively fat disk of mainly hydrogen gas. The crucial issue confronting us is the nature and timing of the dispersal of the gas. Clearly, most of the gas is either incorporated into the proto-Sun or swept out of the solar system, since it is not present now. It is also clear that the gas must remain around long enough to enable the giant planets to form. The cores of these planets are $\sim$10 Earth masses of heavy material and probably formed prior to the accumulation of their hydrogen-rich envelopes (Mizuno, 1980; Stevenson, 1984a). The timescale for this process is not known, but was probably of order 10^6 years. The gas loss is believed to be associated with a T-Tauri phase as the Sun approached the main sequence and probably occurred around an age of 10^6 years or even less (measured from the initial collapse). This phase is not yet well understood (Bertout, 1984), but is probably associated with the establishment of an outer convective zone. Both accretion and strong stellar winds occur during the period of T-Tauri activity.

3. Accumulation of Earth in the Absence of Gas

Many of the issues discussed in this section are also relevant to accumulation in the presence of gas. They are included here to keep the exposition simple. We begin with a disk of planetesimals. At Earth orbit, we need an average of $\sim$10 g/cm^2 in the disk if one Earth mass is spread between 0.8 AU and 1.3 AU from the Sun; this corresponds to 10 km planetesimals spaced $\sim 2 \times 10^9$ cm apart, and initially in circular orbits. Relative motions of neighboring planetesimals due to Keplerian shear are very small ($\sim 10^2$ cm/s). Gravitational interactions in the planetesimal swarm stir the velocity vectors, creating a finite "temperature" (represented by the velocity components associated with eccentric and inclined motions) so that bodies can collide. At this stage, the collisions are very gentle. The *lowest* possible velocity of a collision is of order the escape velocity from the larger of the bodies involved, since the bodies fall into their mutual gravitational well. This is

$$v_{esc} \sim \left(\frac{4\pi G\rho}{3}\right)^{1/2} r \sim 10^3 (r/10 \text{ km}) \text{cm/s}, \tag{5}$$

where ρ is the density and r is the planetesimal radius. If bodies approach each other with velocities of this order or less then the inelasticity of their collisions is likely to cause coagulation, because the rebound velocity is less than the escape velocity.

3.1. The Safronov Steady-State

The simplest picture is one in which the planetesimal swarm is "stirred" by (non-collisional) gravitational encounters and damped by physical collisions, with the latter causing a continuous evolution of the swarm toward progressively larger bodies on average. We think of the swarm as a gas of interacting particles, isotropic in both real and velocity spaces (after the circular Keplerian orbital motion is subtracted), with particle separation small compared to orbital radius and encounter velocities small compared to orbital velocity.

Consider a gravitational encounter between two bodies of mass m that are separated by a distance D at closest approach. In the limit of small deviations of their trajectories, the incremental change in velocity perpendicular to the initial path and caused by their interaction is about

$$\Delta v_T = \tfrac{1}{2} \int_{-\infty}^{\infty} dt \cdot \frac{GmD}{(D^2 + v^2 t^2)^{3/2}} \sim \frac{Gm}{Dv}, \tag{6}$$

where t is time ($t = 0$ is closest approach) and v is the relative velocity. To lowest order, there is no net change to the velocity component parallel to the original trajectory of each particle. The kinetic energy change per encounter is $m(v\Delta v_T + \Delta v_T^2/2)$. However, the *sign* of Δv_T is random, so the time-averaged gain in kinetic energy due to gravitational stirring is $\tfrac{1}{2} m \Delta v_T^2$ per mass, per encounter. Summing over all encounters,

$$\left.\frac{d}{dt}(v^2)\right|_{\text{Gravit}} = \int_{D_{\min}}^{D_{\max}} \left(\frac{Gm}{Dv}\right)^2 nv\cdot 2\pi D\, dD \sim \frac{(Gm)^2 n}{v} \ln\left(\frac{D_{\max}}{D_{\min}}\right), \quad (7)$$

since nv is the flux of encountered bodies per unit area, where n is the planetesimal number density. $D_{\max}$ and $D_{\min}$ are the largest and smallest encounter distances. Clearly $D_{\min} \sim$ planetesimal radius and $D_{\max} \sim$ gravitational sphere of influence, but their precise values are unimportant because they enter into a log function.

To compute collisional damping, we note that if gravitational focusing is ignored (see below) then the mass flux impinging on a given body in the swarm is of order $\pi r^2 mnv$. If most of these collisions lead to coagulation then the reduction in the "incoherent" velocity per collision is of order 50 percent (assuming comparable mass bodies) so that

$$\left.\frac{1}{v^2}\frac{d}{dt}(v^2)\right|_{\text{collisions}} \sim -\pi r_2 nv. \quad (8)$$

From equations (7) and (8) we see that if we assume steady state, i.e.

$$\left.\frac{d}{dt}(v^2)\right|_{\text{Gravit}} + \left.\frac{d}{dt}(v^2)\right|_{\text{coll}} = 0,$$

then it follows that

$$v \simeq \left(\frac{Gm}{r\theta}\right)^{1/2}, \quad (9)$$

where θ is a dimensionless number of order unity that incorporates both uncertainties in the gravitational scattering and uncertainties in the collision physics. It is called the *Safronov number* because he was the first to recognize the importance of the steady state velocity implied by equation (9). There are some subtle distinctions between different possible definitions of what v actually means (see Wetherill, 1980). In Safronov (1969), v is interpreted as the relative velocity of the body with respect to a body moving in a circular orbit at the same instantaneous distance from the Sun. There are other less subtle aspects of the above crude analysis which demand concern however. Do the bodies coagulate efficiently? What is the spectrum of masses (ignored above), and might not the largest masses have runaway growth, because of gravitational focusing? How long does it take to achieve steady-state?

On the question of coagulation, a nice feature of equation (9) is that it predicts impact velocities that are comparable to escape velocities. This means that unless the collisions are remarkably elastic, enough energy is dissipated that coagulation should take place. If impact velocities were much higher then ablation or fragmentation is likely to occur, especially as the velocity becomes high enough that impact stresses become large compared to the strength of rocks. There is an extensive literature of the phenomenology of impact fragmentation and how it depends on material properties and velocity (e.g., Hartmann, 1978; Fujiwara *et al.*, 1978; Kawakami *et al.*, 1983). The Safronov result seems self-consistent in that the velocity it predicts also implies efficient coagulation.

In the presence of a swarm of smaller bodies with relative velocities v at large separation, a larger body of radius r *and* escape velocity $v_{esc} >> v$ has an apparent (gravitational) cross-section $\pi r^2(1 + 2v_{esc}^2/v^2) \propto r^4$ (since $v_{esc}^2 \propto r^2$). This can be enormously larger than the geometrical cross-section. In numerical simulations of the early stages of accretion, Greenberg and coworkers found a tendency for runaway accretion, in which certain embryos grew much faster because of the enhanced cross-section (Greenberg, 1982). The outcome was a rapid development of substantial sized bodies (approximately several hundred kilometers in radius) in nearly circular orbits. A *bimodality* of the mass distribution arose, with substantial mass in these large bodies but also a lot of mass in very small planetesimals. In a number of publications, Wetherill and coworkers have challenged these conclusions, pointing out that the two-body approximation breaks down for very low encounter velocities, restricting the maximum amplitude of gravitational focusing (Wetherill and Cox, 1984, 1985). Most recently, kinetic theory calculations suggest that runaway should occur, if the original swarm is "seeded" with an embryo (Wetherill and Stewart, 1986; Stewart and Wetherill, 1986). Even if there is a runaway at early stages, its consequences may not persist (at least, for any model that eventually can explain the observed planets). Clearly, the larger bodies must also collide with each other, but it may take an extended period of distant encounters before their eccentricities are sufficiently excited to do this.

Assuming no runaway, the collisional evolution is then expected to promote a power-law distribution of masses, at least over a substantial range of masses. A power law can be expressed in the form

$$n(m) \propto m^{-\alpha}, \tag{10}$$

where $n(m)dm$ is the number of bodies with masses between m, $m + dm$. There are a number of theoretical arguments suggesting values of α around 1.8 (e.g., Zvyagina *et al.*, 1973). Two "limiting" values of α are of particular interest. At $\alpha = 5/3$, each decade of mass contributes equally to the total surface area (relevant to calculation of collision rates). At $\alpha = 2$, each decade of mass contributes equally to the total mass. For $\alpha < 2$, most of the mass is in the largest bodies. For $\alpha > 2$ (unlikely), most of the mass is in the smallest bodies. The important point is that for likely values of α, the Safronov steady state still applies except that the relevant *m, r* apply most closely to the largest (or near largest) bodies in the swarm. We will return later to the implications of the mass spectrum for the thermal evolution of the accreting Earth.

Even when the Safronov steady state is an achievable and meaningful state, it takes a finite time to achieve it from some initial condition and it is important to establish this timescale. Numerical simulations by Wetherill (1980) indicate that the timescale for establishing steady state is $\sim 10^6$ yr, about an order of magnitude less than the timescale for which the steady state itself evolves (as progressively larger bodies grow). He determined this, both by starting a swarm "cold" (low v) and letting it heat up to the Safronov value, and by starting a swarm "hot" (high v) and letting it cool down.

3.2. Accumulation Timescales

There has never been a calculation (or coherent set of calculations) that describe the entire accumulation process from planetesimals to completed planets. Ideally, one would want to take the outcome of early stage "kinetics" calculations (in which the total number of planetesimals is immense, and the swarm is treated as a gas) and use this as the input for an orbital simulation calculation. As we saw, the outcome of the early stage evolution is controversial, but everyone agrees that this stage is *rapid* (10^4–10^5 years). In any event, it must be followed by a protracted epoch in which distant encounters excite eccentricities, allowing for the possibility of crossing orbits and collisions. If the Safronov steady state (or something close to it) describes this epoch, then we can crudely estimate the accumulation time of Earth as follows.

Consider a swarm of planetesimals of mass density ρ_s (solid mass per unit volume) in which a particular (largest) embryo of mass M and radius r is growing. The magnitude of the flux of mass it encounters is $\rho_s v$, where v is the encounter velocity at infinity (given by equation 9). The rate of mass increase is then given by

$$\frac{dM}{dt} \simeq \pi r^2(1 + 2\theta)\rho_s v, \tag{11}$$

where allowance is made for gravitational focusing. Now, $dM = 4\pi r^2 \rho_p \, dr$, where ρ_p is the density of the planet, so

$$\frac{dr}{dt} \simeq \frac{\rho_s v}{2\rho_p}(1 + 2\,\theta). \tag{12}$$

Suppose that the final planet mass is M_p and is supplied from a torus of material distributed between $R - \Delta R/2$, $R + \Delta R/2$ radial distance from the Sun, and with a thickness $2\Delta z$ (same cylindrical coordinates as in Figure 11.1). It follows that

$$\rho_s = \frac{M_p - M}{2\pi R\Delta R \cdot 2\Delta z}. \tag{13}$$

But $\Delta z \sim Ri$, where i is the typical orbital inclination. In Safronov steady state, $i \sim v/V_K$, where V_K is the Keplerian orbital velocity. So we finally get

$$\frac{dr}{dt} = \frac{V_K(1 + 2\theta)(M_p - M)}{16\pi R^2 \Delta R \cdot \rho_p}. \tag{14}$$

At early times ($M << M_p$), substitution of $V_K \simeq 30$ km/s, $\theta = 3$, $M_p = 6 \times 10^{27}$ g, $R = 1$ AU $= 1.5 \times 10^{13}$ cm, $\Delta R = 0.5$ AU, $\rho_p = 5.5$ g/cm^3 gives

$$\frac{dr}{dt} \simeq 15 \text{ cm/yr}, \tag{15}$$

which, by coincidence, is similar to the motions of plates on the present surface of Earth. To build a planet of radius 6400 km would take ~40 million years. More precisely, equation (14) with fixed V_K, θ, ΔR, and ρ_p gives

$$\frac{dx}{dt} = \frac{1}{\tau}(1 - x^3), \tag{16}$$

where $\tau \approx 40$ myrs and x is the fraction of the final radius ($x \equiv r/R_\oplus$). For $x << 1$, $x \simeq t/\tau$, and the mass grows as the cube of time. As $x \to 1$, the growth becomes asymptotic, of the form $x = 1 - A\exp(-3t/\tau)$, where A is a constant. In reality, the behavior is more complicated because parameters such as θ, ΔR are not strictly constant.

One feature of this simple model is that it predicts $\tau \sim (1 + 2\theta)^{-1}$ and $M \sim x^3 \sim (1 + 2\theta)^3$, at least at short times. Since this gravitational focusing effect applies only to the largest embryos, Safronov suggested that the second largest body in the swarm would be smaller in mass than the largest body by a factor $(1 + 2\theta)^3 \sim 10^2$ to 10^3. This is an overinterpretation of the Safronov model, however, since the model artificially isolates the evolution of a particular embryo in a narrow zone of radii from the Sun and cannot, therefore, explain why there are only four terrestrial planets rather than, say, one hundred. For that, one needs numerical simulations.

3.3. Numerical Simulations of Late Stage Accretion

The Safronov analytical approach makes approximations and does not embody the full Keplerian dynamics. To date, the only extensive three-dimensional simulations of the "late stages" of accretion are those by Wetherill (1980, 1985), which we now discuss. Other simulations exist, but most are two dimensional (e.g., Lecar and Aarseth, 1986) or severely limited in other ways. By "late stage," one means that substantial ($\gtrsim$1000 km radius) bodies have already been formed, initially in rather low eccentricity orbits. For practical reasons, the simulations typically begin with a "modest" number of bodies (500), although ideally a number like 10^4 or even more would be preferable. The simulation uses a Monte Carlo technique based on Arnold (1965) in which a two-body scattering algorithm is used to calculate mutual planetary perturbations. Probabilistic aspects enter the calculation at several stages: to specify initial orbits, to assess whether a chance encounter occurs between a given pair of bodies in a given time step, and to choose at random (but with appropriate geometric weighting) the particular distance and geometry of an encounter. If, by chance, the encounter distance is less than the gravitational collision radius, then a collision is assumed to take place. The bodies are then assumed to merge, conserving angular momentum. Very distant encounters are treated by an approximate extrapolation rather than by an explicit scattering computation.

Aside from possible subtleties involving resonances, this procedure is believed to treat the gravitational aspect of the problem very well, as confirmed by comparison with direct numerical integration (Wetherill and Cox, 1984, 1985). However, it is much faster than direct orbital integration. The non-gravitational aspects of the simulation (e.g., outcome of impacts or tidally disruptive close encounters, are much less well understood; Wetherill attempts to accommodate these uncertainties by running a range of simulations with different assumptions for these effects. In most respects, these uncertainties do not have a major consequence for the outcomes.

Perhaps the least satisfactory aspects of Wetherill's simulations are the initial conditions. Wetherill chooses initial semi-major axes within a rather modest range of 0.7 to 1.1 AU in most of his recent calculations (Wetherill, 1985). Within this swarm, initial eccentricities were randomly distributed between 0 and 0.05 and initial inclinations were between 0 and 0.025, all consistent with Safronov. Semi-major axes were randomly distributed in accordance with a surface density independent of heliocentric distance, to be consistent with the specific orbital angular momentum and energy of the present system of terrestrial planets. All this seems reasonable, but it is nevertheless strange that our terrestrial planets would be derived solely from bodies that began at orbits lying between 0.7 and 1.1 AU. The probable answer is that a more complete model that includes the giant planets and loss of bodies (inwards, to the Sun and outwards, into the giant planet region and beyond) would not artificially localize the reservoir for the terrestrial planets.

Of course, each simulation is a separate "experiment" with a unique outcome. Many experiments must be carried out to get a good idea of the range of possible outcomes. We seek in these outcomes general features reminiscent of the terrestrial planetary system rather than carbon copies of the system. (Of course, a "carbon copy" would eventually happen if enough simulations were carried out.) The following general features emerge:

1. The accumulation time for over 90 percent of Earth's mass is less than 100 million years and often about the same as Safronov's estimate ($\sim$40 myr). The process is also roughly as Safronov envisaged: excitation of substantial eccentricity by distant encounters, leading eventually to crossing orbits.
2. Despite the high eccentricity of some bodies, the most massive bodies are in generally rather well-defined low eccentricity orbits, because they receive an incoherent sum of perturbations from many smaller bodies. However, Mars or Mercury may not qualify as a massive body in this context. Wetherill finds that there are cases where much of Mars' mass formed at the orbit of Mercury and vice-versa.
3. The number of planets left at the end is about four and usually three to five. Although other outcomes are possible, it is remarkable that about the right number of planets emerges from a plausible initial condition. Moreover, the planets are roughly distributed in a way similar to that observed—although it could be argued that this type of simulation is always likely to give two small "outlying" planets, with most of the mass in between. Since the smallness of Mars may be due to Jupiter perturbations that are not in the simulation, it is consequently unclear whether the simulations identify the cause of the observed mass distribution. Figure 11.2 gives examples of final outcomes, and a comparison to the observed system.
4. Large impacts occurred during the accumulation of Earth. This is not really a new conclusion but received little attention for many years. Up until the mid-1970s and sometimes much later, it was often assumed that most of the mass accreted onto Earth arrived as small bodies (e.g., 10 km planetesimals). This belief probably arose as a misinterpretation of Safronov's results. Hartmann and Davis (1975) and Cameron and Ward (1976) were the first to emphasize the probable important role of very large impactors (bodies the size of Moon or Mars). This has motivated popularity for an impact origin of the Moon (Cameron, 1985; Benz *et al.*, 1986; Stevenson, 1987; see also Hartmann *et al.*, 1986), since it is believed that some of the debris from a giant impact is injected into Earth orbit and becomes source material for the Moon. Figure 11.3 is a histogram repre-

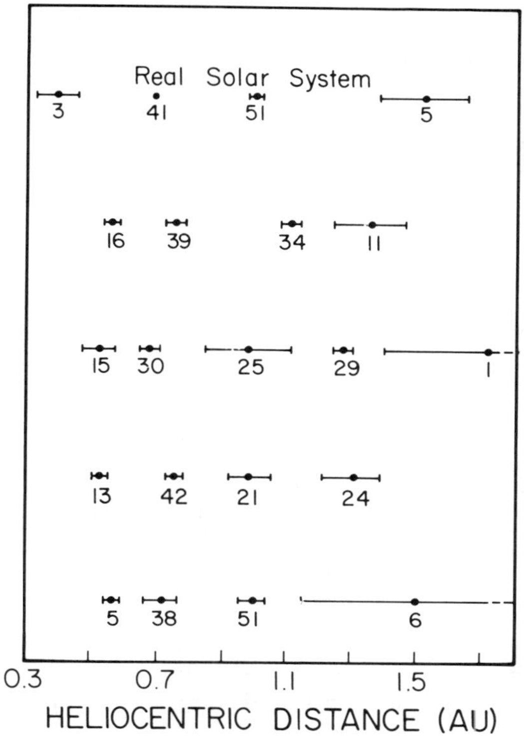

Figure 11.2. Outcomes of Wetherill's 3-D numerical simulations of terrestrial planetary accretion, with the real solar system (top) for comparison. Each point represents the position of a planet, the "error" bar represents the eccentricity of its orbit and the number represents the percentage of total terrestrial planetary mass associated with that planet. The top three simulations are discussed in Wetherill (1980) and the bottom simulation is from Wetherill (1985).

sentation of Wetherill's results, illustrating the number and size of the largest bodies hitting the Earth for a total of five simulations. As we shall see, these impacts have a major importance for the thermal state of the growing Earth. The important conclusion is that most of the mass arrives in *large* bodies. In terms of the mass spectrum of equation (10), this means $\alpha \sim 1.9$. Wetherill's simulations yield similar results, whether he assumes $\alpha \sim 1.9$ as *input* or initializes the calculation with all bodies having the same mass. Even with $\alpha = 1.9$, there are many small bodies, with

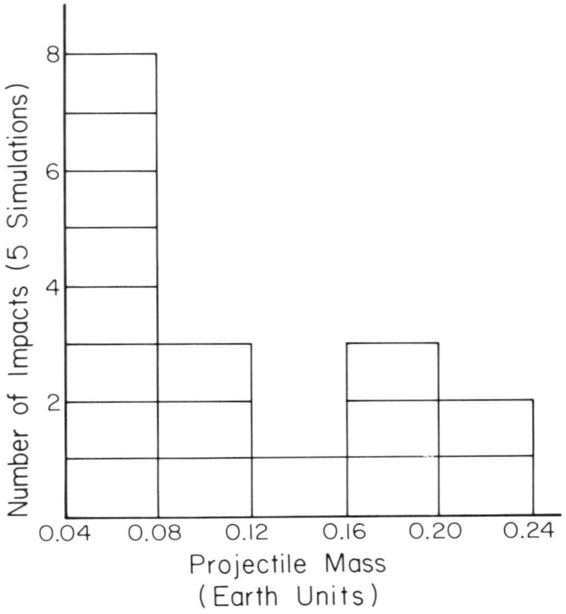

Figure 11.3. Histogram of impactor masses during Earth accumulation for a total of five simulations, as reported by Wetherill (1985). The total mass shown here, divided by five, is ~40 percent of an Earth mass, showing the importance of giant impacts.

significant total mass, and these help to dampen the eccentricity of Earth's orbit.

5. As a related conclusion, the material from which Earth grows has a diverse provenance. The impacting bodies include material that initially coagulated inside the orbit of Mercury or outside the orbit of Mars, as well as more local materials. In view of the time and space dependence of temperatures within the solar nebula, it must be concluded that the source material has diverse condensation temperatures, probably including volatile material. *It is a mistake to suppose that the Earth is composed of material that condensed near Earth orbit*. There may, however, be a trend in the composition of accreting material with time, since material that accretes late has a greater probability of originating from more distant orbits. In particular, it may be volatile rich (as assumed by, for example, Wänke *et al.*, 1984). We will return to this possibility.

3.4. Thermal Regime during Earth Accretion

To appreciate the thermal consequences of earth accumulation, it is instructive to consider some extreme (admittedly unrealistic) limiting cases. Suppose, first, that all the energy of incoming bodies were converted to stored heat. This energy is at least 0.6 GM^2/R, the gravitational self-energy of a constant density body of mass M and radius R. Neglecting latent heat, the temperature rise

$$\Delta T \gtrsim \frac{0.6\, GM}{RC_p} \sim 4 \times 10^4 \text{ K}, \tag{17}$$

assuming a specific heat $C_p \sim 10^7$ erg/g/K. Of course, this value is so large that complete melting and partial vaporization occurs, so latent heat is important and the corresponding ΔT is much lower. But the result amply illustrates that the energy of earth formation is prodigious. This estimate neglects the velocity at infinity relative to the velocity at impact, a reasonable assumption for the Safronov steady state as discussed above.

Suppose we now allow *all* this energy to be available for radiative loss from the surface of an earth that has no atmosphere. We could possibly do this by making the Earth from small particles that can lose their infall energy by prompt infrared radiation. In this case, the radiative flux (in excess of that provided by the Sun) balances the rate at which available gravitational energy is released:

$$4\pi R^2 \sigma (T_e^4 - T_0^4) \simeq \frac{M}{\tau_{acc}}\left[\frac{0.6\, GM}{R} - C_p(T_e - T_0)\right], \tag{18}$$

where T_e is the surface temperature, T_0 is the temperature the surface would have if it were heated by the Sun alone, and τ_{acc} is the accretion time (e.g., the time to accumulate ~90 percent of the mass). In the limit $\tau_{acc} \to 0$, equation (17) is reproduced. In the more realistic limit of large τ, $C_p(T_e - T_0)$ is negligible, and

$$\Delta T \equiv T_e - T_0 \simeq \frac{0.6\, GM^2}{16\pi R^3 \sigma T_0^3 \tau_{acc}} \simeq (23 \text{ K})\left(\frac{10^8 \text{ yr}}{\tau_{acc}}\right), \tag{19}$$

assuming $T_0 = 300$ K and $\Delta T << T_0$. Even at $\tau_{acc} \sim 10^7$ yr, a likely lower bound of Earth formation time based on our previous discussion, the complete equation (18) gives $T_e - T_0 \simeq 70$ K. From this we conclude that black body radiation is an extremely efficient buffer of the *surface* temperature of the growing Earth, at least in the case of a surrounding vacuum.

If equation (18) were rewritten as a differential equation, by replacing M/τ_{acc} by dM/dt and treating T_e, M, R as instantaneous values, then the predicted *internal* temperature of the final earth would be about the same (but somewhat less than) the surface temperature at the end of accretion. This is clearly an unrealistic result, however, since it assumes that all incoming energy is available for prompt radiation. Suppose we assume, instead, that some fraction of the heat from a given impact event is deposited a distance δ below the instantaneous surface. In a time $\tau_{\text{cond}} \sim \delta^2/\kappa$, this heat could escape to the surface and be available for radiation provided the planet has not grown in the meantime. Here, κ is the effective thermal diffusivity. In the time τ_{cond}, the Earth has in fact grown by an amount $\tau_{\text{cond}} \cdot dR/dt$, where $dR/dt \sim R/\tau_{\text{acc}}$. If this is larger than $\sim\delta$ then the heat clearly cannot escape, because the distance it can diffuse in time t increases only as $t^{1/2}$ whereas the distance to the surface is increasing linearly with time. The criterion that the heat cannot efficiently escape is therefore

$$\delta \lesssim \frac{\kappa\tau_{\text{acc}}}{R} \sim (0.5 \text{ km})\left(\frac{\kappa}{10^{-2} \text{ cm}^2/\text{s}}\right)\left(\frac{\tau_{\text{acc}}}{10^8 \text{ yr}}\right). \tag{20}$$

If κ were the value appropriate to solid rock then the critical δ would be very small, much less than the size of the smallest expected planetesimals. This was first realized by Safronov (1969), and led to his prediction of significant energy retention within the growing earth.

The important question, then, is what fraction of the energy in equation (17) is really retained? Clearly, only $\gtrsim$2.5 percent efficiency is needed to guarantee a partially molten Earth. This efficiency factor ϵ can be thought of as the product of two efficiencies: ϵ_d, the fraction of incoming kinetic energy that is converted to heat and is not promptly lost during the impact process (e.g., as radiation from small debris particles or droplets or from a thin ejecta sheet); and ϵ_r, the fraction of this deposited energy that is retained during the time that elapses before another comparable impact in the same region. (Notice that these definitions are rather vague; this is true of the original literature as well, and not just for the convenience of this review.) It is clear that $\epsilon \equiv \epsilon_d\epsilon_r$ depends on a number of considerations, the most important of which are:

1. The physics of individual impact events. We need to understand energy deposition within, below, and adjacent to the crater. This depends on how cratering flow fields scale with impactor size and momentum or energy (e.g., Holsapple and Schmidt, 1982).
2. The mass spectrum of planetesimals. Clearly, large projectiles are more able to deposit energy at depth than small projectiles.
3. The transport of energy from depth to the surface. In equation (20) this is

characterized by a diffusivity, but macroscopic flows (impact stirring, convection) can greatly modify the quantitative inference of this equation or even cast doubt on its validity. For example, melting may cause pervasive volcanism and highly efficient *loss* of energy from the interior to the surface.

4. The accretion rate clearly affects the loss of energy (cf. equation 20) although this is a relatively unimportant factor provided τ_{acc} is neither very short ($\lesssim 10^4$ yr) or extremely long ($\gtrsim 10^9$ yr), criteria that are certainly satisfied. (The lower limit is required to guarantee that the surface never approaches the melting point. The upper limit is required to guarantee that convection does not buffer the temperature below the solidus at all times—this is explained more fully below.)

Of these four factors, by far the most important is the first one. Unfortunately, impact physics is a complicated and imperfectly understood subject, especially when it involves very large impacts or high velocities. Extensive numerical simulations exist for the early stages of the impact process (e.g., O'Keefe and Ahrens, 1982), but these do not directly establish the final disposition of the various internal and kinetic energy components. The most shocked material is ejected as a spray and efficiently loses heat (especially if, as is likely, it is in the form of molten droplets) by prompt radiation in flight or by rapid loss from a relatively thin ejecta sheet. The material most likely to retain significant heat is beneath the floor of the crater. For reasons that are rather obscure (at least to this writer), workers have chosen values of $\epsilon_d \sim 0.2$ (Safronov, 1978; Kaula, 1979, 1980; Coradini *et al.*, 1983; Davies, 1985). An inspection of the computer-simulated flow fields suggests that this value could be too low by a factor of two, but the uncertainties are clearly large. The value of ϵ_d should be relatively insensitive to the size of the impacting body, but may be sensitive to its velocity. At the velocities of interest (~10 km/s), substantial melting is expected and the "debris" should be in the form of droplets (eventually "tektites").

The value of ϵ_r is very difficult to calculate in general, but can be estimated in the limit where only impact stirring and conduction take place (i.e., no convection). In this case, Davies (1985) finds $0.15 < \epsilon_r < 0.7$ but favors $\epsilon_r \sim 0.5$ and $\epsilon \sim 0.1$. The large uncertainty reflects poorly understood aspects of the cratering flow field. Uncertainties in the spectrum of planetesimal sizes are unimportant because almost all the mass (and energy) arrives in bodies whose cratering flow fields are more extensive than the lengthscale δ (equation 20). Likewise, the accretion time is relatively unimportant. However, it is clear that more work is needed.

Assuming a nominal value of $\epsilon \sim 0.1$, we have a sub-surface temperature rise of the Earth of about

$$\Delta T \simeq (5000\ \mathrm{K})\left(\frac{R}{6400\ \mathrm{km}}\right)^2, \tag{21}$$

where R is the instantaneous radius. The temperature must therefore already approach the melting point of silicate phases (or, more correctly, the solidus $T_s \sim 1300$ K) when R is only about 2500–3000 km, and the protoearth is not even the mass of Mars (one tenth of an Earth mass). It seems likely, however, that as this temperature is approached, at least some of the buried energy can be convected outward. We now proceed to estimate this.

Suppose, first, that the Earth's surface participates in the convective cycle (as it does for the ocean basins of the present Earth). According to parameterized descriptions of convection (Tozer, 1967; Turcotte and Schubert, 1982, Ch. 6), the heat flow transported outward is then larger than would be transported by conduction for the same temperature drop by a factor known as the Nusselt number, Nu:

$$\begin{aligned} \mathrm{Nu} &\simeq (\mathrm{Ra}/\mathrm{Ra}_c)^{1/3}, \\ \mathrm{Ra} &\equiv g\alpha\Delta T d^3/\nu(T)\kappa, \\ \mathrm{Ra}_c &\simeq 10^3, \end{aligned} \tag{22}$$

where Ra is the Rayleigh number, Ra_c is its critical value which must be exceeded for convection, g is the gravitational acceleration, α is the coefficient of thermal expansion, ΔT is the temperature drop across the convecting layer of depth d, $\nu(T)$ is the kinematic viscosity of the layer of temperature T, and κ is the thermal diffusivity. We can approximate ΔT by T if the convection penetrates to the surface and the heat flux F is then

$$\begin{aligned} F_{\mathrm{conv}} &\simeq 0.1\, kT^{4/3}(g\alpha/n(T)\kappa)_1^{/3} \\ &\simeq (2500\ \mathrm{erg/cm/s})\left(\frac{T}{1800}\right)^{4/3}\left(\frac{10^{16}\ \mathrm{cm^2/s}}{\nu(T)}\right)^{1/3}, \end{aligned} \tag{23a}$$

assuming a thermal conductivity $k = 2 \times 10^5$ cgs, $g \simeq 10^3$ cm/s^2, $\alpha \simeq 2 \times 10^{-5}$/K. For comparison, the present terrestrial heat flow ~80 erg/cm^2/s is consistent with this equation provided $\nu \sim 10^{21}$ cm^2/s, roughly consistent with postglacial rebound studies (see review by Peltier, 1985). The viscosity of mantle rocks is an extremely strong function of temperature, however. For example, a plausible parameterization for olivine might be (Carter, 1976; Poirier, 1985)

$$\nu(T) \simeq 10^{16} \exp\left[30\left(\frac{T_m}{T} - 1\right)\right] \text{cm}^2/\text{s}, \tag{23b}$$

where T_m is the melting temperature. In a mantle assemblage (e.g., peridotite), *partial* melting probably begins at $T \sim 0.75\, T_m$; $\nu \sim 10^{20}$ cm²/s. We will return to the problem of melting; for the moment, let us concentrate on solid state flow. The thermal energy deposited at depth (i.e., available for convective transport to the surface) can be characterized as an "accretional flux" F_{acc},

$$F_{acc} \simeq \frac{0.6\, \epsilon_d\, GM^2}{4\pi R^3 \tau_{acc}}, \tag{24}$$

where, as usual, the velocity at infinity of incoming projectiles is assumed small. For $\epsilon_d \sim 0.2$,

$$F_{acc} \simeq (2 \times 10^4\ \text{erg/cm}^2/\text{s}) \cdot \left(\frac{10^8\ \text{yr}}{\tau_{acc}}\right). \tag{25a}$$

This is larger than the maximum solid state F_{conv} by about one order of magnitude, even for the longest likely accretional time $\sim 10^8$ yrs. In other words, a viscosity $\nu \sim 10^{13}$ cm²/s would be needed to extract the heat. *This leads to an important conclusion: At least up until melting, convection cannot buffer the internal temperature rise of the accreting Earth.* In other words, equation (21) accurately describes the temperature profile at least up until $\Delta T \sim 1500$ K (radius ~2000–3000 km). The conclusion is even stronger if the convection does not penetrate to the surface since then ΔT in equation (22) is almost an order of magnitude smaller and the same heat flux may require a viscosity several orders of magnitude smaller.

Once melting is initiated, however, heat loss may be prodigious. We have an example of this in our present solar system. Io is tidally heated and releases an average of 10^3 to 2×10^3 erg/cm²/s (Johnson *et al.*, 1984; Ojakangas and Stevenson, 1986). The most remarkable feature of this heat loss is that it is localized in hot spots, perhaps radiative and evaporative loss from sulfur lakes heated by silicate magma chambers (Lunine and Stevenson, 1985), although radiative loss from silicate flows has also been advocated (Carr, 1986). The *local* heat flow at these hot spots is very high, perhaps 10^6–10^7 erg/cm²/s (equivalent to a black-body temperature of 400– 500 K) and persists, at least on the timescale they have been observed (a few years). This tells us that it is possible for a planetary body to lose prodigious amounts of heat by magmatic activity. Even more remarkably, Io does this while apparently remaining mostly solid in most of its volume: it has substantial (several kilometer high) topography.

Since the accretional heat input is small compared to radiative loss from magma, it is possible to envisage a model of episodic volcanism in which a given region of the surface is thought of as the thickening crust of a lava lake that gets punctured or reformed after every interval τ (cf. Stevenson, 1985). In this period, the amount of heat released per unit area is $\sim(\rho C_p \Delta T + L)(\kappa\tau)^{1/2}$, because the conductive layer thickness is $\sim(\kappa\tau)^{1/2}$. This is the standard Stéfan problem (Stéfan, 1891; Turcotte and Schubert, 1982). The resulting heat flow is

$$F_{\text{magmatic}} \sim (5 \times 10^5 \text{ erg/cm}^2\text{/s})\left(\frac{1 \text{ yr}}{\tau}\right)^{1/2}. \tag{25b}$$

If this is larger than F_{acc} then it is presumably possible for the earth to "bleed" magma in steady state with the injected energy, and this magma would freeze; no magma ocean would form at depth. If F_{magmatic} is less than F_{acc}, this means that the input energy cannot be accommodated by magma flowing onto the surface and freezing; a magma ocean must then form beneath the surface. Unfortunately, the appropriate value of τ is rather uncertain. Consider the possibility that it is determined by impact disruption: each incoming body of mass m is responsible for causing the resurfacing of an area $\sim Am^{2/3}$ (i.e., proportional to its cross-sectional area). For high velocity impacts, a plausible assumption is that the affected area is $\sim 10^2$ larger than the physical cross-section of the projectile. The accumulated effect depends on the spectrum of planetesimal masses, $n(m)$, introduced in equation (10). In Table 11.1, we show plausible inventories of the bodies that

Table 11.1. Illustrative inventories of planetesimals required to form the Earth

$\alpha = 2^{\dagger}$	$\alpha = 5/3^{\ddagger}$
1 Mars	5 Mars
10 Moons	3 Moons
10^2 Iapetus	12 Iapetus
.	.
.	.
.	.
10^9 1 km planetesimals	10^5 1 km planetesimals
Total = 1 Earth mass	Total = 1 Earth mass

†Equal amount of mass in each logarithmic mass interval down to 10^{-10} Earth masses.

‡Equal amount of cross-sectional area is contributed by each logarithmic mass interval (lower mass cutoff unimportant for total mass).

accumulated to form Earth, for the limiting cases $\alpha = 5/3$ and $\alpha = 2$, assuming $n(m) \sim m^{-\alpha}$. Since the case $\alpha = 2$ has so many more impacting bodies it might be supposed that the value of τ is much lower in that case. In fact, this is not true since the most massive bodies tend to compensate as α decreases. To be explicit,

$$\frac{1}{\tau} = \int_{m_1}^{m_2} Am^{2/3}n(m)dm/(4\pi R^2\tau_{acc}), \tag{26}$$

assuming $\int_{m_1}^{m_2} n(m)m\,dm$ is the mass of Earth, and m_1, m_2 are the lower and upper mass cut-offs for the mass spectrum, respectively. Assuming $\tau_{acc} \sim 3 \times 10^7$ yrs, $m_1 = 10^{16}$ g, and $m_2 = 10^{27}$ g, one finds that $\tau \sim 10^3$ yrs for $\alpha = 2$ *and* for $\alpha = 5/3$. In these cases, $F_{magmatic} \sim F_{acc}$, and a magma ocean is a marginal possibility at best. It is also possible that early quench crust spontaneously founders because it is more dense than the underlying magma. We conclude that at the very least, the accreting earth should exhibit extensive partial melting and prodigious volcanism. It is possible, but not proven, that the production of magma exceeds the rate at which it can freeze; a magma ocean would then develop. This reviewer is skeptical about magma oceans, at least in the context described here, because melts can migrate very rapidly from planetary interiors at even low degrees of partial melt (Turcotte, 1982; Scott and Stevenson, 1986) and it seems unlikely that the resulting flows are so undisturbed after emplacement that τ can be very long. Of course, one can always construct a model in which a magma ocean exists (e.g., Turcotte and Pflugrath, 1985), by imposing the presence of a "lid" over the ocean.

There are at least two situations in which a magma ocean is inevitable. One involves the production of a melt that is more dense than the solid residue. Since this melt migrates downward, it does not have the opportunity to cool and should accumulate as an ocean. One such "ocean" still exists—the earth's outer core. We discuss core formation in Section 5. Another possibility is that silicate melts may become very dense at high pressure (Rigden *et al.*, 1984) and accumulate in the deep mantle. The second way to make a magma ocean is by giant impact, since then the energy input is so large in a short time that the earth has no possibility of eliminating the heat promptly. The ocean might freeze before the next comparable impact and is thus "transient," but it is nonetheless deep and important to the early earth. Both of the inventories in Table 11.1 admit the possibility of giant impacts. These events have recently received considerable attention because of their possible role in the origin of the moon (Cameron and Ward, 1976; Cameron, 1985;

Hartmann *et al.*, 1986; Stevenson, 1987). Consider, for example, the impact of a Mars-sized projectile at 11 km/s. The total energy is 3×10^{38} ergs; suppose that half of this goes into the Earth as internal (mainly thermal) and gravitational energy (because the Earth "puffs up"). Figure 11.4 shows the resulting post-impact temperature profiles within the Earth for three different assumptions for the distribution of injected energy ranging from uniform to that obtained by assuming a pressure drop-off away from the impact site like r^{-2} (a likely upper bound to the rapidity of the drop-off, e.g., Orphal *et al.*, 1980). These calculations indicate formation of a deep magma ocean with a supercritical near-surface layer that merges continuously with a silicate vapor atmosphere. The cooling time of this system is at least

$$\tau_{\text{cool}} \simeq \frac{\Delta E}{4\pi R^2 \sigma T_e^4}, \tag{27}$$

where T_e is the effective temperature and is governed by the atmospheric level where the opacity begins to be low enough to allow IR photons to

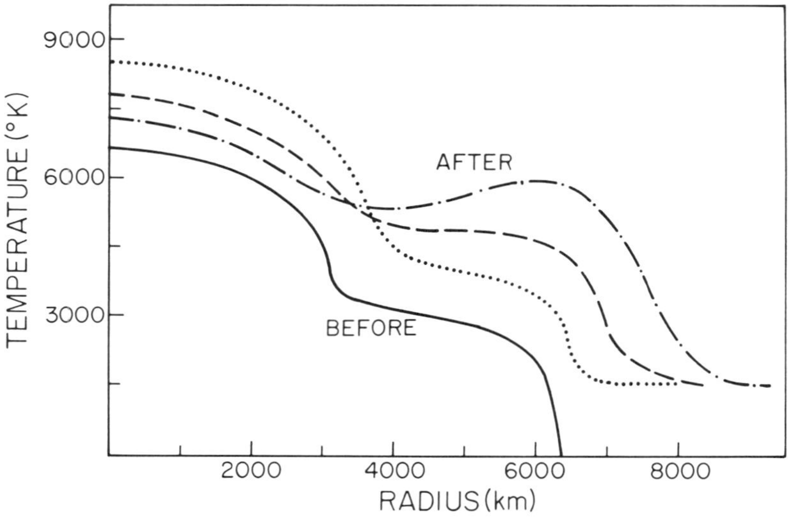

Figure 11.4. Several possible outcomes arising from injecting 1.5×10^{38} ergs into Earth because of a giant impact. The preimpact state (solid line) is assumed to be 50 percent molten because of previous smaller impacts. The three possible post-impact states vary from uniform energy deposition (dotted line) to a concentration in the mantle (dash-dot line). In each case, the planet "puffs up" and an extended transient atmosphere ($T \sim 2000$ K) develops.

escape. Since grain opacity dominates (Thompson and Stevenson, 1986), $T_e \sim 1700$ K and $\tau_{cool} \sim 10^3$–10^4 years. Notice that this is enormous compared to dynamical times (~hours): this is a fundamental distinction between small and large impacts. The ocean may have important geochemical consequences (Anderson, 1981; Hofmeister, 1983; Ohtani, 1985)—we return briefly to this later.

3.5. Impact-Induced Atmospheres

Even when planets accumulate in the absence of gas, an atmosphere may develop, derived from impact induced or thermally induced outgassing of volatiles adsorbed on grains or chemically combined in the planetesimals. By mass, the most important volatiles on the terrestrial planets are H_2O ($\geq 1 \times 10^{24}$ g on Earth) and CO_2 ($\sim 10^{23}$ g on Earth, mostly as carbonates). It is not known when and how these volatiles arrived, but they could have been delivered as hydrated silicates and calcite during primary accretion (i.e., first 10^8 years). Water of hydration is well documented in meteorites. Calcite is present (though not abundant) and is presumably the result of aqueous modification of the parent bodies of meteorites (Kerridge and Bunch, 1979; Macdougall *et al.*, 1984). Actually, the most common form of carbon in meteorites is close to elemental (i.e., reduced, with $C/H >> 1$) and this could also contribute CO_2 in the right postimpact chemical environment. It is important to realize, however, that water and carbon (in some form) can also be supplied by *comets*, scattered into the inner solar system long after the primary accretion. Thus, it is by no means clear that an atmosphere or hydrosphere existed during and immediately after the accumulation of most of Earth's mass. In fact, it is likely that the volatile-rich component of the Earth-forming planetesimal population was very small, since these bodies probably formed beyond or near the orbit of Mars. We argued earlier (based on Wetherill, 1985) that substantial radial mixing of source material is likely, but the quantification of this remains highly uncertain.

Experiments indicate that efficient release of H_2O from serpentine (Lange and Ahrens, 1982) and CO_2 from calcite (Boslough *et al.*, 1982; Lange and Ahrens, 1986) are possible at the shock pressures expected during the impacts accompanying most of Earth accretion. However, the release is not *so* efficient that it occurred on planetesimals too small to retain gravitationally the resulting gas. It is therefore possible that an atmosphere is built up on proto-Earth, induced by impacts. This has been the focus of recent work by Lange and Ahrens (1982, 1986) and by Abe and Matsui (1984; also Matsui and Abe, 1986a, 1986b). In particular, Abe and Matsui have focused on the possibility that the protoatmosphere becomes so massive that the surface of Earth reaches the melting point of silicates and a *surficial* magma ocean develops. Eventually, as accretion continued, this ocean would extend to great depth.

An ideal gas atmosphere with no condensation and heating from below cannot have a steeper temperature gradient than the adiabatic value:

$$\frac{dT}{dr} = \frac{-g}{C_p}. \tag{28}$$

If the mass per unit area of the atmosphere is σ, the thickness of the atmosphere is small compared to the radius, and the adiabatic region extends down from T_e (the photosphere) to T_s (the surface), then

$$T_s \simeq T_e(P_s/P_e)^{\gamma-1/\gamma},$$

$$P_s = \frac{\gamma\sigma kg}{(\gamma - 1)\mu C_p},$$

$$P_e \simeq g/\kappa, \tag{29}$$

where γ is the ratio of specific hearts; P_s and P_e are the respective pressures at the temperature levels T_s, T_e; μ is the mean molecular weight, and κ is the effective transmission opacity. It is assumed that $T_s >> T_e$. If 10^{24} g of water are in the atmosphere then $\sigma \cong 2 \times 10^5$ g/cm^2. For reasonable choices of $\gamma = 1.25$, $T_e = 300$ K, $P_e \sim 10^{-2}$ bars, and $C_p \simeq 5$ k/μ, one finds $T_s \simeq 2000\ K$, $P_s \simeq 200$ bars, and the depth of the atmosphere is ~500 km. This seems to support Abe and Matsui's conclusion that a magma ocean would occur, since the atmosphere is likely to be too opaque for significant radiative loss from great depth. (Actually, this is a complicated issue since heating from the Sun is non-negligible and deposited higher up.) But there is a possible difficulty with such an atmosphere: it can cool by condensation. If we assume that the atmospheric temperature and pressure are dictated by the vapor pressure curve of water, then the temperature gradient is much less steep. In fact, the critical pressure of water (217 bars) occurs at only 647 K; one can easily provide 2×10^5 g/cm^2 of water by a *saturated* atmosphere extending from $T_s \sim 600$ K up to $T_e \sim 300$ K. There could be a very hot (but liquid) ocean below this atmosphere. This state is what meteorologists call a "wet" adiabat.

The situation is somewhat complicated by for CO_2, a situation that Abe and Matsui have not considered in detail. Here, there is no possibility of saturation (the upper part of the Earth's atmosphere being always much too hot for CO_2 condensation), and a Venus-like early Earth seems a definite possibility. If the Earth was ever in this state then it is unclear how it avoided becoming like Venus. Cameron (1983b) has suggested that a giant impact could have "blown away" a primitive CO_2 atmosphere, but the dynamics of this are questionable (Thompson and Stevenson, 1988) and the total reservoirs of CO_2 on Venus and Earth are still comparable today, lessening the rationale for a "special" event in early Earth history. A Venus-like CO_2 atmosphere on Earth would drive T_s to ~800–~1000 K but not cause a magma ocean.

Major uncertainties remain, but a reasonable conclusion at present is that atmospheric blanketing does not *necessarily* lead to a magma ocean on Earth. Our previous discussion of the consequences of energy burial upon impact seems still applicable.

4. Accumulation of Earth in the Presence of Gas

The solar nebula was certainly present during condensation of solids, and probably still present at the time the first planetesimals were built (since this happened rapidly). At least part of the giant planet growth must also have occurred in the presence of gas, but probably not all since otherwise we would expect the giant planets to be closer to "cosmic" (i.e., solar) composition. Nevertheless, astrophysical evidence overwhelmingly indicates that much of the gas must be lost early, perhaps in less than 10^6 years. This is the period of T-Tauri activity (Bertout, 1984) and probable intense UV output of the newly formed Sun (Canuto *et al.*, 1982). It follows that if a theory of terrestrial planet accumulation leads to accretion times much in excess of this, then the gas-rich scenario is likely to be self-inconsistent. As we shall see, this is indeed a problem with the gas-rich theory developed mainly by Hayashi and coworkers (Hayashi, 1981; Hayashi *et al.*, 1985). Accordingly, this theory receives somewhat less attention here than the gas-free case. Even so, the presence of gas at least during the early stages (and perhaps even as far as the formation of Mars-sized bodies) seems plausible and deserves careful attention.

4.1. Aerodynamics in the Solar Nebula

The solar nebula must have had a radial pressure gradient in general, and the orbital motion of a gas element, v_g, accordingly differs from circular Keplerian motion v_K at the same distance r (cylindrical radial coordinate):

$$v_g^2 = v_K^2 + \frac{r}{\rho}\frac{dP}{dr}, \tag{30}$$

where p, ρ are the nebular pressure and density (Whipple, 1973). Since the characteristic lengthscale for changes in P is of order r, and $dP/dr < 0$, it follows that $v_K - v_g \sim c^2/v_K \sim 10^4$ cm/s typically, much smaller than the Keplerian velocity. In the presence of this gas, particles respond in one of two ways depending on their mass. If they are very small then the tangential drag force is large and they tend to move with the gas. But since their orbital motion is slower than Keplerian, there is an unbalanced radial inward force due to solar gravity and they spiral inward. If the particle is so large that the tangential drag force is insufficient to reduce significantly the orbital velocity below the Keplerian value, the particle will still tend to spiral inward because of gradual loss of angular momentum due to this drag. Clearly, the inward spiral is slow both for very small particles (which tend to follow the gas) and very large particles (which do not care about the gas) and a maximum in between. This is illustrated in Figure 11.5, similar to Weidenschilling (1977). For a reasonable nebular density of 10^{-9} g/cm^3, the inward motion can be startlingly rapid (e.g., meter-sized bodies could go 1 AU in 10^3 years) and the *differential* motion of bodies can promote coagulation, as Weidenschilling advocates.

Another feature of gas drag is that it dampens eccentric and inclined motions. In the large body (high Reynold's number) limit, the reduction in kinetic energy due to gas drag is of order the mass of gas encountered divided by the mass of the projectile. If we identify the "random" velocity component as v (in the same way as we did for our discussion of the Safronov steady-state), then

$$\left.\frac{1}{v^2}\frac{d(v^2)}{dt}\right|_{\text{drag}} \sim -\rho_g \pi r^2 v/m, \tag{31}$$

for a body of mass m and radius r. If, in the spirit of Safronov, we seek a steady state with gravitational stirring (equation 7), then

$$v \sim \left(\frac{Gm}{r}\right)^{1/2}\left[\frac{nm}{\rho_g}\ln\left(D_{\text{max}}/D_{\text{min}}\right)\right]^{1/2}, \tag{32}$$

where n is the planetesimal number density, and nm/ρ_g is simply the ratio of total solid mass to total gas mass ($\sim 10^{-3}$). For $m \sim 10^{21}$ g (and $r \sim 50$ km), $v \sim 10^3$ cm/s. Since this velocity is substantially less than the escape velocity or the Safronov value (equation 9), gravitational focusing is more important than in the gas-free case. Nakagawa (1978) finds that 10^{22} g bodies form in

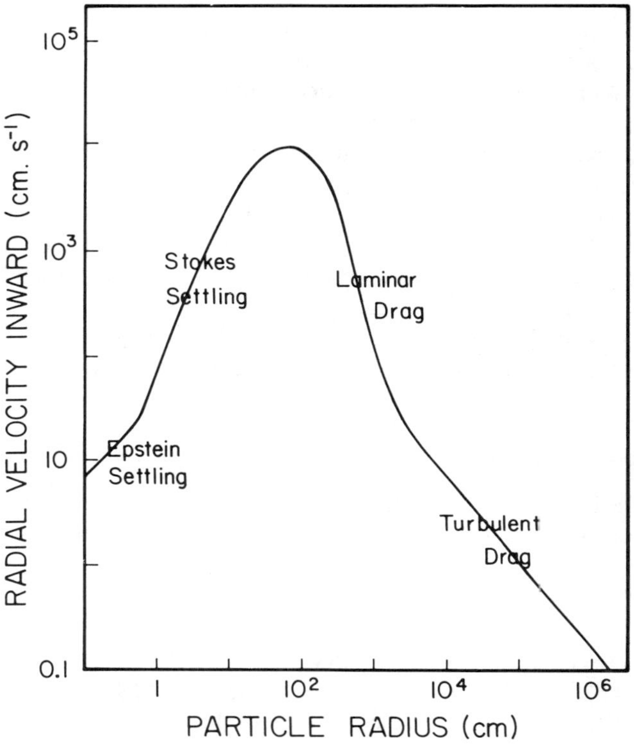

Figure 11.5. Inward radial velocity due to gas drag for a solid particle (density 3 g/cm^3) immersed in a nebula with local density 10^{-9} g/cm^3 and local pressure $p \propto r^{-3}$ (r = distance from the Sun). Four regimes are identified: Epstein settling (mean free path of molecules > grain size), Stokes settling (radial viscous drag), laminar (tangential viscous drag), and turbulent (high Reynolds number tangential drag).

$\sim 10^4$ years. The random velocity of 10^3 cm/s corresponds to eccentricities $\sim v/v_K \sim 10^{-3}$.

4.2. Late Stage Accretion

Once bodies achieve a size for which their escape velocity exceeds the nebular sound speed, a bound atmosphere forms. This occurs at $m \sim 10^{24}$ or 10^{25} g. Hayashi and coworkers propose that if the process analyzed by

Nakagawa (described above) leads to a few embryos of around 10^{25} g, immersed in a swarm of smaller bodies, then the large embryos can grow because of smaller bodies being captured by gas drag within the gravitational sphere of influence of the large body (the Hill sphere). Even with favorable assumptions, this implies a growth time for Earth of $\gtrsim 10^7$ years. This timescale is for the slow inward spiral of 10^{21} g bodies through ~0.1 AU (predicted by equation (30) and Fig. 11.5), since most of the Earth-accreting bodies would have initial orbits substantially more distant than Earth orbit and necessarily rather small eccentricities ($\sim 10^{-3}$). It is important to stress that Earth formation by this mechanism takes much longer than the likely T-Tauri phase of the Sun *even with very favorable assumptions*. It is not clear that even the timescale of 10^7 years is realistic, as Wetherill (1980) discusses.

4.3. Dense Primordial Atmospheres

Within the Hill sphere of a protoplanet immersed in the nebula of gas density ρ_g, the outer layers of gas are isothermal, and the gas density is roughly

$$\rho(r) = \rho_g \exp\left[\frac{Gm}{c^2}\left(\frac{1}{r} - \frac{1}{r_H}\right)\right], \tag{33}$$

where $r_H \sim a(M/M_\odot)^{1/3}$ is the Hill sphere radius, and a is the orbital radius of the planet of mass M. At some level, $r = r_{ph}$ (called the photosphere), the gas is too dense to be transparent and the atmosphere becomes adiabatic. For molecular hydrogen, this occurs at $\rho \sim 10^{-4}$ g/cm^3. For an ideal gas and adiabatic conditions, the temperature rise ΔT between this level and the surface of the solid planet is given by integrating equation (28) with $g = GM/r^2$:

$$\Delta T = GM/RC_p, \tag{34}$$

where R is the physical surface, assuming $R << r_{ph}$. For $C_p \sim 2 \times 10^8$ erg/g/K (for molecular H_2) and Earth radius and mass, $r_{ph} \simeq 2.5 \times 10^9$ cm (comfortably larger than $R = 6.4 \times 10^8$ cm) and $\Delta T = 3000$ K, easily large enough to guarantee a terrestrial magma ocean. This atmosphere would be very massive (Hayashi *et al.*, 1979) and have a profound influence on the evolution, independent of accretion time. Sekiya *et al.* (1981) have analyzed the escape of gas due to the high UV flux of the early Sun and find that the Earth can be stripped of this massive envelope. The fundamental disparity between planetary accumulation time and nebular dissipation time remains, however, and it seems likely that the gaseous envelope is never as massive or hot as equation (34) suggests.

5. The Hadean Earth

The "Hadean" is defined rather arbitrarily as the period of Earth history prior to 3.8×10^9 yr before present. This is the period of "no rock record," although recent discoveries of zircons which date back to 4.2 or even 4.3 byr bp (Froude *et al.*, 1983; Compston and Pidgeon, 1986) indicate that there is some question whether this criterion has any meaning. We return later to this issue; a more useful definition is in terms of process: we wish to consider here that period of Earth's history when the Earth is recovering from the accretion trauma, when impacts may still be important, when radioactive heating is unimportant or only beginning to become important, when primary differentiation (including core formation) takes place, when heat flow and volcanism are prodigious by modern standards (hence the "Hadean" descriptor) and when survivable assemblages of rocks (continents?) are fragmentary or not possible. We discuss in order the processes of core formation, Earth cooling at the liquidus (a magma ocean) through to sub-solidus, the role of continuing impacts, the rapidity of continued outgassing and differentiation, and the stabilization of continents.

5.1. Core Formation

We argued in Section 2 that direct formation of an iron core from condensation in the nebula is implausible, and our discussion in Section 3 indicates further that even though individual planetesimals might differentiate into core and mantle (as some meteorite parent bodies evidently did; Wasson, 1985), the disruptive nature of the accretion process should mix metallic iron with silicate. It is therefore reasonable to suppose that the proto-Earth was undifferentiated, at least during the earliest stages of accumulation. However, we have also seen that energy burial would have heated the subsurface to near the melting point of silicates (coincidentally similar to the melting point of pure iron) when the proto-Earth is perhaps only ~3000 km in radius. At this stage, the Earth would be hot near the outside and cold near the center, because the energy of accretion scales roughly as the square of the radius of the accreting body (equation 21). In these circumstances, how do we make the Earth's core?

Near surface iron, melted by impact, could presumably percolate downward to form a spheroidal layer bounded below by cold, undifferentiated matter. Elsasser (1963) proposed that since this layer is more dense than the underlying material, a Rayleigh-Taylor instability would ensue (cf. Ramberg, 1968, 1981) in which large "blobs" of iron (perhaps hundreds of radius) would grow and migrate downwards by Stokes flow. Rayleigh-Taylor instabilities undoubtedly exist but because the underlying medium has a viscosity that rapidly increases with depth, the preferred "blob" size is only a few kilometers (Stevenson, 1981) and the migration velocity is very slow. The small blob size also prevents a gravitational runaway (as proposed by Turcotte and Emerman, 1983) in which the heat generated by the gravitational energy release of the downgoing blob lubricates its passage. The core can eventually grow by this process—but slowly, over a time of several hundred million years.

Another core formation model, proposed by Vityazev and Mayeva (1976) involves the downward migration of the iron layer, as it eats away at the material below it, assimilating iron in its path and releasing silicate grains that rise up. The problem with this model is that the downward transport of heat needed to lubricate the path is by the slow process of thermal diffusion, and this means that core formation by this method takes $\sim 10^9$ yr.

Of all possible processes, the one that should win is the one that takes least time. One process that is exceptionally rapid is illustrated in Figure 11.6 (Stevenson, 1981), which could be called the spontaneous asymmetry mode. The cold, primordial core can displace sideways *rigidly* in the presence of an iron layer. When it does so, the Earth becomes pear-shaped and large nonhydrostatic stresses develop, causing fracturing of the primordial core (Ste-

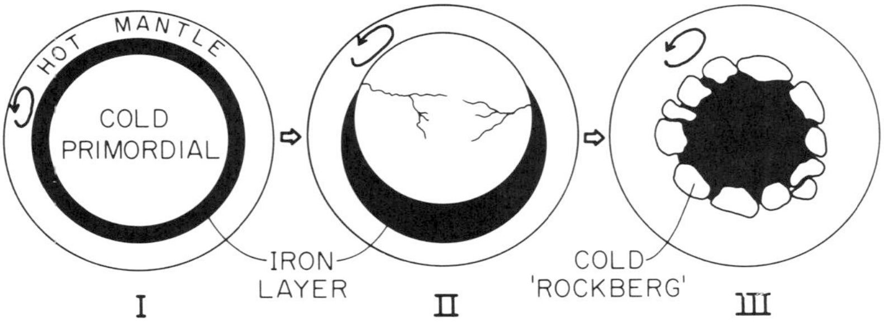

Figure 11.6. Hypothetical scenario for the early stages of core formation. At this stage, the protoearth may be much smaller than the final earth. a, A cold, undifferentiated primordial core is overlain by the denser iron that has accumulated from the partially molten mantle above. b, A spontaneous asymmetry (Stevenson, 1981) develops. Large nonhydrostatic stresses on the primordial core lead to deformation and fracturing. c, The debris from this is distributed in the form of "rockbergs" around the newly formed core. Thermal equilibrium of these rockbergs with the surroundings is by thermal diffusion and takes longer than earth accretion.

venson, 1981; Davies, 1982) and overturn. In principle, this process could be spectacularly rapid (timescale ~ days); but in practice, the timescale is determined by the development of an iron layer (perhaps $\sim 10^6$ yrs). This mechanism may not be as "clean" as Figure 11.6 suggests—for example, iron migrating down cracks may undergo partial freezing because of cooling and increasing pressure. Nevertheless, it seems likely that *core formation was contemporaneous with accretion*. This is consistent with (but not required by) geochemical evidence (e.g., Vidal and Dosso, 1978).

In a sense, the details of the story cartooned in Figure 11.6 are irrelevant since they refer to a very early stage and not to the core growth occurring during most (~90 percent) of Earth accretion. Since the energy of core formation is large, equivalent to a temperature rise of several thousand degrees for the *whole* Earth (Flasar and Birch, 1973; Shaw, 1978) and even substantial for a Mars-sized body (Shaw, 1979), the process is well-lubricated once the initial hurdle of a cold core is cleared. A steady-state can develop (Figure 11.7) in which subsequent arrivals of material are easily accommodated by downgoing blobs whose descent time is very small ($\lesssim 10^6$ yr) compared to the accretion time. These blobs do not equilibrate chemically with the mantle through which they pass, and might therefore preserve a history of low pressure iron-silicate fractionation (Brett, 1984). Most of the mantle is metallic iron-poor in these circumstances; this may be relevant to the outgassing products, as discussed below. Figure 11.8 shows a schematic sequence of temperature profiles for the proto-Earth; before, during, and after core formation (Stevenson, 1983).

As Earth grows, lower mantle pressures may reach a point where an oxygen-bearing metallic iron fluid can form. This suggestion of using oxygen for explaining the low density of the outer core relative to pure iron is an old one, revived by Ringwood (1977) and now receiving substantial experimental support (McCammon *et al.*, 1983a,b; Knittle and Jeanloz, 1986). It is not clear how efficient this process can be, given the expected chemical disequilibrium between downgoing blobs of iron and the deep mantle (Stevenson, 1981). However, the pressure-dependence of partitioning, if allowed to operate, may have important implications for mantle compositional gradients.

5.2. Cooling of a Magma Ocean

We have seen that a steady magma ocean may not develop during Earth accretion, because of the prodigious heat loss possible from a partially

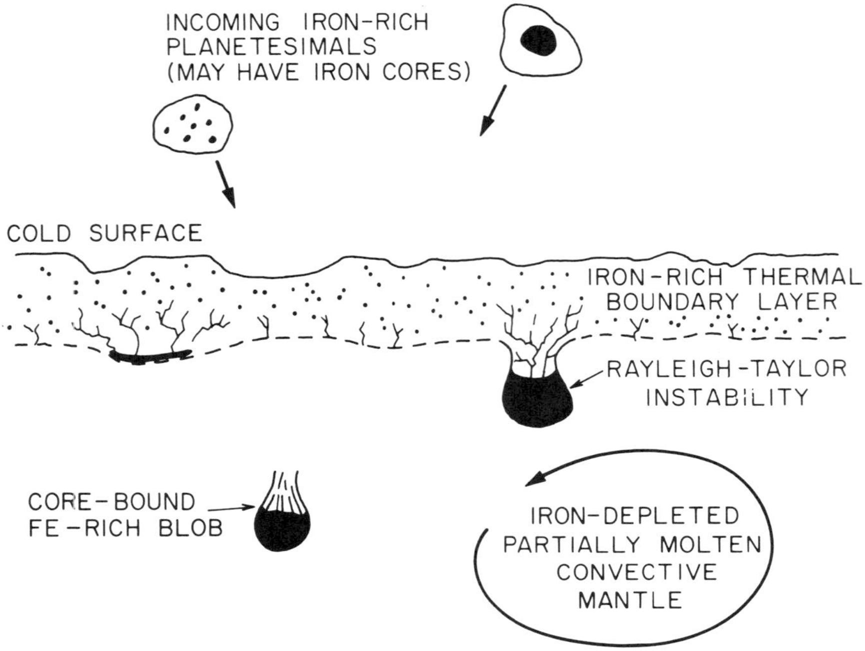

Figure 11.7. Iron migration at a later (steady-state) phase of core formation. Below the thermal boundary, iron percolates into diapirs, which migrate down rapidly, ensuring incomplete chemical equilibrium with the mantle environment.

molten system, and the unlikelihood of an atmosphere so opaque that the surface temperature is pushed up to ~1600 K. However, there is sufficient uncertainty in the parameterization of heat flow that a magma ocean can not be excluded, especially if there were some way to "cap" it with an insulating lid. Moreover, a transient magma ocean seems almost certain following a very large (Mars-sized) impact (cf. Fig. 11.4 and Stevenson, 1987). For these reasons, we examine further the *thermal* aspects of a magma ocean. No attempt is made here to examine the *compositional* evolution. We do not deny its importance (cf. Anderson, 1981; Hofmeister, 1983; Ohtani, 1985) but this is a problem that, in the view of this reviewer, requires much more work before firm conclusions are possible.

The putative lunar magma ocean has received much attention over the last fifteen years (see review by Warren, 1985), motivated mainly by the observed buoyant crust (the anorthositic highlands), explainable as fractional crystal-

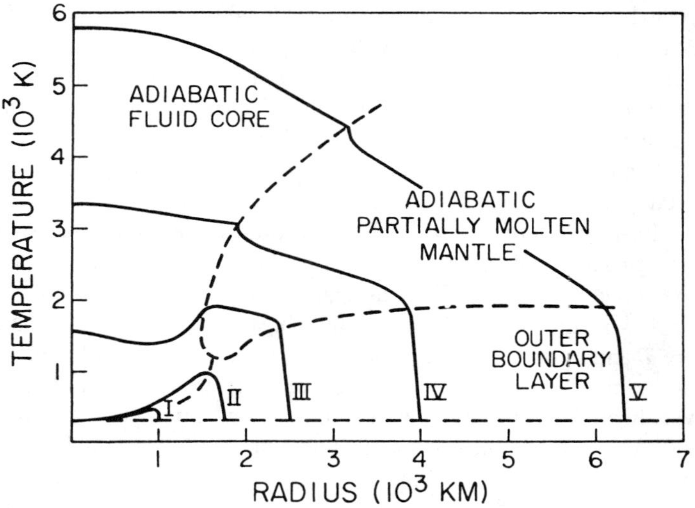

Figure 11.8. Schematic representation of temperature profiles in an accreting Earth. The chronological sequence is from I (outer radius $\simeq$ 1000 km) to V (outer radius $\simeq$ present Earth radius); and the total timescale is $\sim 10^8$ years. In I and II, temperatures are too low for differentiation; but by III, partial separation of core and mantle is achieved. The dashed lines delineate the core, mantle and outer boundary layer. Temperature scale is approximate and excludes the transient traumas of giant impacts.

lization from the ocean. It is generally agreed (but poorly documented) that a similar situation would not have arisen on Earth. The first thing to understand about a magma ocean is that, like any low viscosity system, it has the capability of losing an extraordinary amount of heat. In the high Reynold's number limit ($vl/\nu << 10^3$; v = convective velocity, l = convective length-scale, ν = kinematic viscosity), mixing length theory (Clayton, 1968; Stevenson, 1979) predicts

$$F \simeq 0.1 C_p \delta\rho v / \alpha,$$

$$v \simeq (g l \delta\rho/\rho)^{1/2}, \tag{35}$$

where F is the convective heat flux, $\delta\rho$ is the density contrast (due to thermal expansion) driving the flow, α is the coefficient of thermal expansion, and g is the gravitational acceleration. For reasonable parameter choices,

$$F \simeq 10^{15}(\delta\rho/\rho)^{3/2} \text{ erg/cm}^2\text{-s}, \tag{36}$$

and so the black body radiation from a 1500 K body (corresponding to $F \sim 10^8$ erg/cm^2/s) could be supplied with $\delta\rho/\rho \sim 10^{-5}$, an astonishingly small buoyancy! (It corresponds to temperature fluctuations of only ~0.1 K.) This is in startling contrast to mantle convection, where one is accustomed to large temperature variations even at very low heat flows. The difference arises, of course, from the reduction in viscosity by almost twenty orders of magnitude, relative to present day mantle viscosities.

This illustrative calculation is almost certainly misleading because exposed magma at the surface immediately forms a chill crust and a relatively stagnant, conductive zone. We can analyze this situation as follows: The interior flow is almost isothermal (as the above calculation shows) at temperature T and bounded above by a boundary layer in which the viscosity increases rapidly upwards, because of decreasing temperature and the onset of crystallization. This layer has thickness δ and a temperature drop ΔT across it. Above this again, is a stagnant lid with steady-state thickness D. In the spirit of boundary layer theory (Howard, 1964; see also Lunine and Stevenson, 1985) the correct choice of δ and ΔT should be that which maximizes the local Rayleigh number and yields the critical value of this number for convection (~10^3). This embodies the nonrigorous but physically plausible notion that the convection chooses to be as efficient as possible. It implies

$$F = k\Delta T/\delta,$$

$$\frac{d\mathrm{Ra}(\delta)}{d\delta} = 0 \qquad \text{at } \delta = \delta_c,$$

$$\mathrm{Ra}(\delta) = g\alpha\Delta T\delta^3/\kappa\nu(T - \Delta T/2),$$

$$\mathrm{Ra}(\delta_c) = 10^3, \tag{37}$$

a unique prescription for the preferred δ, ΔT and F. A calculation along these lines was performed for magma chambers on Io (Lunine and Stevenson, 1985) to find the maximum heat extraction possible. Observational data support the idea that Io manages to approach the maximum at localized hotspots; it is possible that Io is a good analog for the primordial Earth. Assuming $\nu \simeq 10^{-6} \exp(3 \times 10^4/T)$ cm^2/s, typical of a tholeiitic basalt for $1500 \lesssim T \lesssim 1700$ (Murase and McBirney, 1973), one finds $F \sim 2 \times 10^6$ erg/cm^2/s and $\delta \sim 50$ cm if $T \sim 1600$ K. An ocean of depth 100 km has of order 10^{17} erg/cm^2 stored in it and would accordingly take ~10^3 years to freeze. This is a highly idealized calculation, but it is not likely to be wrong by many orders of magnitude unless the crust thickens by the addition of buoyant (nonrecyclable) material.

If a crust were to progressively thicken, then the remaining ocean is insulated and the cooling problem is very similar to the classic Stefan problem (Stefan, 1891; Turcotte and Schubert, 1982, p. 168). There can be no doubt that this applies to some problems that seem superficially analogous to the magma ocean problem (e.g., cooling lava lakes at Kilauea). However, impact disruption and large-scale disruption seem likely to invalidate the analogy. If we proceed by ignoring these concerns then, very roughly,

$$F \sim k\Delta T/D \sim \rho(L + C_p\Delta T)\frac{dD}{dt}, \tag{38}$$

where ΔT is now the total temperature drop across the thickening, conducting lid of thickness D, and L is the latent heat of fusion. Not surprisingly,

$$D \sim (\kappa t)^{1/2}, \tag{39}$$

and it takes $\sim 10^8$ yr to freeze a 100 km thick layer. Unless some sort of large scale subduction is invoked, magma oceans can take an extremely long time to freeze if they have a lid (Turcotte and Pflugrath, 1985). But it is difficult to take this model seriously.

5.3. Cooling Through and Below the Solidus

Convection in partially molten media is an unsolved and dynamically difficult problem. We ignore these difficulties here and adopt a parameterized, subsolidus approach to the description of convection, with the hope that it might also encompass mildly supersolidus conditions by choice of a suitably low viscosity. Our parameterization of convection flux F_{conv} has already been introduced (equation 23). The differential equation we wish to understand and solve is

$$4\pi R^2 F_{\text{conv}} = -\frac{dE_{\text{th}}}{dt} + \frac{dE_{\text{acc}}}{dt} + Q_{\text{rad}}, \tag{40}$$

where E_{th} is the internal thermal heat content of Earth, $dE_{\text{acc}}/\text{dt}$ is the addition of heat due to continuing accretion and Q_{rad} is the radiogenic heat production. The presence of three terms on the RHS suggests three regimes which, in chronological order, would be:

1. Accretional input dominates and is balanced by heat output; secular heating or cooling and radiogenic production unimportant.
2. Accretional energy tail-off; secular cooling important for heat output; radiogenic production still unimportant.

3. Negligible impact heating; some secular cooling but planet "buffered" by radiogenic heating.

We can understand these regimes by appropriate parameterization of the various contributions. Consider, first, the impact heating. We have both planetary accumulation timescale estimates (primarily theoretical) and cratering chronologies based on lunar data (Wilhelms, 1984). If we take the latter and scale upward to Earth, allowing for a small gravitational focusing correction, then expressed as a *flux*

$$F_{\text{impact}} \simeq (10^7 \text{ erg/cm}^2\text{/s}) \exp\left(\frac{-t}{3 \times 10^7 \text{ yr}}\right), \tag{41}$$

where we have (somewhat arbitrarily) chosen our origin of time to be at 4.5 byr bp. This formula should *not* be used at small t; it is intended for application at times of two to several hundred million years after formation (i.e., accumulation of most of the mass). In fact, it is a gross overestimate at early times. Recall also, from Section 3, that perhaps only ~10 percent of F_{impact} is subject to deep-seated burial and deserves inclusion in the energy balance of equation (40).

The secular cooling term, $-dE_{\text{th}}/dt$, can be expressed in a convenient non-dimensionalized form by defining a characteristic Kelvin timescale τ_K:

$$\tau_K = \frac{(\text{Heat content of Earth at solidus})}{(\text{Heat flux of earth at } \nu = 10^{16} \text{ cm}^2\text{/s})}$$
$$\simeq 2 \times 10^8 \text{ yr.}$$

If we define $x \equiv t/\tau_K$ and $\theta \equiv T/T_m$, where T_m is the melting point, then equation (40) finally becomes

$$\frac{d\theta}{dx} = -\theta^{4/3} \exp\left[-10\left(\frac{1}{\theta} - 1\right)\right]$$
$$+ 400 \exp(-20x/3)$$
$$+ 0.25 \exp(-x/5). \tag{42}$$

Of course, the coefficients are very approximate, but sufficient for our purpose. The radiogenic term is dominated by ^{235}U and ^{40}K at this stage, and corresponds to a heat flow ~ ten times greater than the present geothermal heat flow. Figure 11.9 shows the general behavior of the solution to equation (42), expressed in terms of heat flow ratioed to its modern value. The curve is smooth but examination confirms that the three regimes appear: At $t \lesssim 0.2 \times 10^9$ yr, the balance is between convective loss and impact energy. For $0.2 \times$

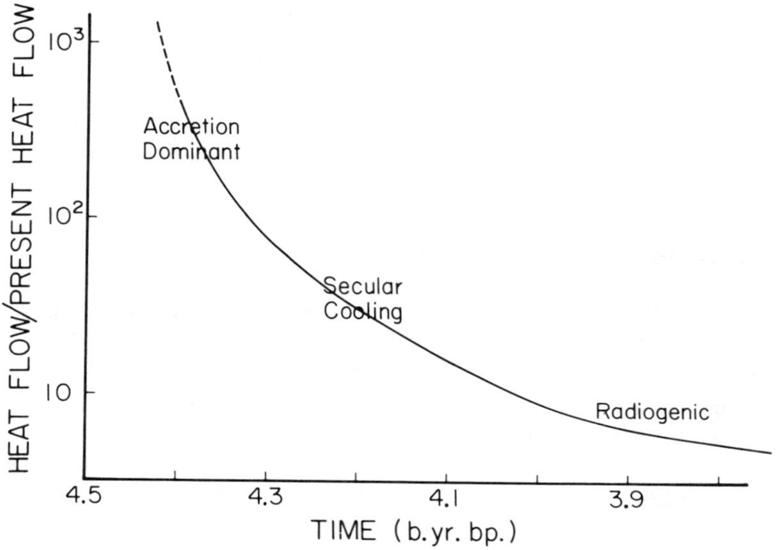

Figure 11.9. Transient cooling of a hot primitive Earth, illustrating the text discussion of the three early regimes: accretional heating, secular cooling, and radiogenic "steady state."

$10^9 \lesssim t \lesssim 0.45 \times 10^9$ yr, the balance is between secular cooling and convective loss. For $t \gtrsim 0.45 \times 10^9$ yr, radiogenic input buffers the cooling. Notice that these energetic arguments say nothing about (and are independent of) the *style* of tectonics and the nature of the convective cells early on, except that it assumes all of the mantle and core heat can be tapped. If the mantle is layered, then the numerical constants in (42) change and so do the timescales for various stages in Figure 11.9, but the basic principles are unaltered. Perhaps the most important result is that the so-called terminal bombardment at 3.8 byr bp inferred from lunar cratering (Wilhelms, 1984) seems to have nothing to do with the so-called "onset" of the rock record on Earth. By 3.8 byr bp the energy input or the crustal recycling due to impacts is pathetic compared to the work being done by the Earth's internal heat engine. Large impacts would have occurred later (e.g., the end of the Cretaceous, Alvarez *et al.*, 1980) but are unimportant for Earth's thermal history.

5.4. Outgassing; Catastrophic vs. Gradual

The history of the Earth's volatiles is a complex subject because one needs to know the details of chemistry and partitioning as well as the large scale physical processes (Fanale, 1971, 1986). This is especially true of H_2O and CO_2, which participate in crustal production and profoundly affect mantle melting relationships (Wyllie, 1978). Moreover, water may be volumetrically extracted by the core forming fluid (Fukai, 1984). We discuss some simple chemical aspects in the next section; here, we focus on "passive" outgassing (i.e. noble gases).

Suppose the mantle is one layer and the outgassing scales as the *mass* flux of material from the interior to the surface boundary layer, caused by convection. Since heat flux is equal to mass flux times heat capacity times temperature difference and since the heat content per unit mass of convected material is not much different now than it was in the early Earth, it follows that the heat flux F is a rough guide to the mass flux. If there were no stirring (i.e. pristine, primitive material is always available for outgassing) then one could define the fraction of outgassing, $f(t)$, relative to present day ($t = t_0, f = 1$) as:

$$f(t) \simeq \int_0^t \mathrm{F}(t)dt \Big/ \int_0^{t_0} \mathrm{F}(t)dt. \tag{42}$$

Inspection of Figure 11.9 indicates that $f(t)$ can already be substantial even at short times, since $F \gtrsim 10F_{\mathrm{now}}$ for $t \lesssim 0.1\ t_0$. In other words, *most of the mantle outgassing may have occurred "catastrophically" as a transient response to accretional heating* (although it may have occurred after nearly all the accretion). Specifically, the model given by equations (42) and (43), augmented by longer-lived heat sources, predicts that $f \cong 0.4$ at 0.25 b.yr, 0.7 at 0.8 b.yr, and 0.9 at 2.0 b.yr. These numbers should not be taken too seriously, but they are crudely consistent with current interpretation of noble gas outgassing, especially the $^{129}Xe/^{130}Xe$ ratio (Thomsen, 1980; Ozima *et al.*, 1986). As before, one could construct more complicated models for a layered mantle, and there is some suggestion that the data require this (Ozima *et al.*, 1986).

5.5. Composition of the Early Atmosphere

We have already discussed an impact-induced atmosphere. The more "traditional" view of the early atmosphere is to be found in the work of Holland (1984; see also Fanale, 1971, 1986; Chang *et al.*, 1983; Cogley and Henderson-Sellers, 1984; Holland *et al.*, 1986) in which outgassing is assumed to determine the composition through three factors:

1. The oxygen fugacity of the mantle source region (e.g., like the present mantle or a metallic iron buffer).
2. The composition of the volatile source, mainly the *H/C* ratio.
3. The release or quench temperature (the temperature below which there is chemical disequilibrium because the reactions are too sluggish).

Based on our earlier discussion, core formation is very rapid so the oxygen buffer may not differ greatly from the present mantle. The *H/C* ratio is uncertain but is plausibly ~25. (If comets contribute, it may be as low as ~10; see the special Halley issue of *Nature*, 15 May 1986). The release temperature could be quite high, perhaps ~1500 K, typical of magmatic sources, but low temperature catalysis can not be excluded. With these assumptions, the conclusion is that the early atmosphere would have been CO_2 rich, containing also N_2 and small amounts (0.1–1 percent) of CH_4. Possible numbers (Chang *et al.*, 1983) are ~35 bars CO_2, ~4 bars CO, 1.5 bars N_2, 0.1 bars CH_4, trace of NH_3, and, of course, a massive hydrosphere (equivalent to a 500 bar atmosphere of H_2O). The small amounts of reduced species may be important for the origin of life. The greenhouse provided by a massive CO_2 atmosphere may have provided a warm surface, offsetting the early low luminosity of the Sun, predicted by stellar evolution (e.g., Newkirk, 1980). However, the CO_2 atmospheric content would have been buffered by the CO_2 cycle, as on the present Earth (Walker, 1977). Moreover, the atmosphere would be subject to impact-induced chemical evolution (Fegley *et al.*, 1986), including production of substantial amounts of HCN.

5.6. Stabilization of Continents

We have argued above that much of the period prior to ~3.8 b.yr.bp was *not* enormously different from later periods. The heat flow was higher but could be accommodated by primarily sub-solidus convection. The mantle temperature was higher than now by a few hundred degrees at most; allowing greater partial melting and the possibility of higher temperature melts (i.e., komatiites; see Nisbett, 1984, 1985). But you have to go back to 4.2 or 4.3 byr bp to encounter a period that was severely different from now (e.g., impact disruption, very extensive melting). Why, then, did it take so long for substantial survivable continental crustal assemblages (cratons) to develop?

There are two very different viewpoints on this. The most popular one seeks to identify one or more aspects of the early environment that are inimical to survival of continental crust (e.g., Hargraves, 1978; Lambert, 1980; Condie, 1980; see review by Ernst, 1983). No exhaustive review of the possibilities is attempted here, but one point should be made clear: contrary to the fond wishes of geochemists, there is *no* compelling evidence that isotopic data (e.g., dePaolo, 1986) places strong constraints on the earliest crustal growth. Thus, "continental growth" curves from this work or from the geological evidence (summarized by Veizer, 1983) do *not* require that continental crust did not exist in substantial amounts early on, they merely require that crust did not survive early on. It is entirely possible to envisage a model in which *enormous* quantities of crust were produced early on and then very efficiently reassimilated.

This brings us to the second, less popular view of continental growth, which simply says that it took a long time for the first survivable craton to "nucleate" and that prior to this, crust suitable for continents existed but was recycled (e.g., subducted). It is possible to construct models of this process (Stevenson, 1984b; Fahnestock and Stevenson, in preparation), borrowing from the language and mathematics of population dynamics. One imagines a population of crustal elements, each being a n-mer (where n is an integer, and a monomer would be a "quantum" of crust; e.g., one typical island arc), with assigned probabilities for subduction and collision with other elements. If a critical size, $N >> 1$, is needed for survival ($N \sim 10$ seems plausible) then one finds that the time to produce the first surviving "craton" is very large compared to the time to make a single island arc, perhaps by a factor of 10^2–10^3. Of course, this picture assumes island arcs are suitable building blocks for continents (e.g., McLennan and Taylor, 1983). The conclusion of the model is that continental growth may be long delayed and the 3.8 byr bp "horizon" may be coincidental, illusory, or irrelevant, probably all three!

Subsequent continental growth has both "secular" (gradual) and "explosive" aspects (Windley, 1984; Schubert and Reymer, 1985; Reymer and Schubert, 1986; dePaolo, 1986) but it seems unlikely that this has a fundamental impact on Earth thermal history. Current evidence points toward thick continental crust, even for the earliest surviving crust (Bickle, 1978; England, 1979; Boyd *et al.*, 1985; Miller and O'Nions, 1985), which implies good insulation. This simply means that the heat flow from the deep interior was channeled out primarily through the ocean basins (e.g., Richter, 1985). Even on the present Earth, heat flow through the continents is small once one takes account of radiogenic heat sources. Henceforth, we ignore the role of continents in the broad view of Earth's thermal history.

6. Later Thermal Evolution of Earth

Previous sections have laid the framework for quantitative assessment of the Earth's thermal history from 3.8 byr bp to now, although some aspects of the convective modeling were introduced in a cavalier way and now require more careful documentation. In this section, we restrict ourselves to "parameterized" convection recipes (e.g., equation 23), and avoid some of the more profound questions of how convection really operates in the mantle; these are dealt with elsewhere in the book. The simpler questions are these:

1. How did the terrestrial heat flux and convective vigor vary through geologic time?
2. How much of the heat flow is attributable to instantaneous (radiogenic) heat production and how much is derived from gradual whole Earth cooling?
3. To what extent do the answers to 1 and 2 depend on the number and nature of convection layers within the Earth?

The main principle guiding most work in this area is that of *self-regulation* first enunciated clearly by Tozer (1965) and then ignored by most for over a decade. Aside from a pioneering paper by McKenzie and Weiss (1975), most activity began in the late 1970s, generated by several groups independently (Sharpe and Peltier, 1978, 1979; Schubert *et al.*, 1980; Davies, 1980) and motivated mainly to address the significance of secular cooling (question 2 above). Later efforts have tended to question the parameterizations used in these thermal histories (McKenzie and Richter, 1981; Christensen, 1985; see review by Jeanloz and Morris, 1986). Some of the controversies remain unresolved, but we will at least explain the issues briefly below.

6.1. Self-regulation and Simple One Layer Models

As envisaged by Tozer, a planet self-regulates by adjusting its mantle viscosity to the value needed to allow convection to eliminate internally generated heat at the rate it is generated. The idea seems attractive because of the very strong dependence of viscosity on temperature. It also implies a very strong buffer: the mantle cools down to a self-regulated steady-state if initially too hot, and heats up if initially too cold. As Figure 11.9 illustrated, cool down from a state hotter than the steady-state can be rapid ($\sim 2 \times 10^8$ years). However, the radiogenic heat production is not constant but decays gradually. This implies that the self-regulated state gradually cools, so that *stored* heat as well as instantaneous heat production contributes to the surface heat flow. Notice that this is simply a consequence of the First Law of Thermodynamics and has nothing to do with the timescale of convective overturn, here assumed to be very small compared to decay times of internal heat sources, or the Kelvin time of the Earth. (The Kelvin time is defined here as total terrestrial heat content divided by the *current* heat loss and is about 1×10^{10} years).

Neglecting the core, and assuming whole mantle convection, the energy balance is given by (cf. equation 40)

$$M_m C_p \frac{\partial T}{\partial t} = M_m Q(t) - 4\pi R^2 F_{\text{conv}}(T), \tag{44}$$

where M_m is the mass of the mantle, assigned a mass averaged temperature T, Q is the heat production per unit mass, and F_{conv} is the convective heat flow. Most workers assume

$$F_{\text{conv}} \cong F_{\text{cond}} \left(\frac{\text{Ra}}{\text{Ra}_{cr}}\right)^{\beta}, \tag{45}$$

where F_{cond} is the heat flux that would be transported by pure conduction for the same temperature drop and layer thickness, Ra is the appropriate Rayleigh number, and Ra_{cr} is the critical value for convective onset ($\sim 10^3$). The exponent $\beta \sim 1/3$ for simple boundary layer analysis, but might be somewhat smaller (see Jeanloz and Morris, 1986). The crucial point concerns how to compute the Rayleigh number. Since mantle convection in Earth penetrates to the surface yet the flow is limited by the viscosity at great depth, it seems reasonable to choose

$$\text{Ra} \equiv g\alpha T d^3/\nu(T)\kappa, \tag{46}$$

where d is the depth of the layer. This leads to the equivalent of equation (23). It follows that $(d \ln F_{\text{conv}})/(d \ln T) \sim 4/3 - 1/3(d \ln \nu/d \ln T) \sim 10$ because $(d \ln \nu)/(d \ln \text{T}) \sim -30$. Since Q has decreased by about a factor of three since the beginning of the Archean (3.8 byr bp), the strict steady-state assumption $M_m Q = 4\pi R^2 F_{\text{conv}}$ would predict a fractional change in temperature of around $0.1 \ln 3 \sim 0.1$, or a couple of hundred degrees. Since the stored heat of the Earth is equivalent to the total geothermal heat flow times 10^{10} years, we see that secular cooling (the LHS of equation 44) could supply at least 10 percent and perhaps as much as 30 percent of the present day heat flow.

Detailed models tend to support this claim (Sharpe and Peltier, 1978, 1979; Schubert *et al.*, 1980; Davies, 1980), but there are some disagreements. Richter and McKenzie (1981) have questioned whether there is any currently acceptable parameterization for convective transport in a medium with a highly temperature dependent viscosity and have suggested that the correct parameterization is closer to that of a constant viscosity medium. This reviewer has been unable to make sense of their arguments. In a more recent and very interesting set of calculations, Christensen (1985) has utilized numerical results for convection in a fluid with a very strongly temperature-

-dependent viscosity, which indicate a very weak dependence of heat flow on Rayleigh number (i.e., very small effective β). Accordingly, he finds the need for greater secular cooling, perhaps 30–50 percent of the geothermal heat flow. Since the parameterization he adopts does not describe the extreme change in viscosity from Earth surface to interior (~20 orders of magnitude) or the existence of subduction, his alternative approach is questionable.

Figure 11.10a and b gives an example of a plausible heat flow and temperature history of Earth, based on the simple models of Schubert *et al.* (1980).

6.2. Multilayered Models

If the mantle is layered then equation (44) is replaced by a set of equations, one for each layer and one or two for the core. Each mantle layer has an associated Rayleigh number, but the low viscosity outer core can be treated as an adiabatic fluid with no significant boundary layers. The general feature of layered models (Spohn and Schubert, 1982; Stevenson et al., 1983) is that deeper layers are less able to lose heat than shallow layers; this delays the loss of heat by secular cooling and can cause either a decrease (early on) or increase (later on) in the contribution of secular cooling to the total heat flow. The *consequences for geochemistry are probably more important than for thermal history*. In particular, the number of layers affects interpretation of isotopic evolution of mantle and crust (e.g., Davies, 1981; DePaolo, 1986).

7. Concluding Comments

We end with an attempt to answer the ten questions posed in the introduction. The reader should be mindful, however, of the admonitions given then, and appropriately skeptical of the "answers" offered below:

1. Since all current quantitative models of Earth formation predict that $>10^7$ years must elapse to accumulate $\gtrsim 90$ percent of the Earth's mass, and since the solar nebula is expected to dissipate earlier than this, the later stages of Earth accretion were probably gas-free (the Safronov scenario).
2. Accumulation was hierarchical; most of the mass was in large bodies. It seems likely that the Earth suffered giant impacts; including bodies as large as Mars.
3. The Earth was mostly formed in $\sim 3 \times 10^7$ yr, but large impacts may have occurred as late as 1×10^8 yr after the initial gravitational collapse responsible for solar system formation.

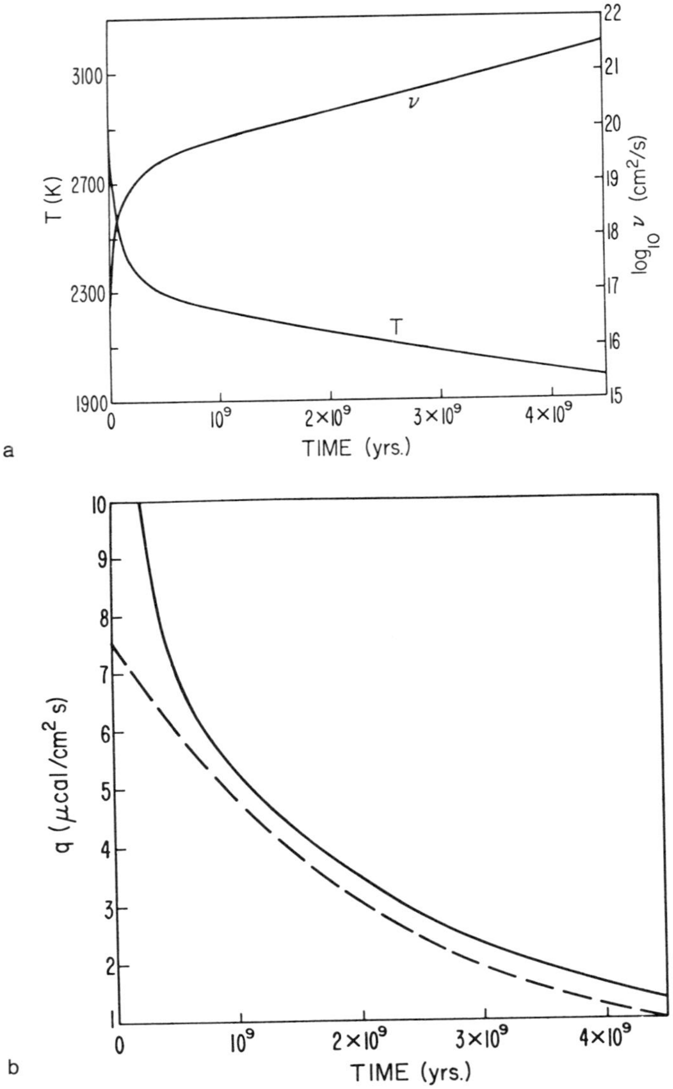

Figure 11.10. Later thermal history of Earth for a simple model (Schubert *et al.*, 1980). In a, the variation of viscosity and temperature with time are shown. In b, the total heat flow (solid line) and instantaneous radiogenic heat production (dashed line) are shown as a function of time.

4. The Earth's core did not form directly from iron-rich nebular condensates, but formed continuously and contemporaneously with accretion by downward aggregation of iron-rich diapirs migrating through the mantle. At least part of the hydrosphere and atmosphere probably formed during primary accretion (first 10^8 years), but significant later editions by comets may also be present.
5. Extensive and pervasive partial melting was unavoidable in the accreting Earth; complete melting may also have occurred during the trauma of a very large ($\geqslant$ Mars-sized) impact event. Mantle differentiation may have been profoundly affected by this extensive melting, but nobody has quantified this yet.
6. Protoearth probably had a substantial H_2O-CO_2 atmosphere, although it may not have been dense enough to drive the surface temperature up to the solidus. Rapid weathering and carbonate production may have buffered surface conditions to a state not much hotter than the present surface temperature.
7. Mantle convection and associated volcanism were far more vigorous on early Earth than now, though not necessarily different in *style* than the present Earth. The thermal regime was first dominated by impacts ($t \lesssim 0.2 \times 10^9$ yr) then by secular cooling (0.2×10^9 yr $\lesssim t \lesssim 0.45 \times 10^9$ yr) then by quasi-steady state with radiogenic heat production ($t \gtrsim 0.45 \times 10^9$ yr). Total crustal production, outgassing, and heat loss were greater during this early period than the integrated effect of all subsequent geologic time.
8. Island arc material and protocontinents should have been formed very early but evidently were recycled. It seems possible that the relative absence of a rock record prior to ~3.8 byr bp is telling us about the "nucleation," i.e., the time delay before survivable continental nuclei (cratons) can form by a stochastic process, rather than about a change in the way the Earth's convective and geochemical engine was operating. A uniformitarian principle could apply here!
9. The general picture for Earth formation should also apply to Venus. However, recent popularity in an impact origin for the Moon leads us to wonder whether Earth may have suffered a more traumatic early history (e.g., a much larger impact event). The Moon is probably *not* a good place to seek an understanding of early Earth processes because its origin may lie in this traumatic event.
10. The best hope for identifying clues of early Earth history lies in understanding the differentiation of a partially or extensively molten early Earth and the extent to which subsequent convective stirring can modify

or eliminate the effects of this chemical differentiation. Much more work is needed on this.

I can think of no better way to end than with a quote from Harold Urey (1952, p. 223), one of the pioneers of studies of early Earth and the solar system:

"The chronology of events given is a complicated one but the true one is certainly even more complicated . . . . And this is a field where judgments in regard to the true and false may require long periods of time. However, the serious student of this subject receives a great reward, for as he must abandon his imperfect hypotheses for others more in accord with the observational data, he finds that the new approximation to the ultimate solution is more satisfying than the old . . . ."

8. Acknowledgment

Partial support for this work came from NASA Planetary Geophysics (NAGW-185) and NSF Experimental and Theoretical Geophysics EAR-8418353.

9. References

Abe, Y. and Matsui, T. *J. Geophys. Res.* C323 (1984).

Alfven, H. and Arrhenius, G. *Evolution of the Solar System*, NASA SP-345, Washington (1976).

Alvarez, L.W., Alvarez, W., Asara, F., and Michel, H.V. *Science* **208**, 1095 (1980).

Anderson, D.L. *Science,* **213**, 82 (1981).

Arnold, J.R. *Astrophys. J.,* **141**, 1536 (1965).

Aumann, H.H., Gillett, F.C., Beichman, C.A., De Jong, T., Houck, J.R., Low, F.J., Neugebauer, G., Walker, R. G., and Wesselius, P.R., *Astrophys. J.* **278**, L23 (1984).

Benz, W., Slattery, W.L., and Cameron, A.G.W. *Lunar Planet. Sci. Conf. Abs. XVII,* **40** (1986).

Bertout, C. *Rep. Prog. Phys.* **47**, 111 (1984).

Bickle, M.J. *Earth Planet. Sci. Lett.* **40**, 301 (1978).

Black, D.C. and Matthews, M.S. (eds.) *Protostars and Planets II*, Tucson, AZ: University of Arizona Press, 1293 pp. (1985).

Boslough, M.B., Ahrens, T.J., Vizgirda, J., Becker, R.H., and Epstein, S. *Earth Planet. Sci. Lett.* **61**, 166 (1982).

Boyd, F.R., Gurney, J.J., and Richardson, S.H. *Nature* **315**, 387 (1985).

Brett, R. *Geochim. Cosmochim. Acta* **48**, 1183 (1984).

Cameron, A.G.W. In T. Gehrels (ed.) *Protostars and Planets,* Tucson, AZ: University of Arizona Press, p. 453 (1978).
Cameron, A.G.W. *Astrophys. Space Sci.* **93**, 295 (1983a).
Cameron, A.G.W. *Icarus* **56**, 195 (1983b).
Cameron, A.G.W. *Icarus* **62**, 319 (1985).
Cameron, A.G.W. and Ward, W.R. *Lunar Sci. Abs. VII*, 120 (1976).
Cameron, A.G.W., DeCampli, W.M., and Bodenheimer, P. *Icarus* **49**, 298 (1982).
Canuto, V.M., Goldman, I., and Hubickyj, O. *Astrophys. J.* **280**, L55 (1984).
Canuto, V.M., Levine, J.S., Augustsson, T.R., and Imhoff, C.L. *Nature* **296**, 816 (1982).
Carr, M.H. *J. Geophys. Res.* **91**, 3521 (1986).
Carter, N.L. *Rev. Geophys. Sp. Phys.* **14**, 301 (1976).
Chang, S., Des Marais, D., Mack, R., Miller, S.L., and Strathearn, G.E. In J.W. Schopf (ed.) *Earth's Earliest Biosphere*, Princeton, NJ: Princeton University Press, 53 (1983).
Clayton, D.D. *Principles of Stellar Evolution and Nucleosynthesis*, New York: McGraw-Hill, p. 256 (1968).
Christensen, U.R. *J. Geophys. Res.* **90**, 2995 (1985).
Cogley, J.G. and Henderson-Sellers, A. *Rev. Geophys. Space Phys.* **22**, 131 (1984).
Compston, W. and Pidgeon, R.T. *Nature* **321**, 766 (1986).
Condie, K.C. *Precambrian Research* **11**, 183 (1980).
Coradini, A., Federico, C., and Lanciano, P. *Phys. Earth Planet. Int.* **31**, 145 (1983).
Davies, G.F. *J. Geophys. Res.* **85**, 7108 (1980).
Davies, G.F. *Nature* **290**, 208 (1981).
Davies, G.F. *Geophys. Res. Lett.* **9**, 1267 (1982).
Davies, G.F. *Icarus* **63**, 45 (1985).
DePaolo, D.J. In M.G. Kivelson (ed.) *The Solar System Observations and Interpretations*, Rubey Vol. IV, Englewood Cliffs, NJ: Prentice-Hall, 94 (1986).
Dermott, S. F., (ed.) *The Origin of the Solar System*, New York: J. Wiley & Sons, 668 pp. (1978).
Edgeworth, K.E. *Mon. Not. Roy. Astron. Soc.* **109**, 600 (1949).
Elsasser, W.M. In J. Geiss, E. Goldberg (eds.) *Earth Science and Meteorites*, Amsterdam: North-Holland, 1 (1963).
England, P.C. *Nature* **277**, 556 (1979).
Ernst, W. G. In J.W. Schopf (ed.) *Earth's Earliest Biosphere*, Princeton, NJ: Princeton University Press, Ch. 3 (1983).
Fanale, F.P. *Chem. Geol.* **8**, 79 (1971).
Fanale, F.P. In M.G. Kivelson (ed.) *The Solar System Observations and Interpretations*, Rubey Vol. IV, Englewood Cliffs, NJ: Prentice-Hall, 136 (1986).
Fegley, B. Jr., Prinn, R.G., Hartman, H. and Watkins, G.H. *Nature* **319**, 305 (1986).
Flasar, F.M. and Birch, F. *J. Geophys. Res.* **78**, 6101 (1973).
Froude, D.O., Ireland, T.R., Kinney, P.D., Williams, I.S., Compston, W., Williams, I.R., and Myers, J.S. *Nature* **304**, 616 (1983).
Fujiwara, A., Kamimoto, G., and Tsukamoto, A. *Nature* **272**, 602 (1978).

Fukai, Y. *Nature* **308**, 174 (1984).
Gehrels, T. (ed.) *Protostars and Planets*, Tucson AZ: University of Arizona Press, 756 pp. (1978).
Goldreich, P. and Ward, W.R. *Astrophys. J.* **183**, 1051 (1973).
Greenberg, R. In A. Brahic (ed.) *Formation of Planetary Systems*, Toulouse: Publ. Lepadues Editions, p. 515 (1982).
Grossman, L. and Larimer, J. *Rev. Geophys. Space Phys.* **12**, 71 (1974).
Hanson, R.B., Jones, B.F., and Lin, D.N.C. *Astrophys. J.* **270**, L27 (1983).
Hargraves, R.B. *Nature* **276**, 459 (1978).
Hartmann, W.K. *Icarus* **33**, 50 (1978).
Hartmann, W.K. and Davis, D.R. *Icarus* **24**, 504 (1975).
Hartmann, W.K., Phillips, R.J., and Taylor, G.J. *Origin of the Moon*, Houston: Lunar & Planetary Institute, (1986).
Hayashi, C. In D. Sugimoto, D.Q. Lamb, and D.N. Schramn (eds): *Fundamental Problems in the Theory of Stellar Evolution, IAU Symp. 93*, Dordrecht, Netherlands: Reidel, p. 113 (1981).
Hayashi, C., Nakazawa, K., and Mizuno, H. *Earth Planet. Sci. Lett.* **43**, 22 (1979).
Hayashi, C., Nakazawa, K., and Nakagawa, Y. In D. Black and M. Matthews (eds.) *Protostars and Planets II*, Tucson, AZ: University of Arizona Press, p. 1100 (1985).
Hofmeister, A.M. *J. Geophys. Res.* **88**, 4963 (1983).
Holland, H.D. *The Chemical Evolution of the Atmosphere and Oceans*, Princeton, NJ: Princeton University Press (1984).
Holland, H.D., Lazar, B., and McCaffrey, M. *Nature* **320**, 27 (1986).
Holsapple, K.A. and Schmidt, R.M. *J. Geophys. Res.* **87**, 1849 (1982).
Howard, L.N. *Proc. 11th Int. Conf. Applied Mech.* Berlin: Springer-Verlag, 1109 (1964).
Jacobs, J.A. *The Earth's Core*, New York: Academic Press (1975).
Jeanloz, R. and Morris, S. *Ann. Rev. Earth Planet. Sci.* **14**, 377 (1986).
Johnson, T.V., Morrison, D., Matson, D.L., Veeder, G., Brown, R.H., and Nelson, R.M. *Science* **226**, 134 (1984).
Kaula, W.M. *J. Geophys. Res.* **84**, 999 (1979).
Kaula, W.M. *Geol. Assoc. Canada Special Paper 20*, 25 (1980).
Kawakami, S., Mizutani, H., Takagi, Y., Kato, M., and Kumazawa, M. *J. Geophys. Res.* **88**, 5806 (1983).
Kerridge, J.F. and Bunch, T.E. In T. Gabriels (ed.) *Asteroids*, Tucson, AZ: University of Arizona Press, 745 (1979).
Knittle, E. and Jeanloz, R. Preprint (1986).
Lambert, R.St.J. *Precambrian Research* **11**, 199 (1980).
Lange, M.A. and Ahrens, T.J. *Icarus* **51**, 96 (1982).
Lange, M.A. and Ahrens, T.J. *Earth Planet. Sci. Lett.* **77**, 409 (1986).
Lecar, M. and Aarseth, S.J. *Astrophys. J.* **305**, 564 (1986).
Lin, C.C. and Shu, F.H. *Proc. Natl. Acad. Sci. U.S.* **55**, 229 (1966).
Lin, D.N.C. and Bodenheimer, P. *Astrophys. J.* **262**, 768 (1982).
Lunine, J.I. and Stevenson, D.J. *Icarus* **64**, 345 (1985).

Lynden-Bell, D. and Pringle, J.E. *Mon. Not. Roy. Astron. Soc.* **168**, 603 (1974).
Macdougall, J.D., Lugmair, G.W., and Kerridge, J.F. *Nature* **307**, 249 (1984).
Matsui, T. and Abe, Y. *Moon and Planets* **34**, 223 (1986a).
Matsui, T. and Abe, Y. *Nature* **319**, 303 (1986b).
McCammon, C.A., Ringwood, A.E., and Jackson, I. *Geophys. J. Roy. Astron. Soc.* **72**, 577 (1983a).
McCammon, C.A., Ringwood, A.E., and Jackson, I. *J. Geophys. Res. Suppl.* **88**, A501 (1983b).
McCarthy, D.W. Jr., Probst, R.G., and Low, F.J. *Astrophys. J. Lett.* **290**, L9 (1985).
McKenzie, D.P. and Richter, F.M. *J. Geophys. Res.* **86**, 11667 (1981).
McKenzie, D.P. and Weiss, N.O. *Geophys. J. Roy. Astron. Soc.* **42**, 131 (1975).
McLennan, S.M. and Taylor, S.R. *Nature* **306**, 169 (1983).
Miller, R.G. and O'Nions, R.K. *Nature* **314**, 325 (1985).
Mizuno, H. *Prog. Theor. Phys.* **64**, 544 (1980).
Murase, T. and McBirney, A.R. *Geol. Soc. Am. Bull.* **84**, 3563 (1973).
Nakagawa, Y. *Prog. Theor. Phys.* **59**, 1834 (1978).
Nakagawa, Y., Nakazawa, K., and Hayashi, C. *Icarus* **45**, 517 (1981).
Newkirk, G.A. In "The Ancient Sun." *Geochim. Cosmochim. Acta Suppl.* **13**, 293 (1980).
Nisbett, E.G. *Nature* **309**, 110 (1984).
Nisbett, E.G. *Nature* **315**, 541 (1985).
O'Keefe, J.D. and Ahrens, T.J. *Geol. Soc. Am. Special Paper 190*, 103 (1982).
Ohtani, E. *Phys. Earth Planet. Int.* **38**, 70 (1985).
Ojakangas, G.W. and Stevenson, D.J. *Icarus* **66**, 341 (1986).
Orphal, D.L., Borden, W.F., Larson, S.A. and Schultz, P.H. *Proc. Lunar Planet. Sci. Conf. XI,* 2309 (1980).
Ozima, M., Podosek, F.A., and Igarashi, G. *Nature* **315**, 471 (1986).
Peltier, W.R. *Nature* **318**, 614 (1985).
Poirier, J-P. *Creep of Crystals*, Cambridge: Cambridge University Press (1985).
Ramberg, H. *Phys. Earth Planet. Int.* **1**, 427 (1968).
Ramberg, H. *Gravity, Deformation and the Earth's Crust*, London: Academic Press (1981).
Reymer, A. and Schubert, G. *Geology* **14**, 299 (1986).
Richter, F.M. *Earth Planet. Sci. Lett.* **73**, 350 (1985).
Rigden, S.M., Ahrens, T.J., and Stolper, E.M. *Science* **226**, 1071 (1984).
Ringwood, A.E. *Geochem. J.* **11**, 111 (1977).
Ringwood, A.E. *Origin of the Earth and Moon*, Berlin: Springer-Verlag, 295 pp. (1979).
Safronov, V.S. *Evolution of the Protoplanetary Cloud and Formation of the Earth and Planets* Moscow: Nauka. In Russian (1969). English transl: NASA TT-F-677 (1972).
Safronov, V.S. *Icarus* **33**, 3 (1978).
Schubert, G. and Reymer, A.P.S. *Nature* **316**, 336 (1985).
Schubert, G., Stevenson, D., and Cassen, P. *J. Geophys. Res.* **85**, 2531 (1980).

Scott, D.R. and Stevenson, D.J. *J. Geophys. Res.*, in press (1986).
Sekiya, M. *Prog. Theor. Phys.* **69**, 1116 (1983).
Sekiya, M., Hayashi, C., and Nakazawa, K. *Prog. Theor. Phys.* **66**, 1301 (1981).
Shakura, N.I. and Sunyaev, R.A. *Astron. Astrophys.* **24**, 337 (1973).
Sharpe, H.N. and Peltier, W.R. *Geophys. Res. Lett.* **5**, 737 (1978).
Sharpe, H.N. and Peltier, W.R. *Geophys. J. Roy. Astron. Soc.* **59**, 171 (1979).
Shaw, G.H. *Phys. Earth Planet. Int.* **16**, 361 (1978).
Shaw, G.H. *Phys. Earth Planet. Int.* **20**, 42 (1979).
Slattery, W.L., DeCampli, W.M., and Cameron, A.G.W. *Moon and Planets* **23**, 381 (1980).
Smith, B.A. and Terrile, R.J. *Science* **226**, 1421 (1984).
Spohn, T. and Schubert, G. *J. Geophys. Res.* **87**, 4682 (1982).
Stéfan, J. *Annalen der Physik und Chem.* 4242269 (1891).
Stevenson, D.J. *Geophys. Astrophys. Fluid Dyn.* **12**, 139 (1979).
Stevenson, D.J. *Science* **214**, 611 (1981).
Stevenson, D.J. *Ann. Rev. Earth Planet. Sci.* **10**, 257 (1982).
Stevenson, D.J. In J.W. Schopf (ed.) *Earth's Earliest Biosphere*, Princeton, NJ: Princeton University Press, Ch. 2 (1983).
Stevenson, D.J. *Lunar Planet. Sci. Abs. XV*, 822 (1984a).
Stevenson, D.J. *G.S.A. Abstracts with Program* **16**, 668 (1984b).
Stevenson, D.J. Lunar Planet. Inst. Technical Report 85-01, 76 (1985).
Stevenson, D.J. *Ann. Rev. Earth Planet. Sci.* **15**, 271 (1987).
Stevenson, D.J., Spohn, T., and Schubert, G. *Icarus* **54**, 466 (1983).
Stewart, G.R. and Wetherill, G.W. *Lunar Planet. Sci. Abs. XVII*, 827 (1986).
Thompson, C. and Stevenson, D.J. *Astrophys. J.*, in press (1988).
Thomsen, L. *J. Geophys. Res.* **85**, 4374 (1980).
Tozer, D.C. *Phil. Trans. R. Soc.* **258A**, 252 (1965).
Turcotte, D.L. *Ann. Rev. Earth Planet. Sci.* **10**, 397 (1982).
Turcotte, D.L. and Emerman, S.H. *J. Geophys. Res. Suppl.* **88** B91 (1983).
Turcotte, D.L. and Pflugrath, J.C. *Proc. 15th Lunar Planet. Sci. Conf. J. Geophys. Res. Suppl. C*, 541 (1985).
Turcotte, D.L. and Schubert, G. *Geodynamics*, J. Wiley & Sons, NY (1982).
Urey, H.C. *The Planets, Their Origin and Development*, New Haven, CT: Yale University Press (1952).
Veizer, J. In J.W. Schopf (ed.) *Earth's Earliest Biosphere*, Princeton, NJ: Princeton University Press, Ch. 10 (1983).
Vidal, P. and Dosso, L. *Geophys. Res. Lett.* **5**, 169 (1978).
Vityazev, A.V. and Mayeva, S.V. *Izv. Acad. Sci. USSR Phys. Solid Earth* **12**, 79 (1976).
Walker, J.C.G. *Evolution of the Atmosphere*, New York: Macmillan, (1977).
Wänke, H., Dreibus, G., and Jagoutz, E. In A. Kröner *et al.* (ed.) *Archean Geochemistry* Berlin: Springer-Verlag, 1 (1984).
Warren, P.H. *Ann. Rev. Earth Planet. Sci.* **13**, 201 (1985).
Wasson, J.T. *Meteorites*, New York: W.H. Freeman (1985).

Weidenschilling, S.J. *Mon. Not. Roy. Astron. Soc.* **180**, 57 (1977).
Weidenschilling, S.J. *Icarus* **44**, 172 (1980).
Wetherill, G.W. *Ann. Rev. Astron. Astrophys.* **18**, 77 (1980).
Wetherill, G.W. *Science* **228**, 877 (1985).
Wetherill, G.W. and Cox, L.P. *Icarus* **60**, 40 (1984).
Wetherill, G.W. and Cox, L.P. *Icarus* **63**, 290 (1985).
Wetherill, G.W. and Stewart, G.R. *Lunar Planet. Sci. Abs. XVII*, 939 (1986).
Whipple, F.L. In C.L. Hemenway, P.M. Millman, and A.F. Cook (eds.) *Evolutionary and Physical Properties of Meteroids*, NASA SP-319, 355 (1973).
Wilhelms, D.E. In *The Geology of the Terrestrial Planets*, NASA SP-469, 107 (1984).
Windley, B.F. *The Evolving Continents*, 2nd Edition, New York: J. Wiley & Sons, (1984).
Wyllie, P.J. *Geophys. Res. Lett.* **5**, 440 (1978).
Yang, J. and Epstein, S. *Geochim. Cosmochim. Acta* **47**, 2199 (1983).
Zvyagina, Y.V., Pechernikova, G.V., and Safronov, V.S. *Astron. Zh.* **50**, 1261 (1973).

Index